Die Grundlehren der mathematischen Wissenschaften

in Einzeldarstellungen
mit besonderer Berücksichtigung
der Anwendungsgebiete

Band 129

Herausgegeben von

J. L. Doob · E. Heinz · F. Hirzebruch · E. Hopf
H. Hopf · W. Maak · S. MacLane
W. Magnus · D. Mumford · F. K. Schmidt · K. Stein

Geschäftsführende Herausgeber

B. Eckmann und B. L. van der Waerden

The Mathematical Apparatus for Quantum-Theories

Based on the Theory of Boolean Lattices

Otton Martin Nikodým

Springer-Verlag Berlin Heidelberg New York 1966

Library of Congress Catalog Card Number 66–27977

Printed in Germany

Titel No. 5112

Dedicated to my wife

Dr. Stanisława Nikodým

Preface

The purpose of this book is to give the theoretical physicist a geometrical, visual and precise mathematical apparatus which would be better adapted to some of their arguments, than the existing and generally applied methods. The theories, presented in this book, are based on the theory of Boolean lattices, whose elements are closed subspaces in the separable and complete Hilbert-Hermite-space.

The first paper, in which the outlines of the said mathematical apparatus is sketched, is that of the author: "Un nouvel appareil mathématique pour la théorie des quanta."[1]

The theory exhibited in this paper has been simplified, generalized and applied to several items of the theory of maximal normal operators in Hilbert-space, especially to the theory of multiplicity of the continuous spectrum and to permutable normal operators, based on a special canonical representation of normal operators and on a general system of coordinates in Hilbert-space, which is well adapted not only to the case of discontinuous spectrum, but also to the continuous one.

The normal operators, which can be roughly characterized as operators with orthogonal eigen-vectors and complex eigen-values, constitute a generalization of hermitean selfadjoint and of unitary operators.

The importance of the methods, sketched in the mentioned paper, has been emphasized in the review in the "Zentralblatt für Mathematik", by the physicist G. Ludwig[2] and later applied by him in his book "Die Grundlagen der Quantenmechanik"[3]. The mentioned theory has

---

[1] Annales de l'Institut Henri Poincaré, tome XI, fasc. II, pages 49—112. The paper constitutes the content of four lectures by the author: February 4, 6, 11 and 13, (1947) at the Institut Henri Poincaré in Paris.

[2] Bd. 37, 1951, p. 278/279.

[3] Berlin/Göttingen/Heidelberg: Springer-Verlag. Die Grundlehren der Mathematischen Wissenschaften 52. (1954), XII + 460 pp. (see the footnote p. 75).

later been simplified, generalized and applied in several papers by the author. The present book can be considered as a systematic synthesis of them all, with suitable preparations, additions and precise proofs.

It contains many new notions, as the notion of "trace", which are defined, studied, and applied.

The author hopes that this book will be useful not only to physicists but also to mathematicians.

Concerning the Boolean-lattice-approach, the following remarks are in order: J. v. NEUMANN has found interesting relations between the logic of propositions and some behaviour of projectors in the Hilbert-space. Now, if we introduce with M. H. STONE suitable and simple operations on closed subspaces of the Hilbert-space, we can perceive that just the Boolean lattices made out of closed subspaces constitute the suitable, useful translation of the relations mentioned, found by v. NEUMANN, and that the Boolean lattices should be chosen as a convenient background for further developments.

An other source can be found in the modern theory of set-function and of general, abstract integration and measure, created by DE LA VALLÉE-POUSSIN, VITALI, HAHN, RADON and especially by M. FRÉCHET who has generalized the LEBESGUEan theory to abstract sets and general denumerably additive, non-negative and bounded measure.

The above few sources have made it possible to construct a geometrical theory of selfadjoint operators and to extend it to normal operators.

We mention that the original approach to the mathematical part of the theory of quanta, based on matrices, is not adequate, as has been shown by J. v. NEUMANN in his paper: [J. reine angew. Math. 161 (1913)]. The matrices have been replaced by operators in Hilbert-space (F. RIESZ, J. v. NEUMANN, M. H. STONE).

Since we do not require that the reader be familiar with the modern abstract theories, we shall start with a sketch of the theory of Boolean lattices.

The reader is supposed to be familiar with

1) basic properties of the structure of Hilbert-space, with basic properties of hermitean selfadjoint operators, the Hilbert spectral-theorem included,
2) with the theory of Lebesgue's measure and integration,

3) with basic notions of abstract topology and

4) with the notion of an ideal in a commutative ring. The reader is also supposed to know the necessity of discrimination between notions having different logical type e.g. between a set and a set of sets.

We apply the usual notations with the following novelty: we shall sometimes use a dot over a letter, say x, to emphasize that x is a variable, e.g. $f(\dot{x})$ means a function of the variable x, and $f(a)$ means the value of the function at the point a, the symbol $A_{\dot{n}}$ will mean the sequence $\{A_1, A_2, \ldots, A_n, \ldots\}$ and $\dot{n}$ the sequence $1, 2, 3, \ldots, n, \ldots$ of natural numbers.

The book contains 29 chapters, which are only partly depending on one another. They are labelled with letters:

A, A I, B, B I, C, C I, D, D I, E, F, G, H,
J, J I, K, L, M, N, P, P I, Q, Q I, R, R I,
S, T, U, W, W I.

We are including the list of contents of these chapters.

We give the list of references labelled with a fat parenthesis **( )**. The list contains not only the papers which are directly applied in the text, but also all those papers, which have influenced the author with some useful ideas.

Though the "Apparatus" is destinated for physicists, it does not contain direct applications to mathematical problems of physics. The author intends to deal with them in subsequent papers or in another book.

I am owing my thanks to Prof. Dr. Helmut Hasse and to Prof. Dr. B. L. van der Waerden for their kind recommendation of my work to the Springer-Verlag.

I wish to thank the U.S.A.-Atomic Energy Commission and the U.S.A.-Office of Ordnance Research for their financial help in my research related to the book, and especially I am owing my thanks to the U. S. A. National Science Foundation for support through several years. I am owing special thanks to that institution whose grants have made possible the final composition of the book.

In addition to that I express my thanks to the French "Fondation Nationale des Recherches Scientifiques," whose financial aid has made possible my research on the "Apparatus" in 1946—1948, and especially to Professor Arnaud Denjoy who kindly arranged that financial aid.

But my most hearty thanks I am owing to my wife Dr. STANISŁAWA NIKODÝM, (also a mathematician) whose help in composing the book, proof reading and typing was very great. Without her efficient help it would have been impossible for me to compose the present book.

Finally, I would like to thank the Springer-Verlag and the printers for a beautiful and very clear setting of a quite difficult text.

Utica, N.Y. USA,
July 10. 1966 Dr. OTTON MARTIN NIKODÝM

List of chapters

A. General tribes (Boolean lattices) 1
A I. Special theorems on Boolean lattices 44
B. Important auxiliaries 130
B I. General theory of traces 145
C. The tribe of figures on the plane 173
C I. The trace-theorem 206
D. The lattice of subspaces of the Hilbert-Hermite space 252
D I. Tribes of spaces 285
E. Double scale of spaces 335
F. Linear operators permutable with a projector 386
G. Some double STIELTJES' and RADON'S integrals 391
H. Maximal normal operator and its canonical representation . . 402
J. Operators $N\dot{f}(\dot{x}) =_{df} \varphi(\dot{x}) \cdot \dot{f}(\dot{x})$ for ordinary functions f . . . 428
J I. Operational calculus on general maximal normal operators . . 450
K. Theorems on normal operators and on related canonical mapping 462
L. Some classical theorems on normal and selfadjoint operators . 502
M. Multiplicity of spectrum of maximal normal operators . . . 521
N. Some items of operational calculus with application to the resolvent and spectrum of normal operators 560
P. Tribe of repartition of functions 586
P I. Permutable normal operators 622
Q. Approximation of somata by complexes 644
Q I. Vector fields on the tribe and their summation 675
R. Quasi-vectors and their summation 689
R I. Summation of quasi-vectors in the separable and complete Hilbert-Hermite-space 708
S. General orthogonal system of coordinates in the separable and complete Hilbert-Hermite-space 718
T. DIRAC'S Delta-function 724
U. Auxiliaries for a deeper study of summation of scalar fields . 760
W. Upper and lower (DARS)-summation of fields of real numbers in a Boolean tribe in the absence of atoms 805
W I. Upper and lower summation in the general case. Complete admissibility. Square summability of fields of numbers . . . 884

References 942

Alphabetical index 947

Chapter A

General tribes (Boolean lattices)

A.1. This first chapter contains the fundamentals of the theory of Boolean lattices.

We start with intuitive remarks; the precise setting will follow later at [A.1.1.].

The Boolean lattices constitute an important item in modern functional analysis, though they did not occur in textbooks of Quantum Mechanics as an essential tool. Nevertheless special Boolean lattices, whose elements are invariant closed subspaces of the Hilbert-space play a quite important role and give a purely geometrical basis to selfadjoint Hermitean operators. The said importance has been shown by the author in (10) and in several subsequent papers by the author.

The algebra of Boolean lattices looks like the algebra of sets, because we have there the operations of addition, multiplication and complementation, obeying just the same formal rules, as in the theory of sets, but with the exception that the relation of belonging of an element to a set, $a \in \alpha$, is not considered at all. We could roughly say, that the elements of the Boolean lattice are "sets without points", though the points may be even available.

Usually people use the term "Boolean algebra", rather than the term Boolean lattice. Now, since in the sequel we shall use not only finite operations, but also the infinite ones, the theory stops to be an algebra. Since the infinite operations are defined as a kind of supremum and infimum of collections of elements, therefore the true basis of the theory is the notion of ordering (partial ordering) (6), (8). Therefore we shall start with this notion and define the lattice as a kind of ordering. In agreement with Russell's and Whitehead's Principia Mathematicae (1), we define the ordering as a *relation* (*correspondence, mapping*), satisfying certain conditions. The *relation* (*correspondence, mapping*) will be understood as a notion which is attached to a *condition* (*propositional function*) (1), *involving two variables*, similarly as the set is a notion attached to a *condition* (*propositional function*), *with one variable only*. Thus e.g. as the condition for numbers $3x - \frac{1}{x} = 5$ with the variable x generates the set $\left\{x \mid 3x - \frac{1}{x} = 5\right\}$, so the condition $3x - y \leqq 4$ generates the relation $\{x, y \mid 3x - y \leqq 4\}$.

We shall not consider the relation $\{x, y \mid w(\cdot\, x, y\, \cdot)\}$ as the set of all ordered couples of numbers (a, b) satisfying the condition $w(\cdot\, a, b\, \cdot)$, but we shall stay with the original Russell's approach.

The statement: a is an element of S, is usually written $a \in S$, where S denotes the set $\{x \mid w(\cdot\, x\, \cdot)\}$. Similarly the statement $v(\cdot\, a, b\, \cdot)$, where a, b are fixed elements, will be (optionally) written $a\, R\, b$, where R denotes the relation $\{x, y \mid v(\cdot\, x, y\, \cdot)\}$ with two variables x, y.

The set of all elements a, satisfying the condition

"there exists b, such that $a\, R\, b$"

is called *domain of* R. We denote it by $\mathsf{C}R$.

The set of all elements b, satisfying the condition

"there exists a, such that $a\, R\, b$"

is called *range of* R. We denote it by $\mathsf{D}R$.

The elements of the set must have the same logical type in order to avoid logical antinomies, but in the relation $a\, R\, b$, the types of a and b may be different.

$$\{x, y \mid w(\cdot\, x, y\, \cdot)\} \quad \text{differs from} \quad \{y, x \mid w(\cdot\, x, y\, \cdot)\}.$$

The relation is said to be *empty*, whenever there does not exist a and b with $w(\cdot\, a, b\, \cdot)$.

Functions are considered as relations.

A lattice will be considered as a special kind of ordering and the Boolean lattice as a special kind of lattice.

We shall use the term "*tribe*" (*Boolean tribe*) to denote a Boolean lattice. The term is borrowed from René de Possel (11). The elements of an ordering will be termed "*somata*" (sing. *soma*); this term is borrowed from Carathéodory (7).

A tribe, if we consider finite operations only, can be reorganized into a commutative ring with unit (*Stone's ring, Boolean algebra*).

We shall state and prove several simple theorems. The proofs will be precise and even sometimes meticulous in order that the reader, who is not familiar with the topic, be acquainted with methods and not spend his time by completing proofs, if they were only sketchy.

A part of the chapter will be devoted to some notions related to the notion of *equality of elements*. This notion is usually considered as something trivial. But we can notice that the equality may differ from the identity.

E. g. In Hilbert-space, whose vectors are square-summable functions $f(x)$, we have the "equality almost everywhere" of functions $f(x)$; hence not the identity.

This fact compels us to make some necessary remarks concerning the so-called "*governing equality*" and the role it plays.

The said discussion of the notion of governing equality enables us to treat the notion of a *subtribe* and *supertribe* of a given tribe with more precision than in many textbooks.

It will enable us to introduce the precise notion of *finitely genuine subtribe of a given tribe* and some variations of this notion. They will be very important in the sequel.

Since the tribe can be reorganized into a ring, we can speak of *ideals in a tribe.* The ideal generates a notion of equality, which reconstructs the tribe into another one. We shall devote to ideals a small part of the chapter.

As the elements of a tribe behave like sets, we can introduce the notion of *measure of somata*, which in turn generates a special ideal and also the notion of the *distance between two somata*. The distance organizes, under simple condition, the tribe into a metric space; hence it yields a topology. In dealing with measure, and generally with functions, we shall put for more clarity a dot upon a letter, say x, to emphasize that x is variable. Thus $f(\dot{x})$ is a function and $f(x)$ the value of the function at x. For more subtile details concerning tribes we refer to our two papers (12), (13).

The part A deals with fundamentals only; more special theorems on tribes will fill up the chapter A I.

Now we are going over to precise definitions and precise proofs of basic theorems.

A.1.1. Ordering. We define it as any correspondence (relation, mapping, application) R, between elements of a manifold, satisfying the following conditions (6):

1) If $x R y$, $y R z$, then $x R z$;

2) If $x R y$, $y R x$, then $x = y$ and conversely;

3) R is not an empty correspondence, i.e. there exist elements x, y with $x R y$.

For an ordering R we have:

$$\text{if} \quad x \in \mathbb{C} R, \quad \text{then} \quad x R x.$$

If R is an ordering, then $\mathbb{C} R = \mathbb{D} R$. When dealing with an ordering, it is convenient to write $\leqq$ instead of R.

A.1.2. Expl.

1) The correspondence $\leqq$, whose domain is the set of all real numbers, is an ordering. The correspondence $(<)$ is not.

2) The correspondence $E \subseteqq F$, of inclusion of sets, whose domain is e.g. the collection of all subsets of the euclidean plane, is an ordering. So is also the relation $E \supseteqq F$.

3) The relation of equality with any domain, is an ordering.

4) The correspondence between propositions p, q, defined as "p implies q", is an ordering.

5) The correspondence between positive integers a, b, defined by "a, b have a common divisor greater than 1", is not an ordering.

A.1.3. Lattice. Let R be an ordering, which we shall write $\leqq$.

Suppose that, given two elements a, b of $\mathrm{Ɑ}\, R$, there exists an element c, such that

α) $a \leqq c$, $b \leqq c$,

β) if $a \leqq c'$, $b \leqq c'$, then $c \leqq c'$.

In the case of existence, c is unique. The element c is something like the supremum of a, b. We shall call c *sum of* a *and* b, (*union* of a, b; *join* of a, b), and write $a + b$, (or $a \cup b$).

The operation will be termed *ordering-addition*.

If for any two elements a, b the join $a + b$ exists, we say that the *ordering admits the sum, union, join of two elements.*

Now suppose that given two elements a, b of $\mathrm{Ɑ}\, R$, there exists a third one d, such that

α′) $d \leqq a$, $d \leqq b$,

β′) if $d' \leqq a$, $d' \leqq b$, then $d' \leqq d$.

In the case of existence, d is unique. We call d *product* of a, b (*intersection of* a, b, *meet* of a *and* b). We shall write $a \cdot b$, $a \cap b$. The operation will be termed *ordering-multiplication*. The product looks like an infimum of two elements. In the case of existence of the product of any two elements, we say that the *ordering admits the product* (*intersection, meet*), *of two elements*.

Lattice. Def. If an ordering admits sums and products of any two elements, the ordering will be termed *lattice*.

We shall use the term *soma* (pl. *somata*) to denote an element of a lattice.

A.1.4. Expl.

6) The ordering 1) in [A.1.2.] is a lattice and we have

$$a \cup b =_{df} \max(a, b);$$

$$a \cap b =_{df} \min(a, b).$$

7) The ordering 2) in [A.1.2.] is a lattice where

$$E + F =_{df} E \cup F \quad \text{(union of sets)},$$

$$E \cdot F =_{df} E \cap F \quad \text{(intersection of sets)}.$$

8) Consider the set A of all closed rectangles on the plane with sides parallel to the axes of a cartesian system of reference. Let us define on A the ordering defined by the relation $a \subseteqq b$ of inclusion of sets. Let us agree to consider the empty set as a rectangle. Under these circumstances the ordering is a lattice.

9) Let us make a similar agreement with the set of all closed circles on the plane. We shall not get any lattices.

10) If in 8) we shall not consider the empty set as a rectangle, the corresponding ordering will not be a lattice.

11) Consider the collection whose all elements are linear subspaces of the euclidean three dimensional space. This means that the elements are: the origin, the whole space, every straight line passing through the origin and every plane passing through the origin.

Define the ordering as the inclusion of sets, confined to the above elements. Under these conditions $a + b$ will be the smallest linear space containing both a and b, and $a \cdot b$ will be the intersection of the sets a, b. (We notice that the join $a + b$ is not the set-union of a and b). The ordering is a lattice.

A.1.4.1. Given a lattice R, let us consider a collection M of its somata, which may be infinite. It may happen that there exists a soma b of R such that

α) if $a \in M$, then $a \leqq b$;

β) if $b' \in \complement R$, and for all $a \in M$ we have $a \leqq b'$, then $b \leqq b'$.

Under these circumstances b is unique, and we say that *the join*

$$\sum_{a \in M} a \tag{1}$$

is meaningful (the *union exists*) and we define (1) as b. It may also happen that there exists a soma $c \in \complement R$ such that

α′) if $a \in M$, then $c \leqq a$,

β′) if $c' \in \complement R$, and for all $a \in M$ we have $c' \leqq a$, then $c' \leqq c$.

Under these circumstances we say that *the meet*

$$\prod_{a \in M} a \tag{2}$$

is *meaningful* (the *meet exists*) and we define (2) as c. If the set M is given by an infinite sequence $a_1, a_2, \ldots, a_n, \ldots$, we may write

$$\sum_{n=1}^{\infty} a_n, \quad \prod_{n=1}^{\infty} a_n$$

respectively. We do similarly if the collection M is finite.

A.1.4.1a. We have defined infinite operations in a lattice (or even in an ordering). A similar definition will be admitted for operations on somata given by an *indexed set* $\{a_i\}$, where the indices make up an abstract or defined non empty set J of some elements.[1] We write $\sum_{i \in J} a_i$, $\prod_{i \in J} a_i$ respectively.

A.1.4.1b. The lattice-addition and multiplication obey the following laws, (8):

$$(3)\quad \begin{aligned} &x \cdot x = x; \quad x + x = x; \quad x \cdot y = y \cdot x; \quad x + (y + z) = (x + y) + z; \\ &x \cdot (y \cdot z) = (x \cdot y) \cdot z; \quad x \cdot (x + y) = x; \quad x + x \cdot y = x. \end{aligned}$$

In addition to that the following are equivalent:

$$(4)\qquad \text{I) } x \leqq y; \quad \text{II) } y = x + y; \quad \text{III) } x \cdot y = x.$$

Conversely, if we have an algebra with operations of addition and multiplication obeying the laws (3), then if we define the ordering ($\leqq$) by means of II) or III), we get a lattice whose operations are just those which are defined in [A.1.3.]. Thus we can reorganize a lattice into an algebra, and conversely, the algebra into a lattice.

A.1.4.2. Let R be a lattice. It may happen that there exists the "smallest" soma in R, which means, an element O, such that $O \leqq a$ for all $a \in \mathrm{C} R$. Such an element is unique, if it exists. We call it *zero* (or *null-element*), *of* R.

It may also happen, that there exists the "greatest" soma, i.e. such soma, denoted by I, that $a \leqq I$ for all $a \in \mathrm{C} R$. Such a soma is unique whenever it exists. We call it *unit of the lattice* R.

A.1.4.3. Let R be a lattice admitting the zero and the unit. Suppose that there exists a correspondence $\mathfrak{C}$, one-to-one with domain and range $\mathrm{C} R$, such that for all $a \in \mathrm{C} R$ we have

α) $a + \mathfrak{C}(a) = I$,

β) $a \cdot \mathfrak{C}(a) = O$,

γ) if $a \leqq b$, then $\mathfrak{C}(b) \leqq \mathfrak{C}(a)$.

If such a correspondences $\mathfrak{C}$ exist and if we have chosen one of them, we shall call the *lattice complementary*, and we call $\mathfrak{C}(a)$ *complement of* a.

Instead of $\mathfrak{C}(a)$ we agree to write co$a$. Thus we have

$$a + \mathrm{co}\,a = I; \quad a \cdot \mathrm{co}\,a = O; \quad \text{if } a \leqq b, \text{ then } \mathrm{co}\,b \leqq \mathrm{co}\,a;$$
$$\mathrm{co}(\mathrm{co}\,a) = a.$$

A lattice may admit several correspondences like $\mathfrak{C}(a)$.

A.1.4.3a. Expl. Take the example 1) in [A.1.4.] and define co$a$ as the whole space if a is the origin; the origin if a is the whole space; the line perpendicular to a, if a is a plane; the plane perpendicular

---

[1] From the logical point of view the indexed set $\{a_i\}$, $(i \in J)$ is a function whose domain is J and the range is a set of elements of a given ordering. Every non empty collection of somata can be indexed by means of ordinal number: this by virtue of the axiom of Zermelo.

to a, if a is a straight line. We get a complementary lattice. Here the complement is defined by means of orthogonality, but if we transform the space by means of a one-to-one linear correspondence, we shall change the orthogonality into something else, which however yields another way of defining the complementation in our lattice.

A.1.4.4. We underscore the fact that the ordering-operations of addition and multiplication, $\sum\limits_{a\in M} a$, $\prod\limits_{a\in M} a$ depend not only on the elements of M, but on the totality of the lattice. If we could increase or diminish the domain of the lattice, the result of the operation can change.

To be more clear, we should write

$$\sum_{a\in M}^{(R)} a,\ \prod_{a\in M}^{(R)} a,\ O^{(R)},\ I^{(R)},\ \mathrm{co}^{(R)} a.$$

If all sums and products exist, we call the lattice *complete.*

A.1.5. We shall give a proof of two important laws for complementary lattices, the so-called *de Morgan laws.*

Theorem. If (R) is a complementary lattice and $M \neq \emptyset$ a collection of somata of R, then the following are equivalent

I) $\sum\limits_{a\in M} a$ exists,

II) $\prod\limits_{a\in M} \mathrm{co}\, a$ exists.

In the case of existence we have

$$\mathrm{co} \sum_{a\in M} a = \prod_{a\in M} \mathrm{co}\, a.$$

We also have the following equivalence of statements:

I′) $\prod\limits_{a\in M} a$ exists,

II′) $\sum\limits_{a\in M} \mathrm{co}\, a$ exists.

In the case of existence we have

$$\mathrm{co} \prod_{a\in M} a = \sum_{a\in M} \mathrm{co}\, a.$$

Proof. To facilitate the reasoning, consider M as an "indexed set" $\{a_i\}$, $i \in J$, where J is a not empty set of some elements; see footnote p. 6. We shall prove the first part of the theorem. Put

(1) $$b =_{df} \sum_{a\in M} a = \sum_i a_i,$$

supposing this exists. By definition of the sum we have

(2) 1) $a_i \leqq b$ for all $i \in J$;

2) if $a_i \leqq b'$ for all i, we have

(3) $$b \leqq b'.$$

From (2) we deduce [A.1.4.3.]

(4) $$\mathrm{co}\, b \leqq \mathrm{co}\, a_i \quad \text{for all} \quad i \in \cdot J.$$

Suppose that

(4') $$\mathrm{co}\, b'' \leqq \mathrm{co}\, a_i \quad \text{for all} \quad i.$$

We have

$$a_i \leqq b'' \quad \text{for all} \quad i;$$

hence, by virtue of (3),

$$b \leqq b''.$$

Consequently

(4'') $$\mathrm{co}\, b'' \leqq \mathrm{co}\, b.$$

Thus we have proved that if

$$\mathrm{co}\, b'' \leqq \mathrm{co}\, a_i \quad \text{for all} \quad i,$$

then we have

(5) $$\mathrm{co}\, b'' \leqq \mathrm{co}\, b.$$

Let $b''' \leqq \mathrm{co}\, a_i$ for all $i \in J$, and put $b'' =_{df} \mathrm{co}\, b'''$. We have $\mathrm{co}\, b'' = b'''$; hence $\mathrm{co}\, b'' \leqq \mathrm{co}\, a_i$ for all $i \in J$. Applying (5) we get $b''' \leqq \mathrm{co}\, b$. Hence, if $b''' \leqq \mathrm{co}\, a_i$ for all i, then $b''' \leqq \mathrm{co}\, b$. This and (4) proves that $\mathrm{co}\, b = \prod_{i \in J} \mathrm{co}\, a_i$. Hence $\prod_i \mathrm{co}\, a_i$, by [A.1.4.1.], exists and is equal to $\mathrm{co}\, b$. By virtue of (1) we get

$$\prod_i \mathrm{co}\, a_i = \mathrm{co} \sum_i a_i, \qquad (i \in J),$$

which constitutes the thesis of the first part of the theorem.[1] Now suppose that $\prod_i \mathrm{co}\, a_i$ exists. Put

$$\mathrm{co}\, b =_{df} \prod_i \mathrm{co}\, a_i,$$

and perform the reasoning similar to the above one. We get the existence of $\sum_i a_i$ and then I).

The second part of the theorem is a consequence of the first. It suffices to replace a by $\mathrm{co}\, a$, and conversely.

A.1.6. Tribes. Let R be a complementary lattice and assume that the distributive law

$$(a + b) \cdot c = a \cdot c + b \cdot c$$

takes place for all somata a, b, c of R. Then we call the lattice *Boolean tribe* (*tribe, Boolean lattice.* The term generally used is *Boolean algebra*). Thus the tribe is defined as a distributive complementary lattice.

[1] This proof is given by S. NIKODÝM.

A tribe is called *trivial* if it is composed of the single soma $O = I$. The most simple non trivial tribe is composed of two different somata O and I.

A.1.6.1. Expl.:

11) The ordering 1) in [A.1.2.] is not a tribe.

12) The ordering 2) in [A.1.2.] is a tribe. Its zero is the empty set and its unit is the whole plane, co$a$ is the set-complement.

13) Let x, y, z be the axes of the cartesian system of coordinates in the euclidean three-dimensional space. Denote by (x, y) the plane passing through the axes x and y and define similarly the symbols (y, z), (z, x). Let (x, y, z) denote the whole space and let O be the set composed of the origin only. We have eight elements: (x), (y), (z), (x, y), (y, z), (z, x), (x, y, z), O. Define the ordering as in [A.1.4], 11), the zero as O, the unit as (x, y, z) and co$a$ as the ortho-complement of a. We have a tribe.

14) We get a tribe by means of an analogous construction in the space with n dimensions ($n = 1, 2, \ldots$).

Remark. Later we shall make a similar construction in the ordinary Hilbert-space with infinite dimensions.

Expl. 15) The expl. 8) in [A.1.4.] with set-complement is not a tribe.

A.2. Boolean algebra. There exists a vast theory of orderings, lattices and tribes, but we shall confine ourselves to quote, without proofs, several laws governing the tribes and we ask the reader, interested in details, to consult special monographs and papers (6), (8), (9), (12), (13). These laws are the following:

A.2.1.

$$a + a = a; \quad a + b = b + a; \quad a + (b + c) = a + b + c;$$

$$a + O = a; \quad a + I = I.$$

$$a \cdot a = a; \quad a \cdot b = b \cdot a; \quad a \cdot (b \cdot c) = (a \cdot b) \cdot c;$$

$$a \cdot O = O; \quad a \cdot I = a.$$

A.2.1 a.

$$(a + b) \cdot c = a\,c + b\,c, \quad a \cdot b + c = (a + c) \cdot (b + c).$$

A.2.2. *De Morgan laws:*

$$\mathrm{co}\,(a + b) = \mathrm{co}\,a \cdot \mathrm{co}\,b,$$

$$\mathrm{co}\,(a \cdot b) = \mathrm{co}\,a + \mathrm{co}\,b.$$

A.2.3.

$$\mathrm{co}\,I = O, \quad \mathrm{co}\,O = I, \quad \mathrm{co}\,(\mathrm{co}\,a) = a.$$

A.2.3a. Def. We define the *subtraction of somata* as follows:

$$a - b \cdot =_{df} \cdot a \cdot \text{co}\, b.$$

A.2.3b. The following laws are valid:

$$a - b = a - a \cdot b, \quad a - a = O,$$
$$a + b = (a - b) + a \cdot b + (b - a),$$
$$a + b = a + (b - a),$$
$$a = a \cdot b + (a - b).$$

A.2.4. Def. Two somata a, b are called *disjoint* whenever $a \cdot b = O$.

A2.5. The following are equivalent:

$$\text{I) } a \leqq b, \quad \text{II) } a \cdot b = a, \quad \text{III) } a + b = b.$$

A.2.6. If $a \leqq b$, $c \leqq d$, then $a\,c \leqq b\,d$ and $a + c \leqq b + d$;

$$a\,b \leqq a, \quad a \leqq a + b, \quad a \leqq a;$$
$$\text{if} \quad b \leqq a, \quad \text{then} \quad a = b + (a - b).$$

The proofs of the statements [A.2.3b.], [A.2.5.], [A.2.6.] are straight forward.

A.2.6.1. We shall need some more complicated formulas, which we shall provide with proofs.

Theorem. If

$$a \geqq b \geqq c,$$

then

$$a - c = (a - b) + (b - c).$$

Proof.[1] We have [A.1.4.3.],

$$\text{co}\, a \leqq \text{co}\, b \leqq \text{co}\, c. \tag{1}$$

By [A.2.3b.] $a = a \cdot b + (a - b)$; hence multiplying both sides by $\text{co}\, c$, we get $a \cdot \text{co}\, c = [(a - b) + a \cdot b] \cdot \text{co}\, c$ and by [A.1.6.]: $a \cdot \text{co}\, c = (a - b) \cdot \text{co}\, c + a\, b \cdot \text{co}\, c$ and, by [A.2.3a.],

$$a \cdot \text{co}\, c = a \cdot \text{co}\, b \cdot \text{co}\, c + a\, b \cdot \text{co}\, c. \tag{2}$$

As by (1) we have: $\text{co}\, b \leqq \text{co}\, c$, and by hypothesis we have $b \leqq a$, therefore we have by [A.2.5.] and by (2):

$$a \cdot \text{co}\, c = a \cdot \text{co}\, b + b \cdot \text{co}\, c.$$

Hence by [A.2.3a.] we get

$$a - c = (a - b) + (b - c) \quad \text{q.e.d.}$$

[1] The given proof is by S. Nikodým.

A.2.6.2. Theorem. If

$$a_1 \geqq a_2 \geqq \cdots \geqq a_n, \qquad (n = 2, 3, \ldots),$$

then

$$a_1 - a_n = (a_1 - a_2) + (a_2 - a_3) + \cdots + (a_{n-1} - a_n).$$

The terms on the right are disjoint.

Proof. We rely on [A.2.6.1.] and apply induction.

A.2.6.3. Theorem.

$$(1) \qquad \sum_{n=1}^{k} a_n = a_1 + [a_2 - a_1] + [a_3 - [a_3 - (a_1 + a_2)] + \cdots + \\ + [a_k - (a_1 + a_2 + \cdots + a_{k-1})],$$

for $k = 1, 2, \ldots$ The terms on the right are disjoint.

Proof. Suppose that for a given k we have (1). We have

$$(2) \qquad \sum_{n=1}^{k} a_n + [a_{k+1} - (a_1 + a_2 + \cdots + a_k)] = b + (a_{k+1} - b),$$

where

$$b =_{df} \sum_{n=1}^{k} a_n.$$

Applying [A.2.3b.], we see that (2) equals

$$b + a_{k+1} = \sum_{n=1}^{k} a_n + a_{k+1} = \sum_{n=1}^{k+1} a_n,$$

which completes the proof.

A.2.6.4. Theorem.

$$\sum_{k=1}^{n} a_k = a_1 + (a_2 - a_1) + (a_3 - a_2) + \cdots + (a_n - a_{n-1}).$$

Proof. Put $b_k =_{df} a_1 + a_2 + \cdots + a_k$. We have $a_k \leqq b_k$, hence $\mathrm{co}\, b_k \leqq \mathrm{co}\, a_k$. As $b_k + a_{k+1} \leqq 1 = b_k + \mathrm{co}\, b_k \leqq b_k + \mathrm{co}\, a_k$, we have

$$\begin{aligned} b_k + a_{k+1} &\leqq (b_k + a_{k+1}) \cdot (b_k + \mathrm{co}\, a_k) \\ &= b_k \cdot b_k + a_{k+1} \cdot b_k + b_k \cdot \mathrm{co}\, a_k + a_{k+1} \cdot \mathrm{co}\, a_k \\ &= b_k + a_{k+1} \cdot \mathrm{co}\, a_k + b_k (a_{k+1} + \mathrm{co}\, a_k) \\ &= b_k + a_{k+1} \cdot \mathrm{co}\, a_k = b_k + (a_{k+1} - a_k). \end{aligned}$$

Hence

$$(1) \qquad b_k + a_{k+1} \leqq b_k + (a_{k+1} - a_k),$$

and we have

$$a_1 + a_2 = a_1 + (a_2 - a_1).$$

Now, suppose that for a given n:

$$(2) \qquad b_n = a_1 + a_2 + \cdots + a_n \leqq a_1 + (a_2 - a_1) + \cdots + (a_n - a_{n-1}).$$

We get from (1)

$$b_n + a_{n+1} \leqq b_n + (a_{n+1} - a_n);$$

hence, from (2),

$$b_n + a_{n+1} \leqq a_1 + (a_2 - a_1) + \cdots + (a_n - a_{n-1}) + (a_{n+1} - a_n),$$

$$\text{i.e.} \quad a_1 + \cdots + a_{n+1} \leqq a_1 + (a_2 - a_1) + \cdots + (a_{n+1} - a_n).$$

The theorem is established, because the converse inequality is clear.

A.2.6.5. Theorem. If for somata a_n, b_n of a tribe, $(n = 1, 2, \ldots)$, we have

$$(1) \qquad \sum_{n=1}^{k} a_n = \sum_{n=1}^{k} b_n \quad \text{for all} \quad k = 1, 2, \ldots,$$

then, in the case of existence of the denumerable sums

$$\sum_{n=1}^{\infty} a_n, \ \sum_{n=1}^{\infty} b_n,$$

we have

$$\sum_{n=1}^{\infty} a_n = \sum_{n=1}^{\infty} b_n.$$

A.2.6.6. Proof. Suppose the existence takes place. Put

$$a =_{df} \sum_{n=1}^{\infty} a_n, \qquad b =_{df} \sum_{n=1}^{\infty} b_n.$$

By the definition of sums we have for all n:

$$a_n \leqq a;$$

hence

$$\sum_{n=1}^{k} a_n \leqq a \quad \text{for every} \quad k = 1, 2, \ldots$$

By (1) we have

$$\sum_{n=1}^{k} b_n \leqq a; \quad \text{hence} \quad b_k \leqq a \quad \text{for every } k;$$

hence, by definition of the sum $\sum_{k=1}^{\infty} b_k$, we get

$$b \leqq a.$$

In a similar way we prove that $a \leqq b$, which completes the proof.

A.2.7. Remark. We see that the above rules of the Boolean algebra are very similar to those which take place in the theory of general sets. But the Boolean algebra is not interested in points of the sets, even if such points are available. It may be called theory of "sets without points".

A.2.8. Remark. One can give the foundation of the theory of tribes by starting with a well selected collection of formal rules [A.2.1.]

to [A.2.3.] and by defining the ordering by [A.2.5.]:

$$a \leqq b \cdot =_{df} \cdot a \cdot b = a, \quad \text{or by} \quad a \leqq b \cdot =_{df} \cdot a + b = b.$$

A.2.9. Remark. If, given a tribe and its algebra, we replace the addition $a + b$ by another one $a \dotplus b$, defined by

$$a \dotplus b =_{df} (a - b) + (b - a),$$

and by keeping the multiplication, the tribe will be reorganized into a commutative ring with unit. We shall call it *Stone's ring* (9).

We have:

$$a \dotplus a = O; \quad a + b = a \dotplus a \cdot b \dotplus b; \quad \operatorname{co} a = I \dotplus a;$$

the equation

$$a \dotplus x = b$$

has the unique solution $a \dotplus b$, so for the corresponding subtraction ($\dot{-}$) we have $b \dot{-} a = b \dotplus a$.

We shall call the operation $a \dotplus b$ *algebraic addition* in discrimination with $a + b$, which we shall call *somatic addition*. The most used term is *symmetric difference*.

A.2.10. Remark. If we have such a ring, we can go back to the Boolean algebra, by defining the somatic addition by

$$a + b \cdot =_{df} \cdot a \dotplus a\, b \dotplus b.$$

A.2.11. Def. A soma a of a tribe is called *atom*, whenever the following conditions are satisfied:

1) $a \neq O$;

2) if $b \leqq a$, then either $b = O$ or $b = a$.

Expl. In the example 2) in [A.1.2.] the ordering is a tribe. Its atoms are sets composed of single points.

A.3. Def. There are tribes which admit all denumerable joins of their somata. Such tribes will be termed *denumerably additive*.

A.3.1. Remark. There are tribes which are not denumerably additive.

Expl. Let T be the class of all finite unions of the sets $(a, b\rangle$ of real numbers, where $0 \leqq a \leqq 1$, $0 \leqq b \leqq 1$,

$$(a, b\rangle =_{df} \{x \mid a < x \leqq b\}.$$

If we order T by means of the relation of inclusion ($\subseteqq$) of sets, we get a tribe which is not denumerably additive.

Indeed, there does not exist the somatic union of the half-open intervals

$$\left(\frac{1}{2p+1}, \frac{1}{2p}\right\rangle, \quad p = 1, 2, \ldots$$

But some denumerable somatic unions may exist; for instance

$$\sum_{n=1}^{\infty} a_n = \sum_{n=1}^{\infty} \left(\frac{1}{n+1}, \frac{1}{n} \right\rangle, \qquad (n = 1, 2, \ldots).$$

This join equals $(0, 1\rangle$.

A.3.2. Def. It may happen that the tribe admits all finite and infinite joins. Such a tribe is called *completely additive.*

A.3.3. Expl. The tribe of all subsets of $(0, 1\rangle$ with inclusion of sets as ordering relation.

A.3.4. Theorem. In a tribe the following are equivalent:

I) All denumerable sums exist;

II) all denumerable products exist.

Proof. Suppose that all sums $\sum_{n=1}^{\infty} a_n$ exist; then there also exist all sums $\sum_{n=1}^{\infty} \operatorname{co} a_n$.

If we take account of DE MORGAN laws [A.1.5.], we get the existence of $\prod_{n=1}^{\infty} a_n$ too. Thus I) implies II). A similar proof is for the implication II $\rightarrow$ I.

A.3.5. Theorem. Let R be a lattice, where all denumerable sums exist.

Then

$$\sum_{n=1}^{\infty} (a_n + b_n) = \sum_{n=1}^{\infty} a_n + \sum_{n=1}^{\infty} b_n .$$

(This is a kind of associative law.)

Proof. It is not difficult to prove the assertion:

$$\sum_{n=1}^{\infty} a_n \leqq \sum_{n=1}^{\infty} (a_n + b_n) . \tag{0}$$

Indeed, we have for all $k = 1, 2, \ldots$

$$a_k + b_k \leqq \sum_{n=1}^{\infty} (a_n + b_n) .$$

Since $a_k \leqq a_k + b_k$, we get for all k

$$a_k \leqq \sum_{n=1}^{\infty} (a_n + b_n) ;$$

hence, by definition of sum,

$$\sum_{n=1}^{\infty} a_n \leqq \sum_{n=1}^{\infty} (a_n + b_n) ,$$

so the assertion (0) is proved.

Similarly we have

$$\sum_{n=1}^{\infty} b_n \leqq \sum_{n=1}^{\infty} (a_n + b_n). \tag{0.1}$$

Adding (0) and (0.1) sidewise, we get

$$\sum_{n=1}^{\infty} a_n + \sum_{n=1}^{\infty} b_n \leqq \sum_{n=1}^{\infty} (a_n + b_n). \tag{1}$$

On the other hand we have

$$a_m \leqq \sum_{n=1}^{\infty} a_n, \qquad b_m \leqq \sum_{n=1}^{\infty} b_n;$$

hence, [A.2.6.],

$$a_m + b_m \leqq \sum_{n=1}^{\infty} a_n + \sum_{n=1}^{\infty} b_n,$$

this for all $m = 1, 2, \ldots$

Applying the definition of the sum, we get

$$\sum_{m=1}^{\infty} (a_m + b_m) \leqq \sum_{n=1}^{\infty} a_n + \sum_{n=1}^{\infty} b_n. \tag{2}$$

From (1) and (2) the theorem follows.

A.3.6. Theorem. If (T) is a tribe denumerably additive, then the "infinite" distributive law takes place:

$$b \cdot \sum_{n=1}^{\infty} a_n = \sum_{n=1}^{\infty} (b \cdot a_n) \quad \text{for all } a_n, b.$$

Proof. We have

$$b \cdot \sum_{n=1}^{\infty} a_n = b \cdot \sum_{n=1}^{\infty} a_n \cdot I = (\text{by [A.1.4.3.)]} = b \cdot \sum_{n=1}^{\infty} a_n \cdot (b + \operatorname{co} b)$$

$$= b \cdot \sum_{n=1}^{\infty} (a_n \cdot b + a_n \cdot \operatorname{co} b),$$

because the finite distributive law is valid in a tribe [A.1.6.].

Applying [A.3.5.], we get

$$b \cdot \sum_{n=1}^{\infty} a_n = b \cdot \left[\sum_{n=1}^{\infty} (a_n b) + \sum_{n=1}^{\infty} (a_n \operatorname{co} b) \right];$$

hence, by the finite distributive law,

$$b \cdot \sum_{n=1}^{\infty} a_n = b \cdot \sum_{n=1}^{\infty} (a_n b) + b \sum_{n=1}^{\infty} (a_n \cdot \operatorname{co} b). \tag{1}$$

Now $a_n \cdot \operatorname{co} b \leqq \operatorname{co} b$ for all n, [A.2.6.], hence, by virtue of the definition of infinite sum:

$$\sum_{n=1}^{\infty} (a_n \cdot \operatorname{co} b) \leqq \operatorname{co} b.$$

Hence

$$b \cdot \sum_{n=1}^{\infty} (a_n \cdot \operatorname{co} b) \leqq b \cdot \operatorname{co} b = 0 .$$

From (1) it follows

$$b \cdot \sum_{n=1}^{\infty} a_n = b \cdot \sum_{n=1}^{\infty} (a_n b) .$$

But $\sum_{n=1}^{\infty} (a_n \cdot b) \leqq b$, because for all n we have $a_n b \leqq b$. Hence

$$b \cdot \sum_{n=1}^{\infty} (a_n \cdot b) = \sum_{n=1}^{\infty} (a_n \cdot b) .$$

It follows that

$$b \cdot \sum_{n=1}^{\infty} a_n = \sum_{n=1}^{\infty} (a_n \cdot b) .$$

A.3.7. Theorem. Let R be a lattice, $a_n \in R$, $(n = 1, 2, \ldots)$, and let

$$b_n =_{df} \sum_{k=1}^{n} a_k .$$

Then the following are equivalent:

I. $\sum_{n=1}^{\infty} a_n$ exists in R.

II. $\sum_{k=1}^{\infty} b_k$ exists in R.

In the case of existence of $\sum_{n=1}^{\infty} a_n$ or of $\sum_{n=1}^{\infty} b_n$ we have:

$$\sum_{n=1}^{\infty} a_n = \sum_{n=1}^{\infty} b_n .$$

Proof. Let I and put

(0) $$b =_{df} \sum_{n=1}^{\infty} a_n .$$

By definition of the lattice-sum, we have

$$a_n \leqq b \quad \text{for all } n .$$

Hence, taking any $k \geqq 1$, we have

$$a_1 \leqq b, \quad a_2 \leqq b, \quad \ldots, \quad a_k \leqq b,$$

which gives, [A.2.6.] and by induction,

$$\sum_{n=1}^{k} a_n \leqq \sum_{n=1}^{k} b = b, \quad \text{i.e., by (0)},$$

(1) $$b_k \leqq b \quad \text{for all } k .$$

Now let $b_k \leqq b'$ for all k. We have

$$\sum_{n=1}^{k} a_n \leqq b' \quad \text{for all } k;$$

hence

$$a_k \leqq b' \quad \text{for all } k.$$

Hence, by definition of the sum $\sum_{n=1}^{\infty} a_n$, we have by (0),

$$b \leqq b'.$$

Thus we have proved that if $b_k \leqq b'$ for all k, then

(2) $$b \leqq b'.$$

If we combine (2) with (1), we get the existence of $\sum_{n=1}^{\infty} b_n$, and we get b as its value.

Now let II, and put

(3) $$c =_{df} \sum_{k=1}^{\infty} b_k.$$

We have $b_k \leqq c$ for all k; hence

$$\sum_{n=1}^{k} a_n \leqq c \quad \text{for all } k;$$

hence

(4) $$a_k \leqq c \quad \text{for all } k.$$

Now suppose, that $a_k \leqq c'$ for all k. Then we have

$$a_1 \leqq c', \quad a_2 \leqq c', \quad \ldots, \quad a_k \leqq c',$$

$$\sum_{n=1}^{k} a_n \leqq c' \quad \text{for all } k, \text{ i.e.}$$

$$b_k \leqq c' \quad \text{for all } k.$$

Hence, by (3), we have $c \leqq c'$.

Since, by (4), $a_k \leqq c$ for all k, and since we have the implication: "if $a_k \leqq c'$ for all k, then $c \leqq c'$," it follows that $\sum_{k=1}^{\infty} a_k$ exists and equals c. The theorem is established.

A.3.8. Theorem. If the tribe (T) is denumerably additive, then the following distributive law takes place for any $b, a_1, a_2, \ldots, a_n, \ldots \in T$:

$$\left(\prod_{n=1}^{\infty} a_n\right) + b = \prod_{n=1}^{\infty} (a_n + b).$$

The proof is based on the theorem [A.3.6.], and on the infinite de Morgan laws [A.1.5.].

A.3.9. Theorems similar to [A.3.6.] and [A.3.8.], but involving non-denumerable addition and multiplication can be proved in an anologous way, so that these more general theorems can be considered corollaries to the proofs of [A.3.6.] and [A.3.8.]. Especially the theorems are useful for completely additive tribes [A.3.2.].

A.4. The role of equality (12). Every mathematical theory possesses a *specific notion of equality of its elements,* denoted usually by the sign (=) and having the formal properties: reflexiveness, symmetry and transitivity. This notion, according to the case, may be introduced axiomatically, or defined in a suitable way. The operations performed on elements and the relations between them are *invariant with respect to this equality.* For instance, in the theory of tribes, we have: if

$$a = a', \quad b = b', \quad c = c',$$

then the sentence

$$a + b = c \quad \text{implies} \quad a' + b' = c';$$

so we are allowed, in any formula, to replace all or some elements by elements equal to them respectively.

For example, if we have $a \leqq b$, we can replace a and b by a' and b' respectively, whenever $a = a'$ and $b = b'$. We call (=) *the governing equality in the given theory.*

A.4.1. If E is a set of elements, and $a \in E$, $a = a'$, then $a' \in E$. The relations have an analogous property. If we have two kinds of elements $a, b, \ldots$ and A, B for which the governing equalities are ($\doteq$) and ($\doteqdot$) respectively, and if S is a correspondence between the elements of the first kind $a, b, \ldots$ and the elements $A, B, \ldots$ of the second kind, we should have:

"if $a \mathsf{S} A$ and $a \doteq a'$, $A \doteqdot A'$, then $a' \mathsf{S} A'$."

We say that S is ($\doteq$)-*invariant in its domain and* ($\doteqdot$)-*invariant in its range.*

To be very strict we should say "(=)-*set*", instead of "set". We also should say ($\doteq$), ($\doteqdot$)-*correspondence,* instead of "correspondence".

A.4.2. The governing equality induces the notion of *the unique element x satisfying the condition* $W(\cdot\, x\, \cdot)$, (1). The statement:

"*there exists a unique element x satisfying the condition* $W(\cdot\, x\, \cdot)$"

means:

"there exists an element x for which $W(\cdot\, x\, \cdot)$ holds true, and if we have $W(\cdot\, x'\, \cdot)$ and $W(\cdot\, x''\, \cdot)$, then $x' = x''$.

Similarly the statement:

"*the unique element x, satisfying the condition* $W(\cdot\, x\, \cdot)$, *has the property* $\varphi(\cdot\, x\, \cdot)$"

means:

"*there exists an element x, such that* $W(\cdot\, x\, \cdot)$ *and* $\varphi(\cdot\, x\, \cdot)$; and if $W(\cdot\, x'\, \cdot)$ and $W(\cdot\, x''\, \cdot)$, then $x' = x''$."

We should say "(=)-unique element" instead of "unique element". In addition to above the notion of equality conditions the cardinal number of a given set of elements. We should say "(=)-cardinal of a set", instead of "cardinal number of the set".

A.4.2a. Expl. The difference between the geometrical free vectors, sliding vectors and fixed vectors is conditioned by various kinds of equality. If we deal with free vectors, there is an infinity of geometrical vectors which are equal to one another, but the set of all vectors geometrically equal to a given vector $\vec{a}$, has the cardinal number $= 1$. If the vectors were "fixed vectors", the cardinal number of the above collection would have the power of continuum.

A.5. In many simple cases the above remarks are not very important, but the situation is different, if we need to introduce many kinds of equalities.

Usually we have in a theory a basic notion of equality, and we call other derived kinds of equality: *equivalences.*

A.5.1. Till now we did not pay attention to the logical difference between an ordering R and its domain.

Def. Now, since we shall have some more subtle reasonings, we shall agree to denote the ordering by (R) and its domain *by* R. Really these two notions are logically different, since R is a set and (R) a relation.

A.6. Homomorphism of tribes (12), (13). Let (T), (T') be two tribes, and let

$$x\,A\,x'$$

be a correspondence which carries the domain $\complement\, T$ onto a subset of $\complement\, T'$ in such a way that the operations and relations for somata of (T) go over to the corresponding operations and relations for somata of (T'). For instance: if $x+y=z$ in (T), $x\,A\,x'$, $y\,A\,y'$, then $x'+y'='\,z'$ in (T'); if $x\leqq y$ in (T), then $x'\leqq'\,y'$ in (T'). [The notions considered in (T') are provided with primes.] In these circumstances we say that A is a *homomorphism from* (T) *into* (T').

If A carries the domain $\complement(T)$ into the whole $\complement(T')$ we say "*homomorphism from* (T) *onto* (T')". The homomorphism should be invariant in $\complement\, A$ with respect to the equality $(=)$ governing in (T), and A should be $(=')$-invariant in $\mathrm{D}\, A$.

If the correspondence is $1 \rightarrow 1$, we call A *isomorphism.* To be very strict we should say "$(=)$, $(=')$-homomorphism" and even "operations and order — $(=)$, $(=')$-homomorphism from (T) into (or onto) (T')".

A.7. Genuine subtribes and supertribes. Let C be a collection of some elements with a relation of equality$(=^C)$ of elements, satisfying the usual conditions of identity. The subsets of C will be also termed collections and we shall speak of elements of C. The collection C will be a kind of substratum to what follows.

Now suppose we have two tribes (T'), (T'') such that T', T'' are subcollections of C, so we can write $T' \subseteqq C$, $T'' \subseteqq C$, and such that $T' \subseteqq T''$. These subcollections are invariant with respect to the equality $(=^C)$; i.e. if $a \in T'$, $a =^C b$, then $b \in T'$ and if $\alpha \in T''$, $\alpha =^C \beta$, then $\beta \in T''$.

Having that background, suppose that in the tribe (T') we have the governing equality $(=')$, and in (T'') the governing equality $(='')$, which may differ from $(=^C)$. We suppose that

$$\text{if} \quad a, b \in T', \quad a =^C b, \quad \text{then} \quad a =' b,$$

and if $\alpha, \beta \in T''$, $\alpha =^C \beta$, then $\alpha ='' \beta$, (but not necessarily conversely).

Under these circumstances, we define the following notion, which will be important in the sequel, **(13)**, **(14)**.

We say that (T') is a *finitely genuine subtribe of* (T'') *or* (T'') is a *finitely genuine supertribe of* (T') whenever the following conditions are satisfied for the somata $a, b, c, \ldots, O', I'$ of (T'):

1^0) $a +' b =' c$ is equivalent to $a +'' b ='' c$;

2^0) $a \cdot' b =' c$ is equivalent to $a \cdot'' b ='' c$;

3^0) $I' ='' I''$;

4^0) $O' ='' O''$.

(Single primes refer to (T') and double primes refer to (T'')).

A.7.1. One can prove that the above four conditions are independent, **(13)**, and that they imply the following rules for somata of (T'):

$$a =' \mathrm{co}' b \quad \text{is equivalent to} \quad a ='' \mathrm{co}'' b,$$

$$a -' b =' c \quad \text{is equivalent to} \quad a -'' b ='' c,$$

$$a \dotplus' b =' c \quad \text{is equivalent to} \quad a \dotplus'' b ='' c,$$

$$a \leqq' b \quad \text{is equivalent to} \quad a \leqq'' b.$$

A.7.2. Def. If in addition to 1^0), 2^0), 3^0) we have not only $T' \subseteqq T''$ but $T' \subseteqq'' T''$, we say that (T') *is a finitely genuine strict subtribe of* (T'') and (T'') *is a finitely genuine strict supertribe of* (T').

A.7.3. Expl. Take as substratum C the collection of all subsets of the interval $(0, 1\rangle =_{df} \{x \mid 0 < x \leqq 1\}$, and denote by $(=^C)$ the equality of these sets, (identity of sets, restricted to C). Now consider all finite unions of sets $\{x \mid a < x \leqq b\}$, $(0 \leqq a, b \leqq 1)$. If we order these collection by the relation of inclusion of sets, we get a tribe (T'), whose governing equality $(=')$ is just $(=^C)$; restricted to T'. Let us call the somata of (T) figures.

On the other hand consider all Lebesgue-measurable subsets of $(0, 1\rangle$ where the ordering is defined by

$$E \leqq'' F \cdot =_{df} \cdot \operatorname{meas}(E - F) = 0.$$

The governing equality will be

$$E ='' F \cdot =_{df} \cdot \operatorname{meas}[(E - F) + (F - E)] = 0.$$

Thus obtained ordering is also a tribe, which will be denoted by (T''). We shall see that (T') is a finitely genuine subtribe of (T'').

To prove that we shall show that the conditions 1^0), 2^0), 3^0), 4^0) in [A.7.] are satisfied. Consider the collection C and its elements, which are subsets of $(0, 1\rangle \cdot E ='' F$ means that E, F do not differ but by a set of measure 0. This is equivalent to the equality

$$E \cup p = F \cup q$$

where p, q are some sets of measure 0.

It follows that if for figures f, g, h we have $f \cup g =' h$ we have also $f \cup g ='' h$. If we have $f \cap g =' h$, then $f \cap g ='' h$. The zero O'' of (T'') is represented by any set of measure 0, and the unit I'' is represented by any set $(0, 1\rangle - p$ where meas $p = 0$. The zero of (T') is $\emptyset$ and its unit is $(0, 1\rangle$.

Consequently we have

$$O' ='' O'', \quad I' ='' I'',$$

and the relations $f +' g =' h$, $f \cdot' g =' h$ for figures imply respectively the relations

$$f +'' g ='' h, \quad f \cdot'' g ='' h.$$

On the other hand, if we have for figures p, q, r, $p +'' q ='' r$, then $p + q = r$, i.e. $p +' q =' r$, follows, because the only figure with measure 0 is the empty set of points. Similarly, if $p \cdot'' q ='' r$, we must have $p \cap q = r$, hence $p \cdot' q =' r$.

Thus we have proved that (T') is a finitely genuine subtribe of (T''). We do not have $T' \subseteqq'' T''$. Indeed this relation requires that T' be invariant with respect to the equality $(='')$. This however is not true, because a set of measure 0, different from O, does not belong to T'. We can change (T') into a tribe (T_1') which would be a strict subtribe of (T''). It suffices to consider all measurable sets E which do not differ from figures but by a set of measure 0. It is, we replace every figure f by the collection of all sets $f - p \cup q$, where p, q are sets of measure 0.

Then we shall have $T_1' \subseteqq'' T''$, so (T_1') is a strictly genuine subtribe of (T''). The tribe (T_1') is isomorphic to (T').

A.7.4. Remark. If (T') is a finitely genuine subtribe of (T''), but not strictly genuine, we can by a simple modification obtain a tribe (T_1') which is isomorphic to (T') and at the same time a finitely strict genuine subtribe of (T'').

Indeed, let $\mathfrak{B}$ be a correspondence which attaches to somata a of (T') all somata α which are $(='')$-equal to a, and belong to (T''). Then $\mathfrak{B}$ can be considered as the $(=')$, $(='')$-isomorphism $a\,\mathfrak{B}\,\alpha$. The tribe $\mathfrak{B}(T') =_{df} T_1'$ is the desired tribe.

A.7.5. Def. If in [A.7.] we replace the conditions 1^0, 2^0 by the following

$$1')\quad \sum_{n=1}^{\infty}{}' a_n =' b \quad \text{is equivalent to} \quad \sum_{n=1}^{\infty}{}'' a_n ='' b,$$

$$2')\quad \prod_{n=1}^{\infty}{}' a_n =' b \quad \text{is equivalent to} \quad \prod_{n=1}^{\infty}{}'' a_n ='' b,$$

for all somata of (T'), we call (T') *denumerably genuine subtribe of* (T''). Such a tribe is called *denumerably genuine strict subtribe of* (T''), whenever $(T') \subsetneqq'' (T'')$.

A.7.5a. Remark. We easily define tribes (T') which are *completely genuine subtribes of* (T''), and such ones which are *completely genuine strict subtribes of* (T'').

A.7.6. Remark. In this work we shall have the need of performing, in a reasonable way, denumerably infinite operations on a finitely additive tribe (T'). Now this can be done if (T') is a finitely genuine subtribe of a denumerably additive tribe (T''). We just use $\sum_{n=1}^{\infty}{}'' \mathfrak{B}(a_n)$ and we shall say that *we took infinite operations from* (T'').

A.9. Ideals in a tribe (12), (13). We know [A.2.9.], [A.2.10.], that a tribe (T) can be reorganized into a commutative ring with unit (Stone's ring). Consequently we can consider ideals in that ring, as we do in any ring in modern algebra.

Def. Therefore we shall call *ideal in* (T) any non empty subset J of T satisfying the following conditions:

1) if $a, b \in J$, then $a \dot{-} b \in J$ ($a \dot{-} b$ means algebraic subtraction in the ring),

2) if $a \in J$, $b \in T$, then $a \cdot b \in J$.

A.9.1. Theorem. For a tribe these conditions are equivalent to the followings ones:

1') if $a, b \in J$, then $a + b \in J$,

2') if $a \in J$, $b \leq a$, then $b \in J$.

Of course, since J is a set of somata, we must have: if $a \in J$, $a = a'$, then $a' \in J$.

We see that the set T is an *ideal in* (T) and also the set, composed of the single soma O, is an ideal. In general we have $O \in J$.

A.9.2. Def. An ideal J in (T) is called *denumerably additive* whenever the following happens:

If

$$a_1, a_2, \ldots, a_n, \ldots \in J,$$

then the sum $\sum_{n=1}^{\infty} a_n$ exists and is a soma belonging to J.

These ideals are important in the case where the tribe (T) is denumerably additive. An ordinary ideal is called sometimes *finitely additive,* to stress that it does not need to be denumerably additive.

A.9.3. Def. An ideal J induces a special ordering of somata of (T), defined by

$$a \leqq^J b \cdot =_{df} \cdot a - b \in J,$$

and also a special equality $(=^J)$ called *equality modulo* J, $a =^J b$, defined by $a \leqq^J b$, $b \leqq^J a$. The relation $a =^J b$ is equivalent to

$$a \dotplus b = (a - b) + (b - a) \in J.$$

A.9.3a. Theorem. The relation $a \leqq b$ implies $a \leqq^J b$. Indeed $a - b = O \in J$. It follows that the relation $a = b$ implies $a =^J b$.

A.9.4. To prove properties of this new equality we mention the following formulas, whose proof can be easily found by drawing two overlapping circles:

$$a = b + (a - b) - (b - a),$$

$$a = b - (b - a) + (a - b),$$

$$a + (a \dotplus b) = b + (a \dotplus b),$$

$$a - (a \dotplus b) = b - (a \dotplus b).$$

A.9.4a. The following are equivalent for somata of (T):

I) $a =^J b$;

II) there exist $p, q \in J$ such that

$$a = b + p - q;$$

III) there exist $p, q \in J$ such that

$$a = b - p + q;$$

IV) there exists $p \in J$ such that

$$a + p = b + p;$$

V) there exists $p \in J$ such that

$$a - p = b - p;$$

VI) there exist $p, p' \in J$, such that

$$a + p = b + p'.$$

A.9.5. Theorem. The correspondence $(\leqq^J)$ is an ordering, and $(=^J)$ obeys the laws of identity.

A.10. Theorem. The ordering $(\leqq^J)$ with domain T is a tribe.

Proof. First we shall prove that the ordering admits joins. Let $a, b \in T$. I say that $a + b$ is the $(\leqq^J)$-*union* of a and b, i.e.

$$a +^J b =^J a + b.$$

We have indeed

$$a \leqq a + b, \quad \text{hence, [A.9.3 a.]} \quad a \leqq^J a + b,$$

and

$$b \leqq a + b, \quad \text{hence} \quad b \leqq^J a + b.$$

Now suppose that

$$a \leqq^J c, \quad b \leqq^J c.$$

We have, [A.9.3.],

$$(1) \qquad a - c \in J \quad \text{and} \quad b - c \in J.$$

To prove that $a + b \leqq^J c$, it suffices to prove that $(a + b) - c \in J$, [A.9.3]. But this follows from

$$(a + b) \operatorname{co} c \in J,$$

and this from

$$a \cdot \operatorname{co} c + b \cdot \operatorname{co} c \in J.$$

This can be written as

$$(a - c) + (b - c) \in J$$

and it follows from (1), [A.9.1.].

A.10a. This being done, we shall show that the ordering $(\leqq^J)$ admits meets. We shall prove that $a \cdot b$ is the *meet* $a \cdot^J b$. We have

$$a \cdot b \leqq a; \quad \text{hence} \quad a \cdot b \leqq^J a;$$

and

$$a \cdot b \leqq b; \quad \text{hence} \quad a \cdot b \leqq^J b,$$

[A.9.3 a]. We must prove that if

$$c \leqq^J a, \quad c \leqq^J b \quad \text{then} \quad c \leqq^J a \cdot b.$$

Let $c \leqq^J a$, $c \leqq^J b$. Hence, [A.9.3.],

$$(2) \qquad c - a \in J, \quad c - b \in J.$$

The relation

$$c \leqq^J a \cdot b$$

is equivalent to $c - a \cdot b \in J$, [A.9.3. Def.]. This is equivalent to

$$c \cdot \operatorname{co}(a \cdot b) \in J;$$

hence to

$$c \operatorname{co} a + c \operatorname{co} b \in J;$$

hence to

$$(c - a) + (c - b) \in J.$$

This however is true because of (2).

Thus we have proved that

$$a +^J b =^J a + b,$$
$$a \cdot^J b =^J a \cdot b,$$

i.e. $(\leqq^J)$ is a lattice.

A.10b. The lattice has the zero O and the unit I.

Indeed we have $O \leqq a$, $a \leqq T$ for any soma a. Hence, [A.9.3a],

$$O \leqq^J a, \quad a \leqq^J I.$$

A.10c. We can show that the lattice is distributive. Indeed we have

$$(a + b) \cdot c = a \cdot c + b \cdot c.$$

Hence, by [A.9.3a.], we have:

$$(a + b) \cdot c =^J a \cdot c + b \cdot c.$$

Now, relying on [A.9.3.] and [A.10a.], we get

$$(a + b) \cdot c =^J (a +^J b) \cdot^J c$$

and

$$a \cdot c + b \cdot c =^J (a \cdot^J c) +^J (b \cdot^J c),$$

so finally:

$$(a +^J b) =^J (a \cdot^J c) +^J (b \cdot^J c).$$

A.10d. Having that we shall prove that the lattice is complementary.

We shall rely on the following remark:

In a distributive lattice with O and I there cannot exist two different notions of complement.

We have

$$a + \operatorname{co} a = I, \quad a \cdot \operatorname{co} a = O.$$

Hence relying on [A.9.3.], [A.10b.] and [A.10a], we get

$$a +^J \operatorname{co} a \cdot =^J I^J = I,$$
$$a \cdot^J \operatorname{co} a \cdot =^J O^J = O.$$

We shall prove that if $a \leqq^J b$, then $\operatorname{co} b \leqq^J \operatorname{co} a$. We have, by definition of $(\leqq^J)$:

(1) $$a - b \in J.$$

We have, by [A.2.3a.]:

$$\operatorname{co} b - \operatorname{co} a = \operatorname{co} b \cdot a = a - b \in J,$$

which gives, [A.9.3]

$$\operatorname{co} b \leqq^J \operatorname{co} a.$$

Since the soma $\operatorname{co} a$ abeys all laws needed for the $(=^J)$-complement, and since there are no two different notions of complement[1], in a distributive lattice with zero and unit, therefore "co" is the complement in the lattice $(T)_J$. Thus, we have proved that $(\leqq^J)$ is a tribe. We shall denote it by (T_J).

A.10.1. Theorem. We have also proved that

$$a +^J b =^J a + b, \quad a \cdot^J b =^J a \cdot b,$$

$$I^J =^J I, \quad O^J =^J O, \quad \operatorname{co}^J a =^J \operatorname{co} a.$$

Thus in performing operations in the new tribe (T_J), we can write these operations as if they were in (T), only the sign $(=)$ must be changed into $(=^J)$ and $(\leqq)$ into $(\leqq^J)$.

A.10.2. Theorem. The operations of addition, multiplication and complementation in the ordering $(\leqq^J)$, i.e. in (T_J) are $(=^J)$-invariant. The following proof may be useful.

Proof. We start with addition. Let $a +^J b =^J c$ and let $a =^J a_1$, $b =^J b_1$, $c =^J c_1$. We shall prove that

$$a_1 +^J b_1 =^J c_1.$$

We have by [A.9.4a.], for some $p, q \in J$,

$$a + p = a_1 + p, \quad b + q = b_1 + q.$$

[1] If the lattice is distributive and has zero and unit, then there cannot exist two different complement-correspondences $\operatorname{co} a$, $\operatorname{co}_1 a$. Indeed, let

(1) $$a + \operatorname{co} a = I, \quad a \cdot \operatorname{co} a = 0,$$

(2) $$a + \operatorname{co}_1 a = I, \quad a \cdot \operatorname{co}_1 a = O.$$

Multiplying the first equation in (1) by $\operatorname{co}_1 a$, we get

$$a \cdot \operatorname{co}_1 a + \operatorname{co} a \cdot \operatorname{co}_1 a = \operatorname{co}_1 a.$$

Hence

$$O + \operatorname{co} a \cdot \operatorname{co}_1 a = \operatorname{co}_1 a;$$

hence

$$\operatorname{co} a \cdot \operatorname{co}_1 a = \operatorname{co}_1 a.$$

It follows

(3) $$\operatorname{co}_1 a \leqq \operatorname{co} a.$$

In a similar way, starting with (2), and multiplying by $\operatorname{co} a$, we get

(4) $$\operatorname{co} a \leqq \operatorname{co}_1 a.$$

From (3) and (4) it follows

$$\operatorname{co} a = \operatorname{co}_1 a,$$

this for every a.

Hence

$$(a+b)+(p+q)=(a_1+b_1)+(p+q).$$

Hence, since $p+q\in J$, we get, by [A.9.4a.],

$$a+b=^J a_1+b_1.$$

Hence by [A.10.]

$$a+^J b=^J a_1+^J b_1.$$

We have $c=^J a+^J b=^J a_1+^J b_1$, $c=^J c_1$. Hence

$$a_1+^J b_1=^J c_1 \quad \text{q.e.d.}$$

A.10.2a. We are going prove the $(=^J)$-invariance of the product. Let $a\cdot^J b=^J c$ and $a=^J a_1$, $b=^J b_1$, $c=^J c_1$.

We have for some somata p, q of J, [A.9.4a],

$$a+p=a_1+p,$$

$$b+q=b_1+q.$$

It follows:

$$(a+p)\cdot(b+q)=(a_1+p)\cdot(b_1+q).$$

Hence

$$a\cdot b+(p\cdot b+a\cdot q+p\cdot q)=a_1\cdot b_1+(p\cdot b_1+a_1\cdot q+p\cdot q).$$

Put

$$r=_{df} p\,b+a\,q+p\,q,$$

$$s=_{df}=p\,b_1+a_1 q+p\,q.$$

We have

(1) $$a\cdot b+r=a_1\cdot b_1+s.$$

Since

$$p\,b\leqq p,\quad a\,q\leqq q,\quad p\,q\leqq q,$$

we get

$$r\leqq p+q\in J.$$

Similarly we prove that $s\in J$.

We have

$$a\cdot b+r=^J a\cdot b,\quad a_1\cdot b_1+s=^J a_1\cdot b_1,$$

which gives, by (1),

(2) $$a\cdot b=^J a_1\cdot b_1.$$

Since $a\cdot^J b=^J a\cdot b$, $a_1\cdot^J b_1=^J a_1\cdot b_1$, [A.10.1.], we get, by (2),

$$a\cdot^J b=^J a_1\cdot^J b_1.$$

We finish the proof as in the foregoing theorem.

A.10.2b. We are going to prove the $(=^J)$-invariance of the operation of complementation. Let

$$b =^J \mathrm{co}^J a, \quad a =^J a_1, \quad b =^J b_1;$$

we shall prove that

$$b_1 =^J \mathrm{co}^J a_1.$$

To do that, apply [A.9.4a]. We get

$$a_1 + p = a + p \quad \text{where} \quad p \in J.$$

Hence

$$\mathrm{co}(a_1 + p) = \mathrm{co}(a + p),$$

$$\mathrm{co}\, a_1 \cdot \mathrm{co}\, p = \mathrm{co}\, a \cdot \mathrm{co}\, p,$$

$$\mathrm{co}\, a_1 - p = \mathrm{co}\, a - p,$$

which gives, [A.9.4a. V)],

$$\mathrm{co}\, a_1 =^J \mathrm{co}\, a.$$

Applying [A.10.1.] we get

(3) $$\mathrm{co}^J a_1 =^J \mathrm{co}^J a.$$

Since

$$\mathrm{co}^J a =^J b \quad \text{and} \quad b =^J b_1,$$

we get

$$\mathrm{co}^J a =^J b_1,$$

hence, by (3),

$$\mathrm{co}^J a_1 =^J b_1 \quad \text{q.e.d.}$$

A.10.2c. Theorem. From what has been proved we can say that the operations $a \dot{+}^J b$ and $a \dot{-}^J b$ are $(=^J)$-equality invariant.

A.10.3. Theorem. We also have the following. The statement

I. $a \leqq^J b$ is equivalent to the statement;

II. there exists $p \in J$ such that

$$a + p \leqq b + p.$$

Proof. Suppose I. i.e. $a \leqq^J b$. It follows

$$a +^J b =^J b.$$

Hence [A.10.1.]

$$a + b =^J b.$$

Hence there exists $p \in J$ such that, [A.9.4a],

$$(a + b) + p = b + p.$$

Hence

$$(a + p) + (b + p) = b + p,$$

which implies

$$a + p \leqq b + p \quad \text{hence II. follows.}$$

Let II., i.e. let $p \in J$ with $a + p \leqq b + p$.

It follows

$$(a + p) + (b + p) = b + p.$$

Hence

$$(a + b) + p = b + p.$$

It follows

$$a + b =^J b;$$

hence

$$a +^J b =^J b$$

and then

$$a \leqq^J b \quad \text{i.e. I.}$$

A.10.4. Theorem. The relation $a \leqq^J b$ is $(=^J)$-invariant.

Proof. Let $a \leqq^J b$ and $a =^J a_1$, $b =^J b_1$; $a \leqq^J b$ is equivalent to

$$a +^J b =^J b,$$

because that statement is equivalent, [A.10.2.], to

$$a_1 +^J b_1 =^J b_1.$$

This however is equivalent to

$$a_1 \leqq^J b_1.$$

A.11. To perform $(\leqq^J)$-operations we use simply the somata of (T) and operations on (T) but changing the signs $(=)$, $(\leqq)$ into $(=^J)$ and $(\leqq^J)$ respectively. In any formula we are allowed to replace somata $a, b, \ldots$ by any others $a_1, b_1, \ldots$ which are $(=^J)$-equal to $a, b, \ldots$ The tribe $(\leqq^J)$ will be termed *tribe* (T) *taken modulo* J and it has already been denoted by (T_J).

A.11.1. If the ideal J is composed of the only soma O, the tribe (T_J) will coincide with (T), and if $J = T$, then the ordering (T_J) will be trivial, because all somata of $(T)_J$ will be $(=^J)$-equal, and equal to O.

A.12. Given an ideal J in (T), we have defined the tribe (T) module J by a change of ordering, and consequently by changing the equality $(=)$-governing in (T) into another kind of equality, viz. $(=^J)$. There is however an other way of using the ideal J in defining a new tribe, viz. by forming so-called *equivalence-classes*. We are going to give a sketch of the corresponding theory. Given a soma a of (T), let us consider the set

$$A =_{df} \{x \mid x \in T,\ x \dotplus a \in J\}.$$

We call A the *J-equivalence class corresponding to* a, and denote it by $[a]$. The following are logically equivalent:

I) $x \in A$, II) $x \dotplus a \in J$.

A.12.1. Theorem. If A is the equivalence class corresponding to a, $a \dotplus b \in J$, then A is also the equivalence class corresponding to b. The equivalence classes corresponding to somata of (T) are either disjoint or identical. The somata of $[a]$ are called *representatives of* $[a]$.

A.12.2. Def. We define for equivalence classes the operations of addition and multiplication, in the following way. If a is a representative of the equivalence class A and b a representative of the equivalence class B, we consider the equivalence class C corresponding to $a + b$, and call C the *J-join of A and B*:

$$[a + b] =_{df} \{x + y \mid x \in A, y \in B\}.$$

One can prove that C does not depend on the choice of representatives a, b of A, B respectively, so we can write

$$[a + b] =_{df} [a] + [b].$$

In a similar way we define *the meet $A \cdot B$ of the equivalence classes*, and for their *complementation*. We prove that just defined operations constitute a Boolean algebra, so we can define the corresponding ordering $(\leqq^J)$ on the set of all equivalence classes. We obtain a new tribe, called *quotient tribe* $(T)/J$.

A.12.3. Theorem. One can prove that the tribes $(T)_J$ and $(T)/J$ are order and operations isomorphic.

We shall confine ourselves to the above sketch, leaving the details to the care of the reader.

A.13. Till now we have considered general ideals in a tribe. Now we are going to study the situation where we have a denumerably additive ideal J in a denumerably additive tribe (T), [A.9.2.].

Theorem. First of all we shall prove that in this case the tribe $(T)_J$ i.e. (T) modulo J is also denumerably additive, i.e for any somata a_n, $(n = 1, 2, \ldots)$ the join $\sum_{n=1}^{\infty}{}^J a_n$ exists. We also shall prove that

$$\sum_{n=1}^{\infty}{}^J a_n =^J \sum_{n=1}^{\infty} a_n.$$

Thus the following are equivalent:

1) $a =^J \sum_{n=1}^{\infty}{}^J a_n$,

2) $a =^J \sum_{n=1}^{\infty} a_n$.

Proof. Put $a =_{df} \sum_{n=1} a_n$. We have $a_n \leqq a$ for every $n = 1, 2, \ldots$ Hence $a_n \leqq^J a$ for all n, [A.9.3.]. Let $a_n \leqq^J b$ for all n. We have

$$a_n - b \in J;$$

hence by denumerable additivity of J, we have

$$\sum_{n=1}^{\infty} (a_n - b) \in J.$$

It follows

$$\sum_{n=1}^{\infty} a_n \operatorname{co} b \in J;$$

hence by the denumerably distributive law, [A.3.6.],

$$\left(\sum_{n=1}^{\infty} a_n\right) \cdot \operatorname{co} b \in J, \quad \sum_{n=1}^{\infty} a_n - b \in J, \quad \sum_{n=1}^{\infty} a_n \leqq^J b;$$

hence

$$a \leqq^J b.$$

It follows that the sum $\sum_{n=1}^{\infty}{}^J a_n$ exists and $(=^J)$-equals to $\sum_{n=1}^{\infty} a_n$.

The remaining part of the thesis follows.

A.13.1. Theorem. If (T) is a denumerably additive tribe and J a denumerably additive ideal in (T), then all meets $\prod_{n=1}^{\infty}{}^J a_n$ exist, and we have

$$\prod_{n=1}^{\infty}{}^J a_n =^J \prod_{n=1}^{\infty} a_n.$$

Proof. We use the forgoing theorem [A.13.] and apply de Morgan rules.

A.13.2. If (T) is a denumerably additive tribe and J a denumerably additive ideal in it, then the denumerable joins are $(=^J)$-invariant.

Proof. Suppose that $a_n =^J b_n$, $(n = 1, 2, \ldots)$.

By [A.9.4a.] there exist $p_n \in J$ such that

$$a_n + p_n = b_n + p_n.$$

Hence

$$\sum_{n=1}^{\infty} (a_n + p_n) = \sum_{n=1}^{\infty} (b_n + p_n).$$

Hence, by [A.3.5.],

$$\sum_{n=1}^{\infty} a_n + \sum_{n=1}^{\infty} p_n = \sum_{n=1}^{\infty} b_n + \sum_{n=1}^{\infty} p_n.$$

The soma $\sum_{n=1}^{\infty} p_n$ belongs to J, since J is a denumerably additive ideal.

Hence, putting $p =_{df} \sum_{n=1}^{\infty} p_n$, we get

$$\sum_{n=1}^{\infty} a_n + p = \sum_{n=1}^{\infty} b_n + p,$$

which gives [A.9.4a.]

$$\sum_{n=1}^{\infty} a_n =^J \sum_{n=1}^{\infty} b_n.$$

Now, by [A.13.] we have

$$\sum_{n=1}^{\infty} a_n =^J \sum_{n=1}^{\infty}{}^J a_n, \quad \sum_{n=1}^{\infty} b_n =^J \sum_{n=1}^{\infty}{}^J b_n.$$

Consequently

$$\sum_{n=1}^{\infty}{}^J a_n =^J \sum_{n=1}^{\infty}{}^J b_n, \quad \text{q.e.d.}$$

A.13.3. Theorem. If (T) is a denumerably additive tribe and J a denumerably additive ideal in it, then the denumerable meets are $(=^J)$-invariant, in (T_J).

Proof. We rely on [A.13.2.], use de Morgan rules and apply the theorem [A.10.2b.] stating that the complementation in (T_J) is $(=^J)$-invariant.

A.14. We shall study the following situation. There is given a tribe (T) and an ideal J in it. In addition to that we have another tribe (S) which is a finitely genuine supertribe of (T). (See def. [A.7.].)

Theorem. Suppose that K is an ideal in (S) and that $J = T \frown K$.*) We shall prove that then the tribe (T_J) is a finitely genuine subtribe of (S_K).

A.14a. Proof. The somata of (T_J) are somata of (T), and conversely.

First we shall prove that, for somata $a, b, c, \ldots$ of (T_J), we have the equivalence of the statements

1) $a +^J b =^J c$,

2) $a +^K b =^K c$.

To prove that, consider 1). By [A.11.] it is equivalent to

$$a + b =^J c.$$

Hence, by [A.9.4a.], this is equivalent to the existence of a soma $p \in J$ such that

$$(1) \qquad (a + b) + p = c + p.$$

*) The somata of T and S are taken from a same collection C of elements [A.7.].

Now, since $J \subseteqq K$, we have $p \in K$. Hence (1) is equivalent to

$$a + b =^K c;$$

hence to

$$a +^K b =^K c.$$

Now we must also prove that the statements

1°) $a \cdot^J b =^J c$,

2°) $a \cdot^K b =^K c$

are equivalent for somata of (T_J).

1°) is equivalent to

$$a \cdot b =^J c.$$

This is equivalent to the existence of $p \in J$ with

$$a \cdot b + p = c + p.$$

Since $J \subseteqq K$, therefore $p \in K$.

Hence 1°) is equivalent to the existence of $p \in K$ with

$$(a \cdot b) + p = c + p.$$

This however, is equivalent to

$$a \cdot b =^K c,$$

hence to

$$a \cdot^K b =^K c.$$

A.14b. To finish the proof we must show that, [A.7.],

(1) $$I^{T_J} =^K I^{S_K},$$

and

(2) $$O^{T_J} =^K O^{S_K}.$$

First let us remind that the somata of (T) and those of (S) are taken from a suitable collection C of elements.

Let us also remark that if we consider the unit I^R of a tribe (R) there may be in general many elements of C which can be denoted by I^R.

Now let x be an element of C which can be denoted I^{T_J} in (T_J). The element x denotes also I^T in (T). Now since (S) is a finitely genuine supertribe of (T), x also denotes I^S in (S), hence also I^{S_K} in (S_k). It follows that

$$I^{T_J} =^{S_K} I^{S_K}.$$

Similarly, we prove that

$$O^{T_J} =^{S_K} O^{S_K}.$$

Thus we have proved that (S_K) is a finitely genuine supertribe of (T_J).

A.14.1. Theorem. If

1. (T) is a finitely additive tribe,
2. J is a finitely additive ideal in (T),
3. (S) is a finitely genuine supertribe of (T),
4. (S) is a denumerably additive tribe,

then there exists in (S) a denumerably additive ideal K, such that

$$J \subseteqq T \frown K.$$

Proof. Let $\mathfrak{B}$ be the isomorphism [A.7.4.] which attaches to somata a of (T) the corresponding somata A of (S):

$$a \smile^{\mathfrak{B}} A = \mathfrak{B}(a).$$

Denote by K the class of all somata P of (S) such that

$$P \leqq^S \sum_{n=1}^{\infty}{}^S \mathfrak{B}(a_n), \quad \text{where} \quad a_n \in J.$$

I say that K is a denumerably additive ideal in (S).

Let $A_1, A_2, \ldots, A_n, \ldots \in K$. Then there exist a_{ik}, $(i, k = 1, 2, \ldots)$, such that, [A.13.1.],

$$A_n \leqq^S \sum_{k=1}^{\infty} \mathfrak{B}(a_{n,k}), \qquad n = 1, 2, \ldots, \qquad \text{where } a_{nk} \in J$$

and where $(\leqq^S)$ is the ordering in (S). Hence, by denumerable additivity,

$$\sum_{n=1}^{\infty} A_n \leqq^S \sum_{n=1}^{\infty} \sum_{k=1}^{\infty} \mathfrak{B}(a_{n,k}) =^S \sum_{n,k=1}^{\infty} \mathfrak{B}(a_{n,k}).$$

Now the set of all $a_{n,k}$ is denumerable; hence the last union $\in K$, and then

$$\sum_{n=1}^{\infty} A_n \in K.$$

Now let $A \in K$ and let $M \leqq^S A$. We shall prove that $M \in K$.

We have for some $a_n \in J$, $(n = 1, 2, \ldots)$,

$$A =^S \sum_{n=1}^{\infty} \mathfrak{B}(a_n).$$

Hence

$$M =^S M A =^S \sum_{n=1}^{\infty} M \cdot \mathfrak{B}(a_n) \leqq^S \sum_{n=1}^{\infty} \mathfrak{B}(a_n).$$

It follows that $M =^S A \cdot M \in K$.

We have proved that K is a denumerably additive ideal in (S). We have $J \subseteqq K$.

Indeed let $a \in J$, then a is also a soma of (S), because S is a finitely genuine supertribe of (T). Hence $a =^S \mathfrak{B}(a)$: Hence $a \leqq^S \mathfrak{B}(a)$, so $a \in K$.

We have proved that $J \subseteqq K$.

Now since $J \subseteqq T$, we get

$$J \subseteqq T \frown K.$$

This completes the proof.

A.15. Measure (13)[1]. Given a tribe (T), let us make correspond to every its soma a a non negative number $\mu(a)$ satisfying the conditions:

1) equality invariance i.e. if $(=)$ is the equality governing in (T), $a = a'$, then $\mu(a) = \mu(a')$,

2) if $a \cdot b = O$, then

$$\mu(a + b) = \mu(a) + \mu(b).$$

Under these circumstances $\mu(\dot{a})$ is called *(finitely additive) measure on* (T).

A.15.1. Def. A measure $\mu(\dot{a})$ is said to be *strictly positive (definite, positive, effective)* whenever the supposition $\mu(a) = 0$ implies $a = O$.

A.15.2. Def. Let (T) be a tribe (which may be finitely additive only). Suppose that if $\{a_{\dot{n}}\}$ is an infinite sequence of disjoint somata of T we have:

$$\text{“if } \sum_{n=1}^{\infty} a_n \text{ exists, then } \mu\left(\sum_{n=1}^{\infty} a_n\right) = \sum_{n=1}^{\infty} \mu(a_n)\text{”.}$$

Then we shall say that $\mu(\dot{a})$ is *denumerably additive on* (T).

A.15.2a. Usually we consider denumerably additive measures in denumerably additive tribes.

A.15.3. Def. A measure $\mu(\dot{a})$ is said to be *trivial* whenever $\mu(I) = 0$. (This implies that $\mu(a) = 0$ for all a.)

The trivial measure is effective [A.15.1.] on the trivial tribe, where $[O = I]$, because $\mu(a) = 0$ implies $a = O$.

A.15.4. Theorem. If (T) is a tribe and $\mu(\dot{a})$ a measure on it, then

$$\mu(O) = 0, \quad [\text{where } O \text{ is the zero of } (T)].$$

Proof. We have

$$\mu(O + O) = \mu(O) + \mu(O),$$

hence

$$\mu(O) = 2\mu(O)$$

which completes the proof.

A.15.5. Theorem. If (T) is a tribe and $\mu(\dot{a})$ a measure on it, then if $a \leqq b$, we have $\mu(a) \leqq \mu(b)$.

---

[1] The measure on a tribe was introduced by the author and by Carathéodory simultaneously and independently from one another (15), (7).

Proof. We have [A.2.6.],

$$b = a + (b - a) \quad \text{and} \quad a \cdot (b - a) = O.$$

[A.2.3 a]. Hence

$$\mu(b) = \mu(a) + \mu(b - a),$$

[A.15.]. Since

$$\mu(b - a) \geqq 0,$$

it follows that

$$\mu(a) \leqq \mu(b) \quad \text{q.e.d.}$$

A.15.6. Theorem. We have for any two somata $a, b \in (T)$

$$\mu(a + b) \leqq \mu(a) + \mu(b).$$

Indeed,

$$a + b = a + (b - a),$$

[A.2.3 b.], and

$$a \cdot (b - a) = O.$$

Hence

$$\mu(a + b) \leqq \mu(a) + \mu(b - a);$$

hence, by [A.15.5.],

$$\mu(a + b) \leqq \mu(a) + \mu(b),$$

because $b - a \leqq b$.

The theorem is proved.

A.15.7. Theorem. If $\mu_1(\dot{a})$, $\mu_2(\dot{a})$ are measures on a tribe (T), then

$$\mu(\dot{a}) =_{df} \mu_1(\dot{a}) + \mu_2(\dot{a})$$

is also a measure on (T).

Proof. Let $a \cdot b = O$. We have

$$\mu(a + b) = \mu_1(a + b) + \mu_2(a + b) = [\mu_1(a) + \mu_1(b)] + [\mu_2(a) + \mu_2(b)]$$
$$= [\mu_1(a) + \mu_2(a)] + [\mu_1(b) + \mu_2(b)] = \mu(a) + \mu(b).$$

On the other hand we have

$$\mu_1(a) \geqq 0, \qquad \mu_2(a) \geqq 0,$$

which gives

$$\mu(a) = \mu_1(a) + \mu_2(a) \geqq 0.$$

Of course $\mu(\dot{a})$ is invariant with respect to the equality governing in (T).

A.15.8. Theorem. If $\mu_1(\dot{a})$, $\mu_2(\dot{a})$ are denumerably additive measures on a denumerably additive tribe (T), then

$$\mu(\dot{a}) =_{df} \mu_1(\dot{a}) + \mu_2(\dot{a})$$

is also a denumerably additive measure on (T).

Proof. Let $a_1, a_2, \ldots, a_n, \ldots$ be disjoint somata of (T). Put

$$a =_{df} a_1 + a_2 + \cdots + a_n + \cdots .$$

We have

$$\mu_1(a) = \mu_1(a_1) + \mu_1(a_2) + \cdots,$$

$$\mu_2(a) = \mu_2(a_1) + \mu_2(a_2) + \cdots,$$

because of denumerable additivity of μ_1 and μ_2.

It follows

$$\mu(a) = \sum_{n=1}^{\infty} \mu_1(a_n) + \sum_{n=1}^{\infty} \mu_2(a_n) = \sum_{n=1}^{\infty} [\mu_1(a_n) + \mu_2(a_n)] = \sum_{n=1}^{\infty} \mu(a_n).$$

In addition to that we have $\mu(a) \geqq 0$, since $\mu_1(a) \geqq 0$ and $\mu_2(a) \geqq 0$ for all $a \in T$.

Of course $\mu(a)$ is invariant with respect to the equality governing in (T).

The theorem is proved.

A.15.9. Expl. Let (T) be the tribe composed of all finite unions of sets $\{x \mid a < x \leqq b\}$ for $0 \leqq a, b \leqq 1$. The Lebesguean measure on them is a finitely additive measure, but not a denumerably additive measure.[1]

A.15.9a. Expl. Let (T) be the tribe of all Lebesgue-measurable subsets of the interval $(0, 1\rangle$. The Lebesguean measure is denumerably additive, but not effective, since there are not empty sets having the measure 0.

A.15.9b. Expl. If we denote by J the denumerably additive ideal, composed of all sets with measure 0, in [A.15.9a.], then the Lebesguean measure on the tribe T/J is denumerably additive and effective.

A.15.9c. For various examples concerning the measure see **(13)**.

A.15.10. Theorem. Let

1) $\mu_1(a), \mu_2(a), \ldots$ be an infinite sequence of finitely additive measures on the tribe (T).

2) Suppose $\sum_{n=1}^{\infty} \mu_n(a)$ be convergent for every $a \in T$. Put

$$\mu(a) =_{df} \sum_{n=1}^{\infty} \mu_n(a),$$

then $\mu(a)$ is also finitely additive.

Proof. Let

$$a \cdot b = O, \quad a, b \in T.$$

We have

$$\mu_n(a + b) = \mu_n(a) + \mu_n(b), \quad \mu_n(a) \geqq 0, \quad \mu_n(b) \geqq 0, \quad (n = 1, 2, \ldots).$$

[1] Because infinite sums in (T) may differ from infinite sums in the tribe of Lebesgue's-measurable sets.

Hence

$$\sum_{i=1}^{n} \mu_i(a+b) = \sum_{i=1}^{n} \mu_i(a) + \sum_{i=1}^{n} \mu_i(b).$$

Both the terms on the right tend to some limits. Those are $\mu(a)$, $\mu(b)$ respectively. It follows that

$$\lim_{n\to\infty} \sum_{i=1}^{n} \mu_i(a+b) = \mu(a) + \mu(b) \quad \text{i.e.}$$

$$\mu(a+b) = \mu(a) + \mu(b).$$

A.15.10.1. The property $\mu(a) \geqq 0$ for all $a \in (T)$ is obvious.

Theorem. Let us suppose that all the measures $\mu_n(a)$ be denumerably additive, and that $\sum_{n=1}^{\infty} \mu_n(a)$ converges for all a, then $\mu(a) =_{df} \sum_{n=1}^{\infty} \mu(a_n)$ is also denumerably additive.

Proof. Let $a_1, a_2, \ldots, a_r, \ldots$ be our infinite sequence of mutually disjoint somata $\in T$. Suppose $a =_{df} \sum_{n=1}^{\infty} a_n \in T$. We have by virtue of what already has been proved:

$$\mu(a_1 + \cdots + a_r) = \mu(a_1) + \cdots + \mu(a_r).$$

Since $a_1 + \cdots + a_r \subseteqq I$, and consequently, [A.15.5.],

$$\mu(a_1 + \cdots + a_r) \leqq \mu(I),$$

we have

$$\mu(a_1) + \cdots + \mu(a_r) \leqq \mu(I) \quad \text{for} \quad r = 2, 3, \ldots$$

and then

(1) $$\sum_{r=1}^{\infty} \mu(a_r)$$

converges.

We have

$$\mu(a_r) = \sum_{n=1}^{\infty} \mu_n(a_r).$$

Hence

$$\sum_{r=1}^{\infty} \mu(a_r) = \sum_{r=1}^{\infty} \sum_{n=1}^{\infty} \mu_n(a_r)$$

exists. The terms on the right side being non negative, we can permute the summations. Hence

$$\sum_{r=1}^{\infty} \mu(a_r) = \sum_{n=1}^{\infty} \sum_{r=1}^{\infty} \mu_n(a_r).$$

Since μ_n was supposed to be a denumerably additive measure, it follows that

$$\sum_{r=1}^{\infty} \mu_n(a_r) = \mu_n(a).$$

Hence

$$\sum_{r=1}^{\infty} \mu(a_r) = \sum_{n=1}^{\infty} \mu_n(a) = \mu(a) \quad \text{q.e.d.}$$

A.15.11. Theorem. If

$$a \leqq b, \quad a, b \in T,$$

then

$$\mu(b - a) = \mu(b) - \mu(a).$$

Proof. Straight forward.

A.15.12. Theorem. If

1. $a_1, a_2, \ldots, a_n, \ldots$ are somata of (T),
2. $a =_{df} \sum_{n=1}^{\infty} a_n \in T$,
3. μ is a measure on (T),

then we have

$$\sum_{n=1}^{\infty} \mu(a_n) \leqq \mu(a).$$

Proof. We have $a_1 + \cdots + a_n \leqq a$. Hence, by [A.15.5.]

$$\mu(a_1 + \cdots + a_n) \leqq \mu(a).$$

By additivity of μ:

$$\mu(a_1) + \cdots + \mu(a_n) \leqq \mu(a).$$

Since all $\mu(a_i) \geqq 0$, the theorem follows.

A.15.13. Theorem. Let (T) be a tribe and $\mu(\dot{a})$ a measure on it. Let J be the set of all somata $a \in T$ for which $\mu(a) = 0$.

Under these conditions J is an ideal in (T).

Proof. Let $a \in J$. We have $\mu(a) = 0$.

Take any $b \leqq a$. We have, [A.15.5.],

$$\mu(b) \leqq \mu(a) = 0, \quad \text{hence} \quad b \in J.$$

Suppose that $a, b \in J$. Then $\mu(a) = \mu(b) = 0$. We have, [A.15.6.],

$$\mu(a + b) \leqq \mu(a) + \mu(b) = 0 + 0 = 0;$$

hence $a + b \in J$. Thus we have proved that if $a, b \in J$, we have $a + b \in J$. These properties prove that J is an ideal in T (see [A.9.1.]).

Measure-topology on a tribe

A.16. Def. Let (G) be a tribe and $\mu(\dot{E})$ a (finitely) additive measure on it. We shall see that the measure induces a *metric-space-topology on* (G). It is constructed by means of the following notion of *"distance" between two somata*:

(1) $$|E, F|_\mu =_{df} \mu(E \dot{+} F) = \mu(E - F) + \mu(F - E),$$

where ($\dot{+}$) means the algebraic addition [A.2.9.]. The notion (1) is equality invariant, i.e., if $E = E_1$, $F = F_1$, then $|E, F|_\mu = |E_1, F_1|_\mu$, where (=) is the governing equality on $(\boldsymbol{G})$.

A.16.1. We shall have some theorems concerning that notion of distance. They are based on properties of the algebraic addition.

We mention the following ones: $E \dot{+} F \leqq E + F$, $E \dot{+} E = O$, the associative and commutative law for the algebraic addition[1].

A.16.2. We have $\mu(E) = |O, E|_\mu$.

A.17.1. $|E, F| = |F, E|$.

A.17.2. $|E, E| = 0$.

A.17.3. $|E, F| \leqq |E, G| + |G, F|$.

A.17.4. $\mu(E \dot{+} F) \leqq \mu(E) + \mu(F)$, $|E, F| \leqq \mu(E) + \mu(F)$.

A.17.5. $(E_1 + F_1) \dot{+} (E_2 + F_2) \leqq (E_1 \dot{+} E_2) + (F_1 \dot{+} F_2)$, which gives by induction:

A.17.6. 1. $(E_1 + \cdots + E_n) \dot{+} (F_1 + \cdots + F_n) \leqq (E_1 \dot{+} F_1) + (E_2 \dot{+} F_2) + \cdots + (E_n \dot{+} F_n)$ for $n \geqq 2$.

A.17.7. $E_1 F_1 \dot{+} E_2 F_2 \leqq (E_1 \dot{+} E_2) + (F_1 \dot{+} F_2)$.

A.17.8. $\operatorname{co} E_1 \dot{+} \operatorname{co} E_2 = E_1 \dot{+} E_2$.

A.17.9. $(E_1 - F_1) \dot{+} (E_2 - F_2) \leqq (E_1 \dot{+} E_2) + (F_1 \dot{+} F_2)$.

A.17.10. $(E_1 E_2 \ldots E_n) \dot{+} (F_1 F_2 \ldots F_n) \leqq (E_1 \dot{+} F_1) + \cdots + (E_n \dot{+} F_n)$.

Proof. By [A.17.8.], $(E_1 \ldots E_n) \dot{+} (F_1 \ldots F_n) = \operatorname{co}(E_1 \ldots E_n) \dot{+} \operatorname{co}(F_1 \ldots F_n) = (\operatorname{co} E_1 + \cdots + \operatorname{co} E_n) \dot{+} (\operatorname{co} F_1 + \cdots + \operatorname{co} F_n)$. Applying [A.17.6.] and afterwards [A.17.8.], we prove the statement. The above lemmas yield proofs for the following properties of the distance of somata of $(\boldsymbol{G})$.

A.17.11. $|E_1 + F_1, E_2 + F_2| \leqq |E_1, E_2| + |F_1, F_2|$.

Proof by [A.17.5.].

A.17.12. $|E_1 + \cdots + E_n, F_1 + \cdots + F_n| \leqq |E_1, F_1| + \cdots + |E_n, F_n|$.

Proof by [A.17.6.].

A.17.13. $|E_1 F_1, E_2 F_2| \leqq |E_1, E_2| + |F_1, F_2|$.

Proof by [A.17.7.].

A.17.14. $|E_1 E_2 \ldots E_n, F_1 F_2 \ldots F_n| \leqq |E_1 F_1| + \cdots + |E_n F_n|$.

Proof by [A.17.10.].

A.17.15. $|E, F| = |\operatorname{co} E, \operatorname{co} F|$.

Proof by [A.17.8.].

---

[1] We shall optionally write $|E, F|$ instead of $|E, F|_\mu$, when no ambiguity can result.

A.17.16. $|E_1 - F_1, E_2 - F_2| \leqq |E_1, E_2| + |F_1, F_2|$.

Proof by [A.17.9.].

A.17.17. $|EH, FH| \leqq |E, F|$.

Proof. By [A.17.13.] we have $|EH, FH| \leqq |E, F| + |H, H|$, from which, by [A.17.2.], the statement follows.

A.17.18. $|E + F, H| \leqq |E, H| + |F, H|$.

Proof. $|E + F, H| = |E + F, H + H| \leqq |E, H| + |F, H|$, by [A.17.11.].

A.18. If $\eta > 0$, $|E, F| < \eta$, then $|\mu E - \mu F| < 2\eta$.

Proof. We have

$$(1) \qquad \begin{aligned} \mu(E) &= \mu(E - F) + \mu(E \cdot F), \\ \mu(F) &= \mu(F - E) + \mu(E \cdot F). \end{aligned}$$

Since $\mu(E - F) + \mu(F - E) < \eta$, we have $\mu(E - F) < \eta$ and $\mu(F - E) < \eta$. Hence for some θ with $|\theta| < 1$ and some θ' with $|\theta'| < 1$ we have $\mu(E - F) = \eta\,\theta$, $\mu(F - E) = \eta\,\theta'$. It follows $\mu(E) - \mu(F) = \eta(\theta - \theta')$; hence $|\mu(E) - \mu(F)| = \eta \cdot |\theta - \theta'| < 2\eta$.

A.18.1. If $|E, F|_\mu < \eta$, then $\mu(E) - \eta \leqq \mu(E \cdot F) \leqq \mu(E)$, $\mu(F) - \eta \leqq \mu(E \cdot F) \leqq \mu(F)$.

Proof. We have $E = EF + (E - F)$, where both terms on the right are disjoint somata. Hence

$$(1) \qquad EF = E - (E - F),$$

which gives, as $E - F \leqq E$, $\mu(E \cdot F) = \mu(E) - \mu(E - F)$. From the hypothesis it follows that $\mu(E - F) < \eta$; hence, from (1) we get $\mu(EF) \geqq \mu(E) - \eta$. The proof of the second thesis is similar.

A.18.2. If $\eta > 0$, $|E_1, F_1| \leqq \eta$, $|E_2, F_2| \leqq \eta$, $E_1 \cdot E_2 = O$, then $\mu(F_1 \cdot F_2) \leqq 2\eta$.

Proof. By [A.17.13.] we have $|E_1 \cdot E_2, F_1 \cdot F_2| \leqq |E_1, F_1| + |E_2, F_2| \leqq 2\eta$. Hence $|O, F_1 F_2| \leqq 2\eta$. The thesis follows by [A.16.2.].

A.18.3. If $E \cdot F = O$, $\eta > 0$, $|E, E'| \leqq \eta$, $|F, F'| \leqq \eta$, then $|E, E' - F'| \leqq 3\eta$, $|F, F' - E'| \leqq 3\eta$, and $E' - F'$, $F' - E'$ are disjoint.

Proof. We have $|E, E' - F'| = |E - O, E' - E'F'| \leqq |E, E'| + |O, E' \cdot F'|$, by [A.17.16.]. By hypothesis, [A.16.2.] and [A.18.2.] it follows that $|E, E' - F'| \leqq 3\eta$. Similarly we get the inequality $|F, F' - E'| \leqq 3\eta$. The second thesis follows from $(E' - F')(F' - E') = E' \operatorname{co} F' \cdot F' \operatorname{co} E' = O$.

A.18.4. Theorem. For somata of G the following are equivalent.

I. μ is an effective measure on (G).

II. $|E, F|_\mu = 0$ implies $E =^G F$.

Proof. Suppose II. be not true. Then there exist E, F such that

(1) $$|E, F| = 0 \quad \text{and} \quad E \neq^G F.$$

Hence

(2) $$\mu(E - F) = \mu(F - E) = 0.$$

I say that either $E - F \neq O$ of $F - E \neq O$.

Indeed, if not, we would have $E - F = O$, $F - E = O$, and then $E \cdot \text{co}F = O$, $F \,\text{co}E = O$. Since $E = E\,F + E\,\text{co}F$, $F = F\,E + F\,\text{co}E$, we would have $E = E\,F$, $F = F\,E$, and then $E = F$, which has been excluded in (1).

If $E - F \neq O$ we get, by (2), $\mu(E - F) = 0$, and if $F - E \neq O$, we get, by (2), $\mu(F - E) = 0$, so μ is not effective. The above arguments show that I implies II. To prove that II implies I, suppose I be not true. There exists $E \neq^G O$, such that $\mu(E) = 0$. Hence, [A.16.2.], $|O, E|_\mu = 0$. Hence, by II, $O =^G E$, which is a contradiction. The theorem is proved.

A.18.5. Suppose that μ is effective. We have the following: $|E, F|_\mu = |F, E|_\mu$, $|E, F|_\mu \leqq |E, G|_\mu + |G, F|_\mu$, and $|E, F|_\mu = 0$ is equivalent to $E = F$. This all by [A.17.1.], [A.17.3.], [A.18.4.], and we see that the notion of distance is (G)-equality invariant.

Hence the notion of distance of somata organizes the tribe (G) into a *Hausdorff-metric space,* hence into a *topology* (2), (3), (4), and especially (15).

A.19. Def. Now let us drop the hypothesis that μ is an effective measure on (G). Introduce for somata of (G) the following new notion of equality $E =^\mu F$, defined by $|E, F|_\mu = 0$.

A.19.1. This notion is invariant with respect to the equality governing in (G). It satisfies the usual axioms of identity. Indeed, we have, by [A.17.2.], $E =^\mu E$. If $E =^\mu F$, then $F =^\mu E$, [A.17.1.]. If $E =^\mu F$ and $F =^\mu G$, then $E =^\mu G$.

Proof. Suppose that $E =^\mu F$ and $F =^\mu G$. We have $\mu(E \dotplus F) = 0$, $\mu(F \dotplus G) = 0$. Now $E \dotplus G = E \dotplus F \dotplus F \dotplus G = (E \dotplus F) \dotplus (F \dotplus G) \leqq \leqq (E \dotplus F) + (F \dotplus G)$. Hence $\mu(E \dotplus G) \leqq \mu(E \dotplus F) + \mu(F \dotplus G) = 0$, so $E =^\mu G$.

A.19.2. The notion of distance $|E, F|_\mu$ is invariant with respect to the μ-equality. Indeed, suppose that $E =^\mu E'$, $F =^\mu F'$. We have $\mu(E' \dotplus F') = \mu[(E \dotplus E') \dotplus (F \dotplus F') \dotplus (E \dotplus F)] \leqq \mu(E \dotplus E') + + \mu(F \dotplus F') + \mu(E \dotplus F)$; hence $\mu(E' \dotplus F') \leqq \mu(E \dotplus F)$. Similarly we get $\mu(E' \dotplus F') \leqq \mu(E' \dotplus F')$, which completes the proof.

A.19.3. The above shows that the notion of distance $|E, F|_\mu$ organizes $(\boldsymbol{G})$ into a Hausdorff-metric topology, in which however not $(=)$ is the governing equality, but $(=^\mu)$.

Indeed the relation $|E, F|_\mu = 0$ is equivalent to $E =^\mu F$.

A.19.4. The notion of measure μ is invariant with respect to $=^\mu$, i.e. if $E =^\mu F$, then $\mu(E) = \mu(F)$. Indeed, we have $\mu(E) = |E, O|_\mu$, $\mu(F) = |F, O|_\mu$. Since the notion of distance is $=^\mu$-invariant, we have $|E, O|_\mu = |F, O|_\mu$, hence $\mu(E) = \mu(F)$.

A.20. The set J of all somata E with $\mu(E) = 0$ is an ideal in the tribe $(\boldsymbol{G})$. If $(\boldsymbol{G})$ is denumerably additive, so is the ideal J, [A.9.2.].

Proof. If $E_1, E_2, \ldots, E_n, \ldots \in J$, then $\mu(E_1 + E_2 + \cdots) = 0$, and then $E_1 + E_2 + \cdots \in J$. On the other hand we have: if $E \in J$ and $E' \leqq E$, then $E' \in J$. Indeed $\mu(E') \leqq \mu(E) = 0$.

A.20.1. The equality $E =^\mu F$, defined in [A.19.] coincides with the equality modulo J in the tribe $(\boldsymbol{G})$, i.e. with $E =^J F$.

We can define $E =^J F$ as $E \dot{-} F \in J$, where $(\dot{-})$ is the algebraic subtraction. Since the subtraction coincides with the addition, $E \dot{-} F \in J$ can be written as $E \dot{+} F \in J$ i.e. $\mu(E \dot{+} F) = 0$. We can apply the general theory of ideals in a tribe, [A.9.] bis [A.14.1.].

A.20.2. Let $(\boldsymbol{G})$ be a denumerably additive tribe. The relation $E \leqq^\mu F$ is defined by $\mu(E - F) = 0$. It organizes $\boldsymbol{G}$ into a denumerably additive Boolean lattice $(\boldsymbol{G}_J)$ with $E =^\mu F$, defined in [A.19.], as the governing equality. $(\boldsymbol{G}_J)$ is just the tribe $(\boldsymbol{G})$, taken module the ideal J of all somata of $(\boldsymbol{G})$ whose μ-measure equals 0. The measure μ is $(=^\mu)$-invariant, denumerably additive and effective on $(\boldsymbol{G}_J)$. The notion of distance $|E, F|_\mu$ is $=^\mu$-invariant. It organizes $(\boldsymbol{G}_J)$ into a Hausdorff-metric space. The author has proved in (16) that this space is complete[1], but it may be non-separable, (17).

A.21. Def. The notion of distance induces in $(\boldsymbol{G})$ the notion of μ limits of an infinite sequence of somata of $(\boldsymbol{G})$. If $E, E_1, E_2, \ldots, E_n, \ldots \in (\boldsymbol{G})$, we say that the sequence E_n *μ-tends* to E, whenever $\lim |E_n, E|_\mu = \lim \mu(E_n \dot{+} E) = 0$. The limit, if it exists, is $(=^\mu)$-unique. We write $E_n \to^\mu E$ or $\lim\limits_{\mu} E_n =^\mu E$.

A.21.1. The notion of limit is $=^\mu$-invariant.

A.21.2. Using methods of the theory of metric-spaces and properties, [A.17.1.]—[A.17.16.], we prove the following: If $E_n \to^\mu E$, $F_n \to^\mu F$, then $\operatorname{co} E_n \to^\mu \operatorname{co} E$; $E_n + F_n \to^\mu E + F$; $E_n \cdot F_n \to^\mu E \cdot F$; $E_n - F_n \to^\mu E - F$ and $E_n \dot{+} F_n \to^\mu E \dot{+} F$.

A.21.3. If $E_n \to^\mu E$, $F_n \to^\mu F$, then $|E_n, F_n|_\mu \to |E, F|_\mu$.

---

[1] The proof will be given in A I.

Proof. By [A.17.3.] we have $0 \leqq |E_n, F_n| \leqq |E_n, E| + |E, F| + |F, F_n|$. Hence, since $|E_n, E| \to 0$, $|F_n, F| \to 0$, [Def. A.21.], we get $\overline{\lim}\,|E_n, F_n| \leqq |E, F|$. We also have $|E, F| \leqq |E, E_n| + |E_n, F_n| + |F_n, F|$; hence $|E, F| \leqq \underline{\lim}\,|E_n, F_n|$. Consequently $|E, F| \leqq \underline{\lim}\,|E_n, F_n| \leqq \overline{\lim}\,|E_n, F_n| \leqq |E, F|$, which completes the proof.

A.21.4. If $E_n \to^{\mu} E$, then $\mu(E) \to \mu(E)$.

Proof. We have $E_n \to^{\mu} E$, $O \to^{\mu} O$, hence by [A.21.3.], $\lim |E_n, O| = |E, O|$ i.e. $\lim \mu\, E_n = \mu\, E$.

Chapter A I

Special theorems on Boolean Lattices

A I.1. In chapter [A.] we have discussed basic properties of tribes. They will be applied in this book. The present chapter [A I.] gives some important auxiliary theorems, especially theorems on extension of tribes and measure. There will be several items: The Wecken's theorem on complete additivity of tribes, Fréchet's condition for denumerable additivity of measure, Borelian extension of tribes, an extension theorem on tribes whose somata are sets, classical measure extension devices, a measure extension device by the author and comparison of measure extension device by Lebesgue, by Caratheodory and by the author, measure topology and somatic topology on a tribe.

Thus the chapter A I. will deal with special theorems on general tribes.

Section 1 of A I.

A I.1.1. Wecken's theorem (25). If

1. (T) is a denumerably additive tribe,
2. (T) admits a denumerably additive measure $\mu(\dot{a})$,
3. this measure $\mu(\dot{a})$ is effective, i.e. if $\mu(a) = 0$, then $a = O$, (the null-element of (T)),

then

1) (T) is completely additive [A.3.2.];

2) if E is any non empty set of somata of (T), then there exists a sequence $a_1, a_2, \ldots, a_n, \ldots$ of elements of E, such that

$$\sum_{a \in E} a = \sum_{n=1}^{\infty} a_n .$$

A I.1.2. Proof (by WECKEN). Since (T) is denumerably additive, every denumerable union of somata of (T) exists. Denote by F the class of all somata b of (T) which can be represented as a sum

$$\sum_{n=1}^{\infty} a_n \quad \text{where} \quad a_n \in E, \qquad (n = 1, 2, \ldots).$$

Since the measure is uniformly bounded, we can put

(1) $$\beta =_{df} \sup_{b \in F} \mu(b).$$

For every natural number n let us find an element $b'_n \in F$, such that

(2) $$\beta - \frac{1}{n} < \mu(b'_n) \leqq \beta.$$

We have for some somata $a_{n,k} \in E$:

(3) $$b'_n = \sum_{k=1}^{\infty} a_{n,k}.$$

Put

(4) $$p =_{df} \sum_{n=1}^{\infty} b'_n = \sum_{n,k} a_{nk}.$$

A I.1.3. We have

(5) $$p \in F,$$

because the union (4) is denumerable.

From (1) it follows

(6) $$\mu(p) \leqq \beta.$$

Since from (4) we have

$$b'_n \leqq p,$$

it follows

$$\mu(b'_n) \leqq \mu(p).$$

But, by (1) and (2) we get

$$\lim_{n \to \infty} \mu(b'_n) = \beta;$$

hence

(7) $$\beta \leqq \mu(p).$$

On the other hand, by (1), we get

$$\mu(p) \leqq \beta,$$

which, together with (7), gives

(8) $$\beta = \mu(p).$$

A I.1.4. Having this, we are going to prove that the infinite lattice-join $\sum_{a \in E}$ exists.

To do that, we will prove the following two items:

1) if $a \in E$, then $a \leqq p$,

2) if, for every $a \in E$, we have $a \leqq p'$, then $p \leqq p'$.

A I.1.5. Item 1). Suppose that a soma

(9) $$a \in E \quad \text{is not included in } p.$$

Put

$$c =_{df} p + a.$$

We have $c \neq p$, for if not, we would have $c = p$, and then $p = p + a$. Hence $a \leqq p$, which contradicts (9). Thus $p \subset c$.

It follows

$$\mu(c) = \mu(p) + \mu(c - p).$$

Since $c \neq p$, we have $\mu(c - p) > 0$, because the measure μ is effective.

Consequently $\mu(c) > \mu(p)$, and then, by (8),

(10) $$\mu(c) > \beta.$$

This however is impossible, because, by (4),

$$c = p + a = \sum_{n,k} a_{n,k} + a \in F,$$

and hence

$$\mu(c) \leqq \beta.$$

This however contradicts (10).

Thus we have proved that every soma $a \in E$ must be contained in p. Thus item 1) is proved.

A I.1.6. Prove of item 2), i.e. if for every $a \in E$ we have $a \leqq p'$, then $p \leqq p'$.

If for every $a \in E$ we have $a \leqq p'$, we also have $\sum_{n,k} a_{n,k} \leqq p'$. Consequently, by (4),

$$p \leqq p'.$$

Thus the item 2) is also proved.

A I.1.7. We have proved that the lattice union $\sum_{a \in E} a$ exists and equals p. In addition to that we have proved that

$$\sum_{a \in E} a = \sum_{n,k} a_{n,k} \quad \text{where} \quad a_{n,k} \in E.$$

The theorem is established.

Section 2 of A I.

A I.2. Relative conditions for denumerability of measure. The problem is this: There is given a finitely additive tribe (A) with finitely additive measure $\mu(\dot{a})$. We like to have a condition which would give the denumerable additivity of the measure. However we like to have infinite unions of somata of (A) taken from a denumerably additive supertribe (A') of (A), [A.7.6.]. The most known condition is Fréchet's condition, (27), given below, but there are also some other ones, which are useful.

A I.2.1. Theorem. If

1) (A) is a non trivial tribe,

2) $\mu(\dot{a})$ is a (finitely) additive, non negative, non trivial (finite) measure on (A),

3) (A') is a denumerably additive tribe,

4) (A') is a finitely genuine supertribe of (A), [A.7.],

then the following six conditions are equivalent:

I. If

$$a_1 \geqq a_2 \geqq \cdots \geqq a_n \geqq \cdots, \qquad a_n \in A,$$

$$\prod_{n=1}^{\infty}{}' a_n =' O',$$

then

$$\lim_{n\to\infty} \mu(a_n) = 0,$$

(*Fréchet's condition*).

II. If

$$a_1 \leqq a_2 \leqq \cdots \leqq a_n \leqq \cdots, \qquad a_n \in A,$$

$$\sum_{n=1}^{\infty}{}' a_n =' a, \qquad a \in A,$$

then

$$\mu(a) \leqq \lim_{n\to\infty} \mu(a_n).$$

III. If

$$a_1, a_2, \ldots, a_n, \ldots \in A, \qquad a \in A,$$

$$a =' \sum_{n=1}^{\infty}{}' a_n,$$

then

$$\mu(a) \leqq \sum_{n=1}^{\infty} \mu(a_n).$$

IV. If

$$a, a_1, a_2, \ldots, a_n, \ldots \in A$$

with a_i disjoint,

$$a \leqq' \sum_{n=1}^{\infty}{}' a_n,$$

then

$$\mu(a) \leqq \sum_{n=1}^{\infty} \mu(a_n),$$

(*Nikodým's condition* (15)).

V. If

$$a_1, a_2, \ldots, a_n, \ldots \in A,$$

a_i are disjoint,

$$a =' \sum_{n=1}^{\infty}{}' a_n, \quad a \in A,$$

then

$$\mu(a) = \sum_{n=1}^{\infty} \mu(a_n).$$

VI. If

$$a_1, a_2, \ldots, a_n, \ldots \in A,$$

a_i are disjoint,

$$I' =' \sum_{n=1}^{\infty}{}' a_n,$$

then

$$\mu(I) = \sum_{n=1}^{\infty} \mu(a_n).$$

A I.2.2. Proof. We easily prove that V, VI are equivalent. Indeed V implies VI.

Now let VI.

Take $a_1, a_2, \ldots, a_n, \ldots$ disjoint, such that

$$a =' \sum_{n=1}^{\infty}{}' a_n, \quad a \in A, \quad \text{and} \quad a_n \in A.$$

Put

$$a_0 =_{df} \mathrm{co}' a =' \mathrm{co}\, a;$$

we have

$$a_0 \in A.$$

We have

$$a_0 +' \sum_{n=1}^{\infty}{}' a_n =' a +' \mathrm{co}\, a =' I',$$

$\sum_{n=0}^{\infty}{}' a_n =' I$, where $a_0, a_1, \ldots$ are disjoint.

From VI it follows:

$$\mu(I) = \sum_{n=1}^{\infty} \mu(a_n) + \mu(a_0).$$

Hence

$$\mu(I) - \mu(a_0) = \sum_{n=1}^{\infty} \mu(a_n),$$

$$\mu(\mathrm{co}\, a_0) = \mu(a) = \sum_{n=1}^{\infty} \mu(a_n),$$

i.e. V holds true.

Now we shall show that IV and V are equivalent.

A I.2.3. Let IV. Take $a_1, a_2, \ldots, a_n, \ldots$ disjoint and $\in A$, with

$$a =' \sum_n^{\infty}{}' a_n, \quad \text{where} \quad a \in A.$$

We have

$$a \leqq' \sum_n{}' a_n.$$

Hence, by IV,

$$\mu(a) \leqq \sum_n \mu(a_n). \tag{1}$$

Now we have

$$\sum_{k=1}^{n} a_k \leqq' a,$$

and then

$$\sum_{k=1}^{n} a_k \leqq a.$$

Hence

$$\mu\left(\sum_{k=1}^{n} a_k\right) \leqq \mu(a).$$

Since the somata a_k are disjoint, it follows

$$\sum_{k=1}^{n} \mu(a_k) \leqq \mu(a).$$

Hence

$$\sum_{k=1}^{\infty} \mu(a_k) \leqq \mu(a). \tag{2}$$

From (1) and (2) it follows

$$\mu(a) = \sum_{k=1}^{\infty} \mu(a_k),$$

i.e. IV implies V.

A I.2.3a. Now let V be true. Take $a_1, a_2, \ldots, a_n, \ldots \in A$, $a \in A$, with a_n disjoint and with $a \leqq' \sum_{n=1}^{\infty}{}' a_n$. We have

$$a =' a \cdot' \sum_{n=1}^{\infty}{}' a_n.$$

Hence

$$a =' \sum_{n=1}^{\infty}{}' a \cdot a_n,$$

where $a \cdot a_n \in A$ and are mutually disjoint.

Hence, by V:

$$\mu(a) = \sum_{n=1}^{\infty} \mu(a\, a_n).$$

Since

$$\mu(a\, a_n) \leqq \mu(a_n),$$

it follows

$$\mu(a) \leqq \sum_{n=1}^{\infty} \mu(a_n).$$

This shows that V implies IV.

So we have proved that IV is equivalent to V.

A I.2.4. Now we shall see that III follows from V. Let V. Take $a_1, a_2, \ldots, a_n, \ldots, a \in A$ with

$$a =' \sum_{n=1}^{\infty}{}' a_n.$$

Put

$$b_1 =_{df} a_1, \quad b_2 =_{df} a_2 - a_1, \quad b_3 =_{df} a_3 - (a_1 + a_2), \quad \ldots,$$
$$b_n = a_n - (a_1 + \cdots + a_{n-1}) \quad \text{for} \quad n > 1.$$

We know that for every $n = 1, 2, \ldots$, [A.2.6.3.],

$$\sum_{k=1}^{n} b_k = \sum_{k=1}^{n} a_k,$$

and that $b_n \in A$. It follows that

$$\sum_{k=1}^{n}{}' b_k =' \sum_{k=1}^{n}{}' a_k,$$

and, by [A.2.6.5.], [A.3.7.], that

$$\sum_{k=1}^{\infty}{}' b_k =' \sum_{k=1}^{\infty}{}' a_k;$$

hence

$$a =' \sum_{k=1}^{\infty}{}' b_k.$$

Since, [A.2.6.3.], b_i are disjoint, it follows from V that

$$\mu(a) = \sum_{k=1}^{\infty} \mu(b_k).$$

But $b_k \leqq a_k$; hence $\mu(b_k) \leqq \mu(a_k)$, and then

$$\mu(a) \leqq \sum_{k=1}^{\infty} \mu(a_k),$$

i.e. III follows from V.

A I.2.4a. We shall prove that III implies V. In the hypothesis of III we have admitted that the somata a_n are any, but only satisfying the condition:

$$a \in A, \quad a_n \in A, \quad a =' \sum_{n=1}^{\infty}{}' a_n.$$

Hence we may admit that a_n are disjoint. By III we get

$$\mu(a) \leqq \sum_{n=1}^{\infty} \mu(a_n). \tag{1}$$

Since $a =' \sum_{n=1}^{\infty}{}' a_n$, we have

$$\sum_{n=1}^{k}{}' a_n \leqq' a.$$

It follows that

$$(2) \qquad \sum_{n=1}^{k} \mu(a_n) \leqq \mu(a), \qquad (k = 1, 2, \ldots).$$

Since the measure is non negative, the relation (2) implies that the series $\sum_{n=1}^{\infty} \mu(a_n)$ converges, and that

$$(3) \qquad \sum_{n=1}^{\infty} \mu(a_n) \leqq \mu(a).$$

Comparing this result with (1) we get

$$\mu(a) = \sum_{n=1}^{\infty} \mu(a_n),$$

so V is proved.

Thus we have proved that III and V are equivalent. Hence we have proved that III, IV, V, VI are equivalent.

A I.2.5. Let II. We shall prove V. Take

$$a_1, a_2, \ldots, a_n, \ldots \in A, \qquad a \in A, \quad \text{with} \quad a =' \sum_n{}' a_n$$

and a_i disjoint.

Since $a_1 + \cdots + a_n \leqq a$, it follows that

$$(1) \qquad \sum_{n=1}^{\infty} \mu(a_n) \leqq \mu(a) \ldots$$

Put

$$c_1 =_{df} a_1, \quad c_2 = a_1 + a_2, \ldots, \quad c_n = a_1 + \cdots + a_n, \ldots \quad (n > 1).$$

We have $c_1 \leqq c_2 \leqq \cdots \leqq c_n \leqq \cdots$ and

$$\sum_{i=1}^{n} c_i = \sum_{i=1}^{n} a_i.$$

Hence, [A.3.7.], [A.2.6.5.],

$$\sum_{i=1}^{\infty}{}' c_i =' \sum_{i=1}^{\infty}{}' a_n =' a.$$

Hence, by II,

$$\mu(a) \leqq \lim_{n \to \infty} \mu(c_n),$$

$$\mu(a) \leqq \lim_n \mu\left(\sum_{k=1}^{n} a_k\right),$$

$$\mu(a) \leqq \lim_n \sum_{k=1}^{n} \mu(a_k) = \sum_{k=1}^{\infty} \mu(a_k),$$

so we have proved that

(2) $$\mu(a) \leqq \sum_{n=1}^{\infty} \mu(a_n).$$

From (1) and (2) we get

$$\mu(a) = \sum_{n=1}^{\infty} \mu(a_n),$$

i.e. II implies V.

A I.2.6. Now let V. We shall prove II. Suppose that

$$a_1 \leqq a_2 \leqq \cdots \leqq a_n \leqq \cdots; \quad a_n, a \in A,$$

$$\sum_{n=1}^{\infty}{}' a_n =' a.$$

Put

$$b_1 =_{df} a_1, \quad b_2 =_{df} a_2 - a_1, \ldots, \quad b_n =_{df} a_n - a_{n-1}, \ldots \quad \text{for} \quad n > 1.$$

We obtain

(3) $$\sum_{k=1}^{n} a_k = \sum_{k=1}^{n} b_k,$$

[A.2.6.4.]. Hence, by [A.2.6.5.], [A.3.7.],

$$a =' \sum_{k=1}^{\infty}{}' a_k =' \sum_{k=1}^{\infty}{}' b_k \quad \text{and } b_k \text{ are disjoint.}$$

Hence, by V,

$$\mu(a) = \sum_{n=1}^{\infty} \mu(b_n),$$

and then

$$\mu(a) = \lim_{k\to\infty} \sum_{n=1}^{k} \mu(b_n);$$

hence by (3)

$$\mu(a) = \lim_{k\to\infty} \mu\left(\sum_{n=1}^{k} b_n\right) = \mu\left(\sum_{n=1}^{k} a_n\right) = \lim_{k\to\infty} \mu(a_k),$$

so II is proved.

Thus we have proved that the conditions II and V are equivalent; hence, if we take into account of what has been proved till now, we can say that the conditions II, III, IV, V, VI are equivalent. It remains to prove the equivalence of II and I.

A I.2.7. Let I. Take

(1) $$a_1 \leqq a_2 \leqq \cdots \leqq a_n \leqq \cdots \quad \text{with} \quad a_n, a \in A,$$

$$\sum_{n}{}' a_n =' a.$$

We get

$$\operatorname{co} a_1 \geqq \operatorname{co} a_2 \geqq \cdots \geqq \operatorname{co} a_n \geqq \cdots,$$

and

$$\prod_{n}{}' \operatorname{co} a_n =' \operatorname{co} a.$$

Hence

$$a \cdot \prod_n{}' \operatorname{co} a_n =' a \cdot \operatorname{co} a = O,$$

$$\prod_n{}' (a \cdot \operatorname{co} a_n) = O,$$

where

$$a \cdot \operatorname{co} a_1 \geqq a \cdot \operatorname{co} a_2 \geqq \cdots,$$

by (1). Hence by I:

$$\lim_{n\to\infty} \mu(a \cdot \operatorname{co} a_n) = 0, \tag{2}$$

and then

$$\lim_{n\to\infty} \mu[I - a \cdot \operatorname{co} a_n] = \lim_{n\to\infty} [\mu(I) - \mu(a \cdot \operatorname{co} a_n)] = \tag{3}$$

$$= \lim_{n\to\infty} \mu(I) - \lim_{n\to\infty} \mu(a \cdot \operatorname{co} a_n) =, \quad \text{by (2)}, \quad = \lim_{n\to\infty} \mu(I) = \mu(I).$$

But

$$I - a \cdot \operatorname{co} a_n = I \cdot \operatorname{co}(a \cdot \operatorname{co} a_n) = \operatorname{co} a + a_n.$$

Hence, by (3),

$$\lim_{n\to\infty} \mu(\operatorname{co} a + a_n) = \mu(I).$$

Since $a_n \leqq a$, we have $a_n \cdot \operatorname{co} a = 0$, and hence, by additivity of μ,

$$\mu(\operatorname{co} a) + \lim \mu(a_n) = \mu(I),$$

i.e.

$$\lim \mu(a_n) = \mu(I) - \mu(\operatorname{co} a) = \mu(I - \operatorname{co} a) = \mu(a).$$

Thus II follows.

A I.2.8. Now let II. We shall prove I. Take:

$$a_1 \geqq a_2 \geqq \cdots \geqq a_n \geqq \cdots \quad \text{with} \quad \prod_n{}' a_n =' O.$$

We get:

$$\operatorname{co} a_1 \leqq \operatorname{co} a_2 \leqq \cdots \quad \text{with} \quad \sum_{n=1}^{\infty}{}' \operatorname{co} a_n =' I.$$

Hence, by II we get

$$\mu(I) \leqq \lim_{n\to\infty} \mu(\operatorname{co} a_n),$$

$$\mu(I) \leqq \lim_{n\to\infty} [\mu(I) - \mu(a_n)],$$

$$\mu(I) \leqq \mu(I) - \lim \mu(a_n).$$

Hence

$$\lim \mu(a_n) \leqq 0, \quad \text{so} \quad \lim \mu(a_n) = 0.$$

Thus II implies I.

Since we have proved that I and II are equivalent, we see that all I—VI are equivalent. The theorem is proved.

Section 3 of A I.

A I.3. Absolute conditions for denumerability of measure. We are going to state and prove the equivalence of some conditions which differ from those, which we would get from [A I.2.], by identifying the tribes (A) and (A'). Indeed, since we shall have only one tribe (A) and no its supertribe, the infinite unions, to be considered, must be taken only with respect to (A). Since (A) will be supposed to be finitely only, infinite sums may not exist.

The theorem to be proved is this:

A I.3.1. Let (A) be a finitely additive tribe and $\mu \geqq 0$ a finitely additive finite measure on (A). The following are equivalent:

I'. μ is a denumerably additive measure on (A), i.e. in the case $a_1, a_2, \ldots, a_n, \ldots \in A$ are disjoint and $\sum_{n=1}^{\infty} a_n$ is significant, we have

$$\mu\left(\sum_{n=1}^{\infty} a_n\right) = \sum_{n=1}^{\infty} \mu(a_n).$$

II'. (FRÉCHET), (27). If $b_1 \geqq b_2 \geqq \cdots \geqq b_n \geqq \cdot\ \cdot$

$$\prod_{n=1}^{\infty} b_n = O,$$

then

$$\lim_{n\to\infty} \mu(b_n) = 0.$$

A I.3.2. Proof. Suppose II'; we shall prove I'.

Let $a_1, a_2, \ldots, a_n, \ldots \in A$, and suppose that the somata a_n are mutually disjoint and have a meaningful union

(0) $$a =_{df} \sum_{n=1}^{\infty} a_n,$$

where $a \in A$, [A.1.4.1.]. It follows

(1) $$\sum_{n=1}^{k} a_n \leqq a \quad \text{for} \quad k = 1, 2, \ldots$$

Put

(2) $$c_k =_{df} a - \sum_{n=1}^{k} a_n, \qquad k = 1, 2, \ldots$$

We have

$$\sum_{n=1}^{k} a_n \leqq \sum_{n=1}^{k+1} a_n;$$

hence

(3) $$c_k \geqq c_{k+1}.$$

A I.3.2a. First we shall prove that the meet $\prod_{n=1}^{\infty} c_n$ is meaningful, and $= O$.

To do that it suffices to prove that

(3a) $$\sum_{k=1}^{\infty} \operatorname{co} c_k$$

is meaningful and $= I$, [A.1.5.]. By (2) we have

(4) $$\operatorname{co} c_k = \operatorname{co}\left[a - \sum_{n=1}^{k} a_n\right] = \operatorname{co}\left[a \cdot \operatorname{co} \sum_{n=1}^{k} a_n\right] = \operatorname{co} a + \sum_{n=1}^{k} a_n,$$

so to prove (3a), we shall prove that the sum

$$\sum_{k=1}^{\infty}\left(\operatorname{co} a + \sum_{n=1}^{k} a_n\right)$$

is meaningful and equal I. According to the definition of union in a lattice we only need to prove that if

(5) $$\operatorname{co} a + \sum_{n=1}^{k} a_n \leqq d \quad \text{for all } k, \text{ then } \quad I \leqq d;$$

because the expression on the left in (5) is $\leqq I$. Let

(5a) $$\operatorname{co} a + \sum_{n=1}^{k} a_n \leqq d \quad \text{for all } k.$$

Since, by (1),

$$\operatorname{co} a \cdot \sum_{n=1}^{k} a_n \leqq \operatorname{co} a \cdot a = O,$$

we have from (5a), $\sum_{n=1}^{k} a_n \leqq d - \operatorname{co} a$, because

$$\sum_{n=1}^{k} a_n = \left(\sum_{n=1}^{k} a_n + \operatorname{co} a\right) - \operatorname{co} a;$$

hence

$$\sum_{n=1}^{\infty} a_n \leqq d - \operatorname{co} a.$$

It follows that

(5b) $$\sum_{u=1}^{\infty} a_n \leqq d,$$

and from (5a) we have

(6) $$\operatorname{co} a \leqq d.$$

Adding (5b) and (6) we get

$$\operatorname{co} a + \sum_{n=1} a_n \leqq d, \quad \text{i.e., by (0),} \quad \operatorname{co} a + a \leqq d,$$

hence $I \leqq d$.

Thus, by (4), we have proved that the sum $\sum_{k=1}^{\infty} \operatorname{co} c_k$ exists and equals I.

Consequently

$$\prod_{k=1}^{\infty} c_k = O.$$

It follows from hypothesis II′, that

(7) $$\lim_{k\to\infty} \mu(c_k) = 0.$$

A I.3.2b. From (2) and (1) we have

$$\mu(c_k) = \mu(a) - \mu\left(\sum_{n=1}^{k} a_n\right),$$

because

$$\sum_{n=1}^{k} a_n \leqq a.$$

Hence, on account of the denumerable additivity of the measure μ, and since the somata a_n are disjoint:

$$\mu(c_k) = \mu(a) - \sum_{n=1}^{k} \mu(a_n).$$

If we let tend k to infinity, and take account of (7), we get

$$\mu(a) = \sum_{n=1}^{k} \mu(a_n),$$

so the condition I′ follows.

A I.3.3. Now let I′, and we shall prove II′. Suppose that $b_1 \geqq b_2 \geqq \cdots \in A$ and suppose that

(0) $$\prod_{n=1}^{\infty} b_n \quad \text{exists and equals } O.$$

Our aim is to prove that $\mu(b_n) \to 0$ for $n \to \infty$. We have

(1) $$\operatorname{co} b_1 \leqq \operatorname{co} b_2 \leqq \cdots \in A.$$

Put

(2) $$a_1 =_{df} \operatorname{co} b_1, \quad a_2 =_{df} \operatorname{co} b_2 - \operatorname{co} b_1, \ldots, \quad a_n =_{df} \operatorname{co} b_n - \operatorname{co} b_{n-1},$$
$$\text{for} \quad n = 2, 3, \ldots$$

We have, [A.2.6.4.],

$$\operatorname{co} b_n = (\operatorname{co} b_n - \operatorname{co} b_{n-1}) + (\operatorname{co} b_{n-1} - \operatorname{co} b_{n-2}) + \cdots +$$
$$+ (\operatorname{co} b_2 - \operatorname{co} b_1) + \operatorname{co} b_1.$$

Hence by (2)

(3) $$\operatorname{co} b_n = \sum_{k=1}^{n} a_k \quad \text{for} \quad n = 2, 3, \ldots,$$

and

$$\operatorname{co} b_1 = \sum_{k=1}^{1} a_k ,$$

so (2) is valid for $n = 1, 2, \ldots$ Now, by (0), we have

$$\sum_{n=1}^{\infty} \operatorname{co} b_n = I ,$$

(by De Morgan laws), and by (2) and (1), we know that all $a_1, a_2, \ldots,$ $a_n, \ldots$ are mutually disjoint. Hence we get, by (3):

$$\sum_{n=1}^{\infty} \left(\sum_{k=1}^{n} a_k \right) = I . \tag{4}$$

Now we can prove that $\sum_{k=1}^{\infty} a_k$ is significant and $= I$. Indeed, we have surely $a_k \leqq I$ for all k.

Let $a_k \leqq c$ for all k. We have

$$\sum_{k=1}^{n} a_k \leqq c \quad \text{for all } n .$$

On account of (4), it follows that $I \leqq \sum_{n=1}^{\infty} c$, i.e. $I = c$, so $\sum_{k=1}^{\infty} a_k$ is significant and $= I$.

A I.3.3a. By hypothesis I′ we have

$$\mu(I) = \sum_{k=1}^{\infty} \mu(a_k) .$$

Hence

$$\lim_{n \to \infty} \left[\mu(I) - \sum_{k=1}^{n} \mu(a_k) \right] = 0 .$$

Since $a_1, \ldots, a_n$ are disjoint, it follows that

$$\lim_{n \to \infty} \mu \left[I - \sum_{k=1}^{n} a_k \right] = 0 .$$

Hence, by (3):

$$\lim_{n \to \infty} \mu [I - \operatorname{co} b_n] = 0 ,$$

i.e.

$$\lim_{n \to \infty} \mu(b_n) = 0 .$$

Thus II is proved, and thus the theorem is established.

A I.3.4. Remark. An obviously equivalent condition is the following similar to the condition of E. Hopf (28):

If $b_1 \geqq b_2 \geqq \cdots \in (A)$, $\prod_{n=1}^{\infty} b_n$ is significant, and $\mu(b_n) \geqq \alpha > 0$ for some α, then $\prod_{n=1}^{\infty} b_n \neq O$.

A I.3.5. Theorem. If (A) is a finitely additive Boolean tribe and $\mu \geqq 0$ is a finite finitely additive measure on (A), then the following are equivalent:

III'. μ is a denumerably additive measure on (A), [A.15.2.].

IV'. If $a \leqq \sum_{n=1}^{\infty} a_n$, with significant sum, then

$$\mu(a) \leqq \sum_{n=1}^{\infty} \mu(a_n) \leqq +\infty.$$

V'. If we have $a \leqq \sum_{n=1}^{\infty} a_n$ with significant sum and converging sum $\sum_{n=1}^{\infty} \mu(a_n)$, then

$$\mu(a) \leqq \sum_{n=1}^{\infty} \mu(a_n).$$

VI'. If $a = \sum_{n=1}^{\infty} a_n$ (with significant sum), then

$$\mu(a) \leqq \sum_{n=1}^{\infty} \mu(a_n).$$

A I.3.6. Proof. First we shall see that IV' → V' → VI' → IV'. Indeed, from IV' follows V'.

Let V' and suppose that $a = \sum_{n=1}^{\infty} a_n$.

Putting

$$b_1 =_{df} a_1, \quad b_2 =_{df} a_2 - a_1, \quad b_3 =_{df} a_3 - (a_1 + a_2), \ldots,$$
$$b_n =_{df} a_n - (a_1 + \cdots + a_{n-1}), \quad \text{for} \quad n = 2, 3, \ldots,$$

we have $b_n \leqq a_n$ and then $\mu(b_n) \leqq \mu(a_n)$. Since, [A.2.6.3.], $\sum_{k=1}^{n} b_k = \sum_{k=1}^{n} a_k$, the sum $\sum_{n=1}^{\infty} b_n$ is significant, [A.3.7.], [A.2.6.5.], and equals a. Since $b_1, b_2, \ldots, b_k, \ldots$ are disjoint, we have

$$\sum_{k=1}^{n} \mu(b_k) = \mu\left(\sum_{k=1}^{n} b_k\right) \leqq \mu(a);$$

hence the series $\sum_{k=1}^{n} \mu(b_k)$ converges. Hence, by V',

$$\mu(a) \leqq \sum_{n=1}^{\infty} \mu(b_n) \leqq \sum_{n=1}^{\infty} \mu(a_n) \quad \text{i.e. VI'}.$$

Let VI' and suppose

$$a \leqq \sum_{n=1}^{\infty} a_n.$$

One can prove, by using the definition of infinite sum, that the distributive property is valid if the sum is significant. Hence

$$a = a \cdot \sum_{n=1}^{\infty} a_n = \sum_{n=1}^{\infty} (a \cdot a_n).$$

Hence, by VI′,

$$\mu(a) \leqq \sum_{n=1}^{\infty} \mu(a\, a_n) \leqq \sum_{n=1}^{\infty} \mu(a_n) \quad \text{i.e. IV}'.$$

Thus IV′, V′ and VI′ are equivalent. We leave III′ to the reader.

Section 4 of A I.

A I.4. Borelian extension of a tribe within a supertribe. Given a collection C of sets, we sometime need to have its extension by performing all finite and denumerable operations on the sets belonging to C, then obtaining a wider collection. This collection will have the property of being closed with respect to these operations. If we order this collection by the relation of ordinary inclusion of sets, we get a tribe called borelian extension of C.

Now this procedure can be applied to collections of general somata. In precise terms we make the following construction:

A I.4.1. Let (G) be a denumerably additive tribe and (F) one of its finitely genuine strict (finitely additive) subtribes, [A.7.2.]. Infinite operation which will be performed on somata of (F) will be taken from (G), [A.7.6.]. We like to have the "smallest" denumerably additive extension of (F) within (G). In precise words, we like to have a denumerably additive tribe (F^b) which is a strictly denumerably genuine subtribe of (G), such that

1) (F) is a finitely genuine strict subtribe of (F^b),

2) If (G') is a strictly denumerably genuine subtribe of (G), such that (F) is a finitely genuine strict subtribe of (G'), then (F^b) is a strictly denumerably genuine subtribe of (G').

Of course, all these tribes will have the same zero and the same unit.

We must prove the unitary existence of the tribe (F^b). In what follows all somatic operations will be taken from (G), and denoted by symbols $+, \cdot, \sum, \prod$, co, $\leqq, =$. We shall have also classes of somata; operations on these classes will be denoted by $\cup, \cap, \subseteqq, =, \bigcap, \bigcup$.

A I.4.2. We shall construct a well ordered sequence of classes of somata of (G), defined as follows:

Put

$$A^1 =_{df} F, \qquad B^1 =_{df} F.$$

A^2 is defined as the collection of all somata of (G), which can be represented as denumerable unions

$$p_1 + p_2 + \cdots + p_n, \ldots,$$

where

$$p_n \in F = A^1 \cup B^1, \quad n = 1, 2, \ldots$$

B^2 is defined as the collection of all somata of (G), which can be represented as co$p$, where $p \in A^2$. Now, suppose that we have already defined all A^β, B^β, where β is any ordinal number less than a given ordinal number $\alpha > 2$. Then we define A^α as the collection of all somata p of (G), which can be represented as denumerable unions

$$p_1 + p_2 + \cdots + p_n, \ldots,$$

where

$$p_n \in \bigcup_{\beta < \alpha} [A^\beta \cup B^\beta],$$

and we define B^α as the collection of all somata of (G) which can be represented as co$p$, where $p \in A^\alpha$.

Now we define the collection F^b of somata of (G) by

$$F^b =_{df} \bigcup_{\alpha} A^\alpha.$$

Where the union is taken for all finite and denumerable ordinals α, i.e. for all $\alpha < \Omega$, where Ω is the smallest non denumerable ordinal.

A I.4.3. We shall prove that if we order the somata of F^b, by taking the ordering of (G), we get a denumerably additive tribe (F^b).

Since all somatic operations are taken from (G) and since (G) is a denumerably additive tribe, it suffices to prove that

1) if $a_1, a_2, \ldots, a_n, \ldots \in F^b$, then

$$\sum_{n=1}^{\infty} a_n \in F^b,$$

2) if $a \in F^b$, then co$a \in F^b$,

3) $O \in F^b$, $\quad I \in F^b$.

A I.4.4. Proof. First we notice that if α_1, α_2 are ordinals $< \Omega$, with $\alpha_1 < \alpha_2$, then

(1) $$A^{\alpha_1} \subseteq A^{\alpha_2}, \quad B^{\alpha_1} \subseteq B^{\alpha_2}.$$

Now suppose that $p_n \in F^b$, $(n = 1, 2, \ldots)$.

There exists an ordinal $\alpha(n) < \Omega$, such that

$$p_n \in A^{\alpha(n)},$$

There also exists an ordinal $\alpha < \Omega$, such that $\alpha(n) \leqq \alpha$ for all $n = 1, 2, \ldots$

Having such α, we get, by (1),

$$A^{\alpha(n)} \subseteq A^\alpha;$$

hence

$$p_n \in A^\alpha \quad \text{for all } n.$$

It follows that

$$\sum_{n=1}^{\infty} p_n \in A^{\alpha+1} \subseteqq \bigcup_{\beta<\Omega} A^\beta \subseteqq F^\beta.$$

Thus we have proved that if $p_n \in F^b$ for $n = 1, 2, \ldots,$ then $\sum_{n=1}^{\infty} p_n \in F^b$.

Let $p \in F^b$. We can find α such that $p \in A^\alpha$. Hence

$$\mathrm{co}\, p \in B^\alpha.$$

It follows that $\mathrm{co}\, p \in A^{\alpha+1} \subseteqq F^b$.

Thus we have proved that if $p \in F^b$, then $\mathrm{co}\, p \in F^b$. Using de Morgan laws, we prove that if $p, q \in F^b$, then $p \cdot q \in F^b$.

Take any soma p of F^b. We have $\mathrm{co}\, p \in F^b$; hence $p + \mathrm{co}\, p \in F^b$, i.e. $I \in F^b$. Since also $p \cdot \mathrm{co}\, p \in F^b$, we get $O \in F^b$. Thus having proved the items 1), 2), 3) in [A I.4.3.], we have proved that (F^b) is denumerably additive tribe, under the ordering considered.

A I.4.5. We have $F^b \subseteqq G$, because $F \subseteqq G$ and (G) is denumerably additive. (F^b) is a denumerably genuine subtribe of (G). The tribe (F^b) is a strictly finitely genuine supertribe of (F); and (G), (F) and (F^b) have the same zero and the same unit.

A I.4.6. Now we are going to prove that the condition 2) in [A I.4.1.] is satisfied. Let (G') be a denumerably additive supertribe of (F), and, at the same time a subtribe of (G), as required in item 2) in [A I.4.1.]. Since $F \subseteqq G'$, and (G') is denumerably additive, we have $A^1 \subseteqq G'$, $B^1 \subseteqq G'$, and then for every $\alpha < \Omega$, we have $A^\alpha \subseteqq G'$, $B^\alpha \subseteqq G'$. Consequently $F^b \subseteqq G'$. Hence (F^b) is a strictly denumerably genuine subtribe of (G'). Now suppose that we have another tribe $(F^b)'$, satisfying the conditions 1), 2) in [A I.4.1.]. Since both these tribes satisfy the same conditions as imposed on (G'), we have $F^{b'} \subseteqq F^b$ and $F^b \subseteqq F^{b'}$, so $F^{b'} = F^b$.

Thus the uniqueness of (F^b) is established.

A I.4.6a. Def. We call (F^b) *the borelian extension of* (F) *within* (G).

A I.4.7. Till now we have admitted that the tribe (F) is a strict subtribe of (G). Now we shall discuss our topic in the case that (F) is not a strict finitely genuine subtribe of (G), but, simply, a finitely genuine one.

Let $\mathfrak{B}$ be the correspondence [A.7.4.] accompanying that situation. Take instead of (F) the tribe $\mathfrak{B}(F)$ which is a strictly finitely genuine subtribe of (G). Then $F_1 = \mathfrak{B}(F)$ and G are in the conditon of our preceeding discussion. We get a tribe $(F_1)^b$, composed of somata belonging to (G). We also call $(F_1)^b$ *the Borelian extension of* (F) *in* (G).

A I.4.8. We can admit a more general point of view, where (F) is a finitely genuine subtribe of (G) through isomorphism. We get a subtribe of (G) which also may be considered as a kind of extension of (F). We call it *Borelian extension of* (F) *in* (G) *through isomorphism.*

Section 5 of A I.

A I.5. Denumerably additive measure-hull of a tribe. It is widely known that the tribe (B) of Borelian subsets of the segment $(0, 1\rangle$ can be extended so as to contain the sets of Lebesgue's measure 0. We get the tribe of all measurable subsets of the segment.

A similar construction can be applied in the general case of an abstract tribe, whose somata are not necessarily sets.

A I.5.1. We consider the following situation. There is given a denumerably additive tribe (M), and a denumerably additive strictly denumerably genuine subtribe (T), [A.7.5 a.], of (M). Suppose that we have on (T) a denumerably additive measure $\mu(\dot{a})$, [A.15.2.], (which we prefer to be not effective), [A.15.1.]. We shall construct a tribe (T_1) which would be an extension of (T) within (M).

We have $T \subseteqq M$, and the operations in (T) can be taken from (M).

A I.5.2. The collection of all somata a of (T), for which $\mu(a) = 0$ is a denumerably additive ideal J in (T), [A.9.2.].

Indeed let $a \in J$ and $b \leqq a$, $b \in T$. Then $\mu(a) = 0$ and $\mu(b) \leqq \mu(a)$. Since the measure is not negative, it follows

$$\mu(b) = 0, \quad \text{and then} \quad b \in J.$$

Now let

$$a_1, a_2, \ldots, a_n, \ldots \in J.$$

We have [A.2.6.3.]

$$a_1 + a_2 + \cdots + a_n = b_1 + b_2 + \cdots + b_n,$$

where

$$b_1 =_{df} a_1, \quad b_2 = a_2 - a_1, \ \ldots, \ b_n =_{df} a_n - (a_1 + a_2 + \cdots + a_{n-1}),$$
$$(n = 2, 3, \ldots).$$

The somata b_n are disjoint.

It follows [A.2.6.5.], since all denumerable operations are performable on T,

$$\sum_{n=1}^{\infty} a_n = \sum_{n=1}^{\infty} b_n, \tag{1}$$

Since the measure μ is denumerably additive, we have

$$\mu\left(\sum_{n=1}^{\infty} b_n\right) = \sum_{n=1}^{\infty} \mu(b_n). \tag{2}$$

Since $a_n \in J$ and $b_n \leqq a_n$, it follows that $\mu(b_n) = 0$. Then from (2), we get

$$\mu\left(\sum_{n=1}^{\infty} b_n\right) = 0;$$

hence, by (1)

$$\mu\left(\sum_{n=1}^{\infty} a_n\right) = 0,$$

which gives

$$\sum_{n=1}^{\infty} a_n \in J.$$

Thus we have proved that J is a denumerably additive ideal in (T).

A I.5.3. Given a soma $a \in J$, consider the class of all somata $P \in M$ such that $P \leqq a$. Let us vary a and take the union J_1, of all classes

$$\{P \mid P \in M,\ P \leqq a\}.$$

We shall prove that J_1 is denumerably additive ideal in (M).

Let $A \in J_1$, and let $A' \in M$ with $A' \leqq A$. Since there exist $a \in J$ with $A \leqq a$, we have also $A' \leqq a$; hence $A' \in J_1$. Now let $A_1, A_2, \ldots, A_n, \ldots \in J_1$. There exist somata

$$a_1, a_2, \ldots, a_n, \ldots \in J,$$

such that

$$A_n \leqq a_n \quad \text{for} \quad n = 1, 2, \ldots$$

Since J is a denumerably additive ideal, we have

$$\sum_{n=1}^{\infty} a_n \in J.$$

Since

$$\sum_{n=1}^{\infty} A_n \leqq \sum_{n=1}^{\infty} a_n,$$

we get

$$\sum_{n=1}^{\infty} A_n \in J_1.$$

The assertion is proved.

A I.5.4. Having that, consider the class T_1 of all somata of M, having the form

(1) $$\alpha = (a + A) - B,$$

where $a \in T$, and $A, B \in J_1$. By [A.9.4a] we can write

$$\alpha + C = a + C,$$

where C is a suitable soma of J_1. We shall prove that the collection T_1 makes up a denumerably additive tribe, where the ordering is taken from M.

A I.5.5. Since all somatic operations are taken from M, it is sufficient to prove that T_1 is closed with respect to complementations and to denumerable unions.

Let $\alpha \in T_1$. There exists $a \in T$, such that

$$\alpha = (a + A) - B \quad \text{for suitable} \quad A, B \in J_1.$$

Hence, [A.9.4a.], we have

$$\alpha \dot{+} a \in J_1.$$

Now

$$\alpha \dot{+} a = \operatorname{co}\alpha \dot{+} \operatorname{co} a \in J_1, \tag{1}$$

[A.17.8.]. Since $\operatorname{co} a \in T$, we get

$$\operatorname{co}\alpha \in T_1.$$

A I.5.6. Let

$$\alpha_1, \alpha_2, \ldots, \alpha_n, \ldots \in T_1$$

and let

$$a_1, a_2, \ldots, a_n, \ldots \in T$$

be such, that

$$\alpha_n \dot{+} a_n \in J_1, \quad (n = 1, 2, \ldots).$$

It follows that for suitable $C_n \in J_1$ we have

$$\alpha_n + C_n = a_n + C_n, \quad (n = 1, 2, \ldots).$$

Hence

$$\sum_{n=1}^{\infty} \alpha_n + \sum_{n=1}^{\infty} C_n = \sum_{n=1}^{\infty} a_n + \sum_{n=1}^{\infty} C_n. \tag{2}$$

Since J_1 is denumerably additive, we get

$$\sum_{n=1}^{\infty} C_n \in J_1.$$

Since (T) is denumerably additive, we have $\sum_{n=1}^{\infty} a_n \in T$. It follows from (2), that $\sum_{n=1}^{\infty} \alpha_n \in T_1$.

Thus we have proved that T_1, if the ordering is taken from M_1, becomes a denumerably additive tribe (T_1).

We have $T \subseteqq T_1 \subseteqq M_1$.

A I.5.7. We have $J_1 \subseteqq T_1$, and J_1 is a denumerably additive ideal in T_1.

Theorem. We have even the identity:

$$J = T \cap J_1. \tag{3}$$

Proof. Let $a \in J$, thus $a \in T$, and since $J \subseteqq J_1$, we have $a \in J_1$.

Thus we get

(4) $$J \subseteqq T \frown J_1.$$

Let $a \in T \frown J_1$. Since $a \in J_1$, there exists a soma $p \in J$, such that

(5) $$a \leqq p,$$

[A I.5.3.]. Since $a \in T$ and J is an ideal in (T), it follows by (5), that $a \in J$.

Thus we have proved that if $a \in T \frown J_1$, then

(6) $$a \in J.$$

This result together with (4) gives the identity (3).

A I.5.8. Having that, we shall prove that if

(7) $$(a + A) - B = (a' + A') - B',$$

where

$$a, a' \in T \quad \text{and} \quad A, B, A', B' \in J_1,$$

then

$$a \dotplus a' \in J.$$

Proof. We have by [A.9.4a], [A.9.3.],

$$(a + A) - B =^{J_1} a,$$

$$(a' + A') - B' =^{J_1} a',$$

where $(=^{J_1})$ is the equality induced by the ideal J_1, [A.9.3.]. Hence from (7) we get $a =^{J_1} a'$; hence

$$a \dotplus a' \in J_1.$$

Now $a, a' \in T$; hence $a \dotplus a' \in T$, (since (T) is a tribe), and then

$$a \dotplus a' \in J_1 \frown T.$$

From [A I.5.7.] it follows that

$$a \dotplus a' \in J. \quad \text{q.e.d.}$$

A I.5.9. Def. Having settled the above auxiliaries, we shall introduce a *measure on the tribe* (T_1).

The definition is this:

If $\alpha \in T_1$, and we put

(8) $$\alpha = a + A - B,$$

where $a \in T$ and $A, B \in J_1$, we define

$$\mu_1(\alpha) =_{df} \mu(a).$$

A I.5.10. We can admit that definition only, if we prove that the number $\mu_1(\alpha)$ does not depend on the representation (8) of α.

Let

$$\alpha = a + A - B = a' + A' - B',$$

where

$$a' \in T, \quad A', B' \in J_1.$$

We have proved that then, [A I.5.8.], $a \dotplus a' \in J$, i.e.

$$a =^J a'.$$

Now, by [A.19.4.] we have

$$\mu(a) = \mu(a'),$$

i.e. the measure is invariant with respect to the equality $(=^{J_1})$. This shows the required independence.

A I.5.11. Theorem. If $\alpha \in J_1$, then $\mu_1(\alpha) = 0$.

Proof. We have

$$\alpha = O + \alpha - O.$$

It follows, by (8), that

$$\mu_1(\alpha) = \mu(0) = 0.$$

A I.5.12. Theorem. If $\alpha \in T$, then $\mu_1(\alpha) = \mu(\alpha)$.

Proof. $\alpha = \alpha + O - O$.

A I.5.13. Theorem. If $\alpha =^{J_1} \beta$, then $\mu_1(\alpha) = \mu_1(\beta)$.

Proof. Let

$$\alpha = a + A' - A'',$$

$$\beta = b + B' - B''$$

be the usual representations.

We get, [A I.5.10.],

$$\alpha =^{J_1} a,$$

$$\beta =^{J_1} b;$$

hence $a =^{J_1} b$; hence $a \dotplus b \in J_1$.

Hence, since $a, b \in T$, we have by [A I.5.8.],

$$a \dotplus b \in J; \quad \text{hence} \quad a =^J b,$$

and then

$\mu(a) = \mu(b)$, [A.19.4.], which gives, [A I.5.9.], $\mu_1(\alpha) = \mu_1(\beta)$.

A I.5.14. We shall prove that the measure μ_1 is denumerably additive on the tribe (T_1).

Proof. Let $\alpha_1, \alpha_2, \ldots, \alpha_n, \ldots$ be disjoint somata of T_1. We have, [A I.5.4.],

(1) $$\alpha_n = a_n + A_n - B_n, \quad (n = 1, 2, \ldots),$$

where

$$a_n \in T, \quad A_n, B_n \in J_1, \quad (n = 1, 2, \ldots).$$

The equation (1) is equivalent to the following:

$$\alpha_n + C_n = a_n + C_n, \quad \text{[A.9.4a.], where} \quad C_n \in J_1.$$

It follows

$$(2) \qquad \sum_{n=1}^{\infty} \alpha_n + \sum_{n=1}^{\infty} C_n = \sum_{n=1}^{\infty} a_n + \sum_{n=1}^{\infty} C_n.$$

Since J_1 is a denumerably additive ideal, [A I.5.2.], we have

$$C =_{df} \sum_{n=1}^{\infty} C_n \in J_1.$$

So we get

$$\sum_{n=1}^{\infty} \alpha_n =^{J_1} \sum_{n=1}^{\infty} a_n,$$

[A.9.4a], and then, [A I.5.13.],

$$(3) \qquad \mu_1\left(\sum_{n=1}^{\infty} \alpha_n\right) = \mu_1\left(\sum_{n=1}^{\infty} a_n\right).$$

Now we have [A.2.6.3.], [A.2.6.5.], [A.3.7.],

$$\sum_{n=1}^{\infty} a_n = a_1 + [a_2 - a_1] + [a_3 - (a_1 + a_2)] + \cdots,$$

where the terms on the right are disjoint.

Since the measure μ is denumerably additive, we get

$$(4) \quad \mu\left(\sum_{n=1}^{\infty} a_n\right) = \mu_1(a_1) + \mu_1(a_2 - a_1) + \mu_1[a_3 - (a_1 + a_2)] + \cdots$$

On the other hand the somata

$$(5) \qquad a_1, a_2 - a_1, \quad a_3 - (a_1 + a_2), \ldots$$

belong respectively to the same J_1-equivalence classes, [A.12.], as

$$(6) \qquad \alpha_1, \alpha_2 - \alpha_1, \alpha_3(\alpha_1 + \alpha_2), \ldots, \quad \text{i.e.}$$

the corresponding elements in the sequences (5) and (6) are $(=^{J_1})$-equal, so they have the equal μ_1-measures respectively, [A I.5.13.], i.e.

$$\mu_1(a_1) = \mu_1(\alpha_1),$$
$$\mu_1(a_2 - a_1) = \mu_1(\alpha_2 - \alpha_1),$$
$$\mu_1[a_3 - (a_1 + a_2)] = \mu_1[\alpha_3 - (\alpha_1 + \alpha_2)],$$
$$\cdots\cdots\cdots\cdots\cdots\cdots\cdots\cdots\cdots$$

This gives, by (4),

$$\mu\left(\sum_{n=1}^{\infty} a_n\right) = \mu_1(\alpha_1) + \mu_1(\alpha_2 - \alpha_1) + \mu_1[\alpha_3 - (\alpha_1 + \alpha_2)] + \cdots$$

Since the somata α_n are disjoint by hypothesis, we have

$$\alpha_2 - \alpha_1 = \alpha_2, \quad \alpha_3 - (\alpha_1 + \alpha_2) = \alpha_3, \ \ldots$$

Consequently, by (3),

$$\mu_1\left(\sum_{n=1}^{\infty} \alpha_n\right) = \mu_1(\alpha_1) + \mu_1(\alpha_2) + \mu_1(\alpha_3) + \cdots,$$

so the denumerable additivity of the measure μ_1 is established.

A I.5.15. The measure $\mu_1(\alpha)$ may be not effective, but if we take the quotient tribe $(T_1)/J_1$, [A.12.], and attach the measure $\mu_1(\alpha)$ to the equivalence class determined by α, the measure will become effective and denumerably additive.

A I.5.16. Expl. As an example we take for M the collection of all subsets of the interval $\langle 0, 1\rangle$. Take for (T) the tribe composed of all borelian subsets of $\langle 0, 1\rangle$. Take for μ the Lebesguean measure on (T). Then (T_1) will be the tribe of all Lebesgue-measurables subsets of $\langle 0, 1\rangle$; J_1 is the collection of all sets of measure 0 and J is the collection of all borelian sets with measure 0.

A I.5.17. Def. In the general case we call the tribe (T_1) provided with the measure μ_1: *denumerably additive measure-hull of* (T). The tribe (T) is a denumerably genuine subtribe of (T_1), but only exceptionally a strict subtribe.

Section 6 of A I.

A I.6. One of the most important problems in measure theory on tribes is the following one: There are given a (finitely additive) tribe (A) provided with finitely additive measure $\mu(\dot{a})$, and another tribe (A') more ample, say a finitely genuine supertribe of (A). We suppose that (A') is denumerably additive and the question is:

Can we find a denumerably additive tribe (B') within (A') and a measure μ' on (B'), such that μ' be a denumerably additive extension of μ on (B')?

The answer is yes, if certain condition is satisfied. To prove this assertion there are three different devices for construction of (B') and of the prolonged measure:

1) *Lebesgue's device,* (23),

2) *Carathéodory's device* (22) and

3) the *author's device* (15), (14).

The two first are the most known, but the third seems to be the most simple.

The first device is based on the notion of *exterior measure* of somata of (A'), obtained by covering them by a denumerable set of somata of (A). This is similar to the known device by LEBESGUE in his theory of measure of sets of real numbers. The second starts with a kind of denumerably *convex measure* $\Gamma(a')$ for all $a' \in A'$ and defines a special notion of measurability of somata of (A').

The third is based on a kind of approximation of somata of (A') by those of (A).

Each of these three theories can be carried through and obtain the tribe (B') and a prolonged denumerably additive measure on (B'). They are also equivalent, i.e. they yield the same tribe (B') and the same prolonged measure.

To save the space we shall not carry through every of these theories. We shall begin with the first and stop after having proved that every soma of (A) is measurable. We shall do similarly with the second method. We shall carry it as long as is neaded for proving that the somata of (A'), considered as measurable by the first approach, are the same as those considered as measurable by the second approach. Measure on (B') will be not yet considered but postponed.

The third method will be carried through completely, i.e. we shall get an extension of the tribe and of the measure.

Now let us exhibit the first method, i.e. the Lebesgue's extension device.

A I.6.1. Lebesgue's extension device. Let (A), (A') be two tribes, where (A) is a finitely genuine subtribe of (A'), [A.7.]. We suppose that (A') is denumerably additive and that $\mu(\dot{a}) \geqq 0$ be a (finite) finitely additive measure on (A), [A.15.].

We shall use primes for notions and relations related to (A'). Since, on (A), infinite operations may not exist at all, we shall use infinite operations, taken from (A'), [A.7.6.].

A I.6.1a. Def. Let $a' \in A'$. By a *covering of* a' we shall understand in this chapter [A I]., any not empty, at most denumerable collection $a_1, a_2, \ldots, a_n, \ldots$ of somata of (A), such that

$$a' \leqq' \sum_n{}' a_n.$$

1) Remark. We can use only infinite coverings, because a finite collection can be completed by using an infinity of somata equal O.

2) Remark. Given an a', we always can find its covering, [A I.6.1 a.],

$$a' \leqq' I + O + O + \cdots.$$

We also we have

$$I' =' I, \quad O' =' O.$$

A I.6.2. Def. Given a covering $\{a_n\}$ of a', we attach to $\sum\limits_{n=1}^{\infty} \mu(a_n)$ the value $+\infty$ in the case of divergence. With this agreement there exists a unique finite, non negative number, defined by the expression

(1) $$\inf_{\{a_n\}} \sum_n \mu(a_n),$$

where the infimum is taken over all coverings $\{a_n\}$ of a'. If we consider only convergent series $\sum_n \mu(a_n)$, we get the same infimum. The number (1) looks like the Lebesgue's outer measure. It will be denoted by $\mu'_e(a')$, and called *outer measure of* a'.

Since I is a covering of a', we have

A I.6.2a. Theorem.

$$0 \leqq \mu'_e(a') \leqq \mu(I).$$

A I.6.2b. Def. We define the *inner measure of* a' as

$$\mu'_i(a') =_{df} \mu'_e(I') - \mu'_e(\mathrm{co}'\, a').$$

A I.6.3. Theorem. If $a' \leqq' b'$, we have

$$\mu'_e(a') \leqq \mu'_e(b').$$

Indeed, every covering of b' is a covering of a'.

A I.6.3a. Theorem.

$$\mu'_i(a') \geqq 0.$$

This follows from [A I.6.2a].

A I.6.3b. Theorem.

$$\mu'_e(O') = 0.$$

Indeed the set composed of the only soma O is a covering of O'.

A I.6.3c. Theorem. If $a \in A$, then

$$\mu'_e(a) \leqq \mu(a).$$

A I.6.4. Remark. It may happen that $\mu(I) > 0$ and $\mu'_e(I') = 0$, and so we can have $\mu'_e(a') = 0$ for every a'. This shows the following example. Denote by $(\alpha, \beta\rangle$ the set of all real numbers x such that $\alpha < x \leqq \beta$. We have $(\alpha, \beta\rangle \neq 0$ if and only if $\alpha < \beta$. The class A of all sets

(1) $$R \cdot \sum_i (\alpha_i, \beta_i\rangle,$$

where the sum is finite, R the set of all rational numbers, and where $0 \leqq \alpha_i, \beta_i \leqq 1$, generates a finitely additive tribe (A), where the ordering relation is that of the inclusion of sets. Denote by A' the class of all subsets of $R \cdot (0, 1\rangle$. This class also generates a tribe (A'). The tribe (A) is a finitely genuine subtribe of (A'), if the ordering relation, in (A'), is also the relation of inclusion of sets. (A') is denumerably additive. Every not empty soma a of (A) can be represented in the form (1), where the half-closed intervals $(\alpha_i, \beta_i\rangle$, $\alpha_i < \beta_i$ are disjoint (in the case there are at least two terms in (1)). If

$$a = R \cdot \sum_i (\alpha_i, \beta_i\rangle$$

is such a representation, define

$$\mu(a) =_{df} \sum_i (\beta_i - \alpha_i),$$

and if $a = O$, put $\mu(a) =_{df} 0$.

We easily see that $\mu(a)$ does not depend on the disjoint representation of a. The measure $\mu(a) \geqq 0$ is effective and finitely additive. Nevertheless, since the set I' is at most denumerable, I' can be covered by denumerable number of segments $(p_i, q_i\rangle$ with $\sum_{i=1}^{\infty} (q_i - p_i)$ as small as we like, so $\mu'_e(I') = 0$ whatever a' may be.

A I.6.5. Theorem. The exterior measure $\mu'_e(a')$ of the soma $a' \in A'$ is the infimum of all numbers $\sum_n \mu(a_n)$, where $\{a_n\}$ is a denumerable covering of a' with disjoint a_n.

Proof. Let a_n, $(n = 1, 2, \ldots)$ be a covering of a' with converging $\sum_{n=1}^{\infty} \mu(a_n)$.

We have

$$a' \leqq \sum_n a_n.$$

Put

$$b_1 =_{df} a_1, \quad b_2 = a_2 - a_1, \quad b_3 = a_3 - (a_1 + a_2), \ldots,$$
$$b_n = a_n - (a_1 + a_2 + \cdots + a_{n-1}), \ldots$$

We have, [A.2.6.3.],

$$\sum_{n=1}^{k} a_n = \sum_{n=1}^{k} b_n,$$

where $b_n \in A$ and where the somata b_n are mutually disjoint. It follows that, [A.2.6.5.], [A.3.7.],

$$(1) \qquad a' \leqq' \sum_{n=1}^{\infty}{}' a_n =' \sum_{n=1}^{\infty}{}' b_n.$$

Since $\sum_{n=1}^{\infty} \mu(a_n)$ converges and since $b_n \leqq a_n$, $n = 1, 2, \ldots$, we get

$$\mu(b_n) \leqq \mu(a_n),$$

and so we get

$$(2) \qquad \sum_{n=1}^{\infty} \mu(b_n) \leqq \sum_{n=1}^{\infty} \mu(a_n),$$

for all coverings $\{a_n\}$ with converging sums of measures.

Denote by p the infimum of sums of measures of elements of disjoint coverings, and by q the infimum of sums of measures of elements of all coverings. We have

$$q \leqq p.$$

But from (2) it follows that

$$p \leqq q.$$

Consequently $p = q$, so the theorem is proved.

A I.6.6. Theorem. If $a \in A$, then $\mu_e'(a)$ is the infimum of all numbers $\sum_n \mu(a_n)$ where $\{a_n\}$ is a covering of a, with disjoint a_n and with

(1) $$a =' \sum_{n=1}' a_n.$$

Proof. Denote by p the infimum of numbers $\sum_n \mu(a_n)$, where $\{a_n\}$ is a disjoint covering of a with (1). We have

(2) $$\mu_e'(a) \leqq p.$$

Now take any disjoint covering $\{b_n\}$ of a and put

$$c_n =_{df} b_n \cdot a.$$

We have

$$\sum_n' c_n =' \sum_n' b_n \cdot a;$$

hence

$$\sum_n' c_n =' a \cdot \sum_n' b_n =' a,$$

because

$$a \leqq' \sum_n' b_n.$$

So $\{c_n\}$ is a disjoint covering of a with $\sum_n' c_n =' a$. Since $\mu(c_n) \leqq \mu(b_n)$, we have

$$p \leqq \sum_n \mu(b_n), \quad \text{for all coverings } \{b_n\}.$$

Hence

(3) $$p \leqq \mu_e'(a).$$

From (2) and (3) the theorems follows.

A I.6.7. Theorem. Let $a' \in A'$. For every $\varepsilon > 0$ there exists a disjoint covering $\{a_n\}$ of a' such that

$$\sum_n \mu(a_n) - \varepsilon \leqq \mu_e'(a') \leqq \sum_n \mu(a_e').$$

This follows from [A I.6.5.].

A I.6.8. Theorem. Let $a' \in A'$. If for every disjoint covering $\{a_n\}$ of a' we have

(1) $$d \leqq \sum_n \mu(a_n),$$

and if for every $\varepsilon > 0$ there exists a covering $\{b_n\}$ of a' such that

$$d \geqq \sum_n \mu(b_n) - \varepsilon, \tag{2}$$

then

$$d = \mu'_e(a').$$

Proof. We have for every covering $\{b_n\}$ of a' the inequality

$$d \leqq \sum_n \mu(b_n). \tag{3}$$

Since (3) and (2) are just the conditions for d, defining the infimum of all $\sum_{n=1}^{\infty} \mu(b_n)$, we get

$$d = \mu'_e(a').$$

A I.6.9. Theorem. For any $a', b' \in A'$ we have

$$\mu'_e(a' + b') \leqq \mu'_e(a') + \mu'_e(b'). \tag{1}$$

Proof. Choose $\varepsilon > 0$. By [A I.6.2.] find coverings $\{a_n\}$, $\{b_n\}$ of a', b' respectively, such that

$$\begin{aligned} &\sum_n \mu(a_n) \leqq \mu'_e(a') + \varepsilon, \\ &\sum_m \mu(b_n) \leqq \mu'_e(b') + \varepsilon. \end{aligned} \tag{2}$$

Since

$$a' \leqq' \sum_n{}' a_n, \qquad b' \leqq' \sum_m{}' b_m,$$

we get

$$a' + b' \leqq' \sum_n{}' a_n + \sum_m{}' b_m.$$

Since the union of sets $\{a_n\}$, $\{b_m\}$ is *a* covering of $a' + b'$, it follows that

$$\mu'_e(a + b') \leqq \sum_n{}' \mu(a_n) + \sum_m{}' \mu(b_m).$$

Hence, by (2),

$$\mu'_e(a' + b') \leqq \mu'_e(a') + \mu'_e(b') + 2\varepsilon.$$

By tending ε toward 0, we get the theorem.

A I.6.9a. Def. We call the property (1) of the above theorem *convexity* or *sub-additivity of* $\mu'_e(\dot{a})$.

A I.6.10. Theorem. The measure μ'_e is denumerably convex on (A'), i.e. if

$$a' =' \sum_{n=1}^{\infty}{}' a'_n, \qquad a', a_n \in A',$$

then

$$\mu'_e(a') \leqq \sum_{n=1}^{\infty} \mu'_e(a'_n),$$

(or $+\infty$ in the case of divergence).

Proof. Take numbers $\varepsilon_n > 0$, $(n = 1, 2, \ldots)$, with

$$\varepsilon_1 + \varepsilon_2 + \cdots + \varepsilon_n + \cdots < \varepsilon.$$

For ε_n and a'_n find a covering

$$a_{n1}, a_{n2}, \ldots, a_{nk}, \ldots \quad \text{of} \quad a'_n,$$

such that, [A I.6.2.],

$$\sum_k \mu(a_{nk}) - \varepsilon_n \leqq \mu'_e(a'_n), \qquad (n = 1, 2, \ldots).$$

We get

(1) $$\sum_{n=1}^{\infty} \sum_k \mu(a_{nk}) - \varepsilon \leqq \sum_{n=1}^{\infty} \mu'_e(a'_n).$$

We also have

$$a' =' \sum_{n=1}^{\infty}{}' a_n \leqq' \sum_{n=1}^{\infty}{}' \sum_k{}' a_{nk}.$$

Since the collection of all a_{nk} constitutes a covering of a', we get

$$\mu'_e(a') \leqq \sum_{n=1}^{\infty} \sum_k \mu(a_{nk});$$

hence, by (1),

$$\mu'_e(a') \leqq \sum_{n=1}^{\infty} \mu'_e(a'_n) + \varepsilon.$$

It follows, through $\varepsilon \to 0$,

$$\mu'_e(a') \leqq \sum_{n=1}^{\infty} \mu'_e(a'_n), \quad \text{q.e.d.}$$

A I.6.11. Theorem. For every $a' \in A'$, we have

$$\mu'_i(a') \leqq \mu'_e(a').$$

Proof. Let $\varepsilon > 0$ and let $\{a_n\}$ be a covering of a' with

(1) $$\mu'_e(a') \leqq \sum_n \mu(a_n) \leqq \mu'_e(a') + \varepsilon.$$

Let $\{b_m\}$ be a covering of $\text{co}' a'$ with

(2) $$\mu'_e(\text{co}' a') \leqq \sum_m \mu(b_m) \leqq \mu'_e(\text{co}' a') + \varepsilon.$$

We have

$$a' + \text{co}' a' =' I' \leqq \sum_n{}' a_n + \sum_m{}' b_m.$$

By [A I.6.9.] we get

$$\mu'_e(I') \leqq \mu'_e(a') + \mu'_e(\text{co}' a');$$

hence, by (1) and (2),

$$\mu'_e(I') \leqq \mu'_e(a') + \mu'_e(\text{co}' a') + 2\varepsilon,$$

$$\mu'_e(I') - \mu'_e(\text{co}' a') \leqq \mu'_e(a') + 2\varepsilon.$$

By virtue of definition [A I.6.2b.] it follows

$$\mu_i'(a') \leqq \mu_e'(a') + 2\varepsilon.$$

If $\varepsilon \to 0$, we get

$$\mu_i'(a') \leqq \mu_e'(a'), \quad \text{q.e.d.}$$

A I.6.12. Theorem. If $a', b' \in A'$, $a' \leqq b'$, then

$$\mu_i'(a') \leqq \mu_i'(b').$$

Proof. By definition [A I.6.2b.], we have

$$\mu_i'(a') = \mu_e'(1') - \mu_e'(\mathrm{co}'\, a'),$$

$$\mu_i'(b') = \mu_e'(1') - \mu_e'(\mathrm{co}'\, b').$$

We subtract:

(1) $$\mu_i'(b') - \mu_i'(a') = \mu_e'(\mathrm{co}'\, a') - \mu_e'(\mathrm{co}'\, b').$$

Since $a' \leqq' b'$, we have $\mathrm{co}'\, b' \leqq \mathrm{co}'\, a'$, and therefore, [A I.6.3.],

$$\mu_e'(\mathrm{co}'\, b') \leqq \mu_e'(\mathrm{co}'\, a').$$

Hence, by (1),

$$\mu_i'(b') - \mu_i'(a') \geqq 0.$$

It follows

$$\mu_i'(a') \leqq \mu_i'(b'), \quad \text{q.e.d.}$$

A I.6.13. We notice that μ_i' is in general no convex measure. E.g. (Choquet): Take for (A) the finitely additive tribe whose somata are finite sums

$$\sum_i (p_i, q_i\rangle,$$

where $0 \leqq p_i$, $q_i \leqq 1$, and where the set-inclusion is taken as the ordering relation. Take for (A') the class of all subsets of $(0, 1\rangle$ with set-inclusion as ordering relation. (A') is denumerably additive. If we take $\mu(a)$ as the ordinary Lebesgue-measure, we have an effective measure on (A); $\mu_e'(a')$ is the Lebesgue-exterior measure and $\mu_i'(a')$ is the interior one. Now there exists a subset c' of $(0, 1\rangle$ such that $\mu_e'(c') = 1$, $\mu_i'(c') = 0$. If we put

$$d' =_{df} \mathrm{co}'\, c',$$

we have

$$\mu_e'(d') = 1, \qquad \mu_i'(d') = 0.$$

Thus

$$\mu_i'(c') + \mu_i'(d') = 0,$$

while

$$\mu_i'(c' +' d') = \mu_i'((0, 1\rangle) = \mathrm{meas}\,(0, 1\rangle = 1.$$

Hence

$$\mu_i'(c' +' d') > \mu_i'(c') + \mu_i'(d'),$$

which shows that μ_i' is not convex.

A I.6.14. Def. A soma $a' \in A'$ is said to be *μ-measurable,* whenever

$$\mu_i'(a') = \mu_e'(a').$$

In the case of measurability of a' we agree to write

$$\mu'(a'), \quad \text{instead of} \quad \mu_e'(a').$$

A I.6.15. Theorem. If $a \in A$, then a is μ-measurable.

Proof. Put

$$b =_{df} \operatorname{co} a =' \operatorname{co}' a.$$

Take $\varepsilon > 0$. Find a disjoint covering $\{c_n\}$ of I', such that

(1) $$\sum_n \mu(c_n) - \varepsilon \leqq \mu_e'(I').$$

Of course we have

$$\sum_n{}' c_n =' I'.$$

Put

$$x_n =_{df} a \cdot c_n, \quad y_n =_{df} b \cdot c_n.$$

We have

$$c_n = a\, c_n + \operatorname{co} a \cdot c_n = a \cdot c_n + b \cdot c_n = x_n + y_n.$$

All x_n, y_n are mutually disjoint, and we have

$$a = \sum_n{}' x_n, \quad b = \sum_m{}' y_m.$$

Since $c_n = x_n + y_n$, we have

$$\mu(c_n) = \mu(x_n) + \mu(y_n),$$

so, by (1),

(2) $$\sum_n \mu(x_n) + \sum_m \mu(y_m) - \varepsilon \leqq \mu_e'(I).$$

Now, since $\{x_n\}$ is a covering of a, and $\{y_n\}$ a covering of b, we have

$$\mu_e'(a) \leqq \sum_n \mu(x_n),$$

$$\mu_e'(b) \leqq \sum_n \mu(y_n).$$

Hence, by (2),

$$\mu_e'(a) + \mu_e'(b) - \varepsilon \leqq \mu_e'(I).$$

Letting tend ε to 0, we get

(3) $$\mu_e'(a) + \mu_e'(b) \leqq \mu_e'(I).$$

On the other hand, by [A I.6.9.],

$$\mu_e'(1) = \mu_e'(a + b) \leqq \mu_e'(a) + \mu_e'(b).$$

Hence, by (3),

$$\mu_e'(a) + \mu_e'(b) = \mu_e'(I),$$

i.e. by Def. [A I.6.2b.],

$$\mu'_e(a) = \mu'_e(1) - \mu'_e(\mathrm{co}\, a) = \mu'_i(a),$$

i.e. a is μ-measurable, q.e.d.

A I.6.16. Theorem. We shall use an analogous reasoning to prove that, if $a, b \in A$, $a \cdot b = O$, then

$$\mu'(a+b) = \mu'(a) + \mu'(b).$$

Proof. Take $\varepsilon > 0$. Find a disjoint covering $\{c_n\}$ of $a+b$ with $\sum\limits_n{}' c_n = a+b$, [A I.6.6.],

(1) $$\sum_n \mu(c_n) - \varepsilon \leqq \mu'_e(a+b) = \mu'(a+b).$$

Put

$$x_n =_{df} a \cdot c_n, \quad y_n =_{df} b \cdot c_n.$$

The x_n, y_m are mutually disjoint, and we have

(2) $$a = \sum_n{}' x_n, \quad b = \sum_n{}' y_n, \quad c_n = x_n + y_n,$$

$$\mu(c_n) = \mu(x_n) + \mu(y_n).$$

Hence, by (1),

(3) $$\sum_n \mu(x_n) + \sum_n \mu(y_n) - \varepsilon \leqq \mu'(a+b).$$

From (2) it follows

$$\mu'(a) \leqq \sum_n \mu(x_n), \quad \mu'(b) \leqq \sum_n \mu(y_n),$$

so, by (3), we get

$$\mu'(a) + \mu'(b) \leqq \mu'(a+b).$$

Since, by [A I.6.9.], the exterior measure is simply convex, we have

$$\mu'(a+b) \leqq \mu'(a) + \mu'(b).$$

This completes the proof.

A I.6.17. Theorem. If $a_1, a_2, \ldots, a_n, \ldots \in A$ are disjoint and

$$\sum_{n=1}^{\infty}{}' a_n =' a \in A,$$

then

$$\mu'(a) = \sum_{n=1}^{\infty} \mu'(a_n).$$

Proof. We have $a_n \leqq' a$; hence $a_n \leqq a$, $(n = 1, 2, \ldots)$. It follows that

$$a_1 + \cdots + a_n \leqq a;$$

and then

$$\mu'(a_1) + \cdots + \mu'(a_n) \leqq \mu'(a).$$

Hence

(1) $$\sum_{n=1}^{\infty} \mu'(a_n) \leqq \mu'(a).$$

Now, since, [A I.6.10.], μ_e' is denumerably convex, we also have:

$$\mu_e'\left(\sum_{n=1}^{\infty}{}' a_n\right) \leqq \sum_{n=1}^{\infty} \mu_e'(a_n), \quad \text{i.e.}$$

(2) $$\mu'(a) \leqq \sum_{n=1}^{\infty} \mu'(a_n).$$

(1) and (2) prove the theorem.

A I.6.18. Remarks and summary. The above device of defining an exterior measure on a given tribe by means of coverings, constitutes a generalization of the classic device, invented by H. LEBESGUE (23) for sets of real numbers, where the denumerable coverings are made out of intervals, thus creating the known theory of measurable sets, of measure, and of a generalized integration of functions.

In our general setting sets are replaced by general somata of a finitely additive tribe (T), and the euclidean measure of segments is replaced by a general finitely additive given measure $\mu(\dot{a})$. What we have made till now, shows that the somata of the tribe (T) are proved to be measurable, and even the defined measure is on (T) denumerably additive, if we are allowed to take infinite unions from a denumerably additive supertribe (T') of (T).

Considering the generalized measure μ_e, it may happen that $\mu_e(a) < \mu(a)$ for some somata a of (T), [A I.6.4. Remark.], so the measure μ_e cannot be considered as a genuine extension of the given measure μ, because we should expect to have $\mu'(a) = \mu(a)$ for all somata of (T), and not only $\mu_e(a) \leqq \mu(a)$.

However it may be shown that if we admit the condition, stating that for all $a \in T$ we have $\mu_e'(a) = \mu(a)$, this will give a necessary and sufficient condition for getting an excellent generalization of the classic Lebesgue's theory, (14).

Though this could be made by continuing our topic, we shall stop it, and prefer to go over to another device of extension of measure, which constitutes a generalization of the known Carathéodory's theory; and we shall use the reasonings taken from the known book by S. SAKS, (22).

Later it will be shown that the generalized Carathéodory's theory can replace the generalized Lebesgue's device, so it will be shown, that these theories are somehow equivalent, even if we do not suppose that $\mu_e(a) = \mu(a)$ for all $a \in T$.

Section 7 of A I.

A I.7. Carathéodory's measure theory. This theory was invented, to give new foundation for the Lebesgue's measure theory of sets. Now, we shall see that it can be as well applied to somata of a tribe, which are not sets. In what follows, we shall imitate the exhibition, given in SAKS's book (22) with some small improvement, as e.g. giving the condition 2^0, which is omitted in SAKS's book and which will be explicitely given below.

The theory is based on the notion of a kind of Lebesgue's exterior measure, which however is given in a generalized form. It is not denumerably additive but denumerably sub-additive. It applies to a single tribe and does not need any supertribe, as the Lebesgue'an theory, [A I.6.1.].

The theory will be later applied to circumstances in [A I.6.].

A I.7.1. We shall consider a non trivial denumerably additive tribe (A), and a non negative real valued function $\Gamma(\dot{a})$, defined over A,*) and satisfying the following conditions:

1^0. $\Gamma(a) \geqq 0$;

2^0. $\Gamma(O) = 0$; **)

3^0. if $a \leqq b$, then $\Gamma(a) \leqq \Gamma(b)$;

4^0. for any infinite sequence

$$a_1, a_2, \ldots, a_n, \ldots$$

of somata of (A) we have

$$\Gamma\left(\sum_{n=1}^{\infty} a_n\right) \leqq \sum_{n=1}^{\infty} \Gamma(a'_n),$$

where, in case of divergence, we replace the right hand side by $+\infty$.

Def. We call the function $\Gamma(\dot{a})$ of the variable soma $\dot{a}$, *Carathéodory's convex measure on* (A).

Remark. Later we shall show that the outer Lebesgue's measure, as considered in [A I.6.2.], is a sort of Carathéodory's convex measure.

A I.7.2. Def. The soma $a \in A$ is said to be *Γ-measurable*, whenever for every $x \in A$, we have

$$\Gamma(x) = \Gamma(x \cdot a) + \Gamma(x \cdot \mathrm{co}\, a).$$

A I.7.3. Theorem. The following are equivalent:

I. a is Γ-measurable;

II. for every $p \leqq a$ and every $q \leqq \mathrm{co}\, a$ we have

$$\Gamma(p + q) = \Gamma(p) + \Gamma(q);$$

III. for every $x \in A$ we have

$$\Gamma(x) \geqq \Gamma(x \cdot a) + \Gamma(x \cdot \mathrm{co}\, a).$$

---

*) We shall consider only the situation, where $\Gamma(a) < \infty$ for all $a \in A$. In SAKS's book the value $+\infty$ is allowed.

**) This condition is unfortunately omitted in SAKS's book.

Proof. Let II. Take any $x \in A$. We have

(1) $$x = x \cdot a + x \cdot \text{co}\, a.$$

Since $x \cdot a \leqq a$, $x \cdot \text{co}\, a \leqq \text{co}\, a$, we have, by II,

$$\Gamma(x \cdot a + x \cdot \text{co}\, a) = \Gamma(x \cdot a) + \Gamma(x \cdot \text{co}\, a),$$

i.e., by (1):

$$\Gamma(x) = \Gamma(x \cdot a) + \Gamma(x \cdot \text{co}\, a). \quad \text{i.e. I.}$$

Thus

(2) $$\text{II} \to \text{I}.$$

Obviously

(3) $$\text{I} \to \text{III}.$$

Suppose III. We have

$$x \cdot a + x \cdot \text{co}\, a + O + O + \cdots = x.$$

Hence, by [A I.7.1., 4⁰],

$$\Gamma(x) \leqq \Gamma(x \cdot a) + \Gamma(x \cdot \text{co}\, a) + \Gamma(O) + \Gamma(O) + \cdots.$$

By [A I.7.1., 2⁰] we have $\Gamma(O) = 0$. Hence

$$\Gamma(x) \leqq \Gamma(x \cdot a) + \Gamma(x \cdot \text{co}\, a),$$

which, together with

$$\Gamma(x) \geqq \Gamma(x \cdot a) + \Gamma(x \cdot \text{co}\, a), \quad \text{gives I}.$$

Thus

(4) $$\text{III} \to \text{I}.$$

Suppose I. Take $p \subseteq a$ and $q \subseteq \text{co}\, a$, and put

$$x =_{df} p + q.$$

We have

$$x \cdot a = (p + q) \cdot a = p\, a + q\, a.$$

But $q \leqq \text{co}\, a$, so $q\, a = 0$. Hence

(5) $$x \cdot a = p\, a = p.$$

In the same way we prove that

(6) $$x \cdot \text{co}\, a = q.$$

Since, by I, for every x, $\Gamma(x) = \Gamma(x \cdot a) + \Gamma(x \cdot \text{co}\, a)$, it follows, by (5), (6), that

$$\Gamma(x) = \Gamma(p) + \Gamma(q),$$

i.e.

$$\Gamma(p + q) = \Gamma(p) + \Gamma(q).$$

Hence

(7) $$\text{I} \to \text{II}.$$

The results (2), (7), (3), (4) prove our theorem.

A I.7.4. Theorem. For any $a, b \in A$ we have

$$\Gamma(a+b) \leqq \Gamma(a) + \Gamma(b).$$

Proof.

$$a + b = a + b + O + O \ldots$$

Hence, by [A I.7.1., 4⁰],

$$\Gamma(a+b) \leqq \Gamma(a) + \Gamma(b) + \Gamma(O) + \Gamma(O) + \cdots$$

The application of [A I.7.1., 2⁰] completes the proof.

A I.7.5. Theorem. The soma O is Γ-measurable.

Proof. Take any $x \in A$. We have

$$\Gamma(x \cdot O) = \Gamma(O) = 0,$$

$$\Gamma(x \cdot \mathrm{co}\, O) = \Gamma(x).$$

Hence $\Gamma(x) = \Gamma(x \cdot O) + \Gamma(x \cdot \mathrm{co}\, O)$ for every $x \in A$; which proves the theorem.

A I.7.6. Theorem. If a is Γ-measurable, so is $\mathrm{co}\, a$.

Proof. We have $a = \mathrm{co}(\mathrm{co}\, a)$. Let $q \leqq \mathrm{co}(\mathrm{co}\, a)$, $p \leqq \mathrm{co}\, a$. We have $q \leqq a$, $p \leqq \mathrm{co}\, a$; hence

$$\Gamma(p+q) = \Gamma(p) + \Gamma(q),$$

which, by [A I.7.3., II], completes the proof.

A I.7.7. Theorem. If $a_1, a_2, \ldots, a_n, \ldots$ are Γ-measurable and disjoint, and

$$a = \sum_{n=1}^{\infty} a_n,$$

then a is also Γ-measurable and

$$\Gamma(a) = \sum_{n=1}^{\infty} \Gamma(a_n).$$

Proof.[1] Put

$$b_n =_{df} \sum_{i=1}^{n} a_i, \quad (n = 1, 2, \ldots).$$

First we shall prove the following lemma: Under hypotheses of the theorem, if $x \in A$, then b_n is Γ-measurable and

$$\Gamma(x) \geqq \sum_{i=1}^{n} \Gamma(x\, a_i) + \Gamma(x \cdot \mathrm{co}\, b_n), \quad n = 1, 2, \ldots \tag{1}$$

The proof will be by induction:

Let $n = 1$. The formula (1) becomes

$$\Gamma(x) \geqq \Gamma(x \cdot a_1) + \Gamma(x \cdot \mathrm{co}\, a_1),$$

which is true, because a_1 is Γ-measurable.

Since $b_1 = a_1$, the soma b_1 is measurable too.

[1] Taken without change from SAKS's book.

Suppose the lemma be true for $n \leqq p$, $(p = 1, 2, \ldots)$. We have

(2) $$\Gamma(x) = \Gamma(x \cdot a_{p+1}) + \Gamma(x \cdot \mathrm{co}\, a_{p+1}).$$

Since b_p is Γ-measurable (by hypothesis), therefore we have

(3) $$\Gamma(x \cdot \mathrm{co}\, a_{p+1}) = \Gamma(x \cdot \mathrm{co}\, a_{p+1} \cdot b_p) + \Gamma(x \cdot \mathrm{co}\, a_{p+1} \cdot \mathrm{co}\, b_p).$$

Now $b_p \cdot a_{p+1} = O$, because $a_1, a_2, \ldots$ are disjoint. Hence

(4) $$b_p \leqq \mathrm{co}\, a_{p+1}, \quad \text{and then} \quad \mathrm{co}\, a_{p+1} \cdot b_p = b_p.$$

Since $b_{p+1} = b_p + a_{p+1}$, we have

(5) $$\mathrm{co}\, b_{p+1} = \mathrm{co}\, b_p \cdot \mathrm{co}\, a_{p+1}.$$

Carrying (4) and (5) into (3), we get

$$\Gamma(x \cdot \mathrm{co}\, a_{p+1}) = \Gamma(x \cdot b_p) + \Gamma(x \cdot \mathrm{co}\, b_{p+1}),$$

and then, from (2),

(6) $$\Gamma(x) = \Gamma(x \cdot a_{p+1}) + \Gamma(x \cdot b_p) + \Gamma(x \cdot \mathrm{co}\, b_{p+1}).$$

Now, we have supposed that for every $y \in A$

$$\Gamma(y) \geqq \sum_{i=1}^{p} \Gamma(y \cdot a_i) + \Gamma(y \cdot \mathrm{co}\, b_p).$$

Hence

$$\Gamma(x \cdot b_p) \geqq \sum_{i=1}^{p} \Gamma(x \cdot b_p \cdot a_i) + \Gamma(x \cdot b_p \cdot \mathrm{co}\, b_p) \geqq \sum_{i=1}^{p} \Gamma(x \cdot a_i) + 0.$$

Hence, from (6),

$$\Gamma(x) \geqq \Gamma(x \cdot a_{p+1}) + \sum_{i=1}^{p} \Gamma(x \cdot a_i) + \Gamma(x \cdot \mathrm{co}\, b_{p+1}),$$

i.e.

(7) $$\Gamma(x) \geqq \sum_{i=1}^{p+1} \Gamma(x \cdot a_i) + \Gamma(x \cdot \mathrm{co}\, b_{p+1}),$$

which constitutes precisely the formula (1) for $n = p + 1$. Now we shall prove that b_{p+1} is Γ-measurable. Since

$$x \cdot b_{p+1} = x \cdot a_1 + \cdots + x \cdot a_{p+1},$$

it follows, by [A I.7.1., 4^0],

$$\Gamma(x \cdot b_{p+1}) \leqq \sum_{i=1}^{p+1} \Gamma(x \cdot a_i).$$

Hence, from (7)

$$\Gamma(x) \geqq \Gamma(x \cdot b_{p+1}) + \Gamma(x \cdot \mathrm{co}\, b_{p+1}),$$

which proves the Γ-measurability of b_{p+1}. Thus, by induction, the lemma is proved, i.e. all $b_1, b_2, \ldots, b_n, \ldots$ are Γ-measurable, and for any $x \in A$, we have

$$\Gamma(x) \geqq \sum_{i=1}^{n} \Gamma(x \cdot a_i) + \Gamma(x \cdot \operatorname{co} b_n), \qquad (n = 1, 2, \ldots). \tag{8}$$

Having this, we notice that

$$\Gamma(x \cdot \operatorname{co} a) \leqq \Gamma(x \cdot \operatorname{co} b_n),$$

because $b_n \leqq a$, and on account of [A I.7.1., 3⁰]. Hence, from (8),

$$\Gamma(x) \geqq \sum_{i=1}^{n} \Gamma(x \cdot a_i) + \Gamma(x \cdot \operatorname{co} a), \qquad (n = 1, 2, \ldots): \tag{9}$$

the clue-inequality in the proof.

It follows that

$$\sum_{i=1}^{\infty} \Gamma(x \cdot a_i)$$

converges, because of [A I.7.1., 1⁰] and we have

$$\Gamma(x) \geqq \sum_{i=1}^{\infty} \Gamma(x \cdot a_i) + \Gamma(x \cdot \operatorname{co} a). \tag{10}$$

Since

$$\Gamma(x \cdot a) = \Gamma\left(x \cdot \sum_{i=1}^{\infty} a_i\right) = \Gamma\left(\sum_{i=1}^{\infty} x \cdot a_i\right) \leqq \sum_{i=1}^{\infty} \Gamma(x \cdot a_i),$$

by [A I.7.1., 4⁰], we get from (10)

$$\Gamma(x) \geqq \Gamma(x \cdot a) + \Gamma(x \cdot \operatorname{co} a),$$

which proves, by [A I.7.2.], the Γ-measurability of a. Now, take a for x in (10), which is valid for any x. We get

$$\Gamma(a) \geqq \sum_{i=1}^{\infty} \Gamma(a \cdot a_i) + \Gamma(a \cdot \operatorname{co} a), \quad \text{i.e.}$$

$$\Gamma(a) \geqq \sum_{i=1}^{\infty} \Gamma(a_i) + 0. \tag{11}$$

Since $a = \sum_{i=1}^{\infty} a_i$, we have by [A I.7.1., 4⁰],

$$\Gamma(a) \leqq \sum_{i=1}^{\infty} \Gamma(a_i).$$

This together with (11) gives

$$\Gamma(a) = \sum_{i=1}^{\infty} \Gamma(a_i).$$

The proof is finished.

A I.7.8. Theorem. If a, b are Γ-measurable, so is $a - b$.

Proof (taken from SAKS's book, **(**22**)**). Let

$$x \leqq a - b, \qquad y \leqq \operatorname{co}(a - b).$$

Put

$$y_1 =_{df} y \cdot b, \quad y_2 =_{df} y \cdot \text{co}\, b.$$

Since b is Γ-measurable, and since $y_1 \leqq b$, $y_2 \leqq \text{co}\, b$, we have, by [A I.7.3., II],

(1) $$\Gamma(y_1 + y_2) = \Gamma(y_1) + \Gamma(y_2).$$

Now, we have

(2) $$x \leqq a$$

and

(3) $$y_2 \leqq \text{co}\, a.$$

Indeed, (2) is obvious, and (3) can be obtained in the following way:

$$y_2 \leqq \text{co}(a - b) \cdot \text{co}\, b = \text{co}(a \cdot \text{co}\, b) \cdot \text{co}\, b = (\text{co}\, a + b) \cdot \text{co}\, b$$
$$= \text{co}\, a \cdot \text{co}\, b \leqq \text{co}\, a.$$

Since a is Γ-measurable, the formulas (2) and (3) imply

(4) $$\Gamma(x + y_2) = \Gamma(x) + \Gamma(y_2).$$

Now, we have

$$y_1 \leqq b, \quad x + y_2 \leqq \text{co}\, b.$$

Hence, b being Γ-measurable,

(5) $$\Gamma(y_1 + (x + y_2)) = \Gamma(y_1) + \Gamma(x + y_2).$$

From (4) and (5) it follows:

$$\Gamma(y_1 + x + y_2) = \Gamma(y_1) + \Gamma(x) + \Gamma(y_2),$$

and then, on account of (1),

$$\Gamma(x + y_1 + y_2) = \Gamma(x) + \Gamma(y_1 + y_2).$$

Since $y_1 + y_2 = y$, we get

$$\Gamma(x + y) = \Gamma(x) + \Gamma(y),$$

and this is true for any $x \leqq a - b$, $y \leqq \text{co}(a - b)$. Hence, by [A I.7.3., II], $a - b$ is Γ-measurable.

A I.7.9. Theorem. If a, b are Γ-measurable, so is $a \cdot b$.

Proof. Since b is Γ-measurable, so is $\text{co}\, b$, by [A I.7.6.], and then, since $a - \text{co}\, b = a \cdot \text{co}\,\text{co}\, b = a \cdot b$, we see that $(a - \text{co}\, b)$ is Γ-measurable, by [A I.7.8.], q.e.d.

A I.7.10. Theorem. If a, b are Γ-measurable, so is $a + b$.

Proof.

(1) $$a + b = \text{co}(\text{co}\, a \cdot \text{co}\, b),$$

and both $\text{co}\, a$ and $\text{co}\, b$ are Γ-measurable. Hence $\text{co}\, a \cdot \text{co}\, b$ is, by [A I.7.9.], Γ-measurable, and then, by (1), $a + b$ too.

A I.7.11. Theorem. If $a_1, a_2, \ldots, a_n$ are Γ-measurable, so is $\sum_{i=1}^{n} a_i$. This follows from [A I.7.10.].

A I.7.12. Theorem. If $a_1, a_2, \ldots, a_n, \ldots$ are Γ-measurable, then

$$a =_{df} \sum_{n=1}^{\infty} a_n$$

is also Γ-measurable.

Proof. We have, by putting

$$b_1 =_{df} a_1, \ \ldots, \ b_n =_{df} a_n - (a_1 + \cdots + a_{n-1}) \quad \text{for} \quad n > 1: \ a = \sum_{n=1}^{\infty} b_n,$$

where b_n are mutually disjoint.

Then [A I.7.7.] completes the proof.

A I.7.13. Theorem. If $\Gamma(a) = 0$, then a is Γ-measurable.

Proof. Let

$$p \leqq a, \quad q \leqq \text{co}\, a.$$

We have

$$0 \leqq \Gamma(p) \leqq \Gamma(a) = 0;$$

hence

(1) $$\Gamma(p) = 0.$$

On the other hand

$$\Gamma(p + q) \leqq \Gamma(p) + \Gamma(q).$$

Hence, by (1),

(2) $$\Gamma(p + q) \leqq \Gamma(q).$$

Since $q \leqq p + q$, we have $\Gamma(q) \leqq \Gamma(p + q)$, and then, by (1) and (2),

$$\Gamma(p + q) = \Gamma(q) = \Gamma(p) + \Gamma(q),$$

which, by [A I.7.3., II], proves that a is Γ-measurable.

A I.7.14. Theorem. The above theorems [A I.7.]—[A I.7.13.] show that the class A_1 of all Γ-measurable somata of (A) gives a denumerably additive Boolean strict subtribe of (A) and a denumerably genuine one. The measure Γ is denumerably additive on (A_1).

AI.7.15. Remarks and summary. The Carathéodory's classic theory deals with sets, but our general approach considers instead somata of a given general denumerably additive tribe (A). The form of definitions and theorems of the generalized theory is just the same, as in the classic theory.

In Lebesgue's theory we admit that the tribe (A) is plunged in a wider one, which is denumerably additive: This just for having the possibility of using denumerable operations. We do not need to have that in the Carathéodory's theory, since the given tribe (A) is supposed to be denumerably additive.

We define in Carathéodory's theory the notion of measurable somata of (A), and the measure is proved to be denumerably additive in the collection of all Carathéodory-measurable somata.

Now we are going to see that the Lebesgue's exterior measure μ_e' is a particular case of Carathéodory's (exterior) measure, so the Carathéodory's theory can be directly applied to Lebesgue's concepts.

We shall also prove that, in this particular case, the Carathéodory's measurability coincides with the Lebesgue's one. Thus we shall also get the proof, that the Lebesgue's measure is denumerably additive on the collection of all Lebesgue's measurable sets.

Section 8 of A I.

A I.8. Confrontation of methods of Lebesgue and of Carathéodory. After having proved some basic theorems on Carathéodory's measurability, we are going back to the measure $\mu_e'(a')$ on (A'), [see A I.6.14.], in order to compare the effectiveness of the two devices of extending the tribe (A).

A I.8.1. First of all we shall prove that $\mu_e'(a')$ is a Carathéodory's convex measure.

Theorem. Let (A) be a finitely additive tribe provided with an additive (finite) measure $\mu(\dot{a})$, and let (A') be a denumerably additive tribe, a finitely genuine supertribe of (A). We consider the outer measure $\mu_e'(a')$ of somata of (A'), as in [A I.6.2.].

Under these conditions $\mu_e'(a')$ is the biggest Carathéodory's convex measure $\Gamma(a')$ on (A'), such that for every $a \in A$ we have

$$\Gamma(a) \leqq \mu(a). \tag{1}$$

Proof. Let $\Gamma(a)$ be a Carathéodory convex measure on (A'), such that (1) is fulfilled. Let $a' \in A'$, and let

$$a' \leqq' \sum_n{}' a_n,$$

where $\{a_n\}$ is a covering of a'. We have

$$a' =' a' \cdot \sum_n{}' a_n =' \sum_n (a' a_n).$$

Hence

$$\Gamma(a') \leqq \sum_n \Gamma(a' a_n) \leqq \sum_n \Gamma(a_n),$$

and, by (1),

$$\Gamma(a') \leqq \sum_n \mu(a_n).$$

This holding for any covering $\{a_n\}$ of a', it follows:

$$\Gamma(a') \leqq \inf_{\{a_n\}} \sum \mu(a_n), \quad \text{i.e.}$$

$$\Gamma(a') \leqq \mu_e'(a'),$$

which completes the proof.

A I.8.1a. We can consider the Carathéodory's measurability of somata of (A'), taken with respect to the Carathéodory's convex measure μ'_e.

Def. If a soma $a' \in A'$ is in this sense measurable, we shall say that *a' is μ'_e C-measurable*. The set of all μ'_e C-measurable somata will be denoted by C^*.

A I.8.2. As we know that $\mu'_e(a')$ is a Carathéodory's convex measure on (A'), we can apply the results got in [A I.7.1]—[A I 7.14.].

The class C^* of all μ'_e C-measurable somata, if ordered as in (A'), is a denumerably additive Boolean tribe: a denumerably genuine strict subtribe of (A'), [A I.7.14.].

Now, we shall prove, by applying a method used in Carathéodory's quoted work (7), the

A I.8.2a. Theorem. If $a \in A$, then a is μ'_e C-measurable.

Proof. Let $x' \in A'$. Put

(1) $$u' =_{df} x' \cdot a, \qquad v' =_{df} x' \cdot \mathrm{co}\, a.$$

Take $\varepsilon > 0$, and a covering of x':

(2) $$x' \leqq \sum_i{}' a_i,$$

with

(2.1) $$\sum_i \mu(a_i) \leqq \mu'_e(x') + \varepsilon,$$

[A I.6.7.]. Put

$$x_i =_{df} a_i \cdot a, \qquad y_i =_{df} a_i \cdot \mathrm{co}\, a, \qquad (i = 1, 2, \ldots).$$

We have, by (1), (2):

$$u' \leqq a \cdot \sum_i{}' a_i = \sum_i{}' a \cdot a_i,$$

$$v' \leqq \mathrm{co}\, a \cdot \sum_i{}' a_i = \sum_i{}' (a_i \cdot \mathrm{co}\, a);$$

hence

(3) $$u' = \sum_i{}' x_i, \qquad v' = \sum_i{}' y_i.$$

Now,

$$a_i = a_i \cdot a + a_i \cdot \mathrm{co}\, a = x_i + y_i, \qquad x_i \cdot y_i = 0.$$

Since μ is additive in (A), we have

$$\mu(a_i) = \mu(x_i) + \mu(y_i).$$

Hence

$$\sum_i \mu(a_i) = \sum_i \mu(x_i) + \sum_i \mu(y_i),$$

and then, by (2.1),

$$\mu'_e(x') + \varepsilon \geqq \sum_i \mu(x_i) + \sum_i \mu(y_i).$$

Hence, from (1), and by [A I.6.10.],

$$\mu_e'(x') + \varepsilon \geqq \mu_e'(u') + \mu_e'(v').$$

Tending ε to 0, we get

$$\mu_e'(x') \geqq \mu_e'(u') + \mu_e'(x'),$$

i.e., by (1):

$$\mu_e'(x') \geqq \mu_e'(x' a) + \mu_e'(x' \cdot \mathrm{co}\, a),$$

for any $x' \in A'$. This proves the $(\mu_e' C)$-measurability of a.

A I.8.3. Theorem. The above theorem [A I.8.2a.] and the general results [A I.7.5.]—[A I.7.14.] show that the class C^*, i.e. the class of all $\mu_e' C$-measurable somata of (A') contains all somata of (A) and all somata which can be obtained from them by applying denumerable somatic operations. Hence C^* contains all somata of (A') which belong to the Borelian extension of (A) within of (A'), [A I.4.6a.], [A I.4.7.]. The class C^* contains also all $a' \in A'$, for which $\mu_e'(a') = 0$, [A I.7.13.]. Hence C^* contains the denumerably additive measure-hull of the above, Borelian extension: [A I.5.17.].

A I.8.3a. Def. Let us agree to write $\mu^*(a')$, instead of $\mu_e'(a')$, in the case of $\mu_e' C$-measurability of the soma $a' \in A'$, i.e. for $a' \in C^*$.

We shall say that a' is *μ^*-measurable* whenever $a' \in C^*$, i.e. when a' is $\mu_e' C$-measurable.

A I.8.4. Theorem. The measure μ^* is, on C^*, denumerably additive, i.e. if $a_1', a_2', \ldots, a_n', \ldots \in C^*$, and are disjoint, then the soma

$$a' =_{df} \sum_{n=1}^{\infty} a_n'$$

also belongs to C^* and we have

$$\mu^*(a') = \sum_{n=1}^{\infty} \mu^*(a_n').$$

Proof. This follows directly from [A I.8.3a.] and [A I.7.7.].

A I.8.4a. Theorem.

$$\mu^*(O') = \mu_e'(O') = \mu(O) = 0.$$

$$\mu^*(I') = \mu_e'(I') \leqq \mu(I).$$

Proof. This is because (A) is a finitely genuine subtribe of (A'), [A.7.].

A I.8.5. Def. The soma $a' \in A'$ is said to be *$\mu_e' L$-measurable,* whenever

$$\mu_i'(a') = \mu_e'(a').$$

This is the same as μ'-measurable, [A I.6.14.]. If a' is $\mu_e' L =$ measurable, its measure will be denoted by $\mu'(a')$, as in [A I.6.14.]. We have

$$\mu'(a') = \mu'_i(a') = \mu'_e(a').$$

A I.8.6. Theorem. If a' is μ'_e C-measurable, then it is μ'_e L-measurable, and we have

$$\mu^*(a') = \mu'(a').$$

Proof. We have

$$a' +' \mathrm{co}' a' =' I'.$$

Since $a' \in C^*$, the soma $\mathrm{co}' a'$ also belongs to C^*, [A I.7.6.]. Since $a' \cdot' \mathrm{co}' a' =' 0'$, we have, [A I.7.7.],

$$\mu^*(a') + \mu^*(\mathrm{co}' a') = \mu^*(I').$$

Hence, [A I.8.3a.], [A I.8.4a],

$$\mu'_e(a') + \mu'_e(\mathrm{co}\, a') = \mu'(I');$$

hence

$$\mu'(I') - \mu'_e(\mathrm{co}\, a') = \mu'_e(a').$$

Hence [A I.6.2b.]

$$\mu'_i(a') = \mu'_e(a').$$

Thus a' is μ'_e L-measurable.

By definition [A I.8.3a.] we have

$$\mu'_e(a') = \mu^*(a'),$$

because a' is μ'_e C-measurable. Hence

$$\mu'(a') = \mu^*(a').$$

The theorem is established.

A I.8.6a. We like to prove the converse of the preceeding theorem. To do that, we shall prove the following.

Lemma. If

1. $a' \in A'$ is μ'_e L-measurable,
2. $\varepsilon > 0$,

then there exists a μ'_e C-measurable soma b' such that

1) $a' \leqq b'$,
2) $\mu'_e a' \leqq \mu^* b' \leqq \mu'_e a' + \varepsilon$.

A I.8.6b. Proof. Let a' be μ'_e L-measurable. In agreement with [Section 6 of A I.], a' can be approximated, as close as we like, by covering made up of somata of (A).

Now we shall find an infinite diminishing sequence of nested coverings approaching a'. These coverings will supply the required b'.

A I.8.6c. To give this idea a precise form, take a positive number ε. Let us take a covering $a_1, a_2, \ldots, a_n, \ldots$ of a', where $a_n \in A$, where the somata are mutually disjoint, and where

$$(1)\qquad \sum_{n=1}^{\infty} \mu(a_n) \leqq \mu'_e(a') + \varepsilon .$$

By [A I.6.7.] such a covering exists.

By [A I.8.2a] a_n is μ'_e C-measurable, and we have

$$(2)\qquad \mu^*(a_n) \leqq \mu(a_n), \quad (n = 1, 2, \ldots).$$

It follows

$$(3)\qquad \sum_{n=1}^{\infty} \mu^*(a_n) \leqq \sum_{n=1}^{\infty} \mu(a_n).$$

Since

$$a' \leqq \sum_{n=1}^{\infty} a_n ,$$

we have, [A I.7.1.],

$$(4)\qquad \mu'_e(a') \leqq \mu'_e\left(\sum_{n=1}^{\infty} a_n\right),$$

because μ'_e is a Carathoédory's convex measure, [A I.8.1.]. We also have, [A I.7.1.],

$$\mu'_e\left(\sum_{n=1}^{\infty} a_n\right) \leqq \sum_{n=1}^{\infty} \mu'_e(a_n).$$

From (4) we get

$$(5)\qquad \mu'_e(a') \leqq \sum_{n=1}^{\infty} \mu'_e(a_n) = \sum_{n=1}^{\infty} \mu^*(a_n),$$

and then we obtain the inequalities:

$$(6)\qquad \mu'_e(a) \leqq \sum_{n=1}^{\infty} \mu^*(a_n) \leqq \sum_{n=1}^{\infty} \mu(a_n) \leqq \mu'_e(a') + \varepsilon ,$$

where the first inequality is taken from (5), the second from (2), and the third from (1). Thus we have proved that in general, if $\{a_n\}$ is a disjoint covering of a' with $a_n \in A$, and satisfying the inequality (1), then the relation (6) takes place.

A I.8.6d. The somata a_n are disjoint, hence we have, by [A I.8.4.],

$$\sum_{n=1}^{\infty} \mu^*(a_n) = \mu^*\left(\sum_{n=1}^{\infty} a_n\right).$$

Put

$$b' =_{df} \sum_{n=1}^{\infty} a_n .$$

Since all $a_n \in C^*$, we have, [A I.7.11.],

$$b' \in C^* .$$

Thus we get from (6):

$$\mu'(a') \leqq \mu^*(b') \leqq \mu'_e(a') + \varepsilon .$$

Since we also have

$$a' \leqq \sum_{n=1}^{\infty} a_n,$$

we get

$$a' \leqq b'.$$

The lemma is established.

A I.8.6e. Having that, we are going to prove that there exists a soma $c' \in C^*$ such that

1) $a' \leqq c'$,

2) $\mu'_e(a') = \mu^*(c')$.

A I.8.6f. We mentioned, [A I.8.6b.], that we need to have a diminishing sequence of nested coverings of a'. To have that, first we shall consider two arbitrary coverings $\{a_n\}$, $\{b_n\}$ of a', each disjoint.

The collection of all somata a_α, b_β is denumerable. It is a covering of a', so we have

$$a' \leqq \sum_{\alpha=1}^{\infty} \sum_{\beta=1}^{\infty} a_\alpha \cdot b_\beta. \tag{7}$$

Indeed, we have

$$\sum_{\alpha=1}^{\infty} \sum_{\beta=1}^{\infty} a_\alpha b_\beta = \left(\sum_{\alpha=1}^{\infty} a_\alpha\right)\left(\sum_{\beta=1}^{\infty} b_\beta\right).$$

Since $a' \leqq \sum_{\alpha=1}^{\infty} a_\alpha$ and $a' \leqq \sum_{\beta=1}^{\infty} b_\beta$, (7) is true. The somata $a_\alpha b_\beta$ belong to A. They are mutually disjoint.

Indeed

$$(a_\alpha b_\beta) \cdot (a_{\alpha'} b_{\beta'}) = (a_\alpha a_{\alpha'}) \cdot (b_\beta b_{\beta'}). \tag{8}$$

Since all a_α are disjoint and all b_β are disjoint, it follows that (8) equals O, whenever $\alpha \neq \alpha'$ or $\beta \neq \beta'$. Thus disjointedness is proved.

Since $a_\alpha b_\beta \leqq a_\alpha$ and $a_\alpha b_\beta \leqq b_\beta$, the covering $\{a_\alpha b_\beta\}$ is nested in the covering $\{a_n\}$ as well as in $\{b_\beta\}$.

A I.8.6g. We have

$$\sum_{\alpha,\beta} \mu(a_\alpha b_\beta) = \sum_{\alpha,\beta} \mu^*(a_\alpha b_\beta),$$

by [A I.8.3.], hence

$$\sum_{\beta,\alpha} \mu^*(a_\alpha b_\beta) = \sum_{\beta} \left(\sum_{\alpha} \mu^*(a_\alpha b_\beta)\right). \tag{9}$$

Now since

$$a_1 b_\beta,\ a_2 b_\beta, \ldots$$

are disjoint somata of C^*, we get

$$\sum_{\alpha} \mu^*(a_\alpha b_\beta) = \mu^*\left(\sum_{\alpha} a_\alpha b_\beta\right),$$

hence

$$\sum_{\beta} \left(\sum_{\alpha} \mu^*(a_\alpha b_\beta)\right) = \sum_{\beta} \mu^*\left(\sum_{\alpha} a_\alpha b_\beta\right) = \sum_{\beta} \mu^*\left(b_\beta \sum_{\alpha} a_\alpha\right) \leqq \sum_{\beta} \mu^*(b_\beta). \tag{10}$$

Since $\mu^*(b_\beta) \leqq \mu(b_\beta)$, [A I.8.1.], we get from (8), (9), (10),

(11) $$\sum_{\alpha, \beta} \mu(a_\alpha b_\beta) \leqq \sum_\beta \mu(b_\beta);$$

hence $\{a_\alpha \cdot b_\beta\}$ is nested in $\{b_\beta\}$.

Similarly we prove that

(11a) $$\sum_{\alpha, \beta} \mu(a_\alpha b_\beta) \leqq \sum_\alpha \mu(a_\alpha).$$

A I.8.6h. Let $\varepsilon > 0$ and $\varepsilon_1 > 0$. Find coverings $\{a_n\}$, $\{b_n\}$ such that

$$\sum_{n=1}^{\infty} \mu(a_n) \leqq \mu'_e(a') + \varepsilon,$$

$$\sum_{n=1}^{\infty} \mu(b_n) \leqq \mu'_e(a') + \varepsilon_1.$$

Then from (11) and (11a) we get

$$\sum_{\alpha, \beta} \mu(a_\alpha b_\beta) \leqq \mu'_e(a') + \varepsilon$$

and

$$\sum_{\alpha, \beta} \mu(a_\alpha b_\beta) \leqq \mu'_e(a') + \varepsilon_1.$$

Hence for the covering $\{a_\alpha b_\beta\}$ we get the inequality

(12) $$\sum_{\alpha, \beta} \mu(a_\alpha b_\beta) \leqq \mu'_e(a') + \min(\varepsilon, \varepsilon_1),$$

and then, by the general remark at the end of [A I.8.6c.],

(13) $$\mu'_e(a) \leqq \sum_{\alpha, \beta = 1}^{\infty} \mu^*(a_\alpha b_\beta) \leqq \sum_{\alpha, \beta = 1}^{\infty} \mu(a_\alpha b_\beta) \leqq \mu'_e(a') + \min(\varepsilon, \varepsilon_1).$$

A I.8.6i. The above argument will be now applied in order to get an infinite sequence of disjoint coverings of a'. Let $\{a_n^{(k)}\}$, $(k = 1, 2, \ldots)$ be a disjoint covering of a', such that

$$\sum_{n=1}^{\infty} \mu(a_n^{(k)}) \leqq \mu'_e(a') + \frac{1}{k}, \qquad k = 1, 2, \ldots$$

Denote by

(14) $$b_n^{(1)}, b_n^{(2)}, \ldots, b_n^{(s)}, \ldots, \qquad (s = 1, 2, \ldots),$$

the coverings

$$\{a_{p(1)}^{(1)}\}, \{a_{p(1)}^{(1)} a_{p(2)}^{(2)}\}, \{a_{p(1)}^{(1)} a_{p(2)}^{(2)} a_{p(3)}^{(3)}\}, \ldots,$$

where $p(1), p(2), \ldots$ are variable indices.

This is a diminishing sequence of nested disjoint coverings, made by using the construction shown in [A I.8.6f.] and [A I.8.6g.].

By (13) we get

$$\sum_{n=1}^{\infty} \mu(b_n^{(s)}) \leqq \mu'_e(a') + \frac{1}{s},$$

and

$$\mu'_e(a') \leqq \sum_{n=1}^{\infty} \mu^*(b_n^{(s)}) \leqq \sum_{n=1}^{\infty} \mu(b_n^{(s)}) \leqq \mu'_e(a') + \frac{1}{s}.$$

We have

$$\sum_{n=1}^{\infty} b_n^{(1)} \geqq \sum_{n=1}^{\infty} b_n^{(2)} \geqq \cdots \geqq \sum_{n=1}^{\infty} b_n^{(s)} \geqq \cdots \geqq a'; \tag{15}$$

hence

$$\mu^*\left(\sum_{n=1}^{\infty} b_n^{(1)}\right) \geqq \mu^*\left(\sum_{n=1}^{\infty} b_n^{(2)}\right) \geqq \cdots \geqq \mu^*\left(\sum_{n=1}^{\infty} b_n^{(s)}\right) \geqq \cdots \geqq \mu'_e(a'). \tag{16}$$

A I.8.6j. It follows that

$$c' =_{df} \prod_{s=1} \sum_{n=1} b_n^{(s)} \geqq a'. \tag{17}$$

We have $c' \in C^*$, [A I.7.6.], [A I.7.7.]. Since the measure μ^* is denumerably additive on C^*, we have

$$\lim_{s\to\infty} \mu^*\left(\sum_{n=1}^{\infty} b_n^{(s)}\right) = \mu^*(c'),$$

and since

$$\mu'_e(a') \leqq \sum_{n=1}^{\infty} \mu(b_n^{(s)}) < \mu'(a') + \frac{1}{s},$$

it follows, by $s \to \infty$,

$$\mu'_e(a') = \lim_{s\to\infty} \sum_{n=1}^{\infty} \mu^*\left(\sum_{n=1}^{\infty} b_n^{(s)}\right);$$

and we get

$$\mu^*(c') = \mu'_e(a'),$$

so the statement [A I.8.6e.] is proved.

A I.8.6k. We have proved a bit more, viz.:

There exists a double sequence $\{b_n^{(s)}\}$ of somata of (A) such that

1) $a' \leqq \prod_{s=1}^{\infty} \sum_{n=1}^{\infty} b_n^{(s)}$,

2) $\sum_{n=1}^{\infty} b_n^{(1)} \geqq \sum_{n=1}^{\infty} b_n^{(2)} \geqq \cdots$,

3) $\mu^*\left(\prod_{s=1}^{\infty} \sum_{n=1}^{\infty} b_n^{(s)}\right) = \mu'_e(a')$,

4) $b_1^{(s)}, b_2^{(s)}, \ldots, b_n^{(s)}, \ldots$ are disjoint.

A I.8.7. Let us apply the obtained results to the soma

$$\operatorname{co} a' = I' - a'.$$

Find a double sequence $\{d_n^s\}$ of somata of (A), such that

1) $\operatorname{co} a' \leqq \prod_{s=1}^{\infty} \sum_{n=1}^{\infty} d_n^s$,

2) $\sum_{n=1}^{\infty} d_n^{(1)} \geqq \sum_{n=1}^{\infty} d_n^{(2)} \geqq \cdots$,

3) $\mu_e'\left(\prod_{s=1}^{\infty} \sum_{n=1}^{\infty} d_n^{(s)}\right) = \mu_e'(\operatorname{co} a')$.

Since (A) is a finitely genuine subtribe of (A'), we have $I' = I$ in (A'). Hence $I' \in C^*$, and then $\operatorname{co} d_n' \in C^*$. We have

$$\mu_e'(\operatorname{co} a') = \mu_e'(I') - \mu_e'(a') = \mu_i'(\operatorname{co} a').$$

From 1) we get by de Morgan law:

$$a' \geqq \sum_{k=1}^{\infty} \prod_{n=1}^{\infty} \operatorname{co} d_n^{(k)}$$

and

$$\mu_i'(a') = \mu^*\left(\sum_{k=1}^{\infty} \prod_{n=1}^{\infty} \operatorname{co} d_n^{(k)}\right).$$

Putting

(18) $$d' =_{df} \sum_{k=1}^{\infty} \prod_{n=1}^{\infty} \operatorname{co} d_n^{(k)},$$

we see that $d' \in C^*$, and we have

$$d' \leqq a', \quad \mu^*(d') = \mu_i'(a').$$

A I.8.7a. Since a' is supposed to be μ' L-measurable, we have

$$\mu_e'(a') = \mu_i'(a').$$

Thus we have

(18a) $$d' \leqq a' \leqq c', \quad \text{(see [A I.8.6j.] and (18))}$$

with

(19) $$\mu^*(d') = \mu_e'(a') = \mu^*(c').$$

From (18a) and (19) we deduce:

$$\mu^*(c' - d') = 0.$$

Since

$$a' - d' \leqq c' - d',$$

we also have

$$\mu^*(a' - d') = 0.$$

Since $a' - d' \in C^*$ and $d' \in C^*$, it follows that

$$a' = (a' - d') + d' \in C^*.$$

Thus we have proved that $a' \in C^*$.

Consequently we have proved the

A I.8.7b. Theorem. If a' is μ'_e L-measurable, then a' is also μ'_e C-measurable, and

$$\mu'(a') = \mu^*(a').$$

A I.8.8. The theorems [A I.8.6.] and [A I.8.7b.] give the following

Theorem. μ'_e L-measurability is the same as μ'_e C-measurability with equality of both measures. Thus we can speak of μ'-measurability in both approaches to measure. We have not yet proved that the μ'_e L-measure is denumerably additive. Now since the measure μ'_e C is so, it follows the denumerable additivity of the μ'_e L-measure, so the general theory of μ'_e L-measurability is completed.

A I.8.9. We cannot say, that the measure $\mu^* = \mu'$ is an extension of the given measure μ in (A). We only know that $\mu^*(a) \leqq \mu(a)$ for all $a \in A$. We even have an example, [A I.6.4.], where μ^* is not an extension of μ. A necessary and sufficient condition for μ' being an extension of μ is of course the equality $\mu^*(a) = \mu(a)$ for all $a \in A$.

A I.8.9a. Theorem. Now we can prove that this condition is equivalent to

$$\mu^*(I') = \mu(I).$$

Indeed let $a \in A$, and suppose that $\mu^*(a) < \mu(a)$. Then we have

$$\mu^*(a) + \mu^*(\mathrm{co}\, a) < \mu(a) + \mu(\mathrm{co}\, a), \quad \text{i.e.} \quad \mu^*(I') < \mu(I).$$

A I.8.10. Theorem. If a' is μ'-measurable, $a' \in A$, then there exists a double sequence $\{a_{ik}\}$ of somata of (A), such that

$$a' \leqq \sum_{i=1}^{\infty} \prod_{k=1}^{\infty} a_{ik}$$

with

$$\mu'(a') = \mu'\left(\sum_{i=1}^{\infty} \prod_{k=1}^{\infty} a_{ik}\right);$$

so

$$\mu'\left(a' \dot{+} \sum_{i=1}^{\infty} \prod_{k=1}^{\infty} a_{ik}\right) = 0;$$

and also there exists a double sequence $\{d_{ik}\}$ of somata of (A), such that

$$\prod_{i=1}^{\infty} \sum_{k=1}^{\infty} d_{ik} \leqq a'$$

with

$$\mu'(a') = \mu'\left(\prod_{i=1}^{\infty} \sum_{k=1}^{\infty} d_{ik}\right),$$

consequently

$$\mu'\left(a' \dot{+} \prod_{i=1}^{\infty} \sum_{k=1}^{\infty} d_{ik}\right) = 0.\text{*}), \text{ [A I.8.6k.]}.$$

Section 9 of A I.

A I.9. Authors device for measure extension. After having discussed the Lebesgue's and the Carathéodory's devices, we shall formulate an other device, by the author, which will be useful in our apparatus for quantum theory.

A I.9.1. Extension problem. Let (T) be a tribe (which does not need to be denumerably additive). Let (T^*) be its finitely genuine supertribe of (T) through isomorphism $\mathfrak{B}$: [A.7.4.]. We suppose that (T^*) is denumerably additive. Let $\mu(a)$ be a measure in (T), which does not need to be denumerably additive (in the sense of [A.15.2.]). Let (T') be the $\mathfrak{B}$-isomorphic image of (T) in (T^*), [A.7.4.]. Let

$$\mu'(a') =_{df} \mu(a)$$

whenever $a' = \mathfrak{B}(a)$.

The function $\mu'(a')$ of the variable soma $a' \in \mathfrak{B}(T)$ is a (finitely additive) measure on (T').

Our problem will consist in finding the possibility and means of extending $\mu'(a')$ in such a way, as to get a denumerably additive strict subtribe (T_1) of (T^*) with denumerably additive measure $\mu^0(a')$, and such that $\mu^0(a') = \mu'(a')$ for all $a' \in T'$. To deal with that problem, we introduce the following auxiliary notion (15):

A I.9.2. Def. If $A, B \in T^*$ and $\sigma > 0$, we shall say that A, B *differ from one another in* (T^*) *by not more than by* σ, if and only if there exists an infinite sequence $a_1, a_2, \ldots, a_n, \ldots \in T$, such that

1) $A \dot{+} B = (A - B) + (B - A) \leqq \sum_{n=1}^{\infty}{}' \mathfrak{B}(a_n)$, (in (T^*)),

2) $\sum_{n=1}^{\infty} \mu(a_n) \leqq \sigma$.

A I.9.3. Remark. Finite covering sequences $\{a_n\}$ are allowed, because we can extend the sequence to an infinite one, by taking an infinite number of elements $= O$. Notice that $(\leqq)$ in 1) is the ordering

---

*) The theorem [A I.8.10.] reminds us a similar one for ordinary Lebesgue's-measurable subsets of the straight line: If E is measurable then there exists a set X of type (G_δ) i.e. denumerable product of open sets, and a set Y of type (F_σ) i.e. denumerable union of closed sets, such that

$$Y \subseteqq E \subseteqq X$$

and where the measure of these three sets is the same.

taken from (T^*). Let us agree to denote notions related to $\mathfrak{B}(T)$ by a prime:

$$a' =_{df} \mathfrak{B}(a) \quad \text{for} \quad a \in T.$$

A I.9.4. Theorem. If A, B do not differ by more than ε_1, and B, C do not differ by more than ε_2, then A, C do not differ by more than $\varepsilon_1 + \varepsilon_2$.

Proof. Let

$$A \dot{+} B \leqq \sum_{n=1}^{\infty} a'_n, \quad \sum_{n=1}^{\infty} \mu(a'_n) \leqq \varepsilon_1,$$

$$B \dot{+} C \leqq \sum_{n=1}^{\infty} b'_n, \quad \sum_{n=1}^{\infty} \mu(b'_n) \leqq \varepsilon_2.$$

It follows

$$A \dot{+} C = (A \dot{+} B) \dot{+} (B + C) \leqq (A \dot{+} B) + (B \dot{+} C) \leqq {}^{*)}$$

$$\leqq \sum_{n=1}^{\infty} a'_n + \sum_{n=1}^{\infty} b'_n \quad \text{in } (T^*).$$

We have

$$\sum_{n=1}^{\infty} \mu(a'_n) + \sum_{n=1}^{\infty} \mu(b'_n) \leqq \varepsilon_1 + \varepsilon_2,$$

which proves the theorem.

A I.9.5. Theorem. If $A, B \in T^*$, and they differ by not more than by $\varepsilon > 0$, then co$A$, co$B$ (in T^*) differ by not more than by ε.

Proof. co$A = I \dot{+} A$, co$B = I \dot{+} B$, [A.2.10.]; hence

$$\text{co}A \dot{+} \text{co}B = (I \dot{+} A) \dot{+} (I \dot{+} B) = A \dot{+} B,$$

which completes the proof.

A I.9.6. Def. Denote by $\boldsymbol{N}$ the class of all somata P of (T^*) such that for every $\varepsilon > 0$ there exists $p \in T$, such that P and $p' = \mathfrak{B}(p)$ differ from one another by not more than ε.

A I.9.7. Theorem. If $a \in T$, then $a' =_{df} \mathfrak{B}(a) \in \boldsymbol{N}$.

Proof. Given $\varepsilon > 0$, the somata $\mathfrak{B}(a)$, $\mathfrak{B}(a)$ do not differ by more than ε, because $a \dot{+} a = O$, $\mu(0) = 0 < \varepsilon$, and $\mathfrak{B}(a) \dot{+} \mathfrak{B}(a) = O$.

A I.9.8. If $M \in \boldsymbol{N}$, then co$M \in \boldsymbol{N}$.

Proof. Let $\varepsilon > 0$. Find $a \in T$ such that M, $\mathfrak{B}(a)$ differ by not more than by ε.

Hence, by [A I.9.5.], co$M$, co$\mathfrak{B}(a) = \mathfrak{B}(\text{co}\,a)$ do not differ by more than ε. This being true for all $\varepsilon > 0$, we get co$M \in \boldsymbol{N}$.

A I.9.9. Theorem. If $M_1, M_2, \ldots, M_n, \ldots \in \boldsymbol{N}$, then

$$\sum_{n=1}^{\infty} M_n \in \boldsymbol{N}.$$

---

*) $M \dot{+} N = (M - N) + (N - M) \leqq M + N$.

Proof. Put

$$M =_{df} \sum_{n=1}^{\infty} M_n.$$

We have $M \in T^*$, because (T^*) is denumerably additive. Let $\varepsilon > 0$. Choose $a_n \in T$, such that M_n, a'_n do not differ by more than $\frac{\varepsilon}{2^n}$, $\left(a'_n =_{df} \mathfrak{B}(a_n)\right)$. Put $A =_{df} \sum_{n=1}^{\infty} a'_n$, we have $A \in T^*$.

We have (with operations in (T^*))

$$\sum_{n=1}^{\infty} M_n - \sum_{n=1}^{\infty} a'_n = \sum_{m=1}^{\infty} M_n \cdot \operatorname{co} \sum_{n=1}^{\infty} a'_n = \sum_{m=1}^{\infty} \left(M_m \cdot \prod_{n=1}^{\infty} \operatorname{co} a'_n\right).$$

But

$$\prod_{n=1}^{\infty} \operatorname{co} a'_n \leqq \operatorname{co} a'_m \quad \text{for} \quad m = 1, 2, \ldots$$

Hence

$$\sum_{n=1}^{\infty} M_n - \sum_{n=1}^{\infty} a'_n \leqq \sum_{m=1}^{\infty} (M_m \cdot \operatorname{co} a'_m),$$

i.e.

$$\sum_{n=1}^{\infty} M_n - \sum_{n=1}^{\infty} a'_n \leqq \sum_{n=1}^{\infty} (M_n - a'_n),$$

i.e.

$$M - A \leqq \sum_{n=1}^{\infty} (M_n - a'_n).$$

In an analogous way we obtain

$$A - M \leqq \sum_{n=1}^{\infty} (a'_n - M_n);$$

consequently

(1) $$M \dotplus A \leqq \sum_{n=1}^{\infty} (M_n \dotplus a'_n).$$

Since M_n and a'_n differ not more than by $\frac{\varepsilon}{2^n}$, there exist a_{n1}, $a_{n2}, \ldots, a_{np}, \ldots \in T$ such that

$$M_n \dotplus a'_n \leqq \sum_{p=1}^{\infty} a'_{np}, \qquad \sum_{p=1}^{\infty} \mu(a'_{np}) \leqq \frac{\varepsilon}{2^n}.$$

Hence, by (1),

$$M \dotplus A \leqq \sum_{n=1}^{\infty} \sum_{p=1}^{\infty} a'_{np},$$

$$\sum_{n=1}^{\infty} \sum_{p=1}^{\infty} \mu(a_{np}) \leqq \frac{\varepsilon}{2^1} + \frac{\varepsilon}{2^2} + \cdots \leqq \varepsilon.$$

Thus we have proved that

(2) M, A do not differ by more than by ε.

A I.9.9a. Now we have

$$A = a_1' + a_2' + \cdots = a_1' + [a_2' - a_1'] + [a_3' - (a_1' + a_2')] + \cdots$$
$$= a_1' + a_2' \operatorname{co} a_1' + a_3' \operatorname{co} a_1' \operatorname{co} a_2' + \cdots$$

Put

$$b_1 =_{df} a,$$
$$b_n =_{df} a_n \cdot \operatorname{co} a_1 \cdot \operatorname{co} a_2 \ldots \operatorname{co} a_{n-1} \qquad (n = 2, 3, \ldots).$$

We have

$$b_n \in T \quad \text{and} \quad b_n \cdot b_m = 0 \text{ *)} \quad \text{for} \quad n \neq m.$$

Besides

$$A = b_1' + b_2' + b_3' + \cdots.$$

Hence

(3) $$\left(A - \sum_{\nu=1}^{n} b_\nu'\right) + \left(\sum_{\nu=1}^{n} b_\nu' - A\right) = A - \sum_{\nu=1}^{n} b_\nu' = \sum_{\nu=n+1}^{\infty} b_\nu',$$

valid for $n = 1, 2, \ldots$

We have

$$b_1 + b_2 + \cdots + b_n \leqq I.$$

Hence

$$\mu(b_1 + b_2 + \cdots + b_n) \leqq \mu(I).$$

By additivity of μ we have

$$\mu(b_1) + \mu(b_2) + \cdots + \mu(b_n) \leqq \mu(I),$$

because the somata b_n are mutually disjoint and belong to (T). Since the infinite series

$$\mu(b_1) + \mu(b_2) + \cdots$$

is bounded and has only non negative terms, it must be convergent. Hence, there exists an index n_0 such that

(4) $$\mu(b_{n_0+1}) + \mu(b_{n_0+2}) + \cdots \leqq \varepsilon.$$

Apply (3) for $n = n_0$. We get

$$A \dotplus \sum_{\nu=1}^{n_0} b_\nu' = \sum_{\nu=n_0+1}^{\infty} b_\nu'.$$

Putting

$$C' =_{df} \sum_{\nu=1}^{n_0} b_\nu',$$

---

*) $b_n \cdot b_{n+p} = a_n \cdot \operatorname{co} a_1 \ldots \operatorname{co} a_{n-1} \cdot a_{n+p} \ \operatorname{co} a_1 \ldots \operatorname{co} a_{n-1} \operatorname{co} a_n \ldots \operatorname{co} a_{n+p-1}$.

we get

(5) $$(A - C') + (C' - A) \leqq \sum_{\nu = n_0 + 1}^{\infty} b'_\nu,$$

with

$$C \in T, \quad b_\nu \in T.$$

The relations (4) and (5) show that A, C' do not differ more than by ε.

Taking (2) into account, and applying [A I.9 4.] we get that M and C' do not differ more than by 2ε. Since $\varepsilon > 0$ was arbitrary and $C \in T$, the theorem is proved.

A I.9.10. Theorem. The foregoing theorems show that if we use operations and order of (T^*), then $\boldsymbol{N}$ is ordered as in (T^*) and is a denumerably additive tribe which is a finitely genuine supertribe of (T) through isomorphism.

The tribe $(\boldsymbol{N})$ is a denumerably genuine strict subtribe of (T^*).

To have more properties of $(\boldsymbol{N})$, we prove the following theorem:

A I.9.11. Theorem. If

1. $M \in \boldsymbol{N}$, $N \in T^*$,
2. for every $\varepsilon > 0$, M does not differ from O^* by more than ε,
3. $N \leqq M$ in (T^*),

then

$$N \in \boldsymbol{N}.$$

Proof. We have

$$O^* - M = O^* - N = O^*;$$

hence

$$(N - O^*) + (O^* - N) = N - O^* = N \leqq M,$$

i.e.

$$N \dotplus O^* \leqq M \dotplus O^*.$$

Since $M \dotplus O^*$ can be enclosed, in T^*, into a denumerable (T^*)-sum of somata of $\mathfrak{B}(T)$ with the sum of measure not exceeding ε, so is with $N \dotplus O^*$.

This proves the theorem.

A I.9.12. Theorem. The null soma O^* and the unit soma I^* of (T^*) belong to $\boldsymbol{N}$.

A I.9.13. Measure on $(\boldsymbol{N})$.

We shall find a measure on $(\boldsymbol{N})$. The following auxiliary notions will be useful.

A I.9.13a. Def. Let $M \in \boldsymbol{N}$. A sequence $a_1, a_2, \ldots, a_n, \ldots \in T$ will be said *approaching* M whenever for every $\varepsilon > 0$ there exists an index λ such that, for every $n > \lambda$, M differs from $a'_n =_{df} \mathfrak{B}(a_n)$ by not more than ε.

A I.9.14. Def. We shall use the following condition for μ, designated by $\mathcal{B}(T, T^*, \mu)$, or simplier $\mathcal{B}(T, T^*)$. Let (T) be a finitely genuine subtribe of the denumerably additive tribe (T^*). Let μ be a measure on (T) and let $p' = \mathfrak{B}(p)$ be the correspondence between somata p of T and the corresponding somata p' of T^*.

Under these circumstances, the condition is:

If $p, q_n \in T$, $(n = 1, 2, \ldots)$,

$$p' \leqq \sum_{n=1}^{\infty}{}' q'_n, \quad \text{in } (T^*),$$

then

$$\mu(p) \leqq \sum_{n=1}^{\infty} \mu(q_n).$$

This condition is just [A I.2.1., IV'].

A I.9.15. Expl. The condition $\mathcal{B}(T, T^*)$ looks very natural, say almost obvious and trivial. Nevertheless the following example shows that the condition may not take place.

Expl. Consider the tribe (A), whose somata are finite unions of half open intervals $(\alpha, \beta\rangle$, where $0 \leqq \alpha, \beta \leqq 1$. Let R be the set of all rational numbers in $J =_{df} (0, 1\rangle$. If f is a soma of A, it can be represented in the form:

$$(1) \qquad f = r_1 + r_2 + \cdots + r_n,$$

where r_i are disjoint half open considered intervals.

Denote by f^0 the set of all rational numbers w, belonging to f, and put $\mu(f^0) =_{df} \sum_{i=1}^{n}$ (length of r_i). The number $\mu(f^0)$ does not depend on the choice of the representation (1) of f.

The sets f^0 constitute a finitely additive tribe (T), if f^0 is ordered by the relation of inclusion of sets.

Let (T^*) be the class of all subsets of R, ordered by the relation of inclusion of sets of points. (T^*) is a finitely genuine strict supertribe of (T).

Let us order the rational numbers of R into an infinite sequence

$$w_1, w_2, \ldots, w_n, \ldots$$

and consider the intervals

$$\Delta_n =_{df} (w_n - \varepsilon_n, w_n + \varepsilon_n\rangle \in A, \quad \text{and the sets } \Delta_n^0 \in T,$$

where $\varepsilon_n > 0$, $\varepsilon_n \to 0$, and where $\sum_{n=1}^{\infty} \varepsilon_n = \frac{1}{4}$. We have $J^0 = R \in T$ and

$$(2) \qquad J^0 \leqq \sum_{n=1}^{\infty} \Delta_n^0.$$

We have $\mu(J^0) = 1$, but

$$\sum_{n=1}^{\infty} \mu(\Delta_n^0) = 2 \sum_{n=1}^{\infty} \varepsilon_n = \tfrac{1}{2}.$$

Thus inequality (2) does not imply $\mu(J^0) \leqq \sum_{n=1}^{\infty} \mu(\Delta_n^0)$, because 1 is not less than $\frac{1}{2}$.

We see that in this example the condition $\mathcal{B}(T, T^*)$ is not satisfied.

A I.9.16. Theorem. If $a_n \in T$, $M \in \boldsymbol{N}$ and $\{a_n\}$ is approaching M, the condition $\mathcal{B}(T, T^*)$ is satisfied,
then $\lim \mu(a_n)$ exists.

Proof. Let $\varepsilon > 0$. By def. [A I.9.13.] there exists λ such that for every $n \geqq \lambda$, $a_n' = \mathfrak{B}(a_n)$ does not differ from M by more than $\varepsilon/2$. Let $n \geqq \lambda$, $m \geqq \lambda$. Since a_n' and a_m' do not differ from M by more than $\varepsilon/2$, therefore a_n' and a_m' do not differ by more than ε, [A I.9.4.]. Hence there exists a sequence $\{b_n\}$, $n = 1, 2, \ldots$, $b_n \in T$, such that

$$a_n' \dotplus a_m' \leqq \sum_{n=1}^{\infty} b_n', \qquad \sum_{n=1}^{\infty} \mu(b_n) \leqq \varepsilon;$$

this for all $n, m \geqq \lambda$.

Since the condition $\mathcal{B}(T, T^*)$ is satisfied, we get

$$\mu(a_n \dotplus a_m) \leqq \sum_{n=1}^{\infty} \mu(b_n) \leqq \varepsilon.$$

By the Theorem [A.18.] it follows

$$|\mu(a_n) - \mu(a_m)| \leqq 2\varepsilon \quad \text{for all} \quad n, m \geqq \lambda.$$

Hence, by the known Cauchy convergence condition, we get the existence of $\lim_{n \to \infty} \mu(a_n)$.

A I.9.17. Theorem. If

1. the condition $\mathcal{B}(T, T^*)$ is admitted, [A I.9.14.],
2. $\{a_n\}$, $\{b_n\}$ are sequences $\in T$, both approaching $M \in T^*$, [A I.9.13.],
3. $M \in \boldsymbol{N}$,

then

$$\lim \mu(a_n) = \lim \mu(b_n).$$

Proof. Let $\varepsilon > 0$. Find an index λ such that for $n \geqq \lambda$, the somata M, a_n' do not differ by more than $\varepsilon/2$, and M and b_n' do not differ by more than $\varepsilon/2$.

Hence a_n', b_n' do not differ by more than ε, [A I.9.4.]. Hence there exists a sequence $a_{n,k} \in T$, $k = 1, 2, \ldots$ such that

$$a_n' \dotplus b_n' \leqq \sum_{k=1}^{\infty} a_{n,k}', \qquad \sum_{k=1}^{\infty} \mu(a_{n,k}) \leqq \varepsilon.$$

The condition $\mathcal{B}(T, T^*)$ gives

$$\mu(a_n \dot{+} b_n) \leqq \sum_{k=1}^{\infty} \mu(a_{n,k}) \leqq \varepsilon .$$

Hence by Theorem [A.18.]

$$(1) \qquad |\mu(a_n) - \mu(b_n)| \leqq 2\varepsilon ,$$

and this for all $n \geqq \lambda$.

Now, by hypothesis and by virtue of theorem [A I.9.16.], the limits $\lim_n \mu(a_n)$ and $\lim_n \mu(b_n)$ both exist.

Put

$$\alpha =_{df} \lim_n \mu(a_n), \quad \beta =_{df} \lim_n \mu(b_n) .$$

Let λ' be such that for all $n \geqq \lambda'$ we have

$$(2) \qquad |\alpha - \mu(a_n)| \leqq \varepsilon , \quad |\beta - \mu(b_n)| \leqq \varepsilon .$$

From (1) and (2) for $n \geqq \max(\lambda, \lambda')$, we get

$$|\alpha - \beta| \leqq 4\varepsilon .$$

This being true for any $\varepsilon > 0$, we get $\alpha = \beta$, which completes the proof.

A I.9.18. Def. The above two theorems show that we can attach to every $M \in \boldsymbol{N}$ a number

$$\mu^0(M) =_{df} \lim \mu(a_n) ,$$

where $\{a_n\}$ approaches M. This number does not depend on the choice of the approaching sequence $\{a_n\}$, but only on M. (The condition $\mathcal{B}(T, T^*)$ is supposed.)

A I.9.19. Theorem. If $a \in T$, then $\mu^0(\mathfrak{B}(a)) = \mu(a)$.

Proof. The sequence $(a, a, \ldots, a, \ldots)$ approaches a', because $a \dot{+} a = O$.

A I.9.20. We see that $\mu^0(M)$ can be considered as extension of the measure $\mu(a)$ through isomorphism $\mathfrak{B}$. We are going to prove that μ^0 is on $\boldsymbol{N}$ denumerably additive, if we admit the condition $\mathcal{B}(T, T^*)$.

A I.9.20a. Lemma. For the somata of any tribe we have

$$a\,b \dot{+} a_1 b_1 \leqq (a \dot{+} a_1) + (b \dot{+} b_1) ,$$

[A.17.7.].

A I.9.21. Lemma. If A, A_1 in (T^*) do not differ by more than ε, and B, B_1 do not differ by more than ε, then $A \cdot B$ do not differ from $A_1 \cdot B_1$ by more than 2ε.

Proof. There exist two sequences $\{a_n\}$, $\{b_n\}$ of somata of (T) such that

$$A \dot{+} A_1 \leqq \sum_{n=1}^{\infty} a'_n, \quad \sum_{n=1}^{\infty} \mu(a_n) \leqq \varepsilon,$$

$$B \dot{+} B_1 \leqq \sum_{n=1}^{\infty} b'_n, \quad \sum_{n=1}^{\infty} \mu(b_n) \leqq \varepsilon.$$

Applying the foregoing lemma [A I.9.20a.], we get

$$A\,B \dot{+} A_1 B_1 \leqq (A \dot{+} A_1) + (B \dot{+} B_1) \leqq \sum_{n=1}^{\infty} a'_n + \sum_{n=1}^{\infty} b'_n,$$

where

$$\sum_{n=1}^{\infty} \mu(a_n) + \sum_{n=1}^{\infty} \mu(b_n) \leqq 2\varepsilon,$$

so the lemma is proved.

A I.9.22. Lemma. If

1. $A, B \in \boldsymbol{N}$,
2. $A \cdot B = O^*$,
3. $a, b \in T$,
4. a', b' differ from A, B respectively not more than by $\varepsilon > 0$,

then there exist $p, q \in T$ such that $p \leqq a$, $q \leqq b$, $p \cdot q = O$, and where p', q' do not differ from a', b' respectively by more than 2ε.

Proof. Put

$$p =_{df} a - b, \quad q =_{df} b - a.$$

We have

$$p \cdot q = O, \quad p, q \in T, \quad p \leqq a, \quad q \leqq b.$$

We also have

$$p \dot{+} a = (a - b) \dot{+} a = a \operatorname{co} b \dot{+} a = a(1 \dot{+} b) \dot{+} a = a \dot{+} a\,b \dot{+} a;$$

hence

(1) $$p \dot{+} a = a \cdot b.$$

Now A, B do not differ from a', b' by more than ε respectively; hence, [A I.9.21.], $A \cdot B$ does not differ from $a' \cdot b'$ by more than 2ε.*) Hence by (1), $p' \dot{+} a'$ does not differ from O^* by more than 2ε.

Hence there exists a sequence $a_n \in T$, $n = 1, 2, \ldots$, such that

$$(p' \dot{+} a') \dot{+} O' \leqq \sum_{n=1}^{\infty} a'_n, \quad \sum_{n=1}^{\infty} \mu(a'_n) \leqq 2\varepsilon.$$

Hence $p' \dot{+} a' \leqq \sum a'_n$, which says that p' and a' do not differ by more than 2ε. A similar result we get for q' and b'. The lemma is established.

---

*) Since $A \cdot B = O^*$, it follows that $a' \cdot b'$ does not differ from O^* by more than 2ε.

A I.9.23. Lemma. If A differs from A_1 by not more than α and B differs from B_1 by not more than β,
then $A + B$ does not differ from $A_1 + B_1$ by more than $\alpha + \beta$.

Proof. There exist sequences $a_1, a_2, \ldots, a_n, \ldots$ and $b_1, b_2, \ldots, b_n, \ldots$ of somata of T such that

$$A \dotplus A_1 \leqq \sum_n a'_n, \quad \sum_n \mu(a_n) \leqq \alpha,$$

$$B \dotplus B_1 \leqq \sum_n b'_n, \quad \sum_n \mu(b_n) \leqq \beta.$$

Hence

$$(A \dotplus A_1) + (B \dotplus B_1) \leqq \sum_n a'_n + \sum_n b'_n.$$

But we have

$$(A + B) \dotplus (A_1 + B_1) \leqq (A \dotplus A_1) + (B \dotplus B_1),$$

[A.17.5.]. It follows that

$$(A + B) \dotplus (A_1 + B_1) \leqq \sum_n a'_n + \sum_n b'_n$$

with

$$\sum_n \mu(a_n) + \sum_n \mu(b_n) \leqq \alpha + \beta.$$

This finishes the proof.

A I.9.24. Theorem. If

1. The condition $\mathcal{B}(T, T^*)$ is admitted,
2. $A, B \in T^*$,
3. $A \cdot B = O^*$,

then

$$\mu^0(A + B) = \mu^0(A) + \mu^0(B).$$

Proof. Let $\{a_n\}$, $\{b_n\}$ be sequences of somata of T, approaching A, B respectively.

a'_n, b'_n do not differ from A, B respectively more than by some number $\varepsilon_n \to 0$, $(\varepsilon_n > 0)$.

By lemma [A I.9.22.], we can find $p_n, q_n \in T$, where $p_n \leqq a_n$, $q_n \leqq b_n$, and such that $p_n \cdot q_n = O$ and where p'_n, q'_n do not differ from a'_n, b'_n by more than $2\varepsilon_n$. Hence, [A I.9.4.], p'_n, q'_n do not differ from A, B respectively by more than $3\varepsilon_n$. Hence p_n, q_n approach A, B respectively. Hence, [A I.9.18.],

(1)
$$\mu^0(A) = \lim \mu(p_n),$$
$$\mu^0(B) = \lim \mu(q_n).$$

Since $p_n \cdot q_n = O$, we have

(2)
$$\mu(p_n + q_n) = \mu(p_n) + \mu(q_n).$$

The sequence $p_n + q_n$ approaches $A + B$.

By [A I.9.18.] we get

$$\mu^0(A+B) = \lim_{n\to\infty} \mu(p_n + q_n).$$

Hence; from (1) and (2),

$$\lim_n \mu(p_n + q_n) = \lim_n \mu(p_n) + \lim_n \mu(q_n),$$

and then

$$\mu^0(A+B) = \mu^0(A) + \mu^0(B), \quad \text{q.e.d.}$$

A I.9.25. Theorem. If we admit the condition $\mathcal{B}(T, T^*)$, then $\mu^0(A)$ defined on $(\boldsymbol{N})$ is denumerably additive, i.e. if $M = M_1 + M_2 + \cdots + M_n + \cdots$, where all M_n are disjoint somata of $(\boldsymbol{N})$, then

$$\mu^0(M) = \mu^0(M_1) + \mu^0(M_2) + \cdots + \mu^0(M_n) + \cdots$$

Proof. Since we know that $(\boldsymbol{N})$ is a denumerably additive tribe, we can apply theorem [A I.3.5.]. Hence to prove that μ^0 is denumerably additive, it suffices to prove that if $A, A_n \in \boldsymbol{N}$, $(n = 1, 2, \ldots)$, and if $A \leqq \sum_{n=1}^{\infty} A_n$, we have $\mu^0(A) \leqq \sum_{n=1}^{\infty} \mu^0(A_n)$, (and $\leqq \infty$ in the case of divergence). Let $M, M_n \in \boldsymbol{N}$, $(n = 1, 2, \ldots)$, and suppose that

$$M \leqq \sum_{n=1}^{\infty} M_n. \tag{1}$$

If $\sum_{n=1}^{\infty} \mu^0(M_n)$ diverges, we have of course,

$$\mu^0(M) \leqq \sum_{n=1}^{\infty} \mu^0(M_n) = +\infty.$$

Suppose that $\sum_{n=1}^{\infty} \mu^0(M_n)$ converges, but that $\mu^0(M)$ is greater than $\sum_{n=1}^{\infty} \mu^0(M_n)$. Hence there exists $\alpha > 0$ such that

$$\mu^0(M) \geqq \alpha + \sum_{n=1}^{\infty} \mu^0(M_n). \tag{2}$$

Notice that, in general, we have for somata A, B of T^*,

$$(A \dotplus B) + A = (A \dotplus B) + B = A + B.^{*)}$$

Take positive numbers $\varepsilon, \varepsilon_n$ such that $\varepsilon = \sum_{n=1}^{\infty} \varepsilon_n$.

---

*) Indeed, [A.2.10],

$$(A \dotplus B) + A = (A + B) \dotplus (A + B) \cdot A \dotplus A = B \dotplus (A \dotplus B) \cdot A$$
$$= B \dotplus (A \dotplus B A) = A \dotplus A B \dotplus B = A + B.$$

Take for M, M_n approaching sequences, [A I.9.13.], and determine for each of them a soma a, a_n respectively such that

$$(2.1)\qquad \begin{aligned} &\mu^0(M) = \mu'(a') \leqq \varepsilon, \\ &|\mu^0(M_n) - \mu'(a'_n)| \leqq \varepsilon_n, \qquad (n = 1, 2, \ldots), \end{aligned}$$

and in addition to that, such that M, M_n do not differ from a, a_n by more than ε, ε_n respectively. There exist sequences b_n, $b_{n,k}$, $(n, k = 1, 2, \ldots)$, such that

$$(3)\qquad \begin{aligned} &M \dotplus a' \leqq \sum_{n=1}^{\infty} b'_n, \quad \text{where} \quad \sum_{n=1}^{\infty} \mu(b_n) \leqq \varepsilon, \\ &M_k + a'_k \leqq \sum_{n=1}^{\infty} b'_{nk}, \quad \text{where} \quad \sum_{n=1}^{\infty} \mu(b_{nk}) \leqq \varepsilon_k, \qquad (k = 1, 2, \ldots). \end{aligned}$$

By (1) we have

$$M \leqq \sum_{n=1}^{\infty} M_n,$$

and it follows, by adding to both sides $M \dotplus a'$:

$$M + (M \dotplus a') \leqq (M \dotplus a') + \sum_{n=1}^{\infty} M_n.$$

By the general rule $[(a \dotplus b) + a = a + b]$ we can write

$$a' + (M \dotplus a') \leqq (M \dotplus a') + \sum_{n=1}^{\infty} [M_n + (M_n \dotplus a)]$$

because

$$M_n \leqq (M_n \dotplus a'_n) + M_n.$$

Hence, by the same mentioned rule, we get:

$$a' \leqq a' + (M \dotplus a') \leqq (M \dotplus a') + \sum_{n=1}^{\infty} [a'_n + (M_n \dotplus a'_n)].$$

Applying (3), we get

$$a' \leqq \sum_{n=1} b'_n + \sum_{n=1} a'_n + \sum_{n=1} (M_n \dotplus a'_n),$$

and

$$a' \leqq \sum_{n=1}^{\infty} b'_n + \sum_{n=1}^{\infty} a'_n + \sum_{n=1}^{\infty} \sum_{k=1}^{\infty} b'_{nk}.$$

The somata a, b_n, b_{nk} belong to T, so we can apply the condition $\mathcal{B}(T, T^*)$. It gives

$$\mu(a) \leqq \sum_{n=1}^{\infty} \mu(b_n) + \sum_{n=1}^{\infty} \mu(a_n) + \sum_{n=1}^{\infty} \sum_{k=1}^{\infty} \mu\, b_{nk}.$$

Hence, by (3),

$$\mu(a) \leqq \varepsilon + \sum_{n=1}^{\infty} \mu(a_n) + \sum_{k=1}^{\infty} \varepsilon_k.$$

So we get, [A I.9.19.],

$$\mu^0(a) \leqq \sum_{n=1}^{\infty} \mu^0(a_n) + 2\varepsilon. \tag{4}$$

Now we take (2.1) into account:

$$|\mu^0(M) - \mu'(a)| \leqq \varepsilon.$$

It follows $\mu^0(M) - \varepsilon \leqq \mu'(a)$, and from (4) we also have

$$\mu^0(M) - \varepsilon \leqq \mu'(a) \leqq \sum_{n=1}^{\infty} \mu'(a_n) + 2\varepsilon. \tag{5}$$

In (2.1) we also have

$$|\mu^0(M_n) - \mu'(a_n)| \leqq \varepsilon_n;$$

hence

$$\mu'(a_n) \leqq \mu^0(M_n) + \varepsilon_n.$$

Summing up for $n = 1, 2, \ldots$ we get

$$\sum_{n=1}^{\infty} \mu'(a_n) \leqq \sum_{n=1}^{\infty} \mu^0(M_n) + \sum_n \varepsilon_n, \quad \text{because} \quad \sum_{n=1}^{\infty} \mu^0(M_n)$$

converges. Hence

$$\sum_{n=1}^{\infty} \mu'(a_n) \leqq \sum_{n=1}^{\infty} \mu^0(M_n) + \varepsilon. \tag{6}$$

Hence, taking (5) into account, we get from (6),

$$\mu^0(M) - \varepsilon \leqq \mu'(a) \leqq \sum_{n=1}^{\infty} \mu^0(M_n) + \varepsilon + 2\varepsilon.$$

It follows that

$$\mu^0(M) \leqq \sum_{n=1}^{\infty} \mu^0(M_n) + 4\varepsilon. \tag{7}$$

Now we may dispose ε by taking, [see (2)],

$$\varepsilon < \frac{\alpha}{8}.$$

We get from (7)

$$\mu^0(M) \leqq \sum_{n=1}^{\infty} \mu^0(M_n) + \frac{\alpha}{2}.$$

But from (2) we have

$$\alpha + \sum_{n=1} \mu^0(M_n) \leqq \mu^0(M) \leqq \sum_{n=1} \mu^0(M_n) + \frac{\alpha}{2};$$

hence $\alpha \leqq \frac{\alpha}{2}$ which is not true. The theorem is established.

A I.9.26. In the subsections [A I.9.]—[A I.9.25.] we have accomplished the following fundamental task: There is given a (finitely additive) tribe (T) with a finitely additive measure μ in it. There is also given (T^*), which is itself denumerably additive and which is a finitely genuine extension of T. We succeeded to get a denumerably additive extension $\boldsymbol{N}$ of T within T^* through isomorphism such that the measure could be extended from T to $\boldsymbol{N}$ so as to become denumerably additive on $\boldsymbol{N}$.

A I.9.27. This task is an abstract and general setting of the known Lebesguean device of extension the ordinary euclidean measure. Indeed consider the tribe T, whose somata f are finite unions of half open sub-intervals of $(0, 1\rangle$, ordered by the relation of inclusion of sets, and define $\mu(f)$ as the ordinary euclidean measure of f. Neither the tribe (T) nor $\mu(f)$ is denumerably additive. Take as (T^*) the tribe composed of all subsets of $(0, 1\rangle$, ordered by the relation of ordinary inclusion of sets. The tribe (T^*) is denumerably (even completely) additive and is a finitely genuine (even strict), supertribe of (T). Thus we are just in the conditions of the mentioned problem of extension the tribe (T) and the measure on it, within the denumerably additive supertribe (T^*). The extension (T^*) is the tribe of all Lebesgue's measurable subsets of $(0, 1\rangle$. The required condition $\mathcal{B}(T, T^*)$ is satisfied. Indeed, if we have for somata of (T), $f, f_1, f_2, \ldots, f_n, \ldots$ the inclusion

$$f' \subseteq' \sum_{n=1}^{\infty} f'_n,$$

where the inclusion and the summation is taken from T^*, and where the primes designate the $\mathfrak{B}$-images of somata of T, it follows that

$$\mu(f) \leqq \sum_{n=1}^{\infty} \mu(f_n).$$

Thus the ordinary euclidean measure μ on (T) is extended to the ordinary Lebesguean measure of the measurable sets, and the extended measure is denumerably additive.

A I.9.28. To perform the above extension in the general case, we have postulated the condition $\mathcal{B}(T, T^*)$ which has shown to be sufficient. Now we can prove that this condition is necessary to have the extension spoken of.

A I.9.29. We shall prove the following theorem: If

1. (T), (T^*) are tribes, the first finitely additive, the second denumerably additive,
2. (T) is a finitely genuine subtribe of (T^*),
3. $\mu(a)$ is a finitely additive measure on (T),
4. there exists a denumerably additive tribe (S'), which is

1) a denumerably genuine strict subtribe of (T^*) and

2) a finitely genuine supertribe of (T),

3) there exists a denumerable additive measure $\mu^0(a)$ on (S), such that for every soma $a \in T$ we have

$$\mu^0(a') = \mu(a), \quad (\text{where } a' = \mathfrak{B}(a)):$$

then the condition $\mathcal{B}(T, T^*)$ is satisfied.

Proof. Let a, a_n be somata of (T) such that

$$a' \leqq' \sum_{n=1}^{\infty} a'_n,$$

where the primes denote images of somata of (T) in (T^*). We have, by denumerable additivity of (T^*), [A.2.6.3.], [A.2.6.5.], [A.3.7.],

$$\sum_{n=1}^{\infty} a'_n = a'_1 + [a'_2 - a'_1] + [a'_3 - (a'_1 + a'_2)] + [a'_4 - (a'_1 + a'_2 + a'_3)] + \cdots$$

Since the measure on (S') is denumerably additive we have, the terms being disjoint,

$$\mu(a') \leqq \mu\left(\sum_{n=1}^{\infty} a'_n\right) = \mu'(a'_1) + \mu'(a'_2 - a'_1) + \mu'[a'_3 - (a'_1 + a'_2)] + \cdots \leqq$$
$$\leqq \mu'(a'_1) + \mu'(a'_2) + \mu'(a'_3) + \cdots$$

which includes the condition $\mathcal{B}(T, T^*)$, because $\mu'(b') = \mu(b)$ whenever $b' = \mathfrak{B}(b)$. The theorem is established.

Section 10 of A I.

A I.10. Comparison of tribe extension devices *L*, *C* and *N*. We have proved in Section 8 of A I. that the Lebesgue's and Carathéodory's extension devices, studied in sections 6 of [A I.] and 7 of [A I.] respectively, are equivalent, if we take for Carathéodory's convex measure the Lebesguean exterior measure. Let us call these devices ***L***, ***C*** respectively. We have noticed that they give an extension of a given tribe, but, in general, not an extension of the measure.

The device ***N***, by the author in Section 9 of [A I.], also gives an extension of the given tribe as ***L***, ***C*** do, but in addition to that it gives an extension of measure, under a suitable condition, denoted in Section 9 of [A I.], by $\mathcal{B}(T, T^*)$.

A I.10.1. We shall prove that, under this condition, all three devices, ***L***, ***C***, ***N***, are equivalent and yield an extension of measurability and of measure, the same for all.

A I.10.1. We shall use μ^0 as the symbol for the measure in the device ***N***, [A I.9.18.] and the symbol μ' for the measure in ***L*** and ***C*** for measurable somata. The symbols T, T^*, μ'_e, μ'_i will have the same meaning as in the Section 6 of [A I.]. We agree to use the terms ***L***-measurable, ***C***-measurable and ***N***-measurable according to the case.

A I.10.2. Theorem. Then, under hypothesis $\mathcal{B}(T, T^*)$, if $\mu^0(a') = 0$, then $\mu_e'(a') = 0$.

Proof. Let $\mu^0(a') = 0$. This implies that $a' \in \boldsymbol{N}$. Hence there exists a sequence approaching a', [A I.9.13.],

$$a_1, a_2, \ldots, a_n, \ldots, a_n \in T.$$

Choose positive numbers $\varepsilon_n > 0$, $k = 1, 2, \ldots$ with $\lim_n \varepsilon_n = 0$, and $\varepsilon_n < \varepsilon$, where ε is given.

For every n find $k(n)$ such that a' differs from $a'_{k(n)}$ by not more than by ε_k, [A I.9.2.].

Put $b'_n = a'_{k(n)}$. The sequence b'_n is also approaching a'. There exist $x_{kn} \in T$, such that $b'_n \dot{+} a' \leqq \sum_{k=1}^{\infty} x_n$, with

(1) $$\sum_{k=1}^{\infty} \mu(x_{kn}) \leqq \varepsilon_n.$$

Since $a' \leqq b'_n + (a' - b'_n)$, we have

(2) $$a' \leqq b'_n + \sum_{k=1}^{\infty} x_{kn}.$$

By hypothesis we have $\lim \mu(b_n) = 0$, [A I.9.18.].

Find an index n_0 such that $\mu(b_{n_0}) \leqq \varepsilon$. From (2) we get, since as $b'_{n_0}, x'_{1n_0}, x'_{2n_0}, \ldots$ is a covering of a', the inequality

$$\mu_e' a' \leqq \mu\, b'_{n_0} + \sum_{k=1}^{\infty} \mu(x_{kn_0}).$$

Hence having, (1)

$$\mu_e' a' \leqq \varepsilon + \varepsilon_{n_0} < 2\varepsilon.$$

Since $\varepsilon > 0$ is arbitrary, we get $\mu_e' a' = 0$, so the theorem is established.

A I.10.3. We are going to prove that, under hypothesis $\mathcal{B}(T, T^*)$, if a' is $\boldsymbol{L}$-measurable, then $a' \in \boldsymbol{N}$, and we have $\mu'(a') = \mu^0(a')$.

Proof. The proof will be similar to that of the theorem [A I.8.7b.] One must replace μ^* by μ^0 and the class C^* by $\boldsymbol{N}$. We shall consider nested and disjoint coverings $\{a_n^{(k)}\}$, $(k = 1, 2, \ldots)$, of a':

$$a' \leqq \sum_{n=1}^{\infty} a_n^{(k)}, \qquad (k = 1, 2, \ldots),$$

$$\sum_{n=1}^{\infty} a_n^{(1)} \geqq \sum_{n=1}^{\infty} a_n^{(2)} \geqq \cdots \geqq \sum_{n=1}^{\infty} a_n^{(k)} \geqq \cdots,$$

such that

$$\mu_e' a' \leqq \sum_{n=1}^{\infty} \mu(a_n^{(k)}) \leqq \mu_e' a' + \frac{1}{k}, \qquad (k = 1, 2, \ldots).$$

Since $a_n^{(k)} \in T$, we have $\mu(a_n^{(k)}) = \mu^0(a_n^{(k)})$, and since the μ^0 measure is denumerably additive, we have, as in the mentioned proof of theorem [A I.8.7b.],

$$\mu_e' a' \leqq \mu^0\left(\sum_{n=1}^{\infty} a_n^{(k)}\right) \leqq \mu_e' a' + \frac{1}{k}.$$

By (1) we get

$$\mu^0\left(\prod_{k=1}^{\infty} \sum_{n=1}^{\infty} a_n^k\right) = \mu_e' a'.$$

Putting

$$p = \prod_{k=1}^{\infty} \sum_{k=1}^{\infty} a_n^k,$$

we get $p \in \boldsymbol{N}$ and

$$\mu^0 p = \mu_e'(a'), \quad a' \leqq p.$$

By de Morgan laws, we get a similar result, stating that there exists some $q \in \boldsymbol{N}$ with $\mu^0(q) = \mu_i'(a')$, and $q \leqq a'$.

Since, by hypothesis, $\mu_i'(a') = \mu_e'(a') = \mu' a'$, it follows that a' is $\boldsymbol{N}$-measurable and we get $\mu^0(a') = \mu_e'(a')$, so the theorem is proved.

A I.10.4. Theorem. Under hypothesis $\mathcal{B}(T, T^*)$, if $a' \in \boldsymbol{N}$ then a' is $\boldsymbol{L}$-measurable with $\mu^0(a') = \mu'(a')$.

Proof. Let $a' \in \boldsymbol{N}$. For every $\varepsilon > 0$ there exists $b \in T$, such that a', b' do not differ by more than by ε, say

(1) $$(a' - b') + (b - a') \leqq \sum_{n=1}^{\infty} x_n$$

with

(2) $$\sum_{n=1}^{\infty} \mu(x_n) \leqq \varepsilon$$

for some $x_n \in T$.

We transform the covering $\{x_n\}$ into another one $\{y_n\}$, where all y_n are disjoint, and where $y_n \leqq x_n$ for $n = 1, 2, \ldots$ It suffices to put

$$y_1 =_{df} x_1, \quad y_2 =_{df} x_2 - x_1, \quad y_3 =_{df} x_3 - (x_1 + x_2), \ldots,$$

$$y_n = x_n - (x_1 + \cdots + x_{n-1}).$$

We have

(3) $$\sum_{n=1}^{\infty} x_n = \sum_{n=1}^{\infty} y_n.$$

Since the somata y_n are disjoint, since $\mu(y_n) = \mu^0(y_n)$, and since the measure μ^0 is denumerably additive, we get from (2)

(4) $$\mu^0\left(\sum_{n=1}^{\infty} y_n\right) \leqq \varepsilon.$$

Since $a', b \in \boldsymbol{N}$, we have $(a'-b)+(b-a') \in \boldsymbol{N}$, so we get from (1), (3) and (4),

(5) $$\mu^0(a'-b)+\mu^0(b-a') \leqq \varepsilon.$$

Choose $\varepsilon_n > 0$ so as to have $\sum_{n=1}^{\infty} \varepsilon_n \leqq \varepsilon$ and find $b_n \in T$, such that

(6) $$\mu^0(a' \dotplus b_n) \leqq \varepsilon_n, \quad (n=1,2,\ldots),$$

according to what just has been proved.

Put

(7) $$c' =_{df} \prod_{k=1}^{\infty} (b'_k + b'_{k+1} + \cdots).$$

Since $b_n \in \boldsymbol{N}$, for every n, we have

(8) $$c' \in \boldsymbol{N}.$$

We also have $c' \in \boldsymbol{L}$, because all $b_k \in \boldsymbol{L}$ and because the tribe of all $\boldsymbol{L}$-measurable somata in denumerably additive.

We have

(9) $$a' - \sum_{i=k}^{\infty} b'_i = a' \cdot \mathrm{co} \sum_{i=k}^{\infty} b'_i = \prod_{i=k}^{\infty} a' \cdot \mathrm{co}\, b'_i = \prod_{i=k}^{\infty} (a'-b'_i) \leqq \sum_{i=k}^{\infty} (a' \dotplus b'_i).$$

We also have

(10) $$\sum_{i=k}^{\infty} b'_i - a' = \sum_{i=k}^{\infty} b'_i \cdot \mathrm{co}\, a' = \sum_{i=k}^{\infty} (b'_i - a') \leqq \sum_{i=k}^{\infty} (b'_i \dotplus a').$$

It follows, from (9) and (10):

$$\sum_{i=k}^{\infty} b'_i \dotplus a' \leqq \sum_{i=k}^{\infty} (b'_i \dotplus a').$$

Since by (6)

$$\sum_{i=k}^{\infty} \mu^0(b'_i \dotplus a') \leqq \sum_{i=k}^{\infty} \varepsilon_i,$$

and since $\sum_{i=k}^{\infty} b'_i \dotplus a' \in \boldsymbol{N}$, we get

(11) $$\mu^0\left(\sum_{i=k}^{\infty} b'_i \dotplus a'\right) \leqq \sum_{i=k}^{\infty} \mu^0(b'_i \dotplus a') \leqq \sum_{i=k}^{\infty} \varepsilon_i \leqq \varepsilon.$$

On the other hand we have from (7) and (8):

$$\mu^0\left(\sum_{i=k}^{\infty} b'_i \dotplus c'\right) \leqq \varepsilon \quad \text{for sufficiently great } k.$$

This, together with (11) gives

$$\mu^0(a' \dotplus c') \leqq \mu^0\left(\sum_{i=k}^{\infty} b'_i \dotplus a'\right) + \mu^0\left(\sum_{i=k}^{\infty} b'_i \dotplus c'\right) \leqq 2\varepsilon.$$

The above inequality being true for every $\varepsilon > 0$, we get

$$\mu^0 (a' \dotplus c') = 0.$$

Applying [A I.10.2.], we get

$$\mu'_e (a' \dotplus c') = 0.$$

It follows that

$$\mu'_e (a' - c') = 0, \quad \mu'_e (c' - a') = 0.$$

Hence

$$(12) \qquad a' - c' \in \boldsymbol{L}, \quad c' - a' \in \boldsymbol{L},$$

where $\boldsymbol{L}$ denotes the set of all $\boldsymbol{L}$-measurable somata of T^*. Since $a' = c' + (a' - c') - (c' - a')$, we get

$$a' \in \boldsymbol{L}, \quad \text{so } a' \text{ is } \boldsymbol{L}\text{-measurable.}$$

Since $\mu'(a' - c') = \mu'(c' - a') = 0$, we get $\mu'(a') = \mu'(c')$, [A.20.2.].

By [A I.10.3.], since c' is $\boldsymbol{L}$-measurable, it is also $\boldsymbol{N}$-measurable, and we get $\mu'(c') = \mu^0(c') = \mu^0(a')$.

Hence $\mu'(a') = \mu^0(a')$.

A I.10.5. Theorem. The above investigations have shown that under the conditions $\mathcal{B}(T, T^*)$ the classes $\boldsymbol{L}$, $\boldsymbol{C}$ and $\boldsymbol{N}$ coincide and that the measure $\mu'(a')$ on them is denumerably additive with $\mu'(a) = \mu(a)$ for every $a \in T$.

We have $T \subseteq \boldsymbol{L}$. Besides, if $\mu'(a') = 0$, $b' \leqq a'$, then $b' \in \boldsymbol{L}$.

If $a' \in \boldsymbol{L}$, then there exists a soma b' belonging to the Borelian extension of T within T', such that

$$\mu'(b' - a') = 0, \quad a' \leqq b'.$$

The soma b' can have the form $\prod\limits_k \sum\limits_i a_{ik}$, $(a_{ik} \in T)$.

If $a' \in \boldsymbol{L}$, then also there exists a soma $c' \in \boldsymbol{L}$, belonging to the Borelian extension of T within T', and such that

$$\mu'(a' - c') = 0, \quad c' \leqq a'.$$

(The soma c' can have the form $\sum\limits_i \prod\limits_k a_{ik}$, $(a_{ik} \in T)$.)

The class J of all $a' \in A'$ for which $\mu'_e(a') = 0$, coincides with the class of all a', for which $\mu'(a') = 0$, and it also coincides with the class of all a', such that, for every $\varepsilon > 0$, a' does not differ from O by more than ε.

J is a denumerably additive ideal in T' and also in $\boldsymbol{L}$. The tribe $(\boldsymbol{L})$ is the smallest denumerably additive denumerably genuine subtribe of (T'), including J and T.

Section 11 of A I.

A I.11.1. Topologies on a denumerably additive tribe. We have proved that, under certain condition, the extension of a finitely additive tribe, provided with a finitely additive measure, to a denumerably additive tribe, having a denumerably additive measure, is the same by all three considered methods $\boldsymbol{L}, \boldsymbol{C}, \boldsymbol{N}$.

A I.11.2. Now we shall be interested in a denumerably additive tribe (T), provided with a denumerably additive measure μ, which may be effective or not effective.

According to the theory, exhibited in Chapter [A], the measure μ generates a denumerably additive ideal J, defined as the set

$$\{a \mid a \in T,\ \mu(a) = 0\}.$$

It also generates a tribe $(T)_J$, where the governing equality $(=^\mu)$ is defined by

$$a =^\mu b \cdot =_{df} \cdot \mu(a \dotplus b) = 0, \quad \text{i.e.} \quad a \dotplus b \in J,$$

and the ordering by

$$a \leqq^\mu b \cdot =_{df} \mu(a - b) = 0, \quad \text{i.e.} \quad a - b \in J.$$

In $(T)_J$ the measure μ is effective.

A I.11.2.1. We have the notion of distance $\|a, b\|_\mu$, defined by

$$\|a, b\|_\mu =_{df} \mu(a \dotplus b), \quad [\text{A.16.}].$$

This notion of distance is (μ)-invariant, and organizes $(T)_J$ into a metric topological space. We call it *μ-measure topology*. In Chapter [A] we have studied the behaviour of somatic and algebraic operation, with respect to the equality in (T) and with respect to the equality $(=^\mu)$ in $(T)_J$.

We recall the basic properties of the distance. They are

$$\|a, b\|_\mu = \|b, a\|_\mu,$$

$$\|a, b\|_\mu \leqq \|a, c\|_\mu + \|c, b\|_\mu,$$

(so-called triangle-law)

$$\|a, b\|_\mu = 0 \quad \text{is equivalent to} \quad a =^\mu b.$$

A I.11.2.2. In this topology the notion of limit — we call it *μ-limit* — of an infinite sequence of somata is defined as follows:

$\lim\limits_{\mu} a_n =^\mu a$ means: "for every $\varepsilon > 0$

there exists ν such that for every $n \geqq \nu$ we have

$$\|a_n, a\|_\mu \leqq \varepsilon".$$

The operations of addition, multiplication, complementation, and algebraic addition are continuous in that topology. This means e.g. that if $\lim_{\mu} a_n =^{\mu} a$, $\lim_{\mu} b_n =^{\mu} b$, then

$$\lim_{\mu} (a_n + b_n) =^{\mu} a + b.$$

The distance $||a, b||_{\nu}$ is also a continuous function in the μ-measure topology.

A I.11.2.3. The author has shown that the μ-measure topology may be not separable (17)[1]. He also has proved that the said topology must be complete (16)[2].

Completeness means that the following two statements are equivalent:

I. The sequence $\{a_n\}$ of somata of (T) possesses a *μ-limit.*

II. For every $\varepsilon > 0$ there exists N, such that for every $n \geqq N$ and every $m \geqq N$ we have $||a_n, a_m|| < \varepsilon$.

(This condition is known under the name: *Cauchy-condition.*)

A I.11.2.4. Theorem. The μ-measure topology, generated in a denumerably additive tribe by a denumerably additive measure, is complete.

A I.11.2.4a. Proof. Suppose I and take $\varepsilon > 0$. Find N such that for all $n \geqq N$ we have

$$||a_n, a||_{\mu} \leqq \frac{\varepsilon}{2},$$

where $a = \lim_{\mu} a_n$.

Let $p \geqq N$, $q \geqq N$. We have

$$||a_p, a_q||_{\mu} \leqq ||a_p, a||_{\mu} + ||a_q, a||_{\mu} \leqq \varepsilon,$$

which completes the proof of II.

Suppose II. Choose positive numbers

$$\varepsilon_n, \varepsilon, \quad (n = 1, 2, \ldots),$$

such that

$$\sum_{n=1}^{\infty} \varepsilon_n \leqq \varepsilon.$$

---

[1] "Separable" means that there exists a denumerable sequence of somata of (T)

(1) $$w_1, w_2, \ldots, w_n, \ldots,$$

such that for every $a \in T$ there exists a subsequence

$$w_{k(1)}, w_{k(2)}, \ldots, w_{k(n)}, \ldots$$

of (1) with $a =^{\mu} \lim_{\mu} w_{k(n)}$.

[2] This theorem has been also found simultanously and independently by N. Aronszajn. His corresponding paper has not been published.

By hypothesis II we can find indices

$$(1) \qquad \nu(1) < \nu(2) < \cdots < \nu(k) < \cdots,$$

such that if

$$n \geqq \nu(k), \quad m \geqq \nu(k),$$

then

$$(2) \qquad \|a_n, a_m\|_\mu \leqq \varepsilon_k.$$

Consider the soma

$$(3) \qquad a =_{df} \sum_{i=1}^{\infty} a_{\nu(i)} \cdot \sum_{i=2}^{\infty} a_{\nu(i)} \cdots \sum_{i=k}^{\infty} a_{\nu(i)} \cdots$$

The product exists because the tribe is denumerably additive. We shall prove that a is just the desired limit of $\{a_n\}$. We have, by [A.6.4.]

$$a_{\nu(k)} + a_{\nu(k+1)} + \cdots = a_{\nu(k)} + [a_{\nu(k+1)} - a_{\nu(k)}] + \cdots + \\ + [a_{\nu(k+l)} - a_{\nu(k+l-1)}] + \cdots$$

It follows by the infinite distributive law [A.3.6.]:

$$\left(\sum_{i=k}^{\infty} (a_{\nu(i)}\right) - a_{\nu(k)} = [a_{\nu(k+1)} - a_{\nu(k)}] \cdot \mathrm{co}\, a_{\nu(k)} + \\ + [a_{\nu(n+2)} - a_{\nu(n+1)}] \cdot \mathrm{co}\, a_{\nu(k)} + \cdots$$

Hence

$$(4) \qquad \mu\left[\sum_{i=k}^{\infty} a_{\nu(i)} - a_{\nu(k)}\right] \leqq \mu(a_{\nu(k+1)} - a_{\nu(k)} + \mu(a_{\nu(k+2)} - a_{\nu(k+1)}) + \cdots$$

By (1) and (2) we have $\mu(a_{\nu(k+1)} - a_{\nu(k)}) \leqq \varepsilon_k$, because $\nu(k)$, and $\nu(k+1)$ are both $\nu(k)$.

Similarly we have $\mu(a_{\nu(k+2)} - a_{\nu(k+1)}) \leqq \varepsilon_{k+1})$, etc.

Thus we deduce from (4) that

$$(5) \qquad \mu\left(\sum_{i=k}^{\infty} a_{\nu(i)} - a_{\nu(k)}\right) \leqq \varepsilon_k + \varepsilon_{k+1} + \cdots$$

As $a_{\nu(k)} \leqq \sum_{i=k}^{\infty} a_{\nu(i)}$, we have $a_{\nu(k)} - \sum_{i=k}^{\infty} a_{\nu(i)} = 0$, which proves together with (5), that

$$(6) \qquad \left\|\sum_{i=1}^{\infty} a_{\nu(i)}, a_{\nu(k)}\right\|_\mu \leqq \varepsilon_k + \varepsilon_{k+1} + \cdots$$

Since the series $\sum_n \varepsilon_n$ converges, we have $\varepsilon_k + \varepsilon_{k+1} + \cdots \to 0$ for $k \to \infty$. Thus we obtain from (6)

$$(7) \qquad \lim_{k\to\infty} \left\|\sum_{i=1}^{\infty} a_{\nu(i)}, a_{\nu(k)}\right\|_\mu = 0.$$

We have from (2)

$$a \leqq \sum_{i=k}^{\infty} a_{\nu(i)}$$

which proves

$$\mu\left(a - \sum_{i=k}^{\infty} a_{\nu(i)}\right) = 0,$$

and

$$\mu\left(\sum_{i=k}^{\infty} a_{\nu(i)} - a\right) = \mu\left(\sum_{i=k}^{\infty} a_{\nu(i)}\right) - \mu(a);$$

so we get

$$\left\|\sum_{i=k}^{\infty} a_{\nu(i)}, a\right\|_{\mu} = \mu\left(\sum_{i=k}^{\infty} a_{\nu(i)}\right) - \mu(a). \tag{8}$$

Now we have by (2),

$$\lim_{k\to\infty} \mu\left(\sum_{i=k}^{\infty} a_{\nu(i)}\right) = \mu(a);$$

so, by (8)

$$\lim_{k} \left\|\sum_{i=k}^{\infty} a_{\nu(i)}, a\right\|_{\mu} = 0. \tag{9}$$

From (7) and (9), by virtue of the triangle-law for distances of somata, we get

$$\lim_{k} \|a_{\nu(k)}, a\|_{\mu} = 0. \tag{10}$$

Having this find σ such that if $n \geqq \nu(\sigma)$, we have

$$\|a_{\nu(\sigma)}, a_n\| \leqq \frac{\varepsilon}{2}, \quad \text{and also} \quad \|a_{\nu(\sigma)}, a\| \leqq \frac{\varepsilon}{2}.$$

These inequalities give, by the triangle-law,

$$\|a_n, a\|_{\mu} \leqq \varepsilon \quad \text{for all} \quad n \geqq \nu(\sigma);$$

so $\lim_{\mu} a_n = a$, i.e. $\{a_n\}$ possesses a μ-limit. The theorem is proved.

A I.11.3. Till now we have only considered a topology, made out of a given Boolean tribe by means of a measure available. Now we shall organize a given Boolean denumerably additive tribe into a kind of topology without the use of any measure (See e.g. DOROTHY MAHARAM, (24)). We shall follow the lines given in (24), but we shall change slightly notations and terminology.

Let us introduce the following definition which gives an immediate generalization of a known notion in the theory of sets.

A I.11.4. Def. Let (B) be a denumerably additive Boolean tribe and $a_1, a_2, \ldots, a_n, \ldots$ an infinite sequence of its somata. We put

$$\operatorname*{Lim\,sup}_n a_n =_{df} \prod_{k=1}^{\infty} \sum_{n=k}^{\infty} a_n,$$

$$\operatorname*{Lim\,inf}_n a_n =_{df} \sum_{k=1}^{\infty} \prod_{n=k}^{\infty} a_n.$$

A I.11.5. Def. Under the same hypothesis we say that $\{a_n\}$ *converges somatically to* a:

$$a = \operatorname*{Lim}_n a_n,$$

whenever

$$\operatorname*{Lim\,sup}_n a_n = \operatorname*{Lim\,inf}_n a_n = a.$$

A I.11.6. We call this *(natural) somatic limit.*

A I.11.7. Lemma. If (B) is denumerably additive Boolean tribe, we have

$$\prod_{n=1}^{\infty} a_n + \prod_{n=1}^{\infty} b_n = \prod_{n,m=1}^{\infty} (a_n + b_m).$$

Proof.

$$\prod_{n=1}^{\infty} a_n + \prod_{m=1}^{\infty} b_m = \prod_{n=1}^{\infty}\left(a_n + \prod_{m=1}^{\infty} b_m\right) = \prod_{n=1}^{\infty} \prod_{m=1}^{\infty} (a_n + b_m)$$

by virtue of the distributive law. This completes the proof.

A I.11.8. Lemma. Under the above condition, if

(1) $$a_1 \geqq a_2 \geqq \cdots,$$
$$b_1 \geqq b_2 \geqq \cdots,$$

then

$$\prod_{n=1}^{\infty} a_n + \prod_{n=1}^{\infty} b_n = \prod_{n=1}^{\infty} (a_n + b_n).$$

Proof. We have

$$\prod_{n=1}^{\infty} a_n \leqq \prod_{n=1}^{\infty} (a_n + b_n), \qquad \prod_{n=1}^{\infty} b_n \leqq \prod_{n=1}^{\infty} (a_n + b_n),$$

(2) $$\prod_{n=1}^{\infty} a_n + \prod_{n=1}^{\infty} b_n \leqq \prod_{n=1}^{\infty} (a_n + b_n).$$

To prove the converse inequality, take an index $p \geqq i$ and keep it fixed for a moment. From (1) it follows that, if $n \geqq p$, $m \geqq p$, then $a_p + b_p \leqq a_n + b_m$. It follows

(3) $$a_p + b_p \leqq \prod_{n, m \leqq p} (a_n + b_m);$$

hence, varying p, we get

$$\prod_{p=1}^{\infty}(a_p + b_p) \leqq \prod_{p=1}^{\infty} \prod_{n,m \leqq p} (a_n + b_m)$$

which gives

$$\prod_{p=1}^{\infty}(a_p + b_p) \leqq \prod_{n,m=1}^{\infty} (a_n + b_m).$$

Applying [A I.11.7.], we get

$$(4) \qquad \prod_{p=1}^{\infty}(a_p + b_p) \leqq \prod_{p=1}^{\infty} a_p + \prod_{p=1}^{\infty} b_p.$$

From (2) and (4) the theorem follows.

A I.11.9. Theorem. We have for any sequence $\{a_n\}$

$$\operatorname{Lim\,inf}_n a_n \leqq \operatorname{Lim\,sup}_n a_n.$$

Proof. By definition,

$$\operatorname{Lim\,inf}_n a_n = \sum_{k=1}^{\infty} \prod_{n=k}^{\infty} a_n, \qquad \operatorname{Lim\,sup}_n a_n = \prod_{k=1}^{\infty} \sum_{n=k}^{\infty} a_n.$$

Consider the equation

$$(1) \qquad \operatorname{co} \sum_{k=1}^{\infty} \prod_{n=k}^{\infty} a_n + \prod_{k=1}^{\infty} \sum_{n=k}^{\infty} a_n = I.$$

This equation is equivalent to

$$(2) \qquad \prod_{k=1}^{\infty} \sum_{n=k}^{\infty} \operatorname{co} a_n + \prod_{k=1}^{\infty} \sum_{n=k}^{\infty} a_n = I.$$

Now we have

$$\sum_{n=1}^{\infty} \operatorname{co} a_n \geqq \sum_{n=2}^{\infty} \operatorname{co} a_n \geqq \sum_{n=3}^{\infty} \operatorname{co} a_n \geqq \cdots,$$

$$\sum_{n=1}^{\infty} a_n \geqq \sum_{n=2}^{\infty} a_n \geqq \sum_{n=3}^{\infty} a_n \geqq \cdots,$$

$$\prod_{k=1}^{\infty} \sum_{n=k}^{\infty} \operatorname{co} a_n + \prod_{k=1}^{\infty} \sum_{n=k}^{\infty} a_n = \prod_{k=1}^{\infty} \left[\left(\sum_{n=k}^{\infty} \operatorname{co} a_n\right) + \left(\sum_{n=k}^{\infty} a_n\right)\right],$$

this by virtue of [A I.11.8.], Now the inequality of the theorem is equivalent to (1), hence to (2) and then to:

$$\prod_{k=1}^{\infty}\left(\sum_{n=k}^{\infty} \operatorname{co} a_n + \sum_{n=k}^{\infty} a_n\right) = I$$

i.e. to

$$\prod_{k=1}^{\infty} \sum_{n=k}^{\infty} (\operatorname{co} a_n + a_n) = I,$$

which is true. The theorem follows.

A I.11.10. Theorem. The following are equivalent:

I. $\operatorname*{Lim\,sup}_n (a_n \dotplus a_0) = O$,

II. $\operatorname*{Lim}_n a_n = a_0$.

Proof. I is equivalent successively to the following statements:

$$\prod_{k=1}^{\infty} \sum_{n=k}^{\infty} (a_n \dotplus a_0) = O,$$

$$\prod_{k=1}^{\infty} \sum_{n=k}^{\infty} [(a_n - a_0) + (a_0 - a_n)] = O,$$

(1) $$\prod_{k=1}^{\infty} \left[\sum_{n=k}^{\infty} (a_n - a_0) + \sum_{n=k}^{\infty} (a_0 - a_n) \right] = O.$$

By [A I.11.8.] the equation (1) is equivalent to

$$\prod_{k=1}^{\infty} \sum_{n=k}^{\infty} (a_n - a_0) + \prod_{k=1}^{\infty} \sum_{n=k}^{\infty} (a_0 - a_n) = O,$$

because

$$\sum_{n=1}^{\infty} (a_n - a_0) \geqq \sum_{n=2}^{\infty} (a_n - a_0) \geqq \cdots$$

and

$$\sum_{n=1}^{\infty} (a_0 - a_n) \geqq \sum_{n=2}^{\infty} (a_0 - a_n) \geqq \cdots$$

Hence (1) is equivalent to the conjunction of the two statements

$$\prod_{k=1}^{\infty} \sum_{n=k}^{\infty} (a_n - a_0) = O, \quad \prod_{k=1}^{\infty} \sum_{n=k}^{\infty} (a_0 - a_n) = O;$$

hence to

$$\prod_{k=1}^{\infty} \sum_{n=k}^{\infty} (a_n \cdot \operatorname{co} a_0) = O, \quad \prod_{k=1}^{\infty} \sum_{n=k}^{\infty} (a_0 \cdot \operatorname{co} a_n) = O, \quad \text{i.e.}$$

$$\operatorname{co} a_0 \cdot \prod_{k=1}^{\infty} \sum_{n=k}^{\infty} a_n = O, \quad a_0 \cdot \prod_{k=1}^{\infty} \sum_{n=k}^{\infty} \operatorname{co} a_n = O.$$

Hence to

$$\Pi\Sigma\, a_n - a_0 = O, \quad a_0 - \Sigma\Pi\, a_n = O,$$

i.e. to

$$\Pi\Sigma\, a_n \leqq a_0, \quad a_0 \leqq \Sigma\Pi\, a_n,$$

i.e. to

(2) $$\operatorname*{Lim\,sup}_n a_n \leqq a_0, \quad a_0 \leqq \operatorname*{Lim\,inf}_n a_n.$$

Thus we have proved that I is equivalent to (2). Let I be true. If follows that

$$\operatorname*{Lim\,sup}_n a_n \leqq \operatorname*{Lim\,inf}_n a_n,$$

and hence, by theorem [A I.11.9.],

$$\operatorname*{Lim\,inf}_n a_n = \operatorname*{Lim\,sup}_n a_n \quad \text{i.e. II.}$$

Let II be true. It follows

$$a_0 = \operatorname*{Lim\,sup}_n a_n, \quad a_0 = \operatorname*{Lim\,inf}_n a.$$

Hence we have (2) and then, by virtue of equivalences proved above, we see that I is true.

The theorem is established.

A I.11.11. Theorem. For a constant sequence $\{a, a, \ldots\}$ we have $\operatorname{Lim} a = a$.

A I.11.11a. Theorem. If $\operatorname{Lim} a_n = a$, then for any partial sequence $a_{\nu(k)}$ we have

$$\operatorname{Lim} a_{\nu(n)} = a.$$

Proof. Since $\operatorname{Lim} a_n = a$, it follows, by [A I.11.10], $\operatorname{Lim\,sup}(a_n \dot{+} a) = O$, i.e.

(1) $$\prod_{k=1}^{\infty} \sum_{n=k}^{\infty} (a_n \dot{+} a) = O.$$

Now we have

$$\sum_{n=k}^{\infty} (a_{\nu(k)} \dot{+} a) \leqq \sum_{n=\nu(k)}^{\infty} (a_n \dot{+} a) \leqq \sum_{n=k}^{\infty} (a_n \dot{+} a),$$

because $\nu(k) \geqq k$.

Hence

$$\prod_{k=1}^{\infty} \sum_{n=k}^{\infty} (a_{\nu(n)} \dot{+} a) \leqq \prod_{k=1}^{\infty} \sum_{n=k}^{\infty} (a_n \dot{+} a).$$

Taking account of (1) we get

$$\prod_{k=1}^{\infty} \sum_{n=k}^{\infty} (a_{\nu(n)} \dot{+} a) = O,$$

i.e.

$$\operatorname{Lim\,sup}(a_{\nu(n)} \dot{+} a) = O,$$

hence, by [A I.11.10.]

$$\operatorname{Lim} a_{\nu(n)} = a, \quad \text{q.e.d.}$$

A I.11.12. By means of the notion of the natural somatic limit the tribe (B) is organized into a limit-topology. We shall show, that in this topology, finite somatic and algebraic operations are continuous. Indeed we have

A I.11.12a. Theorem. If

$$\operatorname{Lim}_n a_n = a_0, \quad \text{then} \quad \operatorname{Lim}_n (\operatorname{co} a_n) = \operatorname{co} a_0.$$

Proof. We have, by def. [A I.11.4.]:

$$\prod_{k=1}^{\infty} \sum_{n=k}^{\infty} a_n = \sum_{k=1}^{\infty} \prod_{n=k}^{\infty} a_n = a_0.$$

From this the theorem follows by de Morgan laws.

A I.11.13. Theorem. If $\operatorname{Lim}_n a_n = a_0$, $\operatorname{Lim}_n b_n = b_0$, then

1) $\operatorname{Lim}_n (a_n + b_n) = a_0 + b_0$,

2) $\operatorname{Lim}_n (a_n \cdot b_n) = a_0 \cdot b_0$,

3) $\operatorname{Lim}_n (a_n \dotplus b_n) = a_0 \dotplus b_0$,

4) $\operatorname{Lim}_n (a_n - b_n) = a_0 - b_0$.

Proof. We have, by theorem [A I.11.10.]:

$$(1) \qquad \prod_{k=1}^{\infty} \sum_{n=k}^{\infty} (a_n \dotplus a_0) = 0, \quad \prod_{k=1}^{\infty} \sum_{n=k}^{\infty} (b_n \dotplus b_0) = 0.$$

We also have

$$\prod_{k=1}^{\infty} \sum_{n=k}^{\infty} [(a_n \dotplus b_n) \dotplus (a_0 \dotplus b_0)] \leqq \prod_{k=1}^{\infty} \sum_{n=k}^{\infty} [(a_n \dotplus a_0) + (b_n \dotplus b_0)] =$$

$$= \prod_{k=1}^{\infty} \left[\sum_{n=k}^{\infty} (a_n \dotplus a_0) + \sum_{n=k}^{\infty} (b_n \dotplus b_0) \right] =, \text{ Lemma [A I.11.8.]},$$

$$= \prod_{k=1}^{\infty} \sum_{n=k}^{\infty} (a_n \dotplus a_0) + \prod_{k=1}^{\infty} \sum_{n=k}^{\infty} (b_n \dotplus b_0) = 0.$$

Hence, by theorem [A I.11.10.],

$$\operatorname{Lim}_n (a_n \dotplus b_n) = a_0 \dotplus b_0,$$

i.e. Thesis 3). Since

$$a_n b_n \dotplus a_0 b_0 = a_n b_n \dotplus a_n b_0 \dotplus a_n b_0 \dotplus a_0 b_0$$

$$= a_n \cdot (b_n \dotplus b_0) \dotplus (a_n \dotplus a_0) \cdot b_0 \leqq (b_n \dotplus b_0) + (a_n \dotplus a_0),$$

we have

$$\prod_{k=1}^{\infty} \sum_{n=k}^{\infty} (a_n b_n + a_0 b_0) \subseteqq \prod_{k=1}^{\infty} \sum_{n=k}^{\infty} [(b_n \dotplus b_0) + (a_n \dotplus a_0)] = , \quad \text{(by [A I.11.8.])},$$

$$= \prod_{k=1}^{\infty} \left[\sum_{n=k}^{\infty} (b_n \dotplus b_0) + \prod_{k=1}^{\infty} \sum_{n=k}^{\infty} (a_n \dotplus a_0) \right].$$

Since, as before the last expression $= O$, the part 2) of the theorem is proved.

Since algebraic operations of addition and multiplication are continuous, it follows that 1) and 4) hold too. Indeed:

$$a_n + b_n = a_n \dotplus a_n b_n \dotplus b_n$$

and

$$a_n - b_n = a_n \cdot \operatorname{co} b_n = a_n (1 \dotplus b_n) = a_n \dotplus a_n \cdot b_n .$$

The theorem is established.

A I.11.14. Remark. The above limit-topology on (B) cannot be considered as a genuine neighborhood topology, if we define *the closure* $\bar{E}$ *of a set* E of somata, as the set of all somata $b \in B$ such that there exists an infinite sequence $\{a_n\} \in B$ with $a_n \in E$, $\operatorname{Lim}_n a_n = b$. It is true that such a closure satisfies the properties:

$$E \subseteqq \bar{E}, \quad \overline{E \cup F} = \bar{E} \cup \bar{F},$$

but $\bar{\bar{E}} = \bar{E}$ fails to be generally true, (see the quoted work of D. MAHARAM p. 155, (24)).

A I.11.15. Theorem. If (B) is denumerably additive Boolean tribe and μ a denumerably additive measure on it, this measure is a continuous function in the natural somatic limit-topology on (B), i.e. if $\operatorname{Lim} a_n = a$, then $\lim \mu(a_n) = \mu(a)$. We also have $\lim \mu(a_n \dotplus a) = 0$.

Proof. Let

$$\operatorname{Lim}_n a_n = a .$$

By theorem [A I.11.10.]

$$\operatorname{Lim}_n \sup (a_n \dotplus a) = O$$

i.e.

$$\prod_{k=1}^{\infty} \sum_{n=k}^{\infty} (a_n \dotplus a) = O .$$

Since

$$\sum_{n=1}^{\infty} (a_n \dotplus a) \geqq \sum_{n=2}^{\infty} (a_n \dotplus a) \geqq \cdots$$

we have, by theorem [A I.2.1.I.],

(1) $$\lim_{k \to \infty} \mu \left(\sum_{n=k}^{\infty} (a_n \dotplus a) \right) = 0 .$$

Since

$$a_k \dotplus a \leqq \sum_{n=k}^{\infty} (a_n \dotplus a),$$

it follows that

$$\mu(a_k \dotplus a) \leqq \mu\left(\sum_{n=k}^{\infty} (a_n \dotplus a)\right),$$

and consequently, by (1),

(2) $$\lim_{k\to\infty} \mu(a_k \dotplus a) = 0.$$

This can be written

$$\lim_{k\to\infty} \|a_k, a\|_\mu = 0;$$

hence,

$$\lim_\mu a_k =^\mu a.$$

By continuity of μ in the μ-measure topology we get finally

$$\lim \mu(a_k) = \mu(a).$$

Thus we have that, if $\mathrm{Lim}\, a_n = a$, then $\lim \mu(a_n) = \mu(a)$.

The theorem [A I.11.15.] is proved.

A I.11.16. Lemma. If

1. $a_1, a_2, \ldots, a_n, \ldots$ is an infinite sequence of somata of a denumerably additive tribe,
2. $a_1 \leqq a_2 \leqq \cdots \leqq a_n \leqq \cdots$,

then

$$\mathrm{Lim}_{n\to\infty}\, a_n = \sum_{n=1}^{\infty} a_n.$$

Proof. By definition [A I.11.5.]:

(1) $$\mathrm{Lim\,sup}_{n\to\infty}\, a_n = (a_1 + a_2 + a_3 + \cdots) \cdot (a_2 + a_3 + \cdots) \cdot (a_3 + a_4 + \cdots).$$

Since $a_1 \leqq a_2$, we have $a_2 = a_1 + a_2$, and since $a_2 \leqq a_3$, we have

$$a_3 = a_1 + a_2 + a_3,$$

etc.

It follows that the factors in the product (1) are all equal. Hence the right hand side in (1) equals $a_1 + a_2 + \cdots$. This completes the proof of the equality $\mathrm{Lim\,sup}\, a_n = a_1 + a_2 + \cdots$

Now we are going to prove that

(2) $$\mathrm{Lim\,inf}_{n\to\infty}\, a_n = a_1 + a_2 + \cdots$$

To do that, consider the expression

(3) $$(a_1 \cdot a_2 \cdot a_3 \ldots) + (a_2 \cdot a_3 \ldots) + (a_3 \ldots) + \cdots$$

Since

$$a_1 \leqq a_2 \leqq a_3 \leqq \cdots,$$

we have

$$a_1 \cdot a_2 \cdot a_3 \cdots = a_1, \qquad a_2 \cdot a_3 \cdots = a_2,$$

etc.

Hence the terms in (3) are equal to $a_1, a_2, a_3, \ldots$ respectively, so the sum (3) equals $a_1 + a_2 + \cdots$. This completes the proof.

A I.11.16a. Lemma. If

1. $a_1, a_2, \ldots, a_n, \ldots$ is an infinite sequence of somata of a denumerably additive tribe;

2. $a_1 \geqq a_2 \geqq \cdots \geqq a_n \geqq \cdots$,

then

$$\operatorname*{Lim}_{n\to\infty} a_n = \prod_{n=1}^{\infty} a_n.$$

Proof. By applying to [A I.11.16.] the de Morgan laws and relying on the theorem [A I.11.12a.], which says that

$$\operatorname*{Lim}_{n\to\infty} (\operatorname{co} a_n) = \operatorname{co}(\operatorname*{Lim}_{n\to\infty} a_n).$$

A I.11.17. Lemma. If

1. $a_1, a_2, \ldots, a_n, \ldots$ is an infinite sequence of somata of a denumerably additive tribe;

2. $\operatorname*{Lim\,inf}_{n\to\infty} a_n =^{\mu} \operatorname*{Lim\,sup}_{n\to\infty} a_n =^{\mu} a$,

then

$$\lim_{n\to\infty} \|a, a_n\|_{\mu} = 0.$$

Proof. Let us remind that all formulas, valid in a tribe (T), are also valid in the tribe (T_J), where J is the ideal composed of all somata with μ-measure equal 0. But we must change only the relation of equality $(=)$ and that inclusion $(\leqq)$ into $(=^{\mu})$ and $(\leqq^{\mu})$ respectively.

We have, by [A I.11.10.],

$$\operatorname*{Lim\,sup}_{n} (a_n \dotplus a) =^{\mu} O,$$

i.e.

$$\operatorname*{Lim\,sup}_{n} [(a_n - a) + (a - a_n)] =^{\mu} O.$$

Now, we have in general, if $A_n \leqq^{\mu} B_n$, $(n = 1, 2, \ldots)$, then

$$\operatorname{Lim\,sup} A_n \leqq^{\mu} \operatorname{Lim\,sup} B_n$$

and

$$\operatorname{Lim\,inf} A_n \leqq^{\mu} \operatorname{Lim\,inf} B_n,$$

which follows directly from the definition of Lim sup and Lim inf. Applying this remark to our case, we get:

$$\operatorname{Lim\,sup}(a_n - a) \leqq^{\mu} \operatorname{Lim\,sup}[(a_n - a) + (a - a_n)],$$

$$\operatorname{Lim\,sup}(a - a_n) \leqq^{\mu} \operatorname{Lim\,sup}[(a_n - a) + (a - a_n)].$$

It follows that

$$\text{Lim sup}\,(a_n - a) =^{\mu} 0$$

and

$$\text{Lim sup}\,(a - a_n) =^{\mu} 0.$$

Hence, by [A I.11.15.], we get

$$\mu(a_n - a) \to 0, \quad \mu(a - a_n) \to 0,$$

which implies

$$\mu(a \dotplus a_n) = 0,$$

i.e.

$$\lim_{n\to\infty} ||a, a_n||_{\mu} = 0, \quad \text{q.e.d.}$$

A I.11.18. Theorem. If

1) (T) is a denumerably additive tribe provided with a denumerably additive measure μ,

2) $\{a_n\}$ is a sequence of somata of (T),

3) $||a_n, a||_{\mu} \to 0$ for $n \to \infty$,

then there exists a partial sequence $\{a_{k(n)}\}$ of $\{a_n\}$, such that

$$\text{Lim sup}\, a_{k(n)} =^{\mu} \text{Lim inf}\, a_{k(n)} =^{\mu} a.$$

A I.11.18a. Proof. To simplify the formulas, we agree to write $=$, $\leqq$ instead of $=^{\mu}$, $\leqq^{\mu}$. Suppose that

$$\text{Lim}_{n\to\infty} ||a_n, a||_{\mu} = 0.$$

We can make $||a, a_n||_{\mu}$ as small, as we like, if we take sufficiently great indices n.

Choose the sequence

$$k(1) < k(2) < \cdots$$

of indices, so that if we put

$$\sigma_n =_{df} ||a_{k(n)}, a||_{\mu},$$

the sum $\sum_{n=1}^{\infty} \sigma_n$ converges.

Put

(0) $$a' =_{df} \text{Lim inf}_{n\to\infty}\, a_{k(n)}, \quad a'' =_{df} \text{Lim sup}_{n\to\infty}\, a_{k(n)},$$

and also

$$A'_n =_{df} a_{k(n)} \cdot a_{k(n+1)} \cdot \cdots,$$

$$A''_n =_{df} a_{k(n)} + a_{k(n+1)} + \cdots, \quad (n = 1, 2, \ldots).$$

We have

$$A'_n \leqq A'_{n+1}, \quad A''_n \geqq A''_{n+1}.$$

Hence by Lemma [A I.11.16.] and [A I.11.16a.], we have

$$a' = \text{Lim}\, A'_n, \quad a'' = \text{Lim}\, A''_n.$$

By virtue of Lemma [A I.11.17.] we obtain:

(1) $$||a', A'_n||_\mu \to 0,$$

(2) $$||a'', A''||_\mu \to 0.$$

A I.11.18b. Now we have

$$||A'_n, a||_\mu = ||a_{k(n)} \cdot a_{k(n+1)} \cdot \cdots, a||_\mu = \mu(a_{k(n)} \cdot a_{k(n+1)} \cdots - a) +$$
$$+ \mu(a - a_{k(n)} \cdot a_{k(n+1)} \cdots) \leqq \mu(a - a_{k(n)}) + \mu(a - a_{k(n+1)}) + \cdots +$$
$$+ \mu(a_{k(n)} - a) + \mu(a_{k(n+1)} - a) + \cdots,$$

this because $\mu(p \cdot q) \leqq \mu(p) + \mu(q)$, and because the formulas involving finite operations can be generalized to formulas involving infinite operations.

It follows that

$$||A'_n, a||_\mu \leqq ||a_{k(n)}, a||_\mu + ||a_{k(n+1)}, a||_\mu + \cdots;$$

hence

$$||A'_n, a||_\mu \leqq \sum_{i=1}^{\infty} \sigma_i.$$

Since $\sum_{i=1}^{\infty} \sigma_i$ converges, we get $\sum_{i=n}^{\infty} \sigma_i \to 0$ for $n \to \infty$; hence

(3) $$||A'_n, a||_\mu \to 0 \quad \text{for} \quad n \to \infty.$$

On the other hand we have

$$||a, A''_n||_\mu = ||a, a_{k(n)} + a_{k(n+1)} + \cdots||_\mu \leqq$$
$$\leqq ||a, a_{k(n)}||_\mu + ||a, a_{k(n+1)}||_\mu + \cdots \leqq \sigma_n + \sigma_{n+1} + \cdots,$$

which gives

(4) $$||a, A''_n||_\mu \to 0 \quad \text{for} \quad n \to \infty.$$

A I.11.18c. From (1) and (3) it follows

$$||a, a'||_\mu \leqq ||a, A'_n||_\mu + ||A'_n, a'||\,\mu \to 0;$$

hence

(5) $$||a, a'||_\mu = 0.$$

Similarly from (2) and (4) it follows

$$||a, a''||_\mu \leqq ||a, A''_n||_\mu + ||A''_n, a''||_\mu \to 0;$$

hence

(6) $$||a, a''|| = 0.$$

Since

$$||a', a''||_\mu \leqq ||a, a'||_\mu + ||a, a''||_\mu,$$

the relations (5) and (6) imply

$$||a', a''||_\mu = 0,$$

i.e., by (0),

$$\operatorname{Lim\,inf} a_{k(n)} =_\mu \operatorname{Lim\,sup} a_{k(n)} =^\mu a, \quad \text{q.e.d.}$$

A I.11.19. Theorem. If

1) $a, a_1, a_2, \ldots, a_n, \ldots$ is an infinite sequence of somata of a denumerably additive tribe (T) and μ a denumerably additive measure on (T), then the following are equivalent:

I. $\lim\limits_{n\to\infty} \|a, a_n\|_\mu = 0$,

II. From every subsequence $\{a_{k(n)}\}$ of $\{a_n\}$ another subsequence $\{a_{kl(n)}\}$ can be extracted, such that

$$\operatorname*{Lim\,sup}_{n\to\infty} a_{kl(n)} =^\mu \operatorname*{Lim\,inf}_{n\to\infty} a_{kl(n)} =^\mu a.$$

A I.11.19a. Proof. Suppose I, i.e. suppose that

$$\operatorname*{Lim}_{n\to\infty} \|a, a_n\|_\mu = 0.$$

Let $\{a_{k(n)}\}$ be a subsequence of $\{a_n\}$, where

$$a_{k(1)} < a_{k(2)} < \cdots < a_{k(n)} < \cdots.$$

We have

$$\lim \|a, a_{k(n)}\|_\mu = 0.$$

Hence, by [A I.11.18.], we can select a subsequence $\{a_{kl(n)}\}$ from $\{a_{k(n)}\}$ such that

$$\operatorname{Lim\,inf} a_{kl(n)} =_\mu \operatorname{Lim\,sup} a_{kl(n)} =^\mu a;$$

so we have proved that I implies II.

Let II, choose any subsequence $\{a_{k(n)}\}$ of $\{a_n\}$, and select in it a subsequence $\{a_{kl(n)}\}$, such that

$$\operatorname*{Lim\,sup}_{n\to\infty} a_{kl(n)} =^\mu \operatorname*{Lim\,inf}_{n\to\infty} a_{kl(n)} =^\mu a.$$

By Lemma [A I.11.17.] we have

(1) $$\operatorname*{Lim}_{n\to\infty} \|a, a_{kl(n)}\|_\mu = 0.$$

A I.11.19b. Let us prove the following general auxiliary theorem.

Theorem. If $\{\alpha_n\}$ is a sequence of non-negative numbers, such that from every its subsequence $\{\alpha_{k(n)}\}$ another sequence $\{\alpha_{kl(n)}\}$ can be extracted such that

$$\lim \alpha_{kl(n)} = 0,$$

then

(1) $$\lim_{n\to\infty} \alpha_n = 0.$$

(2) Indeed, suppose that (1) be not true.

Notice that the sequence $\{\alpha_n\}$ must be bounded, because if not, we would have a partial sequence $\{\alpha_{q(n)}\}$ of $\{\alpha_n\}$ with

$$\alpha_{q(1)} < \alpha_{q(2)} < \cdots < \alpha_{q(n)} < \cdots,$$

and with $\lim \alpha_{q(n)} = +\infty$.

Whatever the partial sequence $\{\alpha_{q\,p(n)}\}$ of $\{\alpha_{q(n)}\}$ may be, we have

$$\lim \alpha_{q\,p(n)} = +\infty,$$

what contradicts the hypothesis of the theorem. Thus $\{\alpha_n\}$ is bounded. Hence, by (2), there exists a subsequence $\{\alpha_{s(n)}\}$ of $\{\alpha_n\}$, having a limit different from 0. But then every partial sequence of $\{\alpha_{s(n)}\}$ would have the same limit, differing from 0. This contradicts the hypothesis of the theorem, that from every subsequence of $\{\alpha_n\}$ another can be extracted with the limit equal to 0.

Thus the auxiliary theorem is true. Consequently

$$\lim_{n\to\infty} \alpha_n = 0.$$

A I.11.19c. Applying the above remark to our case, we get

$$\lim_n \| a_n, a \|_\mu = 0,$$

so the condition I follows. Thus we have proved that II implies I. The theorem is established.

A I.11.20. The above theorem shows the relation between the measure topology and the natural somatic topology.[1]

Chapter B

Important auxiliaries

B.1. The chapter [B] contains some auxiliaries needed in subsequent chapters. We introduce the fundamental notion of the base of a tribe (section 1), of a covering (section 2) and also (section 3) some properties of two-valued measures on a tribe, i.e. those which admits two values only, namely 0 and 1. The main use of these topics will be in Chapter [B I.], C, and in the chapters, dealing with a special kind of integration.

---

[1] The theorem is by the author (18), (1928/29). The theorem in that paper concerns abstract sets, but since the relation ($\in$) of belonging of a point to a set, has not been used in the proof, the proof applies without any change to abstract tribes. See also G. Birkhoff (8), first edition (1939).

Section 1 of B.

B.1.1. Base of a tribe. Let (F) be a finitely additive, non trivial [A.1.6.] tribe. Its somata $f, g, h, \ldots$ will be sometime termed *figures*. Let $\boldsymbol{B}$ be a non empty subset of the domain F of (F), satisfying the following conditions:

1) $O \in \boldsymbol{B}$, $I \in \boldsymbol{B}$, where O, I are the zero, and the unit of (F) respectively;

2) if $a, b \in \boldsymbol{B}$, then $a \cdot b \in \boldsymbol{B}$;

3) if $f \in F$, then there exists a finite number of somata of $\boldsymbol{B}$, whose sum is f.

The somata $a, b, c, \ldots; p, q, \ldots$ of $\boldsymbol{B}$ will be termed *bricks* and the set $\boldsymbol{B}$ will be termed *base of* (F).

B.1.2. We may suppose that the tribe (F) is plunged in a denumerably additive tribe (G), as its finitely genuine subtribe [A.7.]. If so, we shall consider infinite somatic operations on (F), which will be understood, as taken from (G), [A.7.6.].

B.1.3. Def. If we admit [B.1.1.] and [B.1.2.], we shall say that we *are in the conditions* $\boldsymbol{FBG}$.

B.2. Def. According to the case we shall sometimes admit one or the other of the following two conditions, called:

(Hyp Af), and ***(Hyp Ad)***, defined as follows:

(Hyp Af): If $a \in \boldsymbol{B}$, then there exists a finite number of disjoint bricks $a_1, a_2, \ldots, a_n$, $(n \geqq 1)$, such that

$$\operatorname{co} a = a_1 + \cdots + a_n.$$

(Hyp Ad): If $a \in \boldsymbol{B}$, then there exists a denumerably infinite sequence $a_1, a_2, \ldots, a_m, \ldots$ of disjoint bricks, such that

$$\operatorname{co}^G(\mathfrak{B}\, a) =^G \mathfrak{B}(a_1) +^G \mathfrak{B}(a_2) +^G \cdots +^G \mathfrak{B}(a_m) +^G \cdots,$$

where $\mathfrak{B}$ is the correspondence [A.7.4.] attached to the situation where (F) is a finitely genuine subtribe of (G).

B.3. Our condition can be written, by the agreement [B.1.2.], in the simplier form:

$$\operatorname{co} a = a_1 + a_2 + \cdots + \cdots + a_m + \cdots$$

Of course ***(Hyp Af)*** implies ***(Hyp Ad)***. For our main purpose ***(Hyp Af)*** is sufficient. However, since we have further generalizations in mind, we prefer to work under the less restrictive hypothesis ***(Hyp Ad)***. The complication which our approach will imply, is not too great.

B.3.1. Expl. Let F be the tribe, whose somata are $\emptyset$ and finite unions of half-open segments $\langle\alpha, \beta)$ where $0 \leqq \alpha \leqq \beta \leqq 1$. Bricks are just those segments. ***(Hyp Af)*** is satisfied.

B.3.2. Theorem. If

1. $\boldsymbol{F}$ is a finitely additive tribe,
2. $\boldsymbol{B}' \subseteqq \boldsymbol{F}$,
3. $O \in \boldsymbol{B}'$, $I \in \boldsymbol{B}'$,
4. if $a, b \in \boldsymbol{B}'$, then $a \cdot b \in \boldsymbol{B}'$,
5. if $a \in \boldsymbol{B}'$, then $\operatorname{co} a$ is a finite sum of mutually disjoint elements of $\boldsymbol{B}'$,
6. $\boldsymbol{F}$ is the smallest strict subtribe of F containing $\boldsymbol{B}'$,

then

1) every $p \in \boldsymbol{F}$ is a finite sum of mutually disjoint elements of $\boldsymbol{B}'$,
2) every finite sum of mutually disjoint elements of $\boldsymbol{B}'$ belongs to $\boldsymbol{F}$,
3) $\boldsymbol{B}'$ is a base of $\boldsymbol{F}$.

Proof. Denote by $\boldsymbol{F}_1$ the set of all somata of $\boldsymbol{F}$, which are finite disjoint sums of elements of $\boldsymbol{B}'$. We shall first prove that $\boldsymbol{F}_1$ is a strict subtribe of $\boldsymbol{F}$. Let $p \in \boldsymbol{F}_1$. We have $p = a_1 + \cdots + a_n$, where $a_i \in \boldsymbol{B}'$, $n \geqq 1$. Hence $\operatorname{co} p = \operatorname{co}(a_1 + \cdots + a_n) = \operatorname{co} a_1 \ldots \operatorname{co} a_n$. By hyp. 5 we have for $i = 1, 2, \ldots, n$

$$\operatorname{co} a_i = a_{i1} + a_{i2} + \cdots,$$

where the sum is finite and the elements are disjoint. Hence

$$\operatorname{co} p = \sum_{\alpha_1, \ldots, \alpha_n} a_{1\alpha_1} a_{2\alpha_2} \ldots a_{n\alpha_n}.$$

The terms of this sum are, by hyp. 4, elements of $\boldsymbol{B}'$, and they are disjoint. Indeed $a_{1\alpha_1} \ldots a_{n\alpha_1} \cdot a_{1\beta_1} \ldots a_{n\beta_n} = O$ all the time when one of the inequalities $\alpha_1 = \alpha_1 \neq \beta_1, \ldots, \alpha_n \neq \beta_n$ is true; hence different terms are disjoint. Consequently

$$\operatorname{co} p \in \boldsymbol{F}_1.$$

Thus we have proved that if

$$(1) \qquad p \in \boldsymbol{F}_1, \quad \text{then} \quad \operatorname{co} p \in \boldsymbol{F}_1.$$

Now let $p, q \in \boldsymbol{F}_1$. We have

$$p = c_1 + c_2 + \cdots, \qquad q = d_1 + d_2 + \cdots,$$

where the terms of each sum are disjoint elements of $\boldsymbol{B}'$ and the sums are finite.

We get $p \cdot q = \sum_{i,j} c_i d_j$. The terms of this sum are disjoint elements of $\boldsymbol{B}'$. Thus we have proved that if $a, b \in \boldsymbol{F}_1$, then $a \cdot b \in \boldsymbol{F}_1$. This result and (1) imply that $\boldsymbol{F}_1$ is a tribe. Its unit is I and its zero is O. It is a strict subtribe of $\boldsymbol{F}$ and contains $\boldsymbol{B}'$. Hence, by hyp. 6, $\boldsymbol{F} = \boldsymbol{F}_1$. It follows that $\boldsymbol{B}'$ is a base of $\boldsymbol{F}$.

B.3.3. There exists a tribe $\boldsymbol{F}$ and a base $\boldsymbol{B}$ of $\boldsymbol{F}$, such that **(*Hyp Ad*)** is not satisfied — hence **(*Hyp Af*)** neither.

Expl. We shall consider right-half open subintervals $\langle\alpha, \beta)$ of $\langle 0, 1)$. Denote by $a(0)$, b, $a(1)$ the intervals $\langle 0, \frac{1}{3})$, $\langle\frac{1}{3}, \frac{2}{3})$, $\langle\frac{2}{3}, 1)$ respectively, and put $b' =_{df} a(0) \cup b$, $b'' =_{df} b \cup a(1)$. We divide $a(0)$ into three equal parts $a(0, 0)$, $b(0)$, $a(0, 1)$, and we do the same with $a(1)$, getting $a(1, 0)$, $b(1)$, $a(1, 1)$. We put

$$b'(0) =_{df} a(0, 0) \cup b(0), \qquad b''(0) =_{df} b(0) \cup a(0, 1),$$
$$b'(1) =_{df} a(1, 0) \cup b(1), \qquad b''(1) =_{df} b(1) \cup a(1, 1).$$

Suppose we have already difined all $a(\alpha_1, \alpha_2, \ldots, \alpha_r)$ where $\alpha_1, \alpha_2, \ldots, \alpha_r$ are 0 or 1 and r is given. We divide this interval into three equal parts:

$$(1) \qquad a(\alpha_1, \ldots, \alpha_r, 0), \quad b(\alpha_1, \ldots, \alpha_r), \quad a(\alpha_1, \ldots, \alpha_r, 1),$$

and define:

$$(2) \qquad \begin{aligned} b'(\alpha_1, \ldots, \alpha_r) &=_{df} a(\alpha_1, \ldots, \alpha_r, 0) \cup b(\alpha_1, \ldots, \alpha_r), \\ b''(\alpha_1, \ldots, \alpha_r) &=_{df} b(\alpha_1, \ldots, \alpha_r) \cup a(\alpha_1, \ldots, \alpha_r, 1). \end{aligned}$$

Thus we have defined inductively all $a(\alpha_1, \ldots, \alpha_n)$ for $n = 1, 2, \ldots$; (1) defines all $b(\alpha_1, \ldots, \alpha_n)$ and (2) defines all $b'(\alpha_1, \ldots, \alpha_n)$ and $b''(\alpha_1, \ldots, \alpha_n)$. Denote by $\boldsymbol{B}$ the class of sets composed of $\emptyset$, $\langle 0, 1)$, b, b', b'', and all $b(\alpha_1, \ldots, \alpha_n)$, $b'(\alpha_1, \ldots, \alpha_n)$, $b''(\alpha_1, \ldots, \alpha_n)$. Denote by $\boldsymbol{F}$ the smallest tribe of sets which contains $\boldsymbol{B}$. Under the above circumstances $\boldsymbol{B}$ is a base of F, but ***(Hyp Ad)*** is not satisfied.

B.3.4. There exist F and $\boldsymbol{B}$ where ***(Hyp Ad)*** is satisfied, but ***(Hyp Af)*** is not satisfied.

Expl. We consider the segments denoted by b, without or with primes and indices as before, but of all kinds: open, closed, half open on the right and half open on the left. They, and all their endpoints[1] will constitute the base. If we consider the smallest tribe containing that base, we have satisfied ***(Hyp Ad)***, but not ***(Hyp Af)***.

B.3.5. After this preliminary discussion, we shall prove some lemmas under ***(Hyp Ad)*** or ***(Hyp Af)***. To simplify writing of formulas which involve infinite operations, we shall make the following agreement. We shall write $\sum\limits_{n=1}^{\infty} f_n$ and $\prod\limits_{n=1}^{\infty} f_n$ instead of $\sum\limits_{n=1}^{\infty}{}^{G}\, \mathfrak{B}\, f_n$ and $\prod\limits_{n=1}^{\infty}{}^{G}\, \mathfrak{B}\, f_n$ respectively. If special clarity is needed we shall say that the *infinite denumerable operations performed on somata of F are taken from G* [B.1.2.].

B.3.5a. Lemma. Under ***(Hyp Ad)***, if $a_1, \ldots, a_n$ are bricks, $(n \geqq 2)$, then $\operatorname{co}(a_1 + a_2 + \cdots + a_n)$ is the sum of a denumerable number of mutually disjoint bricks, (Infinite summations are taken from G).

Proof. We have $\operatorname{co}(a_1 + \cdots + a_n) = \operatorname{co} a_1 \ldots \operatorname{co} a_n$, and by ***(Hyp Ad)***, $\operatorname{co} a_i = \sum\limits_{s=1}^{\infty} a_{is}$, $(i = 1, \ldots, n)$, where $a_{i1}, a_{i2}, \ldots$ are

[1] We mean single-point-sets.

disjoint bricks. Hence

$$(1)\quad \operatorname{co} a_1 \ldots \operatorname{co} a_n = \sum_{k_1=1}^{\infty} a_{1k_1} \ldots \sum_{k_n=1}^{\infty} a_{nk_n} = \sum_{k_1=1}^{\infty} \ldots \sum_{k_n=1}^{\infty} a_{1k_1} \ldots a_{nk_n}.$$

Now $(a_{1k_1} \ldots a_{nk_n}) \cdot (a_{1j_1} \ldots a_{nj_n}) = O$ all the times, when at least one of the inequalities $k_1 \neq j_1, \ldots, k_n \neq j_n$ holds true. Hence all the terms in (1) are disjoint. They are bricks on account of (1). The lemma is proved.

B.3.5.1. Lemma. Under ***(Hyp Af)***, if $a_1, \ldots, a_n$ are bricks $(n \geqq 2)$, then $\operatorname{co}(a_1 + \cdots + a_n)$ is the sum of a finite number of mutually disjoint bricks.

Proof. Similar to that of Lemma [B.3.5 a.].

B.3.6. Theorem. Under ***(Hyp Ad)***, if $f \in F$, then f is the sum of a denumerable number of mutually disjoint bricks, (infinite summation is taken from G).

Proof. By (1) we have $f = a_1 + \cdots + a_n$, where a_i are bricks. We may suppose $n \geqq 2$, since we can add as many zero-somata as we like. Indeed, O is a brick. Put $g_1 =_{df} a_1$, $g_2 =_{df} a_2 - a_1, \ldots$, $g_n =_{df} a_n - (a_1 + \cdots + a_{n-1})$. We have, [A.2.6.3.];

$$(1)\qquad f = g_1 + \cdots + g_n \quad \text{with} \quad g_1, \ldots, g_n$$

disjoint. Since $g_i = a_i \operatorname{co}(a_1 + \cdots + a_{i-1})$ we have, by [B.3.5 a.] $g_i = a_i \cdot \sum_{s=1}^{\infty} a_{i-1,s}$, where $a_{i-1,1}, a_{i-1,2}, \ldots$ are disjoint bricks. We get

$$g_i = \sum_{s=1}^{\infty} (a_i\, a_{i-1,s}),$$

so, by (1), the lemma is proved.

B.3.6.1. Theorem. Under ***(Hyp Af)***, if $f \in F$, then f is the sum of a finite number of disjoint bricks.

Proof. Similar to that of [B.3.6.].

Section 2 of B.

Coverings

B.4. Def. By a *covering* (*brick-covering*) we shall understand an at most denumerable sum of bricks, (i.e. $\mathfrak{B}$-images of bricks)[1]. Hence O is a covering. We consider F, G, $\boldsymbol{B}$ as in [B.1.1.], [B.1.2.].

---

[1]) There must be at least one brick in the sum, since we do not consider "empty" sums — for reasons given in (12). We can define a covering as $\sum_{n=1}^{\infty} a_n$ where $\{a_n\}$ is an infinite sequence of bricks. Since O is a brick, we always can replace finite sums by infinite ones.

B.4.1. Theorem. If $L \in G$, $L = \mathfrak{B} f_1 + \mathfrak{B} f_2 + \cdots + \mathfrak{B} f_n + \cdots$ where $f_n \in F$, then L can be considered as the sum

$$L = \mathfrak{B} g_1 + \mathfrak{B} g_2 + \cdots + \mathfrak{B} g_n + \cdots$$

where $g_n \in F$ and all g_n are disjoint.

Proof. This follows from [A.2.6.3.].

B.4.2. Theorem. Under ***(Hyp Ad)***, if L is a covering, then it is a denumerable sum of mutually disjoint bricks.

Proof. We have $L = a_1 + a_2 + \cdots + a_n + \cdots$, where a_i are bricks. Put

$$f_1 =_{df} a_1, \quad f_2 =_{df} a_2 - a_1, \quad f_3 =_{df} a_3 - (a_1 + a_2), \ldots$$

We have, by [B.4.1.], $L = f_1 + f_2 + \cdots$, where all f_i are disjoint. On the other hand, since $f_i \in F$, the figure f_i is, by [B.3.6.], a sum of a denumerable number of disjoint bricks. This complete the proof.

B.4.3. Corollary. Under ***(Hyp Ad)***, if L is a covering, $L = f_1 + f_2 + f_3 + \cdots + f_n + \cdots$, where f_n are figures, then there exists an infinite sequence of mutually disjoint bricks $a_1, a_2, \ldots, a_n, \ldots$ such that

1) $L = a_1 + a_2 + a_3 + \cdots$,
2) for every i there exists j such that $a_i \leqq f_j$.

Proof. We start as in the proof of Theorem [B.4.2.], getting

$$L = g_1 + g_2 + \cdots + g_n + \cdots,$$

where $g_1, g_2, \ldots, g_n, \ldots$ are disjoint figures and where

$$g_1 \leqq f_1, \quad g_2 \leqq f_2, \ldots$$

Now, by [B.3.6.], we can decompose each g_n into disjoint bricks

$$g_n = a_{n1} + a_{n2} + \cdots, \qquad (n = 1, 2, \ldots).$$

Since the set $\{a_{ik}\}$, $i, k = 1, 2, \ldots$ is denumerable, it can be represented by an infinite sequence. We have $a_{nk} \leqq g_n \leqq f_n$, so the theorem is proved.

B.5. If $L_1, L_2, \ldots, L_n, \ldots$ are covering, finite or infinite denumerable in number, then $\sum_n L_n$ is also a covering.

B.6. If L_1, L_2 are coverings, so is $L_1 \cdot L_2$.

Indeed, if $L_1 = \sum_n a_n$, $L_2 = \sum_m b_m$, we have $L_1 \cdot L_2 = \sum_n \sum_m a_n b_m$, and $a_n b_m$ is a brick.

Section 3 of B.

Maximal ideals, ultrafilters and two-valued measures on a tribe

B.7. We have already studied ideals in a general tribe, [A.9.]–[A.14.1.], and also noticed [A.20.] that if μ is a finitely additive measure on a tribe (T), then the set

$$\{a \mid \mu(a) = 0\}$$

is an ideal in (T).

However, we need some more information on ideals, especially those which are related to filters and to two-valued measures on a tribe.

B.7a. The classical theory of filters, created by H. CARTAN (29), constituted a new, convenient approach to abstract topology. It deals with sets and defines the filter as a non empty family C_0 of sets, satisfying the conditions:

1⁰. $O \overline{\in} C_0$,

2⁰. if $a \cdot b \in C_0$, then $a \cdot b \in C_0$,

3⁰. if $a \in C_0$, $a \leqq b$, then $b \in C_0$.

Now, it is easy to see that an analogous construction can be introduced in general tribes. We shall do that, with the modification, by which the condition 1⁰ is not admitted. Thus we give the following definition:

Filters and ideals

B.8. Def. Let (T) be a non trivial tribe, [A.1.6.]. By a *filter in* (T) we understand a family C of somata of (T), satisfying the conditions:

2′. if $a, b \in C$, then $a \cdot b \in C$,

3′. if $a \in C$, $a \leqq b$, then $b \in C$.

I indebted to prof. CH. PAUC the following simple, but very instructive remark, that in Boolean lattices filters are very strongly related to ideals, as is seen in the following two theorems:

B.8.1. Theorem. If C is a filter in the tribe (T), then the set

$$J =_{df} \{\mathrm{co}\, a \mid a \in C\}$$

is an ideal in (T).

Proof. Let $p, q \in J$. There exist $a, b \in C$ such that $p = \mathrm{co}\, a$, $q = \mathrm{co}\, b$. Hence $p + q = \mathrm{co}\, a + \mathrm{co}\, b = \mathrm{co}(a \cdot b)$. Since C is a filter, we have $a \cdot b \in C$, and then

$$\mathrm{co}(a \cdot b) \in J.$$

Thus, if

(1) $$p, q \in J, \text{ we have } p + q \in J.$$

Let $p \in J$ and $q \leqq p$. There exist $a \in C$ with $p = \mathrm{co}\, a$. Since $q \leqq p$, we have $\mathrm{co}\, q \geqq \mathrm{co}\, p = a$. Since $a \in C$, we have $\mathrm{co}\, q \in C$, and then $q \in J$.

Thus we have proved that

(2) $$\text{if } p \in J, \quad q \leq p, \text{ then } q \in J.$$

From (1) and (2) it follows that J is an ideal in (T). The converse theorem is also true:

B.8.2. Theorem. If J is an ideal in (T), then the set

$$C' =_{df} \{\text{co}\, p \mid p \in J\}$$

is a filter in (T).

Proof. Let $a, b \in C'$. There exist $p, q \in J$, such that

$$a = \text{co}\, p, \quad b = \text{co}\, q.$$

As J is an ideal, we have $p + q \in J$; hence $\text{co}(p+q) \in C'$; hence $\text{co}\, p \cdot \text{co}\, q \in C'$, and then $a \cdot b \in C'$.

Thus we have proved that if $a, b \in C$, then $a \cdot b \in C$. Let $a \in C$ and let $a \leq b$.

Since $a \in C$, there exists $p \in J$ with $a = \text{co}\, p$. Put $q =_{df} \text{co}\, b$.

Since $a \leq b$, we have $q \leq p$. Since $p \in J$, it follows $q \in J$. Hence $b = \text{co}\, q \in C$. Thus we have proved that if $a \in C$, $a \leq b$, then $b \in C$.

The theorem is established.

B.8.3. Theorem. Let (T) be a not trivial tribe, i.e., $O \neq I$; then there exists a $1 \to 1$ correspondence $\mathfrak{P}$ between ideals J and filters C in (T), defined in [B.8.1.] and [B.8.2.]. Thus

$$J \circlearrowright^{\mathfrak{P}} C, \quad C \circlearrowright^{\mathfrak{P}^{-1}} J$$

means:

$$C = \{a \mid \text{co}\, a \in J\}$$

and

$$J = \{b \mid \text{co}\, b \in C\}$$

respectively. The domain of $\mathfrak{P}$ is the collection of all ideals, and the range the collection of all filters in the given tribe.

B.8.3a. If (T) is the simpliest non trivial tribe, i.e. composed of somata O, I only, where $O \neq I$, then the only ideals in (T) are (O) and T. The only filters are (I) and T. (For a trivial tribe, i.e. where $O = I$, the only filter and the only ideal would be T.) In the sequal we shall deal with non trivial tribes only.

B.8.3b. If we follow the classical patterns of Cartan's filters [B.7a.]; hence admitting the supplementary hypothesis that O is never an element of the filter, then T would be not a filter, and hence the corresponding ideal would differ from T.

In our approach the set T is an ideal and a filter too.

B.8.4. Def. The filter which does not contain O, will be termed *proper filter*, and the ideal which does not contain I, will be termed *proper ideal*.

Hence the proper filter differs from T and the proper ideal also differs from T.

The correspondence $\mathfrak{P}$ carries the collection of all proper ideals onto the collection of all proper filters.

Ultrafilters

B.9. Def. Let (T) be a non trivial tribe. By an *ultrafilter in* (T) we understand any filter C such that

1) $C \neq T$;

2) if C_1 is a filter with $C \subset C_1$, (i.e. equality excluded), then $C_1 = T$.

Thus an ultrafilter is a proper filter (see (29), (31), and also see (32)).

B.9.1. Theorem. The $\mathfrak{P}^{-1}$-correspondent to an ultrafilter C is a maximal ideal J, i.e. a proper ideal in (T) such that

1) $J \neq T$,

2) if J_1 is an ideal in (T) such that if $J \subset J_1$, then $J_1 = T$.

B.9.2. Lemma. If

1. J is an ideal in (T),

2. $p \in T$,

then the set of somata

$$J' =_{df} \{q \dotplus a \,|\, a \in J, q \leqq p\}$$

is also an ideal ($\dotplus$ means algebraic addition, [A.2.9.]).

Proof. First we shall prove that if $x_1, x_2 \in J'$, then

$$x_1 \dotplus x_2 \in J.$$

Since $x_1 \in J'$, there exist $q_1 \leqq p$, $a_1 \in J$, such that

(1) $$x_1 = q_1 \dotplus a_1.$$

Similarly, since $x_2 \in J'$, there exist $q_2 \leqq p$, $a_2 \in J$, such that

(2) $$x_2 = q_2 \dotplus a_2.$$

From (1) and (2) we have

$$x_1 \dotplus x_2 = (q_1 \dotplus a_1) \dotplus (q_2 \dotplus a_2) = (q_1 \dotplus q_2) \dotplus (a_1 \dotplus a_2).$$

Now since $q_1 \leqq p$, $q_2 \leqq p$, it follows that

$$q_1 - q_2 \leqq p \quad \text{and} \quad q_2 - q_1 \leqq p;$$

hence, by adding, we get

(3) $$q' =_{df} q_1 \dotplus q_2 \leqq p.$$

Since $a_1 \in J$, $a_2 \in J$, it follows that

(4) $$a' =_{df} a_1 \dotplus a_2 \in J.$$

Hence $x_1 \dot{+} x_2 = q' \dot{+} a'$, where

$$q' \leqq p \quad \text{and} \quad a' \in J.$$

It follows that

$$x_1 \dot{+} x_2 \in \{q \dot{+} a' \mid a' \in J, q \leqq p\};$$

so

$$x_1 \dot{+} x_2 \in J'.$$

Now we shall prove that if $x \in J'$ and $x' \leqq x$, then $x' \in J'$.

As $x \in J'$, there exists $a \in J$ and $q \leqq p$, such that

$$x = a \dot{+} q.$$

Since $x' \leqq x$, we have $x' = x \cdot x' = (a \dot{+} q) \cdot x' = a\,x' \dot{+} q\,x'$.

We have $a\,x' \leqq a$, so $a\,x' \in J$, because J is an ideal. We also have $q\,x' \leqq p$. Hence

$$x' \in \{q \dot{+} a \mid a \in J, q \leqq p\}.$$

Thus we have proved that J' is an ideal.

There remains to prove that J' includes J and contains p.

Let $x \in J$. We have $x = x \dot{+} O$. Since $O \leqq p$, we have $x \in J'$; hence $J \leqq J'$. We have $p \in J'$. Indeed, $p = O \dot{+} p$. Since $O \in J$ and $p \leqq p$, we have $p \in J'$.

The theorem is established.

B.9.3. Theorem. If J is a proper ideal, and $p \in J$, then $\text{co}\,p \,\bar{\in}\, J$.

Proof. Suppose $\text{co}\,p \in J$. Since J is an ideal, therefore $p + \text{co}\,p \in J$, i.e. $I \in J$, hence $J = T$, which is impossible, because J is a proper ideal.

B.9.4. Theorem. If J is a maximal ideal in (T) and $p \,\bar{\in}\, J$, then $\text{co}\,p \in J$.

Proof. Consider the set of somata

$$J' =_{df} \{q \dot{+} a \mid q \leqq p,\ a \in J,\ q \in T\}.$$

By [B.9.2.] J' is an ideal in (T), which contains J and p. Since $p \,\bar{\in}\, J$, we have $J \subset J'$, and since J is a maximal ideal we have $J' = T$.

Hence $I \in J'$. There exist $q \leqq p$ and $a \in J$ such that

$$I = q \dot{+} a.$$

Since $q \leqq p$, we get $I \leqq q \dot{+} a \leqq p \dot{+} a \leqq I$, hence $p \dot{+} a = I$, which gives $a = I \dot{+} p = \text{co}\,p$, $\text{co}\,p \in J$. q.e.d.

B.9.5. The last two theorems allow to state the following

Theorem. If J is a maximal ideal in (T), then the following are equivalent for every $a \in T$.

I. $a \in J$,

II. $\text{co}\,a \,\bar{\in}\, J$.

B.9.6. Theorem. If

1. J is a proper ideal in (T),
2. for every $a \in T$ we have either $a \in J$ or $\operatorname{co} a \in J$,

then

1) J is a maximal ideal,
2) for every $a \in T$ the alternatives $a \in J$, $\operatorname{co} a \in J$ exclude one another.

Proof. First the exclusion 2) will be proved.

Suppose that for some $a \in J$ we have

$$a \in J \quad \text{and} \quad \operatorname{co} a \in J. \tag{1}$$

Since J is an ideal, we have

$$a + \operatorname{co} a \in J, \quad \text{i.e.} \quad I \in J, \quad \text{which implies} \quad J = T;$$

so J is not a proper ideal.

Thus the statements (1) cannot be simultaneously true. Now suppose that J is not maximal. There exists an ideal J' such that

$$J \subset J' \subset T. \tag{2}$$

There exists a soma p such that $p \neq I$, $p \in J'$, $p \bar{\in} J$.

By hypothesis we have either $p \in J$ or $\operatorname{co} p \in J$. Since $p \bar{\in} J$, we have $\operatorname{co} p \in J$, and then $\operatorname{co} p \in J'$, because $J \subset J'$. But we have also $p \in J'$. Hence $p + \operatorname{co} p \in J'$; hence $I \in J'$, and then $J' = T$, which contradicts (2). The theorem is established.

B.9.7. Theorem. If J is a maximal ideal in (T), then the tribe $T^* =_{df} (T)/J$, whose elements are equivalence classes modulo J, [A.12.], contains only two different equivalence classes: one O^*, whose representatives, [A.12.], are $(=^J)$-equal to O and the other I^* whose representatives are $(=^J)$-equal to I, [A.9.3.].

Thus the tribe $(T)/J$ is composed of two and only two somata: O^*, I^*. Hence the ordering $(T)/J$, (denoted by $\leqq^*$), is defined by $O^* \leqq^* O^*$, $I^* \leqq^* I^*$, $O^* \leqq^* I^*$; the somatic operations are:

$$O^* +^* O^* =^* O^*, \quad O^* +^* I^* =^* I^*, \quad I^* +^* O^* =^* I^*,$$
$$I^* +^* I =^* I^*.$$

Proof. Consider the two following sets of somata of (T),

$$A =_{df} \{x \mid x \in J\},$$
$$B =_{df} \{x \mid x \bar{\in} J\}.$$

We have

$$A \cup B = T.$$

The classes A, B are disjoint.

Indeed, if we had $a \in A \cap B$, we would have $a \in A$; hence $a \in J$ and also $a \in B$; hence $a \bar{\in} J$, which gives a contradiction.

We have $A = \{x \mid x =^J O\}$, because the statement $x =^J O$ is logically equivalent to $x \in J$.

We have $B = \{x \mid x =^J I\}$, because $x =^J I$ means $x \dotplus I \in J$ i.e. $\mathrm{co}\, x \in J$, and then by [B.9.5.] this is equivalent to $x \bar{\in} J$.

Thus A, B are equivalence classes, the first $\{O\}$, the second $\{I\}$.

Since for every ideal J the equivalence classes are either identical or disjoint, and their union is the domain of the tribe, there does not exist any third equivalence class. The remaining theses follow.

Let us notice that the class B is the ultrafilter, $\mathfrak{P}^{-1}$-corresponding to $J = A$.

B.10. We shall prove the following important existence-theorem, (9).

Theorem. (Stone). If (T) is a non trivial tribe,

$$a_0 \in T, \quad a_0 \neq I, \quad a_0 \neq O,$$

then there exists a maximal ideal K in (T), such that $\mathrm{co}\, a_0 \in K$.

B.10a. Proof. Consider the set of somata:

$$(1) \qquad E =_{df} \{x \mid x \leqq \mathrm{co}\, a_0\}.$$

E is an ideal in (T). Indeed, let $a, b \in E$. We have $a \leqq \mathrm{co}\, a_0$, $b \leqq \mathrm{co}\, a_0$; hence $a + b \leqq \mathrm{co}\, a_0$; so $a + b \in E$. Let $a \in E$ and $a' \leqq a$. We have $a \leqq \mathrm{co}\, a_0$; hence $a' \leqq \mathrm{co}\, a_0$; so $a' \in E$. The above proves that E is an ideal. It contains $\mathrm{co}\, a_0$.

B.10.b I say that $a_0 \bar{\in} E$. Indeed, if we had $a_0 \in E$, then, since $\mathrm{co}\, a_0 \in E$, we would have $a_0 + \mathrm{co}\, a_0 \in E$, because E is an ideal. Hence $I \in E$ and then, by (1), $I \leqq \mathrm{co}\, a_0$, which implies $\mathrm{co}\, a_0 = I$, and then $a_0 = O$, which is excluded by hypothesis. Thus E is a proper ideal which contains $\mathrm{co}\, a_0$ and does not contain a_0.

B.10c. Consider the class M of all proper ideals which contain $\mathrm{co}\, a_0$. The class M is not empty. Denote by (M) the ordering of set-inclusion ($\subseteqq$) of ideals belonging to (M). Consider a subchain (N) of (M).[1] We shall prove that (N) has in (M) an upper bound, i.e. that there exists an ideal $J' \in \mathrm{C}\, M$, such that for every $J'' \in \mathrm{C}\, N$ we have $J''(M)\, J'$, i.e. $J'' \subseteqq J$, (set inclusion).

B.10d. Such an ideal J' is given by

$$(1) \qquad J' =_{df} \bigcup_{J \in N} J,$$

[1] By a *chain* we understand a linear ordering (L), i.e. satisfying the condition, that if $a, b \in \mathrm{C}\, L$, then either $a(L)\, b$ or $b(L)\, a$.

(set union). Indeed J' does not contain a_0, since no one of the terms J of the above union does.

J' is an ideal. Indeed, if $a, b \in J'$, then there exist $J_1, J_2 \in N$, such that $a \in J_1$, $b \in J_2$.

Since (M) is a chain, we have either $J_1 \subseteqq J_2$ or $J_2 \subseteqq J_1$.

Consequently there exists in N an ideal J_3, such that $a, b \in J_3$, where $J_3 \in N$, (J_3 is either J_1 or J_2). Since J_3 is an ideal, we have $a + b \in J_3$, hence $a + b \in J'$. Now let $a \in J'$ and let $b \leqq a$. There exists $J_4 \in N$, such that $a \in J_4$. Hence, as J_4 is an ideal in (T), it follows $b \in J_4$; hence $b \in J'$.

Thus we have proved that J' is an ideal. This ideal contains $\mathrm{co}\, a_0$. Indeed, $\mathrm{co}\, a_0$ belongs to every ideal J, where $J \in N$. Since $J \subseteqq J'$, it follows that $\mathrm{co}\, a_0 \in J'$. The ideal J' does not contain a_0, because it contains $\mathrm{co}\, a_0$.

Since $a_0 \overline{\in} J'$, we cannot have $J' = T$. Hence J' is a proper ideal. Consequently $J' \in M$. We have, by (1), $J \subseteqq J'$ for every $J \in N$. Consequently J' is an upper bound of (N) in (M). Thus we have proved that every subchain (N) of (M) possesses in (M) an upper bound.

B.10e. Having that we can apply the *axiom of Zorn*[1].

It gives the existence of a maximal ideal K in the collection of all ideals belonging to M.

B.10f. We shall prove that K is a maximal ideal in (T). Suppose this be not true. Then there exists an ideal K' in (T), such that

$$K \subset K' \subset T. \tag{2}$$

Since $K' \neq T$, therefore K' is a proper ideal in (T).

Now K' cannot contain $\mathrm{co}\, a_0$. Indeed, if we had $\mathrm{co}\, a_0 \in K'$, then K would be not maximal in M, since $K \subset K'$.

But on the other hand, K' contains $\mathrm{co}\, a_0$, because K contains $\mathrm{co}\, a_0$ and $K \subset K'$. Thus we have a contradiction. Consequently there does not exist any ideal K' with $K \subset K' \subset T$. So we have proved that K is a maximal ideal in (T). The theorem is established. The following theorem, which is a small modification of [B.10.] reads:

B.10.1. Theorem. (STONE (9)). If (T) is a non trivial tribe,

$$b_0 \in T, \quad b_0 \neq 1,$$

then there exists a maximal ideal L, such that $b_0 \in L$.

---

[1] This axiom, known under the name "Zorns lemma", is equivalent to the axiom of Zermelo and reads: If (A) is an ordering, and if every subchain (A') of (A) possesses an upper bound in (A), then there exists in (A) a maximal soma, i.e. a soma n such that if $n\,(A)\,n'$, then $n = n'$.

Proof. Put $a_0 = \operatorname{co} b_0$. We have $b_0 = \operatorname{co} a_0$. If we apply [B.10.], we get a maximal ideal L such that $\operatorname{co} b_0 \in L$. Hence, since L is a proper ideal, it follows that $b_0 \bar{\in} L$, because if not, we would have $b_0 + \operatorname{co} b_0 \in L$, and then $L = T$; so L would not be a proper ideal.

Filter-bases

B.11. Def. By a *filter-base in* (T) we shall understand a non empty subset Φ of T such that if

$$a, b \in \Phi,$$

then there exists $c \in T$, such that

$$c \in \Phi, \quad c \leqq a \cdot b, \quad (29).$$

B.11.1. Theorem. If Φ is a filter-base in (T) and we define $C =_{df} \{a \mid \text{there exists } b \in \Phi \text{ with } b \leqq a\}$, then C is a filter.

Def. We call C the *filter generated by the filter-base* Φ.

B.11.2. Theorem. If C is a filter, then it is also a filter-base.

B.11.3. Def. Two filter-bases Φ', Φ'' in the tribe (T) are said *to be equivalent,* if Φ', Φ'' generate the same filter.

B.11.4. Theorem. The equivalence of filter-bases obey the basic rules of identity: reflexiveness, symmetry and transitivity.

We state, on filter-base, some theorems, whose proof we leave to the reader:

B.11.5. Theorem. If Φ', Φ'' are two filter-bases in (T), then the following are logically equivalent:

I. Φ' is equivalent to Φ'',

II. for every $a' \in \Phi'$ there exists $a'' \in \Phi''$, such that $a'' \leqq a'$, and for every $a'' \in \Phi''$ there exists $a' \in \Phi'$ such that $a' \leqq a''$.

B.11.6. Theorem. If C is a filter in (T), $\Phi \subseteqq C$, then the following are equivalent:

I. Φ is a filter-base of C,

II. for every $a \in C$, there exists $b \in \Phi$ such that $b \leqq a$.*)

---

*) The filter-bases are related to so called *directed sets,* invented by MOORE and SMITH. We understand by them any orderings (M), (written $\leqq$), such that for every $a, b \in M$ there exists $c \in M$ such that $a \leqq c$, $b \leqq c$. Instead of the term directed set we prefer to use the term *stream,* and if we have a function, defined on M, we call it *streem-sequence.* The stream defined above, may be termed *up-stream,* and its inverse *down-stream.* Thus filter-bases are down-streams in a given ordering.

The above notion may be compared with a notion introduced in (33) by the author.

Two valued measures on a tribe.

B.12. (TARSKI), (35). **Theorem.** If (T) is non trivial tribe, J a maximal ideal in (T), then if we put for all $a \in (T)$

$$\mu(a) = 0 \quad \text{whenever} \quad a \in J,$$

$$\mu(a) = 1 \quad \text{whenever} \quad a \bar{\in} J,$$

the function $\mu(\dot{a})$ defined over all (T) is an additive measure on (T). It admits two values only.

Proof. We must prove that if $a, b \in (T)$ and $a \cdot b = O$, then

$$\mu(a + b) = \mu(a) + \mu(b).$$

To do that, suppose that $a \in J$, $b \in J$. Then $a + b \in J$. We have $\mu(a) = 0$, $\mu(b) = 0$ and $\mu(a + b) = 0$, so $\mu(a + b) = \mu(a) + \mu(b)$.

Suppose that $a \bar{\in} J$, $b \bar{\in} J$, $a \cdot b = O$. Then

$$\operatorname{co} a \in J, \quad \operatorname{co} b \in J, \quad \text{so} \quad (\operatorname{co} a + \operatorname{co} b) \in J, \quad \operatorname{co}(a \cdot b) \in J.$$

It follows $a \cdot b \bar{\in} J$, so $a \cdot b \neq O$, which is a contradiction. Thus the case considered cannot occur.

Let $a \in J$, $b \bar{\in} J$. We have $\mu(a) = 0$, $\mu(b) = 1$. $a + b$ is a representative of the equivalence-class $O^* +^* I^* =^* I^*$, [B.9.7.], so $a + b \bar{\in} J$. Hence

$$\mu(a + b) = 1, \quad \text{and then} \quad \mu(a + b) = \mu(a) + \mu(b).$$

Thus we have proved that whatever the case may be, μ is additive. The theorem is established.

B.12.1. Theorem. If

1) (T) is a non trivial tribe,

2) $\mu(a)$ is an additive measure on (T) with values $0, 1$,

3) $\mu(I) = 1$,

then there exist a maximal ideal J in (T) such that

$$\mu(a) = 0 \quad \text{whenever} \quad a \in J,$$

and

$$\mu(a) = 1 \quad \text{whenever} \quad a \bar{\in} J.$$

Proof. Let $a \in T$. Put

$$A =_{df} \{x \mid \mu(x) = 0\} \quad \text{and} \quad B =_{df} \{x \mid \mu(x) = 1\}.$$

We have

$$A \cup B = T \quad \text{and} \quad A \cap B = \emptyset.$$

A is an ideal in (T), [A.20.]. It is a proper ideal, because $I \bar{\in} A$. Take any soma $x \in T$ and suppose that $\mu(x) = 1$. Since μ is additive, we get

$$\mu(I - x) = \mu(I) - \mu(x) = 1 - 1 = 0.$$

It follows that $I - x \in A$, i.e. $\operatorname{co} x \in A$. Now suppose that $\mu(x) = 0$. We have $x \in A$.

Thus we have proved that for every soma $x \in T$ we have either $x \in A$ or $\operatorname{co} x \in A$. Applying [B.9.6.], it follows that A is a maximal ideal in (T). The theorem follows.

B.12.2. Remark. The above theorems can be stated in terms of ultrafilters.

B.12.3. We see that maximal ideals, ultrafilters and two-valued measures are related strongly to one another. The existence theorem by Stone, [B.10.], implies that e v e r y n o n t r i v i a l t r i b e a d m i t s a n a d d i t i v e, t w o v a l u e d m e a s u r e, (Tarski, (35)).

Chapter B I

General theory of traces

B I.1. The preceding chapters are busy with discussion of topics, which are rather known in the literature, and which will serve as tool and background for the subsequent, rather new theories.

We shall introduce and study the notions of "trace", which will constitute an important auxiliary for the theory of selfadjoint, unitary, and more generally, for normal operators in Hilbert space, which, roughly speaking, are like hermitean selfadjoint, but with complex eigenvalues.

Before giving a precise definition, we like to make some remarks of intuitional character.

Suppose we would like to define points of the straight line, having given the closed intervals only. Then the points can be defined by means of diminishing sequences of nested intervals:

$$\langle a_1, b_1 \rangle \supseteqq \langle a_2, b_2 \rangle \supseteqq \cdots \supseteqq \langle a_n, b_n \rangle \supseteqq \cdots,$$

whose length tends to 0.

Now, a similar construction can be applied to closed subspaces of the Hilbert space.

The spaces to be considered will be invariant spaces of a given normal operator. Though it may happen that the intersection of spaces of such a nested sequence may be empty, nevertheless those sequences will constitute a mathematically adapted image of something, which could be called "pseudo-eigen spaces", in the case, where the corresponding eigenvalues belong to the continuous spectrum of the operator.

We have introduces such a notion in the paper (37[1]) for the purpose of selfadjoint operators. Now, since we like to deal with normal operators, the said notion has been changed, and generalized and called "trace".

Since we like to have the possibility of further generalizations, we shall introduce the notion on the background of a general Boolean tribe, by using one of its bases [B.1.1.] and by introducing diminishing sequences of bricks [B.1.2.]. Later in the Chapter [C.] and [D I.] we shall apply the general theory to the euclidean plane and to closed subspaces of the Hilbert-space respectively. The notion of the trace will remind that of the filter-base [B.11.].

We shall consider sets of traces and build up a theory of their measurability, which will imitate the pattern of Lebesgue's theory of measurability of sets.

The theory of traces and measurability of sets of traces, if applied to Hilbert space, will enable us to define and apply a "general system" of coordinates in this space, which will be adapted to the continuous spectrum of selfadjoint and even of normal operator as well, as the ordinary, usual orthonormal system of vectors is for the discontinuous spectrum.

B I.1a. The general notion of traces, spoken of, has been given in our C. R. notes (40), (41). It constitutes a modification and generalization of the notion, defined in (37), (38), (39), (40).

B I.1.1. Starting with a precise theory, we consider an abstract, general tribe (F); choose one of its bases $\boldsymbol{B}$, and suppose that (F) is plonged in a finitely genuine supertribe [A.7.], (G) which is denumerably additive and provided with a denumerably additive finite and effective measure. To avoid non essential complications, we admit that (F) is a finitely genuine s t r i c t subtribe of (G), [A.7.2.].

We recall that the measure μ is said to be e f f e c t i v e, whenever $\mu(a) = 0$ implies $a = O$, [A.15.]. We also recall, that the tribe (F) is called s t r i c t subtribe of (G), whenever the ordering (F) is the same as in (G), but only restricted to all somata of F, [A.7.2.], and to their $(=^G)$-equals.

We do not admit neither ***(Hyp Af)*** nor ***(Hyp Ad)***, [B.2.].

B I.2. We shall consider infinite *descending sequences* $a_1 \geqq a_2 \geqq \cdots \geqq \geqq a_n \geqq a_{n+1} \geqq \cdots$ *of bricks*. If $\{a_n\}$, $\{b_n\}$ are two such sequences, we say that $\{a_n\}$ *is included in* $\{b_n\}$: $\{a_n\} \leqq \{b_n\}$, whenever for every

[1] The French term applied is "lieu."

n there exists m such that $a_m \leqq b_n$. We see that the sequence $\{O, O, \ldots\}$ *is included in* every sequence, and every sequence is included in $\{I, I, \ldots\}$. If $\{a_n\} \leqq \{b_n\}$, $\{b_n\} \leqq \{c_n\}$, then $\{a_n\} \leqq \{c_n\}$. We have $\{a_n\} \leqq \{a_n\}$. We say that $\{a_n\}$ *is equivalent to*

$$\{b_n\}; \quad \{a_n\} \sim \{b_n\},$$

$$\text{if} \quad \{a_n\} \leqq \{b_n\} \quad \text{and} \quad \{b_n\} \leqq \{a_n\}.$$

The notion of equivalence obeys the formal rules of identity, and the notion of "being included" is invariant with respect to the equivalence.

If the sequences $\{a_n\}$, $\{b_n\}$ differ only by finite number of elements, they are equivalent.

B I.2.1. Def. A descending sequence $\{a_n\}$ of bricks is said to be *minimal*, if the following conditions are satisfied:

1) $\{a_n\}$ is not equivalent to $\{O, O, \ldots\}$,
2) if $\{b_n\} \leqq \{a_n\}$, then either $\{b_n\} \sim \{O, O, \ldots\}$ or else $\{b_n\} \sim \{a_n\}$.

B I.2.2. Remark. We cannot prove the existence of a minimal sequence without supplementary hypotheses, concerning $\boldsymbol{B}$. Thus, in what follows we shall admit the existence of at least one minimal sequence.

B I.2.3. Def. A saturated class of mutually equivalent minimal sequences of bricks will be termed *trace,* and each of those sequences *representative of that trace.* All elements of a representative of a trace are $\neq O$.

B I.2.3.1. Remark. Notice that there does not exist anything like a "null-trace".

B I.2.4. Def. Two traces x, y are said to be *equal,* $x = y$, if their representatives are equivalent.

B I.2.4.1. This notion of equality obeys the formal laws of identity. It coincides with the identity of classes of sequences, suitably restricted.

B I.2.5. Def. We say that *the brick a covers the trace x*, whenever there exists a representative $\{a_n\}$ of x, such that $a_1 \leqq a$. (Of course it follows that all $a_n \leqq a$.)

B I.3. In [B.] we have defined a *covering* as an at most denumerable sum of bricks. In this chapter we take over that definition

B I.4. Def. Let X be a set of traces (it may be even empty), and L a covering. We say that *X is covered by L* if the following implication is true: "if $x \in X$, then there exists a brick a, such that

1) $a \leqq L$, and
2) x is covered by a".

B I.4.1. The empty set of traces is covered by any covering. Every set X posseses a covering, namely the soma I.

B I.4.2. The following are equivalent:

I. The trace x is covered by the brick a,

II. The set (x), composed of the single trace x, is covered by the covering a.

B I.5. The following lemmas hold true: Let X, Y be sets of traces and L a covering. If $X \subseteqq Y$, and L is a covering of Y, then L is also a covering of X.

B I.5.1. Let L, M be coverings and X a set of traces. If $L \leqq M$, and L is a covering of X, then M is also a covering of X.

B I.5.2. Let $S \neq \emptyset$ be a collection of indices. Suppose, that for every $i \in S$ the set X_i of traces is covered by the covering L. Then $\bigcup_i X_i$, $(i \in S)$ is also covered by L.

B I.5.3. If $X_1, X_2, \ldots$, at most denumerable in number, are covered by $L_1, L_2, \ldots$ respectively, then $\bigcup_n X_n$ is covered by $\sum_n L_n$.

B I.5.4. Theorem. If L, M are coverings of the sets of traces X, Y respectively, then $L \cdot M$ is a covering of $X \frown Y$.

Proof. Let $x \in X \frown Y$. There exist bricks a, b, both covering x and such that $a \leqq L$, $b \leqq M$. Since x is covered by a, there exists a representative $\{a_n\}$ of x such that $a_n \leqq a$, $(n = 1, 2, \ldots)$. Since x is covered by b, there exists a representative $\{b_n\}$ of x with $b_n \leqq b$, $(n = 1, 2, \ldots)$. The sequences $\{a_n\}$, $\{b_n\}$ are equivalent, [B I.2.]. Hence there exists p with $b_p \leqq a_1$. Hence $b_p \leqq a \cdot b \leqq N \cdot M$. Since $\{b_p, b_{p+1}, \ldots\}$ is equivalent to $\{b_1, b_2, \ldots\}$, it is equivalent to $\{a_n\}$. Since $a \cdot b$ is a brick, it follows that x is covered by $L \cdot M$. The lemma is proved.

B I.6. Denote by W the set of all traces. If X is a set of traces, denote by co$X$ the complement of X with respect to W, i.e. $W - X$. We admit following two hypotheses:

Hyp I. If X is a set of traces, L, M are coverings of X and co X respectively, then $L + M = I$.

Hyp II. If a is a brick and X the set of all traces covered by a, then co$a$ is the covering of $\text{co}_W X$. (This means co$X$.)

B I.6.1. Suppose that $\boldsymbol{B}$ has at least one element b differing from O and I. From ***Hyp*** I it follows that there exists at least one trace. Indeed, by [B I.4.1.] the empty set of traces is covered by any covering. If $W = \emptyset$, we would have $W = \text{co}W$. The brick b would cover both W and co$W$, but $b + b \neq I$. From the hypotheses ***Hyp*** I and ***Hyp*** II it follows that, if $a \neq O$ is a brick, but not an atom, there exists a trace covered by a. Indeed, suppose that there does not exist any one, covered by a. Hence, by ***Hyp*** II W is covered by co$a$. There exists a brick b with $0 < b < a$. Now co$W = \emptyset$ is covered by b. Hence, by ***Hyp*** I, $b + \text{co}a = 1$, which is not true. If a is an atom, then

$\{a, a, \ldots\}$ is a minimal sequence, hence representing a trace covered by a.

If F and $\boldsymbol{B}$ are composed of O and unit only, I is an atom.

Thus we have proved, that if $f \in F$ is any figure $\neq O$, then there exists a trace covered by f.

B I.7. The above hypotheses ***Hyp*** I and ***Hyp*** II and the existence of an effective measure on (G) will make possible to develop a theory of measurability of sets of traces. The theorems which will not involve the measure μ explicitely, they will be independent of the choice of the effective measure.

We emphasize that, given the tribe (F), the notion of traces depends on the choice of the base $\boldsymbol{B}$ of F, so we may call them ***B**-traces in F*.

B I.7.1. We shall rely on the following theorem by WECKEN **(25)**: If G is a denumerably additive tribe admitting a denumerably additive, non negative and effective measure, then G is completely additive.

B I.7.2. Def. If X is a set of traces, then by its *outer coat* we shall understand the soma of G:

$$[X]^* = \prod L,$$

where the product is extended over all coverings L of X.

B I.7.3. Def. If X is a set of traces, then by *its inner coat* we shall understand the soma of (G):

$$[X]_* = I - [Y]^* = \mathrm{co}[Y]^*, \quad \text{where} \quad Y = \mathrm{co}_W X.$$

B I.7.4. Theorem. For every set X of traces we have

$$0 \leqq [X]_* \leqq [X]^*.$$

Proof. Put $Y = \mathrm{co}X = W - X$. If L is a covering of X and M a covering of Y, we have, by ***Hyp*** I:

$$L + M = I.$$

Multiplying both sides by $\mathrm{co}M$, we get $L \cdot \mathrm{co}M = \mathrm{co}M$; hence $\mathrm{co}M \leqq L$. This being true for a given L and any M, we get

$$(1) \qquad \sum \mathrm{co}M \leqq L,$$

where the summation is extended over all coverings of Y. The inclusion (1) being valid for any covering L of X, we get

$$\sum \mathrm{co}M \leqq \prod L \quad \text{i.e.} \quad \mathrm{co}\prod M \leqq \prod L,$$

hence

$$\mathrm{co}[Y]^* \leqq [X]^*,$$

and then

$$[X]_* \leqq [X]^*.$$

B I.7.4.1. Theorem. If $X \subseteq Y$ are sets of traces, we have for their outer and inner coats the inclusions:

$$[X]^* \leqq [Y]^*, \quad [X]_* \leqq [Y]_*.$$

B I.7.5. Theorem. If X is a set of traces, then there exists a denumerable sequence of coverings of X:

$$L_1 \geqq L_2 \geqq \cdots \geqq L_n \geqq \cdots,$$

such that $[X]^* = \prod_n L_n$. For such a sequence we have

$$\mu([X]^*) = \lim \mu(L).$$

In addition to that we have $\mu([X]^*) = \inf \mu(M)$, where the infimum is taken for all coverings M of X.

Proof. The coverings are taken from the denumerably additive tribe (G), provided with an effective measure μ.

Denote by K the collection of all coverings L of X. We have, by de Morgan's law [A.1.5.]

$$\operatorname{co} \prod_{L \in K} L = \sum_{L \in K} \operatorname{co} L.$$

Now, by Wecken's theorem [A I.1.1.], there exists a denumerable sequence

$$M_1, M_2, \ldots, M_n, \ldots$$

of coverings of X, such that

$$\sum_{L \in K} \operatorname{co} L = \sum_{n=1} \operatorname{co} M_n.$$

Hence, taking the complement, and applying once more de Morgan laws, we get

$$\prod_{L \in K} L = \prod_{n=1}^{\infty} M_n.$$

Putting

$$M'_n =_{df} M_1 \cdot \cdots \cdot M_n, \quad (n = 1, 2, \ldots),$$

we have

$$\prod_{L \in K} L = \prod_{n=1}^{\infty} M'_n,$$

where all M'_n are coverings of X, and in addition to that

$$M'_1 \geqq M'_2 \geqq \cdots \geqq M'_n \geqq \cdots,$$

so the first part of the theorem is established[1].

B I.7.5a. If $[X]^* = \prod_{n=1}^{\infty} M_n$, where M_n are coverings of X with $M_1 \geqq M_2 \geqq \cdots$, it follows, on account of the denumerable additivity

[1] The proof used in (36) is based on transfinite induction, and is quite complicated.

of the measure μ, that

$$\mu([X]^*) = \lim \mu(M_n).$$

B I.7.5b. To prove the last statement of the thesis, put

$$(11) \qquad \lambda =_{df} \inf \mu(L),$$

where the infimum is taken for all coverings L of X. Take the sequence $\{M_n\}$. We have

$$\lambda \leqq \lim \mu(M_n) = \mu([X]^*).$$

B I.7.5c. Suppose that $\lambda < \mu([X]^*)$. There exists $\varepsilon > 0$ such that $\lambda < \mu([X]^*) - \varepsilon$. By (11) there exists a covering L' such that $\lambda \leqq \mu(L') < \mu([X]^*) - \varepsilon$. Since $M_n \cdot L'$ is also a covering of X, we have

$$(12) \qquad \lambda \leqq \mu(M_n L') \leqq \mu(L') < \mu([X]^*) - \varepsilon.$$

Now, $[X]^* \leqq \prod_{n=1}^{\infty} M_n = [X]^*$; hence $[X]^* = \sum_{n=1}^{\infty} (M_n L')$. In addition to that we have $M_1 L' \geqq M_2 L' \geqq \cdots$, and these somata are coverings of X. It follows, that $\mu([X]^*) = \lim \mu(M_n L')$. On the other hand we get from (12), $\mu([X]^*) < \mu([X]^*) - \varepsilon$, which is false. Thus we have proved that $\lambda = \mu([X]^*)$; and this completes the proof of the whole theorem [B I.7.5.].

B I.8. Measurable sets of traces. Def. A set X of traces is said to be *measurable* whenever its inner coat and its outer coat coincide: $[X]_* = [X]^*$. In this case we speak of *the coat of* X and denote it by $[X]$. We have $[X] = [X]^* = [X]_*$.

B I.8.1. Theorem. If X is a set of traces, $Y = \mathrm{co}_W X$, then the following are equivalent:

I. X is measurable,

II. $[X]^* \cdot [Y]^* = O$,

III. Y is measurable.

In this case we have $[X] = \mathrm{co}[Y]$.

B I.8.2. Theorem. The empty set $\mathit{0}$ of traces is measurable. The total set W of traces is measurable. They are different.

Proof. We have $O \leqq [X]_* \leqq [X]^*$ and $[X]^* = O$. Hence $[X]_* = [X]^*$, so X is measurable. The set $\mathrm{co}_W X$ is also measurable, by [B I.8.1.]. We get $[\mathit{0}] = O$ and $[W] = \mathrm{co}[\mathit{0}] = \mathrm{co}\, O = I$. Since $[\mathit{0}] \neq I$, the sets $[\mathit{0}]$ and $[W]$ are different.

Remark. Soon we shall prove that if a is a brick, then the set of all traces, covered by a, is measurable.

B I.8.3. Def. If X is a measurable set of traces, then the number

$$\mu([X]) = \mu([X]^*) = \mu([X]_*)$$

is called *measure of X and denoted by $\mu(X)$*.

B I.8.4. Theorem. If X is a measurable set of traces, $[X] = \prod_{n=1}^{\infty} L_n$, where $L_1 \geqq L_2 \geqq \cdots$, are coverings of X, then $\mu(X) = \lim \mu(L_n)$.

B I.8.5. Theorem. Let X be a set of traces, $Y = \mathrm{co}_W X$, and let L_n, M_n be coverings of X, Y respectively with

$$L_{n+1} \leqq L_n, \quad M_{n+1} \leqq M_n, \quad (n = 1, 2, \ldots),$$

where

$$[X]^* = \prod_n L_n, \quad [Y]^* = \prod_n M_n.$$

If X is measurable, then

$$\lim \mu(L_n \cdot M_n) = 0.$$

Proof. By theorem [B I.8.1.], $[X]^* \cdot [Y]^* = O$; hence

$$O = \prod_n L_n \cdot \prod M_n = \prod_n (L_n \cdot M_n).$$

We have $L_{n+1} \cdot M_{n+1} \leqq L_n \cdot M_n$ which completes the proof, [A I.2.1; I].

B I.8.6. Theorem. Let X be a set of traces, $Y = \mathrm{co}_W X$. Let L_n, M_n, $(n = 1, 2, \ldots)$, be coverings of X, Y respectively. If $\lim \mu(L_n \cdot M_n) = 0$, then X is measurable, and we have $[X] = \prod_n L_n$.

Proof. Put

$$L'_n = L_1 \ldots L_n, \quad M'_n = M_1 \ldots M_n, \quad (n = 1, 2, \ldots).$$

We have

$$L'_n \cdot M'_n \leqq L_n \cdot M_n,$$

and also

(1) $$\lim \mu(L'_n \cdot M'_n) = 0.$$

Now

$$[X]^* \leqq L'_n, \quad [Y]^* \leqq M'_n;$$

hence

(2) $$[X]^* \cdot [Y]^* \leqq \prod_n (L'_n \cdot M'_n).$$

Since

$$L'_{n+1} \cdot M'_{n+1} \leqq L'_n \cdot M'_n,$$

we get, by (1),

$$\mu(\prod_n (L'_n \cdot M'_n)) = \lim \mu(L'_n \cdot M'_n) = 0,$$

and hence, by the effectiveness of μ, $\prod_n (L'_n \cdot M'_n) = O$, which gives, by (2), $[X]^* \cdot [Y]^* = O$, and then, by virtue of theorem [B I.8.1.], the measurability of X, and of Y.

To prove the second part of the thesis, notice that $\text{co}[X] = [Y] \leqq M_n$; hence

$$L_n \cdot \text{co}[X] \leqq L_n \cdot M_n, \quad \mu(L_n - [X]) \leqq \mu(L_n \cdot M_n).$$

Since $\lim \mu(L_n \cdot M_n) = 0$, it follows

(3) $$\lim \mu(L_n - [X]) = 0.$$

Put

(4) $$p = \prod_n L_n.$$

We have $[X] \leqq p \leqq L_n$; hence $\mu(p - [X]) \leqq \mu(L_n - [X])$. This gives, on account of (3), $p - [X] = O$, i.e. $p \leqq [X]$, and then $p = [X]$. Consequently, by (4), $[X] = \prod_n L_n$, which is the second thesis. The theorem is established.

B I.8.6.1. The measure of the empty set $\emptyset$ of traces is 0; the measure of the total set W of traces is $\mu(I)$.

B I.8.7. Theorem. If a is a brick and X the set of all traces covered by a, then X is measurable. We have $[X] = a$, $\mu([X]) = \mu(a)$.

Proof. By [***Hyp*** II], $\text{co}X$ is covered by $\text{co}a$. Put $L_n = a$, $M_n = \text{co}a$, $(n = 1, 2, \ldots)$. We get, by theorem [B I.8.6.], that X is measurable, $[X] = a$, and $\mu(X) = \mu(a)$.

B I.8.8. Theorem. For every brick $a \neq 0$ there exists a trace covered by a.

Proof. Denoting by X the set of all traces covered by a, suppose that $X = \emptyset$. The complement $\text{co}X$ is covered, by ***Hyp*** II, by $\text{co}a$. Now O is a covering for $\emptyset$. It follows, by ***Hyp*** I, that $\text{co}a + O = I$, which gives $a = O$, thus a contradiction. The theorem is proved, (see [B I.6.1.]).

B I.8.9. Remark. The theory of measure of sets of traces is similar to the Lebesgue's classical theory of measure of point-sets, and it can be developped similarly. We shall apply the original Lebesgue's device (23) in proving that the denumerable union of measurable sets is also measurable. This device can be, however, greatly simplified, owing to the fact, that μ is a denumerably additive measure on G.

B I.8.10. Theorem. If X, Y are measurable sets of traces, then $X \cup Y$ is measurable, and

$$[X \cup Y] = [X] + [Y].$$

Proof. Put $X' = \text{co}X$, $Y' = \text{co}Y$. Let L_n, L'_n, M_n, M'_n be coverings of X, X', Y, Y' respectively with $L_{n+1} \leqq L_n$, $L'_{n+1} \leqq L'_n$, $M_{n+1} \leqq M_n$, $M'_{n+1} \leqq M'_n$, and with

(1) $$[X] = \prod_n L_n, \quad [X'] = \prod_n L'_n, \quad [Y] = \prod_n M_n, \quad [Y'] = \prod_n M'_n.$$

By Lemma [B I.5.3.] and [B I.5.4.]:

(2) $$L_n + M_n, \quad L'_n \cdot M'_n$$

are coverings of $X \cup Y$, $X' \cap Y'$ respectively. We have

(3) $$L_{n+1} + M_{n+1} \leqq L_n + M_n, \quad L'_{n+1} \cdot M'_{n+1} \leqq L'_n \cdot M'_n$$

and

(4) $$X' \cap Y' = \mathrm{co}(X \cup Y).$$

In addition to that,

(5) $$\mu[(L_n + M_n) \cdot L'_n \cdot M'_n] \leqq \mu(L_n \cdot L'_n) + \mu(M_n \cdot M'_n).$$

Since X, Y are measurable, we have, by (1) and theorem [B I.8.5.],

$$\lim \mu(L_n \cdot L'_n) = 0, \quad \lim \mu(M_n \cdot M'_n) = 0;$$

hence, by (5):

$$\lim \mu[(L_n + M_n) \cdot L'_n \cdot M'_n] = 0.$$

If we take account of (3), and apply theorem [B I.8.6.], we obtain the measurability of $X \cup Y$, and in addition to that,

(5.1) $$[X \cup Y] = \prod_n (L_n + M_n).$$

To prove the second part of the thesis, notice that (1) gives $[X] \leqq \prod_n (L_n + M_n)$, $[Y] \leqq \prod_n (L_n + M_n)$, which implies

(6) $$[X] + [Y] \leqq [X \cup Y].$$

On the other hand the relations $L_{n+1} \leqq L_n$, $M_{n+1} \leqq M_n$ yield

$$\prod_{n=1}^{\infty} (L_n + M_n) = \prod_{n=m}^{\infty} (L_n + M_n) \leqq \prod_{n=m}^{\infty} (L_n + M_m) = \prod_{n=m}^{\infty} L_n + M_m.$$

Hence, by (5.1),

$$[X \cup Y] \leqq [X] + M_m, \quad (m = 1, 2, \ldots).$$

It follows

$$[X \cup Y] \leqq \prod_{m=1}^{\infty} ([X] + M_m) = [X] + \prod_{m=1}^{\infty} M_m = [X] + [Y],$$

which together with (6), completes the proof.

B I.8.10.1. Theorem. If X is measurable, then $\mathrm{co}X = W - X$ is also measurable, and

$$[\mathrm{co}X] = \mathrm{co}[X] = \mathit{1} - [X], \quad \mu(\mathrm{co}X) = \mu(\mathit{1}) - \mu(X).$$

Proof. By [B I.8.1.].

B I.8.11. Theorem. If X, Y are measurable sets of traces, then $X \frown Y$, $X - Y$, $X \dotplus Y$ are also measurable, and their coats are $[X] \cdot [Y]$, $[X] - [Y]$, $[X] \dotplus [Y]$ respectively.

Proof. We use de Morgan laws and theorems [B I.8.1.], [B I.8.10.].

B I.8.12. Theorem. If X, Y are measurable sets of traces, $X \subseteqq Y$, then $[X] \leqq [Y]$ and $\mu(X) \leqq \mu(Y)$.

Proof. $X \subseteqq Y$ is equivalent to $X = X \frown Y$. We use theorem [B I.8.11.].

B I.8.13. Theorem. If X, Y are measurable sets of traces, and $X \frown Y = \emptyset$, we have $\mu(X \smile Y) = \mu(X) + \mu(Y)$.

Proof. This follows from theorems [B I.8.11.], [B I.8.10.].

B I.8.14. Theorem. If

1) $X_1, X_2, \ldots, X_n, \ldots$ are measurable sets of traces,
2) $X_i \frown X_j = \emptyset$ for $i \neq j$,
3) $X = \bigcup_{n=1}^{\infty} X_n$,

then

1) X is also measurable,
2) $[X] = \sum_{n=1}^{\infty} [X_n]$,
3) $\mu(X) = \sum_{n=1}^{\infty} \mu(X_n)$.

B I.8.14a. Proof. Let $\delta > 0$. Choose positive numbers $\delta_1 > \delta_2 > \cdots$ such that $\sum_{p=1}^{\infty} \delta_p \leqq \delta$. Fix p, and consider X_p and $X'_p = \mathrm{co}\, X_p$. By theorem [B I.7.5.] and [B I.8.5.], there exist coverings L_n, L'_n of X_p and X'_p respectively, such that

(1) $$L_{n+1} \leqq L_n, \quad L'_{n=1} \leqq L'_n,$$

(2) $$[X_p] = \prod_p L_n, \quad [X'_p] = \prod_p L'_n,$$

and

(3) $$\lim_{n \to \infty} \mu(L_n \cdot L'_n) = 0.$$

Hence there exists $n = N(p)$, such that

(4) $$\mu(L_n \cdot L'_n) \leqq \delta_p \quad \text{for all} \quad n \geqq N(p).$$

B I.8.14b. On the other hand we have for $n = 1, 2, \ldots$ $[X_p] \leqq L_n$, $L_n - [X_p] = L_n \cdot [X'_n] = L_n \cdot \prod_m L'_m$; hence $L_n - [X_p] \leqq L_n \cdot L'_n$ which gives $\mu(L_n - [X_p]) \leqq \mu(L_n \cdot L'_n)$, and then, by (3), $\lim_{n \to \infty} \mu(L_n - [X_p]) = 0$.

Hence there exists $n = M(p)$, such that

(5) $$\mu(L_n - [X_p]) \leqq \delta_p \quad \text{for all} \quad n \geqq M(p).$$

B I.8.14c. Combining (4) and (5), we can say, that for every p there exist coverings M_p, M'_p of X_p, X'_p respectively, such that

(6) $$\mu(M_p \cdot M'_p) \leqq \delta_p, \quad \mu(M_p - [X_p] \leqq \delta_p.$$

Let us fix such coverings.

B I.8.14d. We have $\operatorname{co} X = \bigcap\limits_{n=1}^{\infty} \operatorname{co} X_n \subseteqq \operatorname{co} X_m = X'_m$ for $m = 1, 2, \ldots$ As M'_m is a covering of $\operatorname{co} X'_m$, it is also (by Lemma [B I.5.]), a covering of $\operatorname{co} X$. Hence $M'_1, M'_2, \ldots, M'_m$ is (by Lemma [B I.5.4.] a covering of $\operatorname{co} X$ for any m. On the other hand $M_1 + M_2 + \cdots$ is (by Lemma [B I.5.3.]) a covering of X. We have[1] for a fixed m:

$$(M_1 + M_2 + \cdots + M_n + \cdots) \cdot M'_1 \cdot M'_2 \ldots M'_m = M_1 \cdot (M'_1 \ldots M'_u) + \\ + \cdots + M_m \cdot (M'_1 \ldots M'_m) + M_{m+1} \cdot (M'_1 \ldots M'_m) + \cdots \leqq M_1 \cdot M'_1 + \\ + M_2 M'_2 + \cdots + M_m \cdot M'_m + M_{m+1} + M_{m+2} + \cdots.$$

Since μ is denumerably additive and non negative, it follows

(7) $$\mu[(M_1 + M_2 + \cdots) \cdot M'_1 \cdot M'_2 \ldots M'_m] \leqq \mu(M_1 \cdot M'_1) + \cdots + \\ + \mu(M_m M'_m) + \mu(M_{m+1}) + \mu(M_{m+2}) + \cdots;$$

this inequality holding even if, by chance, the right hand series were divergent. Now, by (6), we have

$$\mu(M_p \cdot M'_p) \leqq \delta_p, \quad (p = 1, \ldots, m)$$

and

$$\mu(M_{m+k} - [X_{m+k}]) \leqq \delta_{m+k}, \quad (k = 1, 2, \ldots);$$

hence, since

$$[X_{m+k}] \leqq M_{m+k},$$

we get, [B I.8.3.]

$$\mu(M_{m+k}) - \mu(X_{m+k}) \leqq \delta_{m+k},$$

i.e.

$$\mu(M_{m+k}) \leqq \mu(X_{m+k}) + \delta_{m+k}, \quad (k = 1, 2, \ldots).$$

Consequently (7) gives:

(8) $$\mu[(M_1 + M_2 + \cdots) \cdot M'_1 \ldots M'_m] \leqq \delta_1 + \cdots + \delta_m + \\ + [\mu(X_{m+1}) + \delta_{m+1}] + [\mu(X_{m+2}) + \delta_{m+2}] + \cdots, \quad (m = 1, 2, \ldots).$$

[1] This is the LEBESGUE's device (23).

B I.8.14e. We shall get help from the denumerable additivity of μ. We have supposed that $X_i \frown X_j = \emptyset$ for $i \neq j$. Hence, by theorem [B I.8.11.] and [B I.8.2.], we also have $[X_i] \cdot [X_k] = O$. It follows that $\mu \sum_{n=1}^{\infty} [X_n] = \sum_{n=1}^{\infty} \mu [X_n]$, and hence, there exists m such that

$$\sum_{i=m+1}^{\infty} \mu(X_i) \leqq \delta.$$

For such an m we get from (8)

$$(8.1) \qquad \mu[(M_1 + M_2 + \cdots + M_m) \cdot (M'_1 \ldots M'_m)] \leqq \delta_1 + \cdots + \delta_m + \\ + \delta_{m+1} + \cdots + \delta \leqq 2\delta.$$

Thus for every $\delta > 0$ there exists m with (8.1). Applying Theorem [B I.8.6.], we deduce the measurability of X.

B I.8.14f. We have, by (6),

$$\mu(M_n - [X_n]) \leqq \delta_n, \qquad (n = 1, 2, \ldots);$$

hence

$$\mu(M_n) \leqq \mu(X_n) + \delta_n.$$

It follows

$$(9) \qquad \sum_{n=1}^{\infty} \mu(M_n) \leqq \sum_{n=1}^{\infty} \mu(X_n) + \delta.$$

Since $\sum_{n=1}^{\infty} M_n$ is a covering of X, (by Lemma [B I.5.3.]), we have

$$\mu(X) \leqq \mu\left(\sum_{n=1}^{\infty} M_n\right);$$

hence, by (9),

$$\mu(X) \leqq \sum_{n=1}^{\infty} \mu(X_n) + \delta.$$

This being true for any $\delta > 0$, we obtain

$$(10) \qquad \mu(X) \leqq \sum_{n=1}^{\infty} \mu(X_n).$$

B I.8.14g. On the other hand we have $[X_n] \leqq [X]$, since $X_n \leqq X$, (Theorem [B I.8.12.]); hence

$$(11) \qquad \sum_{n=1}^{\infty} [X_n] \leqq [X],$$

$$\mu\left(\sum_{n=1}^{\infty} [X_n]\right) \leqq \mu(X).$$

Now, by hypothesis, all $[X_n]$ are disjoint; hence

$$\sum_{n=1}^{\infty} \mu(X_n) \leqq \mu(X). \tag{12}$$

From (10) and (12) we obtain

$$\sum_{n=1}^{\infty} \mu(X_n) = \mu(X). \tag{13}$$

B I.8.14h. From (13) we get

$$\mu(X) - \mu\left(\sum_{n=1}^{\infty} [X_n]\right) = 0;$$

hence

$$\mu([X] - \sum_{n=1}^{\infty} [X_n]) = 0.$$

It follows

$$[X] - \sum_{n=1}^{\infty} [X_n] = O \quad \text{i.e.} \quad [X] \leqq \sum_{n=1}^{\infty} [X_n]. \tag{14}$$

This combined with (11) gives $[X] = \sum_{n=1}^{\infty} [X_n]$; so all items of the thesis are proved.

B I.8.15. Theorem. If $X_1, X_2, \ldots, X_n, \ldots$ are measurable sets of traces, then $X = \bigcup_{n=1}^{\infty} X_n$ is also measurable, and we have $[X] = \sum_{n=1}^{\infty} [X_n]$.

Proof. Put

$$Y_n = \bigcup_{i=1}^{n} X_i, \quad (n = 1, 2, \ldots).$$

We have, [A.2.6.4.]:

$$X_1 \cup X_2 \cup \cdots \cup X_n = Y_1 \cup (Y_2 - Y_1) \cup (Y_3 - Y_2) \cup \cdots \cup (Y_n - Y_{n-1}),$$
$$(n = 2, 3, \ldots).$$

Hence

$$X = Y_1 \cup (Y_2 - Y_1) \cup (Y_3 - Y_2) \cup \cdots. \tag{15}$$

The set X_n being measurable, so are Y_n, (Theorem [B I.8.10.]), hence $Y_{n+1} - Y_n$ are so (Theorem [B I.8.11.]). Since all terms in (15) are disjoint, we can apply Theorem [B I.8.14.], which gives the measurability of X.

In addition to that we have

$$[Y_{n+1} - Y_n] = [Y_{n+1}] - [Y_n],$$

(by Theorem [B I.8.11.]), and $[Y_n] = \sum_{i=1}^{n} [X_i]$, (by Theorem [B I.8.10.]). It follows

$$[X] = \sum_{n=1}^{\infty} [X_n],$$

so the theorem is proved.

B I.8.16. Theorem. If $X_1, X_2, \ldots, X_n, \ldots$ are measurable sets of traces, then

$$X = \bigcap_{n=1}^{\infty} X_n$$

is also measurable, and we have

$$[X] = \prod_{n=1}^{\infty} [X_n].$$

Proof. By de Morgan laws.

B I.9. Def. A set X of traces is called *null-set of traces* if its outer coat is O.

A null set is measurable, since $O \leqq [X]_* \leqq [X]^* = O$; its measure is 0; $\emptyset$ is a null set; W is not a null set. We shall state some theorems, whose proofs we omit.

B I.9.1. If X is a null set, $Y \subseteqq X$, then Y is also a null set.

B I.9.2. If $X_1, X_2, \ldots, X_n, \ldots$ are null sets of traces, so is $\bigcup_{n=1}^{\infty} X_n$.

B I.9.3. For a set X of traces the following are equivalent:

I. X is a null set,

II. $\mu(X) = 0$.

B I.9.4. Theorem. If X is measurable and N a null set, then $X - N$ is measurable, and $[X - N] = [X]$, and $\mu(X - N) = \mu(X)$.

Proof. $X - N$ is measurable on account of [B I.9.] and [B I.8.11.]. We have $(X - N) \cup (N \cap X) = X$, and the sets $X - N$, $N \cap X$ are disjoint. Hence by [B I.8.13.], $\mu(X - N) + \mu(N \cap X) = \mu(X)$. Now $N \cap X \subseteqq N$, hence $\mu(N \cap X) = 0$. It follows $\mu(X - N) = \mu(X)$.

B I.9.5. Theorem. If X is measurable and N is a null-set, then $X \cup N$ is measurable, $[X \cup N] = [X]$ and $\mu(X \cup N) = \mu(X)$.

Proof. $X \cup N$ is measurable, because of [B I.9.] and [B I.8.11.]. We have $X \cup N = X \cup (N - X)$, where X and $N - X$ are disjoint sets. Hence, by [B I.8.13.], $\mu(X \cup N) = \mu(X) + \mu(N - X)$. Since $N - X \subseteqq N$, we have $\mu(N - X) = 0$. It follows:

$$\mu(X \cup N) = \mu(X). \quad \text{q.e.d.}$$

B I.9.6. Consider the class $\boldsymbol{T}$ of all measurable sets of traces. We have $\emptyset \in \boldsymbol{T}$ and $W \in \boldsymbol{T}$. We have $\emptyset \neq W$.

If $X \in \boldsymbol{T}$, then $\operatorname{co} X = W - X \in \boldsymbol{T}$.

If $X_1, X_2, \ldots, X_n, \ldots \in \boldsymbol{T}$, then $\bigcup_n X_n \in \boldsymbol{T}$.

It follows that $\boldsymbol{T}$ is organized into a Boolean tribe with identity of sets as governing equality and inclusion $\subseteqq$ of sets as ordering correspondence. The tribe is denumerably additive. The class $\boldsymbol{J}$ of all null-sets is a denumerably additive ideal in $\boldsymbol{T}$. Hence $\boldsymbol{T}$ can be reorganized into another denumerably additive tribe $\boldsymbol{T}_J$ with $=^J$ as governing equality and $\subseteqq^J$, defined by $X \subseteqq^J Y \cdot =_{df} \cdot X - Y \in \boldsymbol{J}$, as lattice ordering.

B I.9.7. The correspondence $\mathfrak{A}$ which attaches to a variable measurable set X of traces the coat $[X]$ is pluri-one, because if $X =^J Y$, then $[X] = [Y]$, (this follows from [B I.9.4.] and [B I.9.5.]). The correspondence $\mathfrak{A}$ is invariant with respect to the identity of sets in the domain and equality of somata in the range. It preserves finite and denumerable operations, carries the null set into O and the set W into I. It also preserves the measure. $\mathfrak{A}$ is a homomorphism from $\boldsymbol{T}$ into G. The tribe $\mathfrak{A}(\boldsymbol{T})$ is a denumerably genuine, denumerably additive strict subtribe of (G).

B I.9.8. The correspondence $\mathfrak{A}$ is also invariant with respect to $=^J$ in the domain, and as such one it constitutes an isomorphism from $\boldsymbol{T}_J$ into (G), with preservation of measure.

B I.9.9. The tribe $\mathfrak{A}(\boldsymbol{T})$ contains all bricks; hence the tribe (F), and then the borelian extension of (F) within (G). (Remember, that we have supposed that (F) is a strict subtribe of (G), [B I.1.1.].

Theorem. If we suppose that (G) coincides with the Lebesgue's covering extension $(F)^L$ of (F) within (G), then $\mathfrak{A}(\boldsymbol{T})$ coincides with G.

Proof. Straight forward, [A I.6.].

B I.9.10. Theorem. If $A \in G$, then there exists a measurable set X of traces such that $A = [X]$. All sets X, for which $[X] = A$, can be obtained from one of one of then, say X_1, by taking $X_1 - N_1 \cup N_2$, where N_1, N_2 are null sets. If $A \neq 0$, then X is not empty.

Proof. The existence follows from [B I.9.9.]. If $A \neq O$, then $\mu A \neq 0$, because the measure μ is effective on (G). We cannot have $X = \emptyset$ because $\mu X = \mu[A] \neq 0$. The remaining thesis follows from that $\mathfrak{A}$ is $1 \to 1$ from $\boldsymbol{T}_J$ into G.

B I.10. Admit (***Hyp Ad***), [B.2.], ***Hyp*** I and ***Hyp*** II, [B I.6.]. We shall have some remarks on single traces.

Def. By the *neighborhood of the trace* t we shall understand any brick which covers t. Denote by $v(t)$ the set of all neighborhoods of t.

B I.10.1. If $a \in v(t)$ and $b \in v(t)$, then $a \cdot b \neq O$.

Proof. Suppose that $a \cdot b = O$. Since a covers t there exists a representative

(1) $$a = a_1 \geqq a_2 \geqq \cdots \quad \text{of } t.$$

Since b covers t, there exists a representative

(2) $$b = b_1 \geqq b_2 \geqq \cdots \quad \text{of } t.$$

As $a \cdot b = O$ we have $a_n \cdot a_m = O$ for all n, m. Now the sequences $\{a_n\}$, $\{b_n\}$ are equivalent, so we have $\{a_n\} \leqq \{b_n\}$. Hence [B I.2.] there exists m with $a_m \leqq b_1$. Since $a_m \cdot b_1 = O$, it follows that $a_m = O$ and then $a_m = a_{m+1} = \cdots = O$, so $\{a_n\}$ is equivalent to $(O, O, \ldots)$, which is impossible.

B I.10.2. If a sequence $\{a_n\} \sim \{O, O, O, \ldots\}$, then there exists m such that $a_m = a_{m+1} = \cdots = O$.

Indeed, there exists n with $a_n \leqq O$; hence $a_n = O$ and then $a_n = a_{n+1} = \cdots = O$.

B I.10.3. If

$$a'_1 \geqq a'_2 \geqq \cdots \quad \text{is a representative of } t',$$

$$a''_1 \geqq a''_2 \geqq \cdots \quad \text{is a representative of } t'',$$

$$b_n = a'_n a''_n \neq O \quad \text{for all } n, \text{ then } t' = t''.$$

Proof. We get $b_1 \geqq b_2 \geqq \cdots$. Since for every n we have $b_n \leqq a'_n$, there exists m with $b_m \leqq a'_n$; hence $\{b_n\} \leqq \{a'_n\}$; Similarly we get $\{b_n\} \leqq \{a''_n\}$. Since $\{a'_n\}$ is a minimal sequence, we have either $\{b_n\} = \{O, O, \ldots\}$ or $\{a'_n\} \sim \{b_n\}$. The first alternative is impossible; hence $\{a'_n\} \sim \{b_n\}$. Similarly we get $\{a''_n\} \sim \{b_n\}$; hence $\{a'_n\} \sim \{a''_n\}$, so $t' = t''$.

B I.10.4. If $t' \neq t''$, then there exist neighborhoods a' of t', and a'' of t'' such that $a', a'' = O$.

Proof. Let $a_1 \geqq a_2 \geqq \cdots$, $b_1 \geqq b_2 \geqq \cdots$ be representatives of t', t'' respectively. There exists, [B I.10.3.], at least one n with $a_n \cdot b_n = O$. a_n is a neighborhood of t' and b_n is a neighborhood of t''. The theorem is proved.

B I.10.5. If $v(t') = v(t'')$, then $t' = t''$.

Proof. Suppose $t' \neq t''$. By [B I.10.4.] there exist neighborhoods a', a'' ot t', t'' respectively, such that $a' \cdot a'' = O$.

Now $a' \in v(t')$, $a'' \in v(t'')$. Since $v(t') = v(t'')$, we get $a', a'' \in v(t')$. Hence, by [B I.10.1.], $a' \cdot a'' \neq O$, which is contradiction. The theorem is proved.

B I.10.6. From [B I.10.5.] it follows, since $t' = t''$ implies $v(t') = v(t'')$, that the set of all neighborhoods $v(t)$ completely characterizes the trace t. Different traces have different total sets of neighborhoods.

B I.11. Def. We say that the trace t is an *elusive trace* t: if $\mu\, a_n \to 0$, where $\{a_n\}$ is a representative of t. We say that t is a *heavy trace*, whenever $\mu\, a_n \to \alpha > 0$. The following remarks may be interesting:

B I.11.1. If

1) $a_1 \geqq a_2 \geqq \cdots \geqq a_n \geqq \cdots$ is a minimal sequence of bricks,

2) $\lambda > 0$, $\mu(a_n) \geqq \lambda$, then $b =_{df} \prod_n a_n$ is an atom in (G).

To prove it, suppose b is not an atom. We have $\mu b \geqq \lambda$. Hence there exists a decomposition $b = b' + b''$ where $b' \cdot b'' = O$, $\mu b' > 0$, $\mu b'' > 0$. Imitating an argument similar to that applied in the paper [(36), § 1, 1d, e, f] we obtain a brick c, such that $c < b$, $\mu c > 0$. Putting $c_n =_{df} c$ for $n = 1, 2, \ldots$, we get a sequence

$$c_1 \geqq c_2 \geqq \cdots \geqq c_n \geqq \cdots, \quad \text{with} \quad \{c_n\} \leqq \{a_n\},$$

where $\{c_n\}$ is not equivalent to $\{O, O, \ldots\}$.

Since $\{a_n\}$ is a minimal sequence, it follows that $\{c_n\} \sim \{a_n\}$. Hence there exists n with $a_n \leqq c$. This is, however impossible, because $\mu c < \mu b \leqq \mu a_n$. Thus we have proved that b is an atom in (G). In the above proof we have taken into account the circumstance that μ is effective on (G).

B I.11.2. If A is an atom in F^L, then there exists a minimal sequence a_n such that

$$b = \prod a_n, \quad \mu A = \lim \mu a_n > 0.$$

To show it, let A be an atom. There exists an infinite sequence of bricks $a_1 \geqq a_2 \geqq \cdots \geqq a_n \geqq \cdots \geqq A$, such that $\mu(a_n) \to \mu(A) > 0$. We have $A \leqq \prod_{n=1}^{\infty} a_n$. Since the measure is effective, we cannot have $A < \prod_{n=1}^{\infty} a_n$. Thus we have

$$A = \prod_{n=1}^{\infty} a_n.$$

Consider the set $\boldsymbol{U}$ of all traces each of which being covered by all bricks a_n. We shall prove that the set $\boldsymbol{U}$ is not empty. Denote by $\boldsymbol{U}_n$ the set of all traces which are covered by a_n. We have $\mu \boldsymbol{U}_n = \mu a_n > 0$, hence $\boldsymbol{U}_n \neq \emptyset$, [B I.8.7.]. Since $\boldsymbol{U} = \bigcap_n \boldsymbol{U}_n$, and $\boldsymbol{U}_{n+1} \subseteqq \boldsymbol{U}_n$, we have $\mu \boldsymbol{U} = \lim \mu \boldsymbol{U}_n = \lim \mu a_n = \mu A > 0$. Hence $\boldsymbol{U} \neq \emptyset$.

The traces belonging to $\boldsymbol{U}$ may be elusive or not. Denote by $\boldsymbol{U}_e$ the subset of $\boldsymbol{U}$ composed of all elusive traces.

We shall prove $\boldsymbol{U}_e$ is a null-set.

Let $t \in \boldsymbol{U}_e$. Let $c_1 \geqq c_2 \geqq \cdots \geqq c_n \geqq \cdots$ be a representative of t.

Since t is elusive, we have $\lim \mu c_n = 0$. Hence there exists m_0 such that for all $m \geqq m_0$ we have

$$\mu(c_m) < \frac{\mu(A)}{2}. \tag{1}$$

We have $c_m A \leqq A$. Since A is an atom, we have either $c_m \cdot A = O$ or $c_m \cdot A = A$. In the second case we get $A \leqq c_m$, and hence

$\mu(A) \leqq \mu(c_m)$ which contradicts (1). Hence $c_m A = O$, i.e.

(2) $$c_m \leqq \text{co} A .$$

Since t is covered by c_m and by a_n, and $c_m \cdot a_n$ is a brick, it follows that there exists a brick $c_n(t)$ which covers t and is contained in co$A$ and a_n. Hence

(3) $$c_n(t) \leqq a_n - A .$$

Such a brick can be found for every $t \in \boldsymbol{U}_e$ and for every n.

By the axiom of choice we can find for any given n such set of bricks $c_n(t)$. We get

(4) $$\sum_{t \in U_e} c_n(t) \leqq a_n - A .$$

Let L_n be a covering of $a_n - A$ with $\mu(L_n - (a_n - A)) < \frac{1}{n}$. Such a covering exists, [B I.7.5.]. L_n is also a covering of $\boldsymbol{U}_e$. Indeed if $t \in \boldsymbol{U}_e$, $c_n(t)$ is its covering, and $c_n(t) \leqq a_n - A \leqq L_n$. Hence $[\boldsymbol{U}_e]^* \leqq L_n$. Now, since $\lim \mu(a_n - A) = 0$, and $\lim \mu(L_n - (a_n - A)) = 0$, we get $\lim \mu L_n = 0$; hence

$$\mu[\boldsymbol{U}_e]^* = 0,$$

which proves that $\boldsymbol{U}_e$ is a null-set.

Denote by $\boldsymbol{U}_h$ the subset of all heavy traces contained in $\boldsymbol{U}$. Since $\boldsymbol{U} = \boldsymbol{U}_h \cup \boldsymbol{U}_e$, $\boldsymbol{U}_h \cap \boldsymbol{U}_e = \emptyset$ and $\mu \boldsymbol{U}_e = 0$, it follows that $\mu \boldsymbol{U}_h = \mu \boldsymbol{U} = \mu A$. We shall prove that $\boldsymbol{U}_h$ contains only one trace. Suppose that $t', t'' \in \boldsymbol{U}_h$, $t' \neq t''$. Let $b'_1 \geqq b'_2 \geqq \cdots$; $b''_1 \geqq b''_2 \geqq \cdots$ be representatives of t', t'' respectively. Since the trace t' is covered by a_m, there exists n with $b'_n \leqq a_m$. Thus we can find a subsequence $\{b'_{k(n)}\}$ of $\{b'_n\}$ such that $b'_{k(n)} \leqq a_n$ for $n = 1, 2, \ldots$, and with $k(1) \leqq k(2) \leqq \cdots$. Putting $c'_n =_{df} b'_{k(n)}$, we have $c'_1 \geqq c'_2 \geqq \cdots$ and $c'_n \leqq a_n$; $\{c'_n\}$ is a representative of t'. In a similar way we shall find a subsequence $\{c''_n\}$ of $\{b''_n\}$ such that $c''_1 \geqq c''_2 \geqq \cdots \geqq c''_n \geqq \cdots, \leqq a_n$, and where $\{c''_n\}$ is a representative of t''. Since $t' \neq t''$, there exists by [B I.10.3.], an index n_0 such that $c'_{n_0} \cdot c''_{n_0} = O$. The sequences $c'_{n_0} \geqq c'_{n_0+1} \geqq \cdots$, $c''_{n_0} \geqq c''_{n_0+1} \geqq \cdots$ are also representatives of t', t'' respectively. The bricks of the first sequence are disjoint with the bricks of the second one. Since t', t'' are heavy traces, we have

$$\lambda' =_{df} \lim_{n \to \infty} \mu(c'_n) > 0, \qquad \lambda'' =_{df} \lim_{n \to \infty} \mu(c''_n) > 0;$$

hence, by [B I.11.1.], the somata

$$A' =_{df} \prod_{n=1}^{\infty} c'_n, \qquad A'' =_{df} \prod_{n=1}^{\infty} c''_n$$

are atoms in G. They are disjoint. Since

$$c'_n \leqq a_n, \quad c''_n \leqq a_n$$

it follows that

$$A' \leqq \prod_{n=1} a_n = A, \quad A'' \leqq \prod_{n=1} a_n = A.$$

Thus we have

$$A' + A'' = A, \quad \mu A' > 0, \quad \mu A'' > 0,$$

which contradicts the hypothesis that A is an atom. Thus we have proved that $\boldsymbol{U}_e$ has the measure 0 and $\boldsymbol{U}_h$ is composed of a single heavy trace; denote it by t_0. Let $d_1 \geqq d_2 \geqq \cdots$ be a representative of t_0. We can, as before, derive from it another sequence

$$d'_1 \geqq d'_2 \geqq \cdots,$$

representative of t_0, with $d'_n \leqq a_n$, because t_0 is covered by the bricks a_n. We get

$$B =_{df} \prod_{n=1} d'_n \leqq \prod_{n=1} a_n = A, \quad \text{hence} \quad B \leqq A.$$

Since t_0 is heavy, we have $\lim_{n\to\infty} \mu(d'_n) > 0$; hence $\mu B > 0$. Now A is an atom; consequently $B = A$. Thus we have proved that if t_0 is the unique heavy trace covered by bricks a_n, then for every its representative $\{a_n\}$ we have $\prod_{n=1}^{\infty} a_n = A$.

B I.11.3. Corrollar. If b is an atom in (F^L) and $a_1 \geqq a_2 \geqq \cdots$ is a sequence of bricks with $\prod_{n=1}^{\infty} a_n = b$, then the set $\boldsymbol{U}_e$ of all elusive traces which are covered by all a_n is a null-set, and the set $\boldsymbol{U}_h$ composed of all heavy traces which are covered by all a_n, is confined to a single trace.

B I.11.4. Remark. The Theorems [B I.11.1.], [B I.11.2.], and [B I.11.3.] yield information concerning the relation between atoms and heavy traces. They are not the same, of course, but there is a $1 \to 1$ correspondence between them. Concerning the proof of [B I.11.2.], we notice that $\boldsymbol{U}_e$ may be empty.

B I.11.5. Remark. Applying a theorem by STONE on ultrafilters (rather on maximal ideals), we can prove the following:

Every infinite decreasing sequence $a_1 \geqq a_2 \geqq \cdots$ of bricks with $a_n \neq O$, $\lim \mu(a_n) = 0$, contains a minimal sequence b_n, [B I.2.1.], though $\{a_n\}$ may not be minimal.

B I.11.6. Theorem. The set (t) composed of a single elusive trace t is measurable. Its measure is 0.

Proof. Let a_n be a representative of t. We have $\mu a_n \to 0$ for $n \to \infty$. Let X_n be the set of all traces covered by a_n. By [B I.8.7.], X_n is measurable, and we have

$$\mu X_n = \mu a_n.$$

We have

$$(t) \subseteqq \prod_{n=1}^{\infty} X_n.$$

Now

$$\mu \prod_{n=1}^{\infty} X_n = \lim \mu(X_n) = \lim \mu(a_n) = 0.$$

Hence $\prod_{n=1}^{\infty} X_n$ is a null set of traces. Consequently (t) is a null set of traces; hence (t) is measurable.

B I.11.7. Theorem. The set (t) composed of a single heavy trace t is measurable. Its measure is positive.

Proof. Let $a_1 \geqq a_2 \geqq \cdots$ be a representative of (t). We have

$$\lim_{n\to\infty} \mu(a_n) > 0.$$

Let X_n be the set of all traces which are covered by a_n. We have

$$\mu(X_n) = \mu(a_n).$$

We have

(1) $$(t) = \bigcap_{n=1}^{\infty} X_n.$$

The set $\bigcap_{n=1}^{\infty} X_n$ is measurable. By [B I.11.1.] the soma $A =_{df} \prod_{n=1}^{\infty} a_n$ is an atom. Hence, by [B I.11.3. Corrollar.], the set $\boldsymbol{U}_e$ of all elusive traces covered by the bricks a_n is a null set, and the set $\boldsymbol{U}_h$ of all heavy traces covered by all a_n is confined to a single trace.

We have $\boldsymbol{U}_e \cup \boldsymbol{U}_h = \bigcap_{n=1}^{\infty} X_n$; hence $\boldsymbol{U}_h = \bigcap_{n=1}^{\infty} X_n - \boldsymbol{U}_e$. Hence $\boldsymbol{U}_h$ is measurable. From (1) it follows that $(t) = \boldsymbol{U}_h$. Hence (t) is measurable. We have $(t) = \bigcap_{n=1}^{\infty} X_n - \boldsymbol{U}_e$, and then $\mu(t) = \mu \bigcap_{n=1}^{\infty} X_n = \mu \prod_{n=1}^{\infty} a_n = \mu A > 0$. The theorem is proved.

B I.11.8. ***Hyp*** I and ***Hyp*** II are necessary conditions for having a measure theory of sets of traces.

B I.12. Measurable functions of traces and integration. The class $\boldsymbol{T}$ of all measurable sets [B I.9.6.] possesses the properties:

1⁰. if $X \in \boldsymbol{T}$, then $\operatorname{co} X \in \boldsymbol{T}$,

2⁰. if $X_1, X_2, \ldots, X_n, \ldots \in \boldsymbol{T}$,

then $\bigcup_{n=1}^{\infty} X_n \in \boldsymbol{T}$, and there is available a denumerably additive non negative measure $\mu(X)$ defined for all $X \in \boldsymbol{T}$.

These properties enable us to apply Fréchet's theory of measurable functions $f(x)$ of all traces, and in addition to that, to consider Fréchet's integrals $\int f(x)\, d\mu$, (20), (16).[1]

This theory follows the known features of the Lebesgue's integration theory. In our case of number-valued trace-functions we shall confine ourselves to a sketch only, referring for detailed proofs to (16), (26), (22).

B I.12.1. Let $f(x)$ be a real valued function defined almost μ-everywhere (a.e.) in W, [B I.6.]. [This means that $\mu(\text{co}\, \complement f) = O$.] We say that $f(x)$ *fits* $\boldsymbol{T}$ or *that* $f(x)$ *is measurable*, if whatever the real number λ may be, the set $\{x \mid f(x) \leqq \lambda\}$ belongs to $\boldsymbol{T}$, (i.e. is measurable).

B I.12.2. This condition is equivalent to each of the following ones:

1⁰ for every λ the set $\{x \mid f(x) < \lambda\}$ is measurable,
2⁰. for every λ the set $\{x \mid f(x) \geqq \lambda\}$ is measurable,
3⁰. for every λ the set $\{x \mid f(x) > \lambda\}$ is measurable.

B I.12.3. Def. A measurable function $\varphi(x)$ is called *simple* if it is defined almost μ-everywhere and admits an at most denumerable number of values. The following properties hold true:

B I.12.4. If $f(x)$ is measurable, then there exists an infinite sequence

$$\varphi_1(x) \geqq \varphi_2(x) \geqq \cdots \geqq \varphi_n(x) \geqq \cdots$$

of simple functions converging on W a.e. uniformly to $f(x)$, and conversely, a uniform a.e. limit of a sequence $\varphi_1(x) \geqq \varphi_2(x) \geqq \cdots$ of simple function is measurable.

B I.12.5. If $f(x)$ is measurable, so is $|f(x)|$.

B I.12.6. If $f(x)$, $g(x)$ are measurable, so are also $f(x) + g(x)$, $f(x) \cdot g(x)$, $f(x) \cup g(x)$, $f(x) \cap g(x)$. The last two functions are defined as $\max[f(x), g(x)]$, $\min[f(x), g(x)]$ for a.e. point x separately.

B I.12.7. If $\{f_n(x)\}$ is an infinite sequence of measurable functions, and if $\overline{\lim} f_n(x)$, $[\underline{\lim} f(x)]$ is defined a.e., then it is a measurable function.

B I.13. Let $\varphi(x)$ be a simple function, admitting the values $\varphi_1, \varphi_2, \ldots, \varphi_n, \ldots$ on the measurable and disjoint sets $X_1, X_2, \ldots, X_n, \ldots$, respectively with $\mu(\text{co} \bigcup\limits_n X_n) = 0$.

We say that $\varphi(x)$ *is μ-summable,* when the series $\sum\limits_{n=1}^{\infty} \varphi_n\, \mu(X_n)$ converges absolutely.

---

[1] Integrals of functions defined on abstract sets are currently called Lebesgue's integrals, but FRÉCHET (20) was the first who has liberated the integration theory from topological and metrical notions, and this step in that time, was a tremendous progress, and the corresponding idea was far from the intuition of contemporaneous mathematicians.

B I.13.1. The following definition is a generalization of the above one. A measurable function $f(x)$ is said to be *μ-summable,* if there exist two simple μ-summable functions $\varphi(x)$, $\psi(x)$ such that a.e.

$$\psi(x) \leqq f(x) \leqq \varphi(x).$$

If $f(x)$ is summable,

$$\varphi_1(x) \geqq \varphi_2(x) \geqq \cdots$$

is a sequence of μ-summable simple functions tending a.e. uniformly to $f(x)$, and if $\varphi_n(x)$ admits the values $\varphi_{1n}, \varphi_{2n}, \ldots$ on the measurable sets $X_{1n}, X_{2n}, \ldots$ respectively, then $\lim_{n\to\infty} \sum_{k=1}^{\infty} \varphi_{kn}\, \mu(X_{kn})$ exists and does not depend on the choice of $\{\varphi_n(x)\}$. This limit is denoted by $\int f(x)\, d\mu$, and termed *Fréchet's integral of $f(x)$ on W.*

B I.13.2. For a simple function $\varphi(x)$, admitting the values φ_k in the sets X_k we have

$$\int \varphi(x)\, d\mu = \sum_{k=1}^{\infty} \varphi_k\, \mu(X_k).$$

B I.13.3. Remark. The integral $\int f(x)\, d\mu$ can also be defined similarly as did LEBESGUE in his "Leçons sur l'integration", (23), and thus be given an equivalent definition.

B I.13.4. Theorem. (LEBESGUE). If

1. $f_n(x)$ are μ-summable, $(n = 1, 2, \cdots)$,
2. $f_n(x) \leqq f_{n+1}(x)$ a.e.,
3. $f(x) = \lim_{n\to\infty} f_n(x)$ a.e.,

then the following are equivalent:

I. $f(x)$ is μ-summable;

II. $\lim_{n\to\infty} \int f_n(x)\, d\mu = \int f(x)\, d\mu$.

B I.14. The notion of measurability of functions and of their integrals can be extended to complex valued functions. A complex valued function $F(x)$, defined a.e. on W, is said to be *measurable* (or else, *fitting* $\boldsymbol{T}$), if, in the representation $F(x) = f(x) + i\, g(x)$, the real valued functions $f(x)$, $g(x)$ are both measurable.

B I.14.1. Every measurable complex-valued function can be uniformly approximated a.e. by complex-valued simple functions.

B I.15. Def. If $f(x)$, $g(x)$ are both μ-summable, the function $F(x)$ is also termed *μ-summable* and we *define*

$$\int F(x)\, d\mu = \int f(x)\, d\mu + i \int g(x)\, d\mu.$$

The corresponding notions for real valued functions are but a particular case of the above more general notions.

B I.16. The notion of integral can be further generalized by introducing *integrals over measurable subsets of* W. If X is a measurable set of traces, $F(x)$ is a measurable complex valued function. defined a.e. on W, then the function $F_1(x)$ defined by setting $F_1(x) = F(x)$ whenever $x \in X$, and $F_1(x) = 0$ whenever $x \in \operatorname{co} X$, is also measurable.

B I.16.1. Def. We define, in the case of summability of $F_1(x)$, [B I.16.]

$$\int_X F(x)\, d\mu = \int_W F_1(x)\, d\mu,$$

and we say that $F(x)$ *is* μ*-summable on* X.

B I.16.2. The following theorems are true: If $F(x)$ is μ-summable and X is measurable, then $F(x)$ is μ-summable on X.

B I.16.3. If $F(x)$, $G(x)$ are μ-summable on a measurable set X of traces, and α, β are complex numbers, then $\alpha F(x) + \beta G(x)$ is also μ-summable on X, and we have

$$\int_X (\alpha F(x) + \beta G(x))\, d\mu = \alpha \int_X F(x)\, d\mu + \beta \int_X G(x)\, d\mu.$$

B I.16.4. If $F(x)$ is μ-summable on W, $X_1, X_2, \ldots, X_n, \ldots$ are mutually disjoint measurable sets, then if we put $X = \sum_n X_n$, we have

$$\int_X F(x)\, d\mu = \sum_n \int_{X_n} F(x)\, d\mu.$$

This theorem says that the complex valued set-function

$$A(X) =_{df} \int_X F(x)\, d\mu,$$

defined for all $X \in \boldsymbol{T}$, *is denumerably additive on* $\boldsymbol{T}$.

B I.16.5. If X is measurable and $F(x)$ is summable on X, so is $|F(x)|$, and we have

$$\left| \int_X F(x)\, d\mu \right| \leqq \int_X |F(x)|\, d\mu.$$

B I.16.6. (Lebesgue). If

1) $F_n(x)$ are μ-summable functions on a measurable set X,
2) $g(x)$ is a real valued μ-summable function on X,
3) $|F_n(x)| \leqq g(x)$, $(n = 1, 2, \ldots)$, a.e. on X,
4) $\lim_{n\to\infty} F_n(x) = F(x)$ a.e. on X,

then

$$\lim_{n\to\infty} \int_X F_n(x)\, d\mu = \int_X F(x)\, d\mu.$$

B I.16.7. If $F_n(x)$ are μ-summable functions on X, and if $F_n(x)$ tends a.e. on X uniformly to $F(x)$, we have

$$\int_X F(x)\, d\mu = \lim \int_X F_n(x)\, d\mu.$$

B I.17. Def. We direct our attention to *μ-square summable* complex valued functions $F(x)$. We shall state definitions and theorems for functions defined a.e. on W, but the analogous statements will be valid and useful also in the case where $F(x)$ are defined a.e. on a measurable set X of traces. A function $F(x)$ is said to be *μ-square summable if*

1) $|F(x)|^2$ is μ-summable on W,
2) $F(x)$ is measurable on W.

B I.17.1. If $F(x)$, $G(x)$ are μ-square summable on W, so is with $\alpha F(x) + \beta G(x)$, where α, β are complex constants.

B I.17.2. If $F(x)$, $G(x)$ are μ-square summable on W, then $F(x) \cdot G(x)$ is μ-summable on W, and we have the *Cauchy-Schwarz inequality*

$$\left| \int F(x) \cdot G(x)\, d\mu \right|^2 \leqq \int |F(x)|^2\, d\mu \cdot \int |G(x)|^2\, d\mu,$$

and the *Minkowski-inequality*

$$\sqrt{\int |F(x) + G(x)|^2\, d\mu} \leqq \sqrt{\int |F(x)|^2\, d\mu} + \sqrt{\int |G(x)|^2\, d\mu}.$$

B I.17.3. If $F(x)$ is μ-square summable, then there exists a sequence $\{\Phi_n(x)\}$, $(n = 1, 2, \ldots)$ of complex simple μ-square summable functions such that

$$\lim_{n\to\infty} \Phi_n(x) = F(x),$$

a.e. uniformly.

B I.18. Def. Let $\{F_n(x)\}$ be an infinite sequence of μ-square summable functions. We say that *it converges in μ-square mean*, if for every $\varepsilon > 0$ there exists N such that if $n \geqq N$, $m \geqq N$, we have

$$\int |F_n(x) - F_m(x)|^2\, d\mu \leqq \varepsilon.$$

B I.18.1. If $F_n(x)$ *converges in μ-square mean*, then there exists a square μ-summable function $F(x)$, determined uniquely up to a null-set of traces, such that

$$\lim_{n\to\infty} |F_n(x) - F(x)|^2\, d\mu = 0.$$

We say that $\{F_n(x)\}$ converges in μ-square mean toward $F(x)$.

B I.18.2. If $\{F_n(x)\}$ converges in μ-square mean, then there exists $M \geqq 0$ such that

$$\int |F_n(x)|^2\, d\mu \leqq M \quad \text{for} \quad n = 1, 2, \ldots$$

B I.18.3. If $\{F_n(x)\}$ tends to $F(x)$ in μ-square mean, then every subsequence $\{F_{k(n)}\}$ contains another subsequence $\{F_{1(n)}\}$, converging toward $F(x)$ almost μ-everywhere.

B I.18.4. If $\{F_n(x)\}$, $\{G_n(x)\}$ tend in μ-square mean respectively to $F(x)$, $G(x)$, then $\alpha F_n(x) + \beta G_n(x)$ tends in μ-square mean toward

$$\alpha F(x) + \beta G(x),$$

and

$$\lim_{n\to\infty} \int F_n(x)\, G_n(x)\, d\mu = \int F(x)\, G(x)\, d\mu.$$

B I.18.5. If $\{F_n(x)\}$ converges on W in μ-square mean toward $F(x)$, if X is a measurable set of traces, then $F_n(x)$, if restricted to X, converges in μ-square mean toward $F(x)$ restricted to X.

B I.18.6. If $F(x)$, $G(x)$ are μ-square summable functions defined a.e., we say that $F(x)$ *is equivalent* to $G(x)$, $F(x) =_{\text{a.e.}} G(x)$, if the set

$$(\complement F \dotplus \complement G) \cup \{x \mid F(x) \neq G(x)\}$$

is a null set.

B I.18.7. The equivalence possesses the formal properties of the identity. The addition of two functions, the multiplication of a function by a complex number, the integral and the μ-square mean convergence, are invariant with respect to equivalence.

B I.19. Null sets of traces. The theorems on traces show that there exists a correspondence R between measurable sets X of traces and somata of F^b defined by the equivalence of $X R p$ and $p = [X]$. It is a homomorphism whose domain is the class of all measurable sets of traces and whose range is F^b. This relation preserves the measure and transforms finite and infinite unions and intersections into sums and products, and it preserves the complementation. The empty set $\emptyset$ is mapped on O and the total set W on 1.

B I.19.1. The class τ of all measurable sets of traces, if we consider it ordered by the relation of inclusion, make up a denumerably additive tribe $\boldsymbol{T}$, where joins and meets are unions and intersections of sets. F^b is a homomorphic image of $\boldsymbol{T}$ through the relation R.

This relation may not be an isomorphism since there may be several sets of traces which are mapped on O.

B I.19.2. Def. A set X of traces is a *null-set of traces* if its outer coat is O.

A null-set is obviously measurable, since

$$0 \leqq [X]_* \leqq [X]^* = 0;$$

its measure is 0; $\emptyset$ is a null-set.

B I.19.3. If X is a null-set, $Y \subseteqq X$, then Y is also a null-set.

B I.19.4. If $X_1, X_2, \ldots, X_n, \ldots$ are null-sets of traces, so is

$$\bigcup_{n=1}^{\infty} X_n.$$

B I.19.5. For a set X of traces the following are equivalent:

I. X is a null-set.

II. $\mu(X) = 0$.

B I.20. Theorem. *The class J of all null-sets of traces is a denumerably additive ideal in the tribe $\boldsymbol{T}$ of all measurable sets of traces. The tribe $\boldsymbol{T}/J$ is isomorphic with F^b.*

B I.21. Theorem. The following are equivalent:

I. The ideal J of all null-sets possesses at least two different elements, i.e. there exists a not empty set of traces belonging to J.

II. There exists a trace x such that the measure of (x) is $= 0$.

III. There exists at least one descending sequence of bricks

$$a_1 \geqq a_2 \geqq \cdots \text{ with } a_n \neq 0, \text{ where}$$

$$\prod_{n=1}^{\infty} a_p = 0.$$

We omit the proof which is based on the Stone's theorem on the existance of prime ideals, [B.10.], (9).

B I.22. Theorem. Under circumstances given by ***FBG***, [B I.3.] if the condition (***Hyp Af***) is satisfied [B.2.], then ***Hyp*** I also takes place, [B I.6.].

Proof. The ***Hyp*** I reads: if a is a brick and X the set of all traces covered by a, then the set $\mathrm{co}_W X = W - X$, (where W is the set of all traces), is covered by $\mathrm{co}\, a$.

B I.22a. Take a brick a, let τ be a trace, and let $\{R_n\}$ be its representative [B I.2.3.]. We shall prove that if there exists n_0, such that for all $n \geqq n_0$

$$(1) \qquad R_n \cdot \mathrm{co}\, a \neq O,$$

then τ is covered by $\mathrm{co}\, a$.

By ***Af***, the soma $\mathrm{co}\, a$ can be represented as a disjoint union of bricks:

$$\mathrm{co}\, a = a_1 + \cdots + a_k, \qquad (k \geqq 1).$$

Thus

$$R_n \,\mathrm{co}\, a = R_n a_1 + \cdots + R_n a_k \quad \text{for every} \quad n \geqq n_0.$$

By (1) for every $n \geqq n_0$ there exists an index i such that

$$R_n a_i \neq O.$$

Let A_s, $(s = 0, 1, \ldots)$, be the set of all indices i such that

$$R_{n_0+s} \cdot a_i \neq O.$$

The set A_s is not empty whatever $s = 0, 1, \ldots$ may be.

Since

$$R_{n_0} \geqq R_{n_0+1} \geqq \cdots \geqq R_{n_0+s} \geqq \cdots$$

we have

$$A_0 \supseteqq A_1 \supseteqq \cdots \supseteqq A_s \supseteqq \cdots$$

The sets A_s being finite and not empty, there exists an index α, belonging to all $A_0, A_1, \ldots$

Indeed, starting from some index all A_s must be identical. Take such an index α. We have

$$R_{n_0+s} \cdot a_\alpha \neq O \quad \text{for all} \quad s = 0, 1, \ldots$$

Now the somata $R_{n_0+s} \cdot a_\alpha$ are bricks and $\{R_{n_0+s} \cdot a_\alpha\}$ is a diminishing sequence, different from $\{O, O, \ldots\}$ and included, [B I.2.], in $\{R_{n_0+s}\}$, $(s = 1, 2, \ldots)$.

Since $\{R_n\}$ is minimal [B I.2.1.], it follows that

$$\{R_{n_0+s} \cdot a_n\} \quad \text{and} \quad \{R_{n_0+s}\}$$

are equivalent.

Hence, as $\{R_{n_0+s} \cdot a_n\}$ is a representation of τ and

since $R_{n_0+s} \cdot a_n \leqq \operatorname{co} a$, it follows that τ is covered by $\operatorname{co} a$. Thus the assertion is proved.

B I.22b. Now suppose that for all $n \geqq n_0$ we have

$$R_n \cdot a \neq O.$$

Since $R_n \cdot a$ are bricks, $\{R_n \cdot a\}$ is a diminishing sequence, differing from $\{O, O, \ldots\}$, and included in $\{R_n\}$.

Since $\{R_n\}$ is minimal, it follows that

$$\{R_n\} \sim \{a\, R_n\}.$$

Since $a\, R_n \leqq a$, it follows that τ is covered by a.

We have proved that if there exists n_0 such that for all $n \geqq n_0$ we have $a \cdot R_n \neq O$, then τ is covered by a.

B I.22c. Let us notice that τ cannot be covered by a and by $\operatorname{co} a$ simultaneously. Indeed suppose τ is covered by a and by $\operatorname{co} a$; then there exists a representative $\{R_n\}$ of τ such that starting with some index, we have $R_n \leqq a$ and $R_n \leqq \operatorname{co} a$. This gives $R_n \leqq O$ and then $R_n = O$, which is impossible.

B I.22d. Let us suppose that τ is neither covered by a nor by $\operatorname{co} a$. Then we cannot have, starting with some index,

$$R_n\, a \neq O.$$

Similarly we cannot have starting with an index n

$$R_n \cdot \text{co}\, a \neq O.$$

Hence the relations

$$R_n\, a \neq O, \quad R_n \cdot \text{co}\, a \neq O$$

can be valid only for a finite number of indices n.

Hence, starting with some index n_0 we have

$$R_n\, a = R_n\, \text{co}\, a = O.$$

Hence

$$R_n(\text{a} + \text{co}\, a) = O$$

and then

$$R_n = O, \quad \text{so} \quad \{R_n\}$$

cannot be a representative of a trace.

Thus the situation considered cannot take place. From what has been proved it follows that, given a trace τ, it must be either covered by a or by co$a$ disjointedly.

B I.22e. Having that, take a trace $\tau_0 \in W - X$, this τ_0 cannot be covered by a, because X and $W - X$ are disjoint, and X is the set of all traces covered by a. Thus τ_0 must be covered by co$a$.

Chapter C

The tribe of figures on the plane

This chapter gives a detailed study of a special, finitely additive tribe $(\boldsymbol{F})$, whose somata are finite sums of rectangles, on the euclidean plane, and whose sides are parallel to the axes of a given cartesian system of coordinates.

This tribe will be needed in our apparatus for quantum mechanics, since we shall define and study representation of normal operators by operators acting on functions $f(Z)$ of the complex variables Z. This will yield a simple and very handy representation of normal operators. In the present chapter we shall be busy with the study of traces related to $\boldsymbol{F}$, whose base will be chosen as the collection of all mentioned rectangles. As a tool we shall use two-valued measures on $\boldsymbol{F}$, [see Chapter B I.].

The tribe of figures in the euclidean plane

C.1. Let P be the euclidean plane provided with a cartesian system of coordinates x, y. We may consider the collection of all ordered couples (x, y) as the collection of all points of P, or if we like, as the collection of all complex numbers $z = x + iy$.

C.1.1. Def. In this section the greek letters α, β will designate the real numbers and also the symbols $-\infty$, $+\infty$: we shall call them *generalized numbers*.[1] The letters x, y will always, in this chapter, mean ordinary real numbers.

C.1.2. Def. By a *generalized rectangle* we shall understand the set of points

$$R(\alpha_1, \alpha_2; \beta_1, \beta_2) =_{df} \{(x, y) \mid \alpha_1 < x \leqq \alpha_2, \beta_1 < y \leqq \beta_2\}.$$

Since we do not suppose that $\alpha_1 \leqq \alpha_2$, $\beta_1 < \beta_2$, the empty set $\emptyset$ of points is a generalized rectangle. The whole plane P is also a generalized rectangle, viz.

$$R(-\infty, +\infty; -\infty, +\infty).$$

C.1.3. If $x_1 < x_2$, $y_1 < y_2$, then the rectangle $R(x_1, x_2; y_1, y_2)$ contains its right side and its upper side, but it does not contain its left side and does not contain its lower side: this is a half-closed rectangle.

C.1.4. Def. We mention some special kinds of generalized rectangles: the rectangles $R(-\infty, \alpha; -\infty, +\infty)$ are called *vertical half-planes* and the rectangles $R(-\infty, +\infty; -\infty, \beta)$, are called *horizontal half-planes* (P is a horizontal and a vertical half-plane). The rectangles $R(\alpha_1, \alpha_2; -\infty, +\infty)$, if not empty, are called *vertical strips*, and the

---

[1] If $\alpha_1 \leqq \alpha_2 \leqq \cdots \leqq \alpha_n \leqq \cdots$ is a sequence of generalized numbers, we define $\lim \alpha_n$ as the generalized number α such that the following two conditions are satisfied:

1) $\alpha_n \leqq \alpha$ for every n,

2) if $\beta < \alpha$, then there exists an index n_0, such that for every $n \geqq n_0$, we have $\beta \leqq a_n$.

E. g. If all α_n are equal $-\infty$, then $\lim \alpha_n = -\infty$. Indeed we have $\alpha_n \leqq -\infty$ for all n, so the condition 1) is satisfied. The condition 2) also is satisfied because the inequality $\beta < -\infty$ is false, whatever β may be.

A similar definition will be admitted for $\lim \alpha_n$, where $\alpha_1 \geqq \alpha_2 \geqq \cdots \geqq \alpha_n \geqq \cdots$. We get e.g. if $\alpha_n = +\infty$ for all n, then $\lim \alpha_n = +\infty$.

If E is a non empty set of generalized numbers, then $\inf E$ is defined as the generalized number β, such that

1) $\beta \leqq x$ for every $x \in E$,

2) if $\beta < \gamma$, then there exists a number $y \in E$ such that $y < \gamma$.

E. g. If E is composed of the single number $+\infty$, then $\inf E = +\infty$. If it is composed of the single number $-\infty$, then $\inf E = -\infty$. A similar definition we admit for $\sup E$.

rectangles $R(-\infty, +\infty; \beta_1, \beta_2)$, if not empty, are called *horizontal strips*. The rectangles $Q(\alpha, \beta) =_{df} R(-\infty, \alpha; -\infty, \beta)$, even if empty, are called *plane-quarters*.

C.2. Def. The union of a finite number of rectangles will be called *figure*[1].

Theorem. One can prove that every figure is the union of a finite collection of mutually disjoint generalized rectangles.

One also can prove that the complement $\text{co} f =_{df} P - f$ of a figure f is also a figure.

C.2.1. Def. The collection $\boldsymbol{F}$ of all figures, ordered by the relation $(\subseteqq)$ of inclusion of sets, is a finitely additive tribe $(\boldsymbol{F})$: we call it *tribe of figures*. The relation $(\subseteqq)$-confined to $\boldsymbol{F}$, will be in a more suitable way denoted by $(\leqq)$. The operations in this tribe will be written: $+, \cdot, -, \text{co}, \sum, \prod$. The zero of $(\boldsymbol{F})$ will be denoted by O and the unit by I. The zero O is the empty set $\emptyset$ of points, and the unit I is the whole plane P.

C.2.2. Theorem. The collection $\boldsymbol{B}$ of all generalized rectangles is a *base of* $(\boldsymbol{F})$.

Indeed, $O \in \boldsymbol{B}$; $I \in \boldsymbol{B}$; if $a \in \boldsymbol{B}$, $b \in \boldsymbol{B}$, then $a \cdot b \in \boldsymbol{B}$; every soma of $(\boldsymbol{F})$ is the union of finite number of elements of $\boldsymbol{B}$. The elements of $\boldsymbol{B}$ are called *bricks*.

C.2.3. Theorem. ***(Hyp Af)*** is satisfied.

C.2.4. Theorem. $(\boldsymbol{F})$ is the smallest tribe of figures, containing all plane-quarters $Q(\alpha, \beta)$, i.e. it is generated by them. Indeed we have

$$R(\alpha_1, \alpha_2; \beta_1, \beta_2) = \{[Q(\alpha_2, \beta_2) - Q(\alpha_1, \beta_2)] + Q(\alpha_1, \beta_1)\} - Q(\alpha_2, \beta_1).$$

C.3. $(\boldsymbol{F})$ being a tribe and $\boldsymbol{B}$ its base, we can introduce, according to [Chap B I], the *traces*, ($\boldsymbol{B}$-*traces*). We consider *descending sequences of rectangles*:

$$a_1 \geqq a_2 \geqq \cdots \geqq a_n \geqq \cdots,$$

denoted by $a_{\dot{n}}$ or $\{a\}$.*) We say that *the descending sequence* $\{a\}$ *is included in* $\{b\}$:

$$\{a\} \subset \{b\},$$

whenever for every n there exists m such that

$$a_m \leqq b_n.$$

[1] SAKS Integration (22).

*) The dot over n will emphasize that n is considered as variable.

We define the *equivalence* ($\sim$) *of the sequences* $\{a\}$, $\{b\}$

$$\{a\} \sim \{b\},$$

as $\{a\} \subset \{b\}$ and $\{b\} \subset \{a\}$.

A descending sequence $\{a\}$ is said to be *minimal* whenever

1) it is not equivalent to $\{O, O, \ldots\}$ and

2) if $\{b\} \subset \{a\}$, then either $\{b\} \sim \{O, O, \ldots\}$ or else $\{b\} \sim \{a\}$.

Later we shall prove the existence of minimal $\boldsymbol{B}$-sequences in $\boldsymbol{F}$.

C.3.1. By a *trace* (***B**-trace*) we shall understand a saturated set of mutually equivalent minimal descending sequences.

C.4. In the general theory of traces [B I.] we have proved that every trace generates a filter base, such that the corresponding filter is an ultrafilter. We have also seen, that an ultrafilter generates a well determined finitely additive measure on $(\boldsymbol{F})$ admitting two values 0 and 1 only. Conversely such a two-valued measure determines an ultrafilter. To every ultrafilter in $(\boldsymbol{F})$ corresponds a trace.

Information. Though in our case the above takes place, we prefer to give an independent proof, because we shall see at the same time, what the plane traces look like.

C.4.1. Let $\nu(f)$ be a finitely additive measure on $\boldsymbol{F}$, admitting two values 0, 1 only. We suppose that this measure is not trivial, i.e. not equal 0 identically. Thus we have for every two disjoint figures f, g

$$\nu(f+g) = \nu(f) + \nu(g). \tag{1}$$

We have

$$\nu(0) = 0, \quad \nu(P) = 1. \tag{2}$$

If $f \leqq g$, then

$$\nu(f) \leqq \nu(g). \tag{3}$$

(3.0) Two disjoint figures cannot both have the ν-measure 1.

Information. The discussion will be based on considering half-planes and their measure ν. Then we shall proceed to strips and their ν-measure. It will be shown, that the structure of the measure is characterized by what we shall call vertex and characteristic.

C.4.2. Consider the vertical half-planes

$$D_v(\alpha) =_{df} \{(x, y) \mid x \leqq \alpha\}, \quad (-\infty \leqq \alpha \leqq +\infty).$$

Since $D_v(+\infty) = P$, we have $\nu[D_v(+\infty)] = 1$, hence there exists α such that $\nu[D_v(\alpha)] = 1$.

Put, [C.1.1.],

$$\alpha_0 =_{df} \inf\{\alpha \mid \nu[D_v(\alpha)] = 1\}. \tag{3.1}$$

There exists an infinite sequence

$$\alpha_1 \geqq \alpha_2 \geqq \cdots \geqq \alpha_n \geqq \cdots$$

with $\lim \alpha_n = \alpha_0$ and with

(4) $$\nu[D_v(\alpha_n)] = 1, \quad (n = 1, 2, \ldots).$$

If $\alpha_0 < \alpha$, we have for a suitable n

$$\alpha_n \leqq \alpha.$$

Hence, by (3),

$$\nu[D_v(\alpha)] \geqq \nu[D_v(\alpha_n)];$$

hence, by (4),

$$\nu[D_v(\alpha)] = 1.$$

Thus we have proved that if $\alpha_0 < \alpha$, then

(5) $$\nu[D_v(\alpha)] = 1.$$

Let $\alpha < \alpha_0$. We cannot have $\nu[D_v(\alpha)] = 1$, because this would contradict (3.1). Thus we have proved that if $\alpha < \alpha_0$, then

(6) $$\nu[D_v(\alpha)] = 0.$$

C.4.2a. We shall study the vertical strips

$$B_v(\alpha_1 < \alpha_2) =_{df} \{(x, y) \mid \alpha_1 < x \leqq \alpha_2\},$$

[C.1.4.], in order to get some conditions, under which their ν-measure $= 1$. To do that we must consider two cases:

(7) Case I. $$\nu[D_v(\alpha_0)] = 1,$$

(8) Case II. $$\nu[D_v(\alpha_0)] = 0.$$

C.4.2b. Considering the case I, (7), we shall prove the

Lemma. We have

(9) $$-\infty < \alpha_0 \leqq +\infty,$$

and if $\alpha_1 < \alpha_2$, then the following are equivalent:

I′) The strip $B_v(\alpha_1 < \alpha_2)$ contains some strip $B_v(\gamma < \alpha_0)$, where $\gamma < \alpha_0$,

II′) $\nu[B_v(\alpha_1 < \alpha_2)] = 1$.

Proof. We cannot have $\alpha_0 = -\infty$. Indeed, if $\nu[D_v(-\infty)] = 1$, we have

$$\nu[\{(x, y) \mid -\infty < x \leqq -\infty\}] = 1,$$

but the set within the brackets is empty, so its measure $= 0$. This is a contradiction.

Thus we have proved that

$$-\infty < \alpha_0 \leqq +\infty,$$

i.e. (9).

C.4.2c. Suppose I′) and let γ, where $\gamma < \alpha_0$, be such that the strip $B_v(\gamma, \alpha_0)$ is contained in $B_v(\alpha_1 < \alpha_2)$. We have

$$\alpha_1 \leqq \gamma < \alpha_0 \leqq \alpha_2 .$$

We have

$$B_v(\gamma, \alpha_0) = D_v(\alpha_0) - D_v(\gamma),$$

and

$$D_v(\gamma) \leqq D_v(\alpha_0) .$$

Hence

$$\nu[B_v(\gamma, \alpha_0)] = \nu[D_v(\alpha_0)] - \nu[D_v(\gamma)] = 1 - 0 = 1$$

by (6) and (7). Since

$$B_v(\gamma, \alpha_0) \leqq B_v(\alpha_1 < \alpha_2),$$

we get

$$\nu[B_v(\alpha_1 < \alpha_2)] = 1 .$$

(10) Thus II′) follows from I′).

C.4.2d. Now suppose II′), i.e.

$$\nu[B_v(\alpha_1 < \alpha_2)] = 1 .$$

We cannot have $\alpha_2 < \alpha_0$.

Indeed, if $\alpha_2 < \alpha_0$, we have $\alpha_1 < \alpha_0$, and then, by (6)

$$\nu[D_v(\alpha_1)] = \nu[D_v(\alpha_2)] = 0 . \tag{11}$$

Since

$$B_v(\alpha_1 < \alpha_2) = D_v(\alpha_2) - D_v(\alpha_1),$$

we get, by (11),

$$\nu[B_v(\alpha_1 < \alpha_2)] = 0 - 0 = 0,$$

which is impossible.

Hence

$$\alpha_0 \leqq \alpha_2 . \tag{12}$$

We cannot have

$$\alpha_0 \leqq \alpha_1 .$$

Indeed, let $\alpha_0 \leqq \alpha_1$; hence $\alpha_0 \leqq \alpha_1 < \alpha_2$. We have

$$B_v(\alpha_1 < \alpha_2) = D_v(\alpha_2) - D_v(\alpha_1) .$$

We have

$$\nu[D_v(\alpha_2)] = 1 , \quad \nu[D_v(\alpha_1)] = 1 ,$$

by (5) and by (7). Hence

$$\nu[B_v(\alpha_1 < \alpha_2)] = 1 - 1 = 0,$$

which is a contradiction.

Thus we have

$$\alpha_1 < \alpha_0 . \tag{13}$$

If we combine this inequality with (12), we get

$$\alpha_1 < \alpha_0 \leqq \alpha_2 ,$$

so $B_v(\alpha_1 < \alpha_2)$ contains the strip $B_v(\alpha_1, \alpha_0)$, where $\alpha_1 < \alpha_0$.

(13.1) Thus II') implies I').

By (10) and (13.1) the lemma is proved in the case I) i.e. (7).

C.4.2e. Let us go over to the case II) i.e. where

$$\nu[D_v(\alpha_0)] = 0. \tag{8}$$

In that case II) we have $\alpha_0 < +\infty$.

C.4.2f. Indeed, suppose that $\alpha_0 = +\infty$. We have, by (8),

$$0 = \nu[D_v(\alpha_0)] = \nu\{(x, y) \,|\, x \leqq +\infty\} = \nu(P) = 1,$$

which is a contradiction. Thus we have

$$-\infty \leqq \alpha_0 < +\infty. \tag{14}$$

C.4.2g. Lemma. In the case II) considered, the following are equivalent:

I'') the strip $B_v(\alpha_1 < \alpha_2)$ contains a strip $B_v(\alpha_0, \gamma)$ for some γ, where $\alpha_0 < \gamma$;

II'') $\nu[B_v(\alpha_1 < \alpha_2)] = 1$.

Proof. Let $\alpha_0 < \gamma$, and suppose that $B_v(\alpha_1 < \alpha_2)$ contains the strip $B_v(\alpha_0, \gamma)$.

We have

$$\alpha_1 \leqq \alpha_0 < \gamma \leqq \alpha_2 .$$

We have, by (6),

$$\nu[D_v(\alpha_1)] = 0 \tag{15}$$

for $\alpha_1 < \alpha_0$. But, if $\alpha_1 = \alpha_0$, we have the same equality (15). We have $\nu[D_v(\alpha_2)] = 1$, by (15) and (5).

Since

$$B_v(\alpha_1 < \alpha_2) = D_v(\alpha_2) - D_v(\alpha_1),$$

it follows that

$$\nu[B_v(\alpha_1 < \alpha_2)] = \nu[D_v(\alpha_2)] - \nu[D_v(\alpha_1)] = 1 - 0 = 1,$$

so I'') implies II'').

Now suppose II'') i.e.

$$\nu[B_v(\alpha_1 < \alpha_2)] = 1.$$

We cannot have $\alpha_0 < \alpha_1$. Indeed, suppose that $\alpha_0 < \alpha_1$. Then we have $\alpha_0 < \alpha_1 < \alpha_2$. By (5) we have

$$\nu[D_v(\alpha_1)] = 1, \quad \nu[D_v(\alpha_2)] = 1. \tag{16}$$

Since

$$B_v(\alpha_1 < \alpha_2) = D_v(\alpha_2) - D_v(\alpha_1),$$

we get by (16),

$$\nu[B_v(\alpha_1 < \alpha_2)] = 1 - 1 = 0,$$

which is a contradiction.

Thus we have

(17) $$\alpha_1 \leqq \alpha_0.$$

We cannot have

$$\alpha_2 \leqq \alpha_0.$$

Indeed, suppose that

$$\alpha_2 \leqq \alpha_0.$$

Then

$$\alpha_1 < \alpha_2 \leqq \alpha_0.$$

By (6) we have

(18) $$\nu[D_v(\alpha_1)] = 0,$$

and, if $\alpha_2 < \alpha_0$, also $\nu[D_v(\alpha_2)] = 0$; but if $\alpha_2 = \alpha_0$, we have $\nu[D_v(\alpha_2)] = 0$ by (8).

Hence, whatever the case may be, we have

(19) $$\nu[D_v(\alpha_2) = 0.$$

From (18) and (19) we get

$$\nu[B_v(\alpha_1 < \alpha_2)] = \nu[D_v(\alpha_2)] - \nu[D_v(\alpha_1)] = 0 - 0 = 0.$$

Thus II″) implies I″), so the lemma is established.

C.5. We can get similar results in analogous two cases for horizontal strips

$$B_h(\beta_1 < \beta_2).$$

We define

$$\beta_0 =_{df} \inf\{\beta \mid \nu[D_h(\beta)] = 1\},$$

and consider the two possible cases

I_h: $\nu[D_h(\beta_0)] = 1$,

II_h: $\nu[D_h(\beta_0)] = 0$.

In the case I_h we have $-\infty < \beta_0$, and the following are equivalent:

I′: $\nu[B_h(\beta_1 < \beta_2)] = 1$.

II′: The strip $B_h(\beta_1 < \beta_2)$ contains a strip $B_h(\delta < \beta_0)$ for some δ.

In the case II_h we have $\beta_0 < +\infty$ and the following are equivalent

I″. $\nu[B_h(\beta_1 < \beta_2)] = 1$.

II″. The strip $B_h(\beta_1 < \beta_2)$ contains a strip $B_h(\beta_0 < \delta)$ for some δ.

C.6. Given a two valued, not trivial measure $\nu(f)$ with values 0, 1, there are well determined numbers α_0, β_0 having the properties stated in the above lemmas.

We call the couple (α_0, β_0) *vertex of* $\nu(f)$.

Given (α_0, β_0), there are in general four possibilities, which we label

$$(0, 0), \ (1, 0), \ (0, 1), \ (1, 1),$$

and which are respectively defined by:

$$\begin{aligned}
(0, 0)&: \nu[D_v(\alpha_0)] = 0, \quad \nu[D_h(\beta_0)] = 0,\\
(1, 0)&: \nu[D_v(\alpha_0)] = 1, \quad \nu[D_h(\beta_0)] = 0,\\
(0, 1)&: \nu[D_v(\alpha_0)] = 0, \quad \nu[D_h(\beta_0)] = 1,\\
(1, 1)&: \nu[D_v(\alpha_0)] = 1, \quad \nu[D_h(\beta_0)] = 1.
\end{aligned}$$

The labels will be called *characteristic of the measure* ν. Till now we did not prove that all these possibilities can occure really, but let us notice that some possibilities are excluded. Indeed, by Lemma [C.4.2b.] the vertex $(-\infty, \beta)$ cannot be accompanied by the characteristic $(1, 0)$ or $(1, 1)$. The vertex $(+\infty, \beta)$ cannot be accompanied by the characteristic $(0, 0)$ or $(0, 1)$. The vertex $(\alpha, -\infty)$ cannot have neither $(0, 1)$ nor $(1, 1)$, and $(\alpha, +\infty)$ cannot have neither $(0, 0)$ nor $(1, 0)$. Thus

$(-\infty, \beta)$ can have the only characteristics $(0, 0), (0, 1)$,
$(+\infty, \beta)$ can have the only characteristics $(1, 0), (1, 1)$,
$(\alpha, -\infty)$ can have the only characteristics $(0, 0), (1, 0)$,
$(\alpha, +\infty)$ can have the only characteristics $(0, 1), (1, 1)$.

If α, β are ordinary numbers, we call these measures *side-measures at infinity*.

The vertices $(-\infty, +\infty), (-\infty, +\infty), (+\infty, -\infty), (+\infty, +\infty)$ can have only one characteristic, viz. $(0, 0), (0, 1), (1, 0), (1, 1)$ respectively. We call these measures *corner-measures at infinity*.

If both α_0, β_0 are finite, all four characteristic are, in principle, possible.

Later we shall prove that all above cases can really occur.

C.7. Now we are going to study generalized rectangles:

$$R =_{df} R(\alpha_1, \alpha_2; \beta_1, \beta_2), \quad \text{where} \quad \alpha_1 < \alpha_2, \quad \beta_1 < \beta_2.$$

We have

(1) $$R = B_v(\alpha_1 < \alpha_2) \cdot B_h(\beta_1 < \beta_2).$$

If $\nu(R) = 1$, we have

(2) $$\nu[B_v(\alpha_1 < \alpha_2)] = 1 \quad \text{and} \quad \nu[B_h(\beta_1 < \beta_2)] = 1,$$

because if one of these measures were $= 0$, we would have, by [[C.4.1.], (3)], $\nu(R) = 0$.

Conversely if (2), we have

(3) $$\nu(R) = 1.$$

Indeed, if $\nu(R) = 0$, we have

$$\nu[B_v(\alpha_1 < \alpha_2) - R] = 1,$$
$$\nu[B_h(\beta_1 < \beta_2) - R] = 1.$$

Since the figures in brackets are disjoint they cannot have both the measure 1, so we get a contradiction.

C.7.1. Let (α_0, β_0) be the vertex of $\nu(\dot{f})$. There may be four characteristics possible, but we shall confine ourselves e.g. to the characteristic $(1, 0)$, because the others can be treated in an analogous way. By the lemmas, established in [C.5.], [C.4.2g.], the strips with measure 1 are $B_v(\alpha_1 < \alpha_2)$, which contains a strip $B_v(\gamma < \alpha_0)$, and $B_h(\beta_1 < \beta_2)$ which contain a strip $B_h(\beta_0 < \delta)$. Hence the rectangle $R(\alpha_1, \alpha_2; \beta_1, \beta_2)$, having the measure 1, must contain a rectangle

$$R(\gamma, \alpha_0; \beta_0, \delta) \quad \text{with} \quad \gamma < \alpha_0 \quad \text{and} \quad \beta_0 < \delta. \tag{4}$$

Informations. Our discussion of two valued measures has shown that the rectangles R with measure $v(R) = 1$, are well determined by the vertex and characteristic of the measure. They are suggesting to consider special kinds of diminishing sequences with the hope of getting representatives of traces. Now we shall study such sequences independently, without taking account of ν, and we shall see that they really will yield ***B***-traces. The results will be stated in [C.8.], [C.8.1.].

C.7.1a. We shall study the rectangles (4), by taking for γ an infinite sequence of generalized numbers

$$\gamma_1 < \gamma_2 < \cdots < \gamma_n < \cdots \quad \text{with} \quad \lim \gamma_n = \alpha_0, \tag{5}$$

and by taking for δ an infinite sequence of generalized numbers

$$\delta_1 > \delta_2 > \cdots > \delta_n > \cdots \quad \text{with} \quad \lim \delta_n = \beta_0. \tag{6}$$

Thus we have the infinite sequence of rectangles

$$R_n =_{df} R(\gamma_n, \alpha_0; \beta_0, \delta_n), \tag{7}$$

whose measures are $= 1$.

C.7.1b. We shall prove independently and without taking account of ν, that every descending sequence of generalized rectangles (7) with conditions (5), (6), is a minimal sequence.

From (5) and (6) it follows that

$$R_1 \geqq R_2 \geqq \cdots \geqq R_n \geqq \cdots, \tag{8}$$

where all $R_n \neq O$, so R_n is not $(\sim)$-equivalent to

$$\{O, O, \ldots\}. \tag{9}$$

Take any descending sequence of rectangles $R'_{\dot{n}}$ where

(9.1) $$R'_{\dot{n}} \subset R'_{\dot{m}},$$

[C.3.]. By definition, for every n we can find m, such that $R'_m \leqq R_n$. In these formula m can be replaced by any index $p \geqq m$, because R' is a descending sequence. Consequently we can find a sequence of indices

(10) $$k(1) < k(2) < \cdots,$$

such that

(11) $$R'_{k(n)} \leqq R_n, \quad (n = 1, 2, \ldots).$$

$R'_{k(n)}$ is a descending sequence not equivalent to $\{O, O, \ldots\}$. Let

(12) $$R'_{k(n)} = R(\alpha'_n, \alpha''_n; \beta'_n, \beta''_n),$$

where

$$\alpha'_n < \alpha''_n, \quad \beta'_n < \beta''_n.$$

We have, by (11) and (12),

(13) $$\gamma_n \leqq \alpha'_n < \alpha''_n \leqq \alpha_0,$$

(13.1) $$\beta_0 \leqq \beta'_n < \beta''_n \leqq \delta_n.$$

We have

(14) $$(\alpha'_1, \alpha''_1\rangle \geqq (\alpha'_2, \alpha''_2\rangle \geqq \cdots *),$$

and

(15) $$(\beta'_1, \beta''_1\rangle \geqq (\beta'_2, \beta''_2\rangle \geqq \cdots$$

We also have

(16) $$\lim \gamma_n = \lim \alpha'_n = \lim \alpha''_n = \alpha_0, \\ \lim \delta_n = \lim \beta'_n = \lim \beta''_n = \beta_0.$$

C.7.1 c. I say that

$$\alpha''_n = \alpha_0, \quad \beta'_n = \beta_0 \quad \text{for all } n.$$

Indeed, we have by (13), $\alpha''_n \leqq \alpha_0$. Suppose that $\alpha''_{n_0} < \alpha_0$.

Then, by (14), we have

$$\alpha''_{n_0+k} < \alpha''_{n_0} \quad \text{for} \quad k \geqq 0,$$

and then

$$\lim \alpha''_n \leqq \alpha''_{n_0} < \alpha_0$$

which contradicts (16). Thus

(17) $$\alpha''_n = \alpha_0 \quad \text{for all } n.$$

In a similar way we prove that

(18) $$\beta'_n = \beta_0 \quad \text{for all } n.$$

---

*) $(p, q\rangle$ means the half-open interval $\{x \mid p < x \leqq q\}$.

It follows that the sequence (14) is

$$(\alpha'_1, \alpha_0\rangle \geqq (\alpha'_2, \alpha_0\rangle \geqq \cdots,$$

and that in (15) is

$$(\beta_0, \beta''_1\rangle \geqq (\beta_0, \beta''_2\rangle \geqq \cdots$$

It follows that (12) has the form

(19) $$S_n =_{df} R'_{k(n)} = R(\alpha'_n, \alpha_0; \beta_0, \beta''_n), \quad (n = 1, 2, \ldots)$$

where

(20) $$\alpha'_n < \alpha_0, \quad \beta_0 < \beta''_n, \quad \lim \alpha'_n = \alpha_0, \quad \lim \beta''_n = \beta_0.$$

C.7.2. Compare the sequences

$$\gamma_1 < \gamma_2 < \cdots < \alpha_0 \quad \text{and} \quad \alpha'_1 < \alpha'_2 < \cdots$$

They have the same limit α_0.

Hence for every n there exists m_0 such that, if $m \geqq m_0$, we have $\gamma_m > \alpha'_n$.

Similarly, since the sequences

$$\delta_1 > \delta_2 > \cdots \quad \text{and} \quad \beta''_1 > \beta''_2 > \cdots$$

have the same limit β_0, therefore for every n there exists p_0 such that, if $p \geqq p_0$, then

$$\delta_p < \beta''_n.$$

Hence, we can find indices $l(n)$ such that

$$l(1) < l(2) < \cdots,$$

and where

$$\gamma_{l(n)} > \alpha'_n, \quad \delta_{l(n)} < \beta''_n.$$

Hence we get

$$(\gamma_{l(n)}, \alpha_0\rangle \subseteqq (\alpha'_n, \alpha_0\rangle, \quad (\beta_0, \delta_{l(n)}\rangle \subseteqq (\beta_0, \beta''_n\rangle,$$

which implies:

$$R(\gamma_{l(n)}, \alpha_0; \beta_0, \delta_{l(n)}) \subseteqq R(\alpha'_n, \alpha_0; \beta_0, \beta''_n) = R'_{k(n)} = S_n$$

for all n.

Since

$$R(\gamma_{l(n)}, \alpha_0; \beta_0, \delta_{l(n)}) \sim R(\gamma_n, \alpha_0; \beta_0, \delta_n)$$

it follows:

$$R_{\dot{n}} \subset R'_{k(\dot{n})}.$$

But, $R'_{k(\dot{n})} \sim R'_{\dot{n}}$. Hence $R_{\dot{n}} \subset R'_{\dot{n}}$. Taking account of (9.1), we see that $R_{\dot{n}} \sim R'_{\dot{n}}$. Thus we have proved that $R_{\dot{n}}$ is a minimal sequence.

C.8. We have proved the following

Theorem. If

$$\gamma_1 < \gamma_2 < \cdots \quad \text{with} \quad \lim \gamma_n = \alpha_0,$$

and

$$\delta_1 > \delta_2 > \cdots \quad \text{with} \quad \lim \delta_n = \beta_0,$$

then the sequence of rectangles

$$R_n =_{df} R(\gamma_n, \alpha_0; \beta_0, \delta_n)$$

is a minimal descending sequence of bricks of the tribe $(\boldsymbol{F})$. We notice that in the above arguments, (α_0, β_0) could be chosen arbitrarily, without taking account of a two-valued measures.

C.8.1. In the above development we have admitted the symbol $(1, 0)$ as characteristic, but we can see, that for other characteristics $(1, 1)$, $(0, 1)$, $(0, 0)$ the arguments are analogous, whenever the characteristic is admissible for the given (α_0, β_0), (see [C.6.]).

For the chosen sequences

$$\gamma_1 < \gamma_2 < \cdots < \gamma_n \ldots, \qquad \lim \gamma_n = \alpha_0,$$
$$\delta_1 > \delta_2 > \cdots > \delta_n \ldots, \qquad \lim \delta_n = \beta_0,$$

we have got the minimal descending sequence

$$R(\gamma_{\dot{n}}, \alpha_0; \beta_0, \delta_{\dot{n}}). \tag{21}$$

If we take any other sequences

$$\gamma_1' < \gamma_2' < \cdots < \gamma_n' < \cdots, \qquad \lim \gamma_n' = \alpha_0,$$
$$\delta_1' > \delta_2' > \cdots > \delta_n' > \cdots, \qquad \lim \delta_n' = \beta_0,$$

the minimal sequence is

$$R(\gamma_{\dot{n}}', \alpha_0; \beta_0, \delta_{\dot{n}}'). \tag{22}$$

The sequences (21), (22) are equivalent, thus determining a same trace, which we may denote by $\tau[\alpha_0, \beta_0; 1, 0]$.

Similarly we can construct the traces

$$\tau[\alpha_0, \beta_0; 0, 1], \quad \tau[\alpha_0, \beta_0; 1, 1], \quad \tau[\alpha_0, \beta_0; 0, 0].$$

If $\alpha_0 = x_0$, $\beta_0 = y_0$, (finite numbers), there exist all four traces. If one at least of α_0, β_0 is $\pm\infty$, we have two or only one trace. We call them *traces at infinity*. The couple of generalized numbers (α_0, β_0), is called *vertex of the trace*. We can speak of *the characteristic of the trace*.

Two traces having the same vertex, but different characteristics, are called *neighbor-traces*.

We say that a trace is *not at infinity* whenever its vertex is a point of the plane P.

C.8.2. In accordance to the above notation we can also label the vertices of a generalized rectangle:

$$R(\lambda_0, \lambda_1; \mu_0, \mu_1)$$

as follows: the vertices

$$(\lambda_0 \mu_0), (\lambda_0, \mu_1), (\lambda_1 \mu_0), (\lambda_1, \mu_1)$$

may be given the "characteristics"

$$(0, 0), (0, 1), (1, 0), (1, 1).$$

C.8.3. Information. We have proved the existence of many $\boldsymbol{B}$-traces, though we did not prove the existence of a two-valued measure.

Information. Till now we do not know, whether the traces $\tau(\alpha_0, \beta_0; i, j)$ are all possible $\boldsymbol{B}$-traces, or not. But now, since we have proved the existence of $\boldsymbol{B}$-traces, we can proceed to find the corresponding two-valued measure on $(\boldsymbol{F})$: we shall consider an arbitrary $\boldsymbol{B}$-trace and the result will be stated in [C.11.].

C.9. Let τ be any $\boldsymbol{B}$-trace; e.g. one of those, whose existence has been proved. Since we do not know whether the traces $\tau(\alpha_0, \beta_0; i, j)$ are the only ones which exist, we do not make any hypothesis on τ. Let

(1) $$R_1 \geqq R_2 \geqq \cdots \geqq R_n \geqq \cdots,$$

be a representative of τ, where (1) is a minimal descending sequence of bricks.

Starting with (1), we shall construct a two-valued measure with 0 and 1. Let f be a figure, (soma of $\boldsymbol{F}$) and let us attach to it the number $\mu(f) = 1$, whenever there exists n, such that

(2) $$R_n < f,$$

and let us attach to f the number $\mu(f) = 0$ in the remaining case, i.e. whenever there does not exist any n with the property (2).

C.9a. We shall prove that $\mu(f)$ thus defined, is a non trivial measure admitting two values 0, 1 only.

First notice that if f', f'' are two disjoint figures, it is impossible that

(3) $$\mu(f') = \mu(f'') = 1.$$

To prove that, suppose the equality (3) is true. Then there exist n, m such that

$$R_n \leqq f', \quad R_m \leqq f''.$$

It follows that

$$R_n \cdot R_m \leqq f' \cdot f'' = O;$$

but this is impossible, because either $R_n \leqq R_m$, or $R_m \leqq R_n$, and no one of the rectangles considered is empty.

C.9b. We have: if

$$f' \leqq f'' \quad \text{then} \quad \mu(f') \leqq \mu(f'').$$

Indeed, if f' contains a rectangle R_k, so does f'' too, and then $\mu(f') = \mu(f'') = 1$. Consequently $\mu(f') \leqq \mu(f'')$. If f' does not contain any rectangle R_k, we have $\mu(f') = 0$ and then, whatever $\mu(f'')$ may be, $\mu(f') \leqq \mu(f'')$.

The proposition is proved.

C.9c. Let f, g be two disjoint figures, we shall prove that

$$\mu(f+g) = \mu(f) + \mu(g).$$

We have, [C.9b.], putting

$$h =_{df} f + g,$$

the inequalities

(4) $$\mu(f) \leqq \mu(h), \quad \mu(g) \leqq \mu(h).$$

If $\mu(h) = 0$, then both $\mu(f)$, $\mu(g)$ are 0; so

$$\mu(h) = \mu(f) + \mu(g).$$

Suppose that

(5) $$\mu(h) = 1.$$

Then there exists n such that

(6) $$\mu(R_n) = 1$$

and

(7) $$R_n \leqq h = f + g.$$

Hence

(8) $$R_n = R_n h = R_n f + R_n g.$$

We shall consider two cases:

1) $R_n f$ or $R_n g$ are O.

2) $R_n f \neq O$, $R_n g \neq O$.

Let

(9) $$R_n f = O.$$

Then, by (8),

$$R_n = R_n g;$$

hence $R_n \leqq g$, and then $\mu(g) = 1$.

Since f, g are disjoint, it follows by [C.9a.], that $\mu(f) = 0$. Hence we have

(10) $$\mu(f+g) = \mu(f) + \mu(g).$$

We get the same result, if we suppose that $R_n g = O$.

C.9d. Let

(11) $$R_n \neq O \quad \text{and} \quad R_n g \neq O,$$

and take (5).

We know that a figure is either a brick, or it is the finite union of mutually disjoint bricks. Thus we have

(12) $$f = a_1 + a_2 + \cdots + a_s, \quad g = b_1 + b_2 + \cdots + b_t,$$

where

$$s \geqq 1, \quad t \geqq 1,$$

$$R_{n+k} f = a_1 R_{n+k} + a_2 R_{n+k} + \cdots$$

$$R_{n+k} g = b_1 R_{n+k} + b_2 R_{n+k} + \cdots, \quad (k = 1, 2, \ldots).$$

By (11) at least one term $a_i R_{n+k}$ and at least one term $b_j R_{n+k}$ differs from O. Denote by A_{n+k} the set of all indices p, for which $a_p R_{n+k} \neq O$. Similarly denote by R_{n+k} the set of all indices q, for which $b_q R_{n+k} \neq O$. A_{n+k} and B_{n+k} are not empty.

We have

$$R_{n+k} f = \sum_{i \in A_{n+k}} a_i \cdot R_{n+k}, \quad k = 0, 1, \ldots$$

We see that

$$A_k \supseteqq A_{k+1} \supseteqq \cdots$$

Since all these sets are not empty and finite, there exists m_0 such that for all $m \geqq m_0$ we have $A_{m_0} = A_m$.

Thus for these indices m we have

(13) $$R_m f = a_1' R_m + a_2' R_m + \cdots,$$

where $(a_1', a_2', \ldots)$ is a subset of A_n, and where all terms differ from O.

In a similar way we get, for sufficiently great indices m, (we may admit $\geqq m_0$),

(14) $$R_m g = b_1' R_m + b_2' R_m + \cdots$$

where $(b_1', b_2', \ldots)$ is a subset of B_n, and where all terms differ from O.

We have supposed that $f \cdot g = O$. Hence all terms of (13) are disjoint with all terms in (14).

The sequence

(15) $$S_{\dot{n}} =_{df} \{a_1' R_{m_0} \geqq a_1' R_{m_0+1} \geqq \cdots\}$$

is a diminishing sequence, included in $R_{\dot{n}}$. Since $R_{\dot{n}}$ is minimal, we have

$$S_{\dot{n}} \sim R_{\dot{n}}.$$

In a similar way, we prove that, if we define

$$T_{\dot{n}} =_{df} \{b_1' R_{m_0} \geqq b_2' R_{m_0+1} \geqq \cdots\},$$

we get $T_{\dot{n}} \sim R_{\dot{n}}$. It follows: $T_{\dot{n}} \sim S_{\dot{n}}$. Since the elements of $S_{\dot{n}}$ are disjoint with the elements of $T_{\dot{n}}$, none of the relations

$$b'_j R_n \leqq a'_i R_m,$$
$$a'_i R_m \leqq b'_j R_m$$

can take place. Thus we have got a contradiction, showing that the hypothesis (11) cannot take place. Since in remaining cases the additivity of μ has been proved, the additivity of μ is established.

C.10. Starting with $R_{\dot{n}}$ we have constructed a two-valued measure $\mu(\dot{f})$ on $(\boldsymbol{F})$.

We can prove that if, instead of $R_{\dot{n}}$, we take any minimal sequence $R'_{\dot{n}}$, $(\sim)$-equivalent to $R_{\dot{n}}$, we shall get the same measure $\mu(\dot{f})$.

Denote by $\mu'(\dot{f})$ the measure obtained by considering the minimal sequence $R'_{\dot{n}}$. Let g be any figure with $\mu(g) = 1$. Then, by definition of μ, there exists n such that

$$R_n \subseteqq f.$$

As $R_{\dot{n}} \sim R'_{\dot{n}}$, there exists m with $R'_m \subseteqq R_n$. Hence $\mu'(f) = 1$. Thus if $\mu(g) = 1$, we have also $\mu'(g) = 0$. Similarly we prove that if $\mu'(f) = 1$, we get $\mu(f) = 1$.

This proves that the equality $\mu(f) = \mu'(f)$ is valid for all figures f.

C.10.1. We see that the measure $\mu(j)$ depends only on the chosen trace τ. Let us remark, that if we choose for τ any trace $\tau(\alpha_0, \beta_0; i, j)$, we get a well defined measure. Thus we have, at the same time, proved the existence of a two valued non trivial measure on $(\boldsymbol{F})$.

C.10.2. Let us consider the measure $\mu(\dot{f})$, obtained above by means of the chosen trace τ.

By virtue of [C.8.1.] the measure μ generates a well defined trace $\tau' = \tau(\alpha_0, \beta_0; i, j)$ where (i, j) is the characteristic of μ.

We may suppose that e.g. $(i, j) = (1, 0)$, because the remaining case can be treated in an analogous way. The trace τ' possesses a representative

(1) $$R'_{\dot{n}} =_{df} R(\gamma_{\dot{n}}, \alpha_0; \beta_0, \delta_{\dot{n}}), \quad \text{where} \quad \gamma_n < \alpha_0,$$
$$\beta_0 < \delta_n, \quad \lim \gamma_n = \alpha_0, \quad \lim \delta_n = \beta_0, \quad (n = 1, 2, \ldots).$$

A necessary and sufficient condition that a rectangle S has the μ-measure $= 1$, is that there exists n such that

$$R'_n \subseteqq S.$$

On the other hand, a necessary and sufficient condition that $\mu(S) = 1$ is that there exists an index m, such that $R_m \subseteqq S$. The rectangles R_n or R'_n have all the μ-measure 1.

Hence given n, there exists m such that

$$R'_m \leqq R_n,$$

and given p, there exists q such that

$$R_q \leqq R'_p.$$

It follows that $R'_n \sim R_n$, thus defining the same trace. We have proved that

$$\tau = \tau'.$$

C.11. Theorem. All existing $\boldsymbol{B}$-traces are the traces $\tau(\alpha_0, \beta_0; i, j)$ where $i = 0, 1$, $j = 0, 1$, and all two valued (with $0, 1$) non trivial measures are those obtained from these traces. There exists a $1 \to 1$ correspondence between the $\boldsymbol{B}$-traces and the $0, 1$-valued non trivial measures on $(\boldsymbol{F})$.

C.12. If

1) $\tau = \tau(x, y; i, j)$, $(i = 0, 1, j = 0, 1)$ is a $\boldsymbol{B}$-trace not at infinity, i.e. if $(x, y) \in P$,

2) $R_1 \supseteqq R_2 \supseteqq \cdots \supseteqq R_n \supseteqq \cdots$

is representative of τ,

then

1. the diamter of R_n tends to 0,
2. vert $\tau \in \bar{R}_n$, $n = 1, 2, \ldots$ ($\bar{R}_n$ means the closure of R_n),
3. there exists n_0, such that for every $n \geqq n_0$ the point vert τ is a vertex of all the rectangles R_n.

C.12a. If

1. $(x, y) \in P$,
2. $R_1 \supseteqq R_2 \supseteqq \cdots \supseteqq R_n \supseteqq \cdots$

are half-open bounded rectangles, such that

1) the diameter of R_n tends to 0 for $n \to \infty$,

2) they have the common vertex (x, y) for sufficiently great n, then R_n is a representative of a trace $\tau(x, y; i, j)$, which is not at infinity.

C.12.1. If R is a half-open bounded rectangle and $\tau(x, y; i, j)$ is a $\boldsymbol{B}$-trace, covered by R, then its vertex belongs to the closure $\bar{R}$ of R. Every point of $\bar{R}$ is the vertex of a suitable trace which is covered by R.

Proof. We consider all possible traces not at infinity

$$\tau(x, y; 0, 0),\ \tau(x, y; 0, 1),\ \tau(x, y; 1, 0),\ \tau(x, y; 1, 1)$$

with vertex (x, y) in the interior of the rectangle and also the traces with vertices on its boundary.

C.13. After having been oriented in the variety of $\boldsymbol{B}$-traces, we could proceed to measurable sets of traces, in agreement with the general theory exhibited in [B I.]. However we shall not need it now, but only coverings of sets of $\boldsymbol{B}$-traces. As in a general theory, we say that a $\boldsymbol{B}$-trace *is covered by a brick* R, whenever there exists a representative R_n of τ, such that $R_1 \subseteqq R$. Then we have $R_n \subseteqq R$ for all n.

By a covering we shall understand any denumerable union of bricks

$$\bigcup_n a_n, \quad \text{[B I.3.]},$$

where the union is the ordinary set-union of these rectangles, i.e. the sum is taken in the tribe $(\boldsymbol{F}')$ of all borelian subsets of the plane P.

We say that the *trace* τ *is covered by the covering* $L = \bigcup_n a_n$. Whenever there exists a brick b which covers τ and where $b \subseteqq \bigcup_n a_n$ [B I.2.5.].

We say that a covering L is a *covering of the set* X *of* $\boldsymbol{B}$-traces, whenever every trace τ of X is covered by L, [B I.4.].

C.14. Let $\mu(f)$, where f is a variable figure, be a (simply) additive, non negative and not trivial measure on $(\boldsymbol{F})$, which may be not denumerably additive. The tribe $(\boldsymbol{F})$ is a finitely genuine subtribe of the tribe $(\boldsymbol{F}')$ of all borelian subsets of the plane P.

We would like to extend the measure μ to all somata of $(\boldsymbol{F}')$ in such a way that the obtained measure become denumerably additive. We know from [A I.10.] that this is not always possible. The following condition for doing that, is necessary and sufficient, using the relation of ordinary inclusion of sets:

N. If

$$a_1, a_2, \ldots, a_n, \ldots$$

are bricks, with

$$\bigcup_n a_n = P,$$

then

(1) $$\sum_n \mu(a_n) \geqq \mu(P), \quad \text{[A I.2.III.], [A I.9.14.], [A I.9.28.]}.$$

The inequality may be required to take place only in the case of convergence of the sum $\sum_n$, or, if we like it, even without this restriction, then considering ∞ as the value of the sum.

C.14.1. For the purpose of applications we shall consider the following condition **S** for the plane-quarters $Q(\alpha, \beta)$, where $-\infty \leqq \alpha$, $\beta \leqq +\infty$.

S. 1°. If $\alpha_1 \geqq \alpha_2 \geqq \cdots \geqq \alpha_n \geqq \cdots$ tends to α_0, and

$$\beta_1 \geqq \beta_2 \geqq \cdots \geqq \beta_n \geqq \cdots \text{ tends to } \beta_0,$$

then

$$\lim_{n \to \infty} \mu[Q(\alpha_n, \beta_n)] = \mu[Q(\alpha_0, \beta_0)].$$

2^0. If $\alpha_1 \leqq \alpha_2 \leqq \cdots \leqq \alpha_n \leqq \cdots$ tends to $+\infty$, and

$$\beta_1 \leqq \beta_2 \leqq \cdots \leqq \beta_n \leqq \cdots \text{ tends to } +\infty,$$

then

$$\lim_{n\to\infty} \mu[Q(\alpha_n, \beta_n)] = \mu(P).$$

Here $\alpha_n, \alpha_0, \beta_n, \beta_0$ are generalized numbers. We shall prove that the conditions **N** and **S** are logically equivalent.

C.14.2. First we prove that **S** implies **N**.

Proof. Let R_n, $(n = 1, 2, \ldots)$ be generalized rectangles such that

$$P = \bigcup_{n=1}^{\infty} R_n,$$

and let $\varepsilon > 0$; the number ε will be kept fixed in what follows. Choose a natural number N, so that

$$(1) \qquad 0 \leqq \mu(P) - \mu[Q(N, N)] \leqq \varepsilon.$$

This is possible by virtue of the condition 2^0, [C.14.1.]. By Condition 1^0, [C.14.1.], we can find numbers x', y' such that

$$-\infty < x' < N < +\infty,$$

$$-\infty < y' < N < +\infty,$$

$$0 \leqq \mu[Q(x', +\infty)] - \mu[Q(-\infty, +\infty)] \leqq \varepsilon,$$

$$0 \leqq \mu[Q(+\infty, y')] - \mu[Q(+\infty, -\infty)] \leqq \varepsilon.$$

Since

$$Q(-\infty, +\infty) = Q(+\infty, -\infty) = \emptyset,$$

we have

$$(2) \qquad \begin{aligned} 0 &\leqq \mu[Q(x', +\infty)] \leqq \varepsilon, \\ 0 &\leqq \mu[Q(+\infty, y')] \leqq \varepsilon. \end{aligned}$$

We have, [C.2.4.],

$$R(x', N; y', N) = \{[Q(N, N) - Q(x', N)] \cup Q(x', y')\} - Q(N, y'),$$

where

$$Q(x', N) \subseteqq Q(N, N),$$

$$Q(x', y') \cap [Q(N, N) - Q(x', N)] = \emptyset,$$

$$Q(N, y') \subseteqq [Q(N, N) - Q(x', N)] \cup Q(x', y').$$

It follows that

$$\begin{aligned} \mu R(x', N; y', N) &= \mu\{[Q(N, N) - Q(x', N)] \cup Q(x', y')\} - \mu Q(N, y') \\ &= \mu[Q(N, N) - Q(x', N)] + \mu[Q(x', y')] - \mu[Q(N, y')]; \end{aligned}$$

hence

$$(3)\quad \mu R(x', N; y', N) = \mu[Q(N, N)] - \mu[Q(x', N)] + \mu[Q(x', y')] - \\ - \mu[Q(N, y')].$$

We also have

$$Q(x', N) \subseteqq Q(x', +\infty), \quad Q(N, y') \subseteqq Q(+\infty, y'),$$
$$Q(x', y') \subseteqq Q(x', +\infty).$$

By (2) it follows

$$\mu[Q(x', N)] \leqq \varepsilon, \quad \mu[Q(N, y')] \leqq \varepsilon,$$
$$\mu[Q(x', y')] \leqq \varepsilon,$$

which gives, on account of (3),

$$(3.1)\qquad \mu[R(x', N; y', N)] - \mu[Q(N, N)] \leqq 3\varepsilon;$$

hence, by (1),

$$(4)\qquad \mu(P) - \mu[R(x', N; y', N] \leqq 4\varepsilon.$$

C.14.2a. Denoting by a bar the closure of sets, we get

$$(5)\qquad \overline{R(x', N; y', N)} \subseteqq \bigcup_{n=1}^{\infty} R_n.$$

Choose a sequence ε_n of positive numbers, such that

$$(5.1)\qquad \sum_{n=1}^{\infty} \varepsilon_n < \varepsilon,$$

and put

$$R_n =_{df} R(\alpha_n, \beta_n; \gamma_n, \delta_n).$$

By virtue of condition 1^0 there exists $\sigma_n > 0$, such that we have

$$(6)\quad \begin{cases} 0 \leqq \mu[R(\alpha_n, \beta_n + \sigma_n; \gamma_n, \delta_n + \sigma_n)] - \mu(R_n) \leqq \varepsilon_n, \\ \text{whenever} \\ \qquad \beta_n < +\infty, \quad \delta_n < +\infty; \\ 0 \leqq \mu[R(\alpha_n, \beta_n + \sigma_n; \gamma_n, \delta_n)] - \mu(R_n) \leqq \varepsilon_n,, \\ \text{whenever} \\ \qquad \beta_n < +\infty, \quad \delta_n = +\infty; \\ \text{and} \\ 0 \leqq \mu[R(\alpha_n, \beta_n; \gamma_n, \delta_n + \sigma_n)] - \mu(R_n) \leqq \delta_n, \\ \text{whenever} \\ \qquad \beta_n = +\infty, \quad \delta_n < +\infty. \end{cases}$$

The above inequalities can be obtained by applying formulas analogous to (3).

Denote by R'_n the correspondent amplified rectangles R_n, and R if $\beta_n = \delta_n = +\infty$.

In all these cases we have

$$R_n \leqq (R'_n)^o, \qquad (n = 1, 2, \ldots),$$

where the circle o means that we take the set of all points interior of R'_n.

From (5) and (6) it follows:

$$\overline{R(x', N; y', N)} \subseteqq \sum_{n=1}^{\infty} (R'_n)^o.$$

Since the set on the left is bounded and closed, the Heine-Borel theorem can be applied. There exists a finite sequence

$$n(1), n(2), \ldots, n(k), \qquad (k \geqq 1),$$

of indices such that

$$\overline{R(x', N; y', N)} \subseteqq \sum_{i=1}^{k} (R'_{n(i)}).$$

Consequently

$$R(x', N; y', N) \subseteqq \sum_{i=1}^{k} R'_{n(i)}.$$

It follows:

$$\mu[R(x', N; y', N)] \leqq \sum_{i=1}^{k} \mu(R'_{n(i)}). \tag{7}$$

On the other hand we see that (6) implies

$$0 \leqq \mu(R'_{n(i)}) - \mu(R_{n(i)}) \leqq \varepsilon_{n(i)}.$$

Hence we have from (7) and (5.1):

$$\mu[R(x', N; y', N)] \leqq \sum_i \mu(R_{n(i)}) + \varepsilon;$$

hence, from (4), we get

$$\mu(P) \leqq \sum_{i=1}^{\infty} \mu(R_n) + 5\varepsilon.$$

Since this inequality holds true for every $\varepsilon > 0$, we obtain

$$\mu(P) \leqq \sum_{i=1}^{\infty} \mu(R_n),$$

which is just the condition **N**, [C.14.],

C.14.2b. Now we shall prove that, conversely, the condition **N** implies **S**, [C.14.1.].

By the theorems quoted in [C.14.], **N** is a necessary and sufficient condition for the possibility of extension of the measure μ to all somata of $(\boldsymbol{F}')$, in the way that the extended measure be denumerably additive.

Hence the statements 1° and 2° in [C.14.1.] follow. We have proved that the conditions **S** and **N** are equivalent. Hence the condition **S** is necessary and sufficient for the required possibility of extension of measure μ over the borelian extension $(\boldsymbol{F})'$ of $(\boldsymbol{F})^{\cdot}$.

C.15. Let us suppose that the measure $\mu(f)$ on $(\boldsymbol{F})$ satisfies the condition **N** in [C.14.], so that the measure μ can be extended to $(\boldsymbol{F}')$, i.e. there exists a measure $\mu'(E)$ for all borelian sets E, such that

1) if $f \in F$, then $\mu'(f) = \mu(f)$,

2) if $E_1, E_2, \ldots, E_n, \ldots$ are disjoint borelian sets,

then

$$\mu'(\bigcup_{n=1} E_n) = \sum_{n=1}^{\infty} \mu'(E_n).$$

C.15.1. The collection of all borelian sets E with $\mu'(E) = 0$ is a denumerable additive ideal in $(\boldsymbol{F}')$. Denote it by $\boldsymbol{J}'$. Let us reorganize $(\boldsymbol{F}')$ by introducing a new ordering $(\leqq^{J'})$, denoted by

$$A \leqq^{J'} B \cdot =_{df} \cdot B - A \in J'.$$

We get a new denumerably additive tribe $(\boldsymbol{F}_{J'})$ in which the governing equality is

$$A =^{J'} B \cdot =_{df} \cdot (B - A) + (A - B) \in J'.$$

It is equivalent to

$$A =^{J'} B \cdot =_{df} \cdot \mu'(A \dotplus B) = 0.$$

The collection of all figures f with measure $\mu(f) = 0$ is also an ideal $\boldsymbol{J}$, namely an ideal in $(\boldsymbol{F})$, so that it reorganizes $(\boldsymbol{F})$ into a new tribe $(\boldsymbol{F}_J)$, the tribe $(\boldsymbol{F})$ modulo J. We have $\boldsymbol{J} \subseteq \boldsymbol{J}'$.

We shall study the tribe $(\boldsymbol{F}_J)$.

C.16. Information. Since several items will not depend on the quality of $\boldsymbol{J}$, we shall take any general finitely additive ideal in $(\boldsymbol{F})$ which we shall call $\boldsymbol{J}$, and suppose only that $\boldsymbol{J} \neq \boldsymbol{F}$. Later we shall apply the obtained results to the ideal, defined by $\boldsymbol{J} =_{df} \{f \mid \mu f = 0\}$.

First of all, we can apply the general results obtained in [A.] and [B I.].

We define the *ordering in* $(\boldsymbol{F}_J)$ by

$$f =^J g \cdot =_{df} \cdot f - g \in \boldsymbol{J}$$

and the resulting *equality, governing in* $(\boldsymbol{F}_J)$, by

$$f =^J g \cdot =_{df} \cdot f \dotplus g \in \boldsymbol{J}.$$

The tribe $(\boldsymbol{F}_J)$ has a base whose elements (bricks) are the somata of $\boldsymbol{F}$, which are $(=^J)$-equal to the bricks of $\boldsymbol{B}$. In our case the base $\boldsymbol{B}_J$ of $(\boldsymbol{F}_J)$ is composed of generalized rectangles, taken modulo $\boldsymbol{J}$ i.e. the figure f is a soma of $\boldsymbol{B}_J$, whenever there exists a generalized rectangle R with

$$R =^J f.$$

We can take over the topics treated in [B I.] and apply to our case.

We have the $\boldsymbol{B}_J$-descending sequences

$$a_1 \geqq^J a_2 \geqq^J \cdots \geqq^J a_n \geqq^J \cdots$$

of $\boldsymbol{B}_J$-bricks. We have the notion of $(=^J)$-*equality of such sequences*, the notion of $\boldsymbol{J}$-*inclusion of such sequences*:

$$a_{\dot{n}} \subset^J b_{\dot{n}},$$

and of $\boldsymbol{J}$-*equivalence*

$$a_{\dot{n}} \sim^J b_{\dot{n}}.$$

The notion of an $\boldsymbol{J}$-*minimal sequence* is also taken from [B I.] and the same is with $\boldsymbol{B}_J$-*traces*, which are defined as saturated collections of mutually $\boldsymbol{J}$-equivalent $\boldsymbol{J}$-minimal sequences.

C.16.1. Let τ' be a $\boldsymbol{B}_J$-trace, and

$$a_1' \geqq^J a_2' \geqq^J \cdots \geqq^J a_n' \geqq^J \cdots$$

one of its representatives. Let us define the function $\lambda(f')$ defined for all $f' \in \boldsymbol{F}_J$ as follows:

We put $\lambda(f') = 1$, whenever there exists n such that

$$a_n' \leqq^J f',$$

and we put $\lambda(f') = 0$, in the remaining case. Using the same argument as in [C.9.], [C.9a.], [C.9b.], [C.9c.], [C.9d.], we prove that $\lambda(f')$ is an additive measure on $(\boldsymbol{F}_J)$, i.e. if $f' \cdot g' =^J O_J$, then $\lambda(f'+g') = \lambda(f') + \lambda(g')$. The operations in $(\boldsymbol{F}_J)$ are the same as in $(\boldsymbol{F})$, but we must just replace the equality $(=)$ and the inclusion $(\leqq)$ of somata of $(\boldsymbol{F})$ by the $(=^J)$ equality and the $(\leqq^J)$-inclusion. The argument similar to that in [C.10.], shows that the two-valued 0, 1-measure λ does not depend on the choice of the representative $a_{\dot{n}}'$ of τ'. The λ-measure on $(\boldsymbol{F}_J)$ is also a two-valued measure on $(\boldsymbol{F})$.

Indeed, if $f, g \in \boldsymbol{F}$ and $f \cdot g = O$, then

$$f, g \in \boldsymbol{F}_J \quad \text{and} \quad f \cdot g =^J 0_J;$$

hence

$$\lambda(f+g) = \lambda(f) + \lambda(g).$$

C.16.2. Now we know that the two-valued measure on $(\boldsymbol{F})$ generates a $\boldsymbol{B}_J$-trace τ, so we see that to every $\boldsymbol{B}_J$-trace τ' there corresponds a well defined trace $\tau \equiv \tau(\alpha, \beta; i, j)$. The trace τ generates a two-valued measure, which is identical to $\lambda(f)$ on $(\boldsymbol{F})$. This measure being also a measure on $(\boldsymbol{F}_J)$, we see that the correspondence, which carries the trace τ' into τ is $1 \to 1$. Let τ' be a $\boldsymbol{B}_J$-trace, and

(1) $$a_1 \geqq^J a_2 \geqq^J \cdots \geqq^J a_n \geqq^J \cdots$$

one of its representatives.

There exists a representative of τ' of the form

(2) $$b_1 \geqq b_2 \geqq \cdots \geqq b_n \geqq \cdots,$$

where

$$b_n =^J a_n, \quad b_n \in F, \quad n = 1, 2, \ldots$$

Hence the $\boldsymbol{B}_J$-descending sequences (1) and (2) are $(\sim^J)$-equivalent.

C.16.2a. Let τ correspond to τ' according to [C.16.2.], and let

(3) $$A_1 \geqq A_2 \geqq \cdots \geqq A_n \geqq \cdots$$

be a representative of τ.

From (3) it follows

(4) $$A_1 \geqq^J A_2 \geqq^J \cdots \geqq^J A_n \geqq^J \cdots$$

Let $\lambda(f)$ be the two-valued measure generated by (1) by means of the definition [C.16.1.]. This measure is also generated by (2), because (1) and (2) are $(\sim^J)$-equivalent. We have $\lambda(a_n) = 1$ and $\lambda(b_n) = 1$. But since the measures generated by τ' and τ are identical, therefore for every b_n there exists A_m with $A_m \leqq^J b_n$.

Since the same measure is generated by (3), and $\lambda(A_n) = 1$ for all n, therefore for every n there exists m such, that

$$b_m \leqq^J A_n.$$

It follows that the sequences $b_{\dot{n}}$ and $A_{\dot{n}}$ are $(\sim^J)$-equivalent.

Since $b_{\dot{n}}$ is a representative of τ', it follows that $A_{\dot{n}}$ is also a representative of τ'. But $A_{\dot{n}}$ is also a representative of τ. Hence $A_{\dot{n}}$ is a common representative of τ and τ'.

C.16.2b. Let $R_{\dot{n}}$, where $R_1 \geqq R_2 \geqq \cdots \geqq R_n \geqq \cdots$, are rectangles, be a common representative of τ' and τ. Then for all n we have $R_n \overline{\in} \boldsymbol{J}$, because $R_{\dot{n}}$ cannot be $(\sim^J)$-equivalent to the sequence $(O_J, O_J, \ldots)$.

Hence if τ corresponds to τ', and $R_{\dot{n}}$ is a representative of τ, then $R_n \overline{\in} \boldsymbol{J}$ for every n.

C.16.3. Conversely, if $R_{\dot{n}}$ is a representative of a $\boldsymbol{B}$-trace σ, and $R_n \overline{\in} \boldsymbol{J}$ for every n, then $R_{\dot{n}}$ is also a representative of a $\boldsymbol{B}_J$-trace.

C.16.3a. To prove that, let us notice that if λ is the 0, 1-valued measure generated by σ, then we have $\sigma(f) = 0$ for every $f \in \boldsymbol{J}$.

Indeed, if this were not true, there would exist $f \in \boldsymbol{J}$ such that $\sigma(f) = 1$. Hence, if we take a representative $R_{\dot{n}}$ of σ, there would exist an index m such that $R_m \leqq f$. It would follow $R_m \in \boldsymbol{J}$, because $f \in \boldsymbol{J}$. Consequently $\sigma(f) = 0$.

C.16.3b. Having that, we shall prove that if

$$f =^J g, \quad \text{then} \quad \lambda(f) = \lambda(g).$$

To prove this, notice that, [A.9.4a.],

$$f + p - q = g$$

for some $p, q \in \boldsymbol{J}$. We can suppose that $f \cdot p = O$ and $(f + p) \cdot q = O$.

Indeed $f + p = f + (p - f)$; so if we put $p' =_{df} p - f$, we get $p' \in \boldsymbol{J}$ and $f + p = f + p'$ with f, p' disjoint. Then

$$f + p - q = f + p' - q.$$

Now we can have $q \leqq f + p'$. Indeed, put $q' =_{df} q \cdot (f + p')$. Then $q' \in \boldsymbol{J}$ and $q' \leqq f + p'$.

Thus we get the formula $g = f + p' - q'$ with above properties. It follows that λ is an additive measure; hence

$$\lambda(g) = \lambda(f + p') - \lambda(q') = \lambda(f) + \lambda(p') - \lambda(q').$$

Now as $p', q' \in \boldsymbol{J}$, we have $\lambda(p') = \lambda(q') = 0$, by what has been proved before.

Hence $\lambda(g) = \lambda(f)$, and the assertion is proved. Consequently λ is a measure invariant with respect to the equality $(=^J)$ and it yields a $\boldsymbol{B}_J$-trace.

C.17. We have proved the following fundamental theorem on $\boldsymbol{B}_J$-traces.

Theorem. If

1. $(\boldsymbol{F})$ is the tribe of all figures in the plane P, as defined in [C.2.1.],
2. $\boldsymbol{J} \neq \boldsymbol{F}$ is an ideal in $(\boldsymbol{F})$,
3. $\boldsymbol{B}$ is a base of $(\boldsymbol{F})$, made of generalized rectangles of $\boldsymbol{F}$,
4. $\boldsymbol{B}_J$ is the corresponding base of $(\boldsymbol{F}_J)$, i.e. of the tribe $(\boldsymbol{F})$ taken modulo $\boldsymbol{J}$, and defined as the collection of all figures $\in \boldsymbol{F}$, which are $(=^J)$-equal to the bricks of $(\boldsymbol{F})$; then there exists a $1 \to 1$ correspondence between $\boldsymbol{B}_J$-traces and some $\boldsymbol{B}$-traces, such that the domain of the correspondence is composed of all $\boldsymbol{B}_J$-traces, and the region of all $\boldsymbol{B}$-traces, having a representative R_n with $R_n \bar{\in} \boldsymbol{J}$ for all $n = 1, 2, \ldots$

If the correspondence carries the $\boldsymbol{B}_J$-trace τ' into the $\boldsymbol{B}$-trace τ, then τ' and τ have a common representative

$$R_1, R_2, \ldots, R_n, \ldots$$

composed of rectangles.

The $\boldsymbol{B}$-traces which do not correspond to $\boldsymbol{B}_J$-trace, are characterized by the property, stating that if $R_{\dot{n}}$ is a representative of τ, then there exists n_0 such that for every $n \geqq n_0$, $R_n \in \boldsymbol{J}$. They will be termed *non-active* $\boldsymbol{B}$*-traces*.

The $\boldsymbol{B}$-traces τ, which are corresponding to some $\boldsymbol{B}_J$-traces, have the property that if $R_{\dot{n}}$ is a representative of τ, then $R_n \bar{\in} \boldsymbol{J}$. These $\boldsymbol{B}$-traces will be termed *active* or *valid* $\boldsymbol{B}$*-traces*.

C.17.1. Def. If τ' is carried into τ and $\tau = \tau(\alpha_0, \beta_0; i, j)$, where (α_0, β_0) is the vertex of τ and (i, j) its characteristic, we will call (α_0, β_0), the *vertex of* τ', and (i, j) *characteristic of* τ'.

C.18. Information. Having obtained comprehensive information concerning $\boldsymbol{B}$-traces and $\boldsymbol{B}_J$-traces, where $\boldsymbol{J} \neq \boldsymbol{F}$ is any ideal in $(\boldsymbol{F})$, [C.17.], we shall go over to sets of $\boldsymbol{B}_J$-traces, to their measurability and measure, everything according to the general theory of traces, developped in [B I.]. It will be not needed to repeat the whole text in our particular case, so we shall confine ourselves to the recallection of definitions and basic theorems.

Till now the ideal $\boldsymbol{J}$ was arbitrary, but now we must specify it, in reference to [C.14.] where we have introduced a finitely additive measure $\mu(\dot{f})$ on the tribe $(\boldsymbol{F})$. We wanted to extend the measure μ to the whole tribe $(\boldsymbol{F}')$, whose somata are all borelian subsets of the plane P. In this respect we have noticed, that an arbitrary measure μ may be not able to be extended to the tribe $(\boldsymbol{F}')$ so, as to become denumerably additive. Some conditions are needed, and we have found such a condition, called **N** in [C.14.], and another equivalent one, called **S** in [C.14.1.].

Now we admit the condition **S**, which states some properties of plane-quarters, and does extend the measure $\mu(\dot{f})$ to $\mu'(E)$ for all borelian subsets E, of the plane P. The new measure $\mu'(\dot{E})$ satisfies the condition

$$\mu(\dot{f}) = \mu'(\dot{f}) \quad \text{for all} \quad f \in \boldsymbol{F},$$

and the condition of denumerable additivity, i.e. if $E_1, E_2, \ldots, E_n, \ldots$ are disjoint borelian sets, then

$$\mu'\left(\bigcup_{n=1}^{\infty} E_n\right) = \sum_{n=1}^{\infty} \mu'(E_n).$$

In addition to that the measure $\mu'(\dot{E})$ is bounded. The measure μ' generates a denumerably additive ideal $\boldsymbol{J}'$, defined as the collection of all borelian sets E, for which $\mu'(E) = 0$. The measure μ on $(\boldsymbol{F})$ also generates an ideal $\boldsymbol{J}$, which is finitely additive and is composed of all figures $f \in \boldsymbol{F}$ with $\mu(\dot{f}) = 0$.

Of course we have $\boldsymbol{J} \subsetneqq \boldsymbol{J}'$, and even $\boldsymbol{J}' \cap \boldsymbol{F} = \boldsymbol{J}$. We have studied the tribe $(\boldsymbol{F}_J)$ for general $\boldsymbol{J}$. Now we have another tribe $(\boldsymbol{F}')_{J'}$, which is denumerably additive and the measure μ' is effective on it, [A.15.1.]. We shall apply the notions and results given in [C.16], [C.17.], but we shall deal only with an *extensible measure*, by admitting the condition **S**, [C.14.1.].

There is no need to study sets of $\boldsymbol{B}$-traces, but we shall study the sets of $\boldsymbol{B}_J$-traces and the corresponding theory of their measurability.

For this purpose we now have the tribe $(\boldsymbol{F}_J)$, plunged in the denumerably additive supertribe $(\boldsymbol{F}'_{J'})$, with denumerably additive effective measure. We start with coverings and their properties, everything according to the general theory of traces exhibited in [B I.].

C.18.1. Def. We say that the *$\boldsymbol{B}_J$-trace τ is covered by a $\boldsymbol{B}_J$-brick a,* whenever there exists a representative

$$a_1 \geqq^J a_2 \geqq^J \cdots \geqq^J a_n \geqq^J \cdots$$

of τ, such that $a_1 \leqq^J a$.

It follows $a_n \leqq^J a$ for $n = 1, 2, \ldots$

Def. By *a $\boldsymbol{B}_J$-covering* we understand any denumerable union of $\boldsymbol{B}_J$-bricks b_n, $\sum\limits_{n=1}^{\infty} b_n$, where the summation is taken from $(\boldsymbol{F}')_{J'}$.

If

$$(1) \qquad b_n =^J c_n, \quad (n = 1, 2, \ldots),$$

then

$$\sum_{n=1}^{\infty} b_n =^{J'} \sum_{n=1}^{\infty} c_n.$$

By, [A.9.4a.], there exist somata s_n of $\boldsymbol{J}$, such that

$$b_n + s_n = c_n + s_n, \quad (n = 1, 2, \ldots).$$

It follows by using summations from $(\boldsymbol{F}')_{J'}$:

$$\sum_{n=1}^{\infty} b_n + \sum_{n=1}^{\infty} s_n = \sum_{n=1}^{\infty} c_n + \sum_{n=1}^{\infty} s_n.$$

We have $s_n \in \boldsymbol{J}'$, hence $\sum\limits_{n=1}^{\infty} s_n \in \boldsymbol{J}'$, because $\boldsymbol{J}'$ is a denumerably additive ideal in $(\boldsymbol{F}')_{J'}$.

Hence if we put

$$s =_{df} \sum_{n=1}^{\infty} s_n,$$

we get

$$\sum_{n=1}^{\infty} b_n + s = \sum_{n=1}^{\infty} c_n + s;$$

so

$$\sum_{n=1}^{\infty} b_n =^{J'} \sum_{n=1}^{\infty} c_n,$$

i.e. these two coverings differ by a set of μ'-measure 0. Thus in a covering $\sum_n a_n$, we can replace the $\boldsymbol{B}_J$-bricks a_n by others b_n, which are $(=^J)$-equal to a_n.

C.18.2. Def. We say that *a $\boldsymbol{B}_J$-trace is covered by a $\boldsymbol{B}_J$-covering L*, whenever there exists a $\boldsymbol{B}_J$-brick a, which covers τ, and for which $a \leqq^J L$. This notion is invariant with respect to $(=^J)$.

C.18.3. Def. We say that a *set E of $\boldsymbol{B}_J$-traces is $\boldsymbol{B}_J$-covered by the $\boldsymbol{B}_J$-covering L*, whenever for every $\tau \in E$, τ is $\boldsymbol{B}_J$-covered by L.

This notion is invariant with respect to $(=^J)$. To develop the theory of measurability of sets of $\boldsymbol{B}_J$-traces, we need the following two conditions being satisfied [B I.6.]:

C.19. Condition. *Hyp* I. If

1) X, Y are sets of $\boldsymbol{B}_J$-traces, such that $X \cup Y = W^*$, where W^* is the collection of all $\boldsymbol{B}_J$-traces,

2) K, L are $\boldsymbol{B}_J$-coverings of X and Y respectively, then $K \cup L =^J P$, where P is the whole plane.

C.20. Condition. *Hyp* II. If a is a $\boldsymbol{B}_J$-brick and X the set of all $\boldsymbol{B}_J$-traces, which are $\boldsymbol{B}_J$-covered by a, then the set

$$\mathrm{co}_{W'} X = W^* - X \quad \text{is } \boldsymbol{B}_J\text{-covered by} \quad \mathrm{co}_P a =_{df} P - a.$$

C.20.1. Proof of *Hyp* I. We shall prove that the following theorem, which is clearly equivalent to the statement saying that ***Hyp* I** takes place. The theorem is this:

If L is a $\boldsymbol{B}_J$-covering of the set W^* of all $\boldsymbol{B}_J$-traces, then

$$L =^{J'} P. \tag{1}$$

Suppose this be not true. Hence

$$\mu'(P - L) > 0. \tag{2}$$

Let $\mathfrak{M}$ be the mapping of the plane P onto the open square

$$Q =_{df} \left\{(x, y) \mid -\frac{\pi}{2} < x < +\frac{\pi}{2}, \quad -\frac{\pi}{2} < y < +\frac{\pi}{2}\right\},$$

defined as

$$(x, y) \curvearrowright^{\mathfrak{M}} (\mathrm{arc\,tan}\, x,\ \mathrm{arc\,tan}\, y).$$

Borelian subsets of $\mathsf{D}\,\mathfrak{M}$ and those of $\mathsf{G}\,\mathfrak{M} = P$ are in $1 \to 1$ correspondence.

The above transformations carries lines parallel to the axes into parallel, so it transforms rectangles into rectangles. Let us consider the open square, which is the image Q of the whole plane obtained by the transformation $\mathfrak{M}$. We define for every figure in Q, the measure

$$\mu''(f') =_{df} \mu(\mathfrak{M}^{-1} f').$$

From (2) it follows that there exists a point a in the closed square $\bar{Q}$, such that for every its neighborhood U of square shape, we have $\mu''(U) > 0$. If we have a sequence $U_1 \geqq U_2 \geqq \cdots$ of neighborhoods of a with diameter tending to 0, we shall get as $\mathfrak{M}^{-1}$-image a representative of an $\boldsymbol{J}$-trace (which may be even at infinity), such that no U_n has the μ-measure 0.

There may be 1, 2, 3 or 4 traces with vertex $\mathfrak{M}^{-1} a$. Take the case of four traces with vertex a; call then $\tau^1, \tau^2, \tau^3, \tau^4$. The remaining cases are even more simple and can be treated in an analogous way.

Let $R^1_{\dot{n}}, R^2_{\dot{n}}, R^3_{\dot{n}}, R^4_{\dot{n}}$ be their representatives with

$$R^i_n \subseteqq U_n, \quad (n = 1, 2, \ldots).$$

Since $\mu\, U_n > 0$ for all n, therefore for every n there exists an upper index $i = i(n)$ such that

$$\mu''\, R^{i(n)}_n > 0.$$

It follows that there exists an upper index j and a subsequence $k(\dot{n})$ of $\dot{n}$, such that

$$\mu''\, R^j_{k(n)} > 0 \quad \text{for all } n.$$

Thus there exists at least one trace τ^j having a representative, whose no element belongs to $\boldsymbol{J}$. Hence τ^j is a $\boldsymbol{B}_J$-trace which is not covered by L. The contradiction proves the equality (1), hence also ***Hyp*** **I**.

C.20.2. Proof of *Hyp* II.

We know that the ***Hyp*** **II** holds true when the condition **(*Hyp Af*)** is satisfied, [B I.22.]. The condition reads:

If $f' \in \boldsymbol{F}_J$, then there exists a finite number of $\boldsymbol{J}$-disjoint $\boldsymbol{B}_J$-bricks:

$$a'_1, a'_2, \ldots, a'_s$$

with

$$f =^J a'_1 + a'_2 + \cdots + a'_s.$$

Now this is true, because $f' \in \boldsymbol{F}$ too and there are $\boldsymbol{B}$-bricks

$$a_1, a_2, \ldots, a_t,$$

such that

(3) $$f = a_1 + a_2 + \cdots + a_t.$$

From (3) it follows

$$f' =^J a_1 + a_2 + \cdots + a_t,$$

where a_j can be replaced by any figure g_j, where $a_j =^J g_j$. The condition (***Hyp Af***) being satisfied, ***Hyp*** **II** follows.

C.21. Theorem. We shall prove that the tribe $(\boldsymbol{F}_J)$ is a finitely genuine subtribe of $(\boldsymbol{F}')_{J'}$.

Indeed, denote by primes the notions related to $(\boldsymbol{F}')_{J'}$. Let a, b, c be somata of $(\boldsymbol{F}_J)$, then the following are equivalent

1) $a + b =^J c$ and $a +' b =^{J'} c$,

2) $a \cdot b =^J c$ and $a \cdot' b =^{J'} c$. We also have

3) $O =^J O'$, $I =^{J'} I'$.

These relations follow from the fact that $\boldsymbol{J} \subseteqq \boldsymbol{J}'$.

C.22. Having all the notions and properties considered above, we can develop the theory of measurable sets of $\boldsymbol{B}_J$-traces, their coats and measure, just in the way indicated in [B I.]. We shall confine ourselves to the recallection of the definitions and theorems.

C.22.1. Def. Let X be a set of $\boldsymbol{B}_J$-traces. We call *outer-coat of X* the soma of $(\boldsymbol{F}')_{J'}$:

$$[X]^* =_{df} \prod L,$$

where the product is extended over all $\boldsymbol{B}_J$-coverings L of X.

The definition is meaningful, because the measure μ' on $(\boldsymbol{F}')_{J'}$ is not only denumerably additive, but also effective, so, by a theorem by WECKEN [A I., Section 1] all meets are meaningful. As in the general theory, the product $\prod L$ can be replaced by a suitable denumerable product $\prod\limits_{n=1}^{\infty} L_n$, with multiplication taken from the lattice $(\boldsymbol{F}')_{J'}$.

$[X]^*$ is a soma of $(\boldsymbol{F}')_{J'}$, hence a borelian set, taken modulo $\boldsymbol{J}'$.

Let W^* be the collection of all $\boldsymbol{B}_J$-traces.

We define *the inner coat* $[X]_*$ *of* X of $\boldsymbol{B}_J$-traces as

$$[X]_* =_{df} I - [Y]^*,$$

where I is the unit of $(\boldsymbol{F}')_{J'}$ [hence $(=^{J'})$-equal to P], and where

$$Y =^{J'} \mathrm{co}_{W^*} X =^{J'} W^* - X.$$

The inner coat is a borelian set, taken modulo J'. If

$$(1) \qquad [X]_* =^{J'} [X]^*,$$

we say that X is *measurable*, and (1) is written $[X]$, and called *coat of* X. The coat is a borelian set taken modulo $\boldsymbol{J}'$.

C.22.2. We define the *μ^*-measure of a measurable set* X of $\boldsymbol{B}_J$-traces, as

$$\mu^*(X) =_{df} \mu'([X]).$$

If

$$[X] = \prod_{n=1}^{\infty} L_n,$$

where L_n are coverings of X with

$$L_1 \geqq^{J'} L_2 \geqq^{J'} \cdots \geqq^{J'} L_n \geqq^{J'} \cdots,$$

we have

$$\mu^*(X) = \lim_{n\to\infty} \mu'(L_n).$$

C.23. Instead of *measurable* we may also say *μ^*-measurable*. We give a list of theorems and definitions concerning measurable sets of $\boldsymbol{B}_J$-traces, just repeating the correspondent items from the general theory of traces, developped in [B I.].

1) If a is a $\boldsymbol{B}_J$-brick, then the set X of all $\boldsymbol{B}_J$-traces, $\boldsymbol{B}_J$-covered by a, is measurable and we have

$$\mu^* X = \mu' a.$$

(Let us remind that if X is $\boldsymbol{B}_J$-covered by a, and $a =^J b$, then X is also $\boldsymbol{B}_J$-covered by b.)

2) The set composed of a single $\boldsymbol{B}_J$-trace is measurable. Its μ^*-measure may be positive (e.gr. *heavy trace*), and also may be 0, (e.gr. *elusive trace*).

3) The empty set $\emptyset^*$ of $\boldsymbol{B}_J$-traces is measurable, and we have

$$\mu^*(\emptyset^*) = 0.$$

3a) The collection W^* of all $\boldsymbol{B}_J$-traces is measurable, and we have

$$\mu^*(W^*) = \mu'(P).$$

4) If X is a measurable set of $\boldsymbol{B}_J$-traces, so is

$$\mathrm{co}_{W^*} X = W^* - X,$$

and we have

$$[\mathrm{co}_{W^*} X] =^{J'} \mathrm{co}_P [X],$$

$$\mu^*(\mathrm{co}_{W^*} X] = \mu^*(W^*) - \mu^*(X).$$

5) If $X_1, X_2, \ldots, X_n, \ldots$, finite or denumerable in number, are measurable sets of $\boldsymbol{B}_J$-traces, then

$$\bigcup_n X_n$$

is also a measurable set of $\boldsymbol{B}_J$-traces, and we have

$$[\bigcup_n X_n] =^{J'} \bigcup_n [X_n].$$

If the above sets $X_1, X_2, \ldots$ are disjoint (i.e. $X_n \cap X_m = \emptyset^*$ for $n \neq m$), then

$$\mu^*(\bigcup_n X_n) = \sum_n \mu^*(X_n).$$

6) If $X_1, X_2, \ldots, X_n, \ldots$, finite or denumerable in number, are all measurable sets of $\boldsymbol{B}_J$-traces, then so is $\bigcap_n X_n$ and we have

$$[\bigcap_n X_n] =^{J'} \bigcap_n [X_n].$$

If in addition to that,

$$X_1 \geqq X_2 \geqq \cdots \geqq X_n \geqq \cdots,$$

then

$$\mu^*\left(\bigcap_{n=1}^{\infty} X_n\right) = \lim_{n\to\infty} \mu'(X_n).$$

7) If X, Y are measurable sets of $\boldsymbol{B}_J$-traces, then $X - Y$ and $X \dotplus Y$ (algebraic addition, symmetric difference), are also measurable and we have:

$$[X - Y] =^{J'} [X] - [Y],$$
$$[X \dotplus Y] =^{J'} [X] \dotplus [Y],$$

and if $Y \subseteq X$, then

$$\mu^*(X - Y) = \mu^*(X) - \mu^*(Y).$$

8) The measurable sets of $\boldsymbol{B}_J$-traces make up a denumerably additive tribe $(\boldsymbol{F}^*)$ with O^* as zero and W^* as unit. On this tribe the measure μ^* is denumerably additive, but not necessarily effective.

9) In the tribe $(\boldsymbol{F}^*)$ of all measurable sets of $\boldsymbol{B}_J$-traces, the collection $\boldsymbol{J}^*$ of all sets X of $\boldsymbol{B}_j$-traces with $\mu^*(X) = 0$ is a denumerably additive ideal. It reorganizes the tribe $(\boldsymbol{F}^*)$ into the tribe $(\boldsymbol{F}^*)_{J^*}$. The ordering relation $(\leqq^{J^*})$ is defined by

$$X \leqq^{J^*} Y : =_{df} : X - Y \in \boldsymbol{J}^*,$$

and the corresponding equality is

$$X =^{J^*} Y \cdot =_{df} \cdot X \dotplus Y \in \boldsymbol{J}^*.$$

10) The measure μ^* on $(\boldsymbol{F}^*)$ induces the corresponding measure, (also denoted by μ^*) in the tribe $(\boldsymbol{F}^*)_{J^*}$. This measure is effective and denumerably additive.

11) The relation $X \leqq^{J^*} Y$ is equivalent to $\mu^*(X - Y) = 0$, and $X =^{J^*} O$ is equivalent to $X \in \boldsymbol{J}^*$.

11a) Every soma X of $\boldsymbol{J}^*$ is measurable and $\mu^*(X) = 0$. The theorems 4), 5), 6), 7) are true if we replace the signs of $(=)$ and $(\leqq)$ between sets of traces by $(=^{J^*})$ and $(\leqq^{J^*})$.

12) The following are equivalent for measurable sets X, Y of $\boldsymbol{B}_J$-traces

I. $[X] =^J [Y]$.

II. $X =^{J^*} Y$.

C.23.1. The set S of all $\boldsymbol{B}_J$-traces at infinity is measurable and we have

$$\mu^*(S) = 0.$$

Proof. Consider the sequence of rectangles

$$R_n =_{df} \left\{(x, y) \,\middle|\, \begin{matrix} -n < x < +n \\ -n < y < +n \end{matrix}\right\}.$$

We have

$$\sum_{n=1}^{\infty} R_n = P.$$

Denote by ϱ_n the set of all $\boldsymbol{B}_J$-traces covered by R_n. We have

$$[\varrho_n] =^{J'} R_n.$$

Chapter C I

The trace-theorem

The aim of this Chapter [C I.] is an important theorem needed in the theory of representation of maximal normal operators in Hilbert space. This theorem concerns a representation of functions of the variable $\boldsymbol{B}_J$-trace by functions of complex variables. It will be stated at the end of this chapter.

To find and prove the theorem we need the discussion of several auxiliary notions: the correspondence $\mathscr{V}$ between sets of $\boldsymbol{B}_J$-traces and sets of $\boldsymbol{B}$-traces, the correspondence Δ between a set of $\boldsymbol{B}$-traces and its valid parts, the notion of a special measure of sets of $\boldsymbol{B}$-traces, a correspondence denoted by the cross $(\times)$, the notion of distinguished sets of $\boldsymbol{B}$-traces, a notion concerning the vertices of $\boldsymbol{B}$-traces and finally a special property of sets of $\boldsymbol{B}_J$-traces, called Γ-property[1].

C I.24. Since a $\boldsymbol{B}_J$-trace τ' and the corresponding $\boldsymbol{B}$-trace τ possess a same representation, we can many times replace one kind of traces by another one. Though they have some representatives in common, the classes of all representatives of τ and τ' differ. Every representation of τ is also a representation of τ', but not conversely, because there are representatives of τ' which are not representatives of τ.

---

[1] Since this is the continuation of Chapter [C], we shall denote theorems and definitions with labels starting with C I.24.

The $\boldsymbol{B}$-traces which correspond to $\boldsymbol{B}_J$-traces are termed *valid* or active $\boldsymbol{B}$-*traces*, the remaining $\boldsymbol{B}$-traces are termed *non valid* or non active $\boldsymbol{B}$-*traces*, [C.17.].

C I.24.1. Def. Denote by $\mathscr{V}$ the correspondence which attaches to $\boldsymbol{B}_J$-trace τ' the correspondent $\boldsymbol{B}$-trace τ, according to [C.17.]. This correspondence is $1 \to 1$.

Def. Denote by $W^\bullet$ the set of all valid $\boldsymbol{B}$-traces. $W^\bullet$ is the range of the correspondence $\mathscr{V}$ and W^* is the domain of $\mathscr{V}$, where W^* is the set of all $\boldsymbol{B}_J$-traces. We have $\tau' \cup^{\mathscr{V}} \tau$.

Def. The correspondence $\mathscr{V}$ generates another correspondence, also denoted by $\mathscr{V}$, from the sets ξ' of $\boldsymbol{B}_J$-traces onto the sets ξ of corresponding $\boldsymbol{B}$-traces.

In precise terms we define

$$(1) \qquad \mathscr{V}(\xi') = \{\mathscr{V}(\tau') \mid \tau' \in \xi'\}$$

for all sets ξ' of $\boldsymbol{B}_J$-traces.

C I.24.1a. Theorem. The extended correspondence $\mathscr{V}$ is also $1 \to 1$. Its domain is the class of all subsets of W^* and its region is the class of all subsets of $W^\bullet$. Hence $\mathscr{V}^{-1}$ has a meaning, and carries subsets of $W^\bullet$ into subsets of W^*.

C I.24.1b. Theorem. Since the extended correspondence $\mathscr{V}$ is $1 \to 1$, it preserves finite and infinite unions, intersections, and also subtraction, and inclusion of sets.

C I.24.2. Theorem. If θ^* denotes the empty set of $\boldsymbol{B}_J$-traces and θ denotes the empty set of $\boldsymbol{B}$-traces, then

$$\theta = \mathscr{V}(\theta^*).$$

Proof. Indeed, suppose that

$$(1) \qquad \alpha_0 \in \mathscr{V}(\theta^*).$$

There exists a $\boldsymbol{B}_J$-trace α_0', such that $\mathscr{V}(\alpha_0') = \alpha_0$, $\alpha_0' \in \theta^*$. Since the sentence $\alpha_0' \in \theta^*$ is false, (1) cannot be true. Hence $\mathscr{V}(\theta^*) = \theta$, so the assertion is proved.

C I.24.3. Theorem. We have $\mathscr{V}(W^*) = W^\bullet$, where W^* is the set of all $\boldsymbol{B}_J$-traces and $W^\bullet$ the set of all valid $\boldsymbol{B}_J$-traces.

C I.24.4. Theorem. If ξ', η' are sets of $\boldsymbol{B}_J$-traces, then the following are equivalent

I. $\xi' = W^* - \eta' = \mathrm{co}_{W^*}\eta'$,

II. $\mathscr{V}(\xi') = \mathscr{V}(W^*) - \mathscr{V}(\eta') = W^\bullet - \mathscr{V}(\eta') = \mathrm{co}_{W^\bullet}\,\eta'$.

C I.24.5. Theorem. The correspondence $\mathscr{V}$ preserves algebraic additions of sets of traces.

C I.25. Def. Let ξ be a non empty set of $\boldsymbol{B}$-traces. We can decompose it into two parts: a *valid part*, defined as the set of all valid

$\boldsymbol{B}$-traces in ξ, and the *nonvalid part* composed of the remaining $\boldsymbol{B}$-traces in ξ.

The correspondence between the sets ξ of $\boldsymbol{B}$-traces and its valid part will be denoted Δ: we define for any set ξ of $\boldsymbol{B}$-traces,

$$\Delta(\xi) =_{df} \{\alpha \mid \alpha \in \xi, \text{ and there exists a } \boldsymbol{B}_J\text{-trace } \alpha' \text{ such that } \alpha = \mathscr{V}(\alpha')\}.$$

The correspondence Δ is pluri-one, its domain is the class of all sets of $\boldsymbol{B}$-traces and its range the class of all sets of valid $\boldsymbol{B}$-traces.

C I.25.1. Theorem. If ξ is a set of $\boldsymbol{B}$-traces, then

$$\Delta(\xi) = \eta \frown W^{\bullet},$$

where $W^{\bullet}$ is the set of all valid $\boldsymbol{B}$-traces.

Proof. We have

$$\Delta(\xi) \leqq \xi$$

and

$$\Delta(\xi) \leqq W^{\bullet};$$

hence

(1) $$\Delta(\xi) \leqq \xi \frown W^{\bullet}.$$

Conversely, let $\alpha \in \xi \frown W^{\bullet}$. Hence $\alpha \in W^{\bullet}$, so α is a valid trace. Hence there exists a $\boldsymbol{B}_J$-trace α', such that $\alpha = \mathscr{V}(\alpha')$. Since $\alpha \in \xi$, we get, by virtue of definition of Δ,

(2) $$\alpha \in \Delta(\xi).$$

From (1) and (2) the statement follows.

C I.25.2. Theorem. We have

$$\Delta(W^{\bullet}) = W^{\bullet}, \quad \Delta(\Delta(\xi)) = \Delta\xi,$$
$$\Delta(W) = W^{\bullet}, \quad \Delta\theta = \theta,$$

where W is the set of all $\boldsymbol{B}$-traces, $W^{\bullet}$ the set of all valid $\boldsymbol{B}$-traces, and θ the empty set. If $\xi \subseteqq W^{\bullet}$, then $\Delta\xi = \xi$.

C I.25.3. Theorem. The correspondence Δ preserves finite and infinite unions and intersections. Proof by [C I.25.1.].

C I.25.4. Theorem. If ξ is a set of $\boldsymbol{B}$-traces, then

$$\Delta(\mathrm{co}_W\,\xi) = \mathrm{co}_{W^{\bullet}}\,\Delta\xi.$$

Proof. We have, [C I.25.1.],

$$\Delta(\mathrm{co}_W\,\xi) = W^{\bullet} \frown \mathrm{co}_W\,\xi = W^{\bullet} \frown (W - \xi) = (W^{\bullet} \frown W) - (W^{\bullet} \frown \xi)$$
$$= W^{\bullet} - (W^{\bullet} \frown \xi) = W^{\bullet} - \Delta\xi = \mathrm{co}_{W^{\bullet}}\,\Delta\xi.$$

C I.25.5. Theorem. If $\xi \subseteqq W^{\bullet}$, then

$$\Delta(\mathrm{co}_{W^{\bullet}}\,\xi) = \mathrm{co}_{W^{\bullet}}\,\Delta\xi.$$

Proof. We have

$$\Delta(\mathrm{co}_{W^\bullet}\xi) = \Delta(W^\bullet - \xi) =, \text{[C I.25.1.]}, = (W^\bullet - \xi) \cap W^\bullet =$$
$$= (W^\bullet \cap W^\bullet) - (\xi \cap W^\bullet) = W^\bullet - \xi,$$

because $\xi \subseteqq W^\bullet$.

Hence $\Delta(\mathrm{co}_{W^\bullet}\xi) = W^\bullet - \Delta\xi$, because, [C I.25.2.], $\Delta\xi = \xi$. Hence

$$\Delta(\mathrm{co}_{W^\bullet}\xi) = \mathrm{co}_{W^\bullet}\xi, \quad \text{q.e.d.}$$

C I.26. Def. The set X of $\boldsymbol{B}$-traces is said to *be measurable*, whenever the set of $\boldsymbol{B}_J$-traces

$$\mathscr{V}^{-1}\Delta X$$

is measurable (μ^*-measurable), [C.22.1.].

C I.26.1. Theorem. Let X be a set of $\boldsymbol{B}$-traces, (it may contain traces at infinity), [C.23.1.], then the following are equivalent:

I. X is measurable,

II. there exists a μ^*-measurable set Y' of $\boldsymbol{B}_J$-traces such that

$$X = \mathscr{V}(Y') \cap X^0,$$

where X^0 is composed of non valid $\boldsymbol{B}$-traces in X.

Proof. Let I, i.e. let X be a measurable set of $\boldsymbol{B}$-traces. Hence, [C I.26.], the set of $\boldsymbol{B}_J$-traces

(0) $\qquad \mathscr{V}^{-1}\Delta X$ is measurable.

Now we have

(1) $$X = Y \cup X^0,$$

where Y is the valid part of X and X^0 the non valid. We have

(2) $$Y = \Delta X.$$

Put

$$Y' =_{df} \mathscr{V}^{-1} Y.$$

Hence, by (2), $Y' = \mathscr{V}^{-1}\Delta X$; so by (0), Y' is measurable, and we have

$$Y = \mathscr{V}(Y').$$

From (1) we get

$$X = \mathscr{V}(Y') \cup X^0;$$

hence II. Let II, i.e. let

$$X = \mathscr{V}(Y') \cup X^0,$$

(3) $\qquad$ where Y' is a measurable set of $\boldsymbol{B}_J$-traces,

and X^0 is composed of all non valid $\boldsymbol{B}$-traces in X. We have

$$\mathscr{V}^{-1}\Delta X = \mathscr{V}^{-1}\Delta(\mathscr{V}(Y') \cup X^0) = \mathscr{V}^{-1}\Delta\mathscr{V}(Y') =, \text{[C I.25.2.]}, =$$
$$= \mathscr{V}^{-1}\mathscr{V}(Y') = Y',$$

which is measurable, by (3). Hence I, q.e.d.

C I.26.2. Theorem. If a set X of $\boldsymbol{B}$-traces is measurable, then $\varDelta X$ is also measurable.

Proof. Let X be a measurable set of $\boldsymbol{B}$-traces. By [C I.26.1.] there exists a measurable set Y' of $\boldsymbol{B}_J$-traces, such that

$$(1) \qquad X = \mathscr{V}(Y') \cup X^0,$$

where X^0 is the set of all non valid traces in X. Hence

$$(2) \qquad \varDelta X = \varDelta \mathscr{V}(Y'),$$

because we have $\varDelta X^0 = \theta$, (the empty set of $\boldsymbol{B}$-traces), and $\mathscr{V}(Y')$ and X^0 are disjoint. We know that, [C I.25.1.],

$$\varDelta V(Y') = \mathscr{V}(Y') \cap W^\bullet;$$

hence

$$\varDelta X = \mathscr{V}(Y') \cap W^\bullet = \mathscr{V}(Y'),$$

which gives

$$\mathscr{V}^{-1} \varDelta X = \mathscr{V}^{-1} \mathscr{V}(Y') = Y'.$$

Since Y' is measurable, therefore $\varDelta X$ is measurable too, (Def. [C I.26.]).

C I.26.3. Theorem. The set $W^\bullet$ of all valid $\boldsymbol{B}$-traces is measurable.

Proof. We have $\mathscr{V}(W^*) = W^\bullet$, [C I.24.3.], where W^* is the set of all $\boldsymbol{B}_J$-traces.

Since $\varDelta W^\bullet = W^\bullet$, [C I.25.2.], we get

$$\mathscr{V}^{-1} W^\bullet = W^*,$$

and we know that W^* is μ^*-measurable, [C.23.3 a.].

Hence, by Def. [C I.26.], the set $W^\bullet$ is measurable, q.e.d.

C I.26.4. Theorem. The empty set θ of $\boldsymbol{B}$-traces is measurable.

Proof.

$$\mathscr{V}^{-1} \varDelta \theta =, \text{ [C I.25.2.]}, = \mathscr{V}^{-1} \theta =, \text{ [C I.24.2.]}, = \theta^*,$$

where θ^* is the empty set of $\boldsymbol{B}_J$-traces. Since θ^* is measurable, [C.23.3.], it follows, by Def [C I.26.], that θ is measurable, q.e.d.

C I.26.5. Theorem. The set W of all $\boldsymbol{B}$-traces is measurable.

Proof. We have

$$(1) \qquad W = W^\bullet \cup (W - W^\bullet).$$

The set $W^\bullet$ is measurable, [C I.26.3.]. The set $W - W^\bullet$ is composed of all non-valid traces only. Hence from (1), by [C I.26.1.], W is measurable.

C I.26.6. Theorem. Any set ξ, composed of non-valid $\boldsymbol{B}$-traces, is measurable.

Proof.

$$\xi = \theta \cup \xi,$$

hence

$$\xi = \mathscr{V}(\theta^*) \cup \xi,$$

[C I.24.2.]. Since θ^* is measurable, we conclude, by [C I.26.1.], that ξ is measurable.

C I.26.7. Theorem. If the sets of $\boldsymbol{B}$-traces

$$X_1, X_2, \ldots, X_n, \ldots$$

finite or denumerable in number are measurable, so is their union

$$X =_{df} \bigcup_n X_n.$$

Proof. By virtue of theorem [C I.26.1.], there exist measurable sets of $\boldsymbol{B}^J$-traces

$$Y'_1, Y'_2, \ldots, Y'_n, \ldots,$$

and there exist sets of non-valid $\boldsymbol{B}$-traces

$$X^0_1, X^0_2, \ldots, X^0_n, \ldots,$$

such that

$$X_n = \mathscr{V}(Y'_n) \cup X^0_n.$$

Hence

(1) $$\bigcup_n X_n = \bigcup_n \mathscr{V}(Y'_n) \cup \bigcup_n X^0_n.$$

We know that

$$\bigcup_n \mathscr{V}(Y'_n) = \mathscr{V}(\bigcup_n Y'_n).$$

Putting

$$Y' =_{df} \bigcup_n Y'_n, \quad X^0 =_{df} \bigcup_n X^0_n,$$

we see that Y' is measurable, [C.23.], and X^0 is composed of not valid traces.

We have from (1)

$$X = \mathscr{V}(Y') \cup X^0.$$

The theorem [C I.26.1.] completes the proof.

C I.26.8. Theorem. If the sets of $\boldsymbol{B}$-traces

$$X_1, X_2, \ldots, X_n, \ldots,$$

finite or denumerable in number, are measurable, so is their intersection

$$X =_{df} \bigcap_n X_n.$$

Proof. We have, as in the foregoing proof:

$$X_n = \mathscr{V}(Y'_n) \cup X^0_n, \quad (n = 1, 2, \ldots).$$

Hence

$$X =_{df} \bigcap_n X_n = \bigcap_n [\mathscr{V}(Y'_n) \cup X^0_n].$$

The sets $\mathscr{V}(Y'_n)$ are disjoint with the sets X^0_m, because $\mathscr{V}(Y'_n)$ is a set of valid traces, and X^0_m is a set of non valid traces.

Hence

(2) $$X = \bigcap_n \mathscr{V}(Y'_n) \cup \bigcap_n X^0_n = \mathscr{V}(\bigcap_n Y'_n) \cup \bigcap_n X^0_n .$$

Now, by [C.23.], the intersection of a finite or denumerable number of measurable sets of $\boldsymbol{B}_J$-traces is measurable, and $\bigcap_n X^0_n$ is a set composed of non-valid $\boldsymbol{B}$-traces. It follows, by [C I.26.1.], that X is measurable.

C I.26.9. Theorem. If X is a measurable set of $\boldsymbol{B}$-traces, then $\text{co}_W X = W - X$ is also measurable. (W is the set of all $\boldsymbol{B}$-traces.)

Proof. Put

$$Y =_{df} W - X.$$

We have

(0) $$X \cup Y = W,$$

and X, Y are disjoint. We have

(1) $$X = \varDelta X \cup X^0$$
$$Y = \varDelta Y \cup Y^0,$$

where X^0, Y^0 are sets composed of all non-valid traces in X, Y respectively.

Since $X \cup Y = W$, it follows that $X^0 \cup Y^0$ is the set of all not valid traces in W.

We have $\varDelta X \cup \varDelta Y = W^\bullet$, [C I.24.1.], (the set of all valid $\boldsymbol{B}$-traces), because $\varDelta X$, $\varDelta Y$ are the sets of all valid traces in X and Y respectively; hence $\varDelta X \cup \varDelta Y$ is so in

(2) $$X \cup Y = W.$$

Since X is $\boldsymbol{B}$-measurable, $\varDelta X$ is also measurable, [C I.26.2.], i.e. $\mathscr{V}^{-1} \varDelta X$ is μ^*-measurable, [C I.26.]. Since $\mathscr{V}^{-1} W^\bullet$ is also μ^*-measurable, [C I.26.3.], we see, [C.23.], that $\mathscr{V}^{-1} W^\bullet - \mathscr{V}^{-1} \varDelta X$ is also μ^*-measurable. By [C I.24.1 b.] we have $\mathscr{V}^{-1} W^\bullet - \mathscr{V}^{-1} \varDelta X = \mathscr{V}^{-1}(W^\bullet - \varDelta X)$. Hence $\mathscr{V}^{-1}(W^\bullet - X)$ is μ^*-measurable.

Hence $\mathscr{V}\mathscr{V}^{-1}(W^\bullet - \varDelta X) = W^\bullet - \varDelta X$ is measurable. (We rely on [C I.26.1.], where we put $\mathscr{V}^{-1}(W^\bullet - \varDelta X)$ for Y' and θ for X^0.)

Now $W - W^\bullet$ is measurable, since this set is composed of non-valid traces only, [C I.26.6.].

Hence, by [C I.26.7.]

(3) $$(W - W^\bullet) \cup (W^\bullet - \varDelta X)$$

is measurable. Since $\varDelta X \subseteqq W^\bullet \subseteqq W$, the union (3) equals $W - \varDelta X$. Hence, [C I.26.7.], $W - \varDelta X$ is measurable, q.e.d.

C I.26.10. The theorems [C I.26.3.]—[C I.26.9.] show, that the class of all measurable sets of $\boldsymbol{B}$-traces, if ordered by inclusion of sets, make up a denumerably additive tribe $F^\bullet$, where the empty set θ is its zero and the set W of all $\boldsymbol{B}$-traces is its unit.

C I.27. Def. If X is a measurable set of $\boldsymbol{B}$-traces, we put

$$\mu^\bullet(X) =_{df} \mu^*(\mathscr{V}^{-1}\Delta X),$$

where μ^* is the measure of sets of $\boldsymbol{B}_J$-traces, [C.22.2.]. We shall prove that $\mu^\bullet(X)$ is a denumerably additive measure (which however may be not effective).

C I.27.1. Theorem. If X is a measurable set of $\boldsymbol{B}$-traces, then

$$\mu^\bullet X = \mu^\bullet(\Delta X).$$

Proof. We know that ΔX is measurable, [C I.26.2.]. By definition we have:

$$\mu^\bullet X = \mu^*(\mathscr{V}^{-1}\Delta X).$$

Since

$$\Delta(\Delta X) = \Delta X,$$

[C I.25.2.], we get

$$\mu^\bullet(\Delta X) = \mu^*(\mathscr{V}^{-1}\Delta(\Delta X)) = \mu(\mathscr{V}^{-1}\Delta X) = \mu^\bullet(X).$$

C I.27.2. Theorem. If $X_1, X_2, \ldots, X_n, \ldots$, finite or denumerable in number, are all measurable sets of $\boldsymbol{B}$-traces, and if they are mutually disjoint, then

$$\mu^\bullet X = \sum_n \mu^\bullet X_n, \quad \text{where} \quad X =_{df} \bigcup_n X_n.$$

Proof. We have, [C I.25.3.],

$$\Delta X = \bigcup_n \Delta X_n;$$

hence, [C I.24.16.],

$$\mathscr{V}^{-1}\Delta X = \bigcup_n \mathscr{V}^{-1}\Delta X_n.$$

Since the sets X_n are disjoint, and since $\Delta X_n \subseteqq X_n$, therefore ΔX_n are also disjoint.

Hence $\mathscr{V}^{-1}\Delta X_n$ are disjoint.

As $\mathscr{V}^{-1}\Delta X_n$ are measurable sets of $\boldsymbol{B}_J$-traces, and since the measure on these sets is denumerably additive, we have

$$\mu^*(\mathscr{V}^{-1}\Delta X) = \sum_{n=1}^{\infty} \mu^*(\mathscr{V}^{-1}\Delta X_n).$$

Hence, by definition,

$$\mu^\bullet X = \sum_{n=1}^{\infty} \mu^\bullet X_n, \quad \text{q.e.d.}$$

C I.27.3. Theorem. If X is composed of non valid traces only, we have

$$\mu^\bullet X = 0.$$

Proof. We have $\Delta X = \theta$; hence

$$\mathscr{V}^{-1}\Delta X = \theta^*, \quad [\text{C I.24.2.}], \text{ and then } \quad \mu^*(\mathscr{V}^{-1}\Delta X) = 0,$$

[C.23.], which gives

$$\mu^\bullet X = 0.$$

C I.27.4. Theorem. $\mu^\bullet(W) = \mu^*(W^*)$.

Proof. We know that W is measurable, [C I.26.5.]. We have

$$\mu^\bullet W = \mu^*(\mathscr{V}^{-1}\Delta W) = \mu^*(\mathscr{V}^{-1} W) = \mu^* W^*,$$

$$\mu^\bullet W^\bullet = \mu^*(\mathscr{V}^{-1}\Delta W^\bullet), \ [\text{C I.27.}], \ = \mu^*(\mathscr{V}^{-1} W^\bullet) = \mu^* W^*, \ [\text{C I.24.}].$$

C I.27.5. Theorem.

$$\mu^\bullet \theta = 0.$$

Proof.

$$\mu^\bullet \theta = \mu^*(\mathscr{V}^{-1}\Delta\theta) = \mu^*(\mathscr{V}^{-1}\theta) =, \ [\text{C I.24.2.}], \ = \mu^*(\theta^*) = 0, \ [\text{C.23.}].$$

C I.27.6. Theorem. If X is a measurable set of $\boldsymbol{B}$-traces, then

$$\mu^\bullet(\text{co}_W X) = \mu^\bullet W - \mu^\bullet X.$$

Proof. We have, [C I.25.4.],

$$\Delta(\text{co}_W X) = W - \Delta X;$$

hence

(1) $\quad \mathscr{V}^{-1}\Delta(\text{co}_W X) = \mathscr{V}^{-1}(W^\bullet - \Delta X) = \mathscr{V}^{-1} W^\bullet - \mathscr{V}^{-1}\Delta X.$

Since

$$\Delta X \subseteqq W^\bullet,$$

we have

$$\mathscr{V}^{-1}\Delta X \subseteqq \mathscr{V}^{-1} W^\bullet;$$

hence, from (1),

$$\mu^*(\mathscr{V}^{-1}\Delta\, \text{co}_W X) = \mu^*(\mathscr{V}^{-1} W^\bullet) - \mu^*(\mathscr{V}^{-1}\Delta X) = \mu^\bullet W^\bullet - \mu^\bullet X.$$

Now, by [C I.27.4.],

$$\mu^\bullet W^\bullet = \mu^\bullet W.$$

Hence

$$\mu^\bullet(\text{co}_W X) = \mu^\bullet W - \mu^\bullet X, \quad \text{q.e.d.}$$

C I.27.7. Theorem. The measure $\mu^\bullet$ is a denumerably additive measure on the tribe $(\boldsymbol{F}^\bullet)$, [C I.26.10.], where somata are measurable sets of $\boldsymbol{B}$-traces.

Proof. This follows from the forgoing discussion.

C I.27.8. Theorem. The measurable sets of $\boldsymbol{B}$-traces, having the $\mu^\bullet$-measure 0 are those, which are composed of two parts: one of them is the V-image of sets of $\boldsymbol{B}_J$-traces with μ^*-measure 0, the second part is any set of non valid traces. These sets make up in $(\boldsymbol{F}^\bullet)$ a denumerably additive ideal $J^\bullet$.

The ideal $J^\bullet$ reorganizes the tribe $\boldsymbol{F}^\bullet$ into another one $(\boldsymbol{F}^\bullet)_{J^\bullet}$ by means of a new ordering, defined by

$$\xi \leqq^{J^\bullet} \eta \cdot =_{df} : \xi - \eta \in J^\bullet,$$

and consequently gives a new equality

$$\xi =^{J^\bullet} \eta \cdot =_{df} : \eta \dotplus \xi \in J^\bullet.$$

C I.28a. Theorem. The measure $\mu^\bullet$ is invariant with respect to the $(=^{J^\bullet})$-equality.

Proof. This is a consequence of the general theorem [A.19.4.].

C I.28.1. Theorem. The measure $\mu^\bullet$, generated by $J^\bullet$ in the tribe $(\boldsymbol{F}^\bullet)_{J^\bullet}$, is effective.

C I.28.2. Theorem. If the set X' of $\boldsymbol{B}_J$-traces has the μ^*-measure 0, then $X = \mathscr{V}(X')$ has the $\mu^\bullet$-measure 0.

C I.28.3. Theorem. The set Ξ of all $\boldsymbol{B}$-traces, which are not at infinity, is $\mu^\bullet$-measurable, and we have

$$W^\bullet =^{J^\bullet} W =^{J^\bullet} \Xi.$$

Proof. $W = W^\bullet \cup R$, where W is the set of all $\boldsymbol{B}$-traces, $W^\bullet$ is the set of all valid $\boldsymbol{B}$-traces and R is the set of all non-valid $\boldsymbol{B}$-traces.

W, $W^\bullet$ are measurable, [C I.26.5.], [C I.26.3.], and R is also measurable, [C I.26.6.].

We have

(1) $$\mu^\bullet W = \mu^\bullet W^\bullet, \quad \text{[C I.27.4.]}.$$

Hence $\mu^\bullet R = 0$. Since $W^\bullet \subseteqq W$, we get from (1):

(2) $$W =^{J^\bullet} W^\bullet.$$

Now we shall prove that Ξ is measurable.

We have

$$W = \Xi \cup (W - \Xi).$$

The set $W - \Xi$ is the set of all $\boldsymbol{B}$-traces at infinity.

Put

$$W - \Xi = S_v \cup S_n,$$

where S_v is the set of all valid traces, contained in $W - \Xi$, and S_n is the set of all non valid traces in $W - \Xi$. We have $\mu^\bullet S_n = 0$, because S_n is a set of non valid traces, [C I.27.3.].

S_v is the $\mathscr{V}$-correspondent of the set of all $\boldsymbol{B}_J$-traces at infinity.

Since the set S' of all $\boldsymbol{B}_J$-traces at infinity is measurable with $\mu^* S' = 0$, [C.23.1.], therefore any part of S' is also measurable with μ^*-measure equal 0.

It follows that S_v is measurable with $\mu^\bullet$-measure equal 0. Hence $S_v \cup S_n$ is measurable with $\mu^\bullet$-measure 0.

Consequently $W - \Xi$ is measurable and we have

(3) $$\mu^{\bullet}(W - \Xi) = 0.$$

Since W is measurable, it follows that Ξ is measurable, because $\Xi = W - (W - \Xi)$ and $\Xi \subseteqq W$.

Since $W - \Xi \subseteqq W$, we have

$$\mu^{\bullet} \Xi = \mu^{\bullet} W + \mu^{\bullet}(W - \Xi).$$

Hence, by (3),

$$\mu^{\bullet} \Xi = \mu^{\bullet} W.$$

Since $\Xi \subseteqq W$, it follows that $\Xi =^{J^{\bullet}} W$, which together with (2) gives the thesis.

C I.28.4. Theorem. If X' is a set of $\boldsymbol{B}_J$-traces and if $X' \in J^*$, then $\mathscr{V}(X')$ is measurable and $\mathscr{V}(X') =^{J^{\bullet}} \theta$, where θ is the empty set of $\boldsymbol{B}$-traces.

Proof. We have

$$\mu^*(X') = 0,$$

[C.23.11 a.]. Hence X' is measurable.

Put

(1) $$X =_{df} \mathscr{V}(X').$$

We have $\mathscr{V}^{-1} \varDelta(X) \subseteqq \mathscr{V}^{-1} X$, because $\varDelta X \subseteqq X$. Hence, by (1), $\mathscr{V}^{-1} \varDelta(X) \subseteqq X'$. Since $X' \in J^*$, we have

(2) $$\mathscr{V}^{-1} \varDelta(X) \in J^*,$$

and then

$$\mathscr{V}^{-1} \varDelta(X)$$

is measurable.

It follows that, [C I.26.], X is measurable, i.e. $\mathscr{V}(X')$ is measurable.

We have

$$\mu^{\bullet}(X) = \mu^*[\mathscr{V}^{-1}(\varDelta X)],$$

[C I.27. Def.]; hence, by (2),

$$\mu^{\bullet}(X) = 0.$$

Hence $X \in J^{\bullet}$, and then $X =^{J^{\bullet}} \theta$ i.e.

$$\mathscr{V}(X') =^{J^{\bullet}} \theta. \quad \text{q.e.d.}$$

C I.28.5. Theorem. If

1. ξ', η' are measurable sets of $\boldsymbol{B}_J$-traces,
2. $\xi' =^{J^{\bullet}} \eta'$,

then

$$\mathscr{V} \xi' =^{J^{\bullet}} \mathscr{V} \eta'.$$

Proof. By hyp. 2. we have

$$\xi' \dot{+} \xi' \in J^*,$$

(algebraic addition).

Hence

$$(\xi' - \eta') \cup (\eta' - \xi') \in J^*.$$

Hence, by [C I.28.4.]:

$$\mathscr{V}[(\xi' - \eta') \cup (\eta' - \xi')] =^{J^\cdot} \theta.$$

Since $\mathscr{V}$ preserves operations, [C I.24.1 b.], we have

$$(\mathscr{V}\xi' - \mathscr{V}\eta') \cup (\mathscr{V}\eta' - \mathscr{V}\xi') =^{J^\cdot} \theta,$$

i.e.

$$\mathscr{V}\xi' \dot{+} \mathscr{V}\eta' =^{J^\cdot} \theta;$$

hence

$$\mathscr{V}\xi' =^{J} \mathscr{V}\eta', \quad \text{q.e.d.}$$

C I.29. Def. The $\boldsymbol{B}$-trace, (or $\boldsymbol{B}_J$-trace) α_0 is called *ordinary* whenever it is not at infinity, [C.8.1.]. Its vertex, vert α_0 is a point of the plane P. We shall use this notion mainly for $\boldsymbol{B}$-traces.

C I.29a. Theorem. The following properties of α_0 are equivalent:

I. α_0 is an ordinary $\boldsymbol{B}$-trace,

II. vertα_0 is a point of the plane P,

III. α_0 has three different neighbors $\alpha_1, \alpha_2, \alpha_3$, (having the same vertex);

$$\text{vert}\alpha_0 = \text{vert}\alpha_1 = \text{vert}\alpha_2 = \text{vert}\alpha_3.$$

If α_0 is an ordinary trace, so are its neighboring traces.

C I.29.1. Def. A set β of traces is called *ordinary*, whenever the following condition is satisfied:

(1) "if $\alpha \in \xi$, then α is an ordinary trace".

(Let us notice that a measurable set of traces may not be ordinary.)

We mention the following simple properties.

C I.29.2. Theorem. The empty set θ of $\boldsymbol{B}$-traces is ordinary. Indeed, as the proposition $\alpha \in \theta$ is false, the sentence (1) is true.

C I.29.3. Def. We denote by Ξ the set of all ordinary $\boldsymbol{B}$-traces, see [C I.28.3.].

C I.29.4. Theorem. A subset of an ordinary set of $\boldsymbol{B}$-traces is an ordinary set.

Ξ is an ordinary set of $\boldsymbol{B}$-traces.

C I.29.5. Theorem. The union and intersection of a finite or infinite collection of ordinary sets of $\boldsymbol{B}$-traces is ordinary. So is also the set $\Xi - \xi = \text{co}_\Xi \xi$, where ξ is ordinary. The difference and the algebraic sum of two ordinary sets of traces are ordinary.

C I.30. Def. Let ξ be an ordinary set of $\boldsymbol{B}$-traces. Then by $\xi^\times$ we shall understand the set of all traces which either belong to ξ or are neighbors of the traces of ξ. In precise terms we define $\xi^\times$ as the set of $\boldsymbol{B}$-traces:
$\{\alpha \mid \alpha \in \xi$ or α is such that there exists a trace $\beta \in \xi$ with $\operatorname{vert}\beta = \operatorname{vert}\alpha\}$. Since the operation $(\times)$ is defined only for ordinary sets of traces, therefore if we speak of $\xi^\times$, it will imply that ξ is ordinary.

C I.30.1. Theorem. We have $\Xi^\times = \Xi$, and $\theta^\times = \theta$, where θ is the empty set of $\boldsymbol{B}$-traces.

C I.30.2. Theorem. If ξ is an ordinary set of $\boldsymbol{B}$-traces, so is $\xi^\times$.

C I.30.3. Theorem. For every ordinary set ξ of $\boldsymbol{B}$-traces we have

$$\xi \subseteqq \xi^\times .$$

C I.30.4. Theorem. If ξ is an ordinary set of $\boldsymbol{B}$-traces, then

$$(\xi^\times)^\times = \xi^\times .$$

C I.30.5. Theorem. If ξ, η are ordinary sets of $\boldsymbol{B}$-traces and $\xi \subseteqq \eta$, then

$$\xi^\times \subseteqq \eta^\times .$$

C I.30.6. Theorem. If $\xi_1, \xi_2, \ldots, \xi_n, \ldots$, finite or denumerable in number, are all ordinary sets of $\boldsymbol{B}$-traces, and if

$$\xi = \xi_1 \cup \xi_2 \cup \cdots \cup \xi_n \cup \cdots ,$$

then

$$\xi^\times = \xi_1^\times \cup \xi_2^\times \cup \cdots \cup \xi_n^\times \cup \cdots$$

Proof. Since $\xi_n \subseteqq \xi$, we have, by [C I.30.5.],

$$\xi_n^\times \subseteqq \xi^\times, \quad (n = 1, 2, \ldots).$$

Hence

(1) $$\bigcup_n \xi_n^\times \subseteqq \xi^\times .$$

Let $\alpha \in \xi^\times$. Find a trace $\beta \in \xi$ such that $\operatorname{vert}\beta = \operatorname{vert}\alpha$, (Def. [C I.30.]).

Since

$$\xi = \bigcup_n \xi_n^\times, \quad \beta \in \xi ,$$

there exists m such that $\beta \in \xi_m^\times$. By, [C I.30.3.], $\beta \in \xi_m^\times$; hence $\alpha \in \xi_m^\times$, because $\operatorname{vert}\alpha = \operatorname{vert}\beta$.

It follows

$$\alpha \in \bigcup_n \xi_n^\times \quad \text{for all} \quad \alpha \in \xi^\times .$$

Hence

(2) $$\xi^\times \subseteqq \bigcup_n \xi_n^\times$$

which together with (1) gives the thesis.

C I.30.6.1. Remark. Notice that the foregoing theorem is true even for non-denumerable unions of ordinary sets of ***B***-traces.

C I.30.7. Theorem. If

1. $\xi_1, \xi_2, \ldots, \xi_n, \ldots$, infinite denumerable in number, are ordinary sets of ***B***-traces,

2. $\xi_1 \supseteqq \xi_2 \supseteqq \cdots \supseteqq \xi_n \supseteqq \cdots$,

3. $\xi = \bigcap_{n=1}^{\infty} \xi_n$,

then

$$\xi^\times = \bigcap_{n=1}^{\infty} \xi_n^\times.$$

Proof. We have $\xi \subseteqq \xi_n$; hence, [C I.30.5.],

$$\xi^\times \subseteqq \xi_n^\times,$$

which gives, by hyp. 2,

(1) $$\xi^\times \subseteqq \bigcap_{n=1}^{\infty} \xi^\times.$$

Let

(1.1) $$\beta \in \bigcap_{n=1}^{\infty} \xi_n^\times.$$

We have for all n:

$$\beta \in \xi_n^\times.$$

By Def. [C I.30.], there exists α_n such that $\alpha_n \in \xi_n$, $\operatorname{vert}\beta = \operatorname{vert}\alpha_n$.

All traces

(2) $$\alpha_1, \alpha_2, \ldots, \alpha_n, \ldots$$

are either equal β, or are neighbors of β. Hence, since there are only four neighbors at every vertex, there exists a partial sequence

(3) $$\alpha_{k(1)} = \alpha_{k(2)} = \cdots = \alpha_{k(n)} = \cdots$$

of (1). Denote the trace (3) by α. We have

$$\alpha \in \xi_{k(n)}, \quad n = 1, 2, \ldots$$

Consequently

$$\alpha \in \bigcap_{n=1}^{\infty} \xi_{k(n)}, \quad \text{which equals} \quad \bigcap_{m=1}^{\infty} \xi_m,$$

by virtue of hyp. 2.

Since $\operatorname{vert}\alpha = \operatorname{vert}\beta$, we get $\beta \in \xi^\times$, and consequently, by (1.1),

(4) $$\bigcap_{n=1}^{\infty} \xi^\times \subseteqq \xi^\times.$$

From (1) and (4) we obtain the thesis.

C I.30.7.1. Theorem. If

1. $\xi_1, \xi_2, \ldots, \xi_s$ finite in number ($s \geqq 2$) are ordinary sets of ***B***-traces, with

$$\xi_1 \supseteqq \xi_2 \supseteqq \cdots \supseteqq \zeta_s,$$

2. $\xi = \bigcap_{n=1}^{s} \xi_n$,

then

$$\xi^{\times} = \bigcap_{n=1}^{s} \xi_n^{\times}.$$

Proof. We put $\xi_{s+k} =_{df} \xi_s$, for $k = 1, 2, \ldots$ and we are in the condition of the foregoing theorem, [C I.30.7.].

C I.30.7.2. Remark. It is not true that

$$(\xi \frown \eta)^{\times} = \xi^{\times} \frown \eta^{\times}$$

for any ordinary sets ξ, η of $\boldsymbol{B}$-traces. The same can be said of infinite intersections of such sets.

C I.31. Def. By a *distinguished set of $\boldsymbol{B}$-traces* we shall understand any ordinary set ξ of $\boldsymbol{B}$-traces, such that

$$\xi = \xi^{\times},$$

[C I.30. Def.], [C I.29. Def.].

Remark. Since this notion is defined only for ordinary sets of $\boldsymbol{B}$-traces, therefore if we speak of a distinguished set, it implies that the set is ordinary.

C I.31.1. Theorem. The set Ξ of all ordinary $\boldsymbol{B}$-traces is distinguished, [C I.30.1.].

C I.31.2. Theorem. The empty set θ of traces is distinguished, [C I.30.1.].

C I.31.3. Theorem. If ξ is an ordinary set of $\boldsymbol{B}$-traces, then $\xi^{\times}$ is distinguished.

Proof. [C I.30.4.].

C I.31.4. Theorem. If $\xi_1, \xi_2, \ldots, \xi_n, \ldots$ finite or denumerable in number are all distinguished, so is

$$\xi =_{df} \bigcup_n \xi_n.$$

Proof. We have

$$\xi_n = \xi_n^{\times}, \quad (n = 1, 2, \ldots).$$

Hence, [C I.30.6.],

$$\xi^{\times} = (\bigcup_n \xi_n)^{\times} = \bigcup_n \xi_n^{\times} = \bigcup_n \xi_n = \xi.$$

C I.31.5. Theorem. If $\xi_1, \xi_2, \ldots, \xi_n, \ldots$, finite or denumerable in number, are all distinguished, so is with

$$\xi =_{df} \bigcap_n \xi_n.$$

Proof. We have

(1) $$\xi \subseteqq \xi^{\times},$$

[C I.30.3.]. Since $\xi \subseteqq \xi_n$ for all n, we have, [C I.30.5.],

$$\xi^{\times} \subseteqq \xi_n^{\times} = \xi_n, \quad (n = 1, 2, \ldots).$$

Hence

(2) $$\xi^{\times} \subseteqq \bigcap_n \xi_n = \xi.$$

From (1) and (2) we get $\xi = \xi^{\times}$, so the theorem is established.

C I.31.6. Theorem. If ξ is a distinguished set of $\boldsymbol{B}$-traces, then $\mathrm{co}_{\Xi}\xi = \Xi - \xi$ is also distinguished. (Ξ is the set of all ordinary traces, [Def. C I.29.3.].)

Proof. To avoid complication in writing, we shall be allowed to drop sometimes the index Ξ in "co_{Ξ}" and to write "co" instead.

Since $\xi = \xi^{\times}$, we have $\mathrm{co}\,\xi = \mathrm{co}\,(\xi^{\times})$; hence

(1) $$\mathrm{co}\,(\xi^{\times}) \subseteqq (\mathrm{co}\,\xi)^{\times},$$

[C I.30.3.]. Let $\alpha_0 \in (\mathrm{co}\,\xi)^{\times}$, and take the neighbors of α_0:

$$\alpha_1, \alpha_2, \alpha_3,$$

[C I.29.1 a.]. We have

$$\alpha_i \in (\mathrm{co}\,\xi)^{\times}, \quad (i = 0, 1, 2, 3).$$

Hence at least one trace, say α_k, belongs to $\mathrm{co}\,\xi$. I say that no one of the traces $\alpha_0, \alpha_1, \alpha_2, \alpha_3$ belongs to ξ.

To prove that, suppose that $\alpha_1 \in \xi$. Hence, as $\xi = \xi^{\times}$, we have $\alpha_1 \in \xi^{\times}$. Hence all traces $\alpha_0, \alpha_1, \alpha_2, \alpha_3$ belong to $\xi^{\times}$, hence to ξ. Hence no one of these traces belong to $\mathrm{co}\,\xi$, which contradicts the statement: $\alpha_1 \in \xi^{\times}$. Since the traces $\alpha_0, \alpha_1, \alpha_2, \alpha_3$ do not belong to $\xi^{\times}$; they belong all to $\mathrm{co}\,\xi^{\times} = \mathrm{co}\,\xi$. It follows that $\alpha_k \in \mathrm{co}\,(\xi^{\times})$. Consequently

$$(\mathrm{co}\,\xi)^{\times} \subseteqq \mathrm{co}\,(\xi^{\times}).$$

This, together with (1) gives

(2) $$(\mathrm{co}\,\xi)^{\times} = \mathrm{co}\,(\xi^{\times}) = \mathrm{co}\,\xi.$$

Hence $\mathrm{co}\,\xi$ is distinguished, q.e.d.

C I.31.7. Theorem. If ξ is distinguished, then

$$(\mathrm{co}_{\Xi}\xi)^{\times} = \mathrm{co}_{\Xi}(\xi^{\times}).$$

Proof. This follows from (2) in the foregoing proof.

C I.31.8. Theorem. If $\xi_1, \xi_2, \ldots, \xi_n, \ldots$, finite or denumerable in number, are all distinguished, we have

$$\Big(\bigcup_n \xi_n\Big)^{\times} = \bigcup_n (\xi_n^{\times})$$

and

$$\Big(\bigcap_n \xi_n\Big)^{\times} = \bigcap_n (\xi_n^{\times}).$$

Proof. This follows from [C I.31.5.] and [C I.31.7.].

C I.31.9. Theorem. If ξ, η are distinguished sets of $\boldsymbol{B}$-traces, so is $\xi - \eta$ and $\xi \dotplus \eta$.

Proof. [C I.31.5.] and [C I.31.6.].

C I.32. Def. Let ξ be an ordinary set of $\boldsymbol{B}$-traces. By $\operatorname{Vert}\xi$ we shall understand the set of all points of P (complex numbers a) such that there exists a trace α of ξ with

$$a = \operatorname{vert}\alpha,$$

[C I.29a.].

Thus $\operatorname{Vert}\xi$ is the set of all vertices of traces of ξ. If there will be a reasoning concerning $\operatorname{Vert}\xi$, there will be understood that ξ is an ordinary set of $\boldsymbol{B}$-traces.

C I.32.1. Theorem. If θ is the empty set of $\boldsymbol{B}$-traces, then $\operatorname{Vert}\theta$ is the empty set of complex numbers.

Proof. We have

$$\operatorname{Vert}\theta = \{a \,|\, \text{there exists a trace } \alpha \text{ of with } a = \operatorname{vert}\alpha\}.$$

Since the statement on a within braces is false for all a, the set $\operatorname{Vert}\theta$ is empty.

C I.32.2. Theorem. $\operatorname{Vert}\Xi = P$, where P is the set of all complex numbers, [Def. C I.29.3.].

C I.32.3. Theorem. If ξ is an ordinary set of $\boldsymbol{B}$-traces, then

$$\operatorname{Vert}\xi = \operatorname{Vert}(\xi^{\times}).$$

C I.32.4. Theorem. If ξ, η are ordinary sets of $\boldsymbol{B}$-traces, and if $\xi \subseteqq \eta$, then

$$\operatorname{Vert}\xi \subseteqq \operatorname{Vert}\eta.$$

C I.32.5. Theorem. If $\xi_1, \xi_2, \ldots, \xi_n, \ldots$, finite or denumerable in number, are all ordinary sets of $\boldsymbol{B}$-traces, and if

$$\xi = \bigcup_n \xi_n,$$

then

$$\operatorname{Vert}\xi = \bigcup_n \operatorname{Vert}\xi_n.$$

Proof. Since

$$\xi_n \subseteqq \xi, \qquad (n = 1, 2, \ldots),$$

we have, by [C I.32.4.],

$$(1) \qquad \bigcup_n \operatorname{Vert}\xi \subseteqq \operatorname{Vert}\xi.$$

Let $a \in \operatorname{Vert}\xi$. There exists a $\boldsymbol{B}$-trace α such that $\alpha \in \xi$, $a = \operatorname{vert}\alpha$.

Since $\xi = \bigcup_n \xi_n$, $\alpha \in \xi$, there exists m, such that $\alpha \in \xi_m$.

Hence

$$a \in \operatorname{Vert}\xi_m.$$

It follows

$$a \in \bigcup_n \operatorname{Vert}\xi_n,$$

which gives

(2) $$\operatorname{Vert} \xi \subseteqq \bigcup_n \operatorname{Vert} \xi_n .$$

From (1) and (2) the thesis follows.

C I.32.6. Theorem. If

1. $\xi_1, \xi_2, \ldots, \xi_n, \ldots$ is an infinite sequence of ordinary sets of ***B***-traces,

2. $\xi_1 \supseteqq \xi_2 \supseteqq \cdots \supseteqq \xi_n \supseteqq \cdots$,

3. $\xi = \bigcap_{n=1}^{\infty} \xi_n$,

then

$$\operatorname{Vert} \xi = \bigcap_{n=1}^{\infty} \operatorname{Vert} \xi_n .$$

Proof. We have

$$\xi \subseteqq \xi_n , \quad (n = 1, 2, \ldots);$$

hence, [C I.32.4.],

$$\operatorname{Vert} \xi \subseteqq \operatorname{Vert} \xi_n ,$$

which gives

(1) $$\operatorname{Vert} \xi \subseteqq \bigcap_{n=1}^{\infty} \operatorname{Vert} \xi_n .$$

Let $a \in \bigcap_{n=1}^{\infty} \operatorname{Vert} \xi_n$. We have $a \in \xi_n$ for all n. There exists a trace α_n such that

$$\alpha_n \in \xi_n , \quad a = \operatorname{vert} \alpha_n .$$

As all the traces

(2) $$\alpha_1, \alpha_2, \ldots, \alpha_n, \ldots$$

are neighbors, there exists a constant partial sequence

$$\alpha_{s(1)}, \alpha_{s(2)}, \ldots, \alpha_{s(n)}, \ldots$$

of (2). Call these traces α.

We have

$$\alpha \in \xi_{s(n)} , \quad (n = 1, 2, \ldots).$$

Hence, by hyp. 2:

$$\alpha \in \bigcap_{n=1}^{\infty} \xi_{s(n)} = \bigcap_{m=1}^{\infty} \xi_m .$$

Consequently

$$\alpha \in \xi .$$

Hence $a \in \operatorname{Vert} \xi$, which gives

(3) $$\bigcap_{n=1}^{\infty} \operatorname{Vert} \xi_n \subseteqq \operatorname{Vert} \xi .$$

From (1) and (3) the thesis follows.

C I.32.6.1. Remark. It is not true that the thesis takes place if we drop the hyp. 2.

It is neither true that for all ordinary sets ξ, η of $\boldsymbol{B}$-traces:

$$\operatorname{Vert}(\xi \frown \eta) = \operatorname{Vert}\xi \frown \operatorname{Vert}\eta.$$

Expl. Let ξ be composed of a single trace α, and η be composed of the single trace β, where $\alpha \neq \beta$, but $a = \operatorname{vert}\alpha = \operatorname{vert}\beta$.

We have

$$\xi \frown \eta = \emptyset, \quad \text{so} \quad \operatorname{Vert}(\xi \frown \eta) = \emptyset.$$

But

$$\operatorname{Vert}\xi \frown \operatorname{Vert}\eta = \text{set composed of the point } a.$$

C I.32.7. Remark. It is not true that for every ξ we have

$$(1) \qquad \operatorname{Vert}(\operatorname{co}_{\Xi}\xi) = \operatorname{co}_P \operatorname{Vert}\xi.$$

Indeed, suppose that (1) is true for every two ordinary sets X, Y of $\boldsymbol{B}$-traces. We have:

$$X \frown Y = \operatorname{co}_{\Xi}(\operatorname{co}_{\Xi}X \smile \operatorname{co}_{\Xi}Y).$$

Hence, by hypothesis (1):

$$\begin{aligned}\operatorname{Vert}(X \frown Y) &= \operatorname{co}_P \operatorname{Vert}[\operatorname{co}_{\Xi}X \smile \operatorname{co}_{\Xi}Y] = \operatorname{co}_P[\operatorname{Vert}\operatorname{co}_{\Xi}X \smile \operatorname{Vert}\operatorname{co}_{\Xi}Y]\\ &= \operatorname{co}_P[\operatorname{co}_P \operatorname{Vert}X \smile \operatorname{co}_P \operatorname{Vert}Y] = \operatorname{Vert}X \frown \operatorname{Vert}Y,\end{aligned}$$

which is not true for some X, Y, by [C.32.6.1.].

C I.33. Remark. We shall see that the circumstances are simplier, if we confine ourselves to distinguished sets of $\boldsymbol{B}$-traces.

C I.33.1. Theorem. If

1. $\xi_1, \xi_2, \ldots, \xi_n, \ldots$, finite or infinite denumerable in number, are all distinguished, [C I.31. Def.],

2. $\xi = \bigcap_n \xi_n$,

then

$$\operatorname{Vert}\xi = \bigcap_n \operatorname{Vert}\xi_n.$$

Proof. Since $\xi \subseteqq \xi_n$ for all n, we have

$$\operatorname{Vert}\xi \subseteqq \operatorname{Vert}\xi_n;$$

hence

$$(1) \qquad \operatorname{Vert}\xi \subseteqq \bigcap_n \operatorname{Vert}\xi_n.$$

Let

$$(1.1) \qquad a \in \bigcap_n \operatorname{Vert}\xi_n.$$

We have $a \in \operatorname{Vert}\xi_n$ for all n. There exists a trace α_n such that

$$a = \operatorname{vert}\alpha_n, \quad \alpha_n \in \xi_n.$$

Since ξ_n is distinguished, every neighbor β of α_n belongs to ξ_n.

Hence for that β, we have

$$a = \mathrm{vert}\beta, \quad \beta \in \xi_n.$$

Consequently

$$\beta \in \bigcap_n \xi_n = \xi;$$

hence

$$\mathrm{vert}\beta \in \mathrm{Vert}\,\xi.$$

Hence

$$a \in \mathrm{Vert}\,\xi.$$

Thus, by (1.1),

(2) $$\bigcap_n \mathrm{Vert}\,\xi_n \subseteqq \mathrm{Vert}\,\xi.$$

From (1) and (2) the thesis follows.

C I.33.2. Theorem. If ξ is a distinguished set of $\boldsymbol{B}$-traces, then we have

$$\mathrm{Vert}(\mathrm{co}_{\Xi}\,\xi) = \mathrm{co}_P(\mathrm{Vert}\,\xi).$$

Proof. By virtue of theorem [C I.31.6.], the set $\mathrm{co}_{\Xi}\,\xi$ is distinguished.

We have

$$(\mathrm{co}_{\Xi}\,\xi) \frown \xi = \theta,$$

$$(\mathrm{co}_{\Xi}\,\xi) \smile \xi = \Xi,$$

where θ is the empty set of traces.

Applying theorems [C I.32.5.], [C I.32.1.] and [C I.32.2.], we get

$$(\mathrm{Vert}\;\mathrm{co}_{\Xi}\,\xi) \frown \mathrm{Vert}\,\xi = \mathrm{Vert}\,\theta = \theta,$$

$$(\mathrm{Vert}\;\mathrm{co}_{\Xi}\,\xi) \smile \mathrm{Vert}\,\xi = \mathrm{Vert}\,\Xi = P.$$

It follows, [A.1.4.3.], that $\mathrm{Vert}\;\mathrm{co}_{\Xi}\,\xi$ is the complementary set of $\mathrm{Vert}\,\xi$ with respect to P.

Hence

$$\mathrm{Vert}\;\mathrm{co}_{\Xi}\,\xi = \mathrm{co}_P \mathrm{Vert}\,\xi, \quad \text{q.e.d.}$$

C I.33.3. If ξ, η are distinguished sets of $\boldsymbol{B}$-traces, then

$$\mathrm{Vert}(\xi - \eta) = \mathrm{Vert}\,\xi - \mathrm{Vert}\,\eta.$$

Proof. This follows from the relation

$$\xi - \eta = \xi \frown (\mathrm{co}_{\Xi}\,\eta)$$

by application of [C I.33.1.] and [C I.33.2.].

C I.33.4. Lemma. Let L be a $\boldsymbol{B}_J$-covering, and ϱ the set of all $\boldsymbol{B}_J$-traces, $\boldsymbol{B}_J$-covered by L. Then ϱ is measurable, (μ^*-measurable), and for its coat we have

$$[\varrho] =^{J'} L.$$

Proof. We know that L can be represented as the denumerable union of mutually disjoint bricks, [B.4.3.],

$$L = \sum_{n=1}^{\infty} a_n.$$

Indeed, the tribe $\boldsymbol{F}$ of figures on the plane and its base $\boldsymbol{B}$ obey the $(\boldsymbol{Hyp\ Af})$, [B.3.], and consequently the theorem [B.4.3.] holds true.

Denote by α_n the set of all $\boldsymbol{B}_J$-traces covered by a_n. We know, [B I.8.7.], that α_n is measurable and

$$[\alpha_n] =^{J'} a_n.$$

Since $a_n \subseteqq^{J^*} L$, it follows that

$$\alpha_n \subseteqq^{J^*} \varrho,$$

[C.23.], which gives

(1)
$$\sum_{n=1}^{\infty} \alpha_n \subseteqq^{J^*} \varrho.$$

To prove the converse relation, take an arbitrary trace $\tau \in \varrho$. The point vertτ either belongs to the interior of one of the bricks a_n, or else it is lying on the boundary[1] of at most four different bricks a_i. Since τ is then one of the (at most four possible neighbor traces), it follows that there is a well determined brick, which covers τ. Thus $\tau \in \alpha_n$. It follows that

$$\varrho \subseteqq^{J^*} \sum_{n=1}^{\infty} \alpha_n.$$

This, together with (1) gives the equality

(2)
$$\varrho =^{J^*} \sum_{n=1}^{\infty} \alpha_n.$$

We have

$$[\alpha_n] =^{J'} a_n;$$

hence

$$\sum_{n=1}^{\infty} [\alpha_n] =^{J'} \sum_{n=1}^{\infty} a_n =^{J'} L.$$

Now, applying [B I.8.15.], we get

$$\left[\sum_{n=1}^{\infty} \alpha_n\right] =^{J'} L,$$

hence, by (2),

$$[\varrho] =^{J'} L,$$

and ϱ, as the sum (2) of measurable sets of traces, is itself a measurable set of traces. The theorem is established.

[1] Parts of the boundary may be "at infinity".

C I.33.5. Lemma. Suppose that

1) ϱ is a measurable set of $\boldsymbol{B}_J$-traces,

2)

(1) $$L_1 \geqq^J L_2 \geqq^J \cdots \geqq^J L_n \geqq^J \cdots$$

are coverings of ϱ, such that

$$[\varrho] =^{J'} \prod_n L_n =^{J'} \prod_L L,$$

(where intersections are taken from $(\boldsymbol{F}_{J'})$), and where the second product is taken for all possible $\boldsymbol{B}_J$-coverings of ϱ, thus giving the exterior coat of ϱ,

3) $\lambda_1, \lambda_2, \ldots, \lambda_n, \ldots$ are sets of all traces $\boldsymbol{B}_J$-covered by (1) respectively,

then

$$\varrho =^{J^*} \prod_{n=1}^{\infty} \lambda_n.$$

Proof. By the preceding theorem [C I.33.4.],

$$[\lambda_n] =^{J'} L_n.$$

Hence

$$\prod_{n=1}^{\infty} [\lambda_n] =^{J'} \prod_{n=1}^{\infty} L_n.$$

Since ϱ is measurable, we have

$$[\varrho] =^{J'} \prod_{n=1}^{\infty} [\lambda_n];$$

hence, by [B I.8.16.],

$$[\varrho] =^{J'} \left[\prod_{n=1}^{\infty} [\lambda_n\right].$$

Consequently

$$\varrho =^{J^*} \prod_{n=1}^{\infty} \lambda_n. \quad \text{q.e.d.}$$

C I.34. Information. All auxiliaries being settled, we are going to prove a key-theorem which will anable us to get later the canonical representation of maximal normal operators in Hilbert-space. The theorem mentioned concerns sets of μ^*-measurable $\boldsymbol{B}_J$-traces, and it will be extended to functions of the variable $\boldsymbol{B}_J$-trace. We start with recollection of some needed notions:

J' is the denumerably additive ideal in the tribe $(\boldsymbol{F}')$ of all borelian subsets of the plane P. This is the collection of all borelian sets E of points with $\mu' E = 0$, [C.18.].

$J^\bullet$ is the denumerably additive ideal composed of all sets X, [C.27.8.] of $\boldsymbol{B}$-traces having the $\mu^\bullet$-measure equal 0, where the measure is defined by

$$\mu^\bullet(X) = \mu^*(\mathscr{V}^{-1} \Delta X].$$

C I.35. Def. Let ξ' be a measurable (μ^*-measurable), set of $\boldsymbol{B}_J$-traces. It may contain traces at infinity. Let ξ be a measurable ($\mu^\bullet$-measurable), and distinguished set of $\boldsymbol{B}$-traces. We shall say that ξ *is associated with* ξ' whenever the following two conditions are satisfied:

1) $\mathscr{V}\,\xi' =^{J^\bullet} \xi$, $(\xi = \xi^\times)$.

2) $[\xi'] =^{J'} \operatorname{Vert} \xi$.

In addition to that we agree to say thay *the set* ξ' *has the property* Γ, or *satisfies the condition* Γ, whenever there exists a measurable, distinguished set ξ of $\boldsymbol{B}$-traces, associated with ξ'.

Remark. First we shall prove various theorems on sets having the property Γ, and finally we shall prove an existence theorem on the property Γ.

C I.35a. Theorem. If

1. the measurable set of $\boldsymbol{B}_J$-traces ξ' has the property Γ, and ξ is associated to ξ',

2. $\xi' =^{J^*} \eta'$,

then η' also has the property Γ, and ξ is associated to η'.

Proof. We have

$$\mathscr{V}\,\xi' =^{J^\bullet} \xi = \xi^\times, \tag{1}$$

and

$$[\xi'] =^{J'} \operatorname{Vert} \xi. \tag{2}$$

By hypothesis (2) we have

$$\mathscr{V}\,\xi' =^{J^\bullet} \mathscr{V}\,\eta', \quad \text{[C I.28.5.]}, \tag{3}$$

and, by [C.23.12.], we get

$$[\xi'] =^{J'} \operatorname{Vert} \xi.$$

Hence

$$\mathscr{V}\,\eta' =^{J^\bullet} \xi = \xi^\times,$$

and

$$[\eta'] =^{J'} \operatorname{Vert} \xi,$$

so the theorem follows.

C I.35.1. Theorem. If the measurable sets of $\boldsymbol{B}_J$-traces

1.

$$\xi_1', \xi_2', \ldots, \xi_n', \ldots \tag{1}$$

have the property Γ,

2. $\xi' = \bigcup_{n=1}^{\infty} \xi_n'$,

3. $\xi_1, \xi_2, \ldots, \xi_n, \ldots$ are associated with (1) respectively,

4. $\xi =_{df} \bigcup_{n=1}^{\infty} \xi_n$,

then ξ' also has the property Γ and ξ is associated with ξ'.

Proof. By definition, the sets ξ_n are all measurable and distinguished sets of $\boldsymbol{B}$-traces.

We have, by definition of property Γ:

(1) $$\mathscr{V}\,\xi'_n =^{J^\bullet} \xi_n,$$

and

(2) $$[\xi'_n] =^{J'} \mathrm{Vert}\,\xi_n, \qquad (n = 1, 2, \ldots).$$

By [C I.31.4.], ξ is distinguished and, by [C I.26.7.], ξ is measurable.

Since the ideal $J^\bullet$ is denumerably additive, we have

$$\bigcup_n \mathscr{V}\,\xi'_n =^{J^\bullet} \bigcup_n \xi_n.$$

Hence, by [C I.24.1 b.],

$$\mathscr{V}\,(\bigcup_n \xi'_n) =^{J^\bullet} \xi,$$

i.e.

(3) $$\mathscr{V}\,\xi' =^{J^\bullet} \xi.$$

From (2) we deduce, J' being a denumerably additive ideal, that

$$\bigcup_n [\xi'_n] =^{J'} \bigcup_n \mathrm{Vert}\,\xi_n.$$

Since ξ_n are ordinary sets, we have, [C I.32.5.],

$$\bigcup_n [\xi'_n] =^{J'} \mathrm{Vert} \bigcup_n \xi_n.$$

Hence

$$\bigcup_n [\xi'_n] =^{J'} \mathrm{Vert}\,\xi.$$

If we apply a theorem of the theory of coats, [C.23.5.], of measurable sets of $\boldsymbol{B}_J$-traces, we get

$$[\bigcup_n \xi'_n] =^{J'} \mathrm{Vert}\,\xi;$$

hence

(4) $$[\xi'] =^{J'} \mathrm{Vert}\,\xi.$$

From (3) and (4) the theorem follows.

C I.35.2. Theorem. The empty set θ^* of $\boldsymbol{B}_J$-traces has the property Γ, and θ is associated with θ^*.

Proof. The set θ^* is measurable, [C.23., 3)], and we have

(1) $$\mathscr{V}\,\theta^* = \theta,$$

[C I.24.2.]. We also know that θ, i.e. the empty set of $\boldsymbol{B}$-traces is distinguished, [C I.31.2.].

From (1) follows

$$\mathscr{V}\,\theta^* =^{J^\bullet} \theta = \theta^\times.$$

For the coat we have

(2) $$[\theta^*] =^{J'} \mathit{O},$$

i.e. the empty set of points of the plane.

We also have

(3) $$\mathrm{Vert}\,\theta = \mathcal{Q},$$

[C I.32.1.]. From (2) and (3) it follows:

$$[\theta^*] =^{J'} \mathrm{Vert}\,\theta.$$

Since all conditions for the property Γ are satisfied, we deduce that θ^* has the property Γ, and that θ is a set associated with θ^*.

C I.35.3. Theorem. The set W^* of all $\boldsymbol{B}_J$-traces has the property Γ, and the set Ξ of all ordinary $\boldsymbol{B}$-traces is associated with W^*.

Proof. The set W^* is measurable, [C.23.3 a.)]. The set Ξ is ordinary, [C.29.3., Def.], distinguished, [C I.31.1.], and $\mu^\bullet$-measurable [C.28.3.],

$$W^\bullet =^{J^\bullet} W =^{J^\bullet} \Xi;$$

hence

(1) $$\mathcal{V}\, W^* = W^\bullet =^{J^\bullet} \Xi.$$

The coat $[W^*]$ is P and $\mathrm{Vert}\,\Xi = P$, [C I.32.2.]. Hence

(2) $$[W^*] =^{J'} \mathrm{Vert}\,\Xi.$$

(1) and (2) imply the thesis.

C I.35.4. Theorem. If ξ' is a measurable set of $\boldsymbol{B}_J$-traces, having the property Γ, and if ξ is a set of $\boldsymbol{B}$-traces associated with ξ', then

(1) $$\eta' =_{df} \mathrm{co}_{W^*}\, \xi'$$

is also satisfying the condition Γ and

$$\eta =_{df} \mathrm{co}_{\Xi}\, \xi$$

is associated with (1). Here W^* is the collection of all $\boldsymbol{B}_J$-traces and Ξ is the collection of all ordinary $\boldsymbol{B}$-traces, [C I.29.3. Def.].

Proof. By definition [C I.35. Def.], ξ is measurable and distinguished; hence it is an ordinary set of $\boldsymbol{B}$-traces, and we have by hypothesis:

(2) $$\mathcal{V}(\xi') =^{J^\bullet} \xi = \xi^\times,$$

and

(3) $$[\xi'] =^{J'} \mathrm{Vert}\,\xi.$$

From (2) we have

$$W^\bullet - \mathcal{V}(\xi') =^{J^\bullet} W^\bullet - \xi.$$

By [C I.24.3.] we have

$$\mathcal{V}(W^*) - \mathcal{V}(\xi') =^{J^\bullet} W^\bullet - \xi;$$

hence, [C I.24.1 b.],

$$\mathcal{V}(W^* - \xi') =^{J^\bullet} W^\bullet - \xi.$$

We have, [C I.28.3.],

$$W^\bullet =^{J^\bullet} \Xi;$$

hence

$$W^{\bullet} - \xi =^{J^{\bullet}} \Xi - \xi,$$

so we get

$$\mathscr{V}(W^{*} - \xi') =^{J^{\bullet}} \Xi - \xi,$$

which can be written

(4) $$\mathscr{V}(\mathrm{co}_{W^{*}}\xi') =^{J^{\bullet}} \mathrm{co}_{\Xi}\xi.$$

Now, since ξ is measurable by hypothesis, and Ξ is also measurable, [C I.28.3.], therefore, by [C I.26.10.], $\mathrm{co}_{\Xi}\xi = \Xi - \xi$ is measurable.

The set $\mathrm{co}_{\Xi}\xi$ is also distinguished. Indeed, ξ is so by hypothesis; therefore, by [C I.31.6.], $\Xi - \xi$ is distinguished. Thus in (4) we have the first condition for property Γ settled. Let us go over to the second condition.

We have, by hypothesis (3):

$$[\xi'] =^{J'} \mathrm{Vert}\,\xi.$$

It follows

(5) $$\mathrm{co}_P[\xi'] =^{J'} \mathrm{co}_P \mathrm{Vert}\,\xi.$$

Now we have

$$\mathrm{co}_P[\xi'] =^{J'} [\mathrm{co}_{W^{*}}\xi'],$$

[C.23.4.], and

$$\mathrm{co}_P \mathrm{Vert}\,\xi = \mathrm{Vert}(\mathrm{co}_{\Xi}\xi),$$

[C I.33.2.], because ξ is distinguished. Carrying the above into (5) we get

(6) $$[\mathrm{co}_{W^{*}}\xi'] =^{J'} \mathrm{Vert}(\mathrm{co}_{\Xi}\xi).$$

From (4) and (6) the theorem follows.

C I.35.5. Theorem. If $\xi_1', \xi_2', \ldots, \xi_n', \ldots,$ finite or denumerable in number, are measurable sets of $\boldsymbol{B}_J$-traces, all having the property Γ, so is with the set

$$\xi' =_{df} \bigcap_n \xi_n'.$$

If $\xi_1, \xi_2, \ldots, \xi_n, \ldots$ are sets of $\boldsymbol{B}$-traces associated with $\xi_1', \xi_2', \ldots, \xi_n', \ldots$ respectively, then

$$\xi =_{df} \bigcap_n \xi_n$$

is associated with ξ'.

Proof. By hypothesis, ξ_n is measurable and distinguished. Hence, [C I.26.8.], ξ is also measurable, and, [C I.31.5.], distinguished.

By hypothesis we have

$$\mathscr{V}\,\xi_n' =^{J^{\bullet}} \xi_n = \xi_n^{\times};$$

hence, since the ideal $J^{\bullet}$ is denumerably additive,

$$\bigcap_n \mathscr{V}(\xi_n') =^{J^{\bullet}} \bigcap_n \xi_n.$$

Hence, [C I.24.1 b.],

$$\mathscr{V}(\bigcap_n \xi'_n) =^{J^\bullet} \xi,$$

i.e.

(1) $$\mathscr{V}(\xi') =^{J^\bullet} \xi.$$

We also have, by hypothesis,

$$[\xi'_n] =^{J'} \operatorname{Vert} \xi_n.$$

Hence, since

(2) $$\bigcap_n [\xi'_n] =^{J'} \bigcap_n \operatorname{Vert} \xi_n,$$

and since ξ_n is distinguished, we have, [C I.33.1.],

$$\bigcap_n \operatorname{Vert} \xi_n =^{J'} \operatorname{Vert}(\bigcap_n \xi_n) =^{J'} \operatorname{Vert} \xi,$$

and, by [C.23., 6)], we have

$$\bigcap_n [\xi'_n] =^{J'} [\bigcap_n \xi'_n] =^{J'} [\xi'].$$

It follows from (2)

(3) $$[\xi'] =^{J'} \operatorname{Vert} \xi.$$

The relations (1) and (2) yield the thesis.

C I.36. Information. We are going to prove that the set of all $\boldsymbol{B}_J$-traces which are $\boldsymbol{B}_J$-covered by a $\boldsymbol{B}_J$-brick, has the property Γ. As explanation we mention that, if a $\boldsymbol{B}_J$-trace α is covered by a rectangle p, and if $p =^J q$, then α is also covered by q.

To do with easier circumstance, we may consider valid $\boldsymbol{B}$-traces, instead of $\boldsymbol{B}_J$-traces; thus in what follows, we will sometimes deal with valid traces, calling them $\boldsymbol{B}_v$-traces.

Concerning the rectangles on the plane, we shall start with the following particular kind of bounded (non empty) rectangles.

C I.36.1. Def. The measure μ' is denumerably additive and finite on the tribe $(\boldsymbol{F}')$ of borelian subsets of the plane P. Consequently there exists an at most denumerable number of straight lines l, parallel to the x-axis, with $\mu'(l) > 0$. The same can be said of straight lines m, parallel to the y-axis. The remaining lines parallel to the axes will be termed *well selected lines.*

A generalized rectangle half open (in the usual way), non empty and bounded will be termed *well selected,* whenever its sides are lying on well selected lines. For such a rectangle the μ'-measure of its boundary is 0.

C I.36.2. We shall start with a well selected rectangle:

$$R =_{df} \{(x, y) \mid x_1 < x \leqq x_2,\ y_1 < y \leqq y_2\},$$

where

(1) $$x_1 < x_2, \quad y_1 < y_2.$$

We shall prove that the set of all $\boldsymbol{B}_J$-traces $\boldsymbol{B}_J$-covered by R, has the property Γ.

Choose the following sequences of real numbers:

$$x_1^{(1)} < x_1^{(2)} < \cdots < x_1^{(n)} < \cdots < x_1,$$
$$y_1^{(1)} < y_1^{(2)} < \cdots < y_1^{(n)} < \cdots < y_1,$$
$$x_2^{(1)} > x_2^{(2)} > \cdots > x_2^{(n)} > \cdots > x_2,$$
$$y_2^{(1)} > y_2^{(2)} > \cdots > y_2^{(n)} > \cdots > y_2,$$

where

$$\lim x_1^{(n)} = x_1, \quad \lim y_1^{(n)} = y_1,$$
$$\lim x_2^{(n)} = x_2, \quad \lim y_2^{(n)} = y_2.$$

Consider the rectangles:

$$R_n =_{df} \left\{(x, y) \,\Big|\, \begin{matrix} x_1^{(n)} < x \leqq x_2^{(n)} \\ y_1^{(n)} < y \leqq y_2^{(n)} \end{matrix}\right\}.$$

C I.36.2a. We need some preparation. First of all we notice that an open set of points is a covering, [C.13.]. Indeed, let E be an open set. E can be represented as finite or denumerable union of closed bounded rectangles

$$E = \bigcup_n r_n.$$

For every n we can find another closed rectangle r'_n, such that r_n lies in the interior of r'_n, and where $r'_n \subseteqq E$.

Consequently, if we replace r'_n by a half open rectangle r''_n, having the same boundary, we shall have

$$E = \bigcup_n r''_n,$$

so E is a covering, because it is a denumerable union of bricks.

C I.36.2b. Another remark is needed:

If E is an open subset of P, and a a point of E, then all four $\boldsymbol{B}$-traces with vertex a are covered by E.

Indeed, let (x, y) be the coordinates of a, and τ any $\boldsymbol{B}$-trace with vertex a:

$$\tau = \tau(x, y; i, j),$$

where $i = 0, 1$, $j = 0, 1$, [C.8.1.].

Let $S_1 \supseteqq S_2 \supseteqq \cdots \supseteqq S_n \supseteqq \cdots$ be a representative of τ. The diameter of S_n tends to 0, [C.12.], [C.12a.], with $n \to \infty$, and a belongs to the closure of S_n. It follows that starting from some index m, we have $S_n \subseteqq E$, so the assertion is proved.

C I.36.3. Going back to the rectangles R and R_n defined in [C I.36.2.], we notice that

(1)
$$\bar{R} = \bigcap_{n=1}^{\infty} R_n,$$

and

(2)
$$\bar{R}_1 \supset R_1 \supset R_1^\circ \supset \bar{R}_2 \supset R_2 \supset R_2^\circ \supset \ldots,$$

and

$$\bar{R}_n \supset R_n \supset R_n^\circ \supset \bar{R} \supset R \supset R^\circ, \quad (n = 1, 2, \ldots).$$

The bar denotes closure and the small circle denotes the set of all interior points.

C I.36.4. Denote by ϱ, ϱ_n the sets of all ***B***-traces (valid or not), covered by R, R_n respectively.

We shall prove that

$$\varrho = \bigcap_{n=1}^{\infty} \varrho_n.$$

By [C.12.1.], we have

(0)
$$\operatorname{Vert} \varrho = \bar{R}, \quad \text{(closure of } R),$$

and, by (2), we have $\bar{R} \subset R_n^\circ$; hence if $a \in \operatorname{Vert} \varrho$, then, by [C I.36.2b.], all four traces with vertex a are covered by R_n°, hence by R_n. Consequently these traces all belong to the set ϱ_n, i.e. to the set of all ***B***-traces covered by R_n.

It follows that

(1)
$$\varrho^\times \subseteqq \bigcap_{n=1}^{\infty} \varrho_n.$$

Let $\alpha \in \bigcap_{n=1}^{\infty} \varrho_n$; we have

$$\operatorname{vert} \alpha \in \operatorname{Vert} \bigcap_{n=1}^{\infty} \varrho_n.$$

Since $\varrho_1 \supseteqq \varrho_2 \supseteqq \cdots \supseteqq \varrho_n \supseteqq \cdots$, we have, by [C I.32.6.],

$$\operatorname{vert} \alpha \in \bigcap_{n=1}^{\infty} \operatorname{Vert} \varrho_n.$$

By [C.12.1.],

$$\operatorname{Vert} \varrho_n = \bar{R}_n;$$

hence

$$\operatorname{vert} \alpha \in \bigcap_{n=1}^{\infty} \bar{R}_n = \bar{R}.$$

Consequently $\alpha \in \varrho^\times$, [C.12.1.]. Hence

(2)
$$\bigcap_{n=1}^{\infty} \varrho_n \subseteqq \varrho^\times.$$

From (1) and (2) the assertion follows. So we have

(3)
$$\varrho^\times = \bigcap_{n=1}^{\infty} \varrho_n.$$

C I.36.5.

(3a) From (3) it follows that $\varrho^\times$ is measurable,

because, [C I.26.8.], ϱ_n are measurable [C.23.].

We shall prove that $\varrho^\times$ is a set associated to the set ϱ' of all $\boldsymbol{B}_J$-traces, covered by R.

Consider the set of $\boldsymbol{B}$-traces: $\varrho^\times - \varrho$. We have $\varrho \subseteqq \varrho^\times$, hence $\varrho^\times - \varrho \subseteqq \operatorname{co}\varrho$.

Now we know that the traces of $\operatorname{co}\varrho$ are covered by the figure $\operatorname{co}_P R$. The traces belonging to $\varrho^\times$ are also covered by R_n, for any n. Indeed, $\operatorname{Vert}\varrho^\times = \bar{R} \subset R_n^\circ \subset R_n$. Hence, [C I.36.2b.], $\varrho^\times - \varrho$ is covered by $R_n - R$.

Taking into account the hypothesis, stating that R is well selected, [C I.36.1.], we get

(4) $$\bar{R} =^{J'} R.$$

Since

$$\bigcap_{n=1}^{\infty} R_n = \bar{R} =^{J'} R,$$

therefore

$$\bigcap_{n=1}^{\infty} R_n - R =^{J'} \emptyset;$$

hence

$$\bigcap_{n=1}^{\infty} (R_n - R) =^{J'} \emptyset.$$

Hence $\mu'(R_n - R) \to 0$, which gives $\varrho^\times - \varrho =^{J'} \emptyset$, (see [C.27.8.]); hence, as $\varrho \subseteqq \varrho^\times$, we get

(5) $$\varrho^\times =^{J^\bullet} \varrho.$$

C I.36.6. Let ϱ' be the set of all $\boldsymbol{B}_J$-traces ($\boldsymbol{B}_\nu$-traces) covered by R. We have, [C.23.],

$$[\varrho'] =^{J'} R.$$

Now, since by (4) and [C I.36.4., (0)] we have

$$R =^{J'} \bar{R} =^{J'} \operatorname{Vert}\varrho^\times,$$

it follows that

(6) $$[\varrho'] =^{J'} \operatorname{Vert}\varrho^\times.$$

We know that $\varrho^\times$ is measurable, and distinguished, [C I.31.3.], [C I.32.6., (3a)].

Since ϱ is the set of all $\boldsymbol{B}$-traces covered by R, and $\mathscr{V}(\varrho')$ is the set of all valid $\boldsymbol{B}$-traces covered by R, we get by (5)

(7) $$\mathscr{V}(\varrho') =^{J^\bullet} \varrho =^{J^\bullet} \varrho^\times.$$

The statements (6) and (7) prove that $\varrho^\times$ is a set of $\boldsymbol{B}$-traces associated with the set ϱ' of all $\boldsymbol{B}_J$-traces covered by R.

C I.36.7. We have considered the case where the μ'-measure of the boundary of the rectangle R is zero.

Now we are going to consider a more general case, where the left side and the lower side of the rectangle have the μ'-measure 0, but where the other sides may have a positive measure.

In precise terms consider the rectangle

$$R = \{(x, y) \mid x_1 < x \leqq x_2,\ y_1 < y \leqq y_2\},$$

where

$$x_1 < x_2, \quad y_1 < y_2,$$

where the lines $l = \{(x, y) \mid x = x_2\}$ and $m = \{(x, y) \mid y = y_2\}$ may have a positive μ'-measure.

Consider the infinite sequences

$$x^{(1)} > x^{(2)} > \cdots > x^{(n)} > \cdots > x_2,$$

$$y^{(1)} < y^{(2)} < \cdots < y^{(n)} < \cdots < y_2,$$

with

$$\lim_{n\to\infty} x^{(n)} = x_2, \quad \lim_{n\to\infty} y^{(n)} = y_2,$$

and where the rectangles

$$R_n =_{df} \left\{(x, y) \,\middle|\, \begin{array}{l} x_1 < x \leqq x^{(n)} \\ y_1 < y \leqq y^{(n)}, \end{array}\right\}$$

are well selected, [C I.36.1.].

We have

(1) $$\bigcap_{n=1}^{\infty} R_n = R.$$

Let ϱ', ϱ'_n be sets of all $\boldsymbol{B}_J$-traces covered by R, R'_n respectively.

We have

$$\varrho' =^{J^*} \bigcap_{n=1}^{\infty} \varrho'_n.$$

Indeed we have

$$[\varrho'_n] = R_n, \quad [\varrho'] = R.$$

Hence from (1), [C.23.],

$$\bigcap_{n=1}^{\infty} [\varrho'_n] = [\varrho_n], \text{ i.e. } \left[\bigcap_{n=1}^{\infty} \varrho'_n\right] = [\varrho'].$$

It follows that

(2) $$\bigcap_{n=1}^{\infty} \varrho' =^{J^*} \varrho.$$

Since ϱ'_n satisfies the condition Γ, it follows, [C I.35.5.], that $\bigcap_{n=1}^{\infty} \varrho'_n$ satisfies this condition. Hence from (2) we deduce that ϱ' satisfies the condition Γ. Let ϱ be the set of all $\boldsymbol{B}$-traces covered by R and ϱ_n the set of all $\boldsymbol{B}$-traces covered by R_n.

Then by [C I.35.], $\varrho^{\times}$ and $\varrho_n^{\times}$ are associated with ϱ' and ϱ_n' respectively. It follows that $\bigcap_{n=1}^{\infty} \varrho_n^{\times} = \varrho^{\times}$ is associated with ϱ'.

C I.36.8. Having settled the above case, let us proceed to the most general bounded rectangle R.

Consider the sequences

$$x_2 > x^{(1)} > x^{(2)} > \cdots > x^{(n)} > \cdots > x_1,$$

$$y_2 > y^{(1)} > y^{(2)} > \cdots > y^{(n)} > \cdots > y_1,$$

where

$$\lim x^{(n)} = x_1, \qquad \lim y^{(n)} = y_1,$$

and where the lines

$$l^{(n)} =_{df} \{(x, y) \mid x = x^{(n)}\},$$

$$m^{(n)} =_{df} \{(x, y) \mid y = y^{(n)}\},$$

have the μ'-measure 0.

The rectangles

$$R_n =_{df} \left\{(x, y) \,\middle|\, \begin{matrix} x^{(n)} < x \leqq x_2 \\ y^{(n)} < y \leqq y_2, \end{matrix}\right\}$$

belong to the type already considered.

We have

$$R = \bigcup_{n=1}^{\infty} R_n.$$

Denote by ϱ', ϱ_n' the collection of all $\boldsymbol{B}_J$-traces, covered by R, R_n respectively.

Since

$$\varrho =^{J^*} \bigcup_{n=1}^{\infty} \varrho_n',$$

and since the sets ϱ_n' have the property Γ, therefore by [C I.35.1.], ϱ also has that property.

If we denote by ϱ, ϱ_n the set of $\boldsymbol{B}$-traces covered by R, R_n respectively, we get $\varrho^{\times}$ as associate to R. Indeed

$$R = [\varrho'], \qquad R_n = [\varrho_n'].$$

Hence $\varrho =^{J'} \bigcup_{n=1}^{\infty} \varrho_n$. Indeed, the associate to

$$\bigcup_{n=1}^{\infty} \varrho_n \quad \text{is} \quad \bigcup_{n=1}^{\infty} \varrho_n^{\times} = \varrho^{\times}.$$

C I.37. We have proved that if R is a bounded rectangle and ϱ' is the set of all $\boldsymbol{B}_J$-traces covered by R, then ϱ' has the property Γ, and $\varrho^{\times}$ is associated with ϱ', where ϱ' is the set of all $\boldsymbol{B}_J$-traces covered by R.

We also have

$$[\varrho'] =^{J'} R.$$

The above property can be extended to arbitrary $\boldsymbol{B}_J$-coverings. Indeed, let L be a $\boldsymbol{B}_J$-covering, and ξ' the set of all $\boldsymbol{B}_J$-traces covered by L. By [C I.33.4.], the set ξ' is μ^*-measurable and

$$[\xi] =^{J'} L.$$

Let $L =^{J'} \sum_{n=1}^{\infty} R_n$, where R_n are disjoint bounded bricks. If we denote by ξ'_n the set of all $\boldsymbol{B}_J$-traces covered by R_n, we have, [C I.33.4.],

$$\xi' =^{J^*} \bigcup_{n=1}^{\infty} \xi'_n. \tag{1}$$

Since all ξ'_n have the property Γ, it follows that $\bigcup_{n=1}^{\infty} \xi'_n$ has the same property, [C I.35.1.]. Hence by (1) and [C I.35a.], ξ' also has that property. If ξ_n is associated with ξ'_n, then $\bigcup_{n=1}^{\infty} \xi_n$ is associated with ξ'.

In particular, since every not bounded rectangle is a covering, which can be represented as a denumerable union of bounded bricks, we see that every rectangle, bounded or not, has the property Γ.

C I.37.1. Having that, let η' be a μ^*-measurable set of $\boldsymbol{B}_J$-traces. Take, [B I.7.5.], any sequence of coverings of ξ':

$$L_1 \geqq L_2 \geqq \cdots \geqq L_m \geqq \cdots$$

such that

$$\prod_{n=1}^{\infty} L_n = [\xi'].$$

Let $\lambda'_1, \lambda'_2, \ldots, \lambda'_m, \ldots$ be the sets of all $\boldsymbol{B}_J$-traces, covered by L_n respectively.

We have, [C I.33.5.],

$$\prod_{m=1}^{\infty} \xi'_m =^* \lambda'_m.$$

Since all λ'_m have the property Γ, it follows that

$$\prod_{m=1}^{\infty} \lambda'_m$$

has that property. Hence, by [C I.35.5.], ξ' has the property Γ.

C I.38. The following theorem has been proved:

Fundamental trace theorem. If ξ' is a μ^*-measurable set of $\boldsymbol{B}_J$-traces, then it *has the property* Γ. This means the following, [C I.35.]:

There exists a $\mu^\bullet$-measurable and distinguished set ξ of $\boldsymbol{B}$-traces, such that

1) $\mathscr{V}\,\xi' =^{J^\bullet} \xi = \xi^\times$,

2) $[\xi'] =^{J'} \mathrm{Vert}\,\xi$.

(The set ξ is termed *associated with* ξ'.)

Concerning $\mathscr{V}$, $=^{J^\bullet}$, $=^{J'}$, Vert, see [C I.24.1.], [C I.30.], [C I.32.], respectively.

We are going to prove a theorem whose feature reminds [C I.35a.].

C I.38.1. Theorem. If the set ξ of $\boldsymbol{B}$-traces is associated with θ^* (i.e. the empty set of $\boldsymbol{B}_J$-traces),
then

1) $\xi =^{J^\bullet} \theta$ (i.e. the empty set of $\boldsymbol{B}$-traces),

2) $\mathrm{Vert}\,\xi =^{J'} \mathit{O}$ (i.e. the empty set of points of the plane).

Proof. As ξ is associated with θ^*, we have, by definition,

(1) $$\mathscr{V}(\theta^*) =^{J^\bullet} \xi = \xi^\times$$
and
(2) $$[\theta^*] =^{J'} \mathrm{Vert}\,\xi.$$

Now we know that, [C I.35.2.], θ is associated with θ^*. Hence

(3) $$\mathscr{V}(\theta^*) =^{J} \theta,$$

(4) $$[\theta^*] =^{J'} \theta.$$

From (1) and (3) we get
(5) $$\xi =^{J^\bullet} \theta,$$
and from (4) and (2),
$$\mathrm{Vert}\,\xi =^{J'} \mathrm{Vert}\,\theta.$$
Since, by [C I.32.1.],
$$\mathrm{Vert}\,\theta = \mathit{O},$$
we get
(6) $$\mathrm{Vert}\,\xi =^{J'} \mathit{O},$$

(5) and (6) make up the thesis.

C I.38.2. Theorem. If ξ is distinguished, $\mu^\bullet$-measurable and an ordinary set of $\boldsymbol{B}$-traces, such that

(1) $$\xi =^{J^\bullet} \theta,$$
and
(2) $$\mathrm{Vert}\,\xi =^{J'} \mathit{O},$$

then ξ is associated with θ^*.

Proof. We know that θ is associated with θ^*. Hence

(3) $$\mathscr{V}(\theta^*) =^{J^\bullet} \theta,$$

(4) $$[\theta^*] =^{J'} \mathrm{Vert}\,\theta.$$

From (1) and (3) we get

(5) $$\mathscr{V}(\theta^*) =^{J^\bullet} \xi,$$

and from (2) and (4),

(6) $$[\theta^*] =^{J'} \operatorname{Vert}\theta =^{J'} \mathfrak{Q} =^{J'} \operatorname{Vert}\xi.$$

By (5) and (6) the theorem is proved.

Remark. The last two theorems can be stated as follows: If ξ is associated with the set ξ' such that $\mu^* \xi = 0$, then

$$\mu^\bullet \xi = 0 \quad \text{and} \quad \mu'(\operatorname{Vert}\xi) = 0.$$

If $\mu'(\operatorname{Vert}\xi) = 0$, and $\mu^\bullet \xi = 0$, and ξ is associated with ξ', then $\mu^* \xi' = 0$.

C I.38.3. Theorem. If

1) ξ_1, ξ_2 are distinguished, ordinary and $\mu^\bullet$-measurable sets of $\boldsymbol{B}$-traces,

2) $\xi_1 =^{J^\bullet} \xi_2$,

3) $\operatorname{Vert}\xi_1 =^{J'} \operatorname{Vert}\xi_2$,

4) ξ is associated with a μ^*-measurable set ξ' of $\boldsymbol{B}_J$-traces,

then ξ_2 is so.

Proof. Since ξ_1 is associated with ξ', we have

$$\mathscr{V}(\xi') =^{J^\bullet} \xi_1,$$

and

$$[\xi'] =^{J'} \operatorname{Vert}\xi_1.$$

It follows from hypothesis 2) and 3):

$$\mathscr{V}(\xi') =^{J^\bullet} \xi_2,$$

$$[\xi'] =^{J'} \operatorname{Vert}\xi_2,$$

so ξ_2 is associated with the set ξ' of $\boldsymbol{B}_J$-traces, q.e.d.

C I.38.4. Theorem. If ξ_1, ξ_2 are sets of $\boldsymbol{B}$-traces, associated with the set ξ' of $\boldsymbol{B}_J$-traces, then $\xi_1 =^{J^\bullet} \xi_2$ and $\operatorname{Vert}\xi_1 =^{J'} \operatorname{Vert}\xi_2$.

Proof. We have

$$\mathscr{V}(\xi') =^{J^\bullet} \xi_1 \quad \text{and} \quad \mathscr{V}(\xi') =^{J^\bullet} \xi_2;$$

hence

$$\xi_1 =^{J^\bullet} \xi_2.$$

We have

$$[\xi'] =^{J'} \operatorname{Vert}\xi_1,$$

and

$$[\xi'] =^{J'} \operatorname{Vert}\xi_2;$$

hence

$$\operatorname{Vert}\xi_1 =^{J'} \operatorname{Vert}\xi_2,$$

so the theorem is established.

C I.38.4a. Theorem. If ξ_1', ξ_2' have the same set ξ, associated with each of them, then

$$\xi_1' =^{J^*} \xi_2'.$$

Proof. We have

$$[\xi_1'] =^{J'} \operatorname{Vert} \xi,$$

$$[\xi_2'] =^{J'} \operatorname{Vert} \xi.$$

It follows

$$[\xi_1'] =^{J'} [\xi_2'],$$

which gives, [B I.],

$$\xi_1' =^{J^*} \xi_2', \quad \text{q.e.d.}$$

C I.38.5. Lemma. If ξ is $\mu^\bullet$-measurable, then $\Delta\,\xi$ is also $\mu^\bullet$-measurable, and we have

$$\mu^\bullet\,\xi = \mu^\bullet\,\Delta\,\xi.$$

We also have

$$\xi =^{J^\bullet} \Delta\,\xi.$$

Proof. By definition [C I.26.]

$$\mathscr{V}^{-1}\,\Delta\,\xi$$

is μ^*-measurable. Since

$$\Delta\,(\Delta\,\xi) = \Delta\,\xi,$$

the set

$$\mathscr{V}^{-1}\,\Delta\,(\Delta\,\xi)$$

is μ^*-measurable; hence by [C I.26.] $\Delta\,\xi$ is μ^*-measurable.

By definition, [C I.27. Def.], we have

$$\mu^\bullet(\Delta\,\xi) = \mu^*\big(\mathscr{V}^{-1}\,\Delta\,(\Delta\,\xi)\big) = \mu^*\,(\mathscr{V}^{-1}\,\Delta\,\xi) = \mu^\bullet\,\xi.$$

We have $\Delta\,\xi \subseteqq \xi$ and we know that $\xi - \Delta\,\xi$ is a set of non valid traces only. Hence $\xi - \Delta\,\xi$ is $\mu^\bullet$-measurable, [C I.26.6.]. We have $\xi \dot{+} \Delta\,\xi = \xi - \Delta\,\xi \in J^\bullet$, because

$$\mu^\bullet\,(\xi - \Delta\,\xi) = \mu^\bullet\,(\xi) - \mu^\bullet\,(\Delta\,\xi) = 0.$$

Hence

$$\xi =^{J^\bullet} \Delta\,\xi,$$

[C I.28a.].

C I.38.6. Lemma. If ξ, η are $\mu^\bullet$-measurable and

(1) $$\xi =^{J^\bullet} \eta,$$

then

(2) $$\mu^\bullet\,\xi = \mu^\bullet\,\eta.$$

Proof. (1) means that

(3) $$\mu^\bullet\,(\xi \dot{+} \eta) = 0.$$

Now, we have in [A.18.] the theorem, which in the present circumstances reads as follows:

"If

(4) $$\mu^{\bullet}(\xi \dotplus \eta) \leqq \varepsilon, \quad (\varepsilon > 0),$$

then

(5) $$|\mu^{\bullet}\xi - \mu^{\bullet}\eta| \leqq 2\varepsilon."$$

Since (3) is true, the refore (4) is true for all $\varepsilon > 0$. Therefore by (5)

$$|\mu^{\bullet}\xi - \mu^{\bullet}\eta| = 0$$

i.e.

$$\mu^{\bullet}\xi = \mu^{\bullet}\eta.$$

C I.38.6a. Theorem. If η is a distinguished $\mu^{\bullet}$-measurable and ordinary set of $\boldsymbol{B}$-traces, then there exists a μ^*-measurable set ξ' of $\boldsymbol{B}_J$-traces, such that if ξ is any its associated, then

$$\eta =^{J^{\bullet}} \xi.$$

Proof. Put

(1) $$\eta' =_{df} \mathscr{V}^{-1}(\Delta\eta).$$

This set is μ^*-measurable, [C I.26.], because η is $\mu^{\bullet}$-measurable. We have

$$\mu^{\bullet}\eta = \mu^*(\mathscr{V}^{-1}\Delta\eta) = \mu^*\eta'.$$

The set η' is $\mu^{\bullet}$-measurable; hence it possesses a set associated with it, say ξ. We have

(2) $$\mathscr{V}(\eta') =^{J^{\bullet}} \xi.$$

From (1) we get

$$\mathscr{V}(\eta') = \mathscr{V}\mathscr{V}^{-1}(\Delta\eta) = \Delta\eta.$$

From [C I.38.5.] we have

$$\eta =^{J^{\bullet}} \Delta\eta;$$

hence from (2)

$$\eta =^{J^{\bullet}} \xi. \quad \text{q.e.d.}$$

C I.38.7. Remark. The last theorem does not say that every distinguished, $\mu^{\bullet}$-measurable, ordinary set of $\boldsymbol{B}$-traces can be considered as a set associated with some μ^*-measurable set of $\boldsymbol{B}_J$-traces. It only says that if we take a saturated class of mutually $J^{\bullet}$-equivalent sets of such $\boldsymbol{B}$-traces, then among them there exists a set of $\boldsymbol{B}$-traces which is associated to some μ^*-measurable set of $\boldsymbol{B}_J$-traces.

C I.39. Def. It will be convenient to denote by Γ the correspondence which attaches to every μ^*-measurable set ξ' of $\boldsymbol{B}_J$-traces, every set ξ associated to it. We agree to write

$$\xi' \curvearrowright^{\Gamma} \xi.$$

C I.39a. Theorem. If $\xi' \smile^{\Gamma} \xi$, then $\mu^* \xi' = \mu^{\bullet} \xi = \mu'(\text{Vert}\,\xi)$.

Proof. We have

$$\mathscr{V}\, \xi' =^{J^{\bullet}} \xi.$$

Hence, by [C I.38.6.],

$$(1) \qquad \mu^{\bullet}\, \mathscr{V}\, \xi' = \mu^{\bullet}\, \xi.$$

From (1) it follows by [C I.38.5.]:

$$\varDelta(\mathscr{V}\, \xi') =^{J^{\bullet}} \xi,$$

and then

$$\mu^{\bullet}\, \varDelta(\mathscr{V}\, \xi') = \mu^{\bullet}\, \xi.$$

By definition we get

$$(2) \qquad \mu^* [\mathscr{V}^{-1} \varDelta(\mathscr{V}\, \xi')] = \mu^{\bullet}\, \xi.$$

Now since $\mathscr{V}\, \xi'$ is composed of valid traces only, we have

$$\varDelta \mathscr{V}\, \xi' = \mathscr{V}\, \xi'.$$

Hence from (2) we get $\mu^* [\mathscr{V}^{-1} \mathscr{V}\, \xi'] = \mu^{\bullet}\, \xi$, i.e.

$$\mu^*(\xi') = \mu^{\bullet}\, \xi.$$

To prove the second equality of the thesis, we notice that

$$\mu'(\text{Vert}\,\xi) = \mu'[\xi'] = \mu^*\, \xi',$$

so the theorem follows.

C I.39.1. Theorem. The class $\boldsymbol{F}^{\Gamma}$ of all "associated" sets ξ, if ordered by set-inclusion (for sets of $\boldsymbol{B}$-traces) makes up a denumerably additive tribe $(\boldsymbol{F}^{\Gamma})$ with θ as zero and $\varXi$ as unit. [$\varXi$ is the set of all ordinary sets of $\boldsymbol{B}$-traces.]

Proof. Let

$$\xi'_n \smile^{\Gamma} \xi_n, \qquad (n = 1, 2, 3, \ldots),$$

then, by [C I.35.], we have

$$\bigcup_{n=1}^{\infty} \xi'_n \smile^{\Gamma} \bigcup_{n=1}^{\infty} \xi_n.$$

Hence if $\xi_n \in \boldsymbol{F}^{\Gamma}$, $(n = 1, 2, \ldots)$, then $\bigcup_{n=1}^{\infty} \xi_n \in \boldsymbol{F}^{\Gamma}$.

By [C I.35.4.] we have:

$$\text{if} \quad \xi \in \boldsymbol{F}^{\Gamma}, \quad \text{then} \quad \text{co}_{\varXi}\, \xi \in \boldsymbol{F}^{\Gamma}.$$

By [C I.35.3.], we have $\varXi \in \boldsymbol{F}^{\Gamma}$, and by [C I.35.2.], $\theta \in \boldsymbol{F}^{\Gamma}$, so the theorem is proved.

C I.39.2. Def. Consider the class J^{Γ} of all sets ξ belonging to $\boldsymbol{F}^{\Gamma}$, such that

$$\xi =^{J^{\bullet}} \theta, \qquad \text{Vert}\,\xi =^{J'} \mathit{\Omega}.$$

C I.39.3. Theorem. We shall prove that J^{Γ} is a denumerably additive ideal in $(\boldsymbol{F}^{\Gamma})$.

Proof. Let

$$\xi_1, \xi_2, \ldots, \xi_n, \ldots \in J^\Gamma.$$

We have

$$\xi_n =^J \theta, \quad \operatorname{Vert} \xi_n =^{J'} \Theta, \quad (n = 1, 2, \ldots).$$

We have

$$\bigcup \xi_n =^J \theta,$$

because of [C I.28a.]. We also have

$$\bigcup_{n=1}^{\infty} \operatorname{Vert} \xi_n =^{J'} \Theta.$$

It follows, by [C I.33.1.], that

$$\operatorname{Vert} \bigcup_{n=1}^{\infty} \xi_n =^{J'} \Theta.$$

Thus the union of a denumerable number of elements of J^Γ, also belongs to J^Γ.

Now let $\xi \in J^\Gamma$, and let ξ_1 be a subset of ξ with $\xi_1 \in F^\Gamma$. We have

$$\xi =^{J^\bullet} \theta, \quad \operatorname{Vert} \xi =^{J'} \Theta.$$

Since

$$\xi_1 \subseteqq \xi \quad \text{and} \quad \xi \in J^\bullet, \quad \text{therefore} \quad \xi_1 =^{J^\bullet} \theta.$$

On the other hand, since $\xi_1 \subseteqq \xi$, we have $\operatorname{Vert} \xi_1 \subseteqq \operatorname{Vert} \xi$; hence

$$\operatorname{Vert} \xi_1 =^{J'} \Theta.$$

Consequently any subset of a set belonging to J^Γ, belongs to J^Γ. The theorem is proved.

C I.39.3a. Theorem. The following are equivalent:

I. $\xi =^{J^\Gamma} \eta$,

II. $\xi =^{J^\bullet} \eta$, $\operatorname{Vert} \xi =^{J'} \operatorname{Vert} \eta$.

Proof. Straight forward.

C I.39.3b. Theorem. Thus J^Γ is a denumerably additive ideal in the tribe $(\boldsymbol{F}^\Gamma)$ of all associated sets. The ideal J^Γ reorganizes $(\boldsymbol{F}^\Gamma)$ into the tribe $(\boldsymbol{F}^\Gamma_{J^\Gamma})$, modulo J^Γ.

Def. We define $\leqq^{J^\Gamma}$ as usually we do with ideals:

$$\xi_1 \leqq^{J^\Gamma} \xi_2$$

means:

$$\xi_1 - \xi_2 \in J^\Gamma,$$

and

$$\xi_2 =^{J^\Gamma} \xi_2$$

means:

$$\xi_1 \dotplus \xi_2 \in J^\Gamma.$$

C I.39.4. Theorem. If

$$\xi' \smile^{\Gamma} \xi$$

and

$$\xi =^{J^{\Gamma}} \xi_1, \qquad \xi' =^{J^*} \xi_1',$$

then

$$\xi_1' \frown^{\Gamma} \xi_1.$$

Proof. We have $\xi' \smile^{\Gamma} \xi$, and $\xi' =^{J^*} \xi_1'$. Hence, [C I.35 a.], we have $\xi_1' \smile^{\Gamma} \xi$.

Now since $\xi =^{J^{\Gamma}} \xi_1$, we get the equations:

$$\xi =^{J^{\bullet}} \xi_1 \quad \text{and} \quad \operatorname{Vert} \xi =^{J'} \operatorname{Vert} \xi_1.$$

Hence, [C I.39.3 a.],

$$\xi_1' \smile^{\Gamma} \xi_1.$$

C I.40. We shall apply the fundamental trace-theorem [C I.38.] to functions of variable traces.

Def. Let $f(\alpha')$ be a real-number-valued function of the variable $\boldsymbol{B}_J$-trace α', defined on a μ^*-measurable set ξ' of $\boldsymbol{B}_J$-traces.

We say, by definition, that $f(\alpha')$ is μ^*-measurable whenever for every real number p the set

$$\{\alpha' \mid f(\alpha') \leqq p\} \quad \text{is } \mu^*\text{-measurable.}$$

We shall also say that the function $f(\alpha^{\bullet\prime})$ is *fitting the tribe* $(\boldsymbol{F}^*)$ of all μ^*-measurable sets of $\boldsymbol{B}_J$-traces (the dot on α' emphasizes that α' is a variable trace).

We extend this notion to complex-valued functions. We say that such a function is μ^*-measurable whenever its real part and its imaginary part are μ^*-measurable. In presentation the theory of these function we shall follow the pattern of abstract Fréchet's integrals.

We recall that J^* is the ideal in $(\boldsymbol{F}^*)$, composed of all μ^*-measurable sets of $\boldsymbol{B}_J$-traces, with μ^*-measure equal 0.

C I.40.1. Def. We say that $f(\alpha^{\bullet\prime})$ and $g(\alpha^{\bullet\prime})$ are J^*-equal:

$$f(\alpha^{\bullet\prime}) =^{J^*} g(\alpha^{\bullet\prime})$$

whenever

1) $\mathrm{Q} f =^{J^*} \mathrm{Q} g$,
2) the set $\{\alpha' \mid f(\alpha') \neq g(\alpha')\} =^{J^*} \theta^*$.

C I.40.2. The following true theorems are taken over from the general theory of Fréchet's integrals.

1) The J^*-equality of functions obey the ordinary laws of identity.
2) The μ^*-measurable functions are just those which fit the tribe $(\boldsymbol{F}^*)$ of μ^*-measurable sets of $\boldsymbol{B}_J$-traces.

3) If $f(\alpha^{\bullet\prime})$, $g(\alpha^{\bullet\prime})$ are both μ^*-measurable, then

$$f(\alpha^{\bullet\prime})+g(\alpha^{\bullet\prime}),\quad f(\alpha^{\bullet\prime})\cdot g(\alpha^{\bullet\prime}),\quad p\cdot f(\alpha^{\bullet\prime})$$

where p is a complex number, are all μ^*-measurable.

4) The above operations on functions are $(=^{J^*})$-invariant.

5) If $f_n(\alpha^{\bullet\prime})$ are μ^*-measurable, $(n=1,2,\dots)$ and $\lim_n f_n(\alpha^{\bullet\prime})$ exists for all α', excepting perhaps a set of μ^*-measure 0, then the above limit is also a μ^*-measurable function.

6) The operation $\lim_n f_n(\alpha^{\bullet\prime})$ is $(=^{J^*})$-invariant.

C I.41. Def. A function $f(\alpha^{\bullet\prime})$, fitting the tribe $(\boldsymbol{F}^*)$, is said to be simple on the μ^*-measurable set of $\boldsymbol{B}_J$-traces, whenever it admits only a finite or denumerable number of values. Given such a function there exists:

1) a sequence of complex numbers

$$\lambda_1, \lambda_2, \dots, \lambda_n, \dots,$$

all different,

2) a sequence of disjoint sets of $\boldsymbol{B}_J$-traces:

$$\xi', \xi'_1, \xi'_2, \dots, \xi'_n, \dots, \tag{1}$$

such that $\bigcup_n \xi'_n = \xi'$, and such that if $\alpha' \in \xi_n$, then $f(\alpha') = \lambda_n$, $(n=1,2,\dots)$. For such a function all ξ_n are μ^*-measurable sets of $\boldsymbol{B}_J$-traces.

C I.41.1. Given a function as above, consider any sequence

$$\xi, \xi_1, \xi_2, \dots, \xi_n, \dots$$

of distinguished, ordinary and $\mu^\bullet$-measurable sets of $\boldsymbol{B}$-traces such that

$$\xi'_n \cup^{\Gamma} \xi_n, \quad n=1,2,\dots$$

and where

$$\xi' \cup^{\Gamma} \xi.$$

We like to change these sets into other ones which would be disjoint.

To do that, notice that

$$\begin{cases} \xi'_1 = \xi'_1, \\ \xi'_2 = \xi'_2 - \xi'_1, \\ \xi'_3 = \xi'_3 - (\xi'_1 \cup \xi'_2), \\ \dots\dots\dots\dots\dots \\ \xi'_n = \xi'_n - (\xi'_1 \cup \xi'_2 \cup \dots \cup \xi'_{n-1}), \\ \dots\dots\dots\dots\dots\dots\dots\dots \end{cases} \tag{1}$$

because the sets $\xi'_1, \xi'_2, \dots, \xi'_n, \dots$ are mutually disjoint.

By virtue of [C I.35.1.], [C I.35.4.], [C I.35.5.] the sets corresponding to the sets on the right side in (1) are

$$(2)\qquad \begin{cases} \zeta_1 =_{df} \xi_1, \\ \zeta_2 =_{df} \xi_2 - \xi_1, \\ \zeta_3 =_{df} \xi_3 - (\xi_1 \cup \xi_2), \\ \dots\dots\dots\dots\dots\dots \\ \zeta_n =_{df} \xi_n - (\xi_1 \cup \xi_2 \cup \dots \cup \xi_{n-1}), \\ \dots\dots\dots\dots\dots\dots\dots\dots\dots \end{cases}$$

Consequently the Γ-correspondents of

$$\xi', \xi', \dots, \xi', \dots \quad \text{are} \quad \zeta_1, \zeta_2, \dots, \zeta_n, \dots$$

respectively, and we see that the last sets are disjoint. We also have

$$\xi' = \bigcup_n \xi'_n \cup^\Gamma \bigcup_n \zeta_n.$$

Putting

$$\zeta =_{df} \bigcup_n \zeta_n$$

we have

$$\xi' \cup^\Gamma \zeta.$$

Having that consider the function $F(\alpha^\bullet)$ of the variable **B**-trace α, defined on ζ as follows.

$F(\alpha) =_{df} \lambda_n$ whenever $\alpha \in \zeta_n$, $(n = 1, 2, \dots)$.

The function $F(\alpha^\bullet)$ admits the values λ_n only, and on disjoint sets ζ_n associated with the sets ξ_n, which are determined by the function $f(\alpha^\bullet)$.

We call $F(\alpha^\bullet)$ *simple function associated with* $f(\alpha')$, and write

$$f(\alpha^{\bullet\prime}) \cup^\Gamma F(\alpha^\bullet).$$

C I.41.2. We see that, if we change our construction by choosing different η_n, η which are associated with the given ξ'_n, ξ' respectively, we get a function $G(\alpha^\bullet)$, which is equivalent to $F(\alpha^\bullet)$ in the following sense:

$$(3)\qquad \eta_n =^{J^\Gamma} \xi_n, \quad \eta =^{J^\Gamma} \xi.$$

We agree to write

$$F(\alpha^\bullet) =^{J^\Gamma} G(\alpha^\bullet).$$

For real-valued functions, the said equivalence can be expressed as follows: for every real number λ we have

$$\{\alpha \mid F(\alpha) \leqq \lambda\} =^{J^\Gamma} \{a \mid G(\alpha) \leqq \lambda\}.$$

C I.41.3. Theorem. The function $F(\alpha^\bullet)$, associated to $f(\alpha^\bullet)$, has the property:

If $\alpha_1, \alpha_2, \alpha_3$ are neighbour of the trace $\alpha_1 \in \zeta_1$, then

$$F(\alpha_1) = F(\alpha_2) = F(\alpha_3) = F(\alpha).$$

It means that the function $F(\alpha)$ admits four times the same value, namely at each point p of Vert ζ, on four neighboring traces with vertex p. This is true on account of the fact that the sets ζ, ζ_n are all distinguished, [C I.31.1.].

C I.41.4. From the general theory of Lebesgue-Fréchet's integrals we know that if $f(\alpha')$ is a function fitting the tribe $(\boldsymbol{F}^*)$, then there exists an infinite sequence $\varphi_n(\alpha^{\bullet\prime})$ of simple functions, all fitting $(\boldsymbol{F}^*)$, and such that $f(\alpha')$ is the uniform limit of $\varphi_n(\alpha')$:

$$\lim_{n\to\infty} \varphi_n(\alpha') = f(\alpha'). \tag{1}$$

Take a function $f(\alpha')$ and a sequence $\varphi_n(\alpha')$ as above. Let us attach to $\varphi_n(\alpha')$ the function $\Phi_n(\alpha)$, (as it has been made in [C I.41.1.]) associated with φ_n. From (1) it follows that $\Phi_n(\alpha)$ also possesses a uniform limit which is a function $F(\alpha)$, associated to $f(\alpha')$.

In addition to that, if we change the functions $\varphi_n(\alpha')$ into another ones $\psi_n(\alpha')$ with

$$\varphi_n =^{J^*} \psi_n, \quad (n = 1, 2, \ldots),$$

the above procedure gives a function $G(\alpha)$ such that

$$f(\alpha') \circlearrowleft^{\Gamma} G(\alpha), \quad G(\alpha) =^{J^\Gamma} F(\alpha).$$

In this way we attach to every μ^*-measurable function $f(\alpha')$ an associated function $F(\alpha)$, such that if $\alpha_1, \alpha_2, \alpha_3$ are neighbours of α, then

$$F(\alpha) = F(\alpha_1) = F(\alpha_2) = F(\alpha_3).$$

The function $F(\alpha)$ is fitting the tribe $(\boldsymbol{F}^\Gamma)$ of associated sets, [C I.39.1.].

We got the correspondence between μ^*-measurable functions and their associated ones. We denote the correspondence also with Γ, so we write

$$f(\alpha^{\bullet\prime}) \circlearrowleft^{\Gamma} F(\alpha).$$

This correspondence is invariant in its domain with respect to equality $(=^{J^*})$ and in its range with respect to equality $(=^{J^\Gamma})$. The correspondence Γ preserves addition and multiplication of functions.

By [C I.39a.] we have for real valued functions

$$\mu^* \{\alpha' \mid f(\alpha') \leq \lambda\} = \mu^\bullet \{\alpha \mid F(\alpha) \leq \lambda\}$$

and for complex valued ones: if E is an open set of complex numbers, then

$$\mu^* \{\alpha' \mid f(\alpha') \in E\} = \mu^\bullet \{\alpha \mid F(\alpha) \in E\}.$$

C I.42. Consider a function $F(\alpha^\bullet)$ fitting the tribe $(\boldsymbol{F}^\Gamma)$, i.e. an "associated" function of the variable $\boldsymbol{B}$-trace α. We shall attach to it an ordinary function. $\Phi(z)$ of the complex variable z; this is possible owing to the fact that $F(\alpha^\bullet)$ admits the same value on all four neighboring traces. To do that let us start again with simple functions. Let $F(\alpha^\bullet)$ be such a function, which admits the values

$$(1) \qquad \lambda_1, \lambda_2, \ldots, \lambda_n, \ldots$$

on disjoint "associated" sets of $\boldsymbol{B}$-traces

$$\xi_1, \xi_2, \ldots, \xi_n, \ldots,$$

and let attach to it the function $\Phi(z)$ which admits the values (1) on the sets of complex numbers:

$$\operatorname{Vert} \xi_1, \operatorname{Vert} \xi_2, \ldots, \operatorname{Vert} \xi_n, \ldots$$

These sets are disjoint and are μ'-measurable. Applying the method of uniform approximation of functions by simple functions, we get to every "associated" function $F(\alpha)$, i.e. fitting the $(\boldsymbol{F}^\Gamma)$, a function $\Phi(z)$ of the complex variable z, fitting the tribe $(\boldsymbol{F}')$.

This correspondence preserves the measure of sets, i.e. if E is any open set of complex numbers, then

$$\mu^\bullet\{\alpha \mid F(\alpha) \in E\} = \mu'\{z \mid \Phi(z) \in E\}.$$

C I.42.1. Def. Thus we have finally attached to every μ^*-measurable function $f(\alpha')$ of $\boldsymbol{B}_J$-trace a μ'-measurable function $\Phi(z)$ of the complex variable.

Let us call this correspondence $\mathscr{M}$, so we have

$$F(\alpha^\bullet) \smile^{\mathscr{M}} \Phi(z^\bullet)$$

and we shall use the same letter $\mathscr{M}$, to denote the correspondence between sets

$$\xi' \smile^{\mathscr{M}} \operatorname{Vert} \xi, \qquad (\text{where } \xi' \smile^{\Gamma} \xi).$$

C I.43. The generalized fundamental trace-theorem. We can state the following theorem, which was the aim of the Chapter [C I].

Under usual conditions for $(\boldsymbol{F})$, $\boldsymbol{B}$, μ, and if μ is an extensible measure on $(\boldsymbol{F})$, there exists the correspondence $\mathscr{M}$ which carries the μ^*-measurable sets ξ' of $\boldsymbol{B}_J$-traces into μ'-measurable sets Z of complex numbers, and which has the following properties:

1) $\mathscr{M}$ preserves finite and denumerably infinite operations on sets,

2) $\mathscr{M}$ preserves the measurability and measure of measurable sets,

3) the domain of $\mathscr{M}$ is the set $\boldsymbol{F}^*$ of all μ^*-measurable sets of $\boldsymbol{B}_J$-traces. The range of $\mathscr{M}$ is the set $\boldsymbol{F}'$ of all μ'-measurable sets of complex numbers,

4) $\mathcal{M}$ is, in its domain, invariant with respect to the equality $(=^{J^*})$ of μ^*-measurable sets of $\boldsymbol{B}_J$-traces. $\mathcal{M}$ is, in its range, invariant with respect to the equality $(=^{J'})$ of μ'-measurable sets of complex numbers,

5) in this respect $\mathcal{M}$ is a $1 \to 1$ correspondence.

$\mathcal{M}$ induces a correspondence, also denoted by $\mathcal{M}$, between complex valued functions $f(\alpha^{\bullet\prime})$ of the variable $\boldsymbol{B}_J$-trace α', and complex valued functions $\Phi(z^\bullet)$ of the complex variable z.

6) This correspondence preserves measurability of functions, summability and square summability. It preserves also the value of the integrals of functions on measurable sets.

The domain of the correspondence $\mathcal{M}$ is the collection of all μ^*-measurable functions $f(\alpha^{\bullet\prime})$ of the variable $\boldsymbol{B}_J$-trace α', and the range of the generalized $\mathcal{M}$ is the collection of all μ'-measurable functions $\Phi(z^\bullet)$ of the complex variable z.

7) The correspondence is invariant in its domain with respect to the equality $(=^{J^*})$ of functions $f(\alpha^{\bullet\prime})$, and it is invariant, in its range, with respect to the $(=^{J'})$-equality of functions $\Phi(z^\bullet)$.

In this respect the generalized correspondence $\mathcal{M}$ is $1 \to 1$.

8) The correspondence $\mathcal{M}$ preserves operations of addition, subtraction, multiplication and division of functions. It also preserves the limit of a sequence of functions.

C I.44. In this Chapter [C I.] we have considered several tribes and several measures which, however, are generated by a single extensible, finitely additive measure μ on the tribe $\boldsymbol{F}$ of figures on the plane. There are also several corresponding ideals to be taken into account.

To facilitate the readers orientation in the topic of [C I.], we are now going to give the list of notions and notations considered.

The starting point is the tribe $(\boldsymbol{F})$ of all figures of the plane and a finitely additive, finite, non-negative, non trivial and extensible measure μ on $(\boldsymbol{F})$. J is the ideal in $(\boldsymbol{F})$ of all figures f with $\mu(f) = 0$. J is finitely additive. It generates a new ordering in $\boldsymbol{F}$, $f \leqq^J g$ and a new governing equality $f =^J g$, defined by $f - g \in J$ and $f \dotplus g \in J$ respectively. We get the tribe $\boldsymbol{F}$ modulo J, denoted by $(\boldsymbol{F}_J)$.

The tribe $(\boldsymbol{F})$ admits a supertribe $(\boldsymbol{F}')$ composed of all borelian subsets of the plane. It is denumerably additive. The measure μ can be extended to (F') so as to become a denumerably additive measure μ'. The tribe $(\boldsymbol{F}')$ possesses the ideal J' defined as the collection of all borelian sets of P, having the μ'-measure equal 0. The ideal J' is denumerably additive. It generates the tribe $(\boldsymbol{F}')_{J'}$ with equality $=^{J'}$ and ordering $\leqq^{J'}$.

This is the background on which the whole theory has been built up. We have:

1) $(\boldsymbol{F})$ tribe of figures [C.2.1.]. μ measure on $(\boldsymbol{F})$, [C.19.], J ideal on $(\boldsymbol{F})$, [C.15.1.]. $(\boldsymbol{F}_J)$ tribe $\boldsymbol{F}$ taken modulo J, [C.15.]. $=^J$, $\leqq^J$ equality and ordering in $(\boldsymbol{F}_J)$, [C.16.].

2) $(\boldsymbol{F}')$ tribe of all borelian sets, μ' the extended measure μ, in $(\boldsymbol{F}')$. J' ideal in $(\boldsymbol{F}')$, [C.15.1.]. $(\boldsymbol{F}')_{J'}$ tribe $(\boldsymbol{F}')$ taken modulo J'. $=^{J'}$, $\leqq^{J'}$, equality and ordering in $(F')_{J'}$, [C.15.1.].

3) $\boldsymbol{B}$ base in $\boldsymbol{F}$, composed of generalized rectangles, [C.2.2.]. $\boldsymbol{B}$-traces, [C.3.1.]. θ empty set of $\boldsymbol{B}$-traces may be (also denoted by $\theta^\bullet$), [C I.24.2.]. W set of all $\boldsymbol{B}$-traces, [C I.25.2.].

4) $\boldsymbol{B}_J$ base in $(\boldsymbol{F}_J)$, corresponding to $\boldsymbol{B}$, [C.17.]. $\boldsymbol{B}_J$-traces in $(\boldsymbol{F}_J)$. θ^* empty set of $\boldsymbol{B}_J$-traces, [C I.24.2.]. W^* the set of all $\boldsymbol{B}_J$-traces, [C I.24.3.].

5) $\mathscr{V}$ correspondence which attaches to a $\boldsymbol{B}_J$-trace τ' a $\boldsymbol{B}$-trace τ, [C I.24.1.]. This correspondence is generalized to sets of traces, and also denoted by $\mathscr{V}$, so we have

$$\tau' \smile^{\mathscr{V}} \tau \text{ for traces,}$$

$$\xi' \smile^{\mathscr{V}} \xi \text{ for sets of traces.}$$

Valid (or acting) $\boldsymbol{B}$-trace; such that it is the $\mathscr{V}$-correspondent to a $\boldsymbol{B}_J$-trace, non valid $\boldsymbol{B}$-trace, such that there does not exists a $\mathscr{V}^{-1}$-corresponding $\boldsymbol{B}_J$-trace, [C.17.]. $\Delta\,\xi$ is the valid part of the set ξ of $\boldsymbol{B}$-traces. θ (or $\theta^\bullet$) the empty set of valid $\boldsymbol{B}$-traces. $W^\bullet$ the set of all valid $\boldsymbol{B}$-traces.

6) We consider the theory of measurability of sets of $\boldsymbol{B}_J$-traces, just applying the topic of [B I.]. If ξ' is a measurable set of $\boldsymbol{B}_J$-traces we consider the measure of ξ', denoted by $\mu^*(\xi')$. It is induced by $(\boldsymbol{F}')/J'$ according to the general theory exhibited in [B I.].

We call measurable sets of $\boldsymbol{B}_J$-traces μ^*-measurable and the corresponding measure by μ^*. We have

$$\mu^*(\xi') =_{df} \mu'[\xi'],$$

where $[\xi']$ is the coat of ξ'. But we do not consider the standard theory of measurability of $\boldsymbol{B}$-traces according to the general scheme given in [B I.]. We introduce an other kind of measurability of sets of $\boldsymbol{B}$-traces. $(\boldsymbol{F}^*)$ the tribe of all μ^*-measurable sets of $\boldsymbol{B}_J$-traces. J^* the ideal in $(\boldsymbol{F}^*)$ composed of all sets ξ' of $\boldsymbol{B}_J$-traces with

$$\mu^*(\xi') = 0.$$

$=^{J^*}$, $\leqq^{J^*}$ induced equality and ordering in $(F^*)_{J^*}$.

7) Measurable set ξ of $\boldsymbol{B}$-traces ($\mu^\bullet$-measurable), [C I.26.], are such that

$$\mathscr{V}^{-1}\Delta\,\xi \text{ is } \mu^*\text{-measurable,}$$

$$\mu^\bullet(\xi) =_{df} \mu^*(\mathscr{V}^{-1}\Delta\,\xi),$$

a special measure of sets ξ of B-traces, [C I.27.]. $(\boldsymbol{F}^{\bullet})$ the tribe of all $\mu^{\bullet}$-measurable sets of $\boldsymbol{B}_J$-traces. $J^{\bullet}$ the ideal in $(\boldsymbol{F}^{\bullet})$ composed of all sets ξ of $\boldsymbol{B}$-traces with

$$\mu^{\bullet}(\xi) = 0,$$

[C I.28a.]. $(\boldsymbol{F}^{\bullet})_{J^{\bullet}}$ the corresponding tribe $(\boldsymbol{F}^{\bullet})$ taken modulo $J^{\bullet}$. $=^{J^{\bullet}}$, $\leqq^{J^{\bullet}}$ the corresponding equality and, [C I.28.1.], ordering in $(\boldsymbol{F}^{\bullet})_{J^{\bullet}}$.

8) Ξ the set of all traces which are not at infinity, [C I.29.3.]. Ordinary traces: traces which are not at infinity, [C I.29.1.].

9) vert τ, the vertex of the $\boldsymbol{B}$-trace τ [C.8.1.]. Vert ξ is the set of vertices of all traces τ, belonging to the set ξ of $\boldsymbol{B}$-traces, [C I.32.]. $\xi^{\times}$ the set of all neighboring traces of all traces belonging to ξ (which is supposed to be an ordinary set), [C I.30.]. Distinguished set ξ; such that $\xi = \xi^{\times}$ (only for ordinary sets of traces), [C I.31.].

10) Property Γ of μ^{*}-measurable sets of $\boldsymbol{B}_J$-traces, [C I.35.], means: There exists a $\mu^{\bullet}$-measurable distinguished set ξ of $\boldsymbol{B}$-traces, such that

$$\mathscr{V}\ \xi' =^{J^{\bullet}} \xi$$

$[\xi'] =^{J'} \operatorname{Vert}\xi$, where $[\xi']$ is the coat of ξ'.

We say ξ is associated with ξ', [C I.35.].

Correspondence Γ, [C I.31.]. $(\boldsymbol{F}^{\Gamma})$ the tribe of associated sets ξ of $\boldsymbol{B}$-traces ordered by set-inclusion, [C I.39.1.]. θ is its zero, Ξ its unit. J^{Γ} ideal in $(\boldsymbol{F}^{\Gamma})$ defined as the class of all sets $\xi \in \boldsymbol{F}^{\Gamma}$ such that

$$\xi =^{J^{\bullet}} \theta \quad \text{and} \quad \operatorname{Vert}\xi =^{J'} \Theta,$$

[C I.39.2.]. $(\boldsymbol{F}^{J})_{J^{\Gamma}}$ the tribe $(\boldsymbol{F}^{\Gamma})$ taken modulo J^{Γ}. $=^{J^{\Gamma}}$, $\leqq^{J^{\Gamma}}$, equality, and ordering generated in $(\boldsymbol{F}^{\Gamma})$ by J^{Γ}.

The above notions of the equality generate analogous notions of equality for functions.

Chapter D

The lattice of subspaces of the Hilbert-Hermite space[1]

D.1. Def. Let **H** be a separable and complete Hilbert-Hermite-space, which may be of finite or infinite dimension. By a *subspace of* **H** we shall always understand any linear subvariety of **H**, which is closed with respect to the metric topology in **H**.

[1] It is admitted that the reader is familiar with the basic notions and properties related to Hilbert-Hermite space.

Thus the set $(\vec{O})$, composed of the single vector $\vec{O}$ is a subspace, and also the whole space **H** is a subspace. We shall not consider any "empty" space. We shall agree to say "*space*" instead of subspace of **H**.

D.1.1. Def. The space is a set of vectors, hence the relation of inclusion of spaces has a good meaning. We shall write "$\leqq$" to denote the *relation of inclusion of spaces*, e.g. $a \leqq b$. The relation $(\leqq)$ generates the notion of *equality* of spaces, e.g. $a = b$, which may be defined as $a \leqq b$, $b \leqq a$.

The notion of equality of spaces is the mere identity of sets of vectors, restricted to spaces.

D.1.2. We shall prove that the above relation of inclusion is a complete lattice, which admits $(\vec{O})$ as O and **H** as unit I, [A.1.4.2.], and where $(=)$ is the governing equality [A.4.].

D.1.2a. Theorem. First we shall prove the following: If $\{m_k\}$ is a non empty, indexed set of spaces, then their set-intersection $m =_{df} \bigcap_k m_k$ is also a space.

Proof. Let Let $\vec{X}$, $\vec{Y} \in m$. We have, for every k, $\vec{X}$, $\vec{Y} \in m_k$. Since m_k is a linear variety, we have $a\vec{X} + b\vec{Y} \in m_k$ whatever the complex numbers a, b may be. Thus $m = \prod_k m_k$ is a linear variety. Since the intersection of closed sets is closed, it follows that m is closed; hence a space. The assertion is proved.

D.1.2b. This being done, take a non empty class $\{a_\alpha\}$ of spaces and consider the class $\{b_\beta\}$ of all spaces b_β, such that $a_\alpha \subseteqq b_\beta$ for all α. The class $\{b_\beta\}$ is not empty, because the whole Hilbert-space belongs to it. Let b be the set-intersection of all b_β.

By what we have proved, b is a space. We also have $a_\alpha \subseteqq b$ for all α.

I say that if $a_\alpha \leqq b'$ for all α, then $b \leqq b'$.

Indeed, since $a_\alpha \subseteqq b'$ for all α therefore the space b' belongs to the class $\{b_\beta\}$. Hence $\bigcap_\beta b_\beta \subseteqq b'$, i.e. $b \leqq b'$.

Thus, [A.1.4.], we have proved that b is a lattice union of the collection $\{a_\alpha\}$ of spaces. Consequently any non empty set of spaces possesses the ordering-union.

D.1.2c. Let $\{a_\alpha\}$ be a set of spaces as before and let

$$a =_{df} \bigcap_\alpha a_\alpha .$$

We have $a \leqq a_\alpha$ for all α.

Now suppose that $a' \leqq a_\alpha$ for all α. It follows that $a' \subseteqq \bigcap_\alpha a_\alpha = a$, hence

$$a' \leqq a .$$

This proves that a is the ordering-intersection of the spaces $\{a_\alpha\}$.

We have proved that the ordering $\emptyset$ ($\subseteqq$) of inclusion of sets of vectors, if restricted to spaces, is a complete lattice, [A.1.4.4.].

D.1.2d. As $(\vec{O}) \subseteqq a \subseteqq \mathbf{H}$ for any space a, we see that the lattice admits $(\vec{O})$ as lattice-zero O and admits $\mathbf{H}$ as lattice unit I, [A.1.4.2.].

D.1.3. We shall use the signs $+$, $\sum$, $\cdot$, $\prod$ for joins and meets of the lattice. These operations have been introduced by STONE **(26)**. By [A.1.4.1b.] the following rules are valid:

$$a \cdot a = a; \quad a + a = a; \quad a \cdot b = b \cdot a; \quad a + b = b + a;$$

$$a \cdot (b \cdot c) = (a \cdot b) \cdot c; \quad a + (b + c) = (a + b) + c;$$

$$a \cdot (a + b) = a; \quad a + a \cdot b = a.$$

Since the lattice possesses the zero O and the unit I, we have

$$O \leqq a \leqq I; \quad a + I = I; \quad a \cdot I = a; \quad a + O = a; \quad a \cdot O = O.$$

D.1.4. Remark. The lattice union of spaces is not the set-union of spaces; $a + b$ is the smallest space containing a and b, and not the collection of vectors belonging either to a or to b. We have the following theorem:

D.1.5. Theorem. The union $a + b$ of spaces a, b is equal to the closure of the set of vectors:

$$E =_{df} \{\vec{X} + \vec{Y} \mid \vec{X} \in a, \vec{Y} \in b\}. \tag{1}$$

D.1.5a. Proof. First we shall prove that $\bar{E}$ is a space, (where the bar denotes closure in the $\mathbf{H}$-metric topology). Let $Z', Z'' \in \bar{E}$.

There exist infinite sequences of vectors

$$X'_n, Y'_n, X''_n, Y''_n$$

such that

$$X'_n, X''_n \in a, \; Y'_n, Y''_n \in b,$$

and

$$Z' = \lim_n (X'_n + Y'_n),$$

$$Z'' = \lim_n (X''_n + Y''_n).$$

Take any two complex numbers α, β.

We have

$$\begin{aligned} \alpha Z' + \beta Z'' &= \lim [\alpha (X'_n + Y'_n) + \beta (X''_n + Y''_n)] \\ &= \lim [(\alpha X'_n + \beta X''_n) + (\alpha Y'_n + \beta Y''_n)]. \end{aligned} \tag{2}$$

Since

$$\alpha X'_n + \beta X''_n \in a,$$

$$\alpha Y'_n + \beta Y''_n \in b.$$

We have, by (1),

$$R_n =_{df} (\alpha X'_n + \beta X''_n) + (\alpha Y'_n + \beta Y''_n) \in E.$$

From (2) we get

$$\alpha Z' + \beta Z'' = \lim R_n,$$

hence $\alpha Z' + \beta Z'' \in \bar{E}$.

Thus we have proved that $\bar{E}$ is a linear variety. Since $\bar{E}$ is closed, it follows that $\bar{E}$ is a space.

D.1.5b. We have $a \subseteqq \bar{E}$. Indeed

$$a = \{\vec{X} + \vec{O} \mid X \in a\} \subseteqq \bar{E},$$

and similarly for b.

It follows that

(3) $$a + b \subseteqq \bar{E}.$$

D.1.5c. To prove the inclusion inverse to (3), take a vector $Z \in \bar{E}$. There exist sequences of vectors

$$X_n \in a, \qquad Y_n \in b$$

such that

(4) $$Z = \lim_n (X_n + Y_n).$$

We have $X_n + Y_n \in a + b$, because we have the set inclusions

$$X_n \in a \subseteqq a + b,$$
$$Y_n \in b \subseteqq a + b.$$

Since $a + b$ is a space, hence a closed set of vectors, we get from (4),

$$Z \in a + b.$$

Consequently

(5) $$\bar{E} \subseteqq a + b.$$

The relations (3) and (5) prove the theorem.

D.1.6. Theorem. If **H** has finite dimensions, then for all spaces a, b we have

(6) $$a + b = \{\vec{X} + \vec{Y} \mid \vec{X} \in a, \vec{Y} \in b\}.$$

D.1.7. Remark. The equality (6) is not true if **H** has infinite dimensions. STONE has proved this on an example, showing that there exist spaces $a + b$ such that (6) does not hold true.

D.2. Def. Let us consider the notion of orthogonality of vectors $\vec{X}, \vec{Y}$, defined in the general theory of H.-H.-space by the equality:

(1) $$(\vec{X}, \vec{Y}) = 0,$$

where (1) is the scalar product of $\vec{X}$ and $\vec{Y}$. We shall write $\vec{X} \perp \vec{Y}$ to denote the orthogonality of the vectors $\vec{X}, \vec{Y}$.

D.2.1. Def. We say that $\vec{X}$ *is orthogonal to the space* a, whenever

$$\vec{X} \perp \vec{Y} \quad \text{for all} \quad \vec{Y} \in a.$$

We shall write

$$\vec{X} \perp a.$$

One can prove, that if $a = I$, then the only vector $\vec{X}$ orthogonal to a is $\vec{O}$, and if $a = \vec{O}$, then every vector is orthogonal to a.

If $a \neq I$, then there exists at least one vector $\vec{X} \neq \vec{O}$ such that $\vec{X} \perp a$.

D.2.2. Theorem. Let a be a space; then the set of all vectors $\vec{X}$, with $\vec{X} \perp a$ is a space.

D.2.2a. Proof. Let $\vec{X} \perp a$, $\vec{Y} \perp a$. Then, by Def. [D.2.1.], we have

$$\vec{X} \perp \vec{Z}, \quad \vec{Y} \perp \vec{Z} \quad \text{for all} \quad \vec{Z} \in a.$$

Hence

$$(\vec{X}, \vec{Z}) = (\vec{Y}, \vec{Z}) = 0.$$

It follows that

$$(\alpha \vec{X} + \beta \vec{Y}, \vec{Z}) = 0$$

for all complex numbers α, β. Then the set of all vectors, which are orthogonal to a, is a linear manifold b.

D.2.2b. To prove that b is a space, it suffices to prove that b is a closed set of vectors.

Let

$$\vec{X}_n \in b, \quad \lim \vec{X}_n = \vec{X}_0.$$

We have

$$(\vec{X}_n, \vec{Z}) \to (\vec{X}_0, \vec{Z}) \quad \text{for all} \quad \vec{Z} \in a.$$

Since

$$(\vec{X}_n, \vec{Z}) = 0,$$

it follows that

$$(\vec{X}_0, \vec{Z}) = 0,$$

so

$$\vec{X}_0 \in b.$$

Since we know that b is a linear variety, it follows that b is a space. The theorem is established.

D.2.2c. Def. The space b, composed of all vectors orthogonal to the space a, is called the *ortho-complement of* a *and is denoted* co$a$.

We have co$O = I$, co$I = O$.

D.2.3. We can prove that for every space a we have

$$a + \text{co}\,a = I; \quad a \cdot \text{co}\,a = O; \quad \text{co}(\text{co}\,a) = a;$$

$$\text{if} \quad a \leqq b, \quad \text{then} \quad \text{co}\,b \leqq \text{co}\,a.$$

Hence the lattice ($\leqq$) of spaces may be called *complementary* [A.1.4.4.].

This lattice, made complementary by means of the notion of co$a$, will be denoted by $\boldsymbol{L}$. We shall call it *geometrical lattice of spaces.*

D.2.4. The following theorem is known in the general theory of H.-H.-spaces.

Theorem. If a is a space and $\vec{X}$ a vector, then there exist unique vectors

$$\vec{X}_a, \vec{X}_{\mathrm{co}\,a},$$

such that

$$\vec{X} = \vec{X}_a + \vec{X}_{\mathrm{co}\,a},$$

$$\vec{X}_a \perp \vec{X}_{\mathrm{co}\,a}, \quad \vec{X}_a \in a, \quad \vec{X}_{\mathrm{co}\,a} \in \mathrm{co}\,a.$$

D.2.4a. Def. We call $\vec{X}_a$ the *projection of* $\vec{X}$ *on* a; ($\vec{X}_{\mathrm{co}\,a}$ is the projection of $\vec{X}$ on co$a$). We shall write

$$\vec{X}_a = \mathrm{Proj}_a \vec{X} \quad \text{or} \quad \mathrm{Proj}\,(a)\,\vec{X}.$$

We remind that this theorem can be proved by the aid of orthonormal systems of coordinates in $\mathbf{H}$.

D.2.5. The following theorems can be proved:

$$\mathrm{Proj}_a \vec{X} \in a, \quad \|\mathrm{Proj}_a \vec{X}\| \leqq \|\vec{X}\|,^{*)}$$

$$(\mathrm{Proj}_a \vec{X}, \vec{Y}) = (\vec{X}, \mathrm{Proj}_a \vec{Y}),$$

$$\mathrm{Proj}_a (\alpha \vec{X} + \beta \vec{Y}) = \alpha\, \mathrm{Proj}_a \vec{X} + \beta\, \mathrm{Proj}_a \vec{Y}.$$

If $\vec{X}_n \to \vec{X}$, then $\mathrm{Proj}_a \vec{X}_n \to \mathrm{Proj}_a \vec{X}$, where the limit is taken with respect to the metric topology of the H.-H.-space. If $a \leqq b$, then $\mathrm{Proj}_a\, \mathrm{Proj}_b \vec{X} = \mathrm{Proj}_a \vec{X}$.

D.2.6. Def. We say that the spaces a, b *are orthogonal to one another,* $a \perp b$, whenever every vector of a is orthogonal to every vector of b, [D.2.].

D.2.6.1. Theorem. We have: $\mathrm{co}\,a \perp a$, $O \perp a$.

D.2.6.2. Theorem. If $a \perp b$, then $a \cdot b = O$.

D.2.6.3. Theorem. If $a = b + c$, $b \perp c$, $\vec{X} \in a$, then

$$\vec{X} = \mathrm{Proj}_b \vec{X} + \mathrm{Proj}_c \vec{X}.$$

D.2.6.4. Theorem. If $a \perp b$, $a' \leqq a$, $b' \leqq b$, then $a' \perp b'$.

D.2.6.5. Theorem. If $a \perp b$, then

$$\mathrm{Proj}_{a+b} \vec{X} = \mathrm{Proj}_a \vec{X} + \mathrm{Proj}_b \vec{X}.$$

D.2.7. Theorem. If $a' \perp b$, $a'' \perp b$, then $(a' + a'') \perp b$.

Proof. Let $\vec{X} \in a' + a''$.

---

*) $\|\vec{X}\|$ means the *norm of* $\vec{X}$; it is defined as $\sqrt{(\vec{X}, \vec{X})}$. The square root exists, because $(\vec{X}, \vec{X}) \geqq 0$ for all vectors $\vec{X}$.

By [D.1.5.], there exists an infinite sequence

$$\{\vec{X}'_n + \vec{X}''_n\},$$

where

(1) $$\vec{X}'_n \in a', \quad \vec{X}''_n \in a'', \quad \vec{X} = \lim_n (\vec{X}'_n + \vec{X}''_n).$$

Let $Y \in b$.

By hypothesis we have

$$(\vec{X}'_n, \vec{Y}) = (\vec{X}''_n, \vec{Y}) = 0.$$

Hence

$$(\vec{X}'_n + \vec{X}''_n, \vec{Y}) = 0.$$

Since the scalar product is a continuous function, we get from (1)

$$(\vec{X}, \vec{Y}) = 0.$$

This relation is valid for any $\vec{X} \in a' + a''$ and for any $\vec{Y} \in b$. The thesis follows.

D.2.8. Theorem. If $a \perp b$, then $a + b$ coincides with the set

$$\{\vec{X} + \vec{Y} \,|\, \vec{X} \in a, \vec{Y} \in b\}.$$

Proof. Put

$$M =_{df} \{\vec{X} + \vec{Y} \,|\, \vec{X} \in a, \vec{Y} \in b\}.$$

By [D.1.5.] we have

$$\bar{M} = a + b.$$

Hence

(1) $$M \subseteq a + b.$$

To prove the inverse inclusion, take a vector $\vec{Z} \in a + b$. By [D.1.5.] there exists an infinite sequence

$$\{\vec{Z}_n\} = \{\vec{X}_n + \vec{Y}_n\},$$

where

$$\vec{X}_n \in a, \quad \vec{Y}_n \in b, \quad \vec{Z}_n \to \vec{Z}.$$

As $\vec{X}_n \in a$, we have $\vec{X}_n \perp b$, and as $\vec{Y}_n \in b$, we have $\vec{Y}_n \perp a$, which follows from the hypothesis: $a \perp b$.

We have, [D.2.5.],

$$\mathrm{Proj}_a \vec{Z}_n = \mathrm{Proj}_a \vec{X}_n + \mathrm{Proj}_a \vec{Y}_n;$$

hence, as

$$\vec{Y}_n \perp a, \quad \vec{X}_n \in a,$$

we get

$$\mathrm{Proj}_a \vec{Z}_n = \mathrm{Proj}_a \vec{X}_n = \vec{X}_n.$$

Similarly we get

$$\mathrm{Proj}_b \vec{Z}_n = \vec{Y}_n.$$

The operation of projection is continuous and we have $\vec{Z}_n \to \vec{Z}$; hence

(2) $$\left.\begin{array}{l}\mathrm{Proj}_a \vec{Z} = \lim \vec{X}_n,\\ \mathrm{Proj}_b \vec{Z} = \lim \vec{Y}_n\end{array}\right\};$$

hence the limits of $\vec{X}_n$ and of $\vec{Y}_n$ exist.

Put

$$\vec{X} =_{df} \lim \vec{X}_n, \qquad \vec{Y} =_{df} \lim \vec{Y}_n.$$

Since the spaces are closed sets, we get

(3) $$\vec{X} \in a, \qquad \vec{Y} \in b.$$

Now we have

$$\vec{Z} = \lim \vec{Z}_n = \lim (\vec{X}_n + \vec{Y}_n) = \lim \vec{X}_n + \lim \vec{Y}_n = \vec{X} + \vec{Y}.$$

It follows

$$\vec{Z} \in M.$$

Since $\vec{Z}$ is an arbitrary vector belonging to $a + b$, it follows that

$$a + b \subseteqq M.$$

This together with (1) gives

$$a + b = M, \quad \text{q.e.d.}$$

D.2.9. Theorem. If

1. a, b, c are mutually orthogonal spaces,
2. $p =_{df} a + b$, $q =_{df} b + c$,

then

$$p + q = \{\vec{X} + \vec{Y} \mid \vec{X} \in p, \vec{Y} \in q\}.$$

Proof. We have

$$p + q = (a + b) + (b + c) = a + (b + b) + c = a + b + c.$$

Since $a \perp c$, $b \perp c$, we have, [D.2.7.], $a + b \perp c$. By [D.2.8.], $p + q = a + b$ is identical with

(1) $$\{\vec{X} + \vec{Y} \mid \vec{X} \in a + b, \vec{Y} \in c\}.$$

If $\vec{Y} \in c$, we have $\vec{Y} \in q$, because

$$c \leqq b + c = q.$$

Hence the set (1) is contained in the set

(2) $$R =_{df} \{\vec{X} + \vec{Y} \mid \vec{X} \in p, \vec{Y} \in q\}, \quad \text{i.e.} \quad p + q \subseteqq R.$$

But

$$R \subseteqq \bar{R} = p + q,$$

[D.1.5.]. Hence, on account of (2), we get

$$R = p + q, \quad \text{q.e.d.}$$

D.2.10. Theorem. If a, b are spaces, then the following conditions are equivalent:

I. $b \perp a$;

II. $b \leqq \text{co}\, a$;

III. $a \leqq \text{co}\, b$.

D.2.11. Since we have to do with a lattice $\boldsymbol{L}$, with zero and unit, therefore de Morgan's laws take places for spaces: [A.1.5.]:

$$\text{co}\,(a+b) = \text{co}\, a \cdot \text{co}\, b,$$

$$\text{co}\,(a \cdot b) = \text{co}\, a + \text{co}\, b.$$

D.2.12. Theorem. If

$$a \leqq a', \quad b \leqq b',$$

then

$$a + b \leqq a' + b' \quad \text{and} \quad a \cdot b \leqq a' \cdot b'.$$

D.2.13. Theorem. $a \cdot c + b \cdot c \leqq (a+b)\, c$.

Proof. We have $a \leqq a + b$; hence

(1) $$a\, c \leqq (a+b)\, c.$$

Similarly we get

(2) $$b\, c \leqq (a+b)\, c.$$

From (1) and (2), by [D.2.12.], we get the theorem.

D.2.13a. Remark. Let us remark that the inverse inclusion

$$(a+b) \cdot c \leqq a \cdot c + b \cdot c$$

is not always true, even in two-dimensional space.

D.3. Def. By the *difference* $a - b$ *of the spaces* a, b we shall understand the space $a \cdot \text{co}\, b$, [A.2.3a.].

D.3.1. Theorem. If $a \leqq b$, then

$$b = a + (b - a).$$

Proof. Since $a \leqq b$ and $b - a = b \cdot \text{co}\, a \leqq b$, we get

(1) $$a + (b - a) \leqq b.$$

To get the inverse inclusion, take a vector

(2) $$\vec{X} \in b,$$

and put

(2.1) $$\vec{Y} =_{df} \text{Proj}_a \vec{X}.$$

Since

(3) $$a \leqq b \quad \text{and} \quad \vec{Y} \in a, \quad \text{we have} \quad \vec{Y} \in b.$$

On the other hand we have

(4) $$\vec{X} - \vec{Y} = \vec{X} - \text{Proj}_a \vec{X} \in \text{co}\, a,$$

because, by [D.2.4.],

$$\vec{X} = \mathrm{Proj}_a \vec{X} + \mathrm{Proj}_{\mathrm{co}\,a} \vec{X},$$

and

$$\mathrm{Proj}_{\mathrm{co}\,a} \vec{X} \in \mathrm{co}\,a.$$

From (2) and (3) it follows:

$$\vec{X} - \vec{Y} \in b,$$

and from (4) we get

$$\vec{X} - \vec{Y} \in b\,\mathrm{co}\,a = b - a.$$

As $\vec{X} - \vec{Y} \in b - a$, and from (2.1) $\vec{Y} \in a$, we have, by [D.1.5.]:

$$(\vec{X} - \vec{Y}) + \vec{Y} \in (b - a) + a;$$

hence

$$\vec{X} \in (b - a) + a.$$

Taking account of (2), we get

$$b \leqq (b - a) + a,$$

which together with (1) yields the thesis.

D.3.2. Theorem. For any spaces a, b we have

(1) $$a = a \cdot b + (a - a\,b).$$

Proof. We have $a \cdot b \leqq a$. Hence applying the former theorem, [D.3.1.], we get (1).

D.3.2a. Remark. Let us remark that the equality

$$a - b = a - a\,b$$

is not always true.

D.3.3. Theorem. We always have

$$(b - a\,b) \cdot (a - a\,b) = O.$$

Proof. Let

(1) $$\vec{X} \in b - a\,b, \quad \vec{X} \neq \vec{O}.$$

We have

(2) $$\vec{X} \mathrel{\bar{\in}} a\,b,$$

but we do not have $X \in a$. Indeed if we had $X \in a$, we would have, since $\vec{X} \in b$, the relation $\vec{X} \in a \cdot b$, which contradicts (2).

Since $\vec{X} \in a$, we have a fortiori

$$\vec{X} \mathrel{\bar{\in}} a - a\,b.$$

Thus we have proved that if $\vec{X} \neq \vec{O}$ and belongs to $b - a\,b$. Hence it does not belong to $a - a\,b$. It follows that

$$(a - a\,b) \cdot (b - a\,b) = O.$$

D.4. The foregoing discussion has shown that the ordering of spaces is a complete lattice with zero and unit, and that we have made the lattice complementary by introducing the relation of complementation by means of orthogonality of vectors. This lattice, denoted by $\boldsymbol{L}$ and called *geometrical lattice of spaces,* is however not always a Boolean tribe, because the distributive law may not hold true.

However we shall isolate some sublattices which will be tribes. To do that, we are going to introduce the following notion of compatibility of spaces, which will play a great role in the sequel.

D.5. Compatible spaces. Def. Two spaces a, b will be termed *compatible* whenever

$$(a - a\,b) \perp (b - a\,b).$$

D.5.1. Theorem. If $a \leqq b$, then a, b are compatible.

Proof. Indeed $a = a\,b$; hence

$$a - a\,b = a\,b - a\,b = O.$$

Consequently $a - a\,b$ is orthogonal to every space, hence also to $b - a\,b$.

D.5.2. Theorem. If $a \perp b$, then a, b are compatible.

Proof. Indeed, since $a - a\,b \leqq a$, $b - a\,b \leqq b$, and since $a \perp a\,b$, we have [D.2.6.4.]

$$(a - a\,b) \perp (b - a\,b).$$

D.5.3. We know that $a - b$ does not coincide, in general, with $a - a\,b$, but we have:

Theorem. If a, b are compatible, then

$$a - b = a - a\,b.$$

Proof. We have $b \leqq b$ and $\operatorname{co} a \leqq \operatorname{co}(b\,a)$, for $b\,a \leqq a$; hence by [A.2.6.]:

$$b \cdot \operatorname{co} a \leqq b \cdot \operatorname{co}(b\,a),$$

i.e.

(1) $$b - a \leqq b - a\,b.$$

Now we are going to prove the inverse inclusion. Let

(1.1) $$\vec{X} \in b - a\,b.$$

We have

(2) $$\vec{X} \in \operatorname{co}(a\,b), \quad \text{hence} \quad \vec{X} \perp a\,b.$$

Since, by hypothesis

$$(a - a\,b) \perp (b - a\,b),$$

we have

(3) $$\vec{X} \perp a - a\,b.$$

From (2) and (3) we get, [D.2.7.],

$$\vec{X} \perp (a - a\,b) + a\,b;$$

hence, [D.3.2.],

$$X \perp a,$$

which gives

$$\vec{X} \in \mathrm{co}\,a.$$

But from (1.1) we have

$$\vec{X} \in b;$$

hence

$$\vec{X} \in b \cdot \mathrm{co}\,a = b - a.$$

Thus we have proved that

(4) $$b - a\,b \leqq b - a.$$

From (1) and (4) the theorem follows.

D.5.4. Theorem. If a, b are compatible spaces, then

1) $b = a\,b + (b - a)$,

2) $a + b = a\,b + (a - b) + (b - a)$, where all terms are disjoint.

Proof. By [D.3.2.] we have

$$b = a\,b + (b - a\,b);$$

hence, by [D.5.3.],

$$b = a\,b + (b - a),$$

so the first thesis is established. To prove the second thesis, notice that

$$a = a\,b + (a - b),$$

$$b = a\,b + (b - a).$$

Hence

$$a + b = (a\,b + a\,b) + (a - b) + (b - a),$$

i.e.

$$a + b = a\,b + (a - b) + (b - a).$$

The terms are disjoint. Indeed,

$$a\,b \leqq b, \quad a - b \leqq \mathrm{co}\,b,$$

so

$$a\,b \cdot (a - b) = b \cdot \mathrm{co}\,b = O.$$

Similarly we prove that

$$a\,b \cdot (b - a) = O.$$

What concerns the terms $a - b$, $b - a$, they are disjoint, since

$$a - b \leqq a \quad \text{and} \quad b - a \leqq \mathrm{co}\,a.$$

The theorem is proved.

D.6. To go farther, we introduce an auxiliary notion:

Def. Let E be a non empty set of vectors and let a be a space. By *the projection* $\text{Proj}_a E$ *of* E *on the space* a, we shall understand the set of all vectors

$$\text{Proj}_a \vec{X}, \quad \text{where} \quad \vec{X} \in E.$$

D.6.1. Theorem. If a, b are spaces, then

$$\text{Proj}_a b \text{ is linear variety.}$$

Proof. Let $\vec{X}_1, \vec{X}_2 \in \text{Proj}_a b$. There exist vectors $\vec{Y}_1, \vec{Y}_2 \in a$, such that

$$\vec{X}_1 = \text{Proj}_a \vec{Y}_1, \quad \vec{X}_2 = \text{Proj}_a \vec{Y}_2.$$

It follows, [D.2.5.],

$$\lambda \vec{X}_1 + \mu \vec{X}_2 = \text{Proj}_a(\lambda \vec{Y}_1 + \mu \vec{Y}_2).$$

Since $\lambda \vec{Y}_1 + \mu \vec{Y}_2 \in a$, it follows that

$$\lambda \vec{X}_1 + \mu \vec{X}_2 \in \text{Proj}_a b,$$

which completes the proof.

D.6.1a. Remark. The set $\text{Proj}_a b$ is not always a space, because this set may be not closed.

This results from the discussion by Mr. G. JULIA in his C.-R.-notes (Paris 1944). The phenomenon that the projection of a space on an other is not necessarily closed, does not always take place, and never, if **H** has finite dimensions. The phenomenon is in relation to linear varieties which JULIA called "varietés asymptotiques". See also the notes by DIXMIER in the C. R. T. 224, (1947).

D.6.1b. We are giving an example by J. NINOT NOLLA, of two spaces a, b, where $E =_{df} \text{Proj}_a b$ is not a space. Let $\{\vec{Q}_n\}$ be a saturated orthogonal set of vectors in the infinite dimensional separable and complete H.-H.-space. Let $\{\alpha_n\}$ be an infinite sequence of numbers, differing from 0, such that

$$\sum_{n=1}^{\infty} |\alpha_n|^2$$

converges. Let also $\{\theta_n\}$ be an infinite sequence of numbers, where

$$0 < \theta_n < \frac{\pi}{2},$$

and such that

$$\sum_{n=1}^{\infty} \sec^2{}_n \cdot |\alpha_n|^2$$

diverges.

Consider the space a determined by $\{\vec{Q}_{2n-1}\}$, and the space b determined by the sequence

$$\{\cos\theta_n \cdot \vec{Q}_{n-1} + \sin\theta_n \vec{Q}_{2n}\}.$$

Put $E =_{df} \mathrm{Proj}_a b$, and $V =_{df} a - p$, where p is the space determined by the linear variety E.

If $\vec{Y} \in V$, we have $\vec{Y} \in a$, $\vec{Y} \perp b$; hence $\vec{Y} = \vec{O}$.

It follows that $p = a$, and we can prove, by impossible, that the vector

$$\sum_{n=1}^{\infty} \alpha_n \vec{Q}_{2n-1},$$

which belongs to a, does not belong to E.

Of course, this example has been suggested by the example by STONE, mentioned in [D.1.7.].

D.7. We like to find several conditions which would be necessary and sufficient for compatibility of spaces. Therefore we start with some auxiliary theorems.

D.7.1. Lemma. If a, b are spaces, then

$$a \cdot b \subseteqq \mathrm{Proj}_b a,$$

where ($\subseteqq$) is the relation of inclusion of sets of vectors.

Proof. Let

$$\vec{X} \in a \cdot b.$$

To prove that $\vec{X} \in \mathrm{Proj}_b a$, we need to find a vector $\vec{Y} \in a$, whose projection on b be $\vec{X}$.

Now, I say that $\vec{Y} =_{df} \vec{X}$ is such a vector. Indeed, $\vec{Y} \in a \cdot b \leqq a$; hence

$$\vec{Y} \in a.$$

On the other hand $\vec{X} = \mathrm{Proj}_a \vec{X}$, since $\vec{X} \in a$. Consequently $\vec{X} = \mathrm{Proj}_a \vec{Y}$, $\vec{Y} \in b$. The theorem follows.

D.7.2. Theorem. If a, b are spaces, then the following conditions are equivalent:

I. a, b are compatible,

II. $\mathrm{Proj}_b a \subseteqq a \cdot b$.

D.7.2a. Proof. Suppose that the relation II takes place. Let

(1) $$\vec{X} \in a - a\,b, \quad \vec{Y} \in b - a\,b.$$

We shall prove that

$$\vec{X} \perp \vec{Y}.$$

We have

(2) $$\vec{X} \in a;$$

hence

$$\mathrm{Proj}_b \vec{X} \in \mathrm{Proj}_b a.$$

By hypothesis II it follows

(3) $$\mathrm{Proj}_b \vec{X} \in a \cdot b.$$

We have

$$\vec{Y} \in b - a\,b = b \cdot \mathrm{co}\,(a \cdot b)\,;$$

hence

(4) $$\vec{Y} \in b\,.$$

It follows that

(5) $$\mathrm{Proj}_b \vec{Y} = \vec{Y}\,.$$

Consequently we get for scalar products:

(6) $$(\vec{Y}, \vec{X}) = (\mathrm{Proj}_b \vec{Y}, \vec{X}) = (\vec{Y}, \mathrm{Proj}_b \vec{X})\,.$$

Since, from (1),

$$\vec{Y} \in \mathrm{co}\,(a \cdot b)\,,$$

and since, by (3)

$$\mathrm{Proj}_b \vec{X} \in a \cdot b\,,$$

we get

$$(\vec{Y}, \vec{X}) = 0\,,$$

which proves that $\vec{X} \perp \vec{Y}$, i.e.

$$(a - a\,b) \perp (b - a\,b)\,,$$

i.e., [D.5.], the spaces a, b are compatible. We have proved that II $\rightarrow$ I.

D.7.2b. Admit I, i.e.

$$(a - a\,b) \perp (b - a\,b)\,.$$

Let $\vec{X} \in \mathrm{Proj}_b a$. There exists $\vec{Z} \in a$ such that

(7) $$\vec{X} = \mathrm{Proj}_b \vec{Z}\,.$$

The spaces $a - a\,b$ and $a\,b$ are subspaces of a and they are orthogonal. In addition to that we have, [D.3.2.]:

$$(a - a\,b) + a\,b = a\,.$$

Consequently there exists the decomposition of the vector $\vec{Z}$, ([D.2.4.], relativized to the space a):

(7.1) $$\vec{Z} = \vec{Z}' + \vec{Z}''\,,$$

where

(8) $$\vec{Z}' \in a - a\,b\,, \quad \vec{Z}'' \in a\,b\,.$$

Thus we get from (7)

(9) $$\vec{X} = \mathrm{Proj}_b (\vec{Z}' + Z'') = \mathrm{Proj}_b \vec{Z}' + \mathrm{Proj}_b \vec{Z}'' = \mathrm{Proj}_b \vec{Z}' + \vec{Z}''\,,$$

because

$$\vec{Z}'' \in a \cdot b \leq b\,.$$

D.7.2c. Since $\vec{Z}'' \in a \cdot b$, it suffices to prove that

$$\mathrm{Proj}_b \vec{Z}' \in a \cdot b\,.$$

Let $\vec{Y} \in b - a \cdot b$. We have, by (7.1), $\vec{Z}' \in a - a \cdot b$. Since, by hypothesis I, we have $\vec{Y} \perp \vec{Z}'$, we get

(10) $$(\vec{Y}, \vec{Z}') = 0.$$

But we have

$$\vec{Y} \in b - a \cdot b \leqq b,$$

which gives

$$\mathrm{Proj}_b \vec{Y} = \vec{Y};$$

so from (10) we obtain

$$(\mathrm{Proj}_b \vec{Y}, \vec{Z}') = 0.$$

Applying one of the rules [D.2.5.], we get

$$(\vec{Y}, \mathrm{Proj}_b \vec{Z}') = 0,$$

which says that

(11) $$\vec{Y} \perp \mathrm{Proj}_b \vec{Z}'.$$

Since this is true for any $\vec{Y} \in b - a \cdot b$, it follows that

(12) $$(b - a \cdot b) \perp \mathrm{Proj}_b \vec{Z}'.$$

Since

$$b = (b - a \cdot b) + a \cdot b \quad \text{and} \quad b - a \cdot b \perp a \cdot b,$$

we deduce that

(13) $$\mathrm{Proj}_b \vec{Z}' \in a \cdot b.$$

Indeed, we have the decomposition of $\mathrm{Proj}_b \vec{Z}'$ into $\vec{V}_1 + \vec{V}_2$, where $\vec{V}_1 \in b - a \cdot b$, $\vec{V}_2 \in a \cdot b$, which gives

$$\mathrm{Proj}_b \vec{Z}' = \vec{V}_2 \in a \cdot b, \quad \text{as} \quad \mathrm{Proj}_b \vec{Z}' \perp b - a \cdot b.$$

Since $\mathrm{Proj}_b \vec{Z}' \in a \cdot b$ and $\vec{Z}'' \in a \cdot b$, by (8), we get, by (9), $\vec{X} \in a \cdot b$.

As $\vec{X}$ is an arbitrary vector of $\mathrm{Proj}_b a$, it follows that

$$\mathrm{Proj}_b a \subseteqq a \cdot b,$$

i.e. II is proved.

The theorem is established.

D.8. Theorem. For spaces a, b the following conditions are mutually equivalent:

I. a, b are compatible,

II. $\mathrm{Proj}_b a \subseteqq a \cdot b$,

III. $\mathrm{Proj}_b a = a \cdot b$,

IV. $\mathrm{Proj}_a b = \mathrm{Proj}_b a$,

V. $a - b = a - a \cdot b$,

VI. $a - a \cdot b \leqq a - b$,

VII. $a = a \cdot b + a \cdot \mathrm{co}\, b$,

VIII. $a = a \cdot \mathrm{Proj}_b a + a \cdot \mathrm{Proj}_{\mathrm{co}\, b} a$,

IX. a, $\mathrm{co}\, b$ are compatible,

X. $\mathrm{co}\, a$, $\mathrm{co}\, b$ are compatible.

D.8a. Proof. By [D.7.2.] we have

(1) $$\mathrm{I} \rightleftarrows \mathrm{II}.$$

The lemma [D.7.1.] shows that the inclusion

$$a \cdot b \subseteqq \mathrm{Proj}_b a$$

is always true. Hence, if a, b are compatible, we get by II,

$$a \cdot b = \mathrm{Proj}_b a.$$

Thus we have proved that $\mathrm{I} \rightarrow \mathrm{III}$.

Conversely if III, we have II and hence, by what has been proved, we get I.

Thus we have got

(2) $$\mathrm{III} \rightarrow \mathrm{I}, \quad \text{so} \quad \mathrm{I} \rightleftarrows \mathrm{III}.$$

D.8b. Now let I. We get by III:

$$\mathrm{Proj}_b a = a \cdot b,$$

and also

$$\mathrm{Proj}_a b = a \cdot b.$$

It follows

$$\mathrm{Proj}_b a = \mathrm{Proj}_a b,$$

i.e. IV. Conversely, let

(3) $$\mathrm{Proj}_b a = \mathrm{Proj}_a b.$$

We have

$$\mathrm{Proj}_b a \subseteqq b.$$

Similarly

$$\mathrm{Proj}_a b \subseteqq a;$$

hence, by (3),

$$\mathrm{Proj}_b a \subseteqq a \cdot b,$$

i.e. III. Hence, since $\mathrm{III} \rightarrow \mathrm{I}$, we get I.

Thus we have proved that

(4) $$\mathrm{I} \rightleftarrows \mathrm{IV}.$$

From (1), (2) and (4) it follows that all four conditions

(5) I, II, III, IV are equivalent.

D.8c. Let I. By [D.5.3.] we have

$$a - b = a - a\,b,$$

i.e. V. It follows

$$a - a \cdot b \leqq a - b,$$

i.e. IV. Thus we get

(6) $$\mathrm{I} \rightarrow \mathrm{V} \rightarrow \mathrm{VI}.$$

D.8d. Now let VI, i.e.

(7) $$a - a \cdot b \leqq a - b.$$

We always have

$$a - b \leqq a - a \cdot b.$$

Indeed, as $a \cdot b \leqq b$, we have $\mathrm{co}\, b \leqq \mathrm{co}(a \cdot b)$. Hence

$$a \cdot \mathrm{co}\, b \leqq a \cdot \mathrm{co}(a \cdot b),$$

i.e.

(8) $$a - b \leqq a - a \cdot b.$$

From (7) and (8) it follows

$$a - b = a - a \cdot b,$$

i.e. V. Thus we have proved that

(9) $$\mathrm{V} \rightleftarrows \mathrm{VI}.$$

D.8e. Let V, i.e.

(10) $$a - b = a - a \cdot b.$$

We have

$$a - b = a \cdot \mathrm{co}\, b \leqq \mathrm{co}\, b;$$

hence

$$a - b \perp b;$$

hence

$$a - b \perp b - a \cdot b.$$

By (10) it follows

$$a - a \cdot b \perp b - a \cdot b,$$

which, by definition, says that a is compatible with b. Thus we have proved that V → I; hence, by (9), VI → I. Taking account of (6), we get the equivalence of I, V, and VI, and then, by (5), the equivalence of

(11) I, II, III, IV, V and VI.

D.8f. The lemma [D.5.4.] says that I implies

$$a = a \cdot b + (a - b) = a \cdot b + a \cdot \mathrm{co}\, b,$$

i.e. VII. Hence

(1) $$\mathrm{I} \rightarrow \mathrm{VII}.$$

Now suppose VII, i.e.

(2) $$a = a \cdot b + (a - b).$$

We have, by [D.3.2.],

(3) $$a = a \cdot b + (a - a \cdot b).$$

From (2) we see that $a - b$ is the orthocomplement of $a \cdot b$ in a, and from (3) we see that $a - a \cdot b$ is also the orthocomplement of $a\, b$ in a.

Hence

$$a - a\,b = a - b,$$

i.e. V. Hence VII $\to$ V, and then, as V $\rightleftarrows$ I, we get I.

(3.1) Thus we have proved that VII and I are equivalent.

D.8g. Let a, b be compatible. We have, by VII,

$$a = a\,b + a \cdot \mathrm{co}\,b.$$

This can be written:

$$a = a \cdot \mathrm{co}\,b + a \cdot \mathrm{co}\,(\mathrm{co}\,b).$$

This is, however, the condition VII, applied to a and co$b$, i.e.IX. Thus I $\to$ IX.

Conversely, suppose that a, co$b$ are compatible. By VII we get

$$a = a \cdot \mathrm{co}\,b + a \cdot \mathrm{co}\,(\mathrm{co}\,b),$$

i.e.

$$a = a \cdot \mathrm{co}\,b + a\,b.$$

This is the condition VII applied to a and b. Hence a, b are compatible. But, [D.8f.], VII $\to$ I. We have proved the equivalence of I and IX.

D.8h. Let a, b be compatible. From what we already have proved a, co$b$ are compatible; hence by the same condition IX, we have the compatibility of co$b$ and co$a$, i.e. X. Thus we have I $\to$ X. Conversely suppose X, i.e. co$a$, co$b$ are compatible. Applying IX, we get the compatibility of co$a$, and b and then, by IX, the compatibility of a, b.

Thus we have proved the equivalence of I, IX and X.

D.8i. Let VIII, i.e.

(1) $$a = \mathrm{Proj}_b a + \mathrm{Proj}_{\mathrm{co}\,b} a.$$

We have from (1)

$$\mathrm{Proj}_b a \subseteqq a;$$

but we also have

$$\mathrm{Proj}_b a \subseteqq b.$$

It follows

$$\mathrm{Proj}_b a \subseteqq a \cdot b,$$

i.e. II, which implies I. Thus we have got the implication

(2) $$\mathrm{VIII} \to \mathrm{I}.$$

Conversely, suppose that a, b are compatible. We have VII:

$$a = a\,b + (a - b).$$

But we also have, by III:

(3) $$a\,b = \mathrm{Proj}_b a.$$

Since I $\rightarrow$ IX, the spaces a, co$b$ are also compatible; hence we get the formula analogous to (3):

(4) $$a \cdot \mathrm{co}\, b = \mathrm{Proj}_{\mathrm{co}\, b}\, a.$$

From (3) and (4) it follows

$$a\, b + a\, \mathrm{co}\, b = \mathrm{Proj}_b\, a + \mathrm{Proj}_{\mathrm{co}\, b}\, a.$$

Now $a\, b + a \cdot \mathrm{co}\, b = a$, by VII. Consequently

$$a = \mathrm{Proj}_b\, a + \mathrm{Proj}_{\mathrm{co}\, b}\, a,$$

i.e. VIII. We have proved that I $\rightleftarrows$ VIII.

Taking into account all items proved above, we see that all conditions I — X are equivalent. The theorem is established.

D.9. Theorem. The following conditions are mutually equivalent for spaces a, b:

I′. a, b are compatible,

II′. (JULIA). For every vector $\vec{X} \in \mathbf{H}$ we have

$$\mathrm{Proj}_a\, \mathrm{Proj}_b \vec{X} = \mathrm{Proj}_b\, \mathrm{Proj}_a \vec{X}.$$

III′. For every vector $\vec{X} \in \mathbf{H}$ we have

$$\mathrm{Proj}_a\, \mathrm{Proj}_b \vec{X} = \mathrm{Proj}_{a b} \vec{X}.$$

IV′. There exists a space c such that for every $\vec{X} \in \mathbf{H}$:

$$\mathrm{Proj}_a\, \mathrm{Proj}_b \vec{X} = \mathrm{Proj}_c \vec{X}.$$

D.9a. Proof. Let I′. We have by [D.8., VII],

$$b = b\, a + b \cdot \mathrm{co}\, a;$$

hence for every $\vec{X} \in \mathbf{H}$:

$$\mathrm{Proj}_a\, \mathrm{Proj}_b \vec{X} = \mathrm{Proj}_a [\mathrm{Proj}_{b\, a + b\, \mathrm{co}\, a} \vec{X}].$$

Since $b\, a \perp b \cdot \mathrm{co}\, a$, we have

$$\mathrm{Proj}_{b\, a + b\, \mathrm{co}\, a} \vec{X} = \mathrm{Proj}_{b\, a} \vec{X} + \mathrm{Proj}_{b\, \mathrm{co}\, a} \vec{X}.$$

Consequently

(1) $$\mathrm{Proj}_a\, \mathrm{Proj}_b \vec{X} = \mathrm{Proj}_a\, \mathrm{Proj}_{b\, a} \vec{X} + \mathrm{Proj}_a\, \mathrm{Proj}_{b\, \mathrm{co}\, a} \vec{X}.$$

But, as $a\, b \leqq a$, we have, [D.2.5.],

(2) $$\mathrm{Proj}_a\, \mathrm{Proj}_{b\, a} \vec{X} = \mathrm{Proj}_{b\, a} \vec{X}.$$

Since $\mathrm{Proj}_{b\, \mathrm{co}\, a} \vec{X} \in b \cdot \mathrm{co}\, a \leqq \mathrm{co}\, a$, we have

(3) $$\mathrm{Proj}_a\, \mathrm{Proj}_{b\, \mathrm{co}\, a} \vec{X} = \vec{O}.$$

It follows from (1), (2), (3):

$$\mathrm{Proj}_a\, \mathrm{Proj}_b \vec{X} = \mathrm{Proj}_{a b} \vec{X},$$

i.e. III′. An analogous argument gives

$$\mathrm{Proj}_b\,\mathrm{Proj}_a\vec{X} = \mathrm{Proj}_{ab}\vec{X}.$$

Consequently

$$\mathrm{Proj}_a\,\mathrm{Proj}_b\vec{X} = \mathrm{Proj}_b\,\mathrm{Proj}_a\vec{X},$$

i.e. II′. Thus we have proved that

(4) $$\mathrm{I}' \to \mathrm{III}' \to \mathrm{II}'.$$

And we also have

(5) $$\mathrm{III}' \to \mathrm{IV}'.$$

D.9b. Admit IV′. Take a space c, such that for every $\vec{X} \in \mathbf{H}$ we have

(5.1) $$\mathrm{Proj}_a\,\mathrm{Proj}_b\vec{X} = \mathrm{Proj}_c\vec{X}.$$

This is valid for any $\vec{X}$; hence also for any set of vectors, [D.6.]. Take the space b as set of vectors. We get

$$\mathrm{Proj}_a\,\mathrm{Proj}_b b = \mathrm{Proj}_c b;$$

hence

(6) $$\mathrm{Proj}_a b = \mathrm{Proj}_c b.$$

On the other hand we get from (5.1)

$$\mathrm{Proj}_a\,\mathrm{Proj}_b c = \mathrm{Proj}_c c = c,$$

which gives

(7) $$c \subseteqq a.$$

From (5.1) we also get

$$\mathrm{Proj}_a\,\mathrm{Proj}_b(\mathrm{co}\,b) = \mathrm{Proj}_c(\mathrm{co}\,b).$$

As

$$\mathrm{Proj}_b(\mathrm{co}\,b) = O,$$

we get

$$\mathrm{Proj}_c(\mathrm{co}\,b) = O,$$

which implies:

$$\mathrm{co}\,b \perp c;$$

hence

(8) $$c \leqq b.$$

From (7) and (8) we get

(9) $$c \leqq a\,b.$$

But from (6) we have

$$\mathrm{Proj}_a b \subseteqq c.$$

Hence, by (9) we obtain

$$\mathrm{Proj}_a b \subseteqq a \cdot b,$$

i.e. [D.8., II], which says that a, b are compatible. Thus we have proved that IV′ → I′.

If we take account of (4) and (5), we see that the conditions I′, II′, III′, IV′ are equivalent, q.e.d.

D.10. Theorem. If a, b are compatible spaces, then

$$a + b = \{\vec{X} + \vec{Y} \mid \vec{X} \in a, \vec{Y} \in b\}.$$

Proof. We have, [D.5.4.],

$$a = a\,b + (a - b),$$

$$b = a\,b + (b - a).$$

We can apply [D.2.9.], because the spaces $a \cdot b$, $a - b$, $b - a$ are mutually orthogonal. Thus we get the thesis.

D.10.1. Lemma. If

1. $a_1, a_2, \ldots, a_n$, $(n \geqq 2)$ are mutually orthogonal,
2. b is compatible with every a_i, $(i = 1, 2, \ldots, n)$,

then the distributive law

$$(a_1 + \cdots + a_n) \cdot b = a_1\,b + \cdots + a_n\,b$$

takes place.

Proof. We have, [D.5.4.],

$$(1) \qquad a_i = a_i\,b + (a_i - b), \quad (i = 1, \ldots, n),$$

because a_i, b are compatible.

The spaces

$$a_i\,b, \; a_i - b, \quad (i = 1, \ldots, n)$$

are orthogonal to one another.

Take an arbitrary vector $\vec{X}$, belonging to $(a_1 + \cdots + a_n) \cdot b$.

We have, [D.2.6.5.],

$$\vec{X} = (\mathrm{Proj}_{a_1 - b}\vec{X} + \mathrm{Proj}_{a_1 b}\vec{X}) + \cdots + (\mathrm{Proj}_{a_n - b}\vec{X} + \mathrm{Proj}_{a_n b}\vec{X}),$$

since $\vec{X} \in a_1 + \cdots + a_n$, and since, by (1),

$$a_1 + \cdots + a_n = [a_1\,b + (a_1 - b)] + \cdots + [a_n\,b + (a_n - b)].$$

But

$$\mathrm{Proj}_{a_i - b}\vec{X} = \vec{O},$$

because

$$\vec{X} \in b.$$

It follows:

$$(2) \; \vec{X} = \mathrm{Proj}_{a_1 \cdot b}\vec{X} + \cdots + \mathrm{Proj}_{a_n \cdot b}\vec{X} =, \text{[D.2.6.3.]}, = \mathrm{Proj}_{a_1 b + \cdots + a_n b}\vec{X}.$$

The equation (2) gives

$$\vec{X} \in a_1\,b + \cdots + a_n\,b.$$

Thus we have proved that

$$(a_1 + \cdots + a_n) \cdot b \leqq a_1 b + \cdots + a_n b.$$

Taking account of [D.2.13.], we get the thesis.

D.10.2. Theorem. If a, b are compatible with c, then $a + b$ is also compatible with c.

Proof. (S. NIKODÝM). Since a, c are compatible, we have [D.8., VII]

(1) $$a = a c + a \cdot \text{co} c.$$

Similarly we get

(2) $$b = b c + b \cdot \text{co} c.$$

Adding (1) and (2) we get

$$a + b = (a c + a \cdot \text{co} c) + (b c + b \cdot \text{co} c) = (a c + b c) + (a \cdot \text{co} c + b \cdot \text{co} c).$$

But we have, [D.2.13.],

(3) $$a c + b c \leqq (a + b) \cdot c,$$

and by the same reason

(4) $$a \cdot \text{co} c + b \cdot \text{co} c \leqq (a + b) \cdot \text{co} c,$$

Adding (3) and (4) we get

(5) $$a + b \leqq (a + b) \cdot c + (a + b) \cdot \text{co} c;$$

hence, by [D.2.13.],

$$a + b \leqq (a + b)(c + \text{co} c) = (a + b) \cdot I = a + b.$$

It follows that

$$a + b = (a + b) \cdot c + (a + b) \cdot \text{co} c.$$

Hence, [D.8., VII], $a + b$ and c are compatible, q.e.d.

D.10.3. Theorem. If a, b are compatible with c, then $a \cdot b$ is compatible with c.

Proof. Since a, b are compatible with c, it follows, [D.8., IX], that $\text{co} a$, $\text{co} b$ are compatible with c. Hence, by [D.10.2.],

$$\text{co} a + \text{co} b$$

is compatible with c, and then, [D.8., IX],

$$\text{co}(\text{co} a + \text{co} b) = a \cdot b$$

is compatible with c, q.e.d.

D.10.4. Theorem. If a, b are compatible with c, then $a - b$ is compatible with c.

Proof. Since b is compatible with c, so is $\text{co} b$, [D.8., IX]; hence, by [D.10.3.], $a \cdot \text{co} b$ is compatible with c. The thesis follows.

D.10.5. Theorem. The following conditions are equivalent:

I. a, b are compatible,

II. there exist spaces a', b', c mutually orthogonal such that

$$a = a' + c, \quad b = b' + c.$$

(This theorem visualizes well the compatibility of spaces.)

Suppose II, and let a', b', c be mutually orthogonal spaces, such that

$$a = a' + c, \quad b = b' + c.$$

Since

$$a' \perp b', \quad c \perp b',$$

it follows, [D.2.7.],

(1) $$a' + c \perp b';$$

hence $a' + c$ is compatible with b'. Now we see that

(2) $$a' + c$$

is compatible with c, because

$$c \leqq a' + c,$$

[D.5.1.].

From (1) and (2) follows that, [D.10.2.],

$$a' + c \text{ is compatible with } b' + c,$$

i.e. a is compatible with b. Thus I follows.

Suppose I. Let a, b be compatible. We have, by [D.5.4.],

$$a = a\,b + (a - b), \quad b = a\,b + (b - a).$$

The spaces $a\,b$, $a - b$, $b - a$ are mutually orthogonal, so II is proved.

D.10.6. Theorem. If a, b, c are compatible spaces, then

$$(a + b) \cdot c = a \cdot c + b \cdot c.$$

Proof. We have, [D.5.4.],

$$a + b = a \cdot b + (a - b) + (b - a),$$

where the terms on the right are mutually orthogonal. Hence, by [D.10.1.],

$$(a + b) \cdot c = a\,b\,c + (a - b)\,c + (b - a)\,c = [a\,b\,c + (a - b)\,c] + \\ + [a\,b\,c + (b - a)\,c].$$

Since a, b, c are mutually compatible, c is compatible with $a \cdot b$, $a - b$, $b - a$, [D.10.3.], [D.10.4.]. Thus we can apply [D.10.1.], getting

$$(a + b) \cdot c = [a\,b + (a - b)]\,c + [a\,b + (b - a)]\,c,$$

i.e., by [D.5.4.],

$$(a + b) \cdot c = a\,c + b\,c, \quad \text{q.e.d.}$$

D.10.7. Theorem. If $a_1, a_2, \ldots, a_n, b$ are mutually compatible spaces, then

$$(a_1 + a_2 + \cdots + a_n) \cdot b = a_1 b + a_2 b + \cdots + a_n b, \qquad (n = 2, 3, \ldots).$$

Proof. We shall prove this by induction. Suppose that

(1) $$(a_1 + a_2 + \cdots + a_k) \cdot b = a_1 b + a_2 b + \cdots + a_k b.$$

We have

$$A =_{df} (a_1 + a_2 + \cdots + a_{k+1}) \cdot b = [(a_1 + a_2 + \cdots + a_k) + a_{k+1}] \cdot b.$$

Since the distributive law is true for two terms of the sum, [D.10.6.], we get

(2) $$A = (a_1 + a_2 + \cdots + a_k) \cdot b + a_{k+1} \cdot b,$$

because, [D.10.2.], $a_1 + a_2 + \cdots + a_k$ is compatible with b. Since, by hypothesis,

$$(a_1 + \cdots + a_k) \cdot b = a_1 b + \cdots + a_k b,$$

it follows from (2):

$$A = (a_1 b + a_2 b + \cdots + a_k b) + a_{k+1} b,$$

which completes the proof.

D.11. We shall need some auxiliary theorems on infinite collections of spaces in the lattice of all spaces in **H**. They will be useful in farther developping of the theory of compatibility of spaces.

Theorem. If $\{a_n\}$, $(n = 1, 2, \ldots)$ are spaces, then putting

$$b_n =_{df} a_1 + \cdots + a_n,$$

the sum

$$\sum_{n=1}^{\infty} a_n$$

coincides with the set b of all limits of convergent sequences $\{\vec{X}_n\}$, where $\vec{X}_n \in b_n$, and it also coincides with the space spanned by all a_n.

Proof. We have

$$a_i \leqq \sum_{n=1}^{\infty} a_n \quad \text{for all } i;$$

Hence

$$b_m = \sum_{i=1}^{m} a_i \leqq \sum_{n=1}^{\infty} a_n.$$

This being true for all m, we have

(1) $$\sum_{m=1}^{\infty} b_m \leqq \sum_{n=1}^{\infty} a_n.$$

On the other hand, since

$$a_n \leqq b_n,$$

we have

$$a_n \leqq \sum_{i=1}^{\infty} b_i,$$

and this being true for any n, it follows

$$\sum_{n=1}^{\infty} a_n \leqq \sum_{n=1}^{\infty} b_n. \tag{2}$$

From (1) and (2) it follows that

$$\sum_{n=1}^{\infty} a_n = \sum_{n=1}^{\infty} b_n. \tag{3}$$

Hence

$$\sum_{n=1}^{\infty} a_n = \left\{\overline{\bigcup_{n=1}^{\infty} b_n}\right\};$$

and this completes the proof of the first thesis. To have the second thesis, notice that it follows from the definition of

$$\sum_{n=1}^{\infty} a_n.$$

D.11.1. Theorem. Let $a_1, a_2, \ldots, a_\omega, \ldots, a_\alpha, \ldots$ be well ordered sequence of spaces, and define

$$c_\alpha =_{df} \sum_{\gamma \leqq \alpha} a_\gamma \quad \text{for every} \quad \alpha \geqq 1.$$

Then we have:

1) if $\alpha' \leqq \alpha''$, then $c_{\alpha'} \leqq c_{\alpha''}$,
2) $\sum_\alpha a_\alpha = \sum_\alpha c_\alpha$.

Proof. Let $\alpha' < \alpha''$. Put $b =_{df} \sum_\alpha a_\alpha$ and $c =_{df} \sum_\alpha c_\alpha$. We have

$$c_{\alpha'} = \sum_{\gamma \leqq \alpha'} a_\gamma, \qquad c_{\alpha''} = \sum_{\gamma \leqq \alpha''} a_\gamma.$$

Hence

$$c_{\alpha''} = \sum_{\gamma \leqq \alpha'} a_\gamma + \sum_{\alpha' \leqq \gamma \leqq \alpha''} a_\gamma = c_{\alpha'} + \sum_{\alpha' \leqq \gamma \leqq \alpha''} a_\gamma.$$

It follows that

$$c_{\alpha'} \leqq c_{\alpha''}, \quad \text{q.e.d.}$$

To prove the second thesis, we notice that

$$c_\alpha = \sum_{\gamma \leqq \alpha} a_\gamma \leqq \sum_{\gamma \leqq \alpha} a_\gamma + \sum_{\alpha < \gamma} a_\gamma = b.$$

Hence

(1) $$\sum_{\alpha} c_\alpha \leqq b, \quad \text{and then} \quad c \leqq b.$$

Now $\sum_{\alpha} c_\alpha$ is the smallest space c including all c_α. Take any a_β. We have

$$a_\beta \leqq \sum_{\gamma \leqq \beta} a_\gamma = c_\beta.$$

Hence a_β is included in c_β, and then in c. It follows, by the definition of sum of the spaces a_α, that

$$\sum_{\beta} a_\beta \leqq c, \quad \text{i.e.} \quad b \leqq c.$$

Hence, by (1),

$$b = c, \quad \text{q.e.d.}$$

D.11.2. Theorem. If M is a not empty set of spaces and

$$b = \sum_{a \in M} a,$$

then b coincides with the set b' of all limits of converging sequences

$$\vec{x}_1, \vec{x}_2, \ldots, \vec{x}_n, \ldots$$

of vectors such that for every n there exists a finite number of spaces $a_{n1}, \ldots, a_{nk_n} \in M$, where

$$\vec{x}_n \in a_{n1} + \cdots + a_{nk_n}.$$

Proof. First we shall prove that b' is a linear variety. Let $\vec{x}, \vec{y} \in b'$, and let λ be a number.

There exist infinite sequences of finite sums of spaces $\in M$:

$$P_1, P_2, \ldots, P_n, \ldots,$$

$$Q_1, Q_2, \ldots, Q_n, \ldots,$$

and vectors

$$\vec{x}_1, \vec{x}_2, \ldots,$$

$$\vec{y}_1, \vec{y}_2, \ldots,$$

where

$$\vec{x}_n \in \sum_{a \in P_n} a, \quad \vec{y}_n \in \sum_{a \in Q_n} a$$

with

$$\lim_n \vec{x}_n = \vec{x}, \quad \lim_n \vec{y}_n = \vec{y}.$$

We have

$$\vec{x}_n + \vec{y}_n \in \sum_{a \in P_n} a + \sum_{a \in Q_n} a,$$

$$\lambda \vec{x}_n \in \sum_{a \in P_n} a,$$

and

$$\lim_n (\vec{x}_n + \vec{y}_n) = \vec{x} + \vec{y}, \quad \lim_n (\lambda \vec{x}_n) = \lambda \vec{x}.$$

Since

$$\sum_{a \in P_n} a + \sum_{a \in Q_n} a = \sum_{a_n \in P_n + Q_n} a,$$

therefore

$$\vec{x} + \vec{y} \in b'. \quad \text{We also have} \quad \lambda \vec{x} \in b'.$$

Thus b' is a linear variety.

Now we shall prove that b' is a closed set. Let

$$\vec{x}_n \in b', \quad (n = 1, 2, \ldots), \quad \lim_n \vec{x}_n = \vec{x}.$$

It suffices to prove that $\vec{x} \in b'$. We have

$$\vec{x}_k = \lim_n \vec{x}_{n,k} \quad \text{where} \quad \vec{x}_{n,k} \in \sum_{a \in P_{n,k}} a,$$

and where $P_{n,k}$ is a finite subset of M. Take

$$\varepsilon_n > 0 \quad \text{with} \quad \varepsilon_n \to 0.$$

We can find an index $N(n)$ such that

$$\|\vec{x} - \vec{x}_{N(n)}\| \leqq \frac{\varepsilon_n}{2}.$$

Having fixed $N(n)$, we can find $k(n)$ such that

$$\|\vec{x}_{N(n)} - \vec{x}_{N(n), k(n)}\| \leqq \frac{\varepsilon_n}{2}.$$

It follows that

$$\|\vec{x} - \vec{x}_{N(n), k(n)}\| \leqq \varepsilon_n, \quad (n = 1, 2, \ldots).$$

Now

$$\vec{x}_{N(n), k(n)} \in P_{N(n), k(n)}, \quad \text{a finite subset of } M;$$

hence b' is a closed set, and then b' is a space. We clearly have $a_1 \leqq b'$ for every $a_1 \in M$. Indeed, if $\vec{x} \in a_1$, then we have $\vec{x} = \lim_n \vec{x}_n$, where $\vec{x}_n \in \sum_{a \in (a_1)} a$.

Let b'' be a space such that $a \leqq b''$ for every $a \in M$. If $\vec{y} \in b'$, then $\vec{y} = \lim_n \vec{y}_n$ for some vectors $\vec{y}_n$ each of which being included in a finite sum of spaces belonging to M. Therefore $\vec{y}_n \in b''$. Since b'' is closed, it follows that $\vec{y} \subset b''$. Thus $b' \leqq b''$, which proves that

$$b' = \sum_{a \in M} a.$$

The theorem is established.

D.11.3. Theorem. If M is a non empty set of spaces, then

$$\sum_{a \in M} a = \overline{\bigcup_{b \in M'} b},$$

where M' is the set of all finite sums

$$a_1 + \cdots + a_n, \quad (n = 1, 2, \ldots),$$

and where

$$a_1, \ldots, a_n \in M.$$

Proof. Let $\vec{x} \in \overline{\bigcup_{b \in M'} b}$. There exists a sequence $\vec{x}_n$ with $\lim \vec{x}_n = \vec{x}$, such that every $\vec{x}_n$ belongs to $\bigcup_{b \in M'} b$; thus every $\vec{x}_n$ belongs to some b_n, where $b_n \in M'$. Hence by [D.11.2.], $\vec{x} \in \sum_{a \in M} a$. Let $\vec{y} \in \sum_{a \in M} a$. By theorem [D.11.2.], $\vec{y}$ is the limit of some sequence $\{\vec{y}_n\}$ such that $\vec{y}_n$ belongs to some b_n where $b_n \in M'$. Hence $\vec{y}_n \in \bigcup_{b \in M'} b$; hence $\vec{y} \in \overline{\bigcup_{b \in M'} b}$, which completes the proof.

D.12. Theorem. If M is a non empty set of spaces and if $b \perp a$ for every $a \in M$, then

$$b \perp (\sum_{a \in M} a).$$

Proof. By theorem [D.11.3.],

$$\sum_{a \in M} a = \overline{\bigcup_{b \in M'} b},$$

where M' means the same as in theorem [D.11.3.]. Let $\vec{x} \in \sum_{a \in M} a$. There exists a sequence $\vec{x}_n$ with $\vec{x}_n \to \vec{x}$, $\vec{x}_n \in b_n$, where $b_n \in M'$. Let

$$b_n = a_{n,1} + \cdots + a_{n,k_n},$$

where

$$a_{n,i} \in M.$$

Since all $a \in M$ are $\perp b$, it follows that

$$b_n \perp b.$$

Hence $\vec{x}_n \perp b$. Since $\vec{x}_n \to \vec{x}$, we get, [D.11.], $\vec{x} \perp b$. It follows that

$$(\sum_{a \in M} a) \perp b, \quad \text{q.e.d.}$$

D.13. The infinite sum of spaces is defined in a manner not involving any ordering in the set of spaces to be summed. The same holds for infinite product. Thus it is rather trivial to say that the general sum and product are commutative. The associative law is not trivial and it may be stated as follows.

D.13.1.

$$\sum_{\alpha \in A} \sum_{a \in M_\alpha} a = \sum_{a \in \bigcup_{\alpha \in A} M_\alpha} a,$$

$$\prod_{\alpha \in A} \prod_{a \in M_\alpha} a = \prod_{a \in \bigcup_{\alpha \in A} M_\alpha} a.$$

It is supposed that neither A nor any M_α be empty. We leave the proof to the reader.

D.14. Lemma. Let

1⁰. $a_1 \leqq a_2 \leqq \cdots \leqq a_\omega \leqq \cdots \leqq a_\alpha \leqq \cdots$,
2⁰. $b_1 =_{df} a_1$, and
3⁰. $b_\alpha =_{df} a_\alpha - \sum_{\gamma<\alpha} a_\gamma$, for $\alpha > 1$.

Then

1) if $\alpha' \neq \alpha''$, we have $b_{\alpha'} \perp b_{\alpha''}$,
2) $\sum_\alpha a_\alpha = \sum_\alpha b_\alpha$.

Proof. Let

$$\alpha' < \alpha''.$$

We have

$$b_{\alpha'} = a_{\alpha'} - \sum_{\gamma<\alpha'} a_\gamma \leqq a_{\alpha'}, \qquad (\alpha' > 1).$$

Hence

$$b_{\alpha'} \leqq \sum_{\gamma<\alpha''} a_\gamma,$$

and a fortiori,

$$b_{\alpha'} \leqq \operatorname{co} a_{\alpha''} + \sum_{\gamma<\alpha''} a_\gamma,$$

i.e.

$$b_{\alpha'} \leqq \operatorname{co}[a_{\alpha''} \cdot \operatorname{co} \sum_{\gamma<\alpha''} a_\gamma] = \operatorname{co}[a_{\alpha''} - \sum_{\gamma<\alpha''} a_\gamma],$$

i.e.

$$b_{\alpha'} \leqq \operatorname{co} b_{\alpha''},$$

which proves the 1) thesis. To prove the second thesis, notice that for every β

$$b_\beta \leqq a_\beta \leqq \sum_\alpha a_\alpha.$$

Hence

$$\sum_\beta b_\beta \leqq \sum_\alpha a_\alpha. \tag{1}$$

To prove the converse inclusion it suffices to prove, that for every index β:

$$a_\beta \leqq \sum_\alpha b_\alpha.$$

Suppose this proved for all $\beta' < \beta$. We shall prove it for $\beta' = \beta$.

Since $\sum_{\gamma<\beta} a_\gamma \leqq a_\beta$, (by virtue of hypothesis), the space $\sum_{\gamma<\beta} a_\gamma$ is compatible with a_β.

Hence, by theorem [D.8., VII],

$$a_\beta = a_\beta \cdot \sum_{\gamma<\beta} a_\gamma + a_\beta \cdot \operatorname{co} \sum_{\gamma<\beta} a_\gamma = a_\beta \cdot \sum_{\gamma<\beta} a_\gamma + b_\beta. \tag{2}$$

But since for all $\gamma < \beta$ we have supposed, that $a_\gamma \leqq \sum_\alpha b_\alpha$, it follows that

$$\sum_{\gamma<\beta} a \leqq \sum_\alpha b_\alpha .$$

Thus from (2) we get

$$a_\beta \leqq \sum_\alpha b_\alpha + b_\beta \leqq \sum_\alpha b_\alpha .$$

If $\beta = 2$ we have surely $a_\beta \leqq \sum_\alpha b_\alpha$, because

$$a_\beta - a_2 = (a_2 - a_1) + a_2\, a_1 = b_2 + b_1 \leqq \sum_\alpha b_\alpha .$$

Hence we have proved that, whatever the index β might be, we have

$$a_\beta \leqq \sum_\alpha b_\alpha .$$

It follows that

(3) $$\sum_\beta a_\beta \leqq \sum_\alpha b_\alpha .$$

From (1) and (3) the second thesis follows. The theorem is proved.

D.14.1. Lemma. Let $a_1, a_2, \ldots, a_\omega, \ldots, a_\alpha, \ldots$ be any well ordered sequence of spaces. Put

(1) $$c_\alpha =_{df} \sum_{\gamma \leqq \alpha} a_\gamma \quad \text{and} \quad b_1 =_{df} a_1 .$$

Put also

(2) $$b_\alpha =_{df} c_\alpha - \sum_{\gamma<\alpha} a_\gamma \quad \text{for} \quad \alpha > 1 .$$

Then

1) if $\alpha' < \alpha''$, we have $b_{\alpha'} \perp b_{\alpha''}$,
2) $\sum_\alpha a_\alpha = \sum_\alpha b_\alpha$.

Proof. By Lemma [D.14.] we have for every α

(3) $$\sum_{\gamma \leqq \alpha} a_\gamma = \sum_{\gamma \leqq \alpha} c_\gamma ,$$

and

$$\text{if} \quad \alpha' \leqq \alpha'', \quad \text{then} \quad c_{\alpha'} \leqq c_{\alpha''} .$$

By hypothesis (2) and by (3) we get

$$b_\alpha = c_\alpha - \sum_{\gamma<\alpha} c_\gamma \quad \text{for} \quad \alpha > 1 \quad \text{and} \quad b_1 = c_1 .$$

Applying lemma [D.14.], we get:

$$\text{if} \quad \alpha' < \alpha'', \quad \text{then} \quad b_{\alpha'} \perp b_{\alpha''} ,$$

and

$$\sum_\alpha a_\alpha = \sum_\alpha c_\alpha = \sum_\alpha b_\alpha , \quad \text{q.e.d.}$$

D.15. Lemma. Let

1. $\{K\}$ be a non empty class of mutually orthogonal spaces, and
2. let b be a space compatible with every $a \in \{K\}$;

then b is compatible with $\sum_{a \in K} a$.

Proof. Put

$$c =_{df} \sum_{a \in K} a.$$

Since all the spaces of $\{K\}$ are mutually orthogonal, therefore we have for any $\vec{x} \in b$,

$$\text{(1)} \qquad \mathrm{Proj}_c \vec{x} = \sum_{a \in K} \mathrm{Proj}_a \vec{x},$$

where the sum on the right possesses at most a denumerable number of terms $\neq \vec{O}$ and is convergent.

Since b is compatible with a we have, by theorem [D.8.4.]

$$\mathrm{Proj}_a b = \mathrm{Proj}_b a.$$

Hence there exists a vector $\vec{y}_a$ such that

$$\mathrm{Proj}_a \vec{x} = \mathrm{Proj}_b \vec{y}_a.$$

Hence, from (1) we have

$$\mathrm{Proj}_c \vec{x} = \sum_{a \in K} \mathrm{Proj}_b \vec{y}_a.$$

The sum being convergent and at most denumerable, we have,

$$\sum_{a \in K} \mathrm{Proj}_b \vec{y}_a \in b, \quad \text{because} \quad \mathrm{Proj}_b \vec{y}_a \in b,$$

and because b is a closed set. Hence

$$\mathrm{Proj}_c \vec{x} \in b.$$

Hence

$$\mathrm{Proj}_c b \leqq b, \quad \text{and then} \quad \mathrm{Proj}_c b \leqq b \cdot c,$$

which completes the proof because of Theorem [D.8., II].

D.15.1. Lemma. Let $a_1, a_2, \ldots, a_\omega, \ldots, a_\alpha, \ldots$ be a well ordered finite or infinite sequence of spaces and let c be compatible with all spaces a_α.

Put

$$b_\alpha =_{df} \sum_{\gamma \leqq \alpha} a_\gamma - \sum_{\gamma < \alpha} a_\gamma \quad \text{for} \quad \alpha > 1,$$

and

$$b_1 =_{df} a_1;$$

then c is compatible with all b_α.

Proof. First we shall prove that c is compatible with every

$$d_\alpha =_{df} \sum_{\gamma < \alpha} a_\gamma \quad \text{for} \quad \alpha > 1.$$

Suppose this will be true for all $d_{\beta'}$ where $\beta' < \beta$, and where β is fixed for a moment; i.e. $d_{\beta'}$ is compatible with c. Put

$$c_\alpha = \sum_{\gamma \leqq \alpha} a_\gamma .$$

We have

$$b_\alpha = c_\alpha - \sum_{\gamma < \alpha} a_\gamma \quad \text{for} \quad \alpha > 1 .$$

By Lemma [D.14.1.] we have

(0)
$$\sum_{\gamma < \beta} a_\gamma = \sum_{\gamma < \beta} b_\gamma .$$

We have supposed that c is compatible with all $d_{\beta'}$ for $\beta' < \beta$; hence c is compatible with

(1)
$$(d_{\beta'} + a_{\beta'}) - d_{\beta'} .$$

But (1) is equal to

$$\Big(\sum_{\gamma < \beta'} a_\gamma + a_{\beta'}\Big) - \sum_{\gamma < \beta'} a_\gamma = \sum_{\gamma \leqq \beta'} a_\gamma - \sum_{\gamma < \beta'} a_\gamma = b_{\beta'} .$$

Thus c is compatible with all $b_{\beta'}$ for $\beta' < \beta$.

Now, we know that all these $b_{\beta'}$ are mutually orthogonal. It follows, by virtue of Lemma [D.15.], that c is compatible with

$$\sum_{\beta' < \beta} b_{\beta'} ;$$

hence, by (0), c is compatible with

$$\sum_{\gamma < \beta} a_\gamma = d_\beta .$$

Besides we know that c is compatible with $a_1 + a_2$. Thus we have proved, by induction that c is compatible with every d_α, where $\alpha > 1$. Since $b_\alpha = (d_\alpha + a_\alpha) - d_\alpha$, it follows, that c is compatible with every b_α for $\alpha > 1$. Since $b_1 = a_1$ and since c is supposed to be compatible with a_1, it follows that c is compatible with every b_α, q.e.d.

D.15.2. Lemma. Let $a_1, a_2, \ldots, a, \ldots, a_\alpha, \ldots$ be a well ordered sequence of spaces compatible with the space c. Then $\sum_\alpha a_\alpha$ is compatible with c.

Proof. Put

$$b_1 =_{df} a_1 ,$$

$$b_\alpha =_{df} \sum_{\gamma \leqq \alpha} a_\gamma - \sum_{\gamma < \alpha} a_\gamma \quad \text{for} \quad \alpha > 1 .$$

By Lemma [D.15.1.] the space c is compatible with all b_α. Now, by Lemma [D.14.1.],

(1)
$$\sum_\alpha a_\alpha = \sum_\alpha b_\alpha ,$$

and besides, the spaces b_α are orthogonal to one another. It follows, by Lemma [D.15.], that $\sum_\alpha b_\alpha$ is compatible with c, and then, by (1), the theorem is proved.

All the above lemmata allow to prove the following:

D.15.3. Theorem. Let $\{K\}$ be any not empty class of spaces each of which is compatible with b.

Then the spaces

$$\sum_{a \in K} a, \quad \prod_{a \in K} a$$

are also compatible with b.

Proof. Let us arrange all spaces of $\{K\}$ into a well ordered finite or transfinite sequence

$$a_1, a_2, \ldots, a_\omega, \ldots, a_\alpha, \ldots,$$

having the smallest ordinal type, corresponding to the cardinal number of $\{K\}$.

By virtue of Lemma [D.15.1.]

$$\sum_\alpha a_\alpha$$

is compatible with b, which completes the proof for the sum.

Now, we know that the general de Morgan law holds true for any sums and products of spaces.

We have

$$(1) \qquad \prod_\alpha a_\alpha = \operatorname{co} \sum_\alpha \operatorname{co} a_\alpha .$$

Since a_α is compatible with b, it follows by Theorem [D.8., IX], that $\operatorname{co} a_\alpha$ is compatible with b. Hence by what we have already proved

$$\sum_\alpha \operatorname{co} a_\alpha$$

is compatible with b. Hence, using again Theorem [D.8., IX] and relying on (1), we complete the proof.

Chapter D I

Tribes of spaces

D I.1. In the preceding Chapter [D] we have developped the theory of compatibility of spaces, and we have proved that, if the spaces of a set are compatible with one another, all the laws which are valid in a tribe, are valid for them. Especially, the distributive law is valid in the case of compatibility. But it may not take place if the spaces are not compatible. Now we shall isolate some sublattices of the geometrical lattice $\boldsymbol{L}$ of all spaces [D.2.3.], such that they will constitute tribes.

D I.1.1. Def. Let $\boldsymbol{S}$ be a non empty class of spaces, such that

1^0. if $a \in \boldsymbol{S}$, then $\operatorname{co} a \in \boldsymbol{S}$,

2^0. if $a, b \in \boldsymbol{S}$, then $a + b \in \boldsymbol{S}$,

3^0. if $a, b \in \boldsymbol{S}$, and $a \cdot b = O$, then $a \perp b$.

D I.1.2. Theorem. We shall prove that the ordering $\boldsymbol{L}$, if restricted to $\boldsymbol{S}$ and with complementation taken from $\boldsymbol{L}$, is a tribe with $\mathbf{H}$ as unit and $(\vec{O})$ as zero.

We shall prove this in few steps.

D I.1.2a. Theorem. First we shall prove that, if $a, b \in \boldsymbol{S}$, then

$$a \cdot b \in \boldsymbol{S}.$$

Let $a, b \in \boldsymbol{S}$. We have, by hyp. 1^0,

$$\operatorname{co} a, \operatorname{co} b \in \boldsymbol{S};$$

hence, by hyp. 2^0,

$$\operatorname{co} a + \operatorname{co} b \in \boldsymbol{S};$$

hence, by hyp. 1^0,

$$\operatorname{co}(\operatorname{co} a + \operatorname{co} b) \in \boldsymbol{S}.$$

Applying de Morgan law, we get

$$a \cdot b \in \boldsymbol{S}, \quad \text{q.e.d.}$$

D I.1.2b. Theorem. If $a, b \in \boldsymbol{S}$, then $a - b \in \boldsymbol{S}$.

Proof. We have, $\operatorname{co} b \in \boldsymbol{S}$; hence by [D I.1.2a.], $a \operatorname{co} b \in \boldsymbol{S}$, i.e. $a - b \in \boldsymbol{S}$.

D I.1.2c. Theorem. If $a, b \in \boldsymbol{S}$, then a, b are compatible.

Proof. We have, [D.3.3.],

$$(1) \qquad (a - a\,b) \cdot (b - a\,b) = O.$$

Since, by [D I.1.2a.],

$$a \cdot b \in \boldsymbol{S},$$

we have

$$a - a\,b \in \boldsymbol{S} \quad \text{and} \quad b - a\,b \in \boldsymbol{S}.$$

Consequently from (1), by virtue of hyp. 3^0, we have

$$(a - a\,b) \perp (b - a\,b),$$

so [D.5.], a, b are compatible.

D I.1.2d. Theorem. $\boldsymbol{S}$ is a lattice, ordered by the relation of inclusion of spaces, considered as sets of vectors.

Indeed, if $a, b \in \boldsymbol{S}$, we have $a + b \in \boldsymbol{S}$, and $a \cdot b \in \boldsymbol{S}$.

D I.1.2e. Theorem. We have $O_L \in \boldsymbol{S}$ and $I_L \in \boldsymbol{S}$, where O_L and I_L are the zero and the unit of the lattice $\boldsymbol{L}$.

Indeed, there exists a soma $a \in \boldsymbol{S}$, since $\boldsymbol{S}$ is not empty. Hence, by [D I.1.1., 1⁰],

$$\operatorname{co} a \in \boldsymbol{S}.$$

Hence, by [D I.1.1., 2⁰],

$$I_L = a + \operatorname{co} a \in \boldsymbol{S},$$

and, by [D I.1.1a.],

$$O_L = a \cdot \operatorname{co} a \in \boldsymbol{S}.$$

D I.1.2f. Theorem. We see, that in $\boldsymbol{S}$, the distributive law is valid. Indeed, since all spaces of $\boldsymbol{S}$ are compatible, [D I.1.2c.], therefore, by [D.10.6.],

$$(a + b) \cdot c = a\, c + b\, c,$$

whenever $a, b, c \in \boldsymbol{S}$.

Taking account of what has been just proved, we see that $\boldsymbol{S}$, if ordered geometrically, is a distributive, complementary lattice; hence a Boolean tribe.

Def. We shall call it *geometrical tribe of spaces, in* $\mathbf{H}$.

D I.3. Expl. The simplest geometrical tribe in $\mathbf{H}$ is composed of the single space $(\vec{O})$, and of $\mathbf{H}$. A tribe slightly more extensive is the tribe composed of $(\vec{O})$, a, $\operatorname{co} a$, $\mathbf{H}$, where $a \neq O$ and $a \neq \mathbf{H}$.

D I.3.1. Expl. If $\mathbf{H}$ is n-dimensional and

(1) $$\vec{\varphi}_1, \vec{\varphi}_2, \ldots, \vec{\varphi}_n$$

is saturated orthonormal set of vectors, then the collection of all spaces, generated by finite subsets of (1), with additional space $(\vec{O})$, make up a geometrical tribe.

D I.3.2. Expl. Let $\mathbf{H}$ be the space of all square summable, complex valued functions $f(x)$ of the real variable x, varying in $\langle 0, 1\rangle$.

Define the orthogonality of $f(x)$, $g(x)$ by $\int\limits_0^1 \overline{f(x)}\, g(x)\, dx = 0$. $\mathbf{H}$ is an infinite dimensional H.-H.-space.

Let E be a Lebesgue-measurable subset of $\langle 0, 1\rangle$, and define the space a_E as the set of all functions $\in \mathbf{H}$ which vanish for almost all $x \overline{\in} E$. The collection of all spaces a_E is a geometrical tribe in $\mathbf{H}$.

D I.4. Theorem. Let $\boldsymbol{S}$ be a non empty collection of spaces, satisfying the conditions:

1⁰ if $a, b \in \boldsymbol{S}$, then $a + b \in \boldsymbol{S}$,

2⁰ if $a \in \boldsymbol{S}$, then $\operatorname{co} a \in \boldsymbol{S}$.

Then the following are equivalent

I. $\boldsymbol{S}$ if ordered, as in $\boldsymbol{L}$, is a geometrical tribe in $\mathbf{H}$;

II. All spaces of $\boldsymbol{S}$ are mutually compatible.

Proof. From I follows II, by virtue of [D I.1.2c.]. Assume II. Let $a, b \in \boldsymbol{S}$ and suppose that

$$a \cdot b = O. \tag{1}$$

Since a, b are compatible, we have

$$(a - a\,b) \perp (b - a\,b).$$

By (1) we get

$$(a - O) \perp (b - O)$$

i.e.

$$a \perp b. \tag{2}$$

By hyp. 1^0, 2^0 and (2) it follows, [D I.1.1.], [D I.1.2. Theorem], that $\boldsymbol{S}$ is a geometrical tribe in $\mathbf{H}$. The theorem is established.

D I.4.1. Theorem. Let $\boldsymbol{S}$ be non empty collection of spaces, ordered as in the lattice $\boldsymbol{L}$, and satisfying the conditions:

1^0 If $a \in \boldsymbol{S}$, then $\text{co}\,a \in \boldsymbol{S}$.

2^0 If $a, b \in \boldsymbol{S}$, then $a + b \in \boldsymbol{S}$.

Under these circumstances the following are equivalent:

I. S generates a geometrical tribe ($\boldsymbol{S}$) of spaces in $\mathbf{H}$,

II. if $a, b, c \in \boldsymbol{S}$, then $(a + b) \cdot c = a\,c + b\,c$.

Proof. From I the condition II follows, by virtue of [D I.1.2.] or [D I.1.2f.].

Let II. We have

$$a - a \cdot b = a \cdot \text{co}\,(a \cdot b) = a\,(\text{co}\,a + \text{co}\,b),$$

because the de Morgan laws hold true.

But by II:

$$a\,(\text{co}\,a + \text{co}\,b) = a \cdot \text{co}\,a + a \cdot \text{co}\,b = a \cdot \text{co}\,b = a - b.$$

It follows that

$$a - a \cdot b = a - b.$$

This, however, by [D.8., V] proves that a, b are compatible. Applying [D I.4.], we get I. The theorem is established.

D I.5. Remark. Every space a, i.e. subspace of the H.-H.-space, can itself be considered as a H.-H.-space; so we can consider only subspaces of a, which in the case of a tribe of spaces, may be its unit. We shall later consider such subspaces of H.-H., which are subspaces of a given space a. In connection with this remark we shall prove some theorems.

D I.5.1. Lemma. If

1. $\boldsymbol{S}$ is a geometrical tribe of spaces in $\mathbf{H}$,

2. the space p is compatible with all the spaces which are somata of $\boldsymbol{S}$, (*we say p is compatible with $\boldsymbol{S}$*),

then:

$$\text{if} \quad a, b \in \boldsymbol{S}, \quad a\, b\, p = O, \quad \text{then} \quad a\, p \perp b\, p.$$

Proof. We have, [D.5.4.], [D I.4.],

$$b = b\, a + (b - a);$$

hence

$$b\, p = [b\, a + (b - a)]\, p.$$

Since p is compatible with a and b, it is also compatible with $b \cdot a$ and $(b - a)$, [D.10.3.], [D.10.4.]. It follows, by distributive law, [D.10.6.], that

$$b\, p = b\, a\, p + (b - a)\, p.$$

Hence, by virtue of hypothesis $a\, b\, p = O$, we get

(1) $$b\, p = (b - a)\, p.$$

Now we have

$$b\, p - a\, p = b\, p \operatorname{co}(a\, p) = b\, p(\operatorname{co} a + \operatorname{co} p) = b\, p \operatorname{co} a + b\, p \operatorname{co} p$$
$$= b\, p \operatorname{co} a = b \operatorname{co} a \cdot p = (b - a) \cdot p.$$

Hence from (1) we get

(2) $$b\, p = b\, p - a\, p.$$

Hence

(3) $$b\, p \leqq \operatorname{co}(a\, p),$$

which gives, [D.2.10.],

$$b\, p \perp a\, p, \quad \text{q.e.d.}$$

D I.5.2. Theorem. Let $\boldsymbol{S}$ be a geometrical tribe in $\mathbf{H}$, and p a *space compatible with* $\boldsymbol{S}$, i.e. with all spaces of $\boldsymbol{S}$. Under these conditions we have the following:

1) The class $\boldsymbol{P}$ of all spaces $a \cdot p$, where a varies in $\boldsymbol{S}$, generates a tribe, ordered by inclusion of spaces and where the complementation is defined by

$$\operatorname{co}_P t =_{df} p - t,$$

for all $t \in \boldsymbol{P}$. The unit I_P of the tribe $\boldsymbol{P}$ is p and the zero O_P is $(\vec{O})$.

2) The class of all spaces $a \cdot \operatorname{co} p$, where a varies in $\boldsymbol{S}$, also generates a tribe $\boldsymbol{P}'$ of spaces with $(\vec{O})$ as $O_{P'}$ and where the complementation is defined by

$$\operatorname{co}_{P'} s = \operatorname{co} p - s \quad \text{for all} \quad s \in \boldsymbol{P}'.$$

D I.5.3. Def. We call the tribe $\boldsymbol{P}$ *geometrical tribe of spaces in* p, and $\boldsymbol{P}'$ *geometrical tribe of spaces in* co$p$.

Proof of the theorem. Let $s, t \in \boldsymbol{P}$. There exist $a, b \in \boldsymbol{S}$ such that

$$s = a\,p, \quad t = b\,p.$$

By compatibility we get, [D.10.7.],

$$s + t = a\,p + b\,p = (a + b) \cdot p \in \boldsymbol{P}.$$

Let $s \in \boldsymbol{P}$. We have

$$s = a \cdot p \quad \text{for some} \quad a \in \boldsymbol{S}.$$

Hence

$$\mathrm{co}_P\, s = p - s = p - a \cdot p = p - a, \text{ [D.8., V]}, = p \cdot \mathrm{co}\, a.$$

This space belongs to $\boldsymbol{P}$, since co$a \in \boldsymbol{S}$.

We have proved that two conditions for a geometrical tribe are satisfied:

$$\text{if} \quad a, b \in \boldsymbol{P}, \quad \text{then} \quad a + b \in \boldsymbol{P},$$

$$\text{if} \quad a \in \boldsymbol{P} \quad \text{then} \quad \mathrm{co}_P a \in \boldsymbol{P}.$$

There remains to prove that, if $s \cdot t = O$, where $s, t \in \boldsymbol{P}$, then

(1) $$s \perp t.$$

Let $s, t \in \boldsymbol{P}$. We have

$$s = a \cdot p, \quad t = b \cdot p,$$

where

$$a, b \in S.$$

Suppose that $s \cdot t = O$. Then

$$a\,p \cdot b\,p = a\,b\,p = O.$$

Applying the lemma [D I.5.1.], we get

$$a\,p \perp b\,p,$$

which completes the proof of (1).

Since

$$O \leqq a\,p \leqq p \quad \text{for all} \quad a \in \boldsymbol{S},$$

and since

$$p = 1_S \cdot p = 1 \cdot p = p,$$

therefore p is the unit of the tribe $\boldsymbol{P}$.

Since $O \leqq a\,p$ for all $a \in \boldsymbol{S}$, we have $O_P = O$, so O is the zero of $\boldsymbol{P}$.

The theorem is established.

D I.6. Agreement. To facilitate precise wordings, we shall agree to denote a tribe (which is an ordering), e.g., by $(\boldsymbol{T})$. The letter $\boldsymbol{T}$ will denote the domain of the tribe. We shall also call $\boldsymbol{T}$, *tribe-set.*

In this paragraph we shall consider geometrical tribes only, which may be given in the whole space **H**, or only in a given subspace, [D I.5.3.].

D I.6.1. Remark. A geometrical tribe of spaces in **H** may be finitely additive, denumerably additive or even completely additive, [A.3.2.].

D I.6.2. Theorem. The geometrical tribe ($\boldsymbol{T}$) in **H** is denumerably additive, whenever the following condition is satisfied:

If

$$a_1, a_2, \ldots, a_n, \ldots \in \boldsymbol{T},$$

then

$$\sum_{n=1}^{\infty} a_n \in \boldsymbol{T},$$

where $\sum$ denotes lattice-join.

D I.6.3. The tribe ($\boldsymbol{T}$) in **H** is completely additive if and only if the following condition takes place:

If M is any not empty collection of spaces of $\boldsymbol{T}$, then

$$\sum_{a \in M} a \in \boldsymbol{T}.$$

D I.6.4. If the geometrical tribe ($\boldsymbol{T}$) in **H** is denumerably [completely] additive, then, of course, all its spaces are mutually compatible. In addition to that the denumerable [general] de Morgan laws take place, [A.1.5.].

Finally the denumerable (complete) distributive law is valid, i.e.

$$\left(\sum_{n=1}^{\infty} a_n\right) \cdot b = \sum_{n=1}^{\infty} a_n \cdot b,$$

resp.

$$(\sum_{a \in M} a) \cdot b = \sum_{a \in M} a \cdot b$$

for any not empty collection of spaces of $\boldsymbol{T}$, [A.3.6.].

D I.6.5. Remark. The above can be relativized to any space $a \in \boldsymbol{T}$, where the unit becomes a.

D I.6.6. A geometrical tribe of spaces may not be denumerably additive.

Expl. Let **H** be a separable and complete H.-H.-space with infinite dimensions. By vector in **H** we shall understand any infinite sequence $\{x_n\}$ of complex numbers with

$$\sum_{n=1}^{\infty} |x_n|^2 < +\infty.$$

Let $\{\vec{\varphi}_n\}$ be a saturated orthonormal set of vectors. Let R be the set of all rational numbers in the interval $(0, 1\rangle$, given by an infinite sequence $\{r_n\}$*), each number written only once.

Consider the class S of all finite unions

$$(1) \qquad \sum_i (\alpha_i, \beta_i\rangle,$$

where

$$0 \leqq \alpha_i, \beta_i \leqq 1, \quad (i = 1, 2, \ldots).$$

We have $\emptyset \in S$, $(0, 1\rangle \in S$. The subsets A, of $(0, 1\rangle$, if ordered by the relation of inclusion of sets, and where the complementation co$A$ is defined by $(0, 1\rangle - A$, gives a finitely additive tribe (S). If $s \in S$, then $s \cap R$ is a set of all rational numbers, contained in s. Given $s \in S$, define s' as the space in $\mathbf{H}$, spanned by the set of all $\vec{\varphi}_n$, where $r_n \in s \cap R$. Varying s in S we get a collection $\boldsymbol{S}'$ of spaces.

There is a $1 \to 1$ correspondence between the sets s of S and the corresponding spaces s' of $\boldsymbol{S}'$. This correspondence carries complementation into complementation and finite unions into finite unions. $(\boldsymbol{S}')$ is a geometrical tribe in $\mathbf{H}$, isomorphic with the tribe of all $s \cap R$, where $s \in S$. Now the infinite union of intervals:

$$\left(\frac{1}{2}, \frac{1}{3}\right\rangle, \left(\frac{1}{4}, \frac{1}{5}\right\rangle, \ldots, \left(\frac{1}{2n}, \frac{1}{2n+1}\right\rangle, \ldots$$

does not exist in (S), so there does not exist the infinite union of the corresponding spaces in $(\boldsymbol{S}')$.

D I.7. We are going over to some extension-problems of geometrical tribes of spaces.

Def. Let (T), (T') be geometrical tribes of spaces. We say that (T') is an *extension of* (T), (or (T') *contains* (T)), whenever the following takes place:

1) if $a \in T$, then $a \in T'$;

2) the finite operations on somata a, b of (T) are just the corresponding operations taken from (T');

3) (T), (T') both have the same unit $\mathbf{H}$ and the same zero $(\vec{O})$.

Equivalently we can say that, [A.7.2.], (T) is a finitely genuine strict subtribe of (T').

D I.7.1. We shall solve the problem: Given a geometrical tribe (T) in $\mathbf{H}$, and a space p, does there exist a tribe (T'), an extension of (T), which would contain p as one of its somata?

---

*) If a, b are real numbers, put

$$(a, b\rangle =_{df} \{x \mid a < x \leqq b\}.$$

We have $(a, b\rangle \neq \emptyset$ whenever $a < b$.

Def. If such a tribe (T') exists, we shall say that *the space* p *is capable of being adjoint to* (T).

D I.7.2. Theorem. The following is true:

If the space p is capable of being adjoint to (T'), p must be compatible with every space $a \in T$.

Def. We shall say: p must *be compatible with* the tribe (T).

D I.8. Lemma. If the space p is compatible with (T), then we have for all spaces of (T):

1) $\operatorname{co}(a\,p + b \cdot \operatorname{co} p) = \operatorname{co} a \cdot p + \operatorname{co} b \cdot \operatorname{co} p$,

2) for finite somatic sums we have:

$$\sum_i (a_i \cdot p + b_i \cdot \operatorname{co} p) = \sum_i a_i \cdot p + \sum_i b_i \cdot \operatorname{co} p,$$

3) for finite somatic products we have

$$\prod_i (a_i \cdot p + b_i \cdot \operatorname{co} p) = \prod_i a_i \cdot p + \prod_i b_i \cdot \operatorname{co} p,$$

4) $(a \cdot p + b \cdot \operatorname{co} p) - (a' \cdot p + b' \cdot \operatorname{co} p) = (a - a')\,p + (b - b')\operatorname{co} p$,

5) for the algebraic addition we have

$$(a \cdot p + b \cdot \operatorname{co} p) \dotplus (a' \cdot p + b' \cdot \operatorname{co} p) = (a \dotplus a')\,p + (b \dotplus b') \cdot \operatorname{co} p.$$

D I.8a. Proof. To have a possibly simple proof, we shall use Boolean algebra, i.e. we shall consider the commutative and idempotent Stone's ring [A.2.9.], into which a tribe always can be reorganized. The operation in the ring are the algebraic addition, $a \dotplus b = (a - b) + (b - a)$ and the ordinary multiplication $a \cdot b$.

We recall some formulas:

$$a + b = a \dotplus a \cdot b \dotplus b; \quad \operatorname{co} a = I \dotplus a; \quad O \dotplus a = a, \quad a \cdot a = a,$$

$$I \cdot a = a; \quad O \cdot a = O; \quad \text{if} \quad a \cdot b = O, \quad \text{then} \quad a \dotplus b = a + b.$$

Since $(a \cdot p) \cdot (a \cdot \operatorname{co} p) = O$, we have

$$a \cdot p + a \cdot \operatorname{co} p = a \cdot p \dotplus a \cdot \operatorname{co} p.$$

D I.8b. We shall prove the item 5) of the thesis. We have

$$\begin{aligned}(a \cdot p + b \cdot \operatorname{co} p) \dotplus (a' \cdot p + b' \cdot \operatorname{co} p) &= a \cdot p \dotplus b \cdot \operatorname{co} p \dotplus a' \cdot p \dotplus b' \cdot \operatorname{co} p \\ &= (a \cdot p \dotplus a' \cdot p) \dotplus (b \cdot \operatorname{co} p \dotplus b' \cdot \operatorname{co} p) = (a \dotplus a')\,p \dotplus (b \dotplus b')\operatorname{co} p \\ &= (a \dotplus a')\,p + (b \dotplus b')\operatorname{co} p.\end{aligned}$$

D I.8c. To prove 3), we write:

$$\begin{aligned}&(a \cdot p + b \cdot \operatorname{co} p) \cdot (a' \cdot p + b' \cdot \operatorname{co} p) = (a \cdot p \dotplus b \cdot \operatorname{co} p) \cdot (a' \cdot p \dotplus b' \cdot \operatorname{co} p) \\ &= (a \cdot p)\,(a' \cdot p) \dotplus (b \cdot \operatorname{co} p)\,(a' \cdot p) \dotplus (a \cdot p)\,(b' \cdot \operatorname{co} p) \dotplus (b \cdot \operatorname{co} p)\,(b' \cdot \operatorname{co} p) \\ &= a \cdot a' \cdot p + b \cdot b' \cdot \operatorname{co} p = (a\,a')\,p + (b\,b')\operatorname{co} p.\end{aligned}$$

D I.8d. Proof of 1):

$$\mathrm{co}(a \cdot p + b \cdot \mathrm{co}p) = \mathrm{co}(a \cdot p \dotplus b \cdot \mathrm{co}p) = I \dotplus (a \cdot p \dotplus b \cdot \mathrm{co}p)$$

$$= (I \cdot p \dotplus I \cdot \mathrm{co}p) \dotplus (a \cdot p \dotplus b \cdot \mathrm{co}p) = I \cdot p \dotplus a \cdot p \dotplus I \cdot \mathrm{co}p \dotplus b \cdot \mathrm{co}p$$

$$= (I \dotplus a)\, p \dotplus (I \dotplus b)\, \mathrm{co}p = (\mathrm{co}a)\, p \dotplus (\mathrm{co}b)\, \mathrm{co}p = \mathrm{co}a \cdot p + (\mathrm{co}b)\, \mathrm{co}p.$$

By means of [D I.8b., c., d.] we prove straightforward the remaining items.

D I.8.1. Theorem. If p is compatible with (T), then the collection of all spaces

$$a \cdot p + b \cdot \mathrm{co}p, \quad a, b \in T$$

is a geometrical tribe set.

Proof. Denote the set by T'. Let $s, t \in T'$. There exist $a, b, a', b' \in T$, such that

$$s = a \cdot p + b \cdot \mathrm{co}p,$$

$$t = a' \cdot p + b' \cdot \mathrm{co}p.$$

By Lemma [D I.8.],

$$s + t = (a + a')\, p + (b + b')\, \mathrm{co}p;$$

hence

$$s + t \in T'.$$

Let $s \in T'$; we have for some $a, b \in T$:

$$s = a \cdot p + b \cdot \mathrm{co}p.$$

Since, by [D I.8.],

$$\mathrm{co}\, s = (\mathrm{co}a)\, p + (\mathrm{co}b)\, \mathrm{co}p,$$

we have

$$\mathrm{co}\, s \in T'.$$

Now, if

$$s, t, u \in T',$$

we have

$$(s + t)\, u = s \cdot u + t \cdot u,$$

because of the distributive law.

Consequently, by [D I.4.1.], T' is a tribe-set and then (T') is geometrical tribe in $\mathbf{H}$.

D I.8.2. The tribe (T') defined in [D I.8.1.] is the smallest extension of (T) containing p, whenever p is compatible with (T).

Proof. Every tribe (T''), which is an extension of (T) and contains p, must contain the spaces $a \cdot p + b \cdot \mathrm{co}p$, where $a, b \in T$, so (T') is the smallest tribe which contains (T) and p. From what has been proved the following theorem results:

D I.8.3. Theorem. If (T) is a geometrical tribe of spaces, and p a space, then the following are equivalent:

1) p is compatible with (T);

2) p is a space, capable of being adjoint to (T).

In addition to that we have:

Under the conditions 1) or 2), the smallest extension of (T) containing p is the geometrical tribe whose tribe-set is the collection of all spaces $a \cdot p + b \cdot \text{co}p$, where $a, b \in T$.

D I.8.4. Theorem. If

1) (T) is a denumerably (completely) additive tribe of spaces,

2) p is a space compatible with (T),

then the tribe (T') with somata

$$a \cdot p + b \cdot \text{co}p$$

is also denumerably (completely) additive.

Denumerably additive extension of geometrical tribes of spaces

D I.9. We shall attack the following problem:

Given the geometrical tribe set S of spaces, is it possible to find a denumerably additive geometrical tribe (T) including (S)? The answer is positive, and to prove this statement, we shall use some auxiliary theorems, to whose exposition we are now proceeding.

D I.9.1. Lemma. If the spaces a_i, b_k are compatible with one another, $(i, k = 1, 2, \ldots)$, then

$$\sum_{i=1}^{\infty} a_i \cdot \sum_{k=1}^{\infty} b_k = \sum_{i,k=1}^{\infty} a_i b_k .$$

Proof. Putting

$$c = \sum_{k=1}^{\infty} b_k ,$$

we see that all a_i, $(i = 1, 2, \ldots)$, are also compatible with c, by virtue of [D.15.2.]. Hence

$$\left(\sum_{i=1}^{\infty} a_i\right) \cdot c = \sum_{i=1}^{\infty} (a_i \cdot c) . \tag{1}$$

Now

$$a_i \cdot c = a_i \cdot \sum_{k=1}^{\infty} b_k .$$

Hence, by [D.10.7.] we get

$$a_i \cdot c = \sum_{k=1}^{\infty} a_i b_k .$$

Carrying this into (1), we get

$$\sum_{i=1}^{\infty} a_i \cdot c = \sum_{i=1}^{\infty} \sum_{k=1}^{\infty} a_i b_k,$$

which, on account of the associative law for addition of spaces, gives the thesis required. q.e.d.

D I.9.2. If the spaces a_i, b_k, $(i, k = 1, 2, \ldots)$, are compatible with one another, then

$$\prod_{i=1}^{\infty} a_i + \prod_{k=1}^{\infty} b_k = \prod_{i,k=1}^{\infty} (a_i + b_k).$$

Proof. This follows from [D I.9.1.] by de Morgan laws [A.1.5.], which holds true, as we know, for geometrical complementary lattices.

D I.10. Let us introduce the following auxiliary definition. Given the tribe (S) of spaces, let us say that the space p (which may not belong to S), *is of the type* $\sum$, (resp. $\prod$), if there exist $a_1, a_2, \ldots, a_n, \ldots \in S$ such that

$$p = \sum_{n=1}^{\infty} a_n, \quad \left(\text{resp.} \quad p = \prod_{n=1}^{\infty} a_n\right),$$

(**L**-lattice sums and products), [D.2.3.].

The space is said *to be of type* $\prod \sum$ (resp. $\sum \prod$), whenever there exists a sequence $p_1, \ldots, p_n, \ldots$ of spaces of the type $\sum$ (resp. $\prod$) and such that

$$p = \prod_{n=1}^{\infty} p_n, \quad \left(\text{resp.} \quad p = \sum_{n=1}^{\infty} p_n\right).$$

D I.10.1. Lemma. If p, q are of the type $\sum$, then $p + q$ is of the type $\sum$ too.

If p, q are of the type $\prod$, then $p \cdot q$ is also of the type $\prod$.

Proof is straight forward.

D I.10.2. If p, q are of the type $\sum$, then $p \cdot q$ is also of the type $\sum$.

If p, q are of the type $\prod$, then $p + q$ is also of the type $\prod$.

Proof. Let p, q be of the type $\sum$. Then we have the equality:

$$p \cdot q = \sum_{i=1}^{\infty} p_i \cdot \sum_{k=1}^{\infty} q_k,$$

where $p_i \in S$, $q_k \in S$ are some spaces.

Hence, by Lemma [D I.9.1.],

$$p \cdot q = \sum_{i,k=1}^{\infty} p_i q_k.$$

Since $p_i \cdot q_k \in S$, it follows by Def. [D I.10.], that $p \cdot q$ is of type $\sum$.

The proof of the second part is analogous, but based on Lemma [D I.9.2.].

Remark. In the case a_i, b_k belong to a given tribe (S), the above lemmata may be written in the following symbolic manner.

D I.10.2a. Theorem. $\sum \cdot \sum \subseteqq \sum$, i.e. the product of two somate, having the type $\sum$ is also of the type $\sum$.

$\Pi + \Pi \subseteqq \Pi$, i.e. the sum of two somate, having the type Π, is also of the type Π.

In addition to that we have the relations

$$\sum + \sum \subseteqq \sum \quad \text{and} \quad \Pi \cdot \Pi \subseteqq \Pi.$$

D I.10.3. Lemma. If p is of type $\sum \Pi$, then co$p$ is of type $\Pi \sum$, and if p is of type $\Pi \sum$, then co$p$ is of type $\sum \Pi$.

Proof. Let p be of type $\sum \Pi$. There exists a double sequence $a_{ik} \in S$, $i, k = 1, 2, \ldots$, such that

$$p = \sum_{i=1}^{\infty} \prod_{k=1}^{\infty} a_{ik}.$$

It follows that

$$\operatorname{co} p = \operatorname{co}\left(\sum_{i=1}^{\infty} \prod_{k=1}^{\infty} a_{ik}\right) = \prod_{i=1}^{\infty} \operatorname{co} \sum_{k=1}^{\infty} a_{ik} = \prod_{i=1}^{\infty} \sum_{k=1}^{\infty} \operatorname{co} a_{ik}.$$

As $\operatorname{co} a_{ik} \in S$, the first part of the theorem is proved. The proof of the second part is similar.

D I.10.4. Lemma. If p, q are both of the type $\sum \Pi$, then $p + q$ and $p \cdot q$ are also of the type $\sum \Pi$.

If p, q are both of the type $\Pi \sum$, then $p + q$ and $p \cdot q$ are also of the type $\Pi \sum$.

Proof. Let p, q be of the type $\Pi \sum$. Then there exist $a_{ik}, b_{ik} \in S$, $(i = 1, 2, \ldots)$ such that

$$p = \prod_{i=1}^{\infty} \sum_{k=1}^{\infty} a_{ik}, \quad q = \prod_{i=1}^{\infty} \sum_{k=1}^{\infty} b_{ik}.$$

Put

$$p_i =_{df} \sum_{k=1}^{\infty} a_{ik}, \quad (i = 1, 2, \ldots),$$

$$q_i =_{df} \sum_{k=1}^{\infty} b_{ik}, \quad (i = 1, 2, \ldots).$$

We have

$$p = \prod_{i=1}^{\infty} p_i, \quad q = \prod_{j=1}^{\infty} q_j.$$

Hence

(1) $$p \cdot q = \prod_{i=1}^{\infty} p_i \cdot \prod_{j=1}^{\infty} q_j = \prod_{i=1}^{\infty} (p_i \cdot q_i),$$

(by the associative and commutative laws for multiplication of spaces).

We have
$$p_i \cdot q_i = \sum_{k=1}^{\infty} a_{ik} \cdot \sum_{j=1}^{\infty} b_{ij}.$$
Hence, by Lemma [D I.10.1.],
$$p_i \cdot q_i = \sum_{k,j=1}^{\infty} a_{ik} \cdot b_{ij}.$$
Since $a_{ik}, b_{ij} \in S$, and the sum is denumerable,
$$p_i \cdot q_i \text{ is of the type } \sum; \text{ hence } p \cdot q \text{ is } \prod \sum.$$
Now we shall prove that if p, q are of type $\sum \prod$, then $p + q$ is also of the type $\sum \prod$.

To do this, we apply Lemma [D I.10.3.]. The somata co$p$, co$q$ are $\prod \sum$; hence, by what is already proved:
$$\text{co}\,p \cdot \text{co}\,q \quad \text{is of type } \prod \sum.$$
Hence, by Lemma [D I.10.3.],
$$p + q = \text{co}(\text{co}\,p \cdot \text{co}\,q) \quad \text{is of type } \sum \prod.$$

Now we shall prove that, if p, q are of type $\sum \prod$, then $p \cdot q$ is also of type $\sum \prod$.

We have for some somata $a'_{ik}, b'_{ik} \in S$
$$p = \sum_{i=1}^{\infty} \prod_{k=1}^{\infty} a'_{ik}, \quad q = \sum_{i=1}^{\infty} \prod_{k=1}^{\infty} b'_{ik}.$$
Put
$$p'_i =_{df} \prod_{k=1}^{\infty} a'_{ik}, \quad q'_i =_{df} \prod_{k=1}^{\infty} b'_{ik}.$$

Since $b'_{\alpha\beta}$ is compatible with all a'_{ik} it follows, by Theorem [D.15.3.], that $b'_{\alpha\beta}$ is compatible with $p'_i = \prod_{k=1}^{\infty} a'_{ik}$, $i = 1, 2, \ldots$

Thus all p'_i are compatible with all $b'_{\alpha\beta}$. Hence, by the same theorem, p'_i is compatible with
$$\prod_{\beta=1}^{\infty} b'_{\alpha\beta} = q'_\alpha,$$
whatever α may be. Hence all the spaces p'_i are compatible with all q'_j. But p'_α is also compatible with p'_β. Indeed,
$$p'_\alpha = \prod_{k=1}^{\infty} a'_{k\alpha}, \quad p'_\beta = \prod_{k=1}^{\infty} a'_{k\beta}.$$

Since $a'_{i\alpha}$ is compatible with all $a'_{k\beta}$, $(k = 1, 2, \ldots)$, it follows that $a'_{i\alpha}$ is compatible with $\prod_{k=1}^{\infty} a'_{k\beta} = p'_\beta$. This being true for any i, α, β it follows, that p'_β is compatible with

$$\prod_{i=1}^{\infty} a'_{i\alpha} = p'_\alpha.$$

In an analogous manner we prove that all q'_α $(\alpha = 1, 2, \ldots)$ are compatible with one another.

Thus all p'_i, q'_j are compatible with one another. Having this we can apply Lemma [D I.9.1.] which gives

$$\sum_{i=1}^{\infty} p'_i \cdot \sum_{i=1}^{\infty} q'_i = \sum_{i,j=1}^{\infty} p'_i \cdot q'_j.$$

As p'_i, q'_j are of the type Π, it follows that $p'_i \cdot q'_j$ is also of the type Π.

Hence $p \cdot q$ is of the type $\Sigma\,\Pi$.

The remainder of the thesis will be proved by using de Morgan laws. The lemma is established.

D I.10.5. Remark. The above lemma may be written in the following symbolic way:

$$\Sigma\,\Pi + \Sigma\,\Pi \subseteq \Sigma\,\Pi, \quad \Sigma\,\Pi \cdot \Sigma\,\Pi \subseteq \Sigma\,\Pi,$$
$$\Pi\,\Sigma + \Pi\,\Sigma \subseteq \Pi\,\Sigma, \quad \Pi\,\Sigma \cdot \Pi\,\Sigma \subseteq \Pi\,\Sigma.$$

D I.10.6. Theorem. If S is a tribe-set, and S_1 is the class of all spaces which are at the same time of the type $\Pi\,\Sigma$ and $\Sigma\,\Pi$, then S_1 is also a tribe-set; including S (see Def. [D I.7.]).

Proof. First we prove that $S \subseteq S_1$. Take $a \in S$. We have

$$a = \prod_{k=1}^{\infty} \sum_{i=1}^{\infty} a_{ik},$$

where

$$a_{ik} = a \quad \text{for} \quad i, k = 1, 2, \ldots,$$

and

$$a = \sum_{i=1}^{\infty} \prod_{k=1}^{\infty} a_{ik}.$$

Hence a is of type $\Pi\,\Sigma$ and of type $\Sigma\,\Pi$, and then $a \in S_1$.

Let $p \in S_1$. It follows that p is of the type $\Pi\,\Sigma$ and $\Sigma\,\Pi$. Hence, by [D I.10.3.], co$p$ is of the type $\Sigma\,\Pi$ and $\Pi\,\Sigma$, i.e., co$p \in S_1$.

Let $p, q \in S_1$. Since p, q are of the type $\Sigma\,\Pi$ and $\Pi\,\Sigma$, it follows, by Lemma [D I.10.4.], that $p + q$ is also of the type $\Sigma\,\Pi$ and $\Pi\,\Sigma$; hence $p + q \in S_1$.

It remains to prove that if $p, q \in S_1$ then they are compatible. Let

$$p = \sum_{i=1}^{\infty} \prod_{k=1}^{\infty} a_{ik}, \quad q = \sum_{i=1}^{\infty} \prod_{k=1}^{\infty} b_{ik}, \quad \text{where} \quad a_{ik}, b_{ik} \in S.$$

We know that if we put

$$p_i' = \prod_{k=1}^{\infty} a_{ik}, \quad q_i' = \prod_{k=1}^{\infty} b_{ik},$$

then p_i' and q_i' are compatible and compatible with all spaces of S.

Now since q_α' is compatible with all p_i' $(i = 1, 2, \ldots)$, it follows, by [D.15.3.], that

$$q_\alpha' \text{ is compatible with } \sum_{i=1}^{\infty} p_i' = p.$$

This being true for all $\alpha = 1, 2, \ldots$, it follows that p is compatible with

$$\sum_{\alpha=1}^{\infty} q_\alpha' = q.$$

Having this we can conclude that (S_1) is a tribe. The theorem is established.

D I.11. Theorem. Let

$$T_1, T_2, \ldots, T_\omega, \ldots, T_\alpha, \ldots$$

be a finite or transfinite well ordered sequence of tribes-sets in $\mathbf{H}$, such that if $\alpha' \leqq \alpha''$, then

$$T_{\alpha'} \subseteqq T_{\alpha''}.$$

Under these circumstances

$$T =_{df} \bigcup_{\alpha} T_\alpha$$

is also a tribe set on $\mathbf{H}$.

Proof. Let $p \in T$. Then there exists $\alpha \geqq 1$ such that $p \in T_\alpha$. Hence $\text{co}\,p \in T_\alpha$ and then $\text{co}\,p \in T$.

Let $p, q \in T$. There exist α, β, such that $p \in T_\alpha$ and $q \in T_\beta$. We may suppose that $\alpha \leqq \beta$. It follows that $p, q \in T_\beta$, because $T_\alpha \subseteqq T_\beta$. Since T_β is a tribe, it follows that $p + q \in T_\beta \subseteqq T$. Beside p, q are compatible. By application of Theorem [D I.4.] the theorem follows.

D I.11.1. Theorem. If S is a tribe-set of spaces on $\mathbf{H}$, then there exists a denumerably additive tribe (T) in $\mathbf{H}$ such that

$$S \subseteqq T.$$

Proof. Starting with (S), we shall construct a well ordered sequence of tribes (S_α) is the following way. Put

$$S_1 =_{df} S.$$

Suppose we have already defined the tribe (S_j) for all ordinals $\geqq 2$. We consider two cases

1⁰) α has an immediate predecessor $\alpha - 1$,

2⁰) α has no predecessor.

Case 1⁰. We define S_α as the class of all spaces which are of the type $\sum \prod$ and at the same time of the type $\prod \sum$ with respect to $S_{\alpha-1}$.

Case 2⁰. We define

$$S_\alpha =_{df} \bigcup_{j<\alpha} S_j.$$

By Theorem [D I.10.6.] and [D I.11.], (S_α) is a tribe. Now we shall prove that if $\alpha' < \alpha''$, then

$$S_{\alpha'} \subseteqq S_{\alpha''}.$$

Take an ordinal $\alpha \geqq 2$. Suppose that: If $\beta < \alpha$ and $\beta' < \beta'' \leqq \beta$ then

$$S_{\beta'} \subseteqq S_{\beta''}.$$

We shall prove that if $\beta' < \beta'' \leqq \alpha$, then

$$S_{\beta'} \subseteqq S_{\beta''}$$

too. Suppose $\beta'' < \alpha$. We have $\beta' < \beta'' \leqq \beta''$, where $\beta'' < \alpha$. Hence, by hypothesis,

$$S_{\beta'} \subseteqq S_{\beta''}.$$

Now take $\beta'' = \alpha$. We have $\beta' < \beta'' = \alpha$.

In the case α has a predecessor $\alpha - 1$, we have $\beta'' \leqq \alpha - 1$. By hypothesis,

$$S_{\beta''} \subseteqq S_{\alpha-1}.$$

We have

$$S_{\alpha-1} \subseteqq S_\alpha.$$

Hence

$$S_{\beta''} \subseteqq S_\alpha = S_{\beta'}.$$

In the case where α has no predecessor, we have

$$S_\alpha = \bigcup_{\gamma<\alpha} S_\gamma.$$

Hence, since $\beta' < \beta'' = \alpha$, we have

$$S_{\beta'} \subseteqq S_{\beta''}.$$

Thus the proposition is proved.

The above holds for any transfinite sequence of indices. Suppose that for some ordinal α_0 we have

$$S_{\alpha_0} = S_{\alpha_0+1}.$$

In this case for all ordinals $\beta \geqq \alpha_0$, if they are available, we have

$$S_{\alpha_0} = S_\beta .$$

Indeed suppose this be not true for some $\beta_0 \geqq \alpha_0$. Take the smallest $\beta_0 > 0$, for which $S_{\alpha_0} \subset S_{\beta_0}$.

We have $\beta_0 > \alpha_0 + 1$.

In the case $\beta_0 - 1$ is available, we have $\beta_0 - 1 > \alpha_0$; hence

$$\beta_0 - 1 \geqq \alpha_0 + 1 .$$

We have

$$S_{\beta_0 - 1} = S_{\alpha_0} .$$

Since S_{β_0} is made of $S_{\beta_0 - 1}$ by the same construction as S_{α_0+1} is made out of S_{α_0}, we have

$$S_{\beta_0} = S_{\beta_0 - 1} = S_{\alpha_0} ,$$

this is a contradiction.

In the case $\beta_0 - 1$ is not available, we have

$$S_{\beta_0} = \bigcup_{\gamma < \beta_0} S_\gamma .$$

But for all $\gamma < \beta_0$ we have supposed that $S_\gamma = S_{\alpha_0}$. Hence

$$\bigcup_{\gamma < \beta_0} S_\gamma = \bigcup_{\gamma < \beta_0} S_\gamma = S_{\alpha_0} .$$

Hence $S_{\beta_0} = S_{\alpha_0}$, which is the same contradiction as before. Thus the proposition is proved.

Now let $\aleph$ be the cardinal of the set of all subspaces in the given space $\mathbf{H}$. Let $\aleph' > \aleph$, and take the smallest ordinal Ω' corresponding to $\aleph'$.

There surely exist two ordinals $\alpha' < \beta' < \Omega'$ for which

$$S_{\alpha'} = S_{\beta'} ,$$

because if not, the cardinal of the set of different subspaces would be $> \aleph'$. It follows

$$S_{\alpha'} = S_{\alpha'+1} , \quad \text{because} \quad \alpha' + 1 \leqq \beta' ,$$

and

$$S_{\alpha'} \subseteqq S_{\alpha'+1} \subseteqq S_{\beta'} , \qquad S_{\alpha'} \subset S_{\beta'} .$$

Now take the smallest ordinal δ_0 for which

$$S_{\delta_0} = S_{\delta_0 + 1}$$

and consider S_{δ_0}. We have $S \subseteqq S_{\delta_0}$.

We shall prove that S_{δ_0} is a denumerably additive tribe of subspaces in $\mathbf{H}$. Let

$$a_1, a_2, \ldots, a_n, \ldots \in S_{\delta_0} .$$

Put

$$a =_{df} \sum_{n=1}^{\infty} a_n.$$

We have

$$a = \sum_{n=1}^{\infty} a_n \cdot \sum_{n=1}^{\infty} a_n.$$

Hence a is of type $\Pi \Sigma$. We also have

$$a = \sum_{n=1} (a_n \cdot a_n \ldots).$$

Hence a is of type $\Sigma \Pi$. Hence $a \in S_{\delta_0+1}$. Since $S_{\delta_0+1} = S_{\delta_0}$, it follows that

$$a \in S_{\delta_0}.$$

Since (S_{δ_0}) is a tribe and since, therefore, all its spaces are compatible with one another, it follows that (S_{δ_0}) is a denumerably additive tribe on $\mathbf{H}$. Since $S = S_{\delta_0}$, the theorem is established.

D I.11.2. Remark. We shall prove that if S is a tribe-set of spaces in $\mathbf{H}$, it can be extended even to a completely additive tribe of spaces in $\mathbf{H}$. The above construction will be useful for some purposes. It shows some means of extending tribes of spaces.

D I.12. Theorem. If $\mathbf{H}$ is a separable and complete H.-H.-space, and S a denumerably additive tribe in $\mathbf{H}$, then S is completely additive. (The theorem does not hold for non-separable spaces.)

Proof. Let M be a not empty class of spaces $\in S$. Arrange them into a well ordered sequence

$$(1) \qquad a_1, a_2, \ldots, a_\omega, \ldots, a_\alpha, \ldots$$

We make out of (1) the sequence

$$b_1, b_2, \ldots, b_\omega, \ldots, b_\alpha, \ldots,$$

where

$$b_\alpha =_{df} \sum_{\gamma \leqq \alpha} a_\alpha \quad \text{for} \quad \alpha \geqq 1.$$

Thus we have $b_{\alpha'} \leqq b_{\alpha''}$, whenever $\alpha' \leqq \alpha''$.

We have

$$c =_{df} \sum_{a \in M} a = \sum_\alpha a_\alpha = \sum_\alpha b_\alpha.$$

Now, we put

$$c_1 =_{df} b_1,$$

$$c_\alpha =_{df} b_\alpha - \sum_{\gamma < \alpha} b_\alpha.$$

By Theorem [D.14.1.] we know that all c_α are mutually orthogonal and that

$$c = \sum_\alpha c_\alpha.$$

Now there exists an at most denumerable number of spaces $c_\alpha \neq (\vec{O})$, because of separability of **H**.

It follows that there exists a smallest $\alpha_0 < \Omega$, (where Ω is the smallest ordinal corresponding to the first non-denumerable cardinal $\aleph_1$) and such that

$$b_{\alpha_0} = b_\beta \quad \text{for every} \quad \beta \geqq \alpha_0 .$$

Since b_{α_0} is an at most a denumerable sum of sets $\in M$, it follows that $b_{\alpha_0} \in S$. Thus the complete additivity of (S) is proved.

D I.13. A tribe (T) of spaces in the space p is said to *be saturated in* p whenever every tribe (T') in p, which is an extension of (T) in p, coincides with (T).

D I.13.1. There are two questions:

1) Does there exist a saturated tribe in p?

2) if a tribe (T) is not saturated, does there exist an extension (T') of (T) such that (T') is saturated, and if so, whether (T') is unique or not. These two problems will be completely solved.

Radiation scope

D I.14. We shall need a notion, whose special case has been used in the theory of multiplicity of continuous spectrum of selfadjoint operators (26).

D I.14.1. Def. Let (T) be a geometrical tribe in **H**, and E a not empty set of vectors.

By *the radiation scope of* E *on* (T) we shall understand the set

$$\mathcal{M}_T(E)$$

of all vectors $\vec{Y}$ such that, for every $\varepsilon > 0$ there exists a natural number N, a sequence

$$a_1, a_2, \ldots, a_N$$

of spaces of (T), a sequence

$$\lambda_1, \lambda_2, \ldots, \lambda_N$$

of complex numbers, and a sequence

$$\vec{X}_1, \vec{X}_2, \ldots, \vec{X}_N$$

of vectors belonging to E, such that the norm

$$\|\vec{Y} - \sum_{i=1}^{N} \lambda_i \operatorname{Proj}_{a_i} \vec{X}_i\| \leqq \sigma .$$

We can say that $\mathcal{M}_T(E)$ is composed of all vectors $\vec{Y}$ which can be approximated by linear, homogenuous combinations of vectors specified above.

D I.14.1a. Def. In the case E is composed of a single vector $\vec{\xi}$ we agree to write

$$\mathscr{M}_T(\xi) \quad \text{instead of} \quad \mathscr{M}_T((\vec{\xi})).$$

We shall derive some properties of $\mathscr{M}_T(E)$.

D I.14.2. Theorem. The (T)-radiation scope $\mathscr{M}_T(E)$ of E is the smallest space p, spanned by all the vectors

$$\text{Proj}_a \vec{\xi},$$

where $a \in T$ and $\vec{\xi} \in E$.

Proof. First we shall prove that $\mathscr{M}_T(E)$ is a linear manifold. Let $\vec{x}, \vec{y} \in \mathscr{M}_T(E)$. Take $\sigma > 0$.

We have

$$||\vec{x} - \sum_i \lambda_i \text{Proj}_{a_i} \vec{\xi}_i|| \leqq \frac{\sigma}{2},$$

$$||\vec{y} - \sum_j \mu_j \text{Proj}_{b_j} \vec{\eta}_j|| \leqq \frac{\sigma}{2}$$

for some numbers λ_i, μ_j, some vectors $\vec{\xi}_i$, $\vec{\eta}_j \in E$, and some spaces a_i, b_j belonging to T.

We have

$$||(\vec{x} + \vec{y}) - (\sum_i \lambda_i \text{Proj}_{a_i} \vec{\xi}_i + \sum_j \mu_j \text{Proj}_{b_j} \vec{\eta}_j)|| \leqq \sigma,$$

which proves that $\vec{x} + \vec{y} \in \mathscr{M}_T(E)$.

Take $\vec{x} \in \mathscr{M}_T(E)$ and a number α. If $\alpha = 0$, then $\alpha \cdot \vec{x} = \vec{O}$ belongs to $\mathscr{M}_T(E)$, because

$$\vec{O} = \text{Proj}_0 \vec{\xi} \quad \text{if} \quad \vec{\xi} \in E.$$

If $\alpha \neq 0$, we have

$$||\vec{x} - \sum_i \lambda'_i \text{Proj}_{a'_i} \vec{\xi}'_i|| \leqq \frac{\sigma}{|\alpha|}$$

for some linear combination. Hence

$$||\alpha \vec{x} - \sum_\lambda \alpha \lambda'_i \text{Proj}_{a'_i} \vec{\xi}'_i|| \leqq \sigma,$$

which shows that

$$\alpha \vec{x} \in \mathscr{M}_T(E).$$

Thus $\mathscr{M}_T(E)$ is a linear variety.

We shall prove that this set is closed. Let $\vec{x}_n \in \mathscr{M}_T(E)$, where $\vec{x}_n \to \vec{x}$.

For $\sigma > 0$ there is an index n, such that

$$||\vec{x} - \vec{x}_n|| \leqq \frac{\sigma}{2}.$$

Now we have for some linear combination

$$||\vec{x}_n - \sum_i \lambda''_i \text{Proj}_{a''_i} \vec{\xi}''_i|| \leqq \frac{\sigma}{2}.$$

It follows that

$$||\vec{x} - \sum_i \lambda_i'' \operatorname{Proj}_{a_i''} \vec{\xi}_i''|| \leqq \sigma,$$

which gives $\vec{x} \in \mathscr{M}_T(E)$.

Thus $\mathscr{M}_T(E)$ is a space.

Now, we have $\operatorname{Proj}_a \vec{\xi} \in M_T(E)$, whenever $\vec{\xi} \in E$ and $a \in (T)$. It follows that every linear combination

$$\sum_j \mu_j \operatorname{Proj}_{a_j} \vec{\xi}_j, \tag{1}$$

where $a_j \in T$, $\vec{\xi}_j \in E$, belongs to $\mathscr{M}_T(E)$.

The set of all combinations (1) is just the linear variety spanned by the vectors $\operatorname{Proj}_a \vec{\xi}$, where a varies in T and $\vec{\xi}$ varies in E.

Its closure is p. Hence

$$p \subseteqq \mathscr{M}_T(E).$$

But if $\vec{x} \in \mathscr{M}_T(E)$, $\vec{x}$ can be approximated, as close as we like, by linear combinations (1). Hence $\mathscr{M}_T(E) \subseteqq p$. The theorem is proved.

D I.14.3. Theorem. If $E \neq \emptyset$ is a set of vectors, then

$$E \subseteqq \mathscr{M}_T(E).$$

Proof. If $\vec{\xi} \in E$, we have $\vec{\xi} = 1 \cdot \operatorname{Proj}_I \vec{\xi} \in M_T(E)$.

D I.14.4. Theorem. If $E \subseteqq F$, $E \neq \emptyset$, then

$$\mathscr{M}_T(E) \subseteqq \mathscr{M}_T(F).$$

Proof. Denote by α the set of all vectors

$$\operatorname{Proj}_a \vec{\xi}, \quad \text{where} \quad \vec{\xi} \in E, \quad a \in T,$$

and by β the set of all vectors

$$\operatorname{Proj}_a \vec{\eta}, \quad \text{where} \quad \vec{\eta} \in F, \quad a \in T.$$

We have

$$\alpha \subseteqq \beta.$$

Hence the space p, spanned by α, is included in the space q spanned by β.

Since, by [D I.14.2.]

$$p = \mathscr{M}_T(E), \quad q = \mathscr{M}_T(F),$$

it follows that

$$\mathscr{M}_T(E) \subseteqq \mathscr{M}_T(F), \quad \text{q.e.d.}$$

D I.14.5. Theorem. If $E \neq \emptyset$ is a set of vectors, A the linear variety spanned by E, then

$$\mathscr{M}_T(E) = \mathscr{M}_T(A).$$

Proof. We have $E \subseteqq A$; hence, by [D I.14.4.],

$$\mathcal{M}_T(E) \subseteqq \mathcal{M}_T(A). \tag{1}$$

Let $\vec{x} \in M_T(A)$. Take $\sigma > 0$. We have

$$||\vec{x} - \sum_i \lambda_i \operatorname{Proj}_{a_i} \vec{\xi}_i|| \leqq \sigma \tag{2}$$

for some numbers λ_i, some $a_i \in T$ and some $\vec{\xi}_i \in A$. Now $\vec{\xi}_i = \sum_k \mu_{ik} \vec{\eta}_{ik}$, where $\vec{\eta}_{ik} \in E$ and μ_{ik} is a complex number and where the sum is finite. Now

$$\operatorname{Proj}_{a_i} \vec{\xi}_i = \sum_k \mu_{ik} \operatorname{Proj}_{a_i} \vec{\eta}_{ik}.$$

Carrying this into (2), we get an inequality of the same character, as in (2):

$$||\vec{x} - \sum_j \nu_j \operatorname{Proj}_{b_j} \vec{\eta}_j|| \leqq \sigma,$$

with finite sum, and where $b_j \in T$, $\vec{\eta}_j \in E$.

Thus $\vec{x} \in \mathcal{M}_T(E)$. It follows that

$$\mathcal{M}_T(A) \subseteqq \mathcal{M}_T(E),$$

which, on account of (1), completes the proof.

D I.14.6. Theorem. If $E \neq \emptyset$ is a set of vectors, $\bar{E}$ its closure, then

$$\mathcal{M}_T(E) = \mathcal{M}_T(\bar{E}).$$

Proof. We have $E \subseteqq \bar{E}$; hence, [D I.14.4.],

$$\mathcal{M}_T(E) \subseteqq \mathcal{M}_T(\bar{E}). \tag{1}$$

Let $\vec{x} \in \mathcal{M}_T(\bar{E})$. Take $\sigma > 0$.

We have for some linear combination

$$||\vec{x} - \sum_i \lambda_i \operatorname{Proj}_{a_i} \vec{\xi}_i|| \leqq \frac{\sigma}{2}, \quad \text{where} \quad \vec{\xi}_i \in \bar{E}. \tag{2}$$

Now, since $\vec{\xi}_i \in \bar{E}$, there exists a sequence $\vec{\xi}_{i,n} \in E$, tending to $\vec{\xi}_i$ for $n \to \infty$.

We have, by [D.2.5.],

$$\lim_n (\lambda_i \operatorname{Proj}_{a_i} \vec{\xi}_{i,n}) = \lambda_i \operatorname{Proj}_{a_i} \vec{\xi}_i.$$

Find an index m such that

$$||\lambda_i \operatorname{Proj}_{a_i} \vec{\xi}_i - \lambda_i \operatorname{Proj}_{a_i} \vec{\xi}_{i,m}|| \leqq \frac{\sigma}{2N}, \quad (i = 1, \ldots, N),$$

where N is the number of terms of the sum in (2). We get

$$||\sum_i \lambda_i \operatorname{Proj}_{a_i} \vec{\xi}_i - \sum_i \lambda_i \operatorname{Proj}_{a_i} \vec{\xi}_{i,m}|| \leqq \frac{\sigma}{2}. \tag{3}$$

Hence, by (2),

$$||\vec{x} - \sum_i \lambda_i \operatorname{Proj}_{a_i} \vec{\xi}_{i,m}|| \leqq \sigma, \quad \text{where} \quad \vec{\xi}_{i,m} \in E.$$

It follows that

$$\vec{x} \in \mathcal{M}_T(E).$$

Taking into account (1), we have our theorem proved.

D I.14.7. Theorem. If $E \neq \emptyset$, and F is a set everywhere dense in E, then

$$\mathcal{M}_T(E) = \mathcal{M}_T(F).$$

Proof. We have

$$F \subseteq E, \quad E \subseteq \bar{F}; \quad \text{hence} \quad \bar{F} \subseteq \bar{E} \quad \text{and} \quad \bar{E} \subseteq \bar{F}; \quad \text{hence} \quad \bar{E} = \bar{F}.$$

Since, by [D I.14.6.],

$$\mathcal{M}_T(E) = \mathcal{M}_T(\bar{E}), \quad \mathcal{M}_T(F) = \mathcal{M}(\bar{F}),$$

the theorem is proved.

D I.14.8. Theorem. If $E \neq \emptyset$ and p is the space spanned by E, then

$$\mathcal{M}_T(E) = \mathcal{M}_T(p).$$

Proof. Let A be the linear variety spanned by E. By [D I.14.5.],

(1) $$\mathcal{M}_T(E) = \mathcal{M}_T(A).$$

We have

$$\bar{A} = p; \quad \text{hence} \quad \mathcal{M}_T(\bar{A}) = \mathcal{M}_T(p).$$

But

$$\mathcal{M}_T(A) = \mathcal{M}_T(\bar{A});$$

hence

(2) $$\mathcal{M}_T(A) = \mathcal{M}_T(p).$$

From (1) and (2) the theorem follows.

D I.14.9. Theorem. If a is compatible with all spaces of (T), then

$$\mathcal{M}_T(a) = a.$$

Proof. By Theorem [D I.14.2.],

(1) $$a \leqq \mathcal{M}_T(a).$$

To prove the converse inclusion, take a vector $\vec{x} \in \mathcal{M}_T(a)$. The vector $\vec{x}$ is the limit of an infinite sequence of vectors having the form $\sum_i \lambda_i \operatorname{Proj}_{a_i} \vec{\xi}_i$, where $\vec{\xi}_i \in a$, $a_i \in (T)$. Since $\vec{\xi}_i = \operatorname{Proj}_a \vec{\xi}_i$, we have

$$\operatorname{Proj}_{a_i} \vec{\xi}_i = \operatorname{Proj}_{a_i} \operatorname{Proj}_a \vec{\xi}_i.$$

Since a_i and a are compatible [D.5.], we have by virtue of [D.9., II′],

$$\operatorname{Proj}_{a_i} \operatorname{Proj}_a \vec{\xi}_i = \operatorname{Proj}_a \operatorname{Proj}_a \vec{\xi}_i \in a.$$

Hence

$$\sum_i \lambda_i \operatorname{Proj}_{a_i} \vec{\xi}_i \in a.$$

The space a being closed, and $\vec{x}$ being the limit of a sequence of vectors, belonging to a, it follows that $\vec{x} \in a$. Hence

(2) $$\mathcal{M}_T(a) \subseteqq a.$$

From (1) and (2) the theorem follows.

D I.14.10. Theorem. If $E \neq \emptyset$ is a set of vectors, then

$$\mathcal{M}_T(\mathcal{M}_T(E)) = \mathcal{M}_T(E).$$

Proof. Since, by the Theorem [D I.14.2.] we have

$$E \subseteqq \mathcal{M}_T(E),$$

we get by [D I.14.4.]:

(1) $$\mathcal{M}_T(E) \leqq \mathcal{M}_T(\mathcal{M}_T(E)).$$

To prove the converse inclusion, we may use Theorem [D I.14.2.]. Denote by α the set of all vectors $\operatorname{Proj}_a \vec{\xi}$, where $\vec{\xi} \in E$ and $a \in T$. Since $\mathcal{M}_T(E)$ is the space spanned by α, we have, on account of [D I.14.8.],

(2) $$\mathcal{M}_T(\mathcal{M}_T(E)) = \mathcal{M}_T(\alpha).$$

Now, let $\vec{x} \in \mathcal{M}_T(\alpha)$, and take $\sigma > 0$. We have for some linear combination:

$$\|\vec{x} - \sum_i \lambda_i \operatorname{Proj}_{a_i} \vec{\xi}_i\| \leqq \sigma,$$

where

$$a_i \in (T), \quad \vec{\xi}_i \in \alpha.$$

There exist $\vec{\eta}_i \in E$ and $b_i \in T$, such that $\vec{\xi}_i = \operatorname{Proj}_{b_i} \vec{\eta}_i$. Hence

$$\|\vec{x} - \sum_i \lambda_i \operatorname{Proj}_{a_i} \operatorname{Proj}_{b_i} \vec{\eta}_i\| \leqq \sigma,$$

where

$$\vec{\eta}_i \in E, \quad b_i \in T.$$

Since a_i, b_i are compatible, we have, by [D.9., III'],

$$\operatorname{Proj}_{a_i} \operatorname{Proj}_{b_i} \vec{\eta}_i = \operatorname{Proj}_{a_i b_i} \vec{\eta}_i.$$

It follows

$$\|\vec{x} - \sum_i \lambda_i \operatorname{Proj}_{a_i b_i} \vec{\eta}_i\| \leqq \sigma, \quad \text{where} \quad \vec{\eta}_i \in E, \quad \text{and} \quad a_i b_i \in T.$$

Hence

$$\vec{x} \in \mathcal{M}_T(E).$$

Thus we have proved that

(3) $$\mathcal{M}_T(\alpha) \subseteqq \mathcal{M}_T(E).$$

From (1), (2) and (3) we get

$$\mathcal{M}_T(E) = \mathcal{M}_T(\mathcal{M}_T(E)), \quad \text{q.e.d.}$$

D I.14.11. Theorem. If $E \neq \emptyset$ is a set of vectors, then

$$\mathcal{M}_T(E) = \sum_{\vec{x} \in E} \mathcal{M}_T(\vec{x}),$$

(where the sum is somatic i.e. lattice union). If $\vec{x} \in E$, we have $(\vec{x}) \subseteqq E$ and then, by [D I.14.4.],

(1) $$\mathcal{M}_T(\vec{x}) \leqq \mathcal{M}_T(E).$$

The inclusion (1) being true for all $\vec{x} \in E$, it follows, by definition o the general somatic sum of spaces,

(2) $$\sum_{\vec{x} \in E} \mathcal{M}_T(\vec{x}) \leqq \mathcal{M}_T(E).$$

To prove the converse inclusion, let $\vec{y} \in \mathcal{M}_T(E)$ and take $\sigma > 0$. We have for some finite sums

$$|\vec{y} - \sum_i \lambda_i \operatorname{Proj}_{a_i} \vec{\xi}_i| \leqq \sigma,$$

where $a_i \in (T)$ and $\vec{\xi}_i \in E$.

Now, by [D I.14.2.],

$$\lambda_i \operatorname{Proj}_{a_i} \vec{\xi}_i \in \mathcal{M}_T(\vec{\xi}_i).$$

Hence

$$\sum_i \lambda_i \operatorname{Proj}_{a_i} \vec{\xi}_i \in \sum_i \mathcal{M}_T(\vec{\xi}_i) \leqq \sum_{\vec{x} \in E} \mathcal{M}_T(\vec{x}).$$

It follows, by letting σ tend to 0, that $\vec{y}$ is the limit of a sequence of vectors belonging to

$$\sum_{\vec{x} \in E} \mathcal{M}_T(\vec{x}).$$

Since this somatic sum is a space, and then, a closed set, it follows that

$$\vec{y} \in \sum_{\vec{x} \in E} \mathcal{M}_T(\vec{x}).$$

Thus we have proved that

$$\mathcal{M}_T(E) \subseteqq \sum_{\vec{x} \in E} \mathcal{M}_T(\vec{x})$$

which, on account of (2), concludes the proof.

D I.14.12. Theorem. If a is a space compatible with all spaces of (T), then

$$\mathcal{M}_T(a) = \bigcup_{\vec{x} \in a} \mathcal{M}_T(\vec{x}).$$

Proof. We have for $\vec{x} \in a$

$$\mathscr{M}_T(\vec{x}) \leqq \mathscr{M}_T(a).$$

Hence, considering the set-union, we have

$$(1) \qquad \bigcup_{\vec{x} \in a} \mathscr{M}_T(\vec{x}) \subseteqq \mathscr{M}_T(a).$$

On the other hand we have, by [D I.14.9.],

$$(2) \qquad \mathscr{M}_T(a) = \bigcup_{\vec{x} \in \mathscr{M}_T(a)} (\vec{x}) = \bigcup_{\vec{x} \in a} (\vec{x}) \subseteqq \bigcup_{\vec{x} \in a} \mathscr{M}_T(\vec{x}),$$

just by [D I.14.3.]. From (1) and (2), the theorem follows.

D I.14.13. Remark. If we vary the set E and take the spaces $\mathscr{M}_T(E)$, we obtain all possible (T)-radiation scopes. Now, [D I.14.8.] shows, that in order to obtain all (T)-radiation scopes, it suffices to take all $\mathscr{M}_T(p)$ where p are spaces.

Now, given a space p, if we consider an orthonormal system Φ saturated in p, we get, by [D I.14.8.],

$$\mathscr{M}_T(p) = \mathscr{M}_T(\Phi).$$

Now we may ask whether this set Φ of vectors cannot be diminished and, in spite of that, getting the same (T)-radiation scope?

Especially we may ask the following question: can we make correspond to every set E of vectors a single vector $\vec{x}_E$ depending on E such that $\mathscr{M}_T(E)$ be identical with $\mathscr{M}_T(\vec{x}_E)$.

The answer is negative, as is disclosed by the following example in 3-space. Let (T) be the tribe whose elements are $(\vec{O})$, (x, y), (z), 1. Let E be the set composed of two different vectors $\vec{\xi}_1$, $\vec{\xi}_2$, whose plane is neither perpendicular to the z-axis nor to the (x, y)-plane. The (T)-radiation scope of E is the set of all vectors

$$\lambda_1 \operatorname{Proj}_{(xy)} \vec{\xi}_1 + \lambda_2 \operatorname{Proj}_{(xy)} \vec{\xi}_2 + \mu_1 \operatorname{Proj}_z \vec{\xi}_1 + \mu_2 \operatorname{Proj}_z \vec{\xi}_2 + v_1 \vec{\xi}_1 + v_2 \vec{\xi}_2$$

where λ_1, λ_2, μ_1, μ_2, v_1, v_2 are complex numbers.

We see that $\mathscr{M}_T(E) = \mathbf{H}$. Now, if we take any vector $\vec{y}$, its T-radiation scope is O, if $\vec{y} = \vec{O}$; the z-axis if $\vec{y} \neq \vec{O}$ is lying on (z); a one-dimensional variety if $\vec{y} \in (x, y)$; and a plane passing through (z) in the remaining case. Thus never $I = M_T(\vec{y})$.

D I.14.14. Theorem. If $E \neq \emptyset$, $F \neq \emptyset$ are sets of vectors, then

$$\mathscr{M}_T(E \cup F) = \mathscr{M}_T(E) + \mathscr{M}_T(F),$$

where $\cup$ denotes the set-union for sets of vectors.

Proof. We have, by [D I.14.11.],

$$\mathcal{M}_T(E) = \sum_{\vec{x} \in E} \mathcal{M}_T(\vec{x}), \quad \mathcal{M}_T(F) = \sum_{\vec{x} \in F} \mathcal{M}_T(\vec{x}),$$

where the summation is the somatic one. Hence

$$\mathcal{M}_T(E) + \mathcal{M}_T(F) = \sum_{\vec{x} \in E} \mathcal{M}_T(\vec{x}) + \sum_{\vec{x} \in F} \mathcal{M}_T(\vec{x}) = \sum_{\vec{x} \in E \cup F} \mathcal{M}_T(\vec{x})$$

$$= \mathcal{M}_T(E \cup F), \quad \text{q.e.d.}$$

D I.14.15. Theorem. If a, b are spaces, then

$$\mathcal{M}_T(a + b) = \mathcal{M}_T(a) + \mathcal{M}_T(b).$$

Proof. We have, by [D I.14.14.],

(1) $$\mathcal{M}_T(a \cup b) = \mathcal{M}_T(a) + \mathcal{M}_T(b),$$

and, by [D.1.5.],

$$a + b = [a \cup b],$$

where the last expression denotes the space spanned by the set $a \cup b$ of vectors. Using [D I.14.8.], we get

(2) $$\mathcal{M}_T(a \cup b) = \mathcal{M}_T([a \cup b]) = \mathcal{M}_T(a + b).$$

If we compare (1) and (2) we obtain our thesis.

D I.15. Now we are going back to study the problem of extension of a tribe of spaces by adjunction of a new space, which does not belong to the tribe; [D I.7.1.].

D I.15.1. Theorem. If (T) is a tribe of spaces in $\mathbf{H}$ and $E \neq \emptyset$ a set of vectors, then the space

$$\mathcal{M}_T(E)$$

is compatible with all spaces of (T).

Proof. Put

$$p =_{df} \mathcal{M}_T(E).$$

On account of Theorem [D.8., II], the thesis will be proved if we prove that

$$\text{Proj}_a p \subseteqq a \cdot p \quad \text{for every} \quad a \in T.$$

It suffices even to prove that $\text{Proj}_a p \subseteqq p$, because $\text{Proj}_a p \subseteqq a$. Take a vector

(1) $$\vec{y} =_{df} \sum_i \lambda_i \, \text{Proj}_{a_i} \vec{\xi}_i$$

where λ_i are complex numbers, $a_i \in T$, $\vec{\xi}_i \in E$, and the sum is finite. We shall prove that $\text{Proj}_a \vec{y} \in p$. To do this it suffices, on account of (1), to prove that

$$\text{Proj}_a \, \text{Proj}_{a_i} \vec{\xi}_i \in p, \quad (i = 1, 2, \ldots).$$

Now, since a and a_i are compatible, this relation is equivalent to

$$\mathrm{Proj}_{a a_i} \vec{\xi}_i \in p,$$

(see [D.9., III']). We may prove this by proving that, if $b \in T$, $\vec{\xi} \in E$, then we have

$$\mathrm{Proj}_a \vec{\xi} \in p.$$

But this is evident, because $p = \mathscr{M}_T(E)$. Thus we have proved that $\mathrm{Proj}_a \vec{y} \in p$. Now, the vectors, which are represented as finite linear combinations like (1), make up a set everywhere dense in p. Let $\vec{x} \in p$. We have $\vec{x} = \lim_n \vec{y}_n$, where $\vec{y}_n \in E$. We have $\mathrm{Proj}_a \vec{y}_n \in p$. Since $\mathrm{Proj}_a \vec{x} = \lim_n \mathrm{Proj}_a \vec{y}_n$, it follows that $\mathrm{Proj}_a \vec{x} \in p$, because p is closed.

Thus we have proved that

$$\mathrm{Proj}_a \vec{x} \in p \quad \text{for every} \quad \vec{x} \in p;$$

hence $\mathrm{Proj}_a p \subseteqq p$, which completes the proof. The two above theorems allow to state the following.

D I.15.2. Theorem. If (T) is a tribe of space in **H** and p a space, then

a necessary and sufficient condition that p be compatible with all spaces $\in (T)$ is that there exists a set $E \neq \emptyset$ of vectors, such that

$$p = \mathscr{M}_T(E).$$

The above theorems show that the following notions are identical:

D I.15.3.

1) space compatible with every $a \in T$,
2) space capable of being adjoint to (T),
3) (T)-radiation-scope of a non-empty set of vectors.

D I.16. Generating vector. In [D I.14.13.] we have noticed that, generally speaking, if E is a set of vectors ($E \neq \emptyset$) and (T) a tribe of spaces, the radiation-scope $\mathscr{M}_T(E)$ cannot be obtained by replacing E by some single vector. Now we shall see that this is possible in the case where (T) is a saturated tribe, [D I.13.], in **H**. We shall prove the

D I.16.1. Theorem. If

1. is any complete H.-H.-space (not necessarily separable),
2. T a tribe of spaces on **H**,
3. T is saturated,
4. $\vec{x}_1, \vec{x}_2$ are vectors in **H**,

then there always exists a vector $\vec{y}$ such that

$$\mathscr{M}_T((\vec{x}_1) \cup (\vec{x}_2)) = \mathscr{M}_T(\vec{y}).$$

Proof. Put

(1) $\quad p_1 =_{df} \mathscr{M}_T(\vec{x}_1), \quad p_2 =_{df} \mathscr{M}_T(\vec{x}_2), \quad p =_{df} \mathscr{M}_T((\vec{x}_1) \cup (\vec{x}_2)).$

p_1, p_2, p are capable of being adjoint to (T), (see [Def. D I.7.1.]) and [D I.15.3.]. Hence these spaces belong to (T) because (T) is saturated. By [D I.14.3.] we have

(2) $$\vec{x}_2 \in p_2.$$

We can put

$$\vec{x}_2 = \vec{u} + \vec{v}, \quad \text{where} \quad \vec{u} \in p_1 \cdot p_2, \quad \vec{v} \in p_2 - p_1 \cdot p_2.$$

Hence

$$\mathrm{Proj}_{p_1 p_2} \vec{x}_2 = \vec{u}, \quad \mathrm{Proj}_{p_2 - p_1 \cdot p_2} \vec{x}_2 = \vec{v}.$$

By [Theorem D I.1.2c.], p_1, p_2 are compatible and then, by Theorem [D.8., V)]

(2.1) $$p_2 - p_1 \cdot p_2 = p_2 - p_1; \quad \text{hence} \quad \mathrm{Proj}_{p_2 - p_1} \vec{x}_2 = \vec{v}.$$

We have

(3) $$\vec{v} \in p_2 - p_1; \quad \text{hence} \quad \vec{v} \perp p_1.$$

We also have

(4) $$\vec{x}_1 \in p_1, \quad \text{so} \quad \vec{x}_1 \perp p_2 - p_1.$$

Put

(5) $$\vec{y} =_{df} \vec{x}_1 + \vec{v}$$

and

(5.1) $$q =_{df} \mathscr{M}_T(\vec{y}).$$

Our purpose will be a proof of the identity $p = q$. First we shall prove that $p \subseteqq q$.

By Theorem [D I.14.2.], the space q is spanned by all the vectors

(6) $$\mathrm{Proj}_a y,$$

where a varies in (T). Now

$$\mathrm{Proj}_{p_1} \vec{y} = \mathrm{Proj}_{p_1} (\vec{x}_1 + \vec{v}) = \mathrm{Proj}_{p_1} \vec{x}_1 + \mathrm{Proj}_{p_1} \vec{v} = \mathrm{Proj}_{p_1} \vec{x}_1 =,$$

because of (3), $= \vec{x}_1$, because of (4). Hence

(6.1) $$\mathrm{Proj}_{p_1} \vec{y} = \vec{x}_1,$$

$$\mathrm{Proj}_{p_2 - p_1} \vec{y} = \mathrm{Proj}_{p_2 - p_1} (\vec{x}_1 + \vec{v}) = \mathrm{Proj}_{p_2 - p_1} \vec{x}_1 + \mathrm{Proj}_{p_2 - p_1} \vec{v} = \mathrm{Proj}_{p_2 - p_1} \vec{v},$$

because of (4), and $= \vec{v}$, because of (3). Hence

(6.2) $$\mathrm{Proj}_{p_2 - p_1} \vec{y} = \vec{v}.$$

On the other hand we have [by (2.1)]

$$\vec{v} = \mathrm{Proj}_{p_2 - p_1} \vec{x}_2, \quad \text{and then} \quad \mathrm{Proj}_{p_2 - p_1} \vec{y} = \mathrm{Proj}_{p_2 - p_1} \vec{x}_2.$$

Since $p_2 - p_1 \in (T)$, $p_1 \in (T)$, it follows that, for any $a \in T$,

(7) $$\mathrm{Proj}_a \vec{x}_1 \quad \text{and} \quad \mathrm{Proj}_{a (p_2 - p_1)} \vec{x}_2$$

can be found among the vectors (6); hence consequently, they belong to q.

Now we have

$$\mathrm{Proj}_{a\,p_2}\vec{x}_2 = \mathrm{Proj}_a\,\mathrm{Proj}_{p_2}\vec{x}_2 = \mathrm{Proj}_a\vec{x}_2, \quad \text{because} \quad \vec{x}_2 \in p_2.$$

Since, by Theorem [D.8., IV]

$$\mathrm{Proj}_{p_1}p_2 = \mathrm{Proj}_{p_2}p_1,$$

there exists $\vec{\xi} \in p_1$ such that

(8) $$\mathrm{Proj}_{p_1}\vec{x}_2 = \mathrm{Proj}_{p_2}\vec{\xi}.$$

The vector $\vec{\xi}$, as belonging to p_1, is the limit of an infinite sequence of vectors having the form of a finite sum, viz.

$$\sum_i \lambda_i\,\mathrm{Proj}_{a_i}\vec{x}_i,$$

where λ_i are complex numbers and $a_i \in (T)$. Hence, by (8), the vector

$$\mathrm{Proj}_{a\,p_1 p_2}\vec{x}_2$$

is also the limit of a sequence of vectors having the above form. Hence

(9) $$\mathrm{Proj}_{a\,p_2 p_1}\vec{x}_2 \in q, \quad \text{for any} \quad a \in (T).$$

Since by (7)

$$\mathrm{Proj}_{a\,(p_2-p_1)}\vec{x}_2 \in q,$$

it follows from (9) that

$$\mathrm{Proj}_{a\,p_2 p_1}\vec{x}_2 + \mathrm{Proj}_{a\,(p_2-p_1)}\vec{x}_2 \in q.$$

Now $a\,p_2 p_1 \perp a\,(p_2 - p_1)$ and therefore the lefthand side expression equals, [D.2.6.3.],

$$\mathrm{Proj}_{a\,p_2 p_1 + a\,(p_2-p_1)}\vec{x}_2 = \mathrm{Proj}_{a\,[p_2 p_1 + (p_2-p_1)]}\vec{x}_2;$$

hence

$$\mathrm{Proj}_{a\,p_2}\vec{x}_2 = \mathrm{Proj}_a\,\mathrm{Proj}_{p_2}\vec{x}_2 = \mathrm{Proj}_a\vec{x}_2 \in q.$$

Hence, by (7) and the above result, the vectors

(10) $$\mathrm{Proj}_a\vec{x}_1, \quad \mathrm{Proj}_a\vec{x}_2$$

belong to q whatever $a \in (T)$ may be.

Since the vectors (10) span p, it follows that

(11) $$p \subseteqq q.$$

To prove the converse inclusion $q \subseteqq p$, notice that

$$\mathrm{Proj}_{a\,(p_2-p_1)}\vec{x}_2 = \mathrm{Proj}_a\,\mathrm{Proj}_{p_2-p_1}\vec{x}_2 = \mathrm{Proj}\,\vec{v},$$

[this by (2.1)], and then

$$\mathrm{Proj}_a\vec{v} \in a\,(p_2 - p_1) \subseteqq p_2 - p_1 \subseteqq p_2 \subseteqq p,$$

because

$$p_2 = \mathcal{M}_T(\vec{x}_2) \quad \text{and} \quad p = \mathcal{M}_T((\vec{x}_1) \cup (\vec{x}_2)).$$

Besides

$$\mathrm{Proj}_a \vec{x}_2 \in p_2 \subseteqq p.$$

Hence

$$\mathrm{Proj}_a \vec{y} = \mathrm{Proj}_a \vec{v} + \mathrm{Proj}_a \vec{x}_2 \in p.$$

This being true for every $a \in (T)$, it follows, by (5.1), that

(12) $$q \subseteqq p.$$

From (11) and (12) we get $q = p$, i.e.

$$\mathcal{M}_T((\vec{x}_1) \cup (\vec{x}_2)) = \mathcal{M}_T(\vec{y}).$$

The theorem is proved.

D I.16.1.1. Theorem. If

1. the space **H** is separable,
2. (T) is a saturated tribe of spaces in $\mathbf{H} = \mathit{1}$,

then there exists a vector $\vec{x}$ such that

$$\mathcal{M}_T(\vec{x}) = \mathit{1}.$$

Proof. Choose in *1* a denumerable everywhere dense set E of vectors

$$\vec{z}_1, \vec{z}_2, \ldots, \vec{z}_n, \ldots$$

This can be made, for *1* is supposed to be separable. Put

$$p_1 =_{df} \mathcal{M}_T(\vec{z}_1),$$
$$p_2 =_{df} \mathcal{M}_T((\vec{z}_1) \cup (\vec{z}_2)),$$
$$\cdots\cdots\cdots\cdots$$
$$p_n =_{df} \mathcal{M}_T((\vec{z}_1) \cup (\vec{z}_2) \cup \cdots \cup (\vec{z}_n)),$$
$$\cdots\cdots\cdots\cdots$$

By Theorem [D I.14.14.] we have

$$p_n = \mathcal{M}_T(\vec{z}_1) + \mathcal{M}_T(\vec{z}_2) + \cdots + \mathcal{M}_T(\vec{z}_n).$$

Now the Theorem [D I.16.1.] can be stated in the following form, [D I.14.14.]: "for any two vectors $\vec{x}_1$, $\vec{x}_2$, there exists a vector $\vec{y}$ such that

$$\mathcal{M}_T(\vec{x}_1) + \mathcal{M}_T(\vec{x}_2) = \mathcal{M}_T(\vec{y})\text{''}.$$

Using this statement, we can assert the existence of a vector $\vec{\xi}_2$, such that $p_2 = \mathcal{M}(\vec{\xi}_2)$. Hence $p_3 = \mathcal{M}(\vec{\xi}_2) + \mathcal{M}(\vec{z}_3)$. Hence there exists $\vec{\xi}_3$, such that $p_3 = \mathcal{M}(\vec{\xi}_3)$, and so on by induction. If we put $\vec{\xi}_1 =_{df} \vec{z}_1$, we can say that there exists an infinite sequence $\vec{\xi}_1, \vec{\xi}_2, \ldots, \vec{\xi}_n, \ldots$, such that

(1) $$p_n = \mathcal{M}_T(\vec{\xi}_n) \quad \text{for} \quad n = 1, 2, \ldots$$

We have by [D I.14.4.]

(2) $$p_1 \leqq p_2 \leqq \cdots \leqq p_n \leqq \cdots$$

We have $\vec{z}_n \in p_n$, and then

$$E \subseteqq \bigcup_{n=1}^{\infty} p_n.$$

Since E is dense in the space I, we get

$$I = \overline{E} \subseteqq \overline{\bigcup_{n=1}^{\infty} p_n} = \sum_{n=1}^{\infty} p_n,$$

(by Theorem [D.11.3.]). It follows that

(3) $$\sum_{n=1}^{\infty} p_n = I.$$

Two cases can occure:

1^0 case. There exists an index n_0 such that

$$p_{n_0} = p_{n_0+1} = \cdots$$

2^0 case. There exists a partial sequence

$$O \neq p_{\sigma_1} \subset p_{\sigma_2} \subset p_{\sigma_3} \subset \cdots$$

In the first case we have

$$I = \sum_{n=1}^{\infty} p_n = \sum_{n=1}^{n_0} p_n = p_{n_0},$$

and then

$$I = \mathcal{M}_T(\vec{\xi}_{n_0}),$$

which proves the theorem.

Consider the second case. Put

(4) $$q_n =_{df} p_{\sigma_n}, \quad (n = 1, 2, \ldots),$$

and put

(5) $$\vec{x}_n =_{df} \vec{\xi}_{\sigma_n}.$$

Since

(5.1) $$0 \neq q_1 \subset q_2 \subset \cdots \subset q_n \subset \cdots,$$

the spaces

(5.2) $$q_1, q_2 - q_1, q_3 - q_2, \ldots$$

are orthogonal with one another, and they all are $\neq O$. Besides

(5.3) $$I = q_1 + (q_2 - q_1) + \cdots$$

Consider the vectors:

(5.4) $$\begin{cases} \vec{y}_1 =_{df} \vec{x}_1, \\ \vec{y}_2 =_{df} \mathrm{Proj}_{q_2 - q_1} \vec{x}_2, \\ \vec{y}_3 =_{df} \mathrm{Proj}_{q_3 - q_2} \vec{x}_3, \\ \cdots\cdots\cdots\cdots \end{cases}$$

I say that no one $\vec{y}_n$ is $= \vec{O}$. Indeed we have $\vec{y}_1 \neq \vec{O}$. Let v be the smallest index for which $\vec{y}_v = \vec{O}$. We have $v \geqq 2$. As $\vec{x}_v \in q_v$ and $q_v = q_1 + (q_2 - q_1) + \cdots + (q_v - q_{v-1})$ with orthogonal terms, we have

$$\vec{x}_v = \mathrm{Proj}_{q_1} \vec{x}_v + \mathrm{Proj}_{q_2 - q_1} \vec{x}_v + \cdots + \mathrm{Proj}_{q_v - q_{v-1}} \vec{x}_v .$$

Since

$$\mathrm{Proj}_{q_v - q_{v-1}} \vec{x}_v = \vec{y}_v = \vec{O},$$

we have

$$\vec{x}_v = \mathrm{Proj}_{q_1} \vec{x}_v + \cdots + \mathrm{Proj}_{q_{v-1} - q_{v-2}} \vec{x}_v = \mathrm{Proj}_{q_{v-2}} \vec{x}_v .$$

Hence $\vec{x}_v \in q_{v-1}$. It follows, by [D I.14.4.],

(6) $$\mathscr{M}_T(\vec{x}_v) \subseteqq \mathscr{M}_T(q_{v-1}).$$

Since (T) is saturated, the space

$$q_{v-1} = p_{\sigma_{v-1}} = \mathscr{M}_T(\vec{\xi}_{\sigma_{v-1}}),$$

is compatible with all spaces of (T), and belongs itself to (T). Hence, by Theorem [D I.14.9.],

(7) $$\mathscr{M}_T(q_{v-1}) = q_{v-1}.$$

Now, by (5),

$$\mathscr{M}_T(\vec{x}_v) = \mathscr{M}_T(\vec{\xi}_{\sigma_v}) = p_{\sigma_v},$$

and then

$$\mathscr{M}_T(\vec{x}_v) = q_v$$

by (4). Thus, from (6) we get

$$q_v \subseteqq q_{v-1}.$$

But we have $q_{v-1} \subseteqq q_v$ too. It follows that

$$q_{v-1} = q_v,$$

which contradicts the inclusion

(8) $$q_{v-1} \subset q_v,$$

[see (5.1)]. Thus we have proved that all vectors $\vec{y}_n$ $(n = 1, 2, \ldots)$ differ from $\vec{O}$.

Now we shall prove that

$$q_1 = \mathscr{M}_T(\vec{y}_1), \quad q_2 - q_1 = \mathscr{M}_T(\vec{y}_2), \ \ldots, \ q_n - q_{n-1} = \mathscr{M}_T(\vec{y}_n), \ \ldots$$

The first equality is true, because

(8.1) $$\mathscr{M}_T(\vec{y}_1) = \mathscr{M}_T(\vec{x}_1) = \mathscr{M}_T(\vec{\xi}_{\sigma_1}) = p_{\sigma_1} = q_1.$$

Suppose that

$$\vec{x} \in \mathscr{M}_T(\vec{y}_n), \quad (n \geqq 2).$$

There exists an infinite sequence of vectors having the form

(9) $$\sum_i \lambda_i \operatorname{Proj}_{a_i}\vec{y}_n, \quad (a_i \in (T)), \text{ and tending to } \vec{x}.$$

But

$$\vec{y}_n = \operatorname{Proj}_{q_n - q_{n-1}}\vec{x}_n;$$

hence (9) equals

$$\sum_i \lambda_i \operatorname{Proj}_{a_i} \operatorname{Proj}_{q_n - q_{n-1}}\vec{x}_n,$$

and then, on account of compability of $q_n - q_{n-1}$ and a_i, the vector (9) is equal to

$$\operatorname{Proj}_{q_n - q_{n-1}}(\sum_i \lambda_i \operatorname{Proj}_{a_i}\vec{x}_n) \in q_n - q_{n-1}.$$

Hence $\vec{x} \in q_n - q_{n-1}$. Consequently

(10) $$\mathscr{M}_T(\vec{y}_n) \subseteqq q_n - q_{n-1}.$$

Now let

$$\vec{x} \in q_n - q_{n-1}.$$

Since $q_n - q_{n-1} \subseteqq q_n$, therefore there exists $\vec{x}' \in q_n$, such that

(11) $$\vec{x} = \operatorname{Proj}_{q_n - q_{n-1}}\vec{x}'.$$

Since $\vec{x}' \in q_n = \mathscr{M}_T(\vec{x}_n)$, this vector is the limit of a sequence of vectors

$$\sum_j \mu_j \operatorname{Proj}_{b_j}\vec{x}_n, \quad \text{where} \quad b_i \in (T).$$

Hence, by (11), $\vec{x}$ is the limit of the sequence of vectors

(12) $$\operatorname{Proj}_{q_n - q_{n-1}}(\sum_j \mu_j \operatorname{Proj}_{b_j}\vec{x}_n);$$

this because of continuity of the operation of projection. The vector (12) equals

$$\sum_j \mu_j \operatorname{Proj}_{q_n - q_{n-1}} \operatorname{Proj}_{b_j}\vec{x}_n = \sum_j \mu_j \operatorname{Proj}_{b_j} \operatorname{Proj}_{q_n - q_{n-1}}\vec{x}_n;$$

hence, by (5.4), it equals

$$\sum_j \mu_j \operatorname{Proj}_{b_i}\vec{y}_n.$$

Consequently $\vec{x} \in M_T(\vec{y}_n)$. Thus we get

$$q_n - q_{n-1} \subseteqq M_T(\vec{y}_n),$$

and then, by (10),

(13) $$M_T(\vec{y}_n) = q_n - q_{n-1}.$$

This being settled, let us recall that, $\vec{y}_n \neq \vec{O}$, [see (8)]. I say that the vectors $\vec{y}_1, \vec{y}_2, \ldots, \vec{y}_n, \ldots$ are orthogonal with one another. Indeed, by (13) we have

$$\vec{y}_1 \in q_1, \quad \vec{y}_n \in q_n - q_{n-1}$$

for $n \geqq 2$ and, by (5.2), the spaces

$$q_1, q_2 - q_1, \ldots, q_n - q_{n-1}, \ldots$$

are orthogonal with one another. Let $\{w_n\}$ be a sequence of numbers such that

(14) $$\sum_{n=1}^{\infty} |w_n|^2$$

converges. The series $\sum_{n=1}^{\infty} w_n \cdot \frac{\vec{y}_n}{||\vec{y}_n||}$ converges too. Put

(15) $$\vec{y} =_{df} \sum_{n=1}^{\infty} w_n \frac{\vec{y}_n}{||\vec{y}_n||}.$$

We have

(16) $$\text{Proj}_{q_1}\vec{y} = w_1 \frac{\vec{y}_1}{||\vec{y}_1||}, \qquad \text{Proj}_{q_n - q_{n-1}}\vec{y} = w_n \frac{\vec{y}_n}{||\vec{y}_n||},$$

for $n \geqq 2$.

Now we shall prove that

$$\mathcal{M}_T(\vec{y}) = I.$$

Take an arbitrary vector $\vec{\eta} \in I$. Since, by (5.3),

$$q_1 + (q_2 - q_1) + \cdots + (q_n - q_{n-1}) + \cdots = I,$$

we have

(17) $$\vec{\eta} = \text{Proj}_{q_1}\vec{\eta} + \text{Proj}_{q_2 - q_1}\vec{\eta} + \cdots + \text{Proj}_{q_n - q_{n-1}}\vec{\eta} + \cdots$$

Since

$$\text{Proj}_{q_1}\vec{\eta} \in q_1, \qquad \text{Proj}_{q_2 - q_1}\vec{\eta} \in q_n - q_{n-1}$$

for $n \geqq 2$, we have, by (13) and (8.1),

(18) $$\text{Proj}_{q_2}\vec{\eta} \in \mathcal{M}_T(\vec{y}_1), \qquad \text{Proj}_{q_n - q_{n-1}}\vec{\eta} \in \mathcal{M}_T(\vec{y}_n).$$

Take any $\varepsilon > 0$. The sum (17) being convergent, we can find an index N such that

(19) $$||\eta - (\text{Proj}_{q_1}\eta + \cdots + \text{Proj}_{q_N - q_{N-1}}\eta|| \leqq \frac{\varepsilon}{2}.$$

On account of (18) there exists a finite sum

$$\sum_i \lambda_{in} \text{Proj}_{a_{in}}\vec{y}_n, \quad \text{with} \quad a_{in} \in (T),$$

such that

$$||\text{Proj}_{q_1}\vec{\eta} - \sum_i \lambda_{i1} \text{Proj}_{a_{in}}\vec{y}_1|| \leqq \frac{\varepsilon}{2N},$$

$$||\text{Proj}_{q_n - q_{n-1}}\vec{\eta} - \sum_i \lambda_{in} \text{Proj}_{a_{in}}\vec{y}_n|| \leqq \frac{\varepsilon}{2N}.$$

Hence, from (19) we get

$$||\vec{\eta} - (\sum_i \lambda_{i1} \operatorname{Proj}_{a_{i1}} \vec{y}_1 + \cdots + \sum_{in} \lambda_{in} \operatorname{Proj}_{a_{in}} \vec{y}_n|| \leqq \varepsilon .$$

Hence, by (16), we get an inequality of the form

$$||\vec{\eta} - \sum_j k_j \operatorname{Proj}_{a_j} \vec{y}|| \leqq \varepsilon , \quad \text{where} \quad a_j \in T .$$

This proves that

$$\vec{\eta} \in \mathscr{M}_T(\vec{y}) .$$

The vector $\vec{\eta}$ having been chosen arbitrarily in the space I, it follows that

$$I \subseteqq \mathscr{M}_T(\vec{y}) ,$$

so

$$I = \mathscr{M}_T(\vec{y}) . \qquad \text{q.e.d.}$$

The theorem is established.

D I.16.2. Def. Let I be a H.-H.-complete space **H**, which is not supposed to be separable. Let (T) be a tribe of spaces in I. By *a vector generating* **H** *with respect to* (T) we shall understand any vector $\vec{\omega}$ such that

$$I = \mathscr{M}_T(\vec{\omega}) .$$

D I.16.3. Theorem. If

1) I is any complete H.-H.-space,
2) (T) a tribe set of spaces in I,
3) $\vec{\omega}$ a vector generating I with respect to (T),
4) $a \in T$,

then

$$a = \mathscr{M}_T(\operatorname{Proj}_a \vec{\omega}) .$$

Proof. By virtue of Theorem [D I.14.9.],

(1) $$a = M_T(a) .$$

Since $\operatorname{Proj}_a \vec{\omega} \in a$, it follows, by [D I.14.4.], that

$$\mathscr{M}_T(\operatorname{Proj}_a \vec{\omega}) \subseteqq \mathscr{M}_T(a) ,$$

and then, by (1),

(2) $$\mathscr{M}_T(\operatorname{Proj}_a \vec{\omega}) \subseteqq a .$$

Now let

$$\vec{x} \in a .$$

Since $\mathscr{M}_T(\omega) = I$, the vector $\vec{x}$, as belonging to $\mathscr{M}_T(\vec{\omega})$, can be obtained as the limit of a sequence of vectors having the form of a finite sum:

$$\sum_i \lambda_{in} \operatorname{Proj}_{a_{in}} \vec{\omega} , \quad \text{where} \quad a_{in} \in (T) .$$

It follows

$$\vec{x} = \mathrm{Proj}_a \vec{x} = \lim_n (\mathrm{Proj}_a \sum_i \lambda_{in} \mathrm{Proj}_{a_{in}} \vec{\omega}) = \lim_n \sum_i \lambda_{in} \mathrm{Proj}_a \mathrm{Proj}_{a_{in}} \vec{\omega};$$

by [D.9., II] this equals

$$\lim_n \sum_i \lambda_{in} \mathrm{Proj}_{a_{in}} (\mathrm{Proj}_a \vec{\omega}).$$

Hence, by Def. [D I.14.1.],

$$\vec{x} \in \mathcal{M}_T(\mathrm{Proj}_a \vec{\omega}).$$

Thus we have proved that

(3) $$a \subseteqq \mathcal{M}_T(\mathrm{Proj}_a \vec{\omega}).$$

From (2) and (3) the theorem follows.

D I.16.4. Theorem. If

1) I is any complete H.-H.-space,
2) T a tribe-set of spaces in I,
3) there exists a generating vector of I with respect to (T),
4) E is a non empty set of vectors of I,

then there exists a vector $\vec{y}$ such that

$$\mathcal{M}_T(E) = \mathcal{M}_T(\vec{y}).$$

Proof. Let p be the space spanned by E. We have, by the Theorem [D.14.8.],

(1) $$\mathcal{M}_T(E) = \mathcal{M}_T(p) =, \text{(Theorem [D I.14.10.])}, = \mathcal{M}_T(q),$$

where

$$q =_{df} \mathcal{M}_T(p).$$

Let $\vec{\omega}$ be a generating vector, whose existence is supposed. Put

(1.1) $$\vec{y} = \mathrm{Proj}_q \vec{\omega}.$$

Since $\mathrm{Proj}_p \vec{\omega} \in q$, we have

(2) $$\mathcal{M}_T(\vec{y}) \subseteqq \mathcal{M}_T(q) = \mathcal{M}_T(E).$$

Now let $\vec{x} \in \mathcal{M}_T(q)$. We have $\vec{x} \in p \subseteqq q$. Since $\vec{\omega}$ is a generating vector, therefore given an $\varepsilon > 0$, we have

$$||\vec{x} - \sum_i \lambda_i \mathrm{Proj}_{a_i} \vec{\omega}|| \leqq \varepsilon$$

for some λ_i and $a_i \in T$. Hence

$$||\mathrm{Proj}_q \vec{x} - \mathrm{Proj}_q \sum_i \lambda_i \mathrm{Proj}_{a_i} \vec{\omega}|| \leqq \varepsilon.$$

Since q, as a radiation-scope, is compatible with a_i, it follows, by Theorem [D.9., II′]:

(3) $$||\vec{x} - \sum_i \lambda_i \mathrm{Proj}_{a_i} \mathrm{Proj}_q \vec{\omega}|| \leqq \varepsilon,$$

because $\vec{x} \in q$ implies $\mathrm{Proj}_q \vec{x} = \vec{x}$.

From (3) it follows that

$$\vec{x} \in \mathcal{M}_T(\mathrm{Proj}_q \vec{\omega}).$$

Consequently

(4) $$\mathcal{M}_T(E) = \mathcal{M}_T(q) \subseteqq \mathcal{M}_T(\mathrm{Proj}_q \vec{\omega}).$$

From (1), (2) and (4) it follows that

$$\mathcal{M}_T(E) = \mathcal{M}_T(\vec{y}), \quad \text{q.e.d.}$$

D I.17. Saturation of a tribe of spaces. We have discussed the notion of saturated tribe of spaces in Hilbert-space **H**, and the notion of generating vector. The generating vector of **H** with respect to a given geometrical tribe (T) of spaces exists, if and only if the tribe (T) is saturated. Now, there are non saturated tribes, so there comes the problem of possibility of extension of a given tribe to a saturated one. The answer is positive, and the operation of extension of a given tribe will be important in the theory of normal operators.

Though the tribe (T), which is a relation, is a notion different from its domain T, which is a set, nevertheless, to avoid complication in notation, we agree to write some times T to denote both above notions.

Instead of $\mathrm{Proj}_a b$, we shall write occasionally the symbol:

$$(\mathrm{Proj}\, a)\, b.$$

All tribes considered in this section will be geometrical tribes of spaces.

D I.17.1. Theorem. If

1. T is a tribe of spaces,
2. $p_1, p_2, \ldots, p_n, \ldots$ an infinite sequence of spaces,
3. p_n are compatible with every space of (T), $(n = 1, 2, \ldots)$,
4. p_i, p_k are orthogonal $(i \neq k,\ i, k = 1, 2, \ldots)$,

then there exists the smallest tribe T' whose domain contains T and all p_n.

Proof. Put $T_0 =_{df} T$. Since p_1 is compatible with the spaces of T, there exists the smallest tribe T_1, containing T and p_1, [D I.8.4.]. Suppose that we have defined the tribes $T_0, T_1, \ldots, T_s$, $s \geqq 1$, and suppose that for each i, where $1 \leqq i \leqq s$, T_i is the smallest tribe containing T_{i-1} and p_i. We shall prove that p_{s+1} is capable to be adjoint to T_s, [D I.7.1.]. Put $q = \mathcal{M}_{T_s}(p_{s+1})$, [D I.14.1.]. The space q is the space spanned by the vectors $(\mathrm{Proj}\, b)\, \vec{\xi}$ where $b \in T$, $\vec{\xi} \in p_{s+1}$, [D I.14.2.]. Since T_s is the smallest tribe containing T_{s-1} and p_s, any its space b has the form $b = a_1 p_s + a_2 \,\mathrm{co}\, p_s$, where $a_1, a_2 \in T_{s-1}$, [D I.8.4.]. We have

$$(\mathrm{Proj}\, b)\, \vec{\xi} = (\mathrm{Proj}\, a_1 p_s)\, \vec{\xi} + (\mathrm{Proj}\, a_2 \,\mathrm{co}\, p_s)\, \vec{\xi};$$

hence, [D.9., III′],

$$(\operatorname{Proj} b)\,\vec{\xi} = (\operatorname{Proj} a_1)\,(\operatorname{Proj} p_s)\,\vec{\xi} + (\operatorname{Proj} a_2)\,(\operatorname{Proj} \operatorname{co} p_s)\,\vec{\xi},$$

for a_1, a_2, p_s are compatible. But $\vec{\xi} \in p_{s+1}$, $p_{s+1} \perp p_s$; hence $(\operatorname{Proj} p_s)\,\vec{\xi} = \vec{O}$ and $(\operatorname{Proj} \operatorname{co} p_s)\,\vec{\xi} = \vec{\xi}$. It follows that

$$(\operatorname{Proj} b)\,\vec{\xi} = (\operatorname{Proj} a_2)\,\vec{\xi}.$$

Since a_2 may be any space of T_{s-1}, it follows that the set of all vectors $(\operatorname{Proj} b)\,\vec{\xi}$, where b varies in T_s and $\vec{\xi}$ in p_{s+1}, coincides with the set of all vectors $(\operatorname{Proj} a_2)\,\vec{\xi}$, where a_2 varies in T_{s-1} and $\vec{\xi}$ varies in p_{s+1}. Hence $q = \mathcal{M}_{T_{s-1}}(p_{s+1})$. Thus

$$\mathcal{M}_{T_s}(p_{s+1}) = \mathcal{M}_{T_{s-1}}(p_{s+1}).$$

By repeating this argument we get

$$\mathcal{M}_{T_s}(p_{s+1}) = \mathcal{M}_{T_{s-1}}(p_{s+1}) = \cdots = \mathcal{M}_{T_0}(p_{s+1}).$$

Thus we have proved that

$$\mathcal{M}_{T_s}(p_{s+1}) = \mathcal{M}_{T_0}(p_{s+1}) = p_{s+1}.$$

This follows from the fact that if a space c is compatible with the spaces of a tribe Q then $M_Q(c) = c$, [D I.14.9.]. Since p_{s+1} is a radiation scope of I with respect to T_s, the space p_{s+1} is capable of being adjoint to T_s, [D I.15.3.]. Hence there exists the smallest tribe T_{s+1} containing T_s and p_{s+1}. We have $T_s \subseteqq T_{s+1}$. Put $T' = \bigcup_{s=0}^{\infty} T_s$. This is the required tribe.

D I.17.2. Theorem. If

1. T is a denumerably additive tribe of spaces,
2. $p_1, p_2, \ldots, p_n, \ldots$ are spaces, all compatible with the spaces of (T),
3. p_n are orthogonal with one another,
4. $\sum_n p_n = I = \mathbf{H}$,

then

1) there exists the smallest denumerably additive tribe T^* of spaces containing T and all the spaces p_n;

2) T^* is the class of all spaces having the form

$$\sum_i a_i\, p_{\nu(i)}, \tag{1}$$

where the sum is finite or denumerably infinite, and where

$$a_i \in T, \quad \nu(1) \leqq \nu(2) \leqq \cdots$$

Proof. The space T' obtained in the proof of the preceding theorem satisfies the requirements of the thesis 1). To prove item 2), denote by U the set of all spaces (1). We have

$$U \subseteqq T^*.$$

To prove the inverse inclusion, it suffices to prove that U is a denumerably additive tribe. Let $b_n \in U$, $(n = 1, 2, \ldots)$. We have $b_n = \sum_{k=1}^{\infty} a(n\,k)\, p_k$, where $a(n\,k) \in T$. Since $U \subseteqq T^*$, we can apply to the somata of U all formal laws for finite and denumerable operations. We have $\sum_n b_n = \sum_k p_k \sum_n a(n\,k)$ and $\sum_n a(n\,k) \in T$; hence $\sum_n b_n \in U$. Now let $b \in U$, and $b = \sum_k a_k\, p_k$, where $a_k \in T$, and where some of them may be $= O$. We have

$$\text{co}\, b = \prod_k (\text{co}\, a_k + \text{co}\, p_k). \tag{2}$$

Since, by hypothesis $\sum_v p_v = I$, we have

$$\text{co}\, p_k = p_1 + \cdots + p_{k-1} + p_{k+1} + \cdots \in U.$$

On the other hand

$$\text{co}\, a_k + \text{co}\, p_k \in U. \tag{3}$$

We see that to prove that $\text{co}\, b \in U$, it suffices to prove that an infinite product of spaces of U also belongs to U. Let $b_i = \sum_k a(i\,k)\, p_k$, $i = 1, 2, \ldots$ We have $b_j\, p_j = a(i\,j)\, p_j$. Consequently

$$\left(\prod_i b_i\right) p_j = \prod_i (b_i\, p_j) = \prod_i a(i\,j) \cdot p_j. \tag{4}$$

Since $\sum_j p_j = I$, we have

$$\prod_i b_i = \prod_i b_i \cdot \sum_j p_j = \sum_j \left(\prod_i b_i\right) p_j;$$

hence, by (4),

$$\prod_i b_i = \sum_j \left(\prod_i a(i\,j)\right) p_j.$$

Since $\prod_i a(i\,j) \in T$, we see that $\prod_j b_i \in U$. Thus going back to (2) and (3), we see that $\text{co}\, b \in U$. The theorem is established.

D I.17.3. Theorem. If

1. T is a tribe of spaces,
2. p a space compatible with all spaces of T,
3. $\vec{z} \in p$,

then $\mathcal{M}_{T \cdot p}(\vec{z}) = \mathcal{M}_T(\vec{z})$, where $T \cdot p$ is the class of all spaces $a \cdot p$, and where a varies in T.

Proof. $\mathcal{M}_T(\vec{z})$ is the space q spanned by all vectors $(\text{Proj}\, a)\, \vec{z}$, where $a \in T$. Now $\vec{z} \in p$; hence $(\text{Proj}\, p)\, \vec{z} = \vec{z}$. It follows: $(\text{Proj}\, a)\, \vec{z} = (\text{Proj}\, a\, p)\, \vec{z}$, because p and a are compatible. It follows that $q \subseteqq \mathcal{M}_{T \cdot p}(\vec{z})$. Let $b \in T \cdot p$. There exists $a \in T$ with $b = a\,p$. We have:

$$(\operatorname{Proj} b)\,\vec{z} = (\operatorname{Proj} a)\,(\operatorname{Proj} p)\,\vec{z} = (\operatorname{Proj} a)\,\vec{z}.$$

Hence $\mathcal{M}_{Tp}(\vec{z}) \subseteqq q$, which completes the proof.

D I.17.4. Theorem. If

1. T is a denumerably additive tribe of spaces,
2. T is not saturated,

then there exists a vector $\vec{\xi}$ such that $\mathcal{M}_T(\vec{\xi})$ does not belong to T.

Proof. Since T is not saturated, there exists a space p compatible with T but not belonging to T. Let E be a denumerable set of vectors, everywhere dense in p. This is possible to have, since **H** is separable. We can prove that $\mathcal{M}_T(E) = \mathcal{M}_T(\bar{E}) = \mathcal{M}_T(p) = p$. We also have

(1) $$\mathcal{M}_T(E) = \sum_{\vec{x} \in E} \mathcal{M}_T(\vec{x}) = p.$$

Hence there exists $\vec{\xi} \in E$ such that $\mathcal{M}_T(\vec{\xi}) \,\bar{\in}\, T$, because if not, we would have, by (1), $p \in T$. The theorem is proved.

D I.17.5. Theorem. If T is a denumerably additive tribe of spaces in a separable, complete H.-H.-space **H**,
then there exists a finite or denumerable infinite sequence of vectors

(1) $$\vec{\Theta}_1,\ \vec{\Theta}_2,\ \ldots,\ \vec{\Theta}_n,\ \ldots$$

such that, if we put

$$q_n = \mathcal{M}_T(\vec{\Theta}_n),$$

then

1) $q_1, q_2, \ldots$ are orthogonal,

2) $\sum_n q_n = I = \mathbf{H}$,

3) the smallest denumerably additive tribe T^*, containing T and all q_n, is saturated.

Proof. Let us write optionally vectors without arrows. Let

(2) $$x_1,\ x_2,\ \ldots,\ x_\omega,\ \ldots,\ x_\alpha,\ \ldots,$$

be a well ordered sequence, composed of all non null vectors of the space **H**, each taken once only. Put $p_1 = M_T(x_1)$, $\Theta_1 = x_1$. If $p_1 = I$, then x_1 is a generating vector of I with respect to T, and T is a saturated tribe; so the theorem is true with the sequence (1) reduced to a single vector. Suppose $p_1 \neq I$; hence $\operatorname{co} p_1 \neq O$. Consider the tribe $T \cdot \operatorname{co} p_1$. Take a vector belonging to $\operatorname{co} p_1$ and having the smallest index. Denote this vector by z_2. Put

$$p_2 =_{df} \mathcal{M}_{T \cdot \operatorname{co} p_1}(z_2) \subseteqq \operatorname{co} p_1.$$

We have

$$p_2 = \mathcal{M}_T(z_2), \quad \text{and} \quad p_2 \perp p_1.$$

It may happen that $\mathcal{M}_{T\,\mathrm{co}p_1}(z_2) = \mathrm{co}p_1$, and in this case z_2 is a generating vector in $\mathrm{co}p_1$. We put $p_2 =_{df} \mathrm{co}p_1$. We have $p_1 + p_2 = I$, and the process stops, so the theorem is proved. Let α be an ordinal with $2 \leqq \alpha \leqq \Omega$, and suppose that we have already found the well ordered sequences $z_1, z_2, \ldots, z_\omega, \ldots, z_\beta, \ldots$ for all $\beta < \alpha$, and suppose that $p_\beta = \mathcal{M}_T(z_\beta)$. We also suppose that all spaces p_β are orthogonal to one another. Put $q = \sum_{\beta < \alpha} p_\beta$. All spaces p_β are compatible with the spaces of T. It follows that q is also compatible with T. If $q = I$, the process stops. If not, $\mathrm{co}q \neq O$, and hence there exists in (2) a vector with the smallest index and belonging to $\mathrm{co}q$. Denote this vector by z_α. Put $p_\alpha = \mathcal{M}_{T\,\mathrm{co}p}(z_\alpha)$. We have $p_\alpha \subseteqq \mathrm{co}q$; hence p_α is orthogonal to all p_β for $\beta < \alpha$.

Besides $p_\alpha = M_T(z_\alpha)$. In this way we have defined a well ordered sequence $\{z_\alpha\}$ and a corresponding sequence $\{p_\alpha\}$ of mutually orthogonal spaces, compatible with T. Since the vectors in (2) are all different from $\vec{O}$, no space p_α is $= O$. Since $\mathbf{H}$ is separable, the process must stop for an ordinal less than Ω. Hence the number of all vectors z_α is at most denumerable. Let us order them in a single sequence $\Theta_1, \Theta_2, \ldots$ and put $q_n = \mathcal{M}_T(\Theta_n)$. We have $\sum_n q_n = I$. Having this, we have the smallest denumerably additive tribe T^* containing T and all the spaces q_n. We shall prove that T^* is saturated. Put

$$\vec{\omega} = \sum_n \frac{1}{n!} \frac{\vec{\Theta}_n}{\|\Theta_n\|},$$

where the sum is finite or infinite according to the case. We see that

$$(\mathrm{Proj}\, q_i)\, \vec{\omega} = \frac{1}{i!} \frac{\vec{\Theta}_i}{\|\Theta_i\|}.$$

Let $\vec{x} \in \mathbf{H}$. Since the spaces q_n are orthogonal and $\sum_n q_n = I$, we have $x = \sum_i x_i$, where $x_i = (\mathrm{Proj}\, q_i)\, x$. Since $\vec{\Theta}_n$ is a generating vector of $p_n \cdot T$, the vector x_n can be approximated in norm, up to a given positive number, by a finite combination $\sum_i \lambda_{in} (\mathrm{Proj}\, a_{in})\, \vec{\Theta}_n$, where λ_{in} are numbers and $a_{in} \in p_n \cdot T$.

Since x can be approximated in norm up to a given positive number by a finite sum of x_i, it follows that x can be approximated by a finite sum

$$\sum_{is} \mu_{is} (\mathrm{Proj}\, a_{is}\, q_i)\, \vec{\omega},$$

up to a given positive number, in norm. Since $a_{is} q_i \in T^*$, it follows that $x \in \mathcal{M}_{T^*}(\vec{\omega})$, and then that $\vec{\omega}$ is a generating vector of $\mathbf{H}$ with respect to T^*. It follows that T^* is a saturated tribe, so the theorem is proved.

D I.17.6. Def. If T is a denumerably additive tribe of spaces, then every sequence $\vec{\Theta}_1, \vec{\Theta}_2, \ldots$ of vectors, such that the smallest denumerably additive tribe containing T and all the radiation scopes $\mathscr{M}_T(\vec{\Theta}_n)$ is saturated, will be termed *saturating sequence of vectors for* T.

D I.17.7. Def. If T is a denumerably additive non saturated tribe, then an at most countable sequence of spaces $\{p_n\}$ is called *saturating sequence of spaces for* T, whenever p_n are mutually orthogonal, compatible with the spaces of T; with $\sum_n p_n = I$, and where the smallest denumerably additive tribe T^*, containing T and all p_n, is saturated.

D I.17.8. Corollary. If

1. T is a denumerably additive tribe of spaces,
2. $\vec{\Theta}_1, \vec{\Theta}_2, \ldots$ its saturating sequence,

then

$$\vec{\omega} = \sum_n \frac{1}{n!} \frac{\vec{\Theta}_n}{\|\vec{\Theta}_n\|}$$

is a generating vector of **H** with respect to the saturated tribe T^*, obtained from T by adjunction of all $\mathscr{M}_T(\vec{\Theta}_n)$, and by extending thus obtained tribe set to a denumerably additive one.

D I.17.9. Corollary. If

1. T is a non saturated denumerably additive tribe of spaces,
2. $p_1, p_2, \ldots, p_n, \ldots$ its saturating sequence of spaces,
3. T^* the smallest denumerably additive tribe, containing T and all p_n,
4. $\vec{\omega}$ a generating vector of the space **H** with respect to T^*,

then

1) the tribe $p_n T^* = p_n \cdot T$ is a saturated tribe in p_n,

2) $\vec{\Theta}_n =_{df} (\mathrm{Proj}\, p_n)\, \vec{\omega}$ is a generating vector of p_n with respect to $p_n \cdot T$; and we have $p_n = \mathscr{M}_T(\Theta_n)$,

3) if Θ'_n are any vectors such that $p_n = \mathscr{M}_T(\Theta'_n)$, then $\vec{\Theta}'_1, \vec{\Theta}'_2, \ldots$ is a saturating sequence of vectors of T.

D I.18. Measure on a geometrical tribe of spaces. Since the geometrical tribe of spaces in **H** is a particular case of the general abstract tribe, we can introduce, like in [A.] and [A I.], the measure of spaces. The measure on the geometrical tribe will play an important role in the study of normal operators in Hilbert-space.

D I.18.1. Def. We start with some recollections. The definition of an additive measure, given in [A.15.], has been stated as follows: Let (T) be a geometrical tribe. By *a (simply) additive measure on* (T) we understand any function $\mu(\dot{a})$, real valued, finite and defined for all spaces a of (T), and such that, if a, b are orthogonal spaces, then

$$\mu(a + b) = \mu(a) + \mu(b).$$

If $\mu(a) = 0$ for all $a \in T$, the *measure is called trivial*. If (T) is a geometrical tribe, which may be finitely additive only, we say that the *measure* $\mu(\dot{a})$ *is denumerably additive* (compare [A.15.2.]), whenever the following takes place. If $a_1, a_2, \ldots, a_n, \ldots$ are mutually orthogonal spaces and $\sum_{n=1}^{\infty} a_n$ exists in (T), then $\mu\left(\sum_{n=1}^{\infty} a_n\right) = \sum_{n=1}^{\infty} \mu(a_n)$.

(**Remark.** The geometrical sum $\sum_{n=1}^{\infty} a_n$ always exists in the lattice $(\boldsymbol{L})$ of all spaces, but it may not belong to T). The *measure is called effective*, if $\mu(a) = 0$ implies $a = (\vec{O})$.

D I.18.2. Given a geometrical tribe (T), there always exists a simply additive, non trivial measure on (T). Namely there exist measures, admitting the value 0 and 1 only, [B.10.1.] and [B.12.3.].

D I.18.3. The following theorem shows that even a simply additive geometrical tribe admits a denumerably additive, non trivial measure.

Theorem. If

1. (T) is a geometrical tribe (which may be finitely additive only),
2. $\vec{\xi}$ is a vector in $\mathbf{H}$,

then if we put

$$\mu(a) = \|\mathrm{Proj}_a \vec{\xi}\|^2 \quad \text{for all} \quad a \in T,$$

the function $\mu(\dot{a})$ is a denumerably additive measure on (T). If $\vec{\xi} = \vec{O}$, the measure is trivial.

Proof. Let

$$a(1), a(2), \ldots, a(n), \ldots$$

be mutually orthogonal spaces of (T), and suppose that

$$a = \sum_{n=1}^{\infty} a(n) \in T.$$

Let

(1) $$\vec{\xi} \in \mathbf{H} \quad \text{and put} \quad \vec{\eta} =_{df} (\mathrm{Proj}\, a)\, \vec{\xi}.$$

We have $\vec{\eta} \in a$. We have by (1): $\vec{\eta} = \sum_{n=1}^{\infty} [\mathrm{Proj}\, a(n)]\, \vec{\eta}$; hence

(2) $$\|\vec{\eta}\|^2 = \sum_{n=1}^{\infty} \|(\mathrm{Proj}\, a(n)\, \vec{\eta}\|^2.$$

Now,

$$(\mathrm{Proj}\, a(n))\, \vec{\eta} = (\mathrm{Proj}\, a(n)\ \mathrm{Proj}\, a)\, \vec{\xi} = (\mathrm{Proj}\, a(n) \cdot a)\, \vec{\xi},$$

because the spaces $a(n)$ and a are compatible, [D.9., III′]. Hence from (2) we get

$$\|(\mathrm{Proj}\, a)\, \vec{\xi}\|^2 = \sum_{n=1}^{\infty} \|(\mathrm{Proj}\, a(n))\, \vec{\xi}\|^2,$$

i.e.

$$\mu(a) = \sum_{n=1} \mu[a(n)], \quad \text{q.e.d.}$$

D I.18.4. Theorem. If

1. (T) is a finitely additive geometrical tribe of spaces,

2. $\mu(\dot{a})$ is a denumerably additive measure on (T),

3. (T') is the smallest geometrical denumerably additive tribe, containing (T), then the measure $\mu(a)$ can be extended to a denumerably additive measure on (T'), where (T') is the borelian extension of (T) within (T'), [A I.4.]. It is a denumerably additive, finitely genuine, strict supertribe of (T), [A.7.2.].

Proof. We can apply to (T) the Theorem [A I.3.], which says that the following Fréchet's condition takes place:

If $b_1 \geqq b_2 \geqq \cdots \geqq b_n \geqq \cdots$ are spaces of (T) with

$$\prod_{n=1}^{\infty} b_n = 0,$$

then

$$\lim_{n\to\infty} \mu(b_n) = 0.$$

By virtue of Theorem [A I.2.], the condition is equivalent to the following one:

If

$$a_1, a_2, \ldots, a_n, \ldots \in T, \quad a \in T, \quad a = \sum_{n=1}^{\infty} a_n,$$

then

$$\mu(a) \leqq \sum_{n=1}^{\infty} \mu(a_n).$$

This is, however, the condition [A I.9.14.], which, by [A I.9.26.], is necessary and sufficient for the possibility of extension the measure from a tribe to its Lebesguean extension, so that it becomes denumerably additive.

Since the Lebesguean extension contains the Borelian extension, [A I.4.6a.], it follows that the measure $\mu(a)$ can be extended in (T') to a denumerably additive one.

D I.18.5. Remark. In the above proof and the next ones, the following situation makes the reasoning simplier, than in the general case of Boolean tribes: infinite operations are geometrical and do not depend on the totality of spaces in a tribe. They all are operations in the lattice $(\boldsymbol{L})$ of all spaces.

D I.18.6. The Theorem [D I.18.4.] admits the converse:

Theorem. If

1, 3 as before, and if the measure μ can be extended to a denumerably additive one on (T'), the measure μ must be denumerably additive already on (T).

D I.18.7. Theorem. There exists a geometrical tribe (T) and a finitely additive measure on it, which is not denumerably additive.

Expl. It will be similar to the Expl. [D I.6.5.] of a geometrical tribe which is not denumerably additive. Let **H** be a complete, separable Hilbert-space with infinite dimensions, and let

(1) $$\vec{\varphi}_1, \vec{\varphi}_2, \ldots, \vec{\varphi}_n, \ldots$$

be a saturated sequence of orthonormal vectors. Let

(2) $$w_1, w_2, \ldots, w_n, \ldots$$

be all rational numbers, in the half-open interval $(0, 1\rangle$, each of them written only once. Denote by $\mathfrak{G}$ the $1 \to 1$ correspondence which attaches to every w_n the vector $\vec{\varphi}_n$. The correspondence $\mathfrak{G}$ generates another one, also denoted by $\mathfrak{G}$, which attaches to every subset of (2) the subset of (1), composed of the $\mathfrak{G}$-corresponding vectors. If E is a subset of (2) and $\mathfrak{G}(E)$ the corresponding set of vectors, denote by $\boldsymbol{a}(E)$ the space spanned by the set $\mathfrak{G}(E)$. Thus we obtain a collection of spaces $\boldsymbol{a}(E)$ which makes up a denumerably additive geometrical tribe (T') of spaces. The extended correspondence $\mathfrak{G}$ is an isomorphism from the tribe (F') of all subsets of (2), onto (T').

Having that, consider another tribe (F_0), whose somata are finite unions f of half-open subintervals $(\alpha, \beta\rangle$ of $(0, 1\rangle$, and where the ordering is the inclusion of sets of points. (F_0) generates another tribe (F) whose somata are sets of rational numbers

$$f \cap R,$$

where R is the set of all rational numbers, contained in $(0, 1\rangle$. The collection of spaces

$$\mathfrak{G}(f \cap R)$$

is a geometrical tribe (T), which is a finitely additive, finitely genuine, strict subtribe of (T'). Let us define on (T) the measure $\mu[\mathfrak{G}(f \cap R)]$ as follows:

Every soma f of F_0 can be represented as a finite union of mutually disjoint half-open intervals

(3) $$f = \sum_i (\alpha_i, \beta_i\rangle.$$

Put

(4) $$\mu[\mathfrak{G}(f \cap R)] =_{df} \sum_i (\beta_i - \alpha_i).$$

This is just the Lebesguean measure of f. It does not depend on the choice of the representation of f by the formula (3).

Hence the measure μ of the space (4) does not depend on the above representation of f.

The measure (4), is finitely additive. We shall prove that it cannot be extended to a denumerably additive measure on the Borelian extension (T_b), [A I.4.6a.], of (T) within (T').

Suppose this be not true, so the measure can be extended. Then we have:

If

(5) $$a_1, a_2, \ldots, a_n, \ldots \in T$$

and are disjoint and with

(6) $$I = \sum_{n=1}^{\infty} a_n,$$

then

(7) $$\mu(I) = \sum_{n=1}^{\infty} \mu(a_n).$$

(By [A I.2.1.] this is even a necessary and sufficient condition for the possibility of the extension of measure.)

In our case we shall find a sequence (5) satisfying (6) but not satisfying (7). To do that, let M be a nowhere dense, perfect subset of $(0, 1\rangle$, composed of irrational numbers only and with the Lebesguean measure equal $\frac{1}{2}$. All rational numbers (2) are contained in the complementary set

(8) $$(0, 1\rangle - M.$$

Hence for the extended measure μ we have

(9) $$\mu\{\mathfrak{G}[(0, 1) - M]\} = 1.$$

On the other hand the set

$$(0, 1\rangle - M$$

is composed of the denumerable number of disjoint half-open intervals.

Indeed, if (α, β) is an open "free" interval of the perfect set M, we have

$$(\alpha, \beta) = \left(\alpha + \frac{\beta-\alpha}{2},\ \alpha + (\beta-\alpha)\right\rangle \cup \left(\alpha + \frac{\beta-\alpha}{2^2},\ \alpha + \frac{\alpha-\beta}{2}\right\rangle \cup \cdots$$

so a free interval of M is the disjoint union of an denumerable infinity of disjoint half-open intervals.

Hence the set $(0, 1\rangle - M$ is also a denumerable union of disjoint half-open intervals $(0, 1\rangle - M = \sum_j (\gamma_j\, \delta_j\rangle$, whose sum of lengths equals $\frac{1}{2}$, because the Lebesguean measure of M is $\frac{1}{2}$. So we have

$$\tfrac{1}{2} = \mu\left\{\mathfrak{G}\left[\sum_{i=1}^{\infty} (\gamma_i, \delta_i\rangle\right]\right\} = \mu\, \mathfrak{G}[(0, 1\rangle - M] = 1,$$

which is a contradiction. Thus μ cannot be extended in the way considered. Hence, by [D I.18.6.], [D I.18.4.], the measure μ is not denumerably additive on (T).

D I.18.8. Theorem. If

1. **H** is a complete and separable Hilbert-space,

2. (T) is a geometrical, non trivial tribe in **H**, (which may be not denumerably additive), then there exists a denumerably additive effective measure on (T).

Proof. Let

(1) $$\vec{\xi}_1, \vec{\xi}_2, \ldots, \vec{\xi}_n, \ldots$$

be an infinite sequence of vectors, different from $\vec{O}$, making up an everywhere dense set in **H**. Put

(2) $$\mu(a) =_{df} \sum_{n=1}^{\infty} \frac{1}{n!} \frac{\|\mathrm{Proj}_a \vec{\xi}_n\|^2}{\|\vec{\xi}_n\|^2},$$

for all $a \in T$. Since

$$\|\mathrm{Proj}_a \vec{\xi}_n\|^2 \leqq \|\vec{\xi}_n\|^2,$$

the series (2) converges. The function $\|(\mathrm{Proj}\,\dot{a})\, \vec{\xi}_n\|^2$ of a is, [D I.18.3.], a denumerably additive measure on (T). Hence so is also

$$\frac{1}{n!} \frac{\|(\mathrm{Proj}\,\dot{a})\,\vec{\xi}_n\|^2}{\|\vec{\xi}_n\|^2}.$$

It follows, by [A.15.10.], that (2) is also a denumerably additive measure in (T).

It remains to prove the effectiveness of the measure $\mu(\dot{a})$. Suppose that $\mu(\dot{a}) = 0$.

We have

$$\|\mathrm{Proj}_a \vec{\xi}_n\|^2 = 0 \quad \text{for} \quad n = 1, 2, \ldots,$$

since the terms in (2) are all non negative. Hence $\mathrm{Proj}_a \vec{\xi}_n = \vec{O}$ for all n; hence $\vec{\xi}_n \perp a$. Let $\vec{\eta} \in a$. We have $\vec{\eta} \perp \vec{\xi}_n$ for all n. Since the set (1) is everywhere dense in **H**, it follows that $\vec{\eta} = \vec{O}$. Hence a is composed of the only vector $\vec{O}$, i.e. $a = (\vec{O})$ for every $a \in T$, i.e. the tribe is trivial, which contradicts the hypothesis 2. The theorem is established.

D I.18.9. Theorem. If

1. (T) is a geometrical tribe of spaces,

2. $\vec{\omega}$ is a generating vector of I with respect to (T), [D I.16.2.],

3. $\mu(a) =_{df} \|\mathrm{Proj}_a \vec{\omega}\|^2$ for all $a \in T$,

then $\mu(a)$ is a denumerably additive effective measure on (T).

Proof. We already know that $\mu(a)$ is a denumerably additive measure on (T), [D I.18.3.]. To prove the effectiveness, suppose that $\mu(a) = 0$. It follows $\mathrm{Proj}_a \vec{\omega} = \vec{O}$.

Now, by [D I.16.3.], since $\vec{\omega}$ is a generating vector, we have

$$a = \mathscr{M}_T(\mathrm{Proj}_a \vec{\omega}).$$

Hence

$$\mathscr{M}_T(\vec{O}) = a, \quad \text{and consequently} \quad a = O.$$

The theorem is established.

D I.18.10. Theorem. If

1. (T) is a simply additive geometrical tribe of spaces,

2. (T') is the smallest denumerably additive geometrical tribe, which is a finitely genuine, strict supertribe of (T),

3. $a \in T'$,

then there exist double sequences $\{a_{ik}\}$, $\{b_{ik}\}$, $(i, k = 1, 2, \ldots)$, where $a_{ik} \in T$, $b_{ik} \in T$, such that

$$a = \prod_{n=1}^{\infty} \sum_{k=1}^{\infty} a_{ik} = \sum_{n=1}^{\infty} \prod_{k=1}^{\infty} b_{ik}.$$

Proof. We know [D I.17.9.] that every geometrical tribe can be extended to a saturated denumerably additive tribe (T^*). By [D I.16.1.1.] there exists a generating vector $\vec{\omega}$ of I with respect to (T^*).

By [D I.18.9.], the function

$$\mu(a) =_{df} ||\mathrm{Proj}_a \vec{\omega}||^2$$

is a denumerably additive effective measure on (T^*), hence also on (T).

Let us recall the theory of extension of tribe and measure. All three devices, studied in [A I.] give the same denumerably additive extension of a tribe. But they give also the extension of measure to a denumerably additive one, whenever the special condition, [A I.9.14.], called $\mathcal{B}[T, T']$ is satisfied. Now since the extension is already available, the above condition is satisfied, so all three devices $\boldsymbol{L}$, $\boldsymbol{C}$, $\boldsymbol{N}$ yield the same extension of measure.

Let us consider the Lebesgues extension (T_1) of (T) within (T^*). It contains (T'). The measure μ is also a denumerably additive measure on (T').

Since we have here the Lebesgues extension of tribe, we can apply Theorem [A I.8.10.]. Hence if $a \in T'$, there exist two double sequences $\{a_{ik}\}$, $\{b_{ik}\}$ $(i, k = 1, 2, \ldots)$ of T, such that

$$\mu[a \dotplus \sum_{i=1}^{\infty} \prod_{k=1}^{\infty} a_{ik}] = 0,$$

$$\mu[a \dotplus \prod_{i=1}^{\infty} \sum_{k=1}^{\infty} b_{ik}] = 0.$$

Since the measure is effective, it follows that

(1) $$a \dotplus \sum_{i=1}^{\infty} \prod_{k=1}^{\infty} a_{ik} = O, \quad a \dotplus \prod_{i=1}^{\infty} \sum_{k=1}^{\infty} b_{ik} = O.$$

Now, we have, in general, for somata of a tribe:

$$\text{if} \quad a \dotplus b = 0, \quad \text{then} \quad a = b.$$

Indeed, if $a + b = O$, then $(a - b) + (b - a) = O$.
Hence $a - b = O$ and $b - a = O$, i.e.

$$a \cdot \mathrm{co}\, b = O, \qquad b \cdot \mathrm{co}\, a = O.$$

Hence $a \leqq b$ and $b \leqq a$, which gives $a = b$.
Thus from (1) it follows

$$a = \sum_{i=1}^{\infty} \prod_{k=1}^{\infty} a_{ik}, \quad \text{where} \quad a_{ik} \in T,$$

$$a = \prod_{i=1}^{\infty} \sum_{k=1}^{\infty} b_{ik}, \quad \text{where} \quad b_{ik} \in T.$$

The theorem is proved.

D I.18.10a. Theorem. If we consider the proof of the Theorem [D I.11.1.], concerning the extension of a geometrical tribe, we see that the Borelian extension of a geometrical tribe (T) is composed only of spaces, having the type $\sum \prod$ and $\prod \sum$ simultaneously.

D I.18.10b. Remark. Let us notice that the Borelian extension of a geometrical tribe (T) depends only on (T) and not on the supertribe in which (T) is plunged. This is because the operations on spaces depend only on these spaces and not on the totality of somata of T.

D I.18.11. A saturated geometrical tribe is completely additive, [D I.13.].

Proof. Let (T) be a saturated tribe. Let K be a non empty collection of somata of (T). Consider the union:

$$p =_{df} \sum_{a \in K} a, \tag{1}$$

where the addition is geometrical, i.e. taken from the completely additive lattice $(\boldsymbol{L})$ of all spaces. If (1) did not belong to (T), then the tribe composed of (T) and p, will contain (T), and then be identical with (T), because (T) is saturated. Hence $p \in T$. This completes the proof.

Chapter E

Double scale of spaces

In this chapter we shall study a variation of so called "decomposition of identity", a strange and inadequate name for collections E_λ of projectors depending on a real (or complex) parameter λ, and strongly related to selfadjoint or normal operators.

We shall study this topic purely geometrically and independently of any operator.

There are three parts. The first will start with a notion called "scale of spaces", the second will deal with traces related to it, and the third—with representation of vectors by functions.

These topics are concerned with relationship between subsets of the complex plane and closed subspaces of the Hilbert-space.

E.1. Def. Let $s(\dot{Q})$ be a function which attaches to every plane-quarter Q, [C.1.4.], [D.1.], a subspace of the given H.-H.-space $\mathbf{H}$. We make the following assumptions:

I. If Q_1, Q_2 are any plane-quarters, then the spaces $s(Q_1)$, $s(Q_2)$ are compatible, [D.5.] with one another.

II. $s(\Theta) = (\vec{O})$, $s(P) = I = \mathbf{H}$, where Θ is the empty set of points of the plane P and where $\mathbf{H}$ is denoted by I.

III. If $Q_1 \subsetneq Q_2$, then $s(Q_1) \leqq s(Q_2)$, where $\subsetneq$ is the set-inclusion-symbol and $\leqq$ the ordering sign of the lattice $(\boldsymbol{L})$ of all spaces. Of course $\leqq$ means also the set-inclusion of sets of vectors, [D.1.2.].

We call the function satisfying the above conditions: *Double scale of spaces.*

As in [C.1.1.] the greek letters $\alpha, \beta, \ldots$ will denote, in this section, the *generalized numbers*: $-\infty \leqq \alpha, \beta \leqq +\infty$. (In one-dimensional case, where Q denote intervals $(-\infty, \alpha)$, we call $s(\dot{Q})$ simple scale of spaces.)

E.1.1. Extensions. The function $s(\dot{Q})$ will be extended in a suitable way to generalized rectangles and afterwards to figures, [C.2.]. The generalized rectangles were defined, [C.1.2.], by:

$$R(\alpha_1, \alpha_2; \beta_1, \beta_2) =_{df} \{(x, y) \mid \alpha_1 < x \leqq \alpha_2, \beta_1 < y \leqq \beta_2\},$$

where

$$-\infty \leqq \alpha_1, \quad \alpha_2 \leqq +\infty; \quad -\infty \leqq \beta_1, \quad \beta_2 \leqq +\infty.$$

E.1.1a. We have

$$Q(\alpha, \beta) = R(-\infty, \alpha; -\infty, \beta).$$

First the function $s(\dot{Q})$ will be extended to vertical and to horizontal strips, [C.1.4.]:

$$B_v(\alpha_1, \alpha_2) =_{df} R(\alpha_1, \alpha_2; -\infty, +\infty), \quad (\alpha_1 \leqq \alpha_2),$$
$$B_h(\beta_1, \beta_2) =_{df} R(-\infty, +\infty; \beta_1, \beta_2), \quad (\beta_1 \leqq \beta_2).$$

We have the unique representation of strips:

$$(0) \qquad \begin{aligned} B_v(\alpha_1, \alpha_2) &= Q(\alpha_2, +\infty) - Q(\alpha_1, +\infty), \\ B_h(\beta_1, \beta_2) &= Q(+\infty, \beta_2) - Q(+\infty, \beta_1); \end{aligned}$$

hence we can define

$$(1)\qquad \begin{aligned} s[B_v(\alpha_1, \alpha_2)] &=_{df} s[Q(\alpha_2, +\infty)] - s[Q(\alpha_1, +\infty)], \\ s[B_h(\beta_1, \beta_2)] &=_{df} s[Q(+\infty, \beta_2)] - s[Q(+\infty, \beta_1)], \end{aligned}$$

where (—) means subtraction of spaces, [D.3.]. Some strips are plane-quarters too, viz.

$$B_v(-\infty, \alpha) = Q(\alpha, +\infty),$$
$$B_h(-\infty, \beta) = Q(+\infty, \beta),$$

and these are the only strips which are plane-quarters, and the only plane-quarters which are strips.

We see that the values of the function s for these plane-quarters are not only given in [E.1.], but also defined in (1). Now, since

$$B_v(-\infty, \alpha) = Q(\alpha, +\infty) - Q(-\infty, +\infty),$$

the definition (1) gives

$$\begin{aligned} s[B_v(-\infty, \alpha)] &= s[Q(\alpha, +\infty)] - s[Q(-\infty, +\infty)], \\ &= s[Q(\alpha, +\infty)] - s\,Q(\emptyset); \end{aligned}$$

hence, by [E.1., II],

$$s[B_v(-\infty, \alpha)] = s[Q(\alpha, +\infty)].$$

It follows that, for vertical strips, there is no double meaning introduced in our definitions. Similarly, we have the same for horizontal strips. Thus, by introducing the definition (1), we really get a correct extension of the value of the function s, to strips.

E.1.1b. Since, by [E.1., I], $Q(\alpha_2, +\infty)$ and $Q(\alpha_1, +\infty)$ are compatible with all spaces $s(Q)$, the space $s[B_v(\alpha_1, \alpha_2)]$ is also compatible, [D.10.4.], with all these spaces; and similarly do $s[B_h(\beta_1, \beta_2)]$.

Now, since $s[B_v(\alpha_1, \alpha_2)]$ is compatible with $s[Q(+\infty, \beta_2)]$ and with $s[Q(+\infty, \beta)]$, it follows, by virtue of the same [D.10.4.], that $s[B_v(\alpha_1, \alpha_2)]$ is compatible with the difference

$$s[Q(+\infty, \beta_2)] - s[Q(+\infty, \beta_1)];$$

hence $s[B_v(\alpha_1, \alpha_2)]$ is compatible with $s[B_h(\beta_1, \beta_2)]$.

(2) Hence the spaces attached to all plane-quarters and to all strips are mutually compatible.

E.1.1c. We define for rectangles

$$(3)\qquad s[R(\alpha_1, \alpha_2; \beta_1, \beta_2)] =_{df} s[B_v(\alpha_1, \alpha_2)] \cdot s[B_h(\beta_1, \beta_2)],$$

where (·) means the (***L***)-lattice multiplication of spaces (hence their intersection too). This can be done, because the representation of the rectangle

(4) $$R(\alpha_1, \alpha_2; \beta_1, \beta_2) = B_v(\alpha_1, \alpha_2) \frown B_h(\beta_1, \beta_2),$$

by means of strips is unique.

Now, strips are rectangles and some rectangles are strips, viz.:

(5) $$\begin{aligned} B_v(\alpha_1, \alpha_2) &= R(\alpha_1, \alpha_2; -\infty, +\infty), \\ B_h(\beta_1, \beta_2) &= R(-\infty, +\infty; \beta_1, \beta_2). \end{aligned}$$

They are the only ones.

We have

$$R(\alpha_1, \alpha_2; -\infty, +\infty) = B_v(\alpha_1, \alpha_2) \frown B_h(-\infty, +\infty);$$

hence, by Def. (3), we have

$$s[R(\alpha_1, \alpha_2; -\infty, +\infty)] = s[B_v(\alpha_1, \alpha_2)] \cdot s[B_h(-\infty, +\infty)].$$

But, by (0),

$$B_h(-\infty, +\infty) = Q(+\infty, +\infty) - Q(-\infty, +\infty) = P - O = P,$$

and we have $s(P) = I$, by [E.1., II].

Hence the Def. (3) yields

$$s[R(\alpha_1, \alpha_2; -\infty, +\infty)] = s[B_v(\alpha_1, \alpha_2)],$$

so the two definitions are in agreement. Thus the extension of s to rectangles is correct.

Since B_v and B_h are compatible with all $s(Q)$, it follows, [D.10.3.], that $s[R(\alpha_1, \alpha_2; \beta_1, \beta_2)]$ is also compatible with all spaces $s(Q)$.

E.1.1d. Now take $s(R')$, $s(R'')$ where R', R'' are rectangles. Put, using strips

$$R' = B'_v \frown B'_h,$$
$$R'' = B''_v \frown B''_h.$$

We have

$$s(R') = s(B'_v) \cdot s(B'_h),$$
$$s(R'') = s(B''_v) \cdot s(B''_h).$$

Since the spaces $s(R)$ are compatible with all $s(Q)$, therefore $s(R)$ is also compatible with all possible differences $s(Q') - s(Q'')$, [D.10.4.]; hence with all possible $s(B_v)$ and $s(B_h)$. It follows that $s(R')$ is compatible with $s(B''_v)$ and with $s(B''_h)$. Consequently, by [D.10.3.], $s(R')$ is compatible with $s(B''_v) \cdot s(B''_h)$, hence with $s(R'')$.

Thus we have proved that all spaces $s(R)$,

(6) for rectangles R, are compatible with one another.

This includes all plane quarters and strips too.

E.1.1e. Lemma. If

$$\alpha_1 \leqq \alpha_2 \leqq \cdots \leqq \alpha_n \leqq \alpha_{n+1},$$

and we put

$$a_1 =_{df} B_v(\alpha_1, \alpha_2), \ \ldots, \ a_n =_{df} B_v(\alpha_n, \alpha_{n+1}),$$

then

$$s(a_1) + s(a_2) + \cdots + s(a_n) = s[B_v(\alpha_1, \alpha_{n+1})]$$

and the spaces

$$s(a_1), s(a_2), \ldots, s(a_n)$$

are perpendicular to one another. A similar theorem holds true for horizontal strips.

Proof. We have, [E.1.1a., (1)],

$$(1) \quad \begin{cases} s(a_1) = s[Q(\alpha_2, +\infty)] - s[Q(\alpha_1, +\infty)], \\ \cdots\cdots\cdots\cdots\cdots\cdots\cdots\cdots \\ s(a_n) = s[Q(\alpha_{n+1}, +\infty)] - s[Q(\alpha_n, +\infty)], \end{cases}$$

and, [E.1., III],

$$s[Q(\alpha_1, +\infty)] \leqq s[Q(\alpha_2, +\infty)] \leqq \cdots \leqq s[Q(\alpha_n, +\infty)];$$

hence we get, [A.2.6.2.],

$$s[Q(\alpha_{n+1}, +\infty)] - s[Q(\alpha_1, +\infty)] = s(a_1) + s(a_2) + \cdots + s(a_n);$$

hence

$$s[B_v(\alpha_1, \alpha_{n+1})] = s(a_1) + s(a_2) + \cdots + s(a_n),$$

[A.2.6.2.], so the first part of the thesis is established.

To prove the second part of the thesis, notice that, because of (1), the spaces

$$s(a_1), s(a_2), \ldots, s(a_n)$$

are mutually disjoint and compatible.

Hence, [D.5.], these spaces are mutually orthogonal. Thus the second part of the thesis is also proved.

E.1.1f. Lemma. Let

$$\alpha_1 \leqq \alpha_2 \leqq \cdots \leqq \alpha_{n+1} \quad \text{and} \quad \beta' \leqq \beta''.$$

Then we have

$$(1) \quad s[R(\alpha_1, \alpha_{n+1}; \beta', \beta'')] = \sum_{k=1}^{n} s[R(\alpha_k, \alpha_{k+1}; \beta', \beta'')],$$

and the terms at right are mutually orthogonal spaces.

Proof. We have, by the forgoing lemma:

$$s[B_v(\alpha_1, \alpha_{n+1})] = \sum_{k=1}^{n} s[B_v(\alpha_k, \alpha_{k+1})].$$

We multiply by $s[B_h(\beta', \beta'')]$ and apply the distributive law, which is allowed, [D.10.7.], by virtue of compatibility of spaces.

We get:

$$R(\alpha_1, \alpha_n; \beta', \beta'') = \sum_{k=1}^{n} s[B_v(\alpha_k, \alpha_{k+1})] \cdot s[B_h(\beta', \beta'')],$$

so the lemma is proved.

The perpendicularity of the rectangle-spaces in (1) follows from the forgoing Lemma, [E.1.1e.], and from

$$s[R(\alpha_n, \alpha_{n+1}; \beta', \beta'')] \leqq s[B_v(\alpha_k, \alpha_{k+1})],$$

[D.2.6.4.].

E.1.1g. The forgoing Lemma [E.1.1f.] says that we can "slice" the rectangle-space vertically into sub-rectangle-spaces, which are perpendicular to one another. An analogous theorem holds true for horizontal "slicing" of a given rectangle-space. If we combine these two theorems, by slicing in both way, we get the

Theorem. If

$$\alpha' = \alpha_1 \leqq \alpha_2 \leqq \cdots \leqq \alpha_n \leqq \alpha_{n+1} = \alpha'', \qquad (n \geqq 1),$$

$$\beta' = \beta_1 \leqq \beta_2 \leqq \cdots \leqq \beta_m \leqq \beta_{m+1} = \beta'', \qquad (m \geqq 1),$$

then

$$s[R(\alpha', \alpha''; \beta', \beta'')] = \sum_{k=1}^{k=n} \sum_{l=1}^{l=m} s[R(\alpha_k, \alpha_{k+1}; \beta_l, \beta_{l+1})],$$

where all the rectangle-spaces on the right are mutually perpendicular.

E.1.1h. A figure f on the plane is defined as any finite union of rectangles [C.2.]. We can prove that if f is a figure then it can be represented as the union of a finite number of mutually disjoint rectangles.

Indeed, let

$$f = \bigcup_{k=1}^{n} R_k,$$

where

$$R_k = R(\alpha_{k1}, \alpha_{k2}; \beta_{k1}, \beta_{k2}).$$

Consider the half-open interval $(\alpha_{k1}, \alpha_{k2}\rangle$. It may be subdivided into a finite number of disjoint subintervals by mean of vertical lines whose abscisses are among the numbers $\alpha_{s,t}$. These vertical lines will slice the figure f.

We do similarly with the half open vertical intervals $(\beta_{k1}, \beta_{k2}\rangle$, by using horizontal lines. Then the figure f will be decomposed into a finite number of mutually disjoint rectangles.

E.1.1i. Let $f = \bigcup_k R'_k$, where all rectangles R'_k are disjoint, and let $f = \sum_l R''_l$ where all R''_l are disjoint.

We can prove that

$$\sum_k s(R'_k) = \sum_l s(R''_l).$$

The proof uses the same device as before, by which we decompose the spaces $s(R'_k)$ and $s(R''_l)$ into a finite number of rectangle-spaces, which are perpendicular to one another.

The associative law for addition of spaces completes the proof.

E.1.1j. Def. The above result allows to state the definition: If f is a figure, then we put for any decomposition of f into a finite number of mutually disjoint rectangles,

$$f = \bigcup_k R_k,$$

$$s(f) =_{df} \sum_k s(R_k).$$

This notion of $s(f)$ does not depend on the choice of the partition of f into mutually disjoint rectangles.

E.1.1k. If f is a figure, then $s(f)$ is a space compatible with all rectangle-spaces. It follows that if f and g are figures then $s(f)$, $s(g)$ are compatible with one another. Indeed let

$$s(f) = \sum_k s(R'_k)$$

with perpendicular and disjoint R'_k, and

$$s(g) = \sum_l s(R''_l)$$

with perpendicular and disjoint R''_l. The space $s(f)$ is compatible with all $s(R'_k)$, by [D.10.2.].

Take any rectangle-space $s(R)$. It is compatible with all rectangle spaces, hence also with $s(R'_k)$, hence with their sum.

Thus we have proved that $s(f)$ is compatible with absolutely all rectangle-spaces.

Hence $s(f)$ is compatible with $s(R''_l)$ and then, [D.10.2.], with their sum. It follows that $s(f)$ is compatible with $s(g)$.

E.1.1l. Theorem. Thus we have defined the function $s(f)$ for all figures, i.e. for somata of the tribe (F) of figures, and we have proved that for any two figures f, g the spaces $s(f)$, $s(g)$ are compatible with one another.

E.1.2. Our next task is to prove that the correspondence s:

$$f \to s(f)$$

preserves the somatic operations.

First we shall proved that if $f, g \in F$ and $f \frown g = O$, then $s(f) \cdot s(g) = (\vec{O})$.

Proof. Decompose f and g into disjoint rectangles:

$$f = \bigcup_{\alpha} R_{\alpha}, \quad g = \bigcup_{\beta} R'_{\beta}.$$

We have

$$s(f) = \sum_{\alpha} s(R_{\alpha}), \quad s(g) = \sum_{\beta} s(R'_{\beta}).$$

Applying the distributive law, which is valid for compatible spaces, we get

$$s(f) \cdot s(g) = \sum_{\alpha, \beta} s(R_{\alpha}) \cdot s(R'_{\beta}).$$

Since $s(R_{\alpha})$, $s(R'_{\beta})$ are perpendicular to one another, they are disjoint. Hence

$$s(f) \cdot s(g) = (\vec{O}),$$

so the assertion is proved.

E.1.2a. Theorem. Now we shall prove that if

1) $f, g_1, g_2, \ldots, g_n$, $(n \geqq 2)$ are figures,

2) $g_1, g_2, \ldots, g_n$ are disjoint, and

3) $f = \bigcup_{k=1}^{n} g_k$,

then

$$s(f) = \sum_{k=1}^{n} s(g_k).$$

Proof. To prove that, decompose the figures $g_1, g_2, \ldots, g_n$ into disjoint rectangles

$$g_k = \bigcup_{t} g_{kt}, \quad (k = 1, 2, \ldots, n).$$

We have

$$s(g_k) = \sum_{t} s(g_{kt}).$$

Hence

(1)
$$\sum_{k} s(g_k) = \sum_{k} \sum_{t} s(g_{kt})$$

by the associative law for addition of disjoint, compatible spaces.

But the formula (1) constitutes a decomposition of f into disjoint rectangles; therefore

$$s(f) = \sum_{k} \sum_{t} s(g_{kt}).$$

Hence, by (1)

$$s(f) = \sum_{k} s(g_k), \quad \text{q.e.d.}$$

E.1.2b. Theorem. We shall prove that if $f \in F$, then

$$s(P - f) = s(P) - s(f), \quad (P \text{ is the whole plane}),$$

which can be written as

$$s(\mathrm{co}_P f) = \mathrm{co}_I(s(f)), \qquad (I = \mathbf{H}).$$

Proof. Put

$$g =_{df} \mathrm{co}_P f.$$

We have

$$f \frown g = \Theta, \qquad f \smile g = P.$$

Hence, by [E.1.2.] and [E.1.2a.],

$$s(f) \cdot s(g) = (\vec{O}), \qquad s(f) + s(g) = s(P) = I.$$

Hence $s(g)$ is the space-complement of $s(f)$.

E.1.2c. Theorem. If

$$f, g, h \in F \quad \text{and} \quad h = f \smile g,$$

then

$$s(h) = s(f) + s(g).$$

Proof. We have

$$h = (f - g) \smile (f \frown g) \smile (g - f),$$

[A.2.3b.], where all three terms are disjoint figures. We also have

$$f = (f - g) \smile (f \frown g),$$

[A.2.3b.]; hence, by [E.1.2a.],

$$s(f) = s(f - g) + s(f \frown g).$$

Similarly we get

$$s(g) = s(g - f) + s(f \frown g).$$

Hence

$$\begin{aligned} s(f) + s(g) &= s(f - g) + s(f \frown g) + s(f \frown g) + s(g - f) \\ &= s(f - g) + s(f \frown g) + s(g - f) =, \text{ by [E.1.2a.]}, \\ &= s[(f - g) \smile (f \frown g) \smile (g - f) = s(h), \end{aligned}$$

so the assertion is proved.

E.1.2d. Theorem. It follows that: If $f_1, f_2, \ldots, f_n, h$ are any figures with $\bigcup_k f_k = h$, then $\sum_k s(f_k) = s(h)$.

E.1.2e. Theorem. Applying de Morgan-laws [A.1.5.], we get from [E.1.2d.] and [E.1.2b.], that for any figures f_k, $(k = 1, \ldots, n)$

$$s\left(\bigcap_{k=1}^{n} f_k\right) = \prod_{k=1}^{n} s(f_k).$$

E.1.2f. Thus the assertion stated at the beginning of [E.2.] is proved.

We also have for figures

$$s(f - g) = s(f) - s(g),$$

and for the algebraic addition

$$s(f \dot{+} g) = s(f) \dot{+} s(g).$$

E.1.2g. Theorem. We also have:

If $f \subseteqq g$, then $s(f) \leqq s(g)$.

E.1.3. Theorem. From the above discussions it follows that the collection S of all spaces $s(f)$ (where f are figures), ordered by the $\boldsymbol{L}$-lattice-inclusion ($\leqq$), constitutes a finitely additive geometrical tribe. Indeed, we have, [D I.4.],

1) if $a \in S$, then $\text{co}\, a \in S$,
2) if $a, b \in S$, then $a + b \in S$,
3) all spaces of S are compatible with one another.

Applying [D.5.], we see that if $a \cdot b = (\vec{O})$, then a is orthogonal to b. So S, if ordered by inclusion of spaces, with complementation defined as ortho-complementation, is a geometrical tribe of spaces, [D I.1.2f.]. Its zero is $(\vec{O})$ and its unit the whole space $\mathbf{H}$.

E.1.3.1. Theorem. By [E.1. 2a., b., c., d.] the correspondence

$$f \rightarrowtail s(f)$$

is a homomorphism from (F) onto (S), which, however, may not be an isomorphism.

E.1.3.2. Def. We shall call (S) *tribe of spaces determined by the scale* $s(Q)$.

One can prove that (S) is the smallest geometrical tribe, with ordering taken from $(\boldsymbol{L})$ and containing all $s(Q)$. Its complementation is ortho-complementation. We can also call (S) *scale-tribe of spaces.* We have $S = s(F)$, where (F) is the tribe of figures on the plane P.

E.1.4. Theorem. The collection J of all figures f, for which

$$s(f) = (\vec{O})$$

is a (finitely) additive ideal in the tribe (F), [A.9.].

Proof. Indeed, let $f, g \in J$. We have

$$s(f \cup g) = s(f) + s(g),$$

and

$$s(f) = (\vec{O}), \quad s(g) = (\vec{O}).$$

It follows that

$$s(f \cup g) = (\vec{O}) + (\vec{O}) = (\vec{O});$$

hence

$$f \cup g \in J.$$

Let $h \in J$, and $h' \subseteqq h$, $h' \in F$; then

$$s(h') \leqq s(h),$$

[E.1.2f.]. Hence, as

$$s(h) = (\vec{O}),$$

we get

$$s(h') = (\vec{O}), \quad \text{so} \quad h' \in J.$$

The assertion is proved.

E.1.4.1. The ideal J generates in (F) the notion of equality $(=^J)$, defined by

$$f =^J g \cdot =_{df} f \dotplus g \in J,$$

[A.10.], and the notion of ordering

$$f \leqq^J g \cdot =_{df} \cdot f - g \in J,$$

[A.9.3.]. According to [A.10.] this notion reorganizes (F) into the tribe $(F)_J$, i.e. (F) modulo J. Its zero is any soma belonging to J, and its unit any soma belonging to $\operatorname{co} J = I - J$. The tribe $(F)_J$ is finitely additive, like (F). The operations on $(F)_J$ can be performed in the same way, as in (F) only the signs $(\leqq)$, and $(=)$ shall be replaced by $(\leqq^J)$ and $(=^J)$. The operations of addition, subtraction, multiplication, complementation and algebraic addition, and the relations $(=^J)$ and $(\leqq^J)$ are invariant with respect to the equality $(=^J)$, [A.10.1.], [A.10.2.].

E.1.4.2. Theorem. We have defined the correspondence s which is a homomorphism, and which attaches to every figure f the space $s(f)$. But we have more:

$$\text{if} \quad f \leqq^J g, \quad \text{then} \quad s(f) \leqq s(g);$$

$$\text{if} \quad f =^J g, \quad \text{then} \quad s(f) = s(g);$$

$$\text{if} \quad f =^J f', \quad g =^J g', \quad \text{then} \quad s(f \cup g) = s(f' \cup g').$$

In addition to that we have

$$s(f \cap g) = s(f' \cap g'),$$

$$s(f - g) = s(f' - g'),$$

$$s(f \dotplus g) = s(f' \dotplus g'),$$

$$s(\operatorname{co}_P f) = s(\operatorname{co}_P f').$$

We also have:

$$\text{if} \quad s(f) = s(g), \quad \text{then} \quad f =^J g.$$

Indeed, let $s(f) = s(g)$. It follows $s(f) \dotplus s(g) = O$; hence $s(f \dotplus g) = O$, which gives $f \dotplus g \in J$, i.e. $f =^J g$.

E.1.4.3. Def. The above discussion shows that the correspondence

(1) $$f \smile s(f)$$

may be considered as invariant in the domain, with respect to the equality $(=^J)$ and, in the range, with respect to the equality of spaces, [A.4.1.]. Then the correspondence (1) is an isomorphism; denote it by G. It is an isomorphism from $(F)_J$ onto a tribe (S) of spaces.

E.1.4.4. If we consider (1) as a correspondence invariant in the domain, with respect to identity of figures, we get only a homomorphism.

E.1.4.5. Notice that G may be replaced by the correspondence between $(F)/J$ and (S), i.e. by

$$(\text{equivalence class of } f) \smile s(f).$$

E.1.4.6. If the ideal J coincides with F, all figures are $(=^J)$-equal, and (S) is composed of the spaces $(\vec{O})$ and $\mathbf{H}$ only. If J is composed of the empty set only, then the correspondence $f \smile s(f)$ is an isomorphism between figures and spaces.

E.2. The tribe (F) of figures can be extended to the denumerably additive tribe (F'), whose somata are borelian subsets of the plane P. The tribe (S) of spaces also can be extended (geometrically i.e. with ordering relation taken from the lattice $(\boldsymbol{L})$ of all spaces, [D.2.3.]); even to a saturated, completely additive tribe of spaces, [D I.18.11.], [D I.17.2.], [D I.11.1.]. Now we would like to have the function $s(\dot{f})$ extended from figures f to all borelian subsets of the plane.

E.2.1. We shall do that by introducing a measure on the tribes (F) and (S) in a special way which we are now going to determine. Let (T) be a saturated tribe, [D I.18.], completely additive, which is an extension of (S). We know, [D I.16.2.], that there exists a generating vector $\vec{\omega}$ of $\mathbf{H}$ with respect to (T). The function

$$\mu(\dot{a}) = \|\operatorname{Proj}_a \vec{\omega}\|^2 \tag{1}$$

defined for all $a \in T$, is a denumerably additive and effective measure on T, [D I.20.5.]. Hence $\mu(\dot{a})$ is also an effective measure on (S).

We know that, in general, if (A) is a strictly finitely genuine subtribe of a denumerably additive tribe (B) and if $\nu(\dot{a})$ is a finitely additive measure on (A), the measure cannot be always extended to (B) to become denumerably additive. There exists a necessary and sufficient condition, given by e.g. FRÉCHET, [A I.3.1., II′], for the possibility of doing that. This condition is satisfied for the measure (1) in (S), since there is such a prolongation (T) available; but the said condition is not necessarily satisfied by (F).

Having the above in mind, let us take the measure μ defined in (1). Having that, let us define a measure on (F) by

E.2.1a. Def.

$$\mu(f) =_{df} \mu[s(f)] \quad \text{for all} \quad f \in F.$$

(We use the same letter μ on both sides, because the logical type of the measured notions makes a sufficient discrimination.)

E.2.2. Theorem. We shall prove that $\mu(f)$ is finitely additive on (F).

Proof. Let $f \frown g = \Theta$, $f, g \in F$. We have

$$\mu(f \smile g) = \mu[s(f \smile g)] = \mu[s(f) + s(g)]. \tag{1}$$

Since $f \frown g = \Theta$, therefore

$$s(f \frown g) = s(f) \cdot s(g) = s(\Theta) = (\vec{O}).$$

Hence

$$s(f) \cdot s(g) = (\vec{O}).$$

Since the measure μ is additive on S, we have

$$\mu[s(f) + s(g)] = \mu[s(f)] + \mu[s(g)].$$

Hence

$$\mu[s(f \smile g)] = \mu[s(g)] + \mu[s(g)],$$

and then, [E.2.1.],

$$\mu(f \smile g) = \mu(f) + \mu(g), \quad \text{q.e.d.}$$

E.2.3. Theorem. A necessary and sufficient condition that $\mu(f) = 0$ is $f \in J$ [where J is the ideal composed of all figures f, for which $s(f) = (\vec{O})$].

Proof. Let $\mu(f) = 0$. By definition we have

$$\mu[s(f)] = 0,$$

and since the measure μ is effective on (S), we have

$$s(f) = (\vec{O}),$$

it follows that, [E.1.4.], $f \in J$.

Conversely, let $f \in J$. Then, by definition of J, we have $s(f) = (\vec{O})$. Hence $\mu[s(f)] = 0$, i.e. $\mu(f) = 0$, [E.2.1.]. The theorem is established.

E.2.4. Till now, the correspondence $Q \leadsto s(Q)$ for plane-quarters [hence $f \leadsto s(f)$], satisfying the conditions I, II, III in [E.1.], is rather arbitrary. To go further, we impose on s supplementary restrictions, stated in the following two hypotheses:

IV. If for generalized real numbers α_n, α_0, β_n, β_0, [C.1.1.], we have

$$\alpha_1 \geqq \alpha_2 \geqq \cdots \geqq \alpha_n \geqq \cdots; \quad \lim \alpha_n = \alpha_0,$$

$$\beta_1 \geqq \beta_2 \geqq \cdots \geqq \beta_n \geqq \cdots, \quad \lim \beta_n = \beta_0,$$

then

$$\prod_{n=1}^{\infty} s[Q(\alpha_n, \beta_n)] = s[Q(\alpha_0, \beta_0)].$$

V. If

$$\alpha_1 \leqq \alpha_2 \leqq \cdots \leqq \alpha_n \leqq \cdots, \qquad \lim \alpha_n = +\infty,$$

$$\beta_1 \leqq \beta_2 \leqq \cdots \leqq \beta_n \leqq \cdots, \qquad \lim \beta_n = +\infty,$$

then

$$\sum_{n=1}^{\infty} s[Q(\alpha_n, \beta_n)] = s(P).$$

E.2.5. Since the measure μ on (S) is denumerably additive, we have, under the conditions IV and V:

E.3.

$$\lim_{n \to \infty} \mu\{s[Q(\alpha_n, \beta_n)]\} = \mu\{s[Q(\alpha_0, \beta_0)]\},$$

and respectively

$$\lim_{n \to \infty} \mu\{s[Q(\alpha, \beta)]\} = \mu\{s(P)\}.$$

Hence by the definition of measure on (F),

$$(1) \quad \begin{cases} \lim\limits_{n \to \infty} \mu[Q(\alpha_n, \beta_n)] = \mu[Q(\alpha_n, \beta_n)] \\ \text{and respectively} \\ \lim\limits_{n \to \infty} \mu[Q(\alpha_N, \beta_n)] = \mu(P). \end{cases}$$

Now, by [C.14.1.] the relations (1) constitute a necessary and sufficient condition for the possibility of extension the measure μ to a denumerably additive measure on the tribe (F') of all borelian subsets of the plane P.

E.3.1. Till now we have extended the measure to the tribe (F'), but the correspondence $f \smile s(f)$, still waits for the extension. Thus we must define $s(E)$ for all borelian subsets of the plane P.

E.3.2. To perform the said extension of the correspondence $f \frown s(f)$ let us recall the main results of the discussion in Chapter [A I.]. There we have studied [A I., Sections 6, 7, 8, 9, 10] the problem of extension of a finitely additive tribe (A) provided with a finitely additive measure μ, to a supertribe (B) with an extended measure μ' which would be denumerably additive, and where (B) would be a subtribe of a given supertribe (C) of (A).

We have studied in this respect three different devices, denoted $\boldsymbol{L}, \boldsymbol{C}, \boldsymbol{N}$, in [A I., Sect. 6], [A I., Sect. 7] and [A I., Sect. 9] respectively.

The device $\boldsymbol{L}$ is similar to the classical Lebesgue's one, and consists of applying coverings of somata of (C) by somata of (A), thus defining the μ-exterior measure μ'_e of the somata of (C).

The device $\boldsymbol{C}$ is the Carathéodory's device, which uses μ'_e as the Carathéodory's convex measure of somata of (C).

The device $\boldsymbol{N}$ uses a kind of approximation of somata of (C) by somata of (A).

We have compared the devices $\boldsymbol{L}$ and $\boldsymbol{C}$ in [A I., Sect. 8] and proved that both devices give the same extension (B) of (A), and also yield a same measure μ' on (B) which, however, may be not an extension of μ, because if $a \in A$ we may have $\mu(a) \neq \mu'(a)$; really $\mu'(a) \leqq \mu(a)$.

E.3.2.1. We have also compared all three devices and proved that, under certain condition, [A I.9.14.], called $\mathcal{B}$, all three devices yield the same extension (B) of (A) and the same measure μ' on (B), which is a true extension of the measure μ, given on (A). This condition is even necessary for a true extension of the measure μ in (A) to become a denumerably additive measure on (B).

E.3.3. We shall apply the forgoing remarks to the problem of extension the correspondence $f \smile s(f)$.

We are in the circumstances of [E.3.2.]. Indeed we have the tribe (F) of figures with a given finitely additive measure μ in it. The tribe (F) is plunged in the denumerably additive tribe (F^0) of all subsets of the plane P (with inclusion of sets as ordering relation). The measure μ is supposed to satisfy the conditions [E.3.], which are equivalent to the condition N in [C.14.], and which in turn is equivalent to the conditions [A I., Sect. 2], hence to the condition $\mathcal{B}$, mentioned above. Thus if we admit the conditions IV, V for spaces, we have also the condition [E.3.] for sets and measure, so we can apply any one of the devices $\boldsymbol{L}$. $\boldsymbol{C}$, $\boldsymbol{N}$ for extending the measure.

The tribe extended by any one of the above devices will be termed (F^L) and the extended measure μ'.

E.3.4. We know that, starting with (F), we can produce its borelian extension in (F^0), [A I., Sect. 4], which will be denoted by (F^B), and which is the smallest supertribe of (F), containing, with every its soma a, its complement co$a$, and with every infinite sequence $a_1, a_2, \ldots, a_n, \ldots$ of its somata also their union $\sum_n a_n$ and their intersection $\prod_n a_n$. The tribe (F^B) is denumerably additive and it is a denumerably genuine strict subtribe of (F^L). In addition to that (F^L) is identical with the μ'-measure hull of (F^B), [A I., Sect. 5]; and the measure, on the tribes (F), (F'), (F^L), of their common somata, is the same.

The somata of all these tribes will be termed *μ-measurable*, and instead of μ' we shall be allowed to write simplier μ. We have $F \subseteqq F^B \subseteqq F^L$.

E.3.5. Having that we can apply for somata of (F^L) [hence also for somata of (F^B)], the Theorem [A I.8.10.] which states, that if $a \in (F^L)$, then there exists an infinite double sequence $\{f_{ik}\}$ of somata of (F), such that

(1)
$$\mu\left(a \dotplus \bigcap_{i=1}^{\infty} \bigcup_{k=1}^{\infty} f_{ik}\right) = 0.$$

If we use the ideal J composed of all sets b of (F^L) with $\mu(b) = 0$, we can write (1) in the form

(2)
$$a =^J \bigcap_{i=1}^{\infty} \bigcup_{k=1}^{\infty} f_{ik}.$$

E.3.6. This all give us the suggestion that, perhaps, we can produce the prolongation aimed at, by putting for spaces:

(3)
$$s(a) =_{df} \prod_{i=1}^{\infty} \sum_{k=1}^{\infty} s(f_{ik}).$$

But, if we like to follow this way, we must prove that $s(a)$ does not depend on the choice of the sequence $\{f_{ik}\}$ giving the representation (2) of a.

We shall really follow this indicated way, but for technical reason, we shall replace the figures f_{ik} by sets

(4)
$$f_{ik} + p_{ik} - q_{ik},$$

where

$$\mu(p_{ik}) = \mu(q_{ik}) = 0.$$

We shall call (4) J-figures.

E.3.7. We shall start with getting some useful properties of the "J-figures".

Def. By a *J-figure we* shall understand every set E of points of the plane P, such that there exists a figure $f \in F$ with

$$E =^J f.$$

This is equivalent to the following: there exists f and the sets p, q of μ-measure 0, such that

$$E = (f \cup p) - q.$$

E.3.7a. Every J-figure is a μ-measurable set.

Let us study the properties of J-figures.

If α is an J-figure, then $\operatorname{co}\alpha = P - \alpha$ is also an J-figure. Indeed, if $\alpha \dotplus f \in J$, we have

$$\operatorname{co}\alpha \dotplus \operatorname{co} f = \alpha \dotplus f \in J,$$

so $\operatorname{co}\alpha$ is an J-figure, because $\operatorname{co} f$ is a figure.

If α, β are J-figures, so is $\alpha \cup \beta$, and $\alpha \cap \beta$. Indeed, let $\alpha \dotplus f \in J$, $\beta \dotplus g \in J$, where f, g are figures. We have

$$\alpha =^J f, \quad \beta =^J g;$$

hence

$$(\alpha \cup b\rangle =^{\bar{J}} (f \cup g);$$

hence

$$(\alpha \cup \beta) \dot{+} (f \cup g) \in \bar{J}.$$

Since $f \cup g$ is a figure, $\alpha \cup \beta$ is a $\bar{J}$-figure. We also have

$$\alpha \cap \beta =^{\bar{J}} f \cap g;$$

hence

$$(\alpha \cap \beta) \dot{+} (f \cap g) \in \bar{J}.$$

Since $f \cap g$ is a figure, $\alpha \cap \beta$ is a $\bar{J}$-figure. We also see that if α, β are $\bar{J}$-figures, so are $\alpha \dot{+} \beta$, and $\alpha - \beta$.

If f is a figure, then f is also a $\bar{J}$-figure. Every set of μ-measure 0 is a $\bar{J}$-figure.

The above discussion shows that the collection F_1 of all $\bar{J}$-figures, if ordered by set-inclusion, makes up a finitely additive tribe (F_1). (F_1) has $\emptyset$ as zero and P as unit. $\bar{J}$ is an ideal not only in (F), but also in (F_1).

E.4. Def. Let us make correspond to the $\bar{J}$-figure α the space, denoted by $\bar{s}(\alpha)$ and defined by

(1) $$\bar{s}(\alpha) =_{df} s(f),$$

where $\alpha \dot{+} f \in \bar{J}$.

This space $\bar{s}(\alpha)$ does not depend on the choice of f for which $\alpha \dot{+} f \in \bar{J}$, if α is fixed.

Indeed, if $\alpha \dot{+} f \in \bar{J}$ and $\alpha \dot{+} g \in \bar{J}$, then $f =^J g$, so the definition (1) is not ambiguous.

The correspondence $\alpha \leadsto \bar{s}(\alpha)$ is an extension of the correspondence $f \leadsto s(f)$ to $\bar{J}$-figures.

If $\alpha =^{\bar{J}} \beta$, then $\bar{s}(\alpha) = \bar{s}(\beta)$.

The correspondence $\alpha \leadsto \bar{s}(\alpha)$ is a homomorphism.

E.4.1. Lemma. If f_n, $(n = 1, 2, \ldots)$ is a sequence of $\bar{J}$-figures, satisfying the Cauchy-condition: "for every $\varepsilon > 0$ there exists N such that for every $n \geqq N$, $m \geqq N$ we have

$$\mu(f_n \dot{+} f_m) < \varepsilon,$$

then the sequence $\bar{s}(f_n)$ also satisfies a similar Cauchy-condition

$$\mu[\bar{s}(f_n) \dot{+} \bar{s}(f_m)] < \varepsilon \quad \text{for} \quad n, m \geqq N.$$

Proof. The correspondence $f \leadsto \bar{s}(f)$ for $\bar{J}$-figures is a homomorphism which preserves the measure.

Hence if $\mu(f_n \dot{+} f_m) < \varepsilon$, then, since $f_n \dot{+} f_m$ is also a figure, we have

$$\mu\, \bar{s}(f_n \dot{+} f_m) < \varepsilon,$$

hence

$$\mu[\bar{s}(f_n) \dot{+} \bar{s}(f_m)] < \varepsilon,$$

which completes the proof.

E.4.2. We know from [A I.11.2.], that if the sequence $f_{\dot{n}}$ of $\bar{J}$-figures satisfies the Cauchy-condition, then there exists a μ-measurable subset X of P, such that

$$\mu(X \dot{+} f_{\dot{n}}) \to 0,$$

and this set X is $(=^{\bar{J}})$-unique, i.e. if $\mu(X \dot{+} f_n) \to 0$ and $\mu(Y \dot{+} f_n) \to 0$, then $X =^{\bar{J}} Y$.

E.4.2a. If there exists a μ-measurable set X such that

$$\mu(X \dot{+} f_n) \to 0,$$

then $f_{\dot{n}}$ satisfies the Cauchy-condition.

Proof. Let $\varepsilon > 0$. For sufficiently great indices n, m we have

$$\mu(X \dot{+} f_n) \leqq \varepsilon, \quad \mu(X \dot{+} f_m) \leqq \varepsilon.$$

Hence

$$\mu(X \dot{+} f_n) + \mu(X \dot{+} f_m) \leqq 2\varepsilon;$$

hence

$$\mu[(X \dot{+} f_n) \dot{+} (X \dot{+} f_m)] \leqq 2\varepsilon.$$

Hence, since $X \dot{+} X = \emptyset$, we get

$$\mu(f_n \dot{+} f_m) \leqq 2\varepsilon$$

for sufficiently great n, m. This completes the proof.

E.4.3. Lemma. If

1. f_n, g_n, $(n = 1, 2, \ldots)$ are $\bar{J}$-figures,
2. F is a μ-measurable set,
3. $\lim_{n\to\infty} \mu(F \dot{+} f_n) = 0$,
4. $\lim_{n\to\infty} \mu(F \dot{+} g_n) = 0$,
5. Φ, Ψ are spaces determined by the condition

$$\mu(\Phi \dot{+} \bar{s}(f_n)] \to 0, \quad \mu[\Psi \dot{+} \bar{s}(g_n)] \to 0,$$

then

$$\Phi =^{\bar{J}} \Psi.$$

Proof. Choose $\varepsilon > 0$. From hyp. 3 and 4 it follows that, for sufficiently great n, we have

$$\mu(F \dot{+} f_n) \leqq \varepsilon, \quad \mu(F \dot{+} g_n) \leqq \varepsilon.$$

It follows

$$\mu[(F \dot{+} f_n) \dot{+} (F \dot{+} g_n)] \leqq \mu(F \dot{+} f_n) + \mu(F \dot{+} g_n) \leqq 2\varepsilon.$$

Since $F \dot{+} F = \emptyset$, we get

$$\mu(f_n + g_n) \leqq 2\varepsilon$$

for sufficiently great n. Since the correspondence $f \smile \bar{s}(f)$ is a homomorphism, preserving measure, and since $f_n \dot{+} g_n$ is a $\bar{J}$-figure, it follows

$$\mu[s(f_n \dot{+} g_n)] \leqq 2\varepsilon;$$

hence

(1) $$\mu[\bar{s}(f_n) \dot{+} \bar{s}(g_n)] \leqq \varepsilon.$$

Since $f_{\dot{n}}$ and $g_{\dot{n}}$ satisfy the Cauchy-condition, therefore so do $s(f_{\dot{n}})$, $s(g_{\dot{n}})$ by [E.4.1.]. Hence, by [E.4.2.], there exist the spaces Φ_1, Ψ_1, satisfying the condition

$$\mu[\Phi_1 + s(f_n)] \to 0, \qquad \mu[\Psi_1 + s(g_n)] \to 0.$$

Hence by the $\bar{J}$-uniqueness, [E.4.2.] and hyp. 5

$$\Phi_1 =^{\bar{J}} \Phi, \qquad \Psi_1 =^{\bar{J}} \Psi.$$

It follows that for sufficiently great indices we have

(2) $$\mu[\Phi + \bar{s}(f_n)] \leqq \varepsilon, \qquad \mu[\Phi + \bar{s}(g_n)] \leqq \varepsilon.$$

From (1) and (2) follows that

$$\mu[\Phi \dot{+} \bar{s}(f_n)] + \mu[\Psi \dot{+} \bar{s}(g_n)] + \mu[\bar{s}(f_n) \dot{+} \bar{s}(g_n)] \leqq 3\varepsilon$$

for sufficiently great indices n.

Since the measure μ for spaces is additive, we get

$$\mu[\Phi \dot{+} \bar{s}(f_n) \dot{+} \Psi + \bar{s}(g_n) \dot{+} \bar{s}(f_n) \dot{+} \bar{s}(g_n)] \leqq 3\varepsilon;$$

hence

$$\mu(\Phi + \Psi) \leqq 3\varepsilon;$$

this for every $\varepsilon > 0$. Since the measure μ for spaces is effective, we get

$$\Phi \dot{+} \Psi = (\vec{O}), \quad \text{hence} \quad (\Phi - \Psi) + (\Psi - \Phi) = (\vec{O}),$$

which implies

$$\Phi - \Psi = (\vec{O}), \qquad \Psi - \Phi = (\vec{O}).$$

Hence $\Phi \leqq \Psi$ and $\Psi \leqq \Phi$; consequently

$$\Phi = \Psi, \quad \text{q.e.d.}$$

E.4.4. Lemma. If

1. f_n, g_n are $\bar{J}$-figures,
2. $F = f_1 \cup f_2 \cup \cdots \cup f_n \cup \cdots$

   $F = g_1 \cup g_2 \cup \cdots \cup g_n \cup \cdots$
3. $\Phi =_{df} \bar{s}(f_1) + \bar{s}(f_2) + \cdots + \bar{s}(f_n) + \cdots$

   $\Psi =_{df} \bar{s}(g_1) + \bar{s}(g_2) + \cdots + \bar{s}(g_n) + \cdots,$

then $\Phi = \Psi$.

Proof. Put

$$f'_n =_{df} f_1 \cup f_2 \cup \cdots \cup f_n, \quad (n \geqq 1),$$

$$g'_n =_{df} g_1 \cup g_2 \cup \cdots \cup g_n.$$

We have

$$f'_1 \subseteqq f'_2 \subseteqq \cdots \subseteqq f'_n \subseteqq \cdots$$

$$g'_1 \subseteqq g'_2 \subseteqq \cdots \subseteqq g'_n \subseteqq \cdots$$

and

$$\mu(F - f'_n) \to 0, \quad \mu(F - g'_n) \to 0.$$

The sets f'_n, g'_n are $\bar{J}$-figures, and F a subset of P. It follows that

$$\mu(F \dotplus f'_n) \to 0, \quad \mu(F \dotplus g'_n) \to 0,$$

and also

$$\mu[\Phi \dotplus \bar{s}(f'_n)] \to 0, \quad \mu[\Psi \dotplus \bar{s}(g'_n)] \to 0.$$

Since we are in the conditions of Lemma [E.4.3.], the thesis follows.

E.4.5. Let us say that a μ-measurable subset E of P *is of the type* $\sum$, (compare [D I.10.]), whenever it can be represented as the denumerable union of $\bar{J}$-figures:

$$E = \bigcup_n f_n.$$

Similarly we say that E *is of the type* $\prod$ whenever it can be represented as the intersection of a denumerable number of $\bar{J}$-figures:

$$E = \bigcap_n f_n.$$

The Lemmas [E.4.4.], [E.4.5.] give us the suggestion, that to a $\sum$-set $E = \bigcup_n f_n$, we should attach the space $\sum_n \bar{s}(f_n)$.

The space attached in the above way to a $\sum$-set does not depend on the way of sum-representation of the set.

E.4.6. Lemma. If

1. F, G are $\sum$-sets in P,

2. $F \subseteqq^{\bar{J}} G$,

3. Φ, Ψ are spaces attached to F, G respectively by means of their $\sum$-representation,

then

$$\Phi \leqq \Psi.$$

Proof. Let $F = \bigcup_n f_n$, $G = \bigcup_n g_n$, where f_n, g_n are $\bar{J}$-figures, yield $\sum$-representations of F and G.

Since $F \subseteqq G$, we have $F + G = G$. Hence

$$\bigcup_n f_n + \bigcup_n g_n = \bigcup_n g_n,$$

i.e.

$$\bigcup_n (f_n + g_n) = \bigcup_n g_n = G.$$

Now since $f_n + g_n$ is a $\bar{J}$-figure, we have with $\bigcup (f_n + g_n)$, $\bigcup g_n$ two $\sum$-representations of G. Their space images must be equal:

$$\bar{s}[\bigcup (f_n + g_n)] = \bar{s}(g_n).$$

By denumerable additivity of (S) we have

$$\bar{s}(\bigcup f_n) + \bar{s}(\bigcup g_n) = s(\bigcup g_n);$$

hence

$$\Phi + \Psi = \Psi,$$

which gives

$$\Phi \leqq \Psi.$$

The lemma is proved.

E.4.7. Now we shall consider the sets having the type $\Pi \sum$ i.e. the sets having the form

$$\bigcap_{n=1}^{\infty} \bigcup_{m=1}^{\infty} f_{n,m},$$

where $f_{n,m}$ are $\bar{J}$-figures.

We shall prove the Lemma:

Lemma. If

1. F_n, G_n are $\sum$-sets, $(n = 1, 2, \ldots)$,
2. $F_1 \supseteqq^{\bar{J}} F_2 \supseteqq^{\bar{J}} \cdots \supseteqq^{\bar{J}} F_n \supseteqq^{\bar{J}} \cdots$
   $G_1 \supseteqq^{\bar{J}} G_2 \supseteqq^{\bar{J}} \cdots \supseteqq^{\bar{J}} G_n \supseteqq^{\bar{J}} \cdots$
3. $G_n \subseteqq^{\bar{J}} F_n$ for $n = 1, 2, \ldots$
4. $\Gamma = \bigcap_{n=1}^{\infty} F_n$, $\Gamma = \bigcap_{n=1}^{\infty} G_n$,
5. $\Phi =_{df} \prod_{n=1}^{\infty} \bar{s}(F_n)$, $\Psi =_{df} \prod_{n=1}^{\infty} \bar{s}(G_n)$ (where $\bar{s}(F_n)$ and $\bar{s}(G_n)$ are spaces attached to F_n, G_n considered as $\sum$-sets),

then

$$\Phi = \Psi.$$

Proof. We have for spaces:

$$\Phi \dot{+} \Psi = [\Phi \dot{+} s(F_n)] \dot{+} [\Psi \dot{+} s(G_n)] \dot{+} [s(F_n) \dot{+} s(G_n)].$$

Hence:

$$\mu[\Phi \dot{+} \Psi] \leqq \mu[\Phi \dot{+} \bar{s}(F_n)] + \mu[\Psi \dot{+} \bar{s}(G_n)] + \mu[\bar{s}(F_n) \dot{+} \bar{s}(G_n)]. \tag{1}$$

We shall prove that each of the three terms on the right is $\leqq \varepsilon$ for a given $\varepsilon > 0$ and for sufficiently great indices n. By Lemma [E.4.8.] we have

$$\begin{aligned} &\bar{s}(F_1) \supseteqq \bar{s}(F_2) \supseteqq \cdots \supseteqq \bar{s}(F_n) \supseteqq \cdots \\ &\bar{s}(G_1) \supseteqq \bar{s}(G_2) \supseteqq \cdots \supseteqq \bar{s}(G_n) \supseteqq \cdots \end{aligned} \tag{2}$$

and by hyp. 5,

$$(3)\qquad \begin{aligned}&\mu[\bar{s}(F_n) - \Phi] \to 0,\\ &\mu[\bar{s}(G_n) - \Psi] \to 0,\end{aligned}$$

for $n \to \infty$.

Since from 5. and (2)

$$\Phi \leqq \bar{s}(F_n),$$

$$\Psi \leqq \bar{s}(G_n) \quad \text{for all } n,$$

therefore

$$\mu[\Phi - \bar{s}(F_n)] = 0, \qquad \mu[\Psi - \bar{s}(G_n)] = 0.$$

So (3) implies

$$\mu[\bar{s}(F_n) \dotplus \Phi] < \varepsilon,$$

$$\mu[\bar{s}(G_n) \dotplus \Psi] < \varepsilon$$

for sufficiently great indices n.

Hence from (1) it follows

$$(4)\qquad \mu(\Phi \dotplus \Psi) \leqq 2\varepsilon + \mu[\bar{s}(F_n) \dotplus \bar{s}(G_n)].$$

We have $G_n \subseteqq F_n$. Hence, we have

$$(5)\qquad \bar{s}(G_n) \leqq \bar{s}(F_n).$$

From hyp. 4, we have

$$(6)\qquad \begin{aligned}&\mu(F_n - \Gamma) \to 0,\\ &\mu(G_n - \Gamma) \to 0.\end{aligned}$$

Since $\Gamma - F_n = \mathit{O}$, $\Gamma - G_n = \mathit{O}$, (6) can be written

$$\mu(F_n \dotplus \Gamma) \to 0,$$

$$\mu(G_n \dotplus \Gamma) \to 0;$$

hence for sufficiently great indices we have:

$$(7)\qquad \mu(F_n \dotplus \Gamma) \leqq \varepsilon, \qquad \mu(G_n \dotplus \Gamma) \leqq \varepsilon.$$

Now we have

$$F_n \dotplus G_n = F_n \dotplus \Gamma \dotplus G_n \dotplus \Gamma = (F_n \dotplus \Gamma) + (G_n \dotplus \Gamma),$$

which gives

$$\mu(F_n \dotplus G_n) = \mu(F_n \dotplus \Gamma) + \mu(G_n \dotplus \Gamma);$$

hence, by (7)

$$(8)\qquad \mu(F_n \dotplus G_n) \leqq 2\varepsilon.$$

By hyp. 3, we have

$$G_n \subseteqq F_n, \qquad (n = 1, 2, \ldots);$$

hence

$$(9)\qquad G_n \dotplus F_n = F_n - G_n,$$

so (8) can be written:

$$\mu(F_n - G_n) \leqq \varepsilon.$$

Since $G_n \subseteqq^{J} F_n$ (by hyp. 3), we have

(10) $$\mu(F_n - G_n) = \mu(F_n) - \mu(G_n) \leqq \varepsilon.$$

Now we know that

$$\mu(F_n) = \mu(\bar{s}(F_n)),$$

$$\mu(G_n) = \mu(\bar{s}(G_n)),$$

so (10) implies

(11) $$\mu(\bar{s}(F_n)) - \mu(\bar{s}(G_n)) \leqq \varepsilon.$$

Now as $F_n \geqq f_n$, we have, by [E.4.8.],

(12) $$\bar{s}(F_n) \geqq \bar{s}(G_n).$$

Since $G_n - F_n =^{J} \emptyset$, we have

$$\bar{s}(G_n) - \bar{s}(F_n) = 0;$$

hence by (11) we get

$$\mu[\bar{s}(F_n) \dotplus \bar{s}(G_n)] \leqq \varepsilon$$

for sufficiently great n. Carrying this into (4) we get

$$\mu(\Phi + \Psi) \leqq 2\varepsilon + \varepsilon.$$

This being true for all $\varepsilon > 0$, it follows that

$$\mu(\Phi \dotplus \Psi) = 0.$$

Since the measure μ in (S) is effective, we get

$$\Phi \dotplus \Psi = (\vec{O}),$$

and we deduce from it:

$$\Phi = \Psi. \quad \text{q.e.d.}$$

E.4.8. If

1. F_n, G_n are $\sum$-sets on the plane P,

2: $\Gamma = \bigcap\limits_{n=1}^{\infty} F_n = \bigcap\limits_{n=1}^{\infty} G_n$,

3. $\Phi = \prod\limits_{n=1}^{\infty} \bar{s}(F_n)$, $\Psi = \prod\limits_{n=1}^{\infty} \bar{s}(G_n)$,

then

$$\Phi = \Psi.$$

Proof. Put

$$F'_n =_{df} F_1 \frown F_2 \frown \cdots \frown F_n,$$

$$G'_n =_{df} G_1 \frown G_2 \frown \cdots \frown G_n;$$

then we have

$$F'_1 \supseteqq F'_2 \supseteqq \cdots \supseteqq F'_n \supseteqq \cdots,$$

$$G'_1 \supseteqq G'_2 \supseteqq \cdots \supseteqq G'_n \supseteqq \cdots$$

The sets F'_n, G'_n are of the type $\sum$, so we have $\bar{s}(F'_n)$, $\bar{s}(G'_n)$ already defined.

Put

$$H_n =_{df} F'_n \cap G'_n, \quad (n = 1, 2, \ldots).$$

Then we have

$$H_1 \supseteqq H_2 \supseteqq \cdots \supseteqq H_n \supseteqq \cdots$$

and

$$H_n \subseteqq F'_n, \quad H_n \subseteqq G'_n.$$

We have

$$\Gamma = \bigcap_{n=1}^{\infty} F'_n = \bigcap_{n=1}^{\infty} H_n = \bigcap_{n=1}^{\infty} G'_n.$$

Put

$$\Phi =_{df} \prod_{n=1}^{\infty} \bar{s}(F'_n), \quad X =_{df} \prod_{n=1}^{\infty} \bar{s}(H_n), \quad \Psi =_{df} \prod_{n=1}^{\infty} \bar{s}(G'_n).$$

We are in the condition of the preceding lemma, with Φ and X, so we get

$$\Phi = X,$$

and also with Ψ and X, so we get

$$\Psi = X.$$

It follows $\Phi = \Psi$, q.e.d.

E.4.8.1. The above lemma shows, that we can attach to every set of the type

(1) $$\bigcap_{n=1} \bigcup_{m=1} f_{n,m},$$

where $f_{n,m}$ are $\bar{s}$-figures, a definite space

(2) $$\prod_{n=1}^{\infty} \sum_{m=1}^{\infty} s(f_{n,m}).$$

Since if a set has two representations (1), the space (2) will be the same.

E.4.8.2. The tribe (F^L) is the Lebesguean extension of the tribe (F_1) of $\bar{J}$-figures:

$$(F_1)^L = (F)^L.$$

Hence we can apply the Theorem [A I.8.10.], getting the representation

(1) $$a =^{\bar{J}} \bigcap_{i=1}^{\infty} \bigcup_{k=1}^{\infty} f'_{ik}$$

of any set $a \in F^L$, where $\bar{J}$ is the ideal composed of all sets of (F^0), having the measure $= 0$. Now if we define

$$(2) \qquad \bar{s}(a) =_{df} \prod_{i=1}^{\infty} \sum_{k=1}^{\infty} \bar{s}(f'_{ik}),$$

we get a space valued function $\bar{s}(\dot{a})$, which is well defined on all μ-measurable sets a.

We have for every figure f:

$$\bar{s}(f) = s(f).$$

Indeed the ordinary figure f is also a $\bar{J}$-figure and we have the representation (1), if we put $f'_{ik} = f$ for all i and k. Thus (2) yields an extension of the correspondence s to all μ-measurable sets, hence, a fortiori, to all borelian subsets of the plane P.

E.4.8.3. If in (1), instead of $\bar{J}$-figures, we use ordinary figures, the analogous definition (2) will yield the same correspondence $\bar{s}$. Indeed, let:

$$a =^{\bar{J}} \bigcap_{i=1}^{\infty} \bigcup_{k=1}^{\infty} f'_{ik}.$$

Since f'_{ik} is an $\bar{J}$-figure, there exists an ordinary figure f_{ik}, such that

$$f_{ik} =^{\bar{J}} f'_{ik}.$$

Since the ideal $\bar{J}$ is denumerably additive, we get

$$\bigcap_{i=1}^{\infty} \bigcup_{k=1}^{\infty} f'_{ik} =^{\bar{J}} \bigcap_{i=1}^{\infty} \bigcup_{k=1}^{\infty} f_{ik}.$$

Since the representation (2) does not depend on the choice of the representation (1) of a, we get

$$(3) \qquad \bar{s}(a) = \prod_{i=1}^{\infty} \sum_{k=1}^{\infty} s(f_{ik}).$$

By (1), using ordinary figures instead of $\bar{J}$-figures, we obtain the representation of any set belonging to F^L, hence of any borelian subset of the plane.

Thus by defining the extension $\bar{s}$ of s we can confine ourselves to ordinary figures.

Now, we are going to prove that the correspondence $\bar{s}$ preserves somatic operations.

E.4.8.4. Theorem. If a, b are measurable point-sets, then

$$\bar{s}(a \cup b) = \bar{s}(a) + \bar{s}(b).$$

Proof. We shall rely on the following identity, valid for somata of any denumerably additive tribe:

$$\prod_n a_n + \prod_m b_m = \prod_{n,\,m} (a_n + b_m). \tag{1}$$

If h_α are figures, then $\bigcup_{\alpha=1}^{\infty} h_\alpha$ is measurable; hence it belongs to the denumerably additive tribe (F^L).

Hence, if $f_{n\alpha}$, $g_{m\beta}$ are figures, we have, from (1),

$$\bigcap_n \bigcup_\alpha f_{n\alpha} \smile \bigcap_m \bigcup_\beta g_{m\beta} = \bigcap_{n,\,m} [\bigcup_\alpha f_{n\alpha} \smile \bigcup_\beta g_{m\beta}]. \tag{2}$$

Now, the spaces $s(f_{n\alpha})$, $s(g_{m\beta})$ are somata of a geometrical denumerably additive tribe (S) of spaces. Hence so are

$$\bigcup_\alpha s(f_{n\alpha}),\quad \bigcup_\beta s(g_{m\beta}).$$

Hence, applying the general rule (1), we get

$$\prod_n \sum_\alpha s(f_{n\alpha}) + \prod_m \sum_\beta s(g_{m\beta}) = \prod_{n,\,m} [\sum_\alpha s(f_{n\alpha}) + \sum_\beta s(g_{m\beta})]. \tag{3}$$

This is valid for any figures $f_{n\alpha}$, $g_{m\beta}$.

E.4.8.4a. Having that, let

$$a =^{\bar{J}} \bigcap_m \bigcup_\alpha f_{n\alpha},\qquad b =^{\bar{J}} \bigcap_m \bigcup_\beta g_{m\beta}. \tag{4}$$

We have

$$\bar{s}(a) = \prod_n \sum_\alpha s(f_{n\alpha}),$$
$$\bar{s}(b) = \prod_m \sum_\beta s(g_{m\beta}), \tag{5}$$

we get

$$\bar{s}(a) + \bar{s}(b) = \prod_n \sum_\alpha s(f_{n\alpha}) + \prod_m \sum_\beta s(g_{m\beta}). \tag{6}$$

Now

$$a \smile b =^{\bar{J}} \bigcap_n \bigcup_\alpha f_{n\alpha} \smile \bigcap_m \bigcup_\beta g_{m\beta} =^{\bar{J}},\ \text{(by (2))},\ =^{\bar{J}} \bigcap_{n,\,m} [\bigcup_\alpha f_{n\alpha} \smile \bigcup_\beta g_{m\beta}].$$

Since this is a $\prod\sum$-representation of $a \smile b$, we have

$$\bar{s}(a \smile b) = \prod_{n,\,m} [\sum_\alpha s(f_{n\alpha}) + \sum_\beta s(g_{m\beta})]$$

which, by (3), equals

$$\prod_n \sum_\alpha s(f_{n\alpha}) + \prod_m \sum_\beta s(g_{m\beta}) =,\ \text{(by (5))},\ = \bar{s}(a) + \bar{s}(b),$$

so the theorem is proved.

E.4.8.5. Theorem. If a, b are measurable sets, then

$$\bar{s}(a \frown b) = \bar{s}(a) \cdot \bar{s}(b).$$

Proof. We start with the following general formula, valid for somata of any denumerably additive tribe:

$$\prod_{n}^{\infty} a_n \cdot \prod_{m}^{\infty} b_m = \prod_{p=1}^{\infty} (a_p \cdot b_p).$$

Applying this to figures, we get

$$(1) \quad \bigcap_{n=1}^{\infty} \bigcup_{\alpha=1}^{\infty} f_{n\alpha} \frown \bigcap_{m=1}^{\infty} \bigcup_{\beta=1}^{\infty} g_{m\beta} = \bigcap_{p=1}^{\infty} \left(\bigcup_{\alpha=1}^{\infty} f_{p\alpha} \frown \bigcup_{\beta=1}^{\infty} g_{p\beta} \right) = \bigcap_{p=1}^{\infty} \bigcup_{\alpha,\beta}^{\infty} (f_{p\alpha} \frown g_{p\beta}).$$

A similar formula is valid for spaces

$$(2) \quad \prod_{n=1}^{\infty} \sum_{\alpha=1}^{\infty} s(f_{n\alpha}) \cdot \prod_{m=1}^{\infty} \sum_{\beta=1}^{\infty} s(g_{m\beta}) = \prod_{p=1}^{\infty} \sum_{\alpha,\beta}^{\infty} s(f_{p\alpha}) \cdot s(g_{p\beta})$$

$$= \prod_{p=1}^{\infty} \sum_{\alpha,\beta}^{\infty} s(f_{p\alpha} \frown g_{p\beta}).$$

Let

$$a =^{\bar{J}} \bigcap_{n=1}^{\infty} \bigcup_{\alpha=1}^{\infty} f_{n\alpha},$$

$$b =^{\bar{J}} \bigcap_{m=1}^{\infty} \bigcup_{\beta=1}^{\infty} g_{m\beta}.$$

We have

$$\bar{s}(a) \cdot \bar{s}(b) = \prod_{n=1}^{\infty} \sum_{\alpha=1}^{\infty} s(f_{n\alpha}) \cdot \prod_{m=1}^{\infty} \sum_{\beta=1}^{\infty} s(g_{m\beta}) =, \text{ (by (2))},$$

$$= \prod_{p=1}^{\infty} \sum_{\alpha,\beta}^{\infty} s(f_{p\alpha} \frown g_{p\beta}),$$

which equals

$$(3) \quad \bar{s}\left\{ \bigcap_{p=1}^{\infty} \alpha, \beta \bigcup_{\alpha,\beta}^{\infty} (f_{p\alpha} \frown g_{p\beta}), \right.$$

and (3) equals, by (1),

$$\bar{s}\left\{ \bigcap_{n=1}^{\infty} \bigcup_{\alpha=1}^{\infty} f_{n\alpha} \frown \bigcap_{m=1}^{\infty} \bigcup_{\beta=1}^{\infty} g_{m\beta} \right\} = \bar{s}(a \frown b).$$

The theorem is established.

E.4.8.6. Theorem. We have $\bar{s}(\emptyset) = (\vec{O}) = O$ and $\bar{s}(P) = 1$.

E.4.8.7. Theorem. We have for any measurable set a the equality

$$\bar{s}(\mathrm{co}_P\, a) = \mathrm{co}_1 \bar{s}(a).$$

Proof. co$a$ belongs to F^L; hence co$a$ has a $\Pi\Sigma$-representation; hence it gives the space $\bar{s}(\mathrm{co}\,a)$. We have:

$$a \cup \mathrm{co}\,a = P,$$
$$a \cap \mathrm{co}\,a = \emptyset.$$

Hence, by [E.4.8.4.], [E.4.8.5.], and [E.4.8.6.],

$$\text{(1)} \quad \begin{aligned} \bar{s}(a) + \bar{s}(\mathrm{co}\,a) &= \bar{s}(P) = 1, \\ \bar{s}(a) \cdot \bar{s}(\mathrm{co}\,a) &= \bar{s}(\emptyset) = (\vec{O}). \end{aligned}$$

On the other hand we have

$$\text{(2)} \quad \begin{aligned} \bar{s}(a) + \mathrm{co}\,\bar{s}(a) &= 1, \\ \bar{s}(a) \cdot \mathrm{co}\,\bar{s}(a) &= (\vec{O}). \end{aligned}$$

Since the complement is unique in a distributive lattice with zero and unit (see p. 26, footnote), we deduce from (1) and (2), that

$$\bar{s}(\mathrm{co}\,a) = \mathrm{co}\,\bar{s}(a).$$

E.4.8.8. Theorem. We have for measurable sets a, b

$$\bar{s}(a \dot{+} b) = \bar{s}(a) \dot{+} \bar{s}(b).$$

Proof.

$$a \dot{+} b = (a \cap \mathrm{co}\,b) \cup (b \cap \mathrm{co}\,a).$$

Applying preceding theorem, we get the proof.

E.4.8.9. Theorem. If $a, b \in F^L$ and $a \subseteqq b$, then $\bar{s}(a \leqq \bar{s}(b)$.

Proof. $a \subseteqq b$ is equivalent to $a \cap b = a$. This gives $\bar{s}(a \cap b) = \bar{s}(a)$. It follows $\bar{s}(a) \cdot \bar{s}(b) = \bar{s}(a)$; hence $\bar{s}(a) \leqq \bar{s}(b)$, q.e.d.

E.4.9. Theorem. We shall prove that the correspondence $\bar{s}$ preserves the measure of somata i.e.

$$\mu[\bar{s}(a)] = \mu(a)$$

for every measurable soma a.

Proof. We shall use a method borrowed from [A I., Sect. 9], where the extension device ($\boldsymbol{N}$) is discussed.

In agreement with the discussion in the section mentioned, we say that the subsets a, b of P do not differ by more than $\varepsilon > 0$, whenever there exists a covering of $a \dot{+} b$ made of figures $f_1, f_2, \ldots, f_n, \ldots$:

$$a \dot{+} b \subseteqq \bigcup_{n=1}^{\infty} f_n,$$

such that

$$\sum_{n=1}^{\infty} \mu(f_n) \leqq \varepsilon.$$

The μ-measurable sets[1] are defined, in the section mentioned, as sets α, such that for every $\varepsilon > 0$ there exists a figure f which does not differ from α by more than ε. The extended measure $\mu(\alpha)$ of a measurable set α is defined as the limit

$$\lim_{n\to\infty} \mu(f_n),$$

where α does not differ from f_n by more than $\varepsilon_n > 0$, and where $\varepsilon_n \to 0$.

This has been made under certain condition, denoted by $\mathcal{B}$. Now this condition is satisfied, since this condition is necessary and sufficient for the possibility of extending the measure from (F) to (F^L).

The above method is also valid for spaces, since it is general. Having that, take a measurable set a; i.e. $a \in F^L$, and let $\{f_n\}$ be a sequence of figures such that a does not differ from f_n by more than $\varepsilon_n \to 0$. We have

(1)
$$f_n \dotplus a \subseteq \bigcup_{k=1}^{\infty} g_{nk},$$

(where g_{nk} are figures), with

(2)
$$\sum_k \mu(g_{nk}) \leqq \varepsilon_n.$$

From (1) we get

$$\bar{s}(f_n \dotplus a) \leqq \bar{s} \bigcup_{k=1}^{\infty} s(g_{nk}).$$

Since $\bigcup_{k=1}^{\infty} g_{nk}$ is a $\Pi\,\Sigma$-set, we have, by [E.4.8.8.] and [E.4.8.9.]:

$$\bar{s}(f_n) \dotplus \bar{s}(a) \leqq \sum_{k=1}^{\infty} \bar{s}(g_{nk}).$$

Since the measure of a figure f is equal to the measure of $\bar{s}(f)$, we get from (2)

$$\sum_k \mu[s(g_{nk})] \leqq \varepsilon_n.$$

It follows that $\bar{s}(a)$ does not differ from $\bar{s}(f_n)$ by more than ε_n, this for $n = 1, 2, \ldots$ It also follows that

$$\mu\,\bar{s}(a) = \lim \mu(\bar{s}(f_n)) = \lim \mu(f_n) = \mu(a).$$

The theorem is proved.

E.4.10. Theorem. The correspondence $\bar{s}$ preserves denumerably infinite additions.

Proof. Let

1) $a_1, a_2, \ldots, a_n, \ldots$ be measurable sets of points of the plane P, and let

2) $\bigcup_{n=1}^{\infty} a_n =^{J} a$.

---

[1] In [A I. Sect. 9] the class of all measurable sets is denoted by $\boldsymbol{N}$.

First let us suppose that the sets a_n are mutually disjoint. Then we have, by the denumerable additivity of the measure

$$\mu(a) = \sum_{n=1}^{\infty} \mu(a_n).$$

Let $\varepsilon > 0$. There exists N such that, for every $n \geqq N$ we have

$$0 \leqq \mu(a) - \sum_{k=1}^{n} \mu(a_k) < \varepsilon;$$

hence

$$\mu[a - \bigcup_{k=1}^{n} a_k] \leqq \varepsilon.$$

Since

$$\bigcup_{k=1}^{n} a_k \subseteqq a,$$

we get

$$\mu(a \dot{+} \bigcup_{k=1}^{n} a_k) \leqq \varepsilon$$

for all $n \geqq N$. Since the correspondence $\bar{s}$ preserves the measure [E.4.9.], we obtain

$$\mu\, \bar{s}(a \dot{+} \bigcup_{k=1}^{n} a_k) \leqq \varepsilon.$$

Hence we get, [E.4.8.8.],

$$\mu[\bar{s}(a) \dot{+} \sum_{k=1}^{n} \bar{s}(a_k)] \leqq \varepsilon$$

for every $n \geqq N$. We can rely on the following general theorem on somata of a denumerably additive general tribe (T).

If the measure μ on (T) is denumerably additive, and effective, and if $A_1 \leqq A_2 \leqq \cdots \leqq A_n \leqq \cdots$ are somata of (T) such that

$$\mu(A \dot{+} A_n) \to 0 \quad \text{for} \quad n \to \infty,$$

then

$$\sum_{n=1}^{\infty} A_n = A.$$

Now, in our case the spaces

$$A_n =_{df} \sum_{k=1}^{n} \bar{s}(a_k), \qquad A =_{df} \sum_{k=1}^{\infty} \bar{s}(a_k)$$

are in the above situation. Hence we get

$$\sum_{n=1}^{\infty} A_n = \sum_{n=1}^{\infty} \bar{s}(a_n),$$

and then

$$\sum_{n=1}^{\infty} \bar{s}(a_n) = \bar{s}(a).$$

Till now we have considered the case where all a_i are disjoint. Let us take the general case: We have

$$\sum_{i=1}^{n} a_i = a_1 \cup [a_2 - a_1] \cup [a_3 - (a_1 \cup a_2)] \cup \cdots$$

where all terms on the right are disjoint. Thus we have reduced the general case to the special case of disjoint somata, already settled. The theorem is established.

E.4.10a. Theorem. The correspondence $\bar{s}$ preserves denumerably infinite products.

Proof. This follows from [E.4.10.], by applying de Morgan laws and [E.4.8.7.].

E.4.11. Theorem. Suppose that, for the extended correspondence $\bar{s}$, we have for two measurable sets a, b:

(1) $$\bar{s}(a) = \bar{s}(b).$$

We shall prove that

$$\mu(a + b) = 0 \quad \text{i.e.} \quad a =^{J} b.$$

Proof. From (1) we have

$$s(a) + \bar{s}(b) = 0.$$

We cannot have

$$\mu(a \dot{+} b) > 0,$$

because, we would get, by [E.4.9.],

$$\mu\, \bar{s}(a \dot{+} b) > 0;$$

hence

$$\mu[\bar{s}(a) \dot{+} \bar{s}(b)] > 0,$$

which contradicts (1), because the measure of spaces is effective. Thus

$$\mu(a \dot{+} b) = 0 \quad \text{i.e.} \quad a =^{J} b.$$

E.4.11a. Theorem. If a, b are measurable sets and $\mu(a \dot{+} b) = 0$, then

$$\bar{s}(a) = \bar{s}(b).$$

E.4.11b. Theorem. From the above two theorems we deduce that the correspondence

$$a \curvearrowright \bar{s}(a)$$

is $(=^{\bar{J}})$-invariant in the domain and $(=)$-invariant in the range, where $\bar{J}$ is the ideal in (F^L) composed of all sets with μ-measure equal 0.

If we understand the correspondence $\bar{s}$ in this way, we see that it is $1 \to 1$, and then an isometric isomorphism.

E.4.12. The obtained results allow to state the following

Fundamental Theorem.

1. If $s(\dot{Q})$ is a double scale of spaces in the Hilbert-Hermite- (H.-H.)-space $\mathbf{H}$, i.e. a space valued function, defined for all plane-quarters Q and satisfying the conditions:

I. If Q_1, Q_2 are plane quarters, then the spaces $s(Q_1)$, $s(Q_2)$ are compatible;

II. $s(\dot{O}) = (\vec{O}) = O$, $s(P) = I = \mathbf{H}$;

III. $Q_1 \subseteqq Q_2$, implies $s(Q_1) \leqq s(Q_2)$,

then the correspondence

$$Q \smile s(Q)$$

can be extended to all figures f:

$$f \smile s(f)$$

with the properties:

1) it preserves finite somatic operations,

2) the spaces $s(f)$ if ordered by inclusion of spaces, make up a finitely additive geometrical tribe (S); so

$$(F) \smile s(F) = (S),$$

where (F) is the finitely additive tribe composed of all figures on the plane P.

2. Suppose that the scale $s(Q)$ satisfies the following additional conditions:

IV. If for generalized real numbers α_n, α_0, β_n, β_0 we have

$$\alpha_1 \geqq \alpha_2 \geqq \cdots \geqq \alpha_n \geqq \cdots \quad \lim \alpha_n = \alpha_0,$$

$$\beta_1 \geqq \beta_2 \geqq \cdots \geqq \beta_n \geqq \cdots \quad \lim \beta_n = \beta_0,$$

then

$$\prod_{n=1}^{\infty} s[Q(\alpha_n, \beta_n)] = s[Q(\alpha_0, \beta_0)].$$

V. If

$$\alpha_1 \leqq \alpha_2 \leqq \cdots \leqq \alpha_n \leqq \cdots \quad \lim \alpha_n = \infty,$$

$$\beta_1 \leqq \beta_2 \leqq \cdots \leqq \beta_n \leqq \cdots \quad \lim \beta_n = \infty,$$

then

$$\sum_{n=1}^{\infty} s[Q(\alpha_n, \beta_n)] = s(P).$$

Suppose that μ is a denumerably additive effective measure on a saturated supertribe of (S), and define for figures f of (F):

$$\mu(f) =_{df} \mu(s(f));$$

then the measure μ can be extended by the Lebesgue covering device, thus getting a denumerably additive measure μ on the Lebesgue-extended tribe (F^L) of subsets of the plane.

In addition to that the correspondence $s(f)$ can be extended to the correspondence $\bar{s}(a)$ for all sets $a \in (F^L)$, which preserves finite and denumerably infinite somatic operations and ordering, and which also preserves the measure.

The spaces of $\bar{s}(a)$ make up a denumerably additive geometrical tribe (S^L); so that

$$\bar{s}(F^L) = S^L.$$

The measure on S^L is denumerably additive and effective.

3. The correspondence

$$a \smile \bar{s}(a)$$

is invariant in the domain with respect to the equality $(=^{\bar{J}})$, where $\bar{J}$ is the ideal in (F^L), made of all sets with μ-measure 0.

In its range the correspondence is invariant with respect to the equality of spaces.

Remark. One can prove that under the conditions I, II, III the conditions IV, V are necessary and sufficient for the possibility of the extensions stated in the above theorem.

E.5. Space-traces. We have at our disposal the correspondence $a \smile \bar{s}(a)$, which attaches to every set of F^L a space, and which is a homomorphism preserving the ordering, measure μ and also finite and denumerable operations. This correspondence induces an isomorphism from the tribe $(F^L)_{\bar{J}}$ onto the tribe (S^L) of spaces. We have $J \subseteqq \bar{J}$, where J is the ideal in (F) composed of all figures f with $\mu(J) = 0$, and where $\bar{J}$ is the ideal in (F^L) composed of all subsets of P, having the μ-measure 0.

E.5.1. Def. We shall denote by $\mathfrak{S}$ the last isomorphism. We have

$$\mathfrak{S}(F^L)_{\bar{J}} = S^L.$$

We shall also denote by $\mathfrak{S}$ the isomorphism from (F_J) onto the tribe (S) of spaces.

E.5.2. Theorem. The tribe $(F^L)_{\bar{J}}$ is a denumerably additive finitely genuine supertribe of (F_J). The measure on (S^L) is effective and denumerably additive.

E.5.3. Our next discussion will be the study of the said correspondence in relation to traces.

We shall apply to our case the general theory of traces, developped in [B.], [B I.]. We refer also to [C.] and [C I.].

E.5.4. Def. We call *space-bricks* all spaces $s(R)$, where R is a generalized rectangle in P. Especially $s(\Theta) = O$ and $s(P) = I$ are space bricks. Since the measure on (F) is finitely additive only, it may happen that there can exist rectangles R different from Θ, such that $s(R) = O$. They are just the rectangles R with $R \in J$, which is equivalent to $\mu(R) = 0$.

The space-bricks can also be understood as $\mathfrak{S}$-images of $\boldsymbol{B}_J$-bricks.

The $\mathfrak{S}$-images of figures may be called *space-figures*. The space-figures can also be understood as $\mathfrak{S}$-images of $\boldsymbol{B}_J$-figures.

E.5.4.1. Since $\mathfrak{S}$ is an isomorphism, we see that the space-bricks make up a basis of the tribe (S), [B.1.1.]. We denote it $\mathfrak{S}\,\boldsymbol{B}_J$.

Indeed the following conditions are fulfilled:

1) $O \in \mathfrak{S}\,\boldsymbol{B}_J$, $1 \in \mathfrak{S}\,\boldsymbol{B}_J$.
2) if $\alpha, \beta \in \mathfrak{S}\,\boldsymbol{B}_j$, then $\alpha \cdot \beta \in \mathfrak{S}\,\boldsymbol{B}_J$.
3) if φ is a space figure, i.e. $\varphi \in S$,

then there exists a finite number of space-bricks whose sums equal φ.

This is true by virtue of the isomorphism $\mathfrak{S}$ from (F_J) onto (S).

E.5.4.2. Theorem. The tribe (S) is plunged in a denumerably additive tribe (S^L) as its finitely genuine subtribe, and the operations on spaces of (S) can be taken from (S^L), [B.1 a.]. In accordance with [B.1 b.], we can say that we are in the conditions ***FBG***, which should be translated as $(F,\ S\,B,\ S^L)$.

E.5.4.3. Theorem. The hypothesis (***Hyp Af***) holds true for the space, [B.2.].

Indeed if α is a space figure, then $\mathrm{co}_1\alpha = I - \alpha$ can be represented as a finite sum of mutually disjoint space-bricks.

This is true, since $\mathfrak{S}$ is an isomorphism, and in (F_J) with the basis $\boldsymbol{B}_J$, the statement takes place.

E.5.4.4. Def. As in the general case, we can consider *descending sequences of space-bricks*

$$\alpha_1 \geqq \alpha_2 \geqq \cdots \geqq \alpha_n \geqq \cdots$$

[B I.2.]. We say that the descending sequence $\{\alpha_n\}$ *is included in* $\{\beta_n\}$, whenever for every n there exists m such that

$$\alpha_m \leqq \beta_n.$$

We say that $\{\alpha_n\}$ is *equivalent* to $\{\beta_m\}$ whenever $\{\alpha_n\}$ is included in $\{\beta_n\}$, and $\{\beta_n\}$ is included in $\{\alpha_n\}$. We say that the descending sequence $\{\alpha_n\}$ of space-bricks is *minimal*, whenever:

1) $\{\alpha_n\}$ is not equivalent to $\{O, O, \ldots\}$,

2) if $\{\beta_n\}$ is included in $\{\alpha_n\}$, then either $\{\beta_n\}$ is equivalent to $\{\alpha_n\}$ or $\{\beta_n\}$ is equivalent to $\{O, O, \ldots\}$.

E.5.4.5. Def. By a *space-trace* we understand, [B I.2.3.], any saturated class of mutually equivalent minimal descending sequences $\{\alpha_n\}$.

Every $\{\alpha_n\}$ in this collection is called representative of the trace.

E.5.4.6. Theorem. The space-traces are $\mathfrak{S}$-correspondents to $\boldsymbol{B}_J$-traces on the plane; the correspondence is $1 \to 1$.

E.5.4.7. Def. As in [B., Sect. 2], we define *space-coverings* as, [B.4.], an at most denumerable sum of space-bricks. Thus O is a covering.

E.5.4.8. Def. We say that a trace $\boldsymbol{\Phi}$ is covered by the space-brick α, whenever there exists a representative $\{\alpha_n\}$ of $\boldsymbol{\Phi}$ such that

$$\alpha_1 \leqq \alpha,$$

[B I.2.5.].

E.5.4.9. Def. We say that *the set ξ of space traces is covered by a covering λ*, whenever the following is true, [B I.4.]:

If $\tau \in \xi$, then there exists a brick α, such that τ is covered by α, and $\alpha \leqq \lambda$.

E.5.4.10. Theorem. The *hypotheses* ***Hyp*** I *and* ***Hyp*** II hold true for spaces, [B I.6.]. The ***Hyp*** I reads in our case:

If ξ is a set of space-traces, λ, ν are coverings of ξ and co ξ, then $\lambda + \nu = I$.

The ***Hyp*** II reads:

If α is a brick and ξ the set of all space-traces covered by α, [E.5.4.8.], then co α is the covering of co ξ. These hypotheses are true for $\boldsymbol{B}_J$-traces on the plane, hence, by isomorphism $\mathfrak{S}$, they are also true in the space.

E.5.5. Def. Since the measure on (S^L)-is effective, therefore the Wecken's theorem, [A I., Sect. 1], can be applied. Hence the tribe (S^L) is completely additive.

Having that, we define, as in B I., the *coats of sets of space-traces*, as follows, [B I.7.1.], [B I.7.8.].

If ξ is a set of space-traces, then by its *outer coat* $[\xi]^*$, we understand the intersection

$$[\xi]^* = \prod_{\lambda} \lambda,$$

where the product is taken over all coverings of ξ. The outer coat is a space.

By the *inner coat* $[\xi]_*$ of ξ we understand the space

$$[\xi]_* = 1 - [\eta]^*,$$

where $\eta = \operatorname{co}\xi$. (The complement is taken with respect to the set of all space-traces.) The set ξ of traces is called *measurable*, [B I.8.3.], whenever

$$[\xi]^* = [\xi]_*.$$

In that case we speak of the *coat* $[\xi]$ *of* ξ. The coat is a space.

We define the *measure* μ of ξ by

$$\mu(\xi) =_{df} \mu[\xi].$$

E.5.6. In our circumstances all general theorems, concerning measurability of sets of traces, take place, [see B I.].

E.5.7. If $X' \mathfrak{S} \xi$, where X' is a measurable set of $\boldsymbol{B}_J$-traces and ξ a set of space traces, then ξ is a measurable set of space traces, and we have

$$\mu(X') = \mu(\xi).$$

E.5.8. Theorem. If $X' \mathfrak{S} \xi$, where ξ is a μ-measurable set of space-traces, then X' is a μ-measurable[1] set of $\boldsymbol{B}_J$-traces, and we have

$$\mu(X') = \mu(\xi).$$

E.5.9. Def. By a *μ-null set of space-traces* we shall understand a set ξ of space-traces with $\mu(\xi) = 0$. This is equivalent to the condition

$$(\vec{O}) = [\xi]^*, \quad \text{or even} \quad (\vec{O}) = [\xi].$$

Theorem. For such a set ξ we have

$$\mathfrak{S}^{-1}\xi = \Theta^*,$$

where Θ^* is the empty set of $\boldsymbol{B}_J$-traces.

E.5.10. Theorem. Def. The μ-measurable sets of space-traces make up a denumerably additive tribe (S^*) with denumerably additive measure μ on it, which however may be not effective.

E.5.10a. Theorem. The collection K^* of all μ^* null sets is a denumerably additive ideal in (S^*).

The tribe $(S^*)_{K^*}$ is isomorphic with the tribe (S^L), with preservation of measure.

E.5.11. We can consider complex-valued functions $f(\tau)$ of the variable space-trace τ, fitting the tribe (S^L).

---

[1] In C I. we have used the symbol μ^* for the measure of measurable sets of $\boldsymbol{B}_J$-traces.

Since all various measures considered are generated, in a unique way, by the original measure μ, we shall allow us to use the symbol μ for all kinds of measure introduced.

We also can consider integrals

$$\int_{\alpha} f(\tau)\, d\mu$$

taken over a μ-measurable set α of space-traces; they are integrals of the Lebesgue-Fréchet's-type.

Thus the whole Fréchet's theory of abstract integration can be applied.

E.5.12. In [C I.] we have proved that there exists a correspondence M which carries μ-measurable sets of $\boldsymbol{B}_J$-traces into the μ-measurable subsets of the complex plane P,

$$X' \circlearrowright^{\mathsf{M}} Z,$$

this correspondence M preserving ordering, measure and finite and denumerable set operations.

M is a combined correspondence, which first changes the set X' into the associate distinguished set X of $\boldsymbol{B}$-traces, and afterwards changes X onto Vert$X$. See [C I.32.] and [C I.38.].

Now, having the correspondence $\mathfrak{S}$, we see that the correspondence

$$\mathsf{M}\,\mathfrak{S}^{-1}$$

transforms μ-measurable sets of space-traces into μ-measurable sets of complex-numbers, with preservation of order, measure and finite and infinite set-operations.

E.5.13. The correspondence $\mathsf{M}\,\mathfrak{S}^{-1}$ induces a correspondence for functions (also denoted by $\mathsf{M}\,\mathfrak{S}^{-1}$), which transforms μ-square summable functions $f(\tau)$ of a variable space-trace τ into μ-square summable functions $\Phi(z)$ of the complex variable z, with preservation of the values of integrals.

E.6. Single plane representation of vectors by functions. The purpose of the coming discussion is to define a special representation of vectors of $\mathbf{H}$ by μ-square summable functions of the variable space-trace.

We take over the notation of the forgoing subsections of this [Chapter E].

E.6.1. There is given the tribe (S^L) with basis $\mathfrak{S}\,\boldsymbol{B}_J$, provided with a denumerably additive and effective measure, which is defined by

$$\mu(a) = |\mathrm{Proj}_a \vec{\omega}|^2, \quad \text{for all} \quad a \in S^L,$$

where $\vec{\omega}$ is a generating vector of the space $\mathit{1}$ with respect to a saturated supertribe (T) of (S).

We also have the tribe (S^*), whose somata are μ-measurable sets of space-traces.

E.7. We admit the hypothesis:

VI. (S^L) is a saturated tribe of spaces, in addition to already admitted hypotheses I, II, III, IV, V. Thus we admit that $(S^L) = (T)$.

The case, where (S^L) is not saturated, will be later taken into account. Since the measure is effective on (S^L), this tribe is the borelian extension of (S).

If a relation takes place for all space-traces excepting, perhaps, some traces whose set has the μ-measure 0, we shall sometimes use the letters "a.e." which means "*μ-almost everywhere*".

The set of all space-traces will be denoted by Ξ.

E.7.1. First we shall consider (measurable) *step-functions of space-traces*. By a step-function we shall understand a function $\psi(x)$ of the variable trace x of (S^L), such that there exist a finite number of complex numbers

$$\lambda(1), \ldots, \lambda(n), \quad (n \geqq 1)$$

and disjoint measurable sets of traces

$$E(1), \ldots, E(n),$$

with

$$\bigcup_i E(i) = \Xi, \quad \text{[E.7.]},$$

where $\psi(x) = \lambda(i)$ whenever $x \in E(i)$.

E.7.2. Let $\psi(x)$ be a step-function, with numbers $\lambda(i)$ and sets $E(i)$ as before. Denote by $e(i)$ the coat of $E(i)$

$$e(i) =_{df} [E(i)].$$

We shall attach, to $\psi(x)$, the vector in $\mathbf{H}$:

$$\vec{\psi} =_{df} \sum_{i=1}^{n} \lambda(i) \operatorname{Proj}_{e(i)} \vec{\omega}.$$

We shall see that this correspondence between step-functions and vectors does not depend on the choice of $\lambda(i)$ and $E(i)$ characterizing $\psi(x)$.

Denote by $\Omega_E(x)$ the characteristic function of the set E of space-traces, i.e. $\Omega_E(x) = 1$, whenever $x \in E$, and $\Omega_E(x) = 0$, whenever $x \,\bar{\in}\, E$.

We have

$$\psi(x) = \sum_{i=1}^{n} \lambda(i)\, \Omega_{E(i)}(x).$$

We aught to prove that if

$$(1) \qquad \sum_{i=1}^{n} \lambda(i)\, \Omega_{E(i)}(x) = \sum_{j=1}^{m} \lambda'(j) \cdot \Omega_{F(j)}(x),$$

a.e., then

$$(2) \qquad \sum_{i=1}^{n} \lambda(i) \operatorname{Proj}_{e(i)} \vec{\omega} = \sum_{j=1}^{m} \lambda'(j) \cdot \operatorname{Proj}_{f(j)} \vec{\omega},$$

where $e(i) = [E(i)]$, $f(j) = [F(j)]$, and the sets $F(j)$ satisfy analogous conditions as $E(i)$.

From (1) we get

$$\sum_{i=1}^{n}\sum_{j=1}^{m}\lambda(i)\,\Omega_{E(i)\cap F(j)}(x)=\sum_{j=1}^{m}\sum_{i=1}^{n}\lambda'(j)\,\Omega_{F(j)\cap E(i)}(x),$$

a.e.; hence

$$\sum_{i}\sum_{j}[\lambda(i)-\lambda'(j)]\,\Omega_{E(i)\cap F(j)}(x)=0,$$

a.e. It follows that if

$$\mu[E(i)\cap F(j)]>0,$$

we have

$$\lambda(i)=\lambda'(j).$$

Consequently we have

$$\sum_{i}\sum_{j}[\lambda(i)-\lambda'(j)]\cdot\mathrm{Proj}_{e(i)\cdot f(j)}\vec{\omega}=\vec{O},$$

because, if $\mu(E(i)\cap F(j))=0$ we have $e(i)\cdot f(j)=O$, and then

$$\mathrm{Proj}_{e(i)\cdot f(j)}\vec{\omega}=\vec{O}.$$

Thus we have the equality (2) proved.

E.7.3. We shall show that if

$$\vec{p}=\sum_{i=1}^{n}\lambda(i)\,\mathrm{Proj}_{a(i)}\vec{\omega},$$

where $\lambda(i)$ are complex numbers and $a(i)\in S^L$, the vector $\vec{p}$ can be given the form

$$\vec{p}=\sum_{j=1}^{m}\lambda'(j)\,\mathrm{Proj}_{b(j)}\vec{\omega},$$

where $b(1), b(2),\ldots, b(m)\in S^L$ and are disjoint, and $b(1)+\cdots+b(m)=I$.

Consider all products

(1) $\quad a(i_1)\cdot a(i_2),\ldots, a(i_k)\cdot \mathrm{co}\,a(i'_1), \mathrm{co}\,a(i'_2),\ldots, \mathrm{co}\,a(i'_l),$

where $(i_1, i_2,\ldots, i_k)$; $(i'_1, i'_2,\ldots, i'_l)$ constitutes a decomposition of the set $(1, 2,\ldots, n)$ into disjoint parts which can be empty or not.

The spaces (1), which may be denoted by $b(1), b(2),\ldots, b(m)$, are disjoint and every $a(i)$ is a sum of some of them.

Let

$$a(i)=b(k_1)+b(k_2)+\cdots$$

We have

$$\mathrm{Proj}_{a(i)}\vec{\omega}=\mathrm{Proj}_{b(k_1)}\vec{\omega}+\mathrm{Proj}_{b(k_2)}\vec{\omega}+\cdots,$$

so the assertion follows.

E.7.4. Let

$$\vec{p}=\sum_{j=1}^{m}\lambda'(j)\,\mathrm{Proj}_{b(j)}\vec{\omega}$$

as above. Every space $b(j)$ can be considered as the coat of a suitable measurable set $B(j)$ of traces. If

$$\mu(B(j) \dot{+} B'(j)) = 0,$$

we have

$$b(j) = [B(j)] = [B'(j)].$$

Now we see that these sets $B(j)$ can be chosen so as to have them all disjoint, and with

$$B(1) \cup B(2) \cup \cdots = \Xi,$$

where Ξ is the set of all space-traces. To see this we can put

$$\begin{aligned} C(j) &=_{df} B(j) - (B(1) \cup \cdots \cup B(j-1) \cup B(j+1) \cup \cdots) \\ &= B(j) - \{[(B(1) \cap B(j)] \cup \cdots \cup [B(j-1) \cap B(j)] \cup \cdots\}. \end{aligned}$$

It is easy to see that all $C(j)$ are mutually disjoint, and that

$$D =_{df} \Xi - \operatorname{co} \bigcup_j C(j)$$

has the μ-measure $= 0$. We have $[C(j)] = [B(j)]$, and $C(j) \subseteqq B(j)$.

If we change $C(1)$ into $C' =_{df} C(1) \cup D$, we get the required disjoint decomposition of Ξ into measurable sets

$$C', C(2), C(3), \ldots$$

such that their coats are

$$b(1), b(2), b(3), \ldots$$

respectively.

E.7.5. Having this, we see that we have a one-to-one correspondence $\mathcal{G}$ between step functions and the vectors of the form

$$\sum_i \lambda(i) \operatorname{Proj}_{a(i)} \vec{\omega} \tag{1}$$

where $\lambda(i)$ are complex numbers and $a(i) \in S^L$.

The correspondence is one-to-one in the sense that if two step functions $\psi_1(x)$, $\psi_2(x)$ are equal almost μ-everywhere, they correspond to a same vector (1), and conversely, if two step functions correspond to a same vector, they must be equal a.e.

E.7.6. We have: if

$$\psi_1(x)\, \mathcal{G}\, \vec{\psi}_1, \quad \psi_2(x)\, \mathcal{G}\, \vec{\psi}_2,$$

then

$$[\lambda_1 \psi_1(x) + \lambda_2 \psi_2(x)] \cup \mathcal{G} [\lambda_1 \vec{\psi}_1 + \lambda_2 \vec{\psi}_2],$$

and besides for the scalar product we have:

$$(\psi_1(x), \psi_2(x)) = (\vec{\psi}_1, \vec{\psi}_2).$$

Indeed, let

$$\psi_1(x) = \sum_i \lambda'(i)\, \Omega_{E'(i)}(x),$$

$$\psi_2(x) = \sum_j \lambda''(j)\, \Omega_{E''(j)}(x).$$

We have

$$\vec{\psi}_1 = \sum_i \lambda'(i)\, \mathrm{Proj}_{e'(i)}\, \vec{\omega},$$

$$\vec{\psi}_2 = \sum_j \lambda''(j)\, \mathrm{Proj}_{e''(j)}\, \vec{\omega},$$

where

$$e'(i) = [E'(i)] \quad \text{and} \quad e''(j) = [E''(j)].$$

We may suppose that all $E'(i)$ are disjoint, that $\bigcup_i E'(i) = \Xi$, that all $E''(j)$ are disjoint and that $\bigcup_j E''(j) = \Xi$. It follows that the analogous properties hold for $e'(i)$, $e''(j)$. We have

$$\lambda_1 \psi_1(x) + \lambda_2 \psi_2(x) = \sum_{i,j} [\lambda_1 \lambda'(i) + \lambda_2 \lambda''(j)]\, \Omega_{E'(i) \cap E''(j)}(x)$$

and

$$\lambda_1 \vec{\psi}_1 + \lambda_2 \vec{\psi}_2 = \sum_{i,j} [\lambda_1 \lambda'(i) + \lambda_2 \lambda''(j)]\, \mathrm{Proj}_{e'(i) \cdot e''(j)}\, \vec{\omega},$$

so the first part of the assertion is proved.

To prove the second, we write

$$\begin{aligned}
(\psi_1(x), \psi_2(x)) &= \int_{\Xi} \overline{\psi_1(x)}\, \psi_2(x)\, d\mu \\
&= \sum_{i,j} \overline{\lambda'(i)}\, \lambda''(j) \int \overline{\Omega_{E'(i)}(x)} \cdot \Omega_{E''(j)}(x)\, d\mu \\
&= \sum_{i,j} \overline{\lambda'(i)} \cdot \lambda''(j)\, \mu[E'(i) \cap E''(j)] \\
&= \sum_{i,j} \overline{\lambda'(i)}\, \lambda''(j)\, \mu[e'(i) \cdot e''(j)].
\end{aligned}$$

On the other hand we have

$$\begin{aligned}
(\vec{\psi}_1, \vec{\psi}_2) &= \sum_{i,j} \overline{\lambda'(i)}\, \lambda''(j)\, (\mathrm{Proj}_{e'(i)} \vec{\omega}, \mathrm{Proj}_{e''(j)} \vec{\omega}) \\
&= \sum_{i,j} \overline{\lambda'(i)}\, \lambda''(j)\, \| \mathrm{Proj}_{e'(i) \cdot e''(j)} \vec{\omega} \|^2,
\end{aligned}$$

so the second part is proved too.

E.7.7. The correspondence $\mathcal{G}$, [E.7.5.], can be extended to all μ-square summable functions $\xi(x)$ of the variable trace x.

Indeed, let $\xi(x)$ be such a function. We know that there exists an infinite sequence of step-functions $\psi_n(x)$ such that

$$||\xi(x) - \psi_n(x)||^2 = \int ||\xi(x) - \psi_n(x)||^2 \, d\mu \to 0.$$

Let $\psi_n(x) \mathscr{G} \vec{\psi}_n$.

We can see that $\{\vec{\psi}_n\}$ converges. Indeed, for every $\varepsilon > 0$ there exists N such that if $n, m \geqq N$ we have

$$||\psi_n(x) - \psi_m(x)||^2 \leqq \varepsilon.$$

Hence

$$||\vec{\psi}_n - \vec{\psi}_m||^2 \leqq \varepsilon \quad \text{for} \quad n, m \geqq N.$$

Put

$$\vec{\xi} =_{df} \lim_{n\to\infty} \vec{\psi}_n.$$

We easily see that $\vec{\xi}$ does not depend on the choice of the sequence $\{\psi_n(x)\}$ which μ-square-converges to $\xi(x)$. Put $\xi(x) \mathscr{G} \vec{\xi}$, thus extending $\mathscr{G}$ to all square summable functions, because in the case, where $\xi(x)$ is a step function, we shall get accordance with the original definition of $\mathscr{G}$, [E.7.5.]. We see that if $\xi_1(x) = \xi_2(x)$ a.e., they have the same corresponding vector.

The domain of $\mathscr{G}$ is now composed of all μ-square summable functions defined a.e. on Ξ.

It is easy to see, by standard arguments, taken from the theory of ordinary Lebesgue-square-summable functions, that $\mathscr{G}$ preserves addition, multiplication by a complex number and the scalar product.

E.7.8. Now we shall prove that the range of $\mathscr{G}$ is the whole space $\mathbf{H}$. This takes place because $\vec{\omega}$ is a generating vector.

Indeed, let $\vec{\xi} \in \mathbf{H}$. Since $\vec{\omega}$ is a generating vector, therefore for every $\varepsilon > 0$ there exists a vector $\vec{\psi}$ of the form (1) in [E.7.5.] such that

$$||\vec{\xi} - \vec{\psi}||^2 \leqq \varepsilon.$$

Let us take $\varepsilon_n \to 0$, $\varepsilon_n > 0$, and find $\vec{\psi}_n$ such that

(1) $$||\vec{\xi} - \vec{\psi}_n||^2 \leqq \varepsilon_n.$$

We have

(2) $$\lim_{n\to\infty} \vec{\psi}_n = \vec{\xi}.$$

Let $\psi_n(x) \mathscr{G} \vec{\psi}_n$. Since $\{\vec{\psi}_n\}$ is a fundamental sequence, it follows that for every $\varepsilon > 0$ there exists N such that if $n, m \geqq N$, we have

$$||\vec{\psi}_n - \vec{\psi}_m||^2 \leqq \varepsilon.$$

Now we have

$$[\psi_n(x) - \psi_m(x)] \mathscr{G} [\vec{\psi}_n - \vec{\psi}_m]$$

and

$$||\psi_n(x) - \psi_m(x)|| = ||\vec{\psi}_n - \vec{\psi}_m||.$$

Hence

$$||\psi_n(x) - \psi_m(x)||^2 \leqq \varepsilon \quad \text{for} \quad n, m \geqq N.$$

It follows that $\{\psi_n(x)\}$ converges in μ-square mean to some μ-square summable function $\xi(x)$.

We have

$$\lim_{n\to\infty} ||\psi_n(x) - \xi(x)|| = 0.$$

We get

(3) $$\lim_{n\to\infty} ||\vec{\psi}_n - \vec{\xi}'|| = 0,$$

where $\xi(x) \mathscr{G} \vec{\xi}'$.

From (2) and (3) it follows that $\vec{\xi} = \vec{\xi}'$, and hence

$$\xi(x) \mathscr{G} \vec{\xi}.$$

E.7.9. We have proved the following representation theorem, which will be later applied to normal transformations in order to get their canonical representation. We have got the

Theorem. If

1. the double scale of spaces $s(Q)$ satisfies the conditions I⁰, II⁰, III⁰, IV⁰, V⁰ and VI⁰,

2. $\vec{\omega}$ is a generating vector of the space **H** with respect to the saturated tribe (S^L),

3. we define the correspondence $\mathscr{G}$ by $\Omega_E(x) \smile^{\mathscr{G}} \text{Proj}_e\vec{\omega}$ and measure by $\mu(e) = ||\text{Proj}_e\vec{\omega}||^2$, where $\Omega_E(x)$ is the characteristic function of the measurable set E of traces in (S^L), and where $e = [E]$. This correspondence can be extended so as to have:

1) its domain composed of all μ-square summable functions $\xi(\dot{x})$ of the variable trace, defined a.e. on Ξ,

2) its range coinciding with **H**,

3) preservation of addition of functions, their multiplication by complex numbers and of the scalar product.

The correspondence $\mathscr{G}$ is one-to-one, with respect to the equality of vectors and equality a.e. of functions.

E.8. Multiplane representation of vectors by functions. Till now we have dealt with representations of vectors in the case where the tribe (S^L) is saturated.

Now suppose that (S^L) is not saturated. We shall still choose an effective measure by plunging the tribe (S^L) into a saturated tribe (T) and by choosing a generating vector $\vec{\omega}$ of the space I, with respect to (T).

E.8.1. The saturated supertribe (T) may be any, but now we shall select it, by applying the device, exhibited in [D I.17.5.] (see also [D I.17.4.]).

By Theorem [D I.17.10.] there exists a finite or denumerably infinite sequence of mutually orthogonal spaces

$$(1) \qquad p_1, p_2, \ldots, p_n, \ldots,$$

at least two in number, all compatible with the tribe (S^L) and such, that the smallest tribe, containing (S^L) and the spaces (1), be saturated. We have called (1): "*saturating sequence,* [D I.18.8.], of (S^L). We shall denote by $\bar{S}$ the said smallest tribe.

E.8.2. By [D I.17.2.] the tribe $\bar{S}$ is composed of spaces having the form:

$$\sum_i a_i \, p_{v(i)},$$

where the sum is finite or denumerably infinite, where $\{v(i)\}$ is a subsequence of the sequence $1, 2, \ldots, n, \ldots$, and where $a_i \in S^L$.

E.8.3. We shall consider the spaces p_n separately as H.-H.-spaces $\mathbf{H}_n$, which we shall relate to the plane P in just the same way, as we did in the case where (S^L) was saturated. We shall consider geometrical tribes in p_n, which we may denote by I_n.

We consider the plane-quarters Q on the plane P and we attach to them the spaces

$$(2) \qquad s_n(Q) =_{df} s(Q) \cdot p_n,$$

so we have the correspondence

$$(3) \qquad Q \curvearrowright^{s_n} s_n(Q).$$

E.8.4. Theorem. We shall prove that $s_n(\dot{Q})$ is a double scale of spaces in the sense, specified in [E.1.].

E.8.4a. Proof. First we shall prove that if Q', Q'' are plane-quarters, then

$$s_n(Q'),\ s_n(Q'')$$

are compatible with one another.

Indeed, the spaces $s(Q')$, $s(Q'')$ are compatible and each of them is compatible with p_n. It follows, by [D.10.5.], that

$$p_n \cdot s(Q'), \qquad p_n \cdot s(Q'')$$

are also compatible.

E.8.4b. We have

$$s(\emptyset) = (\vec{O}) = O, \qquad s(P) = I.$$

Hence

$$s_n(\emptyset) = s(\emptyset) \cdot p_n = (\vec{O}),$$

$$s_n(P) = s(P) \cdot p_n = I \cdot p_n = p_m = I_n.$$

E.8.4c. If $Q' \subseteqq Q''$, we have

$$s_n(Q') \leqq s_n(Q'').$$

This follows from $s(Q') \leqq s(Q'')$, by multiplying by p_n. The assertion [E.8.4.] is proved.

E.8.5. The correspondence

$$Q \overset{s_n}{\smile} s_n(Q)$$

can be extended to all figures f on the plane

$$f \overset{s_n}{\smile} s_n(f).$$

The correspondence $f \smile s(f)$ has been extended to a wider one, $E \smile \bar{s}(E)$, to all borelian subsets of the plane [Fund. Theor. E.4.12.] owing to the conditions IV. V:

IV. If $\alpha_1 \geqq \alpha_2 \geqq \cdots \to \alpha_0$, $\beta_1 \geqq \beta_2 \geqq \to \cdots \beta_0$, then

$$\prod_{k=1}^{\infty} s[Q(\alpha_k, \beta_k)] = s[Q(\alpha_0, \beta_0)], \tag{1}$$

V. If $\alpha_1 \leqq \alpha_2 \leqq \cdots \to +\infty$, $\beta_1 \leqq \beta_2 \leqq \to +\infty$, then

$$\sum_{k=1}^{\infty} s[Q(\alpha_k, \beta_k)] = I. \tag{2}$$

Now, similar conditions are valid for the space $I_n = p_n$, yielding the extension $E \smile \bar{s}_n(E)$.

E.8.5a. It can be performed by using a measure, defined as follows.

Let $\vec{\omega}$ be a generating vector of the space $\mathbf{H} = I$ with respect to the tribe $\bar{S}$ (which is saturated).

We put

$$\mu(a) =_{df} ||\mathrm{Proj}_a \vec{\omega}||^2 \quad \text{for all} \quad a \in \bar{S}.$$

Applying the Theorem [D I.17.9.] we get

$$p_n \cdot \bar{S} = p_n \cdot S^L,$$

and this is a saturated tribe in p_n. The vector

$$\vec{\omega}_n =_{df} \mathrm{Proj}_{p_n} \vec{\omega}$$

is a generating vector of $p_n = I_n$ with respect to the tribe $p_n \cdot \bar{S}$. We have

$$p_n = \mathscr{M}_{\bar{S}}(\vec{\omega}_n).$$

This is a consequence of the properties similar to I, II, III, [E.1.], which are satisfied in our case, [E.8.4a, 4b, 4c.].

E.8.6. The correspondence s_n may be an isomorphism or not. Whatever the case may be, there exists a collection J_n of all figures f for which $s_n(f) = (\vec{O})$. This collection J_n is a finitely additive ideal in the tribe (F) of figures on the plane, [E.1.4.]. If there is an isomorphism, the ideal J_n is confined to the single figure O (the empty set of points of P).

E.8.7. The spaces $s_n(f)$ where f varies in F, make up a geometrical tribe, which we shall denote by (S_n). The correspondence $(F) \curvearrowright^{s_n} (S_n)$ is a homomorphism, [E.1.3.].

E.8.8. As we mentioned in [E.8.5.], the correspondence S_n can be extended to $\bar{S}_n$ for all borelian sets (and even more), owing to the conditions, obtained from [E.8.5. (1), (2)], by multiplying the formulas by p_n, getting e.g.

$$\prod_{k=1}^{\infty} s_n[Q(\alpha_k, \beta_k)] = s_n[Q(\alpha_0, \beta_0)].$$

We can define on $\bar{S}\, p_n$ the measure:

$$\mu_n(a) =_{df} ||\mathrm{Proj}_a \vec{\omega}_n||^2 \quad \text{for all} \quad a \in \bar{S}\, p_n.$$

The measure μ_n is denumerably additive and effective, [D I.18.9.].

E.8.8b. Having μ_n we define on (F) the measure

$$\mu_n(f) =_{df} \mu_n(s_n f).$$

This measure satisfies the conditions for the possibility of extending it to a denumerably additive measure on borelian sets of the plane, so we can extend it.

At the same time the correspondence

$$f \curvearrowright s_n(f)$$

can be extended to all μ_n-measurable subsets E of P:

$$E \curvearrowright \bar{s}_n(E),$$

and we get

$$\mu_n(E) = \mu_n(\bar{s}_n(E)).$$

E.8.8.1. The extended correspondence $\bar{s}_n$ and the extended measure μ_n generate several tribes, their corresponding ideals and related equalities, correspondences and measures of various objects related to $p_n = I_n$. These objects, originally denoted differently, according to the case, will be now denoted in a more simple way, sinces their character is determined by the type of objects, to which they apply.

E.8.8.2. Def. The correspondence $(F_J) \curvearrowright (S)$, [E.5.1.], if considered as $(=^J)$-invariant in its domain, was $1 \to 1$, and was denoted by $\mathfrak{S}$. Similarly we define $\mathfrak{S}_n$ as the correspondence

$$(F_{J_n}) \curvearrowright \bar{S} \cdot p_n,$$

considered as $(=^{J_n})$-invariant. Hence $\mathfrak{S}_n$ is the correspondence $\bar{s}_n$, slightly modified.

The s_n-image of (F) is a tribe of spaces in $p_n = I_n$ and it depends on n. It is denoted by (S_n).

E.8.8.3. Def. By extension of (S_n) we get the tribe (S_n^L), which is already saturated.

By extension of (F) we get the tribe (F_n^L) of all μ_n-measurable subsets of P.

We get the ideal J_n^L, composed of all subsets of P with μ_n-measure 0.

We have the correspondence

$$(F_n^L)_{J_n^L} \curvearrowright^{s_n} (S_n^L),$$

which if considered as an isomorphism, will be denoted by $\mathfrak{S}_n$.

E.8.8.4. In [C I.] we were interested in $\boldsymbol{B}_J$-traces on P, whose images are space traces in $\mathbf{H}$.

Now we shall consider $\boldsymbol{B}_{J_n}$-traces on P, which are in $1 \to 1$ correspondence with space-traces in $p_n = I_n$.

We consider μ-measurable sets of $\boldsymbol{B}_{J_n}$-traces in $p_n = I_n$. The tribe of all μ_n-measurable sets of $\boldsymbol{B}_{J_n}$-traces will be denoted by (F_n^*) and the corresponding ideal, composed of sets with μ_n-measure 0, by J_n^*. The corresponding equality is

$$(=^{J_n^*}).$$

These sets are related to μ_n-measurable sets of space-traces in $p_n = I_n$. Their tribe will be denoted by (S_n^*) and the corresponding ideal of sets with μ_n-measure 0, by K^*. There is an isometric isomorphism from the tribe

$$(F_n^*)_{J_n^*} \quad \text{onto} \quad (S_n^*)_{K_n^*}.$$

This isomorphism will also be denoted by $\mathfrak{S}_n$.

The correspondence $\mathfrak{S}_n$ carries sets of $\boldsymbol{B}_{J_n}$-traces into sets of s_n-space-traces with preservation of measure μ_n. Hence $\mathfrak{S}_n$ generates a correspondence (also denoted by $\mathfrak{S}_n$), which carries μ_n-square summable functions $f(\sigma)$ of the variable $\boldsymbol{B}_{J_n}$-trace into μ_n-square summable functions $F(\tau)$ of the J_n-space-trace, so we have

$$f(\tau) \curvearrowright^{\mathfrak{S}_n} F(\sigma),$$

where $f(\tau)$ can be replaced by any function $f_1(\tau)$ which is almost μ_n-everywhere equal to $f(\tau)$, and where $F(\sigma)$ can be replaced by any function $F_1(\sigma)$ which is also almost μ_n-everywhere equal to $F(\sigma)$.

E.8.8.5. We know, [C I.5.12.], that there is an isometric isomorphism M_n from the tribe $(F_n^*)_{J_n^*}$ onto the tribe $(F_n^L)_{J_n^L}$ of μ_n-measurable subsets of P, taken modulo J_n^L. Thus we have:

$$(F_n^*)_{J_n^*} \curvearrowright^{\mathsf{M}_n} (F_n^L)_{J_n^L}.$$

The correspondence M_n carries μ_n-measurable sets of $\boldsymbol{B}_{J_n}$-traces into μ_n-measurable subsets of the plane P, with preservation of measures.

M_n generates another correspondence (also denoted M_n), which carries μ_n-square summable functions $f(\tau)$ of the variable $\boldsymbol{B}_{J_n}$-trace into μ_n-square-summable function $\Phi(z)$ of the complex variable z, so we have

$$f(\dot{\tau}) \smile^{\mathsf{M}_n} \Phi(\dot{z}).$$

E.8.8.6. Similarly as in [E.7.9.] there exists a correspondence $\mathscr{G}_n$ which carries μ_n-square summable functions $F(\sigma)$ of the variable I_n-space-trace σ into vectors $\vec{X}$ of $\mathbf{H}_n = I_n = p_n$:

$$F(\sigma) \smile^{\mathscr{G}_n} \vec{X}.$$

In this formula $F(\sigma)$ can be replaced by any function $F_1(\sigma)$ which is almost μ_n-equal to $F(\sigma)$.

If we take [E.8.8.5.] into account, we get the correspondence:

$$(1) \qquad \vec{X} \smile^{\mathscr{G}_n^{-1}} F(\sigma) \smile^{\mathfrak{S}_n^{-1}} f(\tau) \smile^{\mathsf{M}_n} F(\dot{z}),$$

so the correspondence $\mathsf{M}_n\, \mathfrak{S}_n^{-1}\, \mathscr{G}_n^{-1}$ carries vectors of I_n into μ_n-square-summable functions of the complex variable.

This correspondence is in the range invariant with respect to the equality $\left(=^{J_n^L}\right)$ of functions.

E.8.8.7. We shall call the correspondence

$$\mathsf{M}_n\, \mathfrak{S}_n^{-1}\, \mathscr{G}_n^{-1}$$

the C_n-correspondence. Thus we have from (1)

$$\vec{X} \smile^{\mathsf{C}_n^{-1}} f(z).$$

E.8.9. Def. We shall need a denumerable number of e x a c t c o p i e s P_n of the plane P, $(n = 1, 2, \ldots)$ and considering them as disjoint, make their union V. Instead of V we shall also use the symbol $P^\bullet$.

In precise terms, let us consider ordered couples (z, n), where $n = 1, 2, \ldots, n, \ldots$ and z is a complex number. We define the addition and multiplication of these couples:

$$(z', n) + (z'', n) =_{df} (z' + z'', n),$$

$$(z', n) \cdot (z'', n) =_{df} (z' \cdot z'', n).$$

We put

$$P_n =_{df} \{(z, n) \mid z \text{ is a complex number}\}.$$

The correspondence Φ_n from the set P of all complex numbers z onto P_n,

$$z \smile^{\Phi_n} (z, n)$$

is an all-arithmetic operations-isomorphism. We put

$$P^{\bullet} = V =_{df} \bigcup_{n=1}^{\infty} P_n.$$

We call P_n *the n-th-copy of P*, and we call V *multiplane*.

E.8.10. To have a more visual image of the totality of circumstances, generated by various tribes $(p_n \cdot \bar{S})$, it will be convenient to refer these tribes not to P, but to the separate copies P_n of the plane. This will give not only a better outlook, but will generate important correspondences relating the tribe (S^L) to the union V of all copies of P_n.

E.8.10a. Def. Let us agree to denote (in general) by α_n the n-th-copy of α. So if f is a figure on the plane P, then f_n is its copy on P_n. The correspondence $s_n(\dot{E})$, which carries the μ_n-measurable subsets E_n of P_n onto subspaces of (S_n) can also be written:

$$E_n \smile \mathfrak{S}_n(E_n).$$

E.8.11. We provide subsets of P_n with the measure μ_n, getting μ_n-measurable subsets of P_n. The measure μ_n is denumerably additive but it may be non-effective.

Now we generalize this situation to the whole multiplane $V = P^{\bullet}$ as follows:

If $E^{(n)}$ are μ_n-measurable subset of P_n, $n = 1, 2, \ldots$, we consider the union

$$A =_{df} E^{(1)} \cup E^{(2)} \cup \cdots \cup E^{(n)} \cup \cdots \tag{1}$$

and call it *$\mu^{\bullet}$-measurable subset of V*. In precise terms a subset A of V is said to be *$\mu^{\bullet}$-measurable*, whenever for every n the subset

$$E^{(n)} =_{df} A \cap P_n$$

is μ_n-measurable (i.e. the nth-copy of a μ_n-measurable subset of P).

To the $\mu^{\bullet}$-measurable set (1) we attach the $\mu^{\bullet}$-measure, defined as

$$\mu^{\bullet} A = \mu_1(E^{(1)}) + \mu_2(E^{(2)}) + \cdots + \mu_n(E^{(n)}) + \cdots$$

The series always converges, because the measure $\mu_n(a)$ is defined by

$$\mu_n(a) =_{df} ||\mathrm{Proj}_a \vec{\omega}_n||^2 \quad \text{for all} \quad a \in S_n^L.$$

One can prove that $\mu^{\bullet}$ is denumerably additive on the tribe of all $\mu^{\bullet}$-measurable subsets of V. The tribe will be denoted by $(F^{L\bullet})$. It is a denumerably additive tribe, its ordering is the inclusion of sets. The measure $\mu^{\bullet}$ may be not effective.

E.8.12. We have denoted by J_n^L the ideal in the tribe (F_n^L), composed of all sets with μ_n-measure 0.

We denote by $J^{L\bullet}$ the collection of all subsets A of V with $\mu^\bullet(A) = 0$. $J^{L\bullet}$ can be defined as the collection of all sets which can be represented by

$$B_1 \cup B_2 \cup \cdots \cup B_n \cup \cdots$$

where $B_n \in J_n^L$.

E.8.13. We have the correspondence s_n, which attaches to μ_n-measurable subsets of P the space $s_n(E)$ of the tribe (S_n^L) in $p_n = I_n$. Now we extend this correspondence to the correspondence $s^\bullet$ which attaches to the set (1) the space

$$s^\bullet(A) = s_1(E^{(1)}) + s_2(E^{(2)}) + \cdots + s_n(E^{(n)}) + \cdots$$

The correspondence $s^\bullet$ is a homomorphism, preserving finite and denumerable somatic operations and measure. It carries the tribe $(F^{L\bullet})$ onto the tribe (S^L).

E.8.14. If f is a figure, f_n its nth-copy on P_n, then

$$s^\bullet(\bigcup_n f_n) = s(f).$$

E.8.15. Def. We denote by $\boldsymbol{B}^\bullet$ the union of all bases $\boldsymbol{B}_{J_n}$, and call $\boldsymbol{B}^\bullet$ the union of all bases $\boldsymbol{B}_{J_n}$, each placed on the corresponding plane P_n.

E.8.16. Def. By a $\mu^\bullet$-measurable set of $\boldsymbol{B}^\bullet$-traces we understand any set ξ' which can be written:

$$\xi' = \xi_1 \cup \xi_2 \cup \cdots \cup \xi_n \cup \cdots$$

where ξ_n is a μ_n-measurable set of $\boldsymbol{B}_{J_n}$-traces. We define its measure by

$$\mu^\bullet(\xi') = \sum_n \mu_n(\xi_n).$$

The tribe of all $\mu^\bullet$-measurable sets of $\boldsymbol{B}^\bullet$-traces will be denoted by $(F^{*\bullet})$ and the corresponding ideal by $(J^{*\bullet})$.

E.8.17. Def. By a $\mu^\bullet$-measurable set of space-traces we shall understand any set λ, which can be represented as

$$\lambda = \lambda_1 \cup \lambda_2 \cup \cdots \cup \lambda_n \cup \cdots,$$

where λ_n is a μ_n-measurable set of s_n-space-traces.

By its measure we shall understand the number

$$\mu^\bullet(\lambda) = \sum_n \mu_n(\lambda_n).$$

The tribe of all $\mu^\bullet$-measurable sets of space-traces will be denoted by $(S^{*\bullet})$ and the corresponding ideal by $(K^{*\bullet})$.

E.8.17a. Def. Denote by W_n the set of all s_n-space-traces. We put

$$W^\bullet = \bigcup_n W_n.$$

This set of traces differs from the set W of all $\mathbf{H}$-traces, as defined by means of $s(R_n)$, where R_n are rectangles on the plane P. Since

$$s^{\bullet}(\bigcup_n R_n^{(k)}) = s(R^{(k)}),$$

the situation looks like as if the $\mathbf{H}$-traces where decomposed into a finite or denumerable sum of "partial"-traces. Hence there is a difference between $\mu^{\bullet}$-measurable sets of partial traces and μ-measurable sets of $\mathbf{H}$-traces.

E.8.18. We shall amplify the correspondence M_n as follows:

We say that the $\mu^{\bullet}$-measurable set A of points of V is the $\mathsf{M}^{\bullet}$*-image of the $\mu^{\bullet}$-measurable set ξ of $\boldsymbol{B}^{\bullet}$-traces,* whenever A can be written

$$A = A_1 \cup A_2 \cup \cdots \cup A_n \cup \cdots,$$

where A_n is a μ_n-measurable subset of P_n and ξ can be written

$$\xi = \xi_1 \cup \xi_2 \cup \cdots \cup \xi_n \cup \cdots,$$

where

$$\xi_n \circlearrowleft^{\mathsf{M}_n} A_n, \qquad (n = 1, 2, \ldots).$$

We get the correspondence $\mathsf{M}^{\bullet}$, which carries the $\mu^{\bullet}$-measurable sets of $\boldsymbol{B}^{\bullet}$-traces into $\mu^{\bullet}$-measurable subsets of V.

E.8.19. The correspondence $\mathfrak{S}_n$ carries the μ_n-measurable sets of $\boldsymbol{B}_{J_n}$-traces into μ_n-measurable sets of I_n-space-traces.

These correspondences have a joint amplification in the correspondence $\mathfrak{S}^{\bullet}$ which carries $\mu^{\bullet}$-measurable sets of $\boldsymbol{B}^{\bullet}$-trace into $\mu^{\bullet}$-measurable sets of "partial"-space-traces, hence of subsets of $W^{\bullet}$.

It follows that $\mathsf{M}^{\bullet}(\mathfrak{S}^{\bullet})^{-1}$ carries $\mu^{\bullet}$-measurable subsets of $W^{\bullet}$ into $\mu^{\bullet}$-measurable subsets of $V = P^{\bullet}$.

E.8.20. We have a correspondence C_n, which carries vectors $\vec{X}$ of the space I_n into μ_n-square summable functions of the complex variable.

Now let

$$\vec{X} = \vec{X}_1 + \vec{X}_2 + \cdots + \vec{X}_n + \cdots,$$

where

$$\vec{X}_n \in p_n.$$

Thus we define $\mathsf{C}^{\bullet}\vec{X}$ as a function on V, which restricted to P_n, gives the function C_n.

E.8.21. Theorem. Thus $\mathsf{C}^{\bullet}$ carries the vectors $\vec{X}$ of the space $\mathbf{H}$ into $\mu^{\bullet}$-square summable functions of the variable point $P \in V$.

E.8.22. We call $\mathsf{C}^{\bullet}$ *canonical mapping of the space* $\mathbf{H}$. It depends not only on the space-scale $s(\dot{Q})$ but also on the choice of the saturating sequence

$$p_1, p_2, \ldots, p_n, \ldots$$

of spaces and on the choice of the generating vector $\vec{\omega}$ of $\mathbf{H}$.

The mapping **C•** where $\vec{X} \backsim \Phi(\dot{p})$ preserves addition, subtraction of vectors and multiplication by a complex number. It preserves also the norm $\|\vec{X}\|$ of the vector:

$$\|\vec{X}\|^2 = \int_V |\Phi(\dot{p})|^2 \, d\mu^\bullet$$

and also the scalar product of two vectors:

$$(\vec{X}, \vec{Y}) = \int_V \overline{\Phi(\dot{p})} \cdot \Psi(\dot{p}) \, d\mu^\bullet .$$

E.8.22a. The above discussion also covers the case where the tribe (S^L) is saturated: the saturating sequence is confined to the single space S^L.

E.8.23. We terminate the Chapter E., by stating some theorems. To not increase the volume of the book, we leave their proofs to the reader.

E.8.24. If (T) is a geometrical tribe in **H**, and there exists a generating vector $\vec{\omega}$ of **H** with respect to (T), then (T) is saturated in **H**.

E.8.25. If (T) is a geometrical tribe in **H**, and $\mu(\dot{a})$ is a denumerably additive and effective measure on (T), then there exists a vector $\vec{X}$, such that for every $a \in T$ we have

$$\mu(a) = \|\mathrm{Proj}_a \vec{X}\|^2 .$$

The proof of this theorem is based on the known Radon-Nikodým-theorem.

E.8.26. If (T) is a geometrical tribe in **H**, μ an effective measure on (T), then to μ-topology in (T) is separable. (We suppose that the space **H** is separable and complete.)

Chapter F

Linear operators permutable with a projector

F.1. We shall need some auxiliary theorems on projectors $\mathrm{Proj}_a = (\mathrm{Proj}\, a)$, i.e. operators which project orthogonally the whole Hilbert-Hermite-space **H** on a given (closed) subspace a. Some preliminaries, recollections of notions and the meaning of notions will be in order.

By a *linear operation* (*operator*) *in* **H** we shall understand a vector valued function $A(\dot{x})$ of the variable vector x such that for complex numbers λ, μ:

$$A(\lambda x_1 + \mu x_2) = \lambda A(x_1) + \mu A(x_2)$$

for every x_1, x_2 taken in the domain $\mathrm{D}\,A$ of A, where $\mathrm{D}\,A$ is a linear subvariety (manifold) of $\mathbf{H}$; $\mathrm{D}\,A$ may be a closed set or not. The range $\mathrm{D}\,A$ always is a linear subvariety of $\mathbf{H}$. The operation A does not need to be continuous, and $\mathrm{D}\,A$, $\mathrm{D}\,A$ may differ from $\mathbf{H}$. We shall deal with linear operations only. If a is a subspace (i.e. closed linear manifold) of $\mathbf{H}$, we say that B is *a linear operation in* a, if its domain and range are both subsets of a. If A is an operation in a and $b \subseteqq a$, then by $b \upharpoonleft A$ we shall understand the restriction of A to b, i.e. $y = (b \upharpoonleft A)\,x$ if and only if $y = A\,x$, $x \in \mathrm{D}\,A \cap b$. Hence

$$\mathrm{D}(b \upharpoonleft A) = \mathrm{D}\,A \cap b.$$

The *adjoint operation* A^* *of an operation* A *in* a is defined as usually as follows (see e.g. (26)):

$\mathrm{D}\,A^*$ is defined as the set of all vectors $y \in a$ such that there exists y' with

$$(A\,x, y) = (x, y') \quad \text{for all} \quad x \in \mathrm{D}\,A$$

(these are scalar products).

A necessary and sufficient condition that y' be unique, is that $\mathrm{D}\,A$ be everywhere dense in a i.e. the closure $\overline{\mathrm{D}\,A} = a$. Under these circumstances we define $A^*\,y = y'$ for all $y \in \mathrm{D}\,A^*$.

If E, F are sets of vectors $\neq \emptyset$, then by $E \vec{+} F$ we shall understand the set of all vectors $x + y$ where $x \in E$, $y \in F$.

F.1.1. Def. We say that *an operation* A *in* $\mathbf{H}$ *is permutable with* (Proj a), if and only if

$$\mathrm{Proj}_a(A\,x) = A\,(\mathrm{Proj}_a x)$$

for all $x \in \mathrm{D}\,A$.

We give a list of some simple properties of this notion. Their proofs are rather elementary and easy, so we shall confine ourselves to sketching them only.

F.1.2. Theorem. If A is permutable with (Proj a), then

$$\mathrm{Proj}_a \mathrm{D}\,A = \mathrm{D}\,A \cap a.$$

F.1.3. Theorem. If A is permutable with (Proj a), then A is also permutable with (Proj co a) and

$$\mathrm{Proj}_{\mathrm{co}\,a} \mathrm{D}\,A = \mathrm{D}\,A \cap \mathrm{co}\,a$$

(co a means the orthogonal complement of a, (26)).

Proof. We use [F.1.2.] and write

$$A\,(\mathrm{Proj}_{\mathrm{co}\,a} x) = A\,(x - \mathrm{Proj}_a x) = A\,x - \mathrm{Proj}_a A\,x.$$

F.1.4. Theorem. If

1. A is permutable with (Proj a),
2. $\overline{\mathrm{D}\,A} = \mathbf{H}$,

then

$$\mathrm{Proj}_a \text{ᗡ} A = a \cap \text{ᗡ} A$$

is everywhere dense in a, and

$$\mathrm{Proj}_{\mathrm{co}\,a} \text{ᗡ} A = \mathrm{co}\,a \cap \text{ᗡ} A$$

is everywhere dense in $\mathrm{co}\,a$.

Proof. Suppose that $\mathrm{Proj}_a \text{ᗡ} A$ is not everywhere dense in a. Then there exists a sphere $S_a(y_0, r)$ in a with center y_0 and radius $r > 0$ such that

$$S_a(y_0, r) \cap \mathrm{Proj}_a \text{ᗡ} A = \emptyset .$$

It follows that

$$[S_a(y_0, r) \overset{\rightarrow}{+} \mathrm{co}\,a] \cap \text{ᗡ} A = \emptyset ,$$

(where, for sets E, F of vectors, $E \overset{\rightarrow}{+} F$ means the set

$$\{\vec{x} + \vec{y} \,|\, \vec{x} \in E, \vec{y} \in F\}),$$

and then

$$S_{\mathbf{H}}(y_0, r) \cap \text{ᗡ} A = \emptyset ,$$

which is a contradiction. Application of [F.1.2.] and [F.1.3.] completes the proof.

F.1.5. Theorem. If

1. A is permutable with $(\mathrm{Proj}\, a)$,
2. $\overline{\text{ᗡ} A} = \mathbf{H}$,

then

A^* is also permutable with $(\mathrm{Proj}\, a)$ and with $(\mathrm{Proj}\;\mathrm{co}\,a)$

and we have

$$\mathrm{Proj}_a \text{ᗡ} A^* = \text{ᗡ} A^* \cap a ,$$

$$\mathrm{Proj}_{\mathrm{co}\,a} \text{ᗡ} A^* = \text{ᗡ} A^* \cap \mathrm{co}\,a .$$

Proof. By hypothesis 2. the adjoint operation A^* exists [F.1.]. Let $y \in \text{ᗡ} A^*$, and put $y^* = A^* y$. We have

$$(A\,x, y) = (x, y^*) \quad \text{for every} \quad x \in \text{ᗡ} A .$$

Take $z \in \text{ᗡ} A$. We get

$$(A\;\mathrm{Proj}_a z, y) = (\mathrm{Proj}_a z, y^*) .$$

Hence

(1) $$(A\,z, \mathrm{Proj}_a y) = (z, \mathrm{Proj}_a y^*) .$$

It follows

$$\mathrm{Proj}_a y \in \text{ᗡ} A^* .$$

Thus

$$\mathrm{Proj}_a \text{ᗡ} A^* \subseteqq \text{ᗡ} A^* \cap a ;$$

and the converse inclusion is obvious.

From (1) we get

$$\mathrm{Proj}_a y^* = A^* (\mathrm{Proj}_a y),$$

and then

$$\mathrm{Proj}_a A^* y = A^* \,\mathrm{Proj}_a y \quad \text{for all} \quad y \in \mathcal{D} A^*.$$

Theorem [F.1.3.] makes the proof complete.

F.1.6. Def. We shall consider $a \rceil A$, $\mathrm{co}\, a \rceil A$ (see [F.1.]) which we shall denote by A^a, $A^{\mathrm{co}\, a}$ respectively.

F.1.7. Theorem. If A is permutable with $(\mathrm{Proj}\, a)$, then A^a, $A^{\mathrm{co}\, a}$ are linear operations in a and $\mathrm{co}\, a$ respectively, and we have

$$\mathcal{D} A^a = a \cap \mathcal{D} A = \mathrm{Proj}_a \mathcal{D} A,$$

$$\mathcal{D} A^{\mathrm{co}\, a} = \mathrm{co}\, a \cap \mathcal{D} A = \mathrm{Proj}_{\mathrm{co}\, a} A.$$

Proof. The only item requiring a proof is $\mathcal{D} A^a \subseteqq a$, but this follows from the equality

$$A^a x = A x = A \,\mathrm{Proj}_a x \quad \text{for} \quad x \in \mathcal{D} A^a,$$

which gives $A^a x \in a$.

F.1.8. Theorem. If A is permutable with $(\mathrm{Proj}\, a)$, then

$$\mathcal{D} A = \mathcal{D} A^a \overset{\rightarrow}{+} \mathcal{D} A^{\mathrm{co}\, a}.$$

Proof. Let $x \in \mathcal{D} A$, $x_1 = \mathrm{Proj}_a x$, $x_2 = \mathrm{Proj}_{\mathrm{co}\, a} x$. By Theorems [F.1.2.], [F.1.3.], [F.1.7.] we get

$$x_1 \in \mathcal{D} A^a, \qquad x_2 \in \mathcal{D} A^{\mathrm{co}\, a}.$$

This gives

$$\mathcal{D} A \subseteqq \mathcal{D} A^a \overset{\rightarrow}{+} \mathcal{D} A^{\mathrm{co}\, a}.$$

If we take a vector y, belonging to the right hand set, there exist $y_1 \in \mathcal{D} A^a$, $y_2 \in \mathcal{D} A^{\mathrm{co}\, a}$ with $y = y_1 + y_2$.

By using Theorem [F.1.7.], we obtain $y_1, y_2 \in \mathcal{D} A$, which completes the proof.

F.1.9. Theorem. If A is permutable with $(\mathrm{Proj}\, a)$, then

$$A x = A^a x_1 + A^{\mathrm{co}\, a} x_2 \quad \text{for every} \quad x \in \mathcal{D} A,$$

where

$$x_1 = \mathrm{Proj}_a x, \qquad x_2 = \mathrm{Proj}_{\mathrm{co}\, a} x.$$

Proof. We have by Theorem [F.1.7.] and [F.1.3.]:

$$A^a x_1 = \mathrm{Proj}^a A x, \qquad A^{\mathrm{co}\, a} x_2 = \mathrm{Proj}_{\mathrm{co}\, a} A x,$$

which yields the theorem.

F.1.10. Theorem. If

1. A is permutable with $(\mathrm{Proj}\, a)$,
2. $\overline{\mathcal{D} A} = \mathbf{H}$,
3. $(A^*)^a =_{df} a \rceil A^*$, $(A^*)^{\mathrm{co}\, a} =_{df} \mathrm{co}\, a \rceil A^*$,

then

$$\mathrm{D}A^* = \mathrm{D}(A^*)^a \dotplus \mathrm{D}(A^*)^{\mathrm{co}\,a}.$$

Proof. We apply Theorem [F.1.5.] and [F.1.9.].

F.1.11. Theorem. If 1, 2, 3, as before, we have for every $x \in \mathrm{D}A^*$,

$$A^* x = (A^*)^a x_1 + (A^*)^{\mathrm{co}\,a} x_2,$$

where

$$x_1 = \mathrm{Proj}_a x, \quad x_2 = \mathrm{Proj}_{\mathrm{co}\,a} x.$$

Proof. We apply Theorem [F.1.5.] and [F.1.9.].

F.1.12. Our next purpose is to compare $(A^*)^a$ with $(A^a)^*$ and similarly for co$a$. To do this we prove a series of theorems.

F.1.12a. Lemma. If

1. A is permutable with $(\mathrm{Proj}\,a)$,
2. $\overline{\mathrm{D}A} = \mathbf{H}$,

then

1) the adjoint $(A^a)^*$ of A^a in a exists, and similarly for $(A^{\mathrm{co}\,a})^*$,
2) $\mathrm{D}(A^*)^a \subseteqq \mathrm{D}(A^a)^*$, $\mathrm{D}(A^*)^{\mathrm{co}\,a} \subseteqq \mathrm{D}(A^{\mathrm{co}\,a})^*$.

Proof. The existence of adjoints follows from Theorem [F.1.4.] and [F.1.7.]. Let

$$y \in \mathrm{D}(A^*)^a = \mathrm{D}A^* \cap a.$$

We have

$$(A x, y) = (x, A^* y) \quad \text{for all} \quad x \in \mathrm{D}A,$$

and we get

$$(A^a x, y) = (x, A^* y) \quad \text{for all} \quad x \in \mathrm{D}A^a.$$

The first inclusion of 2) follows.

F.1.13. Theorem. If

1. A is permutable with $(\mathrm{Proj}\,a)$,
2. $\overline{\mathrm{D}A} = \mathbf{H}$,

then

1) $\mathrm{D}(A^*)^a = \mathrm{D}(A^a)^*$, $\mathrm{D}(A^*)^{\mathrm{co}\,a} = \mathrm{D}(A^{\mathrm{co}\,a})^*$,
2) $(A^*)^a = (A^a)^*$, $(A^*)^{\mathrm{co}\,a} = (A^{\mathrm{co}\,a})^*$.

Proof. Let $y_1 \in \mathrm{D}(A^a)^*$, $y_2 \in \mathrm{D}(A^{\mathrm{co}\,a})^*$, $y = y_1 + y_2$, and let $x \in \mathrm{D}A$. Decompose x into components x_1, x_2. We have

$$(A^a x_1, y_1) = (x_1, (A^a)^* y_1),$$

$$(A^{\mathrm{co}\,a} x_2, y_2) = (x_2, (A^{\mathrm{co}\,a})^* y_2),$$

which gives

$$(A x, y_1 + y_2) = (x, (A^a)^* y_1 + (A^{\mathrm{co}\,a})^* y_2).$$

Since this is true for all $x \in \mathrm{D}A$, we get

$$(A^a)^* y_1 + (A^{\mathrm{co}\,a})^* y_2 = A^* y.$$

If we put $y_1 = y$, $y_2 = \vec{0}$, we obtain

$$\mathrm{D}(A^a)^* \subseteqq \mathrm{D}\, A^*,$$

and then, [Lemma F.1.12.],

$$\mathrm{D}(A^a)^* = \mathrm{D}(A^*)^a.$$

In addition to this we get $(A^a)^* y = A^* y$, which completes the proof.

F.1.14. Theorem. If

1. A is permutable with (Proj a),
2. $\overline{\mathrm{D}\, A} = \mathbf{H}$,

then

$$A^* y = (A^a)^* y_1 + (A^{\mathrm{co}\, a})^* y_2$$

where

$$y_1 = \mathrm{Proj}_a y, \quad y_2 = \mathrm{Proj}_{\mathrm{co}\, a} y.$$

Proof. We apply Theorem [F.1.13.].

Chapter G

Some double Stieltjes' and Radons integrals

G.1. Up to this date the theory of selfadjoint and also of maximal normal operators uses Stieltjes integrals as essential tool, thus following D. HILBERT's approach to infinite matrices (51). These integrals were afterwards applied to transformations (operations) by F. RIESZ (49), J. v. NEUMANN (50) and H. M. STONE (26). We shall develop a different scheme which, however, must be confronted with the classical one, so we shall need Stieltjes integrals too. Some remarks on these integrals and on Radon's integrals are in order, since there is some misunderstanding in the literature in this respect.

To facilitate the reading of the present paper we start with recollection and remarks concerning Stieltjes integrals in the case of two variables. The case of one variable is commonly known. For details concerning some items we refer to (26).

G.1.1. The classical setting of double Stieltjes integrals has the following scheme: Let $f(x, y)$, $g(x, y)$ be two complex valued functions defined in a rectangle R in the real x, y-plane.

Let $R_1, R_2, \ldots, R_n$ be a finite partition Π of R into non overlapping rectangles, with sides parallel to the axis of a Cartesian system of coordinates. The *Vitali-increment* $\Delta(g; R_i)$ *of* $g(x, y)$ *on*

$$R_i = R_i(x_{i1}, y_{i1}; x_{i2}, y_{i2}), \qquad (x_{i1} < x_{i2}, y_{i1} < y_{i2}),$$

is defined as the number:

$$g(x_{i1}, y_{i1}) + g(x_{i2}, y_{i2}) - g(x_{i2}, y_{i2}) - g(x_{i2}, y_{i1}).$$

The function g is said to be of *bounded Vitali-variation on* R, if $\sum_i |\Delta(g; R_i)|$ is uniformly bounded for all finite partitions Π of R. Choose arbitrary points $(\xi_i, \eta_i) \in R_i$, and consider the sum

$$\sum_i f(\xi_i, \eta_i) \cdot \Delta(g; R_i).$$

If $f(x, y)$ is continuous and bounded on R and $g(x, y)$ of bounded Vitali-variation on R, this sum tends to a limit, when the maximal diameter of R_i tends to 0, this limit is the double Stieltjes integral and is denoted by

$$\iint_R f(x, y)\, dg(v, y).$$

The Stieltjes integral, taken over the whole plane, is defined by the supplementary limit process, by extending R to the whole plane.

G.1.1.1. Remark. For discussion of different notions of functions of bounded variation see **(53)**.

G.1.1.2. Remark. The Stieltjes integral, like the Riemann's integral, *is not a genuine measure-integral,* as the Lebesgue's one (see **(52)**), but one may try to base it on some measure μ. We may replace the not specified rectangles by the half-open ones, as in [C.], to have them disjoint. We define $\mu(R(x, y; x_1, y_1)) = \Delta(g; R)$, and finally, by extending this measure to a finitely additive measure defined for all figures, [C.].

Now, it is not true that this measure can always be extended to all borelian sets, so as to get a denumerably additive measure (see [E.]); hence *the current term Lebesgue-Stieltjes integral is not adequate,* and neither would be "Radon-Stieltjes integral". The regions of the two notions overlap. If the condition, stated in [C I.] is satisfied, we shall obtain a Radon's integral **(54)**.

G.1.1.2a. Remark. The Radon's integral $\int f\, d\mu$ is defined in an analogous way as the Lebesgue's one; the ordinary measure is replaced by general denumerably additive measure $\mu(E)$ defined for *all borelian subsets* of a given euclidean set; its theory is also given in **(16)**.

G.1.2. In view of the spectral theorem for normal operations in the Hilbert-Hermite-space (H.-H.-space) it seems more natural to introduce *vector-valued-measures*, and the corresponding Stieltjes', or Radon's integrals.

We are going to do this, but confining ourselves to a particular case, which will be needed later on.

In close relation to topic of [E.], let $s(Q)$ be a function which attaches to each plane-quarter Q of the plane a subspace of the given H.-H.-space. We make the assumptions taken from [E.1.]:

I⁰ If Q_1, Q_2 are plane-quarters, the spaces $s(Q_1)$, $s(Q_2)$ are compatible with one another;

II⁰ $s(\emptyset) = (\vec{O})$, $s(P) = 1 = \mathbf{H}$;

III⁰ If $Q_1 = Q(\alpha_1; \beta_1), Q_2 = Q(\alpha_2; \beta_2), Q_1 \subseteqq Q_2, (-\infty \leqq \alpha_1 \leqq \alpha_2 \leqq +\infty,$ $-\infty \leqq \beta_1 \leqq \beta_2 \leqq +\infty)$,

then

$$s(Q_1) \subseteqq s(Q_2).$$

G.1.3. Let us fix a bounded figure f^0, [C.], a uniformly continuous complex valued function $M(z)$ of the complex variable $z = x + iy$ defined on f^0, and a vector $X \in \mathbf{H}$.

We know that the function $s(Q)$ can be extended to a well determined function $s(f)$ defined for all figures f, [E.1.4.2.].

We know that for any two figures f_1, f_2 the spaces $s(f_1)$, $s(f_2)$ are compatible and that $s(f)$ is finitely additive, i.e. if $f_1 \frown f_2 = \emptyset$, then $s(f_1 \smile f_2) = s(f_1) + s(f_2)$, [E.].

Let $X(f)$ denote the projection of the vector X on the space $s(f)$. By a *pointed partition of f* we shall understand a set of rectangles

$$R_1, \ldots, R_n, \quad (n \geqq 1;\ \text{half-open}),$$

finite in number, mutually not overlapping, where

$$f^0 = \bigcup_i R_i,$$

and all this together with a sequence of points $z_1, \ldots, z_n$, where $z_i \in R_i$. Let $\Pi = \{R_i, z_i\}$ be a pointed partition of the given figure f^0. Consider the sum of vectors

$$\sum_1^n M(z_i)\, X(R_i), \tag{1}$$

where $X(R_i)$ is the projection of the vector X on the space determined by R_i.

We shall give a sketch of a proof that (1) tends to a well determined limit, if the maximum diameter of the meshes R_i of Π tends to 0.

This limit, a sort of Stieltjes' integral, will be denoted by

$$\int_{f^0} M(z)\, d(\mathrm{Proj}_z X).$$

G.1.3.1. Proof. We have

$$X(R_1) + \cdots + X(R_n) = X(f^0).$$

If f^0 is a rectangle R, we get

$$\begin{aligned} &\|M(z)\, X(R) - \sum_{i=1}^{n} M(z_i) \cdot X(R_i)\|^2 \\ &\quad = \sum_{i=1}^{n} \|XR_i)\|^2 \cdot |f(z) - f(z_i|^2 \leqq \|X(R)\|^2 \end{aligned}$$

(oscillation of M on $R)^2$. This gives for f^0, for its pointed partition $\{R_i, z_i\}$, and its subpartition $\{S_j, u_j\}$:

$$\|\sum_i M(z_i)\, X(R_i) - \sum_j M(u_j)\, X(S_j)\|^2 \leqq \eta^2 \cdot \|X(f^0)\|^2, \tag{2}$$

where η is the maximum oscillation of $M(z)$ on R_i, $(i = 1, 2, \ldots, n)$. It follows that if $\{R_i, z_i\}$, $\{S_i, u_i\}$ are any two pointed partitions of f^0, such that the oscillation on each of these rectangles $\leqq \varepsilon$, then the expression on the left in (2) does not exceed $4\varepsilon^2 \|X(f^0)\|^2$. This completes the proof of the existence of the integral.

As corollaries we obtain the properties

$$\int_{f^0} M(z)\, d(\mathrm{Proj}_z X) \in s(f^0),$$

$$\left\| \sum_{i=1}^{n} M(z_i)\, X(R_i) - \int_{f^0} M(z)\, d(\mathrm{Proj}_z X) \right\|^2 \leqq 4\varepsilon^2 \|X(f^0)\|^2.$$

Since

$$\|\sum_i M(z_i)\, X(R_i)\|^2 = \sum_i |M(z_i)|^2 \cdot \|X(R_i)\|^2.$$

We obtain

$$\left\| \int_{f^0} M(z)\, d(\mathrm{Proj}_z X) \right\|^2 = \int_{f^0} |M(z)|^2\, d\, \|\mathrm{Proj}_z X\|^2, \tag{3}$$

and for the scalar product we get:

$$\left(\int_{f^0} M(z)\, d\, \mathrm{Proj}_z X,\, Y \right) = \int_{f^0} M(z)\, d(\mathrm{Proj}_z X,\, Y) \tag{4}$$

for every $Y \in \mathbf{H}$.

The ordinary Stieltjes' integrals occuring on the right of (3) and (4) are defined as usually by means of the Vitali-variation of the functions

$$g(x, y) = \|\mathrm{Proj}_{s(Q(x, y))} X\|^2$$

and

$$g(x, y) = (\mathrm{Proj}_{s(Q(x,y))} X, Y)$$

respectively.

G.2. Integrals taken over the whole plane. Let us admit that $M(z)$ is a continuous function over the whole plane P. On every bounded rectangle this function is uniformly continuous, so the corresponding vector-valued Stieltjes' integral exists.

Let

$$R(1) \subseteqq R(2) \subseteqq \cdots \quad \text{with} \quad \bigcup_{n=1}^{\infty} R(n) = P.$$

We prove that the following are equivalent:

1⁰. $\lim\limits_{n\to\infty} \int\limits_{R(n)} M(z)\, d\,\mathrm{Proj}_z X$ exists,

2⁰. $\lim\limits_{n\to\infty} \int\limits_{R(n)} |M(z)|^2\, d\,\|\mathrm{Proj}_z\|^2$ exists.

This implies that both limits, in the case of their existence for a single sequence $\{R(n)\}$, also exist, if we take any other sequence $\{S(n)\}$ of rectangles, and that these limits do not depend on this choice. The first limit is denoted by

$$\int\limits_P M(z)\, d(\mathrm{Proj}_z X),$$

and the second is the ordinary Stieltjes integral

$$\int\limits_P |M(z)|^2\, d\,\|\mathrm{Proj}_z\|^2. \tag{5}$$

As a corollary we obtain, in the case of existence of (5),

$$\left\|\int\limits_P M(z)\, d\,\mathrm{Proj}_z X\right\|^2 = \int\limits_P |M(z)|^2 \cdot d\,\|\mathrm{Proj}_z X\|^2,$$

and

$$\left(\int\limits_P M(z)\, (d\,\mathrm{Proj}_z X, Y)\right) = \int\limits_P M(z) \cdot d(\mathrm{Proj}_z X, Y),$$

for every $Y \in \mathbf{H}$.

G.2.1. Remark. If, instead of the whole plane P, we take a not bounded figure f^0, we define $\int\limits_{f^0}$ as $\lim \int\limits_{T(n)}$ as before, where $T(n) = f^0 \cap R(n)$. We obtain similar results.

G.2.2. Theorem. If $\int\limits_P$ exists, then $\int\limits_f$ also exists for any figure f.

G.2.3. Now we shall need integrals taken not only on figures, but also *on arbitrary borelian sets of the plane* P. If we like to preserve the above Stieltjes-scheme, we shall be compelled to use some artificial device for this extension. For instance we may procede as follows.

A way is open through Theorems of (E.], which allows to extend the function $s(f)$ to all borelian sets.

Accordingly to that let us add further assumptions on $s(Q)$, [E.2.4.]:

IV⁰. If

$$\alpha_1 \geqq \alpha_2 \geqq \cdots \geqq \alpha_n \geqq \cdots \to \alpha_0,$$

$$\beta_1 \geqq \beta_2 \geqq \cdots \geqq \beta_n \geqq \cdots \to \beta_0,$$

then

$$\prod_{n=1}^{\infty} s(Q(\alpha_n, \beta_n)) = s(Q(\alpha_0, \beta_0)).$$

V⁰. If

$$\alpha_1 \leqq \alpha_2 \leqq \cdots \to +\infty,$$

$$\beta_1 \leqq \beta_2 \leqq \cdots \to +\infty,$$

then

$$\sum_{n=1}^{\infty} s(Q(\alpha_n, \beta_n)) = I = \mathbf{H},$$

$$(-\infty \leqq \alpha_i, \beta_i \leqq +\infty).$$

By virtue of the quoted theorems the function $s(f)$ can be extended to all borelian sets E of the plane; and any two spaces $s(E_1)$, $s(E_2)$ are compatible. Notice that if

$$\int_P M(z)\, d\,\mathrm{Proj}_z X$$

exists, then *for any figure* f we have

$$\int_f M(z)\, d\,\mathrm{Proj}_z X = \mathrm{Proj}_{s(f)} \int_P M(z)\, d\,\mathrm{Proj}_z X,$$

which can be proved by applying the definitions of $\int_f$ and $\int_P$.

Having this, we can put for a borelian set E:

$$\int_E M(z)\, d\,\mathrm{Proj}_z X =_{df} \mathrm{Proj}_{s(E)} \int_P M(z)\, d\,\mathrm{Proj}_z X,$$

thus getting the required integrals. Now we can define "generalized" Stieltjes integrals

$$\int_E M(z)\, d(\mathrm{Proj}_z X, Y) =_{df} \left(\int_E M(z)\, d\,\mathrm{Proj}_z X, Y\right) \tag{1}$$

and

$$\int_E |M(z)|^2\, d\,\|\mathrm{Proj}_z X\|^2 =_{df} \left\|\int_E M(z)\, d\,\mathrm{Proj}_z X\right\|^2, \tag{2}$$

and prove theorems, whose aim would be the harmony and full accordance with the former definitions and properties.

The above artificial way tries to maintain the Stieltjes integrals. Another artifice would be needed, if we wanted to generalize from continuous functions to borelian functions $M(z)$.

G.2.4. A more natural way for obtaining them all will be given *by dropping the Stieltjes-scheme and by defining instead some Radon's integrals, as follows:*

The integrals (1), (2) are related to number-valued functions

$$(\mathrm{Proj}_{s(Q)}\vec{X}, \vec{Y}) \quad \text{and} \quad \|\mathrm{Proj}_{s(Q)}\vec{X}\|^2$$

defined for all plane quarters Q, and these functions can be extended to

$$(\mathrm{Proj}_{s(f)}\vec{X}, \vec{Y}), \quad \|\mathrm{Proj}_{s(f)}\vec{X}\|^2$$

for all figures f. Both are finitely additive on the Boolean tribe (F) of all figures.

Now the conditions IV^0 and V^0 give the possibility of extending these functions to all borelian sets E.

Indeed, we have from IV^0

$$\lim_{n\to\infty} \mathrm{Proj}_{s(Q(\alpha_n, \beta_n))} X = \mathrm{Proj}_{s(Q(\alpha_0, \beta_0))} X$$

and from V^0

$$\lim_{n\to\infty} \mathrm{Proj}_{s(Q(\alpha_n, \beta_n))} X = \mathrm{Proj}_I X = X$$

respectively, and these relations give

$$\lim_{n\to\infty} (\mathrm{Proj}_{s(Q(\alpha_n, \beta_n))} X, Y) = (\mathrm{Proj}_{s(Q(\alpha_0, \beta_0))} X, Y)$$

and

$$\lim_{n\to\infty} \|\mathrm{Proj}_{s(Q(\alpha_n, \beta_n))} X\|^2 = \|X\|^2$$

respectively, so we can apply [C I.] and get the extension required. Both the set-functions

$$(\mathrm{Proj}_{s(E)} X, Y) \quad \text{and} \quad \|\mathrm{Proj}_{s(E)} X\|^2$$

are denumerably additive and bounded, so they constitute a kind of measure. Having this we can use the original Lebesgue's method (23) (i.e. Radon's) of defining the integrals

$$\int_E M(z)\, d(\mathrm{Proj}_z X, Y), \quad \int_E |M(z)|^2\, d\|\mathrm{Proj}_z X\|^2$$

not only for continuous $M(z)$ but also for borelian ones; in this definitions

$$(\mathrm{Proj}_{s(E)} X, Y), \quad \|\mathrm{Proj}_{s(E)} X\|^2$$

will serve as "measures" of E.

G.2.5. For vector-valued integrals we prefer to choose a shorter way and deal directly with $\int M(z)\, d\,\mathrm{Proj}_z X$, where $M(z)$ is any borelian complex-valued function, thus obtaining a kind of Radon's integrals, based on a *denumerably additive vector-valued measure of borelian sets.* This means that we shall abandon the Stieltjes-scheme. We shall follow the technique, used for real-valued measures in (16) with some modifications and suitable simplifications:

To every borelian set E we attach the vector-valued measure

$$\vec{\mu}(E) = \mathrm{Proj}_{s(E)} X$$

and also auxiliary real-valued measures

$$v(E) = \|\mathrm{Proj}_{s(E)} X\|^2, \quad v_1(E) = (\mathrm{Proj}_{s(E)} X, Y).$$

They are all denumerably additive.

Indeed, let $E_1, E_2, \ldots, E_n, \ldots$ be disjoint borelian sets of points of the plane P. The spaces $s(E_n)$ are mutually disjoint, and hence orthogonal to one another. Put

$$E = \bigcup_{n=1}^{\infty} E_n.$$

We have

$$s(E) = \sum_{n=1}^{\infty} s(E_n)$$

hence

$$(\mathrm{Proj}\, s(E))\, X = \sum_n (\mathrm{Proj}\, s(E_n))\, X.$$

G.2.6. A complex valued function $N(z)$ defined on the plane P is said to be *simple*, if there exists a decomposition $P = \bigcup_n E_n$ into an at most denumerable number of mutually disjoint borelian sets E_n on each of which $N(z)$ is constant. $N(z)$ is defined by $\{E_n, \lambda_n\}$ where λ_n are complex numbers.

We say that $N(z)$ is *integrable*, if the sum

$$\vec{S}(N) = \sum_n \vec{\mu}(E_n) \cdot \lambda_n$$

exists, does not depend on the manner in which $N(z)$ is represented through partition of P. A necessary and sufficient condition for the integrability of $N(z)$ is the convergence of the sum

$$T(N) = \sum_n v(E)\, |\lambda_n|^2.$$

A bounded simple function is always integrable.

We prove that if $N'(z)$, $N''(z)$ are integrable simple functions, so is $N(z) = N'(z) + N''(z)$, and we have

$$\vec{S}(N) = \vec{S}(N') + \vec{S}(N'').$$

G.2.7. The theory of integration may be sketched by the following list of theorems and definitions:

G.2.8. Lemma. If $N'(z)$, $N''(z)$ are integrable simple functions such that

$$|N'(z) - N''(z)| \leq \varepsilon,$$

then

$$||\vec{S}(N') - \vec{S}(N'')|| \leqq \varepsilon \cdot ||X||.$$

G.2.8.1. Def. Let $M(z)$ be a borelian function. We say that $M(z)$ is *summable*, if for every $\varepsilon > 0$ there exists a simple integrable function $N(z)$ such that

$$|M(z) - N(z)| \leqq \varepsilon$$

for all z.

G.2.8.2. Theorem and Def. If

1) $M(z)$ is summable,
2) $\varepsilon_n > 0$, $\varepsilon_n \to 0$,
3) $N_n(z)$ is a simple integrable function, such that

$$|M(z) - N_n(z)| \leqq \varepsilon_n,$$

then $\lim_{n\to\infty} \vec{S}(N_n)$ exists, and is independent of the choice of $\varepsilon_n \to 0$ and of the choice of $N_n(z)$. We put

$$\int M(z)\, d\operatorname{Proj}_z X =_{df} \lim_{n\to\infty} \vec{S}(N_n).$$

G.2.8.3. Theorem. A simple function $N(z)$ is summable if and only if it is integrable, and we have

$$\vec{S}(N) = \int N(z)\, d\operatorname{Proj}_z X.$$

G.2.8.4. Theorem. If $M_n(z)$ are summable, $(n = 1, 2, \ldots)$, $M_n(z) \to M(z)$ uniformly, then $M(z)$ is also summable, and we have

$$\int M(z)\, d\operatorname{Proj}_z X = \lim_{n\to\infty} \int M_n(z)\, d\operatorname{Proj}_z X.$$

G.2.8.5. Theorem. If $M'(z)$, $M''(z)$ are summable, so is $M(z) = M'(z) + M''(z)$, and we have

$$\int M(z)\, d\operatorname{Proj}_z X = \int M'(z) + \int M''(z).$$

If $M(z)$ is summable and λ a complex number, so is $\lambda M(z)$, and we have

$$\int \lambda M(z)\, d\operatorname{Proj}_z X = \lambda \cdot \int M(z).$$

G.2.9. Def. Let E be a borelian set. We say that $M(z)$ is *summable on* E, if the function $M_1(z)$, defined by

$$M_1(z) = 0 \quad \text{for} \quad z \in \operatorname{co} E, \qquad M_1(z) = M(z) \quad \text{for} \quad z \in E,$$

is summable. In this case we define

$$\int_E M(z)\, d\operatorname{Proj}_z X =_{df} \int M_1(z)\, d\operatorname{Proj}_z X.$$

G.2.9.1. Theorem. We have

$$\int_E M(z)\, d\,\mathrm{Proj}_z X \in s(E).$$

G.2.9.2. Theorem. If

$$|M(z) \leqq K \quad \text{for} \quad z \in E,$$

then

$$\left\| \int_E M(z)\, d\,\mathrm{Proj}_z X \right\| \leqq K \cdot \| (\mathrm{Proj}\, s(E))\, X \|.$$

G.2.10. Theorem. The following are equivalent:

I. $\int_E M(z)\, d\,\mathrm{Proj}_z X$ exists,

II. $\int_E |M(z)|^2\, d\, \|\mathrm{Proj}_z X\|^2$ exists (Radon's integral).

In this case we have

$$\left\| \int_E M(z)\, d\,\mathrm{Proj}_z X \right\|^2 = \int_E |M(z)|^2\, d\, \|\mathrm{Proj}_z X\|^2$$

and also

$$\left(\int_E M(z)\, d\,\mathrm{Proj}_z X,\, Y \right) = \int_E M(z)\, d\, (\mathrm{Proj}_z X,\, Y),$$

where on the right we have a Radon's integral.

G.2.11. Theorem. If $E \subseteqq F$, then

$$\int_E M(z)\, d\,\mathrm{Proj}_z X = \mathrm{Proj}_E \left[\int_F M(z)\, d\,\mathrm{Proj}_z X \right].$$

G.2.12. Theorem. If $M(z)$ is summable, then the vector valued function

$$\vec{J}(E) = \int_E M(z)\, d\,\mathrm{Proj}_z X$$

is denumerably additive on the tribe of all borelian sets E.

G.2.13. Theorem. If $M(z)$ is continuous on the whole plane P, then the following are equivalent

1⁰. the Stieltjes integral $\int_P M(z)\, d\,\mathrm{Proj}_z X$, [G.2.4.], exists,

2⁰. $M(z)$ is summable.

In this case the Stieltjes integral $\int_f M(z)\, d\,\mathrm{Proj}_z X$ is equal to the Radon's integral $\int_f M(z)\, d\,\mathrm{Proj}_z X$ for any figure f.

The last theorem shows that we may *confine ourselves to Radon's integrals, and drop the Stieltjes' ones.*

G.2.14. Remarks. The general acceptance of functions whose values are not numbers, but some other mathematical objects, is about 50 years old, though they were already studied explicitely in RUSSELL's and WHITEHEAD's Principia Mathematicae (1). In NAGY's account (44) the integrals $\int M(z)\, d\,\mathrm{Proj}_z X$ are understood as "symbolic" forms only.

Abstract integrals $\int M(z)\, d\,\mathrm{Proj}_z$, whose values are operators, are in (55), where operator-valued functions of a variable interval are considered. See also (56). We do not need operator-valued measures, though they could be also taken into account and dealt with in accordance with the general theory of "functionoids" (57).

G.2.15. Resumé. The above discussion leads to the following setting which will be applied in the sequal.

There will be given a space-valued function $s(\dot{Q})$ defined for all plane-quarters Q, [E.], and where the spaces are (closed) subspaces of a given separable and complete Hilbert-Hermite-space. The function $s(Q)$ will satisfy the conditions I^0, II^0, III^0 of [G.2.] and IV^0, V^0 of [G.2.3.]. $s(Q)$ can be extended to a space valued function $s(\dot{E})$ defined for all borelian sets E of the plane P, so as to become denumerably additive. Let $X \in \mathbf{H}$ be a vector. We consider the vector valued measure

$$\vec{\mu}(E) = \mathrm{Proj}_{s(E)} X.$$

For borelian complex-valued functions $M(z)$ of the complex variable $z = x + iy$ we define vector valued Radon's-integrals

$$\int_F M(z)\, d\,\mathrm{Proj}_z X, \tag{1}$$

where F is a borelian set as in [G.2.7.] — [G.2.12.], where all theorems needed are given. We shall also use ordinary number-valued Radon's integrals

$$\int_F |M(z)|^2\, d\,\|\mathrm{Proj}_z X\|^2, \tag{2}$$

$$\int_F M(z)\, d\,(\mathrm{Proj}_z X,\, Y), \tag{3}$$

as in [G.2.6.], where $Y \in \mathbf{H}$.

The integrals (1), (2), (3) may also be written with a more detailed symbolism:

$$\int_F M(z)\, d\,\mathrm{Proj}_{s(Q(z))} X,$$

$$\int_F |M(z)|^2\, d\,\|\mathrm{Proj}_{s(Q(z))} X\|^2,$$

and

$$\int_F M(z)\, d\,(\mathrm{Proj}_{s(Q(z))} X,\, Y),$$

where $Q(z)$ stands for $Q(x; y)$; $z = x + iy$.

These integrals will be needed for studying the relationship between the known theory of selfadjoint operations and the new one to be developped.

G.2.16. We have discussed integrals in the two-dimensional case. For functions of a single variable, $M(\alpha)$, $(-\infty<\alpha<+\infty)$, the rays $Q(\alpha)=(-\infty,\alpha\rangle$, $(-\infty\leqq\alpha\leqq+\infty)$ aught to be taken instead of plane quarters. The rectangles are replaced by halfopen intervals $R(\alpha,\beta)=(\alpha,\beta\rangle$; figures F are here finite sums of these intervals. If we have a space valued function $s(Q)$ of the variable ray, satisfying the conditions:

If

$$(1)\qquad Q_1\subseteqq Q_2, \quad \text{then} \quad s(Q_1)\subseteqq s(Q_2);$$
$$s(\emptyset)=(\vec{O}), \qquad s(P)=I=\mathbf{H},$$

where P is the whole straight line

$$(-\infty<\lambda<+\infty),$$

no additional compatibility condition is needed, since from (1) the compatibility follows.

The conditions IV⁰, V⁰ is [G.2.3.] are replaced by:

$$(2)\qquad \begin{aligned} &\text{if} \quad \alpha_1\geqq\alpha_2\geqq\cdots\to\alpha_0, &&\text{then} \quad \prod_{n=1}^{\infty} s(Q(\alpha_n))=s(Q(\alpha_0));\\ &\text{if} \quad \alpha_1\leqq\alpha_2\leqq\cdots\to+\infty, &&\text{then} \quad \sum_{n=1}^{\infty} s(Q(\alpha_n))=I=\mathbf{H}. \end{aligned}$$

The extension of $s(Q)$ to all figures is analogous; and the extension to all borelian subsets of the line P is assured by (2).

The vector-valued integral $\int_F M(\lambda)\,d\,\mathrm{Proj}_{s(Q(\lambda))}X$ and the Radon's integrals

$$\int_F |M(\lambda)|^2\,d\,\|\mathrm{Proj}_{s(Q(\lambda))}X\|^2,$$

$$\int_F M(\lambda)\,d(\mathrm{Proj}_{s(Q(\lambda))}X,\,Y)$$

are defined in the same way as in the case of two variables [G.2.15.].

Chapter H

Maximal normal operator and its canonical representation

H.1. Maximal normal operators[1] constitute a generalization of selfadjoint (hermitian) operations. Vagely speaking they may be conceived as operators with mutually orthogonal eigenvectors, but with

[1] Instead of the term *operator*, the following synonyms are used: *operation, transformation*.

complex eigenvalues. Selfadjoint operators and unitary operators are particular cases of maximal normal operators.

H.1.1. We recall some precise definitions taken from Stone's book (26). The operation H in $\mathbf{H}$ is called *selfadjoint*, if $H = H^*$, where H^* is the adjoint of H, [F.1.].

We take as granted the known *spectral theorem* for selfadjoint transformations which reads as follows, (26):

If H is a selfadjoint transformation in $\mathbf{H}$, ($H = H^*$, $\overline{\mathfrak{d} H} = \mathbf{H}$), then there exists a unique function $E(\lambda)$ which attaches to every λ, $(-\infty \leqq \lambda \leqq +\infty)$ a projector, such that

1) $E(\lambda) E(\mu) = E(\lambda)$ for $\lambda \leqq \mu$, $E(\lambda) E(\mu) = E(\mu)$ for $\lambda > \mu$, $E(\lambda + \varepsilon) \to E(\lambda)$, $\varepsilon > 0$, $\varepsilon \to 0$, $E(\lambda)$ tends[1] to the identity-transformations if $\lambda \to +\infty$, and $E(\lambda)$ tends[2] to the null-transformations if $\lambda \to -\infty$;

2) $\vec{X} \in \mathfrak{d} H$, if and only if the Stieltjes integral

$$\int_{-\infty}^{+\infty} \lambda^2 \, d\|E(\lambda) \vec{X}\|^2$$

exists;

3) for every

$$\vec{X} \in \mathfrak{d} H \quad \text{and every} \quad \vec{Y} \in \mathbf{H}$$

we have for the scalar product:

$$(H\vec{X}, \vec{Y}) = \int_{-\infty}^{+\infty} \lambda \, d(E(\lambda) \vec{X}, \vec{Y}),$$

$$\|H X\|^2 = \int_{-\infty}^{+\infty} \lambda^2 \, d\|E(\lambda) \vec{X}\|^2$$

with Stieltjes integrals.

H.2. Theorem. Denote by $t(\lambda)$ the space on which $E(\lambda)$ is projecting. The item 1) in the spectral theorem is equivalent to the following one

1′) There exists a unique space-valued function $t(\lambda)$ defined for all λ, as before, and such that if

$$\lambda' \leqq \lambda'',$$

then

$$t(\lambda') \leqq t(\lambda'');$$

if

$$\lambda_1 \geqq \lambda_2 \geqq \cdots \to \lambda_0,$$

---

[1] This means: for every $\vec{X}$ we have $\|E(\lambda_n)\vec{X} \to \vec{X}\| \to 0$ for $n \to \infty$.

[2] This means: for every $\vec{X}$ we have $\lim_{n\to\infty} \|E(\lambda_n)\vec{X}\| = 0$. $\|\vec{X}\|$ means the norm of $\vec{X}$; it is defined as $\sqrt{(\vec{X}, \vec{X})}$. (see Chapter [D]).

then

$$\prod_{n=1}^{\infty} t(\lambda_n) = t(\lambda_0);$$

if

$$\lambda_1 \geqq \lambda_2 \geqq \cdots \to -\infty,$$

then

$$\prod_{n=1}^{\infty} t(\lambda_n) = (\vec{O});$$

if

$$\lambda_1 \leqq \lambda_2 \leqq \cdots \to +\infty,$$

then

$$\sum_{n=1}^{\infty} t(\lambda_n) = I = \mathbf{H}.$$

H.2.1. Def. We call the function $E(\lambda)$ "*projector-valued spectral scale*", and we call $t(\lambda)$ "*space-valued spectral scale*".

H.2.2. $E(\lambda)$ carries till now the strange name of "decomposition of identity", taken from earlier algebraic theories of matrices.

H.2.3. Remark. Stone's proof, **(**26**)**, of the spectral theorem for normal operators uses theorems of the "Operational Calculus" i.e. theory of operator-valued functions of selfadjoint operators. We prefer to give a sketch of a simplier proof, based on a lemma (see further [H.5 a.]).

H.3. Def. Two selfadjoint operators H', H'' are said to be (*strongly*) *permutable,* if and only if for every λ and μ the projectors $E'(\lambda)$, $E''(\mu)$, taken from their spectral scale, satisfy the condition

$$E'(\lambda)\, E''(\mu)\, \vec{X} = E''(\mu)\, E'(\lambda)\, \vec{X}$$

for every $\vec{X} \in \mathbf{H}$.

H.3a. Remark. This is not equivalent to $H' H'' = H'' H'$, (see **(**43**)**, p. 90), but implies it.

H.3b. Remark. In the above definition we can use the equivalent statement that $t'(\lambda)$, $t''(\lambda)$ be compatible, [D.5.].

H.4. We admit the following definition, which differs slightly from that given by Stone **(**26**)**:

Def. A transformation N in **H** is termed *normal* ("*maximal normal*"), if there exist two strongly permutable selfadjoint transformations H, K in **H** such that

1) $\overline{\mathrm{\mathbf{ɑ}}\, N} = \mathbf{H}$,

2)

(1) $$N \equiv H + i\,K,\text{*)}$$

where i is the imaginary unit. Of course $\mathrm{ɑ}\, N = \mathrm{ɑ}\, H \cap \mathrm{ɑ}\, K$.

---

*) For operators A, B, the symbol $A \equiv B$ will emphasize, that the domains of A and B coincide.

H.5. If A is a selfadjoint operator, then for every $\vec{X} \in \mathrm{Q}A$ and $\vec{Y} \in \mathrm{Q}A$,

$$(A\vec{X}, \vec{Y}) = (\vec{X}, A\vec{Y}).$$

H.5a. Theorem. If H, K are strongly permutable self-adjoint operators, [H.3.], in **H** with $\overline{\mathrm{Q}H \cap \mathrm{Q}K} = \mathbf{H}$, then

$$(H + iK)^* \equiv H - iK,$$
$$(H - iK)^* \equiv H + iK.$$

Proof. We have

$$(1)\quad ((H - iK)\vec{X}, \vec{Y}) = (H\vec{X} - iK\vec{X}, \vec{Y}) = (H\vec{X}, \vec{Y}) - $$
$$- (iK\vec{X}, \vec{Y}) = (H\vec{X}, \vec{Y}) + i(K\vec{X}, \vec{Y})$$

for all $\vec{X} \in \mathrm{Q}H \cap \mathrm{Q}K$.*)

Now (1) equals, for all $\vec{Y} \in \mathrm{Q}H \cap \mathrm{Q}K$, [H.5.],

$$(\vec{X}, H\vec{Y}) + i(\vec{X}, K\vec{Y}) = (\vec{X}, H\vec{Y}) + (\vec{X}, iK\vec{Y})$$
$$= (\vec{X}, H\vec{Y} + iK\vec{Y}) = (\vec{X}, (H + iK)\vec{Y}).$$

It follows that

$$(2)\qquad ((H - iK)\vec{X}, \vec{Y}) = (\vec{X}, (H + iK)\vec{Y}).$$

On the other hand we have for all $X, Y \in \mathrm{Q}H \cap \mathrm{Q}K$:

$$((H - iK)\vec{X}, \vec{Y}) = (\vec{X}, (H - iK)^*\vec{Y}),$$

according to the definition of the adjoint operator, [F.1.]. It follows from (2):

$$(\vec{X}, (H + iK)\vec{Y} = (\vec{X}, (H - iK)^*\vec{Y}));$$

hence

$$(\vec{X}, (H + iK)\vec{Y} - (H - iK)^*\vec{Y}) = 0.$$

Since this equality holds true for every $\vec{X} \in \mathrm{Q}H \cap \mathrm{Q}K$ and since this set is everywhere dense in $\mathfrak{H}$, it follows that

$$(H + iK)\vec{Y} - (H - iK)^*\vec{Y} = \vec{O}.$$

Since this is true for every $\vec{Y} \in \mathrm{Q}H \cap \mathrm{Q}K$, and since $\mathrm{Q}H \cap \mathrm{Q}K$ is the common domain of both operations $H + iK$ and $(H - iK)^*$, we get the second thesis. The proof of the first thesis is similar.

H.5.1. From (1) it follows, by virtue of [H.5.],

$$(2)\qquad N^* \equiv H - iK,$$

and then

$$(3)\qquad N^{**} \equiv H + iK \equiv N.$$

---

*) We admit the rules $(\lambda, \vec{M}, \vec{N}) = \bar{\lambda}(M, N)$, $(\vec{M}, \lambda\vec{N}) = \lambda(\vec{M}, \vec{N})$, as is admitted by physicists.

H.5.1a. We remind that a linear operator A is said to be *closed*, whenever the following condition is satisfied:

If

$$\vec{X}_n \to \vec{X}, \quad A(\vec{X}_n) \to \vec{Y},$$

then

$$\vec{X} \in \daleth A, \quad \text{and} \quad A(\vec{X}) = \vec{Y}.$$

By *the closure* $\tilde{A}$ *of* A we understand the smallest linear closed transformation B, such that for every $\vec{X} \in \daleth A$ we have $A(\vec{X}) = B(\vec{X})$. Not every linear transformation possesses a closure.

H.5.1b. The following theorems are known ((44), p. 30). A^{**} exists, if and only if A possesses a closed prolongation.

H.5.1c. If A is closed and $\overline{\daleth A} = \mathbf{H}$, then $A^{**} = A$.

H.5.2. Theorem. If N is a normal transformation then

$$N = \tilde{N} = N^{**}.$$

H.5.3. Since N^* and N^{**} exist, we have $\overline{\daleth N} = \mathbf{H}$, and $\overline{\daleth N^*} = \mathbf{H}$.

H.5.4. By taking the adjoint in (3) we get $N^{***} = N^*$, i.e.

$$\widetilde{(N^*)} = N^*$$

which proves that N^* is also a closed operation.

H.5.5. From (2) we get

$$\daleth N^* = \daleth H \cap \daleth K = \daleth N.$$

H.5.6. Let $h(\lambda)$, $k(\lambda)$ be the space-valued spectral scales of H, K respectively, [H.2.1.]. We have

$$H\vec{X} = \int_{-\infty}^{+\infty} \lambda \operatorname{Proj}_{h(\lambda)} \vec{X}, \quad K\vec{X} = \int_{-\infty}^{+\infty} \lambda \operatorname{Proj}_{k(\lambda)} \vec{X}, \tag{3.1}$$

[H.1.1.], for every $X \in \daleth N = \daleth H \cap \daleth K$, [H.2.16.], [H.2.3.]. If J denotes a bounded interval $(\alpha, \beta\rangle$, put, in general,

$$h(J) = h(\beta) - h(\alpha), \quad k(J) = k(\beta) - k(\alpha),$$

[A.2.3.]. For any two intervals J', J'' the spaces $h(J')$, $k(J'')$ are compatible with one another, [D.5.].

Having this, fix an interval J, and any partition Π of J into a finite number of disjoint subintervals $J_1, J_2, \ldots, J_n$.

Take $\sigma > 0$. There exists a partition Π sufficiently fine, and points $x(j) \in J_j$, $y(j) \in J_j$ such that, [G.1.2.], [G.2.16.],

$$\left\| \int_J \lambda \, d\operatorname{Proj}_{h(\lambda)} \vec{X} - \sum_j x(j) \operatorname{Proj}_{h(J_j)} \vec{X} \right\| \leq \sigma,$$

$$\left\| \int_J \lambda \, d\operatorname{Proj}_{k(\lambda)} \vec{X} - \sum_j y(j) \operatorname{Proj}_{k(J_j)} \vec{X} \right\| \leq \sigma.$$

Now we have

$$h(J_j) = \sum_l h(J_j) \cdot k(J_l),$$

because of compatibility and orthogonality of spaces.

It follows that

$$\left\| \int_J \lambda \, d\,\mathrm{Proj}_{h(\lambda)} \vec{X} - \sum_{j,l} x(j) \, \mathrm{Proj}_{h(J_j)\,k(J_l)} \vec{X} \right\| \leqq \sigma,$$

and analogously for the other integral.

It follows

$$(4) \quad \left\| \int_J \lambda \, d\,\mathrm{Proj}_{h(\lambda)} \vec{X} + i \int_J \lambda \, d\,\mathrm{Proj}_{k(\lambda)} \vec{X} \right.$$
$$\left. - \sum_{jl} [x(j) + i\,y(l)] \, \mathrm{Proj}_{h(J_j)\,k(J_l)} \vec{X} \right\| \leqq 2\sigma.$$

We extend the definition of $h(x)$, $k(y)$ by putting

$$h(-\infty) = k(-\infty) = (\vec{O}), \quad h(+\infty) = k(+\infty) = 1 = \mathbf{H},$$

and

$$(5) \qquad s(Q(\alpha;\beta)) = h(\alpha) \cdot k(\beta)$$

for

$$-\infty \leqq \alpha, \quad \beta \leqq +\infty,$$

where $Q(\alpha;\beta)$ is a plane quarter [C.1.4.].

We see that all the spaces $Q(\alpha;\beta)$ are compatible with one another, and that they satisfy the conditions I⁰, II⁰, III⁰, [E.1.], and IV⁰, V⁰, [E.2.4.].

In accordance with [G.] we obtain from (4):

$$\int_J \lambda \, d\,\mathrm{Proj}_{h(\lambda)} \vec{X} + i \int_J \lambda \, \mathrm{Proj}_{k(\lambda)} \vec{X} = \iint_{J \times J} z \, d\,\mathrm{Proj}_z \vec{X},$$

where $z = x + iy$ and $J \times J$ denotes the cartesian product of two intervals, hence a square. Since the integrals (3.1) exist, we deduce that

$$H(\vec{X}) + i\,K(\vec{X}) = \int_P z \, d\,\mathrm{Proj}_z \vec{X},$$

where the Radon's integral, [G.2.4.], is extended over the whole plane P.

It follows

$$N(\vec{X}) = \int_P z \, d\,\mathrm{Proj}_z \vec{X} \quad \text{for every} \quad \vec{X} \in \mathfrak{a}\,N.$$

Thus we have proved the following *spectral theorem* for normal operators:

H.5.6a. Theorem. If N is a (maximal) normal operator in $\mathbf{H}$, then there exists a space valued function $s(\dot{Q})$ of the variable plane-quarter Q with the properties, [E.1.], [E.2.9.]:

1) I⁰. If Q_1, Q_2 are any plane-quarters, then the spaces $s(Q_1)$, $s(Q_2)$ are compatible with one another, [D.5.];

II⁰. $s(\emptyset) = (\vec{O})$, $s(P) = 1$;

III⁰. If $Q_1 \subseteq Q_2$, then $s(Q_1) \leq s(Q_2)$;

IV⁰. If for generalized real numbers $\alpha_n, \alpha_0, \beta_n, \beta_0$, [C.1.1.], we have

$$\alpha_1 \geqq \alpha_2 \geqq \cdots \geqq \alpha_n \geqq \cdots \quad \lim \alpha_n = \alpha_0,$$

$$\beta_1 \geqq \beta_2 \geqq \cdots \geqq \beta_n \geqq \cdots \quad \lim \beta_n = \beta_0,$$

then

$$\prod_{n=1}^{\infty} s[Q(\alpha_n, \beta_n)] = s[Q(\alpha_0, \beta_0)];$$

V⁰. If

$$\alpha_1 \leqq \alpha_2 \leqq \cdots \leqq \alpha_n \leqq \cdots \quad \lim \alpha_n = +\infty,$$

$$\beta_1 \leqq \beta_2 \leqq \cdots \leqq \beta_n \leqq \cdots \quad \lim \beta_n = +\infty,$$

then

$$\sum_{n=1}^{\infty} s[Q(\alpha_n, \beta_n)] = s(P),$$

where P is the whole plane.

2) For every $\vec{X} \in \mathrm{D} N$ we have

$$N(\vec{X}) = \int_P z \, d \operatorname{Proj}_z \vec{X}, \tag{6}$$

where z is the complex variable, and where the integral can be considered as vector-valued Stieltjes integral [G.], or as Radon's integral, [G.11.2a.], based on the denumerably additive vector-valued measure $\vec{\mu}(E)$ on all borelian subsets E of the plane, and obtained through extension [E.] of the vector valued function

$$\vec{\mu}(Q) = \operatorname{Proj}_{s(Q)} \vec{X},$$

defined for all plane quarters Q, [G.2.5.].

3) If $N = H + iK$ where H, K are selfadjoint operators in $\mathbf{H}$, then the spectral space-valued scales $h(\lambda)$, $k(\lambda)$ of H and K are

$$h(\lambda) = s(Q(\lambda; +\infty)),$$

$$k(\lambda) = s(Q(+\infty, \lambda)), \quad (-\infty \leqq \lambda \leqq +\infty),$$

so

$$Q(\alpha; \beta) = h(\alpha) \cdot k(\beta).$$

4) A necessary and sufficient condition that the integral (6) exists is that, [G.2.10.],

$$\int_P |z|^2 \, d \|\operatorname{Proj}_z \vec{X}\|^2 < +\infty.$$

Besides, in the case of existence of this integral we have

$$\|N(X)\|^2 = \int_P |z|^2 \, d \|\operatorname{Proj}_z \vec{X}\|^2,$$

and for every $\vec{Y} \in \mathbf{H}$ we have

$$(N(X), \vec{Y}) = \int_P z\, d(\mathrm{Proj}_z \vec{X}, \vec{Y}),$$

where these integrals can be considered as ordinary Stieltjes integrals or as Radon's integrals, which are based on number-valued denumerably additive measures $\mu(E)$, $\nu(E)$ on all borelian sets E, and obtained through extension of the functions of the variable plane quarter Q

$$\mu(Q) = ||\mathrm{Proj}_{s(Q)} \vec{X}||^2, \quad \nu(Q) = (\mathrm{Proj}_{s(Q)} \vec{X}, \vec{Y}),$$

[G.2.5.], respectively.

H.5.7. Def. We call $s(\dot{Q})$ *space-valued double spectral scale of the normal operator* N, and $(\mathrm{Proj}\, s(\dot{Q}))$ *projector-valued double spectral scale* of N.

H.5.8. Remark. On can prove that given the normal operator N, the spectral scale is unique.

H.6. Theorem. If N is a normal transformation, $s(Q)$ is its double spectral scale of spaces, and p a space compatible with all $s(Q)$, then N is permutable with $(\mathrm{Proj}\, p)$, [F.1.1.].

Proof. Let R be a bounded rectangle $\neq \emptyset$. Put

$$M(\vec{X}) =_{df} \int_R z\, d\mathrm{Proj}_z \vec{X} \quad \text{for} \quad \vec{X} \in \mathbb{C}N.$$

Let $R = R(1) \cup \cdots \cup R(n)$ be a partition Π of R into disjoint rectangles $\neq \emptyset$, and $z(i) \in R(i)$, $(i = 1, 2, \ldots)$ some complex numbers.

We have for $\varepsilon > 0$

$$||M(\vec{X}) - \sum_i z_i \,\mathrm{Proj}_{s(R(i))} \vec{X}|| \leqq \varepsilon$$

whenever Π is sufficiently fine.

Hence

$$\|\mathrm{Proj}_p M(\vec{X}) - \sum_i z_i \,\mathrm{Proj}_p \mathrm{Proj}_{s(R(i))} \vec{X}\| \leqq \varepsilon. \tag{1}$$

Since p and $s(R(i))$ are compatible, we have, [D.9.],

$$\mathrm{Proj}_p \mathrm{Proj}_{s(R(i))} \vec{X} = \mathrm{Proj}_{s(R(i))} \vec{X}_p,$$

where

$$\vec{X}_p =_{df} \mathrm{Proj}_p \vec{X}.$$

By letting ε tend to 0, we obtain from (1):

$$\mathrm{Proj}_p M(\vec{X}) = \int_R z\, d\mathrm{Proj}\, \vec{X}_p, \quad \text{for any} \quad R \neq \emptyset. \tag{1.1}$$

Now take an infinite nested sequence $S(n)$ of bounded rectangles with

$$\bigcup_{n=1}^{\infty} S(n) = P$$

and put

(2) $$M(n)(\vec{X}) =_{df} \int_{S(n)} z\, d\,\mathrm{Proj}\,\vec{X} \quad \text{for all} \quad \vec{X} \in \mathsf{a} N.$$

Since

$$\lim_{n\to\infty} M(n)(\vec{X}) = N(\vec{X}) \quad \text{for all} \quad \vec{X} \in \mathsf{a} N,$$

we obtain, by virtue of continuity of the operation of projecting, [D.2.5.],

(3) $$\lim_{n\to\infty} \mathrm{Proj}_p M(n)(\vec{X}) = \mathrm{Proj}_p N(\vec{X}).$$

Applying (1.1) we have

(4) $$\mathrm{Proj}_p M(n)(\vec{X}) = \int_{S(n)} z\, d\,\mathrm{Proj}_p \vec{X} \quad \text{for all} \quad \vec{X} \in \mathsf{a} N.$$

Since the right hand side in (4) tends to $\int_P$, we get, by (3),

$$\mathrm{Proj}_p N(\vec{X}) = \int_P z\, d\,\mathrm{Proj}\,\vec{X}_p$$

i.e.

$$\mathrm{Proj}_p N(\vec{X}) = N\, \mathrm{Proj}_p \vec{X} \quad \text{for all} \quad \vec{X} \in \mathsf{a} N,$$

q.e.d.

H.7. Theorem. If N is a normal transformation and p a space compatible with the spaces of its double spectral scale $s(Q)$, then

1) $p \upharpoonright N$ is a normal transformation in p, [H.1.],

2) $t(Q(\alpha;\beta)) = s(Q(\alpha;\beta)) \cdot p$ is a double spectral scale of $p \upharpoonright N$, [F.1.].

Proof. Let $N = H + iK$ where H, K are strongly permutable selfadjoint operations.

As p is compatible with the spaces $s(Q)$, it is also compatible with $h(\lambda)$, $k(\lambda)$, which are the space-valued spectral scales of H and K respectively. The similar argument as in the proof of Theorem [H.3.6.], applied to the integral representations of H and K, yields the permutability of H and Proj_p, and the permutability of K and Proj_p.

Put

$$N_p = p \upharpoonright N, \quad H_p = p \upharpoonright H, \quad K_p = p \upharpoonright K,$$
$$(N^*)_p = p \uparrow N^*,$$

(compare [F.1.6.]). We have

(1) $$N_p = H_p + i K_p.$$

Since, by [F.1.13.]

$$(N^*)_p = (N_p)^*, \quad H_p = (H^*)_p = (H_p)^*,$$
$$K_p = (K^*)_p = (K_p)^*.$$

The transformations H_p, K_p are selfadjoint in p, and we have

$$(N_p)^* = (N^*)_p = H_p - i K_p. \tag{2}$$

In addition to that H_p and K_p are strongly permutable. Indeed $h(\lambda) \cdot p$, $k(\lambda) \cdot p$ are the spectral scales of H_p and K_p respectively. Since $h(\lambda)$, $k(\mu)$, p are mutually compatible, so is with $h(\lambda) \cdot p$, $k(\mu) \cdot p$. This shows that N_p is a normal (maximal) transformation in p.

To prove the second thesis, consider the sum

$$\sum_j z_j \operatorname{Proj}_{s(R(j))} \vec{X} \quad \text{for} \quad \vec{X} \in \mathfrak{a} N \cap p = \mathfrak{a} N_p,$$

which tends to

$$\int_R z \, d \operatorname{Proj}_{p \cdot s(R(j))} \vec{X}, \qquad R = R(1) \cup R(2) \cup \cdots$$

If $\vec{X} \in p$, this sum can be written

$$\sum_j z_j \operatorname{Proj}_{p \cdot s(R(j))} \vec{X}.$$

If we go to the limit, we get the required integral representation of $N_p(\vec{X})$.

H.8. Theorem. If

1. N is a maximal operator in $\mathbf{H}$,
2. N is Hilbert-bounded i.e. there exists $M > 0$ such that for every $\vec{X} \in \mathfrak{a} N$

$$\|N \vec{X}\| \leq M \cdot \|\vec{X}\|,$$

then

$$\mathfrak{a} N = \mathbf{H}.$$

The proof is based on the spectral theorem.

Trace-representation of normal operators with saturated borelian spectral tribe.

H.9. Let N be a normal transformation in $\mathbf{H}$ and $s(\dot{Q})$ the space-valued double spectral scale of N. The function $s(\dot{Q})$ defined for each plane quarter Q can be extended to a finitely additive function $s(F)$ defined for all figures [E I.] of the plane P. The range $\boldsymbol{S}$ of $s(f)$ is a finitely additive tribe-set of spaces, i.e.

1⁰. if $f, g \in \boldsymbol{S}$, then $f + g \in \boldsymbol{S}$,
2⁰. if $f \in \boldsymbol{S}$, then $\operatorname{co} f \in \boldsymbol{S}$,
3⁰. if $f, g \in \boldsymbol{S}$, $f \cdot g = 0$, then $f \perp g$.

The class $\boldsymbol{S}$, if ordered by the inclusion-relation of spaces, becomes a finitely additive Boolean tribe $(\boldsymbol{S})$.

H.9.1. Def. We call $(\mathbf{S})$ *(simple) spectral tribe of* N.

H.9.2. The space-valued function $s(f)$ can be once more extended to $s(E)$ where E are borelian subsets of the plane P, in such a way that $s(E)$ be denumerably additive, [E.]. (Denoted $\bar{s}(E)$ in [E.].)

This follows from the spectral Theorem [H.5.6a.]. This extension is unique and the range $\boldsymbol{S}^b$ of $s(E)$ is the smallest class of spaces such that

1'. $s(Q) \in \boldsymbol{S}^b$ for every Q;

2'. if $a_1, a_2, \ldots, a_n, \ldots \in \boldsymbol{S}^b$, then $\sum_{n=1}^{\infty} a_n \in \boldsymbol{S}^b$;

3'. if $a \in \boldsymbol{S}^b$, then $\operatorname{co} a \in \boldsymbol{S}^b$;

4'. if $a, b \in \boldsymbol{S}^b$, $a \cdot b = (\vec{O})$, then $a \perp b$.

H.9.3. Def. The class $\boldsymbol{S}^b$ if ordered by the inclusion-relation $a \subseteqq b$ of spaces becomes a denumerably additive Boolean tribe $(\boldsymbol{S}^b)$ which we shall call *borelian spectral tribe of N*.

All spaces of $(\boldsymbol{S}^b)$ are mutually compatible. The tribe $(\boldsymbol{S}^b)$ is isomorphic with the tribe of all borelian sets modulo the ideal composed of all E for which $s(E) = (\vec{O})$, [E.].

H.9.4. Space-trace representation of a normal operator. Let N *be a maximal normal operator*, whose borelian spectral tribe $(\boldsymbol{S}^b)$ is saturated. There exists a generating vector of H with respect to $(\boldsymbol{S}^b)$. We shall prove that such a generating vector can be found in the domain $\mathbb{C} N$ of N. To do this, take a generating vector ξ with $\|\xi\| = 1$. Take an infinite sequence of bounded rectangles

$$Q_1 \subseteqq Q_2 \subseteqq \cdots \subseteqq Q_n \subseteqq \cdots$$

with

$$\bigcup_{n=1}^{\infty} Q_n = P.$$

Put $a_n = s(Q_n)$. We have

$$a_1 \subseteqq a_2 \subseteqq \cdots \qquad \sum_n a_n = I, \qquad a_n \in \boldsymbol{S}.$$

Since a_n is compatible with the spaces of the spectral scale, $N_n =_{df} a_n \rceil N$ is a normal operator in a_n and has the spectral scale $s(\dot{Q}) \cdot a_n$. We also have

(1)
$$N_n \vec{X} = \int_{a_n} z \, d \operatorname{Proj}_z \vec{X}$$

for every $\vec{X} \in \mathbb{C} N_n$. The domain of N_n is the set of all $\vec{X} \in a_n$, for which

$$\int_{a_n} |z|^2 \, d \|\operatorname{Proj}_z \vec{X}\|^2 < +\infty.$$

Since Q_n is bounded, there exists $M_n > 0$ such that for every $z \in Q_n$ we have $|z| \leqq M_n$. We have for every $\vec{X} \in \mathbb{C} N_n$:

(2)
$$\|N_n \vec{X}\|^2 = \int_{a_n} |z|^2 \, d \|\operatorname{Proj}_z \vec{X}\|^2 \leqq M_n^2 \cdot \|(\operatorname{Proj} a_n) \vec{X}\|^2 = M_n^2 \|\vec{X}\|^2.$$

It follows that:

(3) $$\complement N_n = a_n.$$

Thus (1) is valid for all $\vec{X} \in a_n$. Put

(4) $b_0 = a_1,\ b_1 = a_2 - a_1,\ \ldots,\ b_n = a_{n+1} - a_n,\ \ldots,\quad (n = 2, 3, \ldots),$

and

(5) $$\vec{\omega} = \sum_{n=0}^{\infty} \frac{1}{(N+1)\, M_{n+1}} (\operatorname{Proj} b_n)\, \xi.$$

Since

$$\|(\operatorname{Proj} b_n)\, \xi\| \leqq \|\xi\|, \qquad M_n \to \infty,$$

the series (5) converges, so the definition (5) is permissible. Our purpose is to prove that $\vec{\omega} \in \complement N$, and that $\vec{\omega}$ is a generating vector. First we prove that $\vec{\omega} \in \complement N$. Put

$$\vec{\omega}_n = \sum_{k=0}^{n} \frac{1}{(k+1)\, M_{k+1}} (\operatorname{Proj} b_k)\, \xi.$$

We have

(6) $$\lim_{n\to\infty} \vec{\omega}_n = \vec{\omega},$$

and

(7) $$N\vec{\omega}_n = \sum_{k=0}^{n} \frac{1}{(k+1)\, M_{k+1}} N (\operatorname{Proj} b_k)\, \xi,$$

because, by (3)

$$a_{k+1} = \complement N_{k+1} = \complement (a_{k+1} \upharpoonright N) = a_{k+1} \cap \complement N \subseteqq \complement N.$$

From (7) we have

(8) $$N\vec{\omega}_{n+p} - N\vec{\omega}_n = \sum_{k=n+1}^{n+p} \frac{1}{(k+1)\, M_{k+1}} (\operatorname{Proj} b_k)\, \xi.$$

Now the vectors $N(\operatorname{Proj} b_k)\, \xi$, for $k = 0, 1, \ldots$ are orthogonal to one another. Indeed, since b_k is compatible with the space of the spectral scale of N, N and $\operatorname{Proj} b_k$ are permutable. Now

$$(\operatorname{Proj} b_k)\, \xi \in b_k \subseteqq a_{k+1},$$

and then, by (3)

$$(\operatorname{Proj} b_k)\, \xi \in \complement N_{k+1} = a_{k+1} \cap \complement N = \complement N.$$

Hence

(8.1) $N(\operatorname{Proj} b_k)\, \xi = N(\operatorname{Proj} b_k)(\operatorname{Proj} b_k)\, \xi = (\operatorname{Proj} b_k)\, N(\operatorname{Proj} b_k)\, \xi \in b_k;$

and the spaces b_k are orthogonal to one another, because they are disjoint. It follows, from (8), that

(9) $$\|N\vec{\omega}_{n+p} - N\vec{\omega}_n\|^2 = \sum_{k-n+1}^{n+p} \frac{1}{(k+1)^2\, M_{k+1}^2} \|N(\operatorname{Proj} b_k)\, \xi\|^2.$$

Now since $(\mathrm{Proj}\, b_k)\, \vec{\xi} \in b_k \subsetneqq a_{k+1}$ and $a_{k+1} = \complement N_{k+1}$, (by (3)), we have $N(\mathrm{Proj}\, b_k)\, \vec{\xi} = N_{k+1}(\mathrm{Proj}\, b_k)\, \vec{\xi}$, and hence, by (9) and (2),

$$||N\, \vec{\omega}_{n+p} - N\, \vec{\omega}_n||^2 \leqq \sum_{k=n+1}^{n+p} \frac{1}{(k+1)^2 M_{k+1}^2} M_{k+1}^2\, ||(\mathrm{Proj}\, b_k)\, \vec{\xi}\,||^2 \leqq$$

$$\leqq \sum_{k=n+1}^{n+p} \frac{1}{(k+1)^2}\, ||\,\vec{\xi}\,||^2.$$

Since the series $\sum_{n=1} \frac{1}{n^2}$ converges, it follows that the sequence $\{N\, \vec{\omega}_n\}$ converges too. If we take into account (6), we get $N(\vec{\omega}) = \operatorname{Lim}_{n\to\infty} N(\vec{\omega}_n)$ and $\vec{\omega} \in \complement N$, because N is a closed operation. It remains to prove that $\vec{\omega}$ is a generating vector of *1* with respect to $\boldsymbol{S}^b$. The spaces b_k are orthogonal, hence by (8.1),

$$(\mathrm{Proj}\, b_k)\, \vec{\omega} = \frac{1}{(k+1)\, M_{k+1}} (\mathrm{Proj}\, b_k)\, \vec{\xi}.$$

Since $\vec{\xi}$ is a generating vector of *1*, $(\mathrm{Proj}\, b_k)\, \vec{\xi}$ is a generating vector of b_k with respect to the tribe $b_k \cdot \boldsymbol{S}^b$, i.e. $\boldsymbol{S}^b$ restricted to b_k. Hence $(\mathrm{Proj}\, b_k)\, \vec{\omega}$ is a generating vector for $b_k \cdot \boldsymbol{S}^b$. Let $\vec{X} \in 1$. Put $\vec{X}_n = (\mathrm{Proj}\, b_n)\, \vec{X}$. We have $\vec{X} = \vec{X}_1 + \vec{X}_2 + \cdots$ where $\vec{X}_n \in b_n$. Since $(\mathrm{Proj}\, b_k)\, \vec{\omega}$ is a generating vector of b_n with respect to $b_k \cdot (\boldsymbol{S})^b$, the vector $\vec{X}_n$ can be approximated up to $\frac{\varepsilon_n}{2}$ in norm by finite sums $\sum_i \lambda_i (\mathrm{Proj}\, c_i)\, (\mathrm{Proj}\, b_n)\, \vec{\omega}$, where $c_i \in b_n \cdot \boldsymbol{S}^b$. It follows that X can be approximated in norm up to a given $\varepsilon > 0$ by a similar sum:

$$\sum_i \mu_i \cdot (\mathrm{Proj}\, d_i)\, \vec{\omega}, \quad \text{where} \quad d_i \in \boldsymbol{S}^b;$$

so $\vec{\omega}$ is a generating vector of *1* with respect to $\boldsymbol{S}^b$. Thus we have proved the theorem:

H.10.1. Theorem. If

1. N is a (maximal) normal operator in $\mathbf{H}$,
2. $(\boldsymbol{S})^b$ its borelian spectral tribe,
3. $(\boldsymbol{S})^b$ is *saturated in* $\mathbf{H}$,

then there exists a generating vector $\vec{\omega}$ of $\mathbf{H}$ with respect to $\boldsymbol{S}^b$ such that $\vec{\omega} \in \complement N$.

H.10.2. Let $\vec{\omega}$ be such a generating vector with $\vec{\omega} \in \complement N$. Put $\mu(a) = ||\mathrm{Proj}_a \vec{\omega}\,||^2$ for every $a \in \boldsymbol{S}^b$. We know, that $\mu(\dot{a})$ is a denumerably additive, non negative, finite and effective measure on $\boldsymbol{S}^b$. Let N be the normal transformation considered above and denote its $\mathscr{G}^{-1}$-image by $\mathfrak{N}$. This transformation is a normal transformation operating on μ-square summable functions $X(\tau)$ of space-traces. The $\mathscr{G}^{-1}$-image of $\vec{\omega}$ is the function $\Omega(\tau) = 1$, a.e., ([E.7.5.], [E.7.7.]).

Since $\vec{\omega} \in \mathrm{Q}\, N$, we have $\Omega \in \mathrm{Q}\, \mathfrak{N}$. Put $\vec{\varphi} = N(\vec{\omega})$. We have for the corresponding function $\Phi(\tau)$, $\Phi = \mathfrak{N}(\Omega)$, a.e. Now let E be a measurable set of traces with coat e. We have

$$(\mathrm{Proj}_e \vec{\omega})\ \mathscr{G}^{-1}\ \Omega_E(\tau)$$

where Ω_E is the characteristic function of E. Since e is compatible with the spaces of the spectral tribe of N, because e belongs to its borelian extension $(\boldsymbol{S})^b$, we have:

$$\vec{\omega} \in \mathrm{Q}\, N, \qquad \mathrm{Proj}_e N\, \vec{\omega} = N\, \mathrm{Proj}_e \vec{\omega}.$$

Hence

$$\mathfrak{N}\big(\Omega_E(\tau)\big) = \mathrm{Proj}_E\, \mathfrak{N}\, \Omega(\tau)$$

where $\mathrm{Proj}_E f(\tau)$ means the function which admits the value $f(\tau)$ for every $\tau \in E$ and the value 0 for all $\tau \in \mathrm{co}E$. Hence

$$\mathfrak{N}\big(\Omega_E(\tau)\big) = \Phi(\tau) \cdot \Omega_E(\tau)$$

a.e.

If we take a step function $X(\tau) = \sum_i \lambda_i\, \Omega_{E(i)}(\tau)$, we get

$$\mathfrak{N}\big(X(\tau)\big) = \sum_i \lambda_i\, \Omega_{E(i)}(\tau) \cdot \Phi(\tau)$$

a.e.

(1) $$\mathfrak{N}\big(X(\tau)\big) = \Phi(\tau) \cdot X(\tau) \quad \text{a.e.}$$

H.10.2a. Now this formula will be extended over the whole set $\mathrm{Q}\, \mathfrak{N}$. To do this consider the space-brick

$$e(n) = s\big(R(-n, n; -n, n)\big), \quad \text{for} \quad n = 1, 2, \ldots$$

Fix n for a moment and let $E(n)$ be a measurable set of traces with coat $e(n)$.

Since $\vec{\omega}$ is a generating vector, $\big(\mathrm{Proj}\, e(n)\big)\, \vec{\omega}$ is a generating vector of $e(n) \cdot \boldsymbol{S}^b$. Let $\vec{X} \in e(n)$. Take $\sigma_m > 0$, $(m = 1, 2, \ldots)$, where $\sigma_m \to 0$. Let $\vec{X}\ \mathscr{G}^{-1}\ X(\tau)$. There exists a vector

$$\vec{Y}_m = \sum_i \mu_i\, \mathrm{Proj}\,\big(a_i\, e(n)\big)\, \vec{\omega} \in e(n),$$

such that $\|X - Y_m\| \leqq \sigma_m$. We have for the corresponding trace-functions $\|X(\tau) - Y_m(\tau)\| \leqq \sigma_m$.

The operator $N_n = e(n) \upharpoonright N$ is bounded, and hence continuous. It follows that $\lim_{m\to\infty} N(Y_m) = N(X)$. It follows that

$$\lim \mathfrak{N}\big(Y_m(\tau)\big) = \mathfrak{N}\big(X(\tau)\big),$$

i.e.

$$\lim \int_P |\mathfrak{N}\big(X(\tau)\big) - \Phi(\tau) \cdot Y_m(\tau)|^2\, d\mu = 0.$$

Hence there exists a subsequence $\{k(m)\}$ of $\{m\}$ such that

(1) $$\Phi(\tau)\, Y_{k(m)}(\tau) \to \mathfrak{N}(X(\tau))$$

a.e. Since

$$\int |X(\tau) - Y_m(\tau)|^2 \, d\mu \to 0,$$

a new sequence $\{k\,l(m)\}$ can be extracted from $\{k(m)\}$, such that $Y_{k\,l(m)}(\tau) \to X(\tau)$ a.e.; hence

$$\Phi(\tau)\, Y_{k\,l(m)}(\tau) \to \Phi(\tau)\, X(\tau)$$

a.e., and then, since in (1) the limit is unique,

$$\mathfrak{N}(X(\tau)) = \Phi(\tau)\, X(\tau),$$

this formula being valid for all $\vec{X} \in e(n)$, for a given n. Now we shall vary n. Let $\vec{X} \in \mathrm{Q}\, N$. Put

$$\vec{X}(n) = (\mathrm{Proj}\, e(n))\, \vec{X}.$$

We have

(1.0) $$\vec{X} = \lim_{n\to\infty} \vec{X}(n).$$

We shall prove that

$$N(\vec{X}(n)) = \int_{R(n)} z\, d\, \mathrm{Proj}_s(Q(z))\, \vec{X},$$

where

$$R(n) = R(-n, n; -n, n),$$

and

$$N(\vec{X}) = \int_P z\, d\, \mathrm{Proj}_s(Q(z))\, \vec{X}.$$

Since $\vec{X} \in \mathrm{Q}\, N$, we have

$$\int_P |z|^2\, d\|\mathrm{Proj}_s(Q(z))\, \vec{X}\|^2 < +\infty, \qquad \int_P z\, d\, \mathrm{Proj}_s(Q(z))\, \vec{X} = \lim_{n\to\infty} \int_{R(n)},$$

i.e.

$$N(\vec{X}) = \lim_{n\to\infty} N(\vec{X}(n)).$$

If we put

$$X_n(\tau) = X(\tau) \quad \text{whenever} \quad \tau \in E(n),$$

and

$$X_n(\tau) = 0 \qquad \text{whenever} \quad \tau \,\overline{\in}\, E(n),$$

we get

(1.1) $$\mathfrak{N}(X(\tau)) = \lim_{n\to\infty} \mathfrak{N}(X_n(\tau)) = \lim_{n\to\infty} [\Phi(\tau)\, X_n(\tau)].$$

From (1.0) it follows that there exists a subsequence

$$\{q(n)\} \quad \text{of} \quad \{n\} \quad \text{with} \quad X_{q(n)}(\tau) \to X(\tau)$$

a.e. Hence

$$\Phi(\tau)\, X_{q(n)}(\tau) \to \Phi(\tau)\, X(\tau)$$

a.e. By (1.1), another sequence $\{q\,p(n)\}$ can be extracted from $\{q(n)\}$, such that

$$\Phi(\tau)\, X_{q\,p(n)}(\tau) \to \mathfrak{N}(X(\tau))$$

a.e. It follows that

$$\mathfrak{N}(X(\tau)) = \Phi(\tau) \cdot X(\tau)$$

for every $\vec{X} \in \mathsf{d}\, N$. Thus we have proved the following theorem.

H.10.3. Theorem. If

1. N is a maximal normal operator in $\mathbf{H}$,
2. $\boldsymbol{S}^b$ its borelian spectral scale of spaces,
3. $\boldsymbol{S}^b$ is saturated in $\mathbf{H}$,
4. $\vec{\omega}$ is a generating vector of $\mathbf{H}$ with respect to $\boldsymbol{S}^b$, such that $\vec{\omega} \in \mathsf{d}\, N$,
5. $\mu(a) = \|\mathrm{Proj}_a \vec{\omega}\|^2$ for all $a \in \boldsymbol{S}^b$,
6. $\mathcal{G}$ is the correspondence between the characteristic function $\Omega_E(\tau)$ of the measurable set E of traces and $\mathrm{Proj}_e \vec{\omega}$, where e is the coat of E,

then

1) $\mathcal{G}^{-1}$ can be extended to a one-to-one isomorphic and isometric correspondence between all vectors $\vec{X} \in \mathit{1}$ and μ-square summable functions $X(\tau)$ of space-trace τ (with respect to equality of vectors and the $=$-a.e.-equality of functions).

2) The extended $\mathcal{G}^{-1}$ carries $N(\vec{X})$ into the normal transformation

$$\mathfrak{N}(X(\tau)) = \Phi(\tau) \cdot X(\tau)$$

a.e., where $\Phi(\tau)$ is a μ-square summable function and where $\mathsf{d}\, N\, \mathcal{G}^{-1} \mathsf{d}\, \mathfrak{N}$.

H.10.4. We shall calculate the function $\Phi(\tau)$ considered. Take a bounded rectangle

$$R = R(x_1, y_1;\, x_2, y_2) \neq \mathbf{0},$$

put $r = s(R)$, and suppose that $\mu(r) > 0$. Consider the normal transformation $N_1 =_{df} r \upharpoonright N$. We have $\mathsf{d}\, N_1 = r$. We have for $\vec{X} \in r$:

$$N(\vec{X}) = N_1(\vec{X}) = \int_R z\, d\,\mathrm{Proj}_{Q(z)} \vec{X}$$

where $Q(z) = Q(x; y)$, $z = x + i\,y$. Let $z_0 \in R$. We have

$$\|N(\vec{X}) - z_0 \vec{X}\| = \left\| \int_R (z - z_0)\, d\,\mathrm{Proj}_{Q(z)} \vec{X} \right\| \leq \sigma \cdot \|\vec{X}\|,$$

where $\sigma = \sup_{z \in R} |z - z_0|$. If we go over to the $\mathcal{G}^{-1}$-image, we get

$$(0) \qquad \|\Phi(\tau) X(\tau) - z_0 X(\tau)\| \leqq \sigma \|X(\tau)\|,$$

for all $X(\tau)$ vanishing outside the borelian set $E(r)$, whose coat is r.

Put $\vec{X} = \text{Proj}_r \vec{\omega}$. We obtain, from (0), $\|\Phi(\tau) - z_0\| \leqq \sigma$, for every $\tau \in E(r)$; we have $\mu(r) > 0$. Having this, consider an infinite sequence $\{\underset{n}{\Pi}\}$, $(n = 1, 2, \ldots)$ of partitions of the whole plane P into half open equal rectangles $R(n, \alpha)$, such that $\underset{n+1}{\Pi}$ be a subpartition of $\underset{n}{\Pi}$, and that the diagonal of meshes tends to 0. We know that to every trace τ of spaces there corresponds a well defined $\boldsymbol{B}_J$-trace ξ_τ on the plane. If we denote by $R(n, \alpha_\tau)$ the rectangle with covers ξ_τ, the trace τ is covered by $s(R(n, \alpha_\tau))$. The measure μ in $\boldsymbol{S}^b$ induces a measure of borelian sets of the plane, so that $R(n, \alpha_\tau)$ has the measure

$$\mu(R(n, \alpha_\tau)) = \mu(s(R(n, \alpha))) > 0.$$

(Those and only those rectangles $R(n, \alpha_\tau)$ have a positive measure for which there exists τ such that the corresponding trace ξ_τ is covered by this rectangle.)

Let us fix n for a moment, and consider those meshes $R(n, \beta)$ of $\underset{n}{\Pi}$, which have a positive measure. Select in each $R(n, \beta)$ a point $z(n, \beta)$. Denote by δ_n the diameter of the meshes of $\underset{n}{\Pi}$. We get,

$$(1) \qquad \|\Phi(\tau) - \sum_\beta \Omega_{R(n, \beta)}(\tau) \cdot z(n, \beta)\| < \delta_n.$$

This equation is valid with exception of traces τ, whose corresponding plane traces ξ_τ lies on the "border" of the plane. Since these traces make up a set of μ-measure 0, we can say that (1) holds almost μ-everywhere. If $\delta_n \to 0$, the sum $\sum_\beta$ in (1) tends in μ-square mean toward $\Phi(\tau)$. Hence a subsequence $\{k(n)\}$ can be selected in $\{n\}$ such that $\sum_\beta$ tends a.e. to $\Phi(\tau)$. Let τ_0 be a trace at which the convergence takes place. There exists an infinite sequence of rectangles $R(k(n), \beta_n)$ such that τ_0 is covered by $s(R(k(n), \beta_n))$ for $n = 1, 2, \ldots$ We get

$$|\Phi(\tau_0) - z(k(n), \beta_n)| \to 0,$$

which gives

$$(2) \qquad \Phi(\tau_0) = z(\tau_0)$$

where $z(\tau_0)$ is the vertex of the plane $\boldsymbol{B}_J$-trace ξ_{τ_0}, corresponding to τ_0. The equality (2) holds for μ-almost every τ_0. Thus we have proved the

H.10.4.1. Theorem. The function $\Phi(\tau)$ is a.e. equal to the vertex $z(\tau)$ of the plane $\boldsymbol{B}_J$-trace ξ_τ corresponding to τ, [C I.], so we have

$$\mathfrak{N}(X(\tau)) = z(\tau) \cdot X(\tau)$$

a.e. We call this formula: *trace-representation of* $N(\vec{X})$.

The canonical representation of the normal operator N having a saturated borelian spectral tribe S^b

The correspondence $\mathcal{G}^{-1}$ carries isomorphically, isometrically and homeomorphically the vectors $\vec{X}$ of the H.-H.-space **H** into square-summable functions $X(\tau)$ of the variable space-trace τ. The transformation N goes over into the transformation $\mathfrak{N}(F(\tau)) = \Phi(\tau) \cdot F(\tau)$, where $\Phi(\tau)$ is a complex number depending on τ. The generating vector $\vec{\omega}$ of *1* goes over into the constant function $\Omega(\tau) = 1$. Since $\vec{\omega} \in \text{ᗡ} N$, it follows that $\Omega(\tau)$ belongs to ᗡ $\mathfrak{N}$; hence $\Omega(\tau)$ is μ-square summable. There is a $1 \to 1$ correspondence $\mathfrak{S}^{-1}$ between space traces and $\boldsymbol{B}_J$-plane traces, [C I.], this correspondence preserving finite and denumerable set-operations and the measure μ generated by $\vec{\omega}$. There is a $1 \to 1$ correspondence $\mathfrak{S}^{-1}$ between square summable functions $f(\tau)$ and μ-square summable function $f(\alpha)$ of the variable plane $\boldsymbol{B}_J$-trace α, this correspondence being μ-invariant in its domain and in its range. The operator N goes over into $\mathfrak{N}(X(\alpha)) =^\mu \text{vert}\,\alpha \cdot X(\alpha)$. It is an operation-measure and scalar-product-isomorphism, which is μ-invariant in the domain and also in the range. The function vert α is μ-square summable, because $\Phi(\tau)$ is so. Now we can apply the correspondence M which carries the μ-square summable function $X(\dot{\alpha})$ into μ-square summable function $F(\dot{z})$ of the complex variable, [C I.]. The function vert $\dot{\alpha}$ goes over into $\dot{z}$. Thus we get the representation of N in the form $\boldsymbol{N}(F(z) =^\mu z \cdot F(z)$ which is the desired *canonical representation of* $N\vec{X}$. We have proved the following theorem on the transformation $\boldsymbol{C} = \mathsf{M}\,\mathfrak{S}^{-1}\mathcal{G}^{-1}$:

H.10.4.2. Theorem. If

1. N is a maximal normal operator in a separable and complete H.-H.-space **H**.

2. The borelian spectral tribe S^b of N is saturated.

3. $\vec{\omega}$ is a generating vector of **H** with respect to S^b,

then $\vec{\omega}$ generates, by means of the effective, denumerably additive measure μ, defined by

$$\mu(a) = ||\text{Proj}_a \vec{\omega}||^2, \quad (\text{for all } a \in T_b),$$

an isometric, isomorphic and homeomorphic mapping $\boldsymbol{C}$ of **H** onto the space $\mathfrak{H}$ of all μ-quare summable functions $F(z)$ of the complex

27*

variable, taken modulo the ideal of μ-null-sets and defined μ-almost everywhere. The mapping $\boldsymbol{C}$ is called *canonical*. It transforms N into the maximal normal operator $\boldsymbol{N}(F(z)) =^{\mu} z \cdot F(z)$ in $\mathbf{H}$.

Canonical representation of a normal operator N whose borelian spectral tribe S^b is not saturated

H.10.5. Let $(\boldsymbol{S}^b)$ be non saturated. We can find a saturating sequence of spaces, [E.]: $\boldsymbol{H}_1, \boldsymbol{H}_2, \ldots, \boldsymbol{H}_n, \ldots$ (finite or denumerable) of subspaces of $\mathbf{H}$.

This spaces are orthogonal to one another, they are compatible with the spaces of $\boldsymbol{S}^b$ and we have

$$\sum_n \boldsymbol{H}_n = \mathbf{H}, \tag{1}$$

and the smallest tribe $(\boldsymbol{S}^b)^*$ containing $(\boldsymbol{S}^b)$ and all $\boldsymbol{H}_n$ is saturated.

Select a generating vector $\vec{\omega}$ of $\mathbf{H}$ with respect to $(\boldsymbol{S}^b)^*$, such that

$$\vec{\omega} \in \text{ᗡ} N.$$

The vector

$$\vec{\omega}_n = (\operatorname{Proj} \boldsymbol{H}_n)\, \vec{\omega}$$

is a generating vector of $\boldsymbol{H}_n$ with respect to the tribe $\boldsymbol{H}_n \cdot (\boldsymbol{S}^b)$.

If we put

$$N_n =_{df} \boldsymbol{H}_n \upharpoonleft N,$$

we have

$$\vec{\omega}_n \in \text{ᗡ} N_n.$$

The operator N_n is maximal normal in $\boldsymbol{H}_n$, and its borelian spectral tribe is

$$(\boldsymbol{S}^b_n) = \boldsymbol{H}_n \cdot (\boldsymbol{S}^b).$$

This tribe is saturated in $\boldsymbol{H}_n$. Having this, we shall treat each N_n in its space $\boldsymbol{H}_n$ separately, and in accordance with what has been made in the case of the saturated borelian spectral tribe. The measure

$$\mu_n(a) = ||\operatorname{Proj}_a \vec{\omega}_n||^2$$

in the tribe $(\boldsymbol{S}^b_n)$ is effective.

We have

$$\sum_n \mu_n(\boldsymbol{H}_n) = \mu(\mathbf{H}) = ||\vec{\omega}||^2.$$

Let us fix n for a moment and let $\boldsymbol{C}_n$ be the isomorphic and isometric correspondence between the vectors $\vec{X}$ of the space $\boldsymbol{H}_n$ and the μ_n-square summable functions $f(z)$ of the complex variable z, which transforms N_n

into the operator:

(2) $$N_n f(z) =^{\mu_n} z \cdot f(z),$$

where $=^{\mu_n}$ means: equal almost μ_n-everywhere.

H.10.5.1. Take as many disjoint copies $P_1, P_2, \ldots$ of the complex plane as we have made in [E.], and apply the canonical mapping $\boldsymbol{C}_n$ from $\boldsymbol{H}_n$ onto the μ_n-square summable functions $f(z)$ of the complex variable z in P_n.

Denote by V the disjoint union of all these copies P_n (the *multiplane*), as we have defined in [E.] and consider following notions:

If $u \in V$, then by "*Number of u*", *Num u* we understand the ordinary complex number z represented by u. If $E \subseteqq V$, then "*Num E*" denotes the set of all complex z such that there exists $u \in E$ with $z = \mathrm{Num}\, u$. Conversely, if Φ is a set of complex numbers z, then "*Point of Φ*", *Point Φ* means the set of all points $u \in V$ with $\mathrm{Num}\, u \in \Phi$.

H.10.5.2. Let $E \subseteqq V$. We have the disjoint decomposition

$$E = (E \cap P_1) \cup (E \cap P_2) \cap \cdots$$

If $E \cap P_n$ is μ_n-measurable for $n = 1, 2, \ldots$, we shall say that E is *ν-measurable*, and by $\nu(E)$ we shall understand

$$\sum_n \mu_n (E \cap P_n).$$

This measure is denumerably additive. The ν-measurable sets make up a denumerably additive tribe, where the equality is taken modulo sets with ν-measure $= 0$.

A function $f(p)$, defined on V is fitting this tribe (i.e. ν-measurable), if and only if

$$P_n \upharpoonleft f(p), \quad (n = 1, 2, \ldots),$$

is μ_n-measurable. It is ν-square summable, if and only if $P_n \upharpoonleft f(p)$ is μ_n-square summable for $n = 1, 2, \ldots$, and if the sum

$$\sum_n \int |P_n \upharpoonleft f(p)|^2 \, d\mu$$

converges (if there is available an infinity of copies of the complex plane).

H.10.5.3. Having this, let $\vec{X} \in \mathbf{H}$. We have its orthogonal decomposition

$$\vec{X} = \sum_n \vec{X}_n, \quad \vec{X}_n \in \boldsymbol{H}_n.$$

Applying the mapping $\boldsymbol{C}_n$ to $\vec{X}_n$, we get a function $f_n(p)$, where $p \in P_n$. Hence the joint mapping $\boldsymbol{C}$ will transform $\vec{X}$ into a function $f(p)$ on V.

This mapping is an isomorphism with respect to equalities involved. Since

$$\int_{P_n} |f_n(p)|^2 \, d\mu_n$$

exists for all $n = 1, 2, \ldots$, and since this integral equals $\|\vec{X}_n\|^2$, the equality

$$\|X\|^2 = \sum_n \|X_n\|^2$$

implies

$$\sum \int_{P_n} |f_n(p)|^2 \, d\mu_n = \int |f(p)|^2 \, d\nu,$$

and consequently we get

$$\|X\|^2 = \int |f(p)|^2 \, d\nu,$$

so the mapping $\vec{X} \to f(p)$ is isometric. The mapping is "onto", because if $f(p)$ is ν-square summable, so is $P_n \rceil f(p)$ and then

$$\int |f(p)|^2 \, dv = \sum_n \int |f_n(p)|^2 \, d\mu_n.$$

Since $\boldsymbol{C}_n$, extended to functions, is a mapping "onto", it follows that there exists $\vec{X}_n \in \mathbf{H}$ with

$$\vec{X}_n \smile f_n(p).$$

Since the sum

$$\sum_n \|X_n\|^2 = \sum_n \int |f_n(p)|^2 \, d\mu_n$$

exists, we can put

$$\vec{X} =_{df} \sum \vec{X}_n.$$

We get

$$\|X\|^2 = \int |f(p)|^2 \, dv \quad \text{and} \quad \vec{X} \smile^{C} f(p).$$

H.10.5.4. Let a be a space of the tribe $(\boldsymbol{S}^b)^*$. We have

$$a = \sum_n p_n \cdot a_n \quad \text{where} \quad a_n \in \boldsymbol{S}_n^b.$$

Since $\boldsymbol{C}_n$ maps a_n into a μ_n-measurable subset E_n of P_n, it follows that $\boldsymbol{C}$ maps a into a ν-measurable subset E on V.

We have

$$\nu(E) = \sum_n \mu_n(E_n) = \sum_n \|(\mathrm{Proj}\, a_n)\, \vec{\omega}_n\|^2 = \sum_n \|(\mathrm{Proj}\, a_n)\, \vec{\omega}\|^2$$
$$= \|\mathrm{Proj}_a \vec{\omega}\|^2 = \mu(a).$$

Consequently the measure ν can be denoted by the letter μ.

H.10.5.4a. We have: if $a\, \boldsymbol{C}\, E$, then $\mu(a) = \mu(E)$. The measure ν is generated in a usual way by the generating vector $\vec{\omega}$ in $\mathbf{H}$. Conversely if E is a ν-measurable set, and we put

$$E = \bigcup_n (P_n \cap E),$$

the $\boldsymbol{C}_n^{-1}$-image of $P_n \cap E$ is a subspace a_n of $\boldsymbol{H}_n$.

Consequently $\boldsymbol{C}^{-1} E$ is a subspace of $\mathbf{H}$, having the form $\sum_n a_n p_n$; hence it belongs to the saturated tribe $(\boldsymbol{S}^b)^*$.

H.10.6. The mapping $\boldsymbol{C}$ carries N into the operation

$$\boldsymbol{N}(f(p)) = (\mathrm{Num}\, p) \cdot f(p), \tag{1}$$

because each N_n gets the form (1).

The mapping $\boldsymbol{C}$ is not unique, for the tribe $(\boldsymbol{S}^b)$ can be saturated by various saturating sequences $\{\boldsymbol{H}_n\}$ of spaces. In addition to that there are many generating vectors $\vec{\omega}$, and hence many measures μ.

H.10.7. Let us examine the $\boldsymbol{C}$-images of the spaces of $(\boldsymbol{S}^b)^*$ and of the vector $\mathrm{Proj}_a \vec{\omega}$ (i.e. images induced by $\boldsymbol{C}$; they can be said: $\boldsymbol{C}$-images), where a belongs to the saturated tribe $(\boldsymbol{S}^b)^*$. Let $a \in (\boldsymbol{S}^b)^*$, and put $a_n = \boldsymbol{H}_n a$, $(n = 1, 2, \ldots)$. We have $a = \sum_n a_n$. Now we have $a_n = [A_n]_n$ for some μ_n-measurable set A_n of $\boldsymbol{H}_n$-space traces. We have

$$\boldsymbol{H}_n \rceil \boldsymbol{C} = \boldsymbol{C}_n = \mathfrak{M}_n \mathscr{G}_n^{-1},$$

where $\mathfrak{M}_n$ carries A_n onto a μ_n-measurable subset α_n of P_n, this correspondence being determined by the equation $[A_n]_n = \bar{s}_n(\alpha)_n$ i.e. $a_n = \bar{s}_n(\alpha_n)$, [E.]. (Notice that $\mathfrak{M}_n = \mathsf{M}_n \mathfrak{S}_n^{-1}$, [C I.32.], [C I.38.], [E.5.1.].)

Consequently the joint mapping $\boldsymbol{C}$ carries the space

$$a = \sum_n a_n$$

onto the set

$$\bigcup_n \alpha_n,$$

and this mapping is an isomorphism from the tribe $(\boldsymbol{S}^b)^*$ onto the class of all μ-measurable subsets of V, taken with respect to the μ-equality of those sets.

The vector $\mathrm{Proj}(\boldsymbol{H}_n)\, \vec{\omega}$ is a generating vector of $\boldsymbol{H}_n$ with respect to the tribe

$$(\boldsymbol{S}_n^b) = (\boldsymbol{S}^b) \cdot \boldsymbol{H}_n.$$

Hence $\mathscr{G}_n^{-1}$ carries this vector into the function $\Omega_n(\tau) = \mathrm{const} = 1$, defined for all space-traces in $\boldsymbol{H}_n$.

This function is mapped, by $\mathfrak{M}_n$, into the constant function $= 1$, on P_n.

Thus the joint mapping $\boldsymbol{C}$ carries $\vec{\omega}$ into the constant function $= 1$ on the union V of all replicas of the complex plane. If

$$a_n \in \boldsymbol{S}^b_n,$$

then

$$\text{Proj}\,(a_n)\,\vec{\omega} = \text{Proj}\,(a_n)\,\vec{\omega}_n$$

will give the unity-valued constant functions on the set A_n of space-traces in $\boldsymbol{H}_n$, where $[A_n]_n = a_n$. This function is carried through $\mathfrak{M}_n$ into the unity-valued constant function on the subset α_n of P_n, where

$$\bar{s}_n(\alpha_n) = [A_n].$$

Now take an arbitrary space $a \in (\boldsymbol{S}^b_n)^*$. We have

$$a = \sum_n a_n \quad \text{where} \quad a_n \in \boldsymbol{S}^b_n.$$

Consequently the vector $\text{Proj}_a\vec{\omega}$ is carried into the characteristic function of the μ-measurable subset of V:

$$\alpha = \sum_n \boldsymbol{C}(a_n) = \sum_n \alpha_n$$

where

$$\bar{s}_n(\alpha_n) = a_n.$$

Thus we can state the following theorem:

H.10.8. Theorem. The mapping $\boldsymbol{C}$ generates a one-to-one isomorphic correspondence between the spaces of the saturated tribe $(\boldsymbol{S}^b)^*$ and μ-measurable subsets of V. This mapping preserves finite and denumerable lattice-operations. The equality for subsets of V is that of modulo the ideal of sets with μ-measure $= 0$. If $a \in \boldsymbol{H}_n \cdot \boldsymbol{S}^b$, the $\boldsymbol{C}$-image of a is a μ-measurable subset of P_n and conversely.

The mapping $\boldsymbol{C}$ establishes a one-to-one correspondence between the projections $\text{Proj}_a\vec{\omega}$ of the generating vector $\vec{\omega}$ (where $a \in (\boldsymbol{S}^b)^*$), and the characteristic functions of μ-measurable sets α of V.

We have

$$\|\text{Proj}_a\vec{\omega}\|^2 = \mu(\alpha).$$

In particular

$$\mu(P_n) = \|\vec{\omega}_n\|^2 \quad \text{and} \quad \mu(V) = \|\vec{\omega}\|^2.$$

H.10.9. Now consider N_n and its spectral scale $s_n(Q)$, where Q is a variable plane-quarter of the complex plane P. We have

$$s_n(Q) = s(Q) \cdot \boldsymbol{H}_n,$$

where $s(Q)$ is the spectral scale of N.

For every Q, the $\boldsymbol{C}$-image of $s_n(Q)$ is Q, *independently* of $n = 1, 2, \ldots$, and $\boldsymbol{C}_n$ maps the space $a_n = a \cdot \boldsymbol{H}_n$ where $a \in \boldsymbol{S}^b$ (i.e. a space of the borelian spectral tribe of N_n) into a same set of the complex plane P. Hence

$$\operatorname{Num} \boldsymbol{C}(a_n)$$

does not depend on n.

Now, if we consider the joint mapping $\boldsymbol{C}$, and a set $a \in \boldsymbol{S}^b$, we have

$$a = \sum_n a \boldsymbol{H}_n,$$

and hence $\boldsymbol{C}$ maps a onto a same borelian subset on every replica P_n of the complex plane.

We have the theorem

H.10.10. Theorem. If a is a space of the borelian spectral tribe $(\boldsymbol{S}^b)$ of N, then there exists a borelian set α' of complex numbers such that the $\boldsymbol{C}$-image of a is Pointα' i.e. it is composed of exact replicas of α'.

Conversely every borelian set $E \subseteqq V$ such that

$$E = \operatorname{Point}(\operatorname{Num} E),$$

is the $\boldsymbol{C}$-image of some space of the borelian spectral scale of N.

Remark. Of course the μ-measures of $E \cap P_n$ may differ if n varies, and for some n it may be even $= 0$.

H.10.11. Let $\vec{X} \in \mathbf{H}$ and $a \in (\boldsymbol{S}^b)^*$. We like to know the $\boldsymbol{C}$-image of $\operatorname{Proj}_a \vec{X}$. Let $\vec{X}\, \boldsymbol{C}\, f(p)$. We know that $\boldsymbol{C}$ generates a mapping of spaces into point sets on V. Let α be the set corresponding to a. Under these circumstances the $\boldsymbol{C}$-image of $\operatorname{Proj}_a \vec{X}$ will be the function $g(p)$ which is $= f(p)$ almost μ-everywhere on α, but vanishing μ - a.e. on $V - \alpha$. We may denote such a function by

$$[\alpha \rceil f(p)]^V.$$

We have proved the

H.10.12. Theorem. If $a \in (\boldsymbol{S}^b)^*$, $\vec{X} \in \mathbf{H}$, $\boldsymbol{C}$ is a canonical mapping, and α the subset of V, corresponding to a, then the vector $\operatorname{Proj}(a)\, \vec{X}$ goes over to $[\alpha \rceil f(p)]^V$ where $f = \boldsymbol{C} \vec{X}$.

The operator $\operatorname{Proj}(a)\, \vec{X}$ with variable $\vec{X}$ goes over to the operator $(\alpha \rceil f)^V$ with variable f.

The obtained results can be stated in the following.

H.10.13. *Fundamental theorem on (natural) canonical representation of maximal normal operators* in separable and complete H.-H.-spaces:

Theorem. If

1. $\mathbf{H}$ is a separable and complete H.-H.-space,
2. N is a maximal normal operator in $\mathbf{H}$ (with $\mathrm{C} N$ everywhere dense in $\mathbf{H}$),
3. $s(Q)$ its space valued double spectral scale,

4. its borelian spectral tribe $(\boldsymbol{S}^b)$ is not saturated,

5. $\boldsymbol{H}_1, \boldsymbol{H}_2, \ldots$ is a saturating sequence of spaces for $(\boldsymbol{S}^b)$, and $(\boldsymbol{S}^b)^*$ is the corresponding saturated tribe,

6. $\vec{\omega}$ a generating vector of $\mathbf{H}$ with respect to $(\boldsymbol{S}^b)^*$, such that $\vec{\omega} \in \mathrm{d}\, N$,

7. $\mu(a) = \|\mathrm{Proj}\,(a)\,\vec{\omega}\|^2$, defined for all spaces $a \in (\boldsymbol{S}^b)^*$,

8. $N_n = \boldsymbol{H}_n \upharpoonleft N$, $\vec{\omega}_n = \mathrm{Proj}\,(\boldsymbol{H}_n)\,\vec{\omega}$, $\mu_n(a) = \|\mathrm{Proj}\,(a)\,\vec{\omega}_n\|^2$ for all a belonging to $\boldsymbol{S}^b \cdot \boldsymbol{H}_n = \boldsymbol{S}^b_n$,

9. P_n a copy of the complex plane P,

10. $\mu_n(\alpha)$ the denumerably additive measure, defined for all borelian subsets α of P_n by $\mu_n(Q) = \mu_n\, s(Q)$ and afterwards extended in the Lebesguean way, [A I.], to all sets

$$\alpha \frown \alpha_0 \sim \alpha_1, \qquad (\mu_n\text{-measurable sets}),$$

where α_0, α_1 are subsets of borelian sets on P_n with μ_n-measure $= 0$.

11. $\mathfrak{M}_n$ the mapping of μ_n-measurable sets of traces in $\boldsymbol{H}_n$ onto the μ_n-measurable subsets of P_n, defined for borelian sets α by $\mathsf{M}_n\, \mathfrak{S}_n$, so that we have for borelian sets: $A\, \mathsf{M}_n\, \alpha$ is equivalcut to $[A]_n = \bar{s}_n(\alpha)$, (where $[A]$ is the coat of A in $\boldsymbol{H}_n$), and extended to μ_n-measurable subsets of P_n, and afterwards to functions,

12. $\mathscr{G}_n^{-1}$ is the mapping, carrying vectors $\vec{X}$ of $\boldsymbol{H}_n$ into μ_n-square summable functions $f(\tau)$ of the variable space trace in $\boldsymbol{H}_n$, and defined by extension of the relation

$$(\mathrm{Proj}\,(A)\,\vec{\omega}_n)\, \mathscr{G}_n^{-1}(\Omega_a(\tau)),$$

where $[a] = A$, and $\Omega_a(\tau)$ is the characteristic function of the set a,

13. $\boldsymbol{C}_n = \mathfrak{M}_n\, \mathscr{G}_n^{-1}$ is the (natural) canonical mapping of $\boldsymbol{H}_n$ into the space of all μ_n-square summable functions $f(p)$ of the point p, varying in P_n, defined μ_n-a.e.,

14. V is the disjoint union of the planes P_n,

15. The measure $\nu(E) = \sum_n \mu_n(E \frown P_n)$ is defined for the smallest tribe of subsets E of V containing all tribes of μ_n-measurable subsets of P_n, $(n = 1, 2, \ldots)$,

16. $\mathfrak{M}$ is the joint mapping which carries $\bigcup_n A_n$ into $\bigcup_n \mathfrak{M}_n A_n$, where A_n are measurable sets of $\boldsymbol{H}_n$-traces,

17. $\nu(\bigcup_n A_n) = \sum_n \mu_n A_n$ for A_n as above,

18. $\boldsymbol{C}$ is the joint mapping which carries the vectors $\vec{X} = \sum_n \vec{X}_n$, with $\vec{X}_n \in \boldsymbol{H}_n$ into the functions $f(p)$ on V and defined by

$$f(p) = \mathfrak{M}_n\, \mathscr{G}_n^{-1}\, \vec{X}_n = \boldsymbol{C}_n \vec{X}_n \quad \text{for} \quad p \in P_n, \qquad (n = 1, 2, \ldots),$$

then

1) If $a \in (\boldsymbol{S}^b)^*$, $a\ \boldsymbol{C}\ \alpha$, then $\mu(a) = \nu(\alpha)$. If $a \in \boldsymbol{S}_n^b$, $a\ \boldsymbol{C}\ \alpha$, then

$$\mu_n(\alpha) = \nu(\alpha),$$

(thus we can write μ instead of ν),

2) $\mathfrak{M}_n$, $\boldsymbol{C}_n$ are restrictions to $\boldsymbol{H}_n$ of the mappings $\mathfrak{M}$, $\boldsymbol{C}$ respectively.

3) $\mathfrak{M}$ is an isomorphism from the tribe of all sets of traces

$$B = B_1 \cup B_2 \cup \cdots, \tag{1}$$

where B_n is a μ_n-measurable set of $\boldsymbol{H}_n$-traces, onto the set of all μ-measurable point sets of V. This isomorphism is isometric and is taken with respect to μ_n-equalities of the n-th constituents in (1) and with respect to the μ-a.e.-equality of sets in V.

4) $\boldsymbol{C}$ is an isometric isomorphism from the vectors $\vec{X}$ in $\mathbf{H}$ onto μ-square summable functions $f(p)$ defined μ-a.e.-on V.

5) $\boldsymbol{C}$ carries N into the canonical form

$$\boldsymbol{N}(f(p)) =^{\mu} (\text{Num}\, p) \cdot f(p),$$

where the domain of $\boldsymbol{N}$ is composed of all μ-square summable functions defined μ-a.e.-on V, such that $(\text{Num}\, p) \cdot f(p)$ is also μ-square summable. $\boldsymbol{N}$ is a maximal normal operator in the space $\mathfrak{H}$ of all μ-square summable, complex valued, functions defined μ-a.e.-on V; $\boldsymbol{C}$ gives N_n the canonical form

$$\boldsymbol{N}_n(f(p)) =^{\mu} (\text{Num}\, p) \cdot f(p)$$

which is a maximal normal operator in the space of all μ-square summable functions defined μ-a.e.-on V and vanishing a.e. on P_i for all $i \neq n$.

6) Let $[A] \in \boldsymbol{S}^b$ and let

$$A = \bigcup_n (A \cap W_n),$$

where W_n denotes the set of all $\boldsymbol{H}_n$-traces.

Then $\mathfrak{M}$ carries $A \cap W_n$ into a borelian set α_n on the plane P_n, for which $\text{Num}\, \alpha_n$ does not depend on n. Thus $\boldsymbol{C}$ generates an image of A in the form of a borelian subset of the complex plane P.

To the space $[A]$ there corresponds, by influence of $\boldsymbol{C}$, a borelian set $E \subseteqq V$ such that

$$E = \text{Point}(\text{Num}\, E). \tag{2}$$

Conversely to every borelian set E with the property (2) there corresponds a well determined space belonging to $(\boldsymbol{S}^b)$.

7) If $a \in (\boldsymbol{S}^b)$, then $\boldsymbol{C}$ makes correspond to a a borelian subset α of V.

If $a = \sum_n a_n \boldsymbol{H}_n$, where $a_n \in \boldsymbol{S}^b$,

then $\alpha = \bigcup_n \alpha_n$, where α_n corresponds to $a_n \boldsymbol{H}_n$ through the influence of $\boldsymbol{C}_n$.

Conversely if α is a borelian subset of V, then there exists a space $a \in \boldsymbol{S}_n^b$ to which it corresponds.

This correspondence between borelian subsets of V and spaces of $(\boldsymbol{S}^b)^*$ does not depend on the choice of the generating vector $\vec{\omega}$.

8) The generating vector $\vec{\omega}$ goes over through $\boldsymbol{C}$ to a constant function defined a.e. on V.

Proj $(a)\,\vec{\omega}$ for $a \in (\boldsymbol{S}^b)^*$, goes over to the characteristic function of the set α which corresponds to a.

9) If E is a borelian subset of the complex plane, then the set Point E goes over through $\boldsymbol{C}^{-1}$ into the space $\bar{s}(E)$.

H.10.14. Theorem. For a given N the mapping $\boldsymbol{C}$ is well determined by giving

1) the spectral scale $s(Q)$ of N,
2) the saturating sequence $\{\boldsymbol{H}_n\}$,
3) the generating vector $\vec{\omega}$.

Def. We call $\boldsymbol{C}$ *canonical mapping of* $\boldsymbol{H}$.

H.10.15. We know that a selfadjoint operator H is normal. If we consider a canonical mapping $\boldsymbol{C}$ of H on the multiplane, the measure μ is equal 0 on all copies of the subset of P, which is the complement of the real axis.

H.10.16. The unitary operator U is normal. The corresponding measure μ of a canonical mapping can differ from 0 only on the copies of the circle with center at the origin and with radius $= 1$.

H.10.16a. We give the last three theorems without proof to not increase the volume of the book.

Chapter J

Operators $Nf(\dot{x}) =_{df} \varphi(x)\cdot f(\dot{x})$ for ordinary functions f

J.1. The purpose of this Chapter [J.] is to supply preparations for a new theory of functions $F(N)$, where N is a normal operator. These functions are not ordinary compound functions, but they must be conceived as a suitable generalization of iterations of N, viz.

$N^2 = N\,N$, $N^3 = N\,N^2$ etc. Therefore the term "operational calculus" used by STONE (26) seems adequate.

The classical theory of the operational calculus (26) is built up through Radon's integrals. Our method will be based on the trace-representation or canonical representation of normal operators, [H.]. The confrontation of our theory with the classical one will be supplied.

In [H.] we have given the normal operator N the form $\mathfrak{N}(f(\tau)) = \Phi(\tau) \cdot f(\tau)$ where τ is a variable space-trace, and later the canonical form $\boldsymbol{N}(f(p)) = {}^{\mu}\mathrm{Num}\,p \cdot f(p)$ where z is a point varying over the union V of an at most denumerable number of mutually disjoint replicas of the complex plane P, or else varying over a suitable μ-measurable subset of V.

To make the topic simplier we shall rely on this canonical representation. The possibility of having a canonical representation of the most general normal operation is a rather deep result which has required quite complicated proof of some auxiliary Theorems [C I.], [E.], but it can be noticed, that a modification of the method, which shall be developped beneath, would yield an analogous theory of the operational calculus without relying on the canonical representation and by admitting the existence of the trace-representations only.

We shall give detailed arguments for general normal operations and shall only sketch the operational calculus in Hermitean selfadjoint operators, which can be carried through independently of the general theory of normal operators and even in a more simple, though quite analogous way.

The function $F(N)$ where N is a (maximal) normal operator in the given H.-H.-space will be defined through the canonical representation of N. Now, there may be several canonical representations of N; hence the proof of the independence of $F(N)$ of the choice of this canonical representation is essential, and will be supplied.

J.2. Since our starting point will be operations having the form $Nf = \varphi \cdot f$, where f, φ are some functions, the following general topic will be useful:

Let $V \neq \emptyset$ be a set of any elements x, (T) a denumerably additive tribe of subsets a of V, $\mu(a)$ a non negative, denumerably additive measure on (T). The notion of equality of sets a will be that of equality (equivalence) $(\doteq)$-modulo the ideal of all sets a of T, for which $\mu(a) = 0$.

We shall consider complex (or real)-valued functions $f(x)$, fitting (T); i.e. for every halfopen rectangle R on the complex plane, the set

$$\{x \mid f(x) \in R\}$$

belongs to T.

The equality $(\doteq)$ of sets, mentioned above, induces a corresponding equality $f(x) \doteq g(x)$ of functions.

The set of all μ-square summable functions $f(\alpha)$ defined on a set $a \in T$ can be organized, in the known way into a H.-H.-space, by introducing the scalar product.

J.2.1. Remark. The measure μ and (T) may not have the following property:

$$\text{“if} \quad a \in T, \quad b \subseteqq a, \quad \mu(a) = 0, \quad \text{then} \quad b \in T\text{”}.$$

Now it is known, that (T) can be extended to a smallest tribe (T^L) possessing the above property, and μ can be extended to all sets belonging to T^L, so as to remain denumerably additive. This is just the measure-hall extension, [A I.]. The extended tribe is composed of all sets (somata)

$$b = a \cup b_1 \sim b_2, \quad \text{where} \quad a \in T,$$

and b_1, b_2 are subsets of some a_1, a_2, where $a_1, a_2 \in T$, $\mu(a_1) \equiv \mu(a_2) \equiv 0$.

In relation to that we define the new measure as

$$\mu^L(b) = \mu(a).$$

We may call (T^L) the *Lebesgue's μ-extension of the tribe* (T).

The equality $(\dot{=})$ generates a new one $(=^\mu)$ of elements of T^L, defined by

$$\text{“}b =^\mu b' \quad \text{means} \quad \mu^L(b \dot{+} b') = \mu^L(b - b') + \mu^L(b' - b) = 0\text{”}.$$

Lattis operations on (T^L) and μ^L are invariant with respect to $(=^\mu)$.

We refer to sets of T^L by calling them *μ-measurable sets.*

There is an isometric and denumerable lattice-isomorphism between (T) and (T^L) with respect to equalities $(\dot{=})$ and $(=^\mu)$. In the sequal we shall write μ in stead of μ^L, and will not make careful discrimination between (T) and (T^L).

The functions fitting (T^L) will be termed *μ-measurable functions,* and the equality $(=^\mu)$ extended to functions will be termed "*μ-almost everywhere-equality*", (μ-a.e.). The set of all functions $f(x)$, defined on a set $b \in T^L$ almost μ-everywhere and μ-square summable, make up an H.-H.-space. Another space will be obtained if we fix a, and take all functions defined and μ-square summable on sets a' where $a' =^\mu a$. This new space is isometrically isomorphic to the former one, and we may simply identify them.

J.2.2. If E is a μ-measurable subset of V, then μ-square summable functions $g(x)$ defined μ-almost everywhere on E make up also an H.-H.-space which we shall denote by $\boldsymbol{H}(E)$.

If we take such a function $g(x)$, then the $g^V(x)$ or $(g(x))^V$ will mean the function defined as $\dot{=} g(x)$ on E, and $\dot{=} 0$ on $\text{co}E = V - E$. The functions $g^V(x)$ also make up an H.-H.-space which will be denoted by $\boldsymbol{H}^V(E)$. This is a closed subspace of $\boldsymbol{H}$. The spaces $\boldsymbol{H}^V(E)$ and

$\boldsymbol{H}(E)$ are isometrically isomorphic through a correspondence easy to define. The space $\boldsymbol{H}(E)$ is not a subspace of $\boldsymbol{H}$. Usually these two spaces are identified, but we shall not do that, since such an identification would obliterate the clarity of arguments.

We shall use the Russellian symbol

$$E \upharpoonleft f(x)$$

to denote the function $f(x)$ restricted to $E \cap \mathrm{D} f$, (1).

In this section functions and sets will always be considered with respect to the (generalized) equality $(=^{\mu})$, which will often be written simply as $(=)$.

J.3. Lemma. Under the circumstances of [J.2.1.] and [J.2.2.] let $\varphi(x)$ be a complex-valued function fitting (T) and defined a.e. in V.

Then the set of all $f \in \boldsymbol{H}$, such that $f \cdot \varphi \in \boldsymbol{H}$, is everywhere dense in $\boldsymbol{H}$ (in the H.-H.-topology).

Proof. Put

$$a(n) = \{x \mid n \leqq |\varphi(x)| < n+1\} \quad \text{for} \quad n = 0, 1, 2, \ldots$$

Since φ is fitting (T), we have $a(n) \in T$. We also see that $a(i) \cap a(j) = \mathit{O}$ for $i \neq j$ and $\bigcup_{n=1}^{\infty} a(n) = V$. Take the function $f \in \boldsymbol{H}$ and $\varepsilon > 0$. Since $\mu(a)$ and $\int_a |f(x)|^2 \, d\mu$ are denumerably additive functions of $a \in T$, there exists m such that, if we put

$$b = a(1) \cap \cdots \cap a(m), \qquad c = a(m+1) \cap a(m+2) \cap \cdots,$$

we have

(1)
$$\int_c |f(x)|^2 \, d\mu \leqq \varepsilon, \qquad \mu(c) \leqq \varepsilon.$$

Define the function $h \in \boldsymbol{H}$ by putting

$$h(x) = f(x) \quad \text{for} \quad x \in b,$$

and

$$h(x) = \frac{1}{m+k+1} \quad \text{for} \quad x \in a(m+k), \qquad (k = 1, 2, \ldots).$$

We have

(2)
$$h \in \boldsymbol{H}.$$

Indeed

$$\int_{a(m+k)} |h(x)|^2 \, d\mu = \frac{\mu\, a(m+k)}{(m+k+1)^2} \leqq \mu\, a(m+k),$$

hence, by (1),

(3)
$$\int_c |h(x)|^2 \, d\mu \leqq \mu(c) \leqq \varepsilon.$$

We have

(4) $$h \cdot \varphi \in \boldsymbol{H}.$$

Indeed

$$\int\limits_{a(m+k)} |h(x) \quad \varphi(x)|^2 \, d\mu \leqq \int\limits_{a(m+k)} \frac{(m+k+1)^2}{(m+k+1)^2} \, d\mu = \mu\, a(m+k),$$

$$(k = 1, 2, \ldots).$$

Hence

$$\int\limits_{c} |h(x) \cdot \varphi(x)|^2 \, d\mu \leqq \mu(c).$$

Since

$$\int\limits_{b} |h(x) \cdot \varphi(x)|^2 \, d\mu \leqq (m+1)^2 \int\limits_{b} |f(x)|^2 \, dx,$$

it follows that

$$h \cdot \varphi \in \boldsymbol{H}.$$

Now, consider $\|f - h\|$ where $\| \ \|$ is the H.-H.-norm in $\boldsymbol{H}$.

We have

$$\|f - h\|^2 = \int\limits_{b} |f - h|^2 \, d\mu + \int\limits_{c} |f - h|^2 \, d\mu.$$

As $f(x) = h(x)$ on b, the Minkowski's inequality gives

$$\|f - h\| \leqq \sqrt{\int\limits_{c} |f(x)|^2 \, d\mu} + \sqrt{\int\limits_{c} |h(x)|^2 \, d\mu};$$

hence, by (1) and (3),

(5) $$\|f - h\| \leqq 2\sqrt{\varepsilon}.$$

Consequently, for every $f \in \boldsymbol{H}$ and $\varepsilon > 0$ there exists $h \in \boldsymbol{H}$ such that (5) and $h \cdot \varphi \in \boldsymbol{H}$. The lemma is established.

J.4. Theorem. Let $\varphi(x)$ be a complex-valued function, defined a.e. on V and fitting (T). Consider the operation U in $\boldsymbol{H}$, defined by

$$U f \doteq \varphi(x) \cdot f(x),$$

where $\mathrm{D}\,U$ is the set of all $f \in \boldsymbol{H}$, for which $f \cdot \varphi \in \boldsymbol{H}$. Since, by Lemma [J.3.], $\overline{\mathrm{D}\,U} = \boldsymbol{H}$, there exists the adjoint operation U^* of U.

We shall prove that

$$U^* f \doteq \overline{\varphi(x)} \cdot f(x) \text{ and } \quad \mathrm{D}\,U^* = \mathrm{D}\,U,$$

where $\bar{\varphi}$ means the function conjugate to $\varphi(x)$.

Proof. By definition of the adjoint operation, the domain $\mathrm{D}\,U^*$ is the set of all $g \in \boldsymbol{H}$ such that, for every $f \in \mathrm{D}\,U$ there exists $g' \in \boldsymbol{H}$ with the scalar products

$$(U f, g) = (f, g'),$$

i.e.

(1) $$\int_V \overline{\varphi(x) \cdot f(x)} \cdot g(x)\, d\mu = \int_V \overline{f(x)} \cdot g'(x)\, d\mu.$$

This g' is unique and $g' = U^* f$. Now, (1) implies

(2) $$\int_V \overline{f(x)} \cdot [\bar{\varphi}(x) \cdot g(x) - g'(x)]\, d\mu.$$

Let $g \in \mathrm{C}\, U^*$. There exists g' with (2) for all

$$f \in \mathrm{C}\, U.$$

Since $\mathrm{C}\, U$ is everywhere dense in $\boldsymbol{H}$, it follows that

(2.1) $$\bar{\varphi}(x) \cdot g(x) - g'(x) \doteq 0,$$

i.e.

(3) $$g'(x) \doteq \bar{\varphi}(x) \cdot g(x).$$

Hence

$$U^* g \doteq \bar{\varphi}(x) \cdot g(x) \quad \text{for all} \quad g \in \mathrm{C}\, U^*.$$

Since $\bar{\varphi}(x) \cdot g(x)$ is μ-square summable on V, the function $\varphi(x) \cdot g(x)$ is also μ-square summable on V, for $|\varphi \cdot g| = |\bar{\varphi} \cdot g|$. Hence $g \in \mathrm{C}\, U$.

Consequently

(4) $$\mathrm{C}\, U^* \subseteqq \mathrm{C}\, U.$$

Now let $g \in \mathrm{C}\, U$. The function $\varphi \cdot g$ is μ-square summable on V; hence $\bar{\varphi} \cdot g$ is so. Hence if we put $g' = \bar{\varphi} \cdot g$, we get

$$\bar{\varphi} \cdot g - g' \doteq 0.$$

Hence for all $f \in \mathrm{C}\, U$ we obtain

$$\int_V \bar{f} [\bar{\varphi} g - g']\, d\mu = 0.$$

Since $\bar{\varphi} \cdot g$ is μ-square summable and $\bar{f}$ is so, it follows that the integral $\int_V \bar{f}\, \bar{\varphi}\, g\, d\mu$ exists. Hence we get the formula (1) and then

$$(U f, g) = (f, g') \quad \text{for all} \quad W f \in \mathrm{C}\, U.$$

This proves that $g \in \mathrm{C}\, U^*$.

Thus we have $\mathrm{C}\, U \subseteqq \mathrm{C}\, U^*$. If we combine this with (4), we get $\mathrm{C}\, U = \mathrm{C}\, U^*$, which completes the proof.

J.5. Theorem. If $\varphi(x)$ is real-valued and fitting (T), then the operation

$$H f = \varphi(x) \cdot f(x)$$

with $\mathrm{C}\, H$ composed of all $f \in \boldsymbol{H}$, with $\varphi \cdot f \in \boldsymbol{H}$, is a hermitian selfadjoint operator in $\boldsymbol{H}$.

Proof. By Theorem [J.4.], we have

$$H^* f = \bar{\varphi}(x) \cdot f(x), \quad \mathrm{Ɑ} H^* = \mathrm{Ɑ} H.$$

Hence

$$H^* = H, \quad \text{q.e.d.}$$

J.6. Lemma. If under circumstances of [J.2.]

1. $b(x)$ is a complex-valued and μ-summable function, defined μ-a.e. on V,

2. $\varphi(x)$ is a real-valued function, fitting (T) and defined μ-a.e. on V,

3. $b(x) \cdot \varphi(x)$ is μ-summable on V,

4. $E(\lambda) =_{df} \{x \mid \varphi(x) \leqq \lambda\}$ for all $-\infty < \lambda < \infty$,

5. $B(\lambda) = \int_{E(\lambda)} b(x)\, d\mu$ for $-\infty < \lambda < +\infty$,

then

1) $B(\lambda)$ is a complex valued function of bounded variation on $(-\infty, +\infty)$,

2) the Stieltjes integral

$$\int_{-\infty}^{+\infty} \lambda\, dB(\lambda)$$

exists,

3) $\int_{-\infty}^{+\infty} \lambda\, dB(\lambda) = \int_V b(x)\, \varphi(x)\, d\mu$.

Though this lemma is known, we shall supply a proof.

J.6a. Proof. Since φ fits T, we have $E(\lambda) \in T$, and hence, since $b(x)$ is μ-summable on V, it is also μ-summable on $E(\lambda)$, thus the hypothesis 5. is meaningful.

We have for any finite number of disjoint intervals

$$(\alpha(i), \beta(i)\rangle \quad \text{on} \quad (-\infty, +\infty),$$

$$|B(\beta(i)) - B(\alpha(i))| \leqq \int_{E(\beta(i)) - E(\alpha(i))} |b(x)|\, d\mu,$$

and then

$$\sum_i |B(\beta(i)) - B(\alpha(i))| \leqq \int_V |b(x)|\, d\mu,$$

so $B(\lambda)$ is of bounded variation on $(-\infty, +\infty)$. Let

$$\cdots < \lambda_{-1} < \lambda_0 < \lambda_1 < \cdots < \lambda_n < \cdots$$

be a partition of $(-\infty, +\infty)$ with $\sup(\lambda_{n+1} - \lambda_n) \leqq \varepsilon$, where the suprimum is taken for all

$$n = 0, \pm 1, \pm 2, \ldots$$

Choose λ'_n where

$$\lambda_n \leqq \lambda'_n \leqq \lambda_{n+1}, \quad (n = 0, \pm 1, \pm 2, \ldots).$$

Put

$$F(n) = E(\lambda_{n+1}) - E(\lambda_n) \in T.$$

We have

$$(1) \quad \left|\lambda'_n \int_{F(n)} b(x)\,d\mu - \int_{F(n)} \varphi(x)\,b(x)\,d\mu\right| \leqq \int_{F(n)} |\lambda'_n - \varphi(x)|\cdot|b(x)|\,d\mu.$$

Since, on $F(n)$, we have

$$\lambda_n < \varphi(x) \leqq \lambda_{n+1},$$

and

$$\lambda_n \leqq \lambda'_n \leqq \lambda_{n+1},$$

we have

$$|\lambda'_n - \varphi(x)| \leqq \lambda_{n+1} - \lambda_n \leqq \varepsilon \quad \text{for} \quad x \in F(n).$$

It follows that the lefthand side of (1) is $\leqq \varepsilon \int_{F(n)} |b(x)|\,d\mu$. Since $|b(x)|$ is summable on V, and

$$F(i) \frown F(j) = \emptyset \quad \text{for} \quad i \neq j,$$

$$\bigcup_n F(n) = V,$$

it follows, by summation:

$$\left|\sum_{n=-\infty}^{+\infty} \lambda'_n \int_{F(n)} b(x)\,d\mu - \int_V \varphi(x)\,b(x)\,d\mu\right| \leqq \varepsilon \int_V |b(x)|\,d\mu,$$

i.e.

$$\left|\sum_{n=-\infty}^{+\infty} \lambda'_n (B(\lambda_{n+1}) - B(\lambda_n)) - \int_V \varphi(x)\,b(x)\,d\mu\right| \leqq \varepsilon \int_V |b(x)|\,d\mu.$$

Consequently the Stieltjes integral $\int_{-\infty}^{+\infty} \lambda\,dB(\lambda)$ exists and equals $\int_V \varphi(x)\,b(x)\,d\mu$.

J.6.1. Lemma. If

1. $J_n(f)$, $(n = 1, 2, \ldots)$ is a linear continuous complex valued functional with $\mathrm{D}\,J_n = \boldsymbol{H}$, where $\boldsymbol{H}$ is a general H.-H.-space, separable and complete,

2. $J_n(f)$ converges for all $f \in \boldsymbol{H}$,

then

$$J(f) =_{df} \lim_{n\to\infty} J_n(f)$$

is also a linear continuous functional with $\mathrm{D}\,J = \boldsymbol{H}$.

Proof. The theorem is known. See e.g. (44), where the proof is given by a method taken from Osgood (62). It is in accordance with the Osgood-Julia-principle (59). See also (60) and (61).

For the sake of completeness we sketch the proof, taken from (44).

We know that a necessary and sufficient condition for the continuity of a linear functional $J(f)$ with $\mathrm{D}\, J = \boldsymbol{H}$ is its Hilbert-boundedness, i.e.

$$|J(f)| \leqq M \cdot \|f\| \quad \text{where} \quad M > 0$$

is some number. This is, in turn, equivalent to the existence of a sphere in $\boldsymbol{H}$ on which $J(f)$ be bounded.

Hence, to prove the lemma, it suffices to prove that the sequence $\{J_n(f)\}$ is bounded on some sphere.

Suppose that the above sequence is not bounded on any sphere. Hence, whatever the sphere K and $c > 0$ may be, there exists $f \in K$ and an index n, such that

$$|J_n(f)| \geqq c.$$

Take a sphere K and fix f_0 and $n(1)$, so as to have

$$|J_{n(1)}(f_0)| \geqq 1.$$

By continuity we can find a sphere $K_1 \subseteqq K$ with radius $\leqq \frac{1}{2}$,

$$\text{where} \quad |J_{n(1)}(f)| \geqq 1 - \tfrac{1}{2}.$$

Since on K the sequence is not bounded, we can find $f_1 \in K$, an index $n(2) > n(1)$, and a sphere $K_2 \subseteqq K_1$ with radius $\leqq \frac{1}{2^2}$ such that

$$|J_{n(2)}(f)| > 2 - \tfrac{1}{2} \quad \text{in} \quad K_2.$$

By repeating this argument, we get a vector $f \in \bigcap\limits_n K_n$, for which $\{J_n(f)\}$ is not bounded. This completes the proof.

This general theorem gives the possibility of proving the following Lemma on Stieltjes integrals. We refer to [J.6.1.].

J.7. Lemma. If

1. $A(z)$ is a complex-valued and continuous function on the whole complex plane P,

2. $\int\limits_P A(z)\, d(\mathrm{Proj}_z f, g)$ exists for every $g \in \boldsymbol{H}$,

then

$$\int\limits_P |A(z)|^2\, d\|\mathrm{Proj}_z f\|^2$$

also exists.

Proof. Let $R(n)$, $(n = 1, 2, \ldots)$ be nested bounded rectangles in the plane P, such that $\bigcup\limits_n R(n) = P$. Put

$$J_n(g) = \int\limits_{R(n)} A(z)\, d(\mathrm{Proj}_z f, g), \quad (n = 1, 2, \ldots).$$

We have

$$J_n(g) = (f_n, g),$$

where

$$f_n =_{df} \int\limits_{R(n)} A(z)\, \mathrm{Proj}_z f.$$

By hypothesis 2 $\lim_{n\to\infty} J_n(g)$ exists for every g. Hence, by Lemma [J.6.1.],

$$J(g) =_{df} \lim J_n(g)$$

is a linear continuous functional with domain $\boldsymbol{H}$.

Hence, by the known theorem by FRÉCHET (63) (see also (64)), there exists $f_0 \in \boldsymbol{H}$ such that

$$J(g) = (f_0, g) \quad \text{for all} \quad g \in \boldsymbol{H}.$$

Since $(f_n, g) \to (f_0, g)$ for all g, the sequence $\{\|f_n\|\}$ is bounded (see e.g. (44), (46)). Hence

$$\Big\| \int_{R(n)} A \, d\operatorname{Proj}_z f \Big\|^2 = \int_{R(n)} |A|^2 \, d\|\operatorname{Proj}_z f\|^2$$

is bounded.

Since $R(1) \subseteqq R(2) \subseteqq \cdots$, it follows that

$$\lim_{n\to\infty} \int_{R(n)} |A|^2 \, d\|\operatorname{Proj}_z\|^2$$

exists, q.e.d.

As a consequence we obtain the following

J.7.1. Lemma. If

1. $A(z)$ is complex valued and continuous in P,
2. $f \in \boldsymbol{H}$,

then the following are equivalent:

I. $\int_P A(z) \, d(\operatorname{Proj}_z f, g)$ exists for every $g \in \boldsymbol{H}$,

II. $\int_P |A(z)|^2 \, d\|\operatorname{Proj}_z f\|^2$ exists,

III. $\int_P A(z) \, d\operatorname{Proj}_z f$ exists.

J.7.2. An analogous theorem is valid in the case of the straight line, i.e. for functions $A(x)$ of a single real variable x.

J.8. Theorem. If

1. $\varphi(x)$ is a real valued function, defined μ-a.e.-on V and fitting (T),
2. $H f = \varphi \cdot f$ is a self-adjoint hermitean operator with $\mathrm{Q}H$, composed of all $f \in \boldsymbol{H}$, for which $\varphi \cdot f \in \boldsymbol{H}$, (see Theorem [J.5.]),
3. $a(\lambda) =_{df} \{x \mid \varphi(x) \leqq \lambda\}$ for $-\infty < \lambda < +\infty$,

then

$$E(\dot{\lambda}) =_{df} \boldsymbol{H}^V(a(\dot{\lambda})),$$

(see [J.2.]), is the space-valued spectral scale of H.

Proof. Take $g \in \boldsymbol{H}$. We have for $f \in \mathrm{Q}H$: $\operatorname{Proj} E(\lambda) f \doteq f(x)$ for $x \in a(\lambda)$ and $\doteq 0$ for $x \in \operatorname{co} a(\lambda)$.

Hence

$$(\operatorname{Proj} E(\lambda) f, g) = \int_{a(\lambda)} \overline{f(x)} \, g(x) \, d\mu.$$

By Lemma [J.6.] we have

$$(1)\quad \int_{-\infty}^{+\infty} \lambda\, d(\operatorname{Proj} E(\lambda)\, f, g) = \int_V \overline{\varphi(x) f(x)}\, g(x)\, d\mu = \int_V \overline{Hf}\cdot g\, d\mu = (Hf, g).$$

Since this integral exists for all $g \in \boldsymbol{H}$, the Lemma [J.7.] assures the existence of the integral

$$\int_{-\infty}^{+\infty} \lambda^2\, d\|\operatorname{Proj} E(\lambda)\, f\|^2,$$

which equals

$$\left\|\int_{-\infty}^{+\infty} \lambda\, d\operatorname{Proj} E(\lambda)\, f\right\|^2.$$

Now we have

$$(2)\quad \int_{-\infty}^{+\infty} \lambda\, d(\operatorname{Proj} E(\lambda)\, f, g) = \left(\int_{-\infty}^{+\infty} \lambda\, d\operatorname{Proj} E(\lambda)\, f, g\right) \quad \text{for all} \quad g \in \boldsymbol{H}.$$

From (1) and (2) we get

$$Hf = \int_{-\infty}^{+\infty} \lambda\, d\operatorname{Proj} E(\lambda)\, f.$$

We also get

$$\|Hf\|^2 = \int_{-\infty}^{+\infty} \lambda^2\, d\|\operatorname{Proj} E(\lambda)\|^2.$$

Since all other requirements for a spectral scale are also satisfied, the theorem is proved.

J.9. Remark. We remind, that we have admitted the spectral theorem for selfadjoint operators as granted, and even admitted the known fact that the spectral scale of a selfadjoint operator is unique. This implies that $\boldsymbol{H}^V(a(\lambda))$, in [J.8.], is *the* spectral scale of H.

J.10. Theorem. If

1. $\varphi(x)$, $\psi(x)$ are real valued functions fitting (T) and defined μ-a.e.-on V,

2. $Hf = \varphi\cdot f$, $Kf = \psi f$ are corresponding selfadjoint operators in $\boldsymbol{H}$ in agreement with Theorem [J.5.],

then H and K are strongly permutable.

Proof. Put

$$a(\lambda) = \{x \mid \varphi(x) \leqq \lambda\},$$

$$b(v) = \{x \mid \psi(x) \leqq v\},$$

for

$$-\infty < \lambda < +\infty, \qquad -\infty < v < +\infty,$$

and

$$A(\lambda) =_{df} \boldsymbol{H}^V(a(\lambda)), \quad B(v) =_{df} \boldsymbol{H}^V(b(v)).$$

By Theorem [J.8.], $A(\dot{\lambda})$ and $B(\dot{v})$ are the spectral scales of H and K respectively.

We have for $f\in \boldsymbol{H}$:

$$\operatorname{Proj} A(\lambda)\operatorname{Proj} B(v) f \doteq \begin{cases} f(x) & \text{for} \quad x\in a(\lambda)\cap b(v),\\ 0 & \text{for} \quad x\bar{\in} a(\lambda)\cap b(v),\end{cases}$$

$$\operatorname{Proj} B(v)\operatorname{Proj} A(\lambda) f \doteq \begin{cases} f(x) & \text{for} \quad x\in b(v)\cap a(\lambda),\\ 0 & \text{for} \quad x\bar{\in} b(v)\cap a(\lambda),\end{cases}$$

so

$$\operatorname{Proj} A(\lambda)\operatorname{Proj} B(v) = \operatorname{Proj} B(v)\operatorname{Proj} A(\lambda), \quad \text{q.e.d.}$$

J.10.1. Theorem. If

1. $\varphi(x)$ is a complex valued function fitting (T) and defined μ-a.e.-on V,

2. $Uf = \varphi(x)\cdot f(x)$ is the operator in $\boldsymbol{H}$, with $\mathbb{C}U$ composed of all $f\in\boldsymbol{H}$, where $\varphi\cdot f\in\boldsymbol{H}$,

then U is a maximal normal operator in $\boldsymbol{H}$.

Proof. Put $\varphi(x) = \alpha(x) + i\beta(x)$ where $\alpha(x)$, $\beta(x)$ are real-valued. They fit (T) and are defined a.e. on V.

Consider the operators

$$H f = \alpha(x)\cdot f(x),$$
$$K f = \beta(x)\cdot f(x),$$

in accordance with Theorem [J.5.]. They are strongly permutable (by Theorem [J.10.]). It follows that U is a maximal normal operator in $\boldsymbol{H}$, q.e.d.

J.11. Def. We shall say that U, as defined in [J.10.1.], is the operator $Uf = \varphi\cdot f$ with *maximal domain*.

J.12. Remark. If we consider $Uf = \varphi\cdot f$ with the domain which does not embrace all $f\in\boldsymbol{H}$, for which $\varphi\cdot f\in\boldsymbol{H}$, the transformation would be not a maximal normal transformation in $\boldsymbol{H}$.

J.12.3. Theorem. For the operator $Uf=\varphi\cdot f$ of Theorem [J.10.2.], with

$$a(\lambda) =_{df} \hat{x}\{\alpha(x)\leqq\lambda\},$$
$$b(v) =_{df} \hat{x}\{\beta(x)\leqq v\},$$
$$A(\lambda) = \boldsymbol{H}^V(a(\lambda)), \qquad B(v) = \boldsymbol{H}^V(b(v)),$$

(see [J.2.]), and

$$z = \lambda + i v,$$

the function

$$E(z) = A(\lambda)\cdot B(v)$$

is a double space-valued spectral scale of U.

J.13. Remark. Till now we have considered functions defined μ-a.e. on V, but it is easy to see, that if we consider functions defined μ-a.e. on a given subset E of V and fitting (T), analogous theorems can be stated.

J.13.1. Theorem. Let $U f(x) = \varphi(x) \cdot f(x)$ be a maximal normal operator in $\boldsymbol{H}$; so $\varphi(x)$ is defined μ-a.e. in V. Let E be μ-measurable, $E \subseteqq V$.

Put $e =_{df} \boldsymbol{H}^V(E)$ (see [J.2.]). The operation U is permutable with Proj_e, i.e.

$$\mathrm{Proj}_e U f = U \,\mathrm{Proj}_e f \quad \text{for} \quad f \in \mathrm{Ɑ}\, U.$$

The proof is straight forward.

J.13.2. Remark. The set E induces *four operators* generated by U, viz.

1) $U_E(f_1) = (E \upharpoonright \varphi) \cdot f_1$ defined for all $f_1 \in \boldsymbol{H}(E)$ with $(E \upharpoonright \varphi) \cdot f_1 \in \boldsymbol{H}(E)$;

2) $U_E^V(f) = (E \upharpoonright \varphi)^V \cdot f$, defined for all $f \in \boldsymbol{H}^V(E)$, where $(E \upharpoonright \varphi \cdot f) \in \boldsymbol{H}(E)$;

3) $\mathrm{Proj}_e U f$ and

4) $U \,\mathrm{Proj}_e f$ both considered with maximal domains, i.e. for all f for which these expressions are respectively meaningful.

These operators have properties, stated in the following theorem whose proof is easy.

J.14. Theorem. If

1. $E \in T$,
2. $\varphi(x)$ is fitting (T) and is defined μ-a.e. in V,
3. $U f(x) = \varphi(x) \cdot f(x)$ has the maximal domain in $\boldsymbol{H}$,

then

1) the operators U_E, U_E^V, $\mathrm{Proj}_e U$, $U\,\mathrm{Proj}_e$ defined above are maximal normal operators in the spaces $\boldsymbol{H}(E)$, $e = \boldsymbol{H}^V(E)$, $\boldsymbol{H}$, $\boldsymbol{H}$ respectively.

2) U_E is an image of U_E^V through an isometric isomorphism L^{-1} which carries the space $\boldsymbol{H}^V(E)$ onto $\boldsymbol{H}(E)$ and which is defined by

$$f(x)\, L^{-1} f^V(x).$$

3) The domains of U_E^V, $\mathrm{Proj}_e U$, $U\,\mathrm{Proj}_e$ are respectively

$$\mathrm{Ɑ}(e \upharpoonright U) = \mathrm{Proj}(e)\, \mathrm{Ɑ}\, U = \mathrm{Ɑ}\, U \frown e, \; \mathrm{Ɑ}\, U,$$

$$\mathrm{Ɑ}(e \upharpoonright U) \stackrel{\rightarrow}{+} \mathrm{co}\, e.$$

4) $U_E^V = e \upharpoonright U$.

5) $\mathrm{Proj}_e U$ coincides with $e \upharpoonright U$ on $\mathrm{Ɑ}(e \upharpoonright U)$, and $\mathrm{Proj}_e U f = U\,\mathrm{Proj}_e f$ for $f \in \mathrm{Ɑ}\, U$.

6) $U\,\mathrm{Proj}_e$ is $(e \upharpoonright U)$, extended by the operation 0 in $\mathrm{co}\, e$, which carries every vector into $\vec{O}$.

7) U_E^V, $\mathrm{Proj}_e U$, $U\,\mathrm{Proj}_e$, $e \upharpoonleft U$ coincide in

$$\text{ᗡ}\, U \cap e = \mathrm{Proj}\,(e)\, \text{ᗡ}\, U = \text{ᗡ}\,(e \upharpoonleft U).$$

J.15. Lemma. If

1. $\varphi_n(x)$ are functions fitting (T) and are defined μ-a.e. on V, $(n = 1, 2, \ldots)$,

2. $\varphi_n(x)$ converges uniformly μ-a.e. to $\varphi(x)$,

3. $U_n f = \varphi_n \cdot f$, $U f = \varphi \cdot f$ are maximal normal operators in $\boldsymbol{H}$, generated by φ_n, and φ as in [J.10.1.] (i.e. with maximal domains), then

1) there exists n_0 such that for all $n \geqq n_0$

$$\text{ᗡ}\, U = \text{ᗡ}\, U_n,$$

2) for every $f \in \text{ᗡ}\, U$ we have

$$U_n f \to^2 U f$$

(i.e. converges in the μ-square mean).

Proof. Let $f \in U$, i.e. f and $f \cdot \varphi$ are μ-square summable.

We have

$$f \cdot \varphi_n = f \cdot (\varphi_n - \varphi) + f \cdot \varphi.$$

There exists n_0, such that for all $n \geqq n_0$

$$|\varphi_n - \varphi| < 1$$

μ-a.e. As f is μ-square summable, so is $f \cdot (\varphi_n - \varphi)$. It follows that $f \cdot \varphi_n$ is also μ-square summable. Consequently $f \in \text{ᗡ}\, U_n$ for $n \geqq n_0$, and hence

(1) $$\text{ᗡ}\, U \subseteqq \text{ᗡ}\, U_n \quad \text{for} \quad n \geqq n_0.$$

Taking the equality

$$f \cdot \varphi = f(\varphi - \varphi_n) + f\varphi_n,$$

we get in an analogous way, that

(2) $$\text{ᗡ} U_n \subseteqq \text{ᗡ}\, U \quad \text{for} \quad n \geqq n_0.$$

From (1) and (2) the first thesis follows. Let $f \in \text{ᗡ}\, U$. Take $n \geqq n_0$. We have $f \in \text{ᗡ}\, U_n$. Let us confine ourselves only to indices $n \geqq n_0$. Take $\sigma > 0$. Find n_1 such that for all $n \geqq n_1$

$$|\varphi_n(x) - \varphi(x)| \leqq \sigma$$

a.e. we have

$$|\varphi_n f - \varphi f|^2 = |\varphi_n - \varphi|^2 \cdot f^2 \leqq \sigma^2 f^2,$$

a.e. Hence

$$\int_V |\varphi_n f - \varphi f|^2 \, d\mu < \sigma^2 \int_V |f|^2 \, d\mu.$$

If we make σ tend to 0, we get

$$\varphi_n f \to^2 \varphi f,$$

i.e.

$$U_n f \to^2 U f.$$

This completes the proof.

J.15.1. The Lemma [J.15.] holds true, if we replace the space V by any its μ-measurable subset E, and take functions φ_n, φ defined μ-a.e. on E.

From [J.15.1.], if we take account of the isometric isomorphism of the spaces $\boldsymbol{H}(E)$ and $\boldsymbol{H}^V(E)$, we have the following

J.15.2. Lemma. If

1. E is a μ-measurable subset of V,

2. φ_n, φ are functions fitting (T) and are defined μ-a.e. on E $(n = 1, 2, \ldots)$,

3. $U_n f = \varphi_n^V \cdot f$, $U f = \varphi^V f$ are maximal normal operators in $\boldsymbol{H}^V(E)$,

4. $\varphi_n(x)$ converges μ-a.e. uniformly to $\varphi(x)$,

then

1) there exists n_0 such that for all $n \geqq n_0$

$$\text{ᗡ}\, U = \text{ᗡ}\, U_n,$$

2) for every $f \in \text{ᗡ}\, U$ we have

$$U_n f \to^2 U f.$$

J.15.3. Lemma. If

1. $E_1 \subseteqq E_2 \subseteqq \cdots \subseteqq E_n \subseteqq \cdots$ belong to T,

2. $E = \bigcup_{n=1}^{\infty} E_n$,

3. $e_n =_{df} \boldsymbol{H}^V(E_n)$, $e =_{df} \boldsymbol{H}^V(E)$ (see [J.2.]),

4. $\varphi_1(x)$, $\varphi_2(x)$, $\ldots$, $\varphi_n(x)$, $\ldots$ are functions fitting (T), with $\text{ᗡ}\, \varphi_n = E_n$, $(n = 1, 2, \ldots)$,

5. $\varphi(x) = \varphi_n(x)$ for $x \in E_n$, $n = 1, 2, \ldots$ with $\text{ᗡ}\, \varphi = E$,

6. $U_n f = \varphi_n^V(x) \cdot f(x)$ is an operator in e_n with maximal domain, and $U f(x) = \varphi^V(x) \cdot f(x)$ is an operator in e with maximal domain,

7. $M = \bigcup_{n=1}^{\infty} \text{ᗡ}\, U_n$,

then the operator $W g$ defined by

$$W g = \varphi^V(x) \cdot g(x)$$

with domain M exists, and the closure $\tilde{W}$ of W coincides with U (with identity of domains). We may call W "*the hull of* $\{U_n\}$".

J.15.3a. Proof. We have

$$E_n \upharpoonright \varphi = \varphi_n.$$

Consequently

(1) $$(E_n \upharpoonleft \varphi)^V = \varphi_n^V.$$

If

$$f \in \text{ᗡ} U_n,$$

we have

(2) $$\varphi_n^V f = \varphi^V f$$

in V. It follows that

(3) $$\text{ᗡ} U_n \subseteqq \text{ᗡ} U.$$

From (3) and hyp. 7, we get

(4) $$M \subseteqq \text{ᗡ} U.$$

J.15.3b. If $f \in M$, then there exists n, such that

$$f \in \text{ᗡ} U_n.$$

Hence, by (3),

$$f \in \text{ᗡ} U,$$

and consequently $U f$ has a meaning. It follows that the equality

$$W f = \varphi^V f$$

in V is meaningful, so W exists. The first thesis is proved. Thus we have

(5) $$W f = U f \quad \text{for all} \quad f \in M.$$

Since U is a closed operation, it follows that W possesses a closed extension. Hence the closure $\tilde{W}$ exists. We have

(6) $$\tilde{W} \subseteqq U.$$

J.15.3c. To prove the inclusion, inverse to (6), we make the following remark:

Let h be a μ-square summable function on E. Put

$$h_n(x) \doteq h(x) \quad \text{on } E_n, \text{ and } \doteq 0 \quad \text{on } E - E_n.$$

We have

$$h_n(x) \to^2 h(x) \quad \text{on } E.$$

J.15.3d. Let

$$f \in \text{ᗡ} U.$$

Put

$$f_n(x) = f(x) \quad \text{on } E_n, \text{ and } = 0 \quad \text{on } E - E_n.$$

By the Remark [J.15.3c.] we have

$$f_n(x) \to^2 f(x) \quad \text{on } E.$$

Consequently

(6.1) $$f_n^V \to^2 f^V = f \quad \text{on } V.$$

We easily prove that

(7) $$f_n^V(x) \in \text{ᗡ} U_n.$$

Since

$$\mathrm{C}\, U_n \subseteqq M = \mathrm{C}\, W = \mathrm{C}\, \tilde{W},$$

it follows that

$$(8) \qquad f_n^V \in \mathrm{C}\, \tilde{W}.$$

J.15.3e. We shall prove that

$$(9) \qquad \tilde{W} f_n^V \to^2 (E \upharpoonleft \varphi f)^V$$

on V. Indeed, as $f_n^V \in \mathrm{C}\, U_n \subseteqq W$, we have

$$\tilde{W} f_n^V = W f_n^V = \varphi^V \cdot f_n^V.$$

Since $f_n^V \in \mathrm{C}\, U_n$, we have, by (2),

$$(10) \qquad \tilde{W} f_n^V = \varphi^V f_n^V = \varphi_n^V f_n^V = (E_n \upharpoonleft \varphi \cdot f)^V.$$

Now, φf is μ-square summable on E.

If we apply the Remark [J.15.3c.], we get by (10), that

$$\tilde{W} f_n^V \to^2 (E \upharpoonleft \varphi f)^V$$

on V. As

$$f_n^V \in \mathrm{C}\, \tilde{W}, \quad f_n^V \to^2 f^V$$

on V and

$$\tilde{W} f_n^V \to^2 (E \upharpoonleft \varphi f)^V$$

on V, it follows that

$$(E \upharpoonleft \varphi f)^V \in \mathrm{C}\, \tilde{W},$$

and that

$$(11) \qquad \tilde{W} f^V = (E \upharpoonleft \varphi f)^V = U f.$$

Thus we have proved that

$$(12) \qquad U \subseteqq \tilde{W}.$$

From (6) and (12) it follows that $\mathrm{C}\, U = \mathrm{C}\, \tilde{W}$, so (11) completes the proof.

J.15.4. Lemma. If

1. $\varphi_n(x)$ is defined μ-a.e. on V and fitting (T),
2. $\lim_{n\to\infty} \varphi_n(x) \doteq \varphi(x)$, ($\mu$-a.e.),
3. $U f = \varphi \cdot f$, $U_n f = \varphi_n f$, are operations in $\boldsymbol{H}$ with maximal domains,
4. M is the set of all f, for which the sequence $\{U_n f\}$ converges in μ-square mean on V,
5. $W f = \lim^2 U_n f$ with $\mathrm{C}\, W = M$,

then

$$\tilde{W} = U.$$

Proof. Since μ is a denumerably additive measure on (T), the Egoroff-theorem is valid. It reads that, by virtue of Hyp. 1. and 2., for every $\sigma > 0$ there exists a set F, such that $\mu(\mathrm{co}F) \leqq \sigma$, and that on F the sequence $\varphi_n(x)$ converges uniformly.

It follows that there exists a sequence

$$E_1 \subseteqq E_2 \subseteqq \cdots \subseteqq E_m \subseteqq \cdots$$

of μ-measurable subsets of V, such that $\bigcup_{m=1}^{\infty} E_m \doteq V$, and for every m:

(1) $$\lim_{n\to\infty} (E_m \upharpoonright \varphi_n) \doteq (E_m \upharpoonright \varphi)$$

a.e. uniformly on E_n.

Take the operators

$$(U_n)^V_{Em} f = (E_m \upharpoonright \varphi_n)^V \cdot f \quad \text{in} \quad \boldsymbol{H}^V(E_m),$$
$$U^V_{Em} f = (E_m \upharpoonright \varphi)^V f \quad \text{in} \quad \boldsymbol{H}^V(E_m).$$

By (1) we are in the conditions of Lemma [J.15.2.]. Hence, for every $f \in \triangleleft\, U^V_{Em}$ we have

(2) $$(U_n)^V_{Em} f \to^2 U^V_{Em} f$$

on V. Consider the operations

$$U^V_{E_1}, U^V_{E_2}, \ldots$$

We have

$$E_1 \subseteqq E_2 \subseteqq \cdots, \qquad \mathsf{U}_m E_m \doteq V,$$

$U^V_{Em} f = (E_m \upharpoonright \varphi)^V f$ with maximal domain in $\boldsymbol{H}^V(E_m)$, and $U f = \varphi f$ with maximal domain in $\boldsymbol{H}$, so we are in the circumstances of Lemma [J.15.3.]. Consider the operation $W_1 f = \varphi \cdot f$ with domain

$$\triangleleft\, W_1 = \bigcup_{m=1}^{\infty} \triangleleft\, U^V_{Em}.$$

We have

(3) $$\tilde{W}_1 = U.$$

Put

$$e_m = \boldsymbol{H}^V(E_m).$$

If $f \in \triangleleft\, W_1$, we have for some m

$$f \in \triangleleft\, U^V_{Em},$$

and hence, by (2),

$$\lim_{n\to\infty}{}^2 (U_n)^V_{Em} f = U^V_{Em} f$$

on V. On the other hand

$$U^V_{Em} f = U f,$$

because

$$U^V_{Em} = e_m \upharpoonright U \subseteqq U.$$

Hence

$$\lim_{n\to\infty}{}^2 (U_n)^V_{Em} f = U f.$$

Since

$$(U_n)^V_{E_m} \subseteqq U_n,$$

it follows that

$$\lim_{n\to\infty}{}^2 U_n f = U f.$$

Hence

(4) $$W_1 \subseteqq W.$$

We shall prove that

$$W \subseteqq U.$$

Let $f \in W$ and let

$$\lim_{n\to\infty}{}^2 U_n f(x) \doteq g(x).$$

There exists an increasing sequence $k(n)$, $(n = 1, 2, \ldots)$, such that

$$U_{k(n)} f(x)$$

converges μ-a.e. toward $g(x)$.

Since

$$U_{k(n)} f(x) = \varphi_{k(n)} \cdot f(x),$$

it follows that

$$\varphi_{k(n)}(x) \cdot f(x) \to g(x),$$

a.e. Now since

$$\varphi_{k(n)}(x) \dot{\to} \varphi(x),$$

it follows that

$$g(x) \doteq \varphi(x) \cdot f(x).$$

Since $U_n f(x)$ converges in square mean to $g(x)$, the function $g(x)$, and consequently $\varphi \cdot f$, is μ-square summable. Hence $f \in \mathrm{Cl}\, U$.

Thus we get

$$\lim{}^2 U_n f \doteq U f \quad \text{i.e.} \quad W f = U f \quad \text{for all} \quad f \in \mathrm{Cl}\, W.$$

Hence

(5) $$W \subseteqq U.$$

From (4) and (5) we get

$$W_1 \subseteqq W \subseteqq U,$$

and then

$$\tilde{W}_1 \subseteqq \tilde{W} \subseteqq \tilde{U}.$$

Since, by (3),

$$\tilde{W}_1 = U = \tilde{U},$$

it follows that

$$\tilde{W} = U, \quad \text{q.e.d.}$$

J.15.5. The above Lemma [J.8.] will remain true, if we change the space V into a μ-measurable subset E of V. Suppose that:

1. $\varphi_n(x)$ is defined μ-a.e. in a μ-measurable subset E of V, and φ_n is fitting T.

2. $\lim\limits_{n\to\infty} \varphi_n(x) = \varphi(x)$, a.e. in E.

3. $Uf = \varphi^V \cdot f$, $U_n f = \varphi_n^V f$ are operations in $\boldsymbol{H}^V(E)$ with maximal domains.

4. M is the set of all $f \in \boldsymbol{H}^V(E)$, for which $\{U_n f\}$ converges in μ-square mean.

5. $Wf =_{df} \lim^2 U_n f$ with $\complement W = M$.

Then

$$\tilde{W} = U.$$

J.15.6. Let us apply the Lemma [J.15.4.] to the circumstance of Lemma [J.6.1.]. We get the

Lemma. If

1. φ_n are functions fitting (T) and defined μ-a.e. on V $(n = 1, 2, \ldots)$,
2. φ_n converges uniformly μ-a.e. to $\varphi(x)$,
3. $U_n f = \varphi_n f$, $Uf = \varphi f$ are normal operators in $\boldsymbol{H}$ with maximal domains,
4. M is the set of all f, for which $U_n f$ converges in μ-square mean on V,
5. Wg is the operator $\lim^2 U_n g$ with $\complement W = M$,

then

$$\tilde{W} = U.$$

J.15.7. An analogous theorem to [J.15.6.] we obtain, by replacing V by a μ-measurable set E, and also another one, if we change the condition 3 into:

"$U_n f = \varphi_n^V f$, $Uf = \varphi^V f$ are normal operators in $\boldsymbol{H}^V(E)$ with maximal domains",

and 4. into:

"M is the set of all $f \in \boldsymbol{H}^V(E)$, for which $U_n f$ converges in μ-square mean"

J.15.8. Lemma. Suppose that:

1. $\varphi_n(x)$ are fitting (T) and defined μ-a.e. on V $(n = 1, 2, \ldots)$.
2. The sequence $\varphi_n(x)$ converges μ-a.e. on E, and each its partial sequence $\varphi_{k(n)}(x)$ diverges μ-a.e. on co$E$.
3. $U_n f = \varphi_n f$ are operators in $\boldsymbol{H}$ with maximal domains.
4. $\varphi(x) \doteq \lim_{n\to\infty} \varphi_n(x)$ on E.
5. The operator

$$Uf(x) = \varphi^V(x) \cdot f(x)$$

is taken with maximal domain in $\boldsymbol{H}^V(E)$.

6. M is the set of all $f \in \boldsymbol{H}$, for which $U_n f$ converges in μ-square mean.
7. Wf is the operator defined by

$$Wf = \lim_{n\to\infty}{}^2 U_n f \quad \text{with} \quad \complement W = M.$$

Then

$$U = \tilde{W}.$$

Remark. We emphasize Hyp. 3. If we had supposed that $U_n f = \varphi_n f$ are operators in $\boldsymbol{H}^V(E)$, the hyp. 2 would be not needed and the theorem would be a simple consequence of [J.15.7.].

Proof. of [J.15.8.]. Consider the operators $U'_n = (U_n)^V_E$, which are defined in $\boldsymbol{H}^V(E)$ by the equality

$$U'_n = (E \upharpoonright \varphi_n)\mathrm{n} \cdot f;$$

and with maximal domain.

Take the set M' of all $f \in \boldsymbol{H}^V(E)$, for which $U'_n f$ converges in μ-square mean on V.

Take the operation

$$W' f = \lim_{n\to\infty}{}^2 U'_n f \quad \text{with domain } M'.$$

By Lemma [J.15.4.], we have

$$(1) \qquad \tilde{W}' = U$$

with identity of domains, since $(E \upharpoonright \varphi_n)^V \to (E \upharpoonright \varphi)^V$.

J.15.8a. Let $f \in M'$. We have $f \in \boldsymbol{H}^V(E)$ and

$$(2) \qquad \lim_{n\to\infty}{}^2 U'_n f = U f.$$

Hence $f \in \mathrm{C\!\!\!I}\, U'_n$ for all $n \geqq n_0$, where n_0 is a suitable integer depending on f.

Let us consider those indices only. We have

$$U'_n f = (E \upharpoonright \varphi_n)^V \cdot f = \varphi_n f = U_n f.$$

Hence, by (2), the sequence $\{U_n f\}$ converges in μ-square mean on V. Hence $f \in M$.

Thus we have got the inclusion

$$M' \subseteqq M.$$

Since

$$W' f = \lim_{n\to\infty}{}^2 U'_n f = \lim_{n\to\infty}{}^2 U_n f = W f,$$

it follows that

$$W' \subseteqq W,$$

and then

$$\tilde{W}' \subseteqq \tilde{W}.$$

By (1) we obtain

$$(3) \qquad U \subseteqq W.$$

J.15.8b. We shall prove the inverse inclusion.

Let $f \in W$; $\{U_n f\}$ converge in μ-square mean on V; hence $\varphi_n f$ does so.

Let

(3.1) $$\varphi_n f \rightarrow^2 g(x)$$

on V. There exists an increasing sequence $k(n)$ of indices, such that

(4) $$\varphi_{k(n)}(x) \rightarrow g(x)$$

a.e. on V. As

$$\varphi_{k(n)}(x) \rightarrow \varphi(x)$$

a.e. on E, it follows that

(5) $$g(x) \doteq \varphi(x)\, f(x)$$

on E.

J.15.8c. We shall prove that $f(x) \doteq 0$ on $\mathrm{co}E$. It suffices to suppose that $\mu(\mathrm{co}E) > 0$. Suppose that there exists a set $F \subseteqq \mathrm{co}E$ with $\mu F > 0$ and such that $f(x) \neq 0$ on F everywhere.

By (4) we have

$$\varphi_{k(n)}(x)\cdot f(x) \rightarrow g(x)$$

on F a.e. Hence

$$\varphi_{k(n)}(x) \rightarrow \frac{g(x)}{f(x)}$$

on F a.e. But this contradicts the hypothesis that $\varphi_{k(n)}(x)$ diverges μ-a.e. on $\mathrm{co}E$, hence on F. Thus

$$f(x) = 0, \quad \mu\text{-a.e. on } \mathrm{co}E.$$

If we take account of (3.1), we get

$$g(x) \doteq 0 \quad \text{on} \quad \mathrm{co}E.$$

Hence, by (5),

$$g(x) \doteq \varphi^V(x)\, f(x) = U f,$$

and then

$$W f = \varphi^V(x)\, f(x) = U f.$$

This proves that

$$W \subseteqq U,$$

and then

(5) $$\tilde{W} \subseteqq U.$$

From (3) and (5) the theorem follows.

J.16. Lemma. If

1. $E = E_1 \cup E_2 \subseteqq V$,

2. $\varphi_1(x)$ and $\varphi_2(x)$, fitting (T), are defined μ-a.e. on E_1 and E_2 respectively,

3. $\varphi_1(x) \doteq \varphi_2(x)$ on $E_1 \cap E_2$,

4. $U_1 f = \varphi_1^V(x) \cdot f$ with maximal domain in $\boldsymbol{H}^V(E_1)$, and $U_2 f = \varphi_2^V(x) \cdot f$ with maximal domain in $\boldsymbol{H}^V(E_2)$,

5. $\varphi(x) =_{df} \begin{cases} \varphi_1(x) & \text{on } E_1, \\ \varphi_2(x) & \text{on } E_2, \end{cases}$

6. $U f = \varphi^V(x) \cdot f$ with maximal domain in $\boldsymbol{H}^V(E)$,

then

1) U is the smallest linear operator containing both U_1 and U_2,

2) $\boldsymbol{H}^V(E) = \boldsymbol{H}^V(E_1) + \boldsymbol{H}^V(E_2)$.

The proof is easy by considering $U \operatorname{Proj}(e_1)$ and $U \operatorname{Proj}(e_2)$ (see [J.13.2.]), where $e_1 = \boldsymbol{H}^V(E_1)$, $e_2 = \boldsymbol{H}^V(E_2)$. We have

$$U = U \operatorname{Proj}(e_1) + U \operatorname{Proj}(e_2).$$

Chapter J I

Operational calculus on general maximal normal operators

J I. In Chapter [J.] we have dealt with special normal operators; in the present section we shall consider a general (separable and complete) H.-H.-space $\boldsymbol{H}'$ and a maximal normal operator N' in it. Notions related to $\boldsymbol{H}'$ will be provided with primes.

Let (s_b') be the borelian spectral tribe of N', and (T') a denumerably additive tribe of spaces in $\boldsymbol{H}'$, obtained from (s_b') by saturation by means of a saturating sequence of spaces. Let $\mu'(a')$ be a denumerably additive effective measure on (T'), defined by a chosen generating vector.

J I.1. In [H.] we have found a mapping C which carries $\boldsymbol{H}'$ onto the H.-H.-space $\boldsymbol{H}$ composed of functions of the point p, varying on the disjoint union V of several copies of the complex plane P. This mapping C carries vectors of $\boldsymbol{H}'$ into functions which are fitting some denumerably additive tribe (T) of subsets of V provided with a denumerably additive measure μ. The functions of $\boldsymbol{H}$ are μ-square summable and defined μ-a.e. on V. The mapping C is isometric and isomorphic with respect to suitable notions of equalities in the tribes considered. It carries N' into a normal maximal operator N in $\boldsymbol{H}$, having the form

$$N \dot{f}(\dot{p}) =^{\mu} (\operatorname{Num} p) \cdot \dot{f}(\dot{p}),$$

called *canonical*, [H.]. The measure μ is determined by the choice of a saturating sequence of spaces and by the choice of a generating vector. Thus there are several canonical mappings C. Let us choose one of them and keep it fixed.

This remark makes necessary to prove the independence of the important notions, to be introduced, from the choice of C.

J I.2. Let C be a canonical mapping from $\boldsymbol{H}'$ onto $\boldsymbol{H}$.

Let z_0 be a complex number. Take the operator in $\boldsymbol{H}$:

$$M f(p) = z_0 \cdot f(p)$$

with maximal domain (see [J.11.]).

It is maximal normal in $\boldsymbol{H}$, and we have $\mathrm{C} M = \boldsymbol{H}$. If we denote by $I f$ the identity operator $I f = f$ with $\mathrm{C} I = \boldsymbol{H}$, we may agree to write $M f = (z_0 \cdot I) f$. The C^{-1}-image of M is, of course, $(z_0 \cdot I') f'$ where I' is the identity operator in $\boldsymbol{H}'$ with domain $\boldsymbol{H}'$. Thus $C^{-1} M$ is independent of C.

J I.2.1. Consider the operator in $\boldsymbol{H}$:

$$M_1 f(p) = (z - z_0) \cdot f(p) \tag{1}$$

with maximal domain, where $z = \mathrm{Num}\, p$. We have

$$\mathrm{C} M_1 = \mathrm{C} N.$$

(1) is a normal operator in $\boldsymbol{H}$, and we have

$$M_1 f(p) = z \cdot f(p) - z_0 f(p) \quad \text{for} \quad p \in \mathrm{C} N.$$

Hence

$$M_1 f(p) = z \cdot f(p) - z_0 \cdot I f(p),$$

and then

$$C^{-1} M_1 f = C^{-1}(z f) - z_0 C^{-1} I'(f),$$

$$C^{-1} M_1 f(p) = N' f' - z_0 I' f'$$

with domain $\mathrm{C} N'$. This proves, by [J.2.], the independence of $C^{-1} M_1$ of the choice of C. If we put

$$M_1' f' = C^{-1} M_1 f \quad \text{for} \quad f \in \mathrm{C} N,$$

we get

$$M_1' = N' - z_0 I'.$$

M_1' is a maximal normal operator in $\boldsymbol{H}'$, since C is an isometric isomorphism.

J I.2.2. Let $n = 1, 2, \ldots$ and consider the operator in $\boldsymbol{H}$:

$$M_n f(p) = (z - z_0)^n \cdot f(p),$$

with maximal domain. We have

$$M_n' f' = C^{-1} M_n f, \quad \text{where} \quad f' \in C^{-1} \mathrm{C} M_n,$$

is a maximal normal operator in $\boldsymbol{H}'$, because M_n is so in $\boldsymbol{H}$.

We have

(1) $$\text{ᗡ} M_{n+1} \subseteqq \text{ᗡ} M_n,$$

because if $(z - z_0)^{n+1} f(p)$ and $f(p)$ are μ-square summable on V, so is $(z - z_0)^n f(p)$. We have, for $f \in \text{ᗡ} M_{n+1}$,

(2) $$M_{n+1} f(p) = (z - z_0) \cdot M_n f(p),$$

(3) $$M_{n+1} f(p) = z \cdot M_n f(p) - z_0 \cdot M_n f(p).$$

Now suppose that $C^{-1} M_n$ is independent of C; we shall prove that so is $C^{-1} M_{n+1}$.

By (1) we have for $f \in \text{ᗡ} M_{n+1}$,

$$C^{-1}(M_{n+1} f(p)) = C^{-1}(z \cdot M_n f(p)) - z_0 (C^{-1} M_n f(p)).$$

Now $z \cdot M_n f(p)$ is μ-square summable on V, since by (3), and by virtue of (1), $z_0 \cdot M_n f(p)$ is so. It follows that

$$C^{-1}(M_{n+1} f(p)) = N'(C^{-1} M_n f(p)) - z_0 (C^{-1} M_n f(p)).$$

If we take account of the hypothesis and of [J I.2.], we see that $C^{-1}(M_{n+1} f(p))$ does not depend on C. By [J I.2.1.] the independence of $C^{-1} M_n$ of C is proved.

J I.2.3. Def. We agree to write $C^{-1} M_n$ in the form

$$(N' - z_0 I')^n.$$

This is a maximal normal operation in $\boldsymbol{H}'$, obtained by iteration of $N' - z_0 I'$.

J I.2.3a. Remark. Our aim is to define, independently of C, the operation $\frac{1}{N' - z_0 I'}$ by means of a limit process and by a sequence of "polynomials" $P(N')$, which we will define by [J I.2.1.], [J I.2.2.], [J I.2.3.]. This will be done by using tools, got in [J.]. Especially Lemma [J.10.] will be applied. To do this, we must find a sequence of polynomials of the ordinary complex variable z, which would converge in a halfplane toward $\frac{1}{z - z_0}$, but whose each subsequence diverges in the other halfplane. The common boundary of these halfplanes must have the measure $= 0$. To achieve our aims we shall make a remark and prove some inequalities.

J I.3. Let us remark that the number of straight lines l on each replica of P, parallel to the y-axis, with $\mu\, l > 0$ is at most denumerable. Hence the number of such lines l where $\mu(\text{Point}\, l) > 0$ is at most denumerable.

Having this, let C_1, C_2 be two canonical mappings, yielding canonical representations of N. There exists a complex number z_0, such that if l_0 is the straight line, parallel to the y-axis, and passing through z_0, then

$$\mu_1(\text{Point}\, l_0) = 0, \qquad \mu_2(\text{Point}\, l_0) = 0,$$

where μ_1, μ_2 are measures related to C_1 and C_2 respectively. The set of those z_0 is of the 2^0 cathegory in the complex plane.

J I.3.1. Lemma. If

1) $k = 1, 2, \ldots,$

2) $|z| \geqq 1 + \frac{1}{k^3}$,

3) $n \geqq 27k^9$, $(n = 1, 2, \ldots)$,

then

$$|1 + z + \cdots + z^n| \geqq k^2.$$

Proof. Put

$$u_n(z) = 1 + z + \cdots + z^n,$$

where

$$|z| = 1 + a, \qquad a > 0.$$

We have

$$u_n(z) = \frac{z^{n+1} - 1}{z - 1}, \qquad (n = 1, 2, \ldots).$$

As

$$\begin{aligned} |z^{n+1} - 1| &\geqq |z|^{n+1} - 1, \\ &\geqq (1 + a)^{n+1} - 1, \\ &\geqq 1 + (n + 1)\, a - 1, \\ &\geqq (n + 1)\, a, \end{aligned}$$

and

$$|z - 1| \leqq a + 2,$$

we get

$$|u_n(z)| \geqq (n + 1) \cdot \frac{a}{a + 2}.$$

Now let

$$a > \frac{1}{k^3},$$

we have

$$|u_n(z)| > (n + 1) \cdot \frac{\frac{1}{k^3}}{\frac{1}{k^3} + 2},$$

because the function $\frac{a}{a + 2}$ increases in $(0, +\infty)$.

Hence

$$|u_n(z)| \geqq \frac{n + 1}{1 + 2k^3}. \tag{1}$$

If $n \geqq (1 + 2k^3)^3$, then

$$\frac{n + 1}{1 + 2k^3} \geqq (1 + 2k^3)^2 \geqq k^2, \tag{2}$$

We have

$$(2.1) \qquad 27k^9 \geqq (k^3 + 2k^3)^3 \geqq (1 + 2k^3)^3.$$

From (1), (2) and (3) we have: if

$$n \geqq 27k^9, \quad |z| \geqq 1 + \frac{1}{k^3},$$

then

$$|u_n(z)| \geqq k^2, \quad \text{q.e.d.}$$

J I.3.2. We have, for $z_1 \neq z_0$,

$$\frac{1}{z - z_0} = \frac{-1}{z_0 - z_1} \cdot \left[1 + \frac{z - z_1}{z_0 - z_1} + \left(\frac{z - z_1}{z_0 - z_1}\right)^2 + \cdots\right],$$

in the interior of the circle with center z_1 and passing through z_0. In every smaller circle this series converges uniformly.

Putting

$$z_1 = z_0 + k, \qquad (k = 1, 2, \ldots),$$

we get

$$(2.2) \qquad \frac{1}{z - z_0} = \frac{1}{k} \cdot \left[1 + \frac{z + k - z}{k} + \frac{(z_0 + k - z)^2}{k^2} + \cdots\right].$$

Put

$$\varphi_{k,n}(z) = \frac{1}{k}\left[1 + \frac{z_0 + k - z}{k} + \cdots + \frac{(z_0 + k - z)^n}{k^n}\right].$$

We shall apply Lemma [J.13.1.]:

Let

$$|z - (z_0 + k)| \geqq k + \frac{1}{k^2}.$$

We have

$$\left|\frac{z - (z_0 + k)}{k}\right| \geqq 1 + \frac{1}{k^3}.$$

Hence, if $n \geqq 27k^9$, we have

$$(3) \qquad |\varphi_{k,n}(z)| \geqq \frac{1}{k} \cdot k^2 = k.$$

Denote by B_k the circle

$$|z - (z_0 + k)| < k + \frac{1}{k^2}.$$

The segment in B_k, intercepted by the line l_0, has the length

$$2\sqrt{\left(k + \frac{1}{k^2}\right)^2 - k^2}.$$

Since

$$\left(k + \frac{1}{k^2}\right)^2 - k^2 = k^2 + 2\frac{1}{k} + \frac{1}{k^4} - k^2,$$

and since the derivative of this function for $k > 0$ is negative, therefore the length considered decreases if $k \to \infty$.

It follows that, if we have

$$P' = \{z \mid \mathrm{Real}\, z < \mathrm{Real}\, z_0\},$$

then

$$P' \cap \mathrm{co}\, B_k \subsetneqq P' \cap \mathrm{co}\, B_{k+1}, \qquad (k = 1, 2, \ldots).$$

We also have

$$\bigcup_{k=1}^{\infty} (P' \cap \mathrm{co}\, B_k) = P'.$$

It follows that if $z \in P'$ and $n(k)$ is any increasing sequence of indices with $n(k) \geqq 27 k^9$, then the sequence

$$(4) \qquad \varphi_{1,\, n(1)}(z), \quad \varphi_{2,\, n(2)}(z), \ldots$$

diverges and so does any its subsequence.

J I.3.3. Having this, denote by B'_k the circle

$$|z - (z_0 + k)| \leqq k - \frac{1}{2^k}, \qquad (k = 1, 2, \ldots).$$

The series (2.1) converges uniformly in B'_k. Hence there exists $m(k)$ such that, if $n \geqq m(k)$ then

$$(5) \qquad \left| \frac{1}{z - z_0} - \varphi_{k,\, n}(z) \right| \leqq \frac{1}{k} \quad \text{for} \quad z \in B'_k.$$

We have

$$B'_k \subsetneqq B'_{k+1} \quad \text{and} \quad \bigcup_{k=1}^{\infty} B'_k = P'',$$

where

$$(6) \qquad P'' = \{z \mid \mathrm{Real}\, z > \mathrm{Real}\, z_0\}.$$

It follows from (4), (5) and (6) that there exists an increasing sequence $q(k)$ of indices, such that the sequence $\{\varphi_{k,\, q(k)}(z)\}$ diverges everywhere in P', and converges everywhere in P'' toward

$$(7) \qquad \frac{1}{z - z_0}.$$

J I.3.4. Let us go back to [J I.3.]. We have two different canonical representations C_1, C_2 of N, (see [J I.3.]), and they may involve different numbers of replicas of the complex plane. Put $j = 1, 2$.

Define

$$\psi_{k;\, j}(p) = \varphi_{k,\, q(k)}(\mathrm{Num}\, p) \quad \text{for all} \quad p \in V_j.$$

The sequence $\{\psi_{k;\, j}\}$, $(k = 1, 2, \ldots)$ converges everywhere in $\mathrm{Point}_j P''$ toward

$$\frac{1}{\mathrm{Num}\, p - z_0}.$$

It diverges μ_1-almost everywhere and μ_2-almost everywhere in V_j-$\mathrm{Point}_j P''$, and so does any its partial sequence. Indeed $\mu_j(\mathrm{Point}_j\, l_0) = 0$, by [J I.3.].

J I.3.5. We are in the conditions of Lemma [J.15.8.], so we can apply it to $\{\psi_{k;j}(p)\}$.

Indeed, $\psi_{k;j}(p)$ are fitting (T_j) and are defined on V_j, $(k=1, 2, \ldots)$ Let

$$U_{n;j} f(p) = \psi_{n;j}(p) \cdot f(p), \quad j = 1, 2,$$

be operators in $\boldsymbol{H}_j$ with maximal domains, and let

$$U_j f(p) = \left(\frac{1}{z_0 - \operatorname{Num} p}\right)^{V_j} \cdot f(p)$$

be an analogous operator in $\boldsymbol{H}_j(\text{Point}_j P'')$; let

M_j be the set of all $f \in \boldsymbol{H}_j$ for which

$U_{n;j} f$ converges in μ_j-square mean,

and let

$U_j(f)$ be defined on M_j by

$$W_j f = \lim_{n\to\infty}{}^2 U_{n;j} f.$$

By Lemma [J.15.8.] it follows that

$$U_j = \tilde{W}_j.$$

Take the C_j^{-1} images.

Since $\psi_{n,j}$ is a polynomial, therefore, by [J I.2.1.], the operator $C_j^{-1} U_{n;j}$ does not depend on $j = 1, 2$. Since the μ_1 and μ_2-square convergence is transformed by C_j^{-1} into the ordinary H.-H.-convergence of vectors in $\boldsymbol{H}'$, we get

$$C_1^{-1} M_1 = C_2^{-1} M_2.$$

Since the closure of the operation in $\boldsymbol{H}_1$ and $\boldsymbol{H}_2$ corresponds to the H.-H.-closure in $\boldsymbol{H}'$, we get

$$C_1^{-1} U_1 = C_2^{-1} U_2$$

with equality of domains.

J I.3.6. Now we can use the same argument for the lefthand side of the line L_0 by considering the series

$$\frac{1}{z - z_0} = -\frac{1}{z_0 - z_1}\left[1 + \frac{z - z_1}{z_0 - z_1} + \cdots\right]$$

and by taking $z_1 = z_0 - k$, $(k = 1, 2, \ldots)$. Thus we obtain operations

$$U_1^{\times}, U_2^{\times},$$

analogous to U_1, U_2, and they will have the analogous property

$$C_1^{-1} U_1^{\times} = C_2^{-1} U_2^{\times}$$

with equality of domains.

The two halfplanes involved are disjoint.

It follows, by Lemma [J.16.], that the measure-hull Z_j of $C_j^{-1} U_j$ and $C_j^{-1} U_j^{\times}$ does not depend on j, and we have

(7.1) $$Z_j f = \frac{1}{z_0 - \mathrm{Num}\, p} \cdot f(p)$$

with maximal domain in $\boldsymbol{H}_j$.

We have proved that both the C_j^{-1} images of the operation (6) do not depend on j.

J I.3.7. Thus we have proved the following theorem. If

1) C_1, C_2 are two correspondences yielding a canonical representation of N,

2) z_0 is a complex number such that if we denote by l_0 the straight line passing through it and parallel to the y-axis, then

$$\mu_1(\mathrm{Point}_1 l_0) = \mu_2(\mathrm{Point}_2 l_0) = 0,$$

3) $Z_j f = \frac{1}{\mathrm{Num}\, p - z_0} \cdot f(p)$ with maximal domain in $\boldsymbol{H}_j$, $(j = 1,2)$,

then

(7.2) $$C_1^{-1} Z_1 = C_2^{-1} Z_2$$

with identity of domains.

An analogous theorem holds, if the straight line l_0 is parallel to the x-axis and has an analogous property.

J I.3.7a. Def. Under circumstances of [J I.3.7.] we shall denote the operator (7) by

$$\frac{1}{N - z_0 I}.$$

J I.3.8. Take C_1 and C_2 as above and consider a bounded rectangle R on the complex plane P; denote its boundary by R_0. Let R be chosen so as to have the straight lines, which are prolongations of the sides of R_0, satisfying the condition that their μ_1 and μ_2-measures are $= 0$.

By the Theorem [J I.3.7.] we have for every $z_0 \in R_0$ a well defined operator $\frac{1}{N - z_0 I}$, viz. the C_1^{-1}-image, and, at the same time, the C_2^{-1} image of

$$\frac{1}{z_0 - \mathrm{Num}\, p} \cdot f(p)$$

with maximal domain in $\boldsymbol{H}_1$ and $\boldsymbol{H}_2$ respectively. Take the counterclockwise direction on R_0, and consider the integral

$$J(R, z_0) = \frac{1}{2\pi i} \int_{R_0} \frac{dz}{z - z_0}$$

which is equal to 1 for all z_0 lying in the interior of R and equal to 0 for all z_0 lying in the exterior of R.

This integral can be considered as the limit, for $m \to \infty$, of a suitable expression

$$\frac{1}{2\pi i} \cdot \sum_k \frac{s_k}{z_k - z_0} = \chi_m(z_0),$$

where z_k are points on R_0, constituting a partition of the boundary, and s_k the differences of consecutive partition-points.

Now the images C_1^{-1} and C_2^{-1} of the operator

$$U_j f(p) = \chi_m(\mathrm{Point}_j p) \cdot f(p)$$

with maximal domain in $\boldsymbol{H}_j$ are identical for $j = 1, 2$; this on account of [J I.3.7.].

Hence, by Lemma [J.15.6.], the operators

(8) $$W_j f(p) = r(\mathrm{Num}\, p) \quad f(p)$$

with maximal domain in $\boldsymbol{H}_j$, where

$$r(z) = 1 \quad \text{for} \quad z \in R \quad \text{and} \quad \doteq 0 \quad \text{for} \quad z \in \mathrm{co}R,$$

have their both C_j^{-1} images identical.

Thus $C_j^{-1} W_j$ is a well defined normal operator in $\boldsymbol{H}'$.

J I.3.9. Having this, let z_0 be a point such that both lines, passing through $z_0 = x_0 + i\, y_0$ and parallel to the x- and y-axes, have both measures μ_1 and μ_2 equal 0 (see [J I.3.]).

Consider an infinite sequence of rectangles R_n, as in [J I.3.8.], with vertices $z_n = x_n + i\, y_n$, where

$$x_n < x_0, \quad x_n \to -\infty,$$

$$y_n < y_0, \quad y_n \to -\infty.$$

Consider the corresponding operations, for R_n,

$$W_{n;j} f(p) = r_n(\mathrm{Num}\, p) \cdot f(p),$$

as in [J I.3.8.].

The function $r_n(\mathrm{Num}\, p)$ tends, for $n \to \infty$, to the function:

$$\varrho\, z_0(p) = 1$$

for

$$\mathrm{Real\ Num}\, p < x_0, \quad \mathrm{Imag\ Num}\, p < y_0,$$

and $= 0$ almost μ_1-everywhere and almost μ_2-everywhere in the complementary set.

Applying Lemma [J.15.6.], we obtain the operators

$$X_{z_0;j} f(p) = \varrho_{z_0}(p) \cdot f(p)$$

with maximal domains. Their corresponding C_j^{-1} images do not depend on j.

J I.3.10. Now take a completely arbitrary $z_0 = x_0 + i\,y_0$ and the plane quarter

$$\{z \mid \mathrm{Real}\, z \leqq x_0,\ \mathrm{Imag}\, y \leqq y_0\}.$$

There exists a sequence $z_n = x_n + i\,y_n$ of complex numbers, such that

$$x_0 < x_n,\quad y_0 < y_n,$$

$$x_n \to x_0,\quad y_n \to y_0,$$

and where both the measures μ_j of the straight lines, passing through z_n and parallel to the axes, are equal 0.

Consider the operations

$$X_{z_{n;j}} = \varrho_{z_n}(p) \cdot f(p),$$

as in [J I.3.9.].

We have

$$\lim_{n\to\infty} \varrho_{z_n}(p) = \varrho(p) \quad \text{where} \quad \varrho(p) = 1$$

for p, such that

$$\mathrm{Real\ Num}\, p \leqq x_0,\qquad \mathrm{Imag\ Num}\, p \leqq y_0,$$

and equal $= 0$ otherwise.

By Lemma [J.15.6.], the images of the operator

$$Y_{z_{0;j}} f(p) = \varrho(\mathrm{Num}\, p) \cdot f(p)$$

with maximal domain coincide.

We have

$$C_1^{-1}\, Y_{z_0;1} = C_2^{-1}\, Y_{z_0;2}$$

with equality of domains.

The correspondences C_1 and C_2 were arbitrary and so was z_0.

Thus we have proved the following

J I.3.11. Theorem. If

1. N' is a normal operation in $\boldsymbol{H}'$,
2. C a correspondence yielding a canonical representation N of N',
3. $\varphi = Q(x_0, y_0)$ is a plane quarter $\{(x, y) \mid x \leqq x_0, y \leqq y_0\}$,
4. $q(z)$ the function $= 1$ on Q and $= 0$ on co$Q$,
5. $U f(p) = q(\mathrm{Num}\, p) \cdot f(p)$ is an operator with maximal domain in $\boldsymbol{H}$,

then

$C^{-1} U$ is an operator which does not depend on the choice of C. Its domain is the C^{-1}-image of the set of all $f \in \boldsymbol{H}$ for which the function $q(\mathrm{Num}\, p) \cdot f(p)$ is μ-square summable.

J I.4. Theorem. Let $R = R(x_1, y_1; x_2, y_2)$ be the rectangle

$$\{(x, y) \mid x_1 < x \leqq x_2,\ y_1 < y \leqq y_2\}$$

where

$$-\infty < x_1 < x_2 < +\infty, \qquad -\infty < y_1 < y_2 < +\infty.$$

Put

$$\varphi_R(z) = 1 \quad \text{for} \quad z \in R$$

and

$$\varphi_R(z) = 0 \quad \text{for} \quad z \,\overline{\in}\, R.$$

Let C be the correspondence transforming the normal operator N' into a canonical form. Let

$$U\,f(p) = \varphi_R(\operatorname{Num} p) \cdot f(p)$$

be an operator in $\boldsymbol{H}$ with maximal domain. Under these circumstances $C^{-1}U$ is an operator in $\boldsymbol{H}'$, which does not depend on C.

Proof. We have, by putting, in general,

$$Q(x_0 + i\,y_0) = Q(x_0, y_0) = \{(x, y) \,|\, x \leqq x_0, y \leqq y_0\},$$

$$z_0 = x_0 + i\,y_0,$$

and

$$q(z; z_0) = 1 \quad \text{for} \quad z \in Q(z_0)$$

and

$$= 0 \quad \text{for} \quad z \,\overline{\in}\, Q(z_0),$$

the equality

$$\varphi_R(z) = q(z; x_1 + i\,y_1) + q(z; x_2 + i\,y_2) - \\ - q(z; x_1 + i\,y_2) - q(z; x_2 + i\,y_1).$$

From Theorem [J I.3.10.] it easily follows, that

$C^{-1}U$ is independent of C.

J I.4.1. Theorem. Let $\varphi(z)$ be a function, defined everywhere on P and continuous, and let

$$U\,f(p) = \varphi(\operatorname{Num} p) \cdot f(p)$$

be an operator with maximal domain;
then
$C^{-1}U$ is a normal operator in $\boldsymbol{H}$ which does not depend on C.

Proof. Let $R(1) \subsetneqq R(2) \subsetneqq \cdots$ be nested halfopen rectangles whose union is P.

Fix an index n. The function $\varphi(z)$ is uniformly continuous and bounded on $R(n)$.

Put

$$\varphi_n(z) = \varphi(z) \quad \text{for} \quad z \in R(n),$$

and

$$= 0 \quad \text{for} \quad z \,\overline{\in}\, R(n).$$

Let $\delta > 0$. There exists a finite partition Π: $R'(1)$, $R'(2)$, ... of $R(n)$ into equal rectangles, such that if we take $\eta_k \in R'(k)$, we have

$$|\varphi(z) - \varphi(\eta_k)| \leqq \delta \quad \text{on} \quad R'(k).$$

Thus for every $\delta > 0$, $\varphi_n(z)$ can be approximated, up to δ, by a step-function $\xi(z)$ vanishing outside $R(n)$.

By Theorem [J I.4.1.] it follows, that the C^{-1}-image of the operator

$$U_\xi f(p) = \xi(\operatorname{Num} p) \cdot f(p)$$

with maximal domain does not depend on C.

Since

$$\varphi_n(z) = \lim_{\delta \to \infty} \xi(z)$$

on P, it follows, by Lemma [J.15.8.], that the C^{-1}-image of the operator

$$U_n f = \varphi_n(\operatorname{Num} p) \cdot f(p)$$

with maximal domain in $\boldsymbol{H}$ does not depend on C.

Now let $n \to \infty$. We have $\lim_{n\to\infty} \varphi_n(z) = \varphi(z)$. If we apply once more Lemma [J.15.8.] we complete the proof.

J I.4.2. By an easy transfinite induction, and by applying Lemma [J.15.8.] we prove the general

Theorem. If

1. N' is a normal operator in $\boldsymbol{H}'$,

2. C a canonical mapping, yielding a canonical representation of N' in $\boldsymbol{H}$,

3. $\varphi(z)$ a complex valued borelian function of the complex variable z, defined everywhere on the complex plane P,

4. $U f(p) = \varphi(\operatorname{Num} p) \cdot f(p)$ is an operator in $\boldsymbol{H}$ with maximal domain,

then

$C^{-1} U$ does not depend on the choice of C.

J I.4.3. Def. The operator $C^{-1} U$ of Theorem [J I.4.2.] will be denoted by

$$\varphi(N').$$

J I.4.4. Remark. If

$$\varphi(z) = \text{const} = c,$$

then

$$\varphi(N') = c \cdot N', \quad \text{and}$$

if

$$\varphi(z) = z^n,$$

then

$$\varphi(N') = (N')^n;$$

i.e. the n-times iterated N', $(n = 1, 2, \ldots)$.

J I.4.5. The above discussion has yielded a definition of $\varphi(N')$, which is independent of the choice of the canonical correspondence. Thus the φ-operation on N' has an intrinsic meaning; so the operational calculus in normal operators is founded.

Chapter K

Theorems on normal operators and on related canonical mapping

Till now the theory of normal operators and their canonical mappings did not suppose the uniqueness of the double space-valued spectral scale of normal operators.

The proof of uniqueness will be supplied in this Chapter [K.]; but first we shall prove an important theorem, concerning canonical mappings.

K.1. Let N' be a maximal normal operator in the space $\mathbf{H}'$, C a canonical mapping [H.], which carries $\mathbf{H}'$ isomorphically and isometrically into the space $\mathbf{H}$ of all μ-square summable functions $f(p)$ defined a.e. on the disjoint union V of the replicas of the complex plane P, and which carries N' into the operator N where

$$N f(p) =^{\mu} (\mathrm{Num}\, p) \cdot f(p).$$

Let E be a borelian subset of the ordinary complex plane P, and let $\xi(z)$ be the characteristic function of E, i.e. $\xi(z) = 1$ whenever $z \in E$ and $\xi(z) = 0$ whenever $z \bar{\in} E$.

Consider the operator $M' =_{df} \xi(N')$ whose C-image is

$$M f(p) = \xi(\mathrm{Num}\, p) \cdot f(p),$$

(by [J I.4.]).

If f is given, then $M f(p)$ is a function on V, such that

$$M f(p) = f(p), \quad \text{whenever} \quad \mathrm{Num}\, p \in E; \quad \text{hence whenever} \quad p \in \mathrm{Point}\, E,$$

$$M f(p) = 0, \quad \text{whenever} \quad \mathrm{Num}\, p \,\bar{\in}\, E; \quad \text{hence whenever} \quad p \,\bar{\in}\, \mathrm{Point}\, E.$$

Consequently

$$M(f(p)) = [\mathrm{Point}\, E \upharpoonright f(p)]^V.$$

Take the subspace

$$\mathbf{H}^V(\text{Point}\,E)$$

of $\mathbf{H}$ composed of functions which are a.e. equal 0 on

$$\text{co Point}\,E = V - \text{Point}\,E$$

(see [J.]).

We have

$$M(f(p)) = \text{Proj}\,(\mathbf{H}^V(\text{Point}\,E))\,f(p).$$

Its C^{-1}-image is

$$M'(\vec{X}) = \text{Proj}_e \vec{X};$$

hence

(1) $$\xi(N') = \text{Proj}_e,$$

where e is the C^{-1}-image of $\mathbf{H}^V(\text{Point}\,E)$. Since the operation $\xi(N')$ does not depend on the choice of C, [J.], it follows that the space e is independent of C too.

Let us take a canonical mapping C_0, and apply the fundamental representation Theorem. The image of Point$E$ through C_0 is the space of the borelian spectral scale (s_b) of N'.

Thus we get

$$e = s(E),$$

and have the

K.1a. Theorem. If

1. N' is a maximal normal operator in $\mathbf{H}$,

2. C a canonical mapping of $\mathbf{H}$ which gives a canonical representation of N,

3. E a borelian subset of the complex plane P,

4. $\xi(z)$ the characteristic function of E,

then

1) $\xi(N') = \text{Proj}_e$, where e does not depend on C, and where $e = s(E)$, (borelian spectral scale),

2) e is the C^{-1}-image of Point$E$.

K.2. Now we can proceed to the proof of the uniqueness of the double spectral scale of normal operators.

Let $s_1(Q)$, $s_2(Q)$ be two spectral scales of N', and C_1, C_2 canonical mappings of $\mathbf{H}'$ made by means of $s_1(Q)$, $s_2(Q)$ respectively:

$$s_1(Q) \text{ is the } C_1\text{-image of } Q,$$

$$s_2(Q) \text{ is the } C_2\text{-image of } Q.$$

Let $\xi(z)$ be the characteristic function of Q.

We have, by Theorem [K.4.1.],

$$\xi(N') = \text{Proj}_q,$$

where

$$q = s_1(Q) = s_2(Q).$$

Thus the spectral scales s_1 and s_2 coincide.

K.2a. Theorem. If N' is a maximal normal operator, then its double space-valued spectral scale is unique. It follows that the borelian spectral scale is also unique.

The C-image of $s(Q)$ is Q, for any natural canonical mapping C (and also the sets equal Q modulo the ideal of μ-null sets, where μ corresponds to C).

K.2.1. Remark. The proof of the uniqueness of the spectral scale can also be supplied by direct application of the fundamental Theorem [H.13.], without using the operational calculus.

K.3. A remark important for the sequel. Let N' be a maximal normal operator in $\mathbf{H}'$. Its space-valued spectral scale is well determined (see Theorem [K.2.]), and so is (s_b). Hence, if we choose a generating sequence $\mathbf{H}_1'$, $\mathbf{H}_2'$, ... for the borelian spectral tribe (s_b) of N' and select a generating vector $\vec{\omega}$ of $\mathbf{H}'$ with respect to

$$(s_b)^* \quad \text{with} \quad \vec{\omega} \in \complement N',$$

the natural canonical mapping C will be well determined. If (s_b) is saturated, we can consider this as a *particular case* of [H.13.], simply by confining the values of the index n to the single number 1 and by putting $\mathbf{H}_1' = \mathbf{H}'$; so we can *treat jointly* both the cases where (s_b) is saturated or not.

In addition to that, we can always have an *infinite* sequence $\{P_n\}$ of replicas of the complex plane, whatever the case may be. Indeed we can agree to put $\mathbf{H}_n = (\vec{O})$ and $\mu(P_n) = 0$ whenever the n-th copy in our original approach has been lacking in the canonical mapping. We put $V = \bigcup_{n=1}^{\infty} P_n$ and $\mathbf{H} = \mathbf{H}(V)$.

K.3.1. Def. Denote, under [K.3.], by $(\mathscr{T})^P$ the tribe of all borelian subsets of the complex plane P, and by $(\mathscr{T})^V$ the tribe of all borelian subsets of V, both ordered by the relation of inclusion of sets.

K.4. Having this, let us fix the saturating sequence $\{\mathbf{H}_n'\}$, $(n = 1, 2, \ldots)$, but let us vary the generating vector, and see, what will happen to the resulting natural canonical mapping. Let $\vec{\omega} \in \complement N'$, $\vec{\omega}_0 \in \complement N'$ be two generating vectors, and

$$\mu(a) = ||\operatorname{Proj}(a)\,\vec{\omega}\,||^2, \qquad \mu_0(a) = ||\operatorname{Proj}(a)\,\vec{\omega}_0\,||^2$$

the corresponding measures on $(s_b)^*$.

Let C, C_0 be the corresponding natural mappings carrying $\mathbf{H}'$ into $\mathbf{H}$, $\mathbf{H}_0$ respectively, where $\mathbf{H}$ is composed of all μ-square summable functions defined μ-a.e. on V, and $\mathbf{H}_0$ of all μ_0-square summable functions defined μ_0-a.e. on V.

From Theorem [H.10.13.] we know the following:

Let

$$\alpha \in \mathscr{T}^V.$$

We have

$$\alpha = \bigcup_{n=1}^{\infty} (\alpha \frown P_n).$$

Let a_n be the subspace of $\mathbf{H}_n$, which corresponds through C_n^{-1} (see [H.10.13.]), and hence through C^{-1}, to $\alpha \frown P_n$. If we put

$$a = \sum_{n=1}^{\infty} a_n,$$

then a corresponds, through C^{-1}, to α. We have

(1) $$\mu(\alpha) = \mu(a).$$

If we do the same for C_0^{-1} we get the same space a as before, and we have

(2) $$\mu_0(\alpha) = \mu_0(a).$$

Since μ and μ_0 are both effective measures on $(s_b)^*$, it follows that $\mu(\alpha) = 0$ if and only if $\mu_0(\alpha) = 0$. Hence the ideals in $(\mathscr{T})^V$ of sets for which the measure $= 0$, coincide.

Thus we can state the

K.4a. Theorem. If we have

1. N', a normal maximal operator in $\mathbf{H}$,

2. $\mathbf{H}'_n$, $n = 0, 1, \ldots$ its infinite saturating sequence (see [K.3.]),

3. $\vec{\omega}$, $\vec{\omega}_0$ two generating vectors of $\mathbf{H}'$ with respect to the saturated tribe $(s_b)^*$, and μ, μ_0 the corresponding measures on $(s_b)^*$,

4. C, C_0, the resulting natural canonical mappings are giving the corresponding measures on $(\mathscr{T})^V$,

then

the ideals J, J_0 of sets of $(\mathscr{T})^V$, with μ, μ_0-measures $= 0$ respectively, coincide.

K.4.1. Def. The ideal determined by $\{\mathbf{H}'_n\}$, by Theorem [K.4a.], will be denoted by

$$J\{\mathbf{H}'_n\}.$$

K.4.2. Theorem. Under circumstances of Theorem [K.4a.], the function

$$f(p) = \operatorname{Num} p$$

is

$$\mu\text{-square summable on } P.$$

and also

$$\mu_0\text{-square summable on } P.$$

Proof. It suffices to prove it for μ. Let

$$N f(p) = (\mathrm{Num}\, p) \cdot f(p)$$

be the canonical C-image of N'.

The vector $\vec{\omega}$ goes over to the constant function $x(p) = 1$ (see [H.10.13.]). Since $\vec{\omega} \in \mathbb{Q} N'$, we have $x(p) \in \mathbb{Q} N$, and then

$$N x(p) = \mathrm{Num}\, p,$$

so $\mathrm{Num}\, p$ is μ-square summable.

K.4.3. To have a converse Theorem to [K.4a.] we need the following known lemma: (It will be restated later on in a more general form and supplied with an independent proof).

Lemma. If

1. $(\mathscr{T})^V$ is the tribe of all borelian subsets of V,
2. $\mu(\alpha) \geqq 0$ a denumerably additive (finite) measure on $(\mathscr{T})^V$,
3. $\varphi(p) \geqq 0$ is a μ-measurable function, defined μ-a.e. on V,
4. $v(\alpha) = \int\limits_\alpha \varphi(p)\, d\mu$ for all μ-measurable sets $\alpha \subseteq V$,
5. $f(p)$ is v-summable on V,

then the integral

$$\int\limits_V f(p)\, \varphi(p)\, d\mu$$

exists too and equals $\int\limits_V f(p)\, dv$.

K.4.4. Theorem. If

1. N' is a maximal normal operator in $\mathbf{H}'$,
2. $\mathbf{H}'_n$ a saturating sequence of the borelian spectral tribe (s_b),
3. $J(\mathbf{H}'_n)$ the ideal defined in [K.4.1.],
4. $v(\alpha)$ a finite, non negative, denumerably additive measure on $(\mathscr{T})^V$ (see [K.3.1.]),
5. the function $f(p) = \mathrm{Num}\, p$ is v-square summable on V,
6. the ideal composed of all $\alpha \in \mathscr{T}^V$, for which $v(\alpha) = 0$, coincides with $J\{\mathbf{H}'_n\}$,

then there exists a generating vector $\vec{\xi}$ of $\mathbf{H}'$ with respect to $(s_b)^*$, with $\vec{\xi} \in \mathbb{Q} N'$, such that, if we take the canonical mapping D determined by $\mathbf{H}'_n$ and $\vec{\xi}$, then the corresponding measure is $v(\alpha)$ on $(\mathscr{T})^V$.

Proof. Take any canonical mapping C obtained from $\{\mathbf{H}'_n\}$ by the choice of a generating vector. Define

$$v(a) = v(\alpha), \quad \text{for all} \quad \alpha \in \mathscr{T}^V$$

where a is the image of α through C^{-1}. This defines a denumerably additive, non negative (finite), measure on the tribe $(s_b)^*$.

The author has proved [(10), Theorem p. 36] the following general theorem:

K.4.4a. Theorem. "If

1. (T) is a denumerably additive tribe of spaces in the separable and complete H.-H.-space $\mathbf{H}'$,

2. $v(b)$ is a denumerably additive, $\geqq 0$ (finite), measure on (T), then there exists a vector $\vec{\xi} \in \mathbf{H}'$ such that

$$v(b) = ||\mathrm{Proj}\,(b)\,\vec{\xi}||^2 \quad \text{for all} \quad b \in T\text{''}.$$

Applying this to our case, let $\vec{\xi}$ be such a vector in $\mathbf{H}'$, for which

(1) $$v(\alpha) = v(a) = ||\mathrm{Proj}\,(a)\,\vec{\xi}||^2 \quad \text{for} \quad a \in s_b^*; \quad \text{and}$$

let μ be the measure related to C.

Since $J\{\mathbf{H}_n\}$ is the ideal of all μ-null sets as well as the ideal of all v-null sets in $\mathscr{T}^V$, we have $v(\alpha) = 0$ if and only if $\mu(\alpha) = 0$.

Let $a \in s_b^*$, $a \neq 0$. We have $\mu(a) > 0$ because μ is effective on $(s_b)^*$. Let α corresponds to a through C. We have $\mu(\alpha) > 0$. Consequently $v(\alpha) > 0$ and then $v(a) > 0$. Thus $v(a)$ is an effective measure on $(s_b)^*$.

Now we can apply another theorem by the author. It says that:

"If

1. (T) is a saturated tribe of spaces in $\mathbf{H}'$,

2. $v(b)$ an effective measure, $\geqq 0$, denumerably additive (and finite),

3. $\vec{\eta} \in \mathbf{H}'$ such that

$$v(b) = ||\mathrm{Proj}\,(b)\,\vec{\eta}||^2 \quad \text{for} \quad b \in T,$$

then

$\vec{\eta}$ is a generating vector of $\mathbf{H}'$ with respect to (T)."

By virtue of this theorem the vector $\vec{\xi}$ is a generating vector of $\mathbf{H}'$ with respect to $(s_b)^*$.

Let $\vec{\xi}$ be carried, through C, into the μ-square summable function $g(p)$ on V, so that we have

$$v(a) = ||\mathrm{Proj}\,(a)\,\vec{\xi}||^2 \quad \text{for all} \quad a \in s_b^*.$$

Since $\mathrm{Proj}\,(a)\,\vec{\xi}$ goes over, through C, to

$$\mathrm{Proj}\,(\mathbf{H}^V(\alpha))\,g(p) = (\alpha \uparrow g(p))^V,$$

we get

$$v(a) = \int_\alpha |g(p)|^2\,d\mu.$$

We have supposed that $|\mathrm{Num}\, p|^2$ is v-summable on V. By the Lemma [K.4.3.] it follows that

$$\int v\, |\mathrm{Num}\, p|^2\, |g(p)|^2\, d\mu < \infty.$$

Since $(\mathrm{Num}\, p) \cdot g(p)$ and $g(p)$ are both μ-square summable, it follows that $g(p) \in \mathrm{Q}\, N$. It also follows that its C^{-1}-image $\vec{\xi}$ belongs to $\mathrm{Q}\, N'$.

Consequently $\vec{\xi}$ is a generating vector of $\mathbf{H}'$ with respect to $(s_b)^*$ with $\vec{\xi} \in \mathrm{D}\, N'$.

Take the canonical mapping D, determined by $\{\mathbf{H}'_n\}$ and $\vec{\xi}$. By (1) v is just the measure attached to D.

Now let $a \in s_b^*$. The C^{-1}-image of a and the D^{-1}-image of a coincide, by Theorem [H.10.13.].

Hence, by (1), we have

$$v(a) = v(\alpha),$$

whenever α corresponds through D to a. The theorem is established.

K.4.5. Theorem. If we keep $\{\mathbf{H}'_n\}$ fixed and change the generating vector $\vec{\omega} \in \mathrm{Q}\, N'$ into

$$\vec{\omega}_0 = \lambda\, \vec{\omega}$$

where λ is any complex number with $|\lambda| = 1$, then

$$\vec{\omega}_0 \in \mathrm{Q}\, N'$$

too, and this is also a generating vector yielding the same measure.

Proof. Indeed

$$\|\mathrm{Proj}\,(a)\, \vec{\omega}_0\|^2 = \|\lambda\, \mathrm{Proj}\,(a)\, \vec{\omega}\|^2 = |\lambda|^2 \cdot \|\mathrm{Proj}\,(a)\, \vec{\omega}\|^2$$
$$= \|\mathrm{Proj}\,(a)\, \vec{\omega}\|^2.$$

K.4.6. Remark. The converse to [K.4.5.] is not true. There may exist two vectors $\vec{\omega}_0$, $\vec{\omega}$ yielding the same measure μ, but where $\vec{\omega}_0$ is not a multiplum of $\vec{\omega}$.

K.4.7. Remark. The above theorems say that, to a given saturating sequence $\{\mathbf{H}'_n\}$, there exists a well defined ideal $J\{\mathbf{H}_n\}$ in $(\mathscr{T})^V$; but not all measures v with the same null-ideal are eligible for a canonical mapping, but those and only those for which $\mathrm{Num}\, p$ is v-square summable on V. To a given measure there correspond many generating vectors yielding it. What will happen to C, if we change the saturating sequence, will be discussed later on.

K.5. Let N' be a normal operator, $\{\mathbf{H}'_n\}$ a saturating sequence, and C a canonical (natural) mapping with measure μ.

Consider the tribe $(\mathscr{T}^P)$ of all borelian subsets of the complex plane P.

Lemma. If $E \in \mathscr{T}^P$, put

$$\bar{\mu}(E) = \mu(\mathrm{Point}\, E).$$

The set-function $\bar{\mu}$ is denumerably additive, non negative and finite.

Indeed, let $E \frown F = \emptyset$. Then

$$(\text{Point}\, E) \frown (\text{Point}\, F) \neq \emptyset .$$

Since Point E is a borelian subset of V, we get the denumerable additivity. The remaining items are obvious. Thus we have defined on $(\mathscr{T}^P)$ a measure, which may not be effective.

K.5.1. Let us change the measure μ into μ_0, in accordance with [K.4.7.], and define

$$\bar{\mu}_0(E) = \mu_0(\text{Point}\, E)$$

for all $E \in \mathscr{T}^P$.

The following are equivalent:

I. $\bar{\mu}_0(E) = 0$,

II. $\bar{\mu}(E) = 0$.

The proof is straight forward.

Let us change the saturating sequence $\{\mathbf{H}'_n\}$ into $\{G'_n\}$ and take, for the resulting new saturated tribe $(s_b'^*)$ a canonical natural mapping D with measure v. We shall compare $\bar{\mu}(E)$ and $\bar{v}(E)$ for $E \in \mathscr{T}^P$. By Theorem [K.1a.], the space corresponding to E through C^{-1} and D^{-1} is the same, $s(E)$.

We have

$$\bar{\mu}(E) = \mu(\text{Point}\, E) = \mu\, s(e),$$

$$\bar{v}(E) = v(\text{Point}\, E) = v\, s(e),$$

[H.10.13.]. Now both measures μ, v are effective on (s_b).

Consequently, if $\bar{\mu}(E) > 0$ then $\bar{v}(E) > 0$ and conversely. It follows that the ideal of all sets $E \in \mathscr{T}^P$ with $\bar{\mu}(E) = 0$ coincides with the ideal of all sets $E \in \mathscr{T}^P$ with $\bar{v}(E) = 0$. We have the

Theorem. If C is a canonical mapping with measure μ and if we define for all $E \in \mathscr{T}^P$ the function

$$\bar{\mu}(E) = \mu(\text{Point}\, E)$$

then $\bar{\mu}(E)$ is a denumerably additive measure on $(\mathscr{T}^P)$ and the ideal of all sets $E \in \mathscr{T}^P$ with $\bar{\mu}(E) = 0$ does not depend on choice of C.

K.5.1a. Def. Thus we can denote it by $J_{N'}$, since it depends only on N'.

K.5.2. Def. We can construct the measure-hull $(\mathscr{T}^P_L)$ of $\mathscr{T}^P$, by adjunction to $\mathscr{T}^P$ all $J_{N'}$-null sets, [A I., Sect. 5].

We shall call them the *N'-measurable sets* in P, and the functions $f(z)$ fitting $(\mathscr{T}^P_L)$, *N'-measurable functions*. For $E, F \in \mathscr{T}^P_L$ we say that $E =^{N'} F$, whenever $(E - F) \smile (F - E) \in J_{N'}$. The equality $E =^{N'} F$ is the equality modulo $J_{N'}$, on $\mathscr{T}^P_L$. We shall use the expression N'-almost everywhere, N'-a.e. We may also consider the inclusion modulo

$J_{N'}$, denoted by

$$E \subseteqq^{N'} F.$$

K.5.3. Theorem. If $E \in \mathcal{T}_L^P$, then Point$E$ is μ-measurable whatever the canonical mapping with measure μ may be. The proof is easy.

K.5.4. Theorem. If $E \in \mathcal{T}_L^P$, and C is a natural canonical mapping, then the space e, which corresponds through C^{-1} to Point$E$, belongs to s_b and is independent of C.

Proof. Indeed, by [K.5.3.], $F = \text{Point}E$ is μ-measurable.

Since

$$\text{Num}F = E,$$

and hence

$$\text{Point Num}F = \text{Point}E = F,$$

the existence of e and its independence of C is assured by Theorem [H.10.13.].

K.5.5. Theorem. If

1. $E_1, E_2 \in \mathcal{T}_L^P$,
2. $E_1 =^{N'} E_2$,
3. C is a canonical mapping,

then the C^{-1}-image-spaces e_1, e_2 of PointE_1, PointE_2 respectively, coincide.

Proof. Let μ be the measure, related to C. We have

$$(E_1 \dotplus E_2) \in J_{N'}.$$

Hence

$$\mu\,\text{Point}(E_1 \dotplus E_2) = \mu(\text{Point}E_1 \dotplus \text{Point}E_2) = 0;$$

so

$$\text{Point}E_1 =^{\mu} \text{Point}E_2.$$

Consequently the C^{-1}-images of PointE_1, PointE_2 coincide, since C^{-1} is an isomorphism with respect to μ-equality (see [H.10.13.]).

K.5.6. We can prove the converse of Theorem [K.5.3.].

Theorem. If E is a subset of P and Point$E$ is μ-measurable for some natural canonical mapping C with measure μ, then

$$E \in \mathcal{T}_L^P.$$

Proof. Let e be the space $\in s_b^*$, corresponding to $F = \text{Point}E$ through C^{-1}.

Since

$$\text{Point Num}F = F,$$

we have by [H.10.13.] $e \in s_b$ and e is independent of the choice of C. Hence to e there corresponds by [H.10.13.] a borelian subset A of P, such that Point$A$ corresponds, through C, to e.

Since C is one-to-one modulo the ideal of μ-null sets of V, (see [H.10.13.]), it follows that

$$\mu(\text{Point}\, A \dotplus \text{Point}\, E) = 0.$$

Hence

$$\mu \,\text{Point}\,(A \dotplus E) = 0;$$

hence

$$A \dotplus E \in J_{N'},$$

(see [K.5.1.]), and then

$$E \in \mathscr{T}_L^P, \quad \text{q.e.d.}$$

Remark. If $E \in \mathscr{T}_L^P$, there may exist sets $\alpha \subseteqq V$ with

$$\text{Point}\, E =^{\mu} \alpha,$$

but where α has not the property

$$\text{Point Num}\,\alpha = \alpha.$$

K.5.6a. Theorem. If

$$E_1, E_2 \in \mathscr{T}^P$$

and

$$E_1 =^{N'} E_2,$$

then

$$s(E_1) = s(E_2).$$

Proof. Since E_1, E_2 are both borelian, so is

$$E_1 \dotplus E_2 = (E_1 - E_2) \cup (E_2 - E_1).$$

We have, by hypothesis,

$$E_1 \dotplus E_2 \in J_{N'},$$

and then, by [K.5.1.], [K.5.1a.], [K.5.2.],

$$\mu \,\text{Point}\,(E_1 \dotplus E_2) = 0,$$

$$\mu(\text{Point}\, E_1 \dotplus \text{Point}\, E_2) = 0.$$

Hence

$$\text{Point}\, E_1 =^{\mu} \text{Point}\, E_2,$$

and then, since C^{-1} is one-to-one with respect to the μ-equality, the C^{-1}-images of Point E_1, and Point E_2 coincide. Hence, by Theorem [K.1a.], the theorem is proved.

K.5.6b. Def. The above theorem and [K.5.4.] suggest the following definition, which generalizes the function $s(E)$ related to (s_b).

If $F \in \mathscr{T}_L^P$, then by $s(F)$ we shall understand the space $s(E)$ where $E =^{\mu N'} F$ and E is a borelian subset of P.

(This space $s(E)$ does not depend on the choice of the borelian E with $E =^{N'} F$.)

K.5.6c. Theorem. If

$$E_1, E_2 \in \mathscr{T}_L^P$$

and

$$E_1 =^{N'} E_2,$$

then

$$s(E_1) = s(E_2).$$

The easy proof refers to [K.5.6a.], [K.5.6b.].

K.5.6d. Theorem. Relying on the above theorems, it is easily to prove, that the relation $e = s(E)$ is an isomorphism between the $(\mathscr{T}_L^P)$ with equality $=^{N'}$ and the borelian spectral tribe (s_b) of N'.

K.5.7. The $\mu_{N'}$-measurability of sets and the extension of $(\mathscr{T}^P)$ to $(\mathscr{T}_L^P)$ induces analogous notions to complex-valued functions.

Thus if a complex valued function $\varphi(z)$ is $\mu_{N'}$-measurable, then the function

$$f(p) = \varphi(\text{Num}\, p) \tag{1}$$

is also μ-measurable on V, where μ is the measure relative to any canonical mapping.

If $\varphi(z)$ is defined N'-a.e.-on P and is N'-measurable (i.e. fitting $\mathscr{T}_L^P$), then (1) is also μ-measurable on V.

If

$$\varphi_1(z) =^{N'} \varphi_2(z),$$

then the functions

$$f_1(p) = \varphi_1(\text{Num}\, p), \quad f_2(p) = \varphi_2(\text{Num}\, p)$$

are μ-equivalent.

K.6. Till now we have defined the operation φ performed on N' in the only case, where $\varphi(z)$ was a *borelian* function defined *everywhere* on P.

Now we shall extend slightly the notion of $\varphi(N')$. First we shall prove the following: If

1. C is a natural canonical mapping with measure μ,
2. $\varphi(z)$ is a borelian function defined everywhere on P,
3. $\varphi_1(z)$ a N'-measurable function such that

$$\varphi(z) =^{N'} \varphi_1(z),$$

4. U is the operator

$$U f(p) =^{\mu} \varphi(\text{Num}\, p) \cdot f(p)$$

with maximal domain in $\mathbf{H}$, ($\mathbf{H}$ is composed of all μ-square summable functions on V, defined μ-a.e. on V),

5. $U_1 f(p) =^{\mu} \varphi_1(\text{Num}\, p) \cdot f(p)$ with maximal domain in $\mathbf{H}$,

then the C^{-1}-images U', U_1' of U and U_1 respectively, coincide.

Proof. Let $\vec{X} \in \mathbf{\Delta}\, U'$, and let $f(p)$ correspond to $\vec{X}$ through C. We have

(1) $$U f(p) =^{\mu} \varphi(\mathrm{Num}\, p) \cdot f(p);$$

and $f(p)$ and $\varphi(\mathrm{Num}\, p) \cdot f(p)$ are μ-square summable.

Since

$$\varphi(z) =^{N'} \varphi_1(z),$$

we have:

$$\{z \mid z \bar{\in} \mathbf{\Delta}\, \varphi_1 \quad \text{or} \quad \varphi(z) \neq \varphi_1(z)\}$$

is a $\mu_{N'}$-null set.

Hence, by [K.4.5.], [K.4.5.1.],

$$\mathrm{Point}\{z \mid z \bar{\in} \mathbf{\Delta}\, \varphi_1 \quad \text{or} \quad \varphi(z) \neq \varphi_1(z)\}$$

is a μ-null set.

Hence, if we put

$$A(p) = \varphi(\mathrm{Num}\, p), \qquad A_1(p) = \varphi_1(\mathrm{Num}\, p),$$

then the set

$$\{p \mid p \bar{\in} \mathbf{\Delta}\, A(p) \quad \text{or} \quad A(p) \neq A_1(p)\}$$

is a μ-null set, and then

$$A(p) =^{\mu} A_1(p).$$

Consequently

$$\varphi_1(\mathrm{Num}\, p) \cdot f(p) =^{\mu} \varphi(\mathrm{Num}\, p) \cdot f(p),$$

and then, since C is one-to-one with respect to μ-equality, it follows that $\varphi_1(\mathrm{Num}\, p) \cdot f(p)$ is μ-square summable; hence $\vec{X} \in \mathbf{\Delta}\, U_1'$, and in addition to that

$$U' \vec{X} = U_1' \vec{X}.$$

An analogous argument can be used, if we suppose that $\vec{X} \in \mathbf{\Delta}\, U_1'$ so the theorem is established.

K.6.1. Having this in mind, we see that if

$$\varphi_1(z) =^{N'} \varphi_2(z),$$

then the C^{-1}-images of the operators

$$U_1 f(p) =^{\mu} \varphi_1(\mathrm{Num}\, p) \cdot f(p),$$

$$U_2 f(p) =^{\mu} \varphi_2(\mathrm{Num}\, p) \cdot f(p),$$

both taken with maximal domains in $\mathbf{H}$, are identical. This image does not depend on C.

This allows to state the

Def. If $\varphi(z)$ is a N'-measurable function defined N'-a.e. on the complex plane P, then by $\varphi(N')$ we shall understand the operator $\varphi_1(N')$, where $\varphi_1(z)$ is a borelian function defined everywhere on P and such that

$$\varphi_1(z) =^{N'} \varphi(z).$$

Here C is a canonical mapping and μ the measure related to it. $\varphi(N')$ does not depend on C, neither on the choice of φ_1.

K.7. In this subsection we take a space $e \in s_b$ and find the natural canonical mappings of the operator $e \rceil N'$, i.e. N' restricted to e.

We know that its domain is

$$\text{ᗡ} N' \frown e = \operatorname{Proj}(e)\, \text{ᗡ} N',$$

and that

$$\text{ᗡ}(e \rceil N') \subseteqq \text{ᗡ} N'.$$

Let C be the natural canonical mapping of $\mathbf{H}'$ onto $\mathbf{H}$, determined by the saturating sequence $\{\mathbf{H}_n\}$ and the generating vector $\vec{\omega} \in \text{ᗡ} N'$, yielding the measure

$$\mu(a) = \|\operatorname{Proj}(a)\, \vec{\omega}\|^2 \quad \text{for all} \quad a \in (s_b^*).$$

Put

$$N_1' = e \rceil N'.$$

We know that the spectral space-valued scale $s_1(Q)$ of N_1' is

$$s_1(Q) = e \cdot s(Q)$$

where $s(Q)$ is the spectral scale of N'. Hence the borelian spectral tribe: (s_{1b}) of N_1' is $e \cdot (s_b)$, where (s_b) is the borelian spectral tribe of N.

We easily prove that, if we take the sequence

$$(1) \qquad e \cdot \mathbf{H}_1',\ e \cdot \mathbf{H}_2',\ \ldots,\ e\, \mathbf{H}_n',\ \ldots,$$

we obtain a saturating sequence of (s_{1b}).

(Here we may drop those spaces (1) which are $= (\vec{O})$ or not, in agreement with the Remark [K.3.]),

$$\vec{\omega}_1 = \operatorname{Proj}(e)\, \vec{\omega}$$

is a generating vector of the space e with respect to the saturated tribe (s_{1b}^*), with $\vec{\omega}_1 \in \text{ᗡ} N_1$.

We have

$$\vec{\omega}_1 \in \operatorname{Proj}(e)\, \text{ᗡ} N';$$

hence

$$\vec{\omega}_1 \in \text{ᗡ} N_1'.$$

The measure μ_1, generated by $\vec{\omega}_1$, is

$$\mu_1(a) = \|\operatorname{Proj}(a)\, \vec{\omega}_1\|^2$$

for all

$$a \in (s_{1b})^*.$$

Hence

$$\mu_1(a) = \|\operatorname{Proj}(a) \cdot \vec{\omega}\|^2 = \mu(a)$$

for all

$$a \in (s_{1b})^*.$$

Let $e = s(E)$, where E is a borelian subset of P. Let C_1 be the canonical mapping of the space $\mathbf{H}'$ into the space $\mathbf{H}_1$ of all μ_1-square summable functions defined μ_1-a.e. on V, and related to the saturating sequence $\{\mathbf{H}'_n e\}$ of (s_{1b}) and N'_1.

K.7.1. Def. The mapping C_1 will be termed *generated by* C. The measure μ_1 differs from μ in that

$$\mu_1(V - \operatorname{Point} E) = 0;$$

but $\mu_1(\alpha) = \mu(\alpha)$ for all $\alpha \subseteqq \operatorname{Point} E$.

Consequently the μ_1-square summable functions $f_1(p)$ on V can be obtained from μ-square summable functions $f(p)$ on V in the way, that we put $f_1(p) = f(p)$ on $\operatorname{Point} E$ and change the values of $f(p)$ on $\operatorname{co} \operatorname{Point} E$ in any way we like.

If $f_1(p)$ is μ_1-square summable on V, then

$$(\operatorname{Point} E \upharpoonright f_1(p))^V$$

is μ-square summable on V and vice-versa.

If $f(p)$ is μ-square summable on V, then

$$\operatorname{Point} E \upharpoonright f(p) =^{\mu_1} (\operatorname{Point} E \upharpoonright f(p))^V,$$

since the function on the left is defined on $\operatorname{Point} E$, hence μ_1-a.e. on V.

The C-image of N'_1 is the operation

$$U f(p) =^{\mu} (\operatorname{Num} p) \cdot f(p)$$

with maximal domain in $\mathbf{H}^V(\operatorname{Point} E)$.

The C_1-image of N'_1 is the operation

$$U_1 f(p) =^{\mu_1} (\operatorname{Num} p) \cdot f(p)$$

with maximal domain in $\mathbf{H}_1$, i.e. in the space of all μ_1-square summable functions, defined μ_1-a.e.-on V.

K.7.2. Theorem. If

1. $\vec{X} \in e$,
2. $f(p)$, $g(p)$ are its C, C_1-images respectively,

then

$$f(p) =^{\mu_1} g(p).$$

K.7.3. Theorem. If

1. $f(p)$ is μ_1-square summable on V,
2. $(E \upharpoonright g(p))^V =^{\mu} g(p)$,

then

the C^{-1} and C_1^{-1}-images of $g(p)$ are identical vectors in $\mathbf{H}$.

They belong to e.

For the proof we mention that the mapping of vectors is defined as extension of the mapping of the projections of the generating vector onto characteristic functions of point sets. By dealing with characteristic functions we rely on Theorem [K.1.].

K.7.4. Remark. We may, with reference to [K.5.]—[K.5.7.] and [K.6.]—[K.6.1.], consider the N'_1-measurability on $\mathcal{T}^P$, related to $N'_1 = e \upharpoonleft N'$. It has the property that if $F \subseteqq E$, $F \in \mathcal{T}^P$, then $F \in J_{N'_1}$ is equivalent to $F \in J_{N'}$; but $\operatorname{co} E \in J_{N'_1}$, i.e. $\operatorname{co} E$ is an N'_1-null set.

K.8. Till now we have dealt with $\varphi(N')$, where $\varphi(z)$ is N'-measurable and defined N'-a.e. on P.

Def. Now suppose that $\varphi(z)$ is not necessarely defined N'-a.e. on P but N'-a.e. on a N'-measurable subset E of P only. Let C be a canonical mapping related to N', and with measure μ.

Take the space $e = s(E)$, which is the C^{-1}-image of Point E (see [K.5.6b.]).

Consider the function

$$\varphi^P(z),$$

defined by

$$\varphi^P(z) = \varphi(z) \quad \text{for all} \quad z \in E,$$

$$\varphi^P(z) = 0 \quad \text{for all} \quad z \in \operatorname{co} E = P - E.$$

We easily prove that

$$e \upharpoonleft \varphi^P(N')$$

does not depend on C. Thus we define $\varphi(N')$ as

$$e \upharpoonleft \varphi^P(N').$$

K.8.1. We easily prove that if

$$\varphi_1(z) \overset{N'}{=} \varphi_2(z),$$

then

$$\varphi_1(N') = \varphi_2(N'),$$

with identity of domains.

K.8.2. Theorem. If

1. N' is a normal operator in $\mathbf{H}'$,
2. $E \in \mathcal{T}^P_L$,
3. $\varphi(z)$ is N'-measurable and defined N'-a.e. on E,
4. $e = s(E)$,

then

$$\varphi(N') = e \upharpoonleft \varphi^P(N') = \varphi(e \upharpoonleft N'),$$

(see Def. [K.8.]).

Proof. Consider the canonical mapping C of $\mathbf{H}'$ into $\mathbf{H}$, with measure μ, and related to N', and also the canonical mapping C_1 of

$\mathbf{H}'$ into $\mathbf{H}_1$ with measure μ_1 and related to

$$N_1 = e \rceil N$$

and C.

Let

$$X \in \mathrm{D}\, \varphi(N').$$

Let $g(p)$ and $g_1(p)$ be its C and C_1-images respectively. We have

(1) $$g(p) =^{\mu_1} g_1(p), \quad g(p) \in \mathbf{H}^V(E).$$

The C-image of $\varphi(N')$ is the operator

$$U f(p) =^{\mu} \varphi^P(\mathrm{Num}\, p) \cdot f(p)$$

with maximal domain in $\mathbf{H}$.

Hence we have

(2) $$U g(p) =^{\mu} \varphi^P(\mathrm{Num}\, p) \cdot g(p), \quad \text{and}$$

$g(p)$ and

(3) $$\varphi^P(\mathrm{Num}\, p) \cdot g(p)$$

are μ-square summable.

The C_1-image of $\varphi(e \rceil N')$ is the operator

$$U_1 f(p) =^{\mu_1} \varphi(\mathrm{Num}\, p) \cdot f(p)$$

with maximal domain in $\mathbf{H}_1$.

K.8.2a. We shall prove that $g_1(p) \in \mathrm{D}\, U_1$. $g_1(p)$ is μ_1-square summable. By (1) we have

$$\varphi(\mathrm{Num}\, p) \cdot g_1(p) =^{\mu_1} \varphi^P(\mathrm{Num}\, p) \cdot g(p).$$

Hence, by (3), $\varphi(\mathrm{Num}\, p) \cdot g_1(p)$ is μ_1-square summable, and hence $g_1 \in \mathrm{D}\, U_1$. It follows that

$$\vec{X} \in \varphi(e \rceil N').$$

K.8.2b. To prove that

$$\varphi(N')\, \vec{X} = \varphi(e \rceil N\, \vec{X}),$$

we shall prove that the C^{-1}-image of

$$\varphi^P(\mathrm{Num}\, p) \cdot g(p)$$

and the C_1^{-1}-image of

$$\varphi(\mathrm{Num}\, p) \cdot g_1(p)$$

coincide.

The C_1^{-1}-image of

$$\varphi(\mathrm{Num}\, p) \cdot g_1(p)$$

coincides with that of $\varphi(\mathrm{Num}\, p) \cdot g(p)$, because of (1), and the C_1^{-1}-image of the last function coincides with that of

$$\varphi^P(\mathrm{Num}\, p) \cdot g(p).$$

Now, if we apply [K.7.3.], we obtain that the images are identical.

Thus we have proved that if $\vec{X} \in \text{a}\,\varphi(N')$, then

(4) $$\varphi(N')\,\vec{X} = \varphi(e \upharpoonleft N')\,\vec{X}.$$

K.8.2c. Now let

$$\vec{X} \in \text{a}\,\varphi(e \upharpoonleft N').$$

Since $\vec{X} \in e$, a similar reasoning leads to the conclusion that

$$\vec{X} \in \text{a}\,\varphi(N').$$

If we apply (4), we get the complete identity of the operators $\varphi(N')$ and $\varphi(e \upharpoonleft N')$. q.e.d.

K.8.3. Theorem. If

1. N' is a normal transformation in $\mathbf{H}'$,
2. C a natural canonical mapping of $\mathbf{H}'$ into $\mathbf{H}$,
3. $E \in \mathcal{T}_L^P$,
4. $\varphi(z)$ a N'-measurable complex valued function defined N'-a.e. on E,
5. $e = s(E)$,
6. $N_1 f(p) = \ulcorner\varphi(\text{Num}\,p)\urcorner^V \cdot f(p)$ is an operator with maximal domain in $\mathbf{H}$,

then

the C^{-1}-image of N_1 is the normal transformation N_1' in $\mathbf{H}'$, which coincides with $\varphi(N')$ on e, and $N_1'\vec{X} = 0$ for all $\vec{X} \in \text{co}\,e$. Hence $N_1' = \varphi(N')\,\text{Proj}_e$. The proof is easy.

K.9. We are going to prove some convergence theorems, derived from theorems in [J.].

Theorem. If

1. N is a maximal normal operator in $\mathbf{H}$,
2. $(\mathcal{T}_L^P)$ the tribe of subsets of the complex plane P, defined in [K.5.2.] and J_N, the ideal in $(\mathcal{T}_L^P)$ defined in [K.5.1a.],
3. $E \in \mathcal{T}_L^P$,
4. $\varphi_n(z)$, $(n = 1, 2, \ldots)$, N-measurable (see [K.5.7.], [K.5.2.]) functions of the complex variable z, defined N-a.e. on E (see [K.5.7.]),
5. $\varphi_n(z)$ converges uniformly to $\varphi(z)$ on E outside of a N-null set,

then

1) there exists n_0 such that for every $n \geqq n_0$

$$\text{a}\,\varphi_n(N) = \text{a}\,\varphi(N),$$

2) for every $\vec{X} \in \text{a}\,\varphi(N)$ we have

$$\lim_{n\to\infty} \varphi_n(N)\,\vec{X} = \varphi(N)\,\vec{X},$$

(topological limit in $\mathbf{H}$),

3) if M denotes the set of all vectors $\vec{X}$, for which

$$\{\varphi_n(N)\,\vec{X}\}$$

converges, and if we define

$$W\,\vec{X} = \lim_{n\to\infty} \varphi_n(N)\,\vec{X}$$

for all $\vec{X}\in M$, then the operator W thus defined coincides with the operator $\varphi(N)$.

Proof. Take a canonical mapping C which carries **H** into **H**$'$, the space of all μ-square summable functions defined μ-a.e.-on **H**$'$. Put

$$U_n = \varphi_n^P(N), \qquad U = \varphi^P(N).$$

Their C-images are

$$U'_n f(p) = \varphi_n^P(\mathrm{Num}\,p)\cdot f(p),$$

$$U' f(p) = \varphi^P(\mathrm{Num}\,p)\cdot f(p)$$

with maximal domain in **H**$'$.

Now $\varphi_n^P(\mathrm{Num}\,p)$ converges μ-a.e.-uniformly toward $\varphi^P(\mathrm{Num}\,p)$, so we are in the conditions of Lemma [J.15.].

Hence there exists n_0, such that for all $n\geqq n_0$

$$\mathrm{\complement}\,U'_n = \mathrm{\complement}\,U',$$

and for every $f(p)\in \mathrm{\complement}\,U'$, the sequence

$$U'_n f(p)$$

converges in μ-square mean toward $U' f(p)$.

Since C^{-1} is a homeomorphic isomorphism, it follows that

(1) $$\mathrm{\complement}\,U_n = \mathrm{\complement}\,U \quad \text{for all} \quad n\geqq n_0;$$

and if $\vec{X}\in \mathrm{\complement}\,U$, then

(2) $$\lim U_n\vec{X} = U\cdot\vec{X}.$$

Let $e = s(E)$, (see Def. [K.5.6b.]), i.e. e corresponds to Point E through C^{-1}.

We have, by Def. [K.8.],

$$\varphi_n(N) = e \upharpoonright U_n, \qquad \varphi(N) = e \upharpoonright U,$$

and we know that

$$\mathrm{\complement}(e \upharpoonright U_n) = \mathrm{\complement}\,U_n \cap e, \qquad \mathrm{\complement}(e \upharpoonright U) = \mathrm{\complement}\,U \cap e.$$

It follows from (1), that for $n\geqq n_0$

$$\mathrm{\complement}\,\varphi_n(N) = \mathrm{\complement}\,\varphi(N).$$

Let

$$\vec{X}\in \mathrm{\complement}\,\varphi(N).$$

We have

$$\vec{X} \in \text{ᗡ}\, U,$$

because

$$\text{ᗡ}(e \upharpoonleft U) \subseteqq \text{ᗡ}\, U.$$

Hence, by (2),

$$\lim_{n\to\infty} \varphi_n(N)\,\vec{X} = \varphi(N)\,\vec{X}.$$

K.9a. Let us proceed to the proof of the 3rd thesis. Let $\vec{X} \in M$. The sequence $\{\varphi_n(N)\,\vec{X}\}$ converges. Hence, for sufficiently great n, $\vec{X} \in \text{ᗡ}\,\varphi_n(N)$, and then, by thesis 1),

$$\vec{X} \in \text{ᗡ}\,\varphi(N).$$

Thus

$$M \subseteqq \text{ᗡ}\,\varphi(N).$$

Now let $\vec{X} \in \text{ᗡ}\,\varphi(N)$. It follows, by thesis 2), that

$$\lim_n \varphi_n(N)\,\vec{X}$$

exists, and then $\vec{X} \in M$, and so $\text{ᗡ}\,\varphi(N) \subseteqq M$.

Thus we have proved that $M = \text{ᗡ}\,\varphi(N)$.

Now let

$$\vec{X} \in \text{ᗡ}\,\varphi(N) = M.$$

We have

$$\lim \varphi_n(N)\,\vec{X} = \varphi(N)\,\vec{X},$$

by thesis 2), and then

$$W\vec{X} = \varphi(N)\,\vec{X}.$$

It follows that

$$W = \varphi(N).$$

K.9.1. Theorem. If

1. $E_1 \subseteqq^N E_2 \subseteqq^N \cdots \subseteqq^N E_n \subseteqq^N \cdots$ belong to $\mathscr{T}_L^P$,

2. $E =^N \bigcup_{n=1}^{\infty} E_n$,

3. $\varphi_n(z)$ is a complex valued N-measurable function, defined N-a.e.-on E_n, $(n = 1, 2, \ldots)$,

4. $\varphi(z)$ is a complex valued, N-measurable function defined N-a.e.-on E,

5. $\varphi_n(z) =^N \varphi(z)$ on E_n,

6. $M = \bigcup_{n=1}^{\infty} \text{ᗡ}\,\varphi_n(N)$,

then

the operator W with $\text{ᗡ}\,W = M$ and defined by $W\vec{X} = \varphi(N)\,\vec{X}$ exists.

For this operator we have (closure)

$$\tilde{W} = \varphi(N),$$

(with identity of domains).

Proof. Take a natural canonical mapping C for N, transforming $\mathbf{H}$ into the space $\mathbf{H}'$ of all μ-square summable functions $f(p)$ on V.

Let N' be the C-image of N.

It is not difficult to see that we may suppose that

$$E_1 \subseteqq E_2 \subseteqq \cdots \quad \text{and that} \quad E = \bigcup_{n=1}^{\infty} E_n.$$

Put

$$e_n = s(E_n),$$

(see Def. [K.5.6b.]),

$$e = s(E).$$

We have

$$\varphi_n(N) = e_n \upharpoonleft \varphi_n^P(N),$$

$$\varphi(N) = e \upharpoonleft \varphi^P(N),$$

by Def. [K.8.]. The C-images of the operators $\varphi_n^P(N)$, $\varphi^P(N)$ are

$$\varphi_n^P(N'), \quad \varphi^P(N')$$

respectively, i.e. the operators on f:

(1) $$\varphi_n^V(\operatorname{Num} p) \cdot f(p), \quad \varphi^V(\operatorname{Num} p) \cdot f(p),$$

taken with maximal domains in $\mathbf{H}'$.

Hence the C-images S_n, S of $\varphi_n(N)$, $\varphi(N)$ are the operators (1), restricted to

$$\mathbf{H}'^V(\operatorname{Point} E_n), \quad \mathbf{H}'^V(\operatorname{Point} E),$$

respectively.

The sets $\operatorname{Point} E_n$, $\operatorname{Point} E$ are μ-measurable and we have:

$$\operatorname{Point} E_1 \subseteqq \operatorname{Point} E_2 \subseteqq \cdots$$

$$\operatorname{Point} E = \bigcup_{n=1}^{\infty} \operatorname{Point} E_n,$$

$$\varphi_n(\operatorname{Num} p) \text{ are } \mu\text{-measurable},$$

and

$$\varphi(\operatorname{Num} p) = \varphi_n(\operatorname{Num} p) \quad \text{for} \quad p \in \operatorname{Point} E_n.$$

The domain of $S_n = C\,\varphi_n(N)$ is the set of all

$$f(p) \in \mathbf{H}'^V(\operatorname{Point} E_n),$$

for which

$$\varphi_n^V(\operatorname{Num} p) \cdot f(p)$$

is μ-square summable on V. Since the domain of $\varphi_n(N)$ is $e_n \frown \mathbb{C}\,\varphi_n^V(N)$, its C-image is

$$\mathbf{H}'^V(E_n) \frown \mathbb{C}\,\varphi_n^P(N'),$$

hence just the set $\mathbb{C}\,S_n$.

Thus the C-image of M is $M' = \bigcup_{n=1}^{\infty} \mathfrak{D}\, S_n$.

The C-image of the operator W is the operator W', which carries $f \in M'$ into $\varphi^V(N')\, f$ i.e. into

$$\varphi^V(\mathrm{Num}\, p) \cdot f(p).$$

Thus in what concerns the C-images, we are in the conditions of Lemma [J.15.3.].

Consequently

$$\tilde{W}' = S,$$

with equality of domains. Since C is a homeomorphism, it follows that

$$\tilde{W} = \varphi(N)$$

with equality of domains.

K.9.2. Theorem. If

1. N is a maximal normal operator in $\mathbf{H}$,
2. E is an N-measurable subset of the complex plane P,
3. $e = s(E)$,
4. $\varphi_n(z)$, $\varphi(z)$ are N-measurable complex valued functions defined N-a.e.-on E,
5. $\varphi_n(z) \to \varphi(z)$ N-a.e.-on E,
6. M is the set of all $\vec{X} \in e$, for which

$$\lim_{n\to\infty} \varphi_n(N)\, \vec{X}$$

exists (topological limit in $\mathbf{H}$),

7. W is the operator with $\mathfrak{D}\, W = M$ defined by

$$W\vec{X} = \lim_{n\to\infty} \varphi_n(N)\, \vec{X},$$

then

$$\tilde{W} = \varphi(N).$$

Proof. Take a natural canonical mapping C with measure μ carrying $\mathbf{H}$ into $\mathbf{H}'$.

Let

$$e = s(E).$$

The C-image of $\varphi_n(N)$ and $\varphi(N)$ is the operator S_n, S where

$$S_n f(p) = \varphi_n^V(\mathrm{Num}\, p) \cdot f(p),$$
$$S f(p) = \varphi^V(\mathrm{Num}\, p) \cdot f(p),$$

with maximal domains in $\mathbf{H}'^V(\mathrm{Point}\, E)$ respectively. The functions $\varphi_n^V(\mathrm{Num}\, p)$, $\varphi^V(\mathrm{Num}\, p)$ are μ-measurable and we have

$$\varphi_n^V(\mathrm{Num}\, p) \to \varphi^V(\mathrm{Num}\, p)$$

almost μ-everywhere in V.

The C-image of M is the set M' of all functions $f(p) \in \mathbf{H}'^V(\text{Point}\,E)$, for which

$$S_n f(p)$$

converges in μ-square mean. The C-image of W is the operator W' with domain M', defined by

$$W' f = \lim_{n\to\infty}{}^2 S_n f(p)$$

on V. Since we are in the conditions of Lemma [J.15.5.], we have

$$\tilde{W}' = S$$

with equality of domains. It follows, through C^{-1}, that

$$\tilde{W} = \varphi(N)$$

with equality of domains.

K.9.3. Theorem. If

1. N is a maximal normal operator in $\mathbf{H}$,
2. E is an N-measurable subset of the complex plane P,
3. $\varphi_n(z)$ are defined N-a.e. on P and N-measurable,
4. $\varphi_n(z) \to \varphi(z)$ N-a.e. on E,
5. each partial sequence $\varphi_{k(n)}(z)$ diverges N-a.e. on co$E$,
6. M is the set of all vectors $\vec{X} \in \mathbf{H}$, for which

$$\lim \varphi_n(N)\,\vec{X}$$

exists (topological limit),

7. W is the operator with domain M and defined by

$$W\vec{X} = \lim_{n\to\infty} \varphi_n(N)\,\vec{X},$$

then

$$\tilde{W} = \varphi(N)$$

(with identity of domains).

Proof. Easy, by taking a canonical mapping, and by relying on Lemma [J.15.8.].

K.9.4. Theorem. If

1. $E_1, E_2 \in \mathcal{T}_L^P$,
2. $E_1 \cup E_2 =^N E$,
3. $\varphi(z) = \begin{cases} \varphi_1(z) & \text{for } z \in E_1, \\ \varphi_2(z) & \text{for } z \in E_2, \end{cases}$
4. $\varphi_1(z) = \varphi_2(z)$ on $E_1 \cap E_2$,
5. $e_1 = s(E_1)$, $e_2 = s(E_2)$,

then $\varphi(N)$ is the smallest operator such that

$$e_1 \upharpoonright \varphi(N) = \varphi_1(N),$$
$$e_2 \upharpoonright \varphi(N) = \varphi_2(N).$$

The proof, by means of a natural canonical mapping and Lemma [K.8.], is easy.

K.10. We shall study the so called *resolvent operator* $\varphi_\lambda(N)$ of N in **H**, where

$$\varphi_\lambda(z) = \frac{1}{z-\lambda},$$

and where λ is a complex number.

This operator is denoted by

$$\frac{1}{N - \lambda I}$$

where I is the identity operator with domain **H**. We know that if we take a natural canonical mapping C, then N induces on the complex plane an ideal J_N (see [K.5.1a.]) which does not depend on the choice of C. The sets belonging to J_N are called N-null sets.

Now the set (λ) composed of the single point λ may belong to J_N or not.

Suppose $(\lambda) \in J_N$. In this case $\frac{1}{z-\lambda}$ is a function defined N-a.e.-on the complex plane P; so $\frac{1}{N-\lambda I}$ is a maximal normal operator in **H** and its domain is everywhere dense in **H**.

K.10.1. Suppose $(\lambda) \,\overline{\in}\, J_N$. In this case, if μ is the measure accompanying the mapping C, we have $\mu(\text{Point}\,\lambda) > 0$, and the space $e = s((\lambda))$ is not the null-space. The function $\varphi_\lambda(z)$ is *not* defined N-a.e.-on P, so $\frac{1}{N-\lambda I}$ is a normal maximal operator in the subspace $\mathrm{co}\,e$, i.e. in the orthogonal complement of $s((\lambda))$.

Indeed, as $\varphi_\lambda(z)$ is defined in $\mathrm{co}(\lambda)$ only, we have

$$\varphi_\lambda(N) = \mathrm{co}\,e \upharpoonright \varphi_\lambda^P(N).$$

The domain of $\frac{1}{N-\lambda I}$ is everywhere dense in $\mathrm{co}\,e$ (see Def. [K.8.]).

The set A of all points λ, for which $(\lambda) \,\overline{\in}\, J_N$, is at most denumerable.

Indeed, if $\lambda \in A$, then $\mu(\text{Point}\,\lambda) > 0$. If A were uncountable, there would exist a non denumerable infinity of points p_n on V with $\mu((p_n)) \geqq \alpha p_n$ where α is some positive number. It would follow that the measure μ is not finite.

K.10.1a. Def. A complex number λ, for which $(\lambda) \,\overline{\in}\, J_N$ is termed *eigenvalue of* N.

The space $s((\lambda))$ is termed *eigenspace* belonging to the eigenvalue λ.

Thus the number of various eigenvalues is at most infinite denumerable. An eigenspace has the dimension $\geqq 1$, but can be even infinite dimensional.

K.10.2. Def. The complex number λ will be said *regular with respect to N (N-regular)*, if there exists $\varepsilon > 0$, such that the open circle with center λ and radius ε, is an N-null set.

Theorem. The set of all N-regular points is an open subset of the complex plane P.

The proof is obvious.

Let λ be N-regular. Let S be the open circle with center λ such that S is an N-null set.

We have

$$\psi(z) =_{df} \varphi_\lambda(z) = \frac{1}{z-\lambda} =^N \left[(P-S) \upharpoonright \frac{1}{z-\lambda}\right]^P.$$

Consequently the C-image of the operator $\frac{1}{N - \lambda I} = \varphi_\lambda(N)$ is the operator $R(f)$:

$$R f = \psi(\operatorname{Num} p) \cdot f(p)$$

with maximal domain in the space $\mathbf{H}'$ of all μ-square summable functions on V.

Now $\psi(\operatorname{Num} p)$ is bounded on V. Let

$$|\psi(\operatorname{Num} p)| \leqq M.$$

We get

$$\int_V |R(f)|^2 \, d\mu \leqq M^2 \cdot \int_V f^2 \, d\mu;$$

and the domain of R is the whole space $\mathbf{H}'$. If we go over to C^{-1}-images, we find that

$$\frac{1}{N - \lambda I}$$

is a bounded operator in $\mathbf{H}$; hence continuous and its domain is the whole space $\mathbf{H}$.

K.10.3. Suppose that $(\lambda) \in J_N$, but that for every $\varepsilon > 0$, the circle with center λ and radius ε never belongs to the ideal J_N. Denote the circle by $K(\lambda; \varepsilon)$. We have

$$\mu[\operatorname{Point} K(\lambda; \varepsilon)] > 0 \quad \text{for every} \quad \varepsilon > 0.$$

Take the replicas $K_n(\lambda; \varepsilon)$ of $K(\lambda; \varepsilon)$ on the copies P_n of P.

At least for one of them $\mu K_n(\lambda; \varepsilon) > 0$.

It is not difficult to prove the existence of a function $f(p)$ on V which would be μ-square summable on V, but for which

$$\frac{1}{\operatorname{Num} p - \lambda} \cdot f(p)$$

be not μ-square summable on V.

Thus the domain of the C-image of the operator $\frac{1}{N - \lambda I}$ does not contain the whole space $\mathbf{H}'$.

It follows that $\text{Ɑ}\frac{1}{N-\lambda 1}$ does not contain the whole space $\mathbf{H}$. By Theorem [J.3.] it follows that this domain is everywhere dense in $\mathbf{H}$. This operator is not continuous, because if it were, it could be extended to the whole space by continuity, and become a wider normal operator. This would contradict the fact that $\frac{1}{N-\lambda 1}$ is maximal normal in $\mathbf{H}$.

K.10.4. From the above it follows that if the domain of $\frac{1}{N-\lambda 1}$ is the whole space, then it is an N-regular point. The same follows, if we suppose that $\frac{1}{N-\lambda 1}$ is bounded (continuous), and its domain is everywhere dense in $\mathbf{H}$.

Thus we have proved the

K.10.5. Theorem. If

1. N is a maximal normal operator in $\mathbf{H}$,

2. the complex number λ has the property that λ is an eigenvalue,

then the operator

$$\frac{1}{N-\lambda 1}$$

is maximal normal in the space ortho-complementary to the space (eigenspace) $s((\lambda))$. The set of various complex numbers λ of this kind is at most denumerable.

Conversely, if the domain of $\frac{1}{N-\lambda 1}$ is not everywhere dense in $\mathbf{H}$, then λ is an eigenvalue of N.

If $(\lambda) \in J_N$,
then the operator

$$\frac{1}{N-\lambda 1}$$

is maximal normal in $\mathbf{H}$. Hence its domain is everywhere dense in $\mathbf{H}$.

Its domain is $\mathbf{H}$ if and only if it is N-regular point, i.e. if there exists a circle K around λ, which belongs to the ideal J_N. In this case this operator is continuous in $\mathbf{H}$ (Hilbert-bounded in $\mathbf{H}$).

The set of all points of the last kind (N-regular points) is an open subset of the complex plane.

K.10.6. According to Stone's terminology (26) the set of all N-regular points is called *resolvent set of* N.

By the *spectrum of* N we understand the complementary of the resolvent set. In particular the eigenvalues make up the "*discontinuous*" spectrum; the remaining points of the spectrum make up the "*continuous*" spectrum.

K.10.7. Let λ be an eigenvalue. There are two possible cases:
1) there exists a circle K around λ such that the set $K - (\lambda) \in J_N$,
2) no such a circle exists.

In the first case, if we denote by e the eigenspace $s((\lambda))$, then

$$\frac{1}{N - \lambda I}$$

is continuous in co e.

In the second case this operator is discontinuous in co e.

K.10.8. Let λ be any point of the complex plane. We have for $\lambda_0 \neq \lambda$

$$\frac{1}{z-\lambda} = \frac{1}{\lambda_0 - \lambda}\left[1 + \frac{z-\lambda_0}{\lambda-\lambda_0} + \left(\frac{z-\lambda_0}{\lambda-\lambda_0}\right)^2 + \cdots\right],$$

and the series converges for all z for which

(1) $$|z - \lambda_0| < |\lambda - \lambda_0|.$$

Denote this open circle by K.

Put

$$\varphi_n(z) = \frac{1}{\lambda_0 - \lambda}\left[1 + \cdots + \left(\frac{z-\lambda_0}{\lambda-\lambda_0}\right)^n\right], \qquad (n = 1, 2, \ldots),$$

restricted to K. We have

$$\lim \varphi_n(z) = \frac{1}{z-\lambda}$$

for all $z \in K$.

Consider the space

$$e = s(K),$$

and the operators $\varphi_n(N)$.

We have, by [K.8.2.] that

$$-\varphi_n(N) = e \upharpoonright \left(\frac{1}{\lambda-\lambda_0}\, 1 + \frac{N - I\lambda_0}{(\lambda-\lambda_0)^2} + \cdots + \frac{(N - I\lambda_0)^n}{(\lambda-\lambda_0)^{n+1}}\right).$$

We have, by [K.9.2.], that if we denote by M the set of all $\vec{X} \in e$, for which

$$\lim \varphi_n(N)\, \vec{X}$$

exists, and define on M the operator W with domain M by putting

$$W\vec{X} = \lim \varphi_n(N)\, \vec{X}.$$

We obtain the closure of $e \upharpoonright \dfrac{1}{N - \lambda I}$.

K.10.9. Remark. To have a non trivial expansion into a series we must take care of that K contains a part of the spectrum, which, is not an N-null set.

Let us suppose that the boundary of the circle K (see [K.10.7.]) is an N-null set.

Of course this excludes the case, where λ is an eigenvalue.

The sequence $\varphi_n(z)$ has the property that it diverges at every point not belonging to the closure $\bar{K}$ of K; but also each partial sequence $\varphi_{k(n)}(z)$ diverges there. We can apply Theorem [K.9.3.]. Hence $\varphi_{k(n)}$ diverges N-a.e. in $\operatorname{co} K$.

Now, let M_1 denote the set of *all vectors* $\vec{X} \in \mathbf{H}$, for which

$$\lim \varphi_n(N)\,\vec{X}$$

converges. Let W_1 be the operator defined by

$$W_1\,\vec{X} = \lim \varphi_n(N)\,\vec{X},$$

with

$$\mathrm{d}\,M_1 = M_1.$$

Then the closure of W_1 coincides with the operator

$$e \upharpoonright \frac{1}{N - \lambda\,1},$$

where

$$e = s(K).$$

As a consequence we have the following: If K is an N-null set, then the sequence $\varphi_n(N)\,\vec{X}$ converges only if $\vec{X} = \vec{O}$.

Remark. Usual theorems give expansions for the scalar product $\left(\frac{1}{N - \lambda\,1}\vec{X}, \vec{Y}\right)$. The above ones are concerned with strong convergence and not with the weak one.

K.10.10. Consider the range of $(N - \lambda\,1)^{-1}$. Let λ be not an eigenvalue of N. Consider a natural canonical mapping C with measure μ. The C-image of $(N - \lambda\,1)^{-1}$ is the operator

$$R\,f(p) = \frac{1}{\operatorname{Num} p - \lambda} \cdot f(p)$$

with maximal domain in $\mathbf{H}'$.

Let Φ be the set of all μ-square summable functions $g(p)$, such that

$$(\operatorname{Num} p - \lambda) \cdot g(p)$$

is μ-square summable. By Theorem [J.3.], Φ is everywhere dense in $\mathbf{H}'$.

Let $g \in \Phi$. Put

$$f(p) = (\operatorname{Num} p - \lambda) \cdot g(p).$$

The function

$$g(p) = \frac{f p}{\operatorname{Num} p - \lambda}$$

is μ-square summable. Hence $f(p) \in \mathrm{d}\,R$, and then $g(p) \in \mathrm{d}\,R$.

Theorem. If λ is not an eigenvalue, then the range of the resolvent $(N - \lambda\,1)^{-1}$ is everywhere dense in $\mathbf{H}$.

K.10.11. The following theorem resulted from a question kindly put to the author by Mr. N. DUNFORD.

Theorem. If E is the spectrum of N, then for any measure μ related to a natural canonical mapping we have

$$\mu(\mathrm{Point}\, E) = \mu(V),$$

$$\mu(\mathrm{Point\ co}\, E) = 0.$$

If

$$\alpha \subseteqq V, \quad \mu(\alpha) > 0$$

then

$$\mu(\mathrm{Point}\, E \cap \alpha) > 0.$$

Remark. Thus a positive μ-measure is possible only within Point E, where E is the spectrum of N.

K.10.12. We shall investigate the solutions of the equation

$$N\vec{X} - \lambda\vec{X} = \vec{Y}, \tag{1}$$

where N is a maximal normal operator in **H**, λ a complex number, $\vec{Y}$ a given vector and $\vec{X}$ a vector to be found. Let C be a natural canonical mapping of **H** onto the space **H**$'$ of all μ-square summable functions on V.

Let N, Y, X go over into N', $g(p)$, $f(p)$. The equation (1) is equivalent to the following one:

$$N' f(p) - \lambda \cdot f(p) =^{\mu} g(p) \tag{2}$$

on V, so we shall deal with a given $g(p)$ defined μ-a.e. on V, and μ-square summable on V.

K.10.13. *Case, where λ is an eigenvalue of N.* The set (λ) is not an N-null set; hence

$$\mu(\mathrm{Point}\,\lambda) > 0.$$

Let

$$\lambda_1, \lambda_2, \ldots, \lambda_n, \ldots$$

be all points of V for which $\mathrm{Num}\,\lambda_n = \lambda$. Since Point λ is the set of these points, there exists n such that $\mu((\lambda_n)) > 0$.

Let $p_1, p_2, \ldots$ be all those or which $\mu((p_i)) > 0$ and let $q_1, q_2, \ldots$ the remaining ones with $\mu((q_i)) = 0$. Suppose $f(p)$ satisfies (2). There exists a μ-null set α such that

$$(\mathrm{Num}\, p - \lambda) \cdot f(p) = g(p) \tag{3}$$

everywhere on $V - \alpha$. No one of the points $p_1, p_2, \ldots$ belongs to α. If $p \in V - \alpha$, $\mathrm{Num}\, p = \lambda$, the equation (3) gives

$$0 \cdot f(p) = g(p);$$

hence

$$g(p) = 0.$$

Thus we have

(4) $$g(p_1) = g(p_2) = \cdots = 0.$$

Let e be the eigenspace belonging to the eigenvalue λ. The space e is the C^{-1}-image of the set $\bigcup_{n=1}^{\infty}(\lambda_n)$, hence the C^{-1}-image of $\bigcup_i (p_i)$, because q_j do not contribute anything, for their set is an N-null set.

From (4) it follows that $\bigcup_i (p_i) \upharpoonright g = 0$, and then $\operatorname{Proj}(e)\, \vec{Y} = 0$. Consequently $\vec{Y} \in \operatorname{co} e$.

We have proved the

Theorem. In order that the equation $N\vec{X} - \lambda \vec{X} = \vec{Y}$ have a solution where λ is an eigenvalue, it is necessary that $\vec{Y}$ be orthogonal to the eigenspace e belonging to λ.

K.10.13a. Now suppose that $\vec{Y}$ is orthogonal to the eigenspace e; consequently

$$\bigcup_i (p_i) \upharpoonright g(p) = 0.$$

Let $\vec{X}$ be a solution of (1); hence

(5) $$(\operatorname{Num} p - \lambda) \cdot f(p) =^{\mu} g(p)$$

on V.

For

$$p = p_1, p_2, \ldots$$

we have

$$g(p) = 0,$$

and

$$\operatorname{Num} p - \lambda = 0,$$

so for any $f(p)$ the equation (5) is satisfied.

For

$$p = q_1, q_2, \ldots$$

the equation

$$(\operatorname{Num} p - \lambda)\, f(p) = q(p)$$

may be satisfied or not — it does not matter, since

$$\mu(\bigcup_j q_j) = 0.$$

Consider the points p with

$$\operatorname{Num} p \neq \lambda.$$

We get

$$f(p) =^{\mu} \frac{g(p)}{\operatorname{Num} p - \lambda} \quad \text{on} \quad V - \bigcup_{n=1}^{\infty} (\lambda_n).$$

It follows that $f(p)$ is μ-square summable on this set, and then

(6) $$\frac{g(p)}{\operatorname{Num} p - \lambda}$$

is μ-square summable on V. It follows that $g(p)$ belongs to the domain of the operator

$$U\,h(p) = \frac{h(p)}{\operatorname{Num} p - \lambda}$$

with maximal domain in the space of all μ-square summable functions, defined μ-a.e.-in V and vanishing μ-a.e. on $\bigcup_{n=1}^{\infty}(\lambda_n)$.

Its C^{-1}-image is

$$\frac{1}{N - \lambda 1}.$$

Thus

$$\vec{Y} \in \text{ᗡ}\,\frac{1}{N - \lambda 1}.$$

Theorem. In order that the equation $N\vec{X} - \lambda\vec{X} = \vec{Y}$, where λ is an eigenvalue, possesses a solution $\vec{X}$ it is necessary that

$$\vec{Y} \in \mathrm{D}\,\frac{1}{N - \lambda 1}.$$

This set of all those Y may fill out the whole space co$e$, or be only everywhere dense in it, in accordance with Theorem [K.10.5.].

K.10.13b. We shall prove the

Theorem. If

1. λ is an eigenvalue of N and e the corresponding eigenspace,
2. $\vec{Y} \perp e$,
3. $\vec{Y} \in \text{ᗡ}\,\frac{1}{N - \lambda 1}$,
4. $\vec{Z}$ is any vector $\in e$,

then

(7) $$\vec{X} = \vec{Z} + \frac{1}{N - \lambda 1}\vec{Y}$$

is a solution of the equation

(8) $$N\vec{X} - \lambda\vec{X} = \vec{Y}.$$

For a given $\vec{Y}$ there exists no other solution.

Proof. Consider C-images and use former notions. Let $h(p)$ be the C-image of $\vec{Z}$

We have

$$h(p) =^{\mu} 0 \quad \text{on} \quad V - E,$$

where

$$E =^{df} \bigcup_{n=1}^{\infty}(\lambda_n).$$

We also have: $\vec{Y} \in \operatorname{co} e$, hence $g(p) =^{\mu} 0$ on E.

The C-image of (7) is

$$f(p) =^{\mu} h(p) + \frac{g(p)}{\operatorname{Num} p - \lambda} \quad \text{on} \quad \operatorname{co} E,$$

and the equation

$$f(p) =^{\mu} h(p)$$

on E (8) goes over to

$$(9) \qquad (\operatorname{Num} p - \lambda) \cdot f(p) =^{\mu} g(p)$$

on V

Since on E we have $g(p) =^{\mu} 0$, and $\operatorname{Num} p - \lambda = 0$, we have (9) satisfied. On $V - E$ we have $h(p) =^{\mu} 0$ and then (9) is satisfied too

The first part of the thesis is proved.

The second part follows easily from what has been said in [K.10.13 a.].

K.10.14. *Case where λ is not an eigenvalue.* We have

$$(10) \qquad \mu(\operatorname{Point} \lambda) = 0.$$

Taking the C-image we transform (1) into

$$(\operatorname{Num} p - \lambda) \cdot f(p) =^{\mu} g(p).$$

This is equivalent to

$$f(p) =^{\mu} \frac{g(p)}{\operatorname{Num} p - \lambda};$$

because of (10). This allows to state the

Theorem. If λ is not an eigenvalue, then a necessary and sufficient condition that the equation

$$N \vec{X} - \lambda \vec{X} = \vec{Y}$$

possesses a solution is that

$$\vec{y} \in \mathrm{D} \frac{1}{N - \lambda I}.$$

The set of these $\vec{Y}$ is everywhere dense in $\mathbf{H}$, and in accordance with Theorem [K.10.5.].

The solution is unique:

$$\vec{X} = \frac{1}{N - \lambda I} \vec{Y}.$$

Remark. The equation $N x = \lambda X$ does not need to be treated separately, since it is a particular case of (1) in [K.10.12.].

K.11. We like to prove a theorem (J. v. NEUMANN (43)) important for the theory of permutable normal operators.

Some auxiliary theorems on the closure of operators will be in order for this purpose, and also for further development.

We remind that a linear operator T with $\mathrm{D}\, T \subseteqq \mathbf{H}$ is said to be closed, if and only if the hypotheses $x_n \in \mathrm{D}\, T$, $x_n \to x$, $T(x_n) \to y$ imply that $x \in \mathrm{D}\, T$ and that $y = T x$.

We recall the following theorem:

K.11.2. If T possesses a closed extension, then there exists a unique operator T', such that

1) T' is closed,

2) if T'' is a closed extension of T,

then $T'x = T''x$ for all $x \in \mathfrak{D}\, T'$.

This operator T' is termed *closure of* T and denoted $\tilde{T}$.

K.11.3. Lemma. If

1. T is a linear transformation with $\mathfrak{D}\, T \subseteqq \mathbf{H}$,

2. T possesses a closed extension,

then $\tilde{T}$ is the transformation S defined as follows:

1) $x \in \mathfrak{D}\, S$ if and only if there exists $x_n \in \mathfrak{D}\, T$ with $x_n \to x$ and for which $\lim T x_n$ exists.

2) $S(x) = \lim T x_n$ for every sequence $\{x_n\}$ of item 1).

K.11.4. Lemma. If

1. a projector P is permutable with T,

2. T has a closure,

then P is permutable with $\tilde{T}$.

Proof. The hypothesis 1) means that, [F],

$$P(T\,X) = T(P\,X) \quad \text{for all} \quad X \in \mathfrak{D}\, T.$$

Let

$$X \in \mathfrak{D}\, \tilde{T}, \qquad Y = \tilde{T}(X).$$

By [K.11.3.] there exists a sequence

$$X_n \to X, \qquad X_n \in \mathfrak{D}\, T,$$

such that

$$T(X_n) \to Y.$$

We have, since $X_n \in \mathfrak{D}\, T$, the equation

$$P(T\,X_n) = T(P\,X_n). \tag{1}$$

The projector is a continuous operator with domain $\mathbf{H}$.

Hence

$$T\,X_n \to Y$$

implies

$$P(T\,X_n) \to P(Y) = P(T\,X). \tag{2}$$

From (2) and (1) it follows that

$$T(P\,X_n) \to P(\tilde{T}\,X). \tag{3}$$

In addition to that we have

$$P\,X_n \to P\,X, \tag{4}$$

and

$$P\,X_n \in \mathfrak{D}\, T,$$

because, by (1), $T(P\,X_n)$ has a meaning.

From (3) and (4) we get

$$T(P X_n) \to P(\tilde{T} X), \tag{5}$$

$$P X_n \to P X, \tag{6}$$

and

$$P X_n \in \mathrm{D}\, T \subseteqq \mathrm{D}\, \tilde{T}. \tag{7}$$

Since $\tilde{T}$ is closed, (5), (6), (7) imply that

$$\tilde{T}(P X) = P(\tilde{T} X),$$

so the theorem is proved, because X was any vector $\in \mathrm{D}\, \tilde{T}$.

Remark. We did not suppose neither that T is normal maximal, nor that $\mathrm{D}\, T$ is everywhere dense in $\mathbf{H}$.

K.12. Lemma. If

1. $\mathbf{H}$ is an H.-H.-space,
2. U_n are operators with $\mathrm{D}\, U_n \subseteqq \mathbf{H}$,
3. M is the set of all X in $\mathbf{H}$, for which $\{U_n X\}$ converges,
4. $W X$ is the operator with $\mathrm{D}\, W = M$ and defined by

$$W X = \lim_{n \to \infty} U_n X,$$

5. P is a projector in $\mathbf{H}$,
6. each U_n is permutable with P, i.e.

$$P U_n X = U_n P X \quad \text{for all} \quad X \in \mathrm{D}\, U_n, \qquad (n = 1, 2, \ldots),$$

then

$$W \text{ is permutable with } P.$$

Proof. Let $X \in \mathrm{D}\, W = M$.

As $\{U_n X\}$ converges topologically, there exists n_0 such that, for every $n > n_0$, $U_n X$ has a meaning, and hence $X \in \mathrm{D}\, U_n$.

Consider those indices n only. We have

$$U_n P X = P U_n X. \tag{1}$$

Since $U_n X$ converges, in the $\mathbf{H}$-topology, toward $W X$, and since P is a continuous operator with domain $\mathbf{H}$, we get

$$P U_n X \to P W X.$$

Hence, by (1)

$$U_n(P X) \to P W X. \tag{2}$$

Since $U_n(P X)$ converges topologically, we have

$$P X \in M.$$

Hence

$$U_n(P X) \to W(P X). \tag{3}$$

From (2) and (3) it follows:

$$W(P\,X) = P(W\,X)$$

for all $X \in M = \mathfrak{D}\,W$, so the Lemma is proved.

K.12.1. Theorem. If

1. N is a maximal normal operator in $\mathbf{H}$,
2. E is an N-measurable subset of the complex plane P,
3. $\varphi_n(z)$ is defined N-a.e. on P and N-measurable, and $\varphi(z)$ is defined N-a.e. on E,
4. $\lim_{n\to\infty} \varphi_n(z) = \varphi(z)$, N-a.e. on E,
5. every partial sequence $\{\varphi_{k(n)}(z)\}$ diverges N-a.e. on $P - E$,
6. A is a projector in $\mathbf{H}$,
7. $\varphi_n(N)$ is permutable with A,

then

$$\varphi(N) \text{ is also permutable with } A.$$

Proof. Consider the set M of all vectors $X \in \mathbf{H}$, for which $\varphi_n(N)\,X$ converges (topologically); and define the operator W with domain $\mathfrak{D}\,W = M$ by

$$W\,X = \lim \varphi_n(N)\,X.$$

Then, on account of Lemma [K.12.], W is permutable with A. Applying Theorem [K.9.3.] we get

$$\tilde{W} = \varphi(N).$$

Since W is permutable with A, the Lemma [K.11.4.] implies that

$$\varphi(N) \text{ is permutable with } A, \quad \text{q.e.d.}$$

K.12.2. Theorem. If

1. N is a maximal normal operator in $\mathbf{H}$,
2. a a (closed) subspace of $\mathbf{H}$, such that N is permutable with the projector $A = \operatorname{Proj}(a)$,

then a is compatible with all spaces of the borelian spectral tribe (s_b) of N.

Proof. Let z_0 be a complex number, $k = 1, 2, \ldots$ Let l be a straight line on P passing through z_0 and parallel to the imaginary axis and suppose that l is an N-null set. There is an everywhere dense set of such lines l. Put

$$\varphi_{k,n}(z) = \frac{1}{k}\left[1 + \frac{z_0 + k - z}{k} + \cdots + \frac{(z_0 + k - z)^n}{k^n}\right].$$

There exists an increasing sequence $\{q(k)\}$ of indices such that the sequence

$$\{\Phi_k(z)\} = \{\varphi_{k,\,q(k)}(z)\}$$

diverges with all its partial sequences everywhere in the half-plane

$$P' = \{z \mid \text{Real}\, z < \text{Real}\, z_0\}$$

and converges everywhere toward $\frac{1}{z - z_0}$ in the half-plane

$$P'' = \{z \mid \text{Real}\, z > \text{Real}\, z_0\}.$$

We see that $\Phi_k(z)$ converges N-a.e. on P'' and each its partial sequence diverges N-a.e. on $P - P''$.

K.12.2a. We easily prove that, for $n = 1, 2, \ldots, N^n$ is permutable with A.

Suppose it is for n; we shall prove it for $n + 1$. Let

$$X \in \text{ᗡ}\, N^{n+1}.$$

Since

$$N^{n+1} X = N(N^n X),$$

it follows that

$$N^n X \in \text{ᗡ}\, N.$$

N is permutable with A. Hence

$$A(\text{ᗡ}\, N) = \text{ᗡ}\, N \cap A \subseteqq \text{ᗡ}\, N.$$

Consequently

$$A\, N^n X \in \text{ᗡ}\, N;$$

so

(1) $$N(A\, N^n X)$$

has a meaning. We know that $\text{ᗡ}\, N^{n+1} \subseteqq \text{ᗡ}\, N^n$. Hence $X \in \text{ᗡ}\, N^n$, and then, by hypothesis,

$$N^n A\, X = A\, N^n X.$$

By (1) it follows that

$$N(N^n A\, X) = N(A\, N^n X).$$

This gives, by the associative law for operations:

(2) $$(N\, N^n)\,(A\, X) = (N\, A)\,(N^n X).$$

Now

$$N^n X \in \text{ᗡ}\, N.$$

Indeed $N^{n+1} X$ has a meaning and

$$N^{n+1} X = N(N^n X), \quad \text{so} \quad N^n X \in \text{ᗡ}\, N.$$

Consequently, we get from (2):

$$N^{n+1}(A\, X) = (A\, N)\,(N^n X).$$

Hence

$$N^{n+1}(A\, X) = A\,(N\, N^n)\, X = A\, N^{n+1} X,$$

which proves the permutability of N^{n+1} with A. It follows that, for every polynomial $\psi(z)$, the operator $\psi(N)$ is permutable with A. Let us go back to [K.12.1.]. The function $\Phi_k(z)$ is a polynomial, so $\Phi_k(N)$ is permutable with A.

Let us apply the Theorem [K.12.1.].

It gives that, if we put

$$t'(z) = P' \rceil \frac{1}{z - z_0}, \quad \text{where} \quad z_0 \in l,$$

the operator

$$t'(N) \text{ is permutable with } A.$$

An analogous argument will yield the permutability of $t''(z)$ with A, where

$$t''(z) = P'' \rceil \frac{1}{z - z_0}.$$

Since $s(P')$, $s(P'')$ are orthocomplementary spaces in $\mathbf{H}$ (because l is an N-null set), it follows that the operator $\frac{1}{N - z_0 I}$ is permutable with A. Thus we have proved that if z_0 is lying on a straight line l parallel to the imaginary axis, where l is an N-null set, then $\frac{1}{N - z_0 I}$ is permutable with A. In an analogous way we prove the same for a point z_0, which lies on a straight line m parallel to the real axis, and where m is an N-null set.

K.12.2b. Having this, let R be an open rectangle lying on the complex plane, and such that its sides are lying on lines which are N-null sets.

Take on the boundary of R a finite number of points which divide the sides of the rectangle into segments with length $< \varepsilon$, say

$$z_1, z_2, \ldots, z_p = z,$$

where the vertices of R are included, and where these points make up a counterclockwise directed succession of points of the boundary of R. If $\varepsilon \to 0$, the expression

$$\text{(1)} \qquad \varphi_m(W) =_{df} \frac{1}{2\pi i} \sum_{i=2}^{p} (z_i - z_{i-1}) \frac{1}{z_i - W}$$

tends to the integral

$$\text{(2)} \qquad \frac{1}{2\pi i} \int \frac{dz}{z - W}$$

for every W, which does not lie on the boundary of R.

This integral $= 1$ inside and $= 0$ outside of the rectangle.

Since the boundary is an N-null set, (1) tends to 1 in R, and to 0 in co$R$, μ-a.e.

Hence, if we put $\varphi(z) = 1$ in R and $\varphi(z) = 0$ in co$R$, we can apply Theorem [K.9.2.], because

$$\varphi_n(z) \to \varphi(z)$$

μ-a.e.-on P. If we denote by M the set of all vectors X, for which $\lim \varphi_n(N)$ exists, and define the operator WX with $\complement W = M$ as

$$W X = \lim \varphi_n(N)\, X,$$

then, by Lemma [K.12.], W is permutable with A. Now by Theorem [K.9.2.], $\varphi(N) = \tilde{W}$.

Hence, by Lemma [K.11.4.], $\varphi(N)$ is permutable with A. Now $\varphi(N)\, X$ is the operator $\operatorname{Proj} s(R)\, X$. Consequently $\operatorname{Proj} s(R)$ is permutable with A. Hence the space $s(R)$ is compatible with the space $\boldsymbol{a}$.

K.12.2c. We know that if the spaces $\mathfrak{b}_1, \mathfrak{b}_2, \ldots, \mathfrak{b}_n, \ldots$ are compatible with one another and compatible with $\boldsymbol{a}$, then

$$\sum_{n=1}^{\infty} \mathfrak{b}_n \quad \text{and} \quad \prod_{n=1}^{\infty} \mathfrak{b}_n, \quad [D],$$

is compatible with $\boldsymbol{a}$. Since the lines parallel to the axes and being at the same time N-null sets make up everywhere dense sets on the plane, we easily prove that the spaces $s(Q)$, where Q are halfopen plane-quarters, are compatible with $\boldsymbol{a}$. Thus the spaces of the spectral scale of N are compatible with $\boldsymbol{a}$, and then, by the above remark on compatibility, all spaces of (s_b) are compatible with $\boldsymbol{a}$. The Theorem [K.12.2.] is established.

K.13. The operational calculus for hermitian selfadjoint operators can be founded independently of considering general normal operators; and in an analogous and even much more simple way. In [H.] we have spoken of the canonical representation

$$H f(x) \doteq x \cdot f(x)$$

of selfadjoint operators, valid on an at most denumerable number of mutually disjoint replicas of the straight line L of the real variable x.

Just the same method yields a canonical mapping for selfadjoint operators independently of the whole theory of normal operators. The only difference consists of the circumstance that instead of copies of the complex plane, we shall have copies of the straight line.

Denote by V the union of all these replicas involved, and introduce analogous symbols. Num p and Point x, as in [E.]. Thus the theorems analogous hold for selfadjoint operators.

K.13.1. Let the original H.-H.-space be $\mathbf{H}'$ and let C be a correspondence which carries the given selfadjoint operator H' in $\mathbf{H}'$ into the selfadjoint operator H:

$$H f(p) \doteq (\operatorname{Num} p) \cdot f(p)$$

in $\mathbf{H}$. The symbol Num p represents a real number. We can apply all general lemmas stated and proved in [H.]. In a quite similar way we define the operators $(H' - x_0 1')^n$, where $n = 1, 2, \ldots$, and x_0 is a real number.

This operator is the C^{-1}-image of the operator in $\mathbf{H}$:

$$U_n f(p) \doteq (\mathrm{Num}\, p - x_0)^n \cdot f(p),$$

with maximal domain in $\mathbf{H}$. This image does not depend on the choice of C.

K.13.2. Under these "one-dimensional" circumstances a topic analogous to that in [J.], would be inadequate, but we can obtain the theorem of independence, analogous to that in [J I.] in a much more simple way, which will be sketched in what now follows:

Let $x_0 < x_1$ be two real numbers.

The continuous function

$$\varphi(x) = \varphi(x_0, x_1; x) = \begin{cases} 1 & \text{for} \quad x \leqq x_0, \\ 0 & \text{for} \quad x \geqq x_1, \\ \text{linear in } \langle x_0, x_1 \rangle \end{cases}$$

can be up to $\frac{1}{n}$ approximated by a suitable polynomial $\varphi_n(x)$ in the interval

$$(-n, n), \qquad n = 1, 2, \ldots$$

Hence $\varphi_n(x) \to \varphi(x)$ on the whole line L and thus, by Lemma [J.15.4.], the operator

$$U f(p) = \varphi(\mathrm{Num}\, p) \cdot f(p)$$

has its C^{-1}-correspondent operator, independent of C. Now, if we keep x_1 fixed and make x_0 increasing and tending to x_1, we get defined an operator in $\mathbf{H}'$, which is C^{-1}-corresponding to the operator

$$A f(p) = a(\mathrm{Num}\, p) \cdot f(p),$$

where

$$a(x) = \begin{cases} 1 & \text{for} \quad x \leqq x_1, \\ 0 & \text{for} \quad x > x_1. \end{cases}$$

By subtraction we define the operator

$$B f(p) = \xi(\mathrm{Num}\, p) \cdot f(p),$$

where $\xi(x)$ is the characteristic function of the segment

$$\{x \mid \alpha < x \leqq \beta\}.$$

K.13.3. Now in the analogous way, as in [J.], we obtain the C^{-1}-image of the operator $\varphi(\mathrm{Num}\, p) \cdot f(p)$ where φ is any continuous real valued function on $(-\infty, +\infty)$.

This, gives general operators, which are C^{-1}-images of

$$U f(p) = \varphi(\mathrm{Num}\, p) \cdot f(p),$$

where $\varphi(\alpha)$ is *any real valued borelian function*, defined everywhere on L.

We can designate this operator by

$$\varphi(H'),$$

thus, the operational calculus on hermitian selfadjoint operators is founded independently of the theory of normal operators.

K.13.4. Remark. Let us remark, that analogously, as for normal operators, if we deal only *with selfadjoint operators*, the spectral theorem for them is only needed, and even the uniqueness of the corresponding spectral scale is not required. Just the mere existence of at least one spectral scale will do.

K.13.4a. The uniqueness of the spectral scale of hermitian selfadjoint operators can be proved in the same way, as given in [K.1.], [K.2.] because theorems analogous to [K.1.] can be proved for them in the same way.

K.14. We know that a maximal normal operator N' in $\mathbf{H}'$ has the form

$$N' = H' + i K', \tag{1}$$

where H', K' are strongly permutable selfadjoint operators in $\mathbf{H}'$.

Theorem. We shall give a proof that such a decomposition (1) is unique for a given N'.

The spectral scales of H', K' are

$$\begin{aligned} h(\lambda) &= s(Q(\lambda, +\infty)), \\ k(\lambda) &= s(Q(+\infty, \lambda)) \end{aligned} \tag{2}$$

respectively. Since the spectral scale of N' is unique [K.2.], therefore the functions (2) are well determined. The spectral scale determines perfectly the operator.

Thus the uniqueness of the decomposition (1) is proved.

K.14.1. Theorem. If

1. N' is a normal operator in $\mathbf{H}'$,
2. $N' = H' + i K'$ its decomposition into selfadjoint operators H', K',
3. C a canonical mapping of $\mathbf{H}'$ into $\mathbf{H}$ carrying N' into

$$N f(p) = (\text{Num}\, p) \cdot f(p),$$

then

the C-image of H' and K' are

$$H_1 f(p) = (\text{Real Num}\, p) \cdot f(p), \tag{1}$$

$$K_1 f(p) = (\text{Imag Num}\, p) \cdot f(p) \tag{2}$$

with maximal domains (see [J.11.]) respectively.

The simple spectral scales of these operators are

$$h(\lambda) = \mathbf{H}^V[(p \mid \text{Real Num}\, p \leqq \lambda)],$$
$$k(v) = \mathbf{H}^V[(p \mid \text{Imag Num}\, p \leqq v)].$$

If we consider Real$z$, Imag$z$ as functions of z, we can write

$$H' = \text{Real}(N'), \quad K' = \text{Imag}(N').$$

Proof. The operators (1), (2) are selfadjoint in $\mathbf{H}$, (by [J.5.]).

They are strongly permutable, by virtue of Theorem [J.10.]. N is normal in $\mathbf{H}$.

Since

$$\text{Num}\, p = \text{Real Num}\, p + i \,\text{Imag Num}\, p,$$

it follows that

$$N = H_1 + i K_1.$$

As C^{-1} is an isometric isomorphism, it carries H_1 and K_1 into selfadjoint, strongly permutable operations H_1', K_1', and N into

$$H_1' + i K_1'.$$

Hence

$$H' + i K' = H_1' + K_1'.$$

By the Theorem [K.14.] we get: $H' = H_1'$, $K' = K_1'$, and then

$$H = H_1, \quad K = K_1.$$

The forlast theses follow from Theorem [J.8.]. The last thesis follows from Theorem [J I.4.2.]. We easily obtain the following

K.14.2. Theorem. If

1. N' is a maximal normal operator,
2. $\varphi(z)$ a borelian complex-valued function defined for every complex z,
3. $\varphi(z) = \alpha(z) + i\beta(z)$ where α, β are real valued,
4. $\varphi(N') = H' + i K'$ where H', K' are selfadjoint components of $\varphi(N')$,

then

$$H' = \alpha(N'), \quad K' = \beta(N').$$

In particular, if we put

$$R(z) = \text{Real}\, z,$$
$$J(z) = \text{Imag}\, z,$$

we get

$$N' = R(N') + i J(N')$$

and

$$(N')^* = R(N') - i J(N').$$

K.15. The Theorem [K.12.2.] is of course true also for selfadjoint hermitian operators, since every hermitian selfadjoint operator is normal maximal.

As we have mentioned on several occasions, the theory of selfadjoint operators can be carried through independently according to analogous patterns and in a simplified way. The same can be said of the last theorem whose independent proof can be supplied as follows: Having settled the operational calculus for selfadjoint operators H on the basis of the canonical representation

$$\mathscr{H} f(p) = (\operatorname{Num} p) \cdot f(p),$$

where p is the variable point on a disjoint union of an at most denumerable copies of the real straight line, we consider the function $\varphi(x; x_0)$ defined by

$$\varphi(x, x_0) = 1 \quad \text{for} \quad x \leqq x_0,$$

$$\varphi(x, x_0) = 0 \quad \text{for} \quad x > x_0,$$

and where

$$\mu(\operatorname{Point} x_0) = 0.$$

Now

$$\varphi(x, x_0) = \lim_{n\to\infty} \varphi_n(x, x_0)$$

where

$$\varphi_n(x, x_0) = \begin{cases} 1 \quad \text{for} \quad x \leqq x_0, \\ n\left(x + \frac{1}{n} - x_0\right) \quad \text{for} \quad x_0 \leqq x \leqq x_0 + \frac{1}{n}, \\ 0 \quad \text{for} \quad x > x_0 + \frac{1}{n}; \end{cases}$$

and $\varphi_n(x, x_0)$ can be approximated by polynomials.

Using Theorems [K.11.4.], [K.12.], [K.12.1.], we obtain theorems on the permutability of $\varphi(H, x_0)$ with the given projector A, and then also on the compatibility of the corresponding space a with the borelian spectral scale of H.

Chapter L

Some classical theorems on normal and selfadjoint operators

This Chapter [L.] contains proofs of several classical theorems, as e.g. the Radon-integral representation theorem of $f(N)$ for borelian functions $f(\dot{z})$. The proofs differ from those, contained in the Stone's book, and are sometimes simplier. They are framed on the base of our theory. We start with recallections. Variables will sometimes be denoted by a dot, e.g. $f(\dot{z})$. If E, F are sets of vectors, $E \vec{+} F$ will denote the set

$$\{\vec{X} + \vec{Y} \mid \vec{X} \in E, \vec{Y} \in F\}.$$

L.1. (a). A linear operator A is said to be *in the subspace* a of $\boldsymbol{H}$, whenever

$$\text{ᗡ} A \subseteqq a, \quad \text{D} A \subseteqq a.$$

Let

1) A be an operator in $\boldsymbol{H}$,

2) a a space,

3) $\text{Proj}_a = (\text{Proj}\, a)$ the operator in $\boldsymbol{H}$ projecting the whole space $\boldsymbol{H}$ onto the space a. We say that, [F], *A is permutable with* $(\text{Proj}\, a)$, whenever for every $\vec{X} \in \text{ᗡ} A$,

$$\text{Proj}_a (A \vec{X}) = A (\text{Proj}_a \vec{X}).$$

In this case

$$\text{ᗡ}(a \mid A) = a \frown \text{ᗡ} A \neq \text{Proj}_a \text{ᗡ} A.$$

By $a \upharpoonleft A$ we understand the operator A, whose domain is restricted to a, which will mean that

$$\text{ᗡ}(a \upharpoonleft A) = a \frown \text{ᗡ} A;$$

and that if

$$\vec{X} \in \text{ᗡ}(a \upharpoonleft A),$$

then

$$(a \upharpoonleft A) \vec{X} = A \vec{X}.\text{*)}$$

---

*) **Remark.** The conditions

(α) A is permutable with $\text{Proj}\, a$,

(β) $A\, \text{Proj}_a = \text{Proj}_a A$

are not equivalent, though (β) implies (α). Indeed, the condition (β) requires that

$$\text{ᗡ}(A\, \text{Proj}_a) = \text{ᗡ}(\text{Proj}_a A)$$

i.e. the following should be equivalent:

$$(\alpha') \quad \text{Proj}_a X \in \text{ᗡ} A, \qquad (\beta') \quad X \in \text{ᗡ} A.$$

Now, if (α) takes place, we have $(\beta') \to (\alpha')$, hence

$$\text{ᗡ}(\text{Proj}_a A) \subseteqq \text{ᗡ}(A\, \text{Proj}_a).$$

To have the converse implication

$$\text{ᗡ}(A\, \text{Proj}_a) \subseteqq \text{ᗡ}(\text{Proj}_a A),$$

we should have $(\alpha') \to (\beta')$, i.e. if $\text{Proj}_a \vec{X} \in \text{ᗡ} A$, then

$$\vec{X} \in \text{ᗡ} A; \quad \text{i.e, if} \quad \text{Proj}_a \vec{X} \in (\text{ᗡ} A \frown a) = \text{ᗡ}(a \upharpoonleft A),$$

then $\vec{X} \in \text{ᗡ} A$.

Hence if $\vec{Y} \in \text{ᗡ}(a \upharpoonleft A)$, then $\vec{Y} \in \text{ᗡ} A$.

Consequently

$$\vec{Y} \vec{+} \text{co}\, a \subseteqq \text{ᗡ} A.$$

Hence

$$\text{ᗡ}(a \upharpoonleft A) \vec{+} \text{co}\, a \in \text{ᗡ} A.$$

This is not true, because we have

$$\text{ᗡ} A = \text{ᗡ}(a \upharpoonleft A) \vec{+} \text{ᗡ}(\text{co}\, a \upharpoonleft A),$$

and $\text{ᗡ}(\text{co}\, a \upharpoonleft A)$ may differ from $\text{co}\, a$.

Remark. If A is a projector, and if A is permutable with Proj_a, we agree to say that Proj_a is permutable with A.

(b) If N is a normal (maximal) operator in $\boldsymbol{H}$ and E a borelian subset of the complex plane P, then $s(E)$ means the space, defined through extension of the space-valued spectral scale $s(\dot{Q})$, where Q is a variable plane-quarter.

If $M(\dot{z})$ is a complex valued function of the complex variable $\dot{z}$, varying over the whole plane P, and if the integral

$$\int_P M(z)\, d(E(z)\, \vec{X}),$$

[G.], has a meaning for a given vector $\vec{X}$, then

$$\operatorname{Proj}_{s(E)} \int_P M(z)\, d(E(z)\, \vec{X}) = \int_E M(z)\, d(E(z)\, \vec{X}).$$

(c) If

1) N is a normal operator in $\boldsymbol{H}$,

2) p is a space compatible with all $s(Q)$,

then N is permutable with (Proj_a).

(d) We have for every $\vec{X} \in \boldsymbol{H}$:

$$\vec{X} = \int_P d(E(z)\, \vec{X}).$$

L.1.1. Theorem. If

1) A is a linear operator in $\boldsymbol{H}$,

2) (Proj_p) is permutable with A,

then for all $\vec{X} \in \complement(p \upharpoonright A)$ we have

$$\operatorname{Proj}_p(A\, \vec{X}) = (p \upharpoonright A)\, \vec{X}.$$

Proof. Let $\vec{X} \in \complement(p \upharpoonright A)$. We have

(1) $$\vec{X} \in p$$

and

(2) $$\vec{X} \in \complement A.$$

From hyp. 2) it follows by (2),

$$\operatorname{Proj}_p(A\, \vec{X}) = A\, (\operatorname{Proj}_p \vec{X}) =, \text{ by } (1), = A\, \vec{X}.$$

Hence

$$\operatorname{Proj}_p(A\, \vec{X}) = (p \upharpoonright A)\, \vec{X},$$

because

$$(p \upharpoonright A)\, \vec{X} = A\, \vec{X}.$$

The theorem is proved.

L.1.2. Theorem. If

1. N is a maximal normal operator in $\boldsymbol{H}$,

2. E a borelian subset of the complex plane P,

3. $s(E)$ the corresponding space,

then for every

$$\vec{X} \in \complement(s(E) \upharpoonright N)$$

we have

$$(s(E) \upharpoonright N)\,\vec{X} = \int_E z\,d(E(z)\,\vec{X}) = \mathrm{Proj}_{s(E)}(N\,\vec{X}).$$

Proof. Since $s(E)$ is compatible with all spaces $s(Q)$ of the spectral scale $s(\dot{Q})$ of N, we have, [H.], that $(\mathrm{Proj}\, s(E))$ is permutable with N.

Let $\vec{X} \in \mathrm{a}(S(E) \upharpoonright N)$. It follows:

$$\mathrm{Proj}_{s(E)}(N\,\vec{X}) = (s(E) \upharpoonright N)\,\vec{X}. \tag{1}$$

Since $\vec{X} \in \mathrm{a}\, N$, we have

$$\mathrm{Proj}_{s(E)}(N\,\vec{X}) = \mathrm{Proj}_{s(E)} \int_P z\,d[E(z)\,\vec{X}] = \int_E z\,d[E(z)\,\vec{X}],$$

which completes the proof.

L.1.3. Theorem. If

1. $f(\dot{z})$, $g(\dot{z})$ are borelian functions with domain P, and N is as before,
2. λ is a complex number,
3. $h(z) =_{df} \lambda \cdot f(z)$ for all $z \in P$,
4. $k(z) =_{df} f(z) + g(z)$ for all $z \in P$,

then

$$h(N) \equiv \lambda \cdot f(N) \quad \text{and} \quad k(N) \equiv f(N) + g(N).^{*)}$$

Proof. Let C be a canonical mapping, transforming N into the operator $\boldsymbol{N}$, as in [H.], and μ the corresponding measure.

$$\boldsymbol{N}[F(\dot{p})] \doteq (\mathrm{Num}\,\dot{p}) \cdot F(\dot{p}).$$

$[(\doteq)$ stands for $(=^{\mu})]$. C transforms $f(N)$, $g(N)$ into the operators

$$A(F(\dot{p})) \doteq f(\mathrm{Num}\,\dot{p}) \cdot F(\dot{p}),$$
$$B(F(\dot{p})) \doteq g(\mathrm{Num}\,\dot{p}) \cdot F(\dot{p})$$

respectively.

Hence $\lambda f(N)$ and $f(N) + g(N)$ go over into

$$\lambda A(F(\dot{p})) \quad \text{and} \quad A[f(\dot{p})] + B[F(\dot{p})]$$

respectively, i.e. into

$$\lambda f(\mathrm{Num}\,\dot{p}) \cdot F(\dot{p}) \quad \text{and} \quad [f(\mathrm{Num}\,\dot{p}) + g(\mathrm{Num}\,\dot{p})] \cdot F(\dot{p})$$

respectively, i.e. into

$$h(\mathrm{Num}\,\dot{p}) \cdot F(\dot{p}) \quad \text{and} \quad k(\mathrm{Num}\,\dot{p}) \cdot F(\dot{p})$$

respectively. Applying C^{-1}, we get the thesis.

---

* We use, with STONE (26), the sign $(\equiv)$ to emphasize, that the operators are not only equal in their common domain of significance, but also have the same domain.

L.2. Now we intend to get the Radon-integral representation of $f(N)$. To prove it within the frames of our theory, we need some lemmas.

L.2.1. Theorem. If

1. N is a maximal normal operator in $\boldsymbol{H}$,
2. E a borelian subset of the complex plane P,
3. $w_E(z)$ the characteristic function of the set E,

then:

$$\int_P w_E(z)\, d[E(z)\vec{X}] = [w_E(N)]\vec{X} = \mathrm{Proj}_{s(E)}\vec{X} \quad \text{for all} \quad \vec{X} \in \boldsymbol{H}.$$

Proof. Take $\vec{X} \in \boldsymbol{H}$. We have

$$\vec{X} = \int_P d[E(z)\vec{X}].$$

Hence

$$\mathrm{Proj}_{s(E)}\vec{X} = \mathrm{Proj}_{s(E)} \int_P d[E(z)\vec{X}].$$

This expression equals

$$\int_E d[E(z)\vec{X}] = \int_P w_E(z)\, d[E(z)\vec{X}].$$

Since we have

$$(w_{s(E)} N)\vec{X} = \mathrm{Proj}_{s(E)}\vec{X},$$

the theorem is proved.

L.2.2. Lemma. If

1. $a_1, a_2, \ldots, a_n$ $(n \geqq 1)$ are spaces orthogonal to one another,
2. $A_1, A_2, \ldots, A_n$ are operators in $a_1, a_2, \ldots, a_n$ respectively,
3. S is the operator with domain composed of all $\vec{X}$, such that

$$\vec{X} = \sum_{k=1}^{n} \mathrm{Proj}_{a_k}\vec{X}, \qquad \mathrm{Proj}_{a_k}\vec{X} \in \mathrm{\complement}\, A_k,$$

and defined by

$$S\vec{X} = \sum_{k=1}^{n} A_k(\mathrm{Proj}_{a_k}\vec{X}),$$

then S is permutable with all $(\mathrm{Proj}\, a_k)$, $(k = 1, 2, \ldots, n)$.

Proof. It suffices to prove the permutability of S with $(\mathrm{Proj}\, a_1)$. Let $\vec{X} \in \mathrm{\complement}\, S$. We have, by hyp. 3,

$$(1) \quad \mathrm{Proj}_{a_1}(S\vec{X}) = \mathrm{Proj}_{a_1} \sum_{k=1}^{n} A_k(\mathrm{Proj}_{a_k}\vec{X}) = \sum_{k=1}^{n} \mathrm{Proj}_{a_1} A_k \,\mathrm{Proj}_{a_k}\vec{X}.$$

Since A_k is an operator in a_k, we have

$$A_k \,\mathrm{Proj}_a \vec{X} \in a_k,$$

and then, for $k \neq 1$, we have

$$\mathrm{Proj}_{a_1} A_k \,\mathrm{Proj}_{a_k}\vec{X} = \vec{O}.$$

Hence, from (1), we get:

$$\text{Proj}_{a_1}(S\vec{X}) = \text{Proj}_{a_1} A_1 \text{Proj}_{a_1}\vec{X} = A_1 \text{Proj}_{a_1}\vec{X}, \tag{2}$$

since $\mathrm{D} A_1 \subseteqq a_1$.

On the other hand $\text{Proj}_{a_1}\vec{X} \in \mathrm{D}\, S$, because

$$\text{Proj}_{a_k} \text{Proj}_{a_1}\vec{X} = \begin{cases} \text{Proj}_{a_1}\vec{X}, & \text{for } k = 1, \\ \vec{O}, & \text{for } k \neq 1, \end{cases}$$

$$\text{Proj}_{a_1}\vec{X} \in \mathrm{D}\, A_1; \quad \vec{O} \in \mathrm{D}\, A_k \quad \text{for } k \neq 1,$$

and

$$\text{Proj}_{a_1}\vec{X} = \sum_{k=1}^{n} \text{Proj}_{a_k} \text{Proj}_{a_1}\vec{X},$$

since a_k is orthogonal to a_1 for $k \neq 1$.

We have, by hyp. 3,

$$S(\text{Proj}_{a_1}\vec{X}) = \sum_{k=1}^{n} A_k (\text{Proj}_{a_k} \text{Proj}_{a_1}\vec{X}) = A_1 \text{Proj}_{a_1}\vec{X}. \tag{3}$$

From (2) and (3) we get:

$$\text{Proj}_{a_1} S\vec{X} = S\, \text{Proj}_{a_1}\vec{X};$$

hence S and (Proj_{a_1}) are permutable.

Remark. If we suppose that all A_k are closed, we can deduce that S is also closed.

Let us prove this for $n = 2$. Put

$$a_1 = a_1, \quad a_2 = b, \quad A = A_1, \quad B = A_2,$$

and write, in general, $\vec{Y}_a$ instead of $\text{Proj}_a\vec{Y}$. To prove that S is a closed operator, suppose that

$$\vec{X}_1, \vec{X}_2, \ldots, \vec{X}_n, \ldots \in \mathrm{D}\, S, \quad \lim_{n\to\infty} \vec{X}_n = \vec{X}$$

and $\lim_{n\to\infty} S(\vec{X}_n) = \vec{Y}$. We must prove that

$$S(\vec{X}) = \vec{Y}.$$

By continuity of projection, we have

$$\vec{X}_a = \lim_{n\to\infty} \vec{X}_{n,a}, \quad \vec{X}_b = \lim_{n\to\infty} \vec{X}_{n,b}. \tag{1}$$

Since $\vec{X}_n \in \mathrm{D}\, S$ it follows, by hyp. 3, that

$$\vec{X}_{n,a} \in \mathrm{D}\, A, \quad \vec{X}_{n,b} \in \mathrm{D}\, B. \tag{2}$$

We have $\vec{Y}_a = \lim_{n\to\infty} \text{Proj}_a S\vec{X}_n$.

Since $\vec{X}_n \in \mathrm{D}\, S$ and S is permutable with Proj_a, it follows

$$\vec{Y}_a = \text{llm}_{n\to\infty} S\vec{X}_{n,a}.$$

Similarly we get

$$\vec{Y}_b = \lim_{n\to\infty} S\,\vec{X}_{n,b}.$$

Now

$$S\,\vec{X}_{n,a} = A(\vec{X}_{n,a,a}) + B(\vec{X}_{n,a,b}) = A(\vec{X}_{n,a}).$$

Hence

$$(3)\quad \begin{cases} \vec{Y}_a = \lim_{n\to\infty} A(\vec{X}_{n,a}); \\ \text{and similarly we get} \\ \vec{Y}_b = \lim_{n\to\infty} B(\vec{X}_{n,b}). \end{cases}$$

By (1), (2), (3) and on account of the closure property of A and B, we have

$$(4)\qquad \vec{Y}_a = A(\vec{X}_a), \quad \vec{Y}_b = B(\vec{X}_b).$$

Now $\vec{X}_a \in \mathrm{d}A$, $\vec{X}_b \in \mathrm{d}B$; hence, by hyp. 3, $\vec{X} = \vec{X}_a + \vec{X}_b \in \mathrm{d}S$. Since $\vec{Y} = \vec{Y}_a + \vec{Y}_b$, we have, by (4),

$$\vec{Y} = A(\vec{X}_a) + B(\vec{X}_b) = S(\vec{X}).$$

Thus S is closed, q.e.d.

L.2.3. Lemma. If

1. $a_1, a_2, \ldots, a_n, \ldots$ are mutually orthogonal spaces,
2. $A_1, A_2, \ldots, A_n, \ldots$ are closed operators in a_n respectively,
3. $B_n(\vec{X})$ is the operator whose domain is composed of all $\vec{X} \in a_1 + \cdots + a_n$, such that

$$\mathrm{Proj}_{a_k}\vec{X} \in \mathrm{d}A_k,$$

$$\vec{X} = \sum_{k=1}^{n} \mathrm{Proj}_{a_k}\vec{X},$$

and where

$$B_n(\vec{X}) = \sum_{k=1}^{n} A_k(\mathrm{Proj}_{a_k}\vec{X}),$$

4. $M = \bigcup_{n=1}^{\infty} \mathrm{d}B_n$,
5. $W(\vec{X})$ is the operator with domain M and defined by $W(\vec{X}) = B_n(\vec{X})$ whenever $\vec{X} \in \mathrm{d}B_n$,
6. W possesses the closure $\tilde{W}$,

then

1) $\mathrm{d}\tilde{W}$ is the set of all $\vec{Y}$ for which there exist $\vec{Y}_1, \vec{Y}_2, \ldots, \vec{Y}_n, \ldots$ such that $\vec{Y}_n \in \mathrm{d}A_n$, $\vec{Y} = \sum_{n=1}^{\infty} \vec{Y}_n$, and where $\sum_{n=1}^{\infty} A_n(\vec{Y}_n)$ converges.

2) We have for those $\vec{Y}$:

$$\tilde{W}(\vec{Y}) = \sum_{n=1}^{\infty} A_n(\vec{Y}_n).$$

L.2.3a. Proof. Concerning hyp. 5, we shall prove that $W(\vec{X})$ does not depend on the choice of n.

To do that, suppose that

$$\vec{Y} = B_n(\vec{X}).$$

Then, by hyp. 3,

$$\vec{X} = \sum_{k=1}^{n} \vec{X}_{a_k},$$

where

$$\vec{X}_{a_1} \in ᑕ A_1, \ldots, \vec{X}_{a_n} \in ᑕ A_n.$$

Hence

$$\vec{X} \in \sum_{k=1}^{n} \vec{X}_{a_k} + \vec{O} \quad \text{and} \quad \vec{O} \in ᑕ A_{n+1}.$$

It follows

$$\vec{X} \in ᑕ B_{n+1}.$$

We have

$$B_{n+1} \vec{X} = \sum_{k=1}^{n} A_k \vec{X}_{a_k} + A_{k+1} \vec{O} = \vec{Y};$$

hence $\vec{Y} = B_{n+1}(\vec{X})$, which proves that $B_n \subseteqq B_{n+1}$.

The required independence is proved.

L.2.3b. The operator W is permutable with any (Proj_{a_k}). Indeed, let $\vec{X} \in ᑕ W$. By hyp. 5, we have a natural number m, such that

$$\vec{X} \in ᑕ B_m.$$

Hence

$$\vec{X} \in ᑕ B_{m+s}.$$

B_{n+s} is permutable with

$$(\mathrm{Proj}_{a_1}), \ldots, (\mathrm{Proj}_{a_n}), \ldots, (\mathrm{Proj}_{a_{n+s}}).$$

Since n is arbitrary, therefore B_m is permutable with any (Proj_{a_n}).

Hence

(1) $$\mathrm{Proj}_{a_n} B_m \vec{X} = B_m(\mathrm{Proj}_{a_n} \vec{X}).$$

Hence

(2) $$\mathrm{Proj}_{a_n} \vec{X} \in ᑕ B_m \subseteqq M = ᑕ W.$$

By hyp. 5, it follows that

$$B_m \vec{X} = W \vec{X},$$

and by (2), that

$$B_m(\mathrm{Proj}_{a_n} \vec{X}) = W(\mathrm{Proj}_{a_n} \vec{X}).$$

Hence, by (1),

$$\mathrm{Proj}_{a_n} W \vec{X} = W \, \mathrm{Proj}_{a_n} \vec{X}.$$

Thus the permutability is established.

L.2.3c. Notice that if $\vec{X} \in \mathrm{C}\, a_k$, then $W\vec{X} = A_k \vec{X}$.

L.2.3d. By hyp. 6, the closure $\tilde{W}$ of W exists, and we know, that $\mathrm{C}\,\tilde{W}$ is composed of all vectors $\vec{Y}$, such that there exists a sequence $\{\vec{Y}_n\}$, where $\vec{Y}_n \in W$, $\lim_{n\to\infty} \vec{Y}_n = \vec{Y}$ and where $\lim_{n\to\infty} W(\vec{Y}_n)$ exists.

Let $\vec{Y} \in \tilde{W}$. Choose $\vec{Y}_n \in \mathrm{C}\, W$, with $\lim_{n\to\infty} \vec{Y}_n = \vec{Y}$ and existent $\lim_{n\to\infty} W(\vec{Y}_n)$.

Put

(0) $$\vec{Z} = \lim_{n\to\infty} W(\vec{Y}_n).$$

We prove that

$$\vec{Y}_{a_k} \in \mathrm{C}\, A_k \quad \text{and} \quad A_k(\vec{Y}_{a_k}) = \vec{Z}_{a_k},$$

(where in general we put $X_b =_{def} \mathrm{Proj}_b X$).

We have

$$\vec{Y} = \lim_{n\to\infty} \vec{Y}_n;$$

hence, by continuity of projection

(1) $$\vec{Y}_{a_k} = \lim_{n\to\infty} \vec{Y}_{n,\, a_k}.$$

Now we have

(1.1) $$\vec{Y}_{n,\, a_k} \in \mathrm{C}\, A_k, \quad \text{because} \quad \vec{Y}_n \in \mathrm{C}\, B_{s_n} \quad \text{for some } s_n;$$

hence $\vec{Y}_n \in \mathrm{C}\, B_{\max(s_n, k)}$, because $B_s \subseteq B_{\max(s_n, k)}$.

We have, since $\vec{Y}_n \in \mathrm{C}\, W$ and since W is permutable with Proj_{a_k},

(2) $$\mathrm{Proj}_{a_k} W \vec{Y}_n = W\, \mathrm{Proj}_{a_k} \vec{Y}_n = W \vec{Y}_{n,\, a_k} = B_{\max(s_n, k)} \vec{Y}_{n,\, a_k} = A_k \vec{Y}_{n,\, a_k}.$$

Since $W\vec{Y}_n$ tends to the limit $\vec{Z}$, it follows that

$$\mathrm{Proj}_{a_k} W \vec{Y}_n \text{ tends to the limit } \vec{Z}_{a_k},$$

and hence, by (2), that

(3) $$A_k \vec{Y}_{n,\, a_k} \text{ tends to the limit } \vec{Z}_{a_k}.$$

Since A_k is closed, it follows from (1) and (3):

$$\vec{Y}_{a_k} \in \mathrm{C}\, A_k, \quad \text{and} \quad A_k(\vec{Y}_{a_k}) = \vec{Z}_{a_k},$$

so the assertion is proved.

L.2.3e. We shall prove that

$$\vec{Y} = \sum_{k=1}^{\infty} \vec{Y}_{a_k}, \quad \text{and that} \quad \sum_{k=1}^{\infty} A_k \vec{Y}_{a_k} \quad \text{converges.}$$

The first equation follows from the inclusion

$$M \subseteq \sum_{k=1}^{\infty} a_k,$$

(which implies that $\vec{Y} \in \sum_{k=1}^{\infty} a_k$), and from hyp. 1.

We have proved that

(4) $$A_k(\vec{Y}_{a_k}) = \vec{Z}_{a_k}.$$

Now, by (0),

$$\vec{Z} = \lim_{n\to\infty} W(\vec{Y}_n) \in \sum_{k=1}^{\infty} a_n,$$

(because $W(\vec{Y}_n) \in \sum_{k=1}^{\infty} a_k$, which follows from $\mathrm{D} B_m \subseteqq a_1 + \cdots + a_m \subseteqq \sum_{k=1}^{\infty} a_k$).

Hence

$$\vec{Z} = \sum_{k=1} \vec{Z}_{a_k},$$

and then, by (4),

$$\vec{Z} = \sum_{k=1}^{\infty} \vec{Z}_{a_k} = \sum_{k=1}^{\infty} A_k(\vec{Y}_{a_k}).$$

Consequently this sum converges.

L.2.3f. It remains to prove that if

$$\vec{Y}_n \in \mathrm{D} A_n, \quad \vec{Y} = \sum_{n=1}^{\infty} \vec{Y}_n,$$

and if

$$\sum_{n=1}^{\infty} A_n(\vec{Y}_n)$$

converges, then

$$\vec{Y} \in \mathrm{D} \bar{W}.$$

Suppose that

$$\vec{Y}_n \in \mathrm{D} A_n, \quad \vec{Y} = \sum_{n=1}^{\infty} \vec{Y}_n,$$

and that the sum

$$\sum_{n=1}^{\infty} A_n(\vec{Y}_n)$$

converges.

Put

$$\vec{X}_n = \vec{Y}_1 + \cdots + \vec{Y}_n.$$

We have $\vec{X}_n \in \mathrm{D} W$, $\lim \vec{X}_n = \vec{Y}$.

We also have

$$W(\vec{X}_n) = W(\vec{Y}_1) + \cdots + W(\vec{Y}_n) = A_1(\vec{Y}_1) + \cdots + A_n(\vec{Y}_n).$$

It follows that $W(\vec{X}_n)$ exists.

The lemma [L.2.2.] is proved.

L.2.4. Theorem. If

1. E is a borelian subset of the complex plane P,
2. $f(\dot{z})$ a borelian function defined on P,
3. $|f(z)| \leqq M$ for all z,
4. $U(\dot{\vec{X}}) = \int_E f(z)\, d[E(z)\, \dot{\vec{X}}]$,

then $\text{Ɑ}\, U = \boldsymbol{H}$, U is closed (and even continuous) and the operator $s(E) \rceil U$ is also closed (and even continuous).

Proof. We prove the continuity of U, because closeness will follow. The existence of the integral for all $\vec{X}$ is clear. Let $\vec{X}_n \to \vec{X}$. We have

$$||U(\vec{X}_n) - U(\vec{X})||^2 = ||\int_E f(z)\, d[E(z)(\vec{X}_n - \vec{X})||^2 \leqq$$

$$\leqq \int_E |f(z)|^2\, d||E(z)(\vec{X}_n - \vec{X})||^2 \leqq M^2 ||\mathrm{Proj}_E(\vec{X}_n - \vec{X})||^2 \leqq$$

$$\leqq M^2 ||\vec{X}_n - \vec{X}||^2,$$

which tends to 0.

L.3. Theorem. If

1. N is a maximal normal operator in $\boldsymbol{H}$ (with $\overline{\text{Ɑ}\, N} = \boldsymbol{H}$),
2. $E(\dot{z})$ its double projector-valued spectral scale,
3. $f(\dot{z})$ a complex valued borelian function, defined over the whole complex plane P,

then

$$f(N)\, \dot{\vec{X}} = \int_P f(z)\, d[E(z)\, \dot{\vec{X}}], \tag{1}$$

with identity of the domains of significance of both operators.

L.3.1. Proof. The formula (1) will be proved step-wise by proceding to more and more general functions $f(\dot{z})$.

We start with the case where $f(\dot{z})$ is the characteristic function $w_E(\dot{z})$ of a borelian set $E \neq \text{Ѻ}$.

By [L.2.1.] we have

$$w_E(N)\, \vec{X} = \int_P w_E(z)\, d[E(z)\, \vec{X}]$$

for all $\vec{X} \in \boldsymbol{H}$, so the formula (1) is true in our case.

L.3.2. Let $f(z)$ be a step-function (given by the formula):

$$f(z) = f_1\, w_{E_1}(z) + \cdots + f_n\, w_{E_n}(z),$$

defined for all $z \in \boldsymbol{H}$, where $E_1, \ldots, E_n$ are mutually disjoint borelian sets with

$$E_1 + \cdots + E_n = P,$$

and where $f_1, \ldots, f_n$ are complex numbers.

We get

$$f(N)\, X = \sum_{k=1}^{n} f_k \cdot w_{E_k}(N)\, \vec{X} = \sum_{k=1}^{n} f_k \int_P w_{E_k}(z)\, d[E(z)\, \vec{X}] = \int_P f(z)\, d[E(z)\, \vec{X}],$$

for all $\vec{X} \in \boldsymbol{H}$, so the formula (1) is proved in the case considered.

L.3.3. The next step is: bounded borelian real functions $f(z)$ defined on P.

Let

$$|f(z)| \leqq M.$$

Put

$$f_n(z) = \begin{cases} -M, \text{ if } -M \leqq f(z) < -M + \frac{2M}{n}, \\ -M + \frac{2M}{n}, \text{ if } -M + \frac{2M}{n} \leqq f(z) < -M + 2 \cdot \frac{2M}{n}, \\ \dots\dots\dots\dots\dots\dots\dots\dots \\ -M + k \cdot \frac{2M}{n}, \text{ if } -M + k \cdot \frac{2M}{n} \leqq f(z) < -M + \\ \qquad + (k+1) \cdot \frac{2M}{n}. \end{cases}$$

We have

$$|f(z) - f_n(z)| \leqq \frac{2M}{n},$$

so $f_n(z)$ converges uniformly to $f(z)$, the $f_n(z)$ being step-functions.

By what has been already proved, we have

$$\text{(0)} \qquad f_n(N)\vec{X} = \int_P f_n(z)\, d[E(z)\vec{X}] \quad \text{for all} \quad \vec{X} \in \boldsymbol{H}.$$

We also have

$$\text{(1)} \qquad \left\| \int_P f_n(z)\, d[E(z)\vec{X}] - \int_P f(z)\, d[E(z)\vec{X}] \right\|^2$$

$$= \left\| \int_P (f_n - f)\, d[E(z)\vec{X}] \right\|^2 = \int_P |f_n - f|^2\, d\|E(z)\vec{X}\|^2 \leqq$$

$$\leqq \frac{4M^2}{n^2} \int_P d\|E(z)\vec{X}\|^2 = \frac{4M^2}{n^2} \|\vec{X}\|^2 \to 0$$

for $n \to \infty$.

Of course, both integrals (1) exist for every $\vec{X} \in \boldsymbol{H}$. It follows that

$$\text{(2)} \qquad \lim \int_P f_n(z)\, d[E(z)\vec{X}] = \int_P f(z)\, d[E(z)\vec{X}]$$

for all $\vec{X} \in \boldsymbol{H}$.

On the other hand, there exists n_0 such that, for all $n \geqq n_0$ $\mathrm{D} f_n(N) = \mathrm{D} f(N)$, and $\lim f_n(N)\vec{X} = f(N)\vec{X}$ in this domain. Since

$$\mathrm{D} f_n = \boldsymbol{H},$$

we get

$$\text{(3)} \qquad \lim f_n(N)\vec{X} = f(N)\vec{X} \quad \text{for all} \quad \vec{X} \in \boldsymbol{H}.$$

If we take account of (0), (2) and (3), we get

$$f(N)\vec{X} = \int_P f(z)\, d[E(z)\vec{X}] \quad \text{for all} \quad \vec{X} \in \boldsymbol{H}.$$

L.3.4. Now let $f(z)$ be any real valued borelian function, defined on the whole plane P.

Put

$$F_n =_{df} \{z \mid n-1 \leqq |f(z)| < n\} \quad \text{for} \quad n = 1, 2, \ldots$$

These sets are borelian, disjoint, and their union is P. Put

$$E_n = \bigcup_{k=1}^{n} F_k \quad \text{for} \quad n = 1, 2.$$

We have

$$E_1 \subseteq E_2 \subseteq \cdots; \quad \bigcup_{n=1}^{\infty} E_n = P.$$

Define

$$g_n(z) =_{df} \begin{cases} f(z) & \text{for} \quad z \in E_n, \\ 0 & \text{for} \quad z \in \operatorname{co} E_n. \end{cases}$$

We have

$$\lim_{n\to\infty} g_n(z) = f(z) \quad \text{for all} \quad z \in P;$$

and $g_n(z)$ are bounded real valued borelian functions, defined over the whole plane P. Hence

(1) $$g_n(N)\vec{X} = \int_P g_n(z)\, d[E(z)\vec{X}] \quad \text{for all} \quad \vec{X} \in \boldsymbol{H}.$$

Let M be the set of all $\vec{X}$, for which

$$\lim_{n\to\infty} g_n(N)\vec{X}$$

exists, and put

(2) $$W\vec{X} =_{df} \lim_{n\to\infty} g_n(N)\vec{X} \quad \text{for all} \quad \vec{X} \in M;$$

so $\mathrm{C}\, W = M$. We have

(3) $$f(N) = \tilde{W},$$

with identity of domains.

Since $f(N)$ is a maximal normal operator in $\boldsymbol{H}$, it follows that

(4) $\overline{\mathrm{C} f} = \boldsymbol{H}$, and then M is everywhere dense in $\boldsymbol{H}$.

By (1), (2) and (3) we have for all $\vec{X} \in M$:

$$f(N)\vec{X} = \lim_{n\to\infty} \int_P g_n(z)\, d[E(z)\vec{X}].$$

Let α be a borelian subset of P. We get by continuity of projection:

$$\begin{aligned} \operatorname{Proj}_{s(\alpha)} f(N)\vec{X} &= \lim_{n\to\infty} \operatorname{Proj}_{s(\alpha)} \int_P g_n(z)\, d[E(z)\vec{X}] \\ &= \lim_{n\to\infty} \int_\alpha g_n(z)\, d[E(z)\vec{X}]. \end{aligned}$$

Hence

$$(\mathrm{Proj}_{s(\alpha)} f(N)\vec{X}, \vec{X}) = (\lim_{n\to\infty} \int_\alpha g_n(z)\, d\,([E(z)\vec{X}], \vec{X});$$

hence, by continuity of the scalar product

(5) $$(\mathrm{Proj}_{s(\alpha)} f(N)\vec{X}, \vec{X}) = \lim_{n\to\infty} (\int_\alpha g_n(z)\, d\,([E(z), \vec{X}], \vec{X}) =$$

$$= \lim_{n\to\infty} \int_\alpha g_n(z)\, d\,(E(z)\vec{X}, \vec{X}) = \lim_{n\to\infty} \int_\alpha g_n(z)\, d\,||E(z)\vec{X}||^2.$$

The last equation is valid for all $\vec{X} \in M$ and for all borelian sets α. Let us fix $\vec{X}$ for a moment and let us vary α.

Put

(6) $$H_n(\alpha) = \int_\alpha g_n(z)\, d\,||E(z)\vec{X}||^2.$$

This integral is a Radon's-one, with non negative denumerably additive measure $v(\beta) =_{df} ||\mathrm{Proj}_{s(\beta)}\vec{X}||^2$; hence $H_n(\alpha)$ is a denumerably additive finite real valued function of the variable borelian set α. We can apply the theorem by the author (65), *) stating that the limit of an infinite sequence of denumerably additive finite set functions, converging on every soma of a denumerably additive tribe (whose somata are sets), is also finite and denumerably additive.

If we put

$$H(\alpha) =_{df} \lim_{n\to\infty} H_n(\alpha) \quad \text{for all } \alpha, \text{ we see that}$$

$H(\alpha)$ is a denumerably additive finite real valued set-function, defined for all borelian α.

Now if $v(\beta) = 0$, then $H_n(\beta) = 0$, and then $H(\beta) = 0$. Hence, if we apply author's theorem (16), on integral representation of set functions, we deduce the existence of a v-measurable function $g(z)$, defined on P almost v-everywhere, and such that

$$H(\alpha) = \int_\alpha g(z)\, d\, v.$$

This function is v-unique.

Now, since $g_n(z) = f(z)$ for $z \in E_n$, and since for all $\beta \subseteqq E_n$ we have

$$H_n(\beta) = H(\beta),$$

we get, by the v-uniqueness of the function $g(z)$:

$$g(z) =^v g_n(z) \quad \text{for} \quad z \in E_n.$$

*) Mistakenly referred to, many times, as Vitali-Hahn-Saks theorem.

Hence $g(z) =^v f(z)$ for $z \in P$, because $\sum_{n=1}^{\infty} E_n = P$. Thus we have proved, on account of (5), that

$$(6.1) \qquad (\mathrm{Proj}_{s(\alpha)} f(N)\, \vec{X}, \vec{X}) = \int_{s(\alpha)} f(z)\, d\|E(z)\, \vec{X}\|^2$$

for all $\vec{X} \in M$ and all borelian α.

Hence, taking $\alpha = P$, we get

$$(7) \qquad (f(N)\, \vec{X}, \vec{X}) = \int_P f(z)\, d\|E(z)\, \vec{X}\|^2 \quad \text{for all} \quad \vec{X} \in M.$$

Let us remark that, if $\vec{X}, \vec{Y} \in M$, then $\vec{X} + \vec{Y} \in M$ and $\vec{X} + i\, \vec{Y} \in M$. The identity

$$2i\,(f(N)\, \vec{X}, \vec{Y}) = i\,(f(N)\, (\vec{X} + \vec{Y}),\, \vec{X} + \vec{Y}) +$$
$$+ (f(N)\, (\vec{X} + i\, \vec{Y}),\, \vec{X} + i\, \vec{Y}) + (1+i)\, (f(N)\, \vec{X}, \vec{X}) + (1+i)\, (f(N)\, \vec{Y}, \vec{Y})$$

is valid for all $\vec{X}, \vec{Y} \in M$. Thus we get, by (7),

$$2i\,(f(N)\, \vec{X}, \vec{Y}) = i \int_P f(z)\, d\|E(z)\, (\vec{X} + \vec{Y})\|^2 +$$
$$+ \int_P f(z)\, d\|E(z)\, (\vec{X} + i\, \vec{Y}\|^2 + (1+i) \int_P f(z)\, d\|E(z)\, \vec{X}\|^2 +$$
$$+ (1+i) \int_P f(z)\, d\|E(z)\, \vec{Y}\|^2.$$

Hence

$$(8) \qquad (f(N)\, \vec{X}, \vec{Y}) = \int_P f(z)\, d(E(z)\, \vec{X}, E(z)\, \vec{Y})$$
$$= \int_P f(z)\, d(E(z)\, \vec{X}, \vec{Y}) = (\int_P f(z)\, d[E(z)\, \vec{X}], \vec{Y})$$

valid for all $\vec{X}, \vec{Y} \in M$.

Take any $\vec{Z} \in \boldsymbol{H}$. Since, by (4), M is everywhere dense in $\boldsymbol{H}$, there exists an infinite sequence $\{\vec{Z}_n\} \in M$, tending to $\vec{Z}$. Applying (8) to $\vec{Z}_m$, and taking account of continuity of the scalar multiplication, we get

$$(f(N)\, \vec{X}, \vec{Z}) = (\int_P f(z)\, d[E(z)\, \vec{X}], \vec{Z})$$

for all $\vec{Z} \in \boldsymbol{H}$, and $\vec{Z} \in M$.

Hence

$$W(\vec{X}) = f(N)\, \vec{X} = \int_P f(z)\, d[E(z)\, \vec{X}] \quad \text{for all} \quad \vec{X} \in M.$$

Concerning the set M of vectors, we notice that $s(E_n) \subseteqq M$ and $s(F_m) \subseteqq M$. Indeed, if $\vec{X} \in s(E_n)$, then

$$g_n(N)\, \vec{X} = \int_{E_n} g_n(z)\, d[E(z)\, \vec{X}] \quad \text{for} \quad s = 0, 1, 2, \ldots$$

and then

$$\lim g_n(N)\vec{X} = g_n(N)\vec{X};$$

hence

$$\vec{X} \in M.$$

Thus

(9.1) $$\bigcup_{n=1}^{\infty} s(E_n) \subseteqq M.$$

We have for $\vec{X} \in M$:

(10) $$W(\vec{X}) = \int_P f(z)\, d[E(z)\vec{X}] = \sum_{m=1}^{\infty} \int_{F_m} f(z)\, d[E(z)\vec{X}].$$

This is true on account of denumerable additivity of the generalized Radon's integral (with vector-valued measure):

$$\vec{v}(\alpha) =_{df} \mathrm{Proj}_{s(\alpha)}\vec{X}.$$

Put

$$a_m =_{df} s(F_m), \quad m = 1, 2, \ldots$$

These spaces are orthogonal to one another, because F_m are disjoint.

Put

$$A_m =_{df} \int_{F_m} f(z)\, d[E(z)\vec{X}] \quad \text{for all} \quad \vec{X} \in a_m.$$

This integral is meaningful for these $\vec{X}$, because $f(z)$ is bounded on F_m, and A_m is a closed operator in a_m.

Let $B_m(\vec{X})$ be the operator, whose domain is composed of all vectors $\vec{X} \in a_1 + \cdots + a_n$, and defined as

(11) $$B_m\vec{X} =_{df} \int_{E_n} f(z)\, d[E(z)\vec{X}].$$

This integral is meaningful, because $f(z)$ is bounded on E_n. If $\vec{X} \in a_1 + \cdots + a_m$, then $\mathrm{Proj}_{a_k}\vec{X} \in a_k = \mathrm{Q} A_k$, and we have

$$\begin{aligned} B_m\vec{X} &= \int_{F_1} + \cdots + \int_{F_m} f(z)\, d[E(z)\vec{X}] \\ &= A_1(\mathrm{Proj}_{a_1}X) + \cdots + A_m(\mathrm{Proj}_{a_m}X). \end{aligned}$$

Put

(12) $M_1 =_{df} \bigcup_{m=1}^{\infty} \mathrm{Q}\, B_m = \bigcup_{m=1}^{\infty} (a_1 + \cdots + a_m)$. It equals $\bigcup_{m=1}^{\infty} s(E_m) \subseteqq M$, by (9.1).

Let $W_1(\vec{X})$ be the operator with domain M_1, and defined by

$$W_1(\vec{X}) = B_m(\vec{X}), \quad \text{whenever} \quad \vec{X} \in \mathrm{Q}\, B_m.$$

Hence if $\vec{X} \in a_1 + \cdots + a_m$, we have

(13) $$W_1(\vec{X}) = \int_{F_1 + \cdots + F_m} f(z)\, d[E(z)\vec{X}],$$

by (11). Hence $W_1 \subseteqq W$, by (13), (12) and (10).

W_1 possesses the closure, because, by (3), $\tilde{W}$ exists; so we have

$$W_1 \subseteqq W \subseteqq \tilde{W}.$$

We also have

$$\complement W_1 \subseteqq \complement W \subseteqq \complement \tilde{W},$$

and by (12), the set of vectors

$$M_1 = \complement W_1$$

is everywhere dense in $\boldsymbol{H}$, (because $\sum_{m=1}^{\infty} a_n = \boldsymbol{H}$).

It follows that

(14) $$\tilde{W}_1 = \tilde{W}.$$

Hence we have for every $\vec{X}$,

$$\vec{X} = \mathrm{Proj}_{s(F_1)} \vec{X} + \mathrm{Proj}_{s(F_2)} \vec{X} + \cdots$$

and

$$\int_{s(F_k)} f(z)\, d[E(z)\, \vec{X}] = A_k\, \mathrm{Proj}_{s(F_k)} \vec{X}.$$

A necessary and sufficient condition that, for a $\vec{Y}$, the integral

(15) $$\int_P f(z)\, d[E(z)\, \vec{Y}]$$

exists, is that the sum

$$\sum_{k=1}^{\infty} \int_{s(F_k)} f(z)\, d[E(z)\, \vec{Y}]$$

converges. In this case this sum equals (15).

Consequently, a necessary and sufficient condition that

$$\vec{X} \in \complement \tilde{W},$$

is that

$$\int_P f(z)\, d[E(z)\, \vec{Y}]$$

exists. We have

$$\tilde{W}\, \vec{Y} = \sum_{k=1}^{\infty} \int_{s(F_k)} f(z)\, d[E(z)\, \vec{X}].$$

Consequently, by (10), (13) and (3),

$$N f(\vec{X}) = \int_P f(z)\, d[E(z)\, \vec{Y}],$$

with identity of domains. The theorem is proved for all real $f(z)$.

L.3.5. Now if $f(z)$ is complex valued, we write

$$f(\dot{z}) = \mathrm{Real} f(\dot{z}) + i\, \mathrm{Imag} f(\dot{z}),$$

and apply the result obtained in [L.3.4.].

Thus the theorem is proved in its complete generality.

Remark. The Lemma [L.2.3.] in the above proof was used, because we could not rely on the general theorem:

$$\text{“if} \quad \mu_n \to \mu \quad \text{and} \quad \int_P f(z)\, d\mu_n$$

exists, then

$$\int_P f(z)\, d\mu_n \to \int_P f(z)\, d\mu\text{”},$$

simply because this theorem is not true.

L.4. Theorem. If

1. N is a normal maximal operator in $\boldsymbol{H}$,
2. $E(z)$ its double projector-valued spectral scale,
3. $A(\vec{X}) =_{df} \int_P f(z)\, d[E(z)\,\vec{X}]$,
4. $B(\vec{X}) =_{df} \int_P g(z)\, d[E(z)\,\vec{X}]$, where $f(z)$, $g(z)$ are complex valued

borelian functions, defined over the whole plane P, and where the above operations are taken with maximal domains,
then

$$B[A(\vec{X})] = \int_P f(z)\, g(z)\, d[E(z)\,\vec{X}],$$

where both operators are taken with the maximal domains of their significance.

Proof. By the former theorem we have

$$A(\vec{X}) = f(N)\,\vec{X}, \qquad B(\vec{X}) = g(N)\,\vec{X}.$$

Take a canonical mapping C which transforms N into the operator

$$\boldsymbol{N}\,F(p) =^{\mu} (\mathrm{Num}\,p)\cdot F(p).$$

The transformed operators A, B are

$$A\,F(p) = f(\mathrm{Num}\,p)\cdot F(p),$$
$$B\,F(p) = g(\mathrm{Num}\,p)\cdot F(p).$$

Hence the operator $A\,B(\vec{X})$ is the image of $A\,B\big(F(p)\big) = f(\mathrm{Num}\,p)$ $g(\mathrm{Num}\,p)\cdot F(p)$ and then, by the preceding theorem, its C^{-1}-image

$$A\,B(\vec{X}) \quad \text{is} \quad \int_P f(z)\, g(z)\, d[E(z)\,\vec{X}]$$

with identity of domains.

Remark. A consequence of the above theorems is this:

$$A\,B\,\vec{X} = B\,A\,\vec{X}$$

with identity of domains.

L.5. Theorem. If

1. N is a maximal normal operator in $\boldsymbol{H}$,
2. $E(z)$ its double projector-valued spectral scale,

3. $f(z)$, $g(z)$ are borelian complex-valued fonctions, defined everywhere on the complex plane P,

4. $\int_{P} f(z)\, d\|E(z)\vec{X}\|$ and $\int_{P} g(z)\, d\|E(z)\vec{Y}\|$ exist,

then

$$\int_{P} f(z)\, g(z)\, d(E(z)\vec{X}, \vec{Y})$$

also exists, and

$$\left|\int_{P} f(z)\, g(z)\, d(E(z)\vec{X}, \vec{Y})\right|^2 \leqq$$

$$\leqq \int_{P} |f(z)|^2\, d\|E(z)\vec{X}\|^2 \cdot \int_{P} |g(z)|^2\, d\|E(z)\vec{Y}\|^2.$$

Proof. Take a canonical mapping, transforming N into the operator

$$\boldsymbol{N}(F(\dot{p})) \doteq (\mathrm{Num}\, p) \cdot F(p);$$

then the operators

$$A(\vec{X}) = \int_{P} f(z)\, d[E(z)\vec{X}] = f(N)\vec{X},$$

$$B(\vec{Y}) = \int_{P} g(z)\, d[E(z)\vec{Y}] = g(N)\vec{Y},$$

and

$$S(\vec{Z}) =_{df} \int f(z) \cdot g(z)\, d[E(z)\, Z]$$

go into

$$A(F(\dot{p})) \doteq f(\mathrm{Num}\, p) \cdot F(\dot{p}),$$

$$B(G(\dot{p})) \doteq g(\mathrm{Num}\, p) \cdot G(\dot{p}),$$

$$S(H(\dot{p})) \doteq f(\mathrm{Num}\, p)\, g(\mathrm{Num}\, p) \cdot H(\dot{p})$$

respectively. Since

$$\int_{P} |f(z)|^2\, d\|E(z), \vec{X}\|^2 = \int_{P} |f(\mathrm{Num}\, p)|^2\, |F(p)|^2\, d\mu,$$

$$\int_{P} |g(z)|^2\, d\|E(z), \vec{Y}\|^2 = \int_{P} |g(\mathrm{Num}\, p)|^2\, |G(p)|^2\, d\mu,$$

where

$$\vec{X} \smile^{C} F(\dot{p}), \quad \vec{Y} \smile^{C} G(p);$$

and since

$$\int_{P} f(z)\, g(z)\, d(E(z)\vec{X}, \vec{Y}) = \int_{P} f(\mathrm{Num}\, p)\, g(\mathrm{Num}\, p)\, \overline{F(\dot{p})} \cdot G(\dot{p})\, d\mu,$$

we can apply Schwarz-inequality:

$$\left|\int_{P} f(\mathrm{Num}\, p)\, \overline{F(\dot{p})}\, g(\mathrm{Num}\, p)\, G(\dot{p})\, d\mu\right|^2 \leqq$$

$$\leqq \int_{P} |f(\mathrm{Num}\, p)|^2\, |\overline{F(\dot{p})}|^2\, d\mu \cdot \int_{P} |g(\mathrm{Num}\, p)|^2\, |G(\dot{p})|^2\, d\mu,$$

which yields the theorem.

Chapter M

Multiplicity of spectrum of maximal normal operators

M.1. If the normal operator has discontinuous spectrum, the multiplicity of an eigenvalue is easy to define. The case of continuous spectrum is more difficult, but the canonical multiplane-representation of the operator will be helpful. Our theory will define the multiplicity so that it will be not only precise, but intuitionally simple and even visual. We start with some hypotheses and preliminaries.

M.1.1. Let N be a maximal normal operator in a separable and complete H.-H.-space. Let (S^L) be the Lebesguean spectral tribe of N, [E.4.12.]. Since the tribe is geometrical, and admits a denumerably additive and effective measure, (S^L) coincides with borelian extension (S^B) of (S), so we can call this tribe *"Borelian spectral tribe of N"*. Let $\{\mathbf{H}_n\}$, $n = 1, 2, \ldots$ be a saturating sequence of (S^B), [D I.17.7.] of (S^B) and denote by $(\bar{S})$ the smallest geometrical tribe, containing (S^B) and all spaces $\mathbf{H}_n$, $(n = 1, 2, \ldots)$. The tribe $(\bar{S})$ is composed of all spaces, having the form $\sum_i a_i \mathbf{H}_{v(i)}$, where the sum is finite or denumerably infinite, where $v(i)$ is a finite or denumerable subsequence of the sequence of natural numbers $1, 2, 3, \ldots$ and where $a_i \in S^B$, [E.8.2.]. The tribe $(\bar{S})$ is saturated. Let $\vec{\omega}$ be a generating vector of the space $\mathbf{H}$ with respect to $(\bar{S})$. The vector $\vec{\omega}$ generates a measure on the multiplane V, (see [E.8.16.], where this measure was denoted by $\mu^\bullet$). The measure μ is denumerably additive on the tribe $(F^{L\cdot})$, [E.8.12.], of subsets of V. [To simplify the writing we shall use sometimes the symbol (F^L) instead of $(F^{L\cdot})$.]

The saturating sequence and the generating vector induce the canonical mapping $\boldsymbol{C}$ of the space $\mathbf{H}$ onto the space H of all μ-square-summable functions $F(p)$ of the variable point p of V, [E.8.21.], [E.8.22.]. Let us notice that, even in the case of a saturated tribe (S^L), we can use a multiplane canonical mapping, by attaching to some copies P_n of the plane the measure equal 0. This will allow to have a uniform treatment in all cases. For that all we refer to [H.10.13.].

M.1.1.a. We have defined the n-th copy P_n of the plane P, by means of couples (z, n) where $z \in P$ and n is a natural number, [E.8.9.]. Now we shall recall some auxiliary notions related to the multiplane V and introduced in [H.10.5.1.].

M.1.2. Def. If $p \in V$ and $p = (z, n)$, we shall denote by $\operatorname{Num} p$ the complex number z.

M.1.2.1. Def. If $E \subseteqq V$, we denote by Num$E$ the set

$$\{\mathrm{Num}\,p \mid p \in E\}.$$

M.1.2.2. Def. If $F \subseteqq P$, then by Point F we shall understand the set

$$\{p \mid \mathrm{Num}\,p \in F\}, \quad [\text{H.10.5.1.}].$$

We have the following auxiliary theorems:

M.1.3. Theorem. If $\beta \subseteqq P$, then Num Point$\beta = \beta$.

Proof. Let $z \in \mathrm{Num}\,(\mathrm{Point}\,\beta)$. It follows that there exists $p \in \mathrm{Point}\,\beta$, such that

(1) $$z = \mathrm{Num}\,\beta.$$

Since $p \in \mathrm{Point}\,\beta$, it follows that

(2) $$\mathrm{Num}\,p \in \beta.$$

From (1) and (2) we get

$$z \in \beta.$$

Thus

(3) $$\mathrm{Num}\,(\mathrm{Point}\,\beta) \subseteqq \beta.$$

Now, let $z \in \beta$. We shall prove that

(3.1) $$z \in \mathrm{Num}\,(\mathrm{Point}\,\beta).$$

Now (3.1) is equivalent to

$$z \in \{\mathrm{Num}\,p \mid p \in \mathrm{Point}\,\beta\},$$

hence equivalent to the statement: there exists p_1 such that

(3.2) $$p_1 \in \mathrm{Point}\,\beta, \quad z = \mathrm{Num}\,p_1.$$

Now put

$$p_1 =_{df} (z, 1).$$

We have

$$p_1 \in \mathrm{Point}\,z \quad \text{and} \quad \mathrm{Point}\,z \subseteqq \mathrm{Point}\,\beta.$$

Consequently

(3.3) $$p_1 \in \mathrm{Point}\,\beta.$$

Since $p_1 \in \mathrm{Point}\,z$, we have

(3.4) $$z = \mathrm{Num}\,p_1.$$

From (3.3) and (3.4) the statement (3.2) follows. Hence

$$z \in \mathrm{Num}\,(\mathrm{Point}\,\beta).$$

Thus we have

(4) $$\beta \subseteqq \mathrm{Num}\,(\mathrm{Point}\,\beta).$$

From (3) and (4) the theorem follows.

M.1.3a. Remark. It is not true that if $E \subseteqq V$, then

$$\text{Point Num} E = E,$$

but we only have

$$E \subseteqq \text{Point Num} E.$$

M.1.4. Theorem. If $E \subseteqq P_n$, then $(\text{Point Num} E) \cap P_n = E$; and if $p \in P_n$, then $(\text{Point Num} p) \cap P_n = (p)$.

M.1.5. Theorem. If $E, F \subseteqq V$, then $\text{Num}(E \cup F) = \text{Num} E \cup \text{Num} F$; and a similar theorem holds true for any number of subsets of V.

Proof. Let

$$z \in \text{Num} E \cup \text{Num} F.$$

Either $z \in \text{Num} E$ or $z \in \text{Num} F$. If $z \in \text{Num} E$, there exists $p \in E$ such that $z = \text{Num} p$. We have $p \in E \cup F$. Hence there exists $p \in E \cup F$ such that $z = \text{Num} p$. Consequently $p \in \text{Num}(E \cup F)$. Thus we have got

$$\text{Num} E \cup \text{Num} F \subseteqq \text{Num}(E \cup F). \tag{2}$$

From (1) and (2) the theorem follows.

A similar proof holds for

$$\text{Num} \bigcup_\alpha E_\alpha = \bigcup_\alpha \text{Num} E_\alpha,$$

where $\alpha \neq \emptyset$.

M.1.6. Theorem. If $E, F \subseteqq V$, and $E \subseteqq F$, then

$$\text{Num} E \subseteqq \text{Num} F.$$

This follows from the preceding theorem, if we notice that

$$F = E \cup (F - E).$$

M.1.7. Theorem. If $E, F \subseteqq V$, then

$$\text{Num}(E \cap F) \subseteqq \text{Num} E \cap \text{Num} F,$$

(but the equality may be not true).

Proof. Since $E \cap F \subseteqq E$, we get, by Theorem [M.1.6.],

$$\text{Num}(E \cap F) \subseteqq \text{Num} E.$$

Similarly

$$\text{Num}(E \cap F) \subseteqq \text{Num} F;$$

hence

$$\text{Num}(E \cap F) \subseteqq \text{Num} E \cap \text{Num} F, \quad \text{q.e.d.}$$

M.1.7.1. Remark. Complementary sets E, $\text{co} E$ in V may have equal $\text{Num} E$ and $\text{Num}(\text{co} E)$, but the last sets may be disjoint too.

M.1.8. Theorem. If $\alpha \subseteqq P$, then

$$\text{Num}(\text{Point} \alpha \cap P_n) \in \alpha.$$

Proof. We have, [M.1.6.],

$$\text{Num}(\text{Point}\,\alpha \cap P_n) \subseteqq \text{Num Point}\,\alpha;$$

hence, by [M.1.3.],

$$(1) \qquad \text{Num}(\text{Point}\,\alpha \cap P_n) \subseteqq \alpha.$$

Now, let $z \in \alpha$. We have $(z, n) \in P_n$ and $(z, n) \in \text{Point}\,\alpha$; hence

$$(z, n) \in P_n \cap \text{Point}\,\alpha.$$

Hence

$$z = \text{Num}(z, n) \in \text{Num}(P_n \cap \text{Point}\,\alpha).$$

Thus we have proved that

$$(2) \qquad \alpha \subseteqq \text{Num}(P_n \cap \text{Point}\,\alpha).$$

From (1) and (2) the theorem follows.

M.1.9. Theorem. If $E, F \in P_n$ then

$$\text{Num}(E \cap F) = \text{Num}\,E \cap \text{Num}\,F,$$

and the analogous holds for any collection of subsets of P_n

$$\text{Num}(\text{co}\,E) = \text{co}(\text{Num}\,E),$$

$$\text{Num}(E \dotplus F) = \text{Num}\,E \dotplus \text{Num}\,F$$

(algebraic addition of sets). This follows directly from the meaning of a copy.

M.1.10. Theorem. If $\alpha, \beta \in P$, then

$$\text{Point}(\alpha \cup \beta) = \text{Point}\,\alpha \cup \text{Point}\,\beta,$$

and the similar theorem holds true for any not empty class of subsets of P.

Proof. Let

$$p \in \text{Point}(\alpha \cup \beta).$$

Thus

$$\text{Num}\,p \in \alpha \cup \beta;$$

hence either $\text{Num}\,p \in \alpha$ or $\text{Num}\,p \in \beta$. In the first event we have $p \in \text{Point}\,\alpha$, and in the second $p \in \text{Point}\,\beta$. Whatever the case may be, we get

$$p \in \text{Point}\,\alpha \cup \text{Point}\,\beta.$$

Consequently

$$(1) \qquad \text{Point}(\alpha \cup \beta) \subseteqq \text{Point}\,\alpha \cup \text{Point}\,\beta.$$

Now let

$$p \in \text{Point}\,\alpha \cup \text{Point}\,\beta.$$

Hence

$$\text{Num}\,p \in \alpha \quad \text{or} \quad \text{Num}\,p \in \beta.$$

In any case

$$\text{Num}\, p \in \alpha \cup \beta .$$

Hence

$$p \in \text{Point}\,(\alpha \cup \beta) .$$

Consequently

(2) $$\text{Point}\,\alpha \cup \text{Point}\,\beta \subseteqq \text{Point}\,(\alpha \cup \beta) .$$

From (1) and (2) the theorem follows.

M.1.11. Theorem. If $\alpha, \beta \in P$ and $\alpha \subseteqq \beta$, then

$$\text{Point}\,\alpha \subseteqq \text{Point}\,\beta .$$

This follows from [M.1.10.].

M.1.12. Theorem. If $\alpha \in P$, then

$$\text{Point}\,(\text{co}\,\alpha) = \text{Point}\,(P - \alpha) = \text{co}\,(\text{Point}\,\alpha) = V - \text{Point}\,\alpha .$$

Proof. Let

$$p \in \text{Point}\,(\text{co}\,\alpha) .$$

Hence

$$\text{Num}\, p \in \text{co}\,\alpha .$$

Hence $\text{Num}\, p \,\overline{\in}\, \alpha$. Consequently $p \,\overline{\in}\, \text{Point}\,\alpha$, because if we had $p \in \text{Point}\,\alpha$, we would get $\text{Num}\, p \in \alpha$; hence

$$p \in V - \text{Point}\,\alpha .$$

Thus

(1) $$\text{Point}\,(\text{co}\,\alpha) \subseteqq \text{co}\,(\text{Point}\,\alpha) .$$

Now let

$$p \in V - \text{Point}\,\alpha .$$

Hence $p \,\overline{\in}\, \text{Point}\,\alpha$. Hence $\text{Num}\, p \,\overline{\in}\, \alpha$, because if we had $\text{Num}\, p \in \alpha$, we would have $p \in \text{Point}\,\alpha$. Hence $\text{Num}\, p \in \text{co}\,\alpha$, and then $p \in \text{Point}\,(\text{co}\,\alpha)$.

Thus

(2) $$\text{co}\,(\text{Point}\,\alpha) \subseteqq \text{Point}\,(\text{co}\,\alpha) .$$

From (1) and (2) the theorem follows.

M.1.13. Theorem. If $\alpha, \beta \subseteqq P$, then

$$\text{Point}\,(\alpha \cap \beta) = \text{Point}\,\alpha \cap \text{Point}\,\beta ,$$

and similarly for any number of subsets of P.

Proof. This follows from [M.1.10.] and [M.1.12.]. As a consequence we have for the algebraic addition (symmetric difference) of sets:

M.1.13.1. Theorem. If $\alpha, \beta \in P$, then

$$\text{Point}\,(\alpha \dotplus \beta) = \text{Point}\,\alpha \dotplus \text{Point}\,\beta .$$

M.2. In [E.8.11.] we have defined the measure, denoted by $\mu^{\bullet}$ for all sets E, belonging to the tribe $(F^{L^{\bullet}})$, i.e. for all μ-measurable subsets of the multiplane V;

(1) $$\mu^{\bullet} E = \sum_n \mu_n (E \cap P_n).$$

M.2.1. Theorem. If $E \subseteqq P_n$, and E is μ_n-measurable or $\mu^{\bullet}$-measurable, then $\mu^{\bullet} E = \mu_n E$.

M.2.2. Theorem. If $A \in F^L$, i.e. A is a μ-measurable subset of the plane P, then

$$\mu^{\bullet}(\mathrm{Point}\, A) = \mu(A).$$

Hence $\mu^{\bullet}(\mathrm{Point}\, A)$ does not depend on the choice of the saturating sequence $\mathbf{H}_n$, [D I.17.7.].

Proof. We have

(1) $$\mu^{\bullet}(\mathrm{Point}\, A) = \mu^{\bullet} \bigcup_n A_n,$$

where A_n is the n-th-copy of A. The number (1) equals

(2) $$\sum_n \mu^{\bullet}(A_n),$$

because $\mu^{\bullet}$ is a denumerably additive measure. By [M.2.1.], (2) equals

(3) $$\sum_n \mu_n(A_n).$$

Now, the number (3) is equal to

(4) $$\sum_n \mu[s_n(A_n)] =, \text{[E.8.8b.]}, = \mu[\sum_n s(A), \mathbf{H}_n] = \mu[s(A)],$$

because $\sum_n \mathbf{H}_n = 1$, and because μ is a denumerably additive measure.

Now, [E.8.8b., (1)],

$$\mu[s(A)] = \mu(A).$$

It follows that

$$\mu^{\bullet}(\mathrm{Point}\, A) = \mu(A), \quad \text{q.e.d.}$$

Since $\mu(A)$ depends only on the choice of the generating vector $\vec{\omega}$, the number $\mu(\mathrm{Point}\, A)$ does not depend on $\{\mathbf{H}_n\}$, q.e.d.

M.2.3. In [K.] we have proved that the change of the generating vector $\vec{\omega}$ does not influence the μ-measurability of subsets of the plane P. In [M.2.2.] has been said that it does not depend on $\{\mathbf{H}_n\}$ neither. Hence the said μ-measurability depends only on the normal operator N. This allows to speak of *N-measurability of subsets of the plane*. The same can be said of μ-null sets, whose collection depends only on N. Thus instead of the term μ-null set, we can use, more properly, the term *N-null sets*. The sets with positive N-measure will be termed *N-positive sets*. Also if $E \subseteqq P_n$ and $\mu E > 0$, we shall say that E is N-positive.

M.2.4. Theorem. If α is a subset of P, then the following are equivalent:

I. α is an N-null set,

II. Pointα is a $\mu^\bullet$-null set, i.e. $\mu^\bullet(\text{Point}\,\alpha) = 0$.

Proof. This follows from [M.2.2.].

M.2.4.1. Theorem. An analogous equivalence holds true, if we suppose that α is an N-measurable subset of P.

M.2.4.2. If $\alpha \subseteqq P$ then the following are equivalent.

I. α is N-measurable,

II. Pointα is $\mu^\bullet$-measurable.

M.2.5. Theorem. If α is N-measurable, then

$$s(\alpha) = s^\bullet(\text{Point}\,\alpha),$$

Proof. We have Point$\alpha = \bigcup_n \alpha_n$, where α_n is the n-th-copy of α. By [E.8.13.],

$$s^\bullet(\text{Point}\,\alpha) = s^\bullet(\bigcup_n \alpha_n) =, [\text{E.}], = \sum_n s_n(\alpha_n) = \sum_n s(\alpha)\,\mathbf{H}_n$$

$$= s(\alpha) \cdot \sum_n \mathbf{H}_n = s(\alpha) \cdot 1 = s(\alpha).$$

M.2.6. Theorem. If

1. $\alpha \in P$,

2. α is N-positive,

then there exists n, such that

$$\mu^\bullet(\alpha_n) > 0,$$

(α is the n-th-copy of α).

Proof. We have [M.2.2.],

$$(1) \qquad \mu^\bullet(\text{Point}\,\alpha) = \mu(\alpha) > 0,$$

and we also have

$$\mu^\bullet(\text{Point}\,\alpha) = \mu^\bullet(\bigcup_n \alpha_n) = \sum_n \mu^\bullet(\alpha_n).$$

Since $\mu^\bullet(\alpha_n) \geqq 0$, (1) implies that for some n we must have $\mu^\bullet(\alpha_n) > 0$.

M.3. Def. Let α be N-positive; the number (finite or infinite) of indices n for which α_n is $\mu^\bullet$-positive will be termed *upper multiplicity of α and denoted by $\bar{\xi}(\alpha)$*. By [M.2.], we have $+\infty \geqq \bar{\xi}(\alpha) \geqq 1$ for $\alpha \in F^L$, whenever α is N-positive.

M.3.1. Def. If $\bar{\xi}(\alpha) < \infty$ we call α *set of finite upper multiplicity*. If $\bar{\xi}(\alpha) = \infty$ we call α *set of infinite upper multiplicity*.

M.3.2. Theorem. If

1. α, β are N-positive, $(\in F^L)$,

2. $\alpha \subseteqq \beta$,

then

$$\bar{\xi}(\alpha) \leqq \bar{\xi}(\beta).$$

Proof. Let

$$A_{k(1)}, A_{k(2)}, \ldots$$

be all positive copies of α. Let

$$B_{k(1)}, B_{k(2)}, \ldots$$

be all positive copies of β. Since $A_{k(m)} \subseteqq B_{k(m)}$, and we have $\mu(A_{k(m)}) > 0$, it follows $\mu(B_{k(m)}) > 0$; hence β has at least $\bar{\xi}(\alpha)$ positive copies, so

$$\bar{\xi}(\alpha) \leqq \bar{\xi}(\beta), \quad \text{q.e.d.}$$

M.4. Def. Let α be N-positive, $\alpha \in (F^L)$. Consider the set (of numbers):

$$S =_{df} \{\bar{\xi}(\alpha') \mid \alpha' \subseteqq \alpha, \quad \alpha' \in F^L, \quad \alpha' \text{ is } N\text{-positive}\}.$$

The set S has a minimum $\geqq 1$ (which may be even ∞). We call this minimum *lower multiplicity of* α and denote it by $\underline{\xi}(\alpha)$.

M.4.1. Theorem. If α is N-positive, $\alpha \in F^L$, then

$$\underline{\xi}(\alpha) \leqq \bar{\xi}(\alpha).$$

Proof. This is trivial, if $\bar{\xi}(\alpha) = \infty$.

Suppose that

$$\bar{\xi}(\alpha) = m < \infty.$$

For all $\beta \subseteqq \alpha$, where β is an N-positive set, we have, by [M.3.2.],

$$\bar{\xi}(\beta) \leqq \bar{\xi}(\alpha) \leqq m.$$

Hence

$$\min\{\bar{\xi}(\beta) \mid \beta \subseteqq \alpha, \beta\text{-}N\text{-positive}, \beta \in F^L\} \leqq m,$$

i.e. $\underline{\xi}(\alpha) \leqq \bar{\xi}(\alpha)$.

M.4.2. Theorem. If

1. α, β are N-positive, $\alpha, \beta \in F^L$,
2. $\alpha \subseteqq \beta$,

then

$$\underline{\xi}(\alpha) \geqq \underline{\xi}(\beta).$$

Proof. Put

$$S =_{df} \{\bar{\xi}(\alpha') \mid \alpha' \subseteqq \alpha, \alpha' \text{ is } N\text{-positive}, \alpha' \in F^L\},$$

$$T =_{df} \{\bar{\xi}(\beta') \mid \beta' \subseteqq \beta, \beta' \text{ is } N\text{-positive}, \beta' \in F^L\}.$$

If S is composed of the single "number" ∞, the theorem is trivial, since $\min S = \infty$.

Let $p \in S$, and let $p < \infty$. There is an α' with $\alpha' \subseteqq \alpha$, such that α' is N-positive and $p = \bar{\xi}(\alpha')$.

Since $\alpha' \subseteqq \alpha \subseteqq \beta$, we have

$$\bar{\xi}(\alpha') \in T.$$

Hence

$$S \subseteqq T, \quad \text{and then} \quad \min S \geqq \min T,$$

which completes the proof.

M.4.3. Now we shall be interested in the case, where α is of finite upper multiplicity only, so

$$\underline{\xi}(\alpha) \leqq \bar{\xi}(\alpha) < \infty.$$

M.4.4. Theorem. If

1. α, β are N-positive sets, $\alpha, \beta \in F^L$,
2. α, β are both of finite upper multiplicity,

then

$$\alpha \cup \beta \text{ is also of finite upper multiplicity.}$$

Proof. Since α and β have both a finite number of positive copies A_n, B_m, there exists n_0 such that for all $n \geqq n_0$,

$$\mu(A_n) = 0, \quad \mu(B_n) = 0.$$

Hence $\mu(A_N \cup B_n) = 0$.

Since $A_n \cup B_n$ is the n-th copy of $\alpha \cup \beta$, it follows that for $n \geqq n_0$ all copies of $\alpha \cup \beta$ are null sets. Consequently the positive copies must be contained in the planes P_n, where $n < n_0$. Hence $\bar{\xi}(\alpha \cup \beta) < n_0 < \infty$. The theorem is proved.

M.4.5. Theorem. If

1. α, β are N-positive sets $\in F^L$,
2. $\alpha \subseteqq \beta$,
3. β has a finite upper multiplicity,

then α also has a finite upper multiplicity.

Proof. Easy, by [M.3.2.].

M.4.6. Remark. It is not true that there always exists an N-positive set with finite upper multiplicity.

M.5. Def. Let α be an N-positive set, where $\alpha \in F^L$. We say that α *is of constant (or uniform) upper multiplicity* $\leqq +\infty$, whenever the following happens: if β is an N-positive set, such that $\beta \subseteqq \alpha$, $\beta \in F^L$, then

$$\bar{\xi}(\beta) = \bar{\xi}(\alpha).$$

M.5.1. Def. Let α be an N-positive set, $\alpha \in F^L$. We say that α *is of constant (or uniform) lower multiplicity* $\leqq +\infty$, whenever the following takes place: if β is an N-positive set $\in F^L$, such that $\beta \subseteqq \alpha$, then

$$\underline{\xi}(\beta) = \underline{\xi}(\alpha).$$

M.5.2. Theorem. If

1. α is an N-positive set, $\alpha \in F^L$,
2. α is of constant upper multiplicity $p \leqq \infty$,
3. β is an N-positive set, $\beta \in F^L$,
4. $\beta \subseteqq \alpha$,

then
β is also of constant upper multiplicity p.

Proof. This follows directly from Def. [M.5.].

M.5.2.1. Theorem. If

1. α is an N-positive set, $\alpha \in F^L$,
2. α is of constant lower multiplicity $q \leqq \infty$,
3. β is an N-positive set, $\beta \in F^L$,
4. $\beta \subseteqq \alpha$,

then
β is also of constant lower multiplicity q.

Proof directly from definition.

M.5.3. Theorem. If

1. α is an N-positive set, $\alpha \in F^L$,
2. α is of constant upper multiplicity $p \leqq \infty$,

then

1) $\bar{\xi}(\alpha) = \underline{\xi}(\alpha)$,
2) α is of constant lower multiplicity p.

Proof. The set of numbers

$$\{\bar{\xi}(\beta) \mid \beta \subseteqq \alpha, \beta \text{ is } N\text{-positive}\}$$

is composed of the single number p.

Hence $\underline{\xi}(\alpha)$, i.e. the minimum of this set equals p. Hence

$$\underline{\xi}(\alpha) = \bar{\xi}(\alpha). \tag{1}$$

Let β be an N-positive set with $\beta \subseteqq \alpha$.

We have,

$$\underline{\xi}(\alpha) \leqq \underline{\xi}(\beta) \leqq \bar{\xi}(\beta) \leqq \bar{\xi}(\alpha).$$

Hence, by (1),

$$\underline{\xi}(\alpha) = \underline{\xi}(\beta) = p.$$

The theorem is proved.

M.5.4. Theorem. If

1. α is an N-positive set, $\alpha \in F^L$,
2. $\bar{\xi}(\alpha) = \underline{\xi}(\alpha) = p \leqq \infty$,

then α is of constant upper multiplicity p.

Proof. We have for every $\beta \subseteqq \alpha$, where β is N-positive,

$$\underline{\xi}(\alpha) \leqq \underline{\xi}(\beta) \leqq \bar{\xi}(\beta) \leqq \bar{\xi}(\alpha).$$

Hence $\bar{\xi}(\beta) = \bar{\xi}(\alpha)$ for all $\beta \subseteqq \alpha$, where β is N-positive with $\beta \in F^L$.

Hence, by Def. [M.5.], α is a set of constant upper multiplicity p.

M.5.4.1. Remark. It is not true that, if a set is of constant lower multiplicity, then it must be also of constant upper multiplicity.

Expl. Let $\alpha \cap \beta = 0$, where the only copy of α is A_1, and the only copy of β is B_1 (both positive). Then $\bar{\xi}(\alpha \cup \beta) = 2$, $\bar{\xi}(\alpha) = 1$, $\bar{\xi}(\beta) = 1$,

$\underline{\xi}(\alpha \cup \beta) = 1$ and for all

$$\gamma \subseteqq \alpha \cup \beta, \quad \underline{\xi}(\gamma) = 1.$$

M.5.5. Theorem. If $\alpha \in F^L$ and is an N-positive set, then it contains a subset β, which $\in F^L$ and is N-positive, and of constant upper multiplicity. This subset β of α is such that

$$\bar{\xi}(\beta) = \underline{\xi}(\alpha).$$

Proof. There exists the minimum of the set of numbers

$$\{\bar{\xi}(\gamma) \,|\, \gamma \subseteqq \alpha, \text{ is } N\text{-positive}, \ \gamma \in F^L\}.$$

Hence there exists β, such that $\bar{\xi}(\gamma)$ is equal to this minimum, i.e.

$$\bar{\xi}(\beta) = \underline{\xi}(\alpha).$$

Let $\beta' \subseteqq \beta$, where β' is N-positive.

Since $\bar{\xi}(\beta)$ is minimum and $\beta' \subseteqq \beta$, we have

(1) $$\bar{\xi}(\beta) \leqq \bar{\xi}(\beta').$$

On the other hand, we have

(2) $$\bar{\xi}(\beta') \leqq \bar{\xi}(\beta).$$

From (1) and (2), we get

$$\bar{\xi}(\beta') = \bar{\xi}(\beta),$$

for all $\beta' \beta$, where β' is N-positive. Hence β is of constant upper multiplicity.

M.5.6. Theorem. If

1. $\alpha_1, \alpha_2, \ldots, \alpha_n, \ldots$, finite or denumerably infinite in number, are N-positive and $\in F^L$,

2. each α_n has constant lower multiplicity $p \leqq \infty$,

3. $\alpha = \bigcup_n \alpha_n$,

then

α has constant lower multiplicity p.

M.5.6a. Proof. There exists, [M.5.5.], a set $\gamma \subseteqq \alpha$, such that γ is N-positive and $\in F^L$, and has a constant upper multiplicity.

Let

$$q =_{df} \bar{\xi}(\gamma) = \underline{\xi}(\alpha).$$

The set γ has the constant upper multiplicity q. Now at least one of the sets $(\gamma \cap \alpha_N)$ is N-positive, for $\gamma = \sum_n (\gamma \cap \alpha_n)$. Let $(\gamma \cap \alpha_m)$ be N-positive.

Since γ has a constant upper multiplicity q, and $(\gamma \cap \alpha_m) \subseteqq \gamma$, it follows, that $\gamma \cap \alpha_m$ has, [M.5.2.], the constant upper multiplicity q, and then, [M.5.3.], also a constant lower multiplicity q.

Since $\gamma \cap \alpha_m \subseteqq \alpha_m$ and since α_m has the constant lower multiplicity p, it follows that $\gamma \cap \alpha_m$ has also the constant lower multiplicity p. Hence $p = q$. Thus $\underline{\xi}(\alpha) = p$.

M.5.6b. To prove that α has a constant lower multiplicity p, take any set $\alpha' \subseteqq \alpha$, where α' is N-positive. We have

$$\alpha' = \bigcup_n \alpha_n \cap \alpha'.$$

There exists at least one n, for which $\alpha_n \cap \alpha'$ is N-positive, so we can write

$$\alpha' = \bigcup_n \alpha_{k(n)} \cap \alpha',$$

where the summation is confined to all indices $k(n)$, for which $\alpha_{k(n)} \cap \alpha'$, is N-positive.

Now since $\alpha_{k(n)} \cap \alpha'$ is of constant lower multiplicity p, and since

$$\alpha_{k(n)} \cap \alpha' \subseteqq \alpha_{k(n)},$$

it follows, [M.5.2.1.], that

$$\alpha_{k(n)} \cap \alpha'$$

is of constant lower multiplicity p. Thus we are in the situation of [M.5.6a.].

It follows that

$$\underline{\xi}(\alpha') = p.$$

Thus we have proved that α is of constant lower multiplicity p, q.e.d.

M.5.7. Theorem. If

1. $\alpha \in F^L$ is N-positive and of constant upper multiplicity $m < \infty$,
2. $\alpha \subseteqq \beta$,
3. β has a constant upper multiplicity,

then there exist indices

$$k(1) < \cdots < k(m),$$

such that the copies $A_{k(i)}$, $B_{k(i)}$ of α, β, $(i = 1, \ldots, m)$ respectively are all μ-positive; but for any other index l, the copies A_l, B_l are both μ-null sets.

Proof. Let

$$k(1) < \cdots < k(m) \tag{1}$$

be all the indices n, for which A_n is positive.

Since β is of constant upper multiplicity and $\alpha \subseteqq \beta$, [M.5.2.], β has the constant upper multiplicity m. Hence β has, Def. [M.3.], exactly m μ-positive copies. Since $\alpha \subseteqq \beta$, we have for all $i = 1, \ldots, m$

$$A_{k(i)} \subseteqq B_{k(i)};$$

so $B_{k(i)}$ is μ-positive. Their number is m. Hence if l differs from all $k(i)$, the sets B_l and A_l must be null sets. Indeed if not, β would have at least $m+1$ positive copies. The theorem is established.

M.5.7.1. Def. If α is N-positive and $\in F^L$, then by an *index of* α we shall understand every $n = 1, 2, \ldots$, such that the set $(\text{Point}\alpha) \cap P_n$ is μ-positive, (the n-th copy of α).

M.5.7.2. Def. By *the set of indices of* α we understand the set of all indices of α, whenever α is N-positive. If α is an N-null set, then by *the set of indices of* α we shall understand the empty set of natural numbers. In this case $\text{Point}\alpha$ is a μ-null set, so there does not exist any μ-positive copy of α.

M.5.8. Remark. In Theorem [M.5.7.] the set of indices of α and β is

$$\{k(1), \ldots, k(m)\}.$$

M.6. Theorem. If α is an N-positive set and $\in F^L$, and has a constant finite upper multiplicity, then all $\beta \subseteqq \alpha$, where $\beta \in F^L$ and is N-positive, have the same set of indices.

Proof. Let $k(1), \ldots, k(m)$ be the set of indices of α. Since $\beta \subseteqq \alpha$, β is N-positive, it follows that β also has the set of indices

$$k(1), \ldots, k(m).$$

M.6.2. Theorem. If

1. $\alpha \in F^L$ and is N-positive,
2. α has a finite upper multiplicity,
3. for all β where $\alpha \subseteqq \beta \in F^L$ and β-N-positive, the set of indices is the same, (it must be finite),

then α has a constant upper multiplicity.

Proof straight forward.

M.6.3. Let α be an N-positive set, and suppose that

$$\bar{\xi}(\alpha) < \infty. \tag{1}$$

Let β be a subset of α with constant upper multiplicity m. Such a set exists, by [M.5.5.]. We have

$$m < \infty.$$

M.6.3a. Consider the class Γ of all sets β', such that

$$\beta \subseteqq \beta' \subseteqq \alpha,$$

where β' is of constant upper multiplicity.

There exist m indices

$$k(1) < \cdots < k(m), \tag{1}$$

such that the positive copies of β have the indices (1). This is the set of indices of β, Def. [M.5.7.2.]. Consider the N-measure on the plane P, say $\bar{\mu}$.

The set of numbers

$$\{\bar{\mu}(\beta') \mid \beta' \in \Gamma\}$$

is bounded, so it admits the supremum.

Put

$$d = \sup\{\bar{\mu}(\beta') \mid \beta' \in \Gamma\}.$$

For every $n = 1, 2, \ldots$ there exists a $\bar{\mu}$-measurable set $\beta'_n \in \Gamma$, such that

$$d - \frac{1}{n} \leqq \bar{\mu}(\beta'_n) \leqq d.$$

Put

$$\beta''_1 =_{df} \beta'_1, \quad \beta'_n = \beta'_1 + \cdots + \beta'_n \quad \text{for} \quad n > 1.$$

We have

$$\beta \subseteqq \beta''_1 \subseteqq \beta''_2 \subseteqq \cdots \subseteqq \alpha,$$

and

$$\lim_{n\to\infty} \bar{\mu}(\beta''_n) = d = \bar{\mu}(\bar{\beta}),$$

where

$$\bar{\beta} =_{df} \sum_{n=1}^{\infty} \beta''_n.$$

It follows that $\bar{\beta}$ is N-measurable, and N-positive. Since $\beta \subseteqq \beta'_n$, β has constant upper multiplicity. It follows that the constant upper multiplicity of β'_n is m, and β''_n has constant lower multiplicity m. Consequently β''_n has constant lower multiplicity m, and then $\bar{\beta}$ has constant lower multiplicity m.

Since the only positive copies of β'_n are copies with indices (1), so is for β''_n, and then for $\bar{\beta}$.

M.6.3b. I say that: The set $\bar{\beta}$ has constant upper multiplicity m. Indeed, any N-positive subset γ of $\bar{\beta}$ has the only possible positive copies with an index belonging to (1), so $\bar{\xi}(\gamma) \leqq m$. Since $\underline{\xi}(\gamma) = m$ and since $\underline{\xi}(\gamma) \leqq \bar{\xi}(\gamma)$, it follows that

$$\bar{\xi}(\gamma) = m.$$

Thus we have proved that β has the constant upper multiplicity m. The set β is a maximal set, (up to a nullset), containing β, and having a constant upper multiplicity. Indeed, if $\beta_0 \supseteqq \bar{\beta}$, where β_0 is measurable, and if β_0 has also constant upper multiplicity, this multiplicity must be m, by [M.5.7.]. Hence

$$\beta_0 \in \Gamma, \quad \text{so} \quad \bar{\mu}(\beta_0) \leqq \bar{\mu}(\bar{\beta}) = d.$$

On the other hand, since $\beta_0 \supseteqq \bar{\beta}$, we have

$$\bar{\mu}(\bar{\beta}) \leqq \bar{\mu}(\beta_0).$$

Consequently

$$\bar{\mu}(\beta_0) = \bar{\mu}(\bar{\beta}).$$

Now

$$\beta_0 \dotplus \bar{\beta} = (\beta_0 - \bar{\beta}) + (\bar{\beta} - \beta_0) = \beta_0 - \bar{\beta}, \qquad \bar{\beta} \subseteqq \beta_0.$$

Hence

$$\bar{\mu}(\beta_0 \dotplus \beta) = \bar{\mu}(\beta_0) - \bar{\mu}(\beta) = 0,$$

hence

$$\beta_0 =^N \beta .$$

Thus the maximality is proved. We have got the

M.6.4. Theorem. If

1. $\alpha \in F^L$ and is N-positive,
2. $\bar{\xi}(\alpha) < \infty$,
3. $\beta \subseteqq \alpha$, $\beta \in F^L$,
4. β has the finite constant upper multiplicity m,

then there exists a unique, (up to N-null set), N-positive set $\bar{\beta}$ such that:

1) $\beta \subseteqq \bar{\beta} \subseteqq \alpha$, $\bar{\beta} \in F^L$,

2) $\bar{\beta}$ has constant upper multiplicity m,

3) $\bar{\beta}$ is maximal, (up to a N-null set), with respect to 1) and 2),

4) for every $\beta' \subseteqq \bar{\beta}$, where β' is N-positive, the set of indices of β' coincides with the set of indices of β.

M.7. We know from Remark [M.5.4.1.], that if $\alpha \cap \beta = O$ and α has a constant upper multiplicity m and β has a constant upper multiplicity m, then the set $\alpha \cup \beta$ may not have a constant upper multiplicity.

Let α be N-positive with $\bar{\xi}(\alpha) = m < \infty$. We know that there exists an N-positive set β, such that $\beta \subseteqq \alpha$, and that β has a constant upper multiplicity $\leqq m$. This set β can be even supposed to be maximal in this respect, [M.5.7.]. β has a well determined finite sequence

$$k(1) < k(2) < \cdots \tag{1}$$

of indices corresponding to its positive copies.

If we take a similar subset of $\alpha - \beta$, the sequence of the corresponding indices differs from (1), for β is maximal.

M.7a. Consider the class K of all maximal subsets β of α with constant upper multiplicity. For each $\beta \in K$ determine the set of corresponding indices, as above, and take the maximum index n_β. The set of all these n_β must be bounded. Indeed, suppose it is not bounded. Hence there exists an infinite sequence

$$n_{\beta'} < n_{\beta''} < \cdots \to \infty .$$

It follows that α has, among others, the following positive copies

$$A_{n_{\beta'}}, A_{n_{\beta''}}, \ldots$$

Hence α has an infinite number of positive copies, which is excluded by the hypothesis $\bar{\xi}(\alpha) < \infty$.

It follows, that there exists an index n_0, such that: if $\beta \subseteqq \alpha$, $n \geqq n_0$, then β is a null set.

As the collection of all possible finite sequence of natural numbers, which are $< n_0$, is finite, it follows that K is a finite set.

Consequently, if we take a maximal subset γ' in α, choose another one γ'' in $\alpha - \gamma'$, afterwards another one γ''' in $\alpha - (\gamma' \cup \gamma'')$, etc., the process must finish.

Let $\gamma', \gamma'', \ldots, \gamma^{(s)}$ be all obtained maximal sets. We have $\alpha = \gamma' + \gamma'' + \cdots + \gamma^{(s)}$, because if not, this would imply a contradiction. We have proved the theorem:

M.7.1. Theorem. If $\bar{\xi}(\alpha) < \infty$, then there exists a decomposition

$$\alpha = \gamma' \cup \gamma'' \cup \cdots$$

into a finite number of N-positive, N-disjoint sets, such that each of them has a constant finite upper multiplicity, and each of them is maximal in the sense that, if

$$\gamma^{(i)} \subseteqq^N \gamma \subseteqq^N \alpha,$$

and has a constant upper multiplicity, then $\gamma^{(i)} =^N \gamma$. We have $\bar{\xi}(\gamma^{(i)}) = \underline{\xi}(\gamma^{(i)})$. (Of course, $\gamma^{(i)}$ has constant lower multiplicity). Each of these sets $\gamma^{(i)}$ has also a constant lower multiplicity, which equals to the upper one. The sets $\gamma', \gamma'', \ldots$ will be used in the sequal. We may call them *γ-sets.*

M.7.2. Corrollary. If $\bar{\xi}(\alpha) = m < \infty$, then α is the N-disjoint union of sets

$\alpha_{11}, \alpha_{12}, \ldots$ with constant upper multiplicity m_1,

$\alpha_{21}, \alpha_{22}, \ldots$ with constant upper multiplicity m_2,

. . . .

where

$$m_1 < m_2 < \cdots \qquad (m_k \leqq m).$$

$\alpha_{k1} \cup \alpha_{k2} \cup \cdots$ is the greatest subset of α with constant lower multiplicity m_k.

We have $\underline{\xi}(\alpha) = m_1$. All the sets $\alpha_{k1}, \alpha_{k2}, \ldots$ have the constant lower multiplicity m_k. Each set α_1 is maximal, i.e. it cannot be extended to a wider subset of α with constant upper multiplicity. Such a decomposition is unique.

M.7.3. Theorem. The union of any finite collection of mutually disjoint sets, each of which having a finite constant upper multiplicity, is a set with upper multiplicity $< \infty$.

M.8. There may exist N-positive sets with constant upper multiplicity ∞. Let S be the class of all those sets. Each N-positive set $\alpha \in S$ has also a constant lower multiplicity ∞. Considering the N-measure $\bar{\mu}$, and applying an argument similar to that in [M.6.3b.], we prove that there exists a maximal set π_∞ having this property.

This means that if $\alpha \supseteqq \pi_\infty$ and α has a constant upper multiplicity, then $\pi_\infty = \alpha$, (up to an N-null set).

Put

$$\pi_f =_{df} P - \pi_\infty .$$

If $\alpha \subseteqq \pi_f$, we may still have $\xi(\alpha) = \infty$, but there exists a $\beta \subseteqq \alpha$ with $\xi(\beta) < \infty$.

M.8.1. Def. We call π_∞ *infinite part of* P, and π_f *finite part of* P.

M.8.2. Theorem. The finite part π_f of P, if N-positive, can be split into a finite or at most denumerable number of disjoint subsets, each of which having a finite upper multiplicity.

Proof. The set π_f contains a subset α_1, such that $\xi(\alpha_1) < \infty$, because if not, we would have $\pi_f \subseteqq \pi_\infty$, and this is not true, because $\pi_f \cap \pi_\infty = \Theta$, π_f is positive and π_∞ maximal. Take α_1. If $\pi_f - \alpha_1 \neq \Theta$, we can choose an N-positive $\alpha_2 \subseteqq \pi_f - \alpha_1$ with $\xi(\alpha_2) < \infty$. We can proceed in this way, obtaining a well ordering sequence

$$\alpha_1, \alpha_2, \ldots, \alpha_\omega, \ldots, \alpha_\sigma, \ldots$$

of mutually disjoint positive sets. We define for σ:

$$\alpha_\sigma = \pi_f - \bigcup_{\gamma<\sigma} \alpha_\gamma .$$

Since all α_γ are N-positive, there will exist $\sigma < \Omega$, such that $\alpha_\sigma = \Theta$. Consequently π_f is the sum of an at most denumerable number of mutually disjoint N-positive sets, each of which having a finite upper multiplicity.

M.8.3. The sets of positive measure in V may show a very complicated configuration. Therefore it would be in order to change it into an other, more simple and visual. This can be made by changing the saturating sequence $\{\mathbf{H}_n\}$ into another one $\{\mathbf{H}'_n\}$, such that all somata of $(\bar{S})$, $\{\mathbf{H}_n\}$, $\{\mathbf{H}'_n\}$ be compatible with one another.

The sequence $\{\mathbf{H}'_n\}$ will produce a measure-configuration, such that for the positive parts

$$v_1, v_2, \ldots, v_n, \ldots \quad \text{of} \quad P_1, P_2, \ldots, P_n, \ldots$$

we shall have

$$\operatorname{Num} v_1 \geqq \operatorname{Num} v_2 \geqq \cdots \geqq \operatorname{Num} v_n \geqq \cdots$$

We may call this *terrace configuration*.

To perform the mentioned change of the saturating sequence, we shall cut the multiplane V into suitable parts, and transfer parts of one copy of P to another one with conservation of measure. This intuitive remark will have a precise setting. Since the matter is rather complicated, it cannot be settled at once, but we shall need many auxiliaries to give the topic a logically precise form.

First of all we shall study separately the finite part π_f and separately the infinite part π_∞ to get enough material for further use.

M.8.4. The finite part π_f. Let π_f be the finite part of P, and suppose that π_f is N-positive. We see that there exists a decomposition

$$\pi_f = \beta_1 \cup \beta_2 \cup \cdots$$

into an at most denumerable number of N-disjoint, N-positive sets β_n, each of which having a finite upper multiplicity. Since $\xi(\beta_n) < \infty$, we can decompose β_n into a finite number N-disjoint, N-positive sets

$\beta_{n,11}, \beta_{n,12}, \ldots$ each with constant upper multiplicity $m_{n,1}$,

$\beta_{n,21}, \beta_{n,22}, \ldots$ each with constant upper multiplicity $m_{n,2}$,

. . . .

where

$$m_{n,1} < m_{n,2} < \cdots \leqq \xi(\beta_n).$$

They are all maximal in this respect.

$$\beta_{n,11} \cup \beta_{n,12} \cup \cdots$$

is the N-greatest subset of β_n with constant lower multiplicity $m_{n,1}$,

$$\beta_{n,21} \cup \beta_{n,22} \cup \cdots$$

is the N-greatest subset of β_n with constant lower multiplicity $m_{n,2}$,

. . . .

The range of indices of the copies of P, on which the corresponding copies of $\beta_{n,ik}$ are positive, are different, whenever

$$(i, k) \neq (i', k').$$

If β is an N-positive subset of $\beta_{n,i,k}$, then the positive copies of β have the same indices, as $\beta_{n,i,k}$ has. The lower multiplicity of $\beta_{n,i,k}$ is just the number of its positive copies.

Let us vary n, and make the union of all $\beta_{n,i,k}$, which have the same lower multiplicity. We get a set with the same constant lower multiplicity.

Thus π_f will be decomposed into an at most denumerable number of N-disjoint, N-positive sets

$$\pi_f = \gamma_1 \cup \gamma_2 \cup \cdots,$$

where γ_n is decomposed into an at most denumerable number of sets which are N-positive and N-disjoint

$$\gamma_n = \gamma_{n1} \cup \gamma_{n2} \cup \cdots,$$

each of which having the constant upper multiplicity p_n; but they differ in the property, that to

$$\gamma_{n1}, \gamma_{n2}, \ldots$$

there correspond different ranges of m_n indices of the planes, which contain their positive copies.

Thus we have proved the

M.8.5. Theorem. If π_f is N-positive, then there exists a unique decomposition of π_f into an at most denumerable number of N-disjoint, N-positive sets

$$\pi_f = \gamma_1 \cup \gamma_2 \cup \cdots,$$

where γ_n has a constant lower multiplicity m_n, and where

$$1 \leqq m_1 < m_2 < \cdots$$

There exists a unique decomposition of γ_n into an at most denumerable number of N-disjoint, N-positive sets

$$\gamma_n = \gamma_{n1} \cup \gamma_{n2} \cup \cdots,$$

all of them having different sets of indices corresponding to $\mu^\bullet$-positive copies, [E.8.11].

Each γ_{nk} has a constant upper multiplicity m_n and is maximal in this respect.

M.8.6. Theorem. π_f, if N-positive, is the maximal set, such that every its N-positive subset α has a finite lower multiplicity $\underline{\xi}(\alpha)$.

M.8.7. Theorem. π_∞, if N-positive, is the maximal set with constant lower multiplicity ∞.

M.8.8. We shall decompose the set Point π_f in to an at most denumerable number of positive disjoint sets, which will play a role later on. Take the decomposition.

$$\pi_f = \gamma_{11} \cup \gamma_{12} \cup \cdots, \qquad \gamma_{21} \cup \gamma_{22} \cup \cdots, \qquad \gamma_n = \gamma_{n1} \cup \gamma_{n2} \cup \cdots$$

into disjoint N-positive sets, such that γ_{ki} has the constant upper multiplicity $m(k)$, and where γ_{ki} and each positive subset of it has a positive copy.

M.8.9. Denote the n-th copy of γ_{ik} by $(\Gamma_{ik,n})$. Each γ_{ik} has at least one μ-positive copy. Take this one which has the smallest index n. Let us vary i, k and take all the above copies with the smallest index. Denote by A_1 the union of all these sets.

Thus A_1 is a subset of V, defined as the union of all those positive copies of all sets γ_{ki}, which have the smallest index.

Since every γ_{ik} has at least one positive copy, therefore

$$\operatorname{Num} A_1 =^N \pi_f.$$

Each set $\gamma_{k1}, \gamma_{k2}, \ldots$ has m_k copies, which are $\mu^\bullet$-positive.

Hence they have in $V - A_1$ only $m_k - 1$ positive copies whenever $m_k > 1$, and no positive copy, if $m_k = 1$.

If $m_1 = 1$, the sets $\gamma_{11}, \gamma_{12}, \dots$ will have no positive copies in $V - A_1$. Consider the sets γ_{ki}, for which $m_k - 1 > 0$, and define A_2 as the union of those positive copies of γ_{ki} which are in $V - A_1$ and have the smallest index. If $m_1 - 1 > 0$, $m_2 - 1 > 0, \dots$, then $\operatorname{Num} A_2 = \pi_f$. Take the set $\operatorname{Point} \pi_f - A_1 - A_2$. The sets $\gamma_{k1}, \gamma_{k2}, \dots$ have in $\operatorname{Point} \pi_f - A_1 - A_2$ only $m_1 - 2$, $m_2 - 2, \dots$ positive copies: of course if they have some.

Repeating this construction we decompose $\operatorname{Point} \pi_f$ into an at most infinite denumerable number of sets $A_1, A_2, A_3, \dots$, (it may be confined to the first element A_1).

If $m_1 > 1$, then

$$\operatorname{Num} A_1 =^N \operatorname{Num} A_2 =^N \cdots =^N \operatorname{Num} A_{m_1} = \pi_f.$$

If m_2 is available, we have

$$m_1 < m_2.$$

The set $\gamma_1 = \gamma_{11} \cup \gamma_{12} \cup \cdots$ has no positive copy in $V - (A_1 \cup \cdots \cup A_{m_1})$, but all $\gamma_2, \gamma_3, \dots$ have. Hence

$$\operatorname{Num} A_{m_1+1} =^N \cdots =^N \operatorname{Num} A_{m_2} =^N \pi_f - \gamma_1.$$

In general

$$\operatorname{Num} A_{m_k+1} =^N \cdots =^N \operatorname{Num} A_{m_{k+1}} =^N \pi_f - (\gamma_1 \cup \cdots \cup \gamma_k),$$
$$(k = 1, 2, \dots).$$

Hence

$$\operatorname{Num} A_1 \geqq^N \operatorname{Num} A_2 \geqq^N \cdots \quad \text{and} \quad \bigcup_n A_n =^\mu \operatorname{Point} \pi_f.$$

We have proved the following:

M.9. Theorem. If the finite part π_f of P is N-positive, then $\operatorname{Point} \pi_f$ can be decomposed into an at most denumerable number of subsets $A_1 \cup A_2 \cup \cdots$ such that, if we consider the decomposition

$$\pi_f = \gamma_1 \cup \gamma_2 \cup \cdots,$$

into N-disjoint, N-positive sets $\in F^L$, such that γ_n have the constant lower multiplicity m_n, where

$$1 \leqq m_1 < m_2 < \cdots,$$

then

1) $\operatorname{Num} A_1 =^N \cdots =^N \operatorname{Num} A_{m_1} = \pi_f$,
   $\operatorname{Num} A_{m_k+1} =^N \cdots =^N \operatorname{Num} A_{m_{k+1}} =^N \pi_f - (\gamma_1 \cup \dots \cup \gamma_k)$,
2) all $A_1, A_2, \dots$ are disjoint,
3) $\operatorname{Point} \gamma_1 \subseteqq^\mu A_1 \dots A_{m_1}$,
   $\operatorname{Point} \gamma_2 \subseteqq^\mu A_{m_1+1} \cup \cdots \cup A_{m_2}$,
   . . . .
   $\operatorname{Point} \gamma_n \subseteqq^\mu A_{m_{n-1}+1} \cup \cdots \cup A_{m_n}$.

M.9.1. Def. The sets $A_1, A_2, \ldots$ will be called *A-sets*, and will be used later on.

M.9.2. The infinite part π_∞. Let us consider the infinite part π_∞ of P in the case π_∞ is N-positive. We shall decompose Pointπ_∞ into an at most denumerable number of $\bar{\mu}$-measurable sets as follows: First we have to prove the following general

M.9.3. Lemma. If

1. α is a set of positive measure μ, which is denumerably additive;
2. W is a property of measurable subsets of α, such that

a) W is invariant with respect to μ-equality,

b) if $\beta_1, \ldots, \beta_n, \ldots$ are μ-disjoint, finite or denumerable in number and have all the property W,
then $\bigcup_n \beta_n$ has the property W;

c) If $\alpha_1 \subseteqq^\mu \alpha$ and $\mu(\alpha_1) > 0$, then there exists $\alpha_2 \subseteqq^\mu \alpha_1$ with $\mu(\alpha_2) > 0$, having the property W,
then the whole set α has the property W.

Proof. Since $\mu\alpha > 0$, take (by hyp. c) a set α_1 such that $\alpha_1 \subseteqq^\mu \alpha$, $\mu(\alpha_1) > 0$, and where α_1 has the property W. If $\alpha_1 =^\mu \alpha$, the theorem is proved. Suppose that $\alpha_1 \subset^\mu \alpha$. Hence

$$\mu(\alpha - \alpha_1) > 0.$$

Hence, by hypothesis c), there exists $\alpha_2 \subseteqq^\mu \alpha - \alpha_1$, such that $\mu\alpha_2 > 0$, and that α_2 has the property W. We shall continue this process transfinitely. Suppose

1) we have already defined all α_v for $v < \lambda < \Omega$ where $\lambda \geqq 1$,

2) the sets α_v for $v < \lambda$ are all μ-disjoint,

3) each α_v for $v < \lambda$ has the property W.

If $\bigcup_{v<\lambda} \alpha_v =^\mu \alpha$, the theorem is proved by b). Suppose $\bigcup_{v<\lambda} \alpha_v \subset^\mu \alpha$; then $\mu(\bigcup_{v<\lambda} \alpha_v) < \mu(\alpha)$, hence $\mu(\alpha - \bigcup_{v<\lambda} \alpha_v) > 0$.

Hence there exists [hyp. c)], α_λ such that

$$\alpha_\lambda \subseteqq^\mu \alpha - \bigcup_{v<\lambda} \alpha_v, \quad \mu(\alpha_\lambda) > 0,$$

and that α_λ has the property W.

To have the choice well determined, let us arrange all positive μ-measurable sets into a well ordering, and at each step take sets with the smallest index. We get a well ordering

(1) $\qquad \alpha_1, \alpha_2, \ldots, \alpha_\lambda, \ldots,$ composed

of μ-disjoint sets. Since they all have positive measure and are $\subseteqq^\mu \alpha$, it follows, that this well ordering has the ordinal $< \Omega$. Since the number of sets (1) is at most denumerable, therefore by c), the set $\bigcup_\lambda \alpha_\lambda$ is μ-measurable and has the property W. We must have $\bigcup_\lambda \alpha_\lambda =^\mu \alpha$, because if not, (1) would be not finished.

Consequently α has the property W. The lemma is established.

M.9.4. We shall apply the lemma to the following situation: Let α be an N-measurable set, and W the property of subsets β of α, defined as

(1) $$\mu^\bullet(\text{Point}\,\beta \frown P_1) = 0, \quad [\text{E.8.11.}].$$

The hypothesis *b*) is satisfied. Indeed, if $\beta_1, \beta_2, \ldots$ are N-disjoint and have the property W, we have

$$\mu^\bullet(\text{Point}\bigcup_n \beta_n \frown P_1) = \mu^\bullet(\bigcup_n (\text{Point}\,\beta_n \frown P_1) = 0,$$

because

$$\mu^\bullet(\text{Point}\,\beta_n \frown P_1) = 0.$$

The property (1) is N-invariant. To prove this suppose that $\mu^\bullet(\text{Point}\,\beta \frown P_1) = 0$ and let $\beta =^N \beta'$. We have $\mu(\beta \dot{+} \beta') = 0$.

Hence

$$\mu^\bullet[\text{Point}\,(\beta \dot{+} \beta')] = 0,$$

$$\mu^\bullet(\text{Point}\,\beta \dot{+} \text{Point}\,\beta') = 0,$$

and hence

$$\mu^\bullet[(\text{Point}\,\beta \dot{+} \text{Point}\,\beta') \frown P_1] = 0,$$

$$\mu^\bullet[(\text{Point}\,\beta \frown P_1) \dot{+} (\text{Point}\,\beta' \frown P_1)] = 0,$$

i.e.

$$\mu^\bullet(\text{Point}\,\beta' \frown P_1) = \mu^\bullet(\text{Point}\,\beta \frown P_1) = 0.$$

Thus $\mu^\bullet(\text{Point}\,\beta' \frown P_1) = 0$, so the invariance is established. Now suppose that for every $\alpha_1 \subseteqq \alpha$ with $\mu(\alpha_1) > 0$ there exists $\alpha_2 \subseteqq \alpha_1$, with $\bar{\mu}(\alpha_2) > 0$, having the property W. By lemma it follows that α has the property W, i.e.

$$\mu^\bullet(\text{Point}\,\alpha \frown P_1) = 0.$$

M.10. Having this, we proceed to the main argument: There may exist an N-positive, $\in F^L$ subset φ of π_∞ such that every its N-positive subset has a $\mu^\bullet$-positive copy in P_1. If so, we can prove, by using an argument similar to that in [M.6.3b.], that there exists a maximal subset φ of π_∞: having that property. We denote this maximal set by φ_{11}. If there does not exist any set φ like above, we put $\varphi_{11} =_{df} \mathit{O}$. Put $B_{11} =_{df} \text{Point}\,\varphi_{11} \frown P_1$.

Consider the set

$$\pi_\infty - \varphi_{11}.$$

I say that, if γ is N-positive, and $\gamma \subseteqq \pi_\infty - \varphi_{11}$, then

$$\mu^\bullet(\text{Point}\,\gamma \cap P_1) = 0.$$

To prove that, suppose that there is a set $\gamma \subseteqq \pi_\infty - \varphi_{11}$, which is N-positive and where $\mu^\bullet(\text{Point}\,\gamma \cap P_1) > 0$. Put

$$E =_{df} \text{Point}\,\gamma \cap P_1.$$

Hence

(1) $$\mu^\bullet(E) > 0.$$

We have

$$\mu(\gamma) = \sum_n \mu^\bullet(\text{Point}\,\gamma \cap P_n).$$

It follows that

$$\mu(\gamma) \geqq \mu^\bullet(\text{Point}\,\gamma \cap P_1) = \mu^\bullet(E);$$

hence

(2) $$\mu(\gamma) > 0, \quad \text{i.e. } \gamma \text{ is } N\text{-positive.}$$

M.10.1. We shall prove that in γ there exists an N-positive subset δ, such that each N-positive subset δ_1 of δ has the property

$$\mu^\bullet(\text{Point}\,\delta_1 \cap P_1) > 0.$$

Suppose this be not true. Hence in every N-positive subset δ of γ, there exists another N-positive subset δ_1 of δ such that

(3) $$\mu^\bullet(\text{Point}\,\delta_1 \cap P_1) = 0.$$

This is just the special property, considered in the remark [M.9.4.], and dealing with subsets of γ. There we have proved that the conditions a), b) in Lemma [M.9.3.] are satisfied, and now we have also the property c) stated in (3) in the present reasoning. It follows that

$$\mu^\bullet(\text{Point}\,\gamma \cap P_1) = 0$$

which contradicts (1).

Consequently in γ there exists an N-positive subset δ, such that each N-positive subset δ_1 of δ has the property

$$\mu^\bullet(\text{Point}\,\delta_1 \cap P_1) > 0.$$

Now

$$\delta_1 \cap \varphi_{11} \stackrel{N}{=} O.$$

Hence the set $\varphi_{11} \cup \delta_1$ has the property, that each its N-positive subset δ_2 is such, that

$$\mu^\bullet(\operatorname{Point} \delta_2 \cap P_1) > 0;$$

consequently φ_{11} is not maximal.

Thus we have proved that, if γ is N-positive and

$$\gamma \subseteqq \pi_\infty - \varphi_{11},$$

then

$$\mu^\bullet(\operatorname{Point} \gamma \cap P_1) = 0.$$

We have $\operatorname{Num} B_{11} = \varphi_{11}$, [M.10.].

M.10.2. Consider the set

$$\pi_\infty - \varphi_{11}.$$

Suppose that there exists an N-positive subset φ, such that its every N-positive subset has a positive copy in P_2. Such a set φ can be taken maximal; denote it by φ_{12}. If no φ, as above, exists, we put $\varphi_{12} =_{df} \emptyset$. Define

$$B_{12} =_{df} \operatorname{Point} \varphi_{12} \cap P_2.$$

We have

$$\operatorname{Num} B_{12} = \varphi_{12}.$$

If

$$\gamma \subseteqq \pi_\infty - \varphi_{11} - \varphi_{12},$$

then

$$\mu^\bullet(\operatorname{Point} \gamma \cap P_1) = 0,$$

$$\mu^\bullet(\operatorname{Point} \gamma \cap P_2) = 0, \quad \text{and}$$

$\varphi_{11}, \varphi_{12}$ are N-disjoint.

Continuing this process, we get the subsets of P

$$\varphi_{11}, \varphi_{12}, \ldots, \varphi_{1n}, \ldots$$

and the sets

$$B_{11}, B_{12}, \ldots, B_{1n}, \ldots$$

of V, at most denumerable in number and having the properties:

1) $\varphi_{11}, \varphi_{12}, \ldots, \varphi_{1n}, \ldots$ are all N-disjoint, at most denumerable in number.

2) $B_{1n} =^\mu \operatorname{Point} \varphi_{1n} \cap P_n$, $\operatorname{Num} B_{1n} =^N \varphi_{1n}$,

3) the sets B_{1n} are all $\mu^\bullet$-disjoint,

4) $\mu^\bullet(\operatorname{Point} \gamma \cap P_1) = \mu^\bullet(\operatorname{Point} \gamma \cap P_2) = \cdots = \mu^\bullet(\operatorname{Point} \gamma \cap P_n) = 0$

for all

$$\gamma \subseteqq \pi_\infty - (\varphi_{11} \cap \cdots \cap \varphi_{1n}),$$

5) φ_{11} is the maximal subset of π_∞, possessing with all its subsets a positive copy in P_1,

. . . .

φ_{1n} is the maximal subset of $\pi_\infty - (\varphi_{11} \cup \varphi_{12} \cup \cdots \cup \varphi_{1,\, n-1})$, possessing with all its subsets a positive copy in P_n.

M.10.2.1. Def. We shall call the sets B_{1n} *the B-sets*. They will be used in the sequel.

M.10.2.2. Def. The sets φ_{1n} will also be used in the sequel and they will be called *φ-sets*.

M.10.3. We shall prove that

$$\varphi_{11} \cup \varphi_{12} \cup \cdots \cup \varphi_{1n} \cup \cdots =^N \pi_\infty .$$

Suppose that

$$\varphi_{11} \cup \varphi_{12} \cup \cdots \subset^N \pi_\infty .$$

Put

$$\varphi' =_{df} \pi_\infty - \bigcup_n \varphi_{1n} .$$

We have

$$\bar{\mu}(\varphi') > 0 .$$

The set φ' possesses at least one N-positive copy, say with index m. Put

(3.1) $$D =_{df} \operatorname{Point} \varphi' \cap P_m .$$

We have

$$\mu^\bullet D > 0 .$$

Put $\delta =_{df} \operatorname{Num} D$. We get

(4) $$(\operatorname{Point} \delta \cap P_m) = D .$$

I say that the set δ contains at least one N-positive subset δ', such that if $\delta'' \subseteqq \delta'$ and δ'' is N-positive, then

$$\mu^\bullet(\operatorname{Point} \delta'' \cap P_m) > 0 .$$

Suppose that this be not true. Then every N-positive subset δ' contains an N-positive subset δ'' such that

$$\mu^\bullet(\operatorname{Point} \delta'' \cap P_m) = 0 .$$

Applying Lemma [M.9.3.] and [M.9.4.], we obtain

$$\mu^\bullet(\operatorname{Point} \delta \cap P_m) = 0 ,$$

which, by (4), contradicts (3).

Thus the relation

(5) $$\varphi_{11} \cup \varphi_{12} \cup \cdots =^N \pi_\infty$$

is proved.

M.10.4. Put

$$B_1 =_{df} B_{11} \cup B_{12} \cup \cdots \cup B_{1n} \cup \cdots$$

Since

$$\operatorname{Num} B_1 = \bigcup_n \operatorname{Num} B_{1n}$$

and since

$$\operatorname{Num} B_{1n} = \varphi_{1n},$$

it follows, by (5), that

$$\operatorname{Num} B_1 = \pi_\infty .$$

M.10.5. There exists an N-positive set $\beta' \subseteqq \pi_\infty$, such that at least one of its copies is contained in $P - B_1$.

To prove that suppose that there exists an N-positive set β such that each its positive copy is in B_1. Then at least one of the sets

$$\beta \frown \varphi_{11}, \quad \beta \frown \varphi_{12}, \ldots, \quad \beta \frown \varphi_{1n}, \ldots$$

must be positive, since

$$\bigcup_n \varphi_{1n} = \pi_\infty, \quad \text{and } \pi_\infty \text{ is } N\text{-positive.}$$

Suppose $\beta \frown \varphi_{1k}$ is N-positive.

Each copy of $\beta \frown \varphi_{1k}$ is $\subseteqq^\mu B_1$. Hence each copy of $\beta \frown \varphi_{1k}$ is $\subseteqq^\mu B_{1k}$, because $B_1 \frown \operatorname{Point} \varphi_{1k} = B_{1k}$.

Hence $\beta \frown \varphi_{1k}$ possesses only one positive copy but this contradicts the hypothesis that $\beta \subseteqq \pi_\infty$. Hence there does not exist any positive set β, whose every copy is in B_1. Hence, given $\beta \subset^\mu \pi_\infty$, at least one of its copies $E =_{df} \operatorname{Point} \beta \frown P_n$ is not contained in B_1. Since $\beta \subset^\mu \pi_\infty$, we cannot have $B_1 \subseteqq^\mu E$. If $B_1 \frown E =^\mu \emptyset$, put $\beta' =_{df} \beta$; and if $\mu^\bullet(E \sim B_1) > 0$, put $\beta' =_{df} \beta \sim \varphi_n$. The set β' possesses a positive copy in $V - B_1$.

Having that, denote, similarly as in [M.10.2.], by φ_{2p}, $(p = 1, 2, \ldots)$ the maximal subset of π_∞, for which $B_{2p} =_{df} \operatorname{Point} \varphi_{2p} \frown P_p$ is in $V - B_1$. Put $B_2 =_{df} \bigcup_{n=1}^{\infty} B_{2n}$. Of course B_{21} and φ_{21} are μ-empty.

M.10.6. We get

$$\operatorname{Num} B_2 = \pi_\infty .$$

We continue this construction, getting the sets $B_1, B_2, \ldots$. The process must go to infinity, because if not, we would get a contradiction, since we are in π_∞. To go farther we establish the:

M.10.7. Lemma. If $E \subseteqq P_n$, $\mu^\bullet E > 0$, then there exists an N-measurable, N-positive set α such that $\alpha \subseteqq P$, and $(\operatorname{Point} \alpha) \frown P_n =^{\mu^\bullet} E$.

Proof. Since $E \subseteqq P_n$, $\mu E > 0$, it follows that $\mu_n E > 0$, where μ_n is the measure relative to the space $\mathbf{H}_n$. Hence $\boldsymbol{C}^{-1}(E)$ is a space of the borelian spectral tribe of $\mathbf{H}_n \upharpoonright N$; hence

$$\boldsymbol{C}^{-1}(E) = a - \mathbf{H}_n,$$

where a is a space of the borelian spectral tribe of N. Let α be a borelian subset of P such that

$$a = s(\alpha).$$

The set α is N-positive, and we have

$$E =^{\mu} \operatorname{Point}\alpha \frown P_n.$$

The lemma is proved.

M.10.8. We shall prove that

$$\bigcup_{n=1}^{\infty} B_n =^{\mu} \operatorname{Point}\pi_{\infty}.$$

Suppose this be not true. Hence

$$B =_{df} \operatorname{Point}\pi_{\infty} - \bigcup_{n=1}^{\infty} B_n$$

has a positive $\mu^{\bullet}$-measure.

Hence at least one of the $\mu^{\bullet}$-measurable sets

$$B \frown P_1, \quad B \frown P_2, \ldots$$

has a positive $\mu^{\bullet}$-measure.

Suppose this is $D_1' =_{df} B \frown P_{k_1}$, where $k_1 \geqq 1$. We see that $D_1' \frown \bigcup_n B_n$ is a μ-null set.

Applying Lemma [M.10.7.], we find an N-positive set δ_1', such that

$$(\operatorname{Point}\delta_1') \frown P_{k_1} =^{\mu} D_1'.$$

We have the existence of an N-positive subset δ_1 of δ_1', such that for every N-positive β with $\beta \subseteqq \delta_1$, the set $(\operatorname{Point}\beta) \frown P_{k_1}$ is positive. [We may say that δ_1 is a *"uniformly" positive* k_1-th copy, contained in D.]

Put

$$D_1 =_{df} (\operatorname{Point}\delta_1) \frown P_{k_1}.$$

We have

$$\operatorname{Num} D_1 = \delta_1.$$

We also have $\mu^{\bullet} D_1 > 0$ with

$$D_1 \subseteqq D_1'. \tag{1}$$

The set δ_1 may have a positive copy with an index $< k_1$, or not. Suppose it has. Denote by k_2 the next smallest index, such that δ_1 has a k_2-th positive copy:

$$D_2' = (\operatorname{Point}\delta_1) \frown P_{k_2}.$$

This means, that if $k_2 < l < k_1$, then $(\operatorname{Point}\delta_1) \frown P_1$ is a $\mu^{\bullet}$-null set. We continue the process. We find an N-positive subset δ_2 of δ_1 such that

$$(\mathrm{Point}\,\delta_2') \frown P_{k_2} =^{\mu} D_2'.$$

Hence, by Lemma [M.10.7.] we find an N-positive subset δ_2 of δ_2', such that δ_2 has a "uniformly" k_2-th μ-positive copy, contained in D_2'.

Put

$$D_2 =_{df} (\mathrm{Point}\,\delta_2) \frown P_{k2}.$$

We have $\mu^{\bullet} D_2 > 0$. Suppose we have defined k_p, D_p and δ_p, and have

$$D_p = (\mathrm{Point}\,\delta_p) \frown P_{k_p}, \quad \mu D_p > 0,$$

where δ_p is N-positive. The set δ_p may have a positive copy with an index $< k_p$, or not. Suppose it has. Denote by k_{p+1} the next smallest index, such that the k_{p+1}-copy of δ_p is positive:

$$D'_{p+1} =_{df} (\mathrm{Point}\,\delta_p) \frown P_{k_{p+1}}.$$

In the remaining case the process stops with k_p.

This says that, if $k_{p+1} < l < k_p$, then $(\mathrm{Point}\,\delta_p) \frown P_1$ is a null-set.

We find an N-positive set δ'_{p+1}, such that

$$(\mathrm{Point}\,\delta'_{p+1}) \frown P_{k_{p+1}} =^{\mu} D'_{p+1}.$$

We find an N-positive subset δ_{p+1} of δ'_{p+1}, such that δ_{p+1} has a uniformly μ-positive k_{p+1}-th copy, contained in D'_{p+1}.

M.10.9. Put

$$D_{k+1} =_{df} (\mathrm{Point}\,\delta_{p+1}) \frown P_{k_{p+1}}, \quad \mu^{\bullet} D_{k+1} > 0.$$

We get the indices

$$k_1 > k_2 > \cdots \quad (k_p \geqq 1);$$

hence the process must finish after a finite number of steps. We get the indices

$$k_1 > k_2 > \cdots > k_s \quad \text{where} \quad k_s \geqq 1.$$

We have

$$D_{p+1} = (\mathrm{Point}\,\delta_p) \frown P_{k_{p+1}} \quad \text{and} \quad D'_{p+1} =^{\mu} (\mathrm{Point}\,\delta'_{p+1}) \frown P_{k_{p+1}}.$$

Hence

$$\mathrm{Num} D'_{p+1} =^{N} \delta_p \quad \text{and} \quad =^{N} \delta'_{p+1}.$$

Hence

$$\delta_p =^{N} \delta_{p+1}.$$

We also have

$$\delta_{p+1} \subseteqq \delta'_{p+1};$$

hence

$$\delta_{p+1} \subseteqq^{N} \delta_p.$$

Consequently we have

$$\delta_s \subseteqq^N \delta_{s-1} \subseteqq^N \cdots \subseteqq^N \delta_1.$$

Since δ_1 has a uniformly μ-positive k_1-th copy, it follows that δ_s also has a uniformly positive k_1-th copy. Similarly we prove that δ_s has the uniformly positive copies with indices $k_1, k_2, \ldots, k_s$. We also prove that δ_s has a μ-null copy for every l, where $l \leqq k_s$ and where l differs from all the indices $k_1, k_2, \ldots, k_s$.

If $k_s = 1$, we get δ_s; $\varphi_{11} = \varphi_{1k_s}$ and the 1-th copy of δ_s

$$\text{Point}\,\delta_s \frown P_1 \subseteqq^\mu B_{11}.$$

If $k_s = 2$, we have

$$\text{Point}\,\delta_s \frown P_1 =^\mu \emptyset,$$

and then

$$\delta_s \frown \varphi_{11} =^N \emptyset,$$

hence

$$\delta_s \subseteqq^N \pi_\infty - \varphi_{11}.$$

Since δ_2 has a uniformly positive copy in P_2, it follows that

$$\delta_s \subseteqq \varphi_{12} = \varphi_{1k_s},$$

and that

$$\text{Point}\,\delta_s \frown P_2 \subseteqq^\mu B_{12} = B_{1k_s},$$

and

$$\text{Point}\,\delta_s \frown P_1 =^\mu \emptyset.$$

In general: If $k_s = l$, then $\delta_s \subseteqq \varphi_{1l} = \varphi_{1k_1}$, and $\text{Point}\,\delta_s \frown P_l \subseteqq^\mu B_{1l}$, $\text{Point}\,\delta_s \frown P_{l'} =^\mu \emptyset$ for all $l' < l$.

Consequently

$$\text{Point}\,\delta_s \frown (P_1 \smile \cdots \smile P_{k_s}) =^\mu \text{Point}\,\delta_s \frown P_{k_s} \subseteqq B_{1k_s} \subseteqq B_1.$$

M.10.10. We have the inclusion:

$$\text{Point}\,\delta_s \frown P_{k_s+m} \subseteqq P - B_1, \quad \text{for} \quad m = 1, 2, \ldots$$

Indeed

$$\begin{aligned} \text{Point}\,\delta_s \frown P_{k_s+m} \frown B_1 &\subseteqq (\text{Point}\,\varphi_{1k_s} \frown B_1) \frown P_{k_s+m} \\ &= B_{1k_s} \frown P_{k_s+m} \subseteqq P_{k_s} \frown P_{k_s+m} = \emptyset. \end{aligned}$$

Hence $\delta_s \subseteqq^N \varphi_{2k_{s-1}}$, because δ_s has a uniformly positive copy in

$$P_{k_{s-1}} \frown (P - B_1).$$

Hence

$$(\text{Point}\,\delta_s) \frown P_{k_{s-1}} \subseteqq (\text{Point}\,\varphi_{2k_{s-1}}) \frown P_{k_{s-1}} = B_{2k_{s-1}} \subseteqq B_2.$$

Similarly we obtain

$$\text{Point}\,\delta_s \frown P_{k_{s-2}} \subseteqq B_3,$$

etc., and finally:

$$\text{Point}\,\delta_s \cap P_{k_1} \subseteqq B_s.$$

Now, since $D_1' \cap \bigcup_n B_n$ is μ-null set, so is

$$(\text{Point}\,\delta_s \cap P_{k_1}) \cap \bigcup_n B_n,$$

and then

$$(\text{Point}\,\delta_s \cap P_{k_1}) \cap B_s$$

is a $\mu^\bullet$-null set, i.e.

$$\text{Point}\,\delta_s \cap P_{k_1}$$

is a $\mu^\bullet$-null set, which is a contradiction.

M.11. Thus we have proved that

$$\bigcup_{n=1}^{\infty} B_n =^{\mu} \text{Point}\,\pi_\infty.$$

We also know that all B_n are disjoint with one another and that

$$\text{Num}\,B_n =^N \pi_\infty.$$

Let us put $B_n = \Theta$, whenever $\pi_\infty =^N \Theta$.
Let us put $A_n = \Theta$ whenever $\pi_f =^N \Theta$.
We have

$$\text{Point}\,\pi_f =^{\mu} A_1 \cup A_2 \cup \cdots,$$

and all A_n are $\mu^\bullet$-disjoint with all B_m.

We have

$$\bigcup_n A_n \cup \bigcup_n B_m =^{\mu} P.$$

Hence

$$\bigcup_n (A_n \cup B_n) =^{\mu} P,$$

and the sets

$$A_1 \cup B_1,\ A_2 \cup B_2,\ \ldots$$

are μ-disjoint. They are of course $\mu^\bullet$-measurable.

The saturating sequence yielding the terrace configuration of measure

M.11.1. Put, [E], [H.10.5ff.],

$$\mathbf{H}_1' =_{df} \mathbf{C}^{-1}(A_1 \cup B_1), \quad \mathbf{H}_2' =_{df} \mathbf{C}^{-1}(A_2 \cup B_2), \ \ldots$$

The spaces $\mathbf{H}_1', \mathbf{H}_2', \ldots$ are disjoint. We have $\mathbf{H}_n' \in s_b^*$ and

$$\mathbf{H}_1' + \mathbf{H}_2' + \cdots = \mathbf{H}.$$

We shall prove that

(1) $$\mathbf{H}_1', \mathbf{H}_2', \ldots$$

is a saturating sequence of $\mathbf{H}$ with respect to s_b in the sense that the smallest tribe of spaces, containing s_b and the spaces (1), coincides with s_b^*.

M.11.1a. Proof. Take any space $d \in s_b^*$, $d \neq O$. Put

$$D = \boldsymbol{C}\, d, \quad \text{we have} \quad \boldsymbol{C}^{-1} D = d.$$

We have $D = \bigcup_{n=1}^{\infty} (D \frown P_n)$. There exists n, for which $D \frown P_n$ is positive. Take such an n and consider

$$D^{(n)} =_{df} D\, P_n.$$

By Lemma [M.10.7.], there exists an N-positive set $\delta^{(n)} \subseteqq P$, such that

$$D^{(n)} =^{\mu} \operatorname{Point} \delta^{(n)} \frown P_n.$$

Hence

$$\operatorname{Num} D^{(n)} =^{N} \delta^{(n)}.$$

Now I say that

(1) $$D^{(n)} \frown (A_k \smile B_k) =^{\mu} \operatorname{Point} \delta^{(n)} \frown (A_k \smile B_k).$$

Indeed, A_k is the union of all those positive copies of the sets γ_{ik}, which have the smallest index and are lying in

$$P - (A_1 \smile \cdots \smile A_{k-1}).$$

For every such γ_{ik} one copy of γ_{ik} takes part in the composition of A_k.

Consequently, if ε is an N-measurable set, $E \subseteqq^{\mu} A_k$, $\operatorname{Num} E =^{N} \varepsilon$, then

$$E = A_k \frown \operatorname{Point} \varepsilon.$$

Applying that to the sets $\delta^{(n)}$ and $D^{(n)}$, we get

(2) $$D^{(n)} \frown A_k =^{\mu} \operatorname{Point} \delta^{(nk)} \frown A_k.$$

In a similar manner we prove that

(3) $$D^{(n)} \frown B_k =^{\mu} \operatorname{Point} \delta^{(nk)} \frown B_k.$$

From (2) and (3) the equation (1) follows.

Now if we put

$$a^{(n)} =_{df} s(\delta^{(n)\, k}),$$

we get

$$\boldsymbol{C}^{-1} [D^{(n)} \frown (A_k \smile B_k)] = s(\delta^{nk})\, \mathbf{H}'_k,$$

i.e.

$$\boldsymbol{C}^{-1} [D \frown P_n \frown (A_k \smile B_k)] = s(\delta^{nk})\, \mathbf{H}'_k,$$

and then

$$\boldsymbol{C}^{-1} [D \frown (A_k \smile B_k)] = \sum_n s(\delta^{nk})\, \mathbf{H}'_k.$$

Put

$$d^{(k)} =_{df} \sum_n s(\delta^{nk}).$$

We have $d^{(k)} \in s_b$ and

$$\mathbf{C}^{-1}[D \cap (A_k \cup B_k)] = d^{(k)}\,\mathbf{H}'_k.$$

Consequently

$$\mathbf{C}^{-1}[D \cap \bigcup_k (A_k \cup B_k)] = \mathbf{C}^{-1}[D \cap P] = \mathbf{C}^{-1}(D) = \sum_k d^{(k)}\,\mathbf{H}'_k,$$

i.e. $d = \sum_k d^{(k)}\,\mathbf{H}'_k$. Hence $\{\mathbf{H}'_n\}$ is a saturating sequence of D with respect to s_b^*.

M.12. Let us consider the natural canonical mapping $\mathbf{C}'$ of $\mathbf{H}$ into P, obtained by $\{\mathbf{H}'_n\}$ and $\vec{\omega}$. Denote by μ' the corresponding measure.

We have

$$(\mathbf{C}')^{-1}\,P_n = \mathbf{H}'_n \quad \text{and} \quad \mathbf{C}^{-1}(A_n \cup B_n) = \mathbf{H}'_n.$$

Hence

$$(\mathbf{C}')^{-1}\,P_n = \mathbf{C}^{-1}(A_n \cup B_n).$$

Consequently

$$P_n = \mathbf{C}'\,\mathbf{C}^{-1}(A_n \cup B_n)$$

and

$$A_n \cup B_n = \mathbf{C}\,(\mathbf{C}')^{-1}\,P_n.$$

The correspondence $\mathbf{C}$ carries the vectors of the space into functions $F(p)$, defined almost μ-everywhere on P and μ-square summable. This correspondence is invariant with respect to the μ-equality of functions $F(p)$ and with respect to the equality of vectors.

It also carries (in a generalized way) spaces of s_b^* into μ-measurable subsets of V, this correspondence being invariant with respect to the equality of spaces and μ-equality of subsets of V. We also have:

$$\text{if} \quad E = \mathbf{C}\,e, \quad \text{then} \quad \mu(E) = \mu(e) = ||\operatorname{Proj}_e \vec{\omega}||^2.$$

A similar remark can be made for $\mathbf{C}'$. Hence $\mathbf{C}'\,\mathbf{C}^{-1}$ [and its inverse $\mathbf{C}(\mathbf{C}')^{-1}$] are isomorphisms from V onto V, which are invariant with respect to the μ-equality of sets in $\mathrm{Q}(\mathbf{C}'\,\mathbf{C}^{-1}) = P$ and with respect to μ'-equality of sets in $\mathrm{D}(\mathbf{C}'\,\mathbf{C}^{-1}) \cdot \mathbf{C}'\,\mathbf{C}^{-1}$. It carries μ-measurable sets onto μ'-measurable sets, i.e. $E \cup^{\mathbf{C}'\mathbf{C}^{-1}|} E'$.

Let $E \cup^{\mathbf{C}'\mathbf{C}^{-1}} E'$. There exists a space $e \in s_b^*$, such that

$$E \cup^{\mathbf{C}^{-1}} e \cup^{\mathbf{C}'} E'.$$

Hence

$$\mu(E) = \mu(e) = \mu'(E').$$

Thus we have

$$\mu'(E') = \mu[(\mathbf{C}'\,\mathbf{C}^{-1})\,E]$$

for the changed measure. Of course a μ-null set may be μ'-positive and conversely. The mapping $\mathbf{C}'$ places the spaces $\mathbf{C}^{-1}(A_n \cap B_n)$ in P_n.

M.12.1. If we consider the canonical mapping $\mathbf{C}^{-1}$, the N-measurability and N-positiveness of subsets of P will go over into N-measurability and N-positiveness respectively. The same can be said about the lower multiplicity of N-measurable sets of P. The transformation $\boldsymbol{C}'\boldsymbol{C}^{-1}$ does not affect it. Let us consider the sets $\gamma_1, \gamma_2, \ldots \pi_f$ and π_∞. In the mapping $\boldsymbol{C}'$, the set γ_1 has the uniformly positive copies with indices $1, 2, \ldots, m_1$, the set γ_2 has the uniformly positive copies with indices $1, 2, \ldots, m_2$ but no more, and in general:

γ_p has the uniformly positive copies with indices $1, 2, \ldots, m_p$ but no more.

The set π_∞, if N-positive, has the uniformly positive copies with all indices $1, 2, \ldots$

Thus in the mapping, made by $\boldsymbol{C}'$, the set γ_1 has the uniform lower and upper multiplicity m_1; the set γ_2 has the uniform lower and upper multiplicity m_2, etc. The set π_∞ has the uniform lower and upper multiplicity ∞.

Thus the configuration on V, provided by $\boldsymbol{C}'$, has, speaking intuitively, a terrace configuration.

In P_1 we have the largest set E_1 with positive measure, in P_2 the same or next smaller E_2 "above" E_1, as on a "larger basis", and so on.

We call ***C'-terrace mapping*** of $\mathbf{H}$ and the same name will be used for any natural canonical mapping yielding a similar image. The terrace configuration makes the multiplicity visual even in the case of "continuous spectrum".

M.12.2. Def. We say that an N-measurable subset α of P, and also *the space* $s(\alpha)$, *has the uniform multiplicity* $1, 2, \ldots, \infty$, whenever α has the uniform lower multiplicity $1, 2, \ldots, \infty$ respectively.

Remark. This notion looks, as being dependant on the canonical mapping. We notice at once that the terrace configuration and multiplicity depends on s_b^* only.

We are going to show that the terrace configuration i.e. the repartition of lower multiplicity does not depend on s_b^*.

M.13. The purpose of the next discussion is to prove, that the terrace configuration does not depend on the choice of the saturating sequence $\{\mathbf{H}_n\}$ and on the choice of the generating vector $\vec{\omega}$.

Let $\{\mathbf{H}_n\}$, $\{\mathbf{H}'_n\}$ be two saturating sequences, and $\vec{\omega}$, $\vec{\omega}'$ two generating vectors. Denote by $\boldsymbol{C}$ and $\boldsymbol{C}'$ the resulting natural canonical mappings, which carry the given space $\mathbf{H}$ onto the spaces H, H' of respectively μ and μ'-square summable functions $F(p)$ and $F'(p)$, defined almost μ, μ'-everywhere on the union V of copies of the complex plane P. We may suppose that there are an infinite number of copies

$$P_1, P_2, \ldots, P_k, \ldots \quad \text{of } P.$$

M.13.1. The terrace configuration, related to $\boldsymbol{C}$, is characterized by the infinite sequence of N-measurable subsets

(1) $$\alpha_1, \alpha_2, \ldots, \alpha_n, \ldots, \alpha_\omega$$

of P, such that they are all N-disjoint:

$$\bigcup_{k=1}^{\omega} \alpha_k =^N P;$$

α_k has the homogenuous multiplicity k, and α_ω has ∞. We may suppose that $\alpha_k = \emptyset$, whenever an N-positive set α_k with homogenuous multiplicity k is not available.

Let

(2) $$\alpha'_1, \alpha'_2, \ldots, \alpha'_n, \ldots, \alpha'_\omega$$

be the analogous sequence, corresponding to $\boldsymbol{C}'$.

Our aim is to prove that, for $k = 1, 2, \ldots, n, \ldots, \omega$, we have

$$\alpha_k =^N \alpha'_k.$$

M.13.2. A necessary and sufficient condition for the N-equality of the sequences (1), (2) is that

(3) $$\alpha_i \cap \alpha'_k =^N \emptyset \quad \text{for all} \quad i \neq k, \qquad (i, k = 1, 2, \ldots, \omega).$$

This condition is clearly necessary. To prove its sufficiency, suppose (3).

We have

$$P = P \cap P =^N \bigcup_{n=1}^{\omega} \alpha_n \cap \bigcup_{m=1}^{\omega} \alpha'_m =^N \bigcup_{n=1,\, m=1}^{\omega} (\alpha_n \cap \alpha'_m) =^N \bigcup_{n=1}^{\omega} (\alpha_n \cap \alpha'_n).$$

We know that the N-measurability of subsets of P does not depend on the choice of $\{\mathbf{H}_n\}$ and $\vec{\omega}$. Put $\bar{\mu}(\alpha) = \mu(\text{Point}\,\alpha)$.

Since the terms $\alpha_n \cap \alpha'_n$ in the above union are mutually N-disjoint, it follows that

(4) $$\mu(P) = \mu\left[\bigcup_{n=1}^{\omega} (\alpha_n \cap \alpha'_n)\right] = \sum_{n=1}^{\infty} \bar{\mu}(\alpha_n \cap \alpha'_n).$$

I say that

$$\bar{\mu}(\alpha_n \cap \alpha'_n) \leqq \bar{\mu}(\alpha_n) = \bar{\mu}(\alpha'_n) \quad \text{for all} \quad n = 1, 2, \ldots, \omega.$$

If this were not true, there would exist k, such that

$$\bar{\mu}(\alpha'_k \cap \alpha_k) < \bar{\mu}(\alpha_k),$$

or there would exist k such that

$$\bar{\mu}(\alpha'_k \cap \alpha_k) < \bar{\mu}(\alpha'_k).$$

Consider the first alternative:

(5) $$\bar{\mu}(\alpha_k \cap \alpha'_k) < \bar{\mu}(\alpha_k).$$

From (1) we would get, by (5),

$$\bar{\mu}(P) < \sum_{n=1}^{\infty} \bar{\mu}(\alpha_n), \tag{6}$$

because for all n

$$\bar{\mu}(\alpha_n \cap \alpha_n') \leqq \bar{\mu}(\alpha_n).$$

But (6) contradicts the equality

$$\bar{\mu}(P) = \sum_{n=1}^{\infty} \bar{\mu}(\alpha_n).$$

Thus the assertion is proved.

M.13.3. Now suppose that the terrace configurations do not coincide. It follows that there exist n, m where $m < n$, such that

$$\bar{\mu}(\alpha_n \cap \alpha_m') > 0.$$

Put

$$\alpha = \alpha_m \cap \alpha_m'.$$

The set α has the uniform multiplicity n with respect to $\boldsymbol{C}$ and the uniform multiplicity m with respect to $\boldsymbol{C}'$. If $n = \omega$, choose any natural q such that

$$m < q < \omega = \infty.$$

Now, we know that, if the saturating sequence is kept fixed, the terrace configuration does not depend on the generating vector. We also know that, if we change in P the measure μ, so as to preserve the ideal of all sets with μ-measure 0, and assure that the fonction Num$(\dot{p})$ remains μ-square summable, the new measure will be supplied by a suitable new generating vector with maintenance of the saturating sequence $\{\mathbf{H}_n\}$. In the terrace configuration of both $\boldsymbol{C}$ and $\boldsymbol{C}'$ the copies of β, where $\beta \subseteqq \alpha$, and where β is N-positive,

$$\beta^{(1)}, \beta^{(2)}, \ldots, \beta^{(m)}, \ldots, \beta^{(q)}$$

are the same, though they may have different measures μ, μ'; but all are positive. Let us change the measures μ, μ' on P into μ_1, μ_1' on P, so as to have

$$\begin{aligned}
\mu_1(\beta^{(1)}) &= \mu_1(\beta^{(2)}) = \cdots = \mu_1(\beta^{(q)}) = \bar{\mu}(\beta^{(1)}),\\
\mu_1'(\beta^{(1)}) &= \mu_1'(\beta^{(2)}) = \cdots = \mu_1'(\beta^{(q)}) = \bar{\mu}(\beta^{(1)}),\\
\mu_1({}^{(q+s)}) &= \mu(\beta^{(q+s)}) = 0 \quad \text{for} \quad s = 1, 2, \ldots,\\
\mu_1'(\beta^{(m+s)}) &= \mu(\beta^{(m+s)}) = 0 \quad \text{for} \quad s = 1, 2, \ldots,
\end{aligned}$$

and

$$\begin{aligned}
\mu_1(E) &= \mu(E) \quad \text{for all } \mu\text{-measurable} \quad E \subseteqq \text{Point}(\text{co}\,\alpha),\\
\mu_1'(E) &= \mu'(E) \quad \text{for all } \mu'\text{-measurable} \quad E \subseteqq \text{Point}(\text{co}\,\alpha).
\end{aligned}$$

This change can be extended, so as to get a denumerably additive measures on P. With respect to the changed measures the function $\mathrm{Num}(\dot{p})$ will remain μ_1-square summable and μ_1'-square summable, since $\mathrm{Num}(\dot{p})$ is μ-square summable on P_1, and the change affects a finite number of P_k only.

Consequently the change of measure consists of the change of the generating vectors $\vec{\omega}, \vec{\omega}'$ into some other generating vectors $\vec{\omega}_1, \vec{\omega}_1'$, and $\boldsymbol{C}, \boldsymbol{C}'$ into the canonical mappings $\boldsymbol{C}_1, \boldsymbol{C}_1'$ without influencing the terrace configurations.

M.14. To avoid complicated notations, suppose that $\vec{\omega}, \vec{\omega}'$ just have the above properties of $\vec{\omega}_1, \vec{\omega}_1'$ concerning the measure. Thus we have

$$\mu(\beta^{(1)}) = \mu(\beta^{(2)}) = \cdots = \mu(\beta^{(q)}),$$

and

$$\mu'(\beta^{(1)}) = \cdots = \mu'(\beta^{(m)}) = \mu(\beta^{(1)}).$$

M.14.1. We have

$$H \smile^{\boldsymbol{C}^{-1}} \mathbf{H} \smile^{\boldsymbol{C}'} H';$$

hence

$$H \smile^{\boldsymbol{C}'\boldsymbol{C}^{-1}} H'.$$

The correspondence $\boldsymbol{C}'\boldsymbol{C}^{-1}$ is an isometric isomorphism from H onto H'.

We shall prove the following lemma:

If

$$F(\dot{p}) \cdot G(\dot{p}) =^{\mu} 0, \quad F(\dot{p}) = G(\dot{p}) = 0$$

for all

$$\dot{p} \in \mathrm{Point}(\mathrm{co}\,\alpha),$$

$$F(\dot{p}) \smile^{\boldsymbol{C}'\boldsymbol{C}^{-1}} F'(\dot{p}), \quad G(\dot{p}) \smile^{\boldsymbol{C}'\boldsymbol{C}^{-1}} G'(\dot{p}),$$

then

$$\overline{F'(z_1)}\, G'(z_1) + \cdots + \overline{F'(z_m)} \cdot G'(z_m) =^{N} 0$$

for $z \in \alpha$, where $z_1, z_2, \ldots, z_m$ are the successive copies of the complex number z, on $P_1, P_2, \ldots, P_m$.

Proof. Let us remember that, if the space $\boldsymbol{a}$ belongs to the borelian spectral tribe set s_b, then $\boldsymbol{C}$ carries $\boldsymbol{a}$ into the set of all μ-square summable functions $F(\dot{p})$, which vanish almost μ-everywhere on $\mathrm{Point}(\mathrm{co}\,\gamma)$, where γ is a certain N-measurable subset of P. Similarly $\boldsymbol{C}'$ carries $\boldsymbol{a}$ into the set if all μ'-square summable functions $F'(\dot{p})$, which vanish almost μ'-everywhere on $\mathrm{Point}(\mathrm{co}\,\gamma')$, where γ' is a certain N-measurable subset of P, and conversely. Now, we know that $\gamma' =^{N} \gamma$. Hence, if γ is an N-measurable subset of α, then

$$[(\mathrm{Point}\,\gamma) \upharpoonleft F(\dot{p})^{V}] \smile^{\boldsymbol{C}'\boldsymbol{C}^{-1}} [(\mathrm{Point}\,\gamma) \upharpoonleft F'(\dot{p})]^{V},$$

where the upper index means, that the restricted functions should be completed by putting them $=0$ on the remaining points of V. A similar relations holds for $G(\dot{p})$.

We have

$$\int_P \overline{[(\text{Point}\,\gamma) \upharpoonleft F(\dot{p})]} \cdot [(\text{Point}\,\gamma) \upharpoonleft G(\dot{p})]\, d\mu = 0,$$

i.e.

$$\int_{\text{Point}\,\gamma} \overline{F(\dot{p})} \cdot G(\dot{p})\, d\mu = 0.$$

Since $\boldsymbol{C}' \boldsymbol{C}^{-1}$ is isometric, we get

$$\int_{\text{Point}\,\gamma} \overline{F'(\dot{p})} \cdot G'(\dot{p})\, d\mu' = 0. \tag{7}$$

Since there are only m mattering copies of P, viz. $P_1, P_2, \dots, P_m$, and since all others have the μ-measure $=0$, (7) can be written

$$\sum_{k=1}^{m} \int_{\gamma_k} \overline{F'(\dot{z}_k)}\, G'(\dot{z}_k)\, d\mu' = 0,$$

where $z_1, z_2, \dots, z_m$ are copies of the complex number z.

Since the measure μ' on the copies γ_k is the same, it follows that

$$\int_{\gamma} \left[\sum_{k=1}^{m} \overline{F'(\dot{z}_k)} \cdot G'(\dot{z}_k)\right] d\mu' = 0, \tag{8}$$

where μ' is the measure μ assigned on α by putting

$$\mu'(\gamma) =_{df} \mu'(\gamma^{(1)})$$

for every N-measurable subset γ of α.

Since

$$\sum_{k=1}^{m} \overline{F'(\dot{z}_k)} \cdot G'(\dot{z}_k)$$

can be considered as a function of z, defined μ'-almost everywhere on α, and since γ is an arbitrary μ'-measurable subset of α, it follows from (8), that

$$\sum_{k=1}^{m} \overline{F'(\dot{z}_k)} \cdot G'(\dot{z}_k) = 0$$

for almost μ'-all z. Since the class of all N-null subsets of α coincides with the class of all μ'-null subsets of α, we can say that:

$$\sum_{k=1}^{m} \overline{F'(\dot{z}_k)} \cdot G'(\dot{z}_k) = 0$$

almost N-everywhere on α, q.e.d.

M.14.2. Having this consider the functions

$$F_1(\dot{p}), F_2(\dot{p}), \dots, F_q(\dot{p}),$$

defined almost μ-everywhere on V and such that

$$F_k(\dot{p}) = 1 \quad \text{on } P_k \text{ and } = 0 \text{ on } P_l \text{ where } \quad l \neq k,$$

$$(k = 1, 2, \ldots, q;\ l = 1, 2, \ldots, n, \ldots).$$

Let

$$F_k(\dot{p}) \smile^{\mathbf{C}'\mathbf{C}^{-1}} F'_k(\dot{p}).$$

Since if $k_1 \neq k_2$, we have $F_{k_1}(\dot{p}) \cdot F_{k_2}(\dot{p}) = 0$ on P. It follows:

$$\sum_{k=1}^{m} \overline{F'_{k_1}(z_k)} \cdot F_{k_2}(z_k) \overset{N}{=} 0.$$

Denote by $M(k_1, k_2)$ the set of all z, for which

$$\sum_{k=1}^{m} \overline{F'_{k_1}(z_k)} \cdot F_k(z_k) = 0.$$

The set $\alpha - M(k_1, k_2)$ is an N-null set.

M.14.3. The systems of numbers

$$(9) \qquad [F'_k(z_1), F'_k(z_2), \ldots, F'_k(z_m)],$$

will be considered as vectors, forming for a given z an $m =$ dimensional vector space with Hermitean scalar product. We shall show that for N-almost all $z \in \alpha$, the vector (9) is not a null vector.

Indeed, suppose that there exists a N-measurable subset β of α, on which (9) is a null vector, i.e. that

$$F'_k(z_1) = F'_k(z_2) = \cdots = F'_k(z_m) = 0.$$

It follows that

$$F'_k(\dot{p}) = 0 \quad \text{for all } p, \text{ where } \quad \operatorname{Num} \dot{p} \in \beta;$$

hence

$$(\operatorname{Point}\beta) \upharpoonleft F'_k(\dot{p}) = 0 \quad \text{on} \quad V \frown (\operatorname{Point}\beta).$$

Consequently, by the mentioned property of $\mathbf{C}'\mathbf{C}^{-1}$:

$$(\operatorname{Point}\beta) \upharpoonleft F_k(\dot{p}) = 0 \quad \text{almost } \mu\text{-everywhere on} \quad P \frown (\operatorname{Point}\beta).$$

This, however, contradicts the hypothesis that

$$F_k(\dot{p}) = 1 \quad \text{on} \quad P_k \frown (\operatorname{Point}\alpha).$$

M.14.4. Having this, denote by $M(k)$ the set of all z, for which the vector (9) does not vanish. The set $\alpha - M(k)$ is an N-null set.

Put

$$M = \bigcap_{k=1}^{m} M(k) \bigcup_{k_1 \neq k_2} M(k_1, k_2).$$

The set $\alpha - M$ is also an N-null set.

For every $z \in M$ and $k = 1, 2, \ldots, q$

$$[F'_k(z_1), F'_k(z_2), \ldots, F'_k(z_m)]$$

is not a zero vector; and we have

$$\sum_{k=1}^{m} \overline{F'_{k_1}(z_k)} \cdot F'_{k_2}(z_k) = 0$$

for all $z \in M$ and

$$k_1 \neq k_2, \quad (k_1, k_2 = 1, 2, \ldots, q).$$

Since M is not an N-null set, it follows that $M \neq \emptyset$. Take $z \in M$. We have p mutually orthogonal non zero vectors

$$[F'_k(z_1), F'_k(z_2), \ldots, F'_k(z_m)]$$

in the m-dimensional H.-H.-space, where $m < p$. This is impossible. Thus we have proved that the supposition that $m < n$ leads to a contradiction. It follows that $m = n$, so the theorem is proved.

Thus we have proved the following theorem on the invariance of the multiplicity.

M.15. Theorem. If N is a maximal normal operator, $\{\mathbf{H}_n\}$, $\{\mathbf{H}'_n\}$ two saturating sequences of the borelian spectral tribe s_b, $\vec{\omega}$, $\vec{\omega}'$ two generating vectors of $\mathbf{H}$ with respect to the saturated tribes s_b^*, $s_b^{*\prime}$ and $\boldsymbol{C}$, $\boldsymbol{C}'$ the corresponding canonical mappings, $\alpha_1, \alpha_2, \ldots, \alpha_\omega$ and $\alpha'_1, \alpha'_2, \ldots, \alpha'_\omega$ subsets of P with constant multiplicity $1, 2, \ldots, \infty$, then

$$\alpha_1 =^N \alpha'_1, \ldots, \alpha_\omega =^N \alpha'_\omega.$$

Thus the notion of multiplicity depends on N only.

M.15.1. The above invariance theorem considers the image of the normal operator. Now we can percieve its intrinsic meaning.

Let $\alpha_1, \alpha_2, \ldots, \alpha_\omega$ be the sets as before. The spaces

$$s(\alpha_1), s(\alpha_2), \ldots, s(\alpha_\omega)$$

are mutually orthogonal spaces, spanning the whole space $\mathbf{H}$. They are compatible with the spectral tribe, and they do not depend on the choice of the canonical mapping $\boldsymbol{C}$. They belong to the borelian spectral tribe s_b. Now the spaces

$$s(\alpha_k^{(1)}), s(\alpha_k^{(2)}), \ldots, s(\alpha_k^{(k)}), \quad (k = 1, 2, \ldots),$$

where

$$\alpha_k^{(1)}, \alpha_k^{(2)}, \ldots, \alpha_k^{(k)}$$

are the mattering copies of α_k, span $s(\alpha_k)$. They are orthogonal to one another, and if α_k is not an N-null set, they are not reduced to a single vector.

If we take such a generating vector, which attaches to each of the sets $\alpha_k^{(1)}, \alpha_k^{(2)}, \ldots, \alpha_k^{(k)}$ the same measure (as in the forgoing proof of the invariance theorem), we see, that the normal operators

$$(1) \qquad s(\alpha_k^{(1)}) \restriction N, s(\alpha_k^{(2)}) \restriction N, \ldots, s(\alpha_k^{(k)}) \restriction N,$$

have all constant, homogeneous multiplicity 1, and one of them will be obtained from an other by a unitary transformation.

Thus the saturation of the tribe s_b consists in splitting $s(\alpha) \restriction N$ into similar normal and simple operators (1)[1], and the invariance theorem says, that up to a unitary transformation of $s(\alpha) \restriction N$ into itself, the number of those simple operators, into which $s(\alpha) \restriction N$ can be split, depends only on N and α. Those α enumerated as $\alpha_1, \alpha_2, \ldots, \alpha_n, \ldots$ are also well determined by N. If the infinite sequence

$$s(\alpha_\omega^{(1)}), s(\alpha_\omega^{(2)}), \ldots$$

is available, we have the same, because we can attach to any two copies $\alpha_\omega^{(s)}, \alpha_\omega^{(t)}$ the same measure. Thus the notion of multiplicity has in our theory a visual interpretation.

Parts of $s(\alpha)$, having the form $s(\beta)$, where $\beta \subsetneqq \alpha$, behave in the same manner. The behaviour of $s(\gamma)$, where $\gamma \frown \alpha_{k_1} \mathrel{+\!\!\!=^N} \emptyset$ and $\gamma \frown \alpha_{k_2} \mathrel{+\!\!\!=^N} \emptyset$ for two differing k_1, k_2, is different, since $s(\gamma) \restriction N$ cannot be split into mutually similar parts.

M.16. Theorem. Two saturated tribes $s_b^*, s_b^{*\prime}$ are isomorphic in the sense that there exists a unitary transformation T of the space $\mathbf{H}$, which carries in a one-to-one way s_b^* into s_b^* with preservation of finite and denumerable lattice operations.

Proof. This follows immediately from the fact, that the terrace configurations corresponding to s_b^* and $s_b^{*\prime}$ are identical.

Chapter N

On items of operational calculus with application to the resolvent and spectrum of normal operators

We take over the notions defined in Chapter [J I.]. The present is its continuation, or rather a supplement to the theory of the operator $\varphi(N)$, where $\varphi(z)$ is a N-measurable complex-valued function of the complex variable z, defined on an N-measurable subset of the complex

[1] i.e. with saturated borelian spectral tribe.

plane P, and where N is a maximal normal operator in a Hilbert-Hermite-space $\boldsymbol{H}$. Our purpose is to study the most general operators $\psi(\varphi(N))$, $\varphi(\psi(N))$ and $(\varphi\,\psi)\,(N)$ — which are not identical and need a quite subtle reasonings — an dapply the results to the general theory of the eigenvalue problem, of the resolvent and of the spectrum of N. We shall show the great adaptability of our theory to that topic, and we shall get a classification of points of the spectrum of normal maximal operators, which seems to be more subtle, than the usual one, (87).

We shall be quite explicit in arguments, because the topic, spoken of, is just barely-touched in [J I.]. We shall take care of our argument to be valid even in the case, when $\boldsymbol{H}$ is confined to the single vector $\vec{O}$, and also in the case, where the operator N is trivial (denoted by O), i.e. carrying every vector of the space $\boldsymbol{H}$ into the null-vector. Indeed, the above is needed in many circumstances. Our notations differ slightly from those used in [J I.]; they will be, however, explained in the sequal. If we like to emphasize that a letter, say p, is a variable, we shall provede it with a dot: $\dot{p}$.

N.1. Let $\boldsymbol{H}$ be a separable and complete Hilbert-Hermite-space (H.-H.-space), which may be finite or infinite dimentional, or even be confined to the single vector $\vec{O}$. Let N be a maximal normal operator in $\boldsymbol{H}$, which may be even trivial, i.e. carrying every of $\boldsymbol{H}$ into $\vec{O}$. The spectral tribe of N will be denoted by T_N, and its borelian spectral tribe by T_N^B. The tribe T_N^B may be saturated or not.

Whatever the case may be, there exists an infinite sequence $\{\boldsymbol{H}_n\}$ of mutually disjoint and mutually compatible spaces, which are also compatible with all spaces of T_N^B, and such that

1) the tribe obtained from T_N^B by the adjunction of all $\boldsymbol{H}_n$ is saturated,

2) $\sum_{n=1}^{\infty} \boldsymbol{H}_n = \boldsymbol{H}$. The sequence H_n are termed *saturating sequence.*

If we consider the most general case, some of $\boldsymbol{H}_n$ may be confined to the single vector $\vec{O}$. The saturated tribe will be denoted by $\bar{T}_N^B$. Let $\vec{\omega}$ be a generating vector of $\boldsymbol{H}$ with respect to $\bar{T}_N^B$, and such that $\vec{\omega} \in \mathrm{C\!\!\!\!\:\rm{|}}\, N$, (domain of N). The measure $\mu(a)$, defined for all spaces $a \in \bar{T}_N^B$, by

(1) $$\mu(a) =_{df} ||\operatorname{Proj}_a \vec{\omega}||^2,$$

is denumerably additive, finite, non negative and effective[1].

The saturating sequence and the generating vector determine a canonical mapping $\boldsymbol{C}$ of $\boldsymbol{H}$ onto the H.-H.-space $\mathfrak{H}$ whose vectors are complex valued μ-square summable functions $X(\dot{p})$, defined almost μ-everywhere (a.e.) on the disjoint union V of an infinite denumerable

[1] Even if the tribe is trivial i.e. its unit and zero coincide, the measure is effective, because $\mu(a) = 0$ implies $a = \boldsymbol{0}$.

number of exact copies P_n of the complex plane P. The measure μ on V is induced by (1)[1], and the μ-measurable subsets of V and the μ-measurable functions are taken modulo the ideal J_μ of subsets of V with μ-measure equal 0.

We shall use the sign $=^\mu$ to denote the corresponding equality of sets and functions. The canonical mapping is isometric, homeomorphic and isomorphic. It carries the operator $N(\vec{X})$ into the operator

$$\mathfrak{N}(X(\dot{p})) =^\mu (\mathrm{Num}\,\dot{p}) \cdot X(\dot{p}), \tag{2}$$

where $\boldsymbol{C}$ carries $\vec{X}$ into $X(\dot{p})$ and where the domain of (2) is maximal, which means, that $\mathrm{O}\,\mathfrak{N}$ is composed of all functions $X(\dot{p}) \in \mathfrak{H}$, such that $(\mathrm{Num}\,\dot{p})\, X(\dot{p}) \in \mathfrak{H}$. We refer to [H.16.13.].

N.2. Let $a \in \bar{T}_N^B$. The $\boldsymbol{C}$-image of a vector $\vec{X} \in a$ is a function $X(\dot{p})$, which vanishes μ-a.e.-on a μ-measurable subset b of V, the same for all vectors $\vec{X} \in a$. The complement $a =_{df} V - b$ is termed $\hat{C}$-image of a. The correspondence $\hat{C}$ is generated by $\boldsymbol{C}$, and conversely. $\hat{C}$ will be termed the correspondence associate to $\boldsymbol{C}$. It transforms spaces, belonging to the saturated tribe $\bar{T}_N^B$, into μ-measurable subsets of V, taken modulo J_μ. The correspondence $\hat{C}$ is an isometric isomorphism, preserving finite and denumerable somatic operations.

N.3. The measure μ on $\bar{T}_N^B$ induces a denumerably additive measure, on the complex plane P, which corresponds to the measure on T_N^B. The corresponding measurable sets constitute an assemblage, which does not depend on the choice of the canonical correspondence, but only on N. We call the sets *N-measurable*. J_N will denote the ideal of all N-measurable subsets of P with measure 0. The corresponding equality of sets will be denoted $(=^N)$.

In what follows the μ-measurable subsets of V will always be taken modulo J_μ, and the N-measurable subsets of P will be taken modulo J_N, and the same will be for corresponding functions $\varphi(\dot{z})$ and $X(\dot{p})$.

N.4. If $\varphi(\dot{z})$ is an N-measurable, complex valued function, defined N-a.e. on P, then the operator $\varphi(N)$ is defined as the $\mathbf{C}^{-1}$-image of the operator

$$[\varphi(N)]\, X(\dot{p}) =^\mu \cdot \varphi(\mathrm{Num}\,\dot{p}) \cdot X(\dot{p}),$$

taken with maximal domain. $\varphi(N)$ does not depend on the choice of the canonical mapping $\mathbf{C}$.

N.5. It will be useful to denote by $\mathfrak{H}^V(a)$ the H.-H.-space composed of all μ-square summable functions $X(\dot{p})$ defined μ-a.e. in V, but

[1] The n-th copy P_n of P is the set of all ordered couples (z, n) where $n = 1, 2, \ldots$ and z is a complex number. We define $\mathrm{Num}(z, n) =_{df} z$. If $E \subsetneqq P$, then by Point$E$ we understand the set $\bigcup_{n=1}^{\infty} \{(z, n) \mid z \in E\}$. It is a subset of $V = \bigcup_{n=1}^{\infty} P_n$.

vanishing μ-a.e. on $\text{co}_V a = V - a$. Here a is supposed to be a μ-measurable subset of V. Its $\mathbf{C}^{-1}$-image, also denoted by $a = \hat{C}^{-1} a$ is a space of $\bar{T}_N^B$.

N.6. We shall define $\varphi(N)$ in the case, where $\varphi(z)$ is defined only on a N-measurable subset E of P.

Denote by $\varphi^P(\dot{z})$ the function defined on P by

$$\varphi^P(z) = \begin{cases} \varphi(z) & \text{for } z \in E, \\ 0 & \text{for } z \bar{\in} E. \end{cases}$$

We can prove that the operator

$$e \rceil \varphi^P(N),^{*)},$$

where

$$e =_{df} \hat{C}^{-1}(\text{Point} E),$$

does not depend on the choice of $\{\boldsymbol{H}_n\}$, neither on $\vec{\omega}$, i.e. it does not depend on $\mathbf{C}$, but only on N and φ.

We define $\varphi(N) =_{df} e \rceil \varphi^P(N)$. We can prove that

$$\varphi(N) = \varphi(e \rceil N).$$

The above takes place, even if $\boldsymbol{H}$ is confined to the single vector $\vec{O}$ and also if $E =^N \emptyset$. The operator $\varphi(N)$ operates in e only, i.e. it carries some vectors of e into vectors of e. Hence the $\mathbf{C}$-image of $\varphi(N)$ operates on functions $X(p)$ belonging to $\mathfrak{H}^V(\text{Point} E)$.

Remark. (It does not operate on functions defined only on Point E.) Consequently the $\mathbf{C}$-image of $\varphi(N)$ is the operator

$$(1) \qquad [\varphi(N)]\, X(p) =^\mu \varphi^P(\text{Num}\, \dot{p}) \cdot X(\dot{p}),$$

with maximal domain in $\mathfrak{H}^V(\text{Point} E)$. Thus the domain is composed of all functions of $\mathfrak{H}^V(\text{Point} E)$, such that

$$\varphi^P(\text{Num}\, \dot{p})\, X(\dot{p}) \in \mathfrak{H}^V(\text{Point} E).$$

One can also say, that (1) is restricted to $\mathfrak{H}^V(\text{Point} E)$. We shall study these operators.

N.7. Theorem. If

1) E is an N-measurable subset of the plane P,

2) $\varphi(z)$ is defined in E only,

3) $\varphi(z) = 0$ on E,

then $\varphi(N) = e \rceil \mathbf{0}$, where $e = \mathbf{C}^{-1} \mathfrak{H}^V(\text{Point} E)$.

This operator carries every vector of e into $\vec{O}$.

N.8. Theorem. If

1. E_1, E_2 are N-measurable subsets of P,

2. $E_1 \cap E_2 \cdot =^N \cdot \emptyset, \quad E =_{df} E_1 \cup E_2$,

---

*) $e \rceil$ means "restricted to the space e".

3. $\varphi(z)$ is N-measurable and defined only on $E_1 \cup E_2 = E$,
4. $\varphi_1(z) =_{df} E_1 \upharpoonright \varphi(z)$, $\varphi_2(z) =_{df} E_2 \upharpoonright \varphi(z)$,
5. $e_1 =_{df} \mathbf{C}^{-1} \mathfrak{H}^V(\mathrm{Point}\, E_1)$, $e_2 =_{df} \mathbf{C}^{-1} \mathfrak{H}^V(\mathrm{Point}\, E_2)$,

then

1) $\mathsf{D}\, \varphi(N) = \mathsf{D}\, \varphi_1(N) \overrightarrow{+} \mathsf{D}\, \varphi_2(N)$, which means:

$$\mathsf{D}\, \varphi(N) = \{\vec{X}_1 + \vec{X}_2 \mid \vec{X}_1 \in \mathsf{D}\, \varphi_1(N),\ \vec{X}_2 \in \mathsf{D}\, \varphi_2(N)\}.$$

2) $\varphi(N) = \varphi_1(N) \overrightarrow{+} \varphi_2(N)$, which means:
for every

$$\vec{X} \in \mathsf{D}\, \varphi(N)$$

we have

$$\varphi(N)\, \vec{X} = \varphi_1(N)\, \vec{X}_1 + \varphi_2(N)\, \vec{X}_2,$$

where

$$\vec{X}_1 =_{df} \mathrm{Proj}_{e_1} \vec{X}, \qquad \vec{X}_2 =_{df} \mathrm{Proj}_{e_2} \vec{X}.$$

This is the "smallest" linear operator containing $\varphi_1(N)$ and $\varphi_2(N)$. We omit the proof, which is rather straight forward and does not need special artifices.

N.9. Theorem. If

1. E, F are N-measurable subsets of P,
2. $E \subseteqq F$,
3. $\varphi(z)$ is defined only on E.
4. $\Psi(z)$ is defined only on F and is N-measurable,
5. $\varphi(z) = 0$ on E,

then

$$e \upharpoonright \Psi(N) = \Psi(N) + \varphi(N),^{*)}$$

where the addition is the addition of operators.

N.10. Remark. If $e_1, e_2 \in \bar{T}_N^B$, then

$$e_1 \upharpoonright \mathbf{0} + e_2 \upharpoonright \mathbf{0} = (e_1 + e_2) \upharpoonright \mathbf{0},$$
$$e_1 \upharpoonright \mathbf{0} + e_2 \upharpoonright \mathbf{0} = (e_1\, e_2) \upharpoonright \mathbf{0}.$$

It is the smallest linear operator, containing $e_1 \upharpoonright \mathbf{0}$ and $e_2 \upharpoonright \mathbf{0}$.

N.11. Theorem. Suppose that:

1. E_1, E_2 are N-measurable subsets of the plane P.
2. $E_1 \subseteqq E$.
3. $\varphi(\dot{z})$ is N-measurable and defined only on E.
4. $\varphi(z) = 0$ on $E - E_1$.
5. $e =_{df} \mathbf{C}^{-1} \mathfrak{H}^V(\mathrm{Point}\, E)$, $e_1 =_{df} \mathbf{C}^{-1} \mathfrak{H}^V(\mathrm{Point}\, E_1)$.
6. $\varphi_1(\dot{z}) =_{df} E_1 \upharpoonright \varphi(\dot{z})$.

*) According to definition admitted, the domain of $\Psi(N) + \varphi(N)$ is $\mathsf{D}\, \Psi(N) \cap \mathsf{D}\, \varphi(N)$.

Then

1) the domain of $\varphi(N)$ is $\complement\,\varphi_1(N) \overset{\rightarrow}{+} (e - e_1)$.

2) $\varphi_1(N) = e_1 \upharpoonleft \varphi(N)$.

3) $\varphi(N) \equiv \varphi_1(N) \overset{\rightarrow}{+} (e - e_1) \upharpoonleft \mathbf{0}$.

Proof. Since $E_1 \frown (E - E_1) = \emptyset$, we get by [N.8.]:

$$\complement\,\varphi_1(N) \overset{\rightarrow}{+} \complement[(E - E_1) \upharpoonleft \varphi]\, N = \complement\,\varphi(N).$$

Since $\varphi(z) = 0$ on $E - E_1$, we get by [N.7.]

$$\complement[(E - E_1) \upharpoonleft \varphi]\,(N) = e - e_1.$$

Hence

(1) $$\complement\,\varphi(N) = \complement\,\varphi_1(N) \overset{\rightarrow}{+} (e - e_1).$$

By [N.8.] we also have

$$\varphi(N) = \varphi_1(N) \overset{\rightarrow}{+} [(E - E_1) \upharpoonleft \varphi]\,(N).$$

Hence by [N.7.]

(2) $$\varphi(N) = \varphi_1(N) \overset{\rightarrow}{+} (e - e_1) \upharpoonleft \mathbf{0}.$$

To prove item 2), we notice that the $\boldsymbol{C}$-image of $e_1 \upharpoonleft \varphi(N)$ is the restriction to $\mathfrak{H}^V(\text{Point}\,E_1)$ of the $\boldsymbol{C}$-image of the operator $\varphi(N)$. Hence, it is the restriction to $\mathfrak{H}^V(\text{Point}\,E_1)$ of the operator

$$[\varphi(N)]\,X(\dot{p}) \cdot =^{\mu} \cdot\, \varphi(\text{Num}\,\dot{p}) \cdot X(\dot{p}),$$

which is considered with maximal domain in $\mathfrak{H}^V(\text{Point}\,E)$. Now this restriction is identical with the operator, with maximal domain, which carries functions of $\mathfrak{H}^V(\text{Point}\,E_1)$ into $\varphi_1(\text{Num}\,\dot{p}) \cdot X(\dot{p})$. But the last operator is the $\boldsymbol{C}$-image of $(E_1 \upharpoonleft \varphi)\,(N) = \varphi_1(N)$. Consequently $e_1 \upharpoonleft \varphi(N) = \varphi_1(N)$, q.e.d.

N.12. To simplify the symbolism, denote in general by $D^\bullet$ the set Point$\,D$, where D is an N-measurable subset of the complex plane P. In addition to that we shall use the signs "+" and "·" for union and intersection of sets. The equality for N-measurable subsets of P will be $=^N$, and the sets will be taken modulo J_N. The equality for μ-measurable subsets of V will be $=^\mu$, and the sets will be taken modulo J_μ. We shall use the ordinary sign "=" of equality in both cases, unless we would like to emphasise the character of the equality: then we shall write $=^\mu$ or $=^N$. Under these agreements we have for N-measurable sets $\emptyset^\bullet = \emptyset$ on V, $P^\bullet = V$, $(E + F)^\bullet = E^\bullet + F^\bullet$, $(E \cdot F)^\bullet = E^\bullet \cdot F^\bullet$, $(E - F)^\bullet = E^\bullet - F^\bullet$, $(\text{co}_P E)^\bullet = \text{co}_V E^\bullet$. Thus the sets D and $D^\bullet$ are operation-isomorphic.

N.13. Let A, B, A_1, B_1 be N-measurable subsets of P, where $A_1 \subseteqq A$, $B_1 \subseteqq B$. Let $\varphi(\dot{z})$, $\Psi(\dot{z})$ be N-measurable complex valued functions of the complex variable $\dot{z}$, defined only on A, B respectively.

We suppose that

$$\varphi(z) \neq 0 \quad \text{for} \quad z \in A_1, \qquad \Psi(z) \neq 0 \quad \text{for} \quad z \in B_1,$$
$$\varphi(z) = 0 \quad \text{for} \quad z \in A - A_1, \qquad \Psi(z) = 0 \quad \text{for} \quad z \in B - B_1.$$

We shall consider the operators

$$P =_{df} \varphi(N), \qquad Q =_{df} \Psi(N),$$

and shall study the operators

$$R\vec{X} =_{df} Q(P\vec{X}), \quad \text{and} \quad S\vec{X} =_{df} P(Q\vec{X}).$$

N.13.1. The ***C***-image of the operator P is

(1) $$P(X(\dot{p})) = \varphi^P(\operatorname{Num}\dot{p}) \cdot X(\dot{p}),$$

with maximal domain in $\mathfrak{H}^V(A^\bullet)$.

The ***C***-image of the operator Q is

(2) $$Q(X(\dot{p})) = \Psi^P(\operatorname{Num}\dot{p}) \cdot X(\dot{p}),$$

with maximal domain in $\mathfrak{H}^V(B^\bullet)$.

The ***C***-image of R is

$$R(X(\dot{p}) = Q[P(X(\dot{p})].$$

Its domain is composed of all functions $X(\dot{p})$, for which $P(X(\dot{p})) \in \text{ }\mathrm{d}\, Q$. More explicitly, $\mathrm{d}\, R$ is composed of all functions $X(\dot{p})$, for which

(3) $$P(X(\dot{p})) \in \mathfrak{H}^V(B^\bullet),$$

(4) $$\Psi^P(\operatorname{Num}\dot{p}) \cdot P(X(\dot{p}) \in \mathfrak{H}^V(B^\bullet).$$

Since $P(X(\dot{p}))$ should be meaningful, therefore we must have

$$X(\dot{p}) \in \mathfrak{H}^V(A^\bullet), \qquad \varphi^P(\operatorname{Num}\dot{p}) \cdot X(\dot{p}) \in \mathfrak{H}^V(A^\bullet).$$

Hence $\mathrm{d}\, R$ is composed of all functions $X(\dot{p})$ for which the following conditions are satisfied:

(5) $$\begin{cases} \text{(a)}\ X(\dot{p}) \in \mathfrak{H}, \\ \text{(b)}\ \varphi^P(\operatorname{Num}\dot{p}) \cdot X(\dot{p}) \in \mathfrak{H}, \\ \text{(c)}\ \Psi^P(\operatorname{Num}\dot{p})\, \varphi^P(\operatorname{Num}\dot{p})\, X(\dot{p}) \in \mathfrak{H}; \end{cases}$$

and

(6) $$\begin{cases} \text{(a)}\ X(p) =^\mu 0 \ \text{for}\ p \in \operatorname{co} A^\bullet, \\ \text{(b)}\ \varphi^P(\operatorname{Num} p) \cdot X(p) =^\mu 0 \ \text{for}\ p \in \operatorname{co} A, \\ \text{(c)}\ \Psi^P(\operatorname{Num} p)\, \varphi^P(\operatorname{Num} p)\, X(p) =^\mu 0 \ \text{for}\ p \in \operatorname{co} B^\bullet, \\ \text{(d)}\ \varphi^P(\operatorname{Num} p)\, X(p) =^\mu 0 \ \text{for}\ p \in \operatorname{co} B^\bullet. \end{cases}$$

N.13.2. We shall prove that, under hypotheses (13), the conditions (6) are equivalent to the following one:

(7) $X(p) =^{\mu} 0$ for all $p \in [\mathrm{co}(A \cdot B) - (A - A_1)]^{\bullet}$.

Proof. Suppose that (7) holds true. Since

$$\varphi^P(\mathrm{Num}\, p) = 0 \quad \text{in} \quad A^{\bullet} - A_1^{\bullet},$$

it follows that

(8) $$\varphi^P(\mathrm{Num}\, p) \cdot X(p) = 0 \quad \text{in} \quad A^{\bullet} - A_1^{\bullet}.$$

Since $X(p) = 0$ in $\mathrm{co}(A^{\bullet} \cdot B^{\bullet}) - (A^{\bullet} - A_1^{\bullet})$, we have

(9) $$\varphi^P(\mathrm{Num}\, p) \cdot X(p) = 0 \quad \text{in} \quad \mathrm{co}(A^{\bullet} \cdot B^{\bullet}) - (A^{\bullet} - A_1^{\bullet}).$$

From (8) and (9) we have

$$\varphi^P(\mathrm{Num}\, p) \cdot X(p) = 0 \quad \text{in} \quad \mathrm{co}A^{\bullet} + \mathrm{co}B^{\bullet} - (A^{\bullet} - A_1^{\bullet}).$$

Consequently

$$\varphi^P(\mathrm{Num}\, p) \cdot X(p) = 0 \quad \text{in} \quad \mathrm{co}A^{\bullet},$$

i.e. we have (6) (b), and

$$\varphi^P(\mathrm{Num}\, p) \cdot X(p) = 0 \quad \text{in} \quad \mathrm{co}B^{\bullet},$$

i.e. we have (6) (d). Hence

$$\Psi^P(\mathrm{Num}\, p) \cdot \varphi^P(\mathrm{Num}\, p) \cdot X(p) = 0 \quad \text{in} \quad \mathrm{co}B^{\bullet},$$

i.e. we have (6) (c). From hypothesis (7) we have

$$X(p) = 0 \quad \text{for} \quad p \in (\mathrm{co}A^{\bullet} + \mathrm{co}B^{\bullet})\, \mathrm{co}(A^{\bullet}\, \mathrm{co}A_1^{\bullet})$$
$$= (\mathrm{co}A^{\bullet} + \mathrm{co}B^{\bullet}) \cdot (\mathrm{co}A^{\bullet} + A_1^{\bullet}) \supseteqq \mathrm{co}A^{\bullet}.$$

Hence $X(p) = 0$ for $p \in \mathrm{co}A^{\bullet}$, i.e. we have (6) (a).

Thus we have proved that (7) implies (6).

N.13.2.1. Conversely, suppose that (6) holds true. We have $\varphi^P(z) = 0$ on $A - A_1$ and on $\mathrm{co}A$. Hence $\varphi^P(\mathrm{Num}\, p) = 0$ on $\mathrm{co}A^{\bullet}$ and $A^{\bullet} - A_1^{\bullet}$; hence on

(10) $$\mathrm{co}A^{\bullet} + (A^{\bullet} - A_1^{\bullet}) = \mathrm{co}A_1^{\bullet}.$$

It follows that $\varphi^P(\mathrm{Num}\, p) \cdot X(p) = 0$ on $\mathrm{co}A_1^{\bullet}$. On the other hand we have, by (6) (d),

$$\varphi^P(\mathrm{Num}\, p) \cdot X(p) = 0 \quad \text{on} \quad \mathrm{co}B^{\bullet}.$$

Hence

$$\varphi^P(\mathrm{Num}\, p) \cdot X(p) = 0 \quad \text{on} \quad \mathrm{co}(A_1^{\bullet} \cdot B^{\bullet}).$$

We have

(11) $$\mathrm{co}(A_1^{\bullet} \cdot B^{\bullet}) = \mathrm{co}(A_1^{\bullet} \cdot B^{\bullet}) \cdot A_1^{\bullet} + (\mathrm{co}A_1^{\bullet} \cdot B^{\bullet})\, \mathrm{co}A_1^{\bullet}.$$

The function $\varphi^P(\mathrm{Num}\,p)\cdot X(p)$ vanishes on both terms of (11), but, since $\varphi^P(\mathrm{Num}\,p) \neq 0$ on $A_1^\bullet$, therefore $X(p)$ must vanish on $\mathrm{co}\,(A_1^\bullet\cdot B^\bullet)\cdot A_1^\bullet$. By (6) (a) we have

$$X(p) = 0 \quad \text{on} \quad \mathrm{co}\,A^\bullet.$$

Consequently

$$\text{(11a)} \quad X(p) = 0 \quad \text{on} \quad \mathrm{co}\,(A_1^\bullet\cdot B^\bullet)\cdot A_1 + \mathrm{co}\,A^\bullet$$
$$= \mathrm{co}\,A_1^\bullet + (\mathrm{co}\,A_1^\bullet + \mathrm{co}\,B^\bullet)\cdot A_1 = \mathrm{co}\,A_1^\bullet + (A_1^\bullet - B^\bullet).$$

Now we can prove that

$$\text{(12)} \qquad \mathrm{co}\,A_1^\bullet + (A_1^\bullet - B^\bullet) = \mathrm{co}\,(A^\bullet\cdot B^\bullet) - (A^\bullet - A_1^\bullet),$$

just by taking the complements of both sides.

The complementary set is

$$\text{(13)} \qquad A^\bullet\cdot B^\bullet + (A^\bullet - A_1^\bullet)$$

for both. Thus we have proved, that (6) is equivalent to (7).

N.13.2.2. We have

$$A^\bullet B^\bullet + (A^\bullet - A_1^\bullet) = A_1^\bullet\cdot B^\bullet + (A^\bullet - A_1^\bullet),$$

where both terms are disjoint.

The domain of R is contained in $\mathfrak{H}^V[(A_1^\bullet\cdot B^\bullet) + (A^\bullet - A_1^\bullet)]$.

Let us consider the conditions (5) in [N.13.1.]. They can be replaced by similar conditions:

$$\text{(a')} \qquad X(p) \in \mathfrak{H}^V[(A_1^\bullet B^\bullet) + (A^\bullet - A_1^\bullet)],$$

$$\text{(b')} \qquad X(p)\cdot\varphi^P(\mathrm{Num}\,p) \in \mathfrak{H}^V(A_1^\bullet B^\bullet) + \mathfrak{H}^V(A^\bullet - A_1^\bullet),$$

$$\text{(c')} \quad X(p)\cdot\Psi^P(\mathrm{Num}\,p)\,\varphi^P(\mathrm{Num}\,p) \in \mathfrak{H}^V(A_1^\bullet B^\bullet) + \mathfrak{H}^V(A^\bullet - A_1^\bullet).$$

Since

$$\varphi^P(z) = 0 \quad \text{on} \quad A - A_1$$

the condition (b') is equivalent to

$$X(p)\,\varphi^P(\mathrm{Num}\,p) \in \mathfrak{H}^V(A_1^\bullet B^\bullet).$$

Concerning (c') we have

$$A_1 B = A_1 B_1 + A_1(B - B_1),$$

where the terms are disjoint. Hence

$$A_1 B + (A - A_1) = A_1 B_1 + A_1(B - B_1) + (A - A_1).$$

Since

$$\Psi^P = 0 \quad \text{on} \quad B - B_1,$$

and

$$\varphi^P = 0 \quad \text{on} \quad A - A_1$$

the condition (c′) is equivalent to

$$X(p)\,\varphi^P(\mathrm{Num}\,p)\,\Psi^P(\mathrm{Num}\,p) \in \mathfrak{H}^V(A_1^\bullet B_1^\bullet).$$

Thus the conditions (5) and (6) in [N.13.1.] are equivalent to the following ones:

$$(14)\quad \begin{cases} X(\dot p) \in \mathfrak{H}^V[A^\bullet B^\bullet + (A^\bullet - A_1^\bullet)], \\ X(\dot p)\,\varphi^P(\mathrm{Num}\,\dot p) \quad \text{is } \mu\text{-square summable on } A_1^\bullet B^\bullet, \\ X(\dot p)\,\varphi^P(\mathrm{Num}\,\dot p)\,\Psi^P(\mathrm{Num}\,\dot p) \quad \text{is } \mu\text{-square summable on } A_1^\bullet B_1^\bullet. \end{cases}$$

Since the set $A\,B + (A - A_1)$ is the union of the following three disjoint parts:

$$A_1\,B_1 + A_1(B - B_1) + (A - A_1)$$

and

$$A_1\,B = A_1\,B_1 + A_1(B - B_1),$$

we can say, that the domain of R is composed of all functions $X(p)$, belonging to $\mathfrak{H}^V[A^\bullet B^\bullet + (A^\bullet - A_1^\bullet)]$, for which $X(\dot p)\cdot\varphi^P(\mathrm{Num}\,p)$ is μ-square summable on $A_1^\bullet B_1^\bullet$ and on $A_1^\bullet(B^\bullet - B_1^\bullet)$; and

$$X(p)\,\varphi^P(\mathrm{Num}\,p)\,\Psi^P(\mathrm{Num}\,\dot p)$$

is μ-square summable on $A_1^\bullet B_1^\bullet$.

Thus R can be decomposed into three parts:

1) The operator $\mathbf{0}$ in $\mathfrak{H}^V(A^\bullet - A_1^\bullet)$;

2) the operator $\mathbf{0}$ in $\mathfrak{H}^V A_1^\bullet((B^\bullet - B_1^\bullet))$, confined to functions belonging to $\complement[\varphi(N)]$;

3) the operator, carrying $X(p)$ into $\varphi^P(\mathrm{Num}\,\dot p)\,\Psi^P(\mathrm{Num}\,\dot p)\cdot X(p)$, confined to $\mathfrak{H}^V(A_1^\bullet B_1^\bullet)$ and to functions of $\complement[\varphi(N)]$.

The function $\xi(z) =_{df} \varphi(z)\cdot\Psi(z)$ is defined in $A\cdot B$ and does not vanish in $A_1\cdot B_1$. Let us denote the operator $\xi(N)$ by $(\varphi\,\Psi)(N)$. Its $\boldsymbol{C}$-image is the operator

$$[\xi(N)]\,X(\dot p) =^\mu \Psi^P(\mathrm{Num}\,\dot p)\,\varphi^P((\mathrm{Num}\,\dot p)\,X(\dot p)$$

with maximal domain $\mathfrak{H}^V(A^\bullet B^\bullet)$. Having that; and if we take the $\boldsymbol{C}^{-1}$-image, we get the following result, showing the decomposition of R into three parts:

$$(15)\quad \begin{aligned} \Psi(\varphi(N)) &= [a_1\,b_1 \frown \complement\,\varphi(N) \upharpoonright (\Psi\,\varphi)\,(N)] \vec{+}\, a_1(b - b_1) \frown \\ &\qquad \frown \complement\,\varphi(N) \upharpoonright \mathbf{0} \vec{+} (a - a_1) \upharpoonright \mathbf{0}, \\ \complement\,\Psi(\varphi(N)) &= [a_1\,b_1 \frown \complement\,\varphi(N) \frown \complement(\varphi\,\Psi)\,(N)] \vec{+} [a_1(b - b_1) \frown \\ &\qquad \frown \complement\,\varphi(N)] + (a - a_1). \end{aligned}$$

A similar result we have for $S = \varphi(\Psi(N))$:

$$
\begin{aligned}
&\varphi(\Psi(N)) = [a_1 b_1 \cap \complement\,\Psi(N)] \upharpoonleft (\Psi\,\varphi)\,(N) \overset{\to}{+} [b_1(a - a_1) \cap \\
&\qquad\qquad \cap \complement\,\Psi(N)] \upharpoonleft \mathbf{0} \overset{\to}{+} (b - b_1) \upharpoonleft \mathbf{0}, \\
(15\,\mathrm{a})\quad &\complement\,\varphi(|\,(N)) = [a_1 b_1 \cap \complement\,\Psi(N) \cap \complement(\varphi\,\Psi)\,(N)] \overset{\to}{+} [b_1(a - a_1) \cap \\
&\qquad\qquad \cap \complement\,\Psi(N)] \overset{\to}{+} (b - b_1).
\end{aligned}
$$

N.13.3. From the second term in (15) and (15a) we can see, that the operators $\varphi(\Psi(N))$ and $\Psi(\varphi(N))$ may be not maximal normal. We also see, that they differ. The operator $\varphi(\Psi(N))$ is acting in the spaces

$$a_1 b_1,\ a_1(b - b_1) \quad \text{and} \quad (a - a_1),$$

which are disjoint with one another. The operator $\Psi(\varphi(N))$ is acting in the spaces

$$a_1 b_1,\ b_1(a - a_1) \quad \text{and} \quad (b - b_1).$$

Later we shall show that the domains in these spaces are everywhere dense with respect to them.

Now we like to find the common operation-space for both $\varphi(\Psi(N))$ and $\Psi(\varphi(N))$. We shall examine the space

$$[a_1(b - b_1) + (a - a_1)] \cdot [b_1(a - a_1) + (b - b_1)].$$

A simple computation shows, that this space coincides with $a\,b - a_1 b_1$. Consequently the operators $\Psi(\varphi(N))$ and $\varphi(\Psi(N))$ operate in the common space

$$a_1 b_1 + (a\,b - a_1 b_1) = a\,b.$$

N.13.4. Let us see by what amount the operation space of $\Psi(\varphi(N))$ exceeds $a\,b$. The difference is

$$a_1 b + (a - a_1) - a\,b = a - a_1 - b.$$

Thus outside of the common part, with $\varphi(\Psi(N))$, the operation space of $\Psi(\varphi(N))$ has $a - a_1 - b$, as an additional space.

Thus $\varphi(\Psi(N))$ and $\Psi(\varphi(N))$ have the common space $a\,b$ and in addition to that $\Psi(\varphi(N))$ has the additional space $a - a_1 - b$ and $\varphi(\Psi(N))$ has the additional space $b - b_1 - a$.

Of course, all three spaces $a\,b$, $a - a_1 - b$, $b - b_1 - a$ are disjoint. In the additional spaces the operators are $a - a_1 - b \upharpoonleft \mathbf{0}$ and $b - b_1 - a \upharpoonleft \mathbf{0}$ respectively.

Concerning the common operation space $a \cdot b$, the said operators may differ in their domains. We have

$$
\begin{aligned}
(1)\quad &\Psi(\varphi(N)) = a \cdot b \upharpoonleft \Psi(\varphi(N)) \overset{\to}{+} (a - a_1 - b) \upharpoonleft \mathbf{0}, \\
&\varphi(\Psi(N)) = a \cdot b \upharpoonleft \varphi(\Psi(N)) \overset{\to}{+} (b - b_1 - a) \upharpoonleft \mathbf{0}.
\end{aligned}
$$

Indeed, from (15) and (15a) in [N.13.2.2.] we see, that the operators are null in the whole space $a - a_1$ and $b - b_1$ respectively.

To have another formulas for the combined operators, notice that $a \cdot b$ contains the part $(a - a_1)\, b$, where the operator $\Psi(\varphi(N))$ is null in the whole space $(a - a_1)\, b$. Indeed the operator is null on the whole space $a - a_1$, hence also in $(a - a_1)\, b$.

In the remainder $a\, b - (a - a_1)\, b = [a - (a - a_1)]\, b = a_1\, b$, the operator $\Psi(\varphi(N))$ may be split into two, working in the terms of the sum:

$$a_1\, b_1 + a_1 (b - b_1) = a_1\, b .$$

In $a_1 (b - b_1)$ it is a null operator, but not necessarely on the whole space but (see (15) in [N.13.2.2.]), on the everywhere dense set

$$\complement\, \varphi(N) \frown a_1 (b - b_1) .$$

Thus we can say that

$$\begin{aligned} &(2) \qquad \Psi(\varphi(N)) = a_1\, b \upharpoonleft \Psi(\varphi(N)) \overset{\rightarrow}{+} (a - a_1) \upharpoonleft \mathbf{0}, \\ &\qquad\quad\ \varphi(\Psi(N)) = a\, b_1 \upharpoonleft \varphi(\Psi(N)) \overset{\rightarrow}{+} (b - b_1) \upharpoonleft \mathbf{0}, \end{aligned}$$

and we also can say that $\varphi(\Psi(N))$ can be split into two parts, the first of which being $a_1\, b_1 \upharpoonleft \Psi(\varphi(N))$; and the second is a null operator with domain everywhere dense in $a_1 (b - b_1) + (a - a_1)$.

Similar results we have for $\varphi(\Psi(N))$.

N.13.5. We shall prove that our combined operators have their domains everywhere dense in the various spaces of their decomposition.

Lemma. If

1. $f(p)$ is a μ-square summable function on the set a,
2. $a > a_1 > a_2 > \cdots$ with $\mu(a_n) \to 0$,
3. $f_n(p) =_{df} \begin{cases} f(p) & \text{on } a - a_n, \\ 0 & \text{on } a_n, \end{cases}$

then

$$\int_a |f_n(p) - f(p)|^2\, d\mu \to 0 .$$

Proof.

$$\int_a |f_n(p) - f(p)|^2 = \int_{a - a_n} 0\, d\mu + \int_{a_n} |f(p)|^2\, d\mu \to 0,$$

because $\mu(a_n) \to 0$. Indeed the integral is a denumerably additive set function, which is μ-continuous.

Having this, suppose that the set $A_1\, B_1$ has a positive measure, and consider the functions $\varphi(z)$, $\Psi(z)$ as restricted to $A_1 \cdot B_1$. Put

$$E_n =_{df} \{z \,|\, n - 1 \leqq |\varphi(z)| < n\},$$

$$F_n =_{df} \{z \,|\, n - 1 \leqq |\Psi(z)| < n\} \quad \text{for} \quad n = 1, 2, \ldots$$

Since the sets E_n are disjoint and $\sum_{n=1}^{\infty} E_n = A_1 B_1$, it follows that $\mu\left(\sum_{n=k}^{\infty} E_n\right) \to 0$ for $k \to \infty$. Similarly we have $\mu\left(\sum_{n=k}^{\infty} F_n\right) \to 0$ for $k \to \infty$. It follows that

$$\mu D_k \to 0, \quad \text{where} \quad D_k =_{df} \sum_{n \geqq k \text{ or } m \geqq k} E_n F_m .$$

Take any μ-square summable function $Y(\dot{p})$, defined μ-a.e.-on $A_1^{\bullet} B_1^{\bullet}$. Define

$$Y_k(\dot{p}) =_{df} \begin{cases} Y(\dot{p}) & \text{on} \quad \mathrm{co}_V D_k^{\bullet}, \\ 0 & \text{on} \quad D_k^{\bullet}. \end{cases}$$

The functions

$$\varphi(\mathrm{Num}\,\dot{p}) \cdot Y_k(\dot{p}),$$

$$\Psi(\mathrm{Num}\,p) \cdot \varphi(\mathrm{Num}\,p) \cdot Y_k(p),$$

if restricted to $A_1^{\bullet} B_1^{\bullet}$, are μ-square summable; hence the $\boldsymbol{C}^{-1}$-image $\vec{Y}_k$ of $Y_k(\dot{p})$ belongs (by (15), [N.13.2.2.]) to

$$\mathrm{G}[a_1 b_1 \upharpoonright \Psi(\varphi(N))] = a_1 b_1 \cap \mathrm{G}\,\varphi(N) \cap \mathrm{G}(\varphi\,\Psi)\,N.$$

On the other hand, by the lemma, the functions $Y_k(\dot{p})$, if we vary $Y(\dot{p})$, make an everywhere dense set in $\mathfrak{H}^V(A_1^{\bullet} B_1^{\bullet})$. Consequently the set of vectors

$$a_1 b_1 \cap \mathrm{G}\,\varphi(N) \cap \mathrm{G}(\varphi\,\Psi)\,N$$

is everywhere dense in $a_1 b_1$. Similarly we prove that

$$a_1 b_1 \cap \mathrm{G}\,\Psi(N) \cap \mathrm{G}(\varphi\,\Psi)\,N$$

is everywhere dense in $a_1 b_1$. Consequently, in the decomposition formulas (15), (15a), the domains in all three decomposition spaces are everywhere dense, [N.13.2.2.].

The above will be general, if we agree to consider the set composed of the single vector $\vec{O}$ as everywhere dense in the space O.

N.13.6. Consider the decomposition formula (15) in [N.13.2.2.]:

$$(1) \qquad \Psi(\varphi(N)) = [a_1 b_1 \cap \mathrm{G}\,\varphi(N) \cap \mathrm{G}(\Psi\,\varphi)\,(N)] \upharpoonright (\Psi\,\varphi)\,(N) \dotplus$$
$$\dotplus [a_1(b - b_1) \cap \mathrm{G}\,\varphi(N)] \upharpoonright \mathbf{0} \dotplus (a - a_1) \upharpoonright \mathbf{0}.$$

Since the operator $(\Psi\,\varphi)\,N$, restricted to $a_1 b_1$, is maximal normal in $a_1 b_1$, the domain $\mathrm{G}(\varphi\,\Psi)\,(N) \cap a_1 b_1$ is everywhere dense in $a_1 b_1$. Since, by [N.13.5.], the set $a_1 b_1 \cap \mathrm{G}\,\varphi(N) \cap \mathrm{G}(\Psi\,\varphi)\,(N)$ is everywhere dense in $(a_1 b_1)$ and contained in $a_1 b_1 \cap \mathrm{G}(\varphi\,\Psi)\,(N)$, it follows that the operator

$$a_1 b_1 \upharpoonright (\Psi\,\varphi)\,(N)$$

is a maximal normal extension in $a_1 b_1$ of the first term in (1). Indeed $\complement\varphi(N) \frown \complement(\Psi\varphi)(N) \frown a_1 b_1$ is an everywhere dense part of $\complement(\Psi\varphi)(N)\, a_1 b_1$. Since $a_1(b - b_1) \frown \complement\varphi(N)$ is everywhere dense in $a_1(b - b_1)$, the operator $a_1(b - b_1) \upharpoonleft \mathbf{0}$ is a maximal normal extension in $a_1(b - b_1)$ of the second term in (1). Thus though $\Psi(\varphi(N))$ may not be a maximal normal operator, it has a maximal normal extension in its operation space, viz.

$$a_1 b_1 \upharpoonleft (\Psi\varphi)\, N \overset{\rightarrow}{+} a_1(b - b_1) \upharpoonleft \mathbf{0} \overset{\rightarrow}{+} (a - a_1) \upharpoonleft \mathbf{0},$$

i.e.

$$(a_1 b_1) \upharpoonleft (\Psi\varphi)(N) \overset{\rightarrow}{+} [a_1(b - b_1) \overset{\rightarrow}{+} (a - a_1)] \upharpoonleft \mathbf{0}.$$

Similarly the operator $\varphi(\Psi(N))$ has a maximal normal extension

$$(a_1 b_1) \upharpoonleft (\varphi\Psi)(N) \overset{\rightarrow}{+} [b_1(a - a_1) \overset{\rightarrow}{+} (b - b_1)] \upharpoonleft \mathbf{0}.$$

We see that even these maximal normal extensions are not identical. They have only a common part which may be not trivial:

$$(a_1 b_1) \upharpoonleft (\Psi\varphi)(N).$$

Concerning the operator $(\Psi\varphi)(N)$, we have, by [N.11.],

$$(\Psi\varphi)(N) = a_1 b_1 \uparrow (\Psi\varphi)(N) \overset{\rightarrow}{+} (a b - a_1 b_1) \upharpoonleft \mathbf{0}.$$

Hence, the mentioned normal maximal extension of all three operators $\Psi(\varphi(N))$, $\varphi(\Psi(N))$, $(\varphi\Psi)(N)$ differ, though they have a mattering common part.

N.13.7. Remark. In this book we do not consider the problem of uniqueness of maximal normal extension of the combined operators.

N.13.8. Let us consider some special cases.

1) Case. $A = A_1$, $B = B_1$.

The decomposition formulas are reduced to

$$\Psi(\varphi(N)) = [a_1 b_1 \frown \complement\varphi(N)] \upharpoonleft (\Psi\varphi)(N),$$

$$\varphi(\Psi(N)) = [a_1 b_1 \frown \complement\Psi(N)] \upharpoonleft (\Psi\varphi)(N).$$

Their common maximal normal extension in $a_1 b_1$ is $(a_1 b_1) \upharpoonleft (\Psi\varphi)(N)$, and this is the operator $(\Psi\varphi)(N)$. The operators are in $a_1 b_1$.

2) Case. $A = B$, $A_1 = B_1$.

The decomposition formulas are:

$$\Psi(\varphi(N)) = [a_1 \frown \complement\varphi(N)] \upharpoonleft (\Psi\varphi)(N) \overset{\rightarrow}{+} (a - a_1) \upharpoonleft \mathbf{0},$$

$$\varphi(\Psi(N)) = [a_1 \frown \complement\Psi(N)] \upharpoonleft (\varphi\Psi)(N) \overset{\rightarrow}{+} (a - a_1) | \mathbf{0}.$$

The common maximal normal extension is

$$a_1 \upharpoonleft (\Psi\varphi)(N) \overset{\rightarrow}{+} (a - a_1) \upharpoonleft \mathbf{0}$$

which coincides with the operator $(\Psi\,\varphi)\,(N)$. It operates in the space a.

3) Case. $A = A_1 = B = B_1$. We have

$$\Psi(\varphi(N)) = [a \frown \complement\,\varphi(N)] \upharpoonright (\Psi\,\varphi)\,N,$$
$$\varphi(\Psi(N)) = [a \frown \complement\,\Psi(N)] \upharpoonright (\Psi\,\varphi)\,N.$$

The operator $(\varphi\,\Psi)\,N$ is their common maximal normal extension in a. The operators are acting in a.

N.14. Inverse operator. Let $\varphi(z)$ be defined on A only and let $\varphi(z) \neq 0$ on A_1, where $A_1 \subseteqq A$. Both subsets A, A_1 of the plane P are supposed to be N-measurable, as well as $\varphi(z)$.

The function $\Psi(z) =_{df} \dfrac{1}{\varphi(z)}$ is defined on A_1 only, and we have $\Psi(z) \neq 0$ on A_1.

By [N.11.] we have

$$\complement\,\varphi(N) = \complement\,\varphi_1(N) \overrightarrow{+} (a - a_1),$$

where

$$\varphi_1(z) = \begin{cases} \varphi(z) & \text{for} \quad z \in A_1, \\ 0 & \text{for} \quad z \in A - A_1, \end{cases}$$

and

$$a =_{df} \hat{C}^{-1}\,\mathfrak{H}^V(A^\bullet),$$
$$a_1 =_{df} \hat{C}^{-1}\,\mathfrak{H}^V(A_1^\bullet).$$

We also have

$$\varphi_1(N) = a_1 \upharpoonright \varphi(N)$$

and

$$\varphi(N) = \varphi_1(N) \overrightarrow{+} (a - a_1) \upharpoonright \mathbf{0}.$$

For $\Psi(N)$ we have:

$$\frac{1}{\varphi}(N) = a_1 \upharpoonright \frac{1}{\varphi}(N).$$

If $a_1 = O$, then $\dfrac{1}{\varphi}(N) = \mathbf{0}$.

The general decomposition formulas give the following:

$$\frac{1}{\varphi}(\varphi(N)) = [a_1 \frown \complement\,\varphi(N)] \upharpoonright \mathbf{1} \overrightarrow{+} (a - a_1) \upharpoonright \mathbf{0},$$
$$\varphi\left(\frac{1}{\varphi}(N)\right) = \left[a_1 \frown \complement\,\frac{1}{\varphi}(N)\right] \upharpoonright \mathbf{1},$$

where $\mathbf{1}$ is the "identity" operator, i.e. carrying vectors into themselves. The operator

$$a_1 \upharpoonright \mathbf{1} \overrightarrow{+} (a - a_1) \upharpoonright \mathbf{0}$$

is a maximal normal extension of $\dfrac{1}{\varphi}(\varphi(N))$, and $a_1 \upharpoonright \mathbf{1}$ is the maximal normal extension of $\varphi\left(\dfrac{1}{\varphi}(N)\right)$. The operator $a_1 \upharpoonright \mathbf{1}$ is identical with $\left(\varphi \cdot \dfrac{1}{\varphi}\right)(N)$.

A necessary and sufficient condition for the operator $\varphi(N)$ to possess the inverse $[\varphi(N)]^{-1}$ is:

$$\text{“}\varphi(N)\,\vec{X} = \vec{O} \quad \text{implies} \quad \vec{X} = \vec{O}\text{”}.$$

If we take the **C**-image, we get

$$[\varphi(N)]\,X(\dot{p}) = 0 \quad \text{implies} \quad X(\dot{p}) =^{\mu} 0.$$

Hence for vectors in $\mathfrak{H}^V(A^{\bullet})$, the condition is:

$$\text{“}\varphi(\mathrm{Num}\,\dot{p}) \cdot X(\dot{p}) = 0 \quad \text{implies} \quad X(\dot{p}) =^{\mu} 0\text{”}.$$

Hence if $\mu(A - A_1) > 0$, the inverse $[\varphi(N)]^{-1}$ does not exist, though $\frac{1}{\varphi}(N)$ has a meaning.

The operator $a_1 \upharpoonleft \varphi(N)$ possesses the inverse: $a_1 \upharpoonleft \frac{1}{\varphi}(N) = \frac{1}{\varphi}(N)$. If $a = a_1 \neq O$, then $\frac{1}{\varphi}(N)$ is the inverse of $\varphi(N)$.

In this case we get:

$\varphi\left(\frac{1}{\varphi}(N)\right)$ is the operation **1** restricted to the range of $\varphi(N)$,

and

$\frac{1}{\varphi}(\varphi(N))$ is the operation **1** restricted to the domain of $\varphi(N)$.

N.14a. Remark. The operational calculus does not depend on the choice of the canonical mapping; hence it is independent of the saturating sequence $\{\boldsymbol{H}_n\}$ of spaces, whose practical aim is to split the multiple spectrum. Since such a splitting does not matter in the present topic we may ponder about the possibility of simplifying our general theory just by considering the borelian spectral tribe only (which may be not saturated). Such modified approach however, will not change the vectors of $\boldsymbol{H}$ into functions, but, we would have a mapping of functionoids, based on a tribe of spaces, into functionoids based on a tribe of subsets of the plane. The reader interested in such a possibility, may consult (57). We must add, that our admitted approach is useful, for it will give a more natural and finer classification of spectrum, and a better analysis of the operator than the usual one.

N.15. Resolvent of a maximal normal operator. We shall see in what simple and natural way our theory yields general results. We remind that, equality means both μ-equality in V and N-equality in P. Let $\alpha \in P$; and define (α) as the set, composed of the single complex number α. We put

$$a =_{df} \mathfrak{H}^V((\alpha)^{\bullet}).$$

It may happen that $\mu((\alpha)) > 0$. We put

$$\varphi(\dot{z}) =_{df} \dot{z} - \alpha \quad \text{and} \quad \Psi(\dot{z}) =_{df} \frac{1}{\dot{z} - \alpha},$$

and shall study the operators $\varphi(N)$ and $\Psi(N)$. There are two cases:

1) $\mu((\alpha)) > 0$,
2) $\mu((\alpha)) = 0$.

N.15.1. Consider the first case: $\mu(\alpha) > 0$. The function $\Psi(\dot{z})$ is not defined for $z = \alpha$. Put $P_1 =_{df} P - (\alpha)$, and $\varphi_1(\dot{z}) =_{df} P_1 \upharpoonleft \varphi(\dot{z})$. We have $(\alpha) = P - P_1$

$$\operatorname{co}(\alpha) = P_1, \quad \varphi_1(z) \neq 0 \quad \text{on} \quad P_1, \quad \text{and} \quad \operatorname{co} a = \hat{\boldsymbol{C}}^{-1} \mathfrak{H}^V(P_1^\bullet).$$

Referring to [N.14.] we get:

$$\varphi_1(N) = \operatorname{co} a \upharpoonleft \varphi(N), \qquad \varphi(N) = \varphi_1(N) \overrightarrow{+} a \upharpoonleft \mathbf{0},$$

$$\frac{1}{\varphi}(N) = \Psi(N) = \operatorname{co} a \upharpoonleft \frac{1}{\varphi_1}(N);$$

this is the operator inverse to $\varphi_1(N) = \operatorname{co} a \upharpoonleft \varphi(N)$, i.e.

$$(\varphi_1(N))^{-1}.$$

We also have

$$\mathcal{D}\,\varphi(N) = \mathcal{D}\,\varphi_1(N) \overrightarrow{+} a = \mathcal{D}[\operatorname{co} a \upharpoonleft \varphi(N)] \overrightarrow{+} a = [\operatorname{co} a \cap \mathcal{D}\,\varphi(N)] \overrightarrow{+} a,$$

$$\mathcal{D}\,\Psi(N) = \mathcal{D}\left[\operatorname{co} \upharpoonleft \frac{1}{\varphi}(N)\right] = \operatorname{co} a \cap \mathcal{D}\frac{1}{\varphi}(N) = \mathcal{D}\frac{1}{\varphi}(N).$$

Using the commonly admitted symbolism, we can write:

$$\varphi(N) = N - \alpha\,\mathbf{1} = \operatorname{co} a \upharpoonleft (N - \alpha\,\mathbf{1}] \overrightarrow{+} a \upharpoonleft \mathbf{0},$$

$$\Psi(N) = \frac{1}{N - \alpha\,\mathbf{1}} = \operatorname{co} a \upharpoonleft \frac{1}{N - \alpha\,\mathbf{1}},$$

$$\mathcal{D}\,\varphi(N)\,[\operatorname{co} a \cap \mathcal{D}\,N] \overrightarrow{+} a = \text{domain of } \operatorname{co} a \upharpoonleft N \overrightarrow{+} a,$$

$$\mathcal{D}\,\Psi(N) = \text{range of } \operatorname{co} a \upharpoonleft N.$$

We also have

$$\frac{1}{N - \alpha\,\mathbf{1}}(N - \alpha\,\mathbf{1}) = [\mathbf{1} \text{ restricted to the domain of } \operatorname{co} a \upharpoonleft N \overrightarrow{+} a \upharpoonleft \mathbf{0}],$$

$$(N - \alpha\,\mathbf{1})\left(\frac{1}{N - \alpha\,\mathbf{1}}\right) = [\mathbf{1} \text{ restricted to the range of } \operatorname{co} a \upharpoonleft (N - a\,\mathbf{1})].$$

The domain of

$$\frac{1}{N - \alpha\,\mathbf{1}}(N - \alpha\,\mathbf{1}) \quad \text{is} \quad \mathcal{D}(\operatorname{co} a \upharpoonleft N) \overrightarrow{+} a.$$

The domain of

$$(N - \alpha\,\mathbf{1})\left(\frac{1}{N - \alpha\,\mathbf{1}}\right) \text{ is the range of } \operatorname{co} a \upharpoonleft (N - \alpha\,\mathbf{1}).$$

The operator $N - \alpha \mathbf{1}$ has no inverse, but $\mathrm{co}\, a \rceil (N - \alpha \mathbf{1})$ has one, viz. $\frac{1}{N - \alpha \mathbf{1}}$.

If we put

$$N' =_{df} \mathrm{co}\, a \rceil (N - \alpha \mathbf{1}),$$

we have

$N'(N'^{-1}) = (\text{Range of } N') \rceil \mathbf{1}$,

$N'^{\smile}(N) = (\text{Domain of } N') \rceil \mathbf{1}$. ($N'^{\smile}$ is the operator inverse to N'.)

N.15.2. Consider the second case: $\mu((\alpha)) = 0$. Then, with respect to μ-equality, we see from [N.14.], that

$$\frac{1}{N - \alpha \mathbf{1}}$$

is the inverse of the operator $N - \alpha \mathbf{1}$.

The domain of the first is the range of $N - \alpha \mathbf{1}$.

The domain of the second is that of N. Both are everywhere dense in $\boldsymbol{H}$.

The above results are general.

N.15.3. We shall proceed to the spectral analysis of the given operator N. We know, that it is not the measure which matters, but the ideal J_N, made of N-null sets on P. The cases, considered in [N.15.1.] and [N.15.2.] are those, where $(\alpha) \bar{\in} J_N$ and $(\alpha) \in J_N$ respectively.

Various possibilities, which we are now going to consider in relation to J_N, will be termed *global properties* for the operator N.

The point α will be termed *globally elusive* and *regular*[1], whenever there exists $r > 0$, such that

$$\{z \,\|\, z - \alpha| < r\} \in J_N.$$

It will be termed *globally elusive and singular*, whenever $(\alpha) \in J_N$, but for every $r > 0$:

$$\{z \,\|\, z - \alpha| < r\} \bar{\in} J_N.$$

The point α will be termed *globally heavy and regular*, whenever $(\alpha) \bar{\in} J_N$, but there exists $r > 0$, such that

$$\{z \,|\, 0 < |z - \alpha| < r\} \in J_N.$$

It will be termed *globally heavy and singular* whenever $(\alpha) \bar{\in} J_N$, and for every $r > 0$:

$$\{z \,|\, 0 < |z - \alpha| < r\} \bar{\in} J_N.$$

We shall study these four possibilities separately.

---

[1] Called by STONE: „point of the resolvent set".

N.15.4. First we shall study the property of being globally elusive and regular. The set (GER) of all points α of this cathegory is an open set on the plane P. The set may be empty, or even be equal P. The complementary to (GER) will be called the *global spectrum of* N.

Let $\alpha \in$ (GER). Find $r > 0$, such that

$$R =_{df} \{z \,|\, |z| < r\} \,\bar{\in}\, J_N .$$

We have

$$\varphi(z) =_{df} \frac{1}{z-\alpha} =^N \left(P - R \,\rceil\, \frac{1}{z-\alpha}\right) =_{df} \Psi(z).$$

The function $\Psi(\text{Num}\,p)$ is bounded on V. Let $|\Psi(\text{Num}\,p)| \leq M$. We get

$$\int_V |(N)\, X(p)|^2\, d\mu < M^2 \int_V |X(p)|^2\, d\mu .$$

Hence, taking $\boldsymbol{C}^{-1}$ images, we get

$$||\varphi(N)\,\vec{X}|| < ||\vec{X}|| \cdot M$$

for all $\vec{X} \in \boldsymbol{H}$.

Consequently $\varphi(N)$ is a bounded, hence continuous operator with domain $\boldsymbol{H}$. We have proved the

Theorem. If α is a globally elusive regular point, then $\frac{1}{N - \alpha \mathbf{1}}$ is the inverse operator to $N - \alpha \mathbf{1}$. It is bounded, continuous, and defined over all $\boldsymbol{H}$.

N.15.5. Let α be globally elusive and singular. In this case $\frac{1}{N - \alpha \mathbf{1}}$ is also the inverse of $N - \alpha \mathbf{1}$. The domain and range of $\frac{1}{N - \alpha \mathbf{1}}$ are both everywhere dense in $\boldsymbol{H}$. Both are maximal normal operators. We shall prove, that there exists a μ-square summable function $f(p)$, defined μ-a.e. on V, such that $\frac{1}{\text{Num}\,p - \alpha} f(p)$ is not μ-square summable on V.

Denote by $k(r)$ the open circle with radius $r > 0$ and with center at α. Put $b_0 =_{df} 1$. We have

$$\mu\, k(1) > 0. \tag{1}$$

There exists b_1 such that $0 < b_1 < \frac{1}{\sqrt{2}} b_0$ and

$$\mu[k(b_0) - k(b_1)] > 0. \tag{2}$$

Indeed, the supposition that b_1 does not exist, would have yield the following consequence:

$$\mu\left[k(b_0) - k\left(\frac{b_0}{n\sqrt{2}}\right)\right] = 0 \quad \text{for all} \quad n = 2, 3, \ldots$$

Hence

$$\mu \sum_{n=2}^{\infty} \left[k(b_0) - k\left(\frac{b_0}{n\sqrt{2}}\right) \right] = 0.$$

Hence, since $\mu((\alpha)) = 0$, we would get

$$\mu\, k(b_0) = 0,$$

which contradicts (1). Having that, repeat our argument, finding b_2 such that $0 < b_2 < \frac{1}{\sqrt{2}} b_1$,

(3) $$\mu[k(b_1) - k(b_2)] > 0.$$

In general we shall have a number b_n, such that

$$0 < b_n < \frac{1}{\sqrt{2}} b_{n-1}, \quad (n = 2, 3, \ldots),$$

$$\mu[k(b_{n-1}) - k(b_n)] > 0.$$

As

$$\mu[k(b_{n-1}) - k(b_n)] = \mu[k(b_{n-1}) - k(b_n)] = \sum_{i=1}^{\infty} \mu[k(b_{n-1}) - k(b_n)]^{(i)} > 0,$$

where (i) refers to the i-th-copy, which is a subset of P_n, there exists an i_n, such that the i_n-th-copy of

$$k(b_{n-1}) - k(b_n)$$

has a positive measure. Take such an i_n and put

$$L_n =_{df} i_n\text{-th-copy of } \{k(b_{n-1}) - k(b_n)\}.$$

The sets L_n are mutually disjoint.

Put

$$m_n =_{df} \mu L_n.$$

Let us define the function $f(p)$ in V as follows:

$$f(p) = \sqrt{\frac{b_{n-1}^2 - b_n^2}{m_n}}$$

on L_n and $f(p) = 0$ elsewhere. We have

$$\int_{L_n} |f(p)|^2 \, d\mu = b_{n-1}^2 - b_n^2.$$

Consequently

$$\int_V (f(p))^2 \, d\mu = \sum_{n=2}^{\infty} (b_{n-1}^2 - b_n^2) = b_1^2,$$

so $f(p)$ is μ-square-summable on V. We have

$$\int_{L_n} \left| \frac{1}{\operatorname{Num} p - \alpha} f(p) \right|^2 d\mu = \frac{b_{n-1}^2 - b_n^2}{m_n} \int_{L_n} \frac{1}{|\operatorname{Num} p - \alpha|^2} \, d\mu.$$

Since, if $z \in k(b_{n-1}) - k(b_n)$, we have

$$b_n \leqq z - \alpha < b_{n-1}.$$

Hence

$$\frac{1}{b_{n-1}} < \frac{1}{z-\alpha} \leqq \frac{1}{b_n}.$$

Thus we get

$$(1) \qquad \int_{L_n} \left| \frac{f(p)}{\operatorname{Num} p - \alpha} \right|^2 d\mu \geqq \frac{b_{n-1}^2 - b_n^2}{b_{n-1}^2} \geqq \frac{1}{2}.$$

From (1) it follows, that the series

$$\sum_{n=2}^{\infty} \int_{L_n} \left| \frac{f(p)}{\operatorname{Num} p - \alpha} \right|^2 d\mu$$

diverges. Consequently the function $\frac{1}{\operatorname{Num} p - \alpha} f(p)$ is not μ-square summable.

N.15.5a. The above shows, that the domain of $\frac{1}{N - \alpha \mathbf{1}}$ differs from $\boldsymbol{H}$, though it is everywhere dense. It follows that $\frac{1}{N - \alpha \mathbf{1}}$ is discontinuous (not bounded) in $\boldsymbol{H}$.

Theorem. We have proved that if α is a globally elusive and singular complex number, then $\frac{1}{N\alpha - \mathbf{1}}$ is the inverse operator to $N - \alpha \mathbf{1}$. It is maximal normal in $\boldsymbol{H}$, but discontinuous.

N.15.5b. The converse theorem to [N.15.4.] and [N.15.5a.] is true. Thus we have:

Theorem. If α is a globally elusive complex number, then:

1) $N - \alpha \mathbf{1}$ admits the inverse operator $\frac{1}{N - \alpha \mathbf{1}}$, which is maximal normal.

2) The following are equivalent:

I. $\frac{1}{N - \alpha \mathbf{1}}$ is continuous,

II. α is a globally elusive complex number.

3) The following are equivalent:

I. $\frac{1}{N - \alpha \mathbf{1}}$ is discontinuous,

II. α is a globally elusive singular number.

N.15.6. Let us go over to the case where α is a globally heavy number. The results of [N.15.] give the following:

Put

$$a =_{df} \mathfrak{H}^V(\operatorname{Point} \alpha).$$

The operator $\frac{1}{N - \alpha \mathbf{1}}$ is the inverse operator to $(\operatorname{co} a) \upharpoonright N - \alpha \mathbf{1}$. It is a maximal normal operator in $\operatorname{co} a$.

If we apply to co$a \restriction N - \alpha\mathbf{1}$ the result in Theorem [N.15.5b.], we get:

Theorem. If α is a globally heavy complex number, then

1) $N - \alpha\mathbf{1}$ does not admit any inverse operator,

2) $\frac{1}{N - \alpha\mathbf{1}}$ is the inverse of $N - \alpha\mathbf{1}$ restricted to co$a$, where $a = \mathbf{C}^{-1}\mathfrak{H}^V(\text{Point}\,\alpha)$, and where $a \neq 0$ is independent on the canonical mapping.

3) The operator $\frac{1}{N - \alpha\mathbf{1}}$ is maximal normal in co$a$.

4) The following are equivalent:

I. $\frac{1}{N - \alpha\mathbf{1}}$ is continuous (in co$a$),

II. α is a globally heavy regular number.

5) The following are equivalent:

I. $\frac{1}{N - \alpha\mathbf{1}}$ is discontinuous (in co$a$),

II. α is a globally heavy singular number.

N.15.7. Individual spectrums. If the borelian spectral tribe of N is not saturated, then there exists a saturating sequence of spaces $\{\boldsymbol{H}_n\}$ which may be finite or infinite. We call the spaces $\boldsymbol{H}_n$ *saturation compartments.* The operator $\boldsymbol{H}_n \restriction N$ has a simple spectrum only. The results obtained in [N.15.] to [N.15.6.] can be applied to each individual $\boldsymbol{H}_n \restriction N$, so we can have for complex number α and $\boldsymbol{H}_n$ the four possibilities discussed above. This classification of points α is, of course dependent on the choice of the saturating sequence $\{\boldsymbol{H}_n\}$.

Given a point α, which is globally elusive and regular, it will be so in each compartment $\boldsymbol{H}_n$. The same will be, when α is globally heavy and regular. Then α will be so in each compartment $\boldsymbol{H}_n$. A different behaviour show globally singular points.

The following possibilities may take place (both if α is elusive and if α is heavy):

1) α is regular in all compartments $\boldsymbol{H}_n$, though it is globally singular.

2) α is singular in all compartments $\boldsymbol{H}_n$.

3) α is regular in some compartments $\boldsymbol{H}_n$, and singular in other compartments.

In relation to this there is the possibility of defining the "multiplicity of various behaviour". It suggests the problem of independence of these multiplicities of the choice of the saturating sequence. The choice of the generating vector does not influence the ideal J_μ of μ-null sets on V. We leave these problems to the reader.

N.16. The equation $N\vec{X} - \alpha\vec{X} = \vec{Y}$. The equation can be treated by relying on the obtained theorems on the resolvent of N.

Nevertheless, to show the advantage of the canonical mapping, we shall treat it independently.

The canonical transformation $\boldsymbol{C}$ transforms the equation into

$$\mathfrak{N} f(\dot{p}) - \alpha f(\dot{p}) \cdot =^{\mu} \cdot g(\dot{p}) \text{ on } V, \tag{1}$$

where the μ-square summable function $g(\dot{p})$, defined μ-a.e. on V, is given, and $f(\dot{p})$ is to be found.

The equation (1) is equivalent to

$$(\mathfrak{N} - \alpha \mathfrak{J}) f(\dot{p}) =^{\mu} g(\dot{p}), \tag{2}$$

where $\mathfrak{J}$ is the operation which make correspond functions to themselves.

N.16a. Case 1) *α is globally elusive.* The $\boldsymbol{C}$-image of $N - \alpha \mathbf{1}$ is the operator

$$(\mathfrak{N} - \alpha \mathfrak{J}) X(\dot{p}) \cdot =^{\mu} \cdot (\operatorname{Num} \dot{p} - \alpha) \cdot X(\dot{p}).$$

Hence (2) can be written:

$$(\operatorname{Num} \dot{p} - \alpha) \cdot f(\dot{p}) \cdot =^{\mu} \cdot g(\dot{p}). \tag{3}$$

Since $\operatorname{Num} \dot{p} - \alpha \neq^{\mu} 0$, the equation (3) is equivalent to

$$f(\dot{p}) \cdot =^{\mu} \cdot \frac{g(\dot{p})}{\operatorname{Num} \dot{p} - \alpha}. \tag{4}$$

$f(\dot{p})$ is μ-square summable if and only if $\frac{g(\dot{p})}{\operatorname{Num} \dot{p} - \alpha}$ is so; i.e. $g(\dot{p})$ must belong to the domain of the operator which carries $X(\dot{p})$ into $\frac{X(\dot{p})}{\operatorname{Num} \dot{p} - \alpha}$. This is necessary and sufficient. Taking the $\boldsymbol{C}^{-1}$-image, we get the

N.16.1. Theorem. If α is globally elusive (regular or singular), then the following are equivalent:

I. The equation $N \vec{X} - \alpha \vec{X} = \vec{Y}$ has for the given $\vec{Y}$ a solution $\vec{X}$.

II. $\vec{Y} \in \operatorname{\mathsf{d}} \frac{1}{N - \alpha \mathbf{1}}$.

The solution is unique. The set of the admissible vectors $\vec{Y}$ is everywhere dense in $\boldsymbol{H}$. The solution is

$$\vec{X} = \frac{1}{N - \alpha \mathbf{1}} \vec{Y}.$$

If α is globally elusive and regular, there exists $r > 0$, such that $\operatorname{Num} \dot{p} - \alpha > r$ almost μ-everywhere on P. Hence from (4) we get $f(\dot{p}) \leqq^{\mu} g(\dot{p}) \cdot \frac{1}{r}$. Consequently $f(\dot{p})$ is μ-square summable, if and only if $g(\dot{p})$ is so. Thus all $g(\dot{p})$ give solutions of (4).

We have proved the following:

N.16.2. Theorem. If α is globally elusive and regular, then the equation

(5) $$N\vec{X} - \alpha\vec{X} = \vec{Y}$$

admits a solution for every $\vec{Y}$. This solution is $\vec{X} = \frac{1}{N - \alpha\mathbf{1}}\vec{Y}$. The solution is unique.

If α is globally elusive and singular, then the equation does not admit for every $\vec{Y}$ a solution.

Case 2) α *is globally heavy*. We have $\mu(\text{Point}\,\alpha) > 0$. The equation $N\vec{X} - \alpha\vec{X} = \vec{Y}$ goes through $\boldsymbol{C}$ into the equation

$$\mathfrak{N}X(p) - \alpha X(p) \cdot =^{\mu} \cdot Y(p).$$

This equation is equivalent to

(5 a) $$(\text{Num}\,\dot{p} - \alpha) \cdot X(\dot{p}) \cdot =^{\mu} \cdot Y(\dot{p}).$$

Let $\alpha^{(1)}, \alpha^{(2)}, \ldots, \alpha^{(n)}, \ldots$ be the copies of the number α on $P_1, P_2, \ldots, P_n, \ldots$ respectively. There exists at least one n, for which

$$\mu((\alpha^{(n)})) > 0.$$

Denote by $\alpha'_1, \alpha'_2, \ldots$ all copies of α with positive measure, and by $\alpha''_1, \alpha''_2, \ldots$ the remaining copies of α.

Suppose, that the μ-square summable function $X(\dot{p})$ satisfies (5). There exists a μ-null subset E of V, such that

$$(\text{Num}\,\dot{p} - \alpha) \cdot X(\dot{p}) = Y(\dot{p})$$

everywhere on $V - E$. As E is a μ-null set, no one of $\alpha'_1, \alpha'_2, \ldots$ belongs to E. Hence $\alpha'_m \in V - E$ for all m. We get

$$0 \cdot X(\alpha'_m) = Y(\alpha'_m).$$

Hence $Y(\alpha'_m) = 0$ for all m.

Let $a =_{df} \text{Point}(\alpha)$. We have $a \cdot =^{\mu} \cdot \bigcup_m \alpha'_m$, since the set of all α''_m does not matter. Hence

$$a \upharpoonright g(\dot{p}) \cdot =^{\mu} \cdot 0,$$

and then

$$\text{Proj}_a g(\dot{p}) \cdot =^{\mu} \cdot 0.$$

Taking the $\boldsymbol{C}^{-1}$-image, and putting $a = \boldsymbol{C}^{-1}\,\mathfrak{H}^V(\text{Point}\,\alpha)$, we get $\text{Proj}_a \vec{Y} = \vec{O}$. We have proved that:

N.17. Theorem. If

1. α is globally heavy,
2. $a =_{df} \boldsymbol{C}^{-1}\,\mathfrak{H}^V(\text{Point}\,\alpha)$,
3. $\vec{X}$ is a solution of the equation

(6) $$N\vec{X} - \alpha\vec{X} = \vec{Y},$$

then $\vec{Y}$ is orthogonal to a.

Now suppose that a vector $\vec{Y}$ is orthogonal to a, and suppose that $\vec{X}$ is a solution of (6). Put

$$X(\dot{p}) =_{df} \mathbf{C}\,\vec{X}, \qquad Y(\dot{p}) =_{df} \mathbf{C}\,\vec{Y}.$$

By what has been proved, $\vec{Y}$ is orthogonal to a. Hence

$$(\text{Point}\,\alpha) \upharpoonleft Y(\dot{p}) \cdot =^{\mu} \cdot 0.$$

Hence

(7) $$\bigcup_n \alpha^{(n)} \upharpoonleft Y(\dot{p}) \cdot =^{\mu} \cdot 0.$$

Since $X(\dot{p})$ is a solution of the $\mathbf{C}$-transform of (6), we have

(8) $$(\text{Num}\,\dot{p} - \alpha)\, X(\dot{p}) \cdot =^{\mu} \cdot Y(\dot{p}).$$

If $p = \alpha'_m$, then $Y(p) = 0$; this by (7). Hence, for $p = \alpha'_m$, the equation

(9) $$(\text{Num}\,\dot{p} - \alpha)\, X(\dot{p})^{\mu} = Y(\dot{p})$$

is satisfied, because of (7). For $p = \alpha''_m$, the equation (9) may be satisfied or not: it does not matter, because of (7).

Let us consider the points p, such that $\text{Num}\,p \neq \alpha$. We get

$$X(\dot{p}) \cdot =^{\mu} \cdot \frac{Y(\dot{p})}{\text{Num}\,\dot{p} - \alpha} \quad \text{on} \quad V - \bigcup_{n=1}^{\infty} (\alpha^{(n)}).$$

Hence $\dfrac{Y(\dot{p})}{\text{Num}\,\dot{p} - \alpha}$ is μ-square summable on $V - \bigcup_{n=1}^{\infty} (\alpha^{(n)})$. It follows, that $Y(p)$ belongs to the domain of the operator, which carries the function $g(\dot{p})$ into $\dfrac{g(\dot{p})}{\text{Num}\,\dot{p} - \alpha}$, and has the maximal domain. Hence $\vec{Y} \in \text{ᗡ}\,\dfrac{1}{N - \alpha\mathbf{1}}$. Thus we have proved that:

N.18. Theorem. If

1. α is globally heavy,
2. $\vec{Y}$ is orthogonal to $a =_{df} \mathbf{C}^{-1}\, \mathfrak{H}^V(\text{Point}\,\alpha)$,
3. the equation $N\vec{X} - \alpha\vec{X} = \vec{Y}$ possesses a solution $\vec{X}$,

then $\vec{Y} \in \text{ᗡ}\,\dfrac{1}{N - \alpha\mathbf{1}}$.

The set $\text{ᗡ}\,\dfrac{1}{N - \alpha\mathbf{1}}$ coincides with a, whenever α is globally heavy and regular. It is everywhere dense in $\text{co}\,a$, but not coinciding with $\text{co}\,a$, whenever α is globally heavy and singular. Now we shall prove that:

N.19. Theorem. If

1. α is globally heavy,
2. $\vec{Y}$ is orthogonal to $a = \mathbf{C}^{-1}\, \mathfrak{H}^V(\text{Point}\,\alpha)$,
3. $\vec{Y} \in \text{ᗡ}\,\dfrac{1}{N - \alpha\mathbf{1}}$,
4. $\vec{Z}$ is any vector of a,

then

(1) $$\vec{X} =_{df} \vec{Z} + \frac{1}{N - \alpha \mathbf{1}} \vec{Y}$$

is a solution of the equation

(2) $$N \vec{X} - \alpha \vec{X} = \vec{Y}.$$

For a given $\vec{Y}$ the only solution of (2) is (1), where $\vec{Z} \in a$.

Proof. Consider $\boldsymbol{C}$-images, as before, and put $Z(\dot{p}) =_{df} \boldsymbol{C} \vec{Z}$. We have

$$Z(\dot{p}) =^{\mu} 0 \quad \text{on} \quad \hat{\boldsymbol{C}} \boldsymbol{a} = \text{Point}\,\alpha,$$

and

$$^{\bullet}Y(\dot{p}) =^{\mu} 0 \quad \text{on} \quad V - \text{Point}\,\alpha.$$

From (1) we have

$$X(\dot{p}) \cdot =^{\mu} \cdot \frac{1}{\text{Num}\,\dot{p} - \alpha} \cdot Y(\dot{p}) + Z(\dot{p}).$$

Hence

$$X(\dot{p}) \cdot =^{\mu} \cdot \frac{1}{\text{Num}\,\dot{p} - \alpha} \cdot Y(\dot{p}) \quad \text{on} \quad \text{Point}\,\alpha,$$

and

$$X(\dot{p}) \cdot =^{\mu} \cdot Z(\dot{p}) \quad \text{on} \quad V - \text{Point}\,\alpha.$$

Hence

$$X(\dot{p}) \cdot (\text{Num}\,\dot{p} - \alpha) =^{\mu} \cdot Y(\dot{p}) \quad \text{on} \quad \text{Point}\,\alpha$$

and

$$X(\dot{p}) \cdot (\text{Num}\,\dot{p} - \alpha) =^{\mu} Y(\dot{p}) \quad \text{on} \quad V - \text{Point}\,\alpha,$$

because

$$X(\dot{p}) =^{\mu} Z(\dot{p}) =^{\mu} 0 \quad \text{on} \quad V - \text{Point}\,\alpha,$$

and also

$$Y(\dot{p}) =^{\mu} 0 \quad \text{on} \quad V - \text{Point}\,\alpha.$$

Thus, through $\boldsymbol{C}^{-1}$ we get

$$(N - \mathbf{1}\,\alpha)\,\vec{X} = \vec{Y},$$

i.e. $\vec{X}$ satisfies (2). Conversely, if $\vec{X}$ satisfies (2), we get, taking $\boldsymbol{C}$-images:

$$(\text{Num}\,\dot{p} - \alpha) \cdot X(\dot{p}) \cdot =^{\mu} \cdot Y(\dot{p}),$$

we get on $V - \text{Point}\,\alpha$:

$$X(\dot{p}) \cdot =^{\mu} \cdot \frac{Y(\dot{p})}{\text{Num}\,\dot{p} - \alpha};$$

and putting

$$Z(\dot{p}) =_{df} \text{Point}\,\alpha \upharpoonleft X(\dot{p}),$$

we get

$$X(\dot{p}) \cdot =^{\mu} \cdot \frac{Y(\dot{p})}{\text{Num}\,\dot{p} - \alpha} + Z(\dot{p}),$$

hence

$$\vec{X} = \frac{1}{N - \alpha \mathbf{1}} \vec{Y} + \vec{Z},$$

where

$$\vec{Z} \in a.$$

We have proved the

N.20. Theorem. If

1. α is globally heavy,
2. $\vec{X}$ is a solution of the equation

$$N\vec{X} - \alpha\vec{X} = \vec{Y},$$

then 1) $\vec{Y}$ is orthogonal to a, where

$$a = \boldsymbol{C}^{-1} \mathfrak{H}^V(\text{Point}\,\alpha),$$

2) $\vec{Y} \in \text{ᗡ} \frac{1}{N - \alpha \mathbf{1}}$,

3) there exists $\vec{Z} \in a$ such that $\vec{X} = \vec{Z} + \frac{1}{N - \alpha \mathbf{1}} \vec{Y}$,

4) if α is a globally heavy and regular, the following are equivalent:

I. $\vec{Y} \in \text{co}\, a$,

II. the solution exists,

5) if α is globally heavy and singular, the following are equivalent

I. $\vec{Y} \in \text{ᗡ} \frac{1}{N - \alpha \mathbf{1}}$,

II. the solution exists.

In this case, not for all $\vec{Y}$ there exists a solution; but the set of those $\vec{Y}$, for which the solution exists, is dense in $\text{co}\, a$.

Chapter P

Tribe of repartition of functions

In this chapter a special tribe of subsets of the range of a function $f(x)$ is studied in relation to compound functions $f(g(x))$.

This topic will be applied to permutable normal operators in Hilbert-Hermite-space, in solidarity with their canonical representation. Thus this chapter may be considered as auxiliary to the Chapter [P I.], where properties of permutable normal operators are considered.

P.1.1. We start with auxiliary properties of μ-measurable complex valued functions $f(\dot{z})$ of the complex variable z, varying on the complex plane P.

Let $(\mathfrak{B})$ be the tribe of all borelian subsets of P. It is denumerably additive. Let $\mu(E)$ be a non negative, finite, denumerably additive measure on $\mathfrak{B}$ with $\mu(P) > 0$.

Denote by $(\mathfrak{L})$ the tribe of subsets of P, which is the μ-measure-hull of $(\mathfrak{B})$. Its somata will be termed *μ-measurable sets* and the corresponding measure, extended over the said hull, will also be denoted by μ. The extended measure is denumerably additive.

It is sometimes convenient to replace $(\mathfrak{L})$ by the tribe $(\mathfrak{L}_\mu)$ of μ-measurable sets, taken modulo the ideal, composed of all measurable sets with μ-measure equal 0. This means that we introduce on $(\mathfrak{L})$ the special equality $E =^\mu F$ for μ-measurable sets, this equality being defined by

$$\mu(E \dot{+} F) = \mu[(E - F) + (F - E)] = 0.$$

The ordering relation on $\mathfrak{L}$,

$$E \subseteq^\mu F,$$

defined by $\mu(E - F) = 0$, organizes $(\mathfrak{L})$ just into $(\mathfrak{L}_\mu)$, which is also a denumerably additive tribe, and on which μ is a $(=^\mu)$-invariant, effective measure. Every borelian set is μ-measurable.

P.1.2. Def. A complex valued function $f(\dot{z})$, defined on a set $E \neq O$, is said to *be borelian*, if it fits $\mathfrak{B}$, i.e. for every open subset M of the complex plane P the set (which may be empty):

$$f^{-1}(M) =_{df} \{z \mid f(z) \in M\}$$

belongs to $\mathfrak{B}$. Since $f^{-1}(P) = E$, it follows that $E = \mathrm{D}\, f$ must be borelian.

P.1.3. Remark. If the function is borelian, it does not imply, that $\mathrm{D}\, f$ is also a borelian set. It may be a non-borelian set.

P.2. Def. A function $f(\dot{z})$, defined on a set $E \neq O$, is said *to be μ-measurable*, if it fits $\mathfrak{L}$, i.e. if for every open subset M of the complex plane P, the set

$$f^{-1}(M) = \{z \mid f(z) \in M\}$$

belongs to $\mathfrak{L}$. It follows that $E = \mathrm{D}\, f = f^{-1}(P)$ is measurable.

P.2.1. Set-operations on $(\mathfrak{L})$, and also somatic operations on $(\mathfrak{L}_\mu)$ will be denoted by the symbols: $+, \dot{+}, \cdot, -$, co, $\leqq$, $\sum$, $\prod$, and the equality $=^\mu$ on $(\mathfrak{L})$ sometimes written $=$. The dot over a letter, say $\dot{z}$, is written in order to emphazise, that z is a variable. Instead of using the terms *μ-measurable, almost μ-everywhere,* we shall say sometimes simply *measurable* and "*almost everywhere*", written sometimes: a.e.

P.2.2. We recall the following simple properties of inverse images f^{-1}, valid for any kind of function f: Let M_α be an indexed collection of subsets of P. We have

$$f^{-1}(\sum_\alpha M_\alpha) = \sum_\alpha f^{-1}(M_\alpha),$$

$$f^{-1}(\mathrm{co}\, M) = \mathrm{co}\, f^{-1}(M),$$

$$f^{-1}(\prod_\alpha M_\alpha) = \prod_\alpha f^{-1}(M_\alpha).$$

Let M, N be subsets of P. Then we have the following statements: If $M \cdot N = \emptyset$, then $f^{-1}(M) \cdot f^{-1}(N) = \emptyset$, for any M, N:

$$f^{-1}(M - N) = f^{-1}(M) - f^{-1}(N), \quad f^{-1}(M \dot{+} N) = f^{-1}(M) \dot{+} f^{-1}(N).$$

Concerning direct images of functions we have

$$f(\sum_\alpha E_\alpha) = \sum_\alpha f(E_\alpha),$$

and

$$f(\prod_\alpha E_\alpha) \subseteqq \prod_\alpha f(E_\alpha),$$

but the equality may be not true. Notice that

$$f(\mathrm{co}\, E) \text{ may differ from } \mathrm{co}\, f(E)$$

and $E \cdot F = \emptyset$ may not imply

$$f(E) \cdot f(F) = \emptyset.$$

Concerning the μ-equality, we have for $n = 1, 2, \ldots$:

If

$$E_n =^\mu E'_n, \quad \text{then} \quad \sum_n E_n =^\mu \sum_n E'_n \quad \text{and} \quad \prod_n E_n =^\mu \prod_n E'_n.$$

If

$$E =^\mu E', \quad \text{then} \quad \mathrm{co}\, E =^\mu \mathrm{co}\, E'.$$

P.2.3. Def. Two μ-measurable functions $f(\dot{z})$, $g(\dot{z})$ of the complex variable z are *said to be μ-equal*:

written $f(\dot{z}) =^\mu g(\dot{z})$, or $f(\dot{z}) \doteq g(\dot{z})$, or $f(z) = g(z)$ μ-everywhere,

or $f(z) = g(z)$ a.e.,

if and only if

1) $\complement f(\dot{z}) \dot{+} \complement g(\dot{z})$ is a μ-null set,
2) $\{z | f(z) \neq g(z)\}$ is a μ-null set.

The notion of μ-equality of functions is transitive, reflexive and symmetric. We also have:

If

$$f(\dot{z}) =^\mu f'(\dot{z}), \quad g(\dot{z}) =^\mu g'(\dot{z}),$$

then

$$f(\dot{z}) + g(\dot{z}) =^\mu f'(\dot{z}) + g'(\dot{z})$$

and

$$f(\dot{z}) \cdot g(\dot{z}) =^{\mu} f'(\dot{z}) \cdot g'(\dot{z}).$$

In addition to that we have:

If

$$f_n(\dot{z}), \quad (n = 1, 2, \ldots),$$

are measurable and

$$\lim_{n\to\infty} f_n(\dot{z}) = f(\dot{z})$$

a.e., then

$$f(\dot{z})$$

is also measurable.

If

$$f_n(\dot{z}) \doteq f'_n(\dot{z}), \quad f(\dot{z}) \doteq f'(\dot{z}), \quad \lim_{n\to\infty} f_n(\dot{z}) \doteq f(\dot{z})$$

a.e., then

$$\lim_{n\to\infty} f'_n(z) = f'(z)$$

a.e.

By $E \upharpoonleft f(\dot{z})$ we shall denote the function $f(\dot{z})$ restricted to E, i.e. its domain is $E \cdot \mathrm{Q} f$; and if $z \in E \cdot \mathrm{Q} f$, then the value of the restricted function is $f(z)$. ($\mathrm{Q} f$ means the domain of f, and $\mathrm{D} f$ its range.)

P.3. We shall have some auxiliary theorems, concerning relations between sets and functions, and especially concerning borelian sets and borelian functions.

P.3.1. Theorem. If E is a measurable set, then there exists a borelian set F such that

$$E =^{\mu} F.$$

This follows from the fact, that μ-measurable sets are defined as sets, having the form $F + G' - H'$, where F is borelian, and G', H' are subsets of borelian sets having the measure 0.

Remark. In the thesis of this theorem we can add the requirement

$$E \subseteqq F \quad \text{or} \quad F \subseteqq E.$$

P.3.2. Theorem. If $f(\dot{z})$ is a measurable function, then there exists a borelian function $g(\dot{z})$, such that

$$f(\dot{z}) =^{\mu} g(\dot{z}), \quad \mathrm{Q} g \subseteqq \mathrm{Q} f, \quad \text{and} \quad g(z) = f(z)$$

for all $z \in \mathrm{Q} g$.

Proof. First we suppose that $f(\dot{z})$ is *a simple-function* i.e. admitting a finite or denumerable number of values only. Let f admit the values $f_1, f_2, \ldots, f_n, \ldots$, which are all different, on sets:

(1) $E_1, E_2, \ldots, E_n, \ldots$ respectively,

so that the sets are disjoint and their union is $\mathrm{Q} f$. Since $f(z)$ is measurable, so are the sets E_n. Indeed the set (f_n) composed of the single complex number f_n is borelian, hence $f^{-1}(f_n)$ is measurable, and we see that

$$f^{-1}(f_n) = E_n .$$

Applying Theorem [P.3.1.] to the measurable function

$$g_n(\dot{z}) =_{df} E_n \rceil f(\dot{z}),$$

we find a borelian set E'_n, such that $E'_n \subseteqq E_n$, $E'_n =^{\mu} E$.

Put

$$A =_{df} \sum_n E'_n, \quad \text{and} \quad g(z) =_{df} A \rceil f(\dot{z}).$$

We have $g(z) = f(z)$ for all $z \in A$, and

$$A \doteq \sum_n E'_n \doteq \daleth f .$$

Since $\daleth g = A$, it follows that $g(\dot{z}) \doteq f(\dot{z})$, $g(z)$ is borelian and we have $g(z) = f(z)$ for all $z \in \daleth g$.

The theorem is proved for simple functions.

Now let $f(z)$ be an arbitrary measurable function. Take $\varepsilon > 0$. Consider the halfopen rectangles

$$R_{n,m} = \left\{ z \,\middle|\, \begin{array}{l} n\varepsilon < \mathrm{Real}\, z \leqq (n+1)\,\varepsilon \\ m\varepsilon < \mathrm{Imag}\, z \leqq (m+1)\,\varepsilon \end{array} \right\},$$

where $n, m = 0, \pm 1, \pm 2, \ldots$

We have $\sum_{n,m=-\infty}^{+\infty} R_{n,m} = P$, and any two differing rectangles are disjoint.

Find a borelian set D, such that

$$D \doteq \daleth f, \quad D \subseteqq \daleth f .$$

The function $f(\dot{z})$ is μ-equal to the function

$$f'(\dot{z}) =_{df} D \rceil f(\dot{z}).$$

Consider the function $f_\varepsilon(z)$, defined on D, by putting

$$f_\varepsilon(z) = (n+1)\,\varepsilon + i\,(m+1)\,\varepsilon ,$$

whenever

$$f(z) \in R_{n,m}, \quad z \in D .$$

We have

$$\daleth f_\varepsilon(z) = D$$

and

(2) $$|f_\varepsilon(z) - f(z)| \leqq \varepsilon \sqrt{2} \quad \text{for all} \quad z \in D .$$

The function $f_\varepsilon(\dot{z})$ is a simple measurable function. Hence, by what has been already proved, there exists a borelian function $f'_\varepsilon(\dot{z})$

such that

$$f_\varepsilon(z) \doteq f'_\varepsilon(z), \quad \text{ᗡ} f'_\varepsilon(z) \subseteqq \text{ᗡ} f_\varepsilon(z), \quad f'_\varepsilon(z) = f_\varepsilon(z) \quad \text{on} \quad \text{ᗡ} f'_\varepsilon.$$

We have

$$\text{ᗡ} f'_\varepsilon(z) \doteq \text{ᗡ} f_\varepsilon(z);$$

and from (2) it follows that

$$|f'_\varepsilon(z) - f(z)| \leqq \varepsilon \sqrt{2}$$

a.e. on $\text{ᗡ} f'_\delta$. Now put

$$\varepsilon \doteq \frac{1}{k}, \qquad k = 1, 2, \ldots$$

Put

$$D' =_{df} \prod_k \text{ᗡ} f'_{\frac{1}{k}} \doteq \prod_k \text{ᗡ} f_{\frac{1}{k}}.$$

We have

$$\text{ᗡ} f_{\frac{1}{k}} \doteq D \doteq D' \quad \text{for any} \quad k = 1, 2, \ldots,$$

For every k the domain of $D' \rceil f'_{\frac{1}{k}}(\dot{z})$ μ-equals to the domain of $D' \rceil f(z)$.

We have on D':

$$g(z) =_{df} \lim_{k \to \infty} f'_{\frac{1}{k}}(z) = f(z).$$

Consequently, since the limit of a converging sequence of borelian functions, defined on the same borelian set, is a borelian function, it follows that $f(z)$ is μ-equal to the borelian function $g(z)$. We also have

$$\text{ᗡ} g(z) \subseteqq \text{ᗡ} f(z),$$

and

$$g(z) = f(z) \quad \text{for all} \quad z \in \text{ᗡ} g.$$

P.3.3. Theorem. For any function $f(z)$, we have: if

1. $E \subseteqq \mathrm{D} f(\dot{z})$,
2. $f^{-1}(E) = e$,

then

$$f(\text{ᗡ} f - e) = \mathrm{D} f - E.$$

Proof. Suppose that $z_0 \bar{\in} e$. Then $f(z_0)$ cannot belong to E (because if $f(z_0) \in E$, we have $z_0 \in \{z \mid f(z) \in E\}$, hence $z_0 \in e$).

We have

$$e = f^{-1}(E) = \{z \mid f(z) \in E\}.$$

Thus

$$f(z_0) \bar{\in} E.$$

Hence, if $z \in \text{ᗡ} f - e$, then $z \in \text{ᗡ} f$ and $z \bar{\in} E$. Consequently we have $f(z) \in \mathrm{D} f - E$ for every $z \bar{\in} e$, with $z \in \text{ᗡ} f$. Hence, if $z_0 \in \text{ᗡ} f - e$, then $f(z_0) \in \mathrm{D} f - E$.

Hence

(1) $$f(\reflectbox{D}f - e) \subseteqq \mathrm{D}f - E.$$

Suppose that

$$u \in \mathrm{D}f - E.$$

We have $u \overline{\in} E$, and

(2) $$u \in \mathrm{D}f.$$

There exists z_0 such that $u = f(z_0)$. Take z_0. I say that $z_0 \overline{\in} e$. Indeed, suppose that $z_0 \in e$.

Then $f(z_0) \in f(e) = E$, i.e. $u \in E$, which contradicts (2). Thus

$$z_0 \in \reflectbox{D}f - e;$$

and then

$$u = f(z_0) \in f(\reflectbox{D}f - E).$$

Consequently

(3) $$\mathrm{D}f - E \subseteqq f(\reflectbox{D}f - e).$$

From (1) and (3) the theorem follows.

P.3.4. Theorem. If

1. $f(\dot{z}) =^{\mu} g(\dot{z})$,
2. A is a borelian set,

then

$$\{z \mid f(z) \in A\} =^{\mu} \{z \mid g(z) \in A\}.$$

Proof. Put

$$a =_{df} \{z \mid f(z) \in A\},$$

$$b =_{df} \{z \mid g(z) \in A\}.$$

Since $f(\dot{z}) =^{\mu} g(\dot{z})$, we have μ-almost everywhere in $\reflectbox{D}f \cdot \reflectbox{D}g$ the equality

$$f(z) = g(z).$$

Put

$$c =_{df} \{z \mid f(z) = g(z)\}.$$

I say that

$$a\,c = b\,c.$$

Indeed, let $z_0 \in a\,c$. We have $f(z_0) \in A$, because $z_0 \in a$. Since $z_0 \in c$, we have $f(z_0) = g(z_0)$, and then $g(z_0) \in A$, so $z_0 \in b$. Hence $a\,c \subseteqq b\,c$.

Similarly we prove that $b\,c \subseteqq a\,c$.

Consequently

(1) $$a\,c = b\,c.$$

Now, we have $\mu(\reflectbox{D}f \dotplus c) = 0$; hence

$$\mu(a \cdot \reflectbox{D}f \dotplus a\,c) = 0,$$

i.e.

(2) $$\mu(a \dotplus a\,c) = 0.$$

Similarly we get $\mu(b \dot{+} b\,c) = 0$; hence, by (1)

(3) $$\mu(b \dot{+} a\,c) = 0.$$

We have

$$a \dot{+} b = (a \dot{+} a\,c) \dot{+} (b \dot{+} a\,c) \subseteqq (a \dot{+} a\,c) + (b \dot{+} a\,c).$$

Consequently

$$\mu(a \dot{+} b) \leqq \mu(a \dot{+} a\,c) + (b \dot{+} a\,c)] = 0.$$

Hence

$$a =^{\mu} b, \quad \text{q.e.d.}$$

P.3.5. Theorem. If $f(\dot{z})$ is measurable, then the class of all sets a, for which there exists a borelian subset A of P, such that

$$a =^{\mu} \{z \mid f(z) \in A\},$$

make up a denumerably additive tribe, ordered by the correspondence $\subseteqq$. It is a denumerably (even strictly) genuine subtribe of $\mathfrak{L} \cdot \mathrm{C}f$, [P.1.1.].

Proof. Let us vary A, and denote by F the class of all a, such that

$$a =^{\mu} \{z \mid f(z) \in A\}.$$

If $a \in F$ and $a =^{\mu} a'$, then $a' \in F$, so the class F is $=^{\mu}$ invariant.

If $a \in F$, then $\mathrm{co}\, a \in F$. Indeed, find a borelian A, such that

$$a =^{\mu} \{z \mid f(z) \in A\} = f^{-1}(A).$$

Since

$$\mathrm{C}f - f^{-1}(A) = f^{-1}(\mathrm{co}\, A), \qquad [\mathrm{co}\, A =_{df} P - A],$$

we get

$$\mathrm{C}f - a = f^{-1}(\mathrm{co}\, A);$$

hence

$$\mathrm{C}f - a \in F.$$

We have

$$\mathrm{C}f = f^{-1}(P) \in F.$$

If $a_1, a_2, \ldots \in F$, we can find borelian sets $A_1, A_2, \ldots$, such that

$$a_n = f^{-1}(A_n).$$

Since

$$f^{-1}(\sum_n A_n) = \sum_n f^{-1}(A_n),$$

it follows that

$$a_1 + a_2 + \cdots \in F.$$

Thus we have proved, that the class F is a subclass of $\mathfrak{L}$, closed with respect to complementation and finite and denumerable addition. Consequently the ordering correspondence $\subseteqq^{\mu}$ on $\mathfrak{L}_{\mu}$, if restricted to F, organizes F into a denumerably genuine subtribe F of $\mathfrak{L}_{\mu} \cdot \mathrm{C}f$.

P.3.6. Remark. If we restrict ourselves to subsets of $\mathbb{Q}f$, and restrict accordingly the equality $=^{\mu}$ to subsets of $\mathbb{Q}f$ only, we get a tribe, all-operation-isomorphic to F. This remark is related to the fact, that the above topic can be founded by taking a fixed μ-measurable (or borelian) set as the unit of tribes.

P.4. Def. The tribe composed of all somata

$$a =^{\mu} \{z \mid f(z) \in A\},$$

where A is a variable borelian subset of the complex plane P, and taken with respect to the equality $=^{\mu}$, may be termed borelian scale of $f(\dot{z})$, or *borelian tribe of repartition of* $f(\dot{z})$, *repartition-tribe.*

Its unit is $\mathbb{Q}f$, its zero is $\mathbf{O}$, its ordering is $(\subseteq^{\mu})$.

P.4.1. Theorem. If

1. $f(\dot{z})$, $g(\dot{z})$ are μ-measurable,
2. $f(\dot{z}) =^{\mu} g(\dot{z})$,

then the repartition tribes of $f(\dot{z})$, $g(\dot{z})$ coincide.

Proof. Straight forward.

P.4.2. Def. Let $f(\dot{z})$ be a measurable function. The set-valued function of the variable plane-quarter Q,

$$F(Q) =^{\mu} \{z \mid f(z) \in Q\},$$

$$Q =_{df} \left\{ z \,\middle|\, \begin{matrix} \text{Real}\, z \leqq \alpha \\ \text{Imag}\, z \leqq \beta \end{matrix} \right\}, \quad \text{where} \quad \begin{matrix} -\infty \leqq \alpha \leqq +\infty, \\ -\infty \leqq \beta \leqq +\infty, \end{matrix}$$

will be termed *the* (double) *scale of repartition of* $f(\dot{z})$.

P.4.3. Theorem. If $f(\dot{z}) =^{\mu} g(\dot{z})$, and $F(\dot{Q})$, $G(\dot{Q})$ are the respective scales of repartition of f and g, then $F(Q)$ and $G(Q)$ are μ-equal.

Proof. Straight forward, since the domain of the scale of repartition is a part of the tribe of repartition.

P.4.4. Theorem. If

1. $f(\dot{z})$ is measurable,
2. $F(\dot{Q})$ its scale of repartition of $f(\dot{z})$,

then the following properties hold true:

1) $F(\mathbf{O}) =^{\mu} \mathbf{O}$,

2) if $Q \subseteq Q'$, then $F(Q) \subseteq^{\mu} F(Q')$,

3) if $\alpha_1 \geqq \alpha_2 \geqq \cdots \to \alpha_0$, $\beta_1 \geqq \beta_2 \geqq \cdots \to \beta_0$,

$$Q_n = \left\{ z \,\middle|\, \begin{matrix} \text{Real}\, z \leqq \alpha_n \\ \text{Imag}\, z \leqq \beta_n \end{matrix} \right\}, \quad (n = 0, 1, 2 \ldots)$$

then

$$\prod_{n=1}^{\infty} F(Q_n) =^{\mu} F(Q_0),$$

4) if $\alpha_1 \leqq \alpha_2 \leqq \cdots \to +\infty$, $\beta_1 < \beta_2 < \cdots \to +\infty$,

then

$$\sum_{n=1}^{\infty} F(Q_n) =^{\mu} F(P).$$

Proof does not offer difficulties.

P.4.5. Theorem. If

1. $f(\dot{z})$ is μ-measurable,
2. $F(\dot{Q})$ its scale of repartition,
3. (F) its tribe of repartition,

then (F) is the smallest denumerably additive subtribe of $(\mathfrak{L}_\mu)$ containing all somata of $F(Q)$.

Proof. This follows from the fact, that the borelian tribe $\mathfrak{B}$ of subsets of P is generated by finite and denumerable operations, applied to plane-quarters.

P.4.6. Theorem. If

1. $f(\dot{z})$, $g(\dot{z})$ are measurable,
2. $F(Q)$, $G(Q)$ are their scales of repartition,
3. $F(\dot{Q}) =^{\mu} G(\dot{Q})$,

then the tribes of repartition F, G coincide.

Proof. Straight forward.

P.4.6.1. Remark. It is not true, that if the tribe-sets of repartition of two measurable functions coincide, the functions are μ-equal. Ex. various constant functions.

P.4.7. Theorem. If

1. $f(\dot{z})$, $g(\dot{z})$ are measurable,
2. $F(\dot{Q})$, $G(\dot{Q})$ are their scales of repartition,
3. $F(Q) =^{\mu} G(Q)$ for every plane quarter Q,

then

$$f(\dot{z}) =^{\mu} g(\dot{z}).$$

Proof. Suppose this be not true. There exists a measurable set a with $\mu(a) > 0$, such that for all $z \in a$:

$$f(\dot{z}) \neq^{\mu} g(\dot{z}).$$

We have:

$$(1) \qquad \{z \mid f(z) \neq g(z)\} = \{z \mid \operatorname{Real} f(z) \neq \operatorname{Real} g(z)\} \text{ or } \{z \mid \operatorname{Imag} f(z) \neq \operatorname{Imag} g(z)\}.$$

Indeed, if

$$z \in \{z \mid f(z) \neq g(z)\},$$

then

$$f(z) \neq g(z);$$

hence either

$$\operatorname{Real} f(z) \neq \operatorname{Real} g(z) \quad \text{or} \quad \operatorname{Imag} f(z) \neq \operatorname{Imag} g(z).$$

Hence the set on the left in (1) is contained in the set on the right. Conversely, if $\mathrm{Real} f(z) \neq \mathrm{Real} g(z)$ or $\mathrm{Imag} f(z) \neq \mathrm{Imag}(z)$, then we have $f(z) \neq g(z)$, so the set on the right in (1) is contained in the set on the left.

Now both sets on the right are measurable, and then the measure of one of them is positive.

Suppose that the set

$$b =_{df} \{z \mid \mathrm{Real} f(z) \neq \mathrm{Real} g(z)\}$$

has a positive measure. We have

$$b = \{z \mid \mathrm{Real} f(z) < \mathrm{Real} g(z)\} + \{z \mid \mathrm{Real} f(z) > \mathrm{Real} g(z)\}.$$

Both sets at the right side are measurable and one of them, at least, has a positive measure. Suppose that $\mu(c) > 0$, where

$$c = \{z \mid \mathrm{Real} f(z) < \mathrm{Real} g(z)\}.$$

Now, put for rational numbers w_1, w_2, where $w_1 < w_2$:

$$E(w_1, w_2) =_{df} \{z \mid \mathrm{Real} f(z) \leqq w_1 < w_2 \leqq \mathrm{Real} g(z)\}.$$

Since

$$c = \sum_{w_1 < w_2} E(w_1, w_2),$$

at least one of the terms has a positive measure. Suppose

$$d =_{df} \mu E(w_1', w_2') > 0, \quad \text{where} \quad w_1' < w_2'$$

are rational numbers. We have

(1) $$d = \{z \mid \mathrm{Real} f(z) \leqq w_1'\} \cdot \{z \mid \mathrm{Real} g(z) \geqq w_2'\}.$$

Since $F(\dot{Q}) =^{\mu} G(\dot{Q})$ for all plane quarters Q, it follows that, if we put

$$\alpha =_{df} \{u \mid \mathrm{Real}\, u \leqq w_1'\},$$

we get

$$F(\alpha) =^{\mu} G(\alpha).$$

This means that

$$\{z \mid f(z) \in \alpha\} =^{\mu} \{z \mid g(z) \in \alpha\}.$$

Hence

$$\{z \mid \mathrm{Real} f(z) \leqq w_1'\} =^{\mu} \{z \mid \mathrm{Real} g(z) \leqq w_1'\}.$$

Hence from (1) we have

$$d \subseteq^{\mu} \{z \mid \mathrm{Real} g(z) \leqq w_1'\},$$

and

$$d \subseteq^{\mu} \{z \mid \mathrm{Real} g(z) \geqq w_2'\}.$$

Hence

$$\{z \mid \mathrm{Real} g(z) \leqq w_1'\} \cdot \{z \mid \mathrm{Real} g(z) \geqq w_2'\} \neq^{\mu} 0,$$

i.e. there exists z_0 such that

$$w_2' \leqq g(z_0) \leqq w_1',$$

which, gives $w_2' \leqq w_1'$, which contradicts the inequality $w_1' < w_2'$. The theorem is proved.

P.4.8. Theorem. If

1. $f(\dot{z})$, $g(\dot{z})$ are measurable,
2. $F(\dot{Q})$, $G(\dot{Q})$ are their scales of repartition,
3. the set $u_1, u_2, \ldots, u_n, \ldots$ of complex numbers is everywhere dense in the plane P,
4. for every plane quarter Q_n, where

$$Q_n = \left\{ u \,\middle|\, \begin{matrix} \operatorname{Real} u \leqq \operatorname{Real} u_n \\ \operatorname{Imag} u \leqq \operatorname{Imag} u_n \end{matrix} \right\}, \quad \text{we have} \quad F(Q_n) =^{\mu} G(Q_n),$$

then

$$f(z) =^{\mu} g(z).$$

Proof. Straight forward.

P.4.9. Theorem. Let

1. $f(\dot{z})$ be measurable,
2. $F(\dot{Q})$ its scale of repartition,
3. $Q_n = \left\{ u \,\middle|\, \begin{matrix} \operatorname{Real} u \leqq \operatorname{Real} u_n \\ \operatorname{Imag} u \leqq \operatorname{Imag} u_n \end{matrix} \right\}, \quad (n = 1, 2, \ldots)$
4. $u_1, u_2, \ldots, u_n, \ldots$ is everywhere dense in the complex plane P,
5. $g(z) =_{df} \inf\{\operatorname{Real} u_n \mid f(z) \in Q_n\} + i \inf\{\operatorname{Imag} u_n \mid f(z) \in Q_n\}$,

then

$$f(\dot{z}) =^{\mu} g(\dot{z}).$$

Proof. Straight forward.

P.5. Theorem. If

1. $f(\dot{z})$ is measurable,
2. (F) its tribe of repartition,

then the following are equivalent:

I. $f(z)$ is almost μ-everywhere constant on $\complement f$,

II. F is composed of the somata $\emptyset$ and $\complement f$ only.

Proof. It is sufficient to suppose that $\mu \complement f > 0$, because in the event $\mu \complement f = 0$, the theorem is trivial.

If I then II follows obviously.

Suppose II. I say that if E_1, E_2 are disjoint borelian subsets of P, then at least one of the sets

$$e_1 =_{df} f^{-1}(E_1), \quad e_2 =_{df} f^{-1}(E_2)$$

must have the μ-measure 0.

Suppose this be not true i.e. $\mu e_1 > 0$, $\mu e_2 > 0$.

Since $e_1, e_2 \in F$, we must have $e_1 =^\mu \complement f$, $e_2 =^\mu \complement f$; hence $e_1, e_2 \neq 0$ which implies the existence of a point z, where $f(z) \in E_1$ and $f(z) \in E_2$, which is impossible, since $E_1 \cdot E_2 = \emptyset$. Having this, consider the halfopen squares

$$R_{n,m} = \left\{ z \,\middle|\, \begin{matrix} n < \text{Real} z \leqq n+1 \\ m < \text{Imag} z \leqq m+1 \end{matrix} \right\} \quad \text{for all} \quad n, m = 0, \pm 1, \pm 2, \ldots$$

They are disjoint and their union is P. We have

$$f^{-1}(P) =^\mu \sum_{n,m} f^{-1}(R_{n,m}).$$

Now, since the squares are disjoint, there cannot exist two of them, whose inverse image has the positive measure. Since $f^{-1}(P) = \complement f$ and $\mu \complement f \neq 0$, it follows, that there exists one and only one square, say R', such that

$$f^{-1}(R') =^\mu \complement f.$$

Let us partition R' into four equal halfopen squares

$$R'_i, \quad (i = 1, 2, 3, 4).$$

Using the same argument as before, we prove the existence of a single index say i_1 such that

$$f^{-1}(R'_{i_1}) =^\mu \complement f.$$

If we partition again R'_{i_1} into four equal squares $R'_{i_1, i}$ where $i = 1, 2, 3, 4$, we get a single index i_2, such that

$$f^{-1}(R'_{i_1, i_2}) =^\mu \complement f.$$

Continuing, by induction, the above process, we obtain an infinite sequence of nested squares [call them

$$R''_1 \supset R''_2 \supset R''_3 \supset \cdots],$$

such that $\prod_n R''_n$ is composed of a single point u_0, and where

$$f^{-1}(R''_n) =^\mu \complement f \quad \text{for} \quad n = 1, 2, \ldots$$

It follows that:

$$\prod_{n=1}^{\infty} f^{-1}(R''_n) = f^{-1}\left(\prod_{n=1}^{\infty} R''_n\right) = f^{-1}(u_0) =^\mu \complement f.$$

Hence for almost all $z \in \complement f$ we have $f(z) = u_0$. Hence I follows. The theorem is proved.

P.5.1. Theorem. The following are equivalent for a measurable function $f(\dot{z})$:

I. $f(z)$ is μ-equal to a borelian step function,

II. the repartition tribe (F) of $f(\dot{z})$ is composed of a finite number of somata only.

Proof. Let I. Let $f(\dot{z}) =^{\mu} g(\dot{z})$, where $g(\dot{z})$ is a borelian step-function. This means, that there exists a finite number of disjoint borelian sets

$$a_1, a_2, \ldots, a_m, \quad (m \geqq 1),$$

and different complex numbers

(1) $$u_1, u_2, \ldots, u_m, \ldots,$$

such that $f(z) = u_k$, whenever

$$z \in a_k, \quad (k = 1, 2, \ldots),$$

and where

$$\sum_{k=1}^{m} a_k = \complement g =^{\mu} \complement f.$$

If E is a borelian subset of P, and does not contain any point (1), we have $g^{-1}(E) = \emptyset$. In the other case $g^{-1}(E)$ is the union of some sets a_k. Thus all somata of F are either $\emptyset$ or finite unions of the sets a_k. Hence the tribe of $g(\dot{z})$ contains a finite number of somata only. Since $f =^{\mu} g$, it follows that (F) contains a finite number of somata only. Thus I $\rightarrow$ II.

Now suppose II. We may suppose, that the zero of (F) differs from its unit. There exists a finite number of μ-disjoint somata (atoms)

$$a_1, a_2, \ldots, a_m, \quad (m \geqq 1),$$

such that every soma of F is the sum of some of them. Consider the function $a_k \upharpoonleft f(\dot{z})$. Its repartition tribe is $a_k F$; hence it is composed of O and a_k only.

Hence we have $f(\dot{z}) =^{\mu}$ constant on a_k, say u_k. Since a_i, a_k are μ-disjoint, i.e.

$$\mu(a_i \cdot a_k) = 0 \quad \text{for} \quad i \neq k,$$

the set

$$a_0 =_{df} \sum_{\substack{i,k=1 \\ i \neq k}}^{m} a_i \cdot a_k$$

has the measure 0.

Put

$$a'_k =_{df} a_k - a_0, \quad a'_i =_{df} a_i - a_0.$$

We have

$$a'_i \cdot a'_k = (a_i - a_0) \cdot (a_k - a_0) \subseteqq^{\mu} (a_i - a_i \cdot a_k) \cdot (a_k - a_i a_k) \subseteqq$$

$$\subseteqq (a_i - a_k) \cdot (a_k - a_i) = a_i \cdot \operatorname{co} a_k \cdot a_k \cdot \operatorname{co} a_i = \emptyset.$$

Thus we have

$$a'_k =^{\mu} a_k;$$

hence

$$a'_1 + a'_2 + \cdots + a'_m =^{\mu} a_1 + \cdots + a_k =^{\mu} \complement f.$$

There exist borelian subsets a_k'' of a_k' with $a_k'' =^\mu a_k'$.

We have

1) all a_k'' are disjoint,

2) $\sum_{k=1}^{\infty} a_k'' =^\mu \complement f$,

3) we have $f(z) = u_k$ for all $z \in a_k''$.

Consequently, if we define on $\sum_{k=1}^{\infty} a_k''$, the function $\varphi(z)$, by putting

$$\varphi(z) = u_k, \quad \text{whenever} \quad z \in a_k'',$$

we get a borelian step-function. We have

$$\varphi(z) =^\mu f(z).$$

The theorem is proved.

P.6. Theorem. If $f(z)$ is measurable, F its repartition-tribe, then the following are equivalent:

I. F is μ-atomic,

II. f is equivalent to a simple borelian function $g(z)$ i.e. such that there exist disjoint borelian sets a_n with union μ-equal to $\complement f$, on each of which $g(z)$ being constant.

Proof. Similar to the preceding one, by taking account of that the number of disjoint atoms is at most denumerable, because each atom must have a positive measure.

P.7. Let $g(z)$ be measurable. Define for all borelian subsets A of the complex plane P the function

$$v(A) =_{df} \mu(g^{-1} A). \tag{1}$$

This set-function is $\geqq 0$ and denumerably additive; hence it is a denumerably additive measure on the tribe of all borelian subsets of P. It can be considered as extension of the measure $v(f)$ defined for all figures f of the plane. The measure (1) can be extended in the Lebesguean manner, by using to the denumerably additive collection, composed of all borelian sets A for which $v(A) = 0$. We get subsets of P, which we may call „v-measurable". Now to the class of all v-measurable sets surely belongs $\complement g$, because $g^{-1}(\mathrm{D}\, g) = \complement g$, which is borelian, and we have

$$v(\mathrm{D}\, g) = \mu(\complement g).$$

For every v-measurable set A we have

$$g^{-1} A = g^{-1}(A . \mathrm{D}\, g)$$

We recall that if $f(z)$ is a μ-measurable function, then there exists a borelian function $h(z)$ such that

1) $\complement h \subseteqq \complement f$,
2) $h(z) = f(z)$ for all $z \in \complement h$, so $h(z) = \complement h \upharpoonright f(z)$,
3) $f(z) =^{\mu} h(z)$,
4) $\complement h$ is a borelian set, [P.3.2.].

P.7.1. In the sequal we shall use, as a tool a tribe on P, defined by means of a given μ-measurable function $f(\dot{z})$, and by the „measure" v, [P.7.]:

Let $f(z)$ be a μ-measurable function, and A, B be borelian subsets of the complex plane P.

We define

$$A \subseteqq^{v} B \quad \text{by} \quad f^{-1}(A) \subseteqq^{\mu} f^{-1}(B).$$

This is an ordering correspondence.

Indeed, we have

$A \subseteqq^{v} A$,
if $A \subseteqq^{v} B$, $B \subseteqq^{v} C$, then $A \subseteqq^{v} C$.

From hyp. we get $f^{-1}(A) \subseteqq^{\mu} f^{-1}(B)$ and $f^{-1}(B) \subseteqq^{\mu} f^{-1}(C)$; hence $f^{-1}(A) \subseteqq^{\mu} f^{-1}(C)$ i.e. $a \subseteqq^{v} C$.

We define, as usually, $A =^{v} B$ by $A \subseteqq^{v} B$, $B \subseteqq^{v} A$.

P.8. Theorem. The following are equivalent for borelian sets A, B:

I. $A =^{v} B$.
II. $f^{-1}(A) =^{\mu} f^{-1}(B)$.

Indeed, let I. We have $A \subseteqq^{v} B$, $B \subseteqq^{v} A$.

Hence

$$f^{-1}(A) \subseteqq^{\mu} f^{-1}(B), \qquad f^{-1}(B) \subseteqq^{\mu} f^{-1}(A),$$

i.e.

$$f^{-1}(A) =^{\mu} f^{-1}(B),$$

i.e. II.

Let II. We have

$$f^{-1}(A) \subseteqq^{\mu} f^{-1}(B), \qquad f^{-1}(B) \subseteqq^{\mu} f^{-1}(A);$$

hence

$$A \subseteqq^{v} B, \quad B \subseteqq^{v} A \quad \text{i.e.} \quad A =^{v} B$$

i.e. I.

P.8.1. Theorem. The empty $\emptyset$ subset of P is the zero of this ordering and $\mathrm{D} f$ is the unit.

Proof. We have $f^{-1}(\emptyset) = \emptyset$, and then for every borelian subset A of P we have

$$f^{-1}(\emptyset) \subseteqq^{\mu} f^{-1}(A).$$

Hence

$$\emptyset \subseteqq^{v} A.$$

We have $f^{-1}(\mathsf{D}\, f) = \reflectbox{D}\, f$, and for every borelian A we have

$$f^{-1}(A) \subseteqq \reflectbox{D}\, f, \quad \text{hence} \quad f^{-1}(A) \subseteqq^{\mu} \reflectbox{D}\, f.$$

Consequently $A \subseteqq^{v} \mathsf{D}\, f$ for every borelian set A.

P.8.2. Theorem. We have

$$P =^{v} \mathsf{D}\, f.$$

Indeed,

$$f^{-1}(P) = \reflectbox{D}\, f, \quad f^{-1}(\mathsf{D}\, f) = \reflectbox{D}\, f;$$

hence

$$f^{-1}(P) = f^{-1}(\mathsf{D}\, f),$$

and then

$$f^{-1}(P) =^{\mu} f^{-1}(\mathsf{D}\, f),$$

so

$$P =^{v} \mathsf{D}\, f.$$

P.8.3. Theorem. If

1. A, B are borelian sets,
2. $A \subseteqq B$,

then

$$A \subseteqq^{v} B.$$

Proof. We have

$$f^{-1}(A) \subseteqq f^{-1}(B),$$

hence

$$f^{-1}(A) \subseteqq^{\mu} f^{-1}(B),$$

so $A \subseteqq^{v} B$ follows. The converse is, of course, not true, e.g. we have $P \subseteqq^{v} \mathsf{D}\, f$, but it may not be $P \subseteqq \mathsf{D}\, f$.

P.8.4. Theorem. The correspondence $\subseteqq^{v}$ is a denumerably additive Boolean tribe.

Proof. Let $A_1, A_2, \ldots, A_n, \ldots$ be a finite or infinite sequence of borelian sets.

Put

$$A =_{df} \sum_n A_n.$$

Since

$$A_n \subseteqq A,$$

we get, [P.8.3.],

$$A_n \subseteqq^{v} A \quad \text{for all } n.$$

Now suppose that

$$A_n \subseteqq^{v} B \quad \text{for all } n.$$

It follows

$$f^{-1}(A_n) \subseteqq^{\mu} f^{-1}(B) \quad \text{for all } n.$$

Hence

$$\sum_n f^{-1}(A_n) \subseteqq^{\mu} f^{-1}(B),$$

i.e.

$$f^{-1}(\sum_n A_n) \subseteqq^{\mu} f^{-1}(B),$$

i.e.

$$f^{-1}(A) \subseteqq^{\mu} f^{-1}(B).$$

Consequently $A \subseteqq^v B$. This proves the existence of the joint $\bigcup_n^v A_n$, and we have at the same time

$$\bigcup_n^v A_n =^v \sum_n A_n.$$

Similarly we prove that the meet $\bigcap_n^v A_n$ exists and that

$$\bigcap_n^v A_n =^v \prod_n A_n.$$

Now we shall prove that the lattice $\subseteqq^v$ is *complementary*. Let A be a borelian subset of P. We shall prove that, if we put

$$A' = \mathrm{D}f - A,$$

we get

$$A \frown^v A' =^v \emptyset, \qquad A \smile^v A' =^v \mathrm{D}f.$$

By what has been already proved, we have

$$A \frown^v A' =^v A \cdot A' = \emptyset;$$

hence

$$A \frown^v A' =^v \emptyset,$$

$$A \smile^v A' =^v A + A' = A + (\mathrm{D}f - A) = A + \mathrm{D}f.$$

Hence

$$A \smile^v A' =^v A \smile^v \mathrm{D}f =^v A \smile^v P, \qquad A \smile^v P =^v P,$$

because $A \subseteqq^v P$, and then

$$A \smile^v A' =^v \mathrm{D}f,$$

[P.8.2.], so the assertion is proved.

Since $\mathrm{D}f$ is the unit in the ordering $\subseteqq^v$, it follows that $A' =^v \mathrm{co}^v A$; hence A' is the $\subseteqq^v$-complement of A. Thus $(\subseteqq^v)$ is a complementary lattice. We easily prove that co^v is $=^v$ invariant. Hence the lattice subtraction

$$A -^v B =_{df} A \frown^v (\mathrm{co}^v B)$$

is $=^v$ invariant.

We also have

$$A -^v B =^v A - B.$$

It remains to prove the distributive law. Let A, B, C be any borelian sets. We have

$$(A \smile^v B) \frown^v C =^v (A + B) \cdot C =^v A \cdot C + B \cdot C =^v (A \frown^v C) \smile^v (B \frown^v C).$$

The theorem is established.

P.9. We have defined on the tribe $\underline{\underline{C}}^v$ the measure

$$v(A) \text{ by } v(A) =_{df} \mu(f^{-1}(A)), \text{ [P.7.]}.$$

P.9.1. Theorem. The measure $v(A)$ is $=^v$ invariant, is $\geqq 0$ and denumerably additive.

Proof. Obviously we have $v(A) \geqq 0$. Let $A =^v B$. We have

$$f^{-1}(A) =^\mu f^{-1}(B),$$

and then

$$\mu f^{-1}(A) = \mu f^{-1}(B);$$

hence

$$v(A) = v(B),$$

so the $(=^v)$ invariance of the measure v is proved.

Let $A_1, A_2, \ldots, A_n, \ldots$ be a finite or denumerably infinite sequence of borelian subsets of P, and suppose they are $(=^v)$ disjoint, i.e.

$$A_i \frown^v A_j =^v \emptyset \quad \text{for} \quad i \neq j.$$

We have

$$A_i \cdot A_j =^v \emptyset \quad \text{for} \quad i \neq j.$$

Hence

$$f^{-1}(A_i \cdot A_j) = f^{-1}(A_i) \cdot f^{-1}(A_j) =^\mu \emptyset,$$

hence

$$f^{-1}(A_1), f^{-1}(A_2), \ldots$$

are all μ-disjoint with one another.

Since μ is a denumerably additive measure on the tribe of all μ-measurable sets, we have

$$\mu[\sum_n f^{-1}(A_n)] = \sum_n \mu(f^{-1}(A_n)) = \sum_n v(A_n).$$

Since

$$v[\mathbf{U}^v_n A_n] = v[\sum_n A_n],$$

by the $=^v$ invariance of the measure v, and since

$$f^{-1}(\sum_n A_n) = \sum_n f^{-1}(A_n),$$

it follows that

$$v(\mathbf{U}^v_n A_n) = \sum_n v(A_n).$$

The theorem is proved.

P.9.2. Theorem. If A is a borelian set, then the following are equivalent:

I. $v(A) = 0$,

II. $A =^v \emptyset$.

Proof. Suppose II. Since v is v-invariant, it follows that

$$v(A) = v(\mathbf{O}) = \mu[f^{-1}(\mathbf{O})] = \mu(\mathbf{O}) = 0,$$

hence I.

Suppose I, i.e. $v(A) = 0$. It follows that $\mu[f^{-1}(A)] = 0$; hence $f^{-1}(A)$ is a μ-null set, i.e. $f^{-1}(A) =^{\mu} f^{-1}(\mathbf{O})$. It follows that $A =^{v} \mathbf{O}$ i.e. II.

P.9.3. Theorem. If A, B are borelian sets, then the following are equivalent:

I. $A =^{v} B$,

II. $v(A \dot{+} B) = 0$, $[A \dot{+} B =_{df} (A - B) + (B - A)]$.

Proof. Let I. It follows that

$$A \dot{-} B =^{v} \emptyset, \quad B \dot{-} A =^{v} \emptyset,$$

algebraic subtraction, and then

$$(A \dot{-} B) \cup (B \dot{-} A) =^{v} \mathbf{O}.$$

Consequently

$$(A - B) + (B - A) =^{v} \mathbf{O}.$$

Hence

$$A + B =^{v} \mathbf{O}.$$

We get

$$v(A \dot{+} B) = 0,$$

i.e. II. Let II. We get

$$A \dot{+} B =^{v} \mathbf{O}.$$

Hence

$$A - B =^{v} \mathbf{O},$$

which gives

$$A \subseteqq^{v} B,$$

and also

$$B - A =^{v} 0,$$

which gives

$$B \subseteqq^{v} A.$$

Consequently

$$A =^{v} B,$$

i.e. I.

P.9.4. Theorem. The $=^{v}$ equality-invariant set of all borelian sets A with $v(A) = 0$, is a denumerably additive ideal J in the tribe $\subseteqq^{v}$.

Proof easy.

P.9.5. Theorem. The tribe $\subseteqq^{v}$ coincides with the tribe $(\mathfrak{B}_J)$ of all borelian sets, taken modulo J. The measure v is effective in it.

P.9.6. Having this we can complete in Lebesguean manner the tribe $\subseteqq^{v}$ to another one of all v-measurable sets. The equality will be extended to the extended tribe, and the same can be made with the measure v. We get v-measurable subsets of P. We shall maintain the sign $=^v$ for the completed tribe.

P.9.7. Theorem. The following are equivalent for a borelian function f:

I. A is a v-measurable (it may be not borelian, but belongs to the v-hull of $(\mathfrak{B})$.

II. $f^{-1}(A)$ is μ-measurable.

Proof. Let I. There exist borelian sets B', B'', such that

$$B' \subseteqq A \subseteqq B'', \text{ and} \tag{1}$$

$$v(B') = v(B''). \tag{2}$$

From (1) it follows

$$f^{-1}(B') \subseteqq f^{-1}(A) \subseteqq f^{-1}(B''),$$

and from (2)

$$\mu f^{-1}(B') \leqq \mu f^{-1}(B'').$$

Since $f^{-1}(B')$, $f^{-1}(B'')$ are borelian sets, it follows that $f^{-1}(A)$ is μ-measurable. Thus II follows.

Starting with the definition of extended measure:

$$v(A) =_{df} \mu f^{-1}(A)$$

for v-measurable sets A, and imitating [P.8.4.], we get $(\mathfrak{B})$. Thus II implies I.

P.9.8. Theorem. The following are equivalent:

I. A is a v-null, v-measurable set.

II. $f^{-1}(A)$ is a μ-null set.

Proof. Let I. We have $\mu f^{-1}(A) = 0$, so II.

Let II. Put $a = f^{-1}(A)$. Since $\mu f^{-1}(A) = 0$, we have $v(A) = 0$, i.e. I.

P.10. Def. Let us define the correspondence $\mathcal{T}$ as follows:

Let A be a v-measurable set (it may be not borelian) and let a be a μ-measurable set. We define

$$A \mathcal{T} a$$

by: there exists a borelian set A' and a μ-measurable set a' such that:

1) $A' =^v A$,

2) $a =^\mu a'$, $a' = f^{-1}(A')$.

P.10.1. Theorem. The correspondence $\mathcal{T}$ is equality invariant with respect to $=^v$ in the domain $\mathrm{Q}\,\mathcal{T}$ (which comprises all v-measurable subsets of P), and with respect to $=^\mu$ in the range of $\mathcal{T}$, which is composed of μ-measurable sets.

$\mathcal{T}$ is an isomorphism which preserves finite and denumerable operations, inclusion, zero and unit. If

$$A \mathcal{T} a, \quad \text{then} \quad v(A) = \mu(a).$$

Proof straight forward.

P.10.2. Theorem. If $f(\dot{z})$ is measurable with $\mu \, \mathrm{G} f > 0$, then the following are equivalent:

I. The tribe (F) of repartition of $f(\dot{z})$ coincides with $\mathrm{G} f \cdot \mathfrak{L}_\mu$.

II. There exists a set a_0 with $\mu(a_0) = 0$ such that the function $(\mathrm{G} f - a_0) \upharpoonright f(\dot{z})$ is strictly one to one.

III. If a, b are μ-measurable sets $\subseteqq \mathrm{G} f$, $a \cdot b = \mathsf{O}$ (or $=^\mu \mathsf{O}$), then there exists a set c with $\mu\, c = 0$ such that

$f(a - c) \cdot f(b - c) = \emptyset$. [In general $f(p)$ means $\{f(x) \mid x \in p\}$]

Proof. Suppose III. We shall prove II. Divide the whole plane P into halfopen equal squares

(0) $$q_1, q_2, \ldots, q_n, \ldots,$$

which are mutually disjoint.

There exists p_{ik} with $\mu(p_{ik}) = 0$, such that for $i \neq k$:

(1) $$f(q_i - p_{ik}) \cdot f(q_k - p_{ik}) = 0, \quad i, k = 1, 2, \ldots$$

Put

$$p =_{df} \sum_{i,k=1}^{\infty} p_{ik}.$$

We have

$$\mu\, p = 0.$$

We have

$$q_i - p \subseteqq q_i - p_{ik},$$

hence

$$f(q_i - p) \subseteqq f(q_i - p_{ik});$$

and similarly:

$$f(q_k - p) \subseteqq f(q_k - p_{ik}).$$

Hence, by (1),

$$f(q_i - p) \cdot f(q_k - p) \subseteqq f(q_i - p_{ik}) \cdot f(q_k - p_{ik}) = 0.$$

It follows that all sets

$$f(q_1 - p), f(q_2 - p), \ldots, f(p_n - p), \ldots$$

are mutually disjoint.

This is true for any division of P into equal spaces. Put $p^{(1)} =_{df} p$. Now, starting with a given division (0), subdivide each square q_i into four equal squares $q_{i1}, q_{i2}, q_{i3}, q_{i4}$. They form together another division of P into equal squares, so we can find a set $p^{(2)}$ of measure 0, such that all sets

$$f(q_{i_1 i_2} - p^{(2)}), \quad i_1 = 1, 2, \ldots, \quad i_2 = 1, 2, 3, 4,$$

are disjoint.

Continuing the subdivisions, we obtain an infinite sequence of sets $p^{(n)}$ of measure O such that all sets

$$f(q_{i_1 i_2 \ldots i_n} - p^{(n)}),$$

$$(i_1 = 1, 2, \ldots; \quad i_2, \ldots, i_n = 1, 2, 3, 4)$$

are mutually disjoint.

Now consider the function

$$g(\dot{z}) =_{df} (P - p^{(0)}) \upharpoonright f(\dot{z}), \quad \text{where} \quad p^{(0)} =_{df} \sum_{n=1}^{\infty} p^{(n)}.$$

Since $\mu(p^0) = 0$, we have

$$g(\dot{z}) \doteq f(\dot{z}).$$

I say that $g(z)$ is strictly one to one.

Suppose $z_1 \neq z_2$ and $g(z_1) = g(z_2)$. There exists an N, such that in the N-th division of P, the points z_1, z_2 are lying in disjoint meshes, say

$$q_{i_1, i_2, \ldots i_N}, \quad q_{k_1, \ldots, k_N}.$$

Now,

$$z_1 \bar{\in} p^{(N)}, \quad z_2 \bar{\in} p^{(N)},$$

hence

Since the sets

$$g(q_{i_1 \ldots i_N} - p^{(N)}), \quad g(q_{k_1 \ldots k_N} - p^{(N)})$$

are disjoint, we have

$$g(z_1) \neq g(z_2).$$

Thus II is proved.

Now suppose II. We shall prove I. Suppose there exists p, such that $\mu\, p = 0$; and that if we put

$$g(\dot{z}) =_{df} (P - p) \upharpoonright f(\dot{z}),$$

then $g(\dot{z})$ is strictly one to one.

Let a be any measurable subset of P with $\mu\, a \neq 0$. We have

$$a - p =^{\mu} a.$$

Put

(1 a) $$A =_{df} g(a - p).$$

We have

$$g\, g^{-1}(A) = A$$

and

$$a - p \subseteqq g^{-1}\, g(a - p);$$

hence

(2) $$a - p \subseteqq g^{-1}(A).$$

Suppose that $a - p \subset g^{-1}(A)$. Hence there exists $z_1 \in g^{-1}(A)$, such that $z_1 \bar{\in} a - p$. Put $u =_{df} g(z_1)$. We have $u \in A$; hence, but (1a), there exists $z_2 \in a - p$, such that $g(z_2) = u$. Hence there exists $z_1 \bar{\in} a - p$ and $z_2 \in a - p$ with $g(z_1) = g(z_2)$, which contradicts the hypothesis which says, that $g(\dot{z})$ is strictly one to one. Thus we have proved that

(3) $$a - p = g^{-1}(A).$$

Since $\mu(a - p) > 0$ there exists a borelian subset A' of A, such that

$$g^{-1}(A') =^{\mu} g^{-1}(A),$$

which can be proved by using v-measures.

We have

$$g^{-1}(A') \subseteqq g^{-1}(A).$$

Hence, by (3),

$$g^{-1}(A') =^{\mu} a - p =^{\mu} a.$$

Consequently $a \in G$, where (G) is the repartition-tribe of $g(\dot{z})$. Since $g(\dot{z}) =^{\mu} f(\dot{z})$, we have

$$a \in F.$$

Hence every measurable subset of $\mathrm{Q} f$ belongs to F; hence

$$F = \mathfrak{L}_{\mu} \cdot \mathrm{Q} f$$

i.e. I. Now suppose I. We shall prove III. Let $a \cdot b = O$, where a, b are μ-measurable.

There exist borelian sets A, B, such that

$$a' =_{df} f^{-1}(A) =^{\mu} a, \quad b' =_{df} f^{-1}(B) =^{\mu} b.$$

Hence

(1) $$f(a) = A, \quad f(B) = B.$$

Let us apply, the induced measure v on $\mathrm{D} f$, where in general,

$$v(X) = \mu f^{-1}(X).$$

The sets A, B are v-disjoint. Indeed

$$f^{-1}(A \cdot B) = f^{-1}(A) \cdot f^{-1}(B) =^{\mu} a \cdot b =^{\mu} O,$$

and then

$$v(A \cdot B) = \mu(a \cdot b) = 0.$$

Hence

$$A =^{v} A - A\,B = A - B,$$

$$B =^{v} B - A\,B = B - A.$$

We have

(2) $$(A - B) \cdot (B - A) = \emptyset.$$

Put

(3) $a'' =_{df} f^{-1}(A - B) =^{\mu} a, \quad b'' =_{df} f^{-1}(B - A) =^{\mu} b.$

Hence

$$f(a'') = A - B, \quad f(b'') = B - A,$$

and then

(3.0) $$f(a'') \cdot f(b'') = \emptyset.$$

We have

$$a = a\,a'' + (a - a''),$$

where the terms are disjoint.

Hence

(3.1) $$a\,a'' = a - (a - a'').$$

Now

(3.2) $$(a\,a'') \cdot (b - b'') = a\,a''\,b\,\mathrm{co}\,b'' = 0.$$

We get from (3.1), on account of (3.2),

$$a\,a'' = a - (a - a'') - (b - b'') = a\,\mathrm{co}(a - a'')\,\mathrm{co}(b - b'')$$
$$= a\,\mathrm{co}[(a - a'') + (b - b'')] = a - [(a - a'') + (b - b'')].$$

Hence, if we put

$$p =_{df} (a - a'') + (b - b''),$$

we get

(4) $$a\,a'' = a - p, \quad \text{and similarly} \quad b\,b'' = b - p.$$

Now since, by (3),

(5) $$a - a'', \quad b - b''$$

are μ-null sets, therefore p is a null set.

We have, by (4),

$$f(a - p) =^{\mu} f(a\,a'') \subseteqq f(a''),$$
$$f(b - p) =^{\mu} f(b\,b'') \subseteqq f(b'').$$

Hence, by (3.0),

$$f(a - p) \cdot f(b - p) = \emptyset,$$

where p is a μ-null set. Thus we have got III.

Since we have proved that III $\rightarrow$ II, II $\rightarrow$ I and I $\rightarrow$ III, the theorem is established.

P.10.3. Theorem. If

1. $f(z)$ is measurable with borelian $\mathrm{D}\,f$,
2. (F) its borelian tribe of repartition,
3. $\Phi(u)$ a borelian function defined on $\mathrm{D}\,f$,
4. $g(\dot{z}) =_{df} \Phi(f(\dot{z}))$ with $\mathrm{G}\,g = \mathrm{G}\,f$,
5. (G) the borelian tribe of repartition of $g(\dot{z})$,

then (G) is a denumerably additive, denumerably genuine strict subtribe of (F).

Proof. Of course $g(\dot{z})$ is measurable. Indeed, let α be borelian. $A =_{df} \Phi^{-1}(\alpha)$ is a borelian set, hence $a =_{df} f^{-1}(A)$ is a measurable set. Hence $g^{-1}(\alpha) = f^{-1}\Phi^{-1}(\alpha) = f^{-1}(A) = a$ is measurable. Since A is borelian, it follows that $f^{-1}(A) \in F$. Consequently $g^{-1}(\alpha) \in F$. Since if we vary α, we get all sets of G, it follows that $G \subseteqq F$.

Now, if $a \in G$ and $a =^{\mu} a'$, then $a' \in G$, and the ordering is given by the μ-inclusion. Hence G is a partial tribe of F. The zero and unit of G are the same as in F. Since G is a collection of somata of F, which is closed with respect to denumerable addition and the complementation, it follows that G is a denumerably genuine subtribe of F. The theorem is proved.

P.10.4. Theorem. If

1. $f(\dot{z})$ is a measurable function; $\mu(\complement f) > 0$, and $\mathrm{D} f$ borelian,
2. (F) its tribe of repartition,
3. (G) is a strictly denumerably genuine, denumerably additive subtribe of (F),

then there exists a real valued borelian function Φ, such that

1) $\mathrm{D} f = \complement \Phi$,

2) (G) is the borelian tribe of repartition for the function $g(z) =_{df} \Phi f(\dot{z})$ with domain $\complement g = \complement f$.

P.10.4a. Proof. We shall rely on the Theorem [(10). p. 23.], which we shall apply to the tribe (G).

The notion of distance

$$\|a, b\| =_{df} \mu(a \dotplus b)$$

organizes (G) into a metric topology, which is separable. Let $a_1, a_2, \ldots, a_n, \ldots$ be a denumerable everywhere dense set of somata of (G). According to the theorem mentioned, by using finite somatic operations on the somata a_n, we can construct a linear ordering (R) such that all somata of (G) can be represented by denumerable somatic operations executed on somata of (R).

P.10.4b. If $a \in R$, there exists a borelian subset $T(a)$ of $\mathrm{D} f$, such that

$$a \doteq f^{-1}(T(a)).$$

We have introduced the v-measure on borelian subsets of $\mathrm{D} f$, by putting

$$v A = \mu(f^{-1} A),$$

and completed this measure and the corresponding tribe. So we have got v-measurable subsets of $\mathrm{D} f$.

Denote by $=^{v}$ the governing equality of the tribe $T(G)$ of v-measurable sets. This tribe is isomorphic with (G). If $a \mathrel{\dot{\subseteqq}} b$, we have for corresponding sets $A =_{df} T(a)$, $B =_{df} T(b)$ the inclusion:

$$A \subseteq^v B.$$

Consequently the elements, T-corresponding to the elements of the chain R, form a v-chain $v(R)$, which is generated in the v-topology in the whole tribe $T(G)$. The chain (R) is denumerable. Let its all elements be:

$$c_1, c_2, \ldots, c_n, \ldots$$

and let the T-corresponding borelian sets be:

$$C_1, C_2, \ldots, C_n, \ldots$$

If

$$c_n \dot{\subseteq} c_m,$$

we have

$$C_n \subseteq^v C_m, \quad \text{so} \quad C_n - C_m$$

is a v-null set. Put

$$D =_{df} \sum_{n,m}^{\infty} (C_n - C_m), \qquad c_n \dot{\subseteq} c_m.$$

D is a v-null set.

P.10.4c. We shall prove that, if $c_n \subseteq c_m$, then

$$C_n - D \subseteq C_m - D.$$

We have

$$\begin{aligned}(C_n - D) - (C_m - D) &= C_n \operatorname{co} D \cdot \operatorname{co}(C_m \operatorname{co} D) = C_n \operatorname{co} D \cdot (\operatorname{co} C_m + D)\\ &= C_n \operatorname{co} D \operatorname{co} C_m + C_n \cdot \operatorname{co} D \cdot D = C_n \operatorname{co} D \operatorname{co} C_m = (C_n - C_m) \operatorname{co} D\\ &= (C_n - C_m) - D = (C_n - C_m) - \sum_{\substack{n,m\\ C_n \subseteq C_m}} (C_n - C_m) = \emptyset,\end{aligned}$$

Hence $C_n - D \subseteq C_m - D$, so the assertion is proved.

P.10.4d. Put

$$E_n =_{df} C_n - D \quad \text{for} \quad n = 1, 2, \ldots$$

Since D is a null set, we have

$$E_n =^v C_n.$$

The sets E_n form an ascending (R')-chain in the strict sense of set inclusion. All E_n are borelian, since C_n and D are so. We also have:

$$\sum_{n=1}^{\infty} C_n =^v \mathsf{D} f, \quad \sum_{n=1}^{\infty} E_n =^\mu \mathsf{D} f.$$

If we adjoin the borelian set $\mathsf{D} f$ to the chain (R'), this union will be strictly $= \mathsf{D} f$. Thus we can admit, that the equality

$$\sum_{n=1}^{\infty} E_n = \mathsf{D} f$$

is satisfied.

P.10.4e. As the chain (R') is denumerable, there exists an order isomorphism S from (R') onto a suitable subchain $\{\lambda\}$ of the chain of all rational numbers contained in the interval $\langle 0, 1\rangle$. Denote the set of the chain $S(R')$ by Ω. The following are equivalent for $\lambda', \lambda'' \in \Omega$:

I. $\lambda' \leqq \lambda''$,

II. $S^{-1}(\lambda') \subseteqq S^{-1}(\lambda'')$.

P.10.4f. Having this, define the function $\Phi(\dot{u})$ on $\mathrm{D}f$ as follows:

(0) $$\Phi(u) =_{df} \inf\{\lambda \in \Omega,\ u \in S^{-1}(\lambda)\}.$$

Lemma. If

$$\lambda \in \Omega,$$

then

$$S^{-1}(\lambda) = \{u \mid \Phi(u) \leqq \lambda\}.$$

Proof. Let $u_0 \in S^{-1}(\lambda)$. We have

$$\Phi(u_0) = \inf\{\lambda'_{\in\Omega} \mid u_0 \in S^{-1}(\lambda')\}.$$

Since

$$u_0 \in S^{-1}(\lambda),$$

we have

$$\lambda \in \{\lambda'_{\in\Omega} \mid u_0 \in S^{-1}(\lambda')\},$$

and then

$$\Phi(u_0) \leqq \lambda.$$

Hence

$$u_0 \in \{u \mid \Phi(u) \leqq \lambda\}.$$

This being true for all

$$u_0 \in S^{-1}(\lambda),$$

we get

(1) $$S^{-1}(\lambda) \subseteqq \{u \mid \Phi(u) \leqq \lambda\}.$$

Let

(1.1) $$u_0 \in \{u \mid \Phi(u) \leqq \lambda\}.$$

Hence

(2) $$\Phi(u_0) \leqq \lambda.$$

As

$$\Phi(u_0) = \inf\{\lambda' \mid u_0 \in S^{-1}(\lambda')\},$$

it follows from (2), that there exists

$$\lambda' \in \Omega \quad \text{with} \quad \lambda'' \leqq \lambda,$$

$$u_0 \in S^{-1}(\lambda''), \qquad \Phi(u_0) \leqq \lambda'.$$

Since

$$S^{-1}(\lambda'') \subseteqq S^{-1}(\lambda),$$

we get

$$u_0 \in S^{-1}(\lambda).$$

This being true for all u_0, satisfying (1.1), we have proved that

$$\{u|\Phi(u) \leqq \lambda\} \subsetneqq S^{-1}(\lambda). \tag{3}$$

From (1) and (3) the thesis follows.

P.11. Denote by L the number

$$\sup\{\lambda|\lambda \in \Omega\}.$$

Let $\beta < L$. There exists λ, such that $\lambda \in \Omega$, $\beta \leqq \lambda$. Define for every β, where $0 \leqq \beta < L$,

$$\lambda_\beta =_{df} \inf\{\lambda_{\in\Omega}|\,\lambda \geqq \beta\}.$$

The definition is meaningful. We have $\beta \leqq \lambda_\beta$. Indeed there exists a sequence $\lambda_1, \lambda_2, \cdots \lambda_n, \cdots \to \lambda_\beta$, where $\lambda_n \geqq \beta$, we get. Hence

$$\lim \lambda_n \geqq \beta, \quad \text{i.e.} \quad \lambda_\beta \geqq \beta.$$

We have: If

$$0 \leqq \beta' \leqq \beta'' < L,$$

then

$$\lambda_{\beta'} \leqq \lambda_{\beta''}.$$

Proof. We have: if $\lambda \geqq \beta''$, then $\lambda \geqq \beta'$. Hence

$$\{\lambda_{\in\Omega}|\lambda \geqq \beta''\} \subsetneqq \{\lambda_{\in\Omega}|\lambda \geqq \beta'\}.$$

Consequently

$$\inf\{\lambda_{\in\Omega}|\,\lambda \geqq \beta'\} \leqq \inf\{\lambda_{\in\Omega}|\lambda \geqq \beta''\}$$

i.e.

$$\lambda_{\beta'} \leqq \lambda_{\beta''}, \quad \text{q.e.d.}$$

P.11.1. If

$$0 \leqq a < b \leqq 1, \quad (a, b) \cdot \Omega = \emptyset,$$

$$\beta', \beta'' \in (a, b), \quad 0 \leqq \beta', \beta'' < L,$$

then

$$\lambda_{\beta'} = \lambda_{\beta''}.$$

P.11.2. Proof. I say, that for $\lambda \in \Omega$ the following are equivalent:

I. $\lambda \geqq \beta'$,

II. $\lambda \geqq \beta''$.

Indeed, let $\lambda \geqq \beta'$. Since $\lambda \in \Omega$, $\Omega \cdot (a, b) = \emptyset$, it follows that $\lambda \geqq b$.

Hence $\lambda \geqq \beta''$.

Thus $\lambda \geqq \beta'$ implies $\lambda \geqq \beta''$. Similarly we prove, that $\lambda \geqq \beta''$ implies $\lambda \geqq \beta'$.

Hence

$$\inf\{\lambda_{\in\Omega}|\lambda \geqq \beta'\} = \inf\{\lambda_{\in\Omega}|\lambda \geqq \beta''\}.$$

Hence

$$\inf\{\lambda_{\in\Omega}|\,\lambda \geqq \beta'\} = \inf\{\lambda_{\in\Omega}|\,\lambda \geqq \beta''\},$$

i.e. $\lambda_{\beta'} = \lambda_{\beta''}$.

P.11.3. If

$$\lambda_{\beta'} = \lambda_{\beta''}, \quad \beta' < \beta'',$$

then

$$(\beta', \beta'') \cdot \Omega = \emptyset.$$

Proof. Suppose this be not true. Hence there is $\lambda_1 \in \Omega$ available, such that

$$\beta' < \lambda_1 < \beta''.$$

We have

$$\lambda_1 \in \{\lambda_{\in\Omega}|\,\lambda \geqq \beta'\}.$$

Hence

$$\inf\{\lambda_{\in\Omega}|\,\lambda \geqq \beta'\} \leqq \lambda_1,$$

hence

(1) $$\lambda_{\beta'} \leqq \lambda_1.$$

Since $\lambda < \beta'' \leqq \lambda_{\beta''}$, it follows from (1), that $\lambda_{\beta'} < \lambda_{\beta''}$, which is a contradiction.

P.11.4. If $\beta < \lambda_\beta$, then there does not exist any $\lambda \in \Omega$, $\lambda \geqq \beta$, such that $\beta \leqq \lambda < \lambda_\beta$.

Indeed, suppose that

$$\lambda_1 \in \Omega, \quad \beta \leqq \lambda_1 < \lambda_\beta.$$

Since

$$\lambda_\beta = \inf\{\lambda_{\in\Omega}|\,\lambda \geqq \beta\},$$

we must have $\lambda_\beta \leqq \lambda_1$. Hence $\lambda_\beta \leqq \lambda_1 < \lambda_\beta$; hence $\lambda_\beta < \lambda_\beta$, which is a contradiction.

P.11.5. If

$$\lambda_\alpha < \lambda_\beta,$$

then

$$\alpha < \beta.$$

Proof. If we had $\alpha \geqq \beta$, we would have $\lambda_\beta \leqq \lambda_\alpha$, which is not true.

P.11.6. The following are equivalent for $\lambda \in \Omega$:

$$\lambda \geqq \beta, \quad \lambda \geqq \lambda_\beta.$$

Proof. The interval (λ, λ_β) does not contain any $\lambda \in \Omega$.

P.11.7. We have

$$\Phi(u) = \lambda_{\Phi(u)}.$$

Proof. Indeed, by Def. [P.10.4]f.,

$$\Phi(u) = \inf\{\lambda_{\in\Omega}|\,u \in S^{-1}(\lambda')\}.$$

The following are equivalent for $\lambda \in \Omega$:

$$u \in S^{-1}(\lambda), \qquad \Phi(u) \leqq \lambda.$$

Hence

$$\Phi(u) = \inf\{\lambda_{\in\Omega} | \Phi(u) \leqq \lambda\} = \lambda_{\Phi(u)}, \quad \text{q.e.d.}$$

P.11.8. We have

(1)
$$\prod_{\substack{\lambda\in\Omega\\ \lambda\geqq\beta}} S^{-1}(\lambda) = \prod_{\substack{\lambda\in\Omega\\ \lambda\geqq\lambda_\beta}} S^{-1}(\lambda).$$

Proof. We have $\beta \leqq \lambda_\beta$. Hence if $\lambda \geqq \lambda_\beta$, then $\lambda = \beta$. Consequently each factor occuring on the right in (1) also occurs on the left. Hence

(2)
$$\prod_{\substack{\lambda\in\Omega\\ \lambda\geqq\beta}} S^{-1}(\lambda) \subsetneqq \prod_{\substack{\lambda\in\Omega\\ \lambda\geqq\lambda_\beta}} S^{-1}(\lambda).$$

Let $u \in \prod_{\substack{\lambda\in\Omega\\ \lambda>\lambda_\beta}} S^{-1}(\lambda)$. Hence for every $\lambda \in \Omega$, where $\lambda \geqq \lambda_\beta$,

$$u \in S^{-1}(\lambda).$$

Now let $\lambda \in \Omega$, $\lambda \geqq \beta$. We have $\lambda \geqq \lambda_\beta$; consequently

$$u \in S^{-1}(\lambda).$$

Hence, for every $\lambda \in \Omega$, where $\lambda \geqq \beta$ we have $u \in S^{-1}(\lambda)$, i.e.

$$u \in \prod_{\substack{\lambda\in\Omega\\ \lambda\geqq\beta}} S^{-1}(\lambda).$$

Thus

(3)
$$\prod_{\substack{\lambda\in\Omega\\ \lambda>\lambda_\beta}} S^{-1}(\lambda) \subsetneqq \prod_{\substack{\lambda\in\Omega\\ \lambda>\beta}} S^{-1}(\lambda).$$

From (2) and (3) the theorem follows.

P.12. If

$$\beta \leqq \Phi(u) \leqq \lambda_\beta,$$

then

$$\Phi(u) = \lambda_\beta.$$

Proof. Since $\beta \leqq \Phi(u)$, we have $\lambda_\beta \leqq \lambda_{\Phi(u)}$. Hence $\Phi(u) \leqq \lambda_\beta \leqq \lambda_{\Phi(u)}$. Now, as $\Phi(u) = \lambda_{\Phi(u)}$, it follows that $\Phi(u) = \lambda_\beta$.

Remark. It is not true that

$$\prod_{\substack{\lambda\in\Omega\\ \lambda>\beta}} S^{-1}(\lambda) \subsetneqq \{u | \Phi(u) \leqq \beta\}.$$

P.13. For any real β, where $0 \leqq \beta < L$, we have

$$\{u | \Phi(u) \leqq \lambda_\beta\} = \prod_{\substack{\lambda\in\Omega\\ \lambda\geqq\beta}} S^{-1}(\lambda) = \prod_{\substack{\lambda\in\Omega\\ \lambda\geqq\lambda_\beta}} S^{-1}(\lambda).$$

Proof. Let $u_0 \in \prod_{\substack{\lambda \geqq \beta \\ \lambda \in \Omega}} S^{-1}(\lambda)$; hence for every $\lambda \geqq \beta$, $\lambda \in \Omega$ we have $u_0 \in S^{-1}(\lambda)$. Hence

$$\Phi(u_0) \leqq \lambda.$$

Hence

$$\Phi(u_0) \leqq \lambda_\beta,$$

i.e.

$$u_0 \in \{u \mid \Phi(u) \leqq \lambda_\beta\}.$$

Thus

(1)
$$\{u \mid \Phi(u) \leqq \lambda_\beta\} \geqq \prod_{\substack{\lambda \in \Omega \\ \lambda > \beta}} S^{-1}(\lambda).$$

Let

$$u_1 \in \{u \mid \Phi(u) \leqq \lambda_\beta\}.$$

Hence

$$\Phi(u_1) \leqq \lambda_\beta.$$

P.14. Take any

$$\lambda \geqq \lambda_\beta.$$

We get

$$\Phi(u_1) \leqq \lambda;$$

hence

$$u_1 \in S^{-1}(\lambda);$$

hence

$$u_1 \in \prod_{\substack{\lambda \in \Omega \\ \lambda \geqq \lambda_\beta}} S^{-1}(\lambda) = \prod_{\substack{\lambda \in \Omega \\ \lambda \geqq \beta}} S^{-1}(\lambda).$$

Hence

(2)
$$\{u \mid \Phi(u) < \lambda_\beta\} \subseteqq \prod_{\substack{\lambda \in \Omega \\ \lambda > \beta}} S^{-1}(\lambda).$$

From (1) and (2) the theorem follows.

P.15. For every real β we have

$$\{u \mid \Phi(u) < \beta\} = \sum_{\substack{\lambda < \beta \\ \lambda \in \Omega}} S^{-1}(\lambda).$$

Proof.

$$u_0 \in \{u \mid \Phi(u) < \beta\}.$$

We have

$$\Phi(u_0) < \beta,$$

$$\Phi(u_0) = \inf\{\lambda_{\in \Omega} \mid u_0 \in S^{-1}(\lambda)\} < \beta.$$

Hence there exists $\lambda' \in \Omega$, with $\lambda' < \beta$, such that

$$u_0 \in S^{-1}(\lambda'); \quad \text{hence} \quad u_0 \in \sum_{\lambda' < \beta} S^{-1}(\lambda).$$

Hence

(1) $$\{u \mid \Phi(u) < \beta\} \subseteqq \sum_{\lambda' < \beta} S^{-1}(\lambda').$$

Let $u \in \sum_{\substack{\lambda < \beta \\ \lambda \in \Omega}} S^{-1}(\lambda)$. There exists $\lambda' \in \Omega$, such that

$$u \in S^{-1}(\lambda'), \quad \lambda' < \beta;$$

hence

$$\inf \{\lambda \mid u \in S^{-1}(\lambda)\} \leqq \lambda' < \beta,$$

i.e.

$$\Phi(u) < \beta.$$

Hence

(2) $$\sum_{\lambda < \beta} S^{-1}(\lambda) \subseteqq \{u \mid \Phi(u) < \beta\}.$$

From (1) and (2) the theorem follows.

P.16. Now we shall prove that $\Phi(u)$ is borelian. Let β be a real number. We have

(1) $$\{u \mid \Phi(u) < \beta\} = \sum_{\substack{\lambda \in \Omega \\ \lambda < \beta}} S^{-1}(\lambda).$$

Since Ω is a denumerable set, (1) is the union of an at most denumerable number of borelian sets, so it is borelian. Since the collection of the sets

$$\{u \mid \Phi(u) \leqq \beta\}$$

is the scale of repartition of $\Phi(u)$, it follows that all sets of the tribe of repartition are borelian. Consequently Φ is a borelian function.

P.17. We shall prove that the repartition-tribe (G') of the function

$$g(z) =_{df} \Phi(f(z))$$

is a subtribe of (G).

Indeed

$$g^{-1}(-\infty, \beta) = \{z \mid g(z) < \beta\} = \{z \mid \Phi(f(z)) < \beta\}.$$

Since $\Phi(u) < \beta$ is equivalent to

$$u \in \sum_{\substack{\lambda \in \Omega \\ \lambda < \beta}} S^{-1}(\lambda),$$

it follows, that $\Phi(g(z)) < \beta$ is equivalent to

$$f(z) \in \sum_{\substack{\lambda \in \Omega \\ \lambda < \beta}} S^{-1}(\lambda).$$

Hence

$$g^{-1}(-\infty, \beta) = \{z \mid f(z) \in \sum_{\substack{\lambda \in \Omega \\ \lambda < \beta}} S^{-1}(\lambda)\} = f^{-1}\Big(\sum_{\substack{\lambda \in \Omega \\ \lambda < \beta}} S^{-1}(\lambda)\Big) = \sum_{\substack{\lambda \in \Omega \\ \lambda < \beta}} f^{-1}(S^{-1}(\lambda)).$$

Since $S^{-1}(\lambda)$ are sets of the chain (R'), and since the sums have a denumerable number of terms, we may write them $E_{\alpha(n)}$, and get:

$$g^{-1}(-\infty, \beta) = \sum_n f^{-1}(E_{\alpha(n)}).$$

Since

$$f^{-1}(E_{\alpha(n)}) \in G,$$

we get

$$g^{-1}(-\infty, \beta) \in G.$$

Now, as

$$(-\infty, \beta) = \prod_{n=1}^{\infty} \left(-\infty, \beta + \frac{1}{n}\right),$$

we obtain

$$g^{-1}(-\infty, \beta) = \prod_{n=1}^{\infty} g^{-1}\left(-\infty, \beta + \frac{1}{n}\right).$$

Since each factor belongs to G, and G is denumerably additive, we get

$$g^{-1}(-\infty, \beta) \in G.$$

Consequently, all sets of the scale of repartition of $g(z)$ belong to G. It follows, that the tribe-set of repartition G' of $g(z)$ belongs to G, i.e.

$$(1) \qquad G' \subseteqq G.$$

P.17a. Now we shall prove that $G \subseteqq G'$.

To prove this, it is sufficient to show, that there exists in G a set of somata, which also belongs to G', but this set should be such that every denumerably additive tribe, containing them, contains all $\{a_n\}$.

The sets $S^{-1}(\lambda)$ are the sets E_n of the chain (R'). They are all borelian, and

$$f^{-1}(E_n) =^{\mu} c_n,$$

which constitutes the chain (R); hence

$$S^{-1}(\lambda) \in G.$$

Now we can prove that

$$f^{-1}(S^{-1}(\lambda)) \in G'.$$

Indeed, let $\lambda_0 \in \Omega$. We have:

$$\sum_{\substack{\lambda \in \Omega \\ \lambda < \lambda_0}} S^{-1}(\lambda) = \{u \mid \Phi(u) < \lambda_0\},$$

and

$$\sum_{\substack{\lambda \in \Omega \\ \lambda < \lambda_0}} S^{-1}(\lambda) = \{u \mid \Phi(u) \leqq \lambda_0\}.$$

It follows, that

$$S^{-1}(\lambda_0) = \{u \mid \Phi(u) = \lambda_0\}.$$

Hence
$$g^{-1}(\lambda_0) = \{z|g(z) = \lambda_0\} = \{z|\Phi(f(z)) = \lambda_0\}$$
$$= \{z|f(z) \in S^{-1}(\lambda_0)\} = f^{-1}[S^{-1}(\lambda_0)].$$

Since the set $g^{-1}(\lambda_0)$ belongs to G', it follows that
$$f^{-1}(S^{-1}(\lambda_0)) \in G'.$$

Thus we have found a collection of sets:
$$f^{-1}(S^{-1}(\lambda)),$$
where λ vary in Ω, such that they are common to G and G'.

Now if a denumerably additive tribe contains all $f^{-1}(S^{-1}(\lambda))$, i.e. all $f^{-1}(E_n) =^{\mu} c_n$, then it must contain all $\{a_n\}$; and since the set $\{a_n\}$ is μ-topologically dense in G, the tribe must contain all somata of G.

Consequently $G \subseteqq G'$, q.e.d.

The theorem [P.10.4.] is proved.

P.18. Theorem. If

1. $f(\dot{z})$ is measurable with borelian $\mathrm{G}f$ and $\mathrm{D}f$,
2. (F) is the tribe of repartition of $f(z)$,
3. $\Phi(u)$ a borelian $1 \to 1$ function defined on $\mathrm{D}f$,
4. $g(\dot{z}) =_{df} \Phi(f(\dot{z}))$,
5. (G) is the tribe of repartition of $g(\dot{z})$,

then
$$(F) = (G).$$

Proof. We have $G \subseteqq F$ and G is a denumerably genuine subtribe of (F).

Now take $\alpha \subset F$; we shall prove that $\alpha \in G$. Since (F) is the tribe of repartition of $f(\dot{z})$, there exists a borelian set $a \subseteqq \mathrm{D}f$, such that
$$f^{-1}(a) = \alpha.$$

Now let us consider the v-ordering on $\mathrm{D}f$, as in [P.7.3.].

Every borelian subset of $\mathrm{D}f$ is v-measurable. Now, since $\Phi(\dot{u})$ is $1 \to 1$ on $\mathrm{G}\Phi = \mathrm{D}f$, therefore the tribe of repartition of $\Phi(\dot{u})$ contains all borelian subsets of $\mathrm{D}f$. Consequently a belongs to this tribe of repartition. Hence there exists a borelian set A in $\mathrm{D}\Phi$ such that $\Phi^{-1}(A) =^{v} a$. Since Φ is borelian, the set $\Phi^{-1}(A)$ is also borelian. We have
$$\Phi^{-1}(A) =^{v} a;$$
hence
$$f^{-1}[\Phi^{-1}(A)] =^{\mu} f^{-1}(a) = \alpha.$$
Hence
$$g^{-1}(A) =^{\mu} \alpha.$$

Consequently, α belongs to the repartition tribe of $g(\dot{z})$; hence $\alpha \in G$.

Thus we have proved, that

$$F \subseteqq G.$$

It follows that $F = G$, and then, that $(F) = (G)$, q.e.d.

P.19. Theorem. If

1. $f(z)$ is measurable with borelian $\mathsf{D} f$ and $\mathsf{Q} f$,
2. (F) is the repartition tribe of $f(\dot{z})$,
3. $\Phi(u)$ is a borelian function on $\mathsf{D} f$,
4. $g(\dot{z}) =_{df} \Phi[f(\dot{z})]$ for $z \in \mathsf{Q} f$,
5. (G) is the repartition tribe of $g(z)$,
6. $(G) = (F)$,

then

1) $\Phi(u)$ is almost v-one-to-one, where v is the measure introduced on $\mathsf{D} f$ as in [P.7.], [μ-induced measure on $\mathsf{D} f$].

2) There exists a set a with $v(a) = 0$ such that $(\mathsf{D} f - a) \upharpoonleft \Phi(u)$ is strictly $1 \to 1$.

Proof. It suffices to prove that the repartition tribe of $\Phi(\dot{u})$ contains all borelian subsets of

$$\mathsf{Q}\, \Phi = \mathsf{D} f.$$

Now take any borelian subset a of $\mathsf{D} f$. Put

$$\alpha =_{df} f^{-1}(a).$$

Since $(G) = (F)$, we have $\alpha \in G$. Hence there exists a borelian A, such that

$$g^{-1}(A) =^{\mu} \alpha.$$

Hence

$$f^{-1}(\Phi^{-1}(A)) =^{\mu} \alpha.$$

Hence

$$f^{-1}(a) =^{\mu} g^{-1}(\Phi^{-1}(A)).$$

It follows that

$$a =^{v} \Phi^{-1}(A).$$

Hence a belongs to the repartition tribe of Φ.

We have proved that every borelian subset a of $\mathsf{D} f$ belongs to the repartition tribe of Φ. Hence Φ is almost v-one-to-one, i.e. the remainder of the thesis is also proved.

P.20. Theorem. If

1. E is μ-measurable,
2. (G) is a denumerably genuine subtribe of $(\mathfrak{L}_{\mu}) \cdot E$, then there exists a μ-measurable real valued function $g(z)$ defined almost μ-everywhere on E such that (G) is the repartition tribe of $g(\dot{z})$.

Proof. Consider the function $f(z)$, defined by

$$f(z) \doteq z \quad \text{for all} \quad z \in \mathsf{Q} f.$$

The repartition tribe (F) of $f(\dot{z})$ contains all μ-measurable subsets of E.

Since (G) is a denumerably genuine subtribe of (F), and its partial tribe too, there exists a real valued borelian function $\Phi(u)$, defined on $\mathsf{D}\,f$ such that the repartition tribe of $g(\dot{z}) = \Phi(f(\dot{z}))$ is just (G).

Now $g(z) = \Phi(f(z))$, hence the repartition tribe of $g(\dot{z})$ is just (G), q.e.d.

P.21. Theorem. If

1. $f(\dot{z})$, $g(\dot{z})$ are μ-measurable,
2. (F), (G) are their repartition tribes,

then the following are equivalent:

I. $(F) = (G)$,

II. there exists a $1 \to 1$ borelian function Φ defined on a borelian set containing $\mathsf{D}\,f$, such that

$$g(z) =^{\mu} \Phi(f(z)) \quad \text{for all} \quad z \in \mathsf{Q}\,f.$$

In this case we have

$$\mathsf{Q}\,g =^{\mu} \mathsf{Q}\,f.$$

Proof. Suppose I i.e. $(F) = (G)$. It follows that

$$\mathsf{Q}\,f =^{\mu} \mathsf{Q}\,g,$$

because $\mathsf{Q}\,f$ is the unit of (F) and $\mathsf{Q}\,g$ is the unit of (G), so these units are μ-equal. Since (G) is a denumerably genuine subtribe of (F), it follows, that there exists a borelian function $\Phi(u)$, defined on a set, compraising $\mathsf{D}\,f$, such that

$$g(z) =^{\mu} \Phi(f(z)).$$

Now applying Theorem [P.19.], we get that Φ is almost v-one-to-one. So II is proved.

The statement II $\to$ I is easy to prove.

Chapter P I

Permutable normal operators

P I.1. Def. If M, N are maximal normal operators, we shall call them *strongly permutable*, if and only if the spaces of the spectral scale $s(\dot{Q})$ of M are compatible with the spaces of the spectral scale $t(\dot{Q})$ of N. Since compatibility of spaces a, b is equivalent (Julia (58)) to the relation $P_a P_b \vec{X} = P_b P_a \vec{X}$ for all vectors $\vec{X} \in \mathbf{H}$, the name is justified.

P I.1.1. Def. The operators M, N are said to be weakly (formally) *permutable*, if and only if

$$M\,N = N\,M$$

with equality of domains. The last notion offers difficulties in treating it, because (v. NEUMANN, p. 90, (43)), $N \cdot O(\vec{X})$ has $\mathbf{H}$ as its domain, and $O \cdot N(\vec{X})$ has $\mathrm{C}N$ as its domain, so N and O are not weakly permutable.

P I.1.2. Def. A normal operator is said to *be simple,* whenever its borelian spectral tribe is saturated.

P I.1.3. Let N, M be strongly permutable simple maximal normal operators. Their borelian spectral tribes $s_N(Q)$, $s_M(Q)$ are compatible. They are saturated and their supertribes must coincide with their borelian spectral tribes. It follows. that they are identical:

$$s_b =_{df} s_N = s_M.$$

Take a generating vector $\vec{\omega}$ and the corresponding measure μ. The spectral scales $s_N(Q)$, $s_M(Q)$ may be different, but we have for every plane quarter Q

(1) $$s_N(Q) \in s_b, \quad s_M(Q) \in s_b.$$

P I.1.4. Remark. The traces, whose definition is essentially based on the spaces of the spectral scale may be different for M and N. Hence the correspondence $\mathscr{G}_N^{-1}$ between vectors of $\mathbf{H}$ and the μ-square summable functions of traces may differ from $\mathscr{G}_M^{-1}$. The correspondence $\mathfrak{M}_M^*$, which carries the space of μ-square summable functions of space-traces into μ-square summable functions on the plane P, is determined by the relation:

$$\text{“}A\ \mathfrak{M}_M^*\,\alpha \text{ is equivalent to } [A] = s_M(\alpha)\text{”},$$

where α is a borelian subset of the plane P, and A is a measurable set of space-traces, and $[A]$ its coat, which is a space, and a soma of s_b. Now if α is a plane quarter, then $s_M(\alpha)$ is the space of the spectral scale of M; hence to Q there corresponds through $\mathfrak{M}_M^*$ the set of traces of M with [coat $s_M(Q)$]. Thus $\mathfrak{M}_M^*$ may differ from $\mathfrak{M}_N^*$.

The canonical mapping $\mathbf{C}_M$, generated by $\vec{\omega}$ is $\mathfrak{M}_M^*\mathscr{G}_M^{-1}$ and it carries the vectors of $\mathbf{H}$ into μ-square summable functions on P. Hence $\mathbf{C}_M$ may differ from $\mathbf{C}_N$. In addition to that, if we suppose that $\vec{\omega} \in \mathrm{C}\,M$, it does not follow that $\vec{\omega} \in \mathrm{C}\,N$. Having this all in mind we cannot conclude, that the mappings $\mathbf{C}_M$, $\mathbf{C}_N$ coincide because of the coincidence of borelian spectral tribes of both operators M, N.

P I.1.5. Call the corresponding measures μ_M, μ_N. The correspondence

$$Q \smile s_N(Q)$$

defined for all plane quarters Q is the spectral scale of N, can be extended to the correspondence

$$\alpha \smile s_N(\alpha),$$

defined for all borelian sets α of P. This correspondence is $1 \to 1$ with respect to the μ_N-equality of sets α and the equality of spaces. The domain of this correspondence is the class of all borelian sets taken modulo μ_N, and the region is the class s_b of all spaces of the spectral tribe $s_M = s_N$. The μ_N-measure of α is $= 0$, if and only if $s_N(\alpha) = 0$, because $\vec{\omega}$ is a generating vector of **H** with respect to s_N. Now $s_M(\alpha) = 0$ if and only if the μ_M-measure of $\alpha = 0$, because $\vec{\omega}$ is a generating vector of **H** with respect to s_M.

P I.1.5a. Consequently the ideal of all μ_M-null sets of P coincides with the set of all μ_N-null sets, i.e. with the ideal of all μ_N-null sets. Thus the notion of a "N-positive" borelian set is the same, as the notion of M-positive borelian sets.

P I.1.6. The notions of μ_M-equality and μ_N-equality of borelian sets coincide.

Since the correspondences $\alpha \smile s_N(\alpha)$ and $\alpha \smile s_M(\alpha)$ are both one to one with respect to the equality of spaces and almost equality of sets, and since both correspondences transform finite and denumerable set operations, zero and unit into the corresponding finite and denumerable space operations (on compatible spaces) and into the zero-space O and the unit space **H**, it follows that we have an isomorphism E, defined by

$$\alpha \smile s_N(\alpha) \smile s_M^{-1}[s_N(\alpha)]$$

i.e. the isomorphism

$$E =_{df} s_N \,|\, s_M^{-1} = s_M^{-1} s_N,$$

from the class of all borelian sets onto itself, with respect to almost equality for both, and with preservation of zero, unit, and finite and denumerable set operations.

P I.1.7. In the same way we have an isomorphism defined by

$$a \smile s_N^{-1} a \smile s_M(s_N^{-1} a),$$

i.e. the isomorphism

$$E' =_{df} s_N^{-1} \,|\, s_M = s_M s_N^{-1}$$

(given two relations A, B, the symbol $A|B$ means $B A$), from the class s_b onto itself with preservation of O, 1 and all finite and denumerable space-operations. We shall prove some formulas.

We have

$$E = s_M^{-1} s_N, \tag{2}$$

hence

$$s_M E = s_M s_M^{-1} s_N$$

i.e.

(3) $$s_N = s_M E.$$

We have

(4) $$E' = s_M s_N^{-1},$$

hence

$$E' s_N = s_M s_N^{-1} s_N,$$

i.e.

(5) $$s_M = E' s_N.$$

From (3) we get

(6) $$s_N E^{-1} = s_M,$$

and from (5) we get

(7) $$s_N = E'^{-1} s_M.$$

From (3) and (7) we have $s_M E = E'^{-1} s_M$, hence

(8) $$E = s_M^{-1} E'^{-1} s_M,$$

and

(9) $$E'^{-1} = s_M E s_M^{-1}.$$

From (5) and (6) we get $E' s_N = s_N E^{-1}$; hence

(10) $$E' = s_N E^{-1} s_N^{-1},$$

and

(11) $$E^{-1} = s_N^{-1} E' s_N.$$

Thus from (11):

(12) $$E = (s_N^{-1} E' s_N)^{-1},$$

and from (9):

(13) $$E' = (s_M E s_M^{-1})^{-1}.$$

P I.1.8. The measure of the spaces of s_b is defined by

$$\mu(a) = \|\text{Proj}_a \vec{\omega}\|^2.$$

Now, for borelian sets of P we have

$$\mu_M(\alpha) = \mu(s_M(\alpha)),$$

(13 a) $$\mu_N(\alpha) = \mu(s_N(\alpha)) = (\text{by } (3)), = \mu(s_M(E\alpha)).$$

Since $s_M(\alpha)$ may differ from $s_M(E\alpha)$, the measures may be not equal for borelian sets, though the null-sets are the same.

P I.1.9. The correspondence $E^{-1} = s_N^{-1} s_M$ between borelian sets induces a correspondence between measurable functions on P, defined almost everywhere.

P I.1.10. Def. Since the measurable function is determined, up to almost μ-equality, by its scale of repartitions (see [P.]), we can define for measurable functions the correspondence

(14) $f(\dot{z}) \smile^{E^{-1}} g(\dot{z})$ as the extension of $f^{-1}(Q) \smile^{E^{-1}} g^{-1}(Q)$.

If $f(\dot{z}) = \Omega_\alpha(\dot{z})$ is the characteristic function of the set α, we have

$$f^{-1}(A) = \{z \mid f(z) \in A\}.$$

Since

$$f(z) = 1 \quad \text{for} \quad z \in \alpha,$$

and

$$f(z) = 0 \quad \text{for} \quad z \bar{\in} \alpha,$$

we have for every A, such that $1 \in A$, $f^{-1}(A) = \alpha$, and for every A, such that $1 \bar{\in} A$, we have $f^{-1}(A) = \emptyset$. This is especially true for $\alpha = Q$.

Now, if $f(z) \smile^{E^{-1}} g(z)$, we must also have: if $1 \in A$, then $f^{-1}(Q) = 1$ and hence $g^{-1}(A) = E^{-1}(Q)$, and if $0 \in A$, then $f^{-1}(A) = \emptyset$ and $g^{-1}(A) = \emptyset$.

This property has been proved for characteristic function of any set $E^{-1}(\alpha)$.

P I.1.10a. We see that $\alpha \smile^{E^{-1}} \beta$ is equivalent to

$$\Omega_\alpha(\dot{z}) \smile^{E^{-1}} \Omega_\beta(\dot{z}).$$

Thus (14) is a natural modification, for functions, of E^{-1}, which is defined for sets.

P I.1.11. The correspondence E^{-1} for functions is invariant with respect to almost equality of functions, and it is one to one. Its domain and region are composed of all μ-measurable functions, defined a.e. on P.

P I.1.12. Consider the function $f(\dot{z}) =_{df} \dot{z}$. We have $f^{-1}(\alpha) = \alpha$ for all borelian α; hence the tribe of partition of $f(\dot{z})$ is the Lebesguean tribe taken modulo μ. Now let

$$f(\dot{z}) \smile^{E^{-1}} \varphi(\dot{z}).$$

Since E^{-1} is a one-to-one correspondence from $\mathfrak{L}_\mu$ onto $\mathfrak{L}_\mu$ (where $\mathfrak{L}_\mu$ is the tribe of all μ-measurable sets), it follows that the tribe of repartition of $\varphi(\dot{z})$ coincides with $\mathfrak{L}_\mu$.

Consequently, if we apply theorems of [P.], we find that $\varphi(\dot{z})$ is an almost one-to-one function, and it can be admitted as borelian, [P.].

P I.1.13. Now we shall prove, that for every measurable function $F(\dot{z})$:

$$F(\dot{z}) \smile^{E^{-1}} F(\varphi(\dot{z})) =_{df} G(\dot{z}).$$

To prove that, it suffices to prove, that for every plane quarter Q:

$$F^{-1}(Q) \smile^{E^{-1}} G^{-1}(Q).$$

Now

$$G^{-1}(Q) = \{z | G(z) \in Q\} = \{z | F(\varphi(Z)) \in Q\}$$
$$= \{z | \varphi(z) \in F^{-1}(Q)\} = \{z | z \in \varphi^{-1} F^{-1}(Q)\}$$
$$= \{z | \in E^{-1} F^{-1}(Q)\} = E^{-1}[F^{-1}(Q)].$$

Hence

$$G^{-1}(Q) \smile^{E^{-1}} F^{-1}(Q)$$

for all Q, so the assertion is proved.

P I.1.13a. It follows that the correspondence E^{-1} is an isomorphism with respect to the addition and multiplication of functions and multiplication by complex numbers. Indeed, let

$$f(\dot{z}) \smile^{E^{-1}} f'(\dot{z}),$$

and

$$g(\dot{z}) \smile^{E^{-1}} g'(\dot{z}).$$

We have

$$f'(z) = f(\varphi(z)),$$

and

$$g'(\dot{z}) = g(\varphi(\dot{z})).$$

Consequently:

$$f'(z) + g'(z) = f(\varphi(z)) + g(\varphi(z)) = [f(u) + g(u)]_{u=\varphi(z)},$$
$$f'(\dot{z}) \cdot g'(\dot{z}) = f(\varphi(\dot{z})) \cdot g(\varphi(\dot{z})) = [f(u) \cdot g(u)]_{u=\varphi(z)}.$$

Since, if $f(z) = \text{const} = \lambda$, we get $E^{-1} f(z) = \lambda$, it follows, that E^{-1} is a linear transformation of the space of all measurable functions on P, onto itself.

P I.1.14. Theorem. E^{-1} transforms μ_M-summable functions into μ_M-summable functions.

We prove that for simple function. Let $f(\dot{z})$ be a simple function. There exists a finite or denumerably infinite sequence of μ-disjoint measurable sets

$$\alpha_1, \alpha_2, \ldots, \alpha_n, \ldots, \quad \text{where} \quad \sum_n \alpha_n \doteq P,$$

and a corresponding sequence of distinct numbers:

$$p_1, p_2, \ldots, p_n, \ldots, \tag{15}$$

such that

$$f(z) = p_n \quad \text{whenever} \quad z \in \alpha_n.$$

The E^{-1} image of $f(z)$ is the function, which admits on

$$E^{-1}(\alpha_1), E^{-1}(\alpha_2), \ldots, E^{-1}(\alpha_n), \ldots \tag{16}$$

the values (15) respectively.

Since the sets α_n are μ-disjoint, and $\sum_n \alpha_n = P$, therefore (16) are disjoint and their sum $\doteq P$.

Hence $E^{-1} f(z)$ is also a simple function. Suppose that $f(z)$ is μ_M-summable. Then

$$\sum_{n=1}^{\infty} |p_n| \cdot \mu_M(\alpha_n)$$

converges.

Now

$$\sum_{n=1}^{\infty} |p_n| \, \mu_N[E^{-1}(\alpha_n)] = \sum_{n=1}^{\infty} |p_n| \cdot \mu_M[s_N(E^{-1}(\alpha_n))] = (\text{by } (6)) =$$
$$= \sum_{n=1}^{\infty} |p_n| \cdot \mu_M(s_M(\alpha_n)) = \sum_{n=1}^{\infty} |p_n| \cdot \mu_M(\alpha_n),$$

because this sum converges; so $E^{-1} f(z)$ is also μ_N-summable. Similarly we can prove that for simple function: if $E^{-1} f(\dot{z})$ is μ_N-summable, then $f(\dot{z})$ is μ_M-summable.

If $f(\dot{z})$ is a simple function, μ_M-summable, then

$$\int_P f(z)\, d\mu_M = \int_P E^{-1} f(z)\, d\mu_N. \tag{16a}$$

Similar proof, as before.

P I.1.15. Now let $f(\dot{z})$ be a μ-measurable function. There exists an infinite sequence $\{f_n(z)\}$ of simple functions, converging uniformly a.e. to $f(\dot{z})$.

We have

$$\int_P f_n(z)\, d\mu_M \to \int_P f(z)\, d\mu_M. \tag{17}$$

Now since $f_n(z)$ converges uniformly to $f(z)$, therefore $f_n[\Phi(z)]$ converges uniformly to $f[\Phi(z)]$; hence

$$\int_P (f_n \Phi(z)]\, d\mu_N \to \int_P f[\Phi(z)]\, d\mu_N. \tag{18}$$

Since, in (17) and (18) the left hand sides are equal, by virtue of (16a), it follows that the right hand integrals also are equal.

Hence

$$\int_P f(z)\, d\mu_M = \int_P f[\Phi(z)]\, d\mu_N.$$

This implies that the transformation E^{-1} preserves the scalar product, and hence also the norm. Thus E^{-1} is a unitary transformation from the space of all μ_M-square summable functions onto the space of all μ_N-square summable functions.

P I.1.16. The operator N has its **C**-image in the operator

$$\mathsf{C} N = \boldsymbol{N},$$

which carries μ_M-square summable functions $f(\dot{z})$ into $\dot{z}\cdot f(\dot{z})$. The E^{-1} image of $\boldsymbol{N}$ is

$$E^{-1}\boldsymbol{N} = \boldsymbol{M}',$$

which carries the function $f[\Phi(\dot{z})]$ into

$$\Phi(\dot{z})\cdot f[\Phi(\dot{z})]$$

a.e. Now, since $f[\Phi(\dot{z})]$ may be any μ_N-square summable function $g(\dot{z})$, therefore $\boldsymbol{M}'$ carries

$$g(\dot{z}) \quad \text{into} \quad \Phi(\dot{z})\cdot g(\dot{z})$$

a.e. The spectral scale of the operator

$$\boldsymbol{M}' g(\dot{z}) =^{\mu} \Phi(\dot{z})\cdot g(\dot{z})$$

are its spaces composed of functions vanishing a.e. outside

$$\{z \mid \Phi(z) \in Q\},$$

i.e. outside

$$\Phi^{-1}(Q) = E^{-1}(Q).$$

Hence $\boldsymbol{M}'$ is the **C**-image of N. Since $M' = \Phi(M)$, it follows that

$$N = \Phi(M).$$

P I.1.17. There exists a generating vector $\vec{\omega}$, such that

$$\vec{\omega} \in \text{\textsf{a}}\, N \quad \text{and} \quad \vec{\omega} \in \text{\textsf{a}}\, M.$$

We have proved the

P I.1.18. Theorem. If

1. M, N are strongly permutable and simple (maximal) normal operators in **H**,

then there exists a $1 \to 1$ borelian function $\Phi(z)$ such that

$$M = \Phi(N).$$

P I.2. Theorem. If M, N are simple operators (maximal) normal in **H**, then the following are equivalent:

I. M, N are strongly permutable,

II. there exists a borelian $1 \to 1$ function Φ defined on P, such that

$$M = \Phi(N) \qquad (\text{resp. } N = \Phi^{-1}(M)),$$

III. there exists a unitary transformation $\mathfrak{T}$ of the space **H** into itself, such that

1) $M \cup^{\mathfrak{T}} N$,

2) for the borelian tribes we have $s_M \cup^{\mathfrak{T}} s_N$.

IV. The borelian spectral tribes of M, N coincide.

P I.2.1. Lemma. If B is a denumerably additive tribe, admitting an effective, non negative measure μ, whose topology is separable, then there exists a B-soma-valued function $f(Q)$ of the variable plane-quarter Q on the plane P, such that:

1^0 if $Q \subseteqq Q'$, then $f(Q) \subseteqq f(Q')$,

2^0 $\mathrm{D} f$ generates B, i.e. B is the smallest finitely genuine subtribe of B, containing $\mathrm{D} f$,

3^0 if $\alpha_1 \leqq \alpha_2 \leqq \cdots \to +\infty$,
$\beta_1 \leqq \beta_2 \leqq \cdots \to +\infty$,

then

$$\sum_{n=1}^{\infty} f\{x\,y \mid x \leqq \alpha_n,\ y \leqq \beta_n\} = I,$$

4^0 if $\alpha_1 \geqq \alpha_2 \geqq \cdots \to \alpha_0$,
$\beta_1 \geqq \beta_2 \geqq \cdots \to \beta_0$,

then

$$\prod_{n=1}^{\infty} f\{x\,y \mid x \leqq \alpha_n,\ y \leqq \beta_n\} = f\{x\,y \mid x \leqq \alpha_0,\ y \leqq \beta_0\},$$

where

$$-\infty \leqq \alpha_n, \beta_n, \quad \alpha_0, \beta_0 \leqq +\infty.$$

P I.2.2. Proof. First of all take an infinite sequence

(1) $$a_1, a_2, \ldots, a_n, \ldots$$

of somata of B, which constitutes an everywhere dense set in the μ-topology on B. Such a sequence exists, since μ is effective and its topology separable.

Applying [(10), p. 23, Theor.] we construct a set of somata of B

(2) $$c_1, c_2, \ldots, c_n, \ldots$$

which

1) is a chain with the inclusion $\subseteqq$ governing in B, and such that

2) if a subtribe of B contains all somata (2), then it contains also all somata (1). Denote this chain by (C).

In what follows symbols will have special meaning.

P I.2.3. Let as consider the original Peano-curve

(3) $$x = \varphi(t), \quad y = \psi(t), \quad 0 \leqq t \leqq 1,$$

which maps continuously the segment $\langle 0, 1\rangle$ onto the unit square

$$q = \{x, y \mid 0 \leqq x \leqq 1, 0 \leqq y \leqq 1\}.$$

We remind that (3) is obtained in the following way. We partition $\langle 0, 1\rangle$ into 9 equal segments, and the square q into 9 equal squares. We make correspond the subsquares to the subsegments in a one-to-one way (and so that to the adjacent subintervals correspond adjacent subsquares). We continue this process without end, taking the subintervals and the corresponding subsquares.

At any step of the construction, to adjacent intervals there should correspond adjacent squares. In addition to that the following property takes place: if at the n-th step of the construction we take an interval α of lengs $\frac{1}{9^n}$ and the corresponding square A with side $\frac{1}{3^n}$, then on the next step of partition of α into 9 equal intervals and of A into 9 equal squares, to the subsegment placed in the center of α there corresponds the subsquare placed in the center of A.

The intervals obtained through these partitions will be termed *net-intervals*, and the corresponding squares *net-squares*. The points of partitions of $\langle 0, 1\rangle$, 0 and 1 included, constitute a denumerably infinite set termed *net*.

P I.2.4. The correspondence P, which attaches to every net-interval the corresponding square has the properties:

1) if a, b are net-intervals and a, b do not overlap, then P-corresponding squares A, B also do not overlap and conversely,

2) if $a \subsetneqq b$, then $A \subsetneqq B$, and conversely. The correspondence P, from the net-intervals onto the net-squares determines a function $P(x)$, which makes correspond to every point of $\langle 0, 1\rangle$ a point in q, the function $P(x)$ is defined as follows:

Let $x \in \langle 0, 1\rangle$ and $a_1 \supseteqq a_2 \supseteqq \cdots \supseteqq a_n \supseteqq \cdots$ be the net-intervals of the 1-st, 2-d, ..., n-th, ... partition, such that

$$x \in a_n \quad \text{for} \quad n = 1, 2, \ldots$$

Let $A_1 \supseteqq A_2 \supseteqq \cdots \supseteqq A_n \supseteqq \cdots$ be the corresponding net-squares. Since the diameter of A_n tends to 0, the intersection $\prod_{n=1}^{\infty} A_n$ is composed of a single point, say X. We shall make correspond X to x. This correspondence from x to X is pluri-one. Indeed, if $x \in \langle 0, 1\rangle$ and

$$a_1 \supseteqq a_2 \supseteqq \cdots \supseteqq a_n \supseteqq \cdots$$

and

$$b_1 \supseteqq b_2 \supseteqq \cdots \supseteqq b_n \supseteqq \cdots$$

both contain x, then a_n and b_n are not disjoint; so the corresponding squares A_n, B_n are also not disjoint.

Consequently

$$\prod_{n=1}^{\infty} A_n = \prod_{n=1}^{\infty} B_n;$$

hence X is well determined.

The point-valued function $P(x)$ with $\mathrm{G}P = \langle 0, 1\rangle$ is just the Peano-curve.

The function $P(x)$ is not one-to-one, but if we restrict $P(x)$ to the mid-points of the net-intervals, $P(x)$ will become one-to-one.

Notice that the center of a net-interval is never a net-point.

P I.2.5. Denote by s the set of all these centers. If $x \in s$ and x is the center of a net interval α, then $X = P(x)$ is the center of the square A which corresponds through P to α.

The set s is everywhere dense in $\langle 0, 1\rangle$. The set, range of $s \upharpoonright P(\dot{x})$, is also dense in q.

Denote by P' the correspondence which attaches to every half-open net-interval $(\alpha, \beta\rangle$ the half-open square

$$\left\{x\,y \,\middle|\, \begin{matrix} \alpha' < x \leqq \alpha'' \\ \beta' < y \leqq \beta'' \end{matrix}\right\},$$

whenever the net interval $\langle\alpha, \beta\rangle$ goes over, through P, into the square

$$\left\{x\,y \,\middle|\, \begin{matrix} \alpha' \leqq x \leqq \alpha'' \\ \beta' \leqq y \leqq \beta'' \end{matrix}\right\}.$$

Thus we have: if

$$a, b \in \mathsf{C}\!\!\!\!\mathsf{l}\, P', \qquad A = P'\,a, \qquad B = P'\,b,$$

then the following are equivalent:

I. $a \cdot b = \emptyset$,

II. $A \cdot B = \emptyset$,

and the following are equivalent:

I. $a \subseteqq b$,

II. $A \subseteqq B$.

It follows, that if

1) $a_1, \ldots, a_k$ are disjoint and $\in \mathsf{C}\!\!\!\!\mathsf{l}\, P'$,

2) $b_1, \ldots, b_l$ are disjoint and $\in \mathsf{C}\!\!\!\!\mathsf{l}\, P'$,

3) $a_1 + \cdots + a_k = b_1 + \cdots + b_l$,

then

$$P'\,a_1 + \cdots + P'\,a_k = P'\,b_1 + \cdots + P'\,b_l.$$

The above shows, that if a is a finite sum of half-open net intervals

(3.1) $$a = a_1 + \cdots + a_n,$$

then $P'\,a_1 + \cdots + P'\,a_n$ does not depend on the manner of representation of (3) as such a sum of half-open intervals.

P I.2.6. This allows to extend the correspondence P' to the correspondence P'', which attaches to the finite unions $a_1 + \cdots + a_n$ of half-open net-intervals the union of half-open net-squares

$$P'\,a_1 + \cdots + P'\,a_n;$$

which attaches to the empty subset $\emptyset$ of $(0, 1\rangle$ the empty subset of the half-open square

(4) $$\left\{x\,y \,\middle|\, \begin{matrix} 0 < x \leqq 1 \\ 0 < y \leqq 1 \end{matrix}\right\},$$

and which attaches to $(0, 1\rangle$ the square (4).

The correspondence P'' has the properties:

1) if $a, b \in \mathrm{C} P''$, $a \subseteqq b$, then $P'' a \subseteqq P'' b$,

2) $P''(\mathrm{co}\, a) = \mathrm{co}\, P''(a)$,

3) $P''(a_1 + a_2 + \cdots + a_n) = P''(a_1) + P''(a_2) + \cdots + P''(a_n)$ for all $a_1, a_2, \ldots, a_n \in \mathrm{C} P''$, $(n = 1, 2, \ldots)$,

4) if $a P' A$, then $a P'' A$,

5) $\mathrm{D} P''$ is composed of Θ, and of all finite unions of half-open net-squares.

We see, that $\mathrm{C} P''$ and $\mathrm{D} P''$ can be considered as finitely additive tribe-set, and P'' constitutes an operation-isomorphism between them.

Consider the rectangles

$$E\left(\frac{p}{3^n}\right) =_{df} \left\{ x y \,\middle|\, \begin{array}{l} 0 < x \leqq \frac{p}{3^n} \\ 0 < y \leqq 1 \end{array} \right\},$$

where p, n are natural numbers and $1 \leqq p \leqq 3^n$, $n = 1, 2, \ldots$ Take the P''-corresponding sets:

$$e\left(\frac{p}{3^n}\right) =_{df} (P'')^{-1}\, E\left(\frac{p}{3^n}\right).$$

This is a finite union of half-open net-intervals. Similarly put

$$F\left(\frac{p}{3^n}\right) =_{df} \left\{ x y \,\middle|\, \begin{array}{l} 0 < x \leqq 1 \\ 0 < y \leqq \frac{p}{3^n} \end{array} \right\}$$

and

$$f\left(\frac{p}{3^n}\right) =_{df} (P'')^{-1}\, F\left(\frac{p}{3^n}\right).$$

This is also a finite union of half-open set-intervals.

P I.2.7. For any net-points w', w'' the following are equivalent.

$$(4.1) \quad \left\{ \begin{array}{l} \text{I. } w' \leqq w'', \\ \text{II. } F(w') \leqq F(w''), \\ \text{III. } f(w') \leqq f(w''), \\ \text{IV. } E(w') \leqq E(w''), \\ \text{V. } e(w') \leqq E(w''). \end{array} \right.$$

Thus if we vary the net point w by increasing, we get chains w, $F(w)$, $E(w)$, $e(w)$, $f(w)$, ordered by the correspondence of inclusion of sets.

We have supposed, [p. 630], that B has the effective measure μ. Let us make correspond to every $c \in C$ the measure $\mu(c)$. The following are equivalent:

I. $c \subset c'$;

II. $\mu c < \mu c'$.

Let us make correspond to every soma c of the chain C the number $W(c) = \frac{4}{9} + \frac{1}{9}\,\frac{\mu(c)}{\mu(1)}$. We have $\frac{4}{9} \leqq W(c) \leqq \frac{5}{9}$.

The mapping W is $1 \to 1$, and we have $\mathrm{G}W = C$, $\mathrm{D}W = \left\langle \frac{4}{9}, \frac{5}{9} \right\rangle$.

Take a real number t, where $0 \leqq t \leqq 1$, and consider all numbers $r \in \mathrm{D}W$, such that

$$r \leqq t,$$

and take the sum of somata of B:

$$a_t =_{df} \sum_{\substack{r \leqq t \\ r \in \mathrm{D}W}} W^{-1}(r). \tag{5}$$

If there does not exist such an r, we put $a_t =_{df} O$, (the zero soma of B). We see that if $t' \leqq t''$, then $a_{t'} \subseteqq a_{t''}$, a_0 is the zero of B, and a_1 is its unit 1.

P I.2.8. The correspondence which attaches to the number t the soma a_t will be termed A.

We have $\mathrm{G}A = \langle 0, 1 \rangle$, $\mathrm{D}A \subseteqq B$.

This correspondence can be extended to another one A', which attaches to every interval $(t', t''\rangle$, where $0 \leqq t' \leqq t'' \leqq 1$, and t', t'' are points of the net, the soma $a_{t''} - a_{t'}$, and which carries the empty subset $\emptyset$ of $(0, 1\rangle$ into the soma O, and $(0, 1\rangle$ into the unit 1 of B. Now we see that, if

$$t_1 \leqq t_2 \leqq \cdots \leqq t_n$$

are net points (0, 1 included), we have

$$a_{t_n} - a_{t_1} = \sum_{i=2}^{n} (a_{t_i} - a_{t_{i-1}}),$$

we have

$$a_t = a_t - a_0.$$

Having this in mind, we can attach to every finite union of disjoint half-open intervals (with endpoints belonging to the net)

$$\alpha = (t_1', t_2''\rangle + (t_2' t_2''\rangle + \cdots + (t_n' t_n''\rangle \cdots$$

the soma

$$(a_{t_1''} - a_{t_1'}) + (a_{t_2''} - a_{t_2'}) + \cdots + (a_{t_n''} - a_{t_n'});$$

and thus attached soma does not depend on the manner of representing α by a sum of disjoint intervals of the above kind.

P I.2.9. Denote the correspondence given above by A''. We have:

1) if $\alpha \in \mathrm{G}A$, then $A\,\alpha = A'\,\alpha = A''\,\alpha$,

2) if $\alpha \subseteqq \beta$ then $A''\,\alpha \subseteqq A''\,\beta$,

3) if $\alpha = \alpha_1 + \alpha_2 + \cdots + \alpha_n$, then $A''\alpha = A''\alpha_1 + \cdots + A''\alpha_n$, $A''(\mathrm{co}\,\alpha) = \mathrm{co}\,(A''\alpha)$,

4) $A''O = 0$,

5) $A''(0, 1\rangle = 1$ of B.

$\mathrm{C}\, A''$ is the class, composed of $\emptyset$, $(0, 1\rangle$ and all finite sums of half-open intervals $(r', r''\rangle$, where r', r'' are net-points in $(0, 1\rangle$, 0 and 1 included. They make up a tribe. $\mathrm{D}\, A''$ is a finitely additive subtribe-set of B.

A'' is an isomorphism from the tribe of sets onto the above subtribe of B.

P I.2.10. Consider the correspondence

$$H =_{df} A''(P'')^{-1}, \tag{6}$$

which attaches, to the finite unions of half-open net-squares, some somata of B.

P I.2.10a. We shall be especially interested in the above correspondence, restricted to the class of rectangles $E(w)$, where w are net-points, i.e.

$$M =_{df} \{E(w)\} \upharpoonleft A''(P'')^{-1},$$

and to the correspondence (6), restricted to the class of rectangles $F(w)$:

$$N =_{df} \{F(w)\} \upharpoonleft A''(P'')^{-1}.$$

M attaches to every $E(w)$ a soma of B, and we have

$$\text{if} \quad w' \leqq w'', \quad \text{then} \quad M(E(w')) \subseteqq M(E(w'')).$$

Similarly N attaches to every $F(w)$ a soma of B, and we have

$$\text{if} \quad w' \leqq w'', \quad \text{then} \quad N(F(w')) \subseteqq N(F(w'')).$$

P I.2.11. The following is important:

If we know the functions

$$M(E(\dot{w}')) \quad \text{and} \quad N(F(\dot{w}')),$$

we can calculate $H(A)$, where A is any net-square. H is a finite operation-isomorphism.

Denote by $A(\alpha', \beta'; \alpha'', \beta'')$ the rectangle

$$\{x, y \mid \alpha' < x \leqq \alpha'', \beta' < y \leqq \beta''\}.$$

We have

$$A(\alpha', \alpha''; \beta', \beta'') = \{[A(0, 0; \alpha'', \beta'') - A(0, 0; \alpha''\beta')] + \\ + A(0, 0; \alpha', \beta'')\} - A(0, 0; \alpha', \beta').$$

We also have for any rectangle $A(0, 0; \gamma, \delta)$:

$$A(0, 0; \gamma, \delta) = \left\{x\,y \,\middle|\, \begin{matrix} 0 < x \leqq \gamma \\ 0 < x \leqq 1 \end{matrix}\right\} \cdot \left\{x\,y \,\middle|\, \begin{matrix} 0 < x \leqq 1 \\ 0 < y \leqq \delta \end{matrix}\right\}.$$

Hence if γ, δ are net points, then

$$A(0, 0; \gamma, \delta) = E(\gamma) \cdot F(\delta).$$

Hence if α', β', α'', β'' are net points, we have

$$\{x\,y \,|\, \alpha' < x \leqq \alpha'', \beta' < y \leqq \beta''\}$$
$$= \{[E(\alpha'')\,F(\beta'') - E(\alpha'')\,F(\beta')] + E(\alpha')\,F(\beta'')\} - E(\alpha') \cdot F(\beta').$$

Consequently, as H is an isomorphism, we have:

$$H\{x\,y \,|\, \alpha' < x \leqq \beta'', \beta' < y \leqq \beta''\} =$$
$$= \{[H\,E(\alpha'') \cdot H(\beta'') - H\,E(\alpha'') \cdot H\,F(\beta')] + H\,E(\alpha') \cdot H\,F(\beta'')\} -$$
$$- H[E(\alpha')] \cdot H[F(\beta')].$$

To the half open net-squares there correspond through $(P'')^{-1}$ the half open net-intervals. Hence if we have a net-interval a, we can calculate $A''(a)$. It follows that if a partial tribe of B contains all $M\,E(w')$ and all $N\,E(w')$, it also contains all $A''(a)$, where a are half-open net-intervals.

P I.2.11a. Now, if we know the somata $A''(a)$ for all net-intervals, we can calculate the somata a_t for all net-points t.

Indeed, let $t = \frac{p}{q^n}$, we have

$$(0, t\rangle = \left(0, \frac{1}{9^n}\right\rangle + \left(\frac{1}{9^n}, \frac{2}{9^n}\right\rangle + \cdots + \left(\frac{p-1}{9^n}, \frac{p}{9^n}\right\rangle.$$

Hence, since A'' is an isomorphism, we get

$$A''(0, t\rangle = A''\left(0, \frac{1}{9^n}\right\rangle + \cdots + A''\left(\frac{p-1}{9^n}, \frac{1}{9^n}\right\rangle.$$

It follows that if a partial tribe of B contains all $\mathrm{D}\,M$, and $\mathrm{D}\,N$, then it also contains $A''(0, t\rangle$ for all net-points t.

P I.2.11b. Now let $r \in \mathrm{D}\,W$. Since the net-points are everywhere dense, there exists a sequence of net-points

$$w_1 > w_2 > \cdots \to r.$$

We have

$$A''(0, w_n\rangle = a_{w_n} - a_0 = a_{w_n}.$$

We can prove, that

$$W^{-1}\,r = \prod_{n=1}^{\infty} a_{w_n}.$$

Indeed, we have

$$a_{w_n} = \sum_{\substack{r' \leqq w_n \\ r' \in \mathrm{D}\,W}} W^{-1}(r').$$

Since $r \leqq w_n$, it follows that

$$W^{-1}(r) \subseteqq a_{w_n}.$$

Hence

(7) $$W^{-1}(r) \subseteqq \prod_{n=1}^{\infty} a_{w_n}.$$

Suppose that:

(8) $$W^{-1}(r) \subset \prod_{n=1}^{\infty} a_{w_n};$$

hence

$$\mu W^{-1}(r) < \mu \prod_{n=1}^{\infty} a_{w_n}.$$

P I.2.11c. There are two cases:

1) there exists $\varepsilon > 0$ such that the interval

$$(r, r + \varepsilon)$$

does not contain any point of $\mathrm{D}\, W$.

In this case for sufficiently great n, say $n \geqq n_0$, we have

$$w_n \in (r, r + \varepsilon).$$

Hence $a_{w_n} = W^{-1}(r)$ and then $W^{-1}(r) = \prod_{n=n_0}^{\infty} a_{w_n} = \prod_{n=1}^{\infty} a_{w_n}$, which contradicts (8).

P I.2.11d. 2) Such an ε does not exist.

Hence there exists an infinite sequence

$$r_1 > r_2 > \cdots \to r, \quad \text{where } r_m \in \mathrm{D}\, W.$$

For every n there exists m_0, such that for all $m \geqq m_0$

$$r_m < w_n; \quad \text{hence} \quad W^{-1}(r_m) \subseteqq a_{w_n}, \quad \text{i.e.} \quad \mu W^{-1}(r_m) \leqq \mu\, a\, w_n;$$

and for every m there exists n_0 such that, for $n \geqq n_0$;

$$w_n < r_m, \qquad a_{w_n} \subseteqq W^{-1}(r_m),$$

i.e.

$$\mu\, a_{w_n} \leqq \mu\, W^{-1}(r_m).$$

Hence

$$\mu(W^{-1} r_m) \to \mu(W^{-1} r).$$

Hence

$$\mu\left(\prod_{n=1}^{\infty} a_{w_n}\right) = \mu\, W^{-1}(r).$$

It follows that

$$\mu\left[W^{-1}(r) - \prod_{n=1}^{\infty} a_{w_n}\right] = 0,$$

and then, since μ is effective,

$$W^{-1}(r) = \prod_{n=1}^{\infty} a_{w_n},$$

which contradicts (8). Thus we have proved that

$$W^{-1}(r) = \prod_{n=1}^{\infty} a_{w_n}.$$

P I.2.12. Combining this with preceding results, we can say that if a partial tribe of B contains all $\mathsf{D}\,M$ and $\mathsf{D}\,N$, then it must contain all $W^{-1}(r)$ i.e. all elements of the chain C. Consequently it must contain all $\{a_r\}$.

Hence the somata of $\mathsf{D}\,M$, together with those of $\mathsf{D}\,N$, generate the whole tribe B.

P I.2.12a. Having this, consider the chain $\mathbb{D}\,M$, i.e. the somata

$$H\,E(w') = M\,E(w'),$$

where w' are net-points, and the chain of the corresponding measures is

$$\mu\,M\,E(w). \tag{9}$$

They form a denumerable set of numbers. Since the measure is effective, the correspondence $\mathfrak{S}$, defined by

$$\mu\,M\,E(w) \smile M(E\,w),$$

is $1 \to 1$, and has the property, that the following are equivalent:

I. $\mu\,M\,E(w') < \mu\,M\,E(w'')$,

II. $M\,E(w') \subset M\,E(w'')$.

Notice, that to the domain of $\mathfrak{S}$ belong 0 and $\mu(1)$, and to the range the zero-soma and the unit-soma of B.

Let us adjoin to the set $\{\mu\,M\,E(w)\}$ all left hand side points of accumulation, and to the set $\{M\,F(w)\}$ the corresponding products $\prod M(E\,w)$, which, of course, belong to B.

Having this, consider the interval

$$(-1,\ \mu(1) + 1),$$

and attach to every point x in it the soma

$$M'(\alpha) =_{df} \prod_{\substack{x<\alpha \\ \alpha\in(-1,\ \mu(1)+1) \\ \alpha\in\mathsf{O}\mathfrak{S}}} \mathfrak{S}(\alpha).$$

The function $M'(\alpha)$ has the properties

if $\alpha_1 > \alpha_2 > \cdots \to -1$, then $\prod_n M'(\alpha_n) = 0$,

if $\alpha_1 < \alpha_2 < \cdots \to \mu(1) + 1$, then $\sum_n M'(\alpha_n) = 1$,

if $\alpha_1 > \alpha_2 > \cdots \to \alpha_0$, then $M'(\alpha_0) = \prod_n M'(\alpha_n)$.

Similarly we do with the operator N getting $N'(\alpha)$.

Put

$$\xi =_{df} \tan\left[\frac{x - \frac{\mu(1)}{2}}{\mu(1) + 2}\,\frac{\pi}{2}\right].$$

We get

$$\frac{x - \frac{\mu(1)}{2}}{\mu(1) + 2}\,\frac{\pi}{2} = \operatorname{arc\,tan} \xi,$$

$$x = \frac{\mu(1)}{2} + \frac{2}{\pi}(\mu(1) + 2) \operatorname{arc\,tan} \xi.$$

We define

$$M''(\xi) \quad \text{as} \quad M'\left(\frac{\mu(1)}{2} + \frac{2}{\pi}(\mu(1) + 1) \operatorname{arc\,tan} \xi\right),$$

and similarly

$$N''(\eta) \quad \text{as} \quad N'\left[\frac{\mu(1)}{2} + \frac{2}{\pi}(\mu(1) + 1) \operatorname{arc\,tan} \eta\right],$$

and finally we put

$$Q(\xi, \eta) = M''(\xi) \cdot N''(\eta).$$

Thus we obtain a soma valued functions of the variable plane-quarter, which satisfies the conditions of the theorem.

P I.3. Theorem. The space valued double spectral scale $s'[Q(\lambda, v)]$ of the normal maximal operator

$$U F(\dot{z}) =^{\mu} \varphi(\dot{z}) \cdot F(\dot{z})$$

with maximal domain is

(1) $$\mathbf{H}^{H}[d(Q(\dot{\lambda}, \dot{v}))],$$

where

$$d(Q(\lambda, v)) = \{z \mid \varphi(\dot{z}) \in Q(\dot{\lambda}, \dot{v})\}$$

is the set valued double scale of partition of the function $\varphi(\dot{z})$, and where (1) is the collection of all $F(\dot{z}) \in H$, such that

$$F(\dot{z}) =^{\mu} 0 \quad \text{on} \quad \operatorname{co} d(\varphi(\dot{\lambda}, \dot{v})).$$

P I.3.1. Proof. In [H.], theorem [H.5.6a.], (the spectral theorem for maximal normal transformation), we have proved that: if

1. $N = H + iK$ is a maximal normal transformation in $\mathbf{H}$, and, H, K are selfadjoint transformations in $\mathbf{H}$,

2. $s(Q)$ is the double space valued spectral scale of N,

then the simple space valued spectral scales for H and K are respectively:

$$h(\lambda) =_{df} h((-\infty, \lambda\rangle) = s[Q(\lambda; +\infty)], \quad -\infty \leqq \lambda \leqq +\infty,$$

$$k(v) =_{df} k((-\infty, v\rangle) = s[Q(+\infty; v)],$$

and we have

(2) $$s[Q(\lambda, v)] = h(\lambda) \cdot k(v).$$

Let $\varphi(z)$ be a μ-measurable function, defined μ-almost everywhere on P. The operator

$$U F(z) =^{\mu} \varphi(z) \cdot F(z),$$

with maximal domain V, is a maximal normal operator on H.

If we put

$$\varphi(\dot{z}) = \alpha(\dot{z}) + i\,\beta(\dot{z}),$$

where $\alpha(\dot{z})$, $\beta(\dot{z})$ are real valued function $\in H$, then

$$H F(\dot{z}) =^{\mu} \alpha(\dot{z}) \cdot F(\dot{z}),$$
$$K F(\dot{z}) =^{\mu} \beta(\dot{z}) \cdot F(\dot{z})$$

are selfadjoint operators, and

$$U[F(\dot{z})] = H[F(\dot{z})] + i\,K[F(\dot{z})],$$

where H and K are strongly permutable.

Let $s'(Q)$ be the double space-valued spectral scale of U. The spectral scale of H is, by the above theorem

$$h'(\dot{\lambda}) =_{df} h'((-\infty, \dot{\lambda}\rangle) = s'[Q(\dot{\lambda}, +\infty)]$$

and that of K is:

$$k'(v) =_{df} k'((-\infty, v\rangle) = s'[Q(+\infty; v)].$$

Now we have, if we put

$$a'(\lambda) =_{df} \{z \mid \alpha(z) \leqq \lambda\}, \quad b'(v) =_{df} \{z \mid \beta(z) \leqq v\}, \quad \text{the equality}$$
$$h'(\lambda) = \mathbf{H}^{H}(a'(\lambda)), \quad \text{where} \quad \mathbf{H}^{H}(a'(\lambda))$$

is the collection of all $F(\dot{z}) \in H$, such that $F(z) =^{\mu} 0$ on the complement $\operatorname{co} a'(\lambda)$, and $\mathbf{H}^{H}(b'(v))$ is the collection of all $F(\dot{z}) \in H$, such that $F(z) =^{\mu} 0$ on the complement $\operatorname{co} b'(v)$. By (2) we have

$$s[Q(\lambda, v)] = h'(\lambda) \cdot k'(v) = \mathbf{H}^{H}(a'(\lambda)) \cdot \mathbf{H}^{H}(b'(v)).$$

Hence this is the collection of all functions $F(\dot{z}) \in H$, which vanish almost μ-everywhere on

$$\operatorname{co} a'(\lambda) \quad \text{and on} \quad \operatorname{co} b'(v).$$

Now

$$\alpha(\dot{z}) = \operatorname{Real} \varphi(\dot{z}),$$
$$\beta(\dot{z}) = \operatorname{Imag} \varphi(\dot{z});$$

hence

$$a'(\lambda) = \{z \mid \operatorname{Real} \varphi(z) \leqq \lambda\},$$
$$b'(v) = \{z \mid \operatorname{Imag} \varphi(z) \leqq v\}.$$

Hence $s[Q(\lambda, v)]$ is the collection of all $F(\dot{z}) \in H$, which vanish on the set

$$\{z \mid \mathrm{Real}\,\varphi(z) > \lambda\} \quad \text{and on} \quad \{z \mid \mathrm{Imag}\,\varphi(z) > v\}.$$

Hence they vanish on

$$\{z \mid \mathrm{Real}\,\varphi(z) > \lambda\} + \{z \mid \mathrm{Imag}\,\varphi(z) > v\};$$

hence on

$$\mathrm{co}\,\{z \mid \mathrm{Real}\,\varphi(z) \leqq \lambda\} + \mathrm{co}\,\{z \mid \mathrm{Imag}\,\varphi(z) \leqq v\} =$$

$$= \mathrm{co}\left\{z \,\middle|\, \begin{array}{l} \mathrm{Real}\,\varphi(z) \leqq \lambda \\ \mathrm{Imag}\,\varphi(z) \leqq v \end{array}\right\} = \mathrm{co}\,\{z \mid \varphi(z) \in Q(\lambda; v\}.$$

Consequently

$$s'(Q(\lambda, v)) = \mathbf{H}^H\{z \mid \varphi(z) \in Q(\lambda; v\}.$$

Now $\{z \mid \varphi(z) \in Q(\dot{\lambda}, \dot{v}\}$ is the double set-valued scale of partition

$$d[Q(\lambda, v)]$$

of the function $\varphi(z)$. The theorem is proved.

P I.4. Theorem. Let

1. $K \neq \mathrm{O}$ be a collection of maximal normal operators in $\mathbf{H}$,
2. if $N_1, N_2 \in K$, then N_1, N_2 are permutable,

then there exists a simple maximal normal operator M in $\mathbf{H}$ such that for every $N \in K$ there exists a borelian function $\varphi_N(\dot{z})$ defined over the whole plane P, such that

$$N = \varphi_N(M).$$

P I.4.1. Proof. Since all operators $N \in K$ are permutable, the spaces of their borelian tribes are mutually compatible. By arranging K into a well ordered sequence and by successive adjunction we shall find a tribe t', such that for every $N \in K$, the borelian spectral tribe S_N is a denumerably genuine subtribe. Now there exists a saturated tribe t compraising t'. Thus all s_N are denumerably genuine subtribes of t.

The tribe t admits an effective, non negative measure, which generates in it a separable metric space. Hence, by a forgoing theorem, there exists a space-valued function $t(Q)$ of the variable generalized plane quarter Q, having the following properties:

1⁰. for every Q the space $t(Q) \in t$, and for any Q', Q'' the spaces $t(Q'), t(Q'')$ are compatible.

2⁰. If $Q' \subseteqq Q''$ are plane quarters, then

$$t(Q') \subseteqq t(Q'').$$

3^0. $t(\Theta) = (\vec{O})$, $t(P) = 1 = \mathbf{H}$.

4^0. If $\alpha_1 \leqq \alpha_2 \leqq \cdots \to +\infty$,

$\beta_1 \leqq \beta_2 \leqq \cdots \to +\infty$,

then

$$\sum_{n=1}^{\infty} t\left\{x\,y \,\middle|\, \begin{matrix} x \leqq \alpha_n \\ y \leqq \beta_n \end{matrix}\right\} = t(P).$$

5^0. If $\alpha_1 \geqq \alpha_2 \geqq \cdots \to \alpha_0$,

$\beta_1 \geqq \beta_2 \geqq \cdots \to \beta_0$, $\quad (-\infty \leqq \alpha_i, \beta_i \leqq +\infty)$,

then

$$\prod_{n=1}^{\infty} t\left\{x\,y \,\middle|\, \begin{matrix} x \leqq \alpha_n \\ y \leqq \beta_n \end{matrix}\right\} = t\left\{x\,y \,\middle|\, \begin{matrix} x \leqq \alpha_0 \\ y \leqq \beta_0 \end{matrix}\right\}.$$

P I.4.2. This allows to introduce space-traces, and, since the tribe t is saturated in $\mathbf{H}$, also a generating vector $\vec{\omega}$ of $\mathbf{H}$ with respect to t, the corresponding measure μ and consequently a canonical mapping $\mathbf{C}$ of $\mathbf{H}$ onto the space H of all μ-square summable functions $F(z)$ of the complex variable z on the plane P. Consider the operator

$$M\left(F(\dot{z})\right) =^{\mu} \dot{z} \cdot F(\dot{z}) \tag{3}$$

with maximal domain on the complex plane P.

The space valued spectral scale of (3) is the function

$$s'[Q(\lambda, v)] = \mathbf{H}^H[dQ(\lambda, v)],$$

where $d[Q(\lambda, v)]$ is the set-valued repartition-scale of the function

$$\varphi(\dot{z}) =_{df} \dot{z},$$

i.e.

$$d[Q(\dot{\lambda}, \dot{v})] = \{z \,|\, z \in Q(\dot{\lambda}, \dot{v})\} = Q(\dot{\lambda}, \dot{v}).$$

Hence

$$s'[Q(\lambda, v)] = \mathbf{H}^H[Q(\lambda, v)].$$

Since the smallest subtribe of μ-measurable subsets of P, generated by the plane-quarters $Q(\lambda, v)$, is the whole tribe L_μ^P of all μ-measurable subsets, it follows that the spectral tribe m of (3) is saturated, so the operator (3) is simple. The spectral tribe is $\mathbf{C}(t)$.

It follows that the operator in $\mathbf{H}$ $M =_{df} \mathbf{C}^{-1}(U)$ is also simple.

P I.4.3. Now consider any operator $N \in K$. Its spectral tiibe s_N is a denumerably genuine subtribe of t. Hence, if we put $N =_{df} \mathbf{C}(N)$, the spectral tribe $\mathbf{C}(s_N)$ of N is also a denumerably genuine subtribe of the tribe m.

P I.4.4. Consider the correspondence $\mathcal{T}$ from the μ-measurable sets E onto the spaces $\mathbf{H}^H(E)$. It is an isomorphism and $\mathcal{T}^{-1}\,\mathbf{C}(s_N)$ is a denumerably genuine partial subtribe of m. Hence there exists a borelian function $\varphi(\dot{z})$, such that the repartition-tribe of $\varphi(\dot{z})$ is just $\mathcal{T}^{-1}\,\mathbf{C}(s_N)$.

Hence $\mathcal{T}^{-1}\,\mathbf{C}[s_N(Q)]$ is the spectral tribe of $\mathbf{C}(N)$; hence $\mathcal{T}^{-1}\,\mathbf{C}(N)$ is the repartition-tribe of $\varphi(\dot{z})$. Consequently $\mathbf{C}(N)$ is the operator

$$\boldsymbol{N}\,F(\dot{z}) =^{\mu} \varphi(\dot{z}) \cdot F(\dot{z})$$

with maximal domain. Hence

$$N = \varphi(M).$$

The theorem is proved. As a corollary we have the

P I.5. Theorem. If N is a maximal normal operator, then there exists a simple normal operator M and a borelian function $\Phi(\dot{z})$ on P such that

$$N = \Phi(M).$$

Proof. We apply the former theorem in the case where the collection K is composed of the single operator N.

P I.6. Theorem. If

1. N is a normal maximal operator in $\mathbf{H}$,
2. $\Phi(z)$ is a borelian function on P,

then the spectral tribe of $\Phi(N)$ is a denumerably genuine subtribe of the spectral tribe of N.

P I.6.1. Proof. Take a canonical mapping $\mathbf{C}$ which carries N into the operator

$$N[F(\dot{p})] =^{\mu} (\operatorname{Num}\dot{p}) \cdot F(\dot{p}),$$

with maximal domain, on the disjoint union of copies of the complex plane P.

The operator $\mathbf{C}\,\Phi(N)$ is

$$M[F(\dot{p})] =^{\mu} (\operatorname{Num}\dot{p}) \cdot F(\dot{p})$$

with maximal domain. Now, a space of the spectral scale of M is

$$S[Q(\lambda, v)] = S[z \mid \Phi(z) \in Q(\lambda, v)];$$

hence it is a space, which s-corresponds to an N-measurable subset of P. Hence it belongs to the spectral scale of N. Since this set is close with respect to finite and denumerable operations, it is a denumerably genuine partial subtribe of S. Taking the $\mathbf{C}^{-1}$ image we get the theorem.

Remark. $\Phi(z)$ can be supposed N-measurable instead of borelian.

P I.7. Theorem. If

1. N, M are maximal normal operators in $\mathbf{H}$,
2. the spectral tribe s_N of N is a denumerably genuine partial subtribe of the spectral tribe s_M of M,

then there exists a borelian function $\Omega(z)$, such that

$$N = \Omega(M).$$

P I.7.1. Proof. There exists a simple normal operator U, such that

$$N = \Phi(U), \quad M = \Psi(U)$$

where Φ, Ψ are borelian.

Take a canonical mapping C, which carries U into

$$U[F(\dot{z})] =^{\mu} \dot{z} \cdot F(\dot{z}).$$

N and M go over into

$$N\,F(\dot{z}) =^{\mu} \Phi(\dot{z}) \cdot F(\dot{z}),$$

$$M\,F(\dot{z}) =^{\mu} \Psi(\dot{z}) \cdot F(\dot{z}) \quad \text{respectively}.$$

Since the spectral tribes of N and M are related isomorphically to the repartition tribes d_N, d_M of the functions $\Phi(z)$, $\Psi(z)$, and since d_N is a denumerably genuine partial subtribe of d_M, it follows that there exists a borelian function $\Omega(\dot{z})$ such that $\Phi(z) = \Omega[\Psi(z)]$. Hence

$$\Phi\, U\, F(z) = \Omega\, \Psi\, U\, F(z),$$

which gives

$$N\,F(z) = \Omega\, M\, F(z),$$

and then

$$N = \Omega M.$$

P I.8. Theorem. If M, N are (strongly) permutable and Φ, Ψ are borelian functions on P,
then $\Phi(M)$, $\Psi(N)$ are also (strongly) permutable.

P I.8.1. Proof (sketch). There exists a simple normal operator U, such that

$$M = \Phi(U), \quad N = \Psi(U).$$

Hence the spectral tribes of M and N are denumerably genuine subtribes of the spectral scale of U. Hence both these spectral scales belong to the spectral scale of U.

Hence they are compatible with one another.

Hence $\Phi(M)$, $\Psi(N)$ are strongly permutable.

Chapter Q

Approximation of somata by complexes

Q.1. This chapter is of auxiliary capacity and contains definitions of some notions and proofs of their properties, which are needed for foundation of a kind of integration. Especially we introduce the notion of "complexes", which will constitute approximations of somata of tribes, (42).

Q.1.1. We shall deal with a finitely additive abstract tribe $(\boldsymbol{F})$, which is plunged into a denumerably additive tribe $(\boldsymbol{G})$, as its finitely genuine subtribe. We consider a base $\boldsymbol{B}$ of $(\boldsymbol{F})$, [B.1.1.]. We admit the **(*Hyp Ad*)** [B.2.], though the hypothesis **(*Hyp Af*)**, [B.2.], would be sufficient for our purpose. Since we have further generalization in mind, we prefer to admit **(*Hyp Ad*)**, which is less restrictive. The complication which our approach will imply, is not too great.

The infinite operations on somata of $(\boldsymbol{F})$ will be understood, as taken from $(\boldsymbol{G})$. We may even suppose that $(\boldsymbol{G})$ is a strict supertribe of $(\boldsymbol{F})$, [A.7.2.].

As in [B.3.], by a *covering* (brick-covering), we shall understand an at most denumerable sum of bricks, i.e. elements of $\boldsymbol{B}$. We shall consider the null-soma O, also as a covering. We can define a covering as $\sum_{n=1}^{\infty} a_n$, where $a_n \in \boldsymbol{B}$; and the sum may be supposed to be infinite, because we may add terms equal O to the finite sum.

Q.1.2. If $f \in \boldsymbol{F}$, then f is the sum of a denumerable number of mutually disjoint bricks.

Q.2. If L is a covering, then it is a denumerable sum of mutually disjoint bricks.

Q.3. If $L = f_1 + f_2 + \cdots + f_n + \cdots$, where f_n are figures, i.e. $f_n \in \boldsymbol{F}$, then there exists an infinite sequence $a_1, a_2, \ldots, a_n, \ldots$ of mutually disjoint bricks, such that

1) $L = a_1 + a_2 + \cdots + a_n + \cdots$,
2) for every i there exists j such that $a_i \leqq f_j$.

Q.4. As in [B.1.3.], we say that we are in the conditions $\boldsymbol{FBG}$, if we admit the hypotheses in [B.1.1.], [B.1.2.] concerning $(\boldsymbol{F})$, $\boldsymbol{B}$ and $(\boldsymbol{G})$.

Q.5. We shall consider the hypothesis ***Hyp G μ*** which reads: There is on $\boldsymbol{G}$ a non negative measure $\mu(\dot{a})$, which is non trivial, i.e. $\mu(I) > 0$. We shall keep $\mu(\dot{a})$ fixed. This induces in $(\boldsymbol{F})$ a measure, also denoted by $\mu(\dot{a})$.

Q.6. Under ***Hyp G μ*** we define the notion of *distance between two somata a, b of* $(\boldsymbol{G})$:

$$|a, b|_\mu =_{df} \mu(a \dotplus b)$$

(the algebraic addition, defined as

$$a \dotplus b =_{df} \big((a - b) + (b - a)\big).$$

For properties of $|a, b|_\mu$ see [A.16.ff.].

Q.6.1. The measure μ may be effective, i.e. [A.15.1.], if $\mu(a) = 0$, then $a = O$. We mention the following theorem:

Q.7. Theorem. For somata of $\boldsymbol{G}$ the following are equivalent:

I. μ is an effective measure on $\boldsymbol{G}$,
II. $|E, F|_\mu = 0$ implies $E =^G F$.

Proof. Suppose II is not true. Then there exist E, F such that

(1) $$|E, F|_\mu = 0 \quad \text{and} \quad E \neq^G F.$$

Hence

(2) $$\mu(E - F) = \mu(F - E) = 0.$$

I say that either $E - F \neq O$ or $F - E \neq O$. Indeed, if not, we would have $E - F = O$, $F - E = O$, and then $E \cdot \operatorname{co} F = O$, $F \operatorname{co} E = O$. Since $E = E F + E \operatorname{co} F$, $F = F E + F \operatorname{co} E$, we would have $E = E F$, $F = F E$, and then $E = F$, which has been excluded in (1).

If $E - F \neq O$, we get, by (2), $\mu(E - F) = 0$, and if $F - E \neq O$, we get, by (2), $\mu(F - E) = 0$, so μ is not effective. The above arguments show that I implies II. To prove that II implies I, suppose I is not true.

There exists $E \neq^G O$, such that $\mu(E) = 0$. Hence, $|O, E|_\mu = 0$. Hence, by II, $O =^G E$, which is a contradiction. The theorem is proved.

Q.8. Suppose that μ is effecitve. We have the following:

$$|E, F|_\mu = |F, E|_\mu,$$

$$|E, F|_\mu \leqq |E, G|_\mu + |G, F|_\mu, \quad \text{and} \quad |E, F|_\mu = 0$$

is equivalent to $E = F$. This all by [A.17.1.], [A.17.3.], [A.18.4.] and because the notion of distance is **G**-equality invariant.

Thus the notion of distance of somata reorganizes the tribe **G**:

Q.9. The equality $E =^\mu F$, defined as **G** coincides with the equality modulo J in the tribe **G**, i.e. with $E =^J F$.

Proof. We define $E =^J F$ as $E \dot{-} F \in J$, where $\dot{-}$ is the algebraic subtraction in the Stone's ring **G**. Since the subtraction coincides with the addition, $E \dot{-} F \in J$, can be written as $E \dot{+} F \in J$ i.e. $\mu(E \dot{+} F) = 0$.

Q.10. Def. If we introduce on **G** the μ-inclusion, defined by "$E \subseteqq^\mu F$ means $\mu(E - F) = 0$", then $E \subseteqq^\mu F$ will be an ordering, and **G** will be organized into a Borelian lattice with governing equality $=^\mu$.

Q.11. Def. We shall use the following *hypothesis*, denoted by Hyp $\boldsymbol{L\mu}$. It reads: The measure μ, if extended by the Lebesgue's extensions device [A I., Sect. 6.], within **G**, coincides with the measure admitted on **G**. In addition to that the borelian extension $(\boldsymbol{F}^b)$ of **F** [A I., Sect. 4], coincides with **G**. Hence the μ-measure hull [A I., Sect. 5] of $(\boldsymbol{F}^b)$ within **G**, coincides with **G**. We suppose that μ is denumerably additive on **G**. The measure μ may be not effective, but if we denote by J the ideal composed of all somata a of **G** with $\mu(a) = 0$, the measure μ on $(\boldsymbol{G}_J)$ i.e. on the tribe **G** modulo J, is effective. If we admit that μ is already effective on $(\boldsymbol{G})$, this will not affect seriously our discussion.

Q.12. Def. We denote by ***Hyp R*** the collection of hypothesis ***FBG***, ***(Hyp Ad)***, ***Hyp G μ*** and ***Hyp L μ***.

Q.13. Under ***Hyp R***, if $E \in \boldsymbol{G}$ and $\eta > 0$, then there exists an infinite sequence $g_1, g_2, \ldots, g_n, \ldots$ of disjoint figures (i.e. $\in \boldsymbol{F}$), such that

$$\left| E, \sum_{n=1}^{\infty} \mathfrak{B} g_n \right|_{\mu} < \eta, \quad E \subseteqq \sum_{n=1}^{\infty} \mathfrak{B} g_n \quad \text{[A.7.4.], [A.7.6.]}.$$

Proof. By [B.4.3.] there exists a sequence $g_1, g_2, \ldots, g_n, \ldots$ of disjoint figures, such that

$$(1) \qquad E \subseteqq \sum_{n=1}^{\infty} \mathfrak{B} g_n$$

and $0 \leqq \sum_{n=1}^{\infty} \mu(g_n) - \mu(E) < \eta$. Since g_n are disjoint, we have $0 \leqq \mu\left(\sum_{n=1}^{\infty} \mathfrak{B} g_n\right) - \mu(E) < \eta$, and, then, by (1), $\left| E, \sum_{n=1}^{\infty} \mathfrak{B} g_n \right|_{\mu} < \eta$.

Q.13.1. Under hypothesis ***R***, if $E \in \boldsymbol{G}$ and $\eta > 0$, then there exists a figure f, such that $|E, \mathfrak{B} f|_{\mu} < \eta$.

Proof. By [Q.13.] we can find disjoint figures $g_1, g_2, \ldots, g_n, \ldots$, such that $|E, \sum_{n=1}^{\infty} \mathfrak{B} g_n|_{\mu} < \frac{\eta}{2}$, $E \subseteqq \sum_{n=1}^{\infty} \mathfrak{B} g_n$. If we put

$$(1\text{a}) \qquad P = \mathfrak{B} f_1 + \mathfrak{B} f_2 + \cdots,$$

we have

$$(2) \qquad |E, P|_{\mu} < \frac{\eta}{2}.$$

Now, since the series (1a) converges, there is an index n, such that

$$\mu[P - (\mathfrak{B} g_1 + \mathfrak{B} g_2 + \cdots + \mathfrak{B} g_n)] < \frac{\eta}{2}.$$

Hence

$$(3) \qquad \left| P, \mathfrak{B} \sum_{k=1}^{\infty} g_k \right|_{\mu} < \frac{\eta}{2}.$$

The soma $f =_{df} \sum_{k=1}^{n} g_n$ is a figure. Hence from (3) and (2) we get

$$|E, f| < \eta, \quad \text{q.e.d.}$$

Q.13.2. Under ***Hyp R***, if $E \in \boldsymbol{G}$ and $\eta > 0$, there exists an infinite sequence of disjoint bricks $a_1, a_2, \ldots, a_n, \ldots$, such that

$$\left| E, \sum_{k=1}^{\infty} \mathfrak{B} a_k \right|_{\mu} < \eta, \quad E \subseteqq \sum_{k=1}^{\infty} \mathfrak{B} a_k.$$

Proof. By [Q.13.], there is an infinite sequence $\{g_n\}$ of figures with

$$E \subseteqq \sum_{n=1}^{\infty} g_n, \quad \left| E, \sum_{n=1}^{\infty} \mathfrak{B} g_n \right|_{\mu} < \eta.$$

Now, by ***(Hyp Ad)***

$$\mathfrak{B}\, g_n = \mathfrak{B}\, a_{n1} + \mathfrak{B}\, a_{n2} + \cdots,$$

where $a_{n1}, a_{n2}, \ldots$ are disjoint bricks. Since the set $\{a_{nk}\}$, $(n = 1, 2, \ldots)$, $(k = 1, 2, \ldots)$, is denumerable, the theorem follows.

Q.13.3. Under ***Hyp*** $\boldsymbol{R_e}$, if $E \in \boldsymbol{G}$ and $\eta > 0$, there exists a finite number of disjoint bricks $a_1, \ldots, a_n$, $(n \geqq 1)$ with

$$|E, \mathfrak{B}\, a_1 + \cdots + \mathfrak{B}\, a_n|_\mu < \eta.$$

Proof. By [Q.13.1.], we find a figure f, such that

$$(1) \qquad |E, \mathfrak{B}\, f|_\mu < \frac{\eta}{2}.$$

By ***(Hyp Ad)*** we have $\mathfrak{B}\, f = \mathfrak{B}\, a_1 + \mathfrak{B}\, a_2 + \cdots + \mathfrak{B}\, a_n + \cdots$, where $a_1, a_2, \ldots, a_n$, are disjoint bricks. We can find n, such that

$$0 \leqq \mu\, \mathfrak{B}\, f - \mu(\mathfrak{B}\, a_1 + \mathfrak{B}\, a_2 + \cdots + \mathfrak{B}\, a_n) < \frac{\eta}{2};$$

hence

$$(2) \qquad |\mathfrak{B}\, f, \mathfrak{B}\, a_1 + \cdots + \mathfrak{B}\, a_n|_\mu < \frac{\eta}{2}.$$

From (1) and (2) the theorem follows.

Q.13.4. In the sequal we shall use brick-coverings of somata of $\boldsymbol{G}$, defined in [Q.1.1.]. By *a covering of* $E \in \boldsymbol{G}$ we shall understand any covering L, such that $E \subseteqq L$. To simplify notations, we shall write, for figures, f instead of $\mathfrak{B}\, f$, and the same will be for bricks. In the case of infinite sums *we shall take summations from* $\boldsymbol{G}$, as explained in [A.7.6.].

Q.13.5. Remark. It does not seem true that, in the case the measure μ is effective, the ***(Hyp Ad)*** follows.

Q.14. The following two theorems can be proved, under ***Hyp R***: If $E \in \boldsymbol{G}$, then there exists an infinite sequence of coverings of E, $L_1 \geqq L_2 \geqq \cdots \geqq L_n \geqq \cdots$ such that $\mu\, E = \lim \mu(L_n)$ and $\lim^\mu L_n =^\mu E$.

Q.14.1. Under ***Hyp R***, if

1. $E, F \in \boldsymbol{G}$,
2. $L_1 \geqq L_2 \geqq \cdots \geqq L_n \geqq \cdots$ are coverings of E,
3. $M_1 \geqq M_2 \geqq \cdots \geqq M_n \geqq \cdots$ are coverings of F,
4. $E \cdot F =^\mu O$,
5. $|E, L_n|_\mu \to 0$, $|F, M_n|_\mu \to 0$,

then

$$\mu(L_n, M_n) \to 0.$$

Q.14.2. As we mentioned in [A.19.3.], the notion of distance $|E, F|_\mu$ organizes the tribe $\boldsymbol{G}$, with governing equality $=^\mu$, into a metric space, hence into a topology. This topology is necessarily complete[1], but it may be not separable[2].

Later on we shall discuss the condition of separability.

Q.14.3. Hypothesis. To facilitate the discussion we shall often admit that the measure μ is effective on $\boldsymbol{G}$. This hypothesis does not affect much the generality. Indeed, if the measure μ is not effective, we replace $\boldsymbol{G}$ by the same tribe, taken modulo the ideal J, of null-sets in $\boldsymbol{G}$. To get general theorem from those, which were derived under the effectiveness of measure, we only need to change $=, \subseteqq$ into $=^J, \subseteqq^J$, i.e. $=^\mu, \subseteqq^\mu$ respectively. The relation $E \subseteqq^\mu F$ means $\mu(E - F) = 0$.

Q.14.3.1. Def. We denote by ***Hyp R***$_e$ the ***Hyp R*** together with the admission that μ is an effective measure on $\boldsymbol{G}$.

Q.14.4. Theorem. Under ***Hyp R***$_e$, the tribes $\boldsymbol{F}^b$ and $\boldsymbol{F}^L$ coincide.

Q.15. Def. By a *complex* we shall understand a finite (even empty) set of mutually disjoint bricks. A not empty complex P will be denoted by $\{p_1, p_2, \ldots, p_n\}$, $(n \geqq 1)$ or $\{p_i\}$, where p_i are bricks. By *the soma of the complex* we shall understand $p_1 + p_2 + \cdots + p_n$, where $n \geqq 1$, and the soma O, if the complex is empty. We shall write som P, if P denotes the complex.

By the *measure of the complex* P, we shall understand $\mu(\text{som} P)$. We shall write $\mu\{p_i\}$, $\mu(P)$ or $\mu(\text{som} P)$.

If P, Q are complexes and $E \in \boldsymbol{G}$, then by $|P, Q|$, $|E, P|$, $|P, E)$ we shall understand $|\mathfrak{B}\,\text{som} P, \mathfrak{B}\,\text{som} Q|_\mu$, $|E, \mathfrak{B}\,\text{som} P|_\mu$, $|\mathfrak{B}\,\text{som} P, E|_\mu$ respectively.

Q.15.1. Under ***Hyp R***$_e$, if $E \in G$, $\eta > 0$, then there exists a complex P, such that $|E, P|_\mu < \eta$.

Proof. This follows from [Q.13.3.].

Q.15.2. Under ***Hyp R***$_e$, if $E \in \boldsymbol{G}$ and A is a brick-covering of E, $\eta > 0$; then there exists a complex P, such that $|E, P|_\mu < \eta$, $\text{som} P \leqq A$.

Proof. By [Q.15.1.] we find a complex Q with $|E, Q|_\mu < \frac{\eta}{2}$. We have $E \subseteqq A$. There exists a sequence of mutually disjoint bricks $a_1, a_2, \ldots, a_n, \ldots$ with $A = \sum a_n$. Let $Q = \{q_1, q_2, \ldots, q_m\}$, where q_i are disjoint bricks, $(m \geqq 1)$. Consider the bricks $a_n q_i$, for all n and i. They are disjoint. We have $\text{som} Q \cdot A = \sum_{n,i} a_n q_i$, because $\text{som} Q = \sum_i q_i$. We have

$$|E A, \text{som} Q \cdot A| \leqq \frac{\eta}{2},$$

---

[1] This means, that the existence of the limit [A.21.] $\lim^\mu E_n$ for $E_n \in \boldsymbol{G}$ is equivalent to the *Cauchy conditien*: for every $\eta > 0$ there exists n_0, such that if $n \geqq n_0$, $m \geqq n_0$, we have $|E_n, E_m|_\mu \leqq \eta$.

[2] i.e. it may be not true that there exists a denumerable set of somata $\in \boldsymbol{G}$ which is everywhere dense in $\boldsymbol{G}$.

by [A.17.17.], i.e.

$$(1) \qquad |E, \operatorname{som} Q \cdot A| \leqq \frac{\eta}{2}.$$

Let us arrange the bricks $a_n g_i$, into a sequence; denote it by $p_1, p_2, \ldots$ If it is finite, the complex $\{p_1, p_2, \ldots\}$ yields the thesis. If it is infinite, take m such that

$$\left| \sum_{k=1}^{\infty} p_k, \sum_{k=1}^{m} p_k \right| \leqq \frac{\eta}{2},$$

i.e.

$$(2) \qquad \left| \operatorname{som} Q \cdot A, \sum_{k=1}^{m} p_k \right| \leqq \frac{\eta}{2}.$$

From (1) and (2) we get $\left| E, \sum_{k=1}^{m} p_k \right| \leqq \eta$, $\sum_{k=1}^{m} p_k \leqq A$. The complex $P =_{df} \{p_1, \ldots, p_m\}$ yields the thesis.

Q.15.3. Lemma. If

1. $E_1, \ldots, E_n \in \boldsymbol{G}$, $n = 1, 2, \ldots$,
2. $E = E_1 + E_2 + \cdots + E_n + \cdots$,
3. all E_n are disjoint,
4. $\eta =_{df} \sup_{n=1,2\ldots} \mu(E_n)$, $\eta > 0$,

then

1) there exists i with $\mu(E_i) = \eta$,
2) the number of all j, for which $\mu(E_i) = \eta$, is finite.

Proof. Suppose that there does not exist any index i with $\mu(E_i) = \eta$. There exists an infinite sequence of indices $\alpha(1) < \alpha(2) < \cdots$ with $\lim_{i\to\infty} \mu(E_{\alpha(i)}) = \eta$, $\mu(E_{\alpha(i)}) < \eta$. Hence, starting from some k, we have $\mu(E_{\alpha(k)}) > \frac{\eta}{2}$, $\mu(E_{\alpha(k+1)}) > \frac{\eta}{2}$, ... Since $\mu(E) = \mu(E_1) + \mu(E_2) + \cdots$, it follows that $\mu(E) \geqq \mu(E_{\alpha(k)}) + \mu(E_{\alpha(k+1)}) + \cdots$, and then $\mu(E) \geqq m \cdot \frac{\eta}{2}$ for all $m = 1, 2, \ldots$, which is impossible. Hence there exists i with $\mu(E_i) = \eta$. Now the number of those indices i must be finite, because if not, we would have $\mu(E) \geqq m \cdot \eta$ for all $m = 1, 2, \ldots$, which is impossible.

Q.15.3.1. Def. By a *partition of a soma* $E \in \boldsymbol{G}$ we shall understand an at most denumerable sequence of mutually disjoint subsomata of E, $E_1, E_2, \ldots, E_n, \ldots$ with $E = E_1 + E_2 + \cdots$, but where we do not take care of the order in which $E_1, E_2, \ldots$ are written. (For a more precise setting see (45).)

Q.15.3.2. Def. Given two partitions A: $E = E_1 + E_2 + \cdots$ and B: $E = F_1 + F_2 + \cdots$ of a given soma $E \in \boldsymbol{F}$, by *the product* $A \cdot B$

of them we shall understand the partition $E = \sum_{ik} E_i F_k$ where the terms $E_i F_k$ may be arranged, in any way, into a sequence.

Q.15.3.3. Given two partitions $\sum_i E_i$ and $\sum_k F_k$ of E, we say that the *second partition is a subpartition of the first*, whenever for every k with $F_k \neq O$ there exists i with $F_k \leqq E_i$. This index i is unique for a given k.

Q.15.4. If $\sum_i E_i$, $\sum_i F_i, \ldots$ is an infinite sequence of partitions of E, we say that this is a *nested sequence of partitions* of E, whenever every partition, starting from the second, is a subpartition of the preceding one.

Q.15.5. If A, B are partitions of E, then their product is a subpartition of A and of B.

Q.15.6. We shall be interested in partitions, whose elements are either bricks or figures.

Q.15.7. Def. Given a partition E_i or a complex $P\{p_i\}$, by *the net-number* $\mathcal{N}\{E_i\}$, $\mathcal{N} P$ *of the given partition or complex* we shall understand the maximum of the numbers $\mu(E_i)$, $\mu(p_i)$ respectively. If the complex P is empty we define $\mathcal{N} P$ as the number 0.

Q.15.8. If P is a not empty complex $\{p_1, p_2, \ldots, p_n\}$, $n \geqq 1$, its bricks constitute a partition of som$P$. If P' is a complex which constitutes a subpartition of P, then $\mathcal{N} P' \leqq \mathcal{N} P$.

Q.15.9. We shall introduce a kind of integration, which will be based on approximation of the given soma of $\boldsymbol{G}$ by complexes. The integration requires "small particles", i.e. complexes, whose elements should have "small" measures. This is, however, impossible in the case, where $\boldsymbol{G}$ possesses atoms. To master this difficulty, several lemmas will be introduced, concerning special notions of "smallness", when atoms are taken into account. We start with the case, where there are no atoms in $\boldsymbol{G}$.

Q.16. Lemma.

1) Under ***Hyp R***, hence especially ***Hyp L*** $\boldsymbol{\mu}$, and **(*Hyp Ad*)**, we have the following: if

2) the tribe $\boldsymbol{G}$ has no *measure-atoms*[1],

3) a is a brick, $\mu a > 0$,

4) $\eta > 0$,

then there exists a partition of a:

$$a = a_1 + a_2 + \cdots + a_n + \cdots$$

[1] This means that, if $E \in \boldsymbol{G}$ and $\mu E > 0$, then there exist $E_1, E_2 \in \boldsymbol{G}$ with $E_1, E_2 = O$, $\mu(E_1) > 0$, $\mu(E_2) > 0$, $E_1 + E_2 = E$. If μ is effective, measure atoms coincide with ordinary atoms.

into an at most denumerable number of disjoint bricks a_i, such that

$$\max_{n=1,2...} \mu(a_n) < \eta,$$

i.e. the net number $\mathcal{N}\{a_i\}$ of the partition is $< \eta$.

Q.16a. Proof. We may admit that μ is effective, so the measure-atoms coincide with genuine atoms. First of all we recall, that for any partition A: $a = a_1 + a_2 + \cdots + a_n + \cdots$ of a into an at most denumerable number of bricks, the set of all numbers $\mu(a_n)$ admits a maximum [Q.15.3.], which we shall denote by $M(A)$. The theorem says that there exists a decomposition A of a with $M(A) < \eta$. Suppose this be not true. Hence, if we take all possible above decompositions of a, we shall have

(0) $$\alpha =_{df} \sup M(A) \geqq \eta.$$

Hence there exists an infinite sequence $A_1, A_2, \ldots, A_n, \ldots$ of partitions of a, such that

$$M(A_1) \geqq M(A_2) \geqq \cdots \to \alpha \geqq \eta > 0.$$

Take the sequence of partitions $B_1 =_{df} A_1$, $B_2 =_{df} A_1 \cdot A_2$, $B_3 =_{df} A_1 \cdot A_2 \cdot A_3, \ldots$ (the products of partitions [Q.15.2.]). They make up a nested sequence of partitions [Q.15.4.], [Q.15.5.], $B_1, B_2, B_3, \ldots$ where B_{n+1} is a subpartition of $B_1, \ldots, B_n$, $(n = 1, 2, \ldots)$. We have

$$M(B_1) \geqq M(B_2) \geqq \cdots \to \alpha.$$

Q.16b. Denote by λ_n the number of disjoint bricks b in B_n with $\mu(b) \geqq \alpha$. Their number is finite and $\geqq 1$. Indeed, we have $\lambda_n \cdot \alpha \leqq \mu(a)$; hence

(1) $$\lambda_n \leqq \frac{\mu(a)}{\alpha}.$$

If λ_n were $= 0$, all bricks in B_n would have the measure $< \alpha$, which is excluded. In the infinite sequence

$$\lambda_1, \lambda_2, \ldots, \lambda_n, \ldots$$

we have

$$\lambda_1 \leqq \lambda_2 \leqq \cdots \leqq \lambda_n \leqq \cdots,$$

because $\{B_n\}$ is nested.

Hence, by (1), starting from an index n_0, all λ_n are equal. Put $\lambda =_{df} \lambda_{n_0} = \lambda_{n_0+1} = \cdots$ We have $\lambda \geqq 1$. We shall call λ, temporarily in this proof, characteristic-number of the sequence $\{B_n\}$. Thus we have proved that, if the theorem is not true, then there exists a nested sequence of partitions $\{B_n\}$ of a with

$$M(B_1) \geqq M(B_2) \geqq \cdots \to \alpha,$$

and with characteristic number $\lambda \geqq 1$. Consider such a nested sequence with the smallest characteristic number. Denote this number by λ. We have $\lambda \geqq 1$. Denote the sequence of partitions by $\{D_n\}$. We have

(2) $$M(D_1) \geqq M(D_2) \geqq \cdots \to \alpha.$$

Denote the bricks in D_n, whose measure $\geqq \alpha$, by $d_{n1}, d_{n2}, \ldots, d_{n\lambda}$. Since the partitions D_n are nested, we can admit that

$$\begin{array}{l} d_{11} \geqq d_{21} \geqq \cdots \geqq d_{n1} \geqq \cdots \\ d_{12} \geqq d_{22} \geqq \cdots \geqq d_{n2} \geqq \cdots \\ \cdots\cdots\cdots\cdots\cdots\cdots \\ d_{1\lambda} \geqq d_{2\lambda} \geqq \cdots \geqq d_{n\lambda} \geqq \cdots \end{array}$$

Put

$$c_1 =_{df} \prod_{n=1}^{\infty} d_{n1}, \ldots, \qquad c_\lambda \prod_{n=1}^{\infty} =_{df} d_{n\lambda}.$$

These somata may not be bricks, but they $\in \boldsymbol{G}$. We have $\mu c_1 \geqq \alpha, \ldots,$ $\mu c_\lambda \geqq \alpha$.

Q.16c. We shall prove that for at least one index k we have $\mu c_k = \alpha$. Suppose this be not true. Hence for all indices $i = 1, \ldots, \lambda$ we have $\mu c_i > \alpha$. Hence there exists $\delta > 0$ with

(3) $$\mu c_i > \alpha + \delta \quad \text{for all} \quad i = 1, \ldots, \lambda.$$

Consider the partition D_n. There are only λ bricks, for which $\mu(b) \geqq \alpha$, namely $d_{n1}, \ldots, d_{n\lambda}$; for all other bricks b of D_n we have $\mu b < \alpha$. Now the numbers

$$\mu d_{n1}, \ldots, \mu d_{n\lambda} \quad \text{are all} \quad > \alpha + \delta.$$

Hence $M(D_n) > \alpha + \delta$ which contradicts (2).

Q.16d. Take k such that

(4) $$\mu c_k = \alpha.$$

Since $\mu c_k > 0$ and $\boldsymbol{G}$ has no measure-atoms, there exist two somata E, F of $\boldsymbol{G}$ with $\mu E > 0$, $\mu F > 0$, $E \cdot F = O$, $E + F = c_k$. We have, by (4),

(5) $$\mu E + \mu F = \alpha.$$

Take such sets E, F. Take $\eta > 0$ such that

(6) $$\eta < \frac{1}{24} \min[\mu(E), \mu(F)].$$

Since $\mu(E) < \alpha$, $\mu(F) < \alpha$, we have $\eta < \frac{\alpha}{24}$, i.e.

(7) $$\frac{\alpha}{2} - 12\eta > 0.$$

Take $\delta > 0$ such that

(8) $$\delta < \frac{\alpha}{2} - 12\eta$$

and consider n, such that

(9) $$\mu(d_{kn} - c_k) \leqq \delta.$$

Q.16e. Having that, consider figures e, f, such that

$$|E, e|_\mu < \eta, \quad |F, f|_\mu < \eta.$$

Put

$$e' =_{df} e \cdot d_{nk}, \quad f' =_{df} f \cdot d_{nk}.$$

Since

$$E \leqq d_{nk}, \quad F \leqq d_{nk},$$

we have

$$e' \leqq d_{nk}, \quad f' \leqq d_{nk}.$$

We have [A.17.17.],

(10) $$|E, e'|_\mu = |E\, d_{nk}, e\, d_{nk}|_\mu \leqq |E, e| < \eta.$$

Similarly

$$|F, f'|_\mu < \eta.$$

Put

$$e'' =_{df} e' - f', \quad f'' =_{df} f' - e'.$$

We have

$$e'' \leqq d_{nk}, \quad f'' \leqq d_{nk}, \quad e'' \cdot f'' = 0,$$

and, by [A.18.3.],

$$|E, e''|_\mu \leqq 3\eta, \quad |F, e''|_\mu \leqq 3\eta.$$

Hence, by [A.18.],

$$|\mu(e'') - \mu(E)| \leqq 6\eta, \quad |\mu(f'') - \mu(F)| \leqq 6\eta.$$

This gives

(11) $$\begin{aligned} \mu(E) - 6\eta \leqq \mu(e'') \leqq \mu(E) + 6\eta, \\ \mu(F) - 6\eta \leqq \mu(f'') \leqq \mu(F) + 6\eta. \end{aligned}$$

From (6) we get

$$6\eta < \tfrac{1}{2} \min[\mu(E), \mu(F)].$$

Hence

(11.1) $$6\eta < \tfrac{1}{2}\mu(E), \quad 6\eta < \tfrac{1}{2}\mu(F),$$

and then

(12) $$\begin{aligned} \mu(E) - 6\eta > \mu(E) - \tfrac{1}{2}\mu(E) = \tfrac{1}{2}\mu(E), \\ \mu(F) - 6\eta > \mu(F) - \tfrac{1}{2}\mu(F) = \tfrac{1}{2}\mu(F). \end{aligned}$$

Hence, by (11),

(13) $$\begin{aligned} 0 < \tfrac{1}{2}\mu(E) < \mu(e''), \\ 0 < \tfrac{1}{2}\mu(F) < \mu(f''). \end{aligned}$$

Taking the right-hand-side inequalities in (11), we get, by (11.1) and (5)

$$\mu(e'') \leqq \mu(E) + 6\eta \leqq \alpha - \mu(F) + 6\eta \leqq \alpha - \mu(F) + \tfrac{1}{2}\mu(F) \leqq$$
$$\leqq \alpha - \tfrac{1}{2}\mu(F),$$

and similarly $\mu(f'') \leqq \alpha - \frac{1}{2}\mu(E)$. This together with (13) gives

$$0 < \tfrac{1}{2}\mu(E) < \mu(e'') < \alpha - \tfrac{1}{2}\mu(F),$$
(14)
$$0 < \tfrac{1}{2}\mu(F) < \mu(f'') < \alpha - \tfrac{1}{2}\mu(E).$$

We have from (11)

$$\alpha - 12\eta \leqq \mu(e'') + \mu(f'').$$

Hence

$$\mu(d_{nk} - (e''_k + f''_k)) = \mu\, d_{nk} - \mu(e''_k + f''_k) \leqq \mu\, d_{nk} - (\alpha - 12\eta) \leqq$$

by (9),

$$\leqq \mu(c_k) + \delta - (\alpha - 12\eta) \leqq$$

by (4),

$$\leqq \alpha + \delta - (\alpha - 12\eta) = \delta + 12\eta, \quad \text{by (8)}, \quad < \frac{\alpha}{2}.$$

Thus we have

$$\mu(d_{nk} - (e''_k + f''_k)) < \frac{\alpha}{2}$$

and, by (14),

$$\mu(e''_k) < \alpha - \tfrac{1}{2}\mu(F),$$
(15)
$$\mu(f''_k) < \alpha - \tfrac{1}{2}\mu(E).$$

Q.16f. The brick d_{nk} is thus decomposed into three disjoint figures:

$$g =_{df} d_{nk} - (e''_k + f''_k), \qquad e''_k \text{ and } f''_k,$$

all with a measure smaller than α. If we decompose, (see ***(Hyp Ad)***), [Q.1.1.], each of the figures g, e''_k, f''_k into bricks, the measure of every brick will be surely $< \alpha$. If we do this, the partition D_n will be changed into another partition D' of $\boldsymbol{a}$, where the part outside d_{nk} is not changed, and only d_{nk} is replaced by a partition with max measure of bricks less than α.

Consider the sequence $D_1 D', D_2 D', \ldots, D_n D', \ldots$

This is a sequence of nested partitions of $\boldsymbol{a}$ with $M(D_n D') \to \alpha$, but with characteristic number $< \lambda$, or without any one, which is a contradiction, since λ is the minimum characteristic number.

Q.16.1. Under ***HypR_e*** suppose that

1) **G** has no measure-atoms,

2) we have a partition A of a soma of **G** into an at most denumerable number of disjoint bricks,

3) $\eta > 0$.

Then there exists a subpartition B of A into bricks, such that the not-number $\mathcal{N} B < \eta$.

Proof. This follows from Lemma [Q.16.].

Q.17.1. Lemma. Under $\boldsymbol{Hyp R_e}$, if A is an atom of $\boldsymbol{G}$, 2. $E \in \boldsymbol{G}$, than either $A \leqq E$ or $A \leqq \operatorname{co} E$, disjointedly.

Proof. We have $A = A\,E + A \operatorname{co} E$. If $a\,E = O$, then $A = A \operatorname{co} E$, and then $A \leqq \operatorname{co} E$. If $A \operatorname{co} E = O$, we get $A \leqq E$. The remaining case where is $A\,E \neq O$ and $A \operatorname{co} E \neq O$, is impossible. Indeed we would have $\mu(A\,E) > 0$, $\mu(A \operatorname{co} E) > 0$, so A were not an atom.

Q.17.2. Lemma. Under $\boldsymbol{Hyp R_e}$ if

1. A is an atom in $\boldsymbol{G}$,
2. $p_1, p_2, \ldots, p_n, \ldots$ are disjoint different bricks,
3. $A \leqq \sum_n p_n$, then there exists one an only one index n, such that $A \leqq p_n$.

Proof. First of all we cannot have two different indices i, j with $A \cdot p_i \neq O$, $A \cdot p_j \neq O$. Indeed, by Lemma [Q.17.1.], we would have $A \leqq p_i$; hence $A \subsetneqq \operatorname{co} p_j$, because $p_i \leqq \operatorname{co} p_j$, and the bricks $\{p_i\}$ are different. Hence, since $A \leqq p_j$ [Q.17.1.], we would have $A \leqq p_j \cdot \operatorname{co} p_j = O$, i.e. A is not an atom. Now, by hyp. 3., there exists an index n with $A \cdot p_n \neq O$. Hence, [Q.17.1.], we have $A \leqq p_n$. This index n is unique by virtue of what just has been proved. The lemma is established.

Q.17.3. Lemma. Under $\boldsymbol{Hyp R_e}$ if

1. A is an atom in $\boldsymbol{G}$,
2. $E \in \boldsymbol{G}$,
3. $|A, E|_\mu \leqq \eta$, $\eta > 0$,
4. $\eta < \mu(A)$,

then

$$A \leqq E.$$

Proof. We have

$$\mu(E - A) + \mu(A - E) \leqq \eta. \tag{1}$$

Suppose that the inclusion $A \leqq E$ is not true. Then, [Q.17.1.], $A \leqq \operatorname{co} E$; hence $A \cdot E = 0$. Since $A = (A - E) + A\,E$, we get $A = A - E$, and then, by (1), $\mu(E - A) + \mu(A - E) = \mu(E - A) + \mu\,A \leqq \eta$, hence $\mu\,A \leqq \eta$, which contradicts hyp. 4. The lemma is proved.

Q.17.4. Lemma. Under $\boldsymbol{Hyp R_e}$ if A_1, A_2 are two different atoms of $\boldsymbol{G}$, then there exists a partition of 1 into bricks $\{a_n\}$, such that A_1, A_2 are lying in distinct bricks:

$$A_1 \leqq a_i, \quad A_2 \leqq a_j \quad \text{where} \quad a_i \cdot a_j = O.$$

Proof. We have $\mu\,A_1 > 0$, $\mu\,A_2 > 0$. Take $\eta > 0$, such that $\eta < \frac{1}{3} \min[\mu\,A_1, \mu\,A_2]$. Find, [Q.15.1.], complexes P_1, P_2 such that

$|A_1, P_1| < \eta$, $|A_2, P_2| < \eta$. Since $A_1 A_2 = O$, we get [A.18.3.],

$$|A_1, \text{som}\, P_1 - \text{som}\, P_2| \leqq 3\eta, \quad |A_2, \text{som}\, P_2 - \text{som}\, P_1| \leqq 3\eta.$$

We also have $3\eta < \mu(A_1)$, $3\eta < \mu(A_2)$. Applying [Q.17.3.], we get

$$A_1 \leqq \text{som}\, P_1 - \text{som}\, P_2, \quad A_2 \leqq \text{som}\, P_2 - \text{som}\, P_i.$$

Since $\text{som}\, P_1 - \text{som}\, P_2$ and $\text{som}\, P_2 - \text{som}\, P_1$ are figures, they can be partitioned, ***(Hyp Ad)***, into an at most denumerable number of disjoint different bricks. By [Q.17.2.] there exists, in $\text{som}\, P_1 - \text{som}\, P_2$, one brick of the partition, which contains A_1, and in $\text{som}\, P_2 - \text{som}\, P_1$ there exists one brick, containing A_2.

Since $\text{som}\, P_1 - \text{som}\, P_2$ and $\text{som}\, P_2 - \text{som}\, P_1$ are disjoint, the mentioned bricks are also disjoint and different. Having this, decompose the figure $I - (\text{som}\, P_1 - \text{som}\, P_2) + (\text{som}\, P_2 - \text{som}\, P_1)]$ into a denumerable number of disjoint bricks, ***(Hyp Ad)***. Thus we shall have a partition of I into bricks, such that A_1, A_2 are lying in two different bricks of that partition. The lemma is proved.

Q.17.5. Lemma. Under ***Hyp*** $\boldsymbol{R_e}$, if

1) P is a partition of I into bricks,

2) $A_1, A_2, \ldots, A_n, \ldots$ are some or all atoms of $\boldsymbol{G}$, finite or infinite in number,

3) these atoms are lying in distinct bricks of P, (never, two atoms in one brick),

4) Q is a subpartition of P,

then the above atoms are also lying in distinct bricks of Q.

Proof. By [Q.17.2.].

Q.17.6. Lemma. Under ***Hyp*** $\boldsymbol{R_e}$ let $A_1, A_2, \ldots, A_n$, $(n \geqq 2)$ be some different atoms of $\boldsymbol{G}$. (They may be all atoms or not.) Then there exists a partition P of I such that all $A_1, \ldots, A_n$ are lying in distinct bricks of the partition. (Never two atoms in one brick.)

Proof. Consider a not ordered couple (A_i, A_j) where $A_i \neq A_j$. Take, by [Q.17.4.], a partition P_{ij} of I such that A_i and A_j are lying in different bricks of P_{ij}. The product $P' =_{df} \prod_{(i,j)} P_{ij}$, taken for all different above couples of indices, is a partition of I into a denumerable number of bricks, [Q.15.3.2.], and is, [Q.15.5.], a subpartition of all P_{ij}. By [Q.17.5.], the atoms A_i, A_j, $(i \neq j)$ are lying in different bricks of P'. This being true for any couple i, j of indices, the lemma is established.

Q.17.7. Lemma. Admit the ***Hyp*** $\boldsymbol{R_e}$. If

1. $A_1, A_2, \ldots, A_n$, $(n \geqq 1)$ are some, (or all), different atoms of $\boldsymbol{G}$,

2. $\eta > 0$,

then there exists a partition of I into different bricks

(1) $$a_1, a_2, \ldots, a_m, \ldots$$

such that:

1) The atoms $A_1, A_2, \ldots, A_n$ are lying in different bricks (1),

2) if $A_i \leqq a_m$, then $0 \leqq \mu(a_m) - \mu(A_i) < \eta$.

Proof. Relying on [Q.17.6.], find a partition P of I into different bricks such that each atom $A_1, \ldots, A_n$ is lying in a separate brick of P, i.e. two different above atoms are lying in different bricks. Let

(1.1) $$A_1 \leqq a_1, \ldots, A_n \leqq a_n,$$

where $a_1, \ldots, a_n$ are bricks of P.

Take $\eta' > 0$, such that $\eta' < \min[\eta, \mu A_1, \ldots, \mu A_n]$.

By [Q.15.1.], find complexes $P_1, \ldots, P_n$ such that

(2) $$|A_1, P_1| \leqq \eta', \ldots, |A_n, P_n| < \eta'.$$

Since $\eta' < \mu A_i$ and $|A_i, P_i| < \eta'$, we get, by [Q.17.3.],

(3) $$A_i \leqq \operatorname{som} P_i.$$

From (1.1) and (3) we get

(4) $$A_i \leqq a_i \operatorname{som} P_i.$$

From (2) we have, by [A.17.17.],

$$|A_i a_i, a_i \operatorname{som} P_i| \leqq \eta'.$$

Hence, by (1.1),

(4.1) $$|A_i, a_i \operatorname{som} P_i| \leqq \eta'.$$

Hence, by (4),

(5) $$0 < \mu(a_i \operatorname{som} P_i) - \mu(A_i) \leqq \eta'.$$

Having this, replace the brick a_i, $(i = 1, 2, \ldots, n)$, by the two different figures

$$a_i \operatorname{som} P_i, \quad a_i - a_i \operatorname{som} P_i,$$

whose sum is a_i, and partition these figures into bricks. Since a_i belongs to P, we get, in this way, a subpartition P' of P. Since the atoms $A_1, \ldots, A_n$ are lying separate bricks of P, therefore they are also lying, by [Q.17.5.], in separate bricks of P', say in $a'_1, \ldots, a'_n$ respectively.

I say that $a'_1 \leqq a_i \operatorname{som} P_i$.

Indeed, by (4), $A_i \leqq a_i \operatorname{som} P_i$, and, by (1), $A_i \leqq a_i$. The brick a'_i is contained either in $a_i \operatorname{som} P_i$ or in $a_i - \operatorname{som} P_i$. In the second case we would have $A_i \leqq \operatorname{co} \operatorname{som} P_i$ which is a contradiction. Thus

(6) $$a'_i \leqq a_i \operatorname{som} P_i.$$

We have, by (6),

$$\mu(a_i') - \mu(A_i) \leqq \mu(a_i \operatorname{som} P_i - A_i) =, \text{ by (5)}, = \mu(a_i \operatorname{som} P_i) - \\ - \mu(A_i) \leqq \eta' < \eta.$$

Thus we have got a partition P', which satisfies the requirements of the thesis.

Q.18. Def. There exists in $\boldsymbol{G}$ an at most denumerable number of mutually disjoint measure-atoms, say $\beta_1, \beta_2, \ldots, \beta_n, \ldots$ Put $\beta =_{df} \beta_1 + \beta_2 + \cdots$ Let $P = \{p_1, p_2, \ldots, p_n\}$ be a non empty complex. Then by *the reduced net-number of* P, we understand the number

$$\mathcal{N}_R(P) =_{df} \max[\mu(p_1 - \beta), \ldots, \mu(p_n - \beta)].$$

A similar definition we admit for any, at most denumerable partition, into disjoint bricks (see [Q.15.3.]).

Q.18.1. Theorem. If we admit ***Hyp*** $\boldsymbol{R_e}$, then for every $\eta > 0$ there exists a partition of I into bricks such that

$$\mathcal{N}_R(P) < \eta,$$

Def. [Q.18.].

Q.18.1a. Proof. First consider the case, where $\boldsymbol{G}$ is purely atomic, i.e., every soma of $\boldsymbol{G}$, which $\neq O$, is the sum of an at most denumerable number of somata $A_1, A_2, \ldots, A_n, \ldots$ with $\mu(A_n) > 0$ for $n = 1, 2, \ldots$, where all A_n are different measure-atoms. In that case $\beta = I$. We have $p_i - \beta = O$; hence $\mathcal{N}_R(P) = 0$ for any partition P. In that case the theorem is true.

Q.18.1b. Suppose that $\boldsymbol{G}$ has no atoms at all. Then we are in the conditions of [Q.16.1.], applied to I. Hence, if $\eta > 0$, there exists a partition P of I, such that $\mathcal{N}(P) < \eta$. If $P = \{p_1, p_2, \ldots\}$, we have $\max_i \mu(p_i) = \max_i \mu(p_i - \beta) < \eta$. Hence $\mathcal{N}_R(P) < \eta$, so the theorem is true in our case.

Q.18.1c. The case where $\boldsymbol{G}$ is not purely atomic, but has only a finite number of atoms, will constitute a simplified version of arguments, used in the discussion of the case, where we have an infinite number of atoms.

Q.18.1d. Thus we direct our attention to the case where $\boldsymbol{G}$ is not purely atomic, but has a denumerable infinite number of different atoms, say:

$$A_1, A_2, \ldots, A_m, \ldots$$

We may suppose that

(1) $$\mu(A_1) \geqq \mu(A_2) \geqq \cdots \geqq \mu(A_m) \geqq \cdots$$

Let $\eta > 0$. We can find n such that

(2) $$\mu(A_{n+1} + A_{n+2} + \cdots) < \eta.$$

Q.18.1e. Applying [Q.17.1.], take a partition P of I into different bricks $a_1, a_2, \ldots$, such that

(3) $$A_1 \leqq a_1, \ldots, A_n \leqq a_n$$

and

(4) $$\mu(a_i - A_i) < \eta.$$

Putting

(5) $$\beta =_{df} A_1 + A_2 + \cdots + A_n + A_{n+1} + \cdots,$$

we get

(6) $$\mu(a_i - \beta) < \eta \quad \text{for} \quad i = 1, 2, \ldots, n, \ldots$$

We have $A_1 + \cdots + A_n \leqq a_1 + \cdots + a_n$, but there may also exist atoms among

(7) $$A_{n+1}, A_{n+2}, \ldots,$$

which are included in $a_1 + \cdots + a_n$. If all atoms (7) are included in $a_1 + \cdots + a_n$, the arguments, which follow will be simplified. Let

(8) $$A_{k(1)}, A_{k(2)}, \ldots,$$

finite, or infinite in number, be all atoms, taken from (7), which are not included in $a_1 + \cdots + a_n$. Then they must lie in $\text{co}[a_1 + \cdots + a_n]$; this by [Q.17.1.].

Let

(9) $$A_{l(1)}, A_{l(2)}, \ldots$$

be all atoms among (7), which are lying in $\text{co}[a_1 + \cdots + a_n]$. The sets (8) and (9) make up the whole set (7). These sets are disjoint.

Q.18.1f. We have

(10) $$A_{k(1)} + A_{k(2)} + \cdots \leqq \text{co}(a_1 + \cdots + a_n).$$

Find, by [Q.13.2.], a covering B of $A_{k(1)} + A_{k(2)} + \cdots$, such that

(11) $$|A_{k(1)} + A_{k(2)} + \cdots, B|_\mu < \eta.$$

The soma $\text{co}(a_1 + \cdots + a_n)$ is a figure, hence it is also a covering Def. [Q.1.1.]. Hence

(12) $$B' =_{df} \text{co}(a_1 + \cdots + a_n) \cdot B$$

is a covering of $A_{k(1)} + A_{k(2)} + \cdots$, such that

(13) $$|A_{k(1)} + A_{k(2)} + \cdots, B'|_\mu < \eta,$$

and

(14) $$A_{k(1)} + A_{k(2)} + \cdots \leqq B'.$$

Let us partition B' into disjoint different bricks

(15) $$b_1, b_2, \ldots$$

Since, by (2), $\mu(A_{k(1)} + A_{k(2)} + \cdots) < \eta$, we get, by [A.18.], from (13), $\mu B' < 2\eta$, and then $\mu b_k < 2\eta$. It follows that

(16) $$\mu(b_k - \beta) < 2\eta$$

for all bricks (15).

Q.18.1g. Define

(17) $$C =_{df} \text{co}(B' + a_1 + \cdots + a_n).$$

Since

$$A_1 + A_2 + \cdots + A_n \leqq a_1 + \cdots + a_n,$$

$$A_{l(1)} + A_{l(2)} + \cdots \leqq a_1 + \cdots + a_n,$$

and

$$A_{k(1)} + A_{k(2)} + \cdots \leqq B',$$

we have

(18) $$C - \beta = C,$$

i.e. all atoms of G are disjoint with C. We have

(19) $$C \leqq \text{co}(a_1 + \cdots + a_n).$$

If $\text{co}(a_1 + \cdots + a_n) = 0$, the thesis is proved, and so is, if $C = O$. Suppose that $\text{co}|a_1 + \cdots + a_n) \neq 0$ and that $C \neq O$. Consider the tribe

(20) $$\boldsymbol{G}' =_{df} \text{co}(a_1 + \cdots + a_n) \upharpoonleft \boldsymbol{G};$$

its zero is O' and its unit $I' = \text{co}(a_1 + \cdots + a_n)$. The tribe $\boldsymbol{G}'$ is the Lebesgues-μ-covering extension of the tribe

(21) $$\boldsymbol{F}' =_{df} I' \upharpoonleft \boldsymbol{F};$$

the set $\boldsymbol{B}'$ of all somata $a \cdot I'$, where $a \in \boldsymbol{B}$ constitute a base of $\boldsymbol{F}''$. Its bricks are the bricks of $\boldsymbol{B}'$ contained in I'. We have supposed that $C \neq O$. Consider the tribes

(22) $$\boldsymbol{G}'' =_{df} C \upharpoonleft \boldsymbol{G} = C \upharpoonleft \boldsymbol{G}', \qquad \boldsymbol{F}'' =_{df} C \upharpoonleft \boldsymbol{F} = C \upharpoonleft \boldsymbol{F}'.$$

Take account of (19), and notice that C may not be a figure. Denote by $\boldsymbol{B}''$ the set of all somata $a \cdot C$, where $a \in \boldsymbol{B}$. We see that $\boldsymbol{G}''$ is the Lebesgues-μ-covering extension of $\boldsymbol{F}''$; and $\boldsymbol{B}''$ is a base of $\boldsymbol{F}''$.

Q.18.1h. The tribe $\boldsymbol{G}''$ has no atoms, so we can apply Lemma [Q.16.]. By its virtue the $\boldsymbol{B}''$-brick C (which is the unit I'' of $\boldsymbol{G}''$) can be partitioned into a denumerable number of disjoint $\boldsymbol{B}''$-bricks

$$p_1, p_2, \ldots, p_n, \ldots,$$

such that

(23) $$\mu(p_n) < \eta \quad \text{for} \quad n = 1, 2, \ldots$$

Now, $p_n = q_n \cdot C$, where q_n is a $\boldsymbol{B}'$-brick. We have $q_n \in \boldsymbol{G}'$. Since C is the unit of $\boldsymbol{G}''$, we have $C \in \boldsymbol{G}''$, hence $C \in \boldsymbol{G}'$. Consequently

(24) $$p_n \in \boldsymbol{G}'.$$

Q.18.1 i. There exists in $\boldsymbol{G}'$ a covering of p_n, namely $Q_n = \{q_{n1}, q_{n2}, \ldots\}$, such that

(25) $$|Q_n, p_n| < \eta,$$

and the q_{nk} are disjoint $\boldsymbol{F}'$-bricks, and then $\boldsymbol{F}$-figures. By (23), (25), by virtue of [A.18.], we get $\mu Q_n < 2\eta$, and then $\mu(q_{nk}) < 2\eta$ for all indices n, k. Let us decompose every q_{nk}, which is an $\boldsymbol{F}$-figure, into disjoint $\boldsymbol{B}$-bricks $q_{nk} = \{r_{nk1}, r_{nk2}, \ldots\}$.

In this way every p_n is covered by a denumerable number of $\boldsymbol{B}$-bricks r_{nkj}, such that

(26) $$\mu(r_{nkj}) < 2\eta.$$

Thus C is covered by a denumerable number of $\boldsymbol{B}$-bricks. The soma B' is also covered, by [Q.18.1 f.], by a denumerable number of $\boldsymbol{B}$-bricks. The soma $a_1 + \cdots + a_n$ is also covered by the bricks $a_1, \ldots, a_n$. For all those bricks c we have either $\mu(c) < 2\eta$ or $\mu(c - \beta) < 2\eta$, (see (16), (6), (26)). Hence for all of them we have $\mu(c - \beta) < 2\eta$. Thus we have a denumerable number of bricks c, whose sum equals $I \in \boldsymbol{G}$. We have got I decomposed into a denumerable number of disjoint bricks $d_1, d_2, \ldots$ such that each d_n is lying in one of the bricks c. Hence we get for every n:

$$\mu(d_n - \beta) < 2\eta,$$

so the theorem is established.

Q.18.2. Remark. The presence of atoms hinders making partitions with small "meshes", so the notion of reduced net-number helps. Now, if the tribe $\boldsymbol{G}$ is composed of atoms only, the reduced net-number will always be $= 0$, so another kind of net-number shall be introduced — just to cover all possibilities. Therefore we introduce the following definition:

Q.19. Def. Under ***Hyp R$_e$***, if $a \neq O$ is a brick, put

$$\mathcal{N}'(a) =_{df} \mu(a) - \max_{A \leqq a} \mu(A),$$

where the maximum is taken over all atoms A, included in a; of course this happens, if there exists an atom A included in a. If a does not contain any atom, we put $\mathcal{N}'(a) =_{df} \mu(a)$. If P is a complex $\{p_1, \ldots, p_n\}$, $(n \geqq 1)$, then by *the atom-net-number of* P we shall understand the non negative number

$$\mathcal{N}_A(P) =_{df} \max(\mathcal{N}'(p_1), \ldots, \mathcal{N}'(p_n)).$$

If P is an empty complex, we define $\mathcal{N}_A(P) =_{df} 0$.

A similar definition is admitted for partitions into bricks. We can do this, because the maximum spoken of, always exists (compare Lemma [Q.15.3.]).

Q.19.1. Theorem. We admit the ***Hyp*** $\boldsymbol{R_e}$. For every $\eta > 0$ there exists a partition P of I into disjoint bricks, such that

$$\mathcal{N}_A(P) < \eta.$$

Q.19.1a. Proof. If $\boldsymbol{G}$ has no atoms, then the theorem follows from Lemma [Q.16.]. If $\boldsymbol{G}$ has a finite numbers of atoms, the theorem follows from Lemma [Q.17.7.].

Q.19.1b. Suppose that $\boldsymbol{G}$ has an infinite number of atoms. It must be denumerable. Let $A_1, A_2, \ldots, A_n, \ldots$ be all different atoms, arranged so, as to have

$$(1) \qquad \mu(A_1) \geqq \mu(A_2) \geqq \cdots \geqq \mu(A_n) \geqq \cdots$$

This can be done, because the number of atoms whose measure is $\geqq \varepsilon$, where $\varepsilon > 0$, is at most finite. Let $\eta > 0$. Find n such that, if we put $A =_{df} A_{n+1} + A_{n+2} + \cdots$, we have $\mu A < \eta$.

Relying on [Q.17.7.] we can find a partition P of I into disjoint bricks, such that the atoms $A_1, A_2, \ldots, A_n$ are lying in separate bricks, say $a_1, \ldots, a_n$, so that we have

$$A_1 \leqq a_1, \ldots, A_n \leqq a_n,$$

and that

$$\mu(a_i) - \mu(A_i) < \eta \quad \text{for} \quad i = 1, 2, \ldots, n.$$

On account of (1) we have

$$(2) \qquad \mu(a_i) - \max_{A_s \leqq a_i} \mu(A_s) < \eta,$$

because there does not exist any atom in a_i whose measure were $> \mu A_i$. Suppose that $A \neq \mathit{1}$, because if $A = \mathit{1}$, the thesis follows.

Consider the tribe $\boldsymbol{G}' =_{df} \operatorname{co} A \restriction \boldsymbol{G}$. The unit I' of $\boldsymbol{G}'$ is $\operatorname{co} A$; it is a figure. The atoms of $\boldsymbol{G}'$ are $A_{n+1}, A_{n+2}, \ldots$, if any. By [Q.18.1.] we can find a partition $Q = \{q_1, q_2, \ldots\}$ of $\operatorname{co} A = I'$, so that $\max[\mu(q_1 - A'), \mu(q_2 - A'), \ldots] < \eta$, where A' is the sum of all atoms of $\boldsymbol{G}'$. Since $A' \leqq A$, we get $\mu A' < \eta$ and

$$\max[\mu(q_1 - A), \mu(q_2 - A), \ldots] < \eta.$$

The only atoms, contained in q_j, are among those of A'. Since $\mu A' < \eta$, it follows:

$$\mu(q_j) = \mu(q_j - A') + \mu(q_j A') < \eta + \eta = 2\eta.$$

Hence

$$(3) \qquad 0 \leqq \mu(q_j) - \max_{A_s \leqq q_j} \mu(A_s) < \mu(q_j) < 2\eta.$$

Now $\{a_1, a_2, \ldots, a_n, q_1, q_2, \ldots\}$ is a partition of I into bricks. From (2) and (3) it follows, that if b is any brick of this partition, we have $\mu(b) - \max\limits_{A_s \leqq b} \mu(A_s) < 2\eta$, so the theorem is proved.

Q.19.2. If

1. A is a partition of I into bricks,
2. $\mathcal{N}_R A < \eta$,
3. B is a subpartition of A into bricks,

then

$$\mathcal{N}_R(B) < \eta.$$

Proof. Let $b \in B$. There exists $a \in A$ with $b \leqq a$. We have $\mu(a - \beta) < \eta$; consequently $\mu(b - \beta) < \eta$. This being true for any $b \in B$, the lemma follows.

Q.19.3. If

1. A is a partition of I into bricks $a_1, a_2, \ldots$,
2. $\mathcal{N}_A(a_n) < \eta$ for all $n = 1, 2, \ldots$,
3. $B = \{b_1, b_2, \ldots\}$ is a subpartition of A into bricks,

then

$$\mathcal{N}_A(b_n) < \eta \quad \text{for} \quad n = 1, 2, \ldots$$

Proof. Let $b \in B$. There exists $a \in A$ with $b \leqq a$. Now

$$\mu\, a - \max_{A \leqq a} \mu(A) < \eta,$$

where the maximum is taken for all atoms A, which are included in a — if they exist. We have $\mu(a) < \eta$, if no atoms lying in a, are available. Let

(1) $$A_1, A_2, \ldots, A_s,$$

$(s \geqq 0)$ be all different atoms, included in a, with

$$\mu\, A_1 = \cdots = \mu\, A_s = \max_{A \leqq a} \mu(A) =_{df} \delta.$$

Now $b \leqq a$. Concerning the atoms $A_1, \ldots, A_s$, each one of them is either included in b or in $a - b$. Suppose that one at least of (1), say A_k, is in $a - b$. Since $\mu(a) - \delta < \eta$ and since $b \leqq a - A_k$, we have $\mu(b) < \eta$, and then a fortiori $\mu(b) - \max\limits_{A \leqq b} \mu(A) < \eta$.

Now suppose that no one of (1) is in $a - b$. Hence they all are in b. Hence

$$\max_{A \leqq b} \mu(A) = \max_{A \leqq a} \mu(A).$$

Hence

$$\mu(a) - \max_{A \leqq b} \mu(A) < \eta,$$

and then

$$\mu(b) - \max_{A \leqq b} \mu(A) < \eta.$$

The remaining case is, where no atom is included in a; and then no atom is included in b; we get $\mu(b) \leqq \mu(a) < \eta$. The theorem is proved.

Q.20. Theorem. Under ***Hyp R$_e$***, if $E \in \boldsymbol{G}$ and $\eta > 0$, then there exists a complex $P = \{p_1, \ldots, p_n\}$ such that

1) $|E, P|_\mu < \eta$,

2) β is the sum of all atoms of $\boldsymbol{G}$ in the case they exist, and $\beta = O$, if there are no atoms in $\boldsymbol{G}$,

then

$$\mu(p_1 - \beta), \ldots, \mu(p_n - \beta) < \eta \quad \text{i.e.} \quad \mathcal{N}_R(P) < \eta.$$

Proof. First we find a complex P such that $|E, P|_\mu < \frac{\eta}{2}$. The soma co som$P$ is a figure; hence it can be partitioned into an at most denumerable number of disjoint bricks. This partition, together with $P = \{p_1, p_2, \ldots\}$, make up a partition Q of I into an at most denumerable number of disjoint bricks.

By Theorem [Q.18.1.] we can find a partition S of I into disjoint bricks, such that $\mathcal{N}_R(S) < \eta$. Take the product $Q \cdot S$. This is a subpartition of S and of P. Hence, by [Q.19.2.], $\mathcal{N}_R(Q\,S) < \eta$. If we confine that partition to som$P$, we have partitioned som$P$ into a denumerable number of disjoint bricks with reduced net-number $< \eta$. Let this partition of som$P$ be $p_1', p_2', \ldots$ For sufficiently great n we have $|P, p_1' + \cdots + p_n|_\mu < \frac{\eta}{2}$; and if we put $P' =_{df} \{p_1', \ldots, p_n'\}$, we have $|P, P'|_\mu < \frac{\eta}{2}$; hence $|E, P'| < \eta$ and $\mathcal{N}_R(P') < \eta$. Thus the theorem is proved.

Q.20.1. Theorem. Under ***Hyp R$_e$***, if $E \in \boldsymbol{G}$ and $\eta > 0$, then there exists a complex $P = [p_1, \ldots, p_n]$, such that

1) $|E, P|_\mu < \eta$,

2) if $A_1, A_2, \ldots$, are all atoms of $\boldsymbol{G}$,

then

$$\mu(p_i) - \max_{A_j \subsetneqq p_i} A_j < \eta, \quad (i = 1, \ldots, n),$$

and $\mu(p_i) < \eta$, if there does not exist any atom $A_j \leqq p_i$, (i.e. $\mathcal{N}_A(P) < \eta$).

Proof. The proof is similar to that of Theorem [Q.20.]. The difference is that instead of relying on [Q.18.1.], we rely on [Q.19.1.] and [Q.19.3.].

Q.20.2. Under ***Hyp R$_e$*** if $E \in \boldsymbol{G}$ and $\eta > 0$, then there exists a complex P such that

1) $|E, P|_\mu < \eta$,

2) $\mathcal{N}_R(P) < \eta$,

3) $\mathcal{N}_A(P) < \eta$.

Proof. The proof follows the pattern of the two preceding proofs. Take a complex P, such that $|E, P|_\mu < \frac{\eta}{2}$. The soma co som$P$ is a figure; hence it can be partitioned into an at most denumerable number of disjoint bricks, getting a partition Q of I. By Theorem [Q.18.1.] we can find a partition S_1 of I into bricks such that $\mathcal{N}_R(S_1) < \eta$. By Theorem [Q.19.1.] we can find a partition S_2 of I into bricks such that $\mathcal{N}_A(S_2) < \eta$. The product $R =_{df} Q \cdot S_1 \cdot S_2$ is a partition of I into a denumerable number of bricks. By Theorem [Q.19.2.] we have $\mathcal{N}_R(R) < \eta$ and by Theorem [Q.19.3.] we have $\mathcal{N}_A(R) < \eta$. Let us confine the partition R to $\operatorname{som} P$; we get a partition $P' = \{p_1', p_2', \ldots\}$ of $\operatorname{som} P$. For sufficiently great n we have, putting $P'' =_{df} \{p_1', \ldots, p_n'\}$, $|P', P''|_\mu < \frac{\eta}{2}$, which gives $|P'', E|_\mu < \frac{\eta}{2}$. We also have $\mathcal{N}_A(P'') < \eta$, $\mathcal{N}_R(P'') < \eta$. So the theorem is established.

Q.20.3. Theorem. Under ***Hyp R$_e$***, if $E \in \boldsymbol{G}$, A is a brick-covering of E, (hence $E \leqq A$), $\eta > 0$, then there exists a complex P, such that

1) $\operatorname{som} P \leqq A$,
2) $|P, E|_\mu \leqq \eta$,
3) $\mathcal{N}_A(P) \leqq \eta$,
4) $\mathcal{N}_R(P) \leqq \eta$.

Proof. We rely on [Q.14.3.1.] and apply [Q.20.2.].

Q.21. Def. The μ-topology on $\boldsymbol{G}$ is always complete but it may be not separable (see [Q.14.2.]). In farther sections we shall need an important property (see [Q.21.2.]) of complexes approximating somata of $\boldsymbol{G}$, this property being strongly related to separability. Therefore, in what follows in this Chapter Q, we shall admit in addition to the ***Hyp R$_e$***, the *following hypothesis* (***Hyp S***) of separability.

Q.21.1. Def. (*Hyp S*). There exists a denumerable sequence

$$(1) \qquad M_1, M_2, \ldots, M_n, \ldots$$

of somata of $\boldsymbol{G}$ such that for every soma $E \in \boldsymbol{G}$ we can find a subsequence $\{M_{k(n)}\}$ of 1, such that $|M_{k(n)}, E|_\mu \to 0$ for $n \to \infty$. We have the following important theorem:

Q.21.2. Theorem. Under ***Hyp R$_e$*** and (***Hyp S***), if

1) $E \in \boldsymbol{G}$,
2) A is a brick-covering of E,

then there exists an infinite sequence $\{T_n\}$ of complexes, such that

1. $\mathcal{N}_R(T_n) \to 0$, $\mathcal{N}_A(T_n) \to 0$,
2. $\operatorname{som} T_n \leqq A$,
3. $|E, T_n|_\mu \to 0$,
4. for every soma $F \leqq E$ there exists for every n a partial complex R_n of T_n, i.e. $R_n \subseteqq T_n$, such that $|F, R_n|_\mu \to 0$.

Q.21.2a. Proof. Since the μ-topology is separable, there exists a denumerable set of $\boldsymbol{G}$,

$$(1) \qquad M_1, M_2, \ldots, M_n, \ldots,$$

as in [Q.21.1.]. Let $F \in \mathbf{G}$ and $F \leqq E$. We can find a subsequence $\{M_{k(n)}\}$, $(n = 1, 2, \ldots)$ of I such that $|M_{k(n)}, F|_\mu \to 0$. We have by [A.17.17.],

$$|M_{k(n)} \cdot E, F \cdot E| \leqq |M_{k(n)}, F| \to 0;$$

hence

$$|M_{k(n)} \cdot E, F| \to 0,$$

for $n \to \infty$. Thus the set of all $M_n \cdot E$ is everywhere dense in the topology restricted to E.

Q.21.2b. Having this, let $\eta > 0$, and find a complex P_1, such that $\mathrm{som} P_1 \leqq A$, and where

(1.1) $$|E, P_1| \leqq \eta,$$

[Q.15.1.]. Find another complex Q_1 with $\mathrm{som} Q_1 \leqq A$, [Q.15.2.], such that

(1.2) $$|M_1 \cdot E, Q_1| \leqq \eta.$$

Put

(1.3) $$B_1 =_{df} \mathrm{som}\, P_1 \cdot \mathrm{som} Q_1, \quad C_1 =_{df} \mathrm{som}\, P_1 - \mathrm{som} Q_1.$$

B_1, C_1 are figures. We have

(2) $$B_1 C_1 = 0, \quad B_1 + C_1 = \mathrm{som}\, P_1.$$

Let R_1, S_1 be complexes such that, [Q.20.3.],

(3) $$\mathrm{som}\, R_1 \leqq B_1, \quad \mathrm{som}\, S_1 \leqq C_1,$$

(4) $$|R_1, B_1| \leqq \eta, \quad |S_1, C_1| \leqq \eta,$$

$$\mathcal{N}_R(R_1) \leqq \eta, \quad \mathcal{N}_R(S_1) \leqq \eta, \quad \mathcal{N}_A(R_1) \leqq \eta, \quad \mathcal{N}_A(S_1) \leqq \eta.$$

The set

(4.1) $$T_1 =_{df} R_1 \cup S_1$$

is a complex, because of (2) and (3). From (4) it follows

$$|\mathrm{som}\, R_1 + \mathrm{som}\, S_1, B_1 + C_1| \leqq 2\eta,$$

[A.17.11.], i.e.

$$|T_1, B_1 + C_1| \leqq 2\eta.$$

Hence, since $B_1 + C_1 = \mathrm{som}\, P_1$, we have $|T_1, P_1| \leqq 2\eta$; and hence, by (1.1),

(5) $$|E, T_1| \leqq 3\eta.$$

Q.21.2c. We have, (1.2):

$$|M_1 \cdot E, Q_1| < \eta, \quad |E, P_1| \leqq \eta.$$

Hence

$$|M_1 \cdot E \cdot E, \mathrm{som}\, P_1 \cdot \mathrm{som} Q_1| \leqq |M_1 \cdot E \cdot Q_1| + |E, P_1| \leqq 2\eta,$$

[A.17.13.], i.e.

$$|M_1 \cdot E, \operatorname{som} P_1 \cdot \operatorname{som} Q_1| \leqq 2\eta,$$

i.e., (1.3),

$$|M_1 \cdot E, B_1| \leqq 2\eta. \tag{5.1}$$

Since,

$$|M_1 \cdot E, R_1| \leqq |M_1 \cdot E, B_1| + |R_1, B_1|,$$

we get, by (4) and (5.1),

$$|M_1 \cdot E, R_1| \leqq 3\eta. \tag{6}$$

Now R_1 is, by (4.1), a partial complex of T_1.

Thus we have found a complex T_1 such that $\operatorname{som} T_1 \leqq A$,

$$|E, T_1| \leqq 3\eta, \quad \mathcal{N}_A(T) \leqq \eta, \quad \mathcal{N}_R(T) \leqq \eta,$$

and found a partial complex R_1 of T_1 such that $|M_1 \cdot E, R_1| \leqq 3\eta$. This can be done for every $\eta > 0$.

Q.21.2d. Let us consider the somata E, $M_1 E$ and $M_2 E$. We like to find a complex T_2 within A, approximating E, and such that it would contain a partial complex R_2, approximating $M_1 E$, and another one, approximating $M_2 E$.

First find complexes P_2, Q_{21}, Q_{22}, all within A, and such that

$$|P_2, E| < \eta, \quad |M_1 \cdot E, Q_{21}| < \eta, \quad |M_2 \cdot E, Q_{22}| < \eta. \tag{6.1}$$

Let us agree for the sake of simplicity, to use the same symbol for the complex and its soma. Consider the figures

$$P_2 Q_{21} Q_{22}, \; P_2 \bar Q_{21} Q_{22}, \; P_2 Q_{21} \bar Q_{22}, \; P_2 \bar Q_{21} \bar Q_{22} \tag{7}$$

where, in general, $\bar X$ means $\operatorname{co} X$. They are all disjoint figures, which may be also null-figures. We have

$$P_2 = P_2 Q_{21} Q_{22} + P_2 \bar Q_{21} Q_{22} + P_2 Q_{21} \bar Q_{22} + P_2 \bar Q_{21} \bar Q_{22}, \tag{7.1}$$

and

$$\begin{aligned} P_2 Q_{21} &= P_2 Q_{21} Q_{22} + P_2 Q_{21} \bar Q_{22}, \\ P_2 Q_{22} &= P_2 \bar Q_{21} Q_{22} + P_2 Q_{21} Q_{22}. \end{aligned} \tag{7.2}$$

Find complexes

$$R_{12}, \; R_{\bar 1 2}, \; R_{1 \bar 2}, \; R_{\bar 1 \bar 2} \tag{8}$$

contained in the figures

$$P_2 Q_{21} Q_{22}, \; P_2 \bar Q_{21} Q_{22}, \; P_2 Q_{21} \bar Q_{22}, \; P_2 \bar Q_{21} \bar Q_{22}$$

respectively, such that their atom-net-numbers are $< \eta$, their reduced net-numbers $< \eta$, and that

$$\begin{aligned} &|R_{12}, P_2 Q_{21} Q_{22}| < \eta, \quad |R_{\bar 1 2}, P_2 \bar Q_{21} Q_{22}| < \eta, \\ &|R_{1 \bar 2}, P_2 Q_{21} \bar Q_{22}| < \eta, \quad |R_{\bar 1 \bar 2}, P_2 \bar Q_{21} \bar Q_{22}| < \eta. \end{aligned} \tag{8.1}$$

Since the figures (7) are all disjoint, so are the complexes (8). Hence if we put $T_2 =_{df} R_{12} \cup R_{\bar{1}2} \cup R_{1\bar{2}} \cup R_{\bar{1}\bar{2}}$, we get a complex, for which all (8) are partial complexes.

Q.21.2e. Now we have, by (7.1) and [A.17.12.],

$$|P_2, T_2| \leqq |P_2 Q_{21} Q_{22}, R_{12}| + |P_2 \bar{Q}_{21} Q_{22}, R_{\bar{1}2}| + \\ + |P_2 Q_{21} \bar{Q}_{22}, R_{1\bar{2}}| + |P_2 \bar{Q}_{21} \bar{Q}_{22}, R_{\bar{1}\bar{2}}| \leqq 4\eta.$$

Hence, since by (6.1), $|P_2, E| < \eta$, we get

(9) $$|E, T_2| \leqq 5\eta.$$

Now we have, (6.1),

$$|M_1 \cdot E, Q_{21}| < \eta, \quad |M_2 \cdot E, Q_{22}| < \eta, \quad |E, P_2| < \eta.$$

Hence

$$|E \cdot M_1 E, P_2 Q_{21}| < 2\eta$$

and

$$|E \cdot M_2 E, P_2 Q_{22}| < 2\eta.$$

Hence, by (7.2),

(10) $$|M_1 E, P_2 Q_{21} Q_{22} + P_2 Q_{21} \bar{Q}_{22}| < 2\eta, \\ |M_2 E, P_2 \bar{Q}_{21} Q_{22} + P_2 Q_{21} Q_{22}| < 2\eta.$$

Since, by (8.1),

$$|P_2 Q_{21} Q_{22} + P_2 Q_{21} \bar{Q}_{22}, R_{12} \cup R_{1\bar{2}}| < 2\eta, \\ |P_2 \bar{Q}_{21} Q_{22} + P_2 Q_{21} Q_{22}, R_{\bar{1}2} \cup R_{12}| < 2\eta,$$

we get from (10):

(11) $$|M_1 E, R_{12} \cup R_{\bar{1}\bar{2}}| < 4\eta, \quad |M_2 E, R_{\bar{1}2} \cup R_{\bar{1}\bar{2}}| < 4\eta.$$

Since $R_{12} \cup R_{1\bar{2}}$ and $R_{\bar{1}2} \cap R_{12}$ are partial complexes of T_2, we have found a complex T_2 such that $|E, T_2| \leqq 4\eta$, where the atom- and reduced net-number of T_2 are $< \eta$, and which contains a partial complex

$$R_1^{(1)} \text{ with } |M_1 E, R_1^{(1)}| < 4\eta,$$

and which also contains another partial complex

$$R_2^{(1)} \text{ with } |M_2 E, R_2^{(1)}| < 4\eta.$$

This can be done for every $\eta > 0$.

Q.21.2f. Using a similar method, we can prove that, for $\eta > 0$, there exists a complex T_n contained in A, with $|E, T_n| \leqq (2^n + 1)\,\eta$, with the atom- and reduced net-number $< \eta$. The complex T_n is such, that it contains partial complexes:

$$R_1^{(n)}, R_2^{(n)}, \ldots, R_n^{(n)},$$

such that

$$|R_i^{(n)}, M_i E| \leqq 2^n \eta.$$

This is true for every $\eta > 0$. We leave to the reader to perform these arguments in details.

Q.21.2g. Taking $\eta =_{df} \frac{1}{2^n + 1} \cdot \frac{1}{n}$, we can find for every n a complex T_n, such that $\operatorname{som} T_n \leqq A$, $|T_n, E| \leqq \frac{1}{n}$, with reduced and atom-net number $\leqq \frac{1}{n}$; and such that it contains partial complexes $R_i^{(n)}$ $(i = 1, 2, \ldots, n)$, such that

$$|R_i^{(n)}, M_i E| \leqq \frac{1}{n}, \qquad (i = 1, 2, \ldots, n).$$

Q.21.2h. Having the sequence $\{T_n\}$ let $F \leqq E$, and take $\varepsilon > 0$. Find m_0 such that

$$(12) \qquad |F, E M_{m_0}| \leqq \frac{\varepsilon}{2}.$$

Take $n > m_0$ with $\frac{1}{n} \leqq \frac{\varepsilon}{2}$ and with $|T_n, E| \leqq \frac{1}{n}$. There exists a partial complex R_n of T_n such that $|E M_{m_0}, R_n| \leqq \frac{1}{n} \leqq \frac{\varepsilon}{2}$. By (12) we get $|F, R_n| \leqq \frac{\varepsilon}{2} + \frac{\varepsilon}{2} = \varepsilon$, for sufficiently great n. The theorem is established.

Q.21.3. Def. The sequence $\{T_n\}$ having the properties 1., 3., 4. expressed in [Q.21.2.], will be termed *completely distinguished for* E. The property 4. will be termed *property* (***S***) *for* E.

Q.21.3.1. Corrollary. A slightly modified argument yields the following theorem: Under ***Hyp R***$_e$ and (***Hyp S***), if

1. $E \in \boldsymbol{G}$,
2. $A_1, A_2, \ldots, A_n, \ldots$

is an infinite sequence of coverings of E, then there exists a completely distinguished sequence $\{T_n\}$ of complexes such that $\operatorname{som} T_n \leqq A_n$.

Q.21.4. Remark. The validity of theorem [Q.21.2.] implies the hypothesis (***Hyp S***).

Q.21.5. Considering the item [Q.21.2d.], in the proof of [Q.21.2.], we have (6.1): $|P_2, E| < \eta$, $|M_1 E, Q_{21}| < \eta$ and $|M_2 E, Q_{22}| < \eta$, and we have found the complex

$$T_2 = R_{12} \cup R_{\bar{1}2} \cup R_{1\bar{2}} \cup R_{\bar{1}\bar{2}} \leqq \operatorname{som} P_2 \quad \text{with} \quad |T_2, E| \leqq 4\eta,$$

(in [Q.21.2e.]), getting partial complexes $R_{12} \cup R_{1\bar{2}}$ and $R_{\bar{1}2} \cup R_{12}$, approximating $M_1 E$, $M_2 E$ respectively up to 4η.

A similar remark can be said of the general case, where $M_1 E$, $M_2 E, \ldots, M_n E$ are approximated by subcomplexes $R_1^{(n)}, \ldots, R_n^{(n)}$ of a complex T_n, where $\operatorname{som} T_n \leqq P_n$. We get

$$|R_i^{(n)}, M_i E| \leqq 2^n \eta, \quad (i = 1, 2, \ldots, n) \quad \text{and} \quad |E, T_n| \leqq (2^n + 1)\,\eta.$$

Now let $\{Q_n\}$ be a sequence of complexes, such that $|Q_n, E| \to 0$. We can find a subsequence $\{Q_{k(n)}\}$ of $\{Q_n\}$ such that

$$|Q_{k(n)}, E| \leqq \frac{1}{n} \cdot \frac{1}{2^n + 1}.$$

Taking the corresponding $\{T_{k(n)}\}$, we have $|T_{k(n)}, E| \to 0$, and in each $T_{k(n)}$ we can find partial complexes $\{R_i^{n(k)}\}$ with $|R_i^{n(k)}, M_i E| \to 0$, $(i = 1, 2, \ldots, n(k))$. This allows to state the following:

Q.21.5.1. Corrollary. If $|P_n, E| \to 0$, then there exists a subsequence $k(n)$ of indices, and a sequence of complexes $\{T'_n\}$, such that $\operatorname{som} T'_n \leqq \operatorname{som} P_{k(n)}$, and that $\{T'_n\}$ is completely distinguished with respect to E.

Q.21.6. If $\{T_n\}$ has the property (**S**) for $E \in \boldsymbol{G}$, then any subsequence $\{T_{k(n)}\}$ of it also has that property.

Q.21.7. If

1. $E \cdot F = O$,
2. $\{P_n\}$ has the property (**S**) for E,
3. Q_n has the property (**S**) for F,
4. $\operatorname{som} P_n \cdot \operatorname{som} Q_n = O$,

then $\{P_n \cup Q_n\}$ has the property (**S**) for $E + F$.

Proof. Let $H \leqq E + F$, $H \in \boldsymbol{G}$. We have $H \cdot E \leqq E$, $H \cdot F \leqq F$. There exist partial complexes P'_n, Q'_n of P_n, Q_n respectively, such that

$$(1) \qquad |H \cdot E, P'_n| \to 0, \quad |H \cdot E, Q'_n| \to 0.$$

Since $\operatorname{som} P'_n \leqq \operatorname{som} P_n$, $\operatorname{som} Q'_n \leqq \operatorname{som} Q_n$, we have $\operatorname{som} P'_n \cdot \operatorname{som} Q'_n = O$. Hence $P'_n \cup Q'_n$ is a partial complex of $P_n \cup Q_n$. We have $|P_n, E| \to 0$, $|Q_n, F| \to 0$. Hence $|P_n \cup Q_n, E + F| \to 0$, and from (1) we have

$$|H E + H F, P'_n \cup Q'_n| \to 0, \quad \text{i.e.} \quad |H, P'_n \cup Q'_n| \to 0,$$

which completes the proof.

Q.21.8. If

1. $E \in \boldsymbol{G}$,
2. the sequence $\{P_n\}$ of complexes has the property (**S**) with respect to E,
3. $F \leqq E$, $F \in \boldsymbol{G}$,
4. $Q_n \subsetneqq P_n$, $n = 1, 2, \ldots$,
5. $|Q_n, F|_\mu \to 0$ for $n \to \infty$,

then $\{Q_n\}$ has the property (**S**) with respect to F.

Proof. Let $G \in \boldsymbol{G}$, $G \leqq F$. Since $\{P_n\}$ has the property (**S**) with respect to E, and since $G \leqq E$, there exists, for every n, a partial complex

(1) $\qquad S_n$ of P_n with $|S_n, G|_\mu \to 0$ for $n \to \infty$.

We have

(2) $$S_n \subseteqq P_n, \quad Q_n \subseteqq P_n, \quad S_n \frown Q_n \subseteqq Q_n,$$

and $\operatorname{som}(S_n \frown Q_n) = \operatorname{som} S_n \cdot \operatorname{som} Q_n$. By **(*Hyp S*)** and (1) we get

$$|\operatorname{som} S_n \operatorname{som} Q_n, F\, G| = |\operatorname{som} S_n \operatorname{som} Q_n, G| \to 0.$$

Hence

(3) $$|S_n \frown Q_n, G| \to 0.$$

From (2) and (3) the theorem follows.

Q.21.9. If

1. $\{P_n\}$ is a completely distinguished sequence for E,
2. $F \leqq E$,
3. $Q_n \subseteqq P_n$ for all $n = 1, 2, \ldots$,
4. $|Q_n, F| \to 0$,

then $\{Q_n\}$ is a completely distinguished sequence for F.

Proof. Follows from [Q.21.8.].

Q.21.10. If

1. $E \cdot F = O$,
2. $\{P_n\}$ is a completely distinguished sequence for E,
3. $\{Q_n\}$ is a completely distinguished sequence for F,
4. $\operatorname{som} P_n \cdot \operatorname{som} Q_n = O$, $(n = 1, 2, \ldots)$,

then $\{P_n \smile Q_n\}$ is a completely distinguished sequence for $E + F$.

Proof. Follows from [Q.21.7.].

Q.21.11. If

1. $E \in \boldsymbol{G}$,
2. $\{P_n\}$ has the property (***S***) for E, and $|P_n, E| \to 0$ for $n = 1, 2, \ldots$,
3. $\{P'_n\}$ is a complex, whose every brick is contained in a brick of $\{P_n\}$,
4. $|P'_n, P_n| \to 0$ for $n = 1, 2, \ldots$,

then $\{P'_n\}$ also has the property (***S***) for E.

Proof. Let $F \leqq E$. By hyp. 2, we can find a partial complex Q_n of P_n, such that $|F, Q_n| \to 0$ for $n = 1, 2, \ldots$ Denote by Q'_n the maximal partial complex of P'_n, whose bricks are contained in the bricks of Q_n. We shall prove that

(1) $$|Q'_n, F| \to 0.$$

Let $P_n = \{p_{n1}, p_{n2}, \ldots\}$. Suppose that k is such, that p_{nk} contains at least one brick of P'_n. Denote these bricks by $p'_{nk1}, p'_{nk2}, \ldots$ and their sum by p'_{nk}. If p_{nk} does not contain any brick of P'_n, put $p'_{nk} = O$. We have $\operatorname{som} P_n - \operatorname{som} P'_n = \sum_k (p_{nk} - p'_{nk})$. Since the somata $p_{nk} - p'_{nk}$, $(k = 1, 2, \ldots)$ are disjoint, we have

$$\mu(\operatorname{som} P_n - \operatorname{som} P'_n) = \sum_k \mu(p_{nk} - p'_{nk}).$$

If we confine ourselves to those indices k' only, for which $p'_{nk'} \in Q_n$, we shall get

$$\mu(\text{som}\,Q_n - \text{som}\,Q'_n) = \sum_{k'} \mu(p_{nk'} - p'_{nk'}) \leqq \sum_{k} \mu(p_{nk} - p'_{nk}).$$

Since $|P_n, P'_n| \to 0$, we get

(2) $$|Q_n, Q'_n| \to 0.$$

From (1) and (2) we deduce, that $|F, Q'_n| \to 0$, which completes the proof.

Q.21.12. If

1. $E \in \boldsymbol{G}$,

2. the sequence $\{P_n\}$ of complexes has the property $(\boldsymbol{S})$ with respect to E,

then we can find complexes $\{Q_n\}$, such that

1) each brick of Q_n is contained in some brick of P_n,

2) $\{Q_n\}$ is completely distinguished for E.

Proof. We partition every brick p of P_n into a denumerable number of bricks, so as to have the reduced net-number and the atom-net-number tending to 0 for $n \to \infty$. Having this, we take for each brick p a finite number of meshes of partition, sufficiently great, so as to have p approximated with error $< \frac{1}{n \cdot k(n)}$, where $k(n)$ denotes the number of bricks in P_n. The Theorem [Q.21.8.] will complete the proof.

Q.21.13. If

1. $E \in \boldsymbol{G}$,

2. $\{P_n\}$ is a completely distinguished sequence for E,

then there exists a subsequence $k(n)$ of indices and a sequence $\{Q_{k(n)}\}$, such that

1) $\text{som}\,P_{k(n)} \cdot \text{som}\,Q_{k(n)} = O$,

2) $\{P_{k(n)} \cup Q_{k(n)}\}$ is a completely distinguished sequence for I,

3) $Q_{k(n)}$ is completely distinguished for $\text{co}\,E$.

Proof. We have $|P_n, E| \to 0$. Hence

(1) $$|\text{co som}\,P_n, \text{co}\,E| \to 0.$$

The soma $\text{co som}\,P_n$ is a figure, hence a covering. Hence it can be represented as a denumerable sum of mutually disjoint bricks, say $a_{n1}, a_{n2}, \ldots$ Hence there exists a finite number of them, say $b_{n1}, b_{n2}, \ldots$, such that $|a_{n1} + a_{n2} + \cdots, b_{n1} + b_{n2} + \cdots| \leqq \frac{1}{n}$.

It follows from (1), that $|b_{n1} + b_{n2} + \cdots, \text{co}\,E| \to 0$; $R_n =_{df} b_{n1}, b_{n2}, \ldots$ is a complex, for which $|R_n, \text{co}\,E| \to 0$. Hence we can apply [Q.21.5.], by virtue of which there exists a subsequence $k(n)$ of indices, and a complex $Q_{k(n)}$, such that

1) $\operatorname{som} Q_{k(n)} \leqq \operatorname{som} R_{k(n)} \leqq \operatorname{co} \operatorname{som} P_{k(n)}$,
2) $\{Q_{k(n)}\}$ is completely distinguished for $\operatorname{co} E$.

Now $\{P_{k(n)}\}$ is [Q.21.6.] completely distinguished for E, $\{Q_{k(n)}\}$ is completely distinguished for $\operatorname{co} E$, and $\operatorname{som} P_{k(n)} \cdot \operatorname{som} Q_{k(n)} = O$; hence $P_{k(n)} \frown Q_{k(n)} = \mathit{Ø}$. Hence, [Q.21.10.], the sequence $\{P_{k(n)} \smile Q_{k(n)}\}$ is completely distinguished for I. The theorem is established.

Q.21.14. If

1. $p \neq O$ is a figure,
2. P_n is a completely distinguished sequence for I,
3. Q_n is a partial complex of P_n with $|Q_n, p| \to 0$, $Q_n = q_{n1}, q_{n2}, \ldots$ (then, by [Q.21.9.], $\{Q_n\}$ is completely distinguished for p),
4. let a_{ni} and e_{ni} be those among q_{nk}, for which $a_{ni} \leqq p$, and $e_{ni} \cdot p \neq O$, $e_{ni} \cdot \operatorname{co} p \neq 0$ respectively;

then

$$|\sum a_{ni} + \sum e_{ni} p, p| \to 0, \qquad \mu(\sum e_{ni} \operatorname{co} p) \to 0.$$

Proof. Since $|Q_n, p| \to 0$, we have $|p \cdot \operatorname{som} Q_n, p| \to 0$. Now the bricks of Q_n, whose soma of their sum contributes to $p \cdot \operatorname{som} Q_n$, are $a_{n1}, a_{n2}, \ldots$ and $e_{n1}, e_{n2}, \ldots$ Hence $p \cdot \operatorname{som} Q_n = \sum a_{nk} + p \sum e_{nk}$; so the first part of the thesis is proved. Now we have

$$(1) \qquad |\sum_k a_{nk} + \sum_k b_{nk} + \sum_k e_{nk}, p| \to 0,$$

where b_{nk} are all bricks in Q_n, for which $p \cdot b_{nk} = O$. Hence

$$(2) \quad |(\sum_k a_{nk} + \sum_k b_{nk} + \sum_k e_{nk}) \sum_k b_{nk}, p \cdot \sum_k b_{nk}| \to 0, \quad \text{i.e.} \quad |\sum_k b_{nk}, 0| \to 0.$$

By subtraction of (1) and (2), and by [Q.5.14.], we get $|\sum_k a_{nk} + \sum_k e_{nk}, p| \to 0$. Hence

$$|\sum_k a_{nk} + \sum_k e_{nk} p + \sum_k e_{nk} \operatorname{co} p, p| \to 0.$$

By subtraction we get

$$|\sum_k e_{nk} \operatorname{co} p, O| \to 0;$$

hence

$$\mu(\sum_k e_{nk} \cdot \operatorname{co} p) \to 0,$$

which completes the proof.

Chapter Q I

Vector fields on the tribe and their summation

Various kinds of approximations of somata having been settled, we are going to develop the theory of special kinds of integration.

A particular case of the integration has been considered in our paper (10), for the sake of selfadjoint operators in Hilbert-space. The basic tribe has been made out of subsets of the set of all real numbers and the coverings have been just intervals.

Now, we shall admit a general point of view, in order to not only have applications to maximal normal operators, but also for other purposes, which we have in view.

The bulk of difficulties has been settled in the Chapter [Q.], so the general theory of integration will be not more complicated, then in the mentioned linear case.

Q I.1. We shall consider the tribes ***F***, ***G*** and the base ***B***, [B.1.1.], as before under ***hyp. FBG***, [Q.12.]. We also admit the **(*Hyp Ad*)**, [B.2.]. To simplify arguments, we admit that ***F*** is a finitely genuine strict subtribe [A.7.2.] of ***G***, and that μ is an effective[1] denumerably additive measure on ***G***. The tribe ***G*** is supposed to be the μ-figure covering-Lebesgues extension of ***F***, (***Hyp L***), [Q.11.]. The hypothesis of separability of the μ-topology on ***G***, (***Hyp S***), [Q.21.1.], will be especially important (10), (42).

Q I.1.1. Hypothesis. Let V be an F. Riesz-S.-Banach-vector space, complete. Its elements $\vec{x}, \vec{y}, \ldots$ will be termed *vectors*. The norm of $\vec{x}$ will be denoted by $||\vec{x}||$, (46), (47).

Q I.2. Def. By a V-*field on* ***B*** we shall understand any function $\vec{\varphi}(a)$ where a varies over ***B*** and $\vec{\varphi}(a) \in \mathrm{V}$.

Q I.2.1. Def. In [Q.] we have studied infinite sequences $\{P_n\}$ of complexes which have approximated a given soma E of ***G***. They have the following property: $|E, P_n|_\mu \to 0$. We shall call this property *D-property*.

We may subject the sequence $\{P_n\}$ to additional conditions, as $\mathcal{N}_R(P_n) \to 0$, $\mathcal{N}_A(P_n) \to 0$, called (R), (A) *properties*, and if (***Hyp S***) is admitted, the sequence $\{P_n\}$ may have the property, expressed in [Q.21.3.], and called (***S***)-property.

---

[1] The hypothesis that ***F*** is strict subtribe of ***G*** is a non-essential restriction. It can always be obtained by taking $\mathfrak{B}(\boldsymbol{F})$ and $\mathfrak{B}(\boldsymbol{B})$ instead of ***F*** and ***B*** respectively (see [A.7.4.], [A.7.6.]. A similar remark can be made concerning the non-effectiveness of measure [Q.14.3.].

We shall consider various kinds of sequences $\{P_n\}$, but all with (D)-property; we shall call them *distinguished sequences for E* and denote them by (D), (DR), (DA), (DAR), (DS), (DAS), (DRS), (DARS) according to the specific properties admitted. The (DARS)-sequences will be termed *completely distinguished*, as in [Q.21.3.].

Q I.2.1.1. To each of these kind of approximating sequences there will correspond a notion of summation of vector-field, to be soon introduced. The existence of (DRA)-distinguished sequences has been proved in [Q.20.3.], and under hypothesis of separability, [Q.21.], the existence of (DARS)-distinguished sequences has been proved in [Q.21.1.]. If we shall speak in general of a distinguished sequence without specifying its character, we shall say simply "*distinguished*", and use the symbol (D').

Q I.2.1.2. Remark. We do not know whether (DS) does imply (DARS), or not. At present we do not need to be interested in this question.

Q I.2.2. Def. Let $\vec{\varphi}$ be a V-vector field on $\boldsymbol{B}$ and $E \in \boldsymbol{G}$. We call $\vec{\varphi}$ *summable on E with respect to the given kind* (D') *of distinguished sequences,* whenever for every (D')-distinguished sequence $\{P_n\} = \{p_{n1}, p_{n2}, \ldots\}$ for E, the sum $\vec{\varphi}(P_n) =_{df} \sum_i \vec{\varphi}(p_{ni})$ converges, for $n \to \infty$, in the topology of V. We call this limit *sum of the field* $\vec{\varphi}$ *on E* (*with respect to* D') and denote it by $S_E \vec{\varphi}$ or $S_{E(D')} \vec{\varphi}$.

Q I.2.3. Remark. If instead of $\boldsymbol{G}$, we consider the tribe $f \upharpoonright \boldsymbol{G}$ restricted to a given figure, and suppose that $E \leqq f$, the notion of summability in E may change. Thus the notion depends on the totality of the vector-field.

Q I.3. We shall be mainly interested in distinguished sequences (DARS), so the theorems which follow will concern that case. Changes in statements will concern changes of that case. They will be given in remarks.

The sums, introduced above, constitute, some way, a generalization of Weierstrass-Burkill-integrals (22). They are more general than those in (10).

Q I.3a. Theorem. Under above hypotheses and [Q I.1.1.], if

1. $E \in \boldsymbol{G}$,
2. $\vec{\varphi}$ is a V-vector field in $\boldsymbol{B}$,
3. $\vec{\varphi}$ is (DRA)-summable in E,
4. $\eta > 0$,

then there exists $\delta > 0$ such that if $\mathcal{N}_A(P) \leqq \delta$, $|E, P|_\mu \leqq \delta$, where P is a complex, then

$$\|\vec{\varphi}(P) - S_{E,\,(\mathrm{DRA})} \vec{\varphi}\| < \eta .$$

Proof. Suppose the theorem not true. For every $\delta > 0$ there exists a complex P such that $\mathcal{N}_R(P) \leqq \delta$, $\mathcal{N}_A(P) \leqq \delta$ and $|E, P|_\mu \leqq \delta$, but nevertheless

$$\|\vec{\varphi}(P) - S_E \vec{\varphi}\| \geqq \eta.$$

Take $\delta = \frac{1}{n}$, $(n = 1, 2, \ldots)$, and find $P_1, P_2, \ldots, P_n, \ldots$, according to the above. We have

$$(1) \quad \|\vec{\varphi}(P_n) - S_E \vec{\varphi}\| \geqq \frac{1}{n}, \; |E, P_n|_\mu \leqq \frac{1}{n}, \; \mathcal{N}_R(P) \leqq \frac{1}{n}, \; \mathcal{N}_A(P) \leqq \frac{1}{n}.$$

The sequence $\{P_n\}$ is (DRA)-distinguished for E. Hence

$$\lim_{n\to\infty} \vec{\varphi}(P_n) = S_E \vec{\varphi}.$$

This, however, contradicts (1). The theorem is proved.

Q I.3.1. Remark. A similar theorem holds for the following kinds of distinguished sequences (D), (DA), (DR), but the above proof cannot be used for the sequences (DS), (DAS), (DRS), (DARS).

Q I.4. Theorem. Let us admit hypotheses ***Hyp R***$_e$, **(*Hyp S*)** and [Q I.1.1.]. We shall consider (DARS)-summations.

If $E \in \boldsymbol{G}$, $F \leqq E$, $F \in \boldsymbol{G}$, $S_{E,(\mathrm{DARS})}\, \vec{\varphi}$ exists, then $S_{F,(\mathrm{DARS})}\, \vec{\varphi}$ also exists.

Proof. Suppose that $S_F \vec{\varphi}$ does not exist. There exists a completely distinguished sequence for F, [Q.21.3.],

$$(1) \qquad Q_1, Q_2, \ldots, Q_n, \ldots,$$

such that $\vec{\varphi}(Q_n)$ does not tend to any limit. Hence since V is complete, there exists $\eta > 0$ and a subsequence

$$(2) \qquad Q'_1, Q''_1, Q'_2, Q''_2, \ldots, Q'_n, Q''_n, \ldots$$

of (1) such that

$$(3) \qquad \|\vec{\varphi}(Q'_n) - \vec{\varphi}(Q''_n)\| \geqq \eta.$$

The sequence (2) is completely distinguished for F. We have

$$(3.1) \qquad |Q'_n, F| \to 0, \quad |Q''_n, F| \to 0.$$

Hence

$$(4) \qquad |\mathrm{som}\, Q'_n + \mathrm{som}\, Q''_n, F| \to 0.$$

The soma

$$(4.0) \qquad \mathrm{som}\, Q'_n + \mathrm{som}\, Q''_n$$

is a figure. Consider any sequence $P_1, P_2, \ldots, P_n, \ldots$ completely-distinguished for E with

$$(5) \qquad \|\vec{\varphi}(P_N) - S_E \vec{\varphi}\| \to 0.$$

We have

$$|E, P_n| \to 0. \tag{6}$$

From (6) and (4), we get

$$|\operatorname{som} P_n - (\operatorname{som} Q'_n + \operatorname{som} Q''_n), E - F| \to 0.$$

Put

$$A_n = \operatorname{som} P_n - (\operatorname{som} Q'_n + \operatorname{som} Q''_n).$$

The soma A_n is a figure. We have

$$|A_n, E - F| \to 0. \tag{7}$$

By Theorem [Q.21.5.] there exists a subsequence $\{k(n)\}$ of indices, and there exists a completely distinguished sequence $\{s_n\}$ for $E - F$, such that $\operatorname{som} s_n \cdot \operatorname{som} Q'_{k(n)} = O$, $\operatorname{som} s_n \cdot \operatorname{som} Q''_{k(n)} = O$. Hence $s_n \cup Q'_{k(n)}$, $s_n \cup Q''_{k(n)}$ are complexes. Since $\{s_n\}$ is completely distinguished for $E - F$, and since $Q'_{k(n)}$, $Q''_{k(n)}$ are both completely distinguished for F, it follows, [Q.21.7.], that $\{s_n \cup Q'_{k(n)}\}$ and $\{s_n \cup Q''_{k(n)}\}$ are both completely distinguished for E. Hence

$$\vec{\varphi}(s_n) + \vec{\varphi}(Q'_{k(n)}) \to S_E \vec{\varphi},$$

$$\vec{\varphi}(s_n) + \vec{\varphi}(Q''_{k(n)}) \to S_E \vec{\varphi};$$

hence

$$\|\vec{\varphi}(Q'_{k(n)}) - \vec{\varphi}(Q''_{k(n)})\| \to 0,$$

which contradicts (3).

Q I.4.1. Corollaries. The Theorem [Q I.4.] holds true for any of the summations with character (DS), (DAS), (DRS), and the proof is similar. Denoting these categories by I, II, III respectively, we make the following changes in the proof of [Q I.4.], respectively. Instead of the completely distinguished sequence $\{Q_n\}$ in (1), we suppose its character to be (DS) in I, (DAS) in II, (DRS) in III. The sequences $\{Q'_n\}$, $\{Q''_n\}$ have the same character respectively. The sequence $\{P_n\}$ will be supposed to be (DS), (DAS), (DRS) respectively. The remaining arguments will be not changed.

The Theorem [Q I.4.] is also true for the summations of character (D), (DA), (DR), (DAR). Denote them by I′, II′, III′, IV′ respectively. We take $\{Q_n\}$ of the given character. The character of $\{Q'_n\}$, $\{Q''_n\}$ will be the same. Having obtained the relation (7), we shall not need to select a subsequence $k(n)$, but we shall stay with A_n. We shall choose s_n with $\operatorname{som} s_n \leq A_n$ and with properties I′, II′, III′, IV′ respectively. The sequences $\{s_n \cup Q'_n\}$, $\{s_n \cup Q''_n\}$ will be the required distinguished sequences, yielding the final contradiction. Thus we can state the following theorem:

Q I.4.1a. Theorem. Admitting the hypotheses ***Hyp R_e*** and (***Hyp S***), if needed, take any character (D') of summation. If $F \leqq E$ and $S_{E,(D')}\vec{\varphi}$ exists, then $S_{F,(D')}\vec{\varphi}$ exists too.

Q I.5. Lemma. If

1. $\vec{\varphi}, \vec{\varphi}_1, \vec{\varphi}_2, \ldots, \vec{\varphi}_n, \ldots$ are vectors of V,
2. from every sequence $\{\vec{\varphi}_{k(n)}\}$ another subsequence $\{\vec{\varphi}_{kl(n)}\}$ can be extracted, such that $\lim \vec{\varphi}_{kl(n)} = \vec{\varphi}$,

then

$$\lim \vec{\varphi}_n = \vec{\varphi}.$$

Proof. Suppose that $\vec{\varphi}_n$ does not converge. Then, since V is complete, there exists $\eta > 0$ and subsequences $\vec{\varphi}_{s(n)}, \vec{\varphi}_{t(n)}$, such that

1) $t(n-1) < s(n) < t(n) < s(n+1)$, $n \geqq 2$;
2)

$$(1) \qquad \|\vec{\varphi}_{s(n)} - \vec{\varphi}_{t(n)}\| \geqq \eta, \quad (n = 1, 2, \ldots).$$

Extract from $\vec{\varphi}_{s(n)}$ a subsequence

$$(2) \qquad \vec{\varphi}_{s(s'n)} \quad \text{with} \quad \lim \vec{\varphi}_{s(s'n)} = \vec{\varphi}.$$

We get from (1).

$$(3) \qquad \|\vec{\varphi}_{ss'(n)} - \vec{\varphi}_{ts'(n)}\| \geqq \eta \quad \text{for} \quad n = 1, 2, \ldots$$

Now, from $\{\vec{\varphi}_{ts'(n)}\}$ another subsequence $\{\vec{\varphi}_{ts't'(n)}\}$ can be extracted with

$$(4) \qquad \lim \vec{\varphi}_{ts't'(n)} = \vec{\varphi}.$$

From (3) we get

$$(5) \qquad \|\vec{\varphi}_{ss't'(n)} - \vec{\varphi}_{ts't'(n)}\| \geqq \eta.$$

Since, by (2), $\vec{\varphi}_{ss't'(n)} \to \vec{\varphi}$, we get from (4): $0 \geqq \eta$ which is a contradiction.

Hence $\vec{\varphi}_n$ converges. Let $\vec{\Psi} =_{df} \lim \vec{\varphi}_n$. There exists a partial sequence tending to $\vec{\varphi}$. Hence $\vec{\Psi} = \vec{\varphi}$.

Q I.5.1. Theorem. Admit the hypotheses ***Hyp R_e***, [Q I.1.1.], and (***Hyp S***). We shall consider (DARS)-sums. If

1. $E, E_1, E_2, \ldots, E_n, \ldots \in \boldsymbol{G}$,
2. $E_n \leqq E$ for $n = 1, 2, \ldots$,
3. $\mu(E_n) \to 0$,
4. $\vec{\varphi}$ is (DARS)-summable on E,

then

$$S_{E_n,(\mathrm{DARS})}\vec{\varphi} \to \vec{O}.$$

Proof. Let $\{A_n\}$ be a sequence of complexes with

$$(0) \qquad |A_n, E| \to 0.$$

By Theorem [Q I.4.], $\vec{\varphi}$ is summable on E_n. Find, for every $n = 1, 2, \ldots$, a complex P_n such that

(0.1) $$\|\vec{\varphi}(P_n) - S_{E_n}\vec{\varphi}\| \leqq \frac{1}{n},$$

$$\mathcal{N}_A(P_n) \to 0, \quad \mathcal{N}_R(P_n) \to 0, \quad |E_n, P_n| \leqq \frac{1}{n}.$$

Let $\{P_{k(n)}\}$ be a partial sequence of $\{P_n\}$. We have

(1) $$\|\vec{\varphi}(P_{k(n)} - S_{E_{k(n)}}\vec{\varphi}\| \leqq \frac{1}{k(n)} \leqq \frac{1}{n},$$

(2) $$|E_{k(n)} \cdot P_{k(n)}| \leqq \frac{1}{n}, \quad \mathcal{N}_A(P_{k(n)}) \to 0, \quad \mathcal{N}_R(P_{k(n)}) \to 0.$$

Consider the figure

(3) $$Q_{k(n)} =_{df} A_{k(n)} - \operatorname{som} P_{k(n)}.$$

Since $\mu(E_{k(n)}) \to 0$, we have $|E_{k(n)}, O| \to 0$; hence, by (2),

(4) $$|P_{k(n)}| \to 0.$$

It follows from (0) and (4):

$$|A_{k(n)} - \operatorname{som} P_{k(n)}, E - O| \to 0,$$

i.e.

(5) $$|Q_{k(n)}, E| \to 0.$$

By [Q.21.5.] there exists a sequence $l(n)$ of indices $1, 2, \ldots$ and a complex $T_n \leqq Q_{kl(n)}$, such that $\{T_n\}$ has the property (**S**) with respect to E, and in addition to that:

(6) $$\mathcal{N}_R(T_n) \to 0, \quad \mathcal{N}_A(T_n) \to 0.$$

It follows that

(6.1) $$\vec{\varphi}(T_n) \to S_E\vec{\varphi}.$$

Since $\varphi_{kl(n)}$ and $\operatorname{som} P_{kl(n)}$ are disjoint, it follows that $\operatorname{som} T_n$ and $\operatorname{som} P_{kl(n)}$ are disjoint with F.

(7) $$R_n =_{df} T_n \cup P_{kl(n)}$$

is a complex. By (2) and (6), we have

(8) $$\mathcal{N}_A(R_n) \to 0, \quad \mathcal{N}_R(R_n) \to 0.$$

Since $|E, T_n| \to 0$ and, by (4), $|O, P_{kl(n)}| \to 0$, it follows [Q.21.7.], that $|E, T_n \cup P_{kl(n)}| \to 0$ i.e.

(9) $$|E, R_n| \to 0.$$

T_n has the property, that for every $F \in \mathbf{G}$, $F \leqq E$ there exists a partial complex T'_n, $(n = 1, 2, \ldots)$, with $|T'_n, F| \to 0$. Now T'_n is also a partial complex of $R_n = T_n \cup P_{kl(n)}$. Hence, by (8), (9), $\{R_n\}$ has the property ($\boldsymbol{S}$) with respect to E. It follows $\vec{\varphi}(R_n) \to S_E \vec{\varphi}$, i.e.

$$\vec{\varphi}(T_n) + \vec{\varphi}(P_{kl(n)}) \to S_E \vec{\varphi}. \tag{10}$$

Hence, by (6.1), $\vec{\varphi}(P_{kl(n)}) \to \vec{O}$. If we take account of (0.1), we get

$$S_{E_{kl(n)}} \vec{\varphi} \to \vec{O}. \tag{11}$$

Thus we have proved that from every increasing sequence $k(n)$ of natural number another sequence $kl(n)$ can be extracted so as to have (11). It follows by Lemma [Q I.5.], that $S_{E_n} \vec{\varphi} \to \vec{O}$.

Q I.5.2. Corollaries. In Theorem [Q I.5.1.] we have considered (DARS)-sums. Now the theorem holds true for any one of the following summations: (DS), (DAS), (DRS). The proof is almost the same. In these three cases denoted by I, II, III respectively, we shall drop $\mathcal{N}_A(P_n) \to 0$, $\mathcal{N}_R(P_n) \to 0$ in *1*, we shall drop $\mathcal{N}_R(P_n) \to 0$ in II, and $\mathcal{N}_A(P_n) \to 0$ in III.

The theorem holds also true for summations I′, II′, III′, IV′ of the character (D), (DA), (DR) and (DAR) respectively. The proof is even simpler. We omit the conditions $\mathcal{N}_A(P_n) \to 0$, $\mathcal{N}_R(P_n) \to 0$ in I′, omit $\mathcal{N}_R(P_n) \to 0$ in II′, omit $\mathcal{N}_A(P_n) \to 0$ in III′. We do not choose any partial sequence $\{P_{k(n)}\}$, but stay with $\{P_n\}$. So we get, instead of (5), $|Q_n, E| \to 0$. Now, instead of (6), we find $T_n \leqq Q_n$, with $|T_n, E| \to 0$, som $T_n \leqq Q_n$, $\mathcal{N}_R(T_n) \to 0$, $\mathcal{N}_A(T_n) \to 0$. Thus we get $\vec{\varphi}(T_n) \to S_E \vec{\varphi}$, and putting $R_n =_{df} T_n \cup P_n$, we get $\vec{\varphi}(T_n) + \vec{\varphi}(P_n) \to S_E \vec{\varphi}$, and then $\vec{\varphi}(P_n) \to 0$, so the proof can be completed. Thus we can state the general theorem:

Q I.5.3. Corollaries. Admitting hypotheses [Q I.1.1.], ***Hyp R$_e$*** and (***Hyp S***), if needed, consider sums of any kind (D'). If 1., 2. as before and we admit 3′., stating that $\vec{\varphi}$ is (D')-summable, then

$$S_{E_n, (D')} \vec{\varphi} \to \vec{O}.$$

Q I.6. Theorem. Take the hypotheses [Q I.1.1.], ***Hyp R$_e$***, and also (***Hyp S***), if needed. Suppose that $E \in \boldsymbol{G}$. Let (D') be any kind of summation. Suppose, that $S_{E, (D')} \vec{\varphi}$ exists, then

$$S_{O, (D')} \vec{\varphi} \to \vec{O}.$$

Proof. We apply the theorem [Q I.5.1.], taking $E_n = O$. We get $S_{O, (D)} \vec{\varphi} \to \vec{O}$. Hence, since this is a constant sequence, we have

$$S_{O, (D')} \vec{\varphi} = \vec{O}.$$

Q I.6.1. Theorem. If $S_{O, (D')} \vec{\varphi}$ exists, then $S_{O, (D')} \vec{\varphi} = \vec{O}$.
Proof. This follows from [Q I.6.], because $O \leqq O$.
Q I.6.2. Theorem. If $S_{O, (D')} \vec{\varphi}$ exists, then $\vec{\varphi}(O) = O$.

Proof. The sequence $\{O\}, \{O\}, \{O\}, \ldots$ is a sequence of complexes, each of which being composed of the single brick O. This sequence is distinguished of any character considered. Let $\vec{\varphi}_0 = \vec{\varphi}(O)$. We have $\lim \vec{\varphi}(\{O\}) = \vec{\varphi}_0$. Since $S_{O,(D')} \vec{\varphi}$ exists, we have $S_{O,(D')} \vec{\varphi} = \vec{\varphi}_0$. Hence, by [Q I.6.1.], $\vec{\varphi}_0 = \vec{O}$, which proves the theorem.

Q I.6.3. Theorem. If $S_{E,(D')} \vec{\varphi}$ exists for some E, then $\vec{\varphi}(O) = \vec{O}$.

Proof. By [Q I.4.], the sum $S_{O,(D')} \vec{\varphi}$ exists; hence, by [Q I.6.2.], $\vec{\varphi}(O) = \vec{O}$.

Q I.7. Theorem. Consider the hypotheses [Q I.1.1.], $\boldsymbol{Hyp\, R_e}$ and $(\boldsymbol{Hyp\, S})$. We shall consider (DARS)-summation.

If

1. $E_1, E_2 \in \boldsymbol{G}$,
2. $E_1 \cdot E_2 = O$,
3. $S_{E_1+E_2,\,(\mathrm{DARS})} \vec{\varphi}$ exists,

then

$$S_{E_1+E_2,\,(\mathrm{DARS})} \vec{\varphi} = S_{E_1,\,(\mathrm{DARS})} \vec{\varphi} + S_{E_2,\,(\mathrm{DARS})} \vec{\varphi}.$$

Proof. By [Q I.4.], the sums $S_{E_1} \vec{\varphi}$, $S_{E_2} \vec{\varphi}$ exist. By [Q I.14.] we can find coverings $L_{11} \geqq L_{12} \geqq L_{13} \geqq \cdots \geqq L_{1n} \geqq \cdots$ of E, such that $E_1 \leqq L_{1n}$, $(n = 1, 2, \ldots)$, and

$$(1) \qquad \lim \mu(L_{1n}) = \mu(E_1).$$

Similarly we can find coverings $L_{21} \geqq L_{22} \geqq \cdots \geqq L_{2n} \geqq \cdots$ of E_2 with $E_2 \leqq L_{2n}$, $(n = 1, 2, \ldots)$, and

$$(2) \qquad \lim \mu(L_{2n}) = \mu(E_2).$$

Since $E_1 \cdot E_2 = O$, we have, [Q.14.3.],

$$(3) \qquad \mu(L_{1n} \cdot L_{2n}) \to 0.$$

Find complexes $P'_1, P'_2, \ldots, P'_n, \ldots$, such that

$$(4) \qquad \mathrm{som}\, P'_n \leqq L_{1n}, \quad |P'_n, L_{1n}| \to 0,$$

and complexes $P''_1, P''_2, \ldots, P'_n, \ldots$, such that

$$(5) \qquad \mathrm{som}\, P''_n \leqq L_{2n}, \quad |P''_n, L_{2n}| \to 0.$$

Since $\mathrm{som}\, P'_n \leqq L_{1n}$, $\mathrm{som}\, P''_n \leqq L_{2n}$, we have, by (3),

$$(6) \qquad \lim \mu(\mathrm{som}\, P'_n \cdot \mathrm{som}\, P''_n) \to 0.$$

Consider the figures

$$(7) \qquad Q'_n =_{df} \mathrm{som}\, P'_n - \mathrm{som}\, P''_n, \quad Q''_n =_{df} \mathrm{som}\, P''_n - \mathrm{som}\, P'_n.$$

They are disjoint.

By (4) we have $|P'_n, L_{1n}| \to 0$, and by (1), $|L_{1n}, E_1| \to O$; hence $|P'_n, E_1| \to 0$. Since, by (6), $|\mathrm{som}\, P'_n \cdot \mathrm{som}\, P''_n, O| \to 0$, we get, by

[Q I.5.14.], $|\operatorname{som} P'_n - \operatorname{som} P'_n \cdot \operatorname{som} P''_n, E_1 - O| \to 0$, i.e.

(8) $$|Q'_n, E_1| \to 0.$$

In the same way we obtain

(9) $$|Q''_n, E_2| \to 0.$$

Having this, apply Theorem [Q.21.5.], getting a subsequence $k(n)$ of indices and a completely distinguished sequence $\{R'_{k(n)}\}$ for E_1, such that

(9.1) $$\operatorname{som} R'_{k(n)} \leqq Q'_{k(n)}.$$

Since, by (9),

(10) $$|Q''_{k(n)}, E_2| \to 0,$$

we get, by the same [Q.21.5.], a subsequence $\{k\,l(n)\}$ of the indices $\{k(n)\}$, and a completely distinguished sequence $\{R''_{kl(n)}\}$ for E_2, such that

(11) $$R''_{kl(n)} \leqq Q''_{kl(n)}.$$

Since $\operatorname{som} Q'_n$, $\operatorname{som} Q''_n$ are disjoint, so are $\operatorname{som} Q'_{kl(n)}$, $\operatorname{som} Q''_{kl(n)}$, and then, by (9.1) and (11), so are $R'_{kl(n)}$, $R''_{kl(n)}$. By [Q I.21.6.] $R'_{kl(n)}$ is a completely distinguished sequence for E_1, and $R''_{kl(n)}$ is a completely distinguished sequence for E_2. Hence, by [Q.21.10.],

$$\{R_n\} =_{df} \{R'_{kl(n)} \cup R''_{kl(n)}\}$$

is a completely distinguished sequence for $E_1 + E_2$. By hyp. 3, $\lim R_n(\vec{\varphi})$ exists and equals $S_{E_1+E_2}\vec{\varphi}$. Since $S_{E_1}\vec{\varphi}$, $S_{E_2}\vec{\varphi}$ exist, and since $\{R'_{kl(n)}\}$ and $\{R''_{kl(n)}\}$ are completely distinguished sequences for E_1, E_2 respectively, we have

$$\lim R'_{kl(n)} = S_{E_1}\vec{\varphi}, \quad \lim R''_{kl(n)} = S_{E_2}\vec{\varphi}.$$

Since

$$\lim R_n(\vec{\varphi}) = \lim R'_{kl(n)}(\vec{\varphi}) + \lim R''_{kl(n)}(\vec{\varphi}),$$

it follows that

$$S_{E_1+E_2}\vec{\varphi} = S_{E_1}\vec{\varphi} + S_{E_2}\vec{\varphi}. \quad \text{q.e.d.}$$

Q I.7.1. The above theorem is valid for the following kinds of summation (DS), (DAS), (DRS). The proof is almost the same. Indeed, instead of taking completely distinguished sequences $R'_{kl(n)}$, $R''_{kl(n)}$, we take only those of the characters (DS), (D), (DA), (DR) respectively.

The theorem also holds true for any of the following kinds of summations:

(1) (D), (DA), (DR), (DAR).

The proof will be even simpler. Having obtained $\{Q'_n\}$ and $\{Q''_n\}$ and the relations (7) and (8):

$$|E_1, Q'_n| \to 0, \quad |E_2, Q''_n| \to 0$$

we shall not need to consider subsequences, but we shall find R'_n, R''_n, so as to satisfy the corresponding condition (1) with $\text{som}\, R'_n \leqq Q'_n$, and $\text{som}\, R''_n \leqq Q''_n$. Thus we can state the

Q I.7.1a. Corollary. Under hypotheses [Q I.1.1.], ***Hyp R_e*** and (***Hyp S***), if needed, consider any kind (D') of summation.

If $E_1 \cdot E_2 = O$, $S_{E_1+E_2,\,(D')}\,\vec{\varphi}$ exists, then

$$S_{E_1+E_2,\,(D')}\,\vec{\varphi} = S_{E_1,\,(D')}\,\vec{\varphi} + S_{E_2,\,(D')}\,\vec{\varphi}.$$

Q I.8. Theorem. Let us admit the hypotheses [Q I.1.1.], ***Hyp R_e*** and (***Hyp S***). We shall consider (DARS)-summations.

If

1. $E, F \in \boldsymbol{G}$,
2. $E \cdot F = O$,
3. $\vec{\varphi}$ is (DARS)-summable on E and on F,

then $\vec{\varphi}$ is (DARS)-summable on $E + F$; hence, by [Q I.7.]:

$$S_{E+F}\,\vec{\varphi} = S_E\,\vec{\varphi} + S_F\,\vec{\varphi}.$$

Proof. Consider a completely distinguished sequence $\{P_n\}$ of complexes for $E + F$. There exists a partial complex R_n of P_n, $(n = 1, 2, \ldots)$, such that $|R_n, E|_\mu \to 0$. Let $P_n = R_n \cup S_n$ where $R_n \cap S_n = \mathbf{0}$. We have

$$|P_n, E + F| \to 0 \quad \text{and} \quad |R_n, E| \to 0.$$

Hence

$$|P_n - R_n, (E + F) - E| \to 0, \quad \text{i.e.} \quad |S_n, F| \to 0.$$

By virtue of [Q.21.9.], $\{R_n\}$ is a completely distinguished sequence for E, and $\{S_n\}$ a completely distinguished sequence for F. We have

$$\text{(1)} \qquad \lim \vec{\varphi}(R_n) = S_E\,\vec{\varphi}, \quad \lim \vec{\varphi}(S_n) = S_F\,\vec{\varphi}.$$

Since $P_n = R_n \cup S_n$, we have $\vec{\varphi}(P_n) = \vec{\varphi}(R_n) + \vec{\varphi}(S_n)$.

It follows, from (1), that $\lim \vec{\varphi}(P_n)$ exists. Since the completely distinguished sequence $\{P_n\}$ was any one, the field $\vec{\varphi}$ is summable on $E + F$. Applying [Q I.7.], we get the thesis.

Q I.8.1. Corollary. The Theorem [Q I.8.] is valid for summations of the character (DAS), (DRS) and (DS). For proving it is sufficient, in the forgoing proof, to drop the condition $\mathcal{N}_R(P_n) \to 0$, $\mathcal{N}_A(P_n) \to 0$ or both respectively.

Q I.8.2. Remark. The Theorem [Q I.7.] is not true for summations (D), (DA), (DR), (DAR), even if we admit (***Hyp S***). The following example shows it:

Let $\boldsymbol{B}$ be composed of all half-open rectangles

$$\left\{(x, y) \,\middle|\, \begin{Bmatrix} a \leqq x < b \\ c \leqq y < d \end{Bmatrix}\right\}, \quad \text{where} \quad 0 \leqq a, b, c, d \leqq 1.$$

$\boldsymbol{F}$ is defined as the set of finite unions of those rectangles, and $\boldsymbol{G}$ as the tribe of all Lebesgue's measurable subsets of the square

$$I =_{df} \{(x, y) \mid 0 \leqq x < 1, 0 \leqq y < 1\};$$

μ will be the Lebesgue's measure. Define

$$E =_{df} \{(x, y) \mid 0 \leqq x < 1, 0 \leqq y < \tfrac{1}{2}\},$$
$$F =_{df} \{(x, y) \mid 0 \leqq x < 1, \tfrac{1}{2} \leqq y < 1\}.$$

We have $E \cdot F = O$, $E + F = I$. Let $\vec{\varphi}_0 \in V$, $\vec{\varphi}_0 \neq \vec{O}$. If $a = \{(x,y) \mid \alpha \leqq x < \beta, 0 \leqq y < 1\}$, where $0 \leqq \alpha < \beta \leqq 1$, put $\vec{\varphi}(a) = \vec{\varphi}_0$; and for all other bricks b put $\vec{\varphi}(b) = \vec{O}$. The sums $S_E \vec{\varphi} = S_F \vec{\varphi} = \vec{O}$ exist, but $S_I \vec{\varphi}$ does not exist, if we do not use the (**S**)-property for complexes yielding the sum.

Q I.9. Theorem. If

1. $E \in \boldsymbol{G}$,
2. $S_{E,(D')} \vec{\varphi}$ exists,
3. $\{P_n\}$ is a sequence of complexes with $\mu(P_n) \to 0$ for $n \to \infty$,

then

$$\vec{\varphi}(P_n) \to \vec{O}.$$

Proof. $\{P_n\}$ is distinguished of every kind for O. Since $S_{E,(D')} \vec{\varphi}$ exists, therefore, [Q I.6.], $S_{O,(D')} \vec{\varphi} = \vec{O}$. We have $S_{O,(D')} \vec{\varphi} = \lim \vec{\varphi}(P_n)$; hence, $\lim \vec{\varphi}(P_n) = \vec{O}$.

Q I.9.1. Theorem. If

1. $E \in \boldsymbol{G}$,
2. $S_{E,(D')} \vec{\varphi}$ exists,
3. $\alpha > 0$,

then there exists $\beta > 0$ such that, if P is a complex with $\mu P \leqq \beta$, then $\|\vec{\varphi}(P)\| \geqq \alpha$.

Proof. Suppose that the thesis is not true. Then for every $\beta > 0$ we can find a complex P, such that $\mu P \leqq \beta$ and

$$\|\vec{\varphi}(P)\| \geqq \alpha. \tag{1}$$

Putting $\beta = \frac{1}{1}, \frac{1}{2}, \ldots, \frac{1}{n}, \ldots$, and finding the corresponding P_n, we get $\mu(P_n) \to 0$; hence, by (Q I.9.], $\vec{\varphi}(P_n) \to 0$, which contradicts (1).

Q I.9.2. If

1. $E \in \boldsymbol{G}$,
2. $S_{E,(D')} \vec{\varphi}$ exists,
3. $\alpha > 0$,

then there exists $\beta > 0$, such that, if $\mu(F) \leqq \beta$, $F \in \boldsymbol{G}$, then $\|S_{F,(D')} \vec{\varphi}\| \leqq \beta$.

Proof. Similar to that of [Q I.9.1.], through a contradiction.

Q I.10. Theorem. Under hypotheses [Q I.1.1.], ***Hyp R_e*** and (***Hyp S***), if needed, we have, if

1. $E_1, E_2, \ldots, E_n, \ldots, F \in \boldsymbol{G}$,
2. $S_{F,(D')}\vec{\varphi}$ exists, where (D') is any fixed kind of summation,
3. $E =_{df} E_1 + E_2 + \cdots + E_n + \cdots$,
4. $E \leqq F$,
5. $E_1, E_2, \ldots$ are disjoint,

then

$$S_{E,(D')}\vec{\varphi} = \sum_{n=1}^{\infty} S_{E_n,(D')}\vec{\varphi},$$

where $\sum$ is understood as convergent in the V-topology.

Proof. By hyp. 2 and by virtue of [Q I.4.1 a.], the sums $S_{E_n}\vec{\varphi}$, $S_E\vec{\varphi}$ exist, $(n = 1, 2, \ldots)$. Put $A_n =_{df} E - (E_1 + \cdots + E_n)$. We have $E = E_1 + \cdots + E_n + A_n$. By [Q I.4.1 a.] the sum $S_{A_n(D')}\vec{\varphi}$ exists. By [Q I.7.1.]

$$(1) \qquad S_E\vec{\varphi} = S_{E_1}\vec{\varphi} + \cdots + S_{E_n}\vec{\varphi} + S_{A_n}\vec{\varphi}.$$

Since $\boldsymbol{G}$ is denumerably additive, we have $\mu A_n \to 0$; hence, by [Q I.9.2.] $\|S_{A_n}\vec{\varphi}\| \to 0$. Consequently $\|S_E\vec{\varphi} - (S_{E_1}\vec{\varphi} + \cdots + S_{E_n}\vec{\varphi})\| \to 0$, i.e. $S_E\vec{\varphi} = \sum_{n=1} S_{E_n}\vec{\varphi}$, q.e.d.

Q I.10.1. Theorem. Under the same hypotheses as in [Q I.10.], if $E \in \boldsymbol{G}$ and $S_{E,(D')}\vec{\varphi}$ exists.

We put for all $F \leqq E$, $F \in \boldsymbol{G}$, $\vec{K}(F) =_{df} S_F\vec{\varphi}$, then $\vec{K}(F)$ with variable $F \leqq E$ is denumerably additive. Hence $\vec{K}$ is a kind of vector-valued measure (see (15)).

Proof. Follows from [Q I.10.].

Q I.11. Remark. We can prove the following: Under hypotheses [Q I.1.1.], ***Hyp R_e*** and (***Hyp S***), consider any one of the summations: (DS), (DAS), (DRS), (DARS). Suppose that

1. $E_1, E_2, \ldots, E_n, \ldots$ are all disjoint,
2. put $E =_{df} E_1 + E_2 + \cdots + E_n + \cdots$,
3. suppose that $S_E\vec{\varphi}$ exists for all $n = 1, 2, \ldots$,

then $S_E\vec{\varphi}$ exists too, and we have $S_E\vec{\varphi} = \sum_{n=1}^{\infty} S_{E_n}\vec{\varphi}$.

The theorem is not true for the summations (D), (DA), (DR), (DAR), even if (***Hyp S***) is admitted.

Q I.12. Theorem. Under hypotheses [Q I.1.1.], ***Hyp R_e*** and (***Hyp S***), if needed, consider any kind (D') of summations. Let $E_1, E_2, \ldots, E_n, \ldots, E \in \boldsymbol{G}$, $F \in \boldsymbol{G}$. Suppose that $E_n \leqq F$, $E \leqq F$. If

1. $S_F\vec{\varphi}$ exists,
2. $|E_n, E|_\mu \to 0$,

then $S_{E_n}\vec{\varphi} \to S_E\vec{\varphi}$ for $n \to \infty$ in the V-topology.

Proof. We rely on [Q I.4.1 a.], concerning the existence of sums to be now considered. We have $S_E\vec{\varphi} = S_{E-E_n}\vec{\varphi} + S_{E\cdot E_n}\vec{\varphi}$, by [Q I.7.1.]. Since $\mu(E - E_n) \to 0$, we have, by [Q I.5.3.], $\lim S_{E-E_n}\vec{\varphi} = \vec{O}$. Hence

(1) $$\lim S_{E\cdot E_n}\vec{\varphi} = S_E\vec{\varphi}.$$

We have $S_{E_n}\vec{\varphi} = S_{E_n-E}\vec{\varphi} + S_{E_n\cdot E}\vec{\varphi}$, by [Q I.7.1.]. Since $\mu(E_n - E) \to 0$, we have, by [Q I.5.5.], $\lim S_{E_n-E}\vec{\varphi} = 0$. Hence, by (1), we get $\lim S_{E_n}\vec{\varphi} = S_E\vec{\varphi}$, q.e.d.

Q I.12.1. Remark. The Theorem [Q I.12.] says that $S_E\vec{\varphi}$, considered as the function of E, is (μ-limit)-(V-limit)-continuous for any kind of summation considered.

Q I.13. Theorem. Let (D') be any kind of summation of $\vec{\varphi}$. Suppose that $\vec{\varphi}$ is (D')-summable on I. Put $\vec{K}(a) =_{df} S_a\vec{\varphi}$ for every brick a; then $\vec{K}$ is (D')-summable on I, and we have for every $E \in \boldsymbol{G}$, $S_E\vec{K} = S_E\vec{\varphi}$.

Proof. Let $E \in \boldsymbol{G}$. Consider a (D')-distinguished sequence of complexes $\{P_n\}$, where $\{P_n\} = \{p_{n1}, p_{n2}, \ldots\}$, for E. Put $\vec{K}(P_n) =_{df} \sum_i \vec{K}(p_{ni})$. We have $\vec{K}(P_n) = \sum_i S_{p_{ni}}(\vec{\varphi}) = S_{\mathrm{som}\,P_n}\vec{\varphi}$, [Q I.4.1 a.], [Q I.7.1.]. Since $|\mathrm{som}\,P_n, E|_\mu \to 0$, we have

(1) $$S_{\mathrm{som}\,P_n}\vec{\varphi} \to S_E\vec{\varphi}, \quad \text{i.e.} \quad \vec{K}(P_n) \to S_E\vec{\varphi}.$$

This being true for any (D')-distinguished sequence $\{P_n\}$ for E, it follows that $\vec{K}$ is summable on E, and we have

(2) $$\vec{K}(P_n) \to S_E\vec{K}.$$

From (1) and (2), the theorem follows.

Q I.14. Theorem. Admit the hypotheses [Q I.1.1.], ***Hyp R_e*** and ***(Hyp S)***, if needed. We shall consider any kind of (D')-summation. If

1. $\vec{\varphi}$ is (D')-summable on E, where $E \in \boldsymbol{G}$,
2. λ is a number (real or complex, depending on whether V is a real vector space or a complex one),
3. we define for all bricks a: the vector field $\vec{\Psi}(a) =_{df} \lambda\,\vec{\varphi}(a)$;

then $\vec{\Psi}$ is also (D')-summable on E, and

$$S_E\vec{\Psi} = \lambda\, S_E\vec{\varphi}.$$

Q I.15. Under previous hypotheses, if $\vec{\varphi}_1$ and $\vec{\varphi}_2$ are two vector fields, both (D')-summable on E, then the vector-field $\vec{\varphi}_3(a) =_{df} \vec{\varphi}_1(a) + \vec{\varphi}_2(a)$ is also (D')-summable on E, and we have

$$S_E\vec{\varphi}_3 = S_E\vec{\varphi}_1 + S_E\vec{\varphi}_2.$$

Proof. Both [Q I.14.] and [Q I.15.] follow from the linearity of the vector-space V.

Q I.16. The existence of the (D)-sum $S_E \vec{\varphi}$ requires more, than the existence of any other (D')-sum on E. Generally, the "addition" of a letter "increases" the size of summable fields.

Q I.17. Let $\eta > 0$ and let A be an atom of $\boldsymbol{G}$; then, by [Q.15.1.], there exists a complex P such that $|P, A|_\mu < \eta$. If we take $\eta < \frac{1}{2}\mu(A)$, then, [Q.17.3.], we get $A \leqq P$. Hence, by [Q.17.2.] there exists one and only one brick

(1) $$p \text{ of } P \text{ with } A \leqq p.$$

Since $|P, A|_\mu < \eta$, $A \leqq P$, we have

$$\mu P - \mu A < \eta;$$

hence, by (1),

$$\mu p - \mu A < \eta.$$

Thus, if A is an atom of $\boldsymbol{G}$, then for every $\eta > 0$ sufficiently small there exists a brick p, such that $A \leqq p$, $\mu(p - A) < \eta$.

Q I.17.1. Let $\eta_n \to 0$. We can find by [Q I.17.] a brick p_n with $A \leqq p_n$, $\mu(p_n - A) < \eta_n$. If we put $q_n = p_1 \cdots p_n$, ($n = 1, 2, \ldots$), we get bricks

(2) $$q_1 \geqq q_2 \geqq \cdots \geqq q_n \geqq \cdots$$

with

(3) $$A \leqq q_n, \quad \mu(q_n - A) \to 0.$$

Thus, if A is an atom of $\boldsymbol{G}$, there exists an infinite sequence of bricks (2), satisfying the conditions (3).

Q I.17.2. Now, suppose that A is an atom of $\boldsymbol{G}$. Suppose that $\vec{\varphi}$ is (D)-summable on I. Then $\vec{\varphi}$ is (DARS)-summable on the set $\{A\}$, composed of the single soma A, [Q I.4.1a.]. Consider the sequence (2) in [Q I.7.1.] with properties (3). The sequence of complexes

(4) $$\{q_1\}, \{q_2\}, \ldots, \{q_n\}$$

satisfies the ***Hyp R*** and (S) for $\{A\}$, because the only subsoma of A, differing from A is O. Thus (4) is a completely distinguished sequence for $\{A\}$. It follows that

$$S_A \vec{\varphi} = \lim_{n \to \infty} \vec{\varphi}(q_n).$$

Q I.17.3. Let $A_1, A_2, \ldots, A_m, \ldots$ be a finite or infinite sequence of different atoms (some ones or all) of $\boldsymbol{G}$. Suppose that $\vec{\varphi}$ is (DARS)-summable on I. Then, by [Q I.10.], [Q I.7.1.],

$$S_{A_1} \vec{\varphi} + S_{A_2} \vec{\varphi} + \cdots + = S_{A_1 + A_2 + \cdots} \vec{\varphi}.$$

Q I.17.4. Since the hypothesis of existence of the (DARS)-sum on I is less restrictive, than each one concerning (DS), (DAS), (DRS), the above holds true for those summations too. But even if ***(Hyp S)*** is not supposed, our arguments are valid for (DAR)-sums; hence for any one of the sums of character (D), (DA), (DR) too.

Q I.18. Suppose, that $\vec{\varphi}$ is (D)-summable on I. Let a be a brick. Since $\{a\}, \{a\}, \ldots$ is a (D)-distinguished sequence for $\{a\}$, we have

$$S_a \vec{\varphi} = \vec{\varphi}(a).$$

If $a = a_1 + a_2 + \cdots + a_n + \cdots$ is a finite or denumerable sum of disjoint bricks, we have [Q I.10.]:

$$\vec{\varphi}(a) = \vec{\varphi}(a_1) + \vec{\varphi}(a_2) + \cdots + \vec{\varphi}(a_n) + \cdots :$$

a quite strong condition imposed on $\vec{\varphi}(a)$. Dealing with bricks and atoms, we can get examples, showing that the different kinds of summations do not coincide.

Chapter R

Quasi-vectors and their summation

The theory of summation of fields of vectors having been developped in chapter [Q I.], we shall define the notion of quasi-vectors and discuss the summations of their sets. They will play a role in the theory of a special system of coordinates in Hilbert-space, in Chapter [S.].

R.1. We take the hypothesis ***FBG***, [B.3.], concerning the finitely additive tribe ***F***, its basis ***B***, [B.1.1.], and the denumerably additive extension ***G*** of ***F***. The hypothesis ***(Hyp Ad)***, [B.2.], will be admitted. To avoid non-essential complications we shall admit, that ***F*** is a finitely genuine strict subtribe of ***G***, and that the denumerably additive, non negative measure μ on ***G*** is effective. In addition to that we shall admit that ***G*** is the Lebesguean-covering-extension of ***F*** within ***G***. It follows that the borelian extension $\boldsymbol{F}^b$ of ***F*** within ***G*** coincides with ***G***. We shall take over from Chapter [B I.] the theory of ***B***-traces in ***F*** and admit ***Hyp*** **I** and ***Hyp*** **II**, [B I.6.], to have the whole measure theory of sets of traces at our disposal, as it has been developped in Chapter [B I.].

R.2. Let V be a F. Riesz-Banach-normed and complete linear space. Its elements $\vec{x}, \vec{y}, \ldots$ will be termed *vectors*, as in Chapter [Q I.].

R.3. Def. Denote, as in [B I.10.], by $v(x)$ the set of all neighborhoods of the trace x. By a *quasi-vector* $\vec{f}_x(p)$ (or $\vec{f}_x$), *with support* x, we shall understand any vector-valued function, defined for all bricks p belonging to $v(x)$.

If V is the space of real or complex numbers, we shall use also the term *quasi-number with support* x. We know [B I.10.6.], that $v(x)$ determines uniquely the trace; hence *the support* is well determined by a quasi-vector.

R.3.1. Various operations can be performed on quasi-vectors with the same support: Let $\vec{f}_x$, $\vec{g}_x$ be two quasi-vectors. By their sum (difference) $\vec{f}_x + \vec{g}_x$, $(\vec{f}_x - \vec{g}_x)$ we shall understand the function $\vec{h}_x(p)$ defined by $\vec{h}_x(p) =_{df} \vec{f}_x(p) \pm \vec{g}_x(p)$ for all neighborhoods p of x. By $\lambda \vec{f}_x$ we shall understand the function $\vec{h}_x(p)$ defined by $\vec{h}_x(p) = \lambda \vec{f}_x(p)$ for all $p \in v(x)$. The number λ is real or complex according to the character of the space V.

If we have a number-valued functional $F(\vec{x})$ or a vector-valued operator $\vec{F}(\vec{x})$, defined for all $\vec{x} \in V$, we define $F(\vec{f})$, $\vec{F}(\vec{f})$ as the functions

$$h(p) =_{df} F(\vec{f}(p)), \quad \vec{h}(p) =_{df} \vec{F}(\vec{f}(p))$$

for all $p \in v(x)$.

As a particular case we have the *norm of the quasi-vector* $\|\vec{f}_{x_0}\|$ defined as $\|\vec{f}_{x_0}(p)\|$ for all $p \in v(x)$.

R.4. Let $E \neq \emptyset$ be a set of traces. If we have defined, for every $x \in E$, a quasi-vector $\vec{f}_x$ with support x, we shall say that we have *a set of quasi-vectors with support* E, $\{\vec{f}_x | x \in E\}$. The construction can also be considered as a quasi-vector-valued function, defined on E.

R.4.1. We shall be mainly interested in sets of quasi-vectors with support W, i.e. the set of all traces. Such a set of quasi-vectors will be termed *total*.

R.4.2. A set of quasi-vectors with support E can be conceived as a function $\vec{F}(x, p)$ of two variables: x varying in E, and p varying over the whole set $v(x)$. It is not true that, if p is the common neighborhood of two different traces x', x'', we must have $\vec{F}(x', p) = \vec{F}(x'', p)$. Thus to every neighborhood p, which is taken into account, there corresponds a set $T(p)$ of traces $x \in E$, such that $p \in v(x)$. Hence to every p there corresponds a set $\Phi(p)$ of quasi-vectors $\vec{f}_x(p)$, where x varies in $T(p)$.

R.4.3. Thus we have a function $\vec{F}_1(x, p)$ which attaches to every p considered a whole set of vectors

$$\{\vec{F}(x, p) | x \in T(p)\}. \tag{1}$$

Def. If this set is composed of single vector for every $x \in E$, we shall call *the set of quasi-vectors, regular on* E.

R.4.4. Especially, if the set of quasi-vectors is total and regular on W, the set of quasi-vectors yields a vector-field $\vec{\Phi}(p)$, defined for all bricks p, [Q I.2., Def.]. If it is not regular on E, we can select in many ways, for each brick p, a trace $x = a(p)$, and consider the vector $\vec{f}_{a(p)}(p)$, which is well determined by the quasi-vector $\vec{f}_{a(p)}(p)$. If we do that for a total set of quasi-vectors for every brick p, we shall have defined a vector-valued function $\vec{f}_{a(p)}(p)$, thus constituting a vector-field, defined for all bricks. If, in the case of a total set of quasi-vectors, we consider all possible selections of $a(p)$, we shall get various vector-fields, $\vec{\Phi}_a(p)$. We shall call them *selected vector-fields or generated by the given total set of quasi-vectors.*

R.5. Def. We shall go over to the summation of a given set of quasi-vectors with support E, where E is a measurable set of traces [B I.8.]. We refer to [Q I.]. Let $\vec{f}_x$ be a total set of quasi-vectors. Consider one of the vector-fields $\vec{\Phi}_\alpha(\dot{p})$, defined for all bricks p, and generated by the given total quasi-vectors-set $\vec{f}_x$. Suppose that $\vec{\Phi}_\alpha(\dot{p})$ is summable on $[E]$*), in the sense of [Q I.], with respect to a kind (D') of summation. Now, if whatever the choice of $\vec{\Phi}_\alpha(\dot{p})$ may be, the vector-field $\vec{\Phi}_\alpha(\dot{p})$ is (D')-summable on E, and the sum $S_{[E]}\,\vec{\Phi}_\alpha$ has for all choices of $\vec{\Phi}_\alpha$ the same value, we say that the total set $\vec{f}_x$ of quasi-vectors *is (D')-summable on E (over E)*, we denote the sum by $S_E\,\vec{f}_x$, and call it "*sum of $\vec{f}_x$ on E*, or (over E)".

R.5.1. We shall be only interested in (DARS) sums [Q I.2.1.], and admit the ***(Hyp S)***, [Q.21.1.]. We leave the discussion of other kinds of sums to the reader.

R.5.2. For (DARS)-summation we shall prove the theorem: If for all choices of $\vec{\Phi}_\alpha$, the sum $S_{[E]}\,\vec{\Phi}_\alpha$ exists, then all these sums must be equal.

Proof. Let $\vec{\Phi}'(\dot{p})$, $\vec{\Phi}''(\dot{p})$ be different vector-fields, generated by the given total set $\vec{f}_x$ of quasi-vectors [R.4.4.], and suppose, that $\vec{A}' \neq \vec{A}''$, where

$$\vec{A}' =_{df} S_{[E]}\,\vec{\Phi}', \qquad \vec{A}'' =_{df} S_{[E]}\,\vec{\Phi}''.$$

Let $\{P_n\}$ be a completely distinguished sequence of complexes for E. Put

$$P_n =_{df} \{p_{n1}, p_{n2}, \ldots\}, \qquad (n = 1, 2, \ldots).$$

Consider the set of all bricks p_{nk}, $(n = 1, 2, \ldots)$, $(k = 1, 2, \ldots)$. To such a brick p_{nk} there is, possibly, a double choice of the vector, attached to it:

$$\vec{\Phi}'(p_{nk}), \qquad \vec{\Phi}''(p_{nk}).$$

---

*) $[E]$ means the coat of E, [B I.8].

If p_{nk} is an atom, there exists one and only one trace covered by p_{nk}, hence in this case $\vec{\Phi}'(p_{nk}) = \Phi''(p_{nk})$, so the choice is well determined.

We call p_{nk} single choice-brick or double choice brick according to the case, whether $\vec{\Phi}'(p_{nk})$, $\Phi''(p_{nk})$ are equal or different.

Bricks, which are atoms are always of single choice; another bricks may be single-choice bricks, or not.

We shall find a partial sequence $l(n)$ of indices, as follows. We put $l(1) = 1$. We take the choice $\vec{\Phi}'(p_{11})$, $\vec{\Phi}'(p_{12})$, ...

The bricks $p_{11}, p_{12}, \ldots$ are finite in number.

I say that if n is sufficiently great, then not a single brick among $p_{11}, p_{12}, \ldots$ will occure in the complex P_n, excepting perhaps when the brick is an atom. Suppose this be not true. Then there exists an infinite sequence $t(n)$ of indices, such that in every $P_{t(n)}$ there is available at least one of the non atomic bricks p_{1k}. The number of those bricks p_{1k} is finite. Hence there exists a non-atomic brick, say p_{1m}, and a subsequence $\{p_{s(n)}\}$ of $\{p_{t(n)}\}$, such that p_{1m} is a mesh in every complex $P_{s(n)}$, $(n = 1, 2, \ldots)$: $p_{1m} \in P_{(s)n}$, for $n = 1, 2, \ldots$

Since p_{1m} is not an atom, and since μ is effective, there exist somata A, B, such that $\mu A > 0$, $\mu B > 0$, $A \cdot B = O$, $A + B = p_{1m}$. Now $\{P_{s(n)}\}$ is completely distinguished for E, [Q.21.6.]. We have

$$|E, P_{s(n)}| \to 0 \quad \text{for} \quad n \to \infty.$$

Hence

$$|E\, p_{1m}, \operatorname{som} P_{s(n)} \cdot p_{1m}| \to 0 \quad \text{for} \quad n \to \infty, \quad \text{i.e.} \quad |E\, p_{1m}, p_{1m}| \to 0.$$

It follows, as this is a constant sequence, $|E\, p_{1m}, p_{1m}| = 0$, and hence $E\, p_{1m} = p_{1m}$, which gives $p_{1m} \leqq E$.

It follows that $A \leqq E$. Hence there exists a partial complex Q_n of $P_{s(n)}$ with $|Q_n, A| \to 0$, for $n \to \infty$. Hence $|\operatorname{som} Q_n \cdot p_{1m}, A\, p_{1m}| \to 0$

$$(1) \qquad |\operatorname{som} Q_n \cdot p_{1m}, A| \to 0, \quad \text{because} \quad A \leqq p_{1m}.$$

Since Q_n is a partial complex of $P_{s(n)}$, to which p_{1m} belongs as a mesh, we have either $\operatorname{som} Q_n \cdot p_{1m} = O$ or $\operatorname{som} Q_n \cdot p_{1m} = p_{1m}$.

The first alternative cannot occur for an infinite number of indices n, because from (1) we would get $|O, A| \to O$, i.e. $A = O$, which is not true. Hence the relation, spoken of, can occur only for an at most finite number of indices n; hence, for sufficiently great n we have surely

$$\operatorname{som} Q_n \cdot p_{1m} = p_{1m},$$

and hence (1) yields $|p_{1m}, A| \to O$; hence $|p_{1m}, A| = 0$, and then $p_{1m}, = A$, which is not true. The obtained contradiction proves that the supposition, stating that $p_{1m} \in P_{s(n)}$, for $n = 1, 2, \ldots$, is not true.

Hence, our statement that, if n is sufficiently great, then not a single brick among

(2) $$p_{11}, p_{12}, \ldots$$

will occur in $p_{n1}, p_{n2}, \ldots$, excepting, perhaps, when a brick (2) is an atom, is proved. Thus we can find an index $l(2) > l(1)$, such that not a single non-atomic brick of $P_{l(1)}$ will occur in any P_n when $n \geqq l(2)$.

Considering $P_{l(2)}$, we shall repeat our argument, finding an index $l(3)$, such that, if $n \geqq l(3)$, non a single non atomic brick of $P_{l(2)}$ will occur in P_n. By induction we shall find an infinite sequence of indices $l(1) < l(2) < \cdots < l(n) < \cdots$, such that if $k < n$, not a single non-atomic brick, occuring in

(3) $$P_{l(1)}, P_{l(2)}, \ldots, P_{l(k)},$$

will occur in $P_{l(n)}$. Thus if we consider any brick, which occurs in all (3), we see that either this is an atom, and can occur in many complexes, or else it occurs only once in (3). The sequence $P_{l(1)}, P_{l(2)}, \ldots$ is completely distinguished for E.

Consider the vectors

$$\begin{array}{ll} \vec{\Phi}'(p_{l(1),1}), & \vec{\Phi}'(p_{l(1),2}), \ldots, \\ \vec{\Phi}''(p_{l(2),1}), & \vec{\Phi}''(p_{l(2),2}), \ldots, \\ \vec{\Phi}'(p_{l(3),1}), & \vec{\Phi}'(p_{l(3),2}), \ldots, \\ \vec{\Phi}''(p_{l(4),1}), & \vec{\Phi}''(p_{l(4),2}), \ldots, \end{array}$$

which are defined for bricks $p_{l(n),k}$, $(n = 1, 2, \ldots)$ and $(k = 1, 2, \ldots)$. We denote these vectors by:

$$\vec{\Phi}'''(p_{l(n),k}), \quad n = 1, 2, \ldots, \quad k = 1, 2, \ldots$$

They are defined only for bricks $p_{l(n),k}$. There may be some remaining bricks in $\boldsymbol{B}$, namely $b_1, b_2, \ldots$ We put

$$\vec{\Phi}'''(b_n) = \vec{\Phi}'(b_n), \quad n = 1, 2, \ldots$$

Thus $\vec{\Phi}'''(p)$ is defined for all bricks $\in \boldsymbol{B}$ and they constitute a selection of a vector-field generated by the total set $\vec{f}_x$ of quasi-vectors. Now $\vec{\Phi}'''(P_{l(n)})$ has a limit, say $\vec{A}$, by hypothesis, because $\{P_{l(n)}\}$ is completely distinguished for E. We have:

$$\vec{\Phi}'''(P_{l(2n-1)}) \to \vec{A}' \quad \text{and} \quad \vec{\Phi}'''(P_{l(2n)}) \to \vec{A}''.$$

Hence

$$\vec{A}' = \vec{A}, \quad \vec{A}'' = \vec{A}, \quad \text{and then} \quad \vec{A}' = \vec{A}'',$$

which contradicts the hypothesis that $\vec{A}' \neq \vec{A}''$. The theorem is established.

R.5.3. Remark. The theorem [R.5.2.] is true for (DS), (DAS) and (DRS)-summation, but it is not true for other kind of summation.

R.6. The fundamental theorems [R.6.1.] to [R.6.10.] on sums of quasi-vectors, will be given for (DARS)-summations only; hence we admit ***(Hyp S)***, [Q.21.1.]. These theorems are direct consequences of the corresponding theorems in [Q I.].

R.6.1. Considering (DARS)-summations, suppose that

1) $\vec{f}_\tau$ is a total set of quasi-vectors,
2) E, F are measurable sets of traces,
3) $F \subseteqq E$,
4) $\vec{f}_\tau$ is summable on E;

then $\vec{f}_\tau$ is summable on F.

Proof. Let $\vec{\varphi}_\alpha(\dot{p})$ be a selected vector-field, generated by $\vec{f}_\tau$. By hyp. 4 and Def. [R.5.], the sum $S_{[E];(\text{DARS})}\,\vec{\varphi}_\alpha$ exists. Since $[F] \leqq [E]$, $[F] \in \boldsymbol{G}$, it follows, by [Q I.4.], that $S_{[F]}\,\vec{\varphi}_\alpha$ also exists. Applying [Theorem R.5.2.], we get the thesis.

R.6.2. If

1. E_1, E_2 are measurable sets of traces,
2. $E_1 \cap E_2 = \emptyset$,
3. $S_{E_1 \cup E_2}\,\vec{f}_\tau$ exists,

then

$$S_{E_1 \cup E_2}\,\vec{f}_\tau = S_{E_1}\,\vec{f}_\tau + S_{E_2}\,\vec{f}_\tau.$$

Proof. We apply [Def. R.5.], [Q I.7.], and the fact, that $[E_1 \cup E_2] = [E_1] + [E_2]$, [B I.].

R.6.3. If

1. E_1, E_2 are measurable sets of traces,
2. $E_1 \cap E_2 = \emptyset$,
3. $S_{E_1}\,\vec{f}_\tau$ and $S_{E_2}\,\vec{f}_\tau$ both exist,

then $S_{E_1 \cup E_2}\,\vec{f}_\tau$ exists too, and

$$S_{E_1 \cup E_2}\,\vec{f}_\tau = S_{E_1}\,\vec{f}_\tau + S_{E_2}\,\vec{f}_\tau$$

for (DARS)-sums.

Proof. We rely on [Q I.8.].

R.6.4. If

1. E, E_n are measurable sets of traces, $(n = 1, 2, \ldots)$,
2. $\mu E_n \to 0$,
3. $E_n \subseteqq E$,
4. $S_E\,\vec{f}_\tau$ exists,

then

$$\lim S_{E_n}\,\vec{f}_\tau = \vec{O}$$

in the V-topology.

Proof. We rely on [Q I.5.1.] and on the equality $\mu[E_n] = \mu E_n$, [B I.].

R.6.5. If

1. E is a measurable set of traces,
2. $S_E \vec{f}_\tau$ exists,
3. $\alpha > 0$,

then there exists $\beta > 0$, such that if $\mu(F) \leqq \alpha$, where F is a measurable set of traces, then

$$\|S_F \vec{f}_\tau\| \leqq \beta.$$

Proof. We rely on [Q I.9.2.].

R.6.6. If

1. E_n, F are measurable sets of traces, $(n = 1, 2, \ldots)$,
2. E_n are disjoint with one another,
3. $E_n \subseteqq F$,
4. $S_F \vec{f}_\tau$ exists,

then, if we put $E =_{df} \bigcup_{n=1}^{\infty} E_n$, we get

$$S_{E\tau} \vec{f}_\tau = \sum_{n=1}^{\infty} S_{E_n} \vec{f}_\tau;$$

so the vector-valued function $S_G \vec{f}_\tau$ of the variable measurable set $\in \boldsymbol{G}$ of traces, with $G \subseteqq F$, is denumerably additive.

Proof. We rely on [Q I.10.10.1.], and on the equality

$$\left[\bigcup_{n=1}^{\infty} E_n\right] = \sum_{n=1}^{\infty} [E_n],$$

[B I.].

R.6.7. If

1. E_n, F are measurable sets of traces,
2. $E_n \subseteqq F$, $(n = 1, 2, \ldots)$,
3. $|E_n, E|_\mu \to 0$,
4. $S_F \vec{f}_\tau$ exists,

then

$$\lim_{n\to\infty} S_{E_n} \vec{f}_\tau = S_E \vec{f}_\tau$$

in the V-topology.

Proof. We rely on [Q I.12.].

R.6.8. If $S_E \vec{f}_\tau$ exists for a measurable set E of traces, then if $\mu(F) = 0$, we have $S_F \vec{f}_\tau = \vec{O}$.

R.6.8.1. If $S_E \vec{f}_\tau$ exists, and $E =^\mu F$, then $S_E \vec{f}_\tau = S_F \vec{f}_\tau$.

R.6.9. If

1. $S_W \vec{f}_\tau$ exists, (W is the set of all traces),
2. we put for every brick p: $\vec{K}(p) =_{df} S_P \vec{f}_\tau$, where $[P] =_{df} p$,

then for every measurable set E of traces we have

$$S_E \vec{f}_\tau = S_{[E]} \vec{K}(\dot{p}).$$

Proof. We rely on [Q I.13.].

R.6.9.1. Remark. In relation to [R.6.9.], if we define the quasi-vector $\vec{k}_\tau$ by putting $\vec{k}_\tau(v) =_{df} \vec{K}(v)$ for every neighborhood of τ, we get $S_E \vec{f}_\tau = S_E \vec{k}_\tau$ for every measurable set E of traces. Hence the (DARS)-sums of quasi-vectors can be transformed into sums of regular quasi-vector-sets, [R.4.3.].

R.6.10. If $\vec{f}_\tau$ is summable on a measurable set E of traces and λ is a number, (real or complex depending on the character of V), then

$$S_E \lambda \vec{f}_\tau = \lambda S_E \vec{f}_\tau.$$

Proof. [Q I.4.1.].

R.6.11. If $\vec{f}_\tau$, $\vec{g}_\tau$ are both summable on a measurable set E of traces, then

$$S_E(\vec{f}_\tau \pm \vec{g}_\tau) = S_E \vec{f}_\tau \pm S_E \vec{g}_\tau.$$

Proof. We shall represent the set of quasi-vectors $\vec{f}_\tau$ as a function $\vec{f}(\tau, p)$ of two variables: τ and p, where p is a neighborhood of τ. Similarly for $\vec{g}_\tau$. By definition of the sum of two quasi-vectors [R.3.1.], for the quasi-vector

(1) $\vec{h}_\tau =_{df} \vec{f}_\tau + \vec{g}_\tau$, we have $\vec{h}(\tau, p) = \vec{f}(\tau, p) + \vec{g}(\tau, p)$,

whenever p is a neighborhood of τ. Take a choice of a vector-field $\vec{h}(p)$ generated by $\vec{h}_\tau$. It is determined by the choice of the function $\tau = \alpha(p)$, where p is a neighborhood of τ. Then the vector-fields for $\vec{h}_\tau$, $\vec{f}_\tau$, $\vec{g}_\tau$ will be

$$\vec{h}(\alpha(\dot{p}), \dot{p}), \quad \vec{f}(\alpha(\dot{p}), \dot{p}), \quad \vec{g}(\alpha(\dot{p}), \dot{p}),$$

and, by (1), we have

(2) $$\vec{h}(\alpha(p), p) = \vec{f}(\alpha(p), p) + \vec{g}(\alpha(p), p)$$

for all bricks p. We have

$$S_E \vec{f}_\tau = S_{[E]} \vec{f}(\alpha(\dot{p}), \dot{p}), \quad S_E \vec{g}_\tau = S_{[E]} \vec{g}(\alpha(\dot{p}), \dot{p}).$$

Hence by (2)

(3) $$S_E \vec{f}_\tau + S_E \vec{g}_\tau = S_{[E]} \vec{h}(\alpha(\dot{p}), \dot{p}).$$

Thus the sum on the right in (3) exists and has the same value for any choice of the vector-field generated by $\vec{h}_\tau$. Hence [Def. R.5.]

$$S_E \vec{f}_\tau + S_E \vec{g}_\tau = S_E(\vec{f}_\tau + \vec{g}_\tau), \quad \text{q.e.d.}$$

R.7. We shall deal with (DARS)-sums only. Let $\vec{f}_x$, $\vec{g}_x$ be two total sets of quasi-vectors, which are summable on W.

We say that $\vec{f}_x$ *is equivalent to* $\vec{g}_x$: $\vec{f}_x \approx \vec{g}_x$ whenever for every measurable set E of traces we have $S_E \vec{f}_x = S_E \vec{g}_x$.

R.7.1. We have for total, summable sets of quasi-vectors:

1) If $\vec{f}_x \approx \vec{g}_x$, then $\vec{g}_x \approx \vec{f}_x$,

2) $\vec{f}_x \approx \vec{f}_x$,

3) if $\vec{f}_x \approx \vec{g}_x$, $\vec{g}_x \approx \vec{h}_x$,

then $\vec{f}_x \approx \vec{h}_x$.

R.7.2. If $\vec{f}_x \approx \vec{g}_x$ and λ is a number, then $\lambda \vec{f}_x \approx \lambda \vec{g}_x$.

R.7.3. If $\vec{f}_x \approx \vec{g}_x$, $\vec{f}'_x \approx \vec{g}'_x$, then $\vec{f}_x + \vec{f}'_x \approx \vec{g}_x + \vec{g}'_x$.

Proof. We rely on [R.6.11.].

R.7.4. From [R.6.9.1.] it follows, that every total summable set of quasi-vectors is equivalent to a regular total summable set of quasi-vectors.

R.8. An important case of the vector-space V is the space of real number and the space of complex numbers. The vector-fields, in these particular cases, will be termed *scalar fields* and quasi-vectors will be termed *quasi-numbers*.

The function $\mu(\dot{a})$ of the variable brick a is a scalar field, and if for all neighborhoods p of x we define $\mu_x =_{df} \mu(\dot{p})$, we get a real quasi-number. We shall call it *measure-quasi-number*. The total set μ_x is regular.

R.8.1. Quasi numbers f_x, g_x can be multiplied, getting a new quasi-number $f_x \cdot g_x$ defined as the function $f_x(p) \cdot g_x(p)$ for every neighborhood p of x. Given a quasi-vector $\vec{f}_x$ taken from a general Banach-space V, and given a quasi-number a_x, we can multiply them, getting a quasi-vector in V: $a_x \vec{f}_x$, defined in a similar way, as above.

R.9. If M_x is a total set of quasi-numbers, summable on W, then there exists a complex-valued function $F(\dot{x})$ of the trace x variable in W, and such that for every measurable set A of traces, we have

$$S_A M_x = \int_A F(\dot{x})\, d\mu.$$

The function F is μ-unique. The integral is Frechetean, [B I.12.].

Proof. Put $K(A) =_{df} S_A M_x$ for all measurable A. By Theorem [R.6.6.] the set-function $K(A)$ is denumerably additive. If $\mu(A) = 0$, we have $[A] = O$, since the measure μ is effective on $\boldsymbol{G}$. Hence $K(O) = 0$. Consequently $K(A)$ is μ-continuous. Hence, by a known theorem, there exists a μ-unique function $F(x)$, defined μ-a.e. on W, such that

$$\int_A F(x)\, d\mu = S_A M_x,$$

so the theorem is proved.

R.10. Def. Let $f(\dot{x})$ be a complex-number-valued function of the trace x, variable in W. Suppose this function is μ-Fréchet-summable on W. Hence $\int_A f(x)\,d\mu$ exists for every measurable subset of traces. Considering a brick p, and a measurable set P of traces, with $[P] = p$, the integral $\int_P f(x)\,dx$ does not depend on P but on p only. Having fixed x for a moment, and considering all neighborhoods p of x, we put

$$\operatorname{val}_x f =_{df} \frac{1}{\mu(P)} \int_P f(x)\,d\mu .$$

We may call it: *The mean-value quasi-vector of f at x.* This is a quasi-number with support x.

R.11. We shall prove the theorem: Under circumstances [R.1.], if $f(\dot{x})$ is Fréchet-summable on W, then the total set of quasi-numbers $\mu_x \cdot \operatorname{val}_x f$ is summable on W, and we have $S_A \operatorname{val}_x f \cdot \mu_x = \int_A f(x)\,d\mu$ for every measurable set A of traces.

Proof. Put $k_x =_{df} \mu_x \cdot \operatorname{val}_x f$. This quasi-number is defined as the function $k_x(p) =_{df} \mu(\dot{p}) \cdot \frac{1}{\mu(\dot{p})} \int_P f(x)\,d\mu$, defined for all neighborhoods p of x.

Hence $k_x(p) = \int_P f(x)\,d\mu$ where $[P] = p$.

Hence $S_p\, k_x = \int_P f(x)\,d\mu$ for every brick p. Now, let $P' = \{p_1, p_2, \ldots\}$ be a complex. Since the somata p_i are disjoint, we get

$$S_{P'}\, k_x = \int_{P'} f(x)\,d\mu, \tag{1}$$

where $[P'] = \operatorname{som} P$. Now let A be any measurable set of traces and let $\{P_n\}$ be any sequence of complexes with $|P_n, A|_\mu \to 0$. If we denote by P'_n the set of traces with $[P'_n] = \operatorname{som} P_n$, we get

$$\lim_{n\to\infty} \int_{P'_n} f(x)\,d\mu = \int_A f(x)\,d\mu .$$

Hence, by (1), the sequence $\{S_{P'_n}\, k_x\}$ converges to $\int_A f(x)\,d\mu$. On the other hand this sequence converges to $S_E\, k_x$, [R.6.7.]. Consequently $S_E\, k_x = \int_A f(x)\,d\mu$, q.e.d.

R.11.1. Remark. Notice that the quasi-number-set $\operatorname{val}_x f$ is regular, but it may be not summable.

Expl. Let $f = \text{const} = 1$, we get $\operatorname{val}_x f(p) = 1$ for all p. Hence if a complex P has n bricks, the number $f(P) = n$, so it does not tend to any limit.

R.12. We shall be in circumstances [R.1.] till the end of the Chapter [R.], and shall consider only (DARS)-summation. We admit ***(Hyp S)***.

R.13. If $f(\dot{x})$, $g(\dot{x})$ are Fréchet-summable functions on W, then

$$\mu_x \operatorname{val}_x(f+g) \approx \mu_x \operatorname{val}_x f + \mu_x \operatorname{val}_x g,$$

[Def. R.7.].

Proof. Since f and g are Fréchet-summable on W, so is $f+g$. By [R.11.], we have for every measurable set A of traces

$$S_A \operatorname{val}_x f \cdot \mu_x = \int_A f(x)\, d\mu, \tag{1}$$

$$S_A \operatorname{val}_x g \cdot \mu_x = \int_A g(x)\, d\mu, \tag{2}$$

and

$$S_A \operatorname{val}_x(f+g)\, \mu_x = \int_A (f+g)\, d\mu. \tag{3}$$

Since the sum of the right-hand integrals in (1) and (2) equals that of (3), it follows, that the same is for the left-hand sums; so the theorem is proved.

R.14. If $f(\dot{x})$ is Fréchet-summable on W, and λ is a number, then

$$\mu_x \operatorname{val}_x(\lambda f) \approx \lambda\, \mu_x \operatorname{val}_x f.$$

Proof. Similar to the forgoing one, based on [R.11.].

R.15. Theorem. If $f(x)$ is Fréchet-square-summable on W, then

$$\mu_x \operatorname{val}_x |f|^2 \approx \mu_x\, |\operatorname{val}_x f|^2.$$

The proof of this theorem will require few auxiliary theorems and steps. In what follows, till the end of this Chapter [R.], we shall agree to denote by the same letter a measurable set of traces and its coat. This for simplifying the exposition. There will be an ambiguity up to null-sets of traces. These null-sets however, do not matter: see [R.6.8.1.].

Notice that $\mu_x \operatorname{val}_x|f|^2$ is summable, [R.11.], because $|f|^2$ is Fréchet-summable. The summability of the right-hand side expression in the thesis shall be proved.

R.15a. First of all we shall prove the theorem [R.15.] for the functions $f(\dot{x})$ defined as follows: there is a brick $c \neq O$ such that $f(x) = 1$ for $x \in c$, and $f(x) = 0$ for $x \in \operatorname{co} c$.

Let A be a measurable set of traces. Consider a completely distinguished sequence $\{P_n\}$ of complexes for A, and select any subsequence $\{P_{k(n)}\}$ of it. By [Q.21.6.] this is also a completely distinguished sequence for A. Applying [Q.21.13.], find a subsequence $\{P_{kl(n)}\}$ of it, and a sequence $\{Q_n\}$ of complexes for $\operatorname{co} E$ such that $\operatorname{som} Q_n \cdot \operatorname{som} P_{kl(n)} = O$, and where $\{P_{kl(n)} \cup Q_n\}$ is a completely distinguished sequence for I. Put $R_n =_{df} P_{kl(n)} \cup Q_n$, $(n = 1, 2, \ldots)$. The complex $P_{kl(n)}$ is a partial complex of R_n. Denote by g_x, h_x the quasi-numbers $\mu_x \operatorname{val}_x |f|^2$ and $\mu_x |\operatorname{val}_x f|^2$ respectively. For any brick p we have

$$g(p) = \int_{pc} d\mu = \mu(p\,c), \quad h(p) = \frac{|\mu(p\,c)|^2}{\mu(p)}.$$

Hence

$$g(p) - h(p) = \frac{\mu(p\,c) \cdot \mu(p \operatorname{co} c)}{\mu(p)}.$$

Consider the bricks of $P_{kl(n)}$. Denote by

$$a_{n1}, a_{n2}, \ldots; \quad b_{n1}, b_{n2}, \ldots; \quad e_{n1}, e_{n2}, \ldots$$

those of them, for which

$$a_{nk} \leqq c; \quad b_{nk} \cdot c = O; \quad e_{nk}\, c \neq O; \quad e_{nk} \operatorname{co} c \neq O$$

respectively. We have

$$g(P_{kl(n)}) = \sum_k g(a_{nk}) + \sum_k g(e_{nk}),$$

and

$$h(P_{kl(n)}) = \sum_k h(a_{nk}) + \sum_k g(e_{nk}),$$

because $g(b_{nk}) = 0$. Now

$$g(a_{nk}) = \mu(a_{nk}\, c) = \mu(a_{nk}), \quad h(a_{nk}) = \mu(a_{nk}\, c)^2 = \mu(a_{nk}).$$

It follows that

$$g(P_{kl(n)}) - h(P_{kl(n)}) = \sum_k \frac{\mu(e_{nk} \cdot c)\, \mu(e_{nk} \operatorname{co} c)}{\mu(e_{nk})}. \tag{0}$$

We recall that e_{nk} are all those bricks in $P_{kl(n)}$ for which

$$e_{nk} \cdot c \neq 0, \quad e_{nk} \operatorname{co} c \neq 0.$$

Since $\{R_n\}$ is a completely distinguished sequence for 1, there exist a partial complex S_n of R_n, such that $\{S_n\}$ is completely distinguished for $\operatorname{som} c$. We have $|c \cdot S_n| \to 0$. Hence $|\operatorname{co} S, T_n| \to 0$, where $T_n =_{df} R_n \dot{-} S_n$.

We have

$$S_n \cap T_n = \emptyset, \quad S_n \cup T_n = R_n. \tag{1}$$

By [Q.21.9.], T_n is a completely distinguished sequence for $\operatorname{co} c$. Consider all bricks $d_{n1}, d_{n2}, \ldots$ of R_n, for which $d_{ni} c \neq O$, $d_{ni} \cdot \operatorname{co} c \neq O$. They make up two classes: one composed of those bricks, which are in S_n; denote them by

$$d'_{n1}, d'_{n2}, \ldots \tag{2}$$

and the second, composed of those bricks, which are in T_n: denote them by

(3) $$d''_{n1}, d''_{n2}, \dots$$

The classes (2), (3) are disjoint and make up the set $d_{n1}, d_{n2}, \dots$ of bricks. By [Q.21.14.],

$$\mu(\sum d'_{nk} \operatorname{co} c) \to 0 \quad \text{and} \quad \mu(\sum d''_{nk} c) \to 0.$$

Now a brick e_{nk} either belongs to (2) or to (3). Hence all e_{nk} can be devided into two disjoint classes $e'_{n1}, e'_{n2}, \dots$; and $e''_{n1}, e''_{n2}, \dots$, where we have $\mu(\sum e'_{n1} \operatorname{co} c) \to 0$, $\mu(\sum e''_{n1} c) \to 0$.

Having this, resume the formula (0):

$$g(P_{kl(n)}) - h(P_{kl(n)}) = \sum_k \frac{\mu(e_{nk} c) \cdot \mu(e_{nk} \operatorname{co} c)}{\mu(e_{nk})} = \sum_k \frac{\mu(e'_{nk} c) \cdot \mu(e'_{nk} \operatorname{co} c)}{\mu(e'_{nk})} +$$

$$+ \sum_k \frac{\mu(e''_{nk} c) \cdot \mu(e''_{nk} \operatorname{co} c)}{\mu(e''_{nk})} \leqq \sum_k \mu(e' \operatorname{co} c) + \sum_k \mu(e''_{nk} c) \to 0.$$

Since, as we know, [R.11.], the quasi-vector-set $\mu_x \operatorname{val}_x f^2$ is summable, it follows that $\sum_k h(a_{nk})$ tends to $S_A\, \mu_x \operatorname{val}_x |f|^2$.

Thus from every partial sequence $\{P_{k(n)}\}$ another can be extracted $\{P_{lk(n)}\}$ such that

$$\lim P_{lk(n)}(h) = S_A\, \mu_x \operatorname{val}_x |f|^2.$$

Consequently

$$\lim P_n(h) = S_A\, \mu_x \operatorname{val}_x |f|^2.$$

Hence

$$S_A\, \mu_x |\operatorname{val}_x f|^2 = S_A\, \mu_x \operatorname{val}_x |f|^2$$

for every measurable set A of traces. Thus we get

$$\mu_x |\operatorname{val}_x f|^2 \approx \mu_x \operatorname{val}_x |f|^2, \quad \text{q.e.d.}$$

R.15b. If $f(x)$ is the function as in [R.15 a.], and we put $g(x) = \lambda f(x)$, where λ is a number, then the thesis [R.15.] holds for $g(x)$.

R.15c. Lemma. If

1. $[c] \neq 0$ is a figure,
2. $f(x)$, $g(x)$ are μ-square summable on W,
3. $f(x) = 0$ for $x \in c$, $g(x) = 0$ for $x \in \operatorname{co} c = W - c$,
4. A is a measurable set of traces,
5. $P_n = \{p_{ni}, p_{n2}, \dots\}$, $(n = 1, 2, \dots)$, is a completely distinguished sequence for $[A]$,

then

$$A_n =_{df} \sum_k \frac{\int_{p_{nk}} f\, d\mu \cdot \int_{p_{nk}} g\, d\mu}{\mu(p_{nk})} \to 0 \quad \text{for} \quad n \to \infty;$$

(p_{nk} are considered as sets of traces).

Proof. We take over the notations from the preceding proof. $\{P_{k(n)}\}$ is a subsequence of $\{P_n\}$. We build $R_n =_{df} P_{kl(n)} \cup Q_n$.

We find $\{S_n\}$ and $\{T_n\}$ with $|c, S_n| \to 0$, $|\text{co}\,c, T_n| \to 0$. We get the bricks a_{nk}, b_{nk} where $a_{nk} \leqq p$, $b_{nk} \cdot p = O$, and e'_{nk}, e''_{nk}, which all belong to $P_{kl(n)}$. We have $\mu(\sum e'_{nk}\,\text{co}\,c) \to 0$, $\mu(\sum e''_{nk}\, c) \to 0$. Now

$$A_n = \sum_k \frac{\int\limits_{e'_{nk}} f\,d\mu \cdot \int\limits_{e'_{nk}} g\,d\mu}{\mu(e'_{nk})} + \sum_k \frac{\int\limits_{e''_{nk}} f\,d\mu \cdot \int\limits_{e''_{nk}} g\,d\mu}{\mu(e''_{nk})}.$$

Denote the first term by B', the second by B''. We have

$$|B'| \leqq \sum_k \frac{\sqrt{\mu(e'_{nk}) \int\limits_{e'_{nk}} |f|^2\,d\mu} \cdot \sqrt{\mu(e'_{nk}) \int\limits_{e'_{nk}} |g|^2\,d\mu}}{\mu(e_{nk})}$$

$$= \sum_k \sqrt{\int\limits_{e'_{nk}} |f|^2\,d\mu} \cdot \sqrt{\int\limits_{e'_{nk}} |g|^2\,d\mu}.$$

Applying once more the Cauchy-Schwarz-inequality, we get

$$|B'| \leqq \sqrt{\sum_k \int\limits_{e'_{nk}} |f|^2\,d\mu} \cdot \sqrt{\sum_k \int\limits_{e'_{nk}} |g|^2\,d\mu}.$$

Since $f(x) = 0$ for $x \in c$ and $g(x) = 0$ for $x \in \text{co}\,c$, we get

$$|B'| \leqq \sqrt{\sum_k \int\limits_{e'_{nk}\text{co}\,c} |f|^2\,d\mu} \cdot \sqrt{\sum_k \int\limits_{e'_{nk}\,c} |g|^2\,d\mu}.$$

Since $\mu(\sum e'_{nk}\,\text{co}\,c| \to 0$, we get

$$|B'| \leqq \sqrt{\int\limits_{\sum e'_{nk}\,\text{co}\,c} |f|^2\,d\mu} \cdot \sqrt{\sum_k \int\limits_W |g|^2\,d\mu} \to 0.$$

In a similar way we get $|B''| \to 0$. It follows $A_n \to 0$, q.e.d.

R.15d. Lemma. If

1. $f(x)$, $g(x)$ are μ-square summable on W,
2. c is a set, whose coat $c = [c] \in F$,
3. $\mu_x\,|\text{val}_x f|^2 \approx \mu_x\,\text{val}_x |f|^2$, $\mu_x\,|\text{val}_x g|^2 \approx \mu_x\,\text{val}_x |g|^2$,
4. $f(x) = 0$ for $x \in \text{co}\,c$, $g(x) = 0$ for $x \in c$,
5. $h(x) =_{df} f(x) + g(x)$,

then

$$\mu_x\,|\text{val}_x h|^2 \approx \mu_x\,\text{val}_x |h|^2.$$

Proof.

$$|h|^2 = (\bar f + \bar g)(f + g) = \bar f f + \bar g f + \bar f g + \bar g g = |f|^2 + |g|^2 + \bar f g + \bar g f.$$

Hence

$$(1)\qquad \mathrm{val}_x|h|^2 \approx \mathrm{val}_x|f|^2 + \mathrm{val}_x|g|^2 + \mathrm{val}_x|\bar{f}g| + \mathrm{val}_x|\bar{g}f| \approx \approx \mathrm{val}_x|f|^2 + \mathrm{val}_x|g|^2,$$

because $\bar{f}g = \bar{g}f = 0$.

The quasi-number $\mathrm{val}_x(f+g)$ is the function $\frac{1}{\mu(p)}\int\limits_p (f+g)\,d\mu$; hence $|\mathrm{val}_x(f+g)|^2$ is the function:

$$\frac{1}{\mu(p)^2}\cdot\int\limits_p \overline{f+g}\,d\mu\cdot\int\limits_p (f+g)\,d\mu =$$

$$= \frac{1}{\mu(p)^2}\left\{\left|\int\limits_p f\,d\mu\right|^2 + \left|\int\limits_p g\,d\mu\right|^2 + \int\limits_p \bar{f}\,d\mu\cdot\int\limits_p g\,d\mu + \int\limits_p f\,d\mu\cdot\int\limits_p \bar{g}\,d\mu\right\}.$$

Hence

$$(2)\quad |\mathrm{val}_x h|^2 = |\mathrm{val}_x f|^2 + |\mathrm{val}_x g|^2 + \mathrm{val}_x\bar{f}\cdot\mathrm{val}_x g + \mathrm{val}_x f\cdot\mathrm{val}_x\bar{g}.$$

From (1) and (2) we get

$$\mu_x\,\mathrm{val}_x|h|^2 - \mu_x|\mathrm{val}_x h|^2 \approx -\mu_x\,\mathrm{val}_x\bar{f}\,\mathrm{val}_x g - \mu_x\,\mathrm{val}_x f\cdot\mathrm{val}_x\bar{g}.$$

If we apply the Lemma [R.15c.], we get

$$\mu_x\,\mathrm{val}_x|h|^2 \approx \mu_x|\mathrm{val}_x h|^2,\quad \text{q.e.d.}$$

R.15e. If $f(x)$ is a step-function, then $\mu_x|\mathrm{val}_x f|^2 \approx \mu_x\,\mathrm{val}_x|f|^2$.

Proof. [R.15a.], [R.15b.] and [R.15d.].

R.15f. Lemma. If $f(x)$ is square-summable, then there exists an infinite sequence of step-functions $f_1(x), f_2(x), \ldots,$ which tends in μ-square mean to $f(x)$. This is known from the general theory of Fréchet's integrals.

R.15g. If

1. $f_1(x), f_2(x), \ldots, f_n(x), \ldots$ are all μ-square summable in W,
2. $f(x)$ is the square-mean limit of $\{f_n(x)\}$ in W,
3. for every n we have $\mu_x\,\mathrm{val}_x|f_n|^2 \approx \mu_x|\mathrm{val}_x f_n|^2$,

then

$$\mu_x\,\mathrm{val}_x|f|^2 \approx \mu_x|\mathrm{val}_x f|^2.$$

Proof. Suppose that for all functions $f_1(x), f_2(x), \ldots, f_n(x), \ldots,$ which we suppose μ-square summable in W, we have

$$(0)\qquad \mu_x\,\mathrm{val}_x|f_n|^2 \approx \mu_x|\mathrm{val}_x f_n|^2.$$

Suppose that $\lim f_n(x) = f(x)$.

We shall prove that

$$\mu_x\,\mathrm{val}_x|f|^2 \approx \mu_x|\mathrm{val}_x f|^2.$$

Put

$$A_n =_{df} \int_p |f_n|^2\, d\mu, \quad A =_{df} \int_p |f|^2\, d\mu, \quad B_n =_{df} \frac{1}{\mu(p)} \left| \int_p f_n\, d\mu \right|^2,$$

$$B =_{df} \frac{1}{\mu(p)} \left| \int_p f\, d\mu \right|^2,$$

where $p \neq O$ is a brick. We have

$$A_n - A = \int_p |f_n|^2\, d\mu - \int_p |f|^2\, d\mu = \int_p (f_n \bar{f}_n - f\bar{f})\, d\mu =$$

$$= \int_p (f_n \bar{f}_n - f_n \bar{f} + f_n \bar{f} - f\bar{f})\, d\mu = \int_p f_n(\bar{f}_n - \bar{f})\, d\mu + \int_p \bar{f}(f - f)\, d\mu .$$

Hence

(1) $$|A_n - A| \leqq \sqrt{\int_p |f_n|^2\, d\mu} \cdot \sqrt{\int_p |\bar{f}_n - \bar{f}|^2\, d\mu} + \sqrt{\int_p |\bar{f}|^2\, d\mu} \cdot$$

$$\cdot \sqrt{\int_p |f_n - f|^2\, d\mu} \leqq \sqrt{\int_p |\bar{f}_n - \bar{f}|^2\, d\mu} \cdot \left[\sqrt{\int_p |f_n|^2\, d\mu} + \sqrt{\int_p |\bar{f}|^2\, d\mu} \right].$$

We have

$$\mu(p)\,(B_n - B) = \left| \int_p f_n\, d\mu^2 \right| - \left| \int_p f\, d\mu \right|^2$$

$$= \int_p f_n\, d\mu \cdot \int_p \bar{f}_n\, d\mu - \int_p f\, d\mu \cdot \int_p \bar{f}\, d\mu$$

$$= \int_p f_n\, d\mu \cdot \left[\int_p (f_n - f)\, d\mu \right] + \int_p f\, d\mu \cdot \left[\int_p (f_n - f)\, d\mu \right].$$

Hence

$$|B_n - B| \leqq \sqrt{\int_p |f_n|^2\, d\mu} \cdot \sqrt{\int_p |\bar{f}_n - \bar{f}|^2\, d\mu} +$$

$$+ \sqrt{\int_p |\bar{f}|^2\, d\mu} \cdot \sqrt{\int_p |f_n - f|^2\, d\mu} \leqq$$

(2) $$\leqq \left(\sqrt{\int_p |f_n|^2\, d\mu} + \sqrt{\int_p |f|^2\, d\mu} \right) \cdot \sqrt{\int_p |f_n - f|^2\, d\mu} .$$

We have the same estimate in (1) and (2).

Take any measurable set E of traces.

Let $p_\alpha = \{p_{\alpha 1}, p_{\alpha 2}, \ldots\}$, $(\alpha = 1, 2, \ldots)$, be a distinguished sequence for $[E]$. The relation (0) says that

(2.1) $$S_E\, \mu_x\, \mathrm{val}_x |f|^2 = S_E\, \mu_x\, |\mathrm{val}_x f_n|^2 .$$

The left sum is the limit of the sequence

$$M_{n\alpha} = \sum_k \mu(p_{\alpha k}) \cdot \frac{1}{\mu(p_{\alpha k})} \int_{p_{\alpha k}} |f_n|^2\, d\mu = \sum_k \int_{p_{\alpha k}} |f_n|^2\, d\mu \quad \text{for} \quad n \to \infty,$$

while the right one is the limit of the sequence

$$N_{n\alpha}=\sum_k \mu(p_{\alpha k})\cdot\left|\frac{1}{\mu(p_{\alpha k})}\cdot\int\limits_{p_{\alpha k}} f_n\,d\mu\right|^2=\sum_k\frac{1}{\mu(p_{\alpha k})}\left|\int\limits_{p_{\alpha k}} f_n\,d\mu\right|^2 \quad\text{for } n\to\infty.$$

Consider the sums

$$M_\alpha =_{df}\sum_k\int\limits_{p_{\alpha k}}|f|^2\,d\mu,\qquad N=_{df}\sum_k\frac{1}{\mu(p_{\alpha k})}\left|\int\limits_{p_{\alpha k}} f\,d\mu\right|^2.$$

Taking (1) and (2) into account we have

$$\left.\begin{matrix}|M_{n\alpha}-M_\alpha|\\ |N_{n\alpha}-N_\alpha|\end{matrix}\right\}\leqq\sum_k\sqrt{\int\limits_{p_{\alpha k}}|f_n-f|^2\,d\mu}\cdot\left[\sqrt{\int\limits_{p_{\alpha k}}|f_n|^2\,d\mu}+\sqrt{\int\limits_{p_{\alpha k}}|f|^2\,d\mu}\right].$$

Applying once more the Schwarz-Cauchy-lemma, we get

$$\left.\begin{matrix}|M_{n\alpha}-M_\alpha|\\ |N_{n\alpha}-N_\alpha|\end{matrix}\right\}\leqq\sqrt{\sum_k\int\limits_{p_{\alpha k}}|f_n-f|^2\,d\mu}\cdot\sqrt{\sum_k\left(\sqrt{\int\limits_{p_{\alpha k}}|f_n|^2\,d\mu}+\sqrt{\int\limits_{p_{\alpha k}}|f|^2\,d\mu}\right)^2}.$$

Now since, in general, for non negative numbers, x, y, we have

$$(x+y)^2=x^2+2xy+y^2\leqq x^2+y^2+(x^2+y^2)=2(x^2+y^2),$$

we get

$$\left.\begin{matrix}|M_{n\alpha}-M_\alpha|\\ |N_{n\alpha}-N_\alpha|\end{matrix}\right\}\leqq\sqrt{\int\limits_W|f_n-f|^2\,d\mu}\cdot\sqrt{2\left(\int\limits_W|f_n|^2\,d\mu+\int\limits_W|f|^2\,d\mu\right)}.$$

Since the sequence $\left\{\int\limits_W|f_n|^2\,d\mu\right\}$ is bounded, because $\{f_n\}$ is convergent there exists $K>0$, such that

(3) $$\left.\begin{matrix}|M_{n\alpha}-M_\alpha|\\ |N_{n\alpha}-N_\alpha|\end{matrix}\right\}\leqq K\sqrt{\int\limits_W|f_n-f|^2\,d\mu}.$$

We see that this estimate does not depend on α.

Having that, we shall prove that

(4) $$\lim_{n\to\infty}S_E\,\mu_x\,\mathrm{val}_x|f_n|^2=S_E\,\mu_x\,\mathrm{val}_x|f|^2.$$

The sums in (4) exist. We have

(4.1) $$\lim_{n\to\infty}M_{n\alpha}=S_E\,\mu_x\,\mathrm{val}_x|f_n|^2\quad\text{and}\quad\lim_{n\to\infty}M_\alpha=S_E\,\mu_x\,\mathrm{val}_x|f|^2.$$

Hence

$$\lim_{\alpha\to\infty}|M_{n\alpha}-M_\alpha|=|S_E\,\mu_x\,\mathrm{val}_x|f_n|^2-S_E\,\mu_x\,\mathrm{val}_x|f|^2|.$$

By (3) we get, when $\alpha\to\infty$,

$$|S_E\,\mu_x\,\mathrm{val}_x|f_n|^2-S_E\,\mu_x\,\mathrm{val}_x|f|^2|\leqq K\sqrt{\int\limits_W|f_n-f|^2\,d\mu}.$$

Hence, for $n \to \infty$ we get

$$\lim_{n\to\infty} |S_E\, \mu_x \operatorname{val}_x |f_n|^2 - S_E\, \mu_x \operatorname{val}_x |f|^2| = 0,$$

which gives

$$\lim S_E\, \mu_x \operatorname{val}_x |f_n|^2 = S_E\, \mu_x \operatorname{val}_x |f|^2;$$

so (4) is proved.

Now we shall prove that $\lim_{\alpha\to\infty} N_\alpha$ exists. Suppose it does not. We can find $\delta > 0$ and an infinite sequence of indices $\alpha_1' < \alpha_2'' < \alpha_1' < \alpha_2'' < \cdots$, such that $|N_{\alpha_m'} - N_{\alpha_m''}| \geqq \delta$ for all $m = 1, 2, \ldots$ Now we have in general:

$$|N_{n\alpha} - N_{n\beta}| \geqq |N_\alpha - N_\beta| - |(N_{n\alpha} - N_\alpha) + (N_\beta - N_{n\beta})| \geqq$$
$$\geqq |N_\alpha - N_\beta| - |N_{n\alpha} - N_\alpha| - |N_\beta - N_{n\beta}|.$$

Hence

$$|N_{n,\alpha_m'} - N_{n,\alpha_m''}| \geqq |N_{\alpha_m'} - N_{\alpha_m''}| - |N_{n,\alpha_m'} - N_{\alpha_m'}| -$$
$$- |N_{n,\alpha_n''} - N_{\alpha_m''}| \geqq \text{, by (3)}, \geqq \delta - 2K \sqrt{\int_W |f_n - f|^2\, d\mu}\,.$$

If we choose n_0, so as to have

$$2K \sqrt{\int_W |f_n - f|^2\, d\mu} < \frac{\delta}{2},$$

we shall get

$$N_{n_0,\alpha_m'} - N_{n_0,\alpha_m''}| \geqq \frac{\delta}{2}$$

for all m. Consequently the sequence

$$N_{n_0,\alpha_1'},\ N_{n_0,\alpha_1''},\ N_{n_0,\alpha_2'},\ N_{n_0,\alpha_2''},\ \ldots$$

does not converge. Hence $\lim_{\alpha\to\infty} N_{n_0,\alpha}$ does not exist, which contradicts the fact that

$$\lim_{\alpha\to\infty} N_{n_0,\alpha} = S_E\, \mu_x |\operatorname{val}_x f_n|^2.$$

Thus we have proved that the sum $S_E\, \mu_x |\operatorname{val}_x f|^2$ exists; and we have

(5) $$\lim_{\alpha\to\infty} N_\alpha = S_E\, \mu_x |\operatorname{val}_x f|^2;$$

we also have

(6) $$\lim_{\alpha\to\infty} M_\alpha = S_E\, \mu_x \operatorname{val}_x |f|^2.$$

Having that, consider the following circumstances:

$$\lim_{\alpha\to\infty} N_{n\alpha} = S_E\, \mu_x |\operatorname{val}_x f_n|^2, \qquad \lim_{\alpha\to\infty} N_\alpha = S_E\, \mu_x |\operatorname{val}_x f|^2,$$

$$|N_{n\alpha} - N_\alpha| \leqq K \sqrt{\int_W |f_n - f|^2\, d\mu}\,.$$

It follows that

$$|S_E\,\mu_x|\,\mathrm{val}_x f_n|^2 - S_E\,\mu_x|\,\mathrm{val}_x f|^2\,| \leqq K\sqrt{\int_W |f_n - f|^2\,d\mu}\,.$$

Consequently, for $n \to \infty$ we get

$$\lim_{n\to\infty} S_E\,\mu_x\,|\mathrm{val}_x f_n|^2 = S_E\,\mu_x|\,\mathrm{val}_x f|^2.$$

On the other hand we have by (4),

$$\lim_{n\to\infty} S_E\,\mu_x\,\mathrm{val}_x|f_n|^2 = S_E\,\mu_x\,\mathrm{val}_x|f|^2.$$

If we take account of (2.1), i.e.

$$S_E\,\mu_x\,|\mathrm{val}_x f_n|^2 = S_E\,\mu_x\,\mathrm{val}_x|f_n|^2,$$

we get, when $n \to \infty$:

$$S_E\,\mu_x\,\mathrm{val}_x|f|^2 = S_E\,\mu_x\,|\mathrm{val}_x f|^2, \quad \text{i.e.,} \quad \mu_x\,\mathrm{val}_x|f|^2 \approx \mu_x\,|\mathrm{val}_x f|^2.$$

R.15h. The items [R.15a.] to [R.15g.] imply the Theorem [R.15.], so it is proved.

R.16. Theorem. If $f(\dot{x})$, $g(\dot{x})$ are μ-square-summable complex number-valued functions on W, then

$$\mu_x\,\mathrm{val}_x(f\cdot g) \approx \mu_x\,\mathrm{val}_x f\cdot\mathrm{val}_x g.$$

Proof. We recall the formula for numbers

$$\bar{a}\,b = \frac{1}{2}\,\overline{(a+b)}\,(a+b) - \frac{i}{2}\,\overline{(a+b\,i)}\,(a+b\,i) + \frac{i-1}{2}\,\bar{a}\,a + \frac{i-1}{2}\,\bar{b}\,b.$$

Hence we have

$$\bar{f}\,g = \frac{1}{2}\,\overline{(f+g)}\,(f+g) - \frac{i}{2}\,\overline{(f+i\,g)}\,(f+i\,g) + \frac{i-1}{2}\,\bar{f}\,f + \frac{i-1}{2}\,\bar{g}\,g.$$

Hence

$$(1)\quad \int_P \bar{f}\,g\,d\mu = \frac{1}{2}\int_P |f+g|^2\,d\mu - \frac{i}{2}\int_P |f+i\,g|^2\,d\mu + \\ + \frac{i-1}{2}\int_P |f|^2\,d\mu + \frac{i-1}{2}\int_P |g|^2\,d\mu.$$

Putting

$$a =_{df} \int_P f\,d\mu, \qquad b =_{df} \int_P g\,d\mu,$$

we also have

$$(2)\quad \overline{\int_P f\,d\mu}\cdot\int_P g\,d\mu = \frac{1}{2}\left|\int_P f\,d\mu + \int_P g\,d\mu\right|^2 - \\ - \frac{i}{2}\left|\int_P f\,d\mu + i\int_P g\,d\mu\right|^2 + \frac{i-1}{2}\left|\int_P f\,d\mu\right|^2 + \frac{i-1}{2}\left|\int_P g\,d\mu\right|^2.$$

From (1) it follows

$$(3)\quad \mathrm{val}_x(\bar{f}g) = \frac{1}{2}\mathrm{val}_x|f|^2 - \frac{i}{2}\mathrm{val}_x|f+ig|^2 + \frac{i-1}{2}\mathrm{val}_x|f|^2 + \\ + \frac{i-1}{2}\mathrm{val}_x|g|^2,$$

and from (2) we get

$$(4)\quad \mathrm{val}_x f \cdot \mathrm{val}_x g = \frac{1}{2}|\mathrm{val}_x(f+g)|^2 - \frac{i}{2}|\mathrm{val}_x(f+ig)|^2 + \\ + \frac{i-1}{2}\mathrm{val}_x|f|^2 + \frac{i-1}{2}\mathrm{val}_x|g|^2.$$

By theorem [R.12.5.] we have

$$\mu_x \mathrm{val}_x|f|^2 \approx \mu_x|\mathrm{val}_x f|^2,$$

$$\mu_x \mathrm{val}_x|g|^2 \approx \mu_x|\mathrm{val}_x g|^2,$$

$$\mu_x \mathrm{val}_x|f+g|^2 \approx \mu_x|\mathrm{val}_x(f+g)|^2,$$

$$\mu_x \mathrm{val}_x|f+ig|^2 \approx \mu_x|\mathrm{val}_x(f+ig)|^2.$$

Hence, by (3) and (4) we get

$$\mu_x \mathrm{val}_x(\bar{f}g) \approx \mu_x \mathrm{val}_x\bar{f}\cdot\mathrm{val}_x g.$$

Since $\bar{f}$, g are arbitrary square-summable functions, the theorem is proved.

Chapter R I

Summation of quasi-vectors in the separable and complete Hilbert-Hermite space

R I.1. In this Chapter [R I.] we shall apply the theories, developped in the preceding ones, to tribes of (closed) subspaces in a separable and complete Hilbert-Hermite-spaces **H**.

Let G be a denumerably additive tribe of subspaces of **H** with **H** as unit I, and with the space composed of the single vector $\vec{O}$ as zero O, [D.], [D I.], [E.]. The ordering is the set-inclusion of spaces ($\subseteq$), denoted by ($\leqq$), and the complement is ortho-complement, [D.2.2c.]. The relation $E \cdot F = O$ of spaces implies $E \perp F$. The somata of (G) are compatible with one another, [D.5.]. The tribe (G) is geometrical, [D I.1.2f.].

Let (F) be a (finitely) additive tribe of spaces, a finitely genuine strict subtribe of (G), [A.7.2.]. Let B be a base of (F), [B.1.1.], satisfying the condition $(\boldsymbol{Hyp\ Ad})$, [B.2.], and the conditions $\boldsymbol{Hyp}$ **I** and $\boldsymbol{Hyp}$ **II**, [B I.6.].

We suppose that $(\boldsymbol{G})$ is the smallest denumerably additive supertribe of $(\boldsymbol{F})$. There always exists an effective, denumerably additive measure $\mu(E)$ on $(\boldsymbol{G})$, $(\mu(E) \geqq 0)$, [D I.18.8.]. $(\boldsymbol{G})$ is not only the borelian extension of $(\boldsymbol{F})$, [A I., Sect. 4], but it coincides withe μ-Lebesgue's covering extension of $(\boldsymbol{F})$, [A I., Sect. 6].

Thus we are in the conditions $\boldsymbol{Hyp\ R_e}$, [Q.14.3.1.], and also in that of [Q I.1.1.] because **H** is a complex Banach-space.

The μ-topology on $(\boldsymbol{G})$ is separable, (10), so the condition $(\boldsymbol{Hyp\ S})$, [Q.21.1.], is satisfied.

Having this all, we can apply the theory of measurability of sets of traces to space-traces [B I.], [E.], and use all kinds of summation of quasi-vectors, [R.].

R I.2. This general situation becomes specialized when dealing with normal operators in **H**. The notions $\boldsymbol{F}$, $\boldsymbol{B}$, $\boldsymbol{G}$, measure, traces, fields of vectors, quasi-vectors and their summation are built up by means of analogous notions of $\boldsymbol{F}'$, $\boldsymbol{B}'$, $\boldsymbol{G}'$ etc. on the euclidean plane P through a homomorphism, preserving the measure and with validity of $(\boldsymbol{Hyp.\ Af})$ and the conditions $\boldsymbol{Hyp}$ **I** and $\boldsymbol{Hyp}$ **II**, so that the theory of measurability of sets of traces are analogous. The circumstances in **H** are more simple, but those on P are more visual.

By the *rectangle* (*generalized*),

$$R(\alpha_1, \alpha_2; \beta_1, \beta_2),$$

[C.], where $-\infty \leqq \alpha_1, \alpha_2 \leqq +\infty$, $\quad -\infty \leqq \beta_1, \beta_2 \leqq +\infty$, we understand the set of points (x, y) (complex numbers),

$$\{x, y \mid \alpha_1 < x \leqq \alpha_2, \beta_1 < y \leqq \beta_2\}.$$

They are termed *bricks*.

The somata of $\boldsymbol{F}'$ are finite unions of bricks and $\boldsymbol{G}'$ is the tribe of all borelian sets in P; the ordering is the inclusion $\subseteqq$ of sets. The measure μ on $\boldsymbol{F}'$ and $\boldsymbol{G}'$ is transplanted from **H** onto P, and taken by saturation and by means of a generating vector in the space.

R I.2.2. Concerning $\boldsymbol{B}'$-traces in $\boldsymbol{F}'$, (see Chapter C.), we have the following situation: If (x, y) is a point on the plane, then there are four different traces attached to it and with representatives

$$R\left(x-\frac{1}{n}, x; y-\frac{1}{n}, y\right), \quad R\left(x-\frac{1}{n}, x; y, y+\frac{1}{n}\right),$$
$$R\left(x, x+\frac{1}{n}; y-\frac{1}{n}, y\right), \quad R\left(x, x+\frac{1}{n}; y, y+\frac{1}{n}\right), \quad (n = 1, 2, \ldots)$$

respectively. The point (x, y) has been termed *vertex of these traces*, [C.]. In addition to that there are eight "side-traces" at infinity, with representatives, e.g. $R\left(-\infty, -n; y - \frac{1}{n}, y\right)$, $(n = 1, 2, \ldots)$ and four "corner-traces" at infinity, with representatives, e.g. $R(-\infty, -n; n, +\infty)$. These are all existing traces. This has been proved by considering two-valued measures on $\boldsymbol{F}'$, (35), [B.].

R I.2.3. If we consider $\boldsymbol{F}'$ modulo a suitable ideal, the traces will be essentially the same, though some ones may not exist, because some rectangles, seemingly eligible for yielding them, may belong to the ideal, [C.].

R I.2.4. We have the theorem: If

1. $\mu(f)$ is a finitely additive (and finite), non negative measure on $\boldsymbol{F}'$,

2. $\boldsymbol{G}'$ is the collection of all borelian subsets of the plane P, then the following are equivalent:

I. the measure μ can be extended over $\boldsymbol{G}'$, so as to obtain a denumerably additive measure on $\boldsymbol{G}'$, [E.].

II. 1^0. If

$$\alpha_1 \geqq \alpha_2 \geqq \cdots \geqq \alpha_n \geqq \cdots$$

and tends to α_0,

$$\beta_1 \geqq \beta_2 \geqq \cdots \geqq \beta_n \geqq \cdots$$

tends to β_0,

$$(-\infty \leqq \alpha_n, \alpha_0, \beta_n, \beta_0 \leqq +\infty), \quad (n = 1, 2, \ldots),$$

then

$$\mu R(-\infty, \alpha_n; -\infty, \beta_n) \to \mu R(-\infty, \alpha_0; -\infty, \beta_0).$$

2^0. If

$$\alpha_1 \leqq \alpha_2 \leqq \cdots \to +\infty, \quad \beta_1 \leqq \beta_2 \leqq \cdots \to +\infty,$$

then

$$\mu R(-\infty, \alpha_n; -\infty, \beta_n) \to \mu(P).$$

The proof relies essentially on the fact, that the plane P is locally compact.

R I.2.5. Put, in general,

$$Q(\alpha, \beta) =_{df} R(-\infty, \alpha; -\infty, \beta) \quad \text{for} \quad -\infty \leqq \alpha, \beta \leqq +\infty.$$

We have called these rectangles: *plane-quarters*, [E.]. Let s be a correspondence, which attaches to every plane quarter Q a closed subspace $s(Q)$ of $\mathbf{H}$ with conditions:

I^0. If Q_1, Q_2 are two plane-quarters, the $s(Q_1)$, $s(Q_2)$ are compatible,

II^0. $s(\emptyset) = (\vec{O}) = O$, $s(P) = \mathbf{H} = 1$,

III^0. if $Q_1 \subseteqq Q_2$, then $s(Q_1) \leqq s(Q_2)$.

This correspondence can be extended in a unique way to another one, also denoted by s, which attaches to every figure $f \in \boldsymbol{F}'$ a space $s(f)$, the resulting correspondence having the properties:

$$s(f \cap g) = s(f) \cdot s(g), \quad s(f \cup g) = s(f) + s(g).$$

s is a homomorphism from $\boldsymbol{F}'$ onto a tribe $\boldsymbol{F}$ of spaces. The set of all figures f, for which $s(f) = O$, is an ideal in $\boldsymbol{F}'$.

R I.2.6. Now, in order that s could be extended to all borelian sets of the plane, the following conditions are necessary and sufficient:

1) If

$$\alpha_1 \geqq \alpha_2 \geqq \cdots \to \alpha_0, \quad \beta_1 \geqq \beta_2 \geqq \cdots \to \beta_0,$$

then

$$\prod_{n=1}^{\infty} s(Q(\alpha_n \beta_n) = s(Q(\alpha_0 \beta_0));$$

2) If

$$\alpha_1 \leqq \alpha_2 \leqq \cdots \to +\infty, \quad \beta_1 \leqq \beta_2 \leqq \quad \to +\infty,$$

then

$$\sum_{n=1} \cdot s(Q(\alpha_n \beta_n)) = 1 = \mathbf{H}.$$

Let us remark that this situation is present, if we consider a normal maximal operator in $\mathbf{H}$ and consider its spectral scale. Usually in the spectral theory projectors are used. We prefer to consider the spaces themselves, rather than the corresponding projectors (48), (41), (40). The extended correspondence s yields a denumerably additive tribe $(\boldsymbol{G}) = s(\boldsymbol{G}')$ of spaces. The tribe $\boldsymbol{F} = s(\boldsymbol{F}')$ is its finitely genuine strict subtribe, and the s-correspondants of rectangles on P constitute a *base* $\boldsymbol{B}$ *for* $\boldsymbol{F}$.

R I.2.7. The s-images of rectangles R on the plane are closed subspaces in $\mathbf{H}$, and they constitute a base of $\boldsymbol{F}$, so $s(R)$ has been called *bricks of* $\boldsymbol{F}$ ("space-rectangle-bricks"). We have defined space-traces as in the general theory [B I.] of traces—by means of these bricks. Let us remark that, it is not true that to every $\boldsymbol{B}'$-trace in $\boldsymbol{F}'$ there corresponds through s a space-trace; indeed, if $a_1' \geqq a_2' \geqq \cdots$ is a representative of a trace x on the plane, the spaces $s(a_n')$ may be all $= O$, so they do not yield any space trace.

Now, in the above construction we have proved (considering circumstances on the plane), that for the space-traces the hypotheses I and II hold true, so we can apply the theory of measurability of sets of space-traces.

R I.2.8. Remark. A similar construction of the tribes of spaces with a base $\boldsymbol{B}$ can be obtained, if we consider the straight line instead of the plane, and instead of rectangles, we consider half open intervals. This construction is related to the spectral scale of a selfadjoint operator.

The construction by means of half closed arcs on the unit circle will correspond to the spectral theory of unitary operators.

R I.2.9. The tribe ($\boldsymbol{G}$) may be saturated or not. Whatever will be the case, it admits an effective denumerably additive measure, [E.]. Any tribe of spaces can be saturated by a suitable adjunction of spaces.

R I.3. In the sequal we shall pay special attention to tribes of spaces, obtained through the above rectangle or half-open segment-construction, but in the general discussion which we shall now start, we shall consider the general situation.

R I.4. Let us admit that the tribe $\boldsymbol{G}$ of spaces is saturated. Then there exists an isomorphic mapping of the space **H** onto the space of some measurable, complex-valued fonctions of the variable trace. This mapping $\mathcal{G}^{-1}$ is defined in [E.], by means of a generating vector $\vec{\omega}$ of **H** with respect to $\boldsymbol{G}$. The measure μ is defined by

$$\mu(a) =_{df} \|\mathrm{Proj}_a \vec{\omega}\|^2 \quad \text{for all} \quad a \in \boldsymbol{G}.$$

This measure is denumerably additive, non negative and effective. It induces a denumerably additive measure for all measurable sets of traces, also denoted by μ. We shall consider μ-square-summable functions of the variable trace τ, defined almost μ-everywhere on the set W of all traces. These functions will be considered modulo μ-null-sets.

R I.5. Def. Under circumstances set in [R I.1.], a quasi-vector $\vec{\varphi}(\dot{p})$ with support x will be termed *normal*, whenever

$$\vec{\varphi}(p) \leqq p \quad \text{for all} \quad p \in v(x).$$

R I.5.1. Theorem. Under circumstances [R I.1.] if

1. A is a measurable set of traces,
2. $\vec{f}_x$ is a set of quasi-vectors, summable on A,
3. the set is normal,

then the set of quasi-vectors $\|\vec{f}_x\|^2$ is also summable on A, and

$$S_A \|\vec{f}_x\|^2 = \|S_A \vec{f}_x\|^2.$$

Proof. Take any choice of vector-fields, say $\vec{f}(\dot{p})$, generated by $\vec{f}_x$. Let $P_n = \{p_{n1}, p_{n2}, \ldots\}$, $(n = 1, 2, \ldots)$, be any sequence of distinguished complexes for [A.]. We have

$$\vec{\varphi}_n =_{df} \sum_i \vec{f}(p_{ni}) \to S_A \vec{f}_x \tag{1}$$

in the **H**-topology. Now, since the spaces $p_{n1}, p_{n2}, \ldots$ are mutually orthogonal and since $\vec{f}(p_{ni}) \in p_{ni}$, we have

$$\|\vec{\varphi}_n\|^2 = \sum_i \|\vec{f}(p_{ni})\|^2. \tag{2}$$

From (1) it follows that

$$\|\vec{\varphi}_n\|^2 \to S_A \vec{f}_x\|^2. \tag{3}$$

Now, from (2) it follows, that the sequence

$$\{\sum_i ||\vec{f}(p_{ni})||^2\}, \quad (n = 1, 2, \ldots)$$

possesses a limit. The limit does not depend on the choice of the selected field $\vec{f}(\dot{p})$. Hence this limit is $S_A ||\vec{f}_x||^2$. Consequently

$$S_A ||\vec{f}_x||^2 = ||S_A \vec{f}_x||^2. \quad \text{q.e.d.}$$

R I.6. Theorem. Under circumstances [R I.1.] if

1. $\vec{f}_x$ is a total set of quasi-vectors, (D')-summable on W,
2. the set is normal Def. [R I.5.],

then there exists $M > 0$, such that

(1) $$||S_A \vec{f}_x||^2 = S_A ||\vec{f}_x||^2 < M$$

for all measurable sets A of traces.

Proof. By Theorem [R I.5.1.], the sums (1) exist and we have

$$J(A) =_{df} ||S_A \vec{f}_x||^2 = S_A ||\vec{f}_x||^2.$$

Now, the function $J(\dot{A})$ of the variable measurable set A is denumerably additive. Since $J(A) \geqq 0$, it follows, by the known theorem on denumerably additive non negative measures on a denumerably additive tribe, that $J(A)$ is bounded, so the theorem is proved.

R I.7. Theorem. Under circumstances [R I.1.]. If

1. $\vec{f}_x$ and $\vec{g}_x$ are sets of quasi-vectors summable on A,
2. A is a measurable set of traces,
3. the sets $\vec{f}_x$ and $\vec{g}_x$ are normal, Def. [R I.5.],

then the set of quasi-numbers $(\vec{f}_x, \vec{g}_x)$, is also summable on A, and we have $$S_A(\vec{f}_x, \vec{g}_x) = (S_A \vec{f}_x, S_A \vec{g}_x) \quad \text{(scalar products)}.$$

Proof. Let us choose a selection of vector-fields, one for $\vec{f}_x$ another for $\vec{g}_x$. Denote them by $\vec{f}(\dot{p})$, $\vec{g}(\dot{p})$ respectively. They belong both to the same choice $x = \alpha(p)$ of traces, covered by p. (Indeed, we can operate only on quasi-vectors having the same support.)

Let $P_n = \{p_{n1}, p_{n2}, \ldots\}$ be a distinguished sequence of complexes for [A.]. Since $p_{n1}, p_{n2}, \ldots$ are mutually orthogonal spaces, and since

$$\vec{f}(p_{nk}) \in p_{nk}, \quad \vec{g}(p_{nk}) \in p_{nk},$$

it follows, that

(1) $$\left(\sum_k \vec{f}(p_{nk}), \sum_k \vec{g}(p_{nk})\right) = \sum_k \left(\vec{f}(p_{nk}), \vec{g}(p_{nk})\right).$$

We shall prove the existence of the sum $S_A(\vec{f}_x, \vec{g}_x)$. We have

(2) $$(\vec{f}_x, \vec{g}_x) = \frac{1}{2}(\vec{f}_x + \vec{g}_x, \vec{f}_x + \vec{g}_x) - \frac{i}{2}(\vec{f}_x + i\,\vec{g}_x, \vec{f}_x + i\,\vec{g}_x) + \\ + \frac{i-1}{2}(\vec{f}_x, \vec{f}_x) + \frac{i-1}{2}(g_x, g_x).$$

The sums $\vec{f}_x + \vec{g}_x, \vec{f}_x + i\,\vec{g}_x$ of quasi-vectors exist. Hence the sums

$$S_A\,||\vec{f}_x + \vec{g}_x||^2, \quad S_A\,||\vec{f}_x + i\,\vec{g}_x||^2, \quad S_A\,||\vec{f}_x||^2, \quad S_A\,||\vec{g}_x||^2$$

also exist, and then, the (D')-sum

$$S_A(\vec{f}_x, \vec{g}_x)$$

exists. We have

$$\sum_k (\vec{f}(p_{nk}), \vec{g}(p_{nk})) \to S_A(\vec{f}_x, \vec{g}_x).$$

Since

$$\sum_k \vec{f}(p_{nk}) \to S_A\,\vec{f}_x, \quad \text{and} \quad \sum_k \vec{g}(p_{nk}) \to S_A\,\vec{g}_x,$$

it follows from (1), that

$$S_A\,(\vec{f}_x, \vec{g}_x) = (S_A\,\vec{f}_x, S_A\,\vec{g}_x), \quad \text{q.e.d.}$$

R I.8. Under circumstances [R I.1.], if $\vec{f}_x, \vec{g}_x$ are quasi-vectors sets, both (D')-summable on a measurable set A, and they are normal, then

$$|S_A\,(\vec{f}_x, \vec{g}_x)| \leqq ||S_A\,\vec{f}_x|| \cdot ||S_A\,\vec{g}_x||.$$

Proof. We have

$$S_A(\vec{f}_x, \vec{g}_x) = (S_A\,\vec{f}_x, S_A\,\vec{g}_x),$$

and the Cauchy-Schwarz inequality completes the proof.

R I.9. Under circumstances [R I.1.], if

1. $\vec{f}_x, \vec{g}_x$ are quasi-vector sets, both (D')-summable on a measurable set A of traces,

2. $\vec{f}_x, \vec{g}_x$ are normal,

then

$$||S_A(\vec{f}_x + \vec{g}_x)|| \leqq ||S_A\,\vec{f}_x|| + ||S_A\,\vec{g}_x||.$$

Proof. We have

$$S_A(f_x + g_x) = S_A\,f_x + S_A\,g_x,$$

where we have dropped the arrows for the sake of simplicity. Hence

$$\begin{aligned} ||S_A\,(f_x + g_x)||^2 &= (S_A\,f_x + S_A\,g_x, S_A\,f_x + S_A\,g_x) \\ &= ||S_A\,f_x||^2 + (S_A\,f_x, S_A\,g_x) + (S_A\,g_x, S_A\,f_x) + ||S_A\,g_x||^2. \end{aligned}$$

Hence

$$||S_A(f_x + g_x)||^2 \leqq ||S_A\,f_x||^2 + 2|\,(S_A\,f_x, S_A\,g_x)| + ||S_A\,g_x||^2.$$

Taking [R I.8.] into account, we get

$$||S_A(f_x + g_x)||^2 \leqq (||S_A\,f_x|| + ||S_A\,g_x||)^2,$$

which completes the proof.

R I.10. Lemma. If

1. the sequence $a_1, a_2, \ldots, a_n, \ldots \in \boldsymbol{G}$ of spaces μ-tends to a,
2. $\vec{\xi}_n \in a_n$,
3. $\lim \vec{\xi}_n = \vec{\xi}$,

then

$$\vec{\xi} \in a.$$

R I.10a. Proof. Since $\lim^{\mu} a_n = a$, we can extract from $\{a_n\}$ a partial sequence $a_{k(n)}$, such that $a = \prod_{s=1}^{\infty} b_s$, where $b_s =_{df} \sum_{n=s}^{\infty} a_{k(n)}$. We have $b_s \geqq b_{s+1}$. Consider the sequence $\{\vec{\xi}_{k(n)}\}$. We have $\vec{\xi}_{k(s)} \in a_{k(s)}$; hence $\vec{\xi}_{k(s)} \in b_s$ for $s = 1, 2, \ldots$ We have

(2.1) $$\vec{\xi}_{j(s)} = \mathrm{Proj}_a \vec{\xi}_{k(s)} + \mathrm{Proj}_{\mathrm{co}\,a} \vec{\xi}_{k(s)}.$$

Since, by (1),

$$\mathrm{Proj}_{\mathrm{co}\,a} \vec{\xi}_{k(s)} = \mathrm{Proj}_{\mathrm{co}\,a} \mathrm{Proj}_{b_s} \vec{\xi}_{k(s)},$$

we have

$$\vec{\eta} =_{df} \mathrm{Proj}_{\mathrm{co}\,a} \vec{\xi}_{k(s)} = \mathrm{Proj}_{b_s - a} \vec{\xi}_{k(s)} \in b_s - a.$$

Since projecting is a continuous operation, we have

$$\lim_{s\to\infty} \mathrm{Proj}_a \vec{\xi}_{k(s)} = \mathrm{Proj}_a [\lim_{s\to\infty} \vec{\xi}_{k(s)}];$$

hence

$$\lim_{s\to\infty} \mathrm{Proj}_a \vec{\xi}_{k(s)} = \mathrm{Proj}_a \vec{\xi}.$$

From (2) we get, by the passage to limit:

(3) $$\vec{\xi} = \mathrm{Proj}_a \vec{\xi} + \lim \vec{\eta}_s, \quad \text{where} \quad \vec{\eta}_s \in b_s - a.$$

Hence

(3.1) $$\lim \vec{\eta}_s$$

exists. If we put

$$c_s =_{df} b_s - a,$$

we have

$$c_s \geqq c_{s+1}, \quad \text{and} \quad \prod_{s=1}^{\infty} c_s = O, \quad \vec{\eta}_s \in c_s.$$

To prove the Lemma, it is sufficient to prove, that $\lim_{s\to\infty} \vec{\eta}_s \to \vec{O}$. (This because of (3).)

R I.10b. Supposing that this be not true, we have, by (3.1)

$$\vec{\eta} =_{df} \lim_{s\to\infty} \vec{\eta}_s \neq \vec{O}.$$

To disprove that supposition we shall apply the representation in [R I.4.], which is valid for saturated tribes. Now, our tribe $\boldsymbol{G}$ may be not saturated, but it can always be extended (by adjunction of an at most denumerable number of spaces), so as to get a saturated tribe $\boldsymbol{G}_1$. Take an effective measure μ_1 on $\boldsymbol{G}_1$. The topology on $\boldsymbol{G}$ generated by μ_1 will coincide with that one, generated by μ. Having this, we can operate in $\boldsymbol{G}_1$ instead of $\boldsymbol{G}$.

Let us consider the μ-square summable functions of the variable trace x: $H(x)$, $H_s(x)$, which are images of $\vec{\eta}$ and $\vec{\eta}_s$ respectively. Let $E_1, E_2, \ldots, E_s, \ldots$ be μ-measurable sets of traces with supports $c_1, c_2, \ldots, c_s, \ldots$ respectively. We can admit that

$$E_1 \supseteqq E_2 \supseteqq \cdots \supseteqq E_s \supseteqq \cdots$$

(because if not, we can replace E_s by $E_s' =_{df} E_1 \; E_2 \ldots E_s$, whose support is $c_1 \; c_2 \ldots c_s = c_s$). We have

$$\mu(E_n) = \mu(c_n) \to 0. \tag{3.2}$$

The functions $H_s(x)$ can be chosen so as to have

$$H_s(x) = 0 \quad \text{for} \quad x \in \operatorname{co} E_s. \tag{4}$$

Indeed, we have $\vec{\eta}_s \in c_s$; hence $\operatorname{Proj}_{\operatorname{co} c_n} \vec{\eta} = \vec{O}$. Since $\vec{\eta} \neq \vec{O}$, there exists a set E of positive measure such that $H(x) \neq 0$.

Hence, there exists a measurable set F and $\alpha > 0$, such that $\mu(F) > 0$ and $|H(x)| \geqq \alpha$ for all $x \in F$. We have $|H_s(x) - H(s)|^2 \, dx \to 0$.

Since $H_s(x)$ tends in μ-square means to $H(x)$, there exists a partial sequence $H_{l(s)}(x)$, which tends almost μ-everywhere to $H(x)$. By the theorem of EGOROFF, given $\varepsilon > 0$, there exists a set G, such that $\mu(G) > 1 - \varepsilon$, and where $H_{l(s)}$ converges uniformly to $H(x)$.

Hence, if $\varepsilon > 0$ is sufficiently small, we can find a subset F' of F with $\mu(F') > 0$ where $H_{l(s)}(x)$ converges uniformly to $H(x)$. Hence for n sufficiently great,

$$|H_{l(s)}(x)| \geqq \frac{\alpha}{2} \quad \text{on } F' \text{ for} \quad s = 1, 2, \ldots \tag{5}$$

But, by (4),

$$H_s(x) = 0 \quad \text{for} \quad x \in \operatorname{co} E_s;$$

hence

$$\mu\{x \mid H_s(x) \neq 0\} = \mu E_s \to 0, \tag{6}$$

by (3.2). Since

$$\{x \mid H_s(x) \neq 0\} \geqq F' \quad \text{for all} \quad s = 1, 2, \ldots,$$

we have

$$\mu\{x \mid H_s(x) \neq 0\} \geqq \mu F' > 0.$$

But $\mu F' = 0$ by (6), which is a contradiction. The lemma is established.

R I.11. Theorem. Under circumstances [R I.1.] if
1. $\vec{f}_x$ is a set of quasi-vectors in H with support A,
2. $\vec{f}_x$ is normal,
3. A is a measurable set of traces,
4. $\vec{f}_x$ is (D')-summable on A,

then $S_A \vec{f}_x \in [A]$ where $[A]$ is the coat of A.

Proof. Let $\{P_n\}$ be a distinguished sequence of complexes for $[A]$. Let

$$P_n = \{p_{n1}, p_{n2}, \ldots\}.$$

We have $\vec{f}(P_n) \to S_A \vec{f}_x$, and $|P_n, A|_\mu \to 0$, i.e.

$$(1) \qquad \lim \operatorname{som} P_n = [A].$$

Now $\vec{f}(p_{ni}) \in p_{ni}$; hence $\vec{f}(P_n) \in \operatorname{som} P_n$. By Lemma [R I.9.], we get

$$\lim \vec{f}(P_n) = [A].$$

Hence

$$S_A \vec{f}_x \in [A], \quad \text{q.e.d.}$$

R I.12. Theorem. Under circumstances [R I.1.], if
1. $\vec{f}_x$ is a total set of quasi-vectors in H,
2. $\vec{f}_x$ is normal,
3. $\vec{f}_x$ is summable on W,
4. A is a summable set of traces,

then

$$S_A \vec{f}_x = \operatorname{Proj}_{[A]} S_W \vec{f}_x.$$

Proof. Putting $\operatorname{co} A =_{df} W - A$, we have $A \cup (\operatorname{co} A) = W$; $\operatorname{co} A$ is a measurable set of traces.

We have

$$(1) \qquad S_W \vec{f}_x = S_A \vec{f}_x + S_{\operatorname{co} A} \vec{f}_x.$$

Now, by [R I.11.] we have

$$S_A \vec{f}_x \in [A], \qquad S_{\operatorname{co} A} \vec{f}_x \in \operatorname{co}[A].$$

Hence $S_A \vec{f}_x$ is orthogonal to $S_{\operatorname{co} A} \vec{f}_x$. Taking in (1) the projection on the space $[A]$, we get

$$\operatorname{Proj}_{[A]} S_W \vec{f}_x = \operatorname{Proj}_{[A]} S_A \vec{f}_x + \operatorname{Proj}_{[A]} S_{\operatorname{co} A} \vec{f}_x.$$

The second term is the zero-vector. Since $S_A \vec{f}_x \in [A]$ we get

$$\operatorname{Proj}_{[A]} S_W \vec{f}_x = S_A \vec{f}_x, \quad \text{q.e.d.}$$

Chapter S

General orthogonal systems of coordinates in the separable and complete Hilbert-Hermite-space

One of the advantages of the theory of summation, [R I.], is the possibility of defining a system of coordinates (in H.-H.-space), which is more general, than the usual one, composed of a saturated system of mutually orthogonal and normed vectors $\{\vec{\varphi}_n\}$, $n = 1, 2, \ldots$ The usual, just mentioned system of coordinates, is well adapted to the discontinuous spectrum of a normal operator, but it does not work well in the case of a continuous spectrum. The theory, which we are going to sketch, is completely general: it covers in a unified way both possibilities of spectrum.

The background of that theory is a saturated geometrical tribe of spaces and a generating vector [D I.]. Quasi-vectors and quasi-numbers will be applied in relation to space-traces. We start with hypotheses admitted in the coming discussion.

S.1. We admit the hypotheses of [R I.1.]. Thus $(\boldsymbol{F})$, $(\boldsymbol{G})$ are tribes of subspaces of $\boldsymbol{H}$, and $\boldsymbol{B}$ is a base of $(\boldsymbol{F})$, satisfying all conditions required for the theory of measurability of sets of traces and for applying the (DARS)-summations of sets of quasi-vectors.

S.1.1. We suppose that the tribe $\boldsymbol{G}$ is saturated, and we select a generating vector $\vec{\omega}$ of $\boldsymbol{H}$ with respect to $\boldsymbol{G}$. The effective measure on $\boldsymbol{G}$ will be defined by

$$\mu(a) = \|\operatorname{Proj}_a \vec{\omega}\|^2.$$

S.1.2. Def. The set $\boldsymbol{B}$ (yields by extension through $(\boldsymbol{F})$), a saturated tribe $(\boldsymbol{G})$, and $\vec{\omega}$ determine a system of reference $[(\boldsymbol{B}), \vec{\omega}]$ for vectors in $\boldsymbol{H}$. We shall call it *frame (or system) of orthogonal coordinates in* $\boldsymbol{H}$, (37), (38), (39), (40), (41), (42), (48).

S.2. We introduce the following important notions, related to the given frame of coordinates. Let $\vec{X} \in \boldsymbol{H}$, and τ a $\boldsymbol{B}$-trace in $(\boldsymbol{F})$. By the *τ-component of* $\vec{X}$ we shall understand the quasi-vector $\vec{x}_\tau$, with support τ, defined by $\vec{X}(p) =_{df} \operatorname{Proj}_p \vec{X}$ for all neighborhoods p of τ. By a *τ-component-density of* $\vec{X}$ we shall understand the quasi-vector x^*_τ with support τ, defined by

$$\vec{x}^*_\tau(p) =_{df} \frac{\operatorname{Proj}_p \vec{X}}{\mu(p)}$$

for all neighborhoods of τ.

By the *τ-coordinate of* $\vec{X}$, we shall understand the quasi-number x with support p, defined by $x_\tau(p) =_{df} (\mathrm{Proj}_p \vec{\omega}, \vec{X})$ for all neighborhoods. By the *τ-coordinate-density of* $\vec{X}$ we shall understand the quasi-number x_τ^* with support τ and defined by

$$x_\tau^* =_{df} \frac{(\mathrm{Proj}_p \vec{\omega}, \vec{X})}{\mu(p)}.$$

S.3. We have $\mathrm{Proj}_p \vec{X} = \vec{x}^*(p) \cdot \mu(p)$; hence

$$(1) \qquad \vec{x}_\beta = \vec{x}_\beta^* \cdot \mu_\beta.$$

We have $(\mathrm{Proj}_p \vec{\omega}, \vec{X}) = x^*(p) \cdot \mu(p)$; hence

$$(2) \qquad x_\beta = x_\beta^* \cdot \mu_\beta.$$

S.4. Since

$$(\mathrm{Proj}_p \vec{\omega}, \vec{X}) = (\vec{\omega}, \mathrm{Proj}_p \vec{X}),$$

we have

$$\vec{x}_\beta = (\vec{\omega}, \mathrm{Proj}_\beta \vec{X}) = x_\beta^* \cdot \mu_\beta.$$

S.5. The total set of quasi-vectors $\vec{x}_\beta$ is regular, and normal. The same is for $\vec{x}^*$. Indeed

$$\vec{x}_\beta(p) = \mathrm{Proj}_p \vec{X} \in p, \quad \vec{x}_\beta^*(p) \in p,$$

which proves the normality. Since $\mathrm{Proj}_p \vec{X}$ does not depend on the choice of the trace β where $\beta \in p$, the set of all β-components and also the set of β-component-densities is regular.

S.5.1. The total sets of quasi-numbers x_β and x_β^* are also regular.

S.6. The total set of β-components is summable on W, where W is the set of all traces.

Proof. It suffice to prove, that the vector-field $\mathrm{Proj}_p \vec{X}$, defined for all bricks p, is summable on I. Let $P_n = \{p_{n1}, p_{n2}, \ldots\}$ be a distinguished sequence of complexes for I.

We have $\sum\limits_k \mathrm{Proj}_{p_{nk}} \vec{X} = \mathrm{Proj}_{\mathrm{som}\, P_n} \vec{X}$, because all spaces p_{nk} are mutually orthogonal, and because $\mathrm{Proj}_{p_{nk}} \vec{X} \in p_{nk}$. Now, since $\|P_n, I\|_\mu \to 0$, i.e. $\mu\, \mathrm{som}\, P_n \to 1$, we get

$$\lim \mathrm{Proj}_{\mathrm{som}\, P_n} \vec{X} - \vec{X}.$$

This proves the summability of $\vec{x}_\beta$ on W.

S.6.1. If E is a measurable set of traces, then

$$S_E\, \vec{x}_\beta = \mathrm{Proj}_{[E]} \vec{X}, \quad S_W\, \vec{x}_\beta = \vec{X}.$$

Proof. From [S.6.] it follows that $\vec{x}_\beta$ is summable on E.
We have

$$S_E\, \vec{x}_\beta = \mathrm{Proj}_{[E]}\, S_1\, \vec{x}_\beta = \mathrm{Proj}_{[E]} \vec{X}, \quad \text{q.e.d.}$$

S.7. The total set of x_β is summable on W.

Proof. It suffices to prove, that the vector-field $x_\beta(p)$ is summable on *1*. Let $P_n = p_{n1}, p_{n2}, \ldots$ be a distinguished sequence for *1*. We have

$$\sum_k x_\beta(p_{nk}) = \sum_k (\vec{\omega}, \operatorname{Proj}_{nk}\vec{X}) = \sum_p (\vec{\omega}, \operatorname{Proj}_{\operatorname{som} P_n}\vec{X}) \to (\vec{\omega}, \vec{X}).$$

The summability follows.

S.7.1. We have for every measurable set E of traces: if $\vec{X} \in \boldsymbol{H}$, then

$$S_E\, x_\beta = (\vec{\omega}, \operatorname{Proj}_E \vec{X}), \qquad S_W\, x_\beta = (\vec{\omega}, \vec{X}).$$

Proof. The set x_β is summable on E. Taking a distinguished sequence $\{P_n\}$ for E, we obtain, as in the proof of [S.7.],

$$\sum_k (\vec{\omega}, \operatorname{Proj}_{p_{nk}}\vec{X}) = \sum_k (\vec{\omega}, \operatorname{Proj}_{\operatorname{som} P_n}\vec{X}) \to (\vec{\omega}, \operatorname{Proj}_E \vec{X}).$$

S.8. We get

$$S_E\, \vec{x}_\beta\, \mu_\beta^* = S_E\, \vec{x}_\beta^*\, \mu_\beta = S_E\, \vec{x}_\beta = \operatorname{Proj}_{|E|}\vec{X},$$
$$S_E\, x_\beta\, \mu_\beta^* = S_E\, x_\beta^*\, \mu_\beta = S_E\, x_\beta = (\vec{\omega}, \operatorname{Proj}_E \vec{X})$$

for every measurable set E and every $\vec{X} \in \boldsymbol{H}$.

S.9. If $\vec{X} \in H$, λ is a complex number, then

$$(\lambda\,\vec{X})_\beta = \lambda\,\vec{x}_\beta, \qquad (\lambda\,\vec{X})_\beta^* = \lambda\,\vec{x}_\beta^*,$$
$$(\lambda\,X)_\beta = \lambda\,x_\beta, \qquad (\lambda\,X)_\beta^* = \lambda\,x_\beta^*.$$

S.10. If $\vec{X} \in \boldsymbol{H}$, $\vec{Y} \in \boldsymbol{H}$, then

$$(\vec{X} \pm \vec{Y})_\beta = \vec{x}_\beta \pm \vec{y}_\beta, \qquad (\vec{X} \pm \vec{Y})_\beta^* = \vec{x}_\beta^* \pm \vec{y}_\beta^*.$$

Putting $\vec{Z} =_{df} \vec{X} \pm \vec{Y}$, we have

$$z_\beta = x_\beta \pm y_\beta, \qquad z_\beta^* = x_\beta^* \pm y_\beta^*.$$

S.11. If $\vec{X}, \vec{Y} \in H$, then

$$(\vec{X}, \vec{y}_\beta) = (\vec{x}_\beta, \vec{Y}) = (\vec{x}_\beta, \vec{y}_\beta).$$

Proof. We have

$$(\operatorname{Proj}_p\vec{X}, \vec{Y}) = (\vec{X}, \operatorname{Proj}_p\vec{Y}) = (\vec{X}, \operatorname{Proj}_p \operatorname{Proj}_p\vec{Y}) = (\operatorname{Proj}_p\vec{X}, \operatorname{Proj}_p\vec{Y}).$$

S.12. Our next purpose will be a proof of the formula $\vec{x}_\beta \approx \vec{\omega}_\beta \cdot x_\beta^*$. It will be proved by steps expressed in few lemmas.

S.13.1. Lemma. If $\vec{X} = \operatorname{Proj}_a \vec{\omega}$, and a is a brick, then $\vec{x}_\beta \approx \vec{\omega}_\beta \cdot x_\beta^*$.

S.13.1a. Proof. To simplify formulas, denote by $\vec{\omega}$ the vector $\operatorname{Proj}_E \vec{\omega}$ for any $E \in G$, and use the same letter for a measurable set of traces and for its coat, and write $|E|$ instead of μE. We have

$$\vec{x}_\beta = \mathrm{Proj}_p \vec{X}, \quad \vec{\omega}_\beta x^*_\beta = \mathrm{Proj}_p \vec{\omega} \cdot \frac{(\mathrm{Proj}_p \vec{\omega}, \vec{X})}{\mu(p)};$$

hence

$$\vec{x}_\beta = \mathrm{Proj}_p \mathrm{Proj}_a \vec{\omega} = \mathrm{Proj}_{pa} \vec{\omega} = \vec{\omega}_{pa},$$

$$\vec{\omega}_\beta x^*_\beta = \vec{\omega}_p \frac{\vec{\omega}_p, \vec{\omega}_a}{|p|} = \vec{\omega}_p \cdot \frac{|p\,a|}{|p|},$$

where p varies over all neighborhoods of β. Hence

(1) $$\vec{x}_\beta - \vec{\omega}_\beta x^*_\beta = \vec{\omega}_{pa} - \vec{\omega}_p \frac{|p\,a|}{|p|}.$$

S.13.1b. Take any brick $q \neq 0$. We have

$$\vec{A}_{(q)} =_{df} \vec{\omega}_{qa} - \vec{\omega}_q \cdot \frac{|q\,a|}{|q|} = \vec{\omega}_{qa} - (\vec{\omega}_{qa} + \vec{\omega}_{q\,\mathrm{co}\,a}) \cdot \frac{|q\,a|}{|q|}$$

$$= \vec{\omega}_{qa} - \vec{\omega}_{qa} \frac{|q\,a|}{|q|} - \vec{\omega}_{q\,\mathrm{co}\,a} \cdot \frac{|q\,a|}{|q|} = \vec{\omega}_{qa}\left(1 - \frac{|q\,a|}{|q|}\right) - \omega_{q\,\mathrm{co}\,a} \cdot \frac{|q\,a|}{|q|}$$

$$= \vec{\omega}_{qa} \frac{|q\,\mathrm{co}\,a|}{|q|} - \vec{\omega}_{q\,\mathrm{co}\,a} \cdot \frac{|q\,a|}{|q|}.$$

Hence

(1.1) $$A(q) \in q\,a + q\,\mathrm{co}\,a = q.$$

Since the spaces $q\,\mathrm{co}\,a$, $q\,a$ are orthogonal, we get

(2) $$\|\vec{A}(q)\|^2 = |q\,a| \cdot \frac{|q\,\mathrm{co}\,a|^2}{|q|^2} + |q\,\mathrm{co}\,a| \cdot \frac{|q\,a|^2}{|q|^2}$$

$$= \frac{|q\,a| \cdot |q\,\mathrm{co}\,a|}{|q|^2} (|q\,\mathrm{co}\,a| + |q\,a|), \quad \text{i.e.} \quad \|A(q)\|^2 = \frac{|q\,a| \cdot |q\,\mathrm{co}\,a|}{|q|},$$

which is valid for any brick $q \neq 0$.

S.13.1c. Now let E be a measurable set of traces, and $\{P_n\}$ a completely distinguished sequence for E. We shall use arguments, similar to those in the proof of [R.15.]; they rely on [Q.21.6.], [Q.21.13.], [Q.21.9.], [Q.21.14.]. Take a subsequence $\{P_{k(n)}\}$ of $\{P_n\}$, and get a completely distinguished sequence $P_{kl(n)} \cup Q_n$ of I. Considering a partial complex $\{R_n\}$ of $P_{kl(n)} \cup Q_n$, such that $\{R_n\}$ be a completely distinguished sequence for Q, take the bricks

$$e'_{n1}, e'_{n2}, \ldots \quad \text{with} \quad \mu(\sum_k e'_{nk}\,\mathrm{co}\,a) \to 0;$$

and also the bricks

$$e''_{n1}, e''_{n2}, \ldots \quad \text{with} \quad \mu(\sum_k e''_{nk}\,a) \to 0.$$

The bricks belong to $P_{kl(n)}$. We have, putting

$$\vec{A}_\beta =_{df} \vec{x}_\beta - \vec{\omega}_\beta x^*_\beta:$$

$$\|A(P_{kl(n)})\|^2 = \sum_j \|\vec{A}(e'_{nj})\|^2 + \sum_j \|\vec{A}(e''_{nj})\|^2.$$

Hence, by (2)

$$(2.1)\quad \|A(P_{kl(n)})\|^2 = \sum_j \frac{|e'_{nj}a|\cdot|e'_{nj}\,\mathrm{co}\,a|}{e'_{nj}} + \sum_j \frac{|e''_{nj}a|\cdot|e''_{nj}\,\mathrm{co}\,a|}{e''_{nj}} \leqq$$

$$\leqq \sum_j |e'_{nj}\,\mathrm{co}\,a| + \sum_j |e''_{nj}a| = \mu(\sum_j e'_{nj}\,\mathrm{co}\,a) + \mu(\sum_j e''_{nj}a) \to 0.$$

Thus from every partial sequence $\{Q_n\}$ of $\{P_n\}$, another partial sequence $\{Q_{k(n)}\}$ can be extracted, with (2.1).

Hence $\|\vec{A}(P_n)\|^2 - 0$, which gives $\vec{A}(P_n) \to O$, i.e. the quasi-vector $\vec{x}_\beta - \vec{\omega}_\beta x^*_\beta$ is summable over any measurable set E of traces, and we have $S_E(\vec{x}_\beta - \vec{\omega}_\beta x^*_\beta) = \vec{O}$. Since $\vec{x}_\beta$ is summable on E, it follows, that $\vec{x}_\beta - (\vec{x}_\beta - \vec{\omega}_\beta x^*_\beta) = \vec{\omega}_\beta x^*_\beta$ is also summable on E; and we have $S_E \vec{x}_\beta = S_E \vec{\omega}_\beta x^*_\beta$ for all summable E. Hence

$$\vec{x}_\beta \approx \vec{\omega}_\beta \cdot x^*_\beta. \quad \text{q.e.d.}$$

S.13.2. Lemma. If λ is a complex number, $\vec{X} = \lambda\,\mathrm{Proj}_a \vec{\omega}$, where a is a brick, we have $\vec{x}_\beta \approx \vec{\omega}_\beta \cdot x^*_\beta$.

S.13.3. Now we notice that if $\vec{X}, \vec{Y} \in \boldsymbol{H}$, $\vec{Z} =_{df} \xi \vec{X} + \eta \vec{Y}$, and $\vec{x}_\beta \approx \vec{\omega}_\beta \cdot x^*_\beta$, $\vec{y}_\beta \approx \vec{\omega}_\beta \cdot y^*_\beta$, then $z^*_\beta \approx \vec{\omega}_\beta \cdot z^*_\beta$.

Proof. We have for a measurable set E:

$$S_E \vec{x}_\beta = S_E \vec{\omega}_\beta x^*_\beta,$$

$$S_E \vec{y}_\beta = S_E \vec{\omega}_\beta y^*_\beta;$$

hence

$$S_E \vec{z}_\beta = S_E(\vec{\omega}_\beta \cdot \xi x^*_\beta + \vec{\omega}_\beta \cdot \eta y^*_\beta) = S_E \vec{\omega}_\beta z^*_\beta,$$

which completes the proof.

S.13.4. It follows that the theorem is true for any step function $\sum_{i=1}^{n} \lambda_i\,\mathrm{Proj}_{a_i} (n \geqq 1)$, where $a_1, a_2, \ldots a_n$ are disjoint bricks with $\sum_i a_i = 1$, and λ_i complex numbers.

S.13.5. Now we shall prove that if $\vec{X}_n$ are step functions as above, $\vec{X}_n \to \vec{X}$, then the theorem holds true for $\vec{X}$.

Since $\vec{\omega}$ is generating vector and $\boldsymbol{G}$ is the Lebesgue's covering extension of B, therefore for every $\vec{X}$ there exists a sequence of step-functions $\vec{X}_1, \vec{X}_2, \ldots$, which tend in the $\boldsymbol{H}$-topology to $\vec{X}$. Take $\varepsilon > 0$. Find k with $\|\vec{X} - \vec{X}_k\| < \varepsilon$. Put for any brick p:

$$\vec{\Phi}(p) =_{df} \mathrm{Proj}_p \vec{X}, \qquad \vec{\Psi}(p) =_{df} \mathrm{Proj}_p \vec{\omega} \frac{(\mathrm{Proj}_p \vec{\omega}, \vec{X})}{\mu(p)},$$

$$\vec{\varphi}_k(p) =_{df} \mathrm{Proj}_p \vec{X}_k, \qquad \vec{\Psi}_k(p) =_{df} \mathrm{Proj}_p \vec{\omega} \frac{(\mathrm{Proj}_p \vec{\omega}, \vec{X})}{\mu(p)}.$$

Put

$$\vec{\Delta}(p) = \vec{\Phi}(p) - \vec{\Psi}(p), \qquad \vec{\delta}(p) = \vec{\varphi}_k(p) - \vec{\psi}_k(p).$$

First we shall prove some inequalities. Let

$$P_n = \{p_{n1}, p_{n2}, \ldots, p_{ni}, \ldots\}, \quad n = 1, 2, \ldots$$

be complexes, such that

$$\operatorname{som} P_n = 1, \quad \text{and} \quad \mathcal{N}_A P_n \to 0 \quad \text{and} \quad \mathcal{N}_R P_n \to 0.$$

We have

$$\sum_i (\vec{\Phi}(p_{ni}) - \vec{\varphi}(p_{ni})) = \sum_i (\operatorname{Proj}_{p_{ni}} \vec{X} - \operatorname{Proj}_{p_{ni}} \vec{X}_k)$$

$$= \sum_i \operatorname{Proj}_{p_{ni}} (\vec{X} - \vec{X}_k) =, \text{ since the spaces } p_{ni} \text{ are disjoint,}$$

$$= \operatorname{Proj}_{\sum_i p_{ni}} (\vec{X} - \vec{X}_k).$$

Hence

(1) $$\left|\left| \sum_i (\vec{\Phi}(p_{ni}) - \vec{\varphi}(p_{ni}) \right|\right| \leqq ||\vec{X} - \vec{X}_k|| < \varepsilon.$$

Hence

$$\sum_i ||\vec{\Phi}(p_{ni}) - \vec{\varphi}(p_{ni})||^2 < \varepsilon^2.$$

On the other hand we have

$$\left|\left| \sum_i \{\vec{\Psi}(p_{ni}) - \vec{\psi}(p_{ni})\} \right|\right|^2 = \left|\left| \sum_i \frac{\operatorname{Proj}_{p_{ni}} \vec{\omega}}{\mu(p_{ni})} [(\operatorname{Proj}_{p_{ni}} \vec{\omega}, \vec{X}) - (\operatorname{Proj}_{p_{ni}} \vec{\omega}, \vec{X}_k)] \right|\right|^2$$

$$= \left|\left| \sum_i \frac{\operatorname{Proj}_{p_{ni}} \vec{\omega}}{\mu(p_{ni})} \cdot (\operatorname{Proj}_{p_{ni}} \vec{\omega}, \vec{X} - \vec{X}_k) \right|\right|^2 \leqq$$

$$\leqq \left|\left| \sum_i \frac{\operatorname{Proj}_{p_{ni}} \vec{\omega}}{\mu(p_{ni})} (\operatorname{Proj}_{p_{ni}} \vec{\omega}, \operatorname{Proj}_{p_{ni}} (\vec{X} - \vec{X}_k)) \right|\right|^2 \leqq$$

$$\leqq \sum_i \left|\left| \frac{\operatorname{Proj}_{p_{ni}} \vec{\omega}}{\mu(p_{ni})} \cdot (\operatorname{Proj}_{p_{ni}} \vec{\omega}, \operatorname{Proj}_{p_{ni}} (\vec{X} - \vec{X}_k)) \right|\right|^2$$

$$= \sum_i \frac{||\operatorname{Proj}_{p_{ni}} \vec{\omega}||^2}{\mu(p_{ni})} \cdot |(\operatorname{Proj}_{p_{ni}} \vec{\omega}, \operatorname{Proj}_{p_{ni}} (\vec{X} - \vec{X}_k))|^2,$$

and by Cauchy-Schwarz-inequality:

$$\leqq \sum_i \frac{\mu(p_{ni})}{\mu(p_{ni})^2} \cdot ||\operatorname{Proj}_{p_{ni}} \vec{\omega}||^2 \cdot ||\operatorname{Proj}_{p_{ni}} (\vec{X} - \vec{X}_k)||^2 =$$

$$= \sum_i \frac{\mu(p_{ni})}{\mu(p_{ni})^2} \cdot \mu(p_{ni}) \cdot ||\operatorname{Proj}_{p_{ni}} (\vec{X} - \vec{X}_k)||^2 = \sum_i ||\operatorname{Proj}_{p_{ni}} (\vec{X} - \vec{X}_k)||^2 \leqq$$

$$\leqq ||X - X_k||^2 \leqq \varepsilon^2.$$

Hence we have proved, that

(2) $$\sum_i ||\Psi(p_{ni}) - \psi(p_{ni})||^2 \leqq \varepsilon^2.$$

We have

$$\sum_i \|\vec{\Phi}(p_{ni}) - \Psi(p_{ni})\|^2 = \sum_i \|\{\vec{\Phi}(p_{ni}) - \vec{\varphi}(p_{ni})\} + \vec{\varphi}(p_{ni} - \\ - \{\vec{\Psi}(p_{ni}) - \psi(p_{ni}) + \psi(p_{ni})\|^2$$

$$= \sum_i \|\{\vec{\Phi}(p_{ni}) - \vec{\varphi}(p_{ni})\} + \{\vec{\varphi}(p_{ni}) - \vec{\psi}(p_{ni})\} + \{\vec{\Psi}(p_{ni}) - \\ - \vec{\psi}(p_{ni})\}\|^2 \leqq 4\sum_i \|\Phi(p_{ni}) - \vec{\varphi}(p_{ni})\|^2 + 4\sum_i \|\Psi(p_{ni}) - \\ - \psi(p_{ni})\|^2 + 4\sum_i \|\vec{\varphi}(p_{ni}) - \vec{\psi}(p_{ni})\|^2.$$

Hence, by (1) and (2),

$$\sum_i \|\vec{\Phi}(p_{ni}) - \Psi(p_{ni})\|^2 \leqq 4\varepsilon^2 + 4\varepsilon^2 + 4\sum_i \|\vec{\varphi}(p_{ni}) - \psi(p_{ni})\|^2.$$

Now, we know that for $\vec{X}_k$ we have $\vec{x}_{k\beta} \approx \vec{\omega}_\beta \cdot x^*_{k\beta}$. It follows that

$$\sum_i \|\vec{\varphi}(p_{ni}) - \vec{\psi}(p_{ni})\|^2 < \varepsilon.$$

Thus we get

$$\sum_i \|\vec{\Phi}(p_{ni}) - \vec{\Psi}(p_{ni})\|^2 \leqq 12\varepsilon^2,$$

this for all P_n.

Hence

$$\vec{\Phi}(\beta) \approx \Psi(\beta) \quad \text{i.e.} \quad \vec{x}_\beta \approx \omega_\beta \vec{x}^*_\beta, \quad \text{q.e.d.}$$

S.13.6. Theorem. If $\vec{X}$ is any vector in $\boldsymbol{H}$, then

$$\vec{x}_\beta \approx \vec{\omega}_\beta \cdot x^*_\beta.$$

Chapter T

Dirac's Delta-function

This chapter is devoted to a mathematically precise setting of the δ-function of P. A. M. DIRAC, (78). The original definition reads: $\delta(x)$ is the function defined for all real numbers as follows:

$$(1) \qquad \delta(x) = \begin{cases} 0 & \text{for} \quad x \neq 0, \\ +\infty & \text{whenever} \quad x = 0. \end{cases}$$

The function shall have the property

$$\int_{-\infty}^{+\infty} \delta(x)\, dx = 1, \tag{2}$$

and, for every continuous function $f(x)$, we should have

$$f(x) = \int_{-\infty}^{+\infty} \delta(x-y)\, f(y)\, dy. \tag{3}$$

The function $\delta(x)$ has several basic properties, whose list is given at the end of the present Chapter [T.].

From the mathematical point of view, the conditions (1) and (2) can hardly be understood, though the function $\delta(x)$ has shown itself very useful in the hands of able physicists.

Having this in mind, we think that the science cannot disregard the strange notion, but should find a suitable modification of the definition of $\delta(x)$. There were several cases in the history of mathematics, where a notion, introduced in a contradictory way, has got a suitable, logically precise, setting, e.g. complex numbers.

Looking at (1) and (2), we see that $\delta(x)$ cannot be a function, and $\int_{-\infty}^{+\infty}$ cannot be an ordinary integral. In addition to that the formula (3) shows, that the delta-function should be conceived as a function of two variables x, y, like a kernel in integral equations.

There is spread the opinion, that the Dirac's δ-function is the same, as the following set-function $\mu_a(\varepsilon)$, defined for all subsets ε of $(-\infty, +\infty)$, by

$$\mu_a(\varepsilon) = \begin{cases} 0 & \text{whenever} \quad a \,\overline{\in}\, \varepsilon, \\ 1 & \text{whenever} \quad a \in \varepsilon, \end{cases}$$

because for any function $f(x)$ we have

$$f(a) = \int_{-\infty}^{+\infty} f(x)\, \mu_a(\varepsilon).$$

We see that this is not a representation of $f(a)$ by an ordinary measure integral with fixed measure, independent of a.

The following intuitive approach would be more adequate: The function

$$\varphi_n(x) =_{df} \frac{n}{\sqrt{\pi}}\, e^{-(nx)^2}$$

for $n = 1, 2, \ldots$, would be a kind of intuitive "approximation" of the δ-function. Indeed if $n \to \infty$, $x_0 \neq 0$, then $\varphi_n(x_0) \to 0$, and in the proximity of 0, the graph of the function $\varphi_n(x)$ becomes a long and narrow tube. In what follows, we shall not only give a definition of a variation of the δ-function in accordance with (1) and (2), but we shall give proofs of its many basic properties, listed at the end of this chapter.

We shall introduce even some more general notions, having some properties of the δ-function. Our theory is based on the general topics, which were developped in the preceding chapters.

T.1. Def. We admit the hypotheses stated in [R.1.] and [R.2.], concerning the tribes $\boldsymbol{G}$, $\boldsymbol{F}$, the basis $\boldsymbol{B}$ and the linear vector space V.

Let x_0, y_0 be two traces; then by a *quasi-vector with support* (x_0, y_0) we shall understand any function, $\vec{f}(p, q)$, with values taken from V, and defined for all neighborhoods p of the trace x_0, [B I.10.], and for all neighborhoods q of y_0. We shall write $\vec{f}_{x_0, y_0}$, or $\vec{f}(x_0, y_0)$.

T.1.1. Def. We shall consider sets of quasi-vectors $\vec{f}_{x, y}$, where x varies in a measurable set E of traces, and y varies in a measurable set F of traces. We shall call the couple (E, F) *the support of the set of quasi-vectors.*

T.1.2. We have some modifications of these notions.

If x_0 is a trace and q a brick, we can consider the vector valued function $\vec{f}(x_0, q)$ which attaches to every neighborhood p of x_0 a vector of V. We can vary x_0 over a set of traces, and q over a set of bricks, getting a kind of sets of quasi-vectors.

T.1.3. Given a quasi-vector $\vec{f}(x_0, y_0)$, we shall write it $\vec{f}_{x_0}(\dot{q})$, $\vec{f}_{y_0}(\dot{p})$, $\vec{f}(\dot{p}, \dot{q})$, according to whether we like to emphasize the variable neighborhood q of y_0, the variable neighborhood p of x_0, or both variable neighborhoods respectively.

T.2. Def. The following notion of summation will be important: We say, that the set of quasi-vectors $\vec{f}_{x_0, y}$ with support (x_0, E), (where x_0, y are traces, and y varies in E), *is summable on E with respect to y*, whenever for every neighborhood p of x_0, the set of quasi-vectors $\vec{f}_y(p)$ is summable on E with respect to y, i.e., when $\underset{y \in E}{S} \vec{f}_y(p)$ exists for every neighborhood p of x_0.

In the case of summability, we get a quasi-vector

$$\vec{g}_{x_0} =_{df} \vec{g}(p) =_{df} \underset{y \in E}{S} \vec{f}_y(p)$$

with support x_0.

T.3. Def. We are introducing the number-valued function $\Delta(p, q)$ of variable non null bricks p, q, defining it by:

$$\Delta(p, q) =_{df} \begin{cases} 1 & \text{whenever } \; p \cdot q \neq O, \\ 0 & \text{whenever } \; p \cdot q = O. \end{cases} \tag{1}$$

This function generates the following ones:

If τ is a trace, then $\Delta(\tau, q)$ is the quasi number $\Delta(p, q)$ with support τ, defined for all neighborhoods p of τ, by (1). It depends on the parameter q. Similarly $\Delta(p, q)$ will be denoted by $\Delta(p, \tau)$ whenever q varies over all neighborhoods of τ. By $\Delta(\tau, \xi)$, where τ, ξ are two traces, we shall understand the function $\Delta(p, q)$, defined for all neighborhoods p of τ and for all neighborhoods q of ξ.

T.4. We shall take over the topic of [S.2.], to have a system of coordinates in the H.-H.-space $\boldsymbol{H}$. Thus $(\boldsymbol{G})$ is supposed to be, a saturated tribe.

T.4.1. Lemma. For any vector $\vec{X} \in \boldsymbol{H}$ and any spaces $a, b \in \boldsymbol{G}$ we have

$$\|\mathrm{Proj}_a \vec{X} - \mathrm{Proj}_b \vec{X}\|^2 = \|\mathrm{Proj}_{a-b} \vec{X}\|^2 + \|\mathrm{Proj}_{b-a} \vec{X}\|^2.$$

T.4.2. Lemma. If $a_n \in \boldsymbol{G}$, $n = 1, 2, \ldots$, $\vec{X} \in \boldsymbol{H}$ and $\mu(a_n) \to 0$, then

$$\mathrm{Proj}_{a_n} \vec{X} \to \vec{O}$$

in the $\boldsymbol{H}$-topology.

Proof. The number valued function $\|\mathrm{Proj}_a \vec{X}\|^2$ for a variable $a \in \boldsymbol{G}$ is denumerably additive and continuous in the μ-topology in $\boldsymbol{G}$. Hence $\|\mathrm{Proj}_{a_n} \vec{X}\|^2 \to 0$, which gives $\mathrm{Proj}_{a_n} \vec{X} \to \vec{O}$ in the $\boldsymbol{H}$-topology.

T.4.3. Lemma. If $a_n, a \in \boldsymbol{G}$, $\vec{X} \in \boldsymbol{H}$, $a_n \to^{\mu} a$ in the μ-topology in $\boldsymbol{G}$, then $\mathrm{Proj}_{a_n} \vec{X} \to \mathrm{Proj}_a \vec{X}$ in the $\boldsymbol{H}$-topology.

Proof. Since $a_n \to^{\mu} a$, we have $\mu(a_n - a) + \mu(a - a_n) \to 0$. Hence, by [T.4.2.], $\|\mathrm{Proj}_{a_n} \vec{X}\|^2 \to 0$ and $\|\mathrm{Proj}_{a-a_n} \vec{X}\|^2 \to 0$. Hence, by [T.4.1.], we get $\|\mathrm{Proj}_{a_n} \vec{X} - \mathrm{Proj}_a \vec{X}\|^2 \to 0$; hence $\mathrm{Proj}_{a_n} \vec{X} - \mathrm{Proj}_a \vec{X} \to \vec{O}$, which gives the thesis.

T.5. Def. Let $Q_n = \{q_{n1}, q_{n2}, \ldots\}$, $(n = 1, 2, \ldots)$, be a completely distinguished sequence of complexes for I. Given a brick $p \neq O$, let $p_{n1}, p_{n2}, \ldots$ be all those bricks q_{nk} for which $q_{nk} \leqq p$. We get a complex $\{p_{n1}, p_{n2}, \ldots\}$, which may be empty or not. Now if for every $p \neq O$, we have $\lim_{n\to\infty} \mu(\sum_i p_{ni}) = \mu(p)$, we shall call $\{Q_n\}$ *a special sequence for I.*

T.5.1. Remark. We do not know, whether, from all admitted hypotheses it follows, that there exists at least one completely distinguished and special sequence $\{Q_n\}$ for I.

T.5.2. We shall admit the following *hypothesis*: There exists at least one completely distinguished special sequence of complexes for I.

T.5.3. In the case where the base $\boldsymbol{B}$ is composed of spaces, which correspond to half-open rectangles or half-open segments (see [R I.2.2.] to [R I.2.9.]), the hypothesis [T.5.2.] is satisfied.

T.5.4. We shall consider summations of a total quasi-vector-set, defined by means of special sequences. Given a total set of quasi-vectors $\vec{f}_\tau$, we say that $\vec{f}_\tau$ is *"specially" summable over W*, whenever for every completely distinguished and special sequence

$\{Q_n\} = \{q_{n1}, q_{n2}, \ldots\}$, the limit $\lim_{n\to\infty} \sum_k \vec{f}(q_{nk})$ exists and has the same value. The limit will be denoted by $\underset{W}{S^\bullet} \vec{f}_\tau$ and called *"special" sum.*

T.5.5. If [Hyp. T. 5.2.] holds true and $\vec{f}_\tau$ is (DRAS)-summable, then $\underset{W}{S^\bullet} \vec{f}_\tau$ exists too. The converse does not seem to be true.

T.5.6. Lemma. If

1. $\{Q_n\}$ is a completely distinguished and special sequence of complexes for I,

2. f is a figure ($f \in \boldsymbol{F}$), $f \neq O$,

3. $f_{n1}, f_{n2}, \ldots$ are all bricks of Q_n with $f_{nk} \leqq f$,

then we have

$$\lim_{n\to\infty} \mu(\sum_k f_{nk}) = \mu(f).$$

Proof. First we shall prove the lemma under hypothesis that f is a finite sum of disjoint bricks. Let $f = a_1 + \cdots + a_s$, $(s \geqq 2)$. Denote by $q_{n1}^{(k)}, q_{n2}^{(k)}, \ldots$ all different bricks of Q_n, which are contained in a_k, $(k = 1, \ldots, s)$. We have

$$\lim_{n\to\infty} \mu(\sum_j q_{nj}^{(k)}) = \mu(a_k).$$

Since $a_1, \ldots, a_s$ are disjoint, and consequently also the bricks of Q_n, which are inside of them, we get

$$\lim_{n\to\infty} \mu\left(\sum_{k=1}^{s} \sum_j q_{nj}^{(k)}\right) = \mu\left(\sum_{k=1}^{s} a_k\right) = \mu(f). \tag{1}$$

The bricks $q_{nj}^{(k)}$, for fixed n, are certainly bricks, which are inside of f. If there are some other supplementary bricks of Q_n, which are inside f, we have $\sum_{k=1}^{s} \sum_j q_{nj}^{(k)} \leqq \sum_k f_{nk} \leqq f$, where f_{nk} are all bricks of Q_n, which are inside f; hence

$$\lim_{n\to\infty} \mu(\sum_k f_{kn}) = \mu(f).$$

Having that, let us go over to the general figure f. We have $f = a_1 + a_2 + \cdots$ where a_j are disjoint bricks. Take $s > 0$, and find s such that

$$0 \leqq \mu(f) - \sum_{j=1}^{s} \mu(a_j) \leqq \varepsilon. \tag{1.1}$$

Put $f_s =_{df} a_1 + \cdots + a_s$. For such a figure the theorem has been proved. Let $Q_n = \{q_{n1}, q_{n2}, \ldots\}$ be a special sequence for I. Denote by $q'_{n1}, q'_{n2}, \ldots$ all those bricks q_{nk} of Q_n, which are inside f_s. We have

$$\lim_{n\to\infty} \sum_i \mu(q'_{ni}) = \mu f_{s_0}.$$

There exists M, such that for all $n \geqq M$ we have

$$0 \leqq \mu(f) - \mu(f_s) \leqq \varepsilon.$$

We get

(2) $$0 \leqq \mu(f) - \sum_i \mu(q'_{ni}) \leqq 2\varepsilon \quad \text{for} \quad n \geqq M.$$

Now if there are bricks in Q_n, differing from q'_{ni}, which are inside f; their addition will not spoil the inequality (1.1), so we get

(3) $$0 \leqq \mu(f) - \sum_k \mu(f_{nk}) \leqq 2\varepsilon,$$

where f_{nk} are all bricks of Q_n, which are inside f. The inequality (3) is valid for all $n \geqq M$. This completes the proof of the lemma.

T.5.7. Lemma. If

1. $\{Q_n\}$ is a completely distinguished and special sequence of complexes for I,

2. f is a figure,

3. $e_{n1}, e_{n2}, \ldots$ are all bricks of Q_n, for which $e_{nk} \cdot f \neq O$, $e_{nk} \cdot \text{co} f \neq O$,

then we have

$$\lim_{n\to\infty} \mu\Big(\sum_k e_{nk}\Big) = 0.$$

Proof. Let $a_{n1}, a_{n2}, \ldots$ be all bricks of Q_n for which $a_{nk} \leqq p$, and let $b_{n1}, b_{n2}, \ldots$ be all bricks of Q_n for which $b_{nk} \leqq \text{co}\, p$. By Lemma [T.5.6.] we have

$$\lim_{n\to\infty} \mu\Big(\sum_k a_{nk}\Big) = \mu(p), \qquad \lim_{n\to\infty} \mu\Big(\sum_k b_{nk}\Big) = \mu(\text{co}\, p).$$

Since

$$\lim_{n\to\infty} \mu\Big(\sum_k a_{nk} + \sum_k b_{nk} + \sum_k e_{nk}\Big) = \mu(I),$$

it follows that

$$\mu(p) + \mu(\text{co}\, p) + \lim \mu\Big(\sum_k e_{nk}\Big) = \mu(I),$$

which completes the proof.

T.6. Theorem. If

1. Hypothesis [T.5.2.] is admitted,

2. $p \neq O$ is a brick-space,

3. $\vec{X} \in \boldsymbol{H}$,

then, considering special summations, [T.5.4.], we have

(1) $$\text{Proj}_p \vec{X} = \underset{\beta \in W}{S^{\bullet}} \Delta(p, \beta)\, \vec{x}_\beta,$$

Def. [T.3.], (where W is the set of all traces and $\vec{x}_\beta$ is the β-component of $\vec{X}$.)

T.6a. Proof. We shall schedule our argument so, as to put in evidence the reason of admitting Hyp. [T.5.2.].

To simplify print we shall use the alternative symbol $\operatorname{Proj}(p)\vec{X}$ for $\operatorname{Proj}_p \vec{X}$. Let $Q_n = \{q_{n1}, q_{n2}, \ldots\}$ be a completely distinguished and special sequence of complexes for I. Take the partial complex R_n of Q_n with $|R_n, p|_\mu \to 0$. By [Q.21.9.] R_n is a completely distinguished sequence for p. Consider the complex $S_n =_{df} Q_n \sim R_n$ i.e. the complex complementary to R_n in Q_n. We have

$$|\operatorname{som} Q_n - \operatorname{som} R_n, I - p|_\mu \to 0,$$

i.e.

$$|S_n, \operatorname{co} p|_\mu \to 0.$$

Hence, by [Q.21.9.], S_n is a completely distinguished sequence for $\operatorname{co} p$.

Denote by a_i', b_i', e_i' those bricks of R_n for which $a_i' \leqq p$; $b_i' \cdot p = O$; $e_i' \cdot p \neq O$ respectively, and denote by a_i'', b_i'', e_i'' those bricks of S_n, for which $a_i'' \leqq p$; $b_i'' \cdot p = O$; $e_i'' \cdot p \neq O$; $e_i'' \cdot \operatorname{co} p \neq O$ respectively. We have, by [Q.21.14.],

$$(1.1) \qquad |\textstyle\sum a_i' + \sum e_i' p, p| \to 0, \qquad \mu(\sum e_i' \operatorname{co} p) \to 0,$$

$$|\textstyle\sum b_i'' + \sum e_i'' \operatorname{co} p, \operatorname{co} p| \to 0, \qquad \mu(\sum e_i'' p) \to 0$$

for $n \to \infty$.

T.6b. We also have

$$(2.1) \qquad \mu(\textstyle\sum b_i') \to 0, \qquad \mu(\sum a_i'') \to 0.$$

Indeed, from (1.1) we get, by the help of $|\sum e_i' \operatorname{co} p; 0|_\mu \to 0$,

$$|\textstyle\sum a_i' + \sum e_i' p + \sum e_i' \operatorname{co} p; p| \to 0,$$

i.e.,

$$|\textstyle\sum a_i' + \sum e_i'; p| \to 0.$$

Since, on the other hand, we have $|R_n, p| \to 0$, i.e.,

$$|\textstyle\sum a_i' + \sum b_i' + \sum e_i'; p| \to 0,$$

we get, by subtraction,

$$|\textstyle\sum b_i', O| \to 0, \quad \text{i.e.,} \quad \mu(\sum b_i') \to 0.$$

Similarly we prove the second relation in (2.1).

T.6c. We shall build sums which approximate the expression (1). Put

$$(2.2) \qquad \vec{A}_n =_{df} \sum_k \Delta(p, q_{nk}) \operatorname{Proj}(q_{nk}) \vec{X}.$$

The bricks a_i', b_i', e_i', a_i'', b_i'', e_i'' constitute the whole complex Q_n; and they are disjoint. In (2.2) all terms, where $q_{nk} \cdot p = 0$, disappear, and for the others we have $\Delta = 1$.

Thus

$$\vec{A}_n = \sum_i \operatorname{Proj}(a_i') \vec{X} + \sum_i \operatorname{Proj}(a_i'') \vec{X} + \sum \operatorname{Proj}(e_i') \vec{X} + \sum_i \operatorname{Proj}(e_i'') \vec{X}.$$

Since the brick-spaces are orthogonal to one another, we get

$$(3)\quad \vec{A}_n = \mathrm{Proj}\,(\sum_i a'_i + \sum_i e'_i\,p)\,\vec{X} + \mathrm{Proj}\,(\sum_i a''_i)\,\vec{X} + \\ + \mathrm{Proj}\,(\sum_i e'_i\,\mathrm{co}\,p)\,\vec{X} + \mathrm{Proj}\,(\sum_i e''_i\,p)\,\vec{X} + \mathrm{Proj}\,(\sum_i b''_i)\,\vec{X} - \\ - \mathrm{Proj}\,(\sum_i b''_i)\,\vec{X}.$$

In (3) the first term tends to $\mathrm{Proj}_p \vec{X}$, because of (1.1) and by virtue of Lemma [T.4.3.]; the second term tends to $\vec{O}$, because of (2.1) and Lemma [T.4.2.]; the third term tends to $\vec{O}$, because of (2). Concerning the last three terms in (3), their sum can be written as

$$\mathrm{Proj}\,(\sum_i b''_i + \sum_i e''_i\,p)\,\vec{X} - \mathrm{Proj}\,(\sum_i b''_i)\,\vec{X}.$$

Here the first term tends to $\mathrm{Proj}\,(\mathrm{co}\,p)\,\vec{X}$, by (2) and Lemma [T.4.3.]. Hence a necessary and sufficient condition that $\vec{A}_n$ tends to a limit, is that $\mathrm{Proj}\,(\sum_i b_i)\,\vec{X}$ tends to a limit.

T.6d. Till now we did not use at all the conditions under which the sequence $\{Q_n\}$ is a special one. We used only the fact, that it is completely distinguished. From [T.6c.] we get $\lim_{n\to\infty} \vec{A}_n = \vec{X} - \lim_{n\to\infty} \mathrm{Proj}\,(\sum_i b''_i)\,\vec{X}$, whenever at least one of these limits exists. This shows the role of Hyp. [T.5.2.] which we have admitted in the wording of our theorem. Since, by Lemma [T.5.6.], $|\sum_i b''_i, \mathrm{co}\,p| \to 0$ for $n \to \infty$, it follows, by Lemma [T.4.3.],

$$\mathrm{Proj}\,(\sum_i b''_i)\,\vec{X} \to \mathrm{Proj}\,(\mathrm{co}\,p)\,\vec{X}.$$

Hence we get

$$\lim \vec{A}_n = \vec{X} - \mathrm{Proj}_{(\mathrm{co}p)}\,\vec{X} = \mathrm{Proj}_p\,\vec{X}, \quad \text{q.e.d.}$$

T.6.1. Theorem. If

1. Hypothesis [T.5.2.] is admitted,
2. $\vec{X} \in \boldsymbol{H}$,

then considering special summations, [T.5.4.], we have

$$\vec{x}_\alpha = \underset{\beta \in W}{S^\bullet} \Delta(\alpha, \beta)\,\vec{x}_\beta,$$

where $\vec{x}_\alpha$ is the α-component of $\vec{X}$, [S.2.].

Proof. Follows from [T.6.].

T.6.2. Remark. We do not know whether the formula of Theorem [T.6.1.] is true, if we do not use special summations (see Remark [T.5.1.]).

T.7. Theorem. If

1. Hypothesis [T.5.2.] is admitted,
2. $p \neq O$ is a space-brick,

3. $\vec{\omega}$ is a generating vector of the space $\boldsymbol{H}$ with respect to the saturated tribe $\boldsymbol{G}$,

4. $\vec{X} \in \boldsymbol{H}$,

then, considering special sums, we have

$$(\mathrm{Proj}_p \vec{\omega}, \vec{X}) = \underset{\beta \in W}{S^{\bullet}} \Delta(p, \beta)\, x_\beta$$

where $\vec{x}_\beta$ is the β-coordinate of $\vec{X}$, [S.2.].

Proof. The theorem can be proved just by the method, used in the proof of Theorem [T.6.]. We shall give a simplier proof.

Let $Q_n = \{q_{n1}, q_{n2}, \ldots\}$ be a special, completely distinguished sequence of complexes for I. Let a_i, e_i be those bricks of Q_n, for which $a_i \leqq p$; $a_i \cdot p = O$; $e_i \cdot p \neq O$, $e_i \cdot \mathrm{co}\, p \neq O$ respectively. Put

$$A_n =_{df} \sum_k \Delta(p, q_{nk}) \cdot (\mathrm{Proj}\,(q_{nk})\, \vec{\omega}, \vec{X}).$$

We have

$$A_n = \sum_i (\mathrm{Proj}_{a_i} \vec{\omega}, \vec{X}) + \sum_i (\mathrm{Proj}_{e_i} \vec{\omega}, \vec{X}).$$

Since the bricks are orthogonal spaces and since projections are hermitian operators, we get

$$A_n = (\vec{\omega}, \mathrm{Proj}\,(\sum_i a_i)\, \vec{X}) + (\vec{\omega}, \mathrm{Proj}\,(\sum_i e_i)\, \vec{X}).$$

Now, since by Lemma [T.5.7.], $\mu(\sum_i e_i) \to 0$ for $n \to \infty$, we get by Lemma [T.4.2.], $\mathrm{Proj}\,(\sum_i e_i)\, \vec{X} \to \vec{O}$.

Hence

$$(\vec{\omega}, \mathrm{Proj}\,(\sum_i e_i)\, \vec{X}) \to (\vec{\omega}, \vec{O}) = 0.$$

Consequently

$$\lim A_n = (\vec{\omega}; \mathrm{Proj}_p \vec{X}),$$

because

$$|\sum_i a_i, p|_\mu \to 0,$$

which gives

$$\mathrm{Proj}\,(\sum_i a_i)\, \vec{X} \to \mathrm{Proj}_p \vec{X}.$$

The theorem is established.

T.7.1. Theorem. If

1. Hypothesis [T.5.2.] is admitted,
2. $\vec{X} \in \boldsymbol{H}$,

then, for special sum we get

$$x_\alpha = \underset{\beta \in W}{S^{\bullet}} \Delta(\alpha, \beta)\, x_\beta,$$

where x_α is the α-coordinate of $\vec{X}$.

Proof. This follows from [T.7.].

T.8. We define the number-valued function $\Delta'(\dot{p}, \dot{p})$ of the variable bricks p, q, both $\neq O$, as follows:

$$\Delta'(p, q) =_{df} \begin{cases} 0 & \text{whenever } p \cdot q = O, \\ \dfrac{1}{\max(\mu(p), \mu(q))} & \text{whenever } p \cdot q \neq O. \end{cases}$$

T.8.1. Theorem. If

1. Hypothesis [T.5.2.] is admitted,
2. $p \neq O$ is a brick space,
3. $\vec{X} \in \boldsymbol{H}$,

then, using special summation, we have

$$\frac{\text{Proj}_p X}{\mu(p)} = \underset{\beta \in W}{S^\bullet} \Delta'(p, \beta)\, \vec{x}_\beta, \tag{1}$$

where $\vec{x}_\beta$ is the β-component of $\vec{X}$.

Proof. Consider the completely distinguished special sequence $Q_n =_{df} \{q_{n1}, q_{n2}, \ldots\}$ for I. Denote by a_i, b_i, e_i the bricks of Q_n for which we have $a_i \leqq p$; $b_i \cdot p = O$; $e_i p \neq O$, $e_i \operatorname{co} p \neq O$ respectively. For the sum

$$\vec{A}_n =_{df} \sum_k \Delta'(p, q_{nk}) \,\text{Proj}\,(q_{nk})\, \vec{X},$$

which approximates (1), we have:

$$\vec{A}_n = \sum_i \Delta'(p, a_i) \,\text{Proj}\,(a_i)\, \vec{X} + \sum_i \Delta'(p, e_i) \,\text{Proj}\,(e_i)\, \vec{X}.$$

Since $a_i \leqq p$, we have $\Delta'(p, a_i) = \dfrac{1}{\mu(p)}$. Since $\mu(\sum e_i) \to 0$ (by Lemma [T.5.7.]), we have for sufficiently great n: $\mu(e_i) \leqq \mu(p)$ for all i; hence $\Delta'(p, e_i) = \dfrac{1}{\mu(p)}$. Thus we get

$$\vec{A}_n \cdot \mu(p) = \text{Proj}\,(\sum_i a_i)\, \vec{X} + \text{Proj}\,(\sum_i e_i)\, \vec{X}.$$

Now $|\sum_i a_i, p| \to 0$ and $\mu(\sum_i e_i) \to 0$. Hence $\mu(p) \cdot \lim \vec{A}_n = \text{Proj}_p \vec{X}$, so the theorem is proved.

T.8.2. Theorem. If

1. hypothesis [T.5.2.] is admitted,
2. $\vec{X} \in \boldsymbol{H}$,

then, for special summation, we have

$$\vec{x}^*_\alpha = \underset{\beta \in W}{S^\bullet} \Delta'(\alpha, \beta)\, x_\beta,$$

Def. [T.8.], where $\vec{x}^*_\alpha$ is the α-component-density of $\vec{X}$, and $\vec{x}_\beta$ the β-component of $\vec{X}$, [S.2.].

Proof. Relying on [T.8.1.].

T.9. Theorem. If

1. hypothesis [T.5.2.] is admitted,
2. $p \neq O$ is a brick space,
3. $\vec{X} \in \boldsymbol{H}$,

then, for special summation we have

$$\frac{(\mathrm{Proj}_p \vec{\omega}, \vec{X})}{\mu(p)} = \underset{\beta \in W}{S^\bullet} \Delta'(p, \beta) \cdot x_\beta,$$

Def. [T.8.].

Proof. Similar to that of Theorem [T.8.1.].

T.9.1. Theorem. If

1. hypothesis [T.5.2.] is admitted,
2. $\vec{X} \in \boldsymbol{H}$,

then, for special summation we have

$$x_\beta^* = \underset{\beta \in W}{S^\bullet} \Delta'(\alpha, \beta) \cdot x_\beta,$$

where x_α^* is the α-coordinate-density of $\vec{X}$, [S.2.].

Proof. Follows from [T.9.].

T.9.2. The formulas in [T.8.2.] and [T.9.1.] can be written

$$\vec{x}_\alpha^* = \underset{\beta \in W}{S^\bullet} \Delta'(\alpha, \beta)\, \vec{x}_\beta^* \cdot \mu_\beta, \qquad x_\alpha^* = \underset{\beta \in W}{S^\bullet} \Delta'(\alpha, \beta) \cdot x_\beta^* \cdot \mu_\beta,$$

which have the same shape as the Dirac's formula.

Proof. This follows from the equalities:

$$\vec{x}_\beta = \vec{x}_\beta^* \cdot \mu_\beta, \qquad x_\beta = x_\beta^* \cdot \mu_\beta,$$

[S.3.].

T.10. Lemma. If $E \in \boldsymbol{G}$, E does not contain any atom, P_n is a (DR)-distinguished sequence for E, then $\mathcal{N}(P_n) \to 0$.

Proof. Suppose that the thesis is not true. Then there exists a partial sequence $P_{s(n)}$ and $\eta > 0$, such that $\mathcal{N}(P_{s(n)}) \geqq \eta$.

Put $Q_n =_{df} P_{s(n)}$; $\{Q_n\}$ is also a (DR)-distinguished sequence for E. Put $Q_n = \{q_{n1}, q_{n2}, \ldots\}$. Since $\mathcal{N}(Q_n) = \max\{\mu(q_{n1}), \mu(q_{n2}), \ldots\}$, there exists $k(n)$ with $\mu(q_{n,k(n)}) \geqq \eta$. Now, since $\{Q_n\}$ is a (DR)-distinguished sequence for E, it follows that $\mathcal{N}_R(Q_n) \to 0$ i.e.

$$\max_k \{\mu(q_{nk} - \beta)\} \to 0,$$

where

$$\beta = A_1 + A_2 + \cdots + A_m + \cdots$$

is the sum of all atoms of $\boldsymbol{G}$. Hence $\mu(q_{n,k(n)} - \beta) \to 0$. Since $q_{n,k(n)} = (q_{n,k(n)} - \beta) + q_{n,k(n)}\beta$ for sufficiently great n, there exists n_0, such that for all $n \geqq n_0$,

$$\mu(\beta, q_{nk(n)} \geqq \frac{\eta}{2}. \tag{1}$$

Let m_0 be such, that

(2) $$\sum_{n > m_0} \mu A_n < \frac{\eta}{4}.$$

We have from (1)

$$\mu[q_{nk(n)} \sum_{n \leqq m_0} A_n + q_{nk(n)} \sum_{n > m_0} A_n] \geqq \frac{\eta}{2}.$$

Since the two terms are disjoint, we have

$$\mu[q_{nk(n)} \cdot \sum_{n \leqq m_0} A_n] + \mu[q_{nk(n)} \cdot \sum_{n > m_0} A_n] \geqq \frac{\eta}{2}.$$

Here the second term $< \frac{\eta}{4}$. Hence we have

$$\mu[q_{nk(n)} \cdot \sum_{n \leqq m} A_n] \geqq \frac{\eta}{4} \quad \text{for all} \quad n \geqq n_0.$$

Hence at least one atom among $A_1 \ldots A_{m_0}$ must be contained in $q_{nk(n)}$, for all $n \geqq n_0$.

Consequently there exists a partial sequence $l(n)$ and an atom A_i $(i \leqq m)$, such that

$$A_i \subseteqq q_{l(n), k(l(n))} \quad \text{for} \quad n = 1, 2, \ldots$$

Since

$$|E, Q_{l(n)}| \to 0,$$

we have

$$|E \cdot A_i, Q_{l(n)} A_i| \to 0, \quad [E \cdot A_i, A_i| \to 0,$$

i.e.

$$|O, A_i| \to 0,$$

which is impossible.

T.11. Def. We define the function

$$\Delta''(p, q) = \begin{cases} 0 & \text{whenever} \quad p \cdot q = O, \\ \dfrac{1}{\mu(p) + \mu(q)}, & \text{whenever} \quad p \cdot q \neq O. \end{cases}$$

This function generates the functions $\Delta''(\alpha, q)$, $\Delta''(p, \beta)$, $\Delta''(\alpha, \beta)$, where α, β are traces. We shall see that $\Delta''(\alpha, \beta)$ also has some properties of the δ-function.

T.12. Theorem. If

1. hypothesis [T.5.2.] is admitted,
2. $\boldsymbol{G}$ does not contain any atom,
3. $p \neq O$ is brick-space,
4. $\vec{X} \in \boldsymbol{H}$,

then, using special summations, we have

(1) $$\frac{\operatorname{Proj}_p \vec{X}}{\mu(p)} \underset{\beta \in W}{S^{\bullet}} \Delta''(p, \beta)\, \vec{x}_\beta.$$

Proof. Let $Q_n = q_{n1}, q_{n2}, \ldots$ be a completely distinguished and special sequence of complexes for I. Denote by a_i, b_i, e_i the bricks of Q_n with $a_i \leqq p$; $b_i \cdot p = O$; $e_i \cdot p \neq O$, $e_i \operatorname{co} p \neq O$ respectively. Consider the sum, yielding (1):

$$\vec{A}_n =_{df} \sum_k \Delta''(p, q_{nk}) \cdot \operatorname{Proj}_{q_{nk}} \vec{X}.$$

We have

$$(2) \quad \vec{A}_n = \sum_i \frac{1}{\mu(p)} \operatorname{Proj}_{a_i} \vec{X} + \sum_i \frac{1}{\mu(p)} \operatorname{Proj}_{e_i} \vec{X} + \\ + \sum_i \left[\frac{1}{\mu(p) + \mu(a_i)} - \frac{1}{\mu(p)}\right] \operatorname{Proj}_{a_i} \vec{X} + \\ + \sum_i \left[\frac{1}{\mu(p) + \mu(e_i)} - \frac{1}{\mu(p)}\right] \operatorname{Proj}_{e_i} \vec{X}.$$

The sum of two first terms in (2) is

$$\frac{1}{\mu(p)} \operatorname{Proj} \left(\sum_i a_i + \sum_i e_i\right) \vec{X},$$

and tends to $\dfrac{1}{\mu(p)} \operatorname{Proj}_p \vec{X}$, if $n \to \infty$. This follows from the fact, that $|\sum_i a_i, p|_\mu \to 0$ and $\mu \sum_i e_i \to 0$. Concerning the two last terms in (2), they are composed of expressions having the form:

$$(3) \qquad \left[\frac{1}{\mu(p) + \mu(c)} - \frac{1}{\mu(p)}\right] \operatorname{Proj}_c \vec{X},$$

where c is a brick. We have

$$\left\| \left[\frac{1}{\mu(p) + \mu(c)} - \frac{1}{\mu(p)}\right] \operatorname{Proj}_c \vec{X} \right\|^2 \leqq \frac{\mu(c)^2}{\mu(p)^2} \| \operatorname{Proj}_c \vec{X} \|^2.$$

Hence the square of the norm of the sum of the last terms in (2) does not exceed the number

$$(4) \qquad \frac{1}{\mu(p)^4} \sum_i \mu(a_i)^2 \| \operatorname{Proj}_{a_i} \vec{X} \|^2 + \frac{1}{\mu(p)^4} \sum_i \mu(e_i)^2 \cdot \| \operatorname{Proj} \vec{X} \|^2.$$

Take $\varepsilon > 0$. We have, for sufficiently great n, the inequality $\mathcal{N}(Q_n) \leqq \varepsilon$; hence $\mu(a_i) \leqq \varepsilon$, $\mu(e_i) \leqq \varepsilon$.

Thus the expression (4) does not exceed, (for those n), the number

$$\frac{\varepsilon^2}{\mu(p)^4} \left\{\sum_i \| \operatorname{Proj}_{a_i} X \|^2 + \sum_i \| \operatorname{Proj}_{e_i} \vec{X} \|^2\right\} = \\ = \frac{\varepsilon^2}{\mu(p)^4} \| \operatorname{Proj} \left(\sum_i a_i + \sum_i e_i\right) \vec{X} \|^2 \leqq \frac{\varepsilon^2}{\mu(p)^4} \cdot \| \vec{X} \|^2.$$

Consequently the sum of two last terms in (2) tends to $\vec{O}$, so

$$\lim \vec{A}_n = \frac{1}{\mu(p)} \operatorname{Proj}_p \vec{X},$$

what completes the proof.

T.12.1. If

1. hypothesis [T.5.2.] is admitted,
2. $\boldsymbol{G}$ does not contain atoms,
3. $\vec{X} \in \boldsymbol{H}$,

then for special summation we have:

$$\vec{x}_{\alpha}^{*} = \underset{\beta \in W}{S^{\bullet}} \Delta''(\alpha, \beta)\, x_{\beta} = \underset{\beta \in W}{S^{\bullet}} \Delta''(\alpha, \beta)\, \vec{x}_{\beta}^{*}\, \mu_{\beta}.$$

Proof. The theorem follows from [T.12.].

T.13. Theorem. If

1. hypothesis [T.5.2.] is admitted,
2. $\boldsymbol{G}$ does not contain atoms,
3. $p \neq O$ is a brick,
4. $\vec{X} \in \boldsymbol{H}$,

then, using special summation, we have,

$$\frac{(\mathrm{Proj}_p \vec{\omega}, \vec{X})}{\mu(p)} = \underset{\beta \in W}{S^{\bullet}} \Delta''(p, \beta) \cdot x_{\beta}.$$

Proof. Similar to that of Theorem [T.12.].

T.13.1. Theorem. If

1. hypothesis [T.5.2.] is admitted,
2. $\boldsymbol{G}$ does not contain atoms,
3. $\vec{X} \in \boldsymbol{H}$,

then, using special summation, we have

$$x_{\alpha}^{*} = \underset{\beta \in W}{S^{\bullet}} \Delta''(\alpha, \beta)\, x_{\beta} = \underset{\beta \in W}{S^{\bullet}} \Delta''(\alpha, \beta)\, x_{\beta}^{*}\, \mu_{\beta}. \tag{1}$$

(Which resembles the known Dirac's formula.)

Proof. This follows from [T.13.].

T.13.2. Remark. The formula (1) in Theorem [T.13.1.] may be not true, if $\boldsymbol{G}$ has an atom. E.g. take the one-dimensional H.-H.-space. The corresponding $\boldsymbol{G}$ is composed of two somata only, viz. O and I. There exists only one trace, which is a heavy one. In the right hand-side expression in (1), we get $\Delta''(1, 1) = \frac{1}{2\mu(1)}$, so the formula (1) is not true.

T.14. Consider the mapping $\mathscr{G}^{-1}$ of the H.-H.-space $\boldsymbol{H}$ onto the space $\boldsymbol{H}'$ of μ-square summable functions of the variable trace.

Let the $\mathscr{G}^{-1}$-image of the vector $\vec{X}$ be the function $f(\alpha)$. The system of coordinates in $\boldsymbol{H}$ goes over to an analogous system of coordinates in the space H'. The generating vector $\vec{\omega}$ goes over into the characteristic function $\Omega(\alpha)$ of W i.e. $\Omega(\alpha) = 1$ for all α. Then

$$(\mathrm{Proj}_p \vec{\omega}, \vec{X}) = \int \Omega_P(\alpha)\, f(\alpha)\, d\mu = \int_P f(\alpha)\, d\mu,$$

where P is such a set of traces, that the set of μ-square summable functions, vanishing outside P, is just the $\mathcal{G}^{-1}$-image of the space-brick. It follows that, for the $\mathcal{G}^{-1}$-corresponding system of coordinates in $\boldsymbol{H}'$ we have:

$$f_\alpha = \frac{\int\limits_P f(\alpha)\, d\mu}{\mu(P)} = \mathrm{val}_\alpha f$$

(see Def. [R.10.]). Thus the formula (1) in [T.13.1.] will become

$$\mathrm{val}_\alpha f = \underset{\beta \in W}{S^\bullet} \Delta''(\alpha, \beta)\, \mathrm{val}_\beta f\, \mu_\beta,$$

and a similar formula will be obtained from [T.9.2.], by using the Δ'-function.

T.15. Lemma. Under circumstances [B I.1.] we have the following: If

1. $\{p_n\}$ is a representative of the trace τ,
2. q is a brick,
3. $p_n \cdot q \neq O$ for $n = 1, 2, \ldots$,

then q is a neighborhood of τ, Def. [B I.10.].

Proof. Suppose that the thesis be not true; hence q does not cover the trace τ. Hence, whatever the representative $a_1 \geqq a_2 \geqq \cdots$ of τ may be, we always have $a_1 \cdot \mathrm{co}\, q \neq O$. Since $p_k \geqq p_{k+1} \geqq \cdots$ is a representative of τ, $(k = 1, 2, \ldots)$, it follows:

(1) $$p_k \cdot \mathrm{co}\, p \neq O \quad \text{for} \quad k = 1, 2, \ldots$$

We have

(2) $$p_1 q \geqq p_2 q \geqq \cdots \quad \text{all} \quad \neq O.$$

Since $p_1 \geqq p_2 \geqq \cdots$ is a minimal sequence, and

$$\{p_1 q, p_2 q, \ldots\} \leqq \{p_1, p_2, \ldots\},$$

(see [B I.2.1.]), it follows that either

$$\{p_1 q, p_2 q, \ldots\} \sim \{0, 0, \ldots\}$$

or

$$\{p_1 q, p_2 q, \ldots\} \sim \{p_1, p_2, \ldots\}.$$

The first alternative is impossible, because it would imply that $p_n q = O$ for sufficiently great n, [B I.10.2.], so it would lead to contradiction with (2). Hence $\{p_n q\} \sim \{p_n\}$. Consequently, for every n we can find m such that $p_m \leqq p_n q$; hence $p_m \leqq q$, which contradicts (1). The contradiction, thus obtained, proves that q is a neighborhood of τ.

T.15.1. Lemma. Under circumstances [B I.1.] if q is not a neighborhood of the trace τ, then there exists a neighborhood p of τ with $p \cdot q = O$.

Proof. Suppose the thesis be not true. Then for every neighborhood p of τ we have $p \cdot q \neq O$. Hence, if we take a representative $\{p_1 \geqq p_2 \geqq \cdots\}$ of τ, we get $p_k q \neq O$, $(k = 1, 2, \ldots)$. Hence, by the Lemma [T.15.], we see, that q is a neighborhood of τ; which is a contradiction.

T.16. In this chapter, we have defined "special completely distinguished sequences of complexes for I", and have used them in some theorems. To have useful consequences of them, we admit the following general definition, which however, will be later used only in the case where $\boldsymbol{G}$ is a tribe of spaces in H.-H.

Def. Let $\{Q_n\}$ be a completely distinguished and special sequence for I, and s a figure $\neq O$. Consider all bricks a_{nk}, e_{nk} of Q_n, such that $a_{nk} \leqq s$, $e_{nk} \cdot s \neq O$, $e_{nk} \cdot \mathrm{co}\, s \neq O$. We shall consider the two partial complexes $\{a_{n1}, a_{n2}, \ldots\}$ and $\{a_{n1}, a_{n2}, \ldots, e_{n1}, e_{n2}, \ldots\}$ of Q_n. The first will be termed *inner Q_n-coat of s, and denoted by int$(Q_n)\, s$*, the second will be termed *outer Q_n-coat of s and denoted by* ext$(Q_n)\, s$. Now, take any partial complex T_n of Q_n, such that $\mathrm{int}(Q_n)\, s \subseteqq T_n \subseteqq \mathrm{ext}(Q_n)\, s$. If we do that for all n, we get a sequence $\{T_n\}$ of complexes. Any sequence $\{T_n\}$ obtained in the above way will be termed "*special sequence for s, induced by $\{Q_n\}$*. Of course we have

$$\mathrm{som}\,\mathrm{int}(Q_n)\, s \leqq s \leqq \mathrm{som}\,\mathrm{ext}(Q_n)\, s,$$

and

$$\mu[s - \mathrm{som}\,\mathrm{int}(Q_n)\, s] \to 0, \qquad \mu[\mathrm{som}\,\mathrm{ext}(Q_n)\, s - s] \to 0 \quad \text{for} \quad n \to \infty.$$

Hence we get

$$|s, T_n|_\mu \to 0 \quad \text{for} \quad n \to \infty, \quad \text{so} \quad \{T_n\}$$

is a completely distinguished sequence for s, [Q.21.14.].

T.16.1. Def. Continuing the topic [T.16.], let us consider a regular and total set of quasi-vectors $\vec{f}_\alpha$ in the vector-space V. If the sum

$$\vec{f}(T_n) = \sum_k \vec{f}(t_{nk}),$$

where

$$T_n = \{t_{n1}, t_{n2}, \ldots\}$$

is a special sequence for s, induced by $\{Q_n\}$, tends to a limit, which does not depend neither on the choice of $\{T_n\}$ for a given $\{Q_m\}$, nor on the choice of $\{Q_n\}$, we shall say that $\vec{f}_\tau$ *is "specially summable on the figure s"*, and the limit mentioned above will be termed "*special sum of $\vec{f}_\tau$ on the figure s*", and denoted by

$$\underset{s}{S^\bullet} \vec{f}_\alpha .$$

T.16.2. The following theorem is valid. Consider the circumstances of [R.1.2.] and admit Hyp. [T.5.2.].

Theorem. If the special sum $\underset{I}{S^{\bullet}} \vec{f}_{\alpha}$ exists, and s is a figure, then

(1) $$\underset{s}{S^{\bullet}} \vec{f}_{\alpha}$$

also exists.

Proof. Suppose that (1) does not exist. This means that there exists a completely distinguished and special sequence of complexes $\{Q_n\}$ for I, such that if we consider $\operatorname{int}(Q_n)\, s$ and $\operatorname{ext}(Q_n)\, s$, we can find T_n, as in [T.16.], such that

(2) $$\operatorname{int}(Q_n)\, s \subseteqq T_n \subseteqq \operatorname{ext}(Q_n)\, s,$$

and where $\vec{f}(T_n)$ does not tend to any limit. Having that, we can find indices $n_1' < n_1'' < n_2' < n_2'' < \cdots$ such that

(3) $$\|\vec{f}(T_{n_k''}) - \vec{f}(T_{n_k'})\| \geqq a \quad \text{for} \quad k = 1, 2, \ldots,$$

and for some positive number a.

Put

$$P_n^{(1)} =_{df} \operatorname{int}(Q_n)\, s, \quad P_n^{(2)\prime} =_{df} T_{n_k'}, \quad P_n^{(2)\prime\prime} =_{df} T_{n_k''}, \quad P_n^{(3)} =_{df} \operatorname{ext}(Q_n)\, s,$$
$$P_n^{(4)} =_{df} \operatorname{int}(Q_n)\,(\operatorname{co} P_n^{(3)}).$$

We shall see that the complexes

4) $$P_1^{(2)\prime} \cup P_1^{(4)}, \quad P_1^{(2)\prime\prime} \cup P_1^{(4)}, \quad P_2^{(2)\prime} \cup P_2^{(4)}, \quad P_2^{(2)\prime\prime} \cup P_2^{(4)}, \ \ldots$$

make up a completely distinguished and special sequence for I. To prove that, notice that, since Q_n is completely distinguished and special for I, we have

$$\mu \operatorname{som}[P_n^{(3)} - P_n^{(1)}] \to 0$$

and

$$\mu \operatorname{som}[\operatorname{co} P_n^{(3)} - P_n^{(4)}] \to 0,$$

[T.5.6.] and [T.5.7.]. Since

$$\mu \operatorname{som} P_n^{(3)} + \mu \operatorname{co} \operatorname{som} P_n^{(3)} = \mu(I),$$

it follows, that the sum of all bricks of Q_n, which neither belong to $P_n^{(4)}$ nor to $P_n^{(1)}$, has the measure tending to 0 for

(5) $$n \to \infty.$$

Let q be a brick $\neq O$. We need to prove that

$$\mu \operatorname{som} \operatorname{int}(P_n^{(2)\prime} \cup P_n^{(4)})\, q \to \mu\, q$$

and

$$\mu \operatorname{som} \operatorname{int}(P_n^{(2)\prime\prime} \cup P_n^{(4)})\, q \to \mu\, q.$$

Since

$$P_n^{(2)\prime} \cup P_n^{(4)} \subseteqq Q_n$$

and

$$P_n^{(2)''} \cup P_n^{(4)} \subseteqq Q_n,$$

it follows that

$$\operatorname{int}(P_n^{(2)'} \cup P_n^{(4)})\, q \subseteqq \operatorname{int}(Q_n)\, q \to^{\mu} q,$$

and

$$\operatorname{int}(P_n^{(2)''} \cup P_n^{(4)})\, q \subseteqq \operatorname{int}(Q_n)\, q \to^{\mu} q.$$

Now the bricks of $\operatorname{int}(Q_n)\, q$, which do not belong to $P_n^{(2)'} \cup P_n^{(4)}$, are not contained in $P_n^{(1)}$, because $P_n^{(1)} \subseteqq P_n^{(2)'}$; hence, by (5), the measure of the sum of bricks which do not belong to $P_n^{(2)'} \cup P_n^{(4)}$, has the measure tending to 0. Consequently

$$\mu \operatorname{som}(P_n^{(2)'} \cup P_n^{(4)})\, q \to 0.$$

Similarly we have

$$\mu \operatorname{som}(P_n^{(2)''} \cup P_n^{(4)})\, q \to 0.$$

This proves that (4) is a completely distinguished and special sequence for I. By hypothesis both sequences

$$\vec{f}(P_n^{(2)'} \cup P_1^{(4)}), \quad \vec{f}(P_n^{(2)''} \cup P_1^{(4)})$$

tend to the same limit. Hence

$$\vec{f}(P_n^{(2)'}) - \vec{f}(P_n^{(2)''})$$

tends to $\vec{O}$, which contradicts (3). The theorem is proved.

T.16.3. Theorem. We admit the Hyp. [T.5.2.] and circumstances [R.2.]. Then if

1) $\vec{f}_\tau$ is a regular total set of V-quasi-vectors,
2) a, b, c are bricks, $c = a + b$, $a \cdot b = O$,
3) the special sum $\underset{c}{S^\bullet} \vec{f}_\tau$ exists,

then for special sums we have

$$\underset{a}{S^\bullet} \vec{f}_\tau + \underset{b}{S^\bullet} \vec{f}_\tau = \underset{c}{S^\bullet} \vec{f}_\tau.$$

Proof. By [T.16.2.] all these sums exist. Let $\{Q_n\}$ be a completely distinguished and special sequence of complexes for I. Consider the complexes

$$\operatorname{int}(Q_n)\, a, \quad \operatorname{int}(Q_n)\, b \quad \text{and} \quad \operatorname{int}(Q_n)\, c.$$

We have

$$\operatorname{int}(Q_n)\, a \cup \operatorname{int}(Q_n)\, b \subseteqq \operatorname{int}(Q_n)\, c.$$

We have

$$\mu \operatorname{som} \operatorname{int}(Q_n)\, a \to \mu(a), \quad \mu \operatorname{som} \operatorname{int}(Q_n)\, b \to \mu(b),$$

$$\mu \operatorname{som} \operatorname{int}(Q_n)\, c \to \mu(c).$$

Hence, if we put

$$P_n =_{df} \operatorname{int}(Q_n)\, c - [\operatorname{int}(Q_n)\, a \cup \operatorname{int}(Q_n)\, b]$$

we get

$$\lim \mu P_n = \mu(c), \quad P_n \subseteqq \operatorname{int}(Q_n)\, c.$$

Consider the sequence

(1) $$R_n \equiv_{df} \operatorname{int}(Q_n)\, c - P_n = \operatorname{int}(Q_n)\, a \cup \operatorname{int}(Q_n)\, b.$$

This is a sequence induced for c by the sequence $Q_n - P_n$, which is a completely distinguished and special sequence for I. The last can be proved, by using a method, given in the forgoing proof. We have

$$\lim \vec{f}(Q_n) = \underset{c}{S^\bullet} \vec{f}_\tau, \quad \lim \vec{f}[\operatorname{int}(Q_n)\, a] = \underset{a}{S^\bullet} \vec{f}_\tau,$$

and

$$\lim \vec{f}[\operatorname{int}(Q_n)\, b] = \underset{b}{S^\bullet} \vec{f}.$$

Thus, by (1), we get the thesis.

T.16.4. Remark. We do not know, whether the special sum $\underset{a}{S^\bullet} \vec{f}$ can exist without existence of the sum $\underset{a\,(\mathrm{DARS})}{S} \vec{f}$. Our conjecture is "Yes".

T.16.5. If

1. $\{Q_n\}$ is a completely distinguished and special sequence for I,
2. s is a brick $\neq O$,
3. $T_n =_{df} \operatorname{int}(Q_n)\, s$, $n = 1, 2, \ldots$,

then $\{T_n\}$ is a completely distinguished and special sequence in the tribe $s \rceil \boldsymbol{G}$, (restriction to s), in the sense of Def. [T.5.].

T.17. Having that, we are going to get some useful modifications of various forgoing theorems, which involve Δ-functions. To simplify wording, we shall use the same letter for a measurable set of traces and for its coat.

T.17.1. Theorem. If

1. Hyp. [T.5.2.] is admitted,
2. p is a brick,
3. s is a figure with $s \leqq p$,
4. $\vec{X} \in \boldsymbol{H}$,

then

$$\operatorname{Proj}_p \vec{X} = \underset{\beta \in s}{S^\bullet} \Delta(p, \beta)\, \vec{x}_\beta,$$

Def. [T.3.].

Proof. Since $\underset{\beta \in W}{S^\bullet} \Delta(p, \beta)\, \vec{x}_\beta$ exists, [T.6.], therefore, by [T.16.3.],

(1) $$\underset{\beta \in W}{S^\bullet} \Delta(p, \beta)\, \vec{x}_\beta = \underset{\beta \in s}{S^\bullet} + \underset{\beta \in \cos}{S^\bullet}.$$

To evaluate $\underset{\beta \in \cos}{S^\bullet}$, take a completely distinguished and special sequence $\{Q_n\}$ for I and consider $\operatorname{int}(Q_n) \cdot \cos$. We have for its bricks b_{nk}

(2) $$\sum_k \Delta(p, b_{nk}) \operatorname{Proj}(b_{nk})\, \vec{X} = \vec{O}.$$

Since $\operatorname{int}(Q_n)\cos s$ is a special sequence, induced by $\{Q_n\}$, Def. [T.16.], the sum (2) tends to $\underset{\beta\in\cos s}{S^\bullet}$. Now, $\underset{\beta\in\cos s}{S^\bullet}\,\Delta(p,\beta)\,\vec{x}_\beta=\vec{O}$, because $\Delta(p,b_{nk})=0$ for all k. Consequently, by [T.6.] and (1): $\operatorname{Proj}_p\vec{X}=\underset{\beta\in s}{S^\bullet}$, so the theorem is proved.

T.17.2. In a similar way, considering $\operatorname{int}(Q_n)\cdot\cos s$, we can prove similar variants of the Theorems [T.6.1.], [T.7.], [T.7.1.], [T.8.1.], [T.8.2.], [T.9.], [T.9.1.], [T.9.2.], [T.12.], [T.12.1.], [T.13.], [T.13.1.], [T.14.]. In the changed theorems we have the additional hypothesis $\beta\leq s$, where s is a figure, or in theorems involving the trace α we have the hypothesis $\alpha\in s$, and take account of (Lemma [T.15.] and [T.15.1.]). $\underset{\beta\in W}{S^\bullet}$ is replaced by $\underset{\beta\in s}{S^\bullet}$.

We shall refer to these theorems by giving the number e.g. [T.7.1.] and adding [T.17.2.].

T.18. The following remark considers the influence, exercised on various summation formulas, by the change of measure. If $\underset{\tau\in W}{S^\bullet}\vec{f}_\tau=\underset{\tau\in s}{S^\bullet}\vec{f}_\tau$, where s is a figure, we see that the change of the measure μ outside s will not influence the summation-formula.

T.19. A similar remark can be made on functions, which vanish outside a given brick.

T.20. Theorem. If

1. G has no atoms,
2. hyp. [T.5.2.] is admitted,
3. $\vec{f}_\tau$, $\vec{g}_\tau$ are total sets of regular and normal quasi-vectors,
4. $\vec{f}_\tau\approx\vec{g}_\tau$ in the sense of [R.7.],
5. $p\neq O$ is a brick,
6. $\underset{W}{S}\,\Delta''(p,\tau)\,\vec{f}_\tau$ exists,

then

$$\underset{W}{S}\,\Delta''(p,\tau)\,\vec{g}_\tau$$

exists too and equals the sum 6.

Proof. Hypothesis 3 means, that for every measurable set E of traces we have in the (DARS)-summation, Def. [R.5.],

$$\underset{E}{S}\,\vec{f}_\tau=\underset{E}{S}\,\vec{g}_\tau. \tag{0}$$

Consider a completely distinguished and special sequence

$$Q_n=\{q_{n1},q_{n2},\ldots\}$$

of complexes for I. Put $\vec{h}_\tau=_{df}\vec{f}_\tau-\vec{g}_\tau$. Consider the sum

$$\vec{A}_n=_{df}\sum_k\Delta''(p,q_{nk})\cdot\vec{h}(q_{nk}).$$

Denoting by c_k the bricks of $\operatorname{int}(Q_n)\,p$, [T.16.], we get

$$\vec{A}_n = \sum_k \frac{1}{\mu(p) + \mu(c_k)} \cdot \vec{h}(c_k),$$

(1) $$\vec{A}_n = \sum_k \frac{1}{\mu(p)}\,\vec{h}(c_k) + \sum_k \left(\frac{1}{\mu(p) + \mu(c_k)} - \frac{1}{\mu(p)}\right) \vec{h}(c_k).$$

We have

$$\left\| \sum_k \left(\frac{1}{\mu(p) + \mu(c_k)} - \frac{1}{\mu(p)}\right) \cdot \vec{h}(c_k) \right\|^2 \leqq \sum_k \frac{\mu(c_k)^2}{(\mu(p))^4} \, \|\vec{h}(c_k)\|^2.$$

Since $\boldsymbol{G}$ has no atoms, we have $\mathcal{N}(Q_n) \to 0$. Hence for sufficiently great index n we get $\mu(c_k) \leqq \varepsilon$, where ε is any positive number, given in advance. We get

(2) $$\left\| \sum_k \left(\frac{1}{\mu(p) + \mu(c_k)} - \frac{1}{\mu(p)}\right) \vec{h}(c_k) \right\|^2 \leqq \frac{\varepsilon^2}{(\mu(p))^4} \cdot \|\sum_k \vec{h}(c_k)\|^2.$$

As $\vec{f}_\tau$ and $\vec{g}_\tau$ are summable, it follows that $\vec{h}_\tau = \vec{f}_\tau - \vec{g}_\tau$ is summable. Hence [R I.5.1.] $\|\vec{h}_\tau\|^2$ is summable. Since $\lim \|\sum_k h(c_k)\|^2 = \underset{p}{S} \|\vec{h}_\tau\|^2$, it follows that $\|\sum_k \vec{h}(c_k)\|^2$ is bounded. Thus there exists $M > 0$, such that $\|\sum_k \vec{h}(c_k)\|^2 \leqq M$. It follows that the expression on the left in (2) does not exceed $\frac{\varepsilon^2}{(\mu(p))^4} \cdot M$, so it tends to 0 for $\varepsilon \to 0$. The first term in (1) is $\frac{1}{\mu(p)} \sum_k \vec{h}(c_k)$, and tends to $\frac{1}{\mu(p)} \underset{p}{S}\, \vec{h}_\tau = \vec{O}$, by (0). Consequently $\vec{A}_n \to \vec{O}$, which proves the theorem.

T.21. Def. Denote by c_α the quasi-number, defined by

$$f(p) =_{df} c = \text{const}$$

for all neighborhoods of α.

T.21.1. Theorem. If $\boldsymbol{G}$ does not admit atoms, we have

(1) $$I_\alpha = \underset{\beta \in W}{S^\bullet} \Delta''(\alpha, \beta)\, \mu_\beta.$$

Proof. We have for every $\vec{X} \in \boldsymbol{H}$:

$$x^*_\alpha = \underset{\beta \in W}{S^\bullet} \Delta''(\alpha, \beta)\, x^*_\beta\, \mu_\beta.$$

Applying this formula to the generating vector $\vec{\omega}$, we get

$$\omega^*_\alpha = \underset{\beta \in W}{S^\bullet} \Delta''(\alpha, \beta)\, \omega^*_\beta\, \mu_\beta.$$

Now

$$\omega^*_\alpha = \frac{(\mathrm{Proj}_p \vec{\omega}, \vec{\omega})}{\mu(p)} = \frac{\|\mathrm{Proj}_p \vec{\omega}\|^2}{\mu(p)} = I_\alpha.$$

Since

$$\Delta''(\alpha, \beta) \cdot I_\beta = \Delta''(\dot{p}, \dot{q}) \cdot I = \Delta''(\dot{p}, \dot{q}) = \Delta''(\alpha, \beta),$$

we get the formula (1).

T.22. We have defined three functions Δ, Δ', Δ'', which have some properties of the Dirac's-delta-function under very general conditions. We shall terminate this chapter with the study of the genuine δ-function in relation to ordinary functions of a real variable and Lebesguean measure.

T.22.1. To do this we shall consider the space H of all complex-valued Lebesgue-square-summable functions $f(x)$, defined almost everywhere in the half open interval $(A, B\rangle$. The governing equality, $=$, will be that modulo sets of measure 0. We shall have the alternative notation $\vec{f}$ for $f(x)$. The scalar product $(\vec{f}, \vec{g})$ is defined as $\int_0^1 \overline{f(x)}\, g(x)\, dx$, so H is a separable and complete, infinite dimensional H.-H.-space.

Consider the Lebesgue-measurable subsets $(0, 1\rangle$, considered modulo sets of measure 0; so we have $E \doteq F$ whenever

$$\text{meas}(E - F) + \text{meas}(F - E) = 0.$$

The collection of these sets, with ordering defined by

$$E \geqq F \cdot =_{df} \cdot \text{meas}(E - F) = 0,$$

is organized into a Boolean denumerably additive tribe $\boldsymbol{g}$ with effective measure. Consider the collection $\boldsymbol{b}$ of all subsets $(\alpha, \beta\rangle =_{df} \{x \mid \alpha < x \leqq \beta\}$ of $(0, 1\rangle$, where $0 \leqq \alpha, \beta \leqq 1$; and denote by $\boldsymbol{f}$ all finite unions of sets of $\boldsymbol{b}$. We see that $(\boldsymbol{f})$ is a finitely additive tribe, and $\boldsymbol{b}$ its base. The tribe $\boldsymbol{g}$ is a finitely genuine extension of $\boldsymbol{f}$ through the isomorphism from $\boldsymbol{f}$ into $\boldsymbol{g}$, which attaches to every set $(\alpha, \beta\rangle$ the sets $(\alpha, \beta\rangle + E_1 - E_2$, where $\text{meas}\, E_1 = \text{meas}\, E_2 = 0$. The hypothesis, ***(Hyp Af)*** is satisfied; hence, a fortiori ***(Hyp Ad)***. $\boldsymbol{g}$ contains the borelian extension of $\boldsymbol{f}$ within $\boldsymbol{g}$ and $\boldsymbol{g}$ is also the Lebesgue-covering extension [A I.] of $\boldsymbol{f}$ within $\boldsymbol{g}$, where the measure on $\boldsymbol{f}$ is euclidian (hence Lebesguean). Thus [Hyp. $\boldsymbol{R_e}$] is satisfied. The hypothesis [Hyp. ($\boldsymbol{S}$)] of measure-separability of $\boldsymbol{g}$ [A.21.1.] is satisfied. Consequently there exists a completely distinguished sequence of complexes for the soma $I =_{df} (0, 1\rangle$, Def. [Q.21.3.].

If we consider the partitions of $(0, 1\rangle$ into n equal halfopen-segments $(n = 1, 2, \ldots)$, we get a completely distinguished and special sequence for I, Def. [T.5.]; thus Hyp. [T.5.2.] is satisfied. Since $\boldsymbol{g}$ has no atoms, the distinguished sequences of type (D), (DA), (DR), (DAR) for a measurable set E coincide, [Q I.]. The base $\boldsymbol{b}$ in $\boldsymbol{f}$ gives rise to traces, [B I.] (see also [R I.2.8.]). To every point x, where $0 < x < 1$ there correspond two different traces, one x^+ with representative $(x, x + \varepsilon_n\rangle$, where $\varepsilon_n > 0$, $\varepsilon_n \to 0$, $\varepsilon_n \geqq \varepsilon_{n+1}$, $(n = 1, 2, \ldots)$, $(x, x + \varepsilon_n\rangle \subsetneqq (0, 1\rangle$; and another one x^- with representative $(x - \varepsilon_n, x\rangle$. At $x = 0$ we have only one trace 0^+, and at $x = 1$ only one trace 1^-. The point x *will be termed vertex of* x^+ *and of* x^-.

The hypotheses **(*Hyp* I)** and **(*Hyp* II)**, [B I.6.] are satisfied, so the whole theory of measurability of sets of traces, [B I.], is valid. One can prove, that if α is a measurable set of traces in $\boldsymbol{g}$, then α contains with a trace x^+ also x^- excepting perhaps for a set of traces with measure 0. Such a set of traces has the same measure as the set of their vertices; (see [B I.]).

T.22.2. Let us define the correspondence $\mathfrak{M}$ as follows. If E is a measurable subset of $(0, 1\rangle$, consider all square-summable functions $f(x)$ defined a.e. on $(0, 1\rangle$, such that $f(x) = 0$ on the set $\operatorname{co} E$. Denote by e the collection of all those functions. Now the correspondence $\mathfrak{M}$ is defined as that, which attaches e to E. $\mathfrak{M}$ is invariant in its domain with respect to the equality $\doteq$ of sets. The set e is a closed subspace of the Hilbert-Hermite-space H. The correspondence $\mathfrak{M}$ is an ordering and operation-isomorphism from $\boldsymbol{g}$ into a denumerably additive tribe $\boldsymbol{G}$ of subspaces. $\mathfrak{M}$ transforms sets into spaces. Put $\boldsymbol{F} =_{df} \mathfrak{M}\, \boldsymbol{f}$, $\boldsymbol{B} =_{df} \mathfrak{M}\, \boldsymbol{b}$ and define on $\boldsymbol{G}$ the measure by $\mu\, e =_{df} \operatorname{meas} E$. Thus we see that all circumstances in $\boldsymbol{g}$ have their image in $\boldsymbol{G}$. The brick-spaces are the $\mathfrak{M}$-images of half-open sets $(\alpha, \beta\rangle$.

T.22.3. The tribe $\boldsymbol{G}$ is saturated. The function

$$\vec{\omega} =_{df} \omega(x) =_{df} \text{const} = 1$$

is a generating vector of H with respect to $\boldsymbol{G}$, so we can use the whole theory of quasi-vectors, as developped in [R I.], and also the system of coordinates, as defined in [S.]. The generating vector $\vec{\omega}$ and the saturated tribe generate an isomorphic correspondence $\mathcal{G}^{-1}$, [R I.4.], which transforms the vectors $\vec{f}(x)$ of the space of H into square-summable functions $F(\tau)$ of the variable trace. Now, we can prove that $F(x^+) = F(x^-) = f(x)$ for almost all x, so we may use the symbol $\boldsymbol{f}$ instead of F. We can also always suppose that $f(x) = f(x^+) = f(x^-)$ for every x, since null sets do not matter at all.

T.23. The set $\boldsymbol{B}$ of bricks, the saturated tribe $\boldsymbol{G}$ of spaces and the generating vector $\vec{\omega}$ constitute a system of coordinates in H, [S.]. Let $\vec{F} =_{df} f(x)$ be a vector, Φ a space-trace, P its variable neighborhood. Put $\varphi =_{df} \mathfrak{M}^{-1}\, \Phi$ and $p = \mathfrak{M}^{-1}\, P$. The representatives of Φ are descending sequences of spaces and the representatives of φ a descending sequences of half-open segments p, [B I.2.]. Instead of dealing with spaces, we prefer to describe space circumstances by the $\mathfrak{M}^{-1}$-corresponding items in relation to the real axis. Indeed, the space H and its $\mathcal{G}^{-1}$ image are isomorphic and isometric, so we may "identify"—the corresponding items— just for the sake of simplicity.

T.23.1. The Φ-component of $\vec{F}$ is, Def. [S.2.], the quasi-vector

$$\vec{F}_\Phi =_{df} \operatorname{Proj}_P \vec{F}.$$

We may represent it as the function $f_p(x)$ depending on p and defined by

$$f_p(x) =_{df} \begin{cases} 0 & \text{for} \quad x \in \operatorname{co} p, \\ f(x) & \text{for} \quad x \in p. \end{cases}$$

It can be denoted by $\vec{f}_\varphi$, and it looks like an infinitesimal piece of the function taken on p at the trace φ, and completed a.e. outside of p by the values 0. The Φ-component-density of $\vec{F}$ is, Def. [S.2.], the quasi-vector

$$F^*_\Phi =_{df} \frac{\operatorname{Proj}_{\dot{P}} \vec{F}}{\mu(\dot{P})}, \quad \text{where} \quad \mu(P) = \operatorname{meas} p.$$

It may be represented by "infinitesimal" function-pieces:

$$f^*_p(x) =_{df} \begin{cases} 0 & \text{for} \quad x \in \operatorname{co} p, \\ \dfrac{f(x)}{\mu(p)} & \text{for} \quad x \in p. \end{cases}$$

It can be denoted by $\vec{f}^*_\varphi$. The Φ-component of $\vec{F}$ is, Def. [S.2.], the quasi-number

$$F_\Phi =_{df} (\vec{\omega}_P, \vec{F})$$

(scalar product). Since to the vector $\vec{\omega}$ there corresponds the a.e. constant function $\Omega(x) = 1$, and since to the vector $\vec{\omega}_P$ there corresponds the characteristic function $\Omega_P(x)$ of the interval p, we have

$$(\vec{\omega}_P, F) = \int_0^1 \overline{\Omega_p(x)}\, f(x)\, dx = \int_p f(x)\, dx,$$

so F_Φ can be represented by the quasi-number $f^*_\varphi =_{df} \int_p f(x)\, dx$, defined for all half-open segments p, which cover the trace φ.

The Φ-component-density of $\vec{F}$ is, Def. [S.2.], the quasi-number

$$F^*_\Phi =_{df} \frac{(\vec{\omega}_{\dot{P}}, \vec{F})}{\mu(\dot{P})}.$$

Hence it can be represented by the quasi-number

$$f^*_\varphi =_{df} f^*(\dot{p}) = \frac{1}{\operatorname{meas} \dot{p}} \int_{\dot{p}} f(x)\, dx = \operatorname{val}_\varphi f,$$

Def. [R.10.], i.e. by the locally taken mean-value of $f(x)$.

T.23.2. If $f(x)$ is continuous at the point x_0, and we consider the two traces x_0^+, x_0^- (with vertex x_0), and if $(a, b\rangle$ is a variable segment, which covers one or another of this traces, with $\lim (b - a) = 0$,

we get

$$\lim f^*((a, b\rangle) = f(x_0).$$

T.23.3. We have

$$\vec{F}_\Phi \approx \vec{\omega}_\Phi \cdot F^*_\Phi,$$

which can be written as

$$\vec{f}_\varphi \approx \vec{\Omega}_\varphi \cdot \mathrm{val}_\varphi f = \Omega(\dot{p}) \frac{\int\limits_{\dot{p}} f\,dx}{\mathrm{meas}\,\dot{p}}.$$

The symbol $\vec{\Omega}_\varphi$ may be called *characteristic "function" of the trace* φ. If the function $f(x)$ is continuous at x_0, (compare [T.23.2.]), $\vec{f}_\varphi$ can be intuitively conceived as $f(x_0) \cdot \Omega(\dot{p})$.

T.24. We shall need some operations on sets. In general, if E, F are segments, and $\Gamma(x, y)$ is a real-valued function of the real variables x, y, we define

$$\Gamma(E, F) \quad \text{as the set} \quad \{\Gamma(x, y) \mid x \in E, y \in F\}.$$

Thus e.g. we have

$$E - F = \{x - y \mid x \in E, y \in F\},$$

$$a \cdot E = \{a\,x \mid x \in E\}$$

for any number a,

$$E^2 = \{x^2 \mid x \in E\}.$$

We define

$$E - a = \{x - a \mid x \in E\}, \qquad -E =_{df} \{-x \mid x \in E\}.$$

Similarly we define

$$f(E) = \{f(x) \mid x \in E\}.$$

T.24.1. Concerning the "difference" $E - F$ of two sets E, F, the following are equivalent:

I. $\mathsf{O} \in E - F$,

II. $E, F \mathrel{=}\!\!|\!\!\mathrel{=} \mathsf{O}$.

Indeed, let I. There exist $x \in E$, $y \in F$ with $0 = x - y$. Taking such x and y, we get $x = y$; hence $E \cap F \neq \mathsf{O}$.

Let II. There exists $x \in E \cap F$. Since $x - x = 0$, we get $0 \in E - F$.

T.24.2. The "difference" $(a, b\rangle - (c, d\rangle$ is always an open segment. The difference of any two intervals is an interval.

T.24.3. We have for the Lebesgue's measure: $\mathrm{meas}\,E = \mathrm{meas}\,[E + a]$ for any number a.

T.24.4. For intervals E, F of any kind we have

$$\mathrm{meas}\,(E - F) = \mathrm{meas}\,E + \mathrm{meas}\,F.$$

Proof. Let us close the intervals. This will not affect the measure. Let $a < b$, $c < d$ be the extremities of E and F respectively. If

$$0 \leqq x \leqq b, \quad c \leqq y \leqq d,$$

we get

$$a - d \leqq x - y \leqq b - c.$$

Hence

$$\operatorname{meas}(E - F) = (b - c) - (a - d) = (b - a) + (d - c)$$
$$= \operatorname{meas} E + \operatorname{meas} F.$$

T.24.5. If

$$k > 0,$$

then

$$(a, b\rangle\, k = (k\,a, k\,b\rangle$$

and

$$(a, b\rangle\,(-k) = \langle -k\,b, -k\,a).$$

T.24.6. If

$$0 \leqq a - b,$$

then

$$(a, b\rangle^2 = (a^2, b^2\rangle.$$

If

$$a < b \leqq 0,$$

then

$$(a, b\rangle^2 = \langle b^2, a^2).$$

If

$$a < 0 < b,$$

then

$$(a, b\rangle^2$$

is an interval with left extremity 0; its length is $\max(a^2, b^2)$.

T.24.7. The notion of the number-valued function $f(E)$, or $g(E, F)$, where E, F are intervals half closed to the right, induces the notion of symbols $f(\alpha)$, $g(\alpha, F)$, $g(E, \beta)$, $g(\alpha, \beta)$, where α, β are traces. Thus $f(\alpha)$ means the quasi-number $f(\dot{p})$, where p varies over all neighborhoods of α. Similarly $g(\alpha, \beta)$ is the function $g(\dot{p}, \dot{q})$, where p varies over all neighborhoods of α and q over all neighborhoods of β.

T.25. We define the function $\delta(E, F)$ for any intervals E, F as follows:

$$\delta(E, F) =_{df} 0, \quad \text{whenever} \quad E \cap F = \emptyset,$$

$$\delta(E, F) =_{df} \frac{1}{\operatorname{meas}(E - F)} = \frac{1}{\operatorname{meas} E + \operatorname{meas} F}, \quad \text{whenever} \quad E \cap F \neq \emptyset.$$

This function will be proved to be a good version of Dirac's δ-function. Notice that we consider our δ-function as a function of two variables. The function generates the functions $\delta(\alpha, F)$, $\delta(E, \beta)$, $\delta(\alpha, \beta)$ where α, β are traces (see [T.1.2.]).

T.25.0. E.g. $\delta(-\varphi, p)$ will mean the quasi-number $\delta(-q, p)$, where q varies over all neighborhoods of the trace φ. Thus if $q = (a, b\rangle$, then we take $\delta((-b, -a), p)$, where $(a, b\rangle$ is the variable neighborhood of φ.

Concerning $-\varphi$, we do not need to define it. Its use is meaningful only.

T.25.1. We have $\delta(E, F) = \delta(F, E)$; hence $\delta(\alpha, \beta) = \delta(\beta, \alpha)$.

T.25.1.1. $\delta(E, F) = \delta(-E, -F)$.

T.25.2. The function $\delta(E, F)$ of variable intervals can be considered as depending only on $E - F$. Indeed, let $E - F = E_1 - F_1$. If $E \cap F \neq \emptyset$, then, [T.24.1.], $0 \in E - F$; hence $0 \in E_1 - F_1$ and then, [T.24.1.], $E_1 \cap F_1 \neq \emptyset$. Conversely, if $E_1 \cap F_1 \neq \emptyset$, then $E \cap F \neq \emptyset$. Consequently $\delta(E, F) = \delta(E_1, F_1)$.

T.25.3. $\delta(E, F)$ has the translation property, i.e. if a is a number, then $\delta(E + a, F + a) = \delta(E, F)$.

Proof. If $E \cap F \neq \emptyset$, then $0 \in E - F$, [T.24.1.]; hence there exists x with $x \in E, x \in F$. Hence $x + a \in E + a, x + a \in F + a$. Consequently $x + a \in (E + a) \cap (F + a)$, which gives $(E + a) \cap (F + a) \neq \emptyset$. Similarly we prove that, if $(E + a) \cap (F + a) \neq \emptyset$, we get $E \cap F \neq \emptyset$. Thus $\delta(E, F) = 0$ is equivalent to $\delta(E + a, F + a) = 0$. On the other hand we have, the measure being Lebesguean,

$$\operatorname{meas} E = \operatorname{meas}(a + E), \qquad \operatorname{meas} F = \operatorname{meas}(a + F).$$

Hence

$$\frac{1}{\operatorname{meas} E + \operatorname{meas} F} = \frac{1}{\operatorname{meas}(E + a) + \operatorname{meas}(F + a)}.$$

T.25.4. We have for traces

$$b^+ - a = (b - a)^+,$$
$$b^- - a = (b - a)^-.$$

T.25.5. We have for any two traces $a^\pm$, $b^\pm$

$$\delta(a^\pm, b^\pm) = \delta(a^\pm - a, b^\pm - a) = \delta(0^\pm, (b - a)^\pm)$$

and

$$\delta(a^\pm, b^\mp) = \delta(a^\pm - a, b^\mp - a) = \delta(a^\pm, (b - a)^\mp).$$

T.26. If $f(\dot{x})$ is a square-summable function on $(0, 1\rangle$, then

$$\operatorname{val}_\alpha f(\dot{x}) = \underset{\beta \in W}{S^\bullet} \delta(\alpha, \beta) \operatorname{val}_\beta f(\dot{x}) \, |\beta|,$$

where $|\beta|$ denotes the quasi-number of the measure (denoted in [R.7.8.] as μ_β).

The theorem follows from [T.14.]. Indeed, $\Delta''(\alpha, \beta) = \delta(\alpha, \beta)$, and we can take [T.24.4.] into account.

T.26.0. Remark. The Theorem [T.26.] can be considered as a corollary of the proof of Theorem [T.12.], which is more general. Now, if we consider the proof, we can notice that, in our case with the function δ, in the expressions $\delta(q_{ik}, p)$ the interval p can be replaced by the interval with the same extremities, but closed on the left. There will be only at most two intervals q_{ik}, which have a single point in common with the changed p, and their presence or absence will not influence the limit. Another remark is, that considering the above sums, we can drop those q_{ik}, which have no common point with p, as the summation may be carried out even on suitable subsets of W instead on W itself. Later we shall apply this remark in the proof of Theorem [T.28.].

T.26.1. Remark. The Theorem [T.26.] constitutes the main theorem of the Dirac's δ-function. It is the main source of other ones. The theorem can be put into another form

$$\mathrm{val}_{a\pm} f(\dot{x}) = \underset{\beta\in W}{S^\bullet} \delta(0^\pm; \beta - a) \, \mathrm{val}_\beta f(\dot{x}) \, |\beta|$$
$$= \underset{\beta\in W}{S^\bullet} \delta(\beta - a; 0^\pm) \, \mathrm{val}_\beta f(\dot{x}) \, |\beta| .$$

T.26.2. Remark. We have many theorems, having the shape of [T.26.1.], e.g. From [T.12.1.] we get, by [T.23.3.] and [T.20.]:

$$\vec{f}^*_\alpha = \underset{\beta\in W}{S^\bullet} \delta(\alpha, \beta) \, \vec{f}_\beta |\beta| = \underset{\beta\in W}{S^\bullet} \delta(\alpha, \beta) \, \vec{\Omega}_\beta \, \mathrm{val}_\beta \vec{f} |\beta| .$$

T.27. To give precise statement of some theorem on δ-function, we need the following notion *of Dirac-equivalence for quasi-numbers.*

Def. Let $\vec{f}_\alpha$, $\vec{g}_\alpha$ be two quasi-vectors (quasi-numbers) with support α. We say that $\vec{f}_\alpha =^{\lim} \vec{g}_\alpha$ (they *are equal in "limit"*), whenever $\lim \vec{f}(p) = \lim \vec{g}(p)$ for meas $p \to 0$, and p covering the trace α.

T.27.1. Def. Let $\vec{A}(\varphi, \alpha_0)$, $\vec{B}(\varphi, \alpha_0)$ be two functions, where φ is a variable trace and α_0 a constant trace. We say that $\vec{A}(\varphi, \alpha_0) =^D \vec{B}(\varphi, \alpha_0)$, "*A is Dirac-equal to B*", whenever for every continuous function $h(x)$ in $\langle 0, 1)$ we have

$$\underset{\varphi\in W}{S^\bullet} \vec{A}(\varphi, \alpha_0) \, \mathrm{val}_\varphi h(\dot{x}) \, |\varphi| =^{\lim} \underset{\varphi\in W}{S^\bullet} \vec{B}(\varphi, \alpha_0) \, \mathrm{val}_\varphi h(\dot{x}) \, |\varphi| .$$

Both sides are quasi-vectors with support α_0.

Having these notions, we are going to prove some Dirac's-formulas on δ-function.

T.28. Theorem. For traces φ we have

$$\delta(-\varphi, 0^+) =^D \delta(\varphi, 0^+), \qquad \delta(-\varphi, 0^-) =^D \delta(\varphi, 0^-).$$

Proof. The sum

$$\underset{\varphi\in W}{S^\bullet} \delta(-\varphi, 0^+) \, \mathrm{val}_\varphi h \, |\varphi|$$

is understood as the quasi-number with support 0^+:

$$A(\dot{p}) =_{df} \underset{\varphi\in W}{S^\bullet} \delta(-\varphi, \dot{p}) \, \mathrm{val}_\varphi h \, |\varphi| ,$$

which in turn is the limit of

(1)
$$\sum_k \delta(-a_{nk}, p)\ \mathrm{val}_{a_{nk}} h \cdot |a_{nk}|,$$

where $Q_n =_{df} \{a_{n1}, a_{n2}, \dots\}$ is a completely distinguished and special sequence for I. It equals

(2)
$$\sum_k \delta(-p, a_{nk})\ \mathrm{val}\, a_{nk}\, h \cdot |a_{nk}|.$$

Let $p = (\alpha, \beta\rangle$. Then $-p = \langle -\beta, -\alpha)$. Now, among a_{nk} there exists one and at most only one interval, say $(a'_{nk}, a''_{nk}\rangle$, such that $a''_{nk} = -\beta$, and at most one only, such that $a'_{nk} = -\alpha$.

If we drop these two intervals, from every complex Q_n, we get another sequence $\{Q'_n\}$ which is also completely distinguished and special for I. The dropped terms yield a contribution, tending to 0. By Remark [T.26.0.], if we replace in (2) $-p$ by $(-\beta, \alpha\rangle$ and drop the interval mentioned, we get a sum, which tends to

$$\underset{\varphi \in W}{S^\bullet}\ \delta(\varphi, 0^-)\ \mathrm{val}_\varphi h\, |\varphi| = \mathrm{val}_{0^-} h.$$

Thus

$$\underset{\varphi \in W}{S^\bullet}\ \delta(-\varphi, 0^+)\ \mathrm{val}_\varphi h\, |\varphi| = \mathrm{val}_{0^-} h.$$

On the other hand

$$\underset{\varphi \in W}{S^\bullet}\ \delta(\varphi, 0^+)\ \mathrm{val}_\varphi h\, |\varphi| = \mathrm{val}_{0^+} h.$$

Since h is continuous, we have $\mathrm{val}_{0^-} h =^{\lim} \mathrm{val}_{0^+} h$. Thus we have proved that

$$\delta(-\varphi, 0^+) =^D \delta(\varphi, 0^+), \quad \text{q.e.d.}$$

Similarly we can prove the second thesis.

T.29. Theorem. $\delta(\alpha, 0^\pm) \cdot \mathrm{val}_\alpha \dot{x} =^D 0_\alpha$.

Proof. Let $h(x)$ be a continuous function. Consider

$$J =_{df} S^\bullet\ \delta(\alpha, 0^\pm)\ \mathrm{val}_\alpha x \cdot \mathrm{val}_\alpha h(x)\, |\alpha|.$$

Since

$$\mathrm{val}_\alpha \dot{x} \cdot \mathrm{val}_\alpha h(\dot{x}) \approx \mathrm{val}_\alpha (\dot{x}\, h(\dot{x})),$$

we get, by [T.20.],

$$J = S^\bullet\ \delta(\alpha, 0^\pm)\ \mathrm{val}_\alpha (\alpha\, h(x)).$$

Hence, by [T.26.],

(1)
$$J = \mathrm{val}_{0^\pm} (\dot{x}\, h(\dot{x})).$$

On the other hand

$$S^\bullet\ 0_\alpha\ \mathrm{val}_\alpha h(\dot{x})\, |\alpha| = 0_\alpha.$$

By continuity of $h(x)$ we get

$$\lim_{\operatorname{meas} p \to 0} \frac{\int_p h(x)\,dx}{\operatorname{meas} p} = 0, \quad \text{and} \quad \lim 0(p) = 0 \quad \text{(see [T.21.])}.$$

Hence the theorem is proved[1].

T.30. Theorem. If $k > 0$ then

$$\delta(k\beta, 0^{\pm}) =^{D} \frac{1}{k}\delta(\beta, 0^{\pm}).$$

Proof. Let p be a neighborhood of $0^{\pm}$. The sum

$$\underset{\beta \in W}{S^{\bullet}} \delta((k\beta), p) \operatorname{val}_{\beta} f(x) \cdot |\beta|$$

is the limit of the expression

(1) $$A_n =_{df} \sum_i \delta(k \cdot q_{ni}, p) \int_{q_{ni}} f(x)\,dx,$$

where $Q_n = \{q_{n1}, q_{n2}, \ldots\}$ is a completely distinguished and special sequence for I.

T.31. Put $kx = y$; we have

$$\int_{q_{ni}} f(x)\,dx = \frac{1}{k}\int_{p_{ni}} f\left(\frac{y}{k}\right) dy.$$

This expression tends to the expression

(2) $$\frac{1}{k} S^{\bullet} \delta(\beta, p) \operatorname{val}_{\beta} g(y), \quad \text{where} \quad g(y) = f\left(\frac{y}{k}\right);$$

(2) is the quasi-number

$$\frac{1}{k} S^{\bullet} \delta(\beta, 0^{\pm}) \operatorname{val}_{\beta} g(y)\,|\beta| = \operatorname{val}_{0^{\pm}} g(y) = \operatorname{val}_{0^{\pm}} f\left(\frac{x}{k}\right).$$

For continuous functions f we have

$$\lim \operatorname{val}_{0^{\pm}} f\left(\frac{x}{k}\right) = \lim \operatorname{val}_{0^{\pm}} f(x) = f(0).$$

Hence we can write

$$\delta(k\beta, 0^{\pm}) =^{D} \frac{1}{k}\delta(\beta, 0^{\pm}), \quad \text{q.e.d.}$$

T.32. Theorem. If α is a trace, a, b numbers, then

$$\begin{aligned}\underset{\alpha \in W}{S^{\bullet}} \delta(\alpha - a, b^{\pm}) \operatorname{val}_{\alpha} f(\dot{x})\,|\alpha| &= \operatorname{val}_{(a+b^{\pm})} f(\dot{x}) \\ &= \operatorname{val}_{b^{\pm}} f(\dot{x} + a) = \operatorname{val}_{a^{\pm}} f(\dot{x} + b).\end{aligned}$$

Proof. We have by [T.25.5.] and [T.25.4.]

$$\delta(\alpha - a, b^{\pm}) = \delta(\alpha, b^{\pm} + a) = \delta(\alpha, (b + a)^{\pm}).$$

Hence our theorem is equivalent to

$$\underset{\alpha \in W}{S^{\bullet}} \delta(\alpha, (b + a)^{\pm}) \operatorname{val}_{\alpha} f(x)\,|\alpha| = \operatorname{val}_{a+b^{\pm}} f(x).$$

---

[1] See P. A. M. DIRAC 3d edition. The principles of quantum mechanics.

This, however, follows from [T.6b.]. Let $(b', b''\rangle$ be the variable neighborhood of $b^{\pm}$; then $(a+b', a+b''\rangle$ is the variable neighborhood of $(a+b)^{\pm}$; $\mathrm{val}_{(a+b)^{\pm}} f$ is the quasi-number, defined by

$$(1) \qquad \frac{1}{b''-b'} \int_{a+b'}^{a+b''} f(x)\, dx.$$

Put $x = y + a$. Then (1) equals $\int_{b'}^{b''} f(y+a)\, dy$, which defines $\mathrm{val}_{b^{\pm}} f(x+a)$.

T.33. Theorem. If $a > 0$ and τ is a variable trace, then

$$\delta(\tau^2 - a^2, 0^{\pm}) =^{D} \frac{1}{2a} [\delta(\tau - a, 0^{\pm}) + \delta(\tau + a, 0^{\pm})].$$

Proof. The expression $\delta(\tau^2 - a^2, 0^{\pm})$ is defined, [T.26.], as a number-valued function of two intervals p and q, where p is a variable neighborhood of 0^+, and q a variable neighborhood of τ. Consider the sum

$$(1) \qquad \underset{\tau \in W}{S^{\bullet}}\, \delta(\tau^2 - a^2, 0^{\pm})\, \mathrm{val}_\tau f(x) \cdot |\tau|,$$

where $f(x)$ is continuous. (1) is defined as the quasi-number $A(p)$ with support $0^{\pm}$:

$$A(p) = \underset{\tau \in W}{S}\, \delta(\tau^2 - a^2, p)\, \mathrm{val}_\tau f \cdot |\tau|.$$

Now $A(p)$ is the limit of the following sum

$$(2) \qquad \sum_k \delta(q_{nk}^2 - a^2, p)\, \mathrm{val}_{q_{nk}} f \cdot |q_{nk}|,$$

where $Q_n = \{q_{n1}, q_{n2}, \ldots\}$ is a completely distinguished and special sequence of complexes for I.

The intervals q_{nk} are disjoint, so there exists at most one interval — denote it by $(z'\, z''\rangle$ — such that $z' < 0 < z''$.

All other intervals q_{nk} can be devided into two classes: the first will contain all those $(x_k', x_k''\rangle$, for which $0 \leqq x_k' < x_k''$; the second all those $(y_k', y_k''\rangle$, for which $y_k' < y_k'' \leqq 0$. The sum (2) can be written as

$$\sum_k \delta((x_k', x_k''\rangle^2 - a^2, p) \cdot \mathrm{val}_{(x_k', x_k''\rangle} f \cdot (x'' - x') +$$
$$+ \delta((z', z''\rangle^2 - a^2, p) \cdot \mathrm{val}_{(z', z''\rangle} f \cdot (z'' - z') +$$
$$(2.1) \qquad + \sum_k \delta((y_k', y_k''\rangle^2 - a^2, p) \cdot \mathrm{val}_{(y_k', y_k''\rangle} f \cdot (y_k'' - y_k').$$

Hence the expression equals

$$\sum_k \delta(\langle x_k''^2 - a^2, x_k'^2 - a^2\rangle, p) \int_{x_k'}^{x_k''} f(x)\, dx + \delta(\langle 0, \max(z', z'')\rangle, p) \int_{z'}^{z''} f(x)\, dx +$$
$$(3) \qquad + \sum_k \delta(\langle y_k'^2 - a^2, y_k''^2 - a^2\rangle, p) \int_{y_k'}^{y_k''} f(x)\, dx.$$

We shall transform the first term of (3), by changing the variable x as follows: Put $u =_{df} x^2 - a^2$. We have $x^2 = a^2 + u$ and $2x\,dx = du$. Put

$$u'_k =_{df} x'^2_k - a^2, \qquad u''_k =_{df} x''^2_k - a^2;$$

we have, since all x'_k, x''_k are non positive,

$$x'_k = -\sqrt{u'_k + a^2}, \qquad x''_k = -\sqrt{u''_k + a^2}.$$

Hence

(4) $$\int_{x'_k}^{x''_k} f(x)\,dx = -\int_{u'_k}^{u''_k} f(-\sqrt{a^2+u})\frac{du}{-2\sqrt{u+a^2}}.$$

The formula (4) holds true, even if $u''_k = 0$. Consequently the first term in (3) can be written:

$$-\sum_k \delta(\langle u''_k, u'_k), p)\cdot\int_{u'_k}^{u''_k}\frac{f(-\sqrt{a^2+u})}{2\sqrt{a^2+u}}\,du =$$

(5) $$= \sum_k \delta(\langle u''_k, u'_k), p)\cdot \mathrm{val}_{(u''_k,\, u'_k\rangle}\, g_-(u),$$

where

(6) $$g_-(u) =_{df} \frac{f(-\sqrt{a^2+u})}{2\sqrt{a^2+u}} \quad \text{for} \quad u > -a^2.$$

Concerning (5), let us remark, that the transformation

$$x = -\sqrt{u+a^2}, \qquad x^2 - a^2 = u,$$

is one to one and monotonic, non increasing, transforming the interval

(7) $$-\infty < x \leqq 0 \quad \text{into} \quad -\infty < u \leqq -a^2.$$

If we take $\mathrm{meas}\,p < \frac{a^2}{4}$, the interval p will belong to the set (7). In the sum (5), only those terms may not disappear, for which $\langle u''_k\, u'_k)\cdot p \neq \Theta$. Hence we can confine the sum (5) only to terms, for which $\frac{-a^2}{2} < u''_k < u'_k < \frac{a^2}{2}$. In the interval $\left\langle\frac{-a^2}{2}, \frac{+a^2}{2}\right\rangle$ the function $g_-(u)$ is continuous.

If we take account of Remark [T.26.0.], we can change in (5)

$$\delta(\langle u''_k, u'_k), p) \quad \text{into} \quad \delta((u''_k, u'_k\rangle, p).$$

Hence the limit of the sum (5) is the quasi-number $\mathrm{val}_{0\pm}\, g_-(\dot{u})$. Thus we have proved, that the first term in (2.1) tends to

(8) $$\mathrm{val}_{0\pm}\, g_-(\dot{u}),$$

where $g_-(u)$ is defined in (6).

Now, consider the third term in (2.1), hence in (3). We shall use the transformation

(9) $$v = x^2 - a^2, \quad x = \sqrt{v + a^2},$$

since all y'_k, y''_k are non negative. The transformation (9) is $1 \to 1$ and monotonic non decreasing. It transforms the interval

$$0 \leqq x < \infty \quad \text{into} \quad -a^2 \leqq v < \infty .$$

We put

$$v'_k =_{df} y'^2_k - a^2, \quad v''_k =_{df} y''^2_k - a^2 ;$$

and we have

$$y'_k = \sqrt{v'_k + a^2}, \quad y''_k = \sqrt{v''_k + a^2}.$$

Hence

$$\int_{y'_k}^{y''_k} f(x)\, dx = \int_{v'_k}^{v''_k} \frac{f(\sqrt{a^2+v})}{2\sqrt{a^2+v}}\, dv .$$

Hence the third term in (3) equals

(10) $$\sum_k \delta(\langle v'_k, v''_k\rangle, p) \int_{v'_k}^{v''_k} \frac{f(\sqrt{a^2+v})}{2\sqrt{a^2+v}}\, dv .$$

If we put

(11) $$g_+(v) =_{df} \frac{f(\sqrt{a^2+v})}{2\sqrt{a^2+v}},$$

we get, by argument similar to the above one, that (10) tends to

(12) $$\mathrm{val}_{0\pm} g_+(u) .$$

Since the functions g_+ and g_- are both continuous at 0, the limit value of $\mathrm{val}_{0\pm} g_+$ equals $g_+(0) = \frac{f(a)}{2a}$, because $a > 0$.

Similarly we have the limit value of $\mathrm{val}_{0\pm} g_-$ equal to $g_-(0) = \frac{f(-a)}{2a}$. There remains the second term in (3) and (2.1) to be considered:

$$\delta(\langle 0, \max(z' z'')\rangle - a^2, p) \cdot \int_{z'}^{z''} f(x)\, dx , \quad \text{where} \quad z' < 0 < z'' .$$

For p, chosen in $\left\langle -\frac{a^2}{4}, \frac{a^2}{4}\right\rangle$, as we did above, and sufficiently small intervals q_{nk}, the intervals p and $\langle 0, \max(z' z'')\rangle$ do not overlap, so the term considered vanishes. Thus this term does not contribute anything to the limit. We have proved that

$$\underset{\tau\in W}{S^\bullet}\, \delta(\tau^2 - a^2, 0^\pm)\, \mathrm{val} f(x)\, |\tau|$$

has the limit value $\frac{f(a)+f(-a)}{2a}$. Now we have from [T.32.] that

$$S^\bullet\, \delta(\tau - a, 0^\pm)\, \mathrm{val}_\tau f(x)\, |\tau| = \mathrm{val}_{0\pm} f(x+a) =^{\lim} f(a),$$

and

$$S\,\delta(\tau + a, 0^{\pm})\,\mathrm{val}_\tau f(x)\,|\tau| = \mathrm{val}_{0^\pm} f(x - a) =^{\lim} f(-a).$$

Consequently

$$S^\bullet \left[\frac{\delta(\tau - a, 0^{\pm}) + \delta(\tau + a, 0^{\pm})}{2a}\right] \mathrm{val} f(x)\,|\tau| =^{\lim} \frac{f(a) + f(-a)}{2}$$

which completes the proof.

T.34. Theorem. If β is a variable trace and α_0 a fixed trace, then for every continuous function $f(x)$, we have

$$\mathrm{val}_{\alpha_0} f \cdot \delta(\alpha_0, \beta) =^D \mathrm{val}_\beta f \cdot \delta(\alpha_0, \beta).$$

Proof. Take a continuous function $h(x)$. We have

$$\underset{\beta\in W}{S^\bullet} \mathrm{val}_\beta h \cdot \mathrm{val}_{\alpha_0} f\,\delta(\alpha_0\,\beta) = \mathrm{val}_{\alpha_0} f^{*)} \cdot \underset{\beta\in W}{S^\bullet} \mathrm{val}_\beta h \cdot \delta(\alpha_0\,\beta) =$$

(1) $$= \mathrm{val}_{\alpha_0} f \cdot \mathrm{val}_{\alpha_0} h,$$

$$\underset{\beta\in W}{S^\bullet} \mathrm{val}_\beta h \cdot \mathrm{val}_\beta f \cdot \delta(\alpha_0\,\beta) = \underset{\beta\in W}{S^\bullet} \mathrm{val}_\beta (h\,f) \cdot \delta(\alpha_0\,\beta)$$
$$= \mathrm{val}_{\alpha_0} (h\,f).$$

Since f, h are continuous, we have for a variable neighborhood p of α_0 with $\mathrm{meas}\,p \to 0$, the limit value of $\mathrm{val}_p(f)$ is $f(\mathrm{vert}\,\alpha_0)$, that of $\mathrm{val}_p(h)$ is $h(\mathrm{vert}\,\alpha_0)$ and that of $\mathrm{vert}_p(f\,h)$ is $f(\mathrm{vert}\,\alpha_0) \cdot h(\mathrm{vert}\,\alpha_0)$, so the theorem is proved. $\left(\text{By } \mathrm{val}_p f \text{ we understand } \frac{1}{|p|}\underset{p}{S} f(x)\,dx\right)$.

T.34.1. Remark. We can state the theorem as follows

$$\mathrm{val}_{\alpha_0} f \cdot \delta(\alpha_0 - \mathrm{vert}\,\beta, 0^{\pm}) =^D \mathrm{val}_\beta f \cdot \delta(\alpha_0 - \mathrm{vert}\,\beta, 0^{\pm})$$

which looks like

$$f(a)\,\delta(a - b) = f(b)\,\delta(a - b).$$

T.35. Theorem. If γ_0 is a fixed trace and α a variable trace, then

$$\underset{\beta\in W}{S^\bullet} \delta(\alpha, \beta)\,\delta(\beta, \gamma_0)\,|\beta| =^D \delta(\alpha, \gamma_0).$$

Proof. We shall confine ourselves to a sketch of the proof. Let $h(x)$ be a continuous function. We shall compare the expression

(1) $$\underset{\alpha\in W}{S^\bullet} \Big(\underset{\beta\in W}{S^\bullet} \delta(\alpha, \beta)\,\delta(\beta, \gamma_0)\,|\beta|\,\mathrm{val}_\alpha h \cdot |\alpha|,$$

with

(2) $$\underset{\alpha\in W}{S^\bullet} \delta(\alpha, \gamma_0)\,\mathrm{val}_\alpha h\,|\alpha|.$$

Since

$$\underset{\beta\in W}{S^\bullet} \delta(\alpha\,\beta)\,\delta(\beta, p)\,|\beta|$$

---

*) We can extract $\mathrm{val}_\alpha f$ out side of the integral, since we will look for a limit of $\mathrm{val}_p f \underset{\beta\in W}{S^\bullet} \mathrm{val}_\beta h \cdot \delta(p, \beta)$, which is the same, as of $\underset{\beta\in W}{S^\bullet} \mathrm{val}_p f\,\mathrm{val}_\beta h \cdot \delta(p, \beta)$.

is a function $F(\alpha, \gamma_0)$, the expression (1) has the form:

$$\underset{\alpha \in W}{S^\bullet} F(\alpha, \gamma_0) \operatorname{val}_\alpha h \cdot |\alpha|.$$

Hence, by [R.], it is a quasi-number with support γ_0; so we evaluate it, by taking a neighborhood Γ of γ_0, and considering the sum

(3) $$\underset{\alpha \in W}{S^\bullet} F(\alpha, \Gamma) \operatorname{val}_\alpha h |\alpha|,$$

which in turn, is the limit of

(4) $$F_n(\Gamma) =_{df} \sum_k F(\alpha_{nk}, \Gamma) \operatorname{val}_{\alpha_{nk}} h \cdot |\alpha_{nk}|,$$

where $A_n =_{df} \{\alpha_{n1}, \alpha_{n2}, \ldots\}$ is a completely distinguished and special sequence of complexes for I. (4) is a function of Γ.

Now

(5) $$F(\alpha_{nk}, \Gamma) = \underset{\beta \in W}{S^\bullet} \delta(\alpha_{nk} \beta)\, \delta(\beta, \Gamma)\, |\beta|,$$

hence (4) can be written:

(6) $$F_n(\Gamma) = \sum_k \Big[\underset{\beta \in W}{S^\bullet} \delta(\alpha_{nk} \beta)\, \delta(\beta, \Gamma) \cdot |\beta|\Big] \operatorname{val}_{\alpha_{nk}} h \cdot |\alpha_{nk}|.$$

The sum

(7) $$\underset{\beta \in W}{S^\bullet} \delta(\alpha_{nk}, \beta)\, \delta(\beta, \Gamma) \cdot |\beta|,$$

is just the sum of a total set of quasi-numbers; so it is the limit of the sum:

(8) $$G_{n,m}(\Gamma) =_{df} \sum_j \delta(\alpha_{nk} \beta_{mj})\, \delta(\beta_{mj} \Gamma) \cdot |\beta_{mj}|,$$

where $B_m =_{df} \{\beta_{m1}, \beta_{m2}, \ldots\}$ is a completely distinguished and special sequence of complexes for I. We recall that **G** has no atoms; hence for any special sequence of complexes, the net number tends to 0. It follows, that if the intervals Γ and α_{nk}, after closure are disjoint, then, for sufficiently great m, all terms in (8) will vanish, and hence the sum (7) will be equal 0. Hence the sum (6) can be restricted to only those $n\ k$, for which $\bar{\Gamma} \frown \bar{\alpha}_{nk} \neq \Theta$. Hence it can be restricted to ext$(A_n)\bar{\Gamma}$. If $\bar{\Gamma} \frown \bar{\alpha}_{nk} \neq \Theta$, the corresponding terms in (8) will be

(9) $$\frac{1}{|\alpha_{nk}| + |\beta_{mj}|} \cdot \frac{1}{|\beta_{mj}| + |\Gamma|} \cdot |\beta_{mj}|,$$

and the summation in (8) can be taken over all β_{nj}, where $\beta_{nj} \frown \Gamma \neq \emptyset$ and at the same time $\beta_{mj} \frown a_{nk} \neq \emptyset$.

The sum of these terms in (9) can be written:

$$\sum_j |\beta_m| \cdot \frac{1}{|\alpha_{nk}|} \cdot \frac{1}{|\Gamma|} + \sum_j |\beta_{nj}| \left\{\frac{1}{|\alpha_{nk}| + |\beta_{mj}|} \cdot \frac{1}{|\beta_{mj}| + |\Gamma|} - \frac{1}{|\alpha_{nk}| + |\Gamma|}\right\}.$$

To the given n the factor in braces is small for sufficiently great m; and $\sum_j |\beta_{mj}|$ will exceed meas$(\Gamma \frown \alpha_{nk})$ also by less than ε_n for sufficiently great m; so the last term will be small.

Hence (7), i.e. the limit of the expression (8) will be $\frac{|\alpha_{nk}| + \varepsilon_n}{|\Gamma|\,|\alpha_{nk}|}$, where $|\theta| \leqq 1$ and ε_n tends to 0.

It follows that approximately:

$$F_n(\Gamma) = \sum_k \frac{|\alpha_{nk}| + \varepsilon_n}{|\Gamma| \cdot |\alpha_{nk}|} \operatorname{val}_{\alpha_{nk}} h \, |\alpha_{nk}|,$$

where the summation is extended over all α_{nk} with $\bar{\alpha}_{nk} \cap \bar{\Gamma} \neq \emptyset$.

It differs but a little from

(10) $$\sum_k \frac{1}{|\Gamma|} \int_{\alpha_{nk}} h; \quad \text{hence from } \frac{\int_\Gamma h}{|\Gamma|} = \operatorname{val}_{\gamma_0} h.$$

Concerning (δ, γ_0), we have

(11) $$\underset{\alpha \in W}{S} \delta(\alpha, \gamma_0) \operatorname{val}_\alpha h \, |\alpha| = \operatorname{val}_{\gamma_0} h. |$$

From (10) and (11) the theorem follows.

T.36. Remark. We have proved several theorems, concerning the function $\delta(\alpha, \beta)$, where α, β are traces. We believe, that the Dirac's delta-function should be defined as a function of two variables, since it is like an integral-equation-kernel. We believe that our δ-function should replace the genuine δ-function introduced by DIRAC. We compare the following formulas:

1. $\int \delta(x)\, dx = 1$,	1'. $\underset{\beta}{S^\bullet} \delta(\alpha, \beta) \, |\beta| = 1_\alpha$,
2. $f(x) = \int \delta(x - y) f(y)\, dy$,	2'. $\operatorname{val}_\alpha f(\dot{x}) = \underset{\beta}{S^\bullet} \delta(\alpha, \beta) \cdot \cdot \operatorname{val}_\beta f(\dot{x}) \, |\beta|$,
3. $\delta(-x) = \delta(x)$,	3'. $\delta(-\varphi, 0^\pm) =^D \delta(\varphi, 0^\pm)$,
4. $x\,\delta(x) = 0$,	4'. $\delta(\alpha, 0^\pm) \operatorname{val}_\alpha \dot{x} =^D 0_\alpha$,
5. $\delta(a x) = \frac{1}{a} \delta(x)$, $(a > 0)$,	5'. $\delta(k\beta, 0^\pm) =^D \frac{1}{k} \delta(\beta, 0^\pm)$, $(k > 0)$,
6. $\delta(x^2 - a^2) = \frac{1}{2a} [\delta(x - a) + \delta(x + a)]$, $(a > 0)$,	6'. $\delta(\tau^2 - a^2, 0^\pm) =^D \frac{1}{2a} [\delta(\tau - a, 0^\pm) + \delta(\tau + a, 0^\pm)]$, $(a > 0)$,
7. $\int \delta(a - x)\, dx\, \delta(x - b) = \delta(a - b)$,	7'. $\underset{\beta}{S^\bullet} \delta(\alpha\beta) \, |\beta| \, \delta(\beta, \gamma_0) =^D \delta(\alpha, \gamma_0)$,
8. $f(x)\,\delta(x - a) = f(a)\,\delta(x - a)$.	8'. $\operatorname{val}_{\alpha_0} f(\dot{x}) \, \delta(\alpha_0 \beta) =^D \operatorname{val}_\beta f(\dot{x}) \, \delta(\alpha_0 \beta)$.

In addition to that, $\delta(\alpha, \beta)$ behaves like a function of the difference of variables, since $\delta(\alpha, \beta)$ has the translation property.

Concerning the equality $=^D$, it seams to be in agreement with Dirac's remark:

"The meaning of any of these equation is that its two sides give equivalent results as factors of an integral."

Concerning $\mathrm{val}_\alpha f(\dot{x})$—this is an "ideal" average value which physicists approach, by taking the average values from measurement, made with more and more precise instruments.

T.37. We like to remark, that various statements on the genuine function $\delta(\alpha, \beta)$, can be generalized to the case, where bricks are half open rectangles—or even half open hypercubes in n-dimensional space, with Lebesguean measure admitted.

Chapter U

Auxiliaries for a deeper study of summation of scalar fields

In some foregoing chapters, viz. [Q I.], [R.] we have developped a theory of summation of vector fields. The scalar fields constitute a special, but important case of vector fields. For the sake of application it is needed to have more information on scalar fields; therefore, we shall devote this chapter to some auxiliaries for the subsequent Chapters [W.] and [W I.], where the scalar fields are given special attention.

After some recollection, we shall deal with topics: Section 1, admissible vector fields and admissible sets of quasi-numbers; Section 2, the condition of non-overlapping for complexes; Section 3, some theorems on limes superior and inferior for sequences of complexes.

This Chapter [U.] can be considered as continuation of the Chapters [Q.], [Q I.] and [R.].

Remark. In [Q.] we have considered distinguished sequences of complexes for measurable sets of traces. Now, it is in order to explain, that they have to be understood, as distinguished complexes for the coats [E.] of sets of traces E.

We shall often use the following two lemmas. We shall give the proofs, though they can be considered as easy consequences of other theorems.

Lemma α. Let us admit the ***Hyp*** $\boldsymbol{R_e}$ and **(*Hyp S*)**. Now, if

1. E is a measurable set of traces,

2. $\{T_n\}$ is a (DARS)-sequence for E, (i. e. a completely distinguished sequence for $[E]$,

3. F is a measurable subset of E,

4. $\{P_n\}$ is a (DARS)-sequence for F, such that $P_n \subseteqq T_n$, (i. e. the complex P_n is a partial complex of T_n). (Notice that such a sequence $\{P_n\}$ exists, [Q.21.2.]),

5. Q_n is the complementary (in T_n), partial complex of P_n, i.e.

$$P_n \cup Q_n = T_n, \quad P_n \cap Q_n = \emptyset,$$

then Q_n is a (DARS)-sequence for $E - F$.

Proof. Since $\{T_n\}$ is completely distinguished for E, we have

$$\mathcal{N}_R(T_n) \to 0, \quad \mathcal{N}_A(T_n) \to 0,$$

which imply

(1) $$\mathcal{N}_R(Q_n) \to 0, \quad \mathcal{N}_A(Q_n) \to 0.$$

We have

$$|T_n - P_n, E - F| = |\operatorname{som} T_n - \operatorname{som} P_n; [E] - [F]|_\mu \leqq$$
$$\leqq |\operatorname{som} T_n, E| + |\operatorname{som} P_n, [F]| = |T_n, E| + |P_n, F|.$$

Since $|T_n, E|_\mu \to 0$, and $|P_n, F|_\mu \to 0$, it follows that

(2) $$|Q_n, E - F|_\mu \to 0.$$

Since $E - F \subseteqq E$, $Q_n \subseteqq T_n$, and (2) holds true, the sequence Q_n has the property $(\boldsymbol{S})$ with respect to

(3) $$E - F.$$

From (1), (2), and (3) the theorem follows.

Lemma β. If

1. E is a measurable set of traces,

2. $M_1, M_2, \ldots, M_n, \ldots$ a denumerable collection of measurable subsets of E, such that this collection is everywhere dense in $E \upharpoonright \boldsymbol{G}$ with respect to the μ-topology on $\boldsymbol{G}$,

3. $\{P_n\} = \{p_{n1}, p_{n2}, \ldots\}$, $(n = 1, 2, \ldots)$ is a (DAR)-sequence for E,

4. if for every m and n there exists a partial complex $Q_n^{(m)}$ of P_n, such that for every m

$$\lim_{n\to\infty} |Q_n^{(m)}, M_m| = 0,$$

then $\{P_n\}$ is a (DARS)-sequence for E.

Proof. By hyp. 2, there exists a subsequence $\alpha(m)$ of m such, that

$$\lim_{m\to\infty} |M_{\alpha(m)}; F| = 0.$$

We can select a subsequence $\beta(m)$ of m such, that

(1) $$|M_{\alpha\beta(m)}; F| \leqq \frac{1}{m}, \quad \text{for all} \quad m = 1, 2, \ldots$$

For a given m we can find $k(m)$ such, that for all $n \geqq k(m)$, we have (hyp. 4),

(2) $$|M_{\alpha\beta(m)}, Q_n^{\alpha\beta(m)}| \leqq \frac{1}{m},$$

and arrange so, as to have

(3) $$k(1) < k(2) < \cdots$$

We have

$$|F; Q_n^{\alpha\beta(m)}| \leqq |F; M_{\alpha\beta(m)}| + |M_{\alpha\beta(m)}, Q_n^{\alpha\beta(m)}|;$$

hence, by (1) and (2) we have for every m and $n \geqq k(m)$:

$$|F; Q_n^{\alpha\beta(m)}| \leqq \frac{1}{m} + \frac{1}{m}.$$

Hence

$$\lim_{m\to\infty} |F; Q_n^{\alpha\beta(m)}| = 0,$$

$$k(m) \leqq n \leqq k(m+1).$$

The sequence of complexes

(4) $$Q_{k(1)}^{\alpha\beta(1)}, Q_{k(1)+1}^{\alpha\beta(1)}, \ldots, Q_{k(2)-1}^{\alpha\beta(1)}, Q_{k(2)}^{\alpha\beta(2)}, Q_{k(2)+1}^{\alpha\beta(2)}, \ldots,$$

if we complete it, by adjoining at the beginning

$$P_1, P_2, \ldots, P_{k(1)-1},$$

is a (D)-sequence for F, composed of partial complexes of the complexes P_n (for $n = 1, 2, \ldots$).

It follows that (4), if completed as above, is a (DARS)-sequence for F.

Since F is any measurable subset of E, the thesis follows.

U.1. In the present chapter we shall admit the following collection of hypotheses and agreements, which we shall refer to as ***Hyp*** $\boldsymbol{H_1}$:

U.1a. $(\boldsymbol{G})$ is a non trivial, denumerably additive tribe, and $(\boldsymbol{F})$ its finitely genuine (finitely) additive stricts subtribe. $\boldsymbol{B}$ is a basis of $(\boldsymbol{F})$, satisfying the condition ***(Hyp Ad)***, [B.2.]. μ is an effective denumerably additive, non negative measure on $(\boldsymbol{G})$.

U.1b. We consider $\boldsymbol{B}$-traces, and admit [***Hyp*** **I**] and [***Hyp*** **II**], [B I.6.], to have the theory of measurability of sets of traces, [B I.].

U.1c. The tribe $(\boldsymbol{G})$, its measure μ and the tribe $(\boldsymbol{F})$ satisfy the conditions ***Hyp L*** $\boldsymbol{\mu}$ and ***Hyp R***, so $\boldsymbol{G} = \boldsymbol{F}^L$, i.e. $(\boldsymbol{G})$ is the μ-Lebesgue's covering extension of $(\boldsymbol{F})$, [Q.12.].

U.1d. We admit the hypothesis **(*Hyp S*)** from [Q.21.1.]; and we shall consider only the (DARS)-sequences of complexes. We shall call them completely distinguished.

U.1e. In the whole chapter we shall be interested in regular, real, and total sets of quasi-numbers.

U.1f. To avoid non essential complications in formulas, we shall allow ourselves sometimes to use the same symbol E for a measurable set of traces and for its coat. We shall sometimes use the same symbol for a brick p and for the set, composed of p only; and in addition to that, we shall use sometimes the same symbol $P = \{p_1, p_2, \ldots\}$ for a complex and for its soma, i.e. $\operatorname{som} P = p_1 + p_2 + \cdots$. But we shall apply signs $\sum$, $\bigcup$ or $\prod$, $\bigcap$, $+$, $\cup$, $\cdot$, $\cap$, 0, $\emptyset$ to emphasize that we are dealing with collection of bricks or with sum of these bricks, according to the case.

Section 1 of U.

Admissible scalar fields and admissible sets of quasi-numbers

U.2. Theorem. Under $\boldsymbol{H}_1$, if

1. $\{f_x\}$ is a total, regular set of quasi-vectors,
2. there exists a measurable set E of traces, such that $\{f_x\}$ is (DARS)-summable on E,
3. $P_n = \{p_{n1}, p_{n2}, \ldots\}$, $(n = 1, 2, \ldots)$ is a sequence of complexes with $\sum_k \mu(p_{nk}) \to 0$ for $n \to \infty$,

then

$$\lim_{n\to\infty} \sum_k |\vec{f}(p_{nk})| = \vec{0}.$$

Proof. The sequence $\{P_n\}$ is a (DARS)-sequence for the empty set $\emptyset$ of traces. Indeed we have $|P_n, \emptyset|_\mu \to 0$.

We have

$$\mathcal{N}_R(P_n) = \max[\mu(p_{n1} - \beta), \mu(p_{n2} - \beta), \ldots]$$

where β is the sum of all atoms of $\boldsymbol{G}$. Hence, [Q.19.],

$$\mathcal{N}_R(P_n) \leqq \max[\mu(p_{n1}), \mu(p_{n2}), \ldots] \leqq \sum_k \mu(p_{nk}), \quad \text{so} \quad \mathcal{N}_R(P_n) \to 0,$$

$$\mathcal{N}_A(P_n) = \max[\mathcal{N}'(p_{n1}), \mathcal{N}'(p_{n2}), \ldots] \leqq \max_k[\mu(p_{nk}), \mu(p_{n2}), \ldots] \leqq$$
$$\leqq \sum_k \mu(p_{nk}) \to 0.$$

In addition to that, since a subset of $\emptyset$ is $\emptyset$, the sequence $\{P_n\}$ is a (D)-sequence of any subset of $\emptyset$. Thus $\{P_n\}$ is a (DARS)-sequence for Θ. Since $\{\vec{f}_x\}$ is (DARS)-summable in E, it is (DARS)-summable in Θ, and we have

$$\underset{\Theta, \text{(DARS)}}{S} \{\vec{f}_x\} = \vec{O},$$

[Q I.6.]. Be definition of the sum we get

$$\sum_k \vec{f}_x(p_{nk}) \to \vec{O} \quad \text{for} \quad n \to \infty, \quad \text{q.e.d.}$$

U.2.1. Theorem. Under $\boldsymbol{H}_1$, if

1. $\{\vec{f}_x\}$ is a total and regular set of quasi-vectors,
2. there exists a measurable set E of traces, such that $\{\vec{f}_x\}$ is (DARS)-summable on E,

then for every $\varepsilon > 0$ there exists $\delta > 0$ such that, if $P = \{p_1, p_2, \ldots\}$ is a complex with $\sum_n \mu(p_n) \leqq \delta$, we have $\|\vec{f}(p_1) + \vec{f}(p_2) + \cdots\| \leqq \varepsilon$, where $\|\ \|$ denotes the norm of vectors in V.

Proof. Suppose this will not be true. Then we can find $\varepsilon > 0$, and for every $n = 1, 2, \ldots$ a complex $P_n = \{p_{n1}, p_{n2}, \ldots\}$, such that

$$\mu(\operatorname{som} P_n) \leqq \frac{1}{n} \quad \text{and} \quad \|\sum_k \vec{f}_x(p_{nk})\| > \varepsilon.$$

Now, by the preceding lemma, the last norm tends to 0, because $\sum_k \vec{f}(p_{nk}) \to \vec{O}$, so we get a contradiction.

U.2.2. Theorem. Under $\boldsymbol{H}_1$ if

1. $\{\vec{f}_x\}$ is a total and regular set of quasi-vectors (in a complete Banach-space V),
2. there exists a measurable set E, on which $\{\vec{f}_x\}$ is (DARS)-summable,
3. $p_1, p_2, \ldots, p_n, \ldots$ is an infinite sequence of mutually disjoint bricks,

then the series

$$\vec{f}(p_1) + \vec{f}(p_2) + \cdots + \vec{f}(p_n) + \cdots$$

converges in the V-topology.

Proof. Let $\varepsilon > 0$. Find, according to Theorem [U.2.1.], a number $d > 0$, such that if $P = \{p_1, p_2, \ldots\}$ is a complex with $\mu(P) \leqq \delta$, we get $\|\sum_k \vec{f}(p_k)\| \leqq \varepsilon$. Since the series $\mu(p_1) + \mu(p_2) + \cdots + \mu(p_k) + \cdots$ converges, we can find M, such that for $n \geqq M$, and every $s = 1, 2, \ldots,$ we have

$$\mu(p_{n+1}) + \cdots + \mu(p_{n+s}) \leqq \delta.$$

For those indices n and all $s = 1, 2, \ldots,$ we get

$$\|\vec{f}(p_{n+1}) + \cdots + \vec{f}(p_{n+s})\| \leqq \varepsilon.$$

Since V is supposed to be a complete space, the theorem follows.

U.2.3. Theorem. Under $\boldsymbol{H}_1$ if

1. $\{f_x\}$ is a total and regular set of complex quasi-numbers,
2. there exists a measurable set E of traces, on which $\{f_x\}$ is (DARS)-summable,
3. $P_n = \{p_{n1}, p_{n2}, \ldots\}$, $(n = 1, 2, \ldots)$, is a complex with

$$\lim_{n\to\infty} \sum_k \mu(p_{nk}) = 0,$$

then

$$\sum_k |f(p_{nk})| \to 0 \quad \text{for} \quad n \to \infty.$$

Proof. Since for non negative numbers a, b we have $\sqrt{a^2+b^2} \leqq a+b$, therefore, if we put

$$f(p_{nk}) = g(p_{nk}) + i\,h(p_{nk}),$$

where g and h are real, we have

$$\sum_k |f(p_{nk})| \leqq \sum_k |g(p_{nk})| + \sum_k |h(p_{nk})|. \tag{1}$$

Now, consider g_x and P_n. Let $p'_{n1}, p'_{n2}, \ldots$, be all those p_{nk}, for which $g(p_{nk}) \geqq 0$, if any, and let $p''_{n1}, p''_{n2}, \ldots$ be all those p_{nk}, for which $g(p_{nk}) < 0$, if any. We have:

$$\sum_k \mu(p'_{nk}) \leqq \sum_k \mu(p_{nk}), \quad \sum_k \mu(p''_{nk}) \leqq \sum_k \mu(p_{nk});$$

hence

$$\sum_k \mu(p''_{nk}) \to 0 \quad \text{and} \quad \sum_k \mu(p''_{nk}) \to 0.$$

By the forgoing Lemma [U.2.] we have:

$$\Big|\sum_k g(p'_{nk})\Big| \to 0 \quad \text{and} \quad \Big|\sum_k g(p''_{nk})\Big| \to 0,$$

i.e.

$$\sum_k |g(p'_{nk})| \to 0 \quad \text{and} \quad \sum_k |g(p''_{nk})| \to 0. \tag{2}$$

From (2) it follows, that $\sum_k |g(p_{nk})| \to 0$. In a similar way we get $\sum_k |h(p_{nk})| \to 0$. Taking account of (1), we get the theorem:

U.2.4. Theorem. Under $\boldsymbol{H}_1$, if

1. $\{f_x\}$ is a total, regular set of complex quasi-numbers,
2. there exists the measurable set of traces, on which $\{f_x\}$ is (DARS)-summable,

then for every $\varepsilon > 0$ there exists $\delta > 0$, such that, if $P = \{p_1, p_2, \ldots\}$ is a complex with $\sum_k \mu(p_k) \leqq \delta$, then $\sum_k |f(p_k)| \leqq \varepsilon$.

Proof. Indirect, based on Theorem [U.2.3.], and following the pattern of the proof of Theorem [U.2.1.].

U.2.5. Theorem. Under $\boldsymbol{H}_1$, if

1. $\{f_x\}$ is a total and regular set of complex quasi-numbers,
2. there exists a measurable set E of traces, on which $\{f_x\}$ is (DARS)-summable, then for every $\varepsilon > 0$ there exists $\delta > 0$ such, that if $Q = \{E_1, E_2, \ldots\}$ is an infinite sequence of mutually disjoint bricks with $\mu\left(\sum_{n=1}^{\infty} q_n\right) \leqq \delta$,

then

$$\sum_{n=1}^{\infty} |f(q_n)| \leqq \varepsilon.$$

Proof. Choose $\varepsilon > 0$, and find, by Theorem [U.2.4.], a number $\delta > 0$, such that for any complex $P = \{p_1, p_2, \ldots\}$ with $\mu(P) \leqq \delta$, we have

$$\sum_k |f(p_k)| \leqq \varepsilon.$$

Let $Q = \{q_1, q_2, \ldots\}$ be an infinite sequence of mutually disjoint bricks with $\mu\left(\sum_{n=1}^{\infty} q_n\right) \leqq \delta$. If $s = 1, 2, \ldots$, we have $\mu\left(\sum_{n=1}^{s} q_n\right) \leqq \delta$, and then

$$\sum_{n=1}^{s} |f(q_n)| \leqq \varepsilon. \tag{1}$$

Since, by [U.2.2.], the series $\sum_{n=1}^{\infty} |f(q_n)|$ converges, (1) implies

$$\sum_{n=1}^{\infty} |f(q_n)| \leqq \varepsilon.$$

U.2.6. Remark. In all preceding Theorems [U.2.], [U.2.1.] to [U.2.5.] the hypothesis, stating that there exists a measurable set E of traces, on which the given set of quasi-vectors is (DARS)-summable, can be replaced by the hypothesis that the given set is (DARS)-summable on the empty set $\emptyset$ of traces.

U.2.7. Def. Under $\boldsymbol{H}_1$, a scalar field $f(a)$, (a total and regular set of quasi-numbers f_x) will be termed *admissible*, whenever the following happens:

Let $S_n = \{s_{n1}, s_{n2}, \ldots\}$, $(n = 1, 2, \ldots)$ be an infinite sequence of sets of mutually disjoint bricks s_{nk} (for each n respectively). Then from $\sum_k \mu(s_{nk}) \to 0$ for $n \to \infty$, it follows $\sum_k |f(s_{nk})| \to 0$.

U.3. Lemma. Under $\boldsymbol{H}_1$, if

1. $\boldsymbol{G}$ has no atoms,
2. $a \in \boldsymbol{G}$,
3. $a \neq O$,
4. $0 < \alpha < \mu(a)$,

then there exists a soma $b \neq O$, such that

$$b < a, \quad 0 < \mu(b) < \alpha.$$

U.3a. Proof. Suppose the lemma be not true. We notice the following: If $a' \leqq a$, $a' \neq O$, then there exists a soma b', such that $b' < a'$, $b' \neq O$. Indeed, if we suppose, that if $b' < a'$ then $b' = O$, it would follows that a' is an atom.

U.3b. Having this, we shall prove that, if $a' < a$, $a' \neq O$, then there exist at least two disjoint somata a_1', a_2', such that $a_1' + a_2' \leqq a'$, $\mu(a_1') \geqq \alpha$ and $\mu(a_2') \geqq \alpha$. Indeed, we can find a_1' such that $a_1' \neq O$, $a_1' < a'$. By hypothesis, $\mu(a_1') > \alpha$, because $a_1' < a$, $a_1' \neq O$. Having this, put $a'' =_{df} a' - a_1'$. Since $a_1' < a'$, we have $a'' \neq 0$. Since $a'' < a$, we get, by what has been proved, that there exists a_2' such that $a_2' \neq O$, $a_2' < a''$. Since $a_2' \leqq a$, we get, by hypothesis, $\mu(a_2') \geqq \alpha$. Hence $a_1' + a_2' \leqq a$, $a_1' \cdot a_2' = O$, $\mu(a_1') \geqq \alpha$ and $\mu(a_2') \geqq \alpha$. Thus the statement in [U.3b.] is proved.

U.3c. The above preparations made, consider the soma a. By [U.3b.], we can find two disjoint somata a_1, a_2, such that $a_1 + a_2 \leqq a$, $\mu(a_1) \geqq \alpha$ and $\mu(a_2) \geqq \alpha$. It follows that $\mu(a) > 2\alpha$. For a_i, $(i = 1, 2)$, we have the similar. We can find two disjoint somata a_{i1}, a_{i2} such that $a_{i1}, a_{i2} \leqq a_i$, and $\mu(a_{i1}) \geqq \alpha$, $\mu(a_{i2}) \geqq \alpha$.

Hence

$$\mu(a_i) \geqq 2\alpha.$$

This being true for $i = 1, 2$, we get $\mu(a) \geqq 4\alpha$.

Continuing this process by induction, we get

$$\mu(a) \geqq 2^n \cdot \alpha \quad \text{for} \quad n = 1, 2, \ldots,$$

which is not true. The lemma is established.

U.3.1. Lemma. Under $\boldsymbol{H}_1$, if

1. $\boldsymbol{G}$ has no atoms,
2. $a \in \boldsymbol{G}$, $a \neq O$,

then there exist two somata $a_1, a_2 \in \boldsymbol{G}$, such that

1) $a_1 \cdot a_2 = O$,
2) $a_1 + a_2 = a$,
3) $\mu(a_1) = \mu(a_2) = \frac{1}{2}\mu(a)$.

Proof. First we notice, that if $a = a_1 + a_2$, $a_1 \cdot a_2 = O$, $a_1 \neq O$, $a_2 \neq O$, $\mu(a_1) > \mu(a_2)$, then there exist somata b_1, b_2, such that $a = b_1 + b_2$, $b_1 \cdot b_2 = O$, $b_1 \neq O$, $b_2 \neq O$, $\mu(b_1) > \mu(b_2)$, $a_1 > b_1$, $a_2 < b_2$; (which implies that

$$\mu(b_1) - \mu(b_2) < \mu(a_1) - \mu(a_2)).$$

Indeed, put

$$\alpha =_{df} \mu(a_1) - \mu(a_2) > 0.$$

By Lemma [U.3.], there exists a soma $c \neq O$, such that $c < a_1, \mu(c) < \alpha$. Taking such a soma c, and putting $b_1 =_{df} a_1 - c$, $b_2 =_{df} a_2 + c$, we get

$$\mu(b_2) = \mu(a_2 + c) = \mu(a_2) + \mu(c) > \mu(a_2),$$

$$\mu(b_1) = \mu(a_1 - c) = \mu(a_1) - \mu(c) < \mu(a_1),$$

$$\mu(b_1) - \mu(b_2) = [\mu(a_1) - \mu(c)] - [\mu(a_2) + \mu(c)]$$
$$= \mu(a_1) - \mu(a_2) - 2\mu(c) < \mu(a_1) - \mu(a_2).$$

The correctness of our remark is proved.

Having that, suppose that the Lemma, to be proved, is not true. There exist $a_1^{(1)}, a_2^{(1)}$, both $\neq O$, such that $a_1^{(1)} \cdot a_2^{(1)} = O$, $a_1^{(1)} + a_2^{(1)} = a$, because if not, a would be an atom. We have $\mu(a_1^{(1)}) \neq \mu(a_2^{(1)})$; we may admit that $\mu(a_1^{(1)}) > \mu(a_2^{(1)})$. By what has been proved, we can find somata $a_1^{(2)}$, $a_2^{(2)}$, both $\neq O$, such that $\mu(a_1^{(2)}) > \mu(a_2^{(2)})$ $a_1^{(2)} < a_1^{(1)}$, $a_2^{(2)} > a_2^{(1)}$, $\mu(a_1^{(2)}) - \mu(a_2^{(2)}) < \mu(a_1^{(1)}) - \mu(a_2^{(1)})$.

Continuing this process, we can define the well ordered sequences of somata

$$a_1^{(1)} > a_1^{(2)} > \cdots > a_1^{(\gamma)} > \cdots$$

$$a_2^{(1)} < a_2^{(2)} < \cdots < a_2^{(\gamma)} < \cdots$$

with

$$\mu(a_1^{(1)}) - \mu(a_2^{(1)}) > \mu(a_1^{(2)}) - \mu(a_2^{(2)}) > \cdots > \mu(a_1^{(\gamma)}) - \mu(a_2^{(\gamma)}) > \cdots$$

Indeed, suppose we have defined $a_2^{(\beta)}$, $a_1^{(\beta)}$ for all ordinals $\beta < \gamma$, where γ is given. If $\gamma - 1$ exists, we find $a_2^{(\gamma)}$, $a_1^{(\gamma)}$, both $\neq O$ with $a_2^{(\beta)} < a_2^{(\gamma)}$, $a_1^{(\beta)} > a_1^{(\gamma)}$ and

$$\mu(a_1^{(\beta)}) - \mu(a_2^{(\beta)}) > \mu(a_1^{(\gamma)}) - \mu(a_2^{(\gamma)}) \quad \text{for all} \quad \beta < \gamma.$$

If $\gamma - 1$ does not exist, we put

$$a_2^{(\gamma)} =_{df} \sum_{\beta<\gamma} a_2^{(\beta)}, \qquad a_1^{(\gamma)} =_{df} \prod_{\beta<\gamma} a_1^{(\beta)}.$$

These, perhaps non denumerable operations, are performable by virtue of Wecken's theorem, [A I., Sect. 1.]. We get

$$a_1^{(\beta)} > a_1^{(\gamma)}, \qquad a_2^{(\beta)} < a_2^{(\gamma)}$$

for all $\beta < \gamma$. We also have

$$\mu(a_1^{(\beta)}) - \mu(a_2^{(\beta)}) \geqq \mu(a_1^{(\gamma)}) - \mu(a_2^{(\gamma)}) \geqq 0.$$

Now the last sign of equality is excluded, since the lemma has been supposed to be not true. Thus we have

$$\mu(a_1^{(\gamma)}) - \mu(a_2^{(\gamma)}) > 0.$$

Since the process never stops, we get disjoint somata

$$a_2^{(\gamma)} - a_2^{(\gamma+1)} \neq O$$

for all possible ordinals γ.

This is, however, impossible, since there does not exist a non denumerable set of mutually disjoint somata, each having a positive measure.

The lemma is established.

U.3.2. Lemma. Under $\boldsymbol{H}_1$, if

1. $\boldsymbol{G}$ has no atoms,
2. $a \in \boldsymbol{G}$, $a \neq O$,

then there exists a system $a_{i_1, i_2, \ldots, i_k}$ of somata of $\boldsymbol{G}$ (where $k = 1, 2, \ldots, i_1, i_2, \ldots, i_k = 0$, or $= 1$), such that

1) $a > a_{i_1} > a_{i_1 i_2} > \cdots > a_{i_1, i_2, \ldots, i_k} > \cdots$

2) $a_{i_1, i_2, \ldots, i_k} \cdot a_{j_1, j_2, \ldots, j_k} = O$ for $(i_1, i_2, \ldots, i_k) \neq (j_1, j_2, \ldots, j_k)$,

3) for any collection J of indices we have $a_{J,0} + a_{J,1} = a_J$,

4) $\mu(a_J) = \mu(a_K) = \frac{\mu(a)}{2^n}$, whenever J, K are any collections of n indices.

Proof. This follows directly from Lemma [U.3.1.], by iterated decompositions of somata into two parts with equal measure.

U.3.3. Lemma. Under $\boldsymbol{H}_1$, if

1. $\boldsymbol{G}$ has no atoms,
2. E is a measurable set of traces,
3. $\mu(E) > 0$,

then there exists a system $E_{i_1, i_2, \ldots, i_k}$ of measurable sets of traces, having the properties, analogous to those in Lemma [U.3.2.].

Proof. For every soma a of $\boldsymbol{G}$ there exists a measurable set E of traces, whose coat equals a. This is true, because $\boldsymbol{G}$ is the lebesguean extension of $\mathfrak{B}(\boldsymbol{F})$. Now applying Lemma [U.3.2.], we complete the proof.

U.4. Theorem. Under $\boldsymbol{H}_1$, if

1. $\boldsymbol{G}$ has no atoms,
2. $\{f_x\}$ is an admissible, Def. [U.2.7.], set of real (complex) quasi-numbers,
3. E is a measurable set of traces,
4. $P_n = \{p_{n1}, p_{n2}, \ldots\}$, $(n = 1, 2, \ldots)$, is a (DARS)-sequence of complexes for E,

then the sequence

$$M(P_n) =_{df} |f(p_{n1})| + |f(p_{n2})| + \cdots \qquad (n = 1, 2, \ldots)$$

is bounded.

Proof. Suppose that we have a (DARS)-sequence $\{P_n\}$ for E, such that $\{M(P_n)\}$ is not bounded. Then there exists a partial sequence $\{P_{l(n)}\}$ of $\{P_n\}$, such that the sequence $\{M(P_{l(n)})\}$ is not bounded, and in addition to that,

$\mathcal{N}(P_{l(n)}) \leqq 1, \quad |E, P_{l(n)}|_\mu \leqq 1, \quad M(P_{l(n)}) \geqq 1, \quad$ for all $\quad n = 1, 2, \ldots$

Let us remark that, since $\boldsymbol{G}$ has no atoms, the net numbers $\mathcal{N}_A$, $\mathcal{N}_R$ coincide with $\mathcal{N}$. We see, that $\{P_{l(n)}\}$ is also a (DARS)-sequence for E.

By virtue of Lemma [U.3.3.], choose a system of sets E; E_0, E_1; E_{00}, E_{01}, E_{10}, E_{11}; ..., which are measurable subsets of E, having the properties, expressed in that lemma.

There exists a sequence $\{P^0_{l(n)}\}$, whose elements are partial collections $P_{l(n)}$ respectively, such that $\{P^0_{l(n)}\}$ is a (DARS)-sequence for E_0. By [U.9.], if we denote by $P^1_{l(n)}$ the complex complementary in $P_{l(n)}$ for $P^0_{l(n)}$, the sequence $\{P^1_{l(n)}\}$ will be (DARS)-sequence for E_1.

We have

$$P^0_{l(n)} \cap P^1_{l(n)} = \emptyset, \quad P^0_{l(n)} \cup P^1_{l(n)} = P_{l(n)}.$$

Now

$$M(P_{l(n)}) = M(P^0_{l(n)}) + M(P^1_{l(n)}).$$

Since $M(P_{l(n)})$ is not bounded, so must be at least one of the sequences $M(P^0_{l(n)})$, $M(P^1_{l(n)})$. Choose one of them and denote it by $M(P^{(i_1)}_{l(k)})$. Applying a reasoning, similar to the above, we can find a subsequence $\{P^{(i_1)}_{lk(n)}\}$ of $\{P^{(i_1)}_{l(n)}\}$, such that $\{M(P^{(i_1)}_{lk(n)})\}$ is not bounded, and in addition to that, for all $n = 1, 2, \ldots$:

$$\mathcal{N}(P^{(i_1)}_{lk(n)}) \leqq \tfrac{1}{2}, \quad |E_{i_1}; P^{(i_1)}_{lk(n)}| \leqq \tfrac{1}{2}, \quad M(P^{(i_1)}_{lk(n)}) \geqq 1.$$

Repeating the above construction, we shall find a sequence $\{P^{(i_1, i_2)}_{kls(n)}\}$ of complexes, which is a (DARS)-sequence for E_{i_1, i_2}, and where $M(P^{(i_2, i_2)}_{lks(n)})$ is not bounded,

$$\mathcal{N}(P^{(i_1, i_2)}_{lks(n)}) \leqq \frac{1}{2^2}, \quad |E_{i_1, i_2}; P^{(i_1, i_2)}_{lks(n)}| \leqq \frac{1}{2^2}, \quad M(P^{(i_1, i_2)}_{kls(n)}) \geqq 1.$$

We continue infinitely the above process, getting the sequence

$$Q_1 =_{df} P^{(i_1)}_{kl(1)}, \quad Q_2 =_{df} P^{(i_1, i_2)}_{lks(1)}, \ldots,$$

which has the property:

$$(1) \qquad |E_{i_1, \ldots, i_n}; Q_n| \leqq \frac{1}{2^n}, \quad M(Q_n) \geqq 1, \quad \text{for} \quad n = 1, 2, \ldots$$

Since for $n > 3$, we have

$$|E_{i_1, \ldots, i_n}; Q_n| \leqq \frac{1}{2^n} \leqq \frac{1}{2^{n-1}},$$

we get

$$|\mu(Q_n) - \mu(E_{i_1, \ldots, i_n})| < 2 \cdot \frac{1}{2^{n-1}};$$

hence

$$\mu(Q_n) - \mu(E_{i_1, \ldots, i_n}) < \frac{1}{2^{n-2}},$$

$$\mu(Q_n) < \mu(E_{i_1, \ldots, i_n}) + \frac{1}{2^{n-2}},$$

and then

$$\mu(Q_n) < \frac{1}{2^n} + \frac{1}{2^{n-2}} < 2 \cdot \frac{1}{2^{n-2}} = \frac{1}{2^{n-3}},$$

so $\mu(Q_n) \to 0$. It follows, by Theorem [U.2.6.], that $\lim\limits_{n\to\infty} M(Q_n) = 0$, which contradicts the inequality (1). The theorem is established.

U.5. Remark. It is not true that, if $\{f_x\}$ is admissible, then $\{f_x\}$ is (DARS)-summable on $\boldsymbol{W}$, (the set of all traces).

Expl. Let $\boldsymbol{G}$ be the tribe of all Lebesgue-measurable subsets of the half-open interval $(0, 1\rangle$, taken modulo the ideal of null-sets. Let the measure on $\boldsymbol{G}$ be lebesguean. It is effective.

Consider the system of half-open subintervals of $(0, 1\rangle$, $h_{i_1, i_2, \ldots, i_n}$, where $i_1, i_2, \ldots, i_n = 0, 1$; $n = 1, 2, \ldots$; $h_{i_1, i_2, \ldots, i_n, 0}$, $h_{i_1, i_2, \ldots, i_n, 1}$ are two intervals which are obtained by dividing the interval $h_{i_1, i_2, \ldots, i_n}$ into two equal parts, the first being left, the second right. The intervals h_0, h_1 are the intervals $(0, \frac{1}{2}\rangle$, $(\frac{1}{2}, 1\rangle$. The system is just similar, in properties, to the system [U.3.2.]. If $J = i_1, i_2, \ldots, i_n$, then meas h_J equals $\frac{1}{2^n}$.

The above system together with the empty set will constitute the base $\boldsymbol{B}$; and the tribe $\boldsymbol{F}$ will be defined as the tribe of all finite unions of the somata of $\boldsymbol{B}$. We see that $\boldsymbol{G}$ is the borelian extension of $\boldsymbol{F}$ through a suitable isomorphism. $\boldsymbol{G}$ has no atoms.

Let

$$f(h_{i_1, i_2, \ldots, i_n}) =_{df} \begin{cases} \frac{1}{2^n} & \text{whenever } i_n = 0, \\ 0 & \text{whenever } i_n = 1. \end{cases}$$

That scalar field on $\boldsymbol{B}$ generates a regular and total scalar field $\{f(x)\}$. It is admissible. I say that $\{f_x\}$ is not (DARS)-summable on $\boldsymbol{W}$. Indeed, fix the interval h_J for a while, where J is a set of indices 0, 1; and consider its subintervals $h_{J,0}, h_{J,1,0}, h_{J,1,1,0}, \ldots$ They are disjoint and we have

$$\text{meas}[h_{J,0} \cup h_{J,1,0} \cup \cdots] = \frac{1}{2^{n+1}} + \frac{1}{2^{n+2}} + \cdots$$

$$= \frac{1}{2^{n+1}}\left(1 + \frac{1}{2} + \frac{1}{2^2} + \cdots\right) = \frac{1}{2^n}.$$

We have

$$(1) \quad f(h_{J,0}) + f(h_{J,1,0}) + f(h_{J,1,1\,0}) + \cdots = \frac{1}{2^{n+1}} + \frac{1}{2^{n+2}} + \cdots = \frac{1}{2^n}.$$

Hence, if we take the sum in the expression (1), extended over all systems J of n indices $i_1, i_2, \ldots, i_n$, we get $2^n \cdot \frac{1}{2^n} = 1$.

On the other hand the collection Q_n of all these intervals $h_{J,0}, h_{J,0,1}, \ldots$ constitutes a denumerable partition of the interval $(0, 1\rangle$ where the net-number $\mathcal{N}$ is $\frac{1}{2^{n+1}}$. Let us "trim" each its 2^n infinite parts (by cutting away all terms, starting with a suitable sufficiently great index in Q_n), so as to get a finite complex

$$Q'_n = \{q'_{n1}, q'_{n2}, \ldots\} \quad \text{with} \quad |\operatorname{som} Q_n, \operatorname{som} Q'_n| < \varepsilon_n.$$

We shall get

$$\sum_k f(q'_{nk}) \geqq 1 - \varepsilon_n,$$

where $\varepsilon_n > 0$ is chosen; so as to have $\lim \varepsilon_n = 0$. The sequence $\{Q'_n\}$ is a (DARS)-sequence for $\boldsymbol{W}$; and we get

(1.1) $$\lim_{n \to \infty} \sum_k f(q'_{nk}) = 1.$$

On the other hand consider the intervals

(2) $$h_{J,1}, h_{J,0,1}, h_{J,0,0,1}, \ldots,$$

on which f has the value 0. Proceed with that sequence (2) similarly. We get a (DARS)-sequence $R'_n = \{r'_{n1}, r'_{n2}, \ldots\}$ of complexes for $\boldsymbol{W}$, such that

(2.1) $$\sum_k f(r'_{nk}) = 0;$$

hence proceeding to the limit for $n \to \infty$, we get the limit-value 0.

The results (1.1) and (2.1) prove, that the $\{f_x\}$ is not (DARS)-summable.

U.5a. Consider another set of scalar quasi-numbers, generated by the scalar field

$$g(h_{i_1, i_2, \ldots, i_n}) = \begin{cases} 0 & \text{whenever} \quad i_n = 0, \\ \frac{1}{2^n} & \text{whenever} \quad i_n = 1. \end{cases}$$

We get a similar result as before. Indeed g is a "mirror image" of f. So we shall have two (DARS)-sequences for $\boldsymbol{W}$, yielding the value 0 and 1.

U.5b. Now consider the set $\{s_x\}$ of scalar quasi-numbers, defined by

$$s(h_{i_1, i_2, \ldots, i_n}) = \begin{cases} \frac{1}{2^n} & \text{for} \quad i_n = 0, \\ -\frac{1}{2^n} & \text{for} \quad i_n = 1. \end{cases}$$

We have

$$|s(h_{i_1, i_2, \ldots, i_n})| = \frac{1}{2^n} \quad \text{for all } n.$$

Hence

$$|s(h_{i_1, i_2, \ldots, i_n})| = \operatorname{meas} h_{i_1, i_2, \ldots, i_n}.$$

For any (DARS)-sequence $\{T_n\}$ for $\boldsymbol{W}$ we get

$$\sum_k |s(t_{nk})| = \operatorname{meas} \sum_k t_{nk} \to 1 \quad \text{for} \quad n \to \infty, \quad \text{so} \quad S_W |s_x| = 1.$$

U.5c. Since

$$f(q) - g(q) = s(q),$$
$$f(q) + g(q) = |s(q)|,$$

the scalar set $\{s_x\}$ cannot be summable, since if so, the set $\{f_x\}$ would be summable, because

$$f(q) = \tfrac{1}{2}[s(q) + |s(q)|].$$

Thus we have found a total and regular set of quasi-numbers $\{s_x\}$, which is not summable, though $\{|s_x|\}$ is summable.

U.5.1. The above example shows that there is a difference in behaviour of summation and of ordinary Lebesgue's integration. Indeed, if $|f(x)|$ is Lebesgues integrable on $(0, 1\rangle$; so is also $f(x)$.

U.6. Remark. The theorem, inverse to the Theorem [U.4.], is not true.

Expl. Take G, F, B and put $f\left(0, \frac{1}{n}\right\rangle = 1$, and for all other bricks p put $f(p) = 0$. The sums $\sum_k |f(p_{nk})|$ are bounded for every (DARS)-sequence for $\boldsymbol{W}$, but though $\mu\left(0, \frac{1}{n}\right\rangle \to 0$, $f\left(0, \frac{1}{n}\right\rangle$ does not tend to 0; hence f is not admissible.

U.7. Remark. If all the above sums are bounded, and for the complexes $P_1, P_2, \ldots$ the somata $\operatorname{som} P_1, \operatorname{som} P_2, \ldots$ are disjoint, with $\mu(P_n) \to 0$, then $\sum_k |f_{nk}| \to 0$. Proof indirect.

U.8. Theorem. Under hypothesis $\boldsymbol{H}_1$, if

1. $\boldsymbol{G}$ has no atoms,
2. $\{f_x\}$ is a total, regular set of real quasi-numbers,
3. $\{f_x\}$ is admissible,
4. E is a measurable set of traces,
5. we define for every (DARS)-sequence $P_{\dot{n}}$ for E the number

$$\varphi(P_{\dot{n}}) =_{df} \lim_{n\to\infty} \sup \sum_k f(p_{nk}),$$

then the supremum, taken for all P_n,

$$\sup \varphi(P_{\dot{n}}) < +\infty.$$

Proof. By Theorem [U.4.], we know, that for every P_n the sequence $\sum_k |f(p_{nk})|$ is bounded, so $\varphi(P_n) < +\infty$ for all P_n.

Suppose that $\sup \varphi(P_n) = +\infty$; then there exists an infinite sequence of (DARS)-sequences for E:

$$P_n^{(1)}, P_n^{(2)}, \ldots,$$

such that

$$\varphi(P_n^{(\alpha)}) \geqq \alpha, \quad (\alpha = 1, 2, \ldots).$$

Given α, we can find a sequence of indices $\alpha(1) < \alpha(2) < \cdots$, depending on α, such that

$$f(P_{\alpha(1)}^{(\alpha)}), f(P_{\alpha(2)}^{(\alpha)}), \ldots \geqq \frac{\alpha}{2}.\text{*)}$$

$P_{\alpha(1)}^{(\alpha)}, P_{\alpha(2)}^{(\alpha)}, \ldots$ is a subsequence of $P_1^{(\alpha)}, P_2^{(\alpha)}, \ldots$, which is a (DARS)-sequence for E. Hence $P_{\alpha(1)}^{(\alpha)}, P_{\alpha(2)}^{(\alpha)}, \ldots$ is also so for every $\alpha = 1, 2, \ldots$ a (DARS)-sequence for E. Let us denote it by $\bar{P}_1^{\alpha}, \bar{P}_2^{\alpha}, \ldots$ We have

$$(1) \qquad f(\bar{P}_1^{\alpha}) \geqq \frac{\alpha}{2}, \quad f(\bar{P}_2^{\alpha}) \geqq \frac{\alpha}{2}, \ldots$$

Consider a denumerable sequence

$$(2) \qquad M_1, M_2, \ldots, M_n, \ldots$$

of measurable subsets of E, such that, in the μ-topology, restricted to E, the sequence (2) constitutes an everywhere dense set.

Since $\bar{P}_n^{(1)}$ is a (DARS)-sequence for E, therefore there exist a sequence $Q_n^{(1)}$ of partial complexes of $\bar{P}_n^{(1)}$, such that

$$|Q_n^{(1)}, M_1|_\mu \to 0 \quad \text{for} \quad n \to \infty.$$

Choose an index n_1, such that for all $n \geqq n_1$

$$|Q_n^{(1)}, M_1| \leqq \frac{1}{2^1}, \quad \mathcal{N}(\bar{P}_n^{(1)}) \leqq \frac{1}{2^1}.$$

Similarly there exist partial complexes

$$Q_n^{(2)\,1}, Q_n^{(2)\,2} \quad \text{of} \quad \bar{P}_n^{(2)},$$

such that

$$|Q_n^{(2)\,1}, M_1| \to 0, \quad |Q_n^{(2)\,2}, M_2| \to 0 \quad \text{for} \quad n \to \infty.$$

Hence, we can find an index n_2, such that

$$n_2 > n_1 \quad \text{and} \quad |Q_n^{(2)\,1}, M_1| < \frac{1}{2^2}, \quad |Q_n^{(2)\,2}, M_2| < \frac{1}{2^2},$$

and where

$$\mathcal{N}(\bar{P}_n^{(2)}) < \frac{1}{2^2} \quad \text{for all} \quad n > n_2.$$

---

*) $f(P)$ will denote the number $\sum_k f(p_k)$, supposing that $P = \{p_1, p_2, \ldots\}$.

We continue this process by induction, getting an index n_3, such that $n_3 \geqq n_2$, and that for $n \geqq n_3$ we have, for suitable partial complexes

$$Q_n^{(3)\,1},\ Q_n^{(3)\,2},\ Q_n^{(3)\,3} \quad \text{of} \quad \bar{P}_n^{(3)},$$

the inequalities

$$|Q_n^{(3)\,1}, M_1| < \frac{1}{2^3}, \qquad |Q_n^{(3)\,2}, M_2| < \frac{1}{2^3}, \qquad |Q_n^{(3)\,3}, M_3| < \frac{1}{2^3},$$

$$\mathcal{N}(\bar{P}_n^{(3)}) < \frac{1}{2^3}.$$

Consider the sequence of complexes

(3) $$\bar{P}_{n1}^{(1)}\ \bar{P}_{n2}^{(2)}, \ldots, \bar{P}_{n\beta}^{(\beta)}, \ldots, \qquad (\beta = 1, 2, \ldots).$$

We shall prove, that it is a (DARS)-sequence for E. Indeed we have

$$\mathcal{N}(\bar{P}_{n\beta}^{(\beta)}) < \frac{1}{2^\beta}, \qquad (\beta = 1, 2, \ldots),$$

so the net-number tends to zero. Take the set M_δ. We have

$$|Q_{n\delta}^{(\delta)\,\delta}, M_\delta| < \frac{1}{2^\delta}, \qquad |Q_{n_{\delta+1}}^{(\delta+1)\,\delta}, M_\delta| < \frac{1}{2^{\delta+1}}, \ldots,$$

$$|Q_{n_{\delta+s}}^{(\delta+s)\,\delta}, M_\delta| < \frac{1}{2^{\delta+s}}, \qquad (s = 1, 2, \ldots),$$

and

$$Q_{n_\delta}^{(\delta)\,\delta} \subseteqq \bar{P}_{n_\delta}^{(\delta)}, \quad Q_{n_{\delta+1}}^{(\delta+1)\,\delta} \subseteqq \bar{P}_{n_{\delta+1}}^{(\delta+1)}, \ldots, \quad Q_{n_{\delta+s}}^{(\delta+s)\,\delta} \subseteqq \bar{P}_{n_{\delta+s}}^{(\delta+s)}, \ (s = 0, 1, \ldots),$$

Hence

$$|Q_{n_{\delta+s}}^{(\delta+s)\,\delta}, M_\delta| \to 0 \quad \text{for} \quad s \to \infty.$$

Considering the (D)-sequence for M_δ:

$$\bar{P}_{n_1}^{(1)}, \ldots, \bar{P}_{n_{\delta-1}}^{(\delta-1)}, Q_{n_\delta}^{(\delta+s)\,\delta}, Q_{n_{\delta+1}}^{(\delta+1)\,\delta}, \ldots,$$

we deduce, that (3) is a (DARS)-sequence for E. From (1) we have

$$f(\bar{P}_{n_1}^{(1)}) \geqq \frac{1}{2}, \quad f(\bar{P}_{n_2}^{(2)}) \geqq \frac{2}{2}, \ldots, \quad f(\bar{P}_{n_\beta}^{(\beta)}) \geqq \frac{\beta}{2};$$

hence

$$f(\bar{P}_{n_\beta}^{(\beta)}) \to +\infty \quad \text{for} \quad \beta \to \infty.$$

This however contradicts Theorem [U.4.], because the sequence $f(\bar{P}_{n_1}^{(1)}), f(\bar{P}_{n_2}^{(2)}), \ldots$ must be bounded.

U.9. Theorem. Under hypothesis $\boldsymbol{H}_1$, if

1. $\boldsymbol{G}$ has no atoms,
2. $\{f_x\}$ is a total, regular set of real quasi-numbers
3. the set $\{f_x\}$ is admissible,
4. E is a measurable set of traces,
5. we define, for every (DARS)-sequence P_n for E, the number

$$\varphi(P_n) =_{df} \limsup_{n\to\infty} \sum_k f(p_{nk}),$$

6. we define $\Phi(E) =_{df} \sup \varphi(P_n)$, taken for all (DARS)-sequences for E,

then there exists a (DARS)-sequence $\{T_n\}$ for E, such that

$$\lim_{n\to\infty} f(T_n) = \Phi(E).$$

Proof. By the preceding Theorem [U.8.], $\Phi(E) < +\infty$. Let $P_n^{(1)}, P_n^{(2)}, \ldots$ be (DARS)-sequences for E, such that

$$\lim_{\alpha\to\infty} \varphi(P_n^{(\alpha)}) = \Phi(E).$$

Let

(0) $$Q_n^\alpha =_{df} P_n^{s(\alpha)},$$

where $s(\dot\alpha)$ is a subsequence of $\dot\alpha$, such that

(1) $$|\Phi(E) - \varphi(Q_n^\alpha)| \leqq \frac{1}{\alpha}, \quad (\alpha = 1, 2, \ldots).$$

We have

$$\varphi(Q_n^\alpha) = \limsup_{n\to\infty} f(Q_n^\alpha), \quad (\alpha = 1, 2, \ldots).$$

Hence we can find for every α a subsequence $s_\alpha(\dot n)$ of $\dot n$, such that, if we put

(2) $$R_n^\alpha =_{df} Q_{s_\alpha(n)}^\alpha,$$

we have

$$\varphi(Q_n^\alpha) = \lim_{n\to\infty} f(Q_{s_\alpha(n)}^\alpha) = \lim_{n\to\infty} f(R_n^\alpha).$$

Hence we can have a partial sequence $t_\alpha(\dot n)$ of $\dot n$ such that, if we put $T_n^\alpha =_{df} Q_{s_\alpha t_\alpha(n)}^\alpha$, we get

(2.1) $$|\varphi(Q_n^\alpha) - f(T_n^\alpha)| \leqq \frac{1}{n}, \quad (n = 1, 2, \ldots).$$

Taking account of (1), we get

(3) $$|\Phi(E) - f(T_n^\alpha)| \leqq \frac{1}{n} + \frac{1}{\alpha}.$$

By (0), Q_n^α is a (DARS)-sequence for E; hence, by (2), R_n^α, as a subsequence of Q_n^α, is also a (DARS)-sequence for E, whatever $\alpha = 1, 2, \ldots$ may be. Consequently, T_n^α, as a subsequence of Q_n^α, is also a (DARS)-sequence for E.

Let $M_1, M_2, \ldots, M_\lambda, \ldots$ be an everywhere dense set of subsets of E, in the μ-topology restricted to E. Since T_n^1 is a (DARS)-sequence for E, and $M_1 \subseteqq E$, there exist partial complexes $S_n^{1(1)}$, $(n = 1, 2, \ldots)$, of T_n^1, such that $|S_n^{1(1)}, M_1| \to 0$.

Let n_1 be an index such that if $n \geqq n_1$; then

$$|S_n^{1(1)}, M_1| \leqq \tfrac{1}{1}, \quad \mathcal{N}(T_n) \leqq \tfrac{1}{1}.$$

Since T_n^2 is also a (DARS)-sequence for E, there exist partial complexes $S_n^{2(1)}$, $S_n^{2(2)}$ of T_n^2, such that

$$|S_n^{2(1)}, M_1| \to 0, \quad |S_n^{2(2)}, M_2| \to 0 \quad \text{for} \quad n \to \infty.$$

Let us choose $n_2 > n_1$, such that for all $n \geqq n_2$ we have

$$|S_n^{2(1)}, M_1| \leqq \tfrac{1}{2}, \quad |S_n^{2(2)}, M_2| \leqq \tfrac{1}{2}, \quad \mathcal{N}(T_n^{(2)}) \leqq \tfrac{1}{2}.$$

We continue this process, as in the forgoing proof, getting a (D)-sequence

(4) $$T_{n_1}^1, T_{n_2}^2, \ldots$$

for E. The sequence has the property, that for every M_λ there exists a partial complex of T_{nk}^k, $(k = 1, 2, \ldots)$, which μ-tends to M_λ.

This proves that (4) is a (DARS)-sequence for E. Since by (3),

$$|\Phi(E) - f(T_{nk}^k)| \leqq \frac{1}{k} + \frac{1}{n_k} \leqq \frac{2}{k},$$

the theorem is established.

U.9.1. Remark. It is not true, that under hypothesis of the above theorem, the collection of all $f(P)$, where P varies over all complexes, is bounded from above.

Expl. Let $f\left(\frac{1}{n}, 1\right\rangle = n$, $(n = 1, 2, \ldots)$, and for all other intervals $(\alpha, \beta\rangle \subsetneqq (0, 1\rangle$, let $f(\alpha, \beta\rangle = 0$. This scalar field is admissible, but the collection $f(P)$ is not bounded.

U.9.2. Remark. The above proof can be used, almost without change, to establish the following (under $\boldsymbol{H}_1$, and hyp. 1, 2, 3, 4):

If P_n^α, $\alpha = 1, 2, \ldots$ are (DARS)-sequences for E with

$$A_\alpha =_{df} \lim_{n\to\infty} f(P_n^\alpha),$$

and if

$$B = \lim_{n\to\infty} A_\alpha,$$

then there exists (DARS)-sequence Q_n for E, such that

$$B = \lim_{n\to\infty} f(Q_n).$$

This can be applied even for complex scalar fields in circumstances considered.

Section 2 of U.

The condition of non-overlapping for complexes

U.10. Def. Under hypothesis $\boldsymbol{H}_1$, [U.1.], let $P_1, P_2, \ldots$ be a finite or infinite sequence of complexes. We say, that it satisfies the *condition of non-overlapping*, whenever the following is happening: If p, q are bricks, $p \in P_n$, $q \in P_m$, $n \leqq m$, $p \cdot q \neq O$, then $q \leqq p$.

U.10.1. If $P_{\dot{n}}$ satisfies the condition of non-overlapping, and $\alpha(\dot{n})$ is a subsequence of n, then $P_{\alpha(\dot{n})}$ also satisfies the condition of non-overlapping.

If $P_{\dot{n}}$ satisfies the condition of non overlapping, and $Q_n \subseteqq P_n$, $n = 1, 2, \ldots$, then $Q_{\dot{n}}$ also satisfies the condition of non overlapping.

U.10.2. Theorem. Under hypothesis $\boldsymbol{H}_1$, if $E \neq 0$ is a measurable set of traces, $E \leqq L$, where L is a covering, then there exists a (DARS)-sequence P_n for E, such that

1) $P_n \leqq L$, $(n = 1, 2, \ldots)$,*)

2) $P_1, P_2, \ldots, P_n, \ldots$ satisfies the condition of non overlapping.

Proof. We start with two following lemmas:

U.10.2.1. Lemma. If

1. A, B_n, $(n = 1, 2, \ldots)$ are measurable sets of traces,
2. $|A, B_n|_\mu \to 0$,
3. $\mu A > 0$,

then there exists n_1, such that for all $n \geqq n_1$ we have $A \cdot B_n \neq O$.

Proof. Suppose this be not true. Then there exists a subsequence $s(\dot{n})$ of $\dot{n}$, such that $A \cdot B_{s(n)} = O$ for all n. We have for all n:

$$|B_{s(n)}; A|_\mu = \mu(B_{s(n)} - A) + \mu(A - B_{s(n)}) = \mu B_{s(n)} + \mu A \geqq \mu A > 0.$$

Hence $|B_{s(n)}; A|_\mu$ does not tend to 0, which contradicts hyp. 2.

U.10.2.2. Lemma.. If

1. A, B_1, B_2 are measurable sets of traces,
2. $A B_1 = A B_2$,
3. $B_1 \leqq B_2$,

then

$$|A, B_1|_\mu \leqq |A, B_2|_\mu.$$

Proof. We have

$$|A, B_1|_\mu = \mu(A - B_1) + \mu(B_1 - A) = \mu(A - A B_1) + \mu(B_1 - A B_1),$$

$$(1) \qquad |A, B_1|_\mu = \mu(A - A B_2) + \mu(B_1 - A B_2).$$

Now, by hyp. 3, $B_1 - A B_2 \leqq B_2 - A B_2$; hence, from (1),

$$|A, B_1|_\mu \leqq \mu(A - A B_2) + \mu(B_2 - A B_2) = |A, B_2|_\mu, \quad \text{q.e.d.}$$

U.10.2.3. Let $E_1, E_2, \ldots, E_m, \ldots$ be a μ-everywhere dense collection of subsets of E, where

$$(2) \qquad E_m \neq O, \quad (m = 1, 2, \ldots).$$

---

*) P_n is a set of bricks, which in turn are somata of the tribe $\boldsymbol{F}$, L is a soma of the tribe $\boldsymbol{G}$, hence instead of $E \leqq L$, we should write $[E] \leqq L$, where $[E]$ is the coat of the set E of traces. Instead of $P_n \leqq L$ we should write $\operatorname{som} P_n \leqq L$. Nevertheless to avoid unnecessary complication we shall, in the sequal, write, as in the statement of the theorem.

There exists a (DARS)-sequence Q_n for E, such that

(3) $$Q_n \leqq L \quad \text{for} \quad n = 1, 2, \ldots$$

Since $E \neq O$, and $|E, Q_n| \to 0$ for $n \to \infty$, there exists $n(1)$, such that, Lemma [U.10.1.], for all $n \geqq n(1)$,

(4) $$E \cdot Q_n \neq O.$$

Since $E_1 \leqq E$, there exists a partial complex $T_{n,1}$ of Q_n, such that

(4.1) $$|T_{n,1}, E_1| \to 0 \quad \text{for} \quad n \to \infty.$$

Hence we can subject $n(1)$ to the farther condition:

(4.2) $$|T_{n,1}, E_1| \leqq \tfrac{1}{2} \quad \text{for} \quad n \geqq n(1).$$

By Lemma [U.1.], we can find $n(1)$ such that, in addition to the above, for all $n \geqq n(1)$:

(5) $$T_{n,1} \cdot E_1 \neq O.$$

In addition to that we may require, that for $n \geqq n(1)$, we have

(6) $$|Q_n, E| \leqq \tfrac{1}{2}, \quad \mathcal{N}_R(Q_n) \leqq \tfrac{1}{2}, \quad \mathcal{N}_A(Q_n) \leqq \tfrac{1}{2}.$$

Thus we get by (4),

$$E \cdot Q_{n(1)} \neq O;$$

by (5),

$$E_1 \cdot T_{n(1),1} \neq O;$$

by (3),

$$Q_{n(1)} \leqq L;$$

by (4.1),

(7) $$T_{n(1),1} \leqq Q_{n(1)};$$

by (6),

$$|Q_{n(1)}, E| \leqq \tfrac{1}{2}, \quad \mathcal{N}_R Q_{n(1)} \leqq \tfrac{1}{2}, \quad \mathcal{N}_A Q_{n(1)} < \tfrac{1}{2};$$

and by (4.2.),

$$|T_{n(1)}, E_1| \leqq \tfrac{1}{2}.$$

U.10.2.4. Let

(8) $$p^{1'}, p^{2'}, \ldots$$

be all (not empty) bricks of $Q_{n(1)}$, and let $N(1) - 1$ be their number. We have $N(1) \geqq 2$. Since, by (7), $Q_{n(1)} \leqq L$, we have

$$Q_{n(1)} = p^{1'} \cup p^{2'} \cup \cdots \leqq L.$$

Since, by (7), $E \cdot Q_{n(1)} \neq 0$, there exists at least one α', such that $E \cdot p^{\alpha'} \neq O$. Denote all those $p^{\alpha'}$ by

(9) $$p^1, p^2, \ldots$$

Their number is $\geqq 1$ and $\leqq N(1)-1$, but it may be $<N(1)-1$. Thus surely

(10) $$E \cdot p^1 \neq O.$$

Since the bricks of $T_{n(1),1}$ belong to $Q_{n(1)}$, by (7), and since $E_1 \cdot T_{n(1),1} \neq O$, there exists at least one brick in (9), which belongs to $T_{n(1),1}$.

Let $\bar{T}_{n(1),1}$ be the collection of all those bricks of

(11) $$T_{n(1),1},$$

which are among the bricks (9).

We have

(12) $$E_1 \cdot T_{n(1),1} = E_1 \cdot \bar{T}_{n(1),1}.$$

Indeed, since

$$\bar{T}_{n(1),1} \leqq T_{n(1),1},$$

we have

$$E_1 \cdot \bar{T}_{n(1),1} \leqq E_1 \cdot T_{n(1),1}.$$

Now let $x \in E_1 \cdot T_{n(1),1}$. There exists a brick $p^{\alpha'}$, belonging to $T_{n(1),1}$, such that $x \in E_1 \cdot p^{\alpha'}$. Since $E_1 \cdot p^{\alpha'} \neq O$, we have $E \cdot p^{\alpha'} \neq O$, by (2). Hence, by (9), $p^{\alpha'}$ belongs to the collection (9). Hence there exists α, such that $p^{\alpha'} = p^{\alpha}$. Hence $x \in E_1 \cdot p^{\alpha}$, and then $x \in E_1 \cdot T_{n(1),1}$. Consequently

$$E_1 \cdot T_{n(1),1} \leqq E_1 \cdot \bar{T}_{n(1),1},$$

so the assertion (12) is established. Since, by (7),

$$|T_{n(1),1}, E_1| \leqq \tfrac{1}{2},$$

and since $\bar{T}_{n(1),1} \leqq T_{n(1),1}$, the relation (12), by the help of Lemma [U.10.11.], gives

(13) $$|T_{1,1} \cdot E_1| \leqq \tfrac{1}{2},$$

if we put

$$\bar{T}_{1,1} =_{df} \bar{T}_{n(1),1}.$$

Put

(14) $$P_1 = \bigcup_{\alpha \geqq 1} p^{\alpha}.$$

We prove as above, that

$$E \cdot P_1 = E \cdot Q_{n(1)}.$$

Since $P_1 \leqq Q_{n(1)}$, we get from (7), by Lemma [U.11.], the inequality

(15) $$|P_1, E| \leqq \tfrac{1}{2}.$$

We also have, by (7),

(16) $$P_1 \leqq L$$

and

(17) $$\bar{T}_{1,1} \leqq P_1.$$

U.10.2.5. The above constitutes the first step of an inductive construction. To prepare the second step, recall that, by (8) and (7), $p^{1'} + p^{2'} + \cdots \leqq L$, so if we put

$$p^0 =_{df} L - (p^{1'} + p^{2'} + \cdots), \tag{18}$$

we get

$$L = p^{1'} + p^{2'} + \cdots + p^0. \tag{19}$$

It follows that

$$E = \sum_{\alpha'} E\, p^{\alpha'} + E\, p^0. \tag{20}$$

The set p^0 may be empty or not. It may not be a brick, but it is a figure. From (20) we get

$$E = \sum_{\alpha} E\, p^{\alpha} + E\, p^0; \tag{21}$$

and all terms in (20), and also in (21), are disjoint.

U.10.2.6. The above being settled, let us consider the sets $E\, p^{\alpha}$, for which $E\, p^{\alpha} \neq O$, $\alpha \geqq 0$. We have

$$E = \sum_{\alpha \geqq 0} p^{\alpha} \cdot E, \tag{22}$$

where the terms are disjoint. Since p^{α} is a covering of $p^{\alpha} E$, there exists, [U.3b.], a (DARS)-sequence Q_n^{α} for $p^{\alpha} E$ with

$$Q_n^{\alpha} \leqq p^{\alpha}, \quad (n = 1, 2, \ldots), \tag{23}$$

for all α available. If we suppose, that $p^{\alpha} \cdot E \neq O$, $(\alpha \geqq 0)$, we get, by Lemma [U.10.1.], an index $n(2)$, such that for $n \geqq n(2)$,

$$E, Q_n^{\alpha} \neq O, \tag{24}$$

for $\alpha \geqq 0$ available. Since $E_1 \cdot p^{\alpha}$, $E_2 \cdot p^{\alpha}$ are subsets of $E \cdot p^{\alpha}$, there exist partial complexes $T_{n,1}^{\alpha}$, $T_{n,2}^{\alpha}$ of Q_n^{α}, such that

$$|T_{n,1}^{\alpha}, E_1 p^{\alpha}| \to 0, \quad |T_{n,2}^{n}, E_2 p^{\alpha}| \to 0 \quad \text{for} \quad n \to \infty. \tag{25}$$

Hence we can subject $n(2)$ to the following additional conditions:

$$|Q_{n(2)}^{\alpha}, E\, p^{\alpha}| \leqq \frac{1}{N(1) \cdot 2^2}, \quad \mathcal{N}_R(Q_{n(2)}^{\alpha}) \leqq \frac{1}{2^2}, \quad \mathcal{N}_A(Q_{n(2)}^{\alpha}) \leqq \frac{1}{2^2} \tag{26}$$

and

$$|T_{n(2),1}^{\alpha}, E_1 p^{\alpha}| \leqq \frac{1}{N(1) \cdot 2^2}, \quad |T_{n(2),2}^{\alpha}, E_2 p^{\alpha}| \leqq \frac{1}{N(1) \cdot 2^2}.$$

If, by chance $E_1 \cdot p^{\alpha} \neq O$ or $E_2 \cdot p^{\alpha} \neq O$, we can require that

$$T_{n(2),1}^{\alpha} \cdot E_1 \neq O \quad \text{and} \quad T_{n(2),2}^{\alpha} \cdot E_2 \neq O \tag{27}$$

respectively.

U.10.2.7. Now let, — in the case — $E\,Q^{\alpha}_{n(2)} \neq O$, $(\alpha \geqq 0)$,

(27.1) $$p^{\alpha 1'}, p^{\alpha 2'}, \ldots$$

be all (not empty) bricks of

(28) $$Q^{\alpha}_{n(2)}.$$

So, by (23) we have

(28.0) $$Q^{\alpha}_{(n)1} = p^{\alpha 1'} \cup p^{\alpha 2'} \cup \cdots \leqq p^{\alpha}.$$

Since, by (24), $E \cdot Q^{\alpha}_{n(2)} \neq O$, there exists, among the bricks (27.1), at least one $p^{\alpha\beta}$, such that $p^{\alpha\beta'} \cdot E \neq O$. Denote all those bricks by

(28.1) $$p^{\alpha 1}, p^{\alpha 2}, \ldots, \quad (\alpha \geqq 0);$$

so we have $p^{\alpha\beta} \cdot E \neq O$ for all β.

Since $\sum_{\beta'} p^{\alpha\beta'} \cdot E = \sum_{\beta} p^{\alpha\beta} \cdot E$ and $\bigcup_{\beta} p^{\alpha\beta'} \subseteqq \bigcup_{\beta'} p^{\alpha\beta'} \subseteqq Q^{\alpha}_{n(2)}$, therefore, by (26), we get

(29) $$\left|\bigcup_{\beta} p^{\alpha\beta}, E\,p^{\alpha}\right| \leqq \frac{1}{N(1)\cdot 2^2}.$$

Now denote by $\bar{T}^{\alpha}_{n(2),1}$ the complex, composed of all bricks of $T^{\alpha}_{n(2),1}$, which belong to the collection (28.1). Similarly we define $\bar{T}^{\alpha}_{n(2),2}$ as the complex, composed of all bricks of $T^{\alpha}_{n(2),2}$, which belong to (28.1). We have

$$\bar{T}^{\alpha}_{n(2),1} \subseteqq T^{\alpha}_{n(2),1}.$$

Since

$$\bar{T}^{\alpha}_{n(2),1} \cdot E_1 = T^{\alpha}_{n(2),1} \cdot E_1,$$

we get from (26),

(30) $$\left|\bar{T}^{\alpha}_{n(2),1}, E_1\,p^{\alpha}\right| \leqq \frac{1}{N(1)\cdot 2^2}.$$

Similarly we get:

(30.1) $$\left|\bar{T}^{\alpha}_{n(2),2}, E_2\,p^{\alpha}\right| \leqq \frac{1}{N(1)\cdot 2^2}.$$

U.10.2.8. From (29) we obtain:

(31) $$\left|\bigcup_{\alpha\geqq 0}\bigcup_{\beta} p^{\alpha\beta}, \sum_{\alpha\geqq 0} E\,p^{\alpha}\right| \leqq \frac{1}{N(1)\cdot 2^2}\cdot N(1) = \frac{1}{2^2},$$

and similarly, from (30):

(32) $$\left|\bigcup_{\alpha\geqq 0}\bar{T}^{\alpha}_{n(2),1}, \sum_{\alpha\geqq 0} E_1\,p^{\alpha}\right| \leqq \frac{1}{2^2},$$
$$\left|\bigcup_{\alpha\geqq 0}\bar{T}^{\alpha}_{n(2),2}, \sum_{\alpha\geqq 0} E_2\,p^{\alpha}\right| \leqq \frac{1}{2^2}.$$

Hence, by (24), and from the formulas

$$E_1 = \sum_{\alpha \geqq 0} E_1 p^\alpha, \quad E_2 = \sum_{\alpha \geqq 0} E_2 p^\alpha,$$

which we obtain from (24), by multiplying sidewise by E_1, E_2 respectively, we get

$$(33) \quad \Big| \bigcup_{\alpha \geqq 0} \bigcup_\beta p^{\alpha\beta}, E \Big| \leqq \frac{1}{2^2}; \ \Big| \bigcup_{\alpha \geqq 0} \bar{T}^\alpha_{n(2),1}, E_1 \Big| \leqq \frac{1}{2^2}; \ \bigcup_{\alpha \geqq 0} \bar{T}^\alpha_{n(2),2}, E_2 \Big| \leqq \frac{1}{2^2}.$$

We put

$$P_2 =_{df} \bigcup_{\alpha \geqq 0} \bigcup_\beta p^{\alpha\beta} \quad \text{and} \quad \bar{T}_{n(2),1} =_{df} \bigcup_{\alpha \geqq 0} \bar{T}^\alpha_{n(2),1},$$

$$\bar{T}_{n(2),2} =_{df} \bigcup_{\alpha \geqq 0} \bar{T}^\alpha_{n(2),2}.$$

From (23) we get

$$(34) \quad |P_2, E| < \frac{1}{2^2}, \quad |\bar{T}_{2,1}; E_1| < \frac{1}{2^2}, \quad |\bar{T}^\alpha_{2,2}; E_2| \leqq \frac{1}{2^2},$$

where

$$(35) \quad \bar{T}_{2,1} \subseteqq P_2 \quad \text{and} \quad \bar{T}_{2,2} \subseteqq P_2.$$

We also have

$$(36) \quad \begin{cases} \mathcal{N}_A(P_2) \leqq \frac{1}{2^2}, & \mathcal{N}_R(P_2) \leqq \frac{1}{2^2}, \\ \mathcal{N}_A(\bar{T}_{2,1}) \leqq \frac{1}{2^2}, & \mathcal{N}_R(\bar{T}_{2,1}) \leqq \frac{1}{2^2}, \\ \mathcal{N}_A(\bar{T}_{2,2}) < \frac{1}{2^2}, & \mathcal{N}_R(\bar{T}_{2,2}) < \frac{1}{2^2}. \end{cases}$$

To be prepared for the next step, we put

$$p^{\alpha 0} = p^\alpha - (p^{\alpha 1'} + p^{\alpha 2'} + \cdots), \quad \alpha > 0.$$

We get

$$p^\alpha = p^{\alpha 1'} + p^{\alpha 2'} + \cdots + p^{\alpha 0},$$

where the terms are disjoint, because

$$p^{\alpha\beta'} \leqq p^\alpha,$$

by (28.0). We denote by $N(2)$ the number of all sets $p^{\alpha\beta'}$ together with $p^{\alpha 0}$, $(\alpha \geqq 0)$.

U.10.2.9. P_1, P_2 satisfy the condition of non overlapping. Indeed, suppose that $p^{\alpha\beta} \cdot p^\gamma \neq O$, where $\alpha \geqq 0$, $\beta \geqq 1$, $\gamma \geqq 1$. We have $p^{\alpha\beta} \leqq p^\alpha$, and since all p^γ are disjoint, we have $p^{\alpha\beta} < p^\gamma = O$, whenever $\gamma \neq \alpha$. Consequently $\gamma = \alpha \geqq 1$. Hence $p^{\alpha\beta} \leqq p^\gamma$.

U.10.2.10. We are going to perform the general step of the induction. We shall denote sequences of k indices $\alpha_1, \alpha_2, \ldots, \alpha_k$, $(k \geqq 2)$, by a single letter, say A. Suppose we have already constructed all p^A, where $\alpha_k \geqq 0$, and suppose that $E = \sum_A p^A E$ with all terms disjoint, and $p^A E \neq O$.

We also suppose that all p^A are disjoint. We are going to define $p^{A\alpha_{K+1}}$ with $(k+1)$ indices. Our process will imitate that for two indices, but using $E_1, E_2, \ldots, E_{k+1}$. Let $N(k)$ be the number of all p^A with k indices.

U.10.2.11. Consider the sets $p^A \cdot E$. Since p^A is a covering of $E \cdot p^A$, we can find a (DARS)-sequence Q_n^A of complexes for $E\, p^A$, with

$$Q_n^A \leqq p^A \quad \text{for} \quad n = 1, 2, \ldots$$

We can find an index $n(k+1)$, such that for $n \geqq n(k+1)$:

$$Q_n^A \cdot E \neq O.$$

Since $E_1 p^A, E_2 p^A, \ldots, E_{k+1} p^A$ are subsets of $E\, p^A$, there exist partial complexes

$$T_{n,1}^A, T_{n,2}^A, \ldots, T_{n,k+1}^A$$

of Q_n^A, such that

$$|T_{n,i}^A, E_i\, p^A| \to 0, \quad \text{for} \quad n \to \infty, \qquad (i = 1, \ldots, k+1).$$

Hence we can subject $n(k+1)$ to the following additional conditions

$$(1) \qquad |Q_{n(k+1)}^A, E\, p^A| \leqq \frac{1}{N(k) \cdot 2^{k+1}},$$

$$\mathcal{N}_R(Q_{n(k+1)}^A) \leqq \frac{1}{2^{k+1}}, \qquad \mathcal{N}_A(Q_{n(k+1)}^A) \leqq \frac{1}{2^{k+1}}$$

and

$$|T_{n(k+1),i}^A, E_i\, p^A| \leqq \frac{1}{N(k) \cdot 2^{k+1}}, \qquad (i = 1, \ldots, k+1),$$

where $N(k)$ is the number of all A with k indices $\geqq 0$. If, by chance, $E_i\, p^A \neq O$, we can require that

$$T_{n(k+1),i}^A \cdot E_i \neq O, \qquad (i = 1, 2, \ldots, k+1).$$

Now, let in the case

$$E \cdot Q_{n(k+1)}^A \neq O,$$

(where the last index in A is $\geqq 0$),

$$p^{A1'}, p^{A2'}, \ldots$$

be all (not empty) bricks of $Q_{n(k+1)}^A$, so

$$Q_{n(k+1)}^A = p^{A1'} \cup p^{A2'} \cup \cdots \leqq p^A.$$

Denote all the above $p^{A\beta'}$, for which $p^{A\beta'} \cdot E \neq O$, by

$$(2) \qquad p^{A1}, p^{A2}, \ldots;$$

so $E \cdot p^{A\beta} \neq O$ for $\beta \geqq 1$.

Since

$$\sum_{\beta'} p^{A\beta'} \cdot E = \sum_{\beta} p^{A\beta} \cdot E \quad \text{and} \quad \sum_{\beta} p^{A\beta} \subseteqq \sum_{\beta'} p^{A\beta'},$$

we get from (1),

$$|\sum_{\beta} p^{A\beta}, E\, p^A| \leqq \frac{1}{N(k) \cdot 2^{k+1}},$$

for all A with $p^A \cdot E \neq O$, even if the last index in A is $= 0$. Hence we get

$$|\sum_A \sum_\beta p^{A\beta}, \sum_A E\, p^A| \leqq \frac{1}{N(k) \cdot 2^{k+1}} \cdot N(k) = \frac{1}{2^{k+1}}.$$

Now we define $P_{k+1} = \bigcup_A \bigcup_{\beta > 1} p^{A\beta}$, where $\bigcup_A$ is taken over all A considered in (1). Thus $|P_{k+1}, \sum_A E\, p^A| \leqq \frac{1}{2^{k+1}}$. Since $\sum_A E\, p^A = E$, we get $|P_{k+1}, E| \leqq \frac{1}{2^{k+1}}$. Denote by $\bar{T}^A_{n(k+1),i}$ the union of all bricks of $T^A_{n(k+1),i}$, which occur in (2), $(i = 1, \ldots, k+1)$. We have $\bar{T}^A_{n(k+1),i} \leqq$ $\leqq T^A_{n(k+1),i}$. We also have $\bar{T}^A_{n(k+1),i} E_i = T^A_{n(k+1),i} \cdot E_i$. Hence, we get from (1):

$$|\bar{T}^A_{n(k+1),i}, E_i\, p^A| \leqq \frac{1}{N(k) \cdot 2^{k+1}}, \qquad (i = 1, \ldots, k),$$

and then

$$|\bigcup_A \bar{T}^A_{n(k+1),i}, \sum_A E_i\, p^A| \leqq \frac{1}{2^{k+1}}.$$

Now we have $E_i = \sum_A p^A \cdot E_i$; then if we put $\bar{T}_{k+1,i} =_{df} \bigcup_A \bar{T}^A_{n(k+1),i}$, we get $|\bar{T}_{k+1,i}, E_i| \leqq \frac{1}{2^{k+1}}$, where $\bar{T}_{k+1} \subseteqq P_{k+1}$. Till now we have defined $p^{A\beta}$ with $\beta \geqq 1$. Since $p^{A\beta'} \leqq p^A$, we define $p^{A0} =_{df} p^A - \sum_{\beta'} p^{A\beta'}$, so we get $E\, p^A = \sum_{\beta'} p^{A\beta'} E + p^{A0} E$, and then $E\, p^A = \sum_{\beta \geqq 0} p^{A\beta} \cdot E$. Now, as we have, by hypothesis, $E = \sum_A E\, p^A$, we get

$$E = \sum_B p^B \cdot E,$$

where the sum is extended over all B with $k+1$ indices, even those with the last index 0 included. Thus the inductive construction is completed.

U.10.2.12. We have defined an infinite sequence P_n of complexes, with $\mathcal{N}_A P_n \to 0$, $\mathcal{N}_R P_n \to 0$, and $|P_n, E| \to 0$ for $n \to \infty$. We have $P_n \leqq L$. We have also found for every E_i, $i = 1, 2, \ldots$ a subcomplex $\bar{T}_{n,i}$ of P_n, with $|E_i, \bar{T}_{n,i}| \to 0$ for $n \to \infty$.

We have defined $\bar{T}_{n,i}$, starting with the i-th step, but we may complete that sequence by putting

$$\bar{T}_{m,i} =_{df} P_i \quad \text{for} \quad m < i.$$

We see that P_k is a (DARS)-sequence for E.

U.10.2.13. Since we have $p^{A\beta} \leqq p^A$, whatever A may be, it follows, by induction, that $p^{AB} \leqq p^A$ for any sequences A, B of indices.

U.10.2.14. Now we shall prove that the non-overlapping condition is satisfied for $P_1, P_2, \ldots$ Indeed suppose, that $p^A \cdot p^B \neq O$, where B has less indices than A.

Suppose that it follows, that $p^B \leqq p^A$, whenever both B and A have $\leqq k$ indices, $(k \geqq 2)$.

Now let B have $k+1$ indices, $\alpha_1, \alpha_2, \ldots, \alpha_{k+1}$. We have

$$p^B \leqq p^{\alpha_1, \ldots, \alpha_k}.$$

Suppose that $p^B \cdot p^A \neq O$; then $p^{\alpha_1, \ldots, \alpha_k} \cdot p^A \neq O$. If A has k-indices, we must have $A = (\alpha_1, \ldots, \alpha_k)$, because all p^M with k indices are mutually disjoint. Now as A has less than k indices, we have by hypothesis, $p^{\alpha_1, \ldots, \alpha_k} \leqq p^A$. Consequently $p^B \leqq p^A$. The theorem is established.

U.10.3. Remark. By means of a slight modification of the proof of the Theorem [U.3.3.], one can prove the following one:

Under hypothesis $\boldsymbol{H}_1$, if

1. $E \neq O$ is a measurable set of traces,
2. $L_1 \geqq L_2 \geqq \cdots$ are coverings of E,

then there exists a (DARS)-sequence P_n for E, such that

1) $P_n \leqq L_n$, $(n = 1, 2, \ldots)$,
2) $P_1, P_2, \ldots, P_n, \ldots$ satisfies the condition of non overlapping.

U.10.4. Theorem. Under hypothesis $\boldsymbol{H}_1$, if

1. $E \neq O$ is a measurable set of traces,
2. P_n is a (DARS)-sequence for E, satisfying the condition of non-overlapping,
3. $s_1 < s_2 < \cdots < s_k < \cdots$ are natural numbers,
4. $|Q_n, E| \to 0$ for $n \to \infty$,
5. for every $k = 1, 2, \ldots$ the bricks of Q_n are taken from the collection $\bigcup_{n=s_k}^{\infty} P_n$,

then Q_n is a (DARS)-sequence for E.

Proof. Let p be a brick of P_m. Find n', such that for $n \geqq n'$, we have $m < s_n$. It follows that $m < s_{n+t}$ for $t = 0, 1, 2, \ldots$ Let

$$Q_n = q_{n1} \cup q_{n2} \cup \cdots$$

There exists t, such that $Q_n \subseteqq P_{s_n} \cup P_{s_n+1} \cup \cdots \cup P_{s_n+t}$. We have $Q_n \cdot p = q_{n1} p \cup q_{n2} p \cup \cdots$, where the terms of the union are bricks (which may be O). If $q_{nk} \cdot p \neq O$, we have $q_{nk} \leqq p$, by the condition of non-overlapping. Indeed $q_{nk} \in P_{s_n+t'}$, for some $p \in P_m$, and $s_n + t' > m$. Consequently

$$Q_n \cdot p \subseteqq q_{n1} \cup q_{n2} \cup \cdots, \quad \text{i.e.} \quad Q_n \cdot p \subseteqq Q_n$$

for all sufficiently great indices n. This is true, even if $Q_n \cdot p = O$. Now, since $|Q_n, E| \to 0$, we also have

(1) $$|Q_n p, E p| \to 0.$$

Hence there exist complexes T_n, which are partial complexes of Q_n, such that

$$|T_n, E p| \to 0.$$

This is true for sufficiently great n, depending on p, (the complexes T_n may be empty). Now let $p_1 \cup p_2 \cup \cdots$ be any finite union of disjoint bricks of $\bigcup_{k=1}^{\infty} P_k$. Suppose that $p_1 \in P_{m1}$, $p_2 \in P_{m2}, \ldots$

Put $n =_{df} \max(m_1, m_2, \ldots)$, and find n', such that for all $n \geqq n'$, we have $s_n > n$. Take $n \geqq n'$, we have $Q_n \cdot p_k \subseteqq Q_n$, for $k = 1, 2, \ldots$ Hence, since $Q_n p_{k'}$, and $Q_n p_{k''}$ are disjoint for $k' \neq k''$, we see, that $T_n =_{df} Q_n p_1 \cup Q_n p_2 \cup \cdots$ is a complex with $T_n \subseteqq Q_n$.

Since $|p_k E, Q_n p_k| \to 0$ for $n \to \infty$, we get $|\sum_k p_k E, T_n| \to 0$ for $n \to \infty$. To have a sequence T_n, starting with $n = 1$, we can adjoin $(n-1)$ times Q_n to the sequence, at its beginning.

Thus for every finite sum $p_1 + p_2 + \cdots$ of bricks, belonging to $\bigcup_{k=1}^{\infty} P_k$, there exist for suitably great n a partial complex T_n of Q_n with $|\sum_k p_k E, T_n| \to 0$.

Now we shall see, that the collection of the sets $E(p_1 + p_2 + \cdots)$ is μ-everywhere dense in E. Indeed, since P_n is a (DARS)-sequence for E and $E(p_1 + p_2 + \cdots) \leqq E$, there exists for every δ a partial complex S_n of P_n with

$$|E(p_1 + p_2 + \cdots), P_n| \leqq \delta.$$

Having that, we can see that the sequence Q_n is a (DARS)-sequence for E. The theorem is established.

U.10.5. Def. Under ***Hyp*** $\boldsymbol{H}_1$. We shall need to specialize the tribe $\boldsymbol{F}$ and its basis $\boldsymbol{B}$, by admitting the following condition, which we call "boundary condition for bricks"; and denote it by ***Hyp*** $\mathscr{B}$:

If p is a brick and $\varepsilon > 0$, then there exists $\delta > 0$, such that if K is any set of disjoint bricks q with $\mu(q) \leqq \delta$, then

$$\mu \operatorname{som} \{q \mid q \in K, p\, q \neq O, \operatorname{co} p \cdot q \neq O\} \leqq \varepsilon.$$

U.10.5.1. Remark. To admit ***Hyp*** $\mathscr{B}$. the following prerequisite is welcome:

1) $\boldsymbol{G}$ has no atoms, or

2) every atom of $\boldsymbol{G}$ is a brick.

U.10.6. Theorem. Under hypotheses $\boldsymbol{H}_1$ and ***Hyp*** $\mathscr{B}$ and for $\boldsymbol{G}$ without atoms, we have: if

1. $E \neq O$ is a measurable set of traces,
2. Q_n is a (DARS)-sequence for E,

then there exists a (DARS)-sequence P_n for E such that:

1) for every n there exists an index $s(n)$ such that $P_n \subseteqq Q_{s(n)}$,

2) P_n satisfies the condition of non-overlapping.

U.10.6a. Proof. We start the proof with two remarks.

Remark I. Let $c \neq O$ be a brick and let $\varepsilon > 0$. By ***Hyp*** $\mathscr{B}$ there exists $\delta = \delta(c, \varepsilon) > 0$, such that, if K is any not empty collection of bricks p with $\mu(p) \leqq \delta(\varepsilon, c)$, then

$$\mu\{p \mid p \in K, p \cdot c \neq O, p \cdot \text{co}\, c \neq O\} \leqq \varepsilon. \tag{0}$$

Now since $\mathscr{N}(Q_n) \to 0$ for $n \to \infty$, we can find an index $n(c, \varepsilon)$, such that if $n \geqq n(c, \varepsilon)$, then

$$\mathscr{N}(Q_n) \leqq \delta(c, \varepsilon);$$

hence for every brick

$$q_{nk} \in Q_n, \quad (k = 1, 2, \ldots)$$

we have

$$\mu(q_{nk}) \leqq \delta(c, \varepsilon).$$

Since Q_n is a (DARS)-sequence for E, there exists a partial complex R_n of Q_n, such that

$$|E \cdot c, R_n| \to 0 \quad \text{for} \quad n \to \infty. \tag{1}$$

Let $n \geqq n(c, \varepsilon)$, so all bricks of R_n have their measure $\leqq \delta(c, \varepsilon)$. Denote by $r^c_{n1}, r^c_{n2}, \ldots$ those bricks r_{nk} of R_n, for which $r_{nk} \cdot c = O$; denote by $r^i_{n1}, r^i_{n2}, \ldots$ those bricks r_{nk} of R_n, for which $r_{nk} \leqq c$, and by $r^b_{n1}, r^b_{n2}, \ldots$ all those r_{nk}, for which $r_{nk} \cdot c \neq O$ and $r_{nk} \cdot \text{co}\, c \neq O$. Thus we have

$$r^e_{nk} \cdot c = O, \quad r^i_{nk} \leqq c \quad \text{and} \quad r^b_{nk} \cdot c \neq O, \quad r^b_{nk} \cdot \text{co}\, c \neq O. \tag{2}$$

Since, by (1), $|E\, c, R_n| \to 0$, we can find an index $n' = n'(c)$, such that for all $n \geqq n'$:

$$|E\, c, R_n| \leqq \varepsilon;$$

hence

$$\Big|E\, c, \sum_\alpha r^e_{n\alpha} + \sum_\alpha r^e_{n\alpha} + \sum_\alpha r^i_{n\alpha}\Big| \leqq \varepsilon. \tag{3}$$

By (2) we have

$$E\, c \cdot \sum_\alpha r^e_{nk} = O;$$

hence

$$E\, c\,(\sum_\alpha r^i_{n\alpha} + \sum_\alpha r^b_{n\alpha}) = E\, c\,(\sum_\alpha r^i_{n\alpha} + \sum_\alpha r^b_{n\alpha} + \sum_\alpha r^e_{n\alpha}).$$

Since we also have

$$\sum_\alpha r^i_{n\alpha} + \sum_\alpha r^b_{n\alpha} \leqq \sum_\alpha r^i_{n\alpha} + \sum_\alpha r^b_{n\alpha} + \sum_\alpha r^e_{n\alpha},$$

we get from (3), by [U.2.2.],

(4) $$|E\, c,\, \sum_\alpha r^i_{n\alpha} + \sum_\alpha r^b_{n\alpha}| \leqq \varepsilon.$$

Since

$$|E\, c,\, \sum_\alpha r^i_{n\alpha}| = |E\, c - O,\, (\sum_\alpha r^i_{n\alpha} + \sum_\alpha r^b_{n\alpha}) - \sum_\alpha r^b_{n\alpha}|,$$

we have

$$|E\, c,\, \sum_\alpha r^i_{n\alpha}| \leqq |E\, c,\, \sum_\alpha r^i_{n\alpha} + \sum_\alpha r^b_{n\alpha}| + |O,\, \sum_\alpha r^b_{n\alpha}|.$$

We have

$$|O,\, \sum_\alpha r^b_{n\alpha}| = \mu \sum_\alpha r^b_{n\alpha};$$

hence by (4)

(5) $$|E\, c,\, \sum_\alpha r^i_{n\alpha}| \leqq E + \mu \sum_\alpha r^b_{n\alpha}.$$

Now, $r^b_{n1}, r^b_{n2}, \ldots$ is a collection of bricks, such that $\mu(r^b_{nk}) \leqq \delta(c, \varepsilon)$, and $r^b_{nk} \cdot c \neq O$, $r^b_{nk} \cdot \operatorname{co} c \neq O$. Consequently, by (0), $\mu \sum_\alpha r^b_{n\alpha} \leqq \varepsilon$. Taking account of (5), we get

(6) $$|E\, c,\, \sum_\alpha r^i_{n\alpha}| < 2\varepsilon.$$

We also have $\sum_\alpha r^i_{n\alpha} \leqq c$.

Thus we have proved, that for every brick c and every $\varepsilon > 0$, there exists for all sufficiently great indices n, $[n > n(c, \varepsilon)]$, in Q_n a partial complex

$$\bar{R}_n = r^i_{n1}, r^i_{n2}, \ldots,$$

depending on c and ε, with

$$|E\, c,\, \bar{R}_n| \leqq 2\varepsilon, \quad \bar{R}_n \leqq c, \quad \bar{R}_n \subseteqq Q_n.$$

U.10.6b. Let $c_1, c_2, \ldots, c_s$ be a finite collection of mutually disjoint bricks, and let $\varepsilon > 0$. Put

$$n(c_1, c_2, \ldots, c_s; \varepsilon) =_{df} \max\left[n\left(c_i, \frac{\varepsilon}{s}\right), n\left(c_2, \frac{\varepsilon}{s}\right), \ldots, n\left(c_s, \frac{\varepsilon}{s}\right)\right].$$

Then for every $n > n\ (c_1, \ldots, c_s; \varepsilon)$ we can find complexes R^λ_n, $(\lambda = 1, 2, \ldots, s)$, such that

$$\bar{R}^\lambda_n \leqq c_\lambda, \quad \bar{R}^\lambda_n \subseteqq Q_n, \quad |E\, c_\lambda, \bar{R}^\lambda_n| \leqq \frac{2\varepsilon}{s}.$$

Hence we get

$$\left| E \cdot \sum_{\lambda=1}^{s} c_\lambda , \bigcup_{\lambda=1}^{s} \bar{R}_n^\lambda \right| \leqq 2\varepsilon . \tag{7}$$

The complex

(8) $\bigcup_{\lambda=1}^{s} \bar{R}_n^\lambda$ may be denoted by $\bar{R}_n(c_1, \ldots, c_s; \varepsilon)$.

It is a partial complex of Q_n.

U.10.6c. Remark II. Let $f \in F$ be a figure and $\varepsilon > 0$ a number. f can be represented as the denumerable sum of mutually disjoint bricks, say $f = a_1 + a_2 + \cdots + a_l + \cdots$. We can find l, such that $\mu(a_{l+1} + a_{l+2} + \cdots) \leqq \varepsilon$. Hence $\mu[f - (a_1 + \cdots + a_l)] \leqq \varepsilon$, and then

$$|f \cdot E, f(a_1 + \cdots + a_l) \cdot E| \leqq \varepsilon . \tag{9}$$

Let us apply [Remark I] to $a_1, \ldots, a_l$, and ε. We can find an index n', such that for all $n \geqq n'$ we have a partial complex of Q_n: $\bar{R}_n(a_1, \ldots, a_l; \varepsilon)$, such that

$$|E \cdot \sum_{\alpha=1}^{l} a_\alpha , \bar{R}_n(a_1, \ldots, a_l; \varepsilon)| \leqq 2\varepsilon .$$

Hence from (9) we get

$$|f E, \bar{R}_n(a_1, \ldots, a_l; \varepsilon)| \leqq 3\varepsilon .$$

The bricks of $\bar{R}_n(a_1, \ldots, a_l; \varepsilon)$ are contained in the bricks $a_1, \ldots, a_l$. Thus, given a figure f and a number $\varepsilon > 0$, we can find an index n', such that for every $n \geqq n'$ there exists a complex $\bar{R}_n(f, \varepsilon)$, such that

1) $\bar{R}_n(f, \varepsilon)$ is a partial complex of Q_n,
2) $|f E, \bar{R}_n(f, \varepsilon)| \leqq 3\varepsilon$,
3) the bricks of $\bar{R}_n(f, \varepsilon)$ are contained in f.

U.10.6d. The above [Remark II] is valid if, instead of a figure f, we take the sum of any denumerable collection of disjoint bricks. This will be the case, if we take the sum of a denumerable collection of mutually disjoint figures.

U.10.6e. We have $|E, Q_n| \to 0$ for $n \to \infty$. Since $E \neq O$, there exists an index $k(1)$, such that for all $n \geqq k(1)$ we have

$$E \cdot Q_n \neq O, \quad \text{and in addition to that} \quad |E, Q_n| \leqq \frac{1}{2^1}, \quad \mathcal{N}(Q_n) \leqq \frac{1}{2^1} .$$

Let $p^{1'}, p^{2'}, \ldots$ be all (non empty) bricks of $Q_{k(1)}$. There exists at least one number α', such that

$$p^{\alpha'} \cdot E \neq O . \tag{10}$$

Let all bricks $p^{\alpha'}$, satisfying this inequality, be

(11) $p^1, p^2, \ldots$ and put $P_1 =_{df} p^1 \cup p^2 \cup \cdots$

This complex is not empty.

We have

(12) $$E \cdot p^\alpha \neq O \quad \text{for} \quad \alpha = 1, 2, \ldots,$$

and

(13) $$|P_1, E| \leqq \frac{1}{2^1}, \quad \mathcal{N}(P_1) \leqq \frac{1}{2^1}.$$

Put

(14) $$p^0 =_{df} 1 - (p^{1'} + p^2 + \cdots).$$

We get the disjoint sum

(15) $$E = E\,p^1 + E\,p^2 + \cdots + E\,p^0.$$

(16) $p^1, p^2, \ldots$ are bricks, and p^0 is a figure.

U.10.6f. The somata $p^1, p^2, \ldots, p^0$ will constitute the starting point of an inductive construction. The figure p^0 will be taken into account only, when $p^0 \cdot E \neq O$. If $p^0 \cdot E = O$, the corresponding agreements should be omitted. We shall approximate the bricks p^α, by $(\alpha \geqq 1)$, by parts of the complexes Q_n, by applying Remark I, and also we shall approximate p^0 and apply Remark II, getting mutually disjoint bricks $p^{\alpha 1}, p^{\alpha 2}, \ldots, p^{01}, p^{02}, \ldots$, where $p^{\alpha\beta} \leqq p^\alpha$, and $p^{0\beta} \leqq p^0$. The resulting complex will approximate E more closely than $Q_{k(0)}$. Let us proceed to accomplish the general step of the inductive construction.

U.10.6g. Let A denote a sequence of t indices $\alpha_1, \alpha_2, \ldots, \alpha_t$, where $\alpha_1, \alpha_2, \ldots = 0, 1, 2, \ldots$; A may be even empty; but we shall keep t fixed. We admit, that we have already defined the bricks $p^{A1}, p^{A2}, \ldots$ and the figures p^{A0} for all A, composed of t indices $\alpha_1, \alpha_2, \ldots$ Let

(17) $$E \cdot p^{A\beta} \neq O$$

for all A and $\beta \geqq 1$, but not when $A\beta$ is composed of zeros only.

Concerning p^{A0}, we shall consider them only, whenever $p^{A0} \cdot E \neq O$. We admit, that all $p^{A1}, p^{A2}, \ldots, p^{A0}$ are mutually disjoint and that

(18) $$p^{A\beta} \leqq p^A \quad \text{for} \quad \beta \geqq 1, \quad p^{A0} \leqq p^A, \quad p^{A0} = p^A - \sum_{\beta \geqq 1} p^{A\beta}.$$

If A is empty, the above relation is replaced by:

$$p^\beta \leqq 1 \quad \text{for} \quad \beta \geqq 1, \quad p^0 \leqq 1, \quad p^0 = 1 - \sum_{\beta \geqq 1} p^\beta$$

(see [U.10.6a.]). We define

(19) $$P_t =_{df} \bigcup_A \bigcup_{\beta > 1} p^{A\beta},$$

where the first summation is extended over all A with t indices; and suppose that

(20) $$P_t \subseteqq Q_{k(t)}, \quad |E, Q_{k(t)}| \leqq \frac{1}{2^t}, \quad \mathcal{N}(P_t) \leqq \frac{1}{2^t} \quad \text{for} \quad t \geqq 1,$$

where $k(t)$ is an index.

U.10.6h. We put for p with $t+1$ indices 0:

(21) $$p^{0,\ldots,0} =_{df} 1 - \sum_{A\beta} p^{A\beta},$$

where summation is extended over all $A\beta$, differing from $(0, \ldots, 0)$ with $t+1$ zeros. It follows, that all p^B with $(t+1)$ indices are mutually disjoint, and that

(22) $$E = E\,p^{0,\ldots,0} + \sum_{A\beta} E\,p^{A\beta}.$$

Notice, that if A is empty, the equality (22) is reduced to (15) in [U.10.6e.].

U.10.6i. Let $N(t)$ be the number of all $p^{A\beta}$. We have $N(t) \geqq 1$. The sequence Q_n for $n \geqq k(t)$, is a (DARS)-sequence for E. Since $p^{A\beta} \cdot E \leqq E$, and since all sets $p^{A\beta} \cdot E$ are disjoint, we can get disjoint complexes $Q_n^{A\beta}$ such that

(23) $$|p^{A\beta} \cdot E, Q_n^{A\beta}| \to 0, \quad Q_n^{A\beta} \subseteqq Q_n \quad \text{for all} \quad n \geqq k(t).$$

Now, since we have supposed in (17) that $p^{A\beta} \cdot E \neq 0$, we can find an index $k'(t) \geqq k(t)$, such that for all $n \geqq k'(t)$ and all $A\beta$ we have

(24) $$p^{A\beta} \cdot E \neq 0.$$

Now, let us apply to $p^{A\beta}$ the Remark I [U.10.6a.]. We get an index $k(t+1) \geqq k'(t)$, such that for every $n > k(t+1)$, there exists a partial complex $R_n^{A\beta}$, such that

$$|E \cdot p^{A\beta}, R_n^{A\beta}| \leqq \frac{1}{2^{t+1} \cdot N(t)}, \qquad R_n^{A\beta} \leqq p^{A\beta}, \qquad R_n^{A\beta} \subseteqq Q_n^{A\beta} \subseteqq Q_n,$$

and we also have $\mathcal{N}(R_n^{A\beta}) \leqq \frac{1}{2^{t+1}}$; this for all $A\beta$ with $\beta \geqq 1$. Now since

$$p^{A0} = p^A - \sum_{\beta \geqq 1} p^{A\beta}$$

is a figure, we can get a complex R_n^{A0} such that, for sufficiently great n, $[n \geqq k(t+1)]$, we have

$$R_n^{A0} \leqq p^{A0}, \quad R_n^{A0} \subseteqq Q_n, \quad \mathcal{N}(R_n^{A0}) \leqq \frac{1}{2^{t+1}}, \quad |E \cdot p^{A0}, R_n^{A0}| \leqq \frac{1}{2^{t+1} \cdot N(t)}.$$

In addition to that, the index $k(t+1)$ may be so great, that for all $n \geqq k(t+1)$ there exists a complex $R_n^{0,\ldots,0}$ with $(t+1)$ upper indices 0, such that

$$R_n^{0,\ldots,0} \leqq p^{0,\ldots,0}, \quad R_n^{0,\ldots,0} \subseteqq Q_n, \quad \mathcal{N}(R_n^{0,\ldots,0}) \leqq \frac{1}{2^{t+1}\cdot N(t)},$$

$$|E\cdot p^{0,\ldots,0}, R_n^{0,\ldots,0}| \leqq \frac{1}{2^{t+1}\cdot N(t)}.$$

Let us denote by $p^{A\beta 1}, p^{A\beta 2}, \ldots$ all bricks of $R_{k(t+1)}^{A\beta}$. If $A\beta$ differs from the collection of $t+1$ zeros, there exists at least one γ', such that $p^{A\beta\gamma'}\cdot E \neq O$. Denote all $p^{A\beta\gamma'}$, for which $p^{A\beta\gamma'}\cdot E \neq O$, by

$$p^{A\beta 1}, p^{A\beta 2}, \ldots$$

This collection is not empty, provided that $A\beta$ differs from the collection of $(t+1)$ zeros. We define

$$p^{A\beta 0} =_{df} p^{A\beta} - \sum_{\gamma\geqq 1} p^{A\beta\gamma},$$

and also

$$p^{0,\ldots,0} =_{df} 1 - \sum_{A\beta\gamma \neq (0,\ldots,0,0)} p^{A\beta\gamma}.$$

If we put

$$P_{t+1} = \bigcup_{A\beta\gamma,\,\gamma\geqq 1} p^{A\beta\gamma},$$

we get

$$|E, P_{t+1}| \leqq \frac{1}{2^{t+1}}$$

and

$$\mathcal{N}(P_{t+1}) \leqq \frac{1}{2^{t+1}}$$

and

$$P_{t+1} \subseteqq Q_{k(t+1)}.$$

U.10.6j. The above construction, if continued, yields a sequence $P_0, P_1, \ldots, P_t, \ldots$ of partial complexes of $Q_{k(0)}, Q_{k(1)}, \ldots, Q_{k(t)}, \ldots$, which is a (D)-sequence for E. If we apply [Q.21.9.], we see, since $E \leqq E$, that P_t is a (DARS)-sequence for E.

From the construction it follows, that P_t satisfies the condition of non-overlapping, so the theorem is established.

U.10.7. Theorem. Under hypothesis $\boldsymbol{H}_1$, if

1. $\boldsymbol{G}$ has no atoms,
2. $f(a)$ is a field of complex numbers,
3. $f(a)$ is admissible [U.2.7.],
4. $E \neq O$ is a measurable set of traces,
5. $\{p_n\} = \{p_{n1}, p_{n2}, \ldots\}$ is a (DARS)-sequence for E,
6. $A =_{df} \lim_{n\to\infty} \sum_k f(p_{nk})$,
7. hypothesis $\mathscr{B}$ holds true, Def. [U.10.5.],

then there exists a (DARS)-sequence Q_n for E, such that

1) $Q_n \subseteq P_{s(n)}$ for all n, where $s(n)$ is a suitable subsequence of n,

2) Q_n satisfies the condition of non-overlapping [U.10.],

3) if we put $Q_n = \{q_{n1}, q_{n21}, \ldots\}$,

then

$$\lim_{n\to\infty} \sum_k f(q_{nk}) = A.$$

Proof. By Theorem [U.10.6.] we can find a subsequence $s(n)$ of n and a (DARS)-sequence for E, such that

$$Q_n \subseteq P_{s(n)}, \quad (n = 1, 2, \ldots),$$

and that Q_n satisfies the condition of non-overlapping. We have

$$A = \lim_{n\to\infty} \sum_k f(p_{s(n),k}).$$

Since

$$q_{n,1} + q_{n,2} + \cdots \leqq p_{s(n),1} + p_{s(n),2} + \cdots,$$

and

$$\left|\sum_k q_{nk}, E\right| \to 0, \quad \left|\sum_k p_{s(n),k}, E\right| \to 0,$$

it follows that the set of bricks $r_{n,1}, r_{n,2}, \ldots$, which belongs to $P_{s(n)}$ but not to Q_n, has the property: $\mu(r_{n1} + r_{n2} + \cdots) \to 0$. Since $f(a)$ is admissible, we get

(1) $$\sum f(r_{nk}) \to 0.$$

Since

$$\sum f(r_{nk}) = \sum_k f(p_{s(n),n}) - \sum_k f(q_{nk}),$$

it follows from (1), that

$$\lim \sum_k f(q_{nk}) = A.$$

U.10.7.1. We may restrict the notion of $\underset{E}{S} f(a)$, by taking only those (DARS)-sequences into account, which satisfy the condition of non-overlapping, and call it (DARSN)-sum. Now the Theorem [U.10.7.] says that under ***Hyp*** $\mathscr{B}$, this restriction is irrelevent, so if we consider scalar fields, the (DARS)-summability coincides with the (DARSN)-summability (in that case) with the same values of sums.

Section 3 of U.

Some theorems on limes superior and inferior for sequences of complexes

U.20.1. We shall start with abstract sets, though the topic of the present Chapter [U.] will be concerned only with measurable sets of traces. This general approach will however not imply any unnecessary complication.

Let $(\boldsymbol{L})$ be a denumerably additive tribe of sets, with set-inclusion as ordering relation. We recall the following known definitions:

If $E_{\dot{n}}$ is a sequence of sets, we put

$$\overline{\operatorname*{Lim}_{n\to\infty}} E_n .=._{df} \prod_{\alpha=1}^{\infty} (E_\alpha + E_{\alpha+1} + \cdots),$$

(1)

$$\underline{\operatorname*{Lim}_{n\to\infty}} E_n .=._{df} \sum_{\alpha=1}^{\infty} (E_\alpha \cdot E_{\alpha+1} \cdot \cdots).$$

We have

(2) $$\underline{\operatorname*{Lim}_{n\to\infty}} E_n \leqq \overline{\operatorname*{Lim}_{n\to\infty}} E_n.$$

In the case, where both limits are equal, we write $\operatorname*{Lim}_{n\to\infty} E_n$, and call it "*set-limit of* $E_{\dot{n}}$".

U.20.1.1. Now let μ be a denumerably additive non-negative measure on $(\boldsymbol{L})$; and let us reorganize the tribe $(\boldsymbol{L})$, by introducing the ordering "$\leqq^\mu$" defined by $E \leqq^\mu F .=._{df} \mu(E - F) = 0$, by the equality $E =^\mu .=._{df} \mu(E \dot{+} F) = 0$, and by the corresponding lattice operations. Denote by $\boldsymbol{L}_\mu$ the reorganized tribe. The measure μ is effective on $\boldsymbol{L}_\mu$.

If

$$E_n .=.^\mu F_n, \quad (n = 1, 2, \ldots),$$

we get

$$\overline{\operatorname*{Lim}_{n\to\infty}} E_n .=.^\mu \overline{\operatorname*{Lim}_{n\to\infty}} F_n, \qquad \underline{\operatorname*{Lim}_{n\to\infty}} E_n .=.^\mu \underline{\operatorname*{Lim}_{n\to\infty}} F_n,$$

so the operations $\overline{\text{Lim}}$, $\underline{\text{Lim}}$ and Lim are $=^\mu$-invariant. Thus, if we deal with $\boldsymbol{L}$ only, and write the μ-inclusion, μ-equality and μ-operations, by using some times the superscript μ, the definition (1) becomes

$$\overline{\operatorname*{Lim}_{n\to\infty}}^\mu E_n .=.^\mu \prod_{\alpha=1}^{\infty} (E_\alpha + E_{\alpha+1} + \cdots),$$

$$\underline{\operatorname*{Lim}_{n\to\infty}}^\mu E_n .=.^\mu \sum_{\alpha=1}^{\infty} (E_\alpha \cdot E_{\alpha+1} \cdot \cdots).$$

We also have the μ-set-limit $\text{Lim}^\mu E_n$, characterized by

$$\overline{\text{Lim}}^\mu E_n .=.^\mu \underline{\text{Lim}}^\mu E_n.$$

In all cases we have

$$\underline{\text{Lim}}^\mu E_n \leqq^\mu \overline{\text{Lim}}^\mu E_n.$$

U.20.1.2. Dealing only with $\boldsymbol{L}_\mu$ we have the following: $\overline{\operatorname*{Lim}_{n\to\infty}}^\mu E_n$, $\underline{\operatorname*{Lim}_{n\to\infty}}^\mu E_n$ will not change (in the sense of $=^\mu$), if we drop a finite number of elements E_n of the sequence, or if we add a finite number of any sets to the sequence, or else if we change arbitrarily a finite number of sets into any other sets.

U.20.1.3. If $s(\dot{n})$ is a subsequence of $\dot{n}$, we have

$$\underline{\mathrm{Lim}}^{\mu} E_n \leqq \underline{\mathrm{Lim}}^{\mu} E_{s(n)} \leqq \overline{\mathrm{Lim}}^{\mu} E_{s(n)} \leqq \overline{\mathrm{Lim}}^{\mu} E_n .$$

U.20.2. The following theorem by the author (2) holds true. The following are equivalent:

I. $\lim^{\mu} E_n =^{\mu} E$, *the μ-measure limit* defined by $\mu(E \dotplus E_n) \to 0$ for $n \to \infty$.

II. From every subsequence $E_{\alpha(n)}$ of E_n, another subsequence $E_{\alpha\beta(n)}$ can be extracted such that the μ-set limit $\underset{n\to\infty}{\mathrm{Lim}} E_{\alpha\beta(n)} .=.^{\mu} E$.

U.20.2.1. Notice that the μ-set limit $\mathrm{Lim}^{\mu} E_n$ and the μ-measure — topology — limit $\lim^{\mu} E_n$ may not be $=^{\mu}$-equal.

Eg. Let us consider Lebesgue-measurable subsets of the circumference of a circle C with radius 1. Their collection is organized into a denumerably additive tribe taken modulo sets of measure 0. Choose a point A_0 on C and plot, following the counterclockwise direction, and starting from C, the consecutive arcs $(A_0 A_1)$, $(A_1 A_2)$, ..., $(A_n A_{n+1})$, ... with length $1, \frac{1}{2}, \frac{1}{3}, \ldots, \frac{1}{n}, \ldots$ respectively. One can prove that the measure limit

$$\lim_{n\to\infty} (A_n, A_{n+1}) = O, \qquad \underline{\mathrm{Lim}}_{n\to\infty} (A_n, A_{n+1}) = O,$$

but

$$\overline{\mathrm{Lim}}_{n\to\infty} (A_n, A_{n+1}) = C.$$

U.20.2.2. Theorem. We have: if $\mathrm{Lim}^{\mu} E_n = E$, then $\lim^{\mu} E_n = E$, but not conversely.

U.20.3. Let p, q be arbitrary sets of $\boldsymbol{L}$; and define, in general, $\mathrm{co}_q p =_{df} q - p$. This is the relative complement of p (with respect to q). For any not empty collection K of sets we have the de Morgan-laws:

$$\mathrm{co}_q \prod_{p\in K} p = \sum_{p\in K} \mathrm{co}_q p \quad \text{and} \quad \mathrm{co}_q \sum_{p\in K} p = \prod_{p\in K} \mathrm{co}_q p.^{*)}$$

To prove that, denote the ordinary complement of p by $\mathrm{co}\, p =_{df} I - p$, where I is the unit of the tribe $\boldsymbol{L}$.

We have

$$\mathrm{co}_q \prod_{p\in K} p = q - \prod_{p\in K} p = q \cdot \mathrm{co} \prod_{p\in K} p = q \cdot \sum_{p\in K} \mathrm{co}\, p = \sum_{p\in K} (q \cdot \mathrm{co}\, p)$$
$$= \sum_{p\in K} (q - p) = \sum_{p\in K} \mathrm{co}_q p,$$

$$\mathrm{co}_q \sum_{p\in K} p = q - \sum_{p\in K} p = q \cdot \mathrm{co} \sum_{p\in K} p = q \cdot \prod_{p\in K} \mathrm{co}_q p = \prod_{p\in K} (q \cdot \mathrm{co}\, p)$$
$$= \prod_{p\in K} (q - p) = \prod_{p\in K} \mathrm{co}_q p.$$

---

*) The equality governing in $\boldsymbol{L}$ may be any, provided that $\sum_{p\in K}$, $\prod_{p\in K}$ are meaningful. In the case of the tribe L_{μ} all somatic operations are meaningful.

U.20.4. Let E_n be an infinite sequence of sets, and E a set. We have

$$\mathrm{co}_E(\overline{\mathrm{Lim}}\, E_n) = \underline{\mathrm{Lim}}\,(\mathrm{co}_E E_n), \qquad \mathrm{co}_E(\underline{\mathrm{Lim}}\, E_n) = \overline{\mathrm{Lim}}\,(\mathrm{co}_E E_n).$$

The easy proof relies on [U.20.3.].

U.20.5. If E_n, F are sets, then

$$F \cdot \overline{\mathrm{Lim}}\, E_n = \overline{\mathrm{Lim}}\,(F\, E_n), \qquad F \cdot \underline{\mathrm{Lim}}\, E_n = \underline{\mathrm{Lim}}\,(F\, E_n).$$

U.20.6. If $E_n \leqq E$, then

$$\overline{\mathrm{Lim}}\, E_n \leqq E, \quad \text{and} \quad \underline{\mathrm{Lim}}\, E_n \leqq E.$$

U.20.7. Def. Let $\boldsymbol{G}$ be the tribe of measurable sets of traces with $=^\mu$ as governing equality. Let E_n be an infinite sequence of sets of $\boldsymbol{G}$. Let $C \in \boldsymbol{G}$ be a set of $\boldsymbol{G}$. We shall say that C is *regular with respect to* E_n, whenever $\mathrm{Lim}_{n\to\infty}\, C\, E_n$ exists, i.e.

$$\underline{\mathrm{Lim}}_{n\to\infty}\, C\, E_n \;.{=}.^\mu\; \overline{\mathrm{Lim}}_{n\to\infty}\, C \cdot E_n.$$

We shall say that C *is singular with respect to* E_n whenever

$$\underline{\mathrm{Lim}}_{n\to\infty}\, C\, E_n =^\mu O, \qquad \overline{\mathrm{Lim}}_{n\to\infty}\, C\, E_n =^\mu C.$$

U.20.7.1. The null-set is regular and singular, so we should consider only the sets $\neq O$.

U.20.7.2. If C is regular with respect to E_n, $C' \leqq C$, then C' is also regular with respect to E_n.

If C is singular with respect to E_n, $C' \leqq C$, then C' is also singular with respect to E_n.

U.20.8. Theorem. If $A = \underline{\mathrm{Lim}}\, E_n$, $B = \overline{\mathrm{Lim}}\, E_n$, $C = B - A$, then C is the greatest singular set with respect to E_n, and $\mathrm{co}\, C$ is the greatest regular set with respect to E_n. This means that if $C' \leqq C$ then $\underline{\mathrm{Lim}}\,(C'\, E_n) = O$, $\overline{\mathrm{Lim}}\,(C'\, E_n) = C'$; and if $D \in \mathrm{co}\, C$, then

$$\underline{\mathrm{Lim}}\,(D\, E_n) = \overline{\mathrm{Lim}}\,(D\, E_n).$$

Proof. We shall apply several times [Theorem U.20.5.]. We have

$$\underline{\mathrm{Lim}}\,(C'\, E_n) = C'\, \underline{\mathrm{Lim}}\, E_n = C'\, A \leqq C\, A = (B - A) \cdot A = O,$$

$$\overline{\mathrm{Lim}}\,(C'\, E_n) = C'\, \overline{\mathrm{Lim}}\, E_n = C'\, B = C',$$

because $C' \leqq C \leqq B$, so the first part of the thesis is established.

We have

$$\overline{\mathrm{Lim}}\,(D\, E_n) = D \cdot B, \qquad \underline{\mathrm{Lim}}\,(D\, E_n) = D\, A.$$

It follows

$$\overline{\mathrm{Lim}}\,(D\, E_n) - \underline{\mathrm{Lim}}\,(D\, E_n) = D\, B - D\, A = D\,(B - A) = O,$$

because

$$D \in \operatorname{co} C .$$

Hence

$$\operatorname{Lim}(D E_n) \leqq \operatorname{Lim}(D E_n) .$$

Hence

$$\underline{\operatorname{Lim}}(D E_n) = \overline{\operatorname{Lim}}(D E_n) .$$

U.20.9. Lemma. For sets E_n, F_n, $(n = 1, 2, \ldots)$ we have

$$\left| \sum_{n=1}^{\infty} E_n, \sum_{n=1}^{\infty} F_n \right|_{\mu} \leqq \sum_{n=1}^{\infty} |E_n, F_n|_{\mu},$$

$$\left| \prod_{n=1}^{\infty} E_n, \prod_{n=1}^{\infty} F_n \right|_{\mu} \leqq \sum_{n=1}^{\infty} |E_n, F_n|_{\mu},$$

where the sums on the right may be $+\infty$.

Proof.

$$(1) \qquad \left| \sum_{n=1}^{\infty} E_n, \sum_{n=1}^{\infty} F_n \right| = \mu\left(\sum_{n=1}^{\infty} E_n - \sum_{n=1}^{\infty} F_n \right) + \mu\left(\sum_{n=1}^{\infty} F_n - \sum_{n=1}^{\infty} E_n \right).$$

Now

$$(2) \quad \mu\left(\sum_{n=1}^{\infty} E_n - \sum_{n=1}^{\infty} F_n \right) = \mu\left(\sum_{n=1}^{\infty} E_n \cdot \operatorname{co} \sum_{n=1}^{\infty} F_n \right) = \mu\left(\sum_{n=1}^{\infty} E_n \cdot \prod_{n=1}^{\infty} \operatorname{co} F_n \right)$$

$$= \mu \sum_{n=1}^{\infty} \left(E_n \cdot \prod_{n=1}^{\infty} \operatorname{co} F_n \right) \leqq \mu \sum_{n=1}^{\infty} (E_n \cdot \operatorname{co} F_n) = \mu \sum_{n=1}^{\infty} (E_n - F_n) \leqq$$

$$\leqq \sum_{n=1}^{\infty} \mu (E_n - F_n) .$$

Similarly we get

$$(3) \qquad \mu\left(\sum_{n=1}^{\infty} F_n - \sum_{n=1}^{\infty} E_n \right) \leqq \sum_{n=1}^{\infty} \mu (F_n - E_n) .$$

From (2) and (3), since the measure is non negative, we get

$$\mu\left(\sum_{n=1}^{\infty} E_n - \sum_{n=1}^{\infty} F_n \right) + \mu\left(\sum_{n=1}^{\infty} F_n - \sum_{n=1}^{\infty} E_n \right) \leqq \sum_{n=1}^{\infty} [\mu (E_n - F_n) + \mu (F_n - E_n)] .$$

Hence by (1)

$$\left| \sum_{n=1}^{\infty} E_n, \sum_{n=1}^{\infty} F_n \right| \leqq \sum_{n=1}^{\infty} |E_n, F_n| .$$

To prove the second part of the thesis, we rely on the general rule:

$$|a, b|_{\mu} = |\operatorname{co} a, \operatorname{co} b|_{\mu} .$$

Hence

$$\left| \prod_{n=1}^{\infty} E_n, \prod_{n=1}^{\infty} F_n \right| = \left| \operatorname{co} \prod_{n=1}^{\infty} E_n, \operatorname{co} \prod_{n=1}^{\infty} F_n \right| = \left| \sum_{n=1}^{\infty} \operatorname{co} E_n, \operatorname{co} \sum_{n=1}^{\infty} F_n \right| ;$$

hence by the first part of the thesis, already proved,

$$\left|\prod_{n=1}^{\infty} E_n, \prod_{n=1}^{\infty} F_n\right| \leqq \sum_{n=1}^{\infty} |\operatorname{co} E_n, \operatorname{co} F_n| = \sum_{n=1}^{\infty} |E_n, F_n|, \quad \text{q.e.d.}$$

U.20.9.1. Lemma. If

$$|E_n, F_n| < \frac{1}{2^n},$$

then

$$\overline{\operatorname{Lim}} E_n = \overline{\operatorname{Lim}} F_n, \qquad \underline{\operatorname{Lim}} E_n = \underline{\operatorname{Lim}} F_n.$$

Proof. By [U.20.9.] we have

$$(1) \quad |E_\alpha + E_{\alpha+1} + \cdots, F_\alpha + F_{\alpha+1} + \cdots| \leqq |E_\alpha, F_\alpha| +$$
$$+ |E_{\alpha+1}, F_{\alpha+1}| + \cdots \leqq \frac{1}{2^\alpha} + \frac{1}{2^{\alpha+1}} + \cdots + = \frac{1}{2^{\alpha-1}}.$$

Hence, by [U.20.9.] and (1),

$$\left|\prod_{\alpha=k}^{\infty} (E_\alpha + E_{\alpha+1} + \cdots), \prod_{\alpha=k}^{\infty} (F_\alpha + F_{\alpha+1} + \cdots)\right| \leqq$$

$$\leqq \sum_{\alpha=k}^{\infty} |E_\alpha + E_{\alpha+1} + \cdots, F_\alpha + F_{\alpha+1} + \cdots| \leqq \frac{1}{2^{k-1}} + \frac{1}{2^k} + \cdots = \frac{1}{2^{k-2}}.$$

Hence

$$|\overline{\operatorname{Lim}} E_n, \overline{\operatorname{Lim}} F_n| \leqq \frac{1}{2^{k-2}}.$$

If we make $k \to \infty$, we get

$$|\overline{\operatorname{Lim}} E_n, \overline{\operatorname{Lim}} F_n| = 0,$$

i.e. since μ is effective:

$$\overline{\operatorname{Lim}} E_n = \overline{\operatorname{Lim}} F_n.$$

Since

$$|E_n, F_n| \leqq \frac{1}{2^n},$$

it follows

$$|\operatorname{co} E_n, \operatorname{co} F_n| \leqq \frac{1}{2^n}.$$

Hence, by what has been already proved,

$$(2) \qquad \overline{\operatorname{Lim}} \operatorname{co} E_n = \overline{\operatorname{Lim}} \operatorname{co} F_n.$$

Since, by [U.20.4.]

$$\operatorname{co} \overline{\operatorname{Lim}} \operatorname{co} E_n = \underline{\operatorname{Lim}} E_n,$$

$$\operatorname{co} \overline{\operatorname{Lim}} \operatorname{co} F_n = \underline{\operatorname{Lim}} F_n,$$

it follows, from (2),

$$\underline{\mathrm{Lim}}\, E_n = \underline{\mathrm{Lim}}\, F_n,$$

so the theorem is established.

U.20.9.2. Theorem. If $|E_n, F_n| \to 0$ for $n \to \infty$, then from every partial sequence $s(\dot{n})$ of $\dot{n}$ another sequence $st(\dot{n})$ can be extracted such that

$$\overline{\mathrm{Lim}}\, E_{st(n)} = \overline{\mathrm{Lim}}\, F_{st(n)},$$

$$\underline{\mathrm{Lim}}\, E_{st(n)} = \underline{\mathrm{Lim}}\, F_{st(n)}.$$

Proof. We have $|E_{s(n)}, F_{s(n)}| \to 0$. Hence there exists a subsequence $st(n)$ of $s(n)$ such that

$$|E_{st(n)}, F_{st(n)}| \leqq \frac{1}{2^n}.$$

The Lemma [U.20.9.1.] completes the proof.

U.20.9.3. Remark. It is not true that, if $|E_n, F_n|_\mu \to 0$, then $\overline{\mathrm{Lim}}\, E_n =^\mu \overline{\mathrm{Lim}}\, F_n$.

Expl. Take $E_n = (A_n A_{n+1})$ from [U.20.2.1.] and $F_n = O$.

U.20.10. Now we shall use the forgoing lemmas to prove the following Lemma. Under Hyp. $\boldsymbol{H}_1$, if

1. E is a measurable set of traces,
2. P_n is a (D)-sequence of complexes for E. (Hence $|P_n, E|_\mu \to 0$ for $n \to \infty$),
3. $|E, P_n|_\mu \leqq \frac{1}{2^n}$, $(n = 1, 2, \ldots)$,
4. $P_n = P'_n \cup P''_n$, $O = P'_n \cap P''_n$,
5. we define

$$A' = \underline{\mathrm{Lim}}\, P'_n, \quad B' = \overline{\mathrm{Lim}}\, P'_n,$$

$$A'' = \underline{\mathrm{Lim}}\, P''_n, \quad B'' = \overline{\mathrm{Lim}}\, P''_n,$$

then

1) $B' - A' = B'' - A''$. We define $C =_{df} B' - A' = B'' - A''$.

2) C is singular with respect to P'_n, and also singular with respect to P''_n.

3) $E - C$ is regular with respect to P'_n and also with respect to P''_n.

4) $B'' + A' = E$, $B' + A'' = E$.

5) $A' \cdot A'' = O$, $A' \cdot B'' = O$, $B' \cdot A'' = O$, $B' \cdot B'' = C$.

Proof. We have

$$|E, P'_n \cup P''_n| \leqq \frac{1}{2^n}. \tag{1}$$

Multiplying by P''_n, we get

$$|E P''_n, P''_n| \leqq \frac{1}{2^n}. \tag{2}$$

From (1) and (2), we get by subtraction

$$|E - E P''_n, (P'_n \cup P''_n) - P''_n| \leqq |E, P'_n \cup P''_n| + |E P''_n, P''_n| \leqq \frac{1}{2^{n-1}}.$$

Hence

$$|E - E P''_n, P'_n| \leqq \frac{1}{2^{n-1}};$$

hence

(3) $$|E - P''_n, P'_n| \leqq \frac{1}{2^{n-1}}.$$

Similarly we get

(4) $$|E - P'_n, P''_n E| \leqq \frac{1}{2^{n-1}}.$$

Now we can apply to (3) and (4) the Lemma [U.20.9.1.], by denoting P''_n, e.g., by Q''_{n-1}, and by starting with the second element of the sequence), we get

$$\overline{\mathrm{Lim}}\,(E - P''_n) = \overline{\mathrm{Lim}}\,(P'_n E),$$

$$\overline{\mathrm{Lim}}\,(E - P'_n) = \overline{\mathrm{Lim}}\,(P''_n E).$$

Thus we get, (4),

$$\overline{\mathrm{Lim}}\,(\mathrm{co}_E P''_n) = \mathrm{co}_E\, \underline{\mathrm{Lim}}\, P''_n = E - \underline{\mathrm{Lim}}\, P''_n = E - A'' = E\, \overline{\mathrm{Lim}}\, P'_n = \\ = E \cdot B',$$

and similarly:

$$\mathrm{Lim}\,(\mathrm{co}_E P'_n) = E - A' = E \cdot B''.$$

Thus we have proved, that

(5) $$E - A'' = E \cdot B', \quad E - A' = E \cdot B''.$$

[U.20.10a.] We shall prove, that $B'' \subseteq^{\mu} E$.

We have $P''_n \subseteq P_n$, $n = 1, 2, \ldots$; hence $\sum_{\beta=\alpha}^{\infty} P''_\beta \subseteq \sum_{\beta=\alpha}^{\infty} P_\beta$, which is equivalent to

(1) $$\mu\left(\sum_{\beta=\alpha}^{\infty} P''_\beta - \sum_{\beta=\alpha}^{\infty} P_\beta\right) = 0.$$

Now

(2) $$\sum_{\beta=\alpha}^{\infty} P''_\beta \to^{\mu} \lim P'',$$

$$|E, P_n| \leqq \frac{1}{2^n};$$

hence

$$\left|E, \sum_{\beta=\alpha}^{\infty} P_\beta\right| \leqq \sum_{\beta=\alpha}^{\infty} \frac{1}{2^\beta} = \frac{1}{2^{\alpha-1}};$$

hence

(3) $$\sum_{\beta=1}^{\infty} P_n \to^{\mu} E.$$

By (2) and (3), since the measure is μ-continuous, we get from (1)

$$0 = \lim_{\alpha \to \infty} \mu\left(\sum_{\beta=\alpha}^{\infty} P''_\beta - \sum_{\beta=\alpha}^{\infty} P_\beta\right) = \mu(\overline{\lim_{n\to\infty}}\, P''_n, E).$$

Hence

$$\overline{\lim} P''_n \subseteqq^{\mu} E, \quad \text{q.e.d.}$$

In a similar way we find that $B' \leqq^{\mu} E$.

Since $A' \leqq B'$, $A'' \leqq B''$, we get $A' \leqq E$, $A'' \leqq E$. It follows from (1)

(3.1) $$E - A'' = B', \quad E - A' = B''.$$

Since $A'' \leqq E$, $A' \leqq E$, (2) implies

(3.2) $$E = B' + A'', \quad E = B'' + A'$$

with disjoint terms, i.e.

(3.3) $$B' \cdot A'' = O, \quad B'' \cdot A' = O,$$

and

(4) $$A'' = E - B', \quad A' = E - B''.$$

From (3.1) and (4) we have

$$B'' - A'' = (E - A') - (E - B') = E(B' - A'),$$

but since $B' - A' \leqq E$, we have

(5) $$B'' - A'' = B' - A'.$$

Since $A' = E - B''$, we get

$$A' A'' = E \operatorname{co} B'' \cdot A'' \leqq E \operatorname{co} B'' \cdot B'' = O,$$

because $A'' \leqq B''$. Now

$$E = B' + A'' = (B' - A') + A' + A'',$$

i.e.

(6) $$E = A' + C + A''$$

with terms disjoint. We have

$$B' \cdot B'' = (E - A'')(E - A') = E \operatorname{co} A' \operatorname{co} A'' = E \operatorname{co}(A' + A'')$$
$$= E - (A' + A'') = C,$$

by (6).

Concerning the singularity of C and regularity of $E - C$, we can apply [U.8.]. Since $A' = \underline{\mathrm{Lim}}\, P'_n$, $B' = \overline{\mathrm{Lim}}\, P'_n$, $C = B' - A'$, the singularity of C with respect to P'_n, and the regularity of $E - C$ with respect to P'_n follows. Similar argument and (5) shows that C is singular with respect to P''_n, and that $E - C$ is regular with respect to P''_n. The theorem is proved.

U.20.10.1. Remark. We have stated the lemma [U.20.10.] for complexes — this for farther development — but a similar theorem holds true, if we suppose that if P'_n, P''_n, P_n are any sets, where $|P_n, E|_\mu \leqq \frac{1}{2^n}$, $P'_n \cdot P''_n = 0$, $P'_n + P''_n = P_n$.

U.20.11. Lemma. Under ***Hyp H***$_1$, if

1. E is a measurable set of traces,
2. P_n is a (DARS)-sequence of complexes for E,
3. $|E, P_n| \leqq \frac{1}{2^n}$, $(n = 1, 2, \ldots)$,
4. $P_n = P'_n \cup P''_n$, $\emptyset = P'_n \cap P''_n$, $(n = 1, 2, \ldots)$,
5. $P_{\dot{n}}$ satisfies the condition of non-overlapping,
6. $A' =_{df} \underline{\mathrm{Lim}}\, P'_n$, $\quad B' =_{df} \overline{\mathrm{Lim}}\, P'_n$,
   $A'' =_{df} \underline{\mathrm{Lim}}\, P''_n$, $\quad B'' =_{df} \overline{\mathrm{Lim}}\, P''_n$,
7. $C =_{df} B' - A'$,

then there exists a (DARS)-sequence T'_n for C, where

$$T'_n \subseteqq \bigcup_{n=1}^{\infty} P''_n.$$

Proof. Since $C \leqq E$, and P_n is a (DARS)-sequence for E, therefore there exists $Q_{\dot{n}}$, where $Q_n \leqq P_n$, $n = 1, 2, \ldots$, such that $Q_{\dot{n}}$ is a (DARS)-sequence for C.

Put

$$(1) \qquad Q'_n =_{df} Q_n \cap P'_n, \quad Q''_n =_{df} Q_n \cap P''_n.$$

We have

$$Q'_n \cap Q''_n = \emptyset, \quad Q'_n \cup Q''_n = Q_n, \quad (n = 1, 2, \ldots).$$

Take a subsequence $\alpha(\dot{n})$ of $\dot{n}$, such that

$$(2) \qquad |Q_{\alpha(n)}; C| \leqq \frac{1}{2^n}, \quad n = 1, 2, \ldots,$$

$Q_{\alpha(\dot{n})}$ is also a (DARS)-sequence for C; and $Q_{\alpha(n)}$ satisfies the condition of non-overlapping. I say that $\mathrm{Lim}\, Q'_{\alpha(n)} = C$. Indeed, from (2) it follows

$$|Q_{\alpha(n)} \cap P'_{\alpha(n)}, C \cdot P'_{\alpha(n)}| \leqq \frac{1}{2^n}.$$

Hence,

$$\overline{\mathrm{Lim}}\,[Q_{\alpha(n)} \cap P'_{\alpha(n)}] = \overline{\mathrm{Lim}}\,[C \cdot P'_{\alpha(n)}] = C \cdot \overline{\mathrm{Lim}}\, P'_{\alpha(n)} = C;$$

hence

$$\overline{\operatorname{Lim}}\, Q'_{\alpha(n)} = C.$$

Put

$$R'_n =_{df} Q'_{\alpha(n)}, \qquad (n = 1, 2, \ldots);$$

we have

$$\overline{\operatorname{Lim}}\, R'_n = C.$$

Since

$$\overline{\operatorname{Lim}}\, R'_n = \prod_{s=1}^{\infty} (R'_s + R'_{s+1} + \cdots),$$

we can find $s = s(n)$, such that

$$|R'_s + R'_{s+1} + \cdots, C| \leqq \frac{1}{2^n};$$

and then, we can find $t = t(s)$, such that

$$|R'_s + R'_{s+1} + \cdots, R'_s + \cdots + R'_{s+t}| \leqq \frac{1}{2^n}.$$

Hence

$$|R'_s + R'_{s+1} + \cdots + R'_{s+t}, C| \leqq \frac{1}{2^{n-1}}.$$

Now

$$R'_s, R'_{s+1}, \ldots, R'_{s+t}$$

are subcomplexes of

$$Q'_{\alpha(s)}, Q'_{\alpha(s+1)}, \ldots, Q'_{\alpha(s+t)},$$

respectively.

Thus, since $\alpha(s) \to \infty$, we see that, if we put

$$T'_n =_{df} R'_s \cup R'_{s+1} \cup \cdots \cup R'_{s+t},$$

we can say, that T_n is a (DARS)-sequence for C, and that

$$T'_n \subseteqq \bigcup_{n=1}^{\infty} Q'_{\alpha(n)} \subseteqq \bigcup_{n=1}^{\infty} P'_n.$$

Thus we have proved the existence of a (DARS)-sequence of complexes for C and composed of bricks of

$$\bigcup_{n=1}^{\infty} P'_n$$

only.

In a similar way, we prove the existence of a sequence T''_n of complexes, which is a (DARS)-sequence for C, and which is composed of bricks of $\bigcup_{n=1}^{\infty} P''_n$ only.

The theorem is proved.

Chapter W

Upper and lower (DARS)-summation of fields of real numbers in a Boolean tribe in the absence of atoms

All preparations and auxiliaries being ready, [U.], we proceed to a deeper study of summation of fields of real numbers.

W.1. In this chapter it will be convenient to apply the following simplifications in symbolism, which will be, however, considered as optional only.

1) If A_n (also written $\{A_n\}$), denotes a sequence of complexes, then the bricks of A_n will be denoted by $a_{n1}, a_{n2}, \ldots$

2) If $f(a)$ is a field of scalars, then the sum $\sum_k f(a_{nk})$, where A_n is a complex, will be written $f(A_n)$.

3) Measurable sets of traces and their coats will be denoted by the same letter.

4) Instead of $\|a, b\|_\mu$, $=^\mu$, etc. we shall write $|a, b|$, $=$, etc.

5) $\dot{n}$ will denote the sequence of natural numbers $1, 2, 3, \ldots, n, \ldots$; $s(\dot{n})$ its subsequence and we shall write $s\,t(n)$ instead of $s[t(n)]$.

6) We shall say that $A_{\dot{n}}$ is a sequence for E, instead that $A_{\dot{n}}$ is a (DARS)-sequence of E, whenever no misunderstanding will result.

In addition to that we shall denote by ***Hyp H***$_2$ the collection of the following hypotheses: ***H***$_1$, ***Hyp*** $\mathscr{B}$, ***G*** has no atoms.

All considered sets of traces will be supposed to be measurable. As usually, the symbols $\cdot$, $+$, $-$, co, $\leqq$, $\sum$, Π will be applied to sets of traces, and the symbols $\frown$, $\smile$, $\sim$, $-$, Co, $\subseteqq$, $\bigcup$, $\bigcap$, $\emptyset$ to sets of bricks.

Section 1 of W.

Upper and lower summation

W.2. We start with some general theorems, (on sequences of complexes), which will be useful in the sequel.

Theorem. Under ***Hyp H***$_2$, let

1. E, F be sets of traces,
2. $E \cdot F = O$,
3. $P_{\dot{n}}, Q_{\dot{n}}$ be sequences for E, F respectively,

then there exist $s(\dot{n})$, $t(\dot{n})$ and complexes P'_n, Q'_n, such that:

$$P'_n \subseteqq P_{s(n)}, \quad Q'_n \subseteqq Q_{t(n)}, \quad (n = 1, 2, \ldots),$$

$$P'_n \frown Q'_n = \emptyset, \quad (n = 1, 2, \ldots),$$

$P'_{\dot n}$, $Q'_{\dot n}$ are sequences for E, F respectively, $P'_{\dot n} \cup Q'_{\dot n}$ is a sequence for $E + F$. (We can have $s(n) \geqq t(n)$ or $t(n) \geqq s(n)$.)

W.2a. Proof. We can suppose, that

$$(1) \qquad |P_n, E| \leqq \frac{1}{2^{n+3}}, \quad |Q_n, F| \leqq \frac{1}{2^{n+3}}, \quad (n = 1, 2, \ldots).$$

This can be obtained by suitable choice of a subsequence of $\dot n$.

As we have

$$|\operatorname{som} P_\alpha \cdot \operatorname{som} Q_\beta, E \cdot F| \leqq |P_\alpha, E| + |Q_\beta, F|,$$

we get, by (1),

$$(2) \qquad \mu(\operatorname{som} P_\alpha \cdot \operatorname{som} Q_\beta) \leqq \frac{1}{2^{\alpha+3}} + \frac{1}{2^{\beta+3}}.$$

W.2b. Let us fix α, β for a moment. There may exist bricks $p \in P_\alpha$, such that they are subsomata of some bricks of Q_β. Denote the collection, of these p, by $A_{\alpha\beta}$. There may also exist bricks $q \in Q_\beta$, such that they are subsomata of some bricks of P_α: denote the corresponding collection by $B_{\beta\alpha}$. Since

$$\operatorname{som} A_{\alpha\beta} \leqq \operatorname{som} P_\alpha \cdot \operatorname{som} Q_\beta, \qquad \operatorname{som} B_{\beta\alpha} \leqq \operatorname{som} P_\alpha \cdot \operatorname{som} Q_\beta,$$

we have, by (2),

$$\mu(\operatorname{som} A_{\alpha\beta}) \leqq \frac{1}{2^{\alpha+3}} + \frac{1}{2^{\beta+3}}, \qquad \mu(\operatorname{som} B_{\beta\alpha}) \leqq \frac{1}{2^{\alpha+3}} + \frac{1}{2^{\beta+3}}.$$

Hence

$$|P_\alpha \sim A_{\alpha\beta}, E| \leqq |P_\alpha, E| + |A_{\alpha\beta}, 0| \leqq \frac{1}{2^{\alpha+2}} + \frac{1}{2^{\beta+3}},$$

and similarly,

$$(3) \qquad |Q_\beta \sim B_{\beta\alpha}, F| \leqq \frac{1}{2^{\alpha+3}} + \frac{1}{2^{\beta+2}}.$$

W.2c. Let us notice, that if p, q are any two bricks, there are possible only the following four situations:

$$p \leqq q; \quad q \leqq p; \quad p \cdot q = O; \quad p \cdot q \neq O$$

with

$$p \cdot \operatorname{co} q \neq O \quad \text{and} \quad q \cdot \operatorname{co} p \neq O.$$

The complex $P_\alpha \sim A_{\alpha\beta}$ does not possess any brick, which would be a subsoma of some brick of Q_β, and $Q_\beta \sim B_{\beta\alpha}$ does not possess any brick, which would be a subsoma of some brick of P_α.

Hence, if $p \in P_\alpha - A_{\alpha\beta}$, $q \in Q_\beta - B_{\beta\alpha}$, the only possibilities are

1) $p \cdot q = O$,

2) $p \cdot q \neq O$, $p \operatorname{co} q \neq O$, $q \operatorname{co} p \neq O$.

Now, let us keep β fixed and vary α by increasing it. Since $\boldsymbol{G}$ has no atoms, we have $\mathcal{N}(P_\alpha) \to 0$ for $\alpha \to \infty$. Since the **Hyp** $\mathscr{B}$ is admitted, for every $\varepsilon > 0$ we can find $\delta > 0$, such that, if we take any set K of disjoint bricks $\mu(p) \leqq \delta$, then

$$\mu \operatorname{som}\{p \mid p \in K, p_{\beta\lambda} \neq O, p \cdot \operatorname{co} q_{\beta\lambda} \neq O\} \leqq \frac{\varepsilon}{M(\beta)},$$

where $M(\beta)$ is the number of non-null bricks of Q_β, $(\lambda = 1, 2, \ldots)$; and those bricks p exist.

Let us take α_0 such, that for all $\alpha \geqq \alpha_0$, we have $\mathcal{N}(P_\alpha) \leqq \delta$. Then

$$\mu \operatorname{som}\{p \mid p \in P_\alpha, p \cdot q_{\beta\lambda} \neq O, p \cdot \operatorname{co} q_{\beta\lambda} \neq O\} \leqq \frac{\varepsilon}{M(\beta)}, \quad (\lambda = 1, 2, \ldots).$$

Now if we take, in addition to that, α_0 such that for all $\alpha \geqq \alpha_0$,

$$\mathcal{N}(P_\alpha) < \min[\mu(q_{\beta 1}), \mu(q_{\beta 2}), \ldots],$$

then the condition

$$p \in P_\alpha, \quad p \cdot q_{\beta\lambda} \neq O, \quad p \cdot \operatorname{co} q_\beta \neq O, \quad (\lambda = 1, 2, \ldots)$$

is equivalent to

$$p \in P_\alpha, \; p \cdot q_{\beta\lambda} \neq O, \; p \cdot \operatorname{co} q_{\beta\lambda} \neq O, \; \operatorname{co} p \cdot q_{\beta\lambda} \neq O, \; (\lambda = 1, 2, \ldots),$$

since p cannot contain any brick $q_{\beta\lambda}$.

Hence for $\alpha \geqq \alpha_0$:

$$\mu \operatorname{som}\{p \mid p \in P_\alpha, p \cdot q_{\beta\lambda} \neq O, p \cdot \operatorname{co} q_{\beta\lambda} \neq O, \operatorname{co} p \cdot q_{\beta\lambda} \neq O\} \leqq \frac{\varepsilon}{M(\beta)};$$

this for every $\lambda = 1, 2, \ldots$

Hence the measure of the union of all bricks $p \in P_\alpha$, such that there exists λ with $p \cdot q_{\beta\lambda} \neq O$, $p \cdot \operatorname{co} q_{\beta\lambda} \neq O$, $\operatorname{co} p \cdot q_{\beta\lambda} \neq O$, is $\leqq \varepsilon$. Hence, if we denote by $R_{\alpha\beta}$ the above union of bricks, belonging to P_α, we have

(4) $$\mu R_{\alpha\beta} \leqq \varepsilon, \quad \text{for} \quad \alpha \geqq \alpha_0(\beta).$$

Consequently for those α we have: if we consider the bricks $p \in P_\alpha \sim A_{\alpha\beta}$, and the bricks $q \in Q_\beta - B_{\beta\alpha}$, then the set $R_{\alpha\beta}$ of all p, which strictly overlap with some q, has the measure $\leqq \varepsilon$, where ε is given in advance. Consider the complexes

$$P_\alpha \sim A_{\alpha\beta} \sim R_{\alpha\beta}, \quad Q_\beta \sim B_{\beta\alpha}.$$

In the first of them there does not exist any brick p, for which there exists $q \in Q_\beta \sim B_{\beta\alpha}$, where $p \cdot q \neq O$, $p \cdot \operatorname{co} q \neq O$, $q \cdot \operatorname{co} p \neq O$, neither any brick p, for which $q \leqq p$ for some bricks q of $Q_\beta \sim B_{\beta\alpha}$.

The similar is true for $Q_\beta \sim B_{\beta\alpha}$, with respect to $P_\alpha \sim A_{\alpha\beta} \sim R_{\alpha\beta}$, because the possible bricks of P_α, which would partly overlap with some brick of $Q_\beta \sim B_{\beta\alpha}$, are absent in $P_\alpha \sim A_{\alpha\beta} \sim R_{\alpha\beta}$.

Consequently, the complexes $P_\alpha \sim A_{\alpha\beta} \sim R_{\alpha\beta}$, $Q_\beta \sim B_{\beta\alpha}$ are disjoint.

W.2d. We have

$$|P_\alpha \sim A_{\alpha\beta} \sim R_{\alpha\beta}, E| \leqq |P_\alpha \sim A_{\alpha\beta}, E| + |R_{\alpha\beta}, O| \leqq,$$

$$\text{by (4), (3)}, \quad \leqq \frac{1}{2^{\alpha+2}} + \frac{1}{2^{\beta+3}} + \varepsilon. \tag{5}$$

The number ε is still arbitrary. Put

$$\varepsilon =_{df} \frac{1}{2^{\beta+1}},$$

and find the corresponding $\alpha_0 = \alpha_0(\beta)$. If we put

$$P'_\beta =_{df} P_{\alpha_0} \sim A_{\alpha_0\beta} \sim R_{\alpha_0\beta}, \quad Q'_\beta =_{df} Q_\beta \sim B_{\beta\alpha}, \tag{5.1}$$

we get, (3),

$$|P'_\beta, E| \leqq \frac{1}{2^{\alpha_0+2}} + \frac{1}{2^{\beta+3}} + \frac{1}{2^{\beta+1}}, \quad |Q'_\beta, F| \leqq \frac{1}{2^{\alpha_0+3}} + \frac{1}{2^{\beta+2}}.$$

Now, we always can choose α_0 so as to have $\alpha_0 \geqq \beta$. Then we get

$$|P'_\beta, E| \leqq \frac{1}{2^\beta}, \quad |Q'_\beta, F| \leqq \frac{1}{2^\beta}. \tag{6}$$

We also have

$$P'_\beta \subseteqq P_{\alpha_0}, \quad Q'_\beta \subseteqq Q_\beta.$$

Till now β has been fixed. Let us vary it, by increasing, and choose for each β the corresponding α_0. We get the sequences P'_β, Q'_β, $(\beta = 1, 2, \ldots)$. We see that P'_β is a sequence for E and Q'_β is a sequence for F. Hence, $P'_\beta \cup Q'_\beta$ is a sequence for $E + F$. The theorem is established.

W.3. Theorem. We admit the hypothesis $\boldsymbol{H}_2$. If

1. E, F are sets of traces,
2. $E \cdot F = O$,
3. P_n, Q_n are sequences for E, F respectively,
4. $f(\dot{a})$ is a field of scalars,
5. $f(a)$ is admissible,
6. $\lim_{n\to\infty} f(P_n) = A$, $\lim_{n\to\infty} f(Q_n) = B$,

then there exist $s(\dot{n})$, $t(\dot{n})$ and complexes P'_n, Q'_n, such that:

1) $P'_n \subseteqq P_{s(n)}$, $Q'_n \subseteqq Q_{t(n)}$ for $(n = 1, 2, \ldots)$,
2) $P'_n \cap Q'_n = \mathbf{O}$, $(n = 1, 2, \ldots)$,
3) $P'_{\dot{n}}, Q'_{\dot{n}}$ are sequences for E, F respectively,
4) $P'_n \cup Q'_n$ is a sequence for $E + F$,
5) $\lim_{n\to\infty} f(P'_n) = A$, $\lim_{n\to\infty} f(Q'_n) = B$, $\lim_{n\to\infty} f(P'_n \cup Q'_n) = A + B$.

Proof. It is a corrolary of the foregoing proof of Theorem [W.2.]. In that proof we have supposed that

$$|P_n, E| \leqq \frac{1}{2^{n+3}}, \quad |Q_n, F| \leqq \frac{1}{2^{n+3}},$$

which does not take place in the hypotheses of the present theorem. These inequalities, however, can be obtained by selecting a suitable subsequence of $\dot{n}$, say $\lambda(\dot{n})$. We put $\bar{P}_n =_{df} P_{\lambda(n)}$, $\bar{Q}_n =_{df} Q_{\lambda(n)}$; so we have $|\bar{P}_n, E| \leqq \frac{1}{2^{n+3}}$, $|\bar{Q}_n, F| \leqq \frac{1}{2^{n+3}}$. We consider the relations analogous to (5.1):

$$\bar{P}'_\beta = \bar{P}_{\alpha_0} \sim \bar{A}_{\alpha_0 \beta} \sim \bar{R}_{\alpha_0 \beta},$$

where $\bar{A}_{\alpha_0 \beta}$ and $\bar{R}_{\alpha_0 \beta}$ are disjoint partial complexes of $\bar{P}_\alpha$, and

$$\bar{Q}'_\beta = \bar{Q}_\beta \sim \bar{B}_{\beta \alpha_0}, \quad \text{where} \quad \bar{B}_{\beta \alpha_0} \subseteqq \bar{Q}_\beta.$$

Since

$$|\bar{P}'_\beta, E| \to 0, \quad |\bar{Q}'_\beta, F| \to 0, \quad |\bar{P}_\beta, E| \to 0, \quad |\bar{Q}_\beta, F| \to 0 \text{ for } \beta \to \infty,$$

we get

$$\mu(\bar{A}_{\alpha_0 \beta} \cup \bar{R}_{\alpha_0 \beta}) \to 0, \quad \mu(\bar{B}_{\beta \alpha_0}) \to 0 \quad \text{for} \quad \beta \to \infty.$$

By hyp. 5 it follows: $f(\bar{A}_{\alpha_0 \beta} \cup \bar{R}_{\alpha_0 \beta}) \to 0$, $f(\bar{B}_{\beta \alpha_0}) \to 0$. Consequently

$$\lim_{\beta \to \infty} f(\bar{P}'_\beta) = \lim_{\alpha_0 \to \infty} f(\bar{P}_{\alpha_0}),$$

(1)

$$\lim_{\beta \to \infty} f(\bar{Q}'_\beta) = \lim_{\beta \to \infty} f(\bar{Q}_\beta).$$

Since $\bar{P}_n$ is a subsequence of P_n, and $\bar{Q}_n$ is a subsequence of Q_n, it follows that

$$\lim_{n \to \infty} f(\bar{P}_n) = \lim_{n \to \infty} f(P_n), \qquad \lim_{n \to \infty} f(\bar{Q}_n) = \lim_{n \to \infty} f(Q_n).$$

Hence, by (1) and hyp. 6,

$$\lim_{\beta \to \infty} f(\bar{P}'_\beta) = A, \qquad \lim_{\beta \to \infty} f(\bar{Q}'_\beta) = B.$$

Since $\bar{P}'_\beta \cap \bar{Q}'_\beta = \emptyset$, we get

$$\lim_{\beta \to \infty} f(\bar{P}'_\beta \cup \bar{Q}'_\beta) = A + B,$$

and

$$\bar{P}'_\beta \subseteqq \bar{P}_{\alpha_0} = P_{\lambda(\alpha_0)}, \quad \bar{Q}'_\beta \subseteqq \bar{Q}_\beta = Q_{\lambda(\beta)},$$

where $\lambda(\alpha_0) \to \infty$, $\lambda(\beta) \to \infty$. The theorem is proved.

W.4. Def. Let us admit Hyp. $\boldsymbol{H}_1$. Let $f(a)$ be a field of real numbers, and E a measurable set of traces. We suppose, that $\boldsymbol{G}$ has no atoms and that $f(\dot{a})$ is admissible, Def. [U.2.7.]. Take any (DARS)-sequence $P_{\dot{n}}$ for E, and form the sums:

$$f(P_n) = \sum_k f(p_{nk}), \quad (n = 1, 2, \ldots).$$

By [U.8.],

$$\limsup_{n \to \infty} f(P_n), \quad \liminf_{n \to \infty} f(P_n)$$

are finite. If we vary P_n, we can consider

$$\sup_{P_n} \limsup_{n\to\infty} f(P_n), \qquad \inf_{P_n} \liminf_{n\to\infty} f(P_n).$$

By [U.9.] and [U.8.] they are both finite. We call them *upper and lower* (DARS)-*sums of* $f(a)$ *on* E, and denote them by $\bar{S}_E f(a)$, $\underline{S}_E f(a)$ respectively. Our next purpose will be a study of these kinds of integrals.

W.4.1. By [U.8.] we know, that there exist (DARS)-sequences P_n, Q_n for E, such that

$$\bar{S}_E f(a) = \lim_{n\to\infty} f(P_n), \quad \underline{S}_E f(a) = \lim_{n\to\infty} f(Q_n).$$

W.4.2. We have

$$\underline{S}_E f(a) \leqq \bar{S}_E f(a).$$

W.4.3. Theorem. The following are equivalent:

I. $S_E f$ exists,

II. $\underline{S}_E f = \bar{S}_E f$.

For admissible field $f(a)$ of real numbers and under $\boldsymbol{H}_1$ we have:

W.4.4. Theorem.

$$-S_E f = \bar{S}_E(-f), \quad -\bar{S}_E f = \underline{S}_E(-f).$$

Proof. Let Q_n be a sequence for E, such that the limit $\lim_{n\to\infty} f(Q_n)$ is maximum (for all sequences $P_{\dot{n}}$ for E). Hence

(1) $$\lim_{n\to\infty} f(Q_n) = \bar{S}_E f.$$

Then $\lim[(-f)(Q_n)]$ is the minimum for all sequences $Q_{\dot{n}}$ for E. Hence

(2) $$\lim(-f)(Q_n) = \underline{S}_E(-f).$$

Now $\lim_{n\to\infty}(-f)(Q_n) = -\lim_{n\to\infty} f(Q_n)$; hence, by (1), this equals $-\bar{S}_E f$. We also have, by (2),

(3) $$\underline{S}_E(-f) = -\bar{S}_E f.$$

Applying this result to f, replaced by $(-f)$, we get $\underline{S}_E f = -\bar{S}_E(-f)$; hence

(4) $$\bar{S}_E(-f) = -\underline{S}_E f.$$

The formulas (3) and (4) constitute the theses of the theorem, so the theorem is proved.

W.4.5. Theorem. $\underline{S}_0 f = \bar{S}_0 f = 0$ for admissible f and under hypothesis $\boldsymbol{H}_2$.

W.4.6. Theorem. If $\lambda \geqq 0$, then $\bar{S}_E \lambda f = \lambda \bar{S}_E f$, $\underline{S}_E \lambda f = \lambda \underline{S}_E f$ under the same conditions as before.

W.4.7. If $\lambda \leqq 0$, then $\overline{S}_E \lambda f = \lambda \underline{S}_E f$, and $\underline{S}_E \lambda f = \lambda \overline{S}_E f$.

Proof. We have, by [W.4.4.], $\underline{S}_E f = -\overline{S}_E(-f)$. Hence, by [W.4.6.], $\underline{S}_E |\lambda| f = |\lambda| \cdot \underline{S}_E f$; hence $\underline{S}_E(-\lambda f) = -\lambda \cdot \underline{S}_E f$; hence, by [W.4.4.], $-\overline{S}_E \lambda f = -\lambda \underline{S}_E f$; hence $\overline{S}_E \lambda f = \lambda \underline{S}_E f$.

W.4.8. Theorem. If

1. Hypothesis $\boldsymbol{H}_1$ is admitted,
2. E is a measurable set of traces,
3. $f(\dot{a})$, $g(\dot{a})$ are fields of real numbers,
4. $f(\dot{a})$ is admissible,
5. $g(\dot{a})$ is summable on E,
6. there are no atoms,

then

$$\overline{S}_E(f+g) = \overline{S}_E f + \overline{S}_E g, \quad \underline{S}_E(f+g) = \underline{S}_E f + \underline{S}_E g.$$

Proof. Let Q_n be a sequence for E, such that $\lim_{n\to\infty} f(Q_n) = \overline{S}_E f$. We have $\lim_{n\to\infty} g(Q_n) = S_E g$. Hence $\lim_{n\to\infty} (f+g)(Q_n) = \overline{S}_E f + S_E g$. Since $\lim_{n\to\infty} (f+g)(Q_n) \leqq \overline{S}_E(f+g)$, it follows

$$(1) \qquad \overline{S}_E f + S_E g \leqq \overline{S}_E(f+g).$$

On the other hand, let us take a sequence P_n for E, such that

$$(2) \qquad \lim_{n\to\infty} (f+g)(P_n) = \overline{S}_E(f+g).$$

We have

$$(3) \qquad \lim_{n\to\infty} g(P_n) = S_E g.$$

Hence, from (2), $\lim_{n\to\infty} f(P_n)$ exists and equals $\overline{S}_E(f+g) - S_E g$.

We have

$$\lim_{n\to\infty} f(P_n) \leqq \overline{S}_E f.$$

Hence, from (3),

$$\lim g(P_n) + \lim f(P_n) \leqq S_E g + \overline{S}_E f,$$

i.e.

$$\lim (g+f)(P_n) \leqq S_E g + \overline{S}_E f;$$

hence, by (2),

$$(4) \qquad \overline{S}_E(f+g) \leqq S_E g + \overline{S}_E f.$$

From (1) and (4) the theorem for upper sums follows.

Concerning lower sums, we have, [W.4.4.],

$$(5) \qquad \begin{aligned} \underline{S}_E f &= -\overline{S}_E(-f), \\ S_E g &= -S_E(-g). \end{aligned}$$

Hence

$$(6) \qquad \underline{S}_E f + S_E g = -S_E[(-f)+(-g)].$$

Since $(-g)$ is summable, we get, by what already has been proved,

(7) $\bar{S}_E[(-f)+(-g)] = \bar{S}_E(-f) + S_E(-g) = -\underline{S}_E f - S_E g.$

Hence, by (6),

(8) $$-[\underline{S}_E f + S_E g] = -\underline{S}_E(f+g).$$

By (7) and (8) we get

$$\underline{S}_E(f+g) = \underline{S}_E f + S_E g.$$

The theorem is established under ***Hyp* $\boldsymbol{H}_1$**.

W.4.9. Theorem. If

1. $f(a) \leqq g(a)$,
2. E is a measurable set of traces,
3. $f(a)$, $g(a)$ are admissible,
4. $f \leqq g$,
5. there are no atoms,

then

$$\bar{S}_E f \leqq \bar{S}_E g, \quad \text{and} \quad \underline{S}_E f \leqq \underline{S}_E g.$$

Proof. Let P_n be a (DARS)-sequence for E, such that

$$\lim_{n\to\infty} \sum_k f(p_{nk}) = \bar{S}_E f.$$

Since $f(p_{nk}) \leqq g(p_{nk})$, it follows that

$$\sum_k f(p_{nk}) \leqq \sum_k g(p_{nk}).$$

Hence

$$\lim_{n\to\infty} \sum_k f(p_{nk}) \leqq \limsup_{n\to\infty} \sum_k g(p_{nk}).$$

Hence $\bar{S} f \leqq \bar{S} g$.

Let Q_n be a (DARS)-sequence for E, such that

$$\lim_{n\to\infty} \sum_k g(p_{nk}) = \underline{S}_E g.$$

Since $f(p_{nk}) \leqq g(p_{nk})$, it follows that

$$\liminf_{n\to\infty} \sum_k f(p_{nk}) \leqq \underline{S}_E g.$$

Hence $\underline{S}_E f \leqq \underline{S}_E g$. The theorem is established.

W.4.10. Theorem. Admit hypothesis $\boldsymbol{H}_2$. If

1. f, g are fields of real numbers, both admissible,
2. E is a measurable set of traces,

then

$$\bar{S}_E(f+g) \leqq \bar{S}_E f + \bar{S}_E g \quad \text{and} \quad \underline{S}_E(f+g) \geqq \underline{S}_E f + \underline{S}_E g.$$

Proof. Let Q_n be a sequence for E, yielding $\bar{S}_E(f+g)$, i.e.

$$\lim_{n\to\infty} (f+g)(Q_n) = \bar{S}_E(f+g).$$

This can be written

$$\sum_k [f(q_{nk}) + g(q_{nk})] \to \bar{S}_E(f+g).$$

Take a subsequence $\alpha(\dot{n})$ of $\dot{n}$, such that $\lim_{n\to\infty} \sum_k f(q_{\alpha(n)k})$ exists. This can be done, because the sequence $f(Q_n)$ is bounded.

Hence $\lim \sum_k g(q_{\alpha(n)k})$ exists too. Now since

$$\lim_{n\to\infty} \sum_k f(q_{\alpha(n)k}) \leqq \bar{S}_E f,$$

and

$$\lim_{n\to\infty} \sum_k g(q_{\alpha(n)k}) \leqq \bar{S}_E g,$$

it follows, that

$$\lim_{n\to\infty} \sum_k (f+g)\, q_{\alpha(n)k} \leqq \bar{S}_E f + \bar{S}_E g.$$

But the left side is equal to $\bar{S}_E(f+g)$, so the first inequality of the thesis is proved.

To prove the second inequality, take a sequence P_n for E, such that

$$\lim_{n\to\infty} (f+g)(P_n) = \underline{S}_E(f+g).$$

We choose a subsequence $\beta(\dot{n})$ of $\dot{n}$, such that $\lim_{n\to\infty} f(P_n)$ exists. Then $\lim_{n\to\infty} g(P_n)$ exists too. Since $\lim_{n\to\infty} f(P_n) \geqq \underline{S}_E f$, and $\lim_{n\to\infty} g(P_n) \geqq \underline{S}_E g$, it follows that $\underline{S}_E(f+g) \geqq \underline{S}_E f + \underline{S}_E g$; so the second inequality is also established.

W.5. Now we are going to study the upper and lower sums under hypothesis $\boldsymbol{H}_2$, i.e. for admissible scalar fields under ***Hyp*** $\mathscr{B}$ and when $\boldsymbol{G}$ has no atoms.

W.5.1. Theorem. We admit $\boldsymbol{H}_2$. If

1. E, F are sets of traces,
2. $F \leqq E$,
3. T_n is a sequence for E,
4. $f(\dot{a})$ is an admissible field of real numbers,
5. P_n is a sequence for F, such that $P_n \subseteqq T_n$ for $n = 1, 2, \ldots$,
6. $\lim_{n\to\infty} f(T_n) = \bar{S}_E f$,

then

$$\lim_{n\to\infty} f(P_n) = \bar{S}_E f.$$

Proof. Let $s(\dot{n})$ be a subsequence of $\dot{n}$. Consider the corresponding $T_{s(n)}, P_{s(n)}$, and put

$$T_n^{(1)} =_{df} T_{s(n)}, \quad P_n^{(1)} =_{df} P_{s(n)}. \tag{1}$$

They are sequences for E, F respectively.

Put

(2) $$Q_n^{(1)} =_{df} T_n^{(1)} \sim P_n^{(1)} \quad \text{for} \quad n = 1, 2, \ldots$$

$Q_n^{(1)}$ is a sequence for $E - F$. We have $T_n^{(1)} = P_n^{(1)} \cup Q_n^{(1)}$ and $P_n^{(1)} \cap Q_n^{(1)} = \emptyset$. Since

$$\bar{S}_E f = \lim_{n\to\infty} f(T_n),$$

we also have

$$\bar{S}_E f = \lim_{n\to\infty} f(T_n^{(1)}) = \lim_{n\to\infty} [f(P_n^{(1)}) + f(Q_n^{(1)})].$$

By hyp. 4 and [U.9.], the sequences $f(P_{\dot{n}}^{(1)})$, $f(Q_{\dot{n}}^{(1)})$ are both bounded; consequently we can find a subsequence $t(\dot{n})$ of $\dot{n}$, such that

$$\lim_{n\to\infty} f(P_{t(n)}^{(1)}) \quad \text{and} \quad \lim_{n\to\infty} f(Q_{t(n)}^{(1)})$$

exist. Put

(2.1) $$P_n^{(2)} =_{df} P_{t(n)}^{(1)}, \quad Q_n^{(2)} =_{df} Q_{t(n)}^{(1)}, \quad (n = 1, 2, \ldots),$$

and

(3) $$A =_{df} \lim_{n\to\infty} f(P_n^{(2)}), \quad B =_{df} \lim_{n\to\infty} f(Q_n^{(2)}).$$

We have

(4) $$A \leqq \bar{S}_F f.$$

We also have

(5) $$\bar{S}_E f = A + B.$$

By [W.4.1.] there exists a sequence R_n for F, such that

(6) $$\bar{S}_F f = \lim_{n\to\infty} f(R_n).$$

Consider the sequences R_n and $Q_n^{(2)}$. We are in the conditions of Theorem [W.3.]. There exist $\alpha(\dot{n})$ and $\beta(\dot{n})$ and complexes $R_n^{(3)}$ and $Q_n^{(3)}$, such that

$$R_n^{(3)} \subseteqq R_{\alpha(n)}, \quad Q_n^{(3)} \subseteqq Q_{\beta(n)}^{(2)}, \quad R_n^{(3)} \cap Q_n^{(3)} = \emptyset, \quad n = 1, 2, \ldots$$

$R_n^{(3)}$ is a sequence for F; $Q_n^{(3)}$ is a sequence for $E - F$; $R_n^{(3)} \cup Q_n^{(3)}$ is a sequence for E. We have

(7) $$\begin{aligned} \lim_{n\to\infty} f(R_n^{(3)}) &= \lim_{n\to\infty} f(R_n) = \bar{S}_F f, \quad \text{by (6)},\\ \lim_{n\to\infty} f(Q_n^{(3)}) &= \lim_{n\to\infty} f(Q_n^{(2)}) = B, \quad \text{by (3)}. \end{aligned}$$

Since $R_n^{(3)} \cup Q_n^{(3)}$ is a sequence for E, and $R_n^{(3)}$, $Q_n^{(3)}$ are disjoint, it follows that

$$\lim f(R_n^{(3)}) + \lim f(Q_n^{(3)}) = \lim f(R_n^{(3)} \cup Q_n^{(3)}) \leqq \bar{S}_E f.$$

Hence from (7) and (5) we get

$$\bar{S}_F f \leqq \bar{S}_E f = A + B.$$

Hence

(8) $$\bar{S}_F f \leq A.$$

From (4) and (8) it follows that

$$\bar{S}_F f = A.$$

Thus we have proved that

$$\bar{S}_F f = \lim_{n\to\infty} f(P_n^{(2)}).$$

Now by (2.1) and (1),

$$P_n^{(2)} = P_{t(n)}^1 = P_{st(n)},$$

which is a subsequence of P_n. Hence from every subsequence $P_{s(n)}$ of P_n another sequence $P_{st(n)}$ can be extracted, such that

$$\lim_{n\to\infty} f(P_{st(n)}) = \bar{S}_F f.$$

It follows that

$$\lim f(P_n) = \bar{S}_F f, \quad \text{q.e.d.}$$

W.6. Theorem. We admit $\boldsymbol{H}_2$. If

1. E, F are sets of traces,
2. $E \cdot F = O$,
3. $f(a)$ is an admissible field of real numbers,

then

(1) $$\bar{S}_{E+F} f = \bar{S}_E f + \bar{S}_F f.$$

Proof. Let T_n be a sequence for $E + F$, such that

(1.1) $$\lim_{n\to\infty} f(T_n) = \bar{S}_{E+F} f.$$

Since T_n is a (DARS)-sequence, there exists a partial complex P_n of T_n, $(n = 1, 2, \ldots)$, such that P_n is a sequence for E. If we put $Q_n =_{df} T_n \sim P_n$, then Q_n is a sequence for F. Applying Theorem [W.5.] we get

$$\lim_{n\to\infty} f(P_n) = \bar{S}_E f, \qquad \lim_{n\to\infty} f(Q_n) = \bar{S}_F f.$$

Hence, since $P_n \frown Q_n = \mathbf{O}$, we get

(2) $$\lim_{n\to\infty} f(P_n \smile Q_n) = \bar{S}_E f + \bar{S}_F f.$$

Hence, as $P_n \smile Q_n = T_n$, we get

(3) $$\lim_n f(P_n \smile Q_n) = \bar{S}_{E+F} f.$$

From (2) and (3) theorem follows.

W.7. Theorem. Under Hyp $\boldsymbol{H}_2$ theorems, analogous to (5) and (6), hold true for lower sums $\underline{S}_E f$.

Proof. This follows from [W.5.], [W.6.] and [W.4.4.].

W.8. Theorem. We admit $\boldsymbol{H}_2$. If

1. $f(\dot{a})$ is an admissible field of real numbers,
2. $E_1, E_2, \ldots, E_m, \ldots$ are sets of traces,
3. $\mu E_m \to 0$ for $m \to \infty$,

then

$$\bar{S}_{E_m} f \to 0, \quad \underline{S}_{E_m} f \to 0 \quad \text{for} \quad m \to \infty.$$

Proof. There exists a sequence $P_n^{(m)}$ for E_m, such that $\lim_{n\to\infty} f(P_n^{(m)}) = \bar{S}_{E_m} f$. Hence we can find an index $n(m)$, such that

$$|E_m, P_{n(m)}^{(m)}| \leqq \frac{1}{m}, \tag{1}$$

and

$$|\bar{S}_{E_m} f - f(P_{n(m)}^{(m)})| \leqq \frac{1}{m}. \tag{2}$$

Consider the sequence of complexes

$$P_{n(1)}^{1}, P_{n(2)}^{2}, \ldots, P_{n(m)}^{m}, \ldots$$

We have from (1),

$$|\mu E_m - \mu P_{n(m)}^{m}| \leqq \frac{2}{m}.$$

Hence

$$0 \leqq \mu P_{n(m)}^{m} \leqq \mu E_m + \frac{2}{m} \to 0.$$

Since f is admissible, we get

$$\lim_{m\to\infty} f(P_{n(m)}^{m}) = 0.$$

Hence from (2) we get

$$0 \leqq |\bar{S}_{E_m} f| \leqq |f(P_{n(m)}^{m})| + \frac{1}{m} \to 0 \quad \text{for} \quad m \to \infty.$$

The theorem is proved for upper sums. The theorem for lower sums results from [W.4.4.].

W.8.1. Theorem. We admit $\boldsymbol{H}_2$. If

1. $f(a)$ is an admissible field of real numbers,
2. $\varepsilon > 0$,

then there exists $\delta > 0$ such that, if $\mu f \leqq \delta$, then

$$|\bar{S}_F f| \leqq \varepsilon, \quad |\underline{S}_F f| \leqq \varepsilon.$$

Proof. Indirect. Suppose that the first statement is not true; then there exists $\varepsilon > 0$, such that for every $\delta > 0$ we can find a set F with $|\bar{S}_F f| \geqq \varepsilon$, $\mu F \leqq \delta$. Take $\delta = \frac{1}{m}$, $m = 1, 2$, and find a corresponding set F_m with $\mu(F_m) \leqq \frac{1}{m}$, $|\bar{S}_{F_m} f| \geqq \varepsilon$. This contradicts Theorem [W.8.]. The proof for $\underline{S}_F$ relays on [W.4.4.].

W.9. Theorem. If we admit $\boldsymbol{H}_2$ and $f(a)$ is admissible, then the set-function $U(F) =_{df} \overline{S}_F f$, $L(F) =_{df} \underline{S}_F f$ are both denumerably additive functions of the variable measurable set F of traces.

Proof. Let $E_1, E_2, \ldots, E_m, \ldots$ be disjoint sets of traces. Put $E =_{df} \sum_{m=1}^{\infty} E_m$. We have $U(E_n) =_{df} \overline{S}_{E_n} f$, $u(E) = \underline{S}_E f$.

Choose $\alpha > 0$. By Theorem [W.8.1.] we can find $\beta > 0$, such that if $\mu(F) \leqq \beta$, then $|U(F)| \leqq \alpha$. Find $n' = n(\beta)$, such that

$$\mu\left(E - \sum_{n=1}^{n'} E_n\right) \leqq \beta.$$

Hence

$$\left|U(E - \sum_{n=1}^{n'} E_n)\right| \leqq \alpha.$$

Applying Theorem [W.6.], we get

$$\left|U(E) - \sum_{n=1}^{n'} U(E_n)\right| \leqq \alpha.$$

If we put $\alpha = \frac{1}{m}$, $m \to \infty$, and find the corresponding

$$n' = n\left(\frac{1}{m}\right), \quad \text{we get} \quad \left|U(E) - \sum_{n=1}^{n'} U(E_n)\right| \leqq \frac{1}{m};$$

hence

$$U(E) = \sum_{n=1}^{\infty} U(E_n).$$

A similar argument yields the formula $L(E) = \sum_{n=1}^{\infty} L(E_n)$. The theorem is established.

W.10. We have introduced, under Hyp. $\boldsymbol{H}_1$, for a summable given field f of real numbers, and (DARS)-sums, a function $F(x)$ of the variable trace, such that for every measurable set A of traces, we have

$$\int_A F(x)\, d\mu = S_A f.$$

We can do the same, under $\boldsymbol{H}_2$, for upper and lower sum of admissible fields of real numbers.

W.10.1. Theorem. Let us admit Hyp. $\boldsymbol{H}_2$ and let us consider an admissible field $f(a)$ of real numbers.

Then there exist μ-unique functions $\overline{F}(x)$, $\underline{F}(x)$ of the variable trace x, such that these functions are Fréchet-μ-summable, and that

$$\overline{S}_E f(a) = \int_E \overline{F}(x)\, d\mu, \quad \underline{S}_E f(a) = \int_E \underline{F}(x)\, d\mu,$$

for every measurable set E of traces.

Proof. We apply Theorem [W.9.] and use the known theorem, by the author, **(16)**, on the denumerably additive set-functions.

W.10.2. Def. We call $\bar{F}(x)$, $\underline{F}(x)$ *the upper and lower trace-functions attached to the given (admissible) field of real numbers.* The definition is admitted under Hyp. $\boldsymbol{H}_2$.

W.10.3. Theorem. We have $\underline{F}(x) \leqq^{\mu} \bar{F}(x)$.

W.10.4. Theorem. The following are equivalent for admissible fields $f(a)$ of real numbers, and under Hyp. $\boldsymbol{H}_2$:

I. $f(a)$ is (DARS)-summable on E,

II. $\underline{F}(x) =^{\mu} \bar{F}(x)$ almost μ-everywhere on E (where $\bar{F}$, $\underline{F}$ are functions attached to f).

Section 2 of W.

Summation of fields of numbers admitting two values only

W.20.1. We shall be interested in the fields of real numbers $f(a)\,\mu(a)$, where f admits two values only.

If $f(a)$ admits only the values 0 and 1, then $f(a)\,\mu(a)$ is admissible. Indeed, if P_n is any sequence of complexes, then $\sum_k f(p_{nk})\,\mu(p_{nk}) \leqq \leqq \sum_k \mu(p_{nk})$. Hence, if $\mu P_n = \sum_k \mu(p_{nk}) \to 0$, then $(f\,\mu)(P_n) \to 0$. Hence the upper and lower sums of $f\,\mu$ exist.

W.20.2. Lemma. We admit Hyp. $\boldsymbol{H}_2$. If

1. E is a set of traces, $\mu E > 0$,
2. Q_n is a sequence for E,
3. $|E, Q_n| \leqq \frac{1}{2^n}$ for $n = 1, 2, \ldots$,
4. Q_n satisfies the condition of non overlapping (see Def. [U.10.]),
5. $f(a)$ admits the values 0 and 1 only,
6. we put

$$Q_n^0 =_{df} \{q_{nk} \mid q_{nk} \in Q_n,\ f(q_{nk}) = 0\},\quad Q_n^1 =_{df} \{q_{nk} \mid q_{nk} \in Q_n,\ f(q_{nk}) = 1\},$$

7. $A^0 =_{df} \underline{\operatorname{Lim}}_{n\to\infty} \operatorname{som} Q_n^0$, $B^0 =_{df} \overline{\operatorname{Lim}}_{n\to\infty} \operatorname{som} Q_n^0$, $A^1 =_{df} \underline{\operatorname{Lim}}_{n\to\infty} \operatorname{som} Q_n^1$, $B^1 =_{df} \overline{\operatorname{Lim}}_{n\to\infty} \operatorname{som} Q_n^1$,
8. $C =_{df} B^1 - A^1 = B^0 - A^0$ (see Theorem [U.20.10.]),

then

1) $A^0 \leqq B^0 \leqq E$, $A^1 \leqq B^1 \leqq E$, $C \leqq E$;

2) A^0, C, A^1 are disjoint and $A^0 + C + A^1 = E$;

3) $A^0 \cdot A^1 = A^0 B^1 = O$, $B^0 \cdot B^1 = C$, $B^0 = A^0 + C$, $B^1 = A^1 + C$;

4) there exists a sequence T^0_n for C, such that $T^0_n \subseteqq \bigcup_{m=1}^{\infty} Q^0_m$ for all n; there exists a sequence T^1_n for C, such that $T^1_n \subseteqq \bigcup_{m=1} Q^1_m$ for all n;

5) $\overline{S}_C f \mu = \mu(C)$, $\underline{S}_C f \mu = 0$, and if $C^1 \leqq C$, then $\overline{S}_{C'} f \mu = \mu(C')$, $\underline{S}_{C'} f \mu = 0$;

6) $\underline{S}_{A^0} f \mu = 0$, $\overline{S}_{A^1} f \mu = \mu(A^1)$, and if $D^0 \leqq A^0$, $D^1 \leqq A^1$, then $\underline{S}_{D^0} f \mu = 0$, $\overline{S}_{D^1} f \mu = \mu(D^1)$;

7) $\underline{\mathrm{Lim}}_{n\to\infty} (C \cdot \mathrm{som} Q^0_n) = O$, $\overline{\mathrm{Lim}}_{n\to\infty} (C \cdot \mathrm{som} Q^0_n) = C$,

$\underline{\mathrm{Lim}}_{n\to\infty} (C \cdot \mathrm{som} Q^1_n) = O$, $\overline{\mathrm{Lim}}_{n\to\infty} (C \cdot \mathrm{som} Q^1_n) = C$, and if $C' \leqq C$,

then

$\underline{\mathrm{Lim}}_{n\to\infty} (C' \cdot \mathrm{som} Q^0_n) = O$, $\overline{\mathrm{Lim}}_{n\to\infty} (C' \cdot \mathrm{som} Q^0_n) = C'$,

$\underline{\mathrm{Lim}}_{n\to\infty} (C' \cdot \mathrm{som} Q^1_n) = O$, $\overline{\mathrm{Lim}}_{n\to\infty} (C' \cdot \mathrm{som} Q^1_n) = C'$;

8) $$(E - C) \underline{\mathrm{Lim}}_{n\to\infty} \mathrm{som} Q^0_n = (E - C) \overline{\mathrm{Lim}}_{n\to\infty} \mathrm{som} Q^0_n = \mathrm{Lim}_{n\to\infty} (E - C) \mathrm{som} Q^0_n = \lim_{n\to\infty}{}^{\mu} (E - C) \mathrm{som} Q^0_n,$$

$$(E - C) \underline{\mathrm{Lim}}_{n\to\infty} \mathrm{som} Q^1_n = (E - C) \overline{\mathrm{Lim}}_{n\to\infty} \mathrm{som} Q^1_n = \mathrm{Lim}_{n\to\infty} (E - C) \mathrm{som} Q^1_n = \lim_{n\to\infty}{}^{\mu} (E - C) \mathrm{som} Q^1_n,$$

and if $D \leqq E - C$, then

$$D \underline{\mathrm{Lim}}_{n\to\infty} \mathrm{som} Q^0_n = D \overline{\mathrm{Lim}}_{n\to\infty} \mathrm{som} Q^0_n = \mathrm{Lim}_{n\to\infty} (D \,\mathrm{som} Q^0_n) = \lim_{n\to\infty}{}^{\mu} (D \,\mathrm{som} Q^0_n),$$

$$D \underline{\mathrm{Lim}}_{n\to\infty} \mathrm{som} Q^1_n = D \overline{\mathrm{Lim}}_{n\to\infty} \mathrm{som} Q^1_n = \mathrm{Lim}_{n\to\infty} (D \,\mathrm{som} Q^1_n) = \lim_{n\to\infty}{}^{\mu} (D \,\mathrm{som} Q^1_n);$$

9) $$A^0 \underline{\mathrm{Lim}}_{n\to\infty} \mathrm{som} Q^0_n = A^0 \overline{\mathrm{Lim}}_{n\to\infty} \mathrm{som} Q^0_n = \mathrm{Lim}_{n\to\infty} A^0 \mathrm{som} Q^0_n = \lim_{n\to\infty}{}^{\mu} (A^0 \mathrm{som} Q^0_n) = A^0,$$

$$A^1 \underline{\mathrm{Lim}}_{n\to\infty} \mathrm{som} Q^1_n = A^1 \overline{\mathrm{Lim}}_{n\to\infty} \mathrm{som} Q^1_n = \mathrm{Lim}_{n\to\infty} A^1 \mathrm{som} Q^1_n = \lim_{n\to\infty}{}^{\mu} (A^1 \mathrm{som} Q^1_n) = A^1,$$

$$A^0 \underline{\mathrm{Lim}}_{n\to\infty} \mathrm{som} Q^1_n = A^0 \overline{\mathrm{Lim}}_{n\to\infty} \mathrm{som} Q^1_n = \mathrm{Lim}_{n\to\infty} A^0 \mathrm{som} Q^1_n = \lim_{n\to\infty}{}^{\mu} (A^0 \mathrm{som} Q^1_n) = O,$$

$$A^1 \underline{\mathrm{Lim}}_{n\to\infty} \mathrm{som} Q^0_n = A^1 \overline{\mathrm{Lim}}_{n\to\infty} \mathrm{som} Q^0_n = \mathrm{Lim}_{n\to\infty} A^1 \mathrm{som} Q^0_n = \lim_{n\to\infty}{}^{\mu} (A^1 \mathrm{som} Q^0_n) = O,$$

and if $D^0 \leqq A^0$, $D^1 \leqq A^1$, then

$$D^0 \underline{\operatorname{Lim}}_{n\to\infty} \operatorname{som} Q_n^0 = D^0 \overline{\operatorname{Lim}}_{n\to\infty} \operatorname{som} Q_n^0 = \operatorname{Lim}_{n\to\infty} (D^0 \operatorname{som} Q_n^0) =$$
$$= \lim_{n\to\infty}{}^{\mu} (D^0 \operatorname{som} Q_n^0) = D^0,$$

$$D^1 \underline{\operatorname{Lim}}_{n\to\infty} \operatorname{som} Q_n^1 = D^1 \overline{\operatorname{Lim}}_{n\to\infty} \operatorname{som} Q_n^1 = \operatorname{Lim}_{n\to\infty} (D^1 \operatorname{som} Q_n^1) =$$
$$= \lim_{n\to\infty}{}^{\mu} (D^1 \operatorname{som} Q_n^1) = D^1,$$

$$D^0 \underline{\operatorname{Lim}}_{n\to\infty} \operatorname{som} Q_n^1 = D^0 \overline{\operatorname{Lim}}_{n\to\infty} \operatorname{som} Q_n^1 = \operatorname{Lim}_{n\to\infty} (D^0 \operatorname{som} Q_n^1) =$$
$$= \lim_{n\to\infty}{}^{\mu} (D^0 \operatorname{som} Q_n^1) = O,$$

$$D^1 \underline{\operatorname{Lim}}_{n\to\infty} \operatorname{som} Q_n^0 = D^1 \overline{\operatorname{Lim}}_{n\to\infty} \operatorname{som} Q_n^0 = \operatorname{Lim}_{n\to\infty} (D^1 \operatorname{som} Q_n^0) =$$
$$= \lim_{n\to\infty}{}^{\mu} (D^1 \operatorname{som} Q_n^0) = O;$$

10) if M_n^0, M_n^1 are sequences for A^0, A^1 respectively with

$$M_n^0 \subseteqq Q_n, \quad M^1 \subseteqq Q_n, \quad (n = 1, 2, \ldots),$$

then

a) $\operatorname{som} M_n^0\, A^0 \to^{\mu} A^0$, $\operatorname{som} M_n^1 \cdot A^0 \to^{\mu} 0$, $\operatorname{som} M_n^0 \cdot A^1 \to^{\mu} O$, $\operatorname{som} M_n^1 \cdot A^1 \to^{\mu} A^1$, for $n \to \infty$;

b) $Q_n^0 \cap M_n^0$ is a sequence for A^0, and $Q_n^1 \cap M_n^1$ is a sequence for A^1;

c) $\mu(M_n^0 \cap Q_n^0) \to 0$, $\mu(M_n^1 \cap Q_n^1) \to 0$ for $n \to \infty$;

d) $\lim_{n\to\infty} (f\,\mu)(M_n^0) = 0$, $\lim_{n\to\infty} (f\,\mu) = \mu(A^1)$;

e) $\lim_{n\to\infty} (f\,\mu)(M_n^0 \cap Q_n^0) = 0$, $\lim_{n\to\infty} (f\,\mu)(M_n^1 \cap Q_n^1) = \mu(A^1)$. If we put $R^0 =_{df} \{q \mid q \in M_n^0,\, q\,A^0 \neq O\}$, $R^1 =_{df} \{q \mid q \in M_n^1,\, q\,A^1 \neq O\}$, then R^0, R^1 are sequences for A^0, A^1 respectively.

W.20.2.1. Proof. The items 1), 2), 3) result from the Lemma [U.20.10.], because $Q_n^0 \cap Q_n^1 = \emptyset$, $Q_n^0 \cup Q_n^1 = Q_n$, and because Q_n is a sequence for E, with $|E, Q_n| \leqq \frac{1}{2^n}$. By [U.20.11.] there exist two sequences T_n^0, T_n^1 for C, such that $T_n^0 \subseteqq \bigcup_{m=1}^{\infty} Q_m^0$, $T_n^1 \subseteqq \bigcup_{m=1}^{\infty} Q_m^1$ for $n = 1, 2, \ldots$, so the item 4) is proved. The bricks of T_n^0 belong to some $Q_{\alpha_1}^0, Q_{\alpha_2}^0, \ldots$, and those of T_n^1 belong to some $Q_{\beta_1}^1, Q_{\beta_2}^1, \ldots$ Hence for every brick t_{nk}^0 of T_n^0 we have $f(t_{nk}^0) = 0$, and for every brick t_{nk}^1 of T_n^1 we have $f(t_{nk}^1) = 1$. Consequently $(f\,\mu)(T_n^0) = \sum_k f(t_{nk}^0)\,\mu(t_{nk}^0) = 0$, so we have $\lim_{n\to\infty} (f\,\mu)(T_n^0) = 0$, and $(f\,\mu)(T_n^1) = \sum_k f(t_{nk}^1)\,\mu(t_{nk}^1)$ $= \sum_k \mu(t_{nk}^1) = \mu(T_n^1)$; hence $\lim_n (f\,\mu)(T_n^1) = \lim_{n\to\infty} \mu(T_n^1) = \mu(C)$. Having that, take any sequence $P_{\dot n}$ for C. We have

$$0 \leqq (f\,\mu)\,(P_n) \leqq \sum_k \mu(p_{nk}) = \mu(P_n) \to \mu(C).$$

Hence $\mu(C)$ is the greatest number, which can be obtained by $\overline{\lim}\,(f\,\mu)\,P_n$; and 0 the smallest which can be obtained by $\underline{\lim}\,(f\,\mu)\,P_n$. Consequently, [W.20.2.] and [W.4.1.], we have

$$(1)\qquad \overline{S}_C\, f\,\mu = \mu(C),\quad \underline{S}_C\, f\,\mu = 0,$$

which is a part of item 5).

W.20.2.1 a. Let us take $C' \leqq C$. Since T_n^0, T_n^1 are (DARS)-sequences, there exist partial complexes R_n^0, R_n^1 of T_n^0, T_n^1 respectively, which are both sequences for C'. We get

$$\lim\,(f\,\mu)\,(R_n^0) = \underline{S}_{C'}\, f\,\mu,$$

and

$$\lim\,(f\,\mu)\,(R_n^1) = \overline{S}_{C'}\, f\,\mu.$$

Now

$$(f\,\mu)\,R_n^0 = 0 \quad\text{and}\quad (f\,\mu)\,(R_n^1) = \mu\,R_n^1 \to \mu\,C'.$$

Consequently

$$\underline{S}_{C'}\, f\,\mu = 0,\quad \overline{S}_{C'}\, f\,\mu = \mu(C');$$

thus the second part of item 5) is proved. The item 5) is thus established.

W.20.2.1 b. By [U.20.10.], C is singular with respect to $\operatorname{som} Q_n^0$ and $\operatorname{som} Q_n^1$. This means, Def. [U.20.7.], that

$$\underline{\operatorname{Lim}}_{n\to\infty}\,(C\cdot \operatorname{som} Q_n^0) = O,\quad \overline{\operatorname{Lim}}_{n\to\infty}\,(C\cdot \operatorname{som} Q_n^0) = C,$$

$$\underline{\operatorname{Lim}}_{n\to\infty}\,(C\cdot \operatorname{som} Q_n^1) = O,\quad \overline{\operatorname{Lim}}_{n\to\infty}\,(C\cdot \operatorname{som} Q_n^1) = C.$$

By [U.20.7.2.] every measurable subset C' of C is also singular with respect to both Q_n^0 and Q_n^1. Hence

$$\underline{\operatorname{Lim}}_{n\to\infty}\,(C'\cdot \operatorname{som} Q_n^0) = O,\quad \overline{\operatorname{Lim}}_{n\to\infty}\,(C'\cdot \operatorname{som} Q_n^0) = C',$$

$$\underline{\operatorname{Lim}}_{n\to\infty}\,(C'\cdot \operatorname{som} Q_n^1) = O,\quad \overline{\operatorname{Lim}}_{n\to\infty}\,(C'\cdot \operatorname{som} Q_n^1) = C'.$$

Thus item 7) is established.

W.20.2.1 c. By [U.20.10.], $E - C$ is regular with respect to $\operatorname{som} Q_n^0$ and to $\operatorname{som} Q_n^1$. This means, Def. [U.20.7.], that

$$(1)\qquad \underline{\operatorname{Lim}}_{n\to\infty}\,[(E-C)\cdot \operatorname{som} Q_n^0] = \overline{\operatorname{Lim}}_{n\to\infty}\,[(E-C)\,\operatorname{som} Q_n^0],$$

and

$$(2)\qquad \underline{\operatorname{Lim}}_{n\to\infty}\,[(E-C)\cdot \operatorname{som} Q_n^1] = \overline{\operatorname{Lim}}_{n\to\infty}\,[(E-C)\,\operatorname{som} Q_n^1].$$

Applying [U.20.5.], we see that (1) and (2) are respectively equal to

$$(E - C) \cdot \underline{\mathrm{Lim}}_{n\to\infty} \mathrm{som} Q_n^0 = (E - C)\, \overline{\mathrm{Lim}}_{n\to\infty} \mathrm{som} Q_n^0,$$

$$(E - C) \cdot \underline{\mathrm{Lim}}_{n\to\infty} \mathrm{som} Q_n^1 = (E - C)\, \overline{\mathrm{Lim}}_{n\to\infty} \mathrm{som} Q_n^1.$$

By [U.20.2.2.], since both $\underline{\mathrm{Lim}}$ and $\overline{\mathrm{Lim}}$ are equal, Lim exist and $= \lim^\mu$. Consequently the expressions (1) and (2) are equal to $\lim^\mu[(E-C)\,\mathrm{som} Q_n^0]$, $\lim^\mu[(E - C)\,\mathrm{som} Q_n^1]$ respectively. Thus the first part of item 8) is proved. To get the second part, we apply [U.20.7.2.], stating that if $D \leqq E - C$, then D is also regular. Hence we get:

$$\underline{\mathrm{Lim}}_{n\to\infty} [D \cdot \mathrm{som} Q_n^0] = \overline{\mathrm{Lim}}_{n\to\infty} [D \cdot \mathrm{som} Q_n^0] = \lim_{n\to\infty}{}^\mu [D \cdot \mathrm{som} Q_n^0].$$

Similarly

$$\underline{\mathrm{Lim}}_{n\to\infty} [D \cdot \mathrm{som} Q_n^1] = \overline{\mathrm{Lim}}_{n\to\infty} [D \cdot \mathrm{som} Q_n^1] = \lim_{n\to\infty}{}^\mu [D \cdot \mathrm{som} Q_n^1].$$

By [U.20.5.] the above expressions are equal respectively to

$$D\, \underline{\mathrm{Lim}}_{n\to\infty} \mathrm{som} Q_n^0 = D\, \overline{\mathrm{Lim}}_{n\to\infty} \mathrm{som} Q_n^0, \qquad D\, \underline{\mathrm{Lim}}_{n\to\infty} \mathrm{som} Q_n^1 = D\, \overline{\mathrm{Lim}}_{n\to\infty} \mathrm{som} Q_n^1.$$

The item 8) is proved.

W.20.2.1 d. We have $A^0 \leqq E - C$. Hence, by what has been already proved in [W.20.2.1 b.] we have

$$(3) \qquad \underline{\mathrm{Lim}}_{n\to\infty} (A^0\, \mathrm{som} Q_n^0) = \overline{\mathrm{Lim}}_{n\to\infty} (A^0\, \mathrm{som} Q_n^0) = \mathrm{Lim}_{n\to\infty} (A^0\, \mathrm{som} Q_n^0)$$
$$= \lim_{n\to\infty}{}^\mu (A^0\, \mathrm{som} Q_n^0).$$

The above set equals,

$$A^0\, \underline{\mathrm{Lim}}_{n\to\infty} \mathrm{som} Q_n^0.$$

Hence, it equals $A^0 \cdot A^0 = A^0$, so the first relation of item 9) is established. Since $A' \leqq E - C$, we get, by a similar argument

$$(4) \quad \underline{\mathrm{Lim}}_{n\to\infty} (A^1\, \mathrm{som} Q_n^1) = \overline{\mathrm{Lim}}_{n\to\infty} (A^1\, \mathrm{som} Q_n^1) = \mathrm{Lim}_{n\to\infty} (A^1\, \mathrm{som} Q_n^1)$$
$$= \lim_{n\to\infty}{}^\mu (A^1\, \mathrm{som} Q_n^1) = A^1\, \mathrm{Lim}_{n\to\infty} \mathrm{som} Q_n^1 = A^1 \cdot A^1 = A^1,$$

by hyp. 7. By [U.20.5.] the expressions (3) and (4) are equal to

$$A^0\, \underline{\mathrm{Lim}}_{n\to\infty} \mathrm{som} Q_n^0 = A^0\, \overline{\mathrm{Lim}}_{n\to\infty} \mathrm{som} Q_n^0, \qquad A^1\, \underline{\mathrm{Lim}}_{n\to\infty} \mathrm{som} Q_n^1 = A^1\, \overline{\mathrm{Lim}}_{n\to\infty} \mathrm{som} Q_n^1,$$

so the first two equations of item 9) are proved.

W.20.2.1e. We have

$$A^0 \underline{\operatorname{Lim}}_{n\to\infty} \operatorname{som} Q_n^1 = A^0 \overline{\operatorname{Lim}}_{n\to\infty} \operatorname{som} Q_n^1 = \lim_{n\to\infty}{}^{\mu} (A^0 \operatorname{som} Q_n^1);$$

hence, by hyp. 7, this set equals $A^0 \cdot A^1$; hence, by the proved item 1), it equals O. Hence we get

$$A^0 \underline{\operatorname{Lim}}_{n\to\infty} \operatorname{som} Q_n^1 = A^0 \overline{\operatorname{Lim}}_{n\to\infty} \operatorname{som} Q_n^1 = \operatorname{Lim}_{n\to\infty} (A^0 \operatorname{som} Q_n^1) = \lim_{n\to\infty}{}^{\mu} (A^0 \operatorname{som} Q_n^1) = O;$$

and similarly we prove the last equation of item 9).

W.20.2.1f. Now, let $D^0 \leqq A^0$, $D^1 \leqq A^1$. We proceed, as before, replacing A^0 by D^0, and we get, by the proved relations in [W.20.2.1c.] and by hyp. 7, and [U.20.5.]:

$$\underline{\operatorname{Lim}}_{n\to\infty} (D^0 \operatorname{som} Q_n^0) = \overline{\operatorname{Lim}}_{n\to\infty} (D^0 \operatorname{som} Q_n^0) = \lim_{n\to\infty}{}^{\mu} (D^0 \operatorname{som} Q_n^0) = D^0 \underline{\operatorname{Lim}}_{n\to\infty} \operatorname{som} Q_n^0$$

$$= D^0 \overline{\operatorname{Lim}}_{n\to\infty} \operatorname{som} Q_n^0 = D^0 \cdot A^0 = D^0;$$

and similarly we prove the remaining equalities. We shall also rely on the equation $A^0 \cdot A^1 = O$. Thus the item 9) is settled.

W.20.2.1g. Since M_n^0, M_n^1 are sequences for A^0, A^1 respectively, we have

(1) $$|M_n^0, A^0| \to 0, \quad |M_n^1, A^1| \to 0.$$

It follows, that $|\operatorname{som} M_n^0 A^0, A^0 A^0| \to 0$; hence $\operatorname{som} M_n^0 \cdot A^0 \to \mu A^0$, $|\operatorname{som} M_n^0 \cdot A^1, A^0 A^1| \to 0$; hence, since $A^0 A^1 = O$, $\mu(\operatorname{som} M_n^0 \cdot A^1) \to 0$. Similarly we prove, that $\mu(\operatorname{som} M_n^1 \cdot A^0) \to 0$ and that $\operatorname{som} M_n^1 \cdot A^1 \to^{\mu} A^1$. The item 10a) is proved.

W.20.2.1h. Lemma. In general, for somata: if $a_n \to^{\mu} a$, $\lim^{\mu} b_n = O$, then $a_n - b_n \to^{\mu} a$.

W.20.2.1i. We have $M_n^0 \subseteqq Q_n$ and $Q_n = Q_n^0 \cup Q_n^1$. Hence

(2) $$M_n^0 = (Q_n^0 \cap M_n^0) \cup (Q_n^1 \cap M_n^0),$$

because $M_n^0 \subseteqq Q_n^0 \cup Q_n^1$. The terms in (2) are disjoint complexes. We have

$$\mu(Q_n^1 \cap M_n^0) = \mu(\operatorname{som} Q_n^1 \cdot \operatorname{som} M_n^0) = \mu(\operatorname{som} Q_n^1 \cdot A^0) +$$
$$+ \mu[\operatorname{som} Q_n^1 \cdot (\operatorname{som} M_n^0 - A^0)] - \mu[\operatorname{som} Q_n^1 \cdot (A^0 - \operatorname{som} M_n^0)].$$

Now, since $\operatorname{som} M_n^0 \to^{\mu} A^0$, we have

$$\mu(A^0 - \operatorname{som} M_n^0) \to 0, \quad \mu(\operatorname{som} M_n^0 - A^0) \to 0.$$

Consequently, by [W.20.2.1e.],

(3) $$\mu(Q_n^1 \cap M_n^0) \to 0, \quad \text{because} \quad \mu(\operatorname{som} Q_n^1 \cdot A^0) \to 0.$$

Since $|M_n^0, A^0| \to 0$ and since, by (2), $Q_n^0 \frown M_n^0 = M_n^0 \sim (Q_n^1 \frown M_n^0)$, it follows, by (3), that $|Q_n^0 \frown M_n^0, A^0| \to 0$. In a similar way we prove that

(4) $$|Q_n^1 \frown M_n^1, A^1| \to 0.$$

Since M_n^0 is a (DARS)-sequence for A^0, and $Q_n^0 \frown M_1^0 \subseteq M_1^0$, with (4), it follows that $Q_n^0 \frown M_n^0$ is a (DARS)-sequence for A^0. Similarly we prove that $Q_n^1 \frown M_n^1$ is a (DARS)-sequence for A^1. Thus item 10b) is established.

W.20.2.1j. We have

$$M_n^0 \sim Q_n^0 = M_n^0 \sim (Q_n^0 \frown M_n^0).$$

Since $\operatorname{som} M_n^0 \to^\mu A^0$, $\operatorname{som}(Q_n^0 \frown M_n^0) \to^\mu A^0$, it follows that

$$\operatorname{som}[M_n^0 \sim (Q_n^0 \frown M_n^0)] \to^\mu A^0 - A^0 = O.$$

Hence $\mu(M_n^0 \sim Q_n^0) \to 0$. Similarly we prove that $\mu(M_n^1 \sim Q_n^1) \to 0$. The item 10c) is established.

W.20.2.1k. Proof of item 10d). We have

(5) $$(f\,\mu)\,(M_N^0) = (f\,\mu)\,[(M_n^0 \frown Q_n^0) \smile (M_n^0 \sim Q_n^0)] =$$
$$= (f\,\mu)\,(M_n^0 \frown Q_n^0) + (f\,\mu)\,(M_n^0 \sim Q_n^0).$$

Since $f\,\mu$ is admissible, and by [W.20.2.1j.] $\mu(M_n^0 \sim Q_n^0) \to 0$, it follows that $(f\,\mu)\,(M_n^0 \sim Q_n^0) \to 0$. Concerning the first term in (5), we have $(f\,\mu)\,(M_n^0 \frown Q_n^0) = 0$, because for all bricks $q \in M_n^0 \frown Q_n^0$ we have $f(q) = 0$. Hence

(6) $$\lim_{n\to\infty} (f\,\mu)\,(M_n^0) = 0.$$

We have

$$(f\,\mu)\,(M_n^1) = (f\,\mu)\,[(M_n^1 \frown Q_n^1) \smile (M_n^1 \sim Q_n^1)]$$
$$= (f\,\mu)\,[M_n^1 \frown Q_n^1] + (f\,\mu)\,[M_n^1 \sim Q_n^1].$$

By [W.20.2.1j.] we have $\lim \mu(M_n^1 \sim Q_n^1) = 0$; hence, $f\,\mu$ being admissible,

$$(f\,\mu)\,(M_n^1 \sim Q_n^1) \to 0.$$

Now $(f\,\mu)\,(M_n^1 \frown Q_n^1) = \mu(M_n^1 \frown Q_n^1)$, because for all bricks $q \in M_n^1 \frown Q_n^1$ we have $f(q) = 1$.

Now, $\mu(M_n^1 \frown Q_n^1) \to \mu(A^1)$, because, by [W.20.2.1i.], $M_n^1 \frown Q_n^1$ is a (DARS)-sequence for A^1. Consequently

(7) $$\lim (f\,\mu) \cdot (M_n^1) = \mu\,A^1.$$

Thus the item 10d) is settled.

W.20.2.1 l. By [W.20.2.1 i.] $Q_n^0 \frown M_n^0$ is a sequence for A^0, and by [W.20.2.1 k.], $Q_n^1 \frown M_n^1$ is a sequence for A^1. We have

$$M_n^0 = (M_n^0 \frown Q_n^0) + (M_n^0 \sim Q_n^0),$$

where the terms are disjoint. It follows

$$(f\,\mu)\,(M_n^0) = (f\,\mu)\,(M_n^0 \frown Q_n^0) + (f\,\mu)\,(M_n^0 \sim Q_n^0). \tag{7.1}$$

Since, [W.20.2.1 j.], $\lim_{n\to\infty} (M_n^0 \sim Q_n^0) = O$, and since $f\,\mu$ is admissible, it follows that $(f\,\mu)\,(M_n^0 \sim Q_n^0) \to 0$. Consequently, from (1), (6) and [W.20.2.1 h.], $(f\,\mu)\,(M_n^0 \frown Q_n^0) \to 0$. We have $M_n^1 = (M_n^1 \frown Q_n^1) \smile (M_n^1 \sim Q_n^1)$ where the terms are disjoint. Since $\mu\,(M_n^1 \sim Q_n^1) \to 0$, it follows [W.20.2.1 h.], that $\lim (f\,\mu)\,(M_n^1) = \lim (f\,\mu)\,(M_n^1 \frown Q_n^1) = \mu\,(A^1)$. The item [W.20.2.1 e.] is established.

W.20.2.1 m. Since M_n^0 is a sequence for A^0, for which, by (6), $\lim (f\,\mu)\,(M_n^0) = 0$, and 0 is the lowest limit of $(f\,\mu)\,(P_n)$, where P_n are sequences for A^0, it follows that

$$\underline{S}_{A^0} f\,\mu = 0. \tag{8}$$

Since M_n^1 is a sequence for A^1, for which, by (7), $\lim (f\,\mu)\,(M_n^1) = \mu\,(A^1)$, and $\mu\,(A^1)$ is the greatest amount, which could be yielded by sequences for A^1, it follows that $\overline{S}_{A^1} f\,\mu = \mu\,(A^1)$. Thus the first part of item 6) is established.

W.20.2.1 n. Let $D^0 \leqq A^0$, $D^1 \leqq A^1$. If R_n is a sequence for D^1, then

$$(f\,\mu)\,(R_n) \leqq \sum_k f(r_{nk})\,\mu\,(r_{nk}) \leqq \sum_k \mu\,(r_{nk}) \to \mu\,D^1.$$

Hence

$$\overline{S}_{D^1} f\,\mu \leqq \mu\,D^1.$$

We have

$$\overline{S}_{D^1} f\,\mu + \overline{S}_{A^1 - D^1} f\,\mu = \overline{S}_{A^1} f\,\mu.$$

We cannot have

$$\overline{S}_{D^1} f\,\mu < \mu\,D^1,$$

because we would get

$$\overline{S}_{A^1} f\,\mu < \mu\,D^1 + \mu\,(A^1 - D^1) = \mu\,(A^1),$$

which contradicts (8) in [W.20.2.1 m.]. Thus we have

$$\overline{S}_{D^1} f\,\mu = \mu\,D^1.$$

We have

$$\underline{S}_{D^0} f\,\mu + \underline{S}_{A^0 - D^0} f\,\mu = \underline{S}_{A^0} f\,\mu = 0.$$

Hence, since both terms are non negative, it follows that $\underline{S}_{D^0} f\,\mu = 0$. Thus item 6) is proved.

W.20.2.1 p. To prove the only remaining item 10f), we shall use the following lemma:

Lemma. If

1. $|M_n, A| \to 0$,

2. M_n^A is the complex, composed of all bricks $q \in M_n$, for which $q \cdot A \neq O$,

then

$$|M_n^A, A| \to 0.$$

Proof. We have, putting

$$M_n^B =_{df} M_n \sim M_n^A,$$

$$A - \operatorname{som} M_n = A \cdot \operatorname{co} \operatorname{som} M_n = A \operatorname{co} \operatorname{som} (M_n^A \cup M_n^B),$$

where

$$A \cdot \operatorname{som} M_n^B = O.$$

Hence

$$A - \operatorname{som} M_n = A \cdot \operatorname{co} \operatorname{som} M_n^A \cdot \operatorname{co} \operatorname{som} M_n^B = A \operatorname{co} \operatorname{som} M_n^A \cdot A \operatorname{co} \operatorname{som} M_n^B$$
$$= (A - \operatorname{som} M_n^A) \cdot (A - \operatorname{som} M_n^B).$$

But

$$A - \operatorname{som} M_n^B = A - \operatorname{som} M_n^B \cdot A = A, \quad \text{since} \quad A \cdot \operatorname{som} M_n^B = O.$$

Hence

(9) $$A - \operatorname{som} M_n = (A - \operatorname{som} M_n^A) \cdot A = A - \operatorname{som} M_n^A.$$

On the other hand

(10) $$\operatorname{som} M_n^A - A \leqq \operatorname{som} M_n - A, \quad \text{since} \quad \operatorname{som} M_n^A \leqq \operatorname{som} M_n.$$

From (9) and (10) we get

$$(\operatorname{som} M_n^A - A) + (A - \operatorname{som} M_n^A) \leqq (\operatorname{som} M_n - A) + (A - \operatorname{som} M_n).$$

Since $\mu(\operatorname{som} M_n^A \dotplus A) \to 0$, it follows that $|\operatorname{som} M_n^A \dotplus A| \to 0$, so the lemma is proved.

W.20.2.1 q. By Lemma [W.20.2.1 p.] we have $|R_n^0, A^0| \to 0$. Since $R_n^0 \subseteqq M_n^0$, we see, that R_n^0 is a (DARS)-sequence for A^0.

Similarly, R_n^1 is a (DARS)-sequence for A^1. So 10f) is proved. All items of the theorem [W.20.2.] have been proved.

W.20.2.2. In the preceeding Lemma [W.20.2.1 p] we have admitted the hypotheses 3 and 4, which we would like to get rid of. This can be done. We shall prove the following:

Lemma. Let us admit Hyp. $\boldsymbol{H}_2$. If

1. E is a set of traces,

2. Q_n is a sequence for E,

3. $f(a)$ admits the values 0, 1 only,

then from Q_n a subsequence $Q_{\alpha(\dot{n})}$ can be extracted, such that, if we put

$$Q^0_{\alpha(n)} =_{df} \{\beta \mid \beta \in Q_{\alpha(n)}, f(\beta) = 0\},$$

$$Q'_{\alpha(n)} =_{df} \{\beta \mid \beta \in Q_{\alpha(n)}, f(\beta) = 1\} \quad \text{for} \quad n = 1, 2, \ldots,$$

and if we put

$$A^0 =_{df} \underline{\operatorname{Lim}}_{n\to\infty} Q^0_{\alpha(n)}, \qquad B^0 =_{df} \overline{\operatorname{Lim}}_{n\to\infty} Q^0_{\alpha(n)},$$

$$A' =_{df} \underline{\operatorname{Lim}}_{n\to\infty} Q'_{\alpha(n)}, \qquad B' =_{df} \overline{\operatorname{Lim}}_{n\to\infty} Q'_{\alpha(n)},$$

$C =_{df} B' - A' = B^0 - A^0$, then the items of the thesis in Lemma [W.20.2.1p], 1), 2), 3), 5), 6) hold true. We also have:

4′) there exists a sequence $T^0_{\dot{n}}$ for C, such that $T^0_n \subseteqq \bigcup_{m=1}^{\infty} Q^0_m$ for all n, and there exists a sequence $T'_{\dot{n}}$ for C, such that $T'_n \subseteqq \bigcup_{m=1}^{\infty} Q'_m$ for all n.

We have the relations 7), 8), 9), where Q^0_n, Q'_n are replaced by $Q^0_{\alpha(n)}$, $Q'_{\alpha(n)}$. If M^0_n, M'_n are sequences for A^0, A' respectively with $M^0_n \subseteqq Q_{\alpha(n)}$, $M'_n \subseteqq Q_{\alpha(n)}$, then the items 10a, 10b, 10c, 10d, 10e hold true, if we replace Q^0_n, Q'_n by $Q^0_{\alpha(n)}$, $Q'_{\alpha(n)}$ respectively. The item 10f) holds true unchanged.

W.20.2.2a. Proof. Let $Q_{\dot{n}}$ be a sequence for E. There exist $r(n)$, such that

$$|Q_{r(n)}, E| \leqq \frac{1}{2^{n+1}} \quad \text{for} \quad n = 1, 2, \ldots;$$

$Q_{r(\dot{n})}$ is a sequence for E.

We may remark, that for every subsequence $\lambda(\dot{n})$ of $\dot{n}$ we also have, since

(1) $\qquad \lambda(n) \geqq n, \quad$ the inequality $\quad |Q_{r\lambda(n)}, E| \leqq \frac{1}{2^{n+1}}$.

Applying Theorem [U.10.6.], we can find $s(\dot{n})$ and a sequence $P_{rs(n)}$ for E, such that

(1.1) $\qquad P_{rs(n)} \subseteqq Q_{rs(n)}, \quad n = 1, 2, \ldots,$

and where $P_{rs(n)}$ satisfies the condition of non overlapping. Since $P_{rs(n)}$ is a sequence for E, we can find $t(\dot{n})$ such that

(2) $\qquad |P_{rst(n)}, E| \leqq \frac{1}{2^{n+1}}$.

The sequence $P_{rst(n)}$ also satisfies the condition of non-overlapping (see [U.10.1.1.]). Since, by (1),

(2.1) $\qquad |Q_{rst(n)}, E| \leqq \frac{1}{2^{n+1}}$,

it follows from (2):

(3) $$|P_{rst(n)}, Q_{rst(n)}| \leqq \frac{1}{2^n},$$

and from (1.1):

(4) $$P_{rst(n)} \subseteqq Q_{rst(n)}.$$

By (4) we get from (3)

(5) $$\mu(Q_{rst(n)} \sim P_{rst(n)}) \leqq \frac{1}{2^n}, \quad (n = 1, 2, \ldots).$$

Now

(6) $$P_{rst(\dot{n})} \text{ and } Q_{rst(\dot{n})}$$

are both sequences for E. With (2), (6), and since $P_{rst(n)}$ satisfies the condition of non overlapping, we are in the conditions of Lemma [W.20.2.1 p] for the sequence $P_{rst(n)}$ for E. However, we like to have the items of [W.20.2.] valid for $Q_{rst(n)}$.

W.20.2.2b. Put $\alpha(n) =_{df} r\,s\,t(n)$ for $n = 1, 2, \ldots$ Before going farther, we prove that

(7) $$Q^0_{\alpha(n)} \sim P^0_{\alpha(n)} \subseteqq Q_{\alpha(n)} \sim P_{\alpha(n)} \quad \text{and} \quad Q'_{\alpha(n)} \sim P'_{\alpha(n)} \subseteqq Q_{\alpha(n)} \sim P_{\alpha(n)},$$

where

(8) $$Q^0_{\alpha(n)} =_{df} \{q_{\alpha(n),k} | q_{\alpha(n),k} \in Q_{\alpha(n),k}, f(q_{\alpha(n),k}) = 0\},$$
$$Q'_{\alpha(n),k} =_{df} \{q_{\alpha(n),k} | q_{\alpha(n),k} \in Q_{\alpha(n),k}, f(q_{\alpha(n),k}) = 1\},$$

(9) $$P^0_{\alpha(n),k} =_{df} \{p_{\alpha(n),k} | p_{\alpha(n),k} \in P_{\alpha(n),k}, f(p_{\alpha(n),k}) = 0\},$$
$$P'_{\alpha(n),k} =_{df} \{p_{\alpha(n),k} | p_{\alpha(n),k} \in P_{\alpha(n),k}, f(p_{\alpha(n),k}) = 1\}.$$

In the following proof of (7), we shall omit the indices $\alpha(n)$ and use somatic symbols for operation — this for simplifying the symbolism.

Since $P^0 \leqq P \leqq Q$, by (4), and since $Q = Q^0 + Q'$, we have

$$P^0 = P^0(Q^0 + Q') = P^0 Q^0 + P^0 Q'.$$

Hence, as $P^0 \cdot Q' = O$, we get

(10) $$P^0 \leqq Q^0.$$

Similarly we get

(11) $$P' \leqq Q'.$$

Now, we have $P Q^0 = (P^0 + P') Q^0 = P^0 Q^0 + P' Q^0 = P^0 Q^0$, because $P' Q^0 = O$. Hence $P Q^0 = P^0 Q^0$, and then, since $P^0 \leqq Q^0$, also

(12) $$P^0 = P Q^0.$$

We also have $Q^0 \leqq Q$; hence

(13) $$Q^0 = Q Q^0.$$

From (12) and (13), it follows

$$Q^0 - P^0 = Q\,Q^0 - P\,Q^0 = (Q - P)\,Q^0 \leqq Q - P.$$

In a similar way we prove that $Q' - P' < Q - P$, so (7) is proved.

W.20.2.2c. Since by (5),

$$\mu(Q_{\alpha(n)} - P_{\alpha(n)}) \leqq \frac{1}{2^n},$$

we get from (7),

$$\mu(Q^0_{\alpha(n)} \sim P^0_{\alpha(n)}) \leqq \frac{1}{2^n}, \quad \mu(Q'_{\alpha(n)} \sim P'_{\alpha(n)}) \leqq \frac{1}{2^n}.$$

Having this, we apply Lemma [U.20.9.1.], getting

$$(13.1)\quad \begin{aligned} &\underline{\mathrm{Lim}}\,\mathrm{som}\,Q^0_{\alpha(n)} = \underline{\mathrm{Lim}}\,\mathrm{som}\,P^0_{\alpha(n)}, \quad \overline{\mathrm{Lim}}\,\mathrm{som}\,Q^0_{\alpha(n)} = \overline{\mathrm{Lim}}\,\mathrm{som}\,P^0_{\alpha(n)},\\ &\underline{\mathrm{Lim}}\,\mathrm{som}\,Q'_{\alpha(n)} = \underline{\mathrm{Lim}}\,\mathrm{som}\,P'_{\alpha(n)}, \quad \overline{\mathrm{Lim}}\,\mathrm{som}\,Q'_{\alpha(n)} = \overline{\mathrm{Lim}}\,\mathrm{som}\,P'_{\alpha(n)}, \end{aligned}$$

which implies

$$(14)\quad \begin{aligned} A^0 &= \underline{\mathrm{Lim}}\,\mathrm{som}\,Q^0_{\alpha(n)}, \quad B^0 = \overline{\mathrm{Lim}}\,\mathrm{som}\,Q^0_{\alpha(n)},\\ A' &= \underline{\mathrm{Lim}}\,\mathrm{som}\,Q'_{\alpha(n)}, \quad B' = \overline{\mathrm{Lim}}\,\mathrm{som}\,Q'_{\alpha(n)}. \end{aligned}$$

W.20.2.2d. Let us peruse the proofs of various items of Lemma [W.20.2.]. All of them hold true, if we consider the sequences $P_{\alpha(n)}$, $P^0_{\alpha(n)}$, $P'_{\alpha(n)}$. We must prove that these sequences can be replaced by $Q_{\alpha(n)}$, $Q^0_{\alpha(n)}$, $Q'_{\alpha(n)}$ respectively.

W.20.2.2e. The items 1), 2), 3) result from (14).

W.20.2.2f. The item 4) has the form

$$T^0_n \subseteqq \bigcup_{m=1}^{\infty} P^0_{\alpha(n)}, \quad T'_n \subseteqq \bigcup_{m=1}^{\infty} P'_{\alpha(n)},$$

but since

$$P^0_{\alpha(n)} \subseteqq Q^0_{\alpha(n)}, \quad P'_{\alpha(n)} \subseteqq Q'_{\alpha(n)},$$

we have

$$T^0_n \subseteqq \bigcup_{m=1}^{\infty} P^0_{\alpha(m)} \subseteqq \bigcup_{m=1}^{\infty} Q^0_{\alpha(m)} \subseteqq \bigcup_{m=1}^{\infty} Q^0_m,$$

and similarly $T'_n \subseteqq \bigcup_{m=1}^{\infty} Q'_m$; hence (4′) is proved.

W.20.2.2g. Item 5) does not involve any sequences of complexes, so it holds true.

W.20.2.2h. The same for 6).

W.20.2.2i. We have $\underline{\mathrm{Lim}}_{n\to\infty}(C\,\mathrm{som}\,P^0_{\alpha(n)}) = O$; hence, by [U.20.5.], $C\,\underline{\mathrm{Lim}}_{n\to\infty}\,\mathrm{som}\,P^0_{\alpha(n)} = O$. Since, by (3), $|P^0_{\alpha(n)}, Q^0_{\alpha(n)}| \leqq \frac{1}{2^n}$, $(n = 1, 2, \ldots)$, we get $\underline{\mathrm{Lim}}\,\mathrm{som}\,P^0_{\alpha(n)} = \underline{\mathrm{Lim}}\,\mathrm{som}\,Q^0_{\alpha(n)}$, and then $C \cdot \underline{\mathrm{Lim}}_{n\to\infty}\,\mathrm{som}\,Q^0_{\alpha(n)} = O$, which gives $\underline{\text{Lim}}_{n\to\infty}(C \operatorname{som} Q^0_{\alpha(n)}) = O$. Similarly we prove the analogous remaining equalities 7) of [W.20.2.], with Q^0_n, Q'_n changed into $Q^0_{\alpha(n)}$, $Q'_{\alpha(n)}$.

W.20.2.2j. Items, analogous to 8) and 9) in [W.20.2.], result in a similar way.

W.20.2.2k. Concerning items, analogous to [W.20.2., 10)] they are true, if we suppose that $M^0_n \subseteqq P_{\alpha(n)}$, $M'_n \subseteqq P_{\alpha(n)}$, and that M^0_n, M'_n are sequences for A^0, A' respectively. Now, let us suppose, that $\bar{M}^0_n \subseteqq Q_{\alpha(n)}$, $\bar{M}'_n \subseteqq Q_{\alpha(n)}$, and that $\bar{M}^0_n$, $\bar{M}'_n$ are sequences for A^0, A' respectively.

Looking at [W.20.2.1g.] in the proof of Lemma [W.20.2.], we see that no change is needed, if M^0_n, M'_n are replaced by $\bar{M}^0_n$, $\bar{M}'_n$. Thus the item 10a) is true in our case.

W.20.2.2l. The same can be said of 10b), 10c), 10d), 10e), so Lemma [W.20.2.] is established.

W.20.2.3. Def. If Q_n is a sequence for E, then any subsequence $Q_{\alpha(n)}$ of Q_n, for which the whole thesis of Lemma [W.20.2.] holds true, will be termed *α-subsequence of* Q_n.

The last Lemma [W.20.2.] says, that from every sequence Q_n for E, an α-subsequence for E can be extracted.

W.20.3. Theorem. Let us admit Hyp. $\boldsymbol{H}_2$. If

1. E is a set of traces,
2. $f(a)$ is a field of real numbers, admitting the values 0, 1 only,
3. $f(a)\,\mu(a)$ is (DARS)-summable on E,

then there exist well determined sets A^0, A' of traces such that

1) $S_{A^0} f\mu = 0$, $S_{A'} f\mu = \mu(A')$;

2) if Q_n is any sequence for E, and we put

$$Q^0_n =_{df} \{q \mid q \in Q_n,\ f(q) = 0\}, \qquad Q'_n =_{df} \{q \mid q \in Q_n,\ f(q) = 1\},$$

then

1a) Q^0_n is a sequence for A^0; Q'_n is a sequence for A';

2b) $A^0 + A' = E$, $A^0 \cdot A' = O$;

3) if $\Phi(x)$ defined on E, is the function of traces attached to $f(a)$, then

$$\Phi(x) =^{\mu} 0 \quad \text{for} \quad x \in A^0 \quad \text{and} \quad \Phi(x) =^{\mu} 1 \quad \text{for} \quad x \in A'.$$

Proof. Consider a subsequence $Q_{s(n)}$ of Q_n. By Lemma [W.20.2.2.] there exists an α-subsequence $Q_{s\alpha(n)}$ of $Q_{s(n)}$ (see [W.20.2.3.]). Consider the corresponding sequences $Q^0_{s\alpha(n)}$, $Q'_{s\alpha(n)}$ and the sets A^0, A', C, B^0, B', where $C = B^0 - A^0 = B' - A'$.

Notice, that $\mu C = 0$, because if not, we would have, [W.20.2.2.],

$$\overline{S}_C f\mu = \mu(C) > 0, \qquad \underline{S}_C f\mu = 0.$$

Hence $f\mu$ would be not summable on C. This is, however, impossible, because $f\mu$ is supposed to be summable on E, and we have $C \leqq E$.

Since $C = O$, it follows that $A^0 = B^0$, $A' = B'$, i.e.,

$$(1)\qquad \begin{aligned} A^0 &= \underline{\operatorname{Lim}}_{n\to\infty} \operatorname{som} Q^0_{s\alpha(n)} = \overline{\operatorname{Lim}}_{n\to\infty} \operatorname{som} Q^0_{s\alpha(n)}, \\ A' &= \underline{\operatorname{Lim}}_{n\to\infty} \operatorname{som} Q'_{s\alpha(n)} = \overline{\operatorname{Lim}}_{n\to\infty} \operatorname{som} Q'_{s\alpha(n)}. \end{aligned}$$

Now let us choose another subsequence $Q_{t(n)}$ of Q_n, and let us do the same with it. Find an α-subsequence $Q_{t\beta(n)}$ of $Q_{t(n)}$, and consider the corresponding sets A^0_1, A'_1, C_1, B^0_1, B'_1, where $C_1 = B^0_1 - A^0_1 = B'_1 - A'_1$. Since $C_1 =^\mu O$, it follows that

$$A^0_1 = B^0_1, \quad A'_1 = B'_1.$$

We have

$$E = A^0 + A', \quad A^0 \cdot A' = O,$$

and

$$E = A^0_1 + A'_1, \quad A^0_1 \cdot A'_1 = O.$$

Consider the sets

$$A^0 A^0_1,\ A^0 A'_1,\ A' A^0_1,\ A' A'_1.$$

I say that

$$A^0 \cdot A'_1 = O.$$

Indeed, suppose that $A^0 A'_1 \neq O$; then, by [W.20.2.], $\underline{S}_{A^0 A'_1} f \mu = 0$, because $A^0_1 A'_1 \leqq A^0$, and $\overline{S}_{A^0 A'_1} f \mu = \mu(A^0 A'_1) > 0$. We have $\mu(A^0 A'_1) > 0$, because the measure μ is effective on E. Hence $\underline{S}_{A^0 A'_1} f \mu < \overline{S}_{A^0 A'_1} f \mu$, which is impossible, because $f \mu$ is (DARS)-summable on E, hence also on $A^0 A'_1$. Thus $A^0 A''_1 = O$. In a similar way we prove that $A' A^0_1 = O$. We have $A^0 = A^0 \cdot (A^0_1 + A'_1)$, because $A^0_1 + A'_1 = E$ and $A^0 \leqq E$. Hence $A^0 = A^0 A^0_1 + A^0 A'_1$. Since $A^0 A'_1 = O$, it follows that $A^0 = A^0 A^0_1$; hence $A^0 \leqq A^0_1$.

In a similar way we prove that $A^0_1 \leqq A^0$. Consequently $A^0 = A^0_1$. It follows $A' = A'_1$. Hence we get

$$\begin{aligned} A^0 &= \underline{\operatorname{Lim}}_{n\to\infty} \operatorname{som} Q^0_{t\beta(n)} = \overline{\operatorname{Lim}}_{n\to\infty} \operatorname{som} Q^0_{t\beta(n)}, \\ A' &= \underline{\operatorname{Lim}}_{n\to\infty} \operatorname{som} Q'_{t\beta(n)} = \overline{\operatorname{Lim}}_{n\to\infty} \operatorname{som} Q'_{t\beta(n)}. \end{aligned}$$

Thus from every subsequence $Q^0_{k(n)}$ of Q^0_n another subsequence $Q^0_{kl(n)}$ can be extracted, such that

$$\underline{\operatorname{Lim}}_{n\to\infty} \operatorname{som} Q^0_{kl(n)} = \overline{\operatorname{Lim}}_{n\to\infty} \operatorname{som} Q^0_{kl(n)} = A^0.$$

Hence, by the theorem by the author (2), we get

$$\lim_{n\to\infty}{}^{\mu} \operatorname{som} Q^0_n = A^0.$$

In a similar way we prove that

$$\lim_{n\to\infty}{}^{\mu} \operatorname{som} Q'_n = A'.$$

It follows that Q^0_n is a sequence for A^0, and Q'_n is a sequence for A'. To prove the item 3) of our thesis, put

$$\Phi(x) =_{df} 0 \quad \text{for} \quad x \in A^0, \qquad \Phi(x) =_{df} 1 \quad \text{for} \quad x \in A'.$$

We have for any subset F of E: $\int\limits_F \Phi(x)\, d\mu = \mu(A'F)$. Since also $S_F f\mu = \mu(A'F)$ for every F, and since the function attached to $f\mu$ is unique, it follows that $\Phi(x)$ is the function attached to $f\mu$. The theorem is proved.

W.20.3.1. Remark. It is not true, that every field $f(\dot{a})\,\mu(\dot{a})$ of real numbers, where $f(a) = 0$ or 1, is (DARS)-summable. Indeed, it is not true that for every sequence N_n of complexes $\lim\limits_{n\to\infty}{}^{\mu} N_n$ exists. An explicit example is this:

Let $\boldsymbol{G}$ be the tribe of all Lebesgue's measurable subsets in $(0, 1\rangle$ taken modulo null sets, and let $\boldsymbol{B}$ be composed of $\emptyset$, $(0, 1\rangle$, and of all intervals $\left(\frac{p}{2^n}, \frac{p+1}{2^n}\right\rangle$, where $p = 0, 1, \ldots, 2^{n-1}$, and $n = 1, 2, \ldots$ Put $f\left[\left(\frac{p}{2^n}, \frac{p+1}{2^n}\right\rangle\right] = 0$ or 1, according to whether n is even or odd. Let μ denote the Lebesgue'an measure. We have on $(0, 1\rangle$ $\overline{S} f\mu = 1$, $\underline{S} f\mu = 0$.

W.20.4. Let us consider the case, where $f(a)\,\mu(a)$ is not (DARS)-summable on a set E of traces.

Lemma. We admit Hyp. $\boldsymbol{H}_2$. If

1. E is a set of traces,
2. $f(a)$ is a field of real numbers, admitting the values $0, 1$ only,
3. $F \leqq E$, $\mu F > 0$,
4. $f(a)\,\mu(a)$ is not (DARS)-summable on any subset F' of F with $\mu F' > 0$,

then

$$\overline{S}_{F'} f\mu = \mu(F'), \quad \underline{S}_{F'} f\mu = 0$$

on any set F', where $F' \leqq F$.

W.20.4a. Proof. Let Q_n be a sequence for F, such that

$$\lim_{n\to\infty} (f\mu)\cdot(Q_n) = \overline{S}_F f\mu,$$

(see [W.4.1.]). Take any α-sequence $Q_{s(\dot{n})}$ of $Q_{\dot{n}}$, Def. [W.20.2.3.]. We have

$$\lim_{n\to\infty} (f\mu)\cdot(Q_{s(n)}) = \overline{S}_F f\mu. \tag{1}$$

Consider the sets A^0, A', C, generated by $Q_{s(n)}$. These sets are disjoint, and $A^0 + C + A' = F$.

W.20.4b. There are two cases to be considered:

1) $\mu\, C = 0$,

2) $\mu\, C > 0$.

In the second case, $\mu\, C > 0$, we have, by [W.20.2.],

$$\overline{S}_C\, f\, \mu = \mu(C), \quad \mu\, C > 0.$$

W.20.4c. Take the first case, $\mu\, C = 0$. We have

$$A^0 + A' = F. \tag{1.1}$$

Define as usually,

$$Q^0_{s(n)} =_{df} \{q \mid q \in Q_{s(n)},\, f(q) = 0\},$$

and take any sequence M^0_n for A^0, such that $M^0_n \subseteqq Q^0_{s(n)}$. By [W.20.2.], $Q^0_{s(n)} \frown M^0_n$ is a sequence for A^0, and we have

$$(f\, \mu)\,(M^0_n \frown Q^0_{s(n)}) \to 0. \tag{2}$$

Now $Q^0_{s(n)} \frown M^0_n$ is a partial complex of $Q_{s(n)}$, and $Q^0_{s(n)} \frown M^0_n$ is a sequence for A^0. If we consider (1) and apply Theorem [W.5.1.], we get

$$\lim\,(f\, \mu)\,(Q^0_{s(n)} \frown M^0_n) = \overline{S}_{A^0}\, f\, \mu.$$

Hence, by (1), $\overline{S}_{A^0}\, f\, \mu = 0$. Since $0 \leqq \underline{S}_{A^0}\, f\, \mu \leqq \overline{S}_{A^0}\, f\, \mu = 0$, we see, that $f\, \mu$ is summable on A^0. This is, however, impossible, whenever $\mu\, A^0 > 0$, because of Hyp. 4. Thus $\mu\, A^0 = 0$. It follows, by (1.1), that $A' = F$. By [W.20.2.] we have $\overline{S}_{A'}\, f\, \mu = \mu(A')$; we also have $\mu(A') > 0$; this by Hyp. 3. Since $A' \leqq F$, we see that, whatever the case may be, there always exists a set D, where $\mu\, D > 0$, $D < F$, and such that

$$\overline{S}_D\, f\, \mu = \mu(D). \tag{3}$$

W.20.4d. We shall prove that

$$\overline{S}_F\, f\, \mu = \mu(F). \tag{4}$$

If, in (3), $D_1 = F$, the statement (4) is proved. If $D_1 < F$, consider the set $D_2 =_{df} F = D_1$. We have $\mu\, D_2 > 0$, $D_2 \leqq F$, and; by Hyp. 4, the field $f\, \mu$ is not summable on any part of D_2 with positive measure. Hence we can apply to D_1 our preceeding argument, and then continue the process. Thus we shall define a well ordering of disjoint subsets of F, every one with positive measure:

$$D_1, D_2, \ldots, D_\omega, D_{\omega+1}, \ldots, D_\lambda, \ldots,$$

as follows:

Suppose that $\lambda < \Omega$, and that we have already found all sets D_r with $r < \lambda$. If $\lambda - 1$ exists, and $\sum\limits_{r<\lambda} D_r = F$, the process stops. But if

$\sum_{r<\lambda} D_r < F$, we consider the set $F - \sum_{r<\lambda} D_r$, and apply our argument, getting a set D_λ, such that $\mu D_\lambda > 0$, and such that $f\mu$ is not summable on any part of D_λ with positive measure (we also use [W.9.]). If $\lambda - 1$ does not exist, and $\sum_{r<\lambda} D_r = F$, the process stops. If this sum is $< F$, we find, as before, a set D_λ, where $\mu D_\lambda > 0$ and where $D_\lambda \leqq F - \sum_{r<\lambda} D_r$.

Since the measure is bounded, the process can not be performed for all $\lambda < \Omega$. Hence it must stop somewhere for an ordinal, which is less that Ω, say for an ordinal λ'. We get

$$\sum_{r \leqq \lambda'} D_r = F.$$

Now, the upper sum is a denumerably additive set-function. Hence, since all D_r are disjoint, and their cardinal is $\leqq \aleph_0$, we see that

$$\overline{S}_F f\mu = \mu F. \tag{5}$$

Thus a part of the thesis is established.

W.20.4e. Now, we are going to prove that

$$\underline{S}_F f\mu = 0.$$

Let P_n be a sequence for F, such that

$$\lim_{n\to\infty} (f\mu)(P_n) = \underline{S}_F f\mu.$$

Take any α-sequence $P_{t(n)}$ of P_n. We have

$$\lim_{n\to\infty} (f\mu)(P_{t(n)}) = \underline{S}_F f\mu. \tag{6}$$

Consider the sets A_1^0, A_1^1, C_1, generated by $P_{t(n)}$. These sets are disjoint, and we have $A_1^0 + C_1 + A_1^1 = F$. We consider, as before, two cases:

1) $\mu C_1 = 0$,
2) $\mu C_1 > 0$.

In the second case we have, by [W.20.2.],

$$\underline{S}_{C_1} f\mu = 0, \quad \mu C_1 > 0. \tag{7}$$

In the first case we have

$$A_1^0 + A_1^1 = F. \tag{8}$$

Define

$$P_{t(n)}^1 =_{df} \{p \mid p \in P_{t(n)}, f(p) = 1\},$$

and take any sequence M_n^1 for A_1^0, such that $M_n^1 \subseteqq P_{t(n)}$. By [W.20.2.], $P_{t(n)}^1 \frown M_n^1$ is a sequence for A_1^1. Hence we have

$$(f\mu)(M_n^1 \frown P_{t(n)}^1) \to \mu(A_1^1). \tag{9}$$

Now $M_n^1 \frown P_{t(n)}^1$ is a partial complex of $P_{t(n)}$, and $M_n^1 \frown P_{t(n)}^1$ is a sequence for A_1^1. Hence, by (6),

$$\lim (f\,\mu)\,(M_n^1 \frown P_{t(n)}^1) = \underline{S}_{A_1^1} f\,\mu,$$

and then, by (9),

$$\underline{S}_{A_1^1} f\,\mu = \mu(A_1^1).$$

Since we have, [W.20.2.], $\overline{S}_{A_1^1} f\,\mu = \mu(A_1^1)$, it follows that $\underline{S}_{A_1^1} = \overline{S}_{A_1^1}$, which contradicts Hyp. 4, whenever $\mu\,A_1^1 > 0$. Hence $\mu\,A_1^1 = 0$ and then, by (8), $A_1^0 = F$. By [W.20.2.] we have

$$\underline{S}_{A_1^0} f\,\mu = 0, \quad \mu\,A_1^0 > 0. \tag{10}$$

From (9) and (10) it follows that, whatever the case may be, there exists a subset D_1 of F with $\mu\,D_1 > 0$ and $\underline{S}_{D_1} f\,\mu = 0$. Now, we apply transfinite induction, similar to that exhibited before, getting disjoint sets $D_1, D_2, \ldots, D_\omega, \ldots, D_\alpha, \ldots$; and we use an argument similar to that in [W.20.4d.]. Thus we get

$$\underline{S}_F f\,\mu = 0.$$

The theorem is proved.

W.20.5. Theorem. Let us admit Hyp. $\boldsymbol{H}_2$. If

1. E is a set of traces, $\mu\,E > 0$,
2. $f(a)$ a field of real numbers, admitting the values 0, 1 only,
3. $f(a)$ is not summable on E,

then there exist well determined sets A^0, A', C, such that

1) they are disjoint; with $A^0 + A' + C = E$;
2) $f(a)\,\mu(a)$ is summable on A^0 with $S_{A^0} f\,\mu = 0$;
3) $f(a)\,\mu(a)$ is summable on A' with $S_{A'} f\,\mu = \mu(A')$;
4) $f(a)\,\mu(a)$, (if $\mu\,C > 0$) is not summable on C, and we have

$$\underline{S}_C f\,\mu = 0, \quad \overline{S}_C f\,\mu = \mu(C);$$

5) we have for the functions $\underline{\Phi}(x)$, $\overline{\Phi}(x)$ of traces, attached to $f\,\mu(a)$, the relations:

$$\underline{\Phi}(x) =^{\mu} 0 \quad \text{on} \quad A^0 + C, \quad \text{and} \quad \underline{\Phi}(x) =^{\mu} 1 \quad \text{on} \quad A',$$

$$\overline{\Phi}(x) =^{\mu} 0 \quad \text{on} \quad A^0, \quad \text{and} \quad \overline{\Phi}(x) =^{\mu} 1 \quad \text{on} \quad C + A'.$$

Proof. The set E is the sum of two disjoint sets, such that $E = E' + E''$, $\mu\,E'' > 0$, $f\,\mu$ is summable on E', and not summable on any part of E'' with positive measure.

Indeed, we have the functions of traces $\overline{\Phi}(x)$, $\underline{\Phi}(x)$ with

$$\int_F \overline{\Phi}(x)\,d\mu = \overline{S}_F f\,\mu, \quad \int_F \underline{\Phi}(x)\,d\mu = \underline{S}_F f\,\mu,$$

for any measurable subset F of E.

We have $\underline{S}_F f\mu \leqq \overline{S}_F f\mu$; hence $\underline{\Phi}(x) \leqq^\mu \overline{\Phi}(x)$ for all $x \in E$. Since both $\underline{\Phi}(x)$, $\overline{\Phi}(x)$ are measurable, therefore the set $E =_{df} \{x \mid \underline{\Phi}(x) = \overline{\Phi}(x)\}$ is measurable. Hence

$$E'' =_{df} E - E' = \{x \mid \underline{\Phi}(x) <^\mu \overline{\Phi}(x)\}$$

is also measurable. We have $E' + E'' = E$, $E' \cdot E'' = O$. We have, putting $\Phi(x) = \underline{\Phi}(x) = \overline{\Phi}(x)$ for all $x \in E'$, $\int\limits_{E'} \Phi(x)\, d\mu = \overline{S}_{E'} f\mu = \underline{S}_{E'} f\mu$, so $f\mu$ is summable on E'. We also have $\int\limits_F \underline{\Phi}(x)\, d\mu < \int\limits_F \overline{\Phi}(x)\, d\mu$ for all $F \leqq E''$ with $\mu F > 0$. Hence $\underline{S}_F f\mu < \overline{S}_F f\mu$; so $f\mu$ is not summable on any subset F of E with $\mu(F) > 0$. Having these sets E', E'', we can apply Theorem [W.20.4.] to E''. We get $\overline{S}_{E''} f\mu = \mu(E'')$, $\underline{S}_{E''} f\mu = 0$. Concerning E', we can apply Theorem [W.20.3.]. We get two subsets A^0, A' of E', such that $A^0 \cdot A' = 0$, $A^0 + A' = E'$, where $S_{A^0} f\mu = 0$, $S_{A'} f\mu = \mu(A')$; and even for all $D^0 \leqq A^0$, $D' \leqq A'$ we have $S_{D^0} f\mu = 0$, $S_{D^0} f\mu = \mu(D')$.

Thus, if we put $C = E''$, we get the parts 1), 2), 3), 4) of the thesis, so the theorem is proved, since the sets A^0, A', C are well determined, hence unique.

W.20.6. The Theorems [W.20.3.] and [W.20.5.] imply the following one:

Theorem. Under Hyp. $\boldsymbol{H}_2$, if E is a set of traces, $f(\dot{a})$ a field of real numbers, admitting only the values 0 and 1, then there exists a unique partition of E into three disjoint measurable sets of traces A^0, $A^{0'}$, A' such that

1) on A^0 the field $f\mu$ is summable with $S_{A^0} f\mu = 0$,

2) on A' the field $f\mu$ is summable with $S_{A^0} f\mu = \mu(A')$,

3) on $A^{0'}$ we have $\overline{S}_{A^{0'}} f\mu = \mu(A^{0'})$, $\underline{S}_{A^{0'}} f\mu = 0$. Some of these sets A^0, A', $A^{0'}$ and even all may be null-sets. Notice that the above is true, even if $\mu E = 0$,

4) concerning the functions $\overline{\Phi}(x)$, $\underline{\Phi}(x)$, attached to $f(a)\mu(a)$, we have

$$\overline{\Phi}(x) =^\mu 0 \quad \text{on} \quad A^0, \qquad \overline{\Phi}(x) =^\mu 1 \quad \text{on} \quad A^{0'} + A',$$

$$\underline{\Phi}(x) =^\mu 0 \quad \text{on} \quad A^0 + A^{0'}, \qquad \underline{\Phi}(x) =^\mu 1 \quad \text{on} \quad A',$$

whenever the corresponding sets are not μ-null sets.

W.20.6.1. Def. The sets A^0, A', $A^{0'}$ having the properties stated above may be called *characteristic sets of* $f(a)$ on E; especially A^0, A', $A^{0'}$ may be termed the (0)-*set, the* (1)-*set* and $A^{0'}$ *the* (0 1)-*set.*

W.20.6.2. Theorem. Under the above conditions, if $F \leqq E$, then the characteristic sets on F are

$$A^0 F, A' F, A^{0'} F.$$

Proof. We have, [W.20.3.]:

$$S_{FA^0} f \mu = 0, \quad S_{FA'} f \mu = \mu(A' F),$$

$$\underline{S}_{FA^{0'}} f \mu = 0, \quad \overline{S}_{FA^{0'}} f \mu = \mu(F A^{0'}),$$

so the thesis is true.

W.20.7. We shall consider fields $f(a)$ of real numbers, admitting two values only, viz. M, N where $M \neq N$.

Theorem. If

1. Hyp. $\boldsymbol{H}_2$ is admitted,
2. E is a set of traces,
3. $f(a)$ is a field of real number, admitting the values M, N only, $M \neq N$,

then there exist unique sets A^0, $A^{0'}$, A' such that $f \mu$ is summable on A^0 and on A' with

1) $S_{D^0} f \mu = \mu(D^0) \cdot M$ for all $D^0 \leqq A^0$,
$S_{D'} f \mu = \mu(D') \cdot N$ for all $D' \leqq A'$,

2) $\overline{S}_C f \mu = \max(M, N) \cdot \mu C$, $\underline{S}_C f \mu = \min(M, N) \cdot \mu C$, for all $C \leqq A^{0'}$,

3) if $\underline{\Phi}(x)$ are functions of the variable trace x, attached to $f \mu$, $\overline{\Phi}(x)$,

we have

$$\overline{\Phi}(x) = \underline{\Phi}(x) = M \quad \text{for} \quad x \in A^0, \qquad \overline{\Phi}(x) = \underline{\Phi}(x) = N \quad \text{for} \quad x \in A',$$

$$\overline{\Phi}(x) = \max(M, N) \quad \text{for} \quad x \in A^{0'}, \qquad \underline{\Phi}(x) = \min(M, N) \quad \text{for} \quad x \in A^{0'}.$$

W.20.7.1. Def. We may call

A^0 the M-characteristic set,
A' the N-characteristic set,
$A^{0'}$ the (M, N)-characteristic set of $f(\dot{a})$.

W.20.7a. Proof. Put

$$(1) \qquad g(a) =_{df} \begin{cases} 0 & \text{whenever} \quad f(a) = M, \\ 1 & \text{whenever} \quad f(a) = N. \end{cases}$$

The statement (1) is equivalent to the following:

$$(2) \qquad f(a) = g(a) \cdot (N - M) + M$$

for all a. Let A^0, $A^{0'}$, A' be the characteristic sets for $g(a)$. Since g is summable on A^0 and A', it follows that

$$f(a) = g(a)(N - M) + M$$

is summable on A^0 and A'. We have for

$$D^0 \leqq A^0, \quad D' \leqq A', \quad \text{the equation}$$

$$S_{D^0} f \mu = S_{D^0}[(g(N-M)+M)]\,\mu = (N-M)\,S_{D^0}\,g\,\mu + M\,S_{D^0}\,\mu$$
$$= M\,S_{D^0}\,\mu = M\,\mu\,D^0,$$

and

$$S_{D'} f \mu = S_{D'}(g(N-M)+M)\,\mu = (N-M)\,S_{D'}\,g\,\mu + M\,S_{D'}\,\mu$$
$$= (N-M)\,\mu\,D' + M\,\mu\,D' = N\,\mu\,D'$$

W.20.7b. Concerning $A^{0'}$, we have for $C \leqq A^{0'}$:

$$\overline{S}_C f \mu = \overline{S}_C[(N-M)\cdot g + M]\,\mu = \overline{S}_C(N-M)\,g + M\,\overline{S}_C\,\mu$$
$$= \overline{S}_C(N-M)\,g + M\,\mu(C).$$

Suppose that $M < N$; then

$$\overline{S}_C f \mu = (N-M)\,\overline{S}_C\,g + M\,\mu(C) = (N-M)\,\mu\,C + M\,\mu\,C = N\,\mu\,C$$

and

$$\underline{S}_C f \mu = (N-M)\,\underline{S}_C\,g\,\mu + M\,\mu(C) = M\,\mu(C).$$

W.20.7c. Now, suppose that $M > N$. We have

$$\overline{S}_C f \mu = \overline{S}_C(N-M)\,g\,\mu + M\,\mu(C) = (N-M)\,\underline{S}_C\,g\,\mu + M\,\mu(C)$$
$$= M\,\mu(C),$$

$$\underline{S}_C f \mu = \underline{S}_C(N-M)\,g\,\mu + M\,\mu(C) = (N-M)\,\overline{S}_C\,g\,\mu + M\,\mu(C)$$
$$= (N-M)\,\mu(C) + M\,\mu(C) = N\,\mu(C).$$

W.20.7d. Hence in all cases

$$\overline{S}_C f \mu = \max(M, N)\cdot\mu\,C, \qquad \underline{S}_C f \mu = \min(M, N)\cdot\mu\,C.$$

The theorem is established.

W.20.7e. Remark. The above formulas are also true if $M = N$. In that case $f(a) = \text{const} = M$, and $f\mu$ is summable. We get

$$S_C f \mu = \max(M, N)\,\mu\,C = \min(M, N)\,\mu\,C.$$

W.20.7f. To prove the remaining part of the thesis, put

$$\overline{\Phi}(x) = \underline{\Phi}(x) = M \quad \text{for} \quad x \in A^0, \qquad \overline{\Phi}(x) = \underline{\Phi}(x) = N \quad \text{for} \quad x \in A',$$

and

$$\overline{\Phi}(x) =_{df} \max(M, N) \;\text{ for }\; x \in A^{0'}, \qquad \underline{\Phi}(x) =_{df} \min(M, N) \;\text{ for }\; x \in A^{0'}.$$

We prove that, for all sets D we get

$$\overline{S}_D f \mu = \int_D \overline{\Phi}(x)\,d\mu, \qquad \underline{S}_D f \mu = \int_D \underline{\Phi}(x)\,d\mu,$$

by considering the decomposition $D = D\,A^0 + D\,A' + D\,A^{0'}$, and by relying on the fact, that the functions attached to $f\mu$ are unique.

Section 3 of W.

Theorems on (DARS)-summation of fields of numbers

W.30.1. Remark. We start with the following remarks, which will be important for some items to be settled. If we peruse the theorem and proof of Lemma [U.20.11.], we see that they deal with an arbitrary (DARS)-sequence $P_{\dot n}$ for a measurable set E of traces, and suppose that P_n is decomposed into two disjoint complexes P'_n, P''_n; and we deal only with bricks belonging to P_n, $(n = 1, 2, \ldots)$. This theorem is applied in the proof of Lemma [W.20.2.], where is also considered a general single (DARS)-sequence Q_n for E, satisfying some conditions. There is also considered a field $f(\dot a)$ of real numbers, which admits two values only, viz. 0 and 1. Now, this function serves for a decomposition of the collection $\boldsymbol{B}$ of all bricks into disjoint parts L^0, L', and the induced decomposition of Q_n into disjoint parts Q_n^0, Q'_n. Now, all our arguments will be essentially not changed, if we replace the values 0, 1 of $f(a)$ by M and N, where $M \neq N$. Even more: they will be essentially not changed, if we admit a more general field $f(a)$, which admits not only the values M, N, but also some other values too. We must only admit that the bricks Q_n belong to the set $L^M \cup L^N$, where L^M, L^N are the collection of all bricks of $\boldsymbol{B}$, and where $f(a)$ is either equal to M or to N. The items of the thesis of the above lemma, 1), 2), 3), 4), will be not changed. However, concerning item 5), hence concerning the sums $\overline{S}_C$, $\underline{S}_C$, they should be replaced by

$$\sup_R \limsup_{n\to\infty} \sum_k f(r_{nk})\, \mu(r_{nk}) \tag{1}$$

and by $\inf \liminf \sum$, where R are (DARS)-sequences for C, whose bricks belong to $L^N \cup L^M$ only. Similar change should be made in stating item 6). All other items will hold true without any change. Let us consider Lemma [W.20.2.]. The difference between it and Lemma [W.20.6.2.] is that, instead of Q_n, we deal with a suitable its subsequence $Q_{\alpha(n)}$.

Here we do not need to require that $f(a)$ admits the values 0 and 1 only; we may admit that $f(a)$ admits the values M, N, where $M \neq N$, and some other values too. We must however consider (DARS)-sequences for E, whose bricks belong to $L^M \cup L^N$. Similarly for Theorem [W.20.4.], where $\overline{S}$, $\underline{S}$ would be replaced by expressions (1); and analogously for Theorem [W.20.5.], [W.20.6.], [W.20.7.]. If we do like that, we can state the following general theorem:

W.30.2. Theorem. We admit $\boldsymbol{H}_2$. If

1) L^0, L' are sets of bricks such that $L^0 \cap L' = \emptyset$,
2) E is a measurable set of traces,

3) there exists a (DARS)-sequence for E, whose bricks belong to $L^0 \cup L'$,

and there exist well determined measurable sets A^0, C, A' of traces, such that

1⁰) A^0, C, A' are disjoint,

2⁰) $A^0 + C + A' = E$,

3⁰) if Q_n is any (DARS)-sequence for $E - C$, whose bricks belong to $L^0 \cup L'$, and we put

$$Q_n^0 =_{df} Q_n \cap L^0, \quad Q'_n =_{df} Q_n \cap L', \quad (n = 1, 2, \ldots),$$

then Q_n^0 is a (DARS)-sequence for A^0, and Q'_n is a (DARS)-sequence for A',

4) there exist (DARS)-sequences T_n^0, T'_n with $T_n^0 \subseteqq L^0$, $T'_n \subseteqq L'$, $(n = 1, 2, \ldots)$ both for C, (some of A^0, A', C may be empty),

5) the set C is the μ-maximal set of traces with the property 4),

6) A^0 is the μ-maximal set of traces, such that for every (DARS)-sequence Q_n for $E - C$, whose bricks belong to L^0, the sequence $Q_n \cap L^0$ is a (DARS)-sequence for A^0; and similarly for A'.

W.30.2.1. Remark. The items 4), 5), 6) of the thesis can be stated in the following way: Concerning A^0, there exists a (DARS)-sequence for A^0, whose bricks belong to L^0. But if $\bar{A}^0 \leqq A^0$ with $\mu \bar{A}^0 > 0$, then there does not exist any (DARS)-sequence for $\bar{A}^0$, made up of considered bricks, which do not belong to L^0. Similarly for A'.

Concerning $A^{0'} =_{df} C$, there exists a (DARS)-sequence for $A^{0'}$, made up of bricks of L^0, and also there exists a (DARS)-sequence for $A^{0'}$, made up of bricks of L'. But if $\bar{A}^{0'} \leqq A^{0'}$ with $\mu \bar{A}^{0'} > 0$, then there does not exist any (DARS)-sequence, made up of bricks considered, which do not belong to $L^0 \cup L'$, (of course, because such bricks are not considered).

From the above it follows, that the sets A^0, $A^{0'}$, A' are μ-maximal with respect to their above properties. This means that, if $\underline{A}^0 \geqq A^0$, $\underline{A}^{0'} \geqq A^{0'}$, $\underline{A}' \geqq A'$, and if $\underline{A}^0$, $\underline{A}^{0'}$, $\underline{A}'$ have respectively the properties mentioned above, then

$$\mu(\underline{A}^0 - A^0) = 0, \quad \mu(\underline{A}^{0'} - A^{0'}) = 0, \quad \mu(\underline{A}' - A') = 0.$$

This follows from the fact, that A^0, $A^{0'}$, A' are disjoint and that their sum equals E.

W.30.3. Theorem. We admit Hyp. $\boldsymbol{H}_2$. Suppose that

1) $L^0, L^1, \ldots, L^m$, $(m \geqq 1)$ are not empty disjoint subcollections of the collection $\boldsymbol{B}$ of all bricks,

2) E is a measurable set of traces,

3) there exists a (DARS)-sequence Q_n for E, whose bricks belong to $L^0 \cup L^1 \cup \cdots \cup L^m$,

4) considering only the bricks, which belong to $L^0 \cup L^1 \cup \cdots \cup L^m$, there exist well determined subsets $A^{i_1, i_2, \ldots, i_k}$, $(k > 1;\ i_1, i_2, \ldots, i_k$

are numbers $0, 1, \ldots, m$; the permutations of the indices do not matter, i.e., $i_1, \ldots, i_k$ is considered as a subset of the set $0, 1, \ldots, m$, and all $i_1, \ldots, i_k$ are different), such that

1^0) they are all mutually disjoint,

2^0) their sum equals E,

3a) for every $\alpha = 1, \ldots, k$ there exists a (DARS)-sequence for $A^{i_1, i_2, \ldots, i_k}$, made up of bricks, belonging to L^{i_α},

3b) if $A \leqq A^{i_1, \ldots, i_k}$, $\mu A > 0$, then there does not exist any (DARS)-sequence for A, made up of bricks, which does not belong to $L^{i_1} \cup \cdots \cup L^{i_l}$,

4) if 1') Q_n is a (DARS)-sequence for A, where $A \leqq A^{i_1, \ldots, i_k}$, and where the bricks of Q_n belong to $L^0 \cup \cdots \cup L^m$.

2') $Q_n = Q'_n \cup Q''_n$, where $Q'_n \subseteqq L^{i_1} \cup \cdots \cup L^{i_k}$ and $Q''_n \subseteqq L^{j_1} \cup L^{j_2} \cup \cdots \cup L^{j_l}$, where $j_1, j_2, \ldots, j_l$ is the set of indices complementary to $i_1, \ldots, i_k$, then

$$\mu Q''_n \to 0 \quad \text{for} \quad n \to \infty.$$

W.30.3a. Proof. Suppose the theorem is true for $m = p \geqq 1$; we shall prove it for $m = p + 1$. Let $L^0, L^1, \ldots, L^{p+1}$ be subcollections of $\boldsymbol{B}$, as in Hyp. 1. The only bricks we shall consider, will belong to $L^0 \cup \cdots \cup L^{p+1}$. Put $M =_{df} L^1 \cup \cdots \cup L^{p+1}$. Then $L^0 \cap M = \emptyset$. We can apply Theorem [W.30.2.] and Remark [W.30.2.1.]. We find sets A^0, C, B', such that A^0, C, B' are disjoint, and $A^0 + C + B' = E$.

There exists a (DARS)-sequence for A^0, made of bricks of L^0 but, if $A \leqq A^0$, $\mu A > 0$, there does not exist any sequence for A, whose bricks belong to M. There exists a (DARS)-sequence for B', made of bricks of M, but if $B \leqq B'$, $\mu B > 0$, there does not exist any sequence for B, whose bricks belong to L^0. Concerning C, there exists a (DARS)-sequence for C, made up of bricks of L^0 and there exists a (DARS)-sequence for C, made up of bricks of $L^1 \cup \cdots \cup L^{p+1}$ ($L^1 =_{df} L'$).

W.30.3b. Since there is a (DARS)-sequence for B', made of bricks of M, we can apply the hypothesis of the proof. There exist sets of traces:

$$A^{i_1}, A^{i_1, i_2}, \ldots, A^{i_1, i_2, \ldots, i_{k+1}},$$

where $i_\alpha \geqq 1$, such that $A^{i_1, i_2, \ldots, i_k}$ admits a (DARS)-sequence, whose bricks belong to L^{i_1}; another one, whose bricks belong to L^{i_2}, etc., and a sequence, whose bricks belong to L^{p+1}. But, if $A \leqq A^{i_1, \ldots, i_k}$ with $\mu A > 0$, then there does not exist any (DARS)-sequence for A, whose bricks belong to $L^{j_1} \cup L^{j_2} \cup \cdots \cup L^{j_l}$, where $(j_1, j_2, \ldots, j_l)$ is the subset of the set $(1, 2, \ldots, p + 1)$, complementary to $(i_1, \ldots, i_k)$.

W.30.3c. The set C admits a (DARS)-sequence, made up of bricks of $L^1 \cup \cdots \cup L^{p+1} = M$. Hence, by hypothesis in the proof, we can find the sets, which we shall denote by

$$A^{0, i_1}, A^{0, i_1, i_2}, \ldots, A^{0, i_1, \ldots, i_{p+1}},$$

having the following properties:

$$A^{0,i_1,\ldots,i_k}$$

admits a (DARS)-sequence made of bricks of L^{i_1}, another one made of L^{i_2}, etc. and of L^{p+1}. But if $F \leqq A^{0,i_1,\ldots,i_k}$, with $\mu F > 0$, then there does not exist any sequence, made up of bricks of $L^{j_1} \cup L^{j_2} \cup \cdots \cup L^{j}$.

W.30.3d. We have defined the sets A^0, $A^{i_1,\ldots,i_k}$, and $A^{0,i_1,\ldots,i_k}$. They all are disjoint. Indeed, $A^{i_1,\ldots,i_k} \leqq B'$, $A^{0,i_1,\ldots,i_k} \leqq C$ and we know, that all A^0, B', C are disjoint. In addition to that all $A^{i_1,\ldots,i_k}$ are disjoint with one another, and so are $A^{0,i_1,\ldots,i_k}$ — this by hypothesis. We have

$$A^0 + \sum_{i_1,\ldots,i_k} A^{i_1,\ldots,i_k} + \sum_{i_1,\ldots,i_k} A^{0,i_1,\ldots,i_k} = E.$$

Indeed, $A^0 + C + B' = E$, and, by hypothesis $B' = \sum_{i_1,\ldots,i_k} A^{i_1,\ldots,i_k}$,

$$C = \sum_{i_1,\ldots,i_k} A^{0,i_1,\ldots,i_k}.$$

W.30.3e. $A^{i_1,\ldots,i_k}$ admits, by hypothesis, (DARS)-sequences, made up of bricks of L^{i_1}, of bricks of L^{i_2}, etc., and of bricks of L^{i_k} respectively.

$A^{0,i_1,\ldots,i_k}$ admits, by hypothesis, (DARS)-sequences, made up of bricks of L^{i_1}, of bricks of L^{i_2}, ..., and of bricks of $L^{i_{p+1}}$. But, Theorem [W.30.2.], Remark [W.30.2.1.], it also admits a (DARS)-sequence, made up of bricks of L^0; this because $A^{0,i_1,\ldots,i_k} \leqq C$ and C does so.

W.30.3f. We see, that if $A < A^0$, $\mu A > 0$, it does not admit, Theorem [W.30.2.], Remark [W.30.2.1.], any (DARS)-sequence, made up of bricks of $M = L' \cup \cdots \cup L^{p+1}$. Let $D \leqq A^{i_1,\ldots,i_k}$, $\mu D > 0$. By hypothesis, D does not admit any (DARS)-sequence made up of bricks of

$$L^{j_1} \cup L^{j_2} \cup \cdots \cup L^{j_l}.$$

By Theorem [W.30.2.], Remark [W.30.2.1.], D does not admit any (DARS)-sequence, made up of bricks of L^0. Suppose that D admits a (DARS)-sequence, made up of bricks of $L^0 \cup (L^{j_1} \cup \cdots \cup L^{j_l})$. We apply to D the Theorem [W.30.2.], Remark [W.30.2.1.]. Then D can be decomposed into disjoint sets D^0, X, D', where D^0 admits a (DARS)-sequence, made up of bricks of L^0. If $\mu D^0 > 0$, this is impossible, since $D^0 < D$, and D does not admit any sequence, made up of bricks of L^0. Hence $\mu D^0 = 0$. The set D' admits a (DARS)-sequence, made up of bricks of $L^{j_1} \cup \cdots \cup L^{j_l}$, which is impossible if $\mu D' > 0$. So we have $\mu D' = 0$. Hence we have the μ-equality $X = D$. the set X admits a (DARS)-sequence, made up of bricks of A^0, which is impossible, if $\mu X = 0$, i.e. $\mu D > 0$. So $\mu D = 0$, but this is a contradiction, because $\mu D > 0$. Hence D does not admit any (DARS)-sequence, made up of bricks of $L^0 \cup L^{j_1} \cup \cdots \cup L^{j_l}$.

W.30.3g. In a similar way we prove, that if $F < A^{0, i_1, \ldots, i_k}$, $\mu F > 0$, then it does not admit any (DARS)-sequence, whose bricks belong to $L^{j_1} \cup \cdots \cup L^{j_{p+1}}$. Thus we have proved the theses 1), 2), 3a), 3b). To prove 4), suppose that Q_n is a sequence for A, where $A \leqq A^{i_1, \ldots, i_k}$ and where

$$Q_n \subseteqq L^0 \cup \cdots \cup L^m.$$

Put

(1) $Q'_n =_{df} Q_n \cap (L^{i_1} \cup \cdots \cup L^{i_k})$, $\quad Q''_n =_{df} Q_n \cap (L^{j_1} \cup \cdots \cup L^{j_l})$,

where $(j_1, \ldots, j_l)$ is a subset of $(0, 1, \ldots, m)$, complementary to $(i_1, \ldots, i_k)$.

Since $Q'_n \cup Q''_n = Q_n$, $Q'_n \cap Q''_n = \emptyset$, we can apply Lemma [W.20.2.], modified by the Remark [W.30.1.]. Take a subsequence $Q_{\alpha(n)}$ of Q_n. We can extract out of it another subsequence $Q_{\alpha\beta(n)}$ such that, if we put

$$F' = \underline{\operatorname{Lim}} Q'_{\alpha\beta(n)}, \quad G' = \overline{\operatorname{Lim}} Q'_{\alpha\beta(n)}, \quad F'' = \underline{\operatorname{Lim}} Q''_{\alpha\beta(n)},$$

$$G'' = \overline{\operatorname{Lim}} Q''_{\alpha\beta(n)}, \quad H =_{df} G' - F' = G'' - F'',$$

then there exists a sequence S''_n for H, such that $S''_n \subseteqq \bigcup_{m=1}^{\infty} Q''_m$. Since $Q''_m \subseteqq L^{j_1} \cup \cdots \cup L^{j_l}$, we must have $\mu H = 0$, because if $\mu H > 0$, we would have a contradiction with 3b), as $A \equiv A^{i_1, \ldots, i_k}$. As $F' \leqq G'$, $F'' \leqq G''$, we get $F' = G'$, $F'' = G''$. Hence $A = F' + F''$, and we have

$$F' = \underline{\operatorname{Lim}} Q'_{\alpha\beta(n)} = \overline{\operatorname{Lim}} Q'_{\alpha\beta(n)} = \operatorname{Lim} Q'_{\alpha\beta(n)},$$

$$F'' = \underline{\operatorname{Lim}} Q''_{\alpha\beta(n)} = \overline{\operatorname{Lim}} Q''_{\alpha\beta(n)} = \operatorname{Lim} Q''_{\alpha\beta(n)}.$$

It follows that

$$(2) \qquad |Q''_{\alpha\beta(n)}, F''| \to 0,$$

and then $Q''_{\alpha\beta(n)}$ is a (DARS)-sequence for F''.

Now, by (1), $Q''_{\alpha\beta(n)} \subseteqq L^{j_1} \cup \cdots \cup L^{j_l}$; and we also have $F'' \leqq A \leqq \leqq A^{i_1, \ldots, i_k}$. Hence if we had $\mu F'' > 0$, this would contradict item 3b). Hence $\mu F'' = 0$. Consequently from (2) it follows that $\mu Q''_{\alpha\beta(n)} \to 0$. Thus from every subsequence $\alpha(\dot{n})$ of $\dot{n}$, another $\alpha\beta(\dot{n})$ can be extracted, such that $\mu Q''_{\alpha\beta(n)} \to 0$. Consequently $\mu Q''_n \to 0$, so the item 4) is established. The whole theorem is proved.

W.30.3.1. Remark. We can prove that $A^{i_1, \ldots, i_k}$ (of the thesis) is μ-maximal with respect to the properties 3c) and 3b). The proof is indirect, and based on the properties 1) and 2) of the thesis.

W.30.3.2. Def. We call $A^{i_1, \ldots, i_k}$ the *characteristic sets for* $L^0, \ldots, L^m$ and E.

W.30.4. Theorem. We admit Hyp. $\boldsymbol{H}_2$. Suppose that $f(a)$ admits a finite number of values only, say

$$M_0 < M_1 < \cdots < M_p, \quad (p \geqq 1).$$

Let E be a measurable set of traces. Define $L^k = \{a \mid f(a) = M_k\}$, $(k = 0, \ldots, p)$, and let $\{A^{i_1, \ldots, i_k}\}$ be the assemblage of all characteristic sets for $L^0, \ldots, L^p$. Let $F \leqq E$.

Then

$$(\alpha) \qquad \bar{S}_F f \mu = \sum_k \sum_{i_1, \ldots, i_k} \max[M_{i_1}, \ldots, M_{i_k}] \mu(A^{i_1, \ldots, i_k} \cdot F),$$

$$(\beta) \qquad \underline{S}_F f \mu = \sum_k \sum_{i_1, \ldots, i_k} \min[M_{i_1}, \ldots, M_{i_k}] \mu(A^{i_1, \ldots, i_k}, F).$$

Especially we have:

$$\bar{S}_{A^{i_1, \ldots, i_k}} f \mu = \max[M_{i_1}, \ldots, M_{i_k}] \mu(A^{i_1, \ldots, i_k}),$$

$$\underline{S}_{A^{i_1, \ldots, i_k}} f \mu = \min[M_{i_1}, \ldots, M_{i_k}] \mu(A^{i_1, \ldots, i_k}).$$

Proof. Let F be a set of traces. Put $A =_{df} A^{i_1, \ldots, i_k}$. There exists i, such that $M_i = \max\{M_{i_1}, \ldots, M_{i_k}\}$. Let P_n be a (DARS)-sequence for $A \cdot F$, such that

$$\lim_{n \to \infty} (f \mu)(P_n) = \bar{S}_{AF} f \mu.$$

We decompose P_n into two disjoint complexes P'_n, P''_n, where $P'_n \subseteqq L^{i_1, \ldots, i_k}$, $P''_n \subseteqq L^{j_1, \ldots, j_l}$, where $(j_1, \ldots, j_l)$ is a subset of $(0, 1, \ldots, p)$, complementary to $(i_1, \ldots, i_k)$. By Theorem [W.30.3.] we have $\mu P''_n \to 0$; hence P'_n is a (DARS)-sequence for $A \cdot F$. We have $\lim (f \mu)(P'_n) = \bar{S}_{AF} f \mu$. We also have

$$(f \mu)(P'_n) = \sum_k f(p'_{nk}) \mu f(p'_{nk}) \leqq \sum_k M_i \mu(p'_{nk}) \to M_i \mu(A \cdot F).$$

It follows that

$$(1) \qquad \bar{S}_{AF} f \mu < M_i \mu(A F).$$

Now, since $A F$ admits a (DARS)-sequence Q_n, made up of bricks of M_i, it follows that

$$\bar{S}_{AF} f \mu \geqq \limsup_{n \to \infty} \sum_k f(q_{nk}) \mu(q_{nk}) = M_i \limsup_{n \to \infty} \sum_k \mu(q_{nk}).$$

But, since

$$\sum_k \mu(q_{nk}) = \mu(Q_n) \mu(A F),$$

it follows that

$$(2) \qquad \bar{S}_{AF} f \mu \geqq M_i \mu(A F).$$

From (1) and (2) it follows that $\bar{S}_{AF} f \mu = M_i \mu(A F)$. Now we can apply [W.20.6.], getting the formula (α). In a similar way we prove (β).

W.30.4.1. Theorem. We admit Hyp. $\boldsymbol{H}_2$. Let $f(a)$ be a field of real numbers, admitting a finite number of values only, say $M_0 < M_1 < \cdots < M_p$, $(p \geqq 1)$. Let E be a measurable set of traces, and let

$$L^i =_{df} \{a \mid f(a) = M_i\}, \qquad (i = 0, 1, \ldots, p).$$

Let $\{A^{i_1, \ldots, i_k}\}$ be the system of characteristic sets for $L^0, \ldots, L^p$ and E. Then the following are equivalent:

I. $f(a)$ is (DARS)-summable on E,

II. all $A^{i_1, \ldots, i_k}$, with more than one index, have the measure 0; so $E =^{\mu} A^0 + \cdots + A^p$.

In this case, if we take any (DARS)-sequence Q_n for E, and define $Q_n^i =_{df} Q_n \cup L^i$, $(i = 1, \ldots, p)$, $(n = 1, 2, \ldots)$, then $A^i = \lim\limits_{n \to \infty} \mu Q_n^i$ is even a (DARS)-sequence for A^i. We also have

$$(1) \qquad S_E f \mu = \sum_{i=0}^{p} M_i \, \mu(A^i F) \quad \text{for all} \quad F \leqq E.$$

Proof. Suppose I. Then no one $A^{i_1, \ldots, i_k}$ with more than one index, can have a positive measure. Indeed $\bar{S}$ and $\underline{S}$ on this set would differ, since all $M^0, \ldots, M^p$ are different. Thus II. follows.

Suppose II., then since $A^0 + \cdots + A^p = E$, all sets $A^{i_1, \ldots, i_k}$ have the measure 0; so the terms in Theorem [W.30.4., (α), (β)], where more, than one index is considered, can be dropped. There remain terms, corresponding to $A^0, \ldots, A^p$, and then the maxima and minima written therein reduce to $M_0, \ldots, M_p$, so $\bar{S} = \underline{S}$ for $F = E$. Hence we have also $\bar{S}_F = \underline{S}_F$ for any $F \leqq E$. Thus the equivalence is proved. The formula (1) follows.

Let Q_n be any (DARS)-sequence for E. Define Q_n^i, $(i = 0, 1, \ldots, p)$, as in the statement of the thesis. Consider the set A^0, (the other sets will be treated analogously). By Theorem [W.30.4.], if we decompose Q_n into the parts Q_n^0 and P_n, where $P_n \subseteqq L^2 \cup \cdots \cup L^p$, $Q_n^0 \subseteqq L^0$, we get $\mu P_n \to 0$ for $n \to \infty$. Hence $|Q_n^0, A^0| \to 0$. Q_n^0 is a (DARS)-sequence for A^0. The theorem is proved.

W.30.5. Theorem. We admit $\boldsymbol{H}_2$. If

1. E is a set of traces,
2. Q_n is a (DARS)-sequence for E,

then there exists a subsequence $k(\dot{n})$ of $\dot{n}$ and a (DARS)-sequence T_n for $1 - E$, such that

1) $Q_{k(\dot{n})} \cap T_{\dot{n}} = \emptyset$ for $n = 1, 2, \ldots$,

2) $Q_{k(\dot{n})} \cup T_{\dot{n}}$ is a (DARS)-sequence for 1.

Proof. Choose an index n, and put $A_n =_{df} 1 - \mathrm{som} Q_n$. This set is a figure, hence a denumerable sum of bricks; hence it is a covering of itself. There exists a complex S_n, such that $|S_n, A_n| \leqq \frac{1}{n}$, and $\operatorname{som} S_n \leqq A_n$.

Since $|Q_n, E| = |1 - \operatorname{som} Q_n, 1 - E| = |A_n, 1 - E|$, it follows that $\lim |S_n, 1 - E| \to 0$. There exists $k(n)$ and a sequence T_n of complexes, such that

$$\operatorname{som} T_n \leqq \operatorname{som} S_{k(n)},$$

and where T_n is a (DARS)-sequence for $1 - E$. We have

$$\operatorname{som} T_n \leqq \operatorname{som} S_{k(n)} \leqq A_{k(n)};$$

hence $= 1 - \operatorname{som} Q_{k(n)}$. We have $T_n \frown Q_{k(n)} = \mathit{O}$. As T_n is a (DARS)-sequence for $1 - E$, and $Q_{k(n)}$ is a (DARS)-sequence for E with $T_n \frown Q_{k(n)} = \mathit{O}$ for $n = 1, 2, \ldots$, it follows, that $T_n \smile Q_{k(n)}$ is a (DARS)-sequence for 1. The theorem is proved.

W.30.5.1. Theorem. We admit hypothesis $\boldsymbol{H}_2$. If

1. E, F are sets of traces,
2. $F \leqq E$,
3. Q_n is a (DARS)-sequence for F,

then there exists a subsequence $k(\dot{n})$ of $\dot{n}$ and a (DARS)-sequence $R_{\dot{n}}$ for $E - F$, such that

$$Q_{k(n)} \frown R_n = \mathit{O}, \quad Q_{k(n)} \smile R_n = E$$

is a (DARS)-sequence for E.

Proof. We rely on Theorem [W.30.3.]. Let $k(\dot{n})$ be a subsequence of $\dot{n}$, and T_n a sequence for $1 - F$, such that $Q_{k(n)} \frown T_n = \mathit{O}$. The sequence $Q_{k(n)} \smile T_n$ is a (DARS)-sequence for

(1) $$1 - F.$$

From (1) it follows, that there exists a (DARS)-sequence R_n for $(1 - F)\, E = E - F$, with $R_n \subsetneqq T_n$. Since $Q_{k(n)} \frown R_n = \mathit{O}$, we see, that $Q_{k(n)} \smile R_n$ is a (DARS)-sequence for E. The theorem is proved.

Remark. The above two theorems are extension-theorems for (DARS)-sequences. They allow to complete a (DARS)-sequence for a given set to another of a larger set, by adjoining complexes disjoint to the given ones in the sequence.

W.30.5.2. Theorem. We admit hypothesis $\boldsymbol{H}_2$. If

1. L^0, L' are sets of bricks, such that $L^0 \frown L' = \mathit{O}$, $L^0 \smile L' = \mathbf{B}$, i.e., the set of all bricks,
2. E is a measurable set of traces,

then there exist well determined measurable sets A^0, C, A' of traces, such that

1) A^0, C, A' are disjoint, $A^0 + C + A' = E$ (some of A^0, C, A' may be empty).

2) If Q_n is any (DARS)-sequence for $E - C$, and we put

$$Q_n^0 =_{df} Q_n \frown L^0, \quad Q_n' =_{df} Q_n \frown L', \quad (n = 1, 2, \ldots),$$

then Q_n^0 is a (DARS)-sequence for A^0, and $Q'_{\dot n}$ is a (DARS)-sequence for A';

3) there exist (DARS)-sequences $T_n^0 \subseqq L^0$, $T'_n \subseqq L'$, both for C;

4) the set C is the μ-maximal set of traces satisfying 3) (i.e. there does not exist any set C', with $C' > C$ with the property 3);

5) if $\bar{A}^0$ is a measurable subset of A^0, then for every (DARS)-sequence $P_{\dot n}$ of E, there exists $\bar{P}_n \subseqq P_n \cap L^0$, such that $\bar{P}_{\dot n}$ is a (DARS)-sequence for A^0;

5.1) A^0 is the μ-maximal set with property 5), i.e. if $\bar{\bar{A}}^0$ is a measurable subset of E, such that for every (DARS)-sequence P_n for E, there exists $\bar{\bar{P}}_n \subseqq P_n \cap L^0$, then $\bar{\bar{A}}^0 \leqq^\mu A^0$.

6) If $\bar{A}'$ is a measurable subset of A', then for every (DARS)-sequence R_n for E, there exists an $\bar{R}_n \subseqq R_n \cap L'$, such that $\bar{R}_n$ is a (DARS)-sequence for $\bar{A}'$;

6.1) A' is the μ-maximal set with property 6), i.e., if $\bar{\bar{A}}'$ is a measurable subset of E, such that for every (DARS)-sequence P_n for E there exists $\bar{\bar{P}}_n \subseqq P_n \cap L'$, such that P_n is a (DARS)-sequence for $\bar{\bar{A}}'$, then $\bar{\bar{A}}' \leqq^\mu A'$.

W.30.5.2a. Proof. We consider the auxiliary field of real numbers $\varphi(\dot a)$, such that $\varphi(a) = 0$, whenever $a \in L^0$, and $\varphi(a) = 1$, whenever $a \in L'$. Then if we apply Theorem [W.30.3.] to $E - C$, we get the characteristic sets A^0, C, A', with properties 1), 2), and 3) of the thesis.

W.30.5.2b. To prove 4), suppose that $C' \geqq C$ with $\mu C' > \mu C$, and with property 3). Then on $C' - C$ the field $\varphi(\dot a)$ is not summable. But $C' - C \leqq E - C = A^0 + A'$; hence $\varphi(a)$ is summable on $C' - C$. This is a contradiction.

W.30.5.2c. Let $A \leqq A^0$. Take any sequence Q_n for E. As this is a (DARS)-sequence, there exists Q'_n, such that $Q'_n \subseqq Q_n$, $(n = 1, 2, \ldots)$, and where Q'_n is a (DARS)-sequence for $A^0 + A' = E - C$. Hence, by the proved item 2) of the thesis, $Q'_n \cap L^0$ is a sequence for A^0. Now since this is a (DARS)-sequence, there exist Q''_n, where $Q''_n \subseqq Q'_n \cap L^0$, $(n = 1, 2, \ldots)$, and where Q''_n is a sequence for A. We have $Q''_n \subseqq Q_n \cap L^0$, so item 5) is proved. In a similar way we prove item 6) of the thesis.

W.30.5.2d. Now, let $A \leqq C$, with $\mu A > 0$, and suppose that given any (DARS)-sequence $Q_{\dot n}$ for E, we can find $Q'_n \subseqq Q_n \cap L^0$ such that Q'_n is a sequence for A. Take the sequence T'_n for A, with $T'_n \subseqq L'$. By [W.30.5.1.], there exists a subsequence $s(\dot n)$ of $\dot n$ and a (DARS)-sequence $R_{\dot n}$ for $E - A$, such that $T'_{s(n)} \cap R_n = \emptyset$, and that $T'_{s(n)} \cup R_n$ is a (DARS)-sequence for E.

Suppose that there exists a sequence R'_n for A, such that $R'_n \subseqq (T'_{s(n)} \cup R_n) \cap L^0$. Since $T'_{s(n)} \cap L^0 = \emptyset$, (because $T'_{s(n)} \subseqq L'$), we have

$$R'_n \subseteqq (R_n \frown L^0);$$

hence

(1) $$R'_n = R'_n \frown R_n \frown L^0.$$

Since $\operatorname{som} R'_n \to A$, $\operatorname{som} R_n \to E - A$, it follows that

$$\operatorname{som} R'_n \cdot \operatorname{som} R_n \to A \cdot (E - A) = O;$$

hence $\operatorname{som}(R'_n \frown R_n \frown L^0) \to O$. Consequently, by (1), $\operatorname{som} R'_n \to O$. Hence $\mu(\operatorname{som} R'_n) \to 0$. Since R'_n is a sequence for A, it follows $\mu(\operatorname{som} R'_n) \to \mu(A) > 0$. This is a contradiction. It follows that if $A \leqq C$, and if for every (DARS)-sequence Q_n for E there exists $Q'_n \subseteqq Q_n \frown L^0$, such that Q'_n is a (DARS)-sequence for A, then $\mu A = 0$.

W.30.5.2e. Now, let $A \leqq A'$, and suppose that for every (DARS)-sequence Q_n for E there exists $Q'_n \subseteqq Q_n \frown L^0$, such that Q'_n is a (DARS)-sequence for A. Take Q_n and find Q'_n as above. We have

$$\sum_k \varphi(q'_{nk})\, \mu(q'_{nk}) \to S_A\, \varphi\, \mu,$$

because $\varphi\,\mu$ is summable on A', hence on A. Since $q'_{nk} \in Q_n$, we have $q'_{nk} \in L^0$, and then $\varphi(q'_{nk}) = 0$. Consequently $S_A\, \varphi\, \mu = 0$. On the other hand, since $A \leqq A'$, we have $S_A\, \varphi\, \mu = \mu(A)$. It follows that $\mu(A) = 0$. Hence, if $A \leqq A'$, and if for every (DARS)-sequence Q_n for E there exists $Q'_n \subseteqq Q_n \frown L^0$, such that Q'_n is a (DARS)-sequence for A, then $\mu A = 0$.

W.30.5.2f. Now, take any subset A of E, and suppose that, for every (DARS)-sequence Q_n for E, there exists Q'_n such that $Q'_n \subseteqq Q_n \frown L^0$ and where Q'_n is a (DARS)-sequence for A.

We have

$$A = (A^0 + C + A')\, A = A^0 A + C A + A' A.$$

Since Q'_n is a (DARS)-sequence for E, there exists R_n, such that $R_n \subseteqq Q'_n \subseteqq Q_n \frown L^0$, where R_n is a (DARS)-sequence for $C A$. It follows by item 8d), already settled, that $\mu(C A) = 0$. Similarly, there exists R'_n with $R'_n \subseteqq Q'_n \frown L^0$, such that R'_n is a (DARS)-sequence for $A' A$. Since Q_n is arbitrary, it follows by item 8e) already settled, that $A' A =^{\mu} O$. Consequently $A = A^0 \cdot A$; hence $A \leqq A^0$, so the item 5.1) is proved. Similarly we prove item 6.1). The theorem is established.

W.30.6. Theorem. We admit the Hyp. $\boldsymbol{H}_2$. If

1. $f(\dot{a})$ is a field of real numbers,
2. E is a measurable set of traces,
3. $M < N$ are real numbers,
4. we have either $f(a) \leqq M$ or $f(a) \geqq N$ for all a,

5. $f(\dot{a})\,\mu(\dot{a})$ is (DARS)-summable on E,
6. we define

$$f^{+}(a) =_{df} \begin{cases} f(a) & \text{whenever} \quad f(a) \geqq N, \\ 0 & \text{whenever} \quad f(a) \leqq M, \end{cases}$$

$$f^{-}(a) =_{df} \begin{cases} 0 & \text{whenever} \quad f(a) \geqq N, \\ f(a) & \text{whenever} \quad f(a) \leqq M, \end{cases}$$

then 1⁰) both $f^{+}(\dot{a})\,\mu(\dot{a})$ and $f^{-}(\dot{a})\,\mu(\dot{a})$ are (DARS)-summable on E, and we have

$$S_F f\mu = S_F f^{+}\mu + S_F f^{-}\mu \quad \text{for all} \quad F \leqq E.$$

2⁰) There exist a well determined measurable sets of traces A^{-}, A^{+}, such that

$$A^{-}\cdot A^{+} = O, \quad A^{-} + A^{+} = E, \quad S_{A^-} f\mu = S_{A^-} f^{-}\mu, \quad S_{A^+} f\mu = S_{A^+} f^{+}\mu;$$

3⁰) if Q_n is a (DARS)-sequence for E, and Q_n^{+} is the set of all bricks q of Q_n, for which $f(q) \geqq N$, and Q_n^{-} is the set of all bricks q of Q_n, for which

$$f(q) \leqq M, \quad (n = 1, 2, \ldots),$$

then

$$Q_n^{+} \text{ is a (DARS)-sequence for } A^{+},$$

$$Q_n^{-} \text{ is a (DARS)-sequence for } A^{-};$$

4⁰) if P_n is a (DARS)-sequence for A^{-}, and we put $P_n = P_n^{+} + P_n^{-}$, where P_n^{+}, P_n^{-} are defined analogously to Q_n^{+}, Q_n^{-}, then $\mu P_n^{+} \to 0$. The similar statement is valid for A^{+}.

W.30.6a. Proof. Denote by L^0, L' the sets of bricks $\{p \mid f(p) \leqq M\}$, $\{p \mid f(p) \geqq N\}$ respectively, and define the auxiliary field $\varphi(a)$ by putting

$$\varphi(a) = 0, \quad \text{whenever} \quad a \in L^0, \qquad \text{and} \qquad \varphi(a) = 1, \quad \text{whenever} \quad a \in L'.$$

Since we are in the conditions of [W.20.6.], we have the characteristic disjoint sets A^0, C, A', such that $A^0 + C + A' = E$. I say that $\mu C = 0$. Suppose this be not true, then; [W.20.6.],

$$\overline{S}_C \varphi\mu = \mu(C), \qquad \underline{S}_C \varphi\mu = 0.$$

Consider a sequence T_n for C, for which

$$\lim_{n\to\infty} \sum_k \varphi(t_{nk})\,\mu(t_{nk}) = \overline{S}_C \varphi\mu. \tag{1}$$

Define

$$T_n^0 =_{df} T_n \frown L^0, \qquad T_n' =_{df} T_n \frown L'.$$

We have

$$T_n^0 \frown T_n' = \emptyset, \qquad T_n^0 \smile T_n' = T_n.$$

From (1) we get

$$\lim_{n\to\infty} \Big[\sum_k \varphi(t_{nk}^0)\,\mu(t_{nk}^0) + \sum_l \varphi(t_{nl}')\,\mu(t_{nl}')\Big] = \overline{S}_C \varphi\mu.$$

Since

$$\varphi(t^0_{nk}) = 0, \quad \varphi(t'_{nk}) = 1,$$

we get

$$\lim_{n\to\infty} \sum_k \mu(t'_{nk}) = \mu(C).$$

Since

$$\sum_k \mu(t_{nk}) \to \mu(C),$$

and

$$\sum_k \mu(t'_{nk}) \to \mu(C),$$

and since $T'_n \subseteqq T_n$, it follows that

$$\lim_{n\to\infty} \sum_k \mu(t^0_{nk}) = 0.$$

Since $f\mu$ is summable on E, therefore $f\mu$ is admissible. Hence

(2) $$\lim_{n\to\infty} \sum_k f(t^0_{nk})\,\mu(t^0_{nk}) = 0.$$

By hypothesis, $f\mu$ is summable on E; hence it is so on C. Since $T_{\dot{n}}$ is a sequence for C, it follows

$$\lim_{n\to\infty} \sum_k f(t_{nk})\,\mu(t_{nk}) \to S_C f\mu.$$

Hence, by (2),

$$\lim \sum_k f(t'_{nk})\,\mu(t'_{nk}) = S_C f\mu.$$

Consequently, since $f(t'_{nk}) \geqq N$, we get

(2.1) $$S_C f\mu \geqq N\mu(C).$$

W.30.6b. We have, [W.20.6.], $\underline{S}_C \varphi\mu = 0$. Consider a sequence $R_{\dot{n}}$ for C, for which

(3) $$\lim \sum_k \varphi(r_{nk})\,\mu(r_{nk}) = \underline{S}_C \varphi\mu = 0.$$

Put

$$R^0_n =_{df} R_n \cap L^0, \quad R'_n =_{df} R_n \cap L'.$$

We have

$$R^0_n \cup R'_n = R_n, \quad R^0_n \cap R'_n = \emptyset.$$

From (3) we have

(3.1) $$\sum_k \varphi(r^0_{nk})\,\mu(r^0_{nk}) + \sum_k \varphi(r'_{nk})\,\mu(r'_{nk}) \to 0.$$

Since

$$\varphi(r^0_{nk}) = 0, \quad \varphi(r'_{nk}) = 1,$$

we get

$$\sum_k \mu(r'_{nk}) \to 0.$$

Since $f\mu$ is summable, it is admissible; hence

$$\sum_k f(r^1_{nk})\,\mu(r^1_{nk}) \to 0. \tag{3.2}$$

Since $R_{\dot n}$ is a (DARS)-sequence for C, we have

$$\sum_k f(r_{nk})\,\mu(r_{nk}) \to S_C f\mu,$$

because $f\mu$ is summable on C. It follows from (3.2) that

$$\sum_k f(r^0_{nk})\,\mu(r^0_{nk}) \to S_C f\mu, \quad \sum \mu(r^0_{nk}) \to \mu C.$$

Since $f(r^0_{nk}) \leqq M$, we get $S_C f\mu < M\mu(C)$. Hence, $M\mu(C) \geqq N\mu(C)$, i.e. $M \geqq N$, because $\mu C > 0$. This is a contradiction, because $M < N$. Consequently $\mu C = 0$.

W.30.6c. It follows that $E = A^0 + A'$, $A^0 \cdot A' = O$. We are going to prove that $S_{A^0} f\mu = S_{A^0} f^+\mu$. To do this, let $P_{\dot n}$ be a (DARS)-sequence for A^0. Take any subsequence $P_{s(n)}$ of P_n. The sequence $P_{s(n)}$ is a sequence for A^0. By Theorem [W.30.5.1.] there exists a subsequence $t(\dot n)$ of $\dot n$ and a sequence $R_{\dot n}$ for A', such that

$$P_{st(n)} \frown R_n = \Theta, \quad \text{and} \quad Q_n =_{df} P_{st(n)} \smile R_n$$

is a sequence for E. Put

$$\bar P_n =_{df} P_{st(n)}, \quad \text{so} \quad Q_n = \bar P_n \smile R_n.$$

Define

$$P^0_n =_{df} \bar P_n \frown L^0, \quad P'_n =_{df} \bar P_n \frown L',$$

$$R^0_n =_{df} R_n \frown L^0, \quad R'_n =_{df} R_n \frown L', \quad Q^0_n =_{df} Q_n \frown L^0, \; Q'_n =_{df} Q_n \frown L'.$$

We have

$$\bar P_n = R^0_n \smile R'_n, \quad R_n = R^0_n \smile R'_n, \quad Q_n = Q^0_n \smile Q'_n.$$

By [W.20.6.], $\varphi(\dot a)$ is summable on E, and we know that $Q_{\dot n}$ is a sequence for E. Hence, by [W.20.3.], Q^0_n is a sequence for A^0, and $Q'_{\dot n}$ is a sequence for A'. It follows that

$$\mu(\mathrm{som}\,Q^0_n + A^0) \to 0;$$

hence

$$\mu[\mathrm{som}\,P'_n(\mathrm{som}\,Q^0_n + A^0)] \to 0;$$

hence

$$\mu[\mathrm{som}\,\bar P'_n\,\mathrm{som}\,Q^0_n + \mathrm{som}\,\bar P'_n \cdot A^0] \to 0.$$

Now $\bar P'_n \frown Q^0_n = \Theta$; hence $\mathrm{som}\,\bar P'_n \cdot \mathrm{som}\,Q^0_n = \Theta$. Hence

$$\mu(\mathrm{som}\,\bar P'_n + \mathrm{som}\,A^0) \to 0. \tag{1}$$

We also have

$$|\bar P'_n \smile \bar P^0_n, A^0| \to 0. \tag{2}$$

From (1) and (2) we get

$$|\bar P'_n \smile \bar P^0_n \sim \bar P'_n, A^0| \to 0;$$

hence, since $\bar{P}_n^0 \cap \bar{P}_n' = \emptyset$, we get

(3) $$|\bar{P}_n^0, A^0| \to 0.$$

Hence $\bar{P}_n^0$ is a sequence for A^0, since $\bar{P}_n$ is so, and $\bar{P}_n^0 \subseteqq \bar{P}_n$. Since $f\,\mu$ is summable on A^0, (as it is on E), we have $(f\,\mu)\,(\bar{P}_n^0) \to S_{A^0}\, f\,\mu$. Now, $(f\,\mu)\,(\bar{P}_n^0) = (f\,\mu)\,(f^-\,\mu)\,(\bar{P}_n)$, because the only bricks for which the terms which may differ from 0, are those of $\bar{P}_n^0$. Consequently

$$\lim_{n\to\infty} (f^-\,\mu)\,(\bar{P}_n) = S_{A^0}\, f\,\mu.$$

We recall that

$$\bar{P}_n = P_{st(n)}, \quad (n = 1, 2, \ldots).$$

Hence we can say that, if P_n is any sequence for A^0, then from every subsequence $s(\dot{n})$ of $\dot{n}$, another $s\,t(n)$ can be extracted, such that

$$\lim_{n\to\infty} (f^-\,\mu)\,(P_{st(n)}) = S_{A^0}\, f\,\mu.$$

It follows that

$$\lim_{n\to\infty} (f^-\,\mu)\,(P_n) = S_{A^0}\, f\,\mu.$$

This being valid for any sequence P_n for E, we get

$$S_E\, f^-\,\mu = S_{A^0}\, f\,\mu.$$

In a similar way we obtain

$$S_E\, f^+\,\mu = S_{A'}\, f\,\mu.$$

It follows that

$$S_E\, f\,\mu = S_E\, f^-\,\mu + S_E\, f^+\,\mu.$$

A modification of our argument gives, with the help of [W.20.6.2.], the equality

$$S_F\, f\,\mu = S_F\, f^-\,\mu + S_F\, f^+\,\mu,$$

valid for every $F \leqq E$.

W.30.6d. To prove the thesis 3), take any sequence Q_n for E. We have

$$Q_n^- = Q_n \cap L^0, \quad Q_n^+ = Q_n \cap L'.$$

If we take account of the equality $\mu\, C = 0$ in [W.30.6b.] of the present proof, and if we rely on [W.20.3.], we prove that Q_n^-, Q_n^+ are (DARS)-sequences for A^-, A^+ respectively.

W.30.6e. The proof of item 4) may be based on [W.30.3.].

W.30.7. Theorem. Under hypothesis $\boldsymbol{H}_2$, if

1) E is a measurable set of traces,

2) $f(a)$ is a field of real numbers, admitting a finite number of values only, say $M_1, M_2, \ldots, M_p$, $(p \geqq 2)$,

3) $f(a)$ is (DARS)-summable on E,

4) A', $A^2, \ldots, A^p$ are characteristic sets belonging to the sets of bricks:

$$L^i =_{df} \{q \mid f(q) = M_i\}, \quad (i = 1, \ldots, p),$$

and to E, then if we put, for $i = 1, \ldots, p$,

$$f^i(a) =_{df} \begin{cases} 0 & \text{whenever} \quad f(a) \neq M_i, \\ f(a) & \text{if} \quad f(a) = M_i, \end{cases}$$

then $f^i \mu$ is summable on E, and

$$S_E f^i \mu = S_{A^i} f \mu, \quad (i = 1, \ldots, p).$$

[Of course we have $f(a) = \sum_{i=1}^{p} f^i(a)$.]

Proof. We shall prove that $S_E f^1 \mu = S_{A'} f \mu$, the proofs for $A^2, \ldots, A^p$ being similar.

Let Q_n be a sequence for E. Decompose Q_n into disjoint complexes

$$Q_n = Q'_n \cup \cdots \cup Q^p_n, \quad \text{where} \quad Q^k_n \subseteq L^k.$$

We know, Theorem [W.30.4.1.], that Q^k_n is a sequence for A^k. We have

$$(f' \mu)(Q_n) = (f' \mu)(Q'_n) + \cdots + (f' \mu)(Q^p_n).$$

Since

$$f'(q^i_{nk}) = 0 \quad \text{for} \quad i = 2, 3, \ldots,$$

we get

(1) $$(f' \mu)(Q_n) = (f' \mu)(Q'_n).$$

On the other hand

$$(f^i \mu)(Q'_n] = 0, \quad \text{for} \quad i = 2, 3, \ldots,$$

because

$$f^i(a) = 0, \quad \text{if} \quad a \in L';$$

hence also if $a \in Q'_n \subseteq L'$. Consequently

(2) $$(f' \mu)(Q'_n) = (f \mu)(Q'_n).$$

From (1) and (2) we get

$$(f' \mu)(Q_n) = (f \mu)(Q'_n).$$

Since Q'_n is a sequence for A', and $f \mu$ is summable on A', it follows that

$$\lim_{n \to \infty} (f \mu)(Q'_n) = S_{A'} f \mu.$$

Hence for all sequences Q_n for E we get

$$\lim (f' \mu)(Q_n) = S_{A'} f \mu,$$

so the equality $S_E f' \mu = S_{A'} f \mu$ is proved. The theorem is established.

W.30.7.1. We leave to the reader the examination of fields $f(a)$, which admit a denumerable number of values, whose set has no accumulation point.

W.30.8. Theorem. We admit hypothesis $\boldsymbol{H}_2$. If

1. $f(\dot{a})$, $g(\dot{a})$ are fields of real numbers, both admitting the values $0, 1$ only,

2. $f(\dot{a})$, $g(\dot{a})$ are (DARS)-summable on a measurable set E of traces, then

1) $h(a) =_{df} f(a) \cdot g(a)$ is also (DARS)-summable on E;

2) if A^0, A', are the characteristic sets attached to $f(a)$, and B^0, B' are the characteristic sets attached to $g(a)$, then $C^0 =_{df} A^0 B^0 + A^0 B' + A' B^0$, $C' =_{df} A' B'$ are the characteristic sets attached to $h(a)$;

3) $S_{C^0} h \mu = 0$, $S_{C'} h \mu = \mu(A' B')$;

4) if $F(x)$, $G(x)$, $H(x)$ are functions (of the variable trace x) attached to $f\mu$, $g\mu$ and $fg\mu$ respectively, then $H(x) =^{\mu} F(x) \cdot G(x)$.

W.30.8a. Proof. Since $h(a)$ admits the values 0 and 1 only, we can speak of the characteristic sets C^0, C, C', attached to it, [W.20.5.]. We shall prove that $\mu C = 0$, which will imply, Theorem [W.30.4.1.], that $h(a)$ is summable. Suppose that $\mu C > 0$. Define

$$X^0 =_{df} \{p \mid f(p) = 0,\}, \qquad X' =_{df} \{p \mid f(p) = 1\},$$

$$Y^0 =_{df} \{q \mid g(q) = 0\}, \qquad Y' =_{df} \{q \mid g(q) = 1\}.$$

We have $\overline{S}_C f g \mu = \mu(C)$, $\underline{S}_C f g \mu = 0$. Let $T_{\dot{n}}$ be a sequence for C with $\lim (fg\mu)(T_n) = \overline{S}_C f g \mu$. By Theorem [W.30.5.1.] there exists a subsequence $s(\dot{n})$ of $\dot{n}$ and a sequence $T^*_{\dot{n}}$ for $E - C$, such that $T_{s(n)} \frown T^*_n = \emptyset$, and $P_n =_{df} T_{s(n)} \smile T^*_n$ is a sequence for E. We have

$$T^*_n \subseteqq P_n, \quad T_{s(n)} \subseteqq P_n, \quad \lim_{n \to \infty} (fg\mu)(T_{s(n)}) = \overline{S}_C f g \mu.$$

W.30.8b. Since we shall now deal only with bricks of P_n, we can simplify the symbolism by dropping the index n, dropping the "som" where it should be written, and by using symbols $+$, $\cdot$, $-$, co for operations on sets of bricks. Since P is a sequence for E, there exist sequences Q^0, Q' for A^0, A' respectively, where $Q^0 + Q' = P$, $Q^0 \cdot Q' = O$. We define $P^0 =_{df} P \cdot X^0$, $P' =_{df} P \cdot X'$. Since, by Theorem [W.30.4.1.], P^0, P' are sequences for A^0, A' respectively, therefore $\bar{Q}^0 =_{df} P^0 \cdot Q^0$, $\bar{Q}' =_{df} P' \cdot Q'$ are also sequences for A^0, A'. We have $\bar{Q}^0 \leqq X^0$, $\bar{Q}' \leqq X'$, and $\bar{Q}^0 + \bar{Q}'$ is a sequence for E. We have $\bar{Q}^0 \cdot \bar{Q}' = O$. Now put $R =_{df} \bar{Q}^0 + \bar{Q}'$, which is a sequence for E. There exist sequences S^0, S' for B^0, B' respectively, where $S^0 + S' = R$, $S^0 \cdot S' = O$. On the other hand, by [W.30.4.1.], $R^0 =_{df} R \cdot Y^0$, $R' =_{df} R \cdot Y'$ are also sequences for B^0, B' respectively. Hence $\bar{R}^0 =_{df} S^0 R^0$, $\bar{R}' =_{df} S' R'$ are sequences for B^0, B' respectively. We have $\bar{R}^0 \cdot \bar{R}' = O$; and $\bar{R}^0 + \bar{R}'$ is a sequence for E; $\bar{R}^0 \leqq Y^0$; $\bar{R}' \leqq Y'$.

W.30.8c. Since $\bar{R}^i$ is a sequence for B^i, and $\bar{Q}^j$ is a sequence for A^j, $(i, j = 0, 1)$, it follows that $\bar{R}^i \cdot \bar{Q}^j$ is a sequence for $B^i A^j$. We have $\bar{R}^i \cdot \bar{Q}^j \leqq X^i \cdot Y^j$, $(i, j = 0, 1)$. The sequences $\bar{Q}^0 \cdot \bar{R}^0$, $\bar{Q}' \cdot \bar{R}^0$, $\bar{Q}^0 \cdot \bar{R}'$, $\bar{Q}' \cdot \bar{R}'$ are disjoint, and their bricks belong to $X^0 Y^0$, $X' Y^0$, $X^0 Y'$, $X' Y'$ respectively.

W.30.8d. Now, concerning T, since $\lim_{n\to\infty} (f g \mu)(T_n) = \mu(C)$, we have

$$\lim_{n\to\infty} \sum_k{}' f(t_{nk})\, g(t_{nk}) = (t_{nk}) = \mu(C), \tag{1}$$

where the prime will say, that only those bricks t_{nk} are taken into account, for which $g(t_{nk}) = 1$, $f(t_{nk}) = 1$, i.e., the bricks of $X' Y'$. Denote the subcomplex of T, composed of those bricks, by $\bar{T}$. We have $\bar{T} \to C$ and

$$(\bar{Q}^0 \bar{R}^0 + \bar{Q}^0 \bar{R}' + \bar{Q}' \bar{R}^0) \to A^0 B^0 + A^0 B' + A' B^0.$$

Since

$$\bar{T} \cdot (\bar{Q}^0 \bar{R}^0 + \bar{Q}^0 \bar{R}' + \bar{Q}' \bar{R}^0) = O,$$

it follows that

$$C \cdot (A^0 B^0 + A^0 B' + A' B^0) = O.$$

Hence

$$C \leqq A' B', \quad \text{as} \quad C \leqq E. \tag{2}$$

W.30.8e. We shall repeat our argument by selecting a sequence S_n such, that

$$\lim_{n\to\infty} (f g \mu)(S_n) = S_C\, f g \mu = 0. \tag{3}$$

We get analogous sequences for $A^0 B^0$, $A^0 B'$, $A' B^0$, $A' B'$, which we shall denote by $\bar{Q}_1^0 \cdot \bar{R}_1^0$, $\bar{Q}_1' \cdot \bar{R}_1^0$, $\bar{Q}_1^0 \cdot \bar{R}_1'$, $\bar{Q}_1' \cdot \bar{R}_1'$. They are composed of bricks belonging to $X^0 Y^0$, $X' Y^0$, $X^0 Y'$, $X' Y'$ respectively.

By (3) the collection of bricks, in S_n, belonging to $X' Y'$, must be such, that their joint measure tends to 0. Hence $S_n \bar{Q}_1' \bar{R}_1' \to 0$. Since $S_n \to C$, $\bar{Q}_1' \bar{R}_1' \to A' B'$, it follows that $C \cdot A' B' = O$.

Since, by (2), $C \leqq A' B'$, it follows that $\mu C = 0$, which is in contradiction with $\mu C > 0$.

Since $C = O$, the characteristic decomposition of E for $f g$, is $E = C^0 + C'$, where $f g \mu$ is summable on C^0 with value 0, and on C' with value $\mu(C')$. Since $f g \mu$ is summable on subsets of E: $A^0 B^0$, $A^0 B'$, $A' B^0$, $A' B'$, and since these sets are approximated by complexes $Q^1 R^1$, whose bricks belong respectively to

$$X^0 Y^0, X^0 Y', X' Y^0, X' Y',$$

we have

$$(f \mu)(\bar{Q}^0 \bar{R}^0) \to 0, \quad (f \mu)(\bar{Q}^0 \bar{R}') \to 0, \quad (f \mu)(\bar{Q}' \bar{R}^0) \to 0,$$

$$(f \mu)(\bar{Q}' \bar{R}') \to \mu(A' B').$$

We get $C^0 = A^0 B^0 + A^0 B' + A' B'$, $C' = A' B'$. By direct computation we get for the functions F, G, H, attached to $f\mu$, $g\mu$, $fg\mu$: $F(x)\,G(x) = H(x)$. The theorem is proved.

W.30.9. Theorem. Under hypothesis $\boldsymbol{H}_2$. If

1. $f(a)$, $g(a)$ admit a finite number of values only,
2. E is a set of traces,
3. $f\mu$, $g\mu$ are summable on E,

then so is $fg\mu$, and we get for the functions F, G, H, attached to $f\mu$, $g\mu$, $fg\mu$ respectively, the μ-equality:

$$F(x)\cdot G(x) =^{\mu} H(x).$$

Proof. Let $f(a)$ admit the values

$$M_1, M_2, \ldots, M_p, \qquad p \geqq 1,$$

and $g(a)$ the values

$$N_1, N_2, \ldots, N_q, \qquad q \geqq 1.$$

Define

$$f_k(a) = \begin{cases} 0 & \text{if } f(a) \neq M_k, \\ 1 & \text{if } f(a) = M_k, \end{cases} \qquad (k = 1, \ldots, p),$$

and

$$g_l(a) = \begin{cases} 0 & \text{if } g(a) \neq M_l, \\ 1 & \text{if } g(a) = M_l, \end{cases} \qquad (l = 1, \ldots, q).$$

We have

$$f(a) = \sum_{k=1}^{p} f_k(a)\cdot M_k, \qquad g(a) = \sum_{l=1}^{q} g_l(a)\cdot N_l.$$

The fields $f_k\mu$, $g_l\mu$ are summable, by virtue of Theorem [W.30.7.]. We have

$$f(a)\,g(a)\,\mu(a) = \sum_{k=1}^{p}\sum_{l=1}^{q} f_k(a)\cdot g_l(a)\cdot M_k\cdot N_l\cdot\mu(a).$$

By Theorem [W.30.8.] all terms in this sum are summable; hence $fg\mu$ is summable. Let $F_k(x)$, $G_l(x)$ be the functions of traces, attached to $f_k\mu$, $g_l\mu$, respectively. Then, by Theorem [W.30.8.], the function, attached to $f_k g_l \mu$, is $F_k G_l$; hence the function attached to $M_k N_l f_k g_l \mu$, is $M_k N_l F_k G_l$. Hence, by additivity, the function attached to $fg\mu$ is

$$\sum_{k=1}^{p}\sum_{l=1}^{q} M_k F_k\cdot N_l G_l = \left(\sum_{k=1}^{p} M_k F_k\right)\left(\sum_{l=1}^{q} N_l G_l\right) = F\cdot G.$$

The theorem is established.

W.30.10. Till now we have studied fields, admitting a finite number of values only. To deal with more general fields, we introduce the following notion:

W.30.10.1. Def. Let us admit the hypothesis $\boldsymbol{H}_2$. Let $f(a)$ be an admissible field of real numbers, and E a measurable set of traces. Then [U.9.], [W.4.], $\overline{S}_E f$ and $\underline{S}_E f$ exist. We shall call the difference $\overline{S}_E f - \underline{S}_E f$ *sum-oscillation of* $f(a)$ *on the set* E, and introduce the notation,

$$\Delta_E f(a) =_{df} \overline{S}_E f(a) - \underline{S}_E f(a).$$

W.30.11. Theorem. Under hypothesis $\boldsymbol{H}_2$, if

1. $f(a)\,\mu(a)$, $g(a)\,\mu(a)$ are admissible fields of real numbers,
2. $|f(a) - g(a)| \leqq \alpha$ for all bricks a,
3. E is a measurable set of traces,

then

$$|\Delta_E f\mu - \Delta_E g\mu| \leqq 2\alpha\,\mu(E).$$

Proof. We have for all a:

$$g(a) - \alpha \leqq f(a) \leqq g(a) + \alpha;$$

hence, by [W.4.8.],

$$\overline{S}_E(g-\alpha)\,\mu \leqq \overline{S}_E f\mu \leqq \overline{S}_E(g+\alpha)\,\mu,$$

and

$$\underline{S}_E(g-\alpha)\,\mu \leqq \underline{S}_E f\mu \leqq \underline{S}_E(g+\alpha)\,\mu.$$

By [W.4.7.] it follows

(1) $$\overline{S}_E g\mu - \alpha\,\mu(E) \leqq \overline{S}_E f\mu \leqq \overline{S}_E g\mu + \alpha\,\mu(E),$$

and

(2) $$\underline{S}_E g\mu - \alpha\,\mu(E) \leqq \underline{S}_E f\mu \leqq \underline{S}_E g\mu + \alpha\,\mu(E).$$

From (2) we get

$$-\underline{S}_E g\mu + \alpha\,\mu(E) \geqq -\underline{S}_E f\mu \geqq -\underline{S}_E g\mu - \alpha\,\mu(E);$$

hence

(3) $$-\underline{S}_E g\mu - \alpha\,\mu(E) \leqq -\underline{S}_E f\mu \leqq -\underline{S}_E g\mu + \alpha\,\mu(E).$$

Adding sidewise (1) and (3), we obtain:

$$\Delta_E g\mu - 2\alpha\,\mu(E) \leqq \Delta_E f\mu \leqq \Delta_E g\mu + 2\alpha\,\mu(E);$$

hence

$$|\Delta_E f\mu - \Delta_E g\mu| \leqq 2\alpha\,\mu(E), \quad \text{q.e.d.}$$

W.30.11.1. Theorem. Under hypothesis $\boldsymbol{H}_2$, if

1. $f(\dot{a})$, $g(\dot{a})$ are fields of real numbers,
2. E is a measurable set of traces,
3. $f(\dot{a})\,\mu(\dot{a})$ is (DARS)-summable on E,
4. $g(a)\,\mu(a)$ is admissible,
5. $|f(a) - g(a)| \leqq \alpha$, for all a,

then

$$0 \leqq \Delta_E g\mu \leqq 2\alpha\,\mu(E).$$

Proof. This follows directly from [W.30.11.], because $\Delta_E f\mu = 0$.

W.30.11.2. Lemma. Under hypothesis $\boldsymbol{H}_2$, if

1. $f_n(a)$, $(n = 1, 2, \ldots)$, $f_n(a)$ are fields of real numbers,
2. E is a measurable set of traces,
3. $f_n(\dot{a})\,\mu(\dot{a})$, $f(\dot{a})\,\mu(\dot{a})$ are admissible,
4. $\Delta_E f_n\,\mu \to 0$ for $n \to \infty$,
5. $f_n(a)$ tends uniformly to $f(a)$,

then

$$f\,\mu \text{ is (DARS)-summable on } E.$$

Proof. Let $\alpha > 0$. There exists N, such that for all $n \geqq N$ and all a:

$$|f - f_n| \leqq \alpha.$$

We have by Theorem [W.30.11.]:

$$|\Delta_E f\,\mu - \Delta_E f_n\,\mu| \leqq 2\alpha\,\mu(E) \quad \text{for} \quad n \geqq N.$$

Hence

$$\Delta_E f_n\,\mu \to \Delta_E f\,\mu \quad \text{for} \quad n \to \infty,$$

and then $\Delta_E f\,\mu = 0$, i.e. $f\,\mu$ is (DARS)-summable on E.

W.30.11.3. Lemma. Under hypothesis $\boldsymbol{H}_2$, if

1. f_n, f, E are as before, and 1., 2., 3., in [W.30.11.2.] are satisfied,
2. $f_n(a)$ tends uniformly to $f(a)$,
3. $f\,\mu$ is (DARS)-summable on E,

then

$$\Delta_E f_n\,\mu \to 0 \quad \text{for} \quad n \to \infty.$$

Proof. Let $\alpha > 0$. Then there exists N such that, for all $n \geqq N$, $|f(a) - f_n(a)| \leqq \alpha$ for all a. Hence, by Theorem [W.30.11.],

$$|\Delta_E f\,\mu - \Delta_E f_n\,\mu| \leqq \alpha \cdot \mu(E) \quad \text{for} \quad n \geqq N.$$

It follows

$$\Delta_E f_n\,\mu \leqq \alpha\,\mu(E) \quad \text{for} \quad n \geqq N.$$

Consequently

$$\Delta_E f_n\,\mu \to 0 \quad \text{for} \quad n \to \infty.$$

The lemma is proved.

W.30.11.4. Theorem. Let us admit the hypothesis $\boldsymbol{H}_2$. If

1. f, g are fields of real numbers, both admissible,
2. E is a measurable set of traces,

then

$$\Delta_E(f + g) \leqq \Delta_E f + \Delta_E\,g.$$

Proof. By [W.4.10.] we have

$$\overline{S}_E(f + g) \leqq \overline{S}_E\,f + \overline{S}_E\,g, \tag{1}$$

and

$$\underline{S}_E(f + g) \geqq \underline{S}_E\,f + \underline{S}_E\,g.$$

Hence

(2) $$\underline{S}_E(f+g) \leqq -\underline{S}_E f - \underline{S}_E g.$$

Adding (1) and (2), we get

$$\Delta_E(f+g) \leqq \Delta_E f + \Delta_E g. \quad \text{q.e.d.}$$

W.30.12. Lemma. We admit the hypothesis $\boldsymbol{H}_2$. If

1. $f(a)$ admits a finite number of values only,
2. $0 \leqq f(a) \leqq 1$ for all a,
3. E is a measurable set of traces,

then

$$\Delta_E f^2(a)\,\mu(a) \leqq 2\Delta_E f(a)\,\mu(a).$$

Proof. Let the values M_i, admitted by $f(a)$, be

$$0 \leqq M_1 < M_2 < \cdots < M_p \leqq 1.$$

Consider the characteristic sets $A^{i_1\ldots i_k}$, as in [U.30.4.1.]. We have

$$\overline{S}_{A^{i_1\ldots i_k}} f\,\mu = \max(M_{i_1}, \ldots, M_{i_k}) \cdot \mu(A^{i_1\ldots i_k}),$$

$$\underline{S}_{A^{i_1\ldots i_k}} f\,\mu = \min(M_{i_1}, \ldots, M_{i_k}) \cdot \mu(A^{i_1\ldots i_k}).$$

For some α and β we have

$$M_\alpha = \operatorname{Max}(M_{i_1}, \ldots, M_{i_k}), \qquad M_\beta = \operatorname{Min}(M_{i_1}, \ldots, M_{i_k});$$

hence

$$\Delta_{A^{i_1\ldots i_k}} f\,\mu = (M_\alpha - M_\beta) \cdot \mu(A^{i_1\ldots i_k}).$$

Since $0 \leqq M_1^2 \leqq M_2^2 \leqq \cdots \leqq M_p^2 \leqq 1$, and since the characteristic sets for f^2 are the same as for f, we get

$$\Delta_{A^{i_1\ldots i_k}} f^2\,\mu = (M_\alpha^2 - M_\beta^2) \cdot \mu(A^{i_1\ldots i_k}).$$

Now

$$M_\alpha^2 - M_\beta^2 = (M_\alpha - M_\beta) \cdot (M_\alpha + M_\beta) \leqq 2(M_\alpha - M_\beta),$$

because $0 \leqq f(a) \leqq 1$. It follows

$$\Delta_{A^{i_1\ldots i_k}} f^2\,\mu \leqq 2(M_\alpha - M_\beta) \cdot \mu(A^{i_1\ldots i_k}) = 2\Delta_{A^{i_1\ldots i_k}} f\,\mu.$$

If we take the sum over all subsets $(i_1, \ldots, i_k)$ of $(1, 2, \ldots, p)$, we obtain

$$\Delta_E f^2\,\mu \leqq 2\Delta_E f\,\mu, \quad \text{q.e.d.}$$

W.30.13. Lemma. Under hypothesis $\boldsymbol{H}_2$. If

1. $f(a)$ is a field of real numbers,
2. E is a measurable set of traces,
3. $0 \leqq f(a) < 1$,
4. $f(a)\,\mu(a)$ is (DARS)-summable on E,

then

$[f(a)]^2\,\mu(a)$ is also (DARS)-summable on E.

Proof. Define the fields for $n = 1, 2, \ldots$:

$$f_n(a) =_{df} \begin{cases} \frac{\lambda}{n} & \text{whenever } \frac{\lambda}{n} \leqq f(a) < \frac{\lambda+1}{n}, \\ \text{for} & \lambda = 0, 1, \ldots, n-1. \end{cases}$$

Thus $f_n(a)$ admits the values $0, \frac{1}{n}, \frac{2}{n}, \ldots, \frac{n-1}{n}$ only.

If

$$\frac{\lambda}{n} \leqq f(a) < \frac{\lambda+1}{n},$$

then

$$0 \leqq f(a) - f_n(a) < \frac{\lambda+1}{n} - \frac{\lambda}{n} = \frac{1}{n},$$

so

$$0 \leqq f(a) - f_n(a) < \frac{1}{n}$$

for all a and $n = 1, 2, \ldots$ Consequently $f_n(a)$ tends uniformly to $f(a)$ for all a. Since $f\mu$ is summable on E, it follows, by [W.30.11.3.], that $\lim_{n\to\infty} \Delta_E f_n \mu = 0$. Since, by [W.30.12.], $\Delta_E f_n^2 \mu \leqq 2\Delta_E f_n \mu$, it follows that

$$\lim_{n\to\infty} \Delta_E f_n^2 \mu = 0.$$

By [W.30.11.2.] it follows, that $f_n^2 \mu$ is summable on E, q.e.d.

W.30.14. Theorem. Under hypothesis $\boldsymbol{H}_2$, if

1. $f(a)$ is a field of real numbers,
2. E is a measurable set of traces,
3. $f(a)$ is bounded (i.e. there exists $M > 0$, such that $|f(a)| \leqq M$ for all a),
4. $f(a)\mu(a)$ is (DARS)-summable on E,

then

$$f^2(a)\mu(a) \text{ is also (DARS)-summable on } E.$$

Proof. Let $|f(a)| \leqq M$ for all bricks a, where $M > 0$. Then

$$-M \leqq f(a) \leqq M, \quad 0 \leqq f(a) + M \leqq 2M < 2M + 1, \quad 0 \leqq \frac{f(a)+M}{2M+1} < 1.$$

If we put

$$g(a) =_{df} \frac{f(a)+M}{2M+1},$$

we have

$$0 \leqq g(a) < 1$$

and

$$f(a) = (2M+1)\,g(a) - M.$$

Hence

$$f^2 = (2M+1)^2 g^2 + M^2 - 2M(2M+1)\,g.$$

Since $g\,\mu$ is summable on E, so is

$$[-2M(2M+1)\cdot g + M^2]\,\mu,$$

by Lemma [W.30.13.], $g^2\,\mu$ is summable on E; hence also $(2M+1)\,g^2\,\mu$. Consequently, $f^2\,\mu$ is summable on E. The theorem is proved.

W.30.15. Theorem. Under hypothesis $\boldsymbol{H}_2$, if

1. $f(\dot{a})$, $g(\dot{a})$ are fields of real numbers,
2. E is a measurable set of traces,
3. $f(\dot{a})$, $g(\dot{a})$ are both bounded,
4. $f\,\mu$, $g\,\mu$ are both (DARS)-summable on E,

then

$$f(a)\cdot g(a)\cdot \mu(a) \text{ is also (DARS)-summable on } E.$$

Proof. We have

$$f g = \tfrac{1}{4}[(f+g)^2 - (f-g)^2] \quad \text{for all } a.$$

Since $f\,\mu$, $g\,\mu$ are summable, so is $f+g$ and $f-g$. Since these fields are bounded, therefore, by Theorem [W.30.14.], $(f+g)^2\,\mu$, $(f-g)^2\,\mu$ are summable; hence so is their difference. This completes the proof.

W.30.16. Theorem. If

1. the hypothesis $\boldsymbol{H}_2$ is admitted,
2. $f(a)$ is admissible,
3. E is a measurable set of traces with $\mu E > 0$,
4. for every subset F of E we have $S_F f = 0$,

then

$$\bar{S}_E |f| = 0, \quad \text{hence also} \quad \bar{S}_F |f| = 0 \quad \text{for all} \quad F \leqq E.$$

W.30.16a. Define the fields $f^+(\dot{a})$, $f^-(\dot{a})$ as follows:

$$(1)\qquad \begin{cases} f^+(a) = \begin{cases} f(a) & \text{whenever } f(a) > 0, \\ 0 & \text{whenever } f(a) \leqq 0, \end{cases} \\ f^-(a) = \begin{cases} 0 & \text{whenever } f(a) > 0, \\ f(a) & \text{whenever } f(a) \leqq 0. \end{cases} \end{cases}$$

We have for every a:

$$(2)\qquad f(a) = f^+(a) + f^-(a).$$

If $Q = \{q_1, q_2, \ldots\}$ is a complex, we denote by q_i^+, q_i^-, q_i^0 those bricks of $q_i \in Q$, for which $f(q_i) > 0$, < 0, $= 0$ respectively. The corresponding subsets of Q will be denoted by Q^+, Q^-, Q^0 respectively. The complexes Q^+, Q^-, Q^0 are mutually disjoint, and

$$(3)\qquad Q = \bigcup_i q_i^+ \cup \bigcup_i q_i^- \cup \bigcup_i q_i^0;$$

some of these complexes may be empty.

It follows that

(4) $$g(Q) = g(Q^+) + g(Q^-) + g(Q^0) \quad \text{for any field } g(\dot{a}).$$

W.30.16b. We have

(5) $$f^+(q_i^-) = f^-(q_i^+) = f^+(q_i^0) = f^-(q_i^0) = 0.$$

We also have

(6) $$f(q_i^+) = f^+(q_i^+) = |f(q_i^+)|, \quad f(q_i^-) = f^-(q_i^-) = -|f(q_i^-)|.$$

W.30.16c. Let Q_n be a (DARS)-sequence for F, where $F \leqq E$, and such that

(7) $$\lim_{n\to\infty} \sum_k |f(q_{nk})| = \bar{S}_F |f|.$$

We have by (5) and (6):

(8) $$\sum_k \{|f(q_{nk}^+)| + |f(q_{nk}^-)|\} = \sum_k \{f(q_{nk}^+) - f(q_{nk}^-)\} \to \bar{S}_F |f|.$$

On the other hand, since $f(a)$ is summable on F, with sum equal 0, we have

(9) $$\sum_k \{f(q_{nk}^+) + f(q_{nk}^-)\} \to 0.$$

From (8) and (9) we get by addition and subtraction:

(10) $$2\sum_k f(q_{nk}^+) \to \bar{S}_F |f|, \quad 2\sum_k f(q_{nk}^-) \to -\bar{S}_F |f|;$$

hence, by (6),

(11) $$2\sum_k f^+(q_{nk}^+) \to \bar{S}_F |f|, \quad 2\sum_k f^-(q_{nk}^-) \to -\bar{S}_F |f|.$$

From (11) and (5) we have

$$\sum_k \{f^+(q_{nk}^+) + f^+(q_{nk}^-) + f^+(q_{nk}^0)\} \to \tfrac{1}{2} \bar{S}_F |f|,$$

i.e.

(12) $$\sum_k f^+(q_{nk}) \leftrightarrow \tfrac{1}{2} \bar{S}_P |f|.$$

Now, since $Q_{\dot{n}}$ is a sequence for F, it follows that

$$\lim \sum_k f^+(q_{nk}) \leqq \bar{S}_F f^+.$$

Consequently

(13) $$\tfrac{1}{2} \bar{S}_F |f| \leqq \bar{S}_F f^+.$$

From (11) and (5) we have

(13.0) $$\sum_k \{f^-(q_{nk}^+) + f^-(q_{nk}^-) + f^-(q_{nk}^0)\} \to -\tfrac{1}{2} \bar{S}_F |f|,$$

i.e.

$$\sum_k f^-(q_{nk}) \to -\tfrac{1}{2} \bar{S}_F |f|.$$

Since Q_n is a sequence for F, it follows that

$$\lim \sum_k f^-(q_{nk}) \geqq \underline{S}_F f^-.$$

Consequently

(13.1) $$-\tfrac{1}{2} \bar{S}_F |f| \geqq \underline{S}_F f^-.$$

W.30.16d. Let P_n be a (DARS)-sequence for F, such that

(14) $$\lim_{n\to\infty} f^+(P_n) = \bar{S}_F f^+.$$

We have, by (4),

$$f^+(p^+_{nk}) + f^+(p^-_{nk}) + f^+(p^0_{nk}) \to \bar{S}_F f^+;$$

hence, by (5) and (6),

$$f^+(p^+_{nk}) \to \bar{S}_F f^+.$$

Hence, by (6),

(15) $$f(P^+_n) \to \bar{S}_F f^+.$$

We have, by hypothesis, $f(P_n) \to 0$, i.e.

$$f(P^+_n) + f(P^-_n) \to 0.$$

Consequently

(16) $$f(P^-_n) \to -\bar{S}_F f^+.$$

From (15) and (16) we get

$$f(P^+_n) - f(P^-_n) \to 2\bar{S}_F f^+;$$

hence, by (16),

(16.1) $$\sum_k |f(p^+_{nk})| + \sum_k |f(p^-_{nk})| = \sum_k |f(p_{nk})| \to 2\bar{S}_F f^+.$$

Consequently

(17) $$2\bar{S}_F f^+ \leqq \bar{S}_F |f|.$$

From (13) and (14) it follows

(18) $$2\bar{S}_F f^+ = \bar{S}_F |f|.$$

W.30.16e. We have supposed that

$$\lim \sum_k |f(q_{nk})| = \bar{S}_F |f|,$$

and we have got, (12),

$$\lim \sum_k f^+(q_{nk}) = \tfrac{1}{2} \bar{S}_F |f| =, \text{ by } (15), = \bar{S}_F f^+.$$

This says that, the same (DARS)-sequence for

(18.1) E which yields $\bar{S}_F |f|$, also yields $\bar{S}_F f^+$.

On the other hand, we have supposed, (14), that

$$\lim_{n\to\infty} \sum_k f^+(p_{nk}) = \bar{S}_F f^+$$

and proved that, (16.1)

$$\lim_{n\to\infty}\sum_k |f(p_{nk})| = 2\bar{S}_F f^+;$$

hence, by (18),

$$\lim_{n\to\infty}\sum_k f(p_{nk}) = \bar{S}_F |f|.$$

Thus every sequence for

(18.2) F which yields $\bar{S}_F f^+$, also yields $\bar{S}_F |f|$.

From (18.1), (18.2) it follows, that the same sequences for F yield

(18.3) $\bar{S}_F |f|$ and $\bar{S}_F f^+$.

W.30.16f. Let R_n be a (DARS)-sequence for F, such that

(19) $$\lim_{n\to\infty} f^-(R_n) = \underline{S}_F f^-.$$

We have

$$\sum_k \{f^-(r^+_{nk}) + f^-(r^-_{nk}) + f^-(r^0_{nk})\} \to \underline{S} f^-.$$

Since

$$f^-(r^+_{nk}) = 0, \quad f^-(r^0_{nk}) = 0,$$

it follows

(20) $$\sum_k f^-(r^-_{nk}) \to \underline{S}_F f^-;$$

hence, by (6),

(21) $$\sum_k f(r^-_{nk}) \to \underline{S}_F f^-.$$

On the other hand we have, by hypothesis,

$$\sum_k f(r^+_{nk}) + \sum_k f(r^-_{nk}) \to 0;$$

hence

(22) $$\sum_k f(r^+_{nk}) \to -\underline{S}_F f^-.$$

From (21) and (22), we get

(22.1) $$\sum_k |f(r_{nk})| = \sum_k \{|f(r^+_{nk})| + |f(r^-_{nk})|\}$$

$$= \sum_k f(r^+_{nk}) - \sum_k f(r^-_{nk}) \to -\underline{S}_F f^- - \underline{S}_F f^- = -2\underline{S}_F f^-.$$

Hence

$$-2\underline{S}_F f^- \leqq \bar{S}_F |f|;$$

hence

(23) $$-\tfrac{1}{2}\bar{S}_F |f| \leqq \underline{S}_F f^-.$$

Consequently, from (13.1) we get

(24) $$\underline{S}_F f^- = -\tfrac{1}{2}\bar{S}_F |f|.$$

We have supposed that

$$\lim \sum_k |f(q_{nk})| = \overline{S}_F |f|,$$

and proved that, (13.1),

$$\lim \sum_k f^-(q_{nk}) = -\tfrac{1}{2}\overline{S}_F |f|,$$

i.e., by (24),

$$\lim \sum_k f^-(q_{nk}) = \underline{S}_F f^-.$$

Hence every (DARS)-sequence, which yields $\overline{S}_F |f|$, yields also $\underline{S}_F f^-$. Later we have supposed that, (19),

$$\lim \sum_k f^-(r_{nk}) = \underline{S}_F f^-,$$

and have proved that, (22.1),

$$\lim \sum_k |f(r_{nk})| = -2\underline{S}_F f^-,$$

i.e. by (24),

$$\lim \sum_k |f(r_{nk})| = \overline{S}_F |f|.$$

Hence every (DARS)-sequence, which yields $\underline{S}_F f^-$ also yields $\overline{S}_F |f|$. If we compare this result with (18.1), we can say, that the following lemma is true:

W.30.16g. Lemma. The (DARS)-sequences for F yield, at the same time,

(25) $$\underline{S}_F f^-, \ \overline{S}_F f^+ \ \text{and} \ \overline{S}_F |f|.$$

We have

(26) $$\underline{S}_F f^- = -\tfrac{1}{2}\overline{S}_F |f| \ \text{and} \ S_F f^+ = \tfrac{1}{2}\overline{S}_F |f|.$$

The statements (25), (26) are true for all

$$F \leqq E.$$

W.30.16h. Suppose that

(27) $$M =_{df} \overline{S}_E |f| > 0.$$

Let $\Phi(x)$ be the function of a trace x, attached to $\overline{S}_A |f|$, where A is a variable subset of E. We have $\overline{S}_A |f| = \int\limits_A \Phi(x)\, d\mu$ for all A. We have $\Phi(x) \geqq 0$ for all x.

Denote by E' the set of all x, for which $\Phi(x) > 0$; then $\overline{S}_{E'} |f| > 0$, and for every $F < E'$ we have $\overline{S}\mathrm{n}\, |f| > 0$, and then, by [W.30.16g.],

(27.1) $$\overline{S}\mathrm{n}\, f^+ = -\underline{S}\mathrm{n}\, f^- > 0.$$

W.30.16i. We may assume that $E' = E$. Let Q'_n be a (DARS)-sequence for E, such that

$$\lim_{n\to\infty} \sum_k |f(q'_{nk})| = \overline{S}_E |f|.$$

Since the hypothesis $\boldsymbol{H}_2$ is admitted, we can rely on Theorem [U.10.7.]. Hence there exists a subsequence $s(\dot{n})$ of $\dot{n}$ and a (DARS)-sequence Q_n for E, such that $Q_n \subseteqq Q'_{s(n)}$, $(n = 1, 2, \ldots)$, that Q_n satisfies the condition of non overlapping, and that

$$\lim_{n\to\infty} \sum_k |f(q_{nk})| = \bar{S}_E |f|. \tag{28}$$

Take such a sequence Q_n. By the lemma proved above, we have

$$\begin{gathered}\lim_{n\to\infty} \sum_k |f(q_{nk})| = M > 0, \qquad \lim_{n\to\infty} \sum_k f^+(q_{nk}) = \bar{S}_E f^+ = \frac{M}{2},\\ \lim_{n\to\infty} \sum_k f^-(q_{nk}) = \underline{S}_E f^- = -\frac{M}{2}.\end{gathered} \tag{29}$$

Take a positive number ε, such that

$$0 < \varepsilon < \frac{M}{32}. \tag{30}$$

Find n_1, such that for all $n \geqq n_1$ we have

$$\begin{gathered}\sum_k |f(q_{nk})| = M + \varepsilon\,\Theta_n, \qquad \sum_k f^+(q_{nk}) = \frac{M}{2} + \varepsilon\,\Theta_n^+,\\ \sum_k f^-(q_{nk}) = -\frac{M}{2} + \varepsilon\,\Theta_n^-,\end{gathered} \tag{31}$$

where $|\Theta_n|, |\Theta_n^+|, |\Theta_n^-| \leqq 1$. Since Q_n is a sequence for E, there exists n_2, such that $n_2 \geqq n_1$ and that for all $n \geqq n_2$:

$$|Q_n, E| \leqq \varepsilon. \tag{32}$$

W.30.16j. Lemma. If

$$|a, b| \leqq \varepsilon,$$

then

$$|a\,c, b\,c| \leqq \varepsilon.$$

Indeed,

$$|a\,c, b\,c| = \mu(a\,c \dotplus b\,c) = \mu[c(a \dotplus b)] \leqq \mu(a \dotplus b) = |a, b| \leqq \varepsilon.$$

W.30.16k. Thus from (32) we get

$$|Q_n E, E| \leqq \varepsilon, \qquad |Q_n E, Q_n| \leqq \varepsilon. \tag{33}$$

The complexes Q_n^+, Q_n^- are disjoint, and $Q_n = Q_n^+ \cup Q_n^-$. We have

$$\operatorname{som} Q_n E = \operatorname{som} Q_n^+ \cdot E + \operatorname{som} Q_n^- \cdot E. \tag{34}$$

Hence from (33):

$$|\operatorname{som} Q_n^+ E, \operatorname{som} Q_n^+| \leqq \varepsilon, \qquad |\operatorname{som} Q_n^- E, \operatorname{som} Q_n^-| \leqq \varepsilon. \tag{34.1}$$

From (34), by Theorem [W.6.], we have

$$\bar{S}_{\operatorname{som} Q_n E} |f| = \bar{S}_{\operatorname{som} Q_n^+ E} |f| + \bar{S}_{\operatorname{som} Q_n^- E} |f|. \tag{35}$$

We have by (33)

$$|\operatorname{som} Q_n E, E| \to 0;$$

hence

$$\lim_{n\to\infty} \bar{S}_{Q_n E}|f| = \bar{S}_E|f| = M.$$

This follows, by Fréchet's theorem, (see (16)), because the set-function $\bar{S}_A|f|$ of the variable set $A \leqq E$ is denumerably additive: Theorem [W.9.]. Find $n_3 \geqq n_2$, such that for all $n \geqq n_3$,

(36) $$\bar{S}_{Q_n E}|f| = M + \varepsilon \cdot Q'_n,$$

where $|\Theta'_n| \leqq 1$. Hence we have for all $n \geqq n_3$:

$$\bar{S}_{\mathrm{som}\, Q_n^+ \cdot E}|f| + \bar{S}_{\mathrm{som}\, Q_n^- \cdot E}|f| = M + \varepsilon\, \Theta'_n.$$

Put $N =_{df} n_3$. We have by (30),

$$\bar{S}_{\mathrm{som}\, Q_N^+ \cdot E}|f| + \bar{S}_{\mathrm{som}\, Q_N^- \cdot E}|f| \geqq \tfrac{7}{8} M.$$

Hence one of the numbers

(37) $\bar{S}_{\mathrm{som}\, Q_N^+ \cdot E}|f|$, $\bar{S}_{\mathrm{som}\, Q_N^- \cdot E}|f|$ must be $\geqq \frac{7}{16}$, hence $\geqq \frac{M}{4}$.

W.30.161. Suppose that

(38) $$\bar{S}_{\mathrm{som}\, Q_N^- E}|f| \geqq \frac{M}{4}.$$

Put

(39) $$E' =_{df} \mathrm{som} Q_N^- \cdot E.$$

We have

(40) $$\bar{S}_{E'}|f| \geqq \frac{M}{4},$$

and from (34.1),

(41) $$|E', \mathrm{som} Q_N^-| \leqq \varepsilon;$$

hence

(42) $$E' \leqq \mathrm{som} Q_N^-, \quad E' \cdot \mathrm{som} Q_N^+ = 0, \quad E' \leqq E.$$

Since $Q_{\dot n}$ is a (DARS)-sequence for E, and since $E' \leqq E$, there exists a partial complex P_n of Q_n, such that $P_{\dot n}$ is a (DARS)-sequence for E'. Hence $\{P_{N+1}, P_{N+2}, \ldots\}$ is also a (DARS)-sequence for E'. Now, since $Q_{\dot n}$ satisfies the condition of non overlapping, (see Def. [U.10.1.]), therefore, if $q \in P_n$, $(n > N)$, then either q is disjoint with $\mathrm{som} Q_N^+$, or q is contained in one of the bricks of Q_N^+. By Lemma [W.20.2.1.], since $|P_n, E'| \to 0$ for $n \to \infty$, therefore if P'_n is the complex, composed of all bricks $p \in P_n$, for which $E' \cdot p \neq O$, therefore $|P'_n, E'| \to 0$, and hence, P'_n is (DARS)-sequence for E'.

We have

(43) $$P'_n \frown Q_N^+ = \emptyset \quad \text{for} \quad n > N.$$

In addition to that we have

(44) $$P'_n \subseteqq P_n \subseteqq Q_n \quad \text{for} \quad n > N.$$

W.30.16m. Since P'_n is a (DARS)-sequence for E', there exists $n_3 > N$, such that for all $n \geqq n_3$ we have $|P'_n, E'| \leqq \varepsilon$. Hence $|P'_n, E \cdot \mathrm{som} Q_N^-| \leqq \varepsilon$. As by (34.1), $|Q_N^+, E \cdot \mathrm{som} Q_N^+| \leqq \varepsilon$, it follows, by (43), $|Q_N^+ \cup P'_n, E\, \mathrm{som} Q_N^+ + E\, \mathrm{som} Q_N^-| \leqq 2\varepsilon$; hence by (34) we get: $|Q_N \cup P'_n, E\, Q_N| \leqq 3\varepsilon$. Since, by (33), $|E\, Q_N, E| \leqq \varepsilon$, we get

$$|Q_N^+ \cup P'_n, E| \leqq 3\varepsilon. \tag{45}$$

W.30.16n. Since P'_n is a (DARS)-sequence for E', and since by (44), $P'_n \subseteqq Q_n$, we get $\lim \sum_k f^+(p'_{nk}) = \overline{S}_{E'} f^+$, because by Lemma [W.30.16g.], the sequence, yielding $\overline{S}_{E'} |f|$, also yields $\overline{S}_{E'} f^+$. Hence there exists $n_4 > n_3$, such that for all $n > n_4$ we have

$$\sum_k f^+(p'_{nk}) = \overline{S}_{E'} f^+ + \varepsilon \cdot \Theta_2, \quad \text{where} \quad |\Theta_2| \leqq 1$$

and depends on n.

Put $N' = n_4$. We have $N' > N$, and we have

$$\sum_k f^+(p'_{N',k}) = \overline{S}_{E'} f^+ + \varepsilon \cdot \Theta'_2, \quad |\Theta'_2| \leqq 1. \tag{46}$$

Now, by (39), $E' = \mathrm{som} Q_N^- E$; hence, by (38), $\overline{S}_{E'} |f| \geqq \frac{M}{4}$; consequently, by Lemma [W.30.16g.], $\overline{S}_{E'} f^+ \geqq \frac{M}{8}$.

Consequently, by (46)

$$\sum_k f^+(p'_{N'k}) \geqq \frac{M}{8} + \varepsilon \cdot \Theta'_2. \tag{47}$$

Let us take account of (31):

$$\sum_k f^+(q^+_{Nk}) = \frac{M}{2} + \varepsilon \cdot \Theta^+_N. \tag{48}$$

From (47) and (48) we get

$$\sum_k f^+(q^+_{Nk}) + \sum_k f^+(p'_{N'k}) \geqq \frac{M}{8} + \frac{M}{2} + 2\varepsilon \cdot \bar{\Theta}, \quad \text{where} \quad |\bar{\Theta}| \leqq 1.$$

Hence, by (30),

$$\sum_k f^+(q^+_{Nk}) + \sum_k f^+(p'_{N'k}) > \frac{M}{8} + \frac{M}{2} - \frac{M}{16} = \frac{M}{2} + \frac{M}{16}.$$

If we put $S =_{df} Q_N^+ \cup P'_{N'}$, we get

$$\sum_k f^+(s_k) \geqq \frac{M}{2} + \frac{M}{16}, \tag{49}$$

where, by (45),

$$|S, E| \leqq 3\varepsilon. \tag{50}$$

W.30.16o. We have proved that, in the case (38), given $\varepsilon > 0$, with $0 < \varepsilon < \frac{M}{32}$, there exist two natural numbers N, N', where $N' > N$, and complexes Q_n^+ and $P'_{N'}$, both subcomplexes of some Q_n, such that, if we put $S = Q_N^+ \cup P'_{N'}$, then $|S, E| \leqq 3\varepsilon$ and

$$\sum_k f^+(s_k) \geqq \frac{M}{2} + \frac{M}{16}.$$

W.30.16p. Now consider the second case, where

$$\text{(51)} \qquad \bar{S}_{\operatorname{som} Q_N^+ E} |f| \geqq \frac{M}{4}.$$

The arguments will be similar to those, already exhibited. Put $E'' =_{df} \operatorname{som} Q_N^+ \cdot E$. We have

$$\text{(52)} \qquad \bar{S}_{E''} |f| \geqq \frac{M}{4}.$$

We have

$$|E'', \operatorname{som} Q_N^+| \leqq \varepsilon, \qquad E'' \leqq \operatorname{som} Q_N^+, \qquad E'' \cdot \operatorname{som} Q_N^- = O, \qquad E'' \leqq E.$$

Since $Q_{\dot n}$ is a (DARS)-sequence for E, and since $E'' \leqq E$, there exists a partial complex R_n of Q_n, such that $R_{\dot n}$ is a (DARS)-sequence for E''. Hence $R_{N+1}, R_{N+2}, \ldots$ is also a (DARS)-sequence for E''. Now, since Q_n satisfies the condition of non overlapping, therefore if $r \in R_n$, $(n > N)$, then either r is disjoint with $\operatorname{som} Q_N^-$, or r is contained in one of the bricks of Q_N^-. Hence if R'_n is the complex composed of all bricks $r \in R_n$, for which $E'' \cdot r \neq O$, then $|R'_n, E''| \to 0$, and R'_n is a (DARS)-sequence for E''. We have $R'_n \frown Q_N^- = \emptyset$, for $n > N$. In addition to that $R'_n \subseteqq R_n \subseteqq Q_n$ for $n > N$.

Since $R'_{\dot n}$ is a (DARS)-sequence for E'', there exists $n'_3 > N$, such that for all $n > n'_3$ we have $|R'_n, E''| \leqq \varepsilon$. Hence

$$|R'_n, E \operatorname{som} Q_N^+| \leqq \varepsilon.$$

As

$$|Q_N^-, E \operatorname{som} Q_N^+| \leqq \varepsilon,$$

it follows that

$$|Q_N^- \cup R'_n, E \operatorname{som} Q_N^+ + E \operatorname{som} Q_N^-| \leqq 2\varepsilon.$$

Hence

$$|Q_N^- \cup R'_N, E\, Q_N| \leqq 2\varepsilon.$$

Since

$$|E\, Q_N, E| \leqq \varepsilon,$$

we get

$$|Q_n^- \cup R'_n, E| \leqq 3\varepsilon.$$

Since $R'_{\dot n}$ is a (DARS)-sequence for E'', and $R'_n \subseteqq Q_n$, we get

$$\lim \sum_k f^-(r'_{nk}) = \underline{S}_{E''} f^-,$$

because the sequence, yielding $\bar{S}_{E''}|f|$, also yields $\underline{S}_{E''}f^-$. Hence there exists $n'_4 > n'_3$, such that for all $n \geqq n'_4$,

$$\sum_k f^-(r'_{nk}) = \underline{S}_{E''}f^- + \varepsilon \cdot \Theta'_2,$$

where $|\Theta'_2| \leqq 1$ and depends on n. Put $N'' =_{df} n'_4$. We have

$$N'' > N \quad \text{and} \quad \sum_k f^-(r'_{N'',k}) = \underline{S}_{E''}f^- + \varepsilon\, \Theta''_2,$$

where $|\Theta''_2| \leqq 1$.

Now

$$E'' = \text{som} Q^+_N E \quad \text{and} \quad \bar{S}_{E''}|f| \geqq \frac{M}{4};$$

hence

$$\underline{S}_{E''}f^- \leqq -\frac{M}{8}.$$

Consequently

$$\sum_k f^-(r'_{N''k}) \leqq -\frac{M}{8} + \varepsilon\, \Theta''_2.$$

By (31) we have

$$\sum_k f^-(q^-_{nk}) = -\frac{M}{2} + \varepsilon\, \Theta^-_N.$$

Hence

$$\sum_k f^-(r^{-\prime}_{N''k}) + \sum_k f^-(q^-_{nk}) \leqq -\frac{M}{8} - \frac{M}{2} + 2\varepsilon\, \bar{\Theta}_1,$$

where $|\bar{\Theta}_1| \leqq 1$. Hence

$$\sum_k f^-(r'_{N''k}) + \sum_k f^-(q_{nk}) \leqq -\frac{M}{2} - \frac{M}{16}.$$

If we put

$$T =_{df} Q^-_N \cup R'_{N''},$$

we get

$$\sum_k f^-(t_k) \leqq -\frac{M}{2} - \frac{M}{16},$$

and at the same time, we have $|T, E| \leqq 3\varepsilon$. Thus we have proved that in the case (51), given $\varepsilon > 0$ with $0 < \varepsilon < \frac{M}{32}$, there exist two numbers $N'' > N$, and complexes Q^-_N and $R'_{N'}$, both subcomplexes of $\text{som}Q_n$, such that if we put

$T = Q^-_N \cup R'_{N''}$, then $|T, E| \leqq 3\varepsilon$ and $\sum_k f^-(t_k) \leqq -\frac{M}{2} - \frac{M}{16}$.

W.30.16q. Let us put $\varepsilon = \frac{M}{64m}$, $(m = 1, 2, \ldots)$ and find the corresponding N, which we denote by $N(m)$. There are two cases for each m: either there exists an infinity of m, for which (37) holds true, i.e.

$$\bar{S}_{\text{som}\, Q^+_{N(m)} E}|f| \geqq \frac{M}{4}$$

holds true, or else there exists an infinity of indices m for which

$$\overline{S}_{\operatorname{som} Q^-_{N(m)} E} |f| \geqq \frac{M}{4}$$

holds true.

Both cases will be treated similarly. If the first case occurs, we have an infinite subsequence $s(m)$ of m, such that $\overline{S}_{\operatorname{som} Q^-_{Ns(m)} E} |f| \geqq \frac{M}{4}$. Then we have the corresponding $N'(m) > N s(m)$, and complexes $Q^+_{Ns(m)}$ and $P'_{N'(m)}$, such that if we put

$$S_m = Q^+_{Ns(m)} \cup P'_{N'(m)},$$

then

$$(53) \qquad |S_m, E| \leqq 3 \cdot \frac{M}{64m},$$

and

$$(54) \qquad \sum_k f^+(s_{mk}) \geqq \frac{M}{2} + \frac{M}{16}.$$

Since

$$Q^+_{Ns(m)} \cup P'_{N'(m)} \subseteqq \bigcup_{n=1}^{\infty} Q_n,$$

we get by virtue of Theorem [U.10.4.], that S_m is a (DARS)-sequence for E. We have

$$\limsup \sum_k f^+(s_{mk}) \leqq \overline{S}_E f^+;$$

hence, by (54),

$$\overline{S}_E f^+ > \frac{M}{2} + \frac{M}{16}, \quad \text{i.e.} \quad \frac{M}{2} > \frac{M}{2} + \frac{M}{16},$$

which is a contradiction. If the second case occurs an infinite number of times, then

$$\overline{S}_{\operatorname{som} Q^+_{N(m)}} |f| \geqq \frac{M}{4},$$

and we get the conclusion

$$\underline{S}_E f^- < -\frac{M}{2} - \frac{M}{16},$$

which gives

$$-\frac{M}{2} < -\frac{M}{2} - \frac{M}{16},$$

which is also contradiction. The theorem is established.

W.30.17. Theorem. We admit the hypothesis $\boldsymbol{H}_2$. If

1. $f(\dot{a})$ is a field of real numbers,
2. E is a measurable set of traces,
3. $f(a)$ is (DARS)-summable on E,

then

$$g(a) =_{df} |f(a)| \text{ is also (DARS)-summable on } E.$$

W.30.17a. Proof. Define

$$L^0 =_{df} \{p \mid f(p) < 0\}, \quad L' =_{df} \{p \mid f(p) \geqq 0\}.$$

$L^0 \cup L'$ is the collection of all bricks, and $L^0 \cap L' = \emptyset$. Let A^0, C, A' be the characteristic sets, belonging to L^0, L' and E. They are measurable, and disjoint. We also have

$$A^0 + C + A' = E.$$

There exist (DARS)-sequences T_n^0, T_n' for C, where $T_n^0 \subseteqq L^0$, $T_n' \subseteqq L'$ for $n = 1, 2, \ldots$ The field f is summable on E; hence also on any summable subset $\bar{C}$ of C. Since T_n^0, T_n' are (DARS)-sequences for $\bar{C}$, there exist partial complexes $\bar{T}_n^0$, $\bar{T}_n'$ of T_n^0, T_n' respectively, $(n = 1, 2, \ldots)$, such that $\bar{T}_n^0$, $\bar{T}_n'$ are both (DARS)-sequences for $\bar{C}$. Hence

$$\lim_{n\to\infty} \sum_k f(\bar{t}_{nk}^0) = \lim \sum_k f(\bar{t}_{nk}') = S_{\bar{C}} f.$$

Since

$$f(\bar{t}_{nk}^0) < 0, \quad f(\bar{t}_{nk}') \geqq 0,$$

it follows that

$$S_{\bar{C}} f \leqq 0 \quad \text{and} \quad S_{\bar{C}} f \geqq 0.$$

Consequently

$$S_C f = 0 \quad \text{for all} \quad \bar{C} \leqq C.$$

The Theorem [W.30.16.] allows to state that

(1) $$S_C |f| = 0;$$

hence $|f|$ is (DARS)-summable on C.

W.30.17b. Take a (DARS)-sequence Q_n for $E - C$, such that

$$\lim_{n\to\infty} \sum_k |f(q_{nk})| = \bar{S}_{E-C} |f|.$$

Put

$$Q_n^0 =_{df} Q_n \cap L_n^0, \quad Q_n' =_{df} Q_n \cap L_n' \quad \text{for} \quad (n = 1, 2, \ldots).$$

We know that Q_n^0, Q_n' are (DARS)-sequences for A^0, A' respectively. We have:

(2) $$\lim_{n\to\infty} \sum_k |f(q_{nk}^0)| = \bar{S}_{A^0} |f|, \qquad \lim_{n\to\infty} \sum_k |f(q_{nk}')| = \bar{S}_{A'} |f|.$$

On the other hand,

(3) $$\lim_{n\to\infty} \sum_k |f(q_{nk}^0)| = -\lim_{n\to\infty} \sum_k f(q_{nk}^0), \quad \text{because} \quad f(q_{nk}^0) < 0,$$

$$\lim_{n\to\infty} \sum_k |f(q_{nk}')| = \lim_{n\to\infty} \sum_k f(q_{nk}'), \quad \text{because} \quad f(q_{nk}') \geqq 0.$$

Since f is summable on A^0 and A', and Q_n^0, Q_n' are (DARS)-sequences for A^0, A' respectively, it follows from (2) and (3):

$$\overline{S}_{A^0}|f| = -S_{A^0} f, \quad \overline{S}_{A'}|f| = S_{A'} f. \tag{4}$$

Now, let P_n be a (DARS)-sequence for $E - C$, such that

$$\lim_{n\to\infty} \sum_k f(p_{nk}) = \underline{S}_{E-C}|f|.$$

If we put

$$P_n^0 =_{df} P_n \cap L^0, \quad P_n' =_{df} P_n \cap L', \quad (n = 1, 2, \ldots),$$

we have:

$$\begin{aligned} &\lim_{n\to\infty} \sum_k |f(p_{nk}^0)| = \underline{S}_{A^0}|f|, \\ &\lim_{n\to\infty} \sum_k |f(p_{nk}')| = \underline{S}_{A'}|f|. \end{aligned} \tag{5}$$

On the other hand

$$\lim_{n\to\infty} \sum_k |f(p_{nk}^0)| = \lim_{n\to\infty} \sum_k f(p_{nk}^0) = -S_{A^0} f,$$

$$\lim_{n\to\infty} \sum_k |f(p_{nk}')| = \lim_{n\to\infty} \sum_k f(p_{nk}') = S_{A'} f.$$

Hence from (5)

$$\underline{S}_{A^0}|f| = -S_{A^0} f, \quad \underline{S}_{A'}|f| = S_{A'} f. \tag{6}$$

If we compare (4) and (6), we get

$$\overline{S}_{A^0}|f| = \underline{S}_{A^0}|f|, \quad \overline{S}_{A'}|f| = \underline{S}_{A'}|f|,$$

so

(7) $|f|$ is (DARS)-summable on A^0 and A'.

From (7) and (1) it follows, that $|f|$ is (DARS)-summable on E, q.e.d.

W.30.18. Theorem. Under the hypotheses of Theorem [W.30.17.], if we define

$$f^-(a) =_{df} \begin{cases} f(a), & \text{if } f(a) < 0, \\ 0, & \text{if } f(a) \geqq 0. \end{cases}$$

$$f^+(a) =_{df} \begin{cases} 0, & \text{if } f(a) < 0, \\ f(a), & \text{if } f(a) \geqq 0, \end{cases}$$

then $f^+(a)$ and $f^-(a)$ are both (DARS)-summable on E.

Proof. We have

$$f^+(a) + f^-(a) = f(a)$$

and

$$f^+(a) - f^-(a) = |f(a)|$$

for all a. Hence

$$f^+(a) = \tfrac{1}{2}[f(a) + |f(a)|],$$

$$f^-(a) = \tfrac{1}{2}[f(a) - |f(a)|]$$

for all a. Applying Theorem [W.30.17.], we deduce that the thesis is true.

W.30.19. Theorem. We admit Hyp. $\boldsymbol{H}_2$. If

1. $f(\dot{a})$ is a field of real numbers,
2. E is a measurable set of traces,
3. f is a (DARS)-summable on E,
4. λ is a real number,
5. we define

$$f_\lambda(a) =_{df} \begin{cases} f(a) & \text{whenever} \quad f(a) \geqq \lambda, \\ 0 & \text{otherwise}, \end{cases}$$

then

$$f_\lambda(a) \text{ is (DARS)-summable on } E.$$

Proof. It is slightly similar to that of Theorem [W.30.17]. Define

$$(1) \qquad L^0 =_{df} \{a \mid f(a) < \lambda\}, \qquad L' =_{df} \{a \mid f(a) \geqq \lambda\}.$$

We have: $L^0 \cup L'$ is the collection $\boldsymbol{B}$ of all bricks, and $L^0 \cap L' = \emptyset$. Let A^0, C, A' be the characteristic sets for L^0, L' and E. Let us remark, that the theorem is true if $\lambda = 0$, (see Theorem [W.30.18.]). Let $\lambda \neq 0$. I say that $\mu(C) = 0$. Indeed, suppose that $\mu(C) > 0$. Let T_n^0, T_n' be the (DARS)-sequences for C, such that $T_n^0 \subseteqq L^0$, $T_n' \subseteqq L'$.

Since f is summable on E, it is also summable on C. Hence

$$\lim_{n\to\infty} \sum_k f(t_{nk}^0) = \lim_{n\to\infty} \sum_k f(t_{nk}') = S_C f.$$

Now, as

$$f(t_{nk}^0) < \lambda, \quad f(t_{nk}') \geqq \lambda,$$

we get $\sum_k \lambda \leqq S_C f$. Hence $S_C f = \sum_k \lambda$, which tends to infinity $+\infty$, or $-\infty$, depending on whether $\lambda > 0$ or $\lambda < 0$. This is impossible. Thus $\mu C = 0$. Hence $E =^\mu A^0 + A'$. The field f is (DARS)-summable on A^0 and on A'. Let P_n be a (DARS)-sequence for E. Put

$$P_n^0 =_{df} P_n \cap L^0, \quad P_n' = P_n \cap L'.$$

We have

$$\sum_k f_\lambda(p_{nk}) = \sum_k f_\lambda(p_{nk}^0) + \sum_k f_\lambda(p_{nk}').$$

Now $f_\lambda(p_{nk}^0) = 0$, because $p_{nk}^0 \in L^0$, $f_\lambda(p_{nk}') = f(p_{nk}')$; hence $\sum_k f_\lambda(p_{nk}) = \sum_k f(p_{nk})$. Since P_n' is a (DARS)-sequence for A', and since f is summable on A', it follows that

$$\lim_{n\to\infty} \sum_k f_\lambda(p_{nk}) = S_{A'} f.$$

Since the limit exists, and is the same for every (DARS)-sequence P_n for E, it follows that f_λ is (DARS)-summable on E, q.e.d.

W.30.19.1. Theorem. Under $\boldsymbol{H}_2$ and the hypotheses 1., 2., 3., 4., of Theorem [W.30.19.], if we define

$$f'_\lambda(a) =_{df} \begin{cases} f(a) & \text{whenever } f(a) > \lambda, \\ 0 & \text{otherwise}, \end{cases}$$

$$g_\lambda(a) =_{df} \begin{cases} 0 & \text{whenever } f(a) \geqq \lambda, \\ f(a) & \text{otherwise}, \end{cases}$$

$$g'_\lambda(a) =_{df} \begin{cases} 0 & \text{whenever } f(a) > \lambda, \\ f(a) & \text{otherwise}, \end{cases}$$

then the fields $f'_\lambda(\dot a)$, $g_\lambda(\dot a)$, $g'_\lambda(\dot a)$ are all (DARS)-summable.

Proof. We rely on Theorem [W.30.19.], use $-f(\dot a)$, and also the relations

$$f_\lambda(a) + g_\lambda(a) = f(a), \quad f'_\lambda(a) + g'_\lambda(a) = f(a).$$

W.30.19.2. Remark. Under hypotheses 1., 2., 3., 4., 5. of Theorem [W.30.19.] it is not true that, if $f\,\mu$ is summable, then the field $f_\lambda(a)\,\mu(a)$ also is summable.

Expl. Let $\boldsymbol{G}$ be the set of all Lebesgue's measurable subsets of $(0, 3\rangle$, and taken modulo null-sets. Let μ be the Lebesguean measure, and let the base $\boldsymbol{B}$ be composed of 0, $(0, 3\rangle$ and of all its subintervals $\left(\frac{p}{2^n}, \frac{p+1}{2^n}\right)$. Put $f(a) = 0$ for all $a = \left(\frac{p}{2^n}, \frac{p+1}{2^n}\right\rangle$, which are in $(0, 1\rangle$. Put $f(a) = 1$ for all $a = \left(\frac{p}{2^n}, \frac{p+1}{2^n}\right\rangle$, which are in $(2, 3\rangle$. Concerning the bricks being in $(1, 2\rangle$, put $f(a) = \frac{n}{n+1}$ if n is odd, and $f(a) = \frac{n+1}{n}$ if n is even. The field $f(a)\,\mu(a)$ is (DARS)-summable, because it is so on $(0, 1\rangle$, $(1, 2\rangle$ and $(2, 3\rangle$. But define

$$f_1(a) =_{df} \begin{cases} f(a) & \text{whenever } f(a) \geqq 1, \\ 0 & \text{whenever } f(a) < 1. \end{cases}$$

We have for $a \in (0, 1\rangle : f_1(a) = 0$ for $a \leqq (2, 3\rangle$, and we have $f(a) = 1$; hence $f_1(a) = 1$; for $a \in (1, 2\rangle$ and n even we have $f_1(a) = f(a) = \frac{n+1}{n}$. For $a \in (2, 3\rangle$ and n odd, we have $f_1(a) = 0$. We have

$$\overline{S}_{(1,2\rangle} f_1\,\mu = 1, \quad \underline{S}_{(1,2\rangle} f_1\,\mu = 0;$$

hence $f_1\,\mu$ is not summable. We have

$$S_{(0,1\rangle} f_1\,\mu = 0, \quad S_{(2,3\rangle} f_1\,\mu = 1.$$

W.30.20. Theorem. We admit hypothesis $\boldsymbol{H}_2$. If

1. $f(a)$ is a field of real numbers,
2. E is a measurable set of traces,
3. f is (DARS)-summable on E,

4. λ is a real number,
5. we define

$$f^{\lambda}(a) =_{df} \begin{cases} f(a) & \text{whenever } f(a) < \lambda, \\ \lambda & \text{whenever } f(a) \geqq \lambda, \end{cases}$$

then

$$f^{\lambda}(a)\,\mu(a) \text{ is (DARS)-summable on } E.$$

Proof. Since the field $\mu(a)$ is summable, therefore the field $(f - \lambda)\,\mu$ is also summable. Hence, by virtue of Theorem [W.30.18.], the field $f_1(a)\,\mu(a)$, where

$$f_1(a) =_{df} \begin{cases} (f - \lambda) & \text{for } f - \lambda < 0, \\ 0 & \text{for } f - \lambda \geqq 0, \end{cases}$$

is summable.

Hence $f_1(a)\,\mu(a) + \lambda\,\mu(a)$ is also summable.

Now

$$f_1(a) = \begin{cases} f - \lambda & \text{for } f < \lambda, \\ 0 & \text{for } f \geqq \lambda; \end{cases}$$

hence

$$f_1(a) + \lambda = \begin{cases} f & \text{for } f < \lambda, \\ \lambda & \text{for } f \geqq \lambda; \end{cases}$$

hence $f_1(a) + \lambda = f^{\lambda}$. Thus $f^{\lambda}\,\mu$ is (DARS)-summable on E.

W.30.21. Theorem. We admit the hypothesis $\boldsymbol{H}_2$. If

1. $f(a)$ is a field of real numbers,
2. E is a measurable set of traces,
3. $f(a)\,\mu(a)$ is (DARS)-summable on E,
4. $f(a) \geqq 0$ for all a,
5. $\alpha > 0$ is a number,
6. we define

$$L^0_{\alpha} =_{df} \{a \mid f(a) < \alpha\}, \quad L'_{\alpha} =_{df} \{a \mid f(a) \geqq \alpha\},$$

7. $f_{\alpha}(a) =_{df} \begin{cases} f(a) & \text{if } f(a) < \alpha, \\ \alpha & \text{if } f(a) \geqq \alpha, \end{cases}$

8. A^0_{α}, C_{α}, A'_{α} are characteristic sets for L^0_{α}, L'_{α} and E,

then

1) $f(\dot{a})\,\mu(\dot{a})$ is summable on A^0_{α},
2) if $\alpha \to \infty$, then $A^0_{\alpha} \to^{\mu} E$, and if $\alpha \leqq \beta$, then $A^0_{\alpha} \leqq A^0_{\beta}$,
3) $f_{\alpha}(\dot{a})\,\mu(\dot{a})$ is summable on E,
4) $\lim\limits_{\alpha \to \infty} S_{A^0_{\alpha}}\, f\,\mu = S_E\, f\,\mu$,
5) $\lim\limits_{\alpha \to \infty} S_E\, f_{\alpha}\,\mu = S_E\, f\,\mu$,
6) $S_E\, f_{\alpha}\,\mu = S_{A^0_{\alpha}}\, f\,\mu + \mu(A^0_{\alpha} + C_{\alpha})$,
7) $S_{A^0_{\alpha}}\, f\,\mu = S_{A^0_{\alpha}}\, f_{\alpha}\,\mu$.

W.30.21 a. Proof. For every $\alpha \geqq 0$ we define

$$L^0_\alpha =_{df} \{a \mid f(a) < \alpha\}, \quad L'_\alpha =_{df} \{a \mid f(a) \geqq \alpha\}.$$

We have $L^0_\alpha \cup L'_\alpha = \boldsymbol{B}$, (the base), $L^0_\alpha \cap L'_\alpha = \emptyset$.

Let A^0_α, C_α, A'_α be the characteristic sets for L^0_α, L'_α and E.

W.30.21 b. First of all we shall prove that, if $\alpha \leqq \beta$, then $A^0_\alpha \leqq A^0_\beta$. We have $L^0_\alpha \subseteqq L^0_\beta$, because $f(a) < \alpha$ implies $f(a) < \beta$.

Suppose that

(1) $$\mu(A^0_\alpha - A^0_\beta) > 0.$$

Take a (DARS)-sequence Q_n for E, and put $Q^0_n =_{df} Q_n \cap L^0_\alpha$.

There exists a sequence P^0_n for A^0_n, such that

$$P_n \subseteqq Q_n \cap L^0_n, \quad (n = 1, 2, \ldots).$$

We have

(2) $$P_n \subseteqq Q_n \cap L^0_\beta, \quad \text{since} \quad L^0_\alpha \subseteqq L^0_\beta.$$

There exists a (DARS)-sequence R_n for A^0_β, such that $R_n \subseteqq Q_n \cap L^0_\beta$. Hence we can find a sequence S_n for $A^0_\beta - A^0_\alpha$, where $S_n \subseteqq R_n$; hence

(2.1) $$S_n \subseteqq Q_n \cap L^0_\beta.$$

We have

$$\mathrm{som}\, S_n \to^\mu A^0_\beta - A^0_\alpha, \quad \mathrm{som}\, P_n \to^\mu A^0_\alpha.$$

Since A^0_α, $A^0_\beta - A^0_\alpha$ are disjoint, it follows

$$\mathrm{som}\, S_n \cdot \mathrm{som}\, P_n \to^\mu O, \quad \mathrm{som}\,(S_n \cap P_n) \to^\mu O.$$

Hence

$$S_n \sim (S_n \cap P_n) \to^\mu A^0_\beta - A^0_\alpha, \quad P_n \sim (S_n \cap P_n) \to^\mu A^0_\alpha,$$

i.e.

(2.2) $$S_n \sim P_n \to^\mu A^0_\beta - A^0_\alpha, \qquad P_n \sim S_n \to^\mu A^0_\alpha.$$

As S_n is a (DARS)-sequence for $A^0_\beta - A^0_\alpha$, it follows from (2), that $S_n \sim P_n$ is (DARS)-sequence for

$$A^0_\beta - A^0_\alpha.$$

As P_n is a (DARS)-sequence for A^0_α, it follows, in the similar way, that $P_n \sim S_n$ is a (DARS)-sequence for A^0_α. Hence

$$T_n =_{df} (S_n - P_n) \cup (P_n - S_n)$$

is a (DARS)-sequence for

(3) $$A^0_\alpha + (A^0_\beta - A^0_\alpha).$$

In addition to that, by (1.1),

$$S_n \sim P_n \subseteqq S_n \subseteqq Q_n \cap L^0_\beta,$$

and by (2),

$$P_n \sim S_n \subseteqq P_n \subseteqq Q_n \cap L^0_\beta.$$

Hence

(4) $$T_n \subseteqq Q_n \frown L^0_\beta.$$

Now we have

$$A^0_\alpha + (A^0_\beta - A^0_\alpha) = A^0_\alpha + A^0_\beta = A^0_\beta + (A^0_\alpha - A^0_\beta),$$

where A^0_β, $A^0_\alpha - A^0_\beta$ are disjoint. Since, by (1), $\mu(A^0_\alpha - A^0_\beta) > 0$, it follows that

(5) $$\mu[A^0_\beta + (A^0_\alpha - A^0_\beta)] > \mu A^0_\beta.$$

Thus for every sequence Q_n for E there exists a (DARS)-sequence T_n for $A^0_\beta + (A^0_\alpha - A^0_\beta)$ with

(6) $$T_n \subseteqq Q_n \frown A^0_\beta.$$

Now, by Theorem [W.30.5.2.], A^0_β is the set with maximal measure, such that for every sequence Q_n for E there exists a partial sequence of $Q_n \frown L^0_\beta$ for A^0_β. But (6) shows, that this is not true, because of (5) and (6). The contradiction shows that

$$\mu(A^0_\alpha - A^0_\beta) = 0, \quad \text{i.e.} \quad A^0_\alpha \leqq^\mu A^0_\beta, \quad \text{q.e.d.}$$

W.30.21c. We shall prove that there exists a (DARS)-sequence for $C_\alpha + A'_\alpha$, which is composed of bricks belonging to L'_α. There exists a sequence P_n for C_α with $P_n \subseteqq L'_\alpha$. Now, by Theorem [W.30.5.1.], there exists a subsequence $s(\dot{n})$ of $\dot{n}$ and a (DARS)-sequence M_n for A'_α, such that $P_{s(n)} \smile M_n$ is a sequence for $C_\alpha + A'_\alpha$, and where

$$P_{s(n)} \frown M_n = \emptyset \quad \text{for} \quad n = 1, 2, \ldots$$

Now, by Theorem [W.30.3.], the bricks of M_n, which belong to L^0_α make up a set, whose joint measure tends to 0 if $n \to \infty$. If we drop these bricks in all M_n, we get a new (DARS)-sequence M'_n for A'_n, whose all bricks belong to L'_α. Consequently $P_{s(n)} \frown M'_n$ is a (DARS)-sequence for $C_\alpha + A'_\alpha$, such that

$$P_{s(n)} \frown M'_n \subseteqq L'_\alpha.$$

In this argument no hypothesis, concerning summability has been used. Thus we can apply [W.30.21c.] in the proof of the next theorem.

W.30.21d. Now we shall rely on the hypothesis: $f\mu$ is summable on E. It follows that $S_{A_\alpha} f\mu$ and $S_{C_\alpha + A'_\alpha} f\mu$ exist. Let U_n be a (DARS)-sequence for $C_\alpha + A'_\alpha$, such that $U_n \subseteqq L'_\alpha$ for all $n = 1, 2, \ldots$ (by item [W.30.21c.]). We have $\sum_k f(u_{kn})\,\mu(u_{kn}) \geqq \sum_k \alpha\,\mu(u_{kn})$, because $f(u_{kn}) \geqq \alpha$ for $u_{kn} \in L'_\alpha$, and for $a \in L'_\alpha$ we have $f(a) \geqq \alpha$. It follows that

$$S_{C_\alpha + A'_\alpha} f\mu \geqq \alpha\,\mu(C_\alpha + A'_\alpha);$$

hence, since $f \geqq 0$, we get

$$\alpha \cdot \mu(C_\alpha + A'_\alpha) \leqq S_E f\mu.$$

If $\alpha \to \infty$, we get

(6.0)
$$\lim_{\alpha \to \infty} \mu(C_\alpha + A'_\alpha) = 0,$$

and then, by the denumerable additivity of the (DARS)-sums,

$$\lim_{\alpha \to \infty} S_{C_\alpha + A'_\alpha} f \mu = 0.$$

Since

$$S_E f \mu = S_{A^0_\alpha} f \mu + S_{C_\alpha + A'_\alpha} f \mu,$$

it follows that

(6.1)
$$S_E f \mu = \lim_{\alpha \to \infty} S_{A^0_\alpha} f \mu.$$

We have

$$\lim{}^\mu A^0_\alpha = E;$$

hence

$$A^0_\alpha + (C_\alpha + A'_\alpha) = E.$$

Thus the thesis 1), 2) and 4) are established.

W.30.21e. Now we are going to prove the remaining thesis. By Theorem [W.30.20.] the field $f_\alpha(\dot{a})$ is (DARS)-summable. We shall prove that

(7)
$$S_{A^0_\alpha} f \mu = S_{A^0_\alpha} f_\alpha \mu.$$

Both sums exist. To evaluate them, we may take any (DARS)-sequence for A^0_α. Take a sequence Q_n whose bricks $\in L^0_\alpha$. We have

$$\sum_k f(q_{nk}) \mu(q_{nk}) = \sum_k f_\alpha(q_{nk}) \mu(q_{nk}),$$

because $q_{nk} \in L^0_\alpha$; and then

$$f_\alpha(q_{nk}) = f(q_{nk}).$$

The equality (7) follows.

W.30.21f. Now consider $S_{C_\alpha + A'_\alpha} f_\alpha \mu$. Take, [W.30.21d.], a sequence U_n for $C_\alpha + A'_\alpha$ with $U_n \subseteqq L'_\alpha$. We have

$$\sum_k f_\alpha(u_{nk}) \mu(u_{nk}) = \sum_k \alpha \cdot \mu(u_{nk}),$$

because

$$u_{nk} \in L'_\alpha \quad \text{and then} \quad f_\alpha(u_{nk}) = \alpha.$$

Hence

(8)
$$S_{C_\alpha + A'_\alpha} f_\alpha \mu = \alpha \cdot \mu(C_\alpha + A'_\alpha).$$

Now, we have

$$\lim \mu(C_\alpha + A'_\alpha) = 0.$$

Since

$$S_E f_\alpha \mu = S_{A^0_\alpha} f_\alpha \mu + S_{C_\alpha + A'_\alpha} f_\alpha \mu = S_{A^0_\alpha} f \mu + \alpha \cdot \mu(C_\alpha + A'_\alpha),$$

it follows that

$$\lim_{\alpha \to \infty} S_E f_\alpha \mu = \lim S_{A^0_\alpha} f \mu.$$

Hence, by (6.1):

$$\lim_{\alpha\to\infty} S_E f_\alpha \mu = S_E f \mu,$$

so the whole theorem is proved.

W.30.22. Theorem. We admit the hypothesis $\boldsymbol{H}_2$. If

1. $f(a)$ is a field of real numbers,
2. $f(a) \geqq 0$ for all a,
3. E is a measurable set of traces,
4. we define L^0_α, L'_α and $f_\alpha(a)$, A^0_α, C_α, A'_α, as in the hypotheses of Theorem [W.30.21.],
5. $f(a)$ is admissible,
6. there exists a sequence

$$0 < \alpha(1) < \alpha(2) < \cdots < \alpha(m) < \cdots \to \infty$$

such that $S_{A^0_{\alpha(m)}} f \mu$ exists for $m = 1, 2, \ldots,$

then

$$f \mu \text{ is summable on } E.$$

Proof. By item [W.30.21c.], in the proof of Theorem [W.30.21.], there exists a (DARS)-sequence $U_{\dot{n}}$ for

(1) $C_{\alpha(m)} + A'_{\alpha(m)}$, such that $U_n \subseteqq L'_\alpha$, $(n = 1, 2, \ldots)$.

Since $f(a)\,\mu(a)$ is admissible, we can speak of the upper and lower sum of $f\mu$. We have for a fixed m:

(2) $$\overline{S}_E f \mu = \overline{S}_{A^0_{\alpha(m)}} f \mu + S_{C_{\alpha(m)}+A'_{\alpha(m)}} f \mu.$$

Now, by (1)

$$\sum_k f(u_{nk})\,\mu(u_{nk}) \geqq \alpha(m) \cdot \sum_k \mu(u_{nk}).$$

Hence

$$\overline{S}_E f \mu \geqq \overline{S}_{C_{\alpha(m)}+A'_{\alpha(m)}} f \mu \geqq \alpha(m) \cdot \mu(C_{\alpha(m)} + A'_{\alpha(m)}).$$

This being true for all m, we get by $m \to \infty$ and $\alpha(m) > 0$,

$$\lim_{m\to\infty} \mu(C_{\alpha(m)} + A'_{\alpha(m)}) = 0.$$

Since $\overline{S}_A$ is denumerably additive, it follows

$$\lim_{m\to\infty} \overline{S}_{C_{\alpha(m)}+A'_{\alpha(m)}} f \mu = 0.$$

Consequently from (2) we get

(3) $$\overline{S}_E f \mu = \lim_{m\to\infty} \overline{S}_{A^0_{\alpha(m)}} f \mu.$$

Now, we also have

$$\underline{S}_E f \mu = \underline{S}_{A^0_{\alpha(m)}} f \mu + \underline{S}_{C_{\alpha(m)}+A^0_{\alpha(m)}} f \mu;$$

hence by the same reason, as before, we get

$$\lim_{m\to\infty} \underline{S}_{C_{\alpha(m)}+A'_{\alpha(m)}} f\,\mu = 0,$$

and then

(4) $$\underline{S}_E f\,\mu = \lim \underline{S}_{A^0_{\alpha(m)}} f\,\mu.$$

Since, by hypothesis, μ is summable on $A^0_{\alpha(m)}$, we get from (3) and (4)

$$\overline{S}_E f\,\mu = \lim S_{A^0_{\alpha(m)}} f\,\mu = \underline{S}_E f\,\mu.$$

Consequently $f\,\mu$ is (DARS)-summable on E, q.e.d.

W.30.22.1. Theorem. Let us admit the hypothesis $\boldsymbol{H}_2$ and the hypotheses 1., 2., 3., 4., 5., but instead of 6., let us take 6.1:

$$S_E f_{\alpha(m)}\,\mu \text{ exist for all } m = 1, 2, \ldots:$$

then

$$f\,\mu \text{ is (DARS)-summable on } E.$$

Proof. Let us fix m for a moment, and put $\alpha =_{df} \alpha(m)$. Let Q_n be a (DARS)-sequence for A^0_α. Put

$$Q^0_n =_{df} Q_n \cap L^0, \quad Q'_n =_{df} Q_n \cap L'.$$

We have

$$Q_n = Q^0_n \cup Q'_u, \quad Q^0_n \cap Q'_n = \emptyset.$$

We have

$$\sum_k f(q_{nk})\,\mu(q_{nk}) = \sum_k f(q^0_{nk})\,\mu(q^0_{nk}) + \sum_k f(q'_{nk})\,\mu(q'_{nk}).$$

Now by Theorem [W.30.3.], $\sum_k \mu(q'_{nk}) \to 0$ for $n \to \infty$. Since $f\,\mu$ is admissible, it follows that

(1) $$\sum_k f(q'_{nk})\,\mu(q'_{nk}) \to 0 \quad \text{for} \quad n \to \infty.$$

Hence Q^0_n is a (DARS)-sequence for A^0_α, and we have

$$S_{A^0_\alpha} f_\alpha\,\mu = \lim \sum_k f_\alpha(q^0_{nk})\,\mu(q^0_{nk}).$$

Now, since $q^0_{nk} \in L^0$, we have $f_\alpha(q^0_{nk}) = f(q^0_{nk})$. Hence

$$\lim \sum_n f(q^0_{nk})\,\mu(q^0_{nk}) = S_{A^0} f_\alpha\,\mu.$$

Hence, by (1)

$$\sum_k f(q_{nk})\,\mu(q_{nk}) = \sum_k f(q^0_{nk})\,\mu(q^0_{nk}) + \sum_k f(q'_{nk})\,\mu(q'_{kn}) \to S_{A^0_\alpha} f_\alpha\,\mu.$$

Thus for every (DARS)-sequence for A^0_α the limit: $\lim_{k\to\infty} \sum_k f(q_{nk})\,\mu(q_{nk})$ exists and is the same.

Hence $S_{A^0_\alpha} f\,\mu$ exists. This being true for all $\alpha = \alpha(m)$, it follows. by Theorem [W.30.22.], that $f\,\mu$ is summable on E.

W.30.23. Theorem. We admit the hypothesis $\boldsymbol{H}_2$. If

1. E is a measurable set of traces,
2. $f(a)\,\mu(a)$ and $g(a)\,\mu(a)$ are both (DARS)-summable on E,
3. $f(a)\,g(a)\,\mu(a)$ is admissible,
4. $f(a) \geqq 1$, $g(a) \geqq 1$ for all a,

then

$$f(a)\,g(a)\,\mu(a)$$

is (DARS)-summable on E.

Proof. In general, given any field of scalars $h(a)$ and a number λ, we define

$$h_\lambda(a) =_{df} \begin{cases} f(a) & \text{whenever} \quad f(a) < \lambda, \\ \lambda & \text{whenever} \quad f(a) \geqq \lambda. \end{cases}$$

We shall prove that for $\lambda > 1$ we have

(0) $$(f_\lambda\, g_\lambda)_\lambda = (f\,g)_\lambda.$$

Put

$$k =_{df} f \cdot g, \quad l =_{df} f_\lambda\, g_\lambda,$$

and define

$$L =_{df} \{p \mid f(p) < \lambda,\ g(p) < \lambda\}.$$

Let $a \in L$. Then $f(a) < \lambda$ and $g(a) < \lambda$. Hence

$$f_\lambda(a) = f(a), \quad g_\lambda(a) = g(a);$$

hence

$$l(a) = f_\lambda(a)\,g_\lambda(a) = f(a)\,g(a) = k(a),$$

and then

(1) $$l_\lambda(a) = k_\lambda(a) \quad \text{for all} \quad a \in L.$$

Let $a \,\overline{\in}\, L$. Then either $f(a) \geqq \lambda$ or $g(a) \geqq \lambda$. If $f(a) \geqq \lambda$, then whatever $g(a)$ may be, we have $f(a) \cdot g(a) \geqq \lambda$, because $g(a) \geqq 1$; hence $k(a) \geqq \lambda$ and

(2) $$k_\lambda(a) = \lambda.$$

We have $f_\lambda(a) = \lambda$; hence, whatever g_λ may be, we have

$$f_\lambda(a) \cdot g_\lambda(a) \geqq \lambda;$$

hence $l(a) \geqq \lambda$, and then

(3) $$l_\lambda(a) = \lambda.$$

From (2) and (3) it follows: $l_\lambda(a) = k_\lambda(a)$, for all $a \,\overline{\in}\, L$, provided that $f(a) \geqq \lambda$.

Now, if instead, we suppose that $g(a) \geqq \lambda$, we get similarly the same equality $l_\lambda(a) = k_\lambda(a)$. Hence this equality holds true for all $a \,\overline{\in}\, L$. When we take (1) into account, we get $l_\lambda(a) = k_\lambda(a)$ for all a without exception. Thus (0) is proved.

Having that, we see, by Theorem [W.30.20.] that $f_\lambda(a)\,\mu(a)$, $g_\lambda(a)\,\mu(a)$ are both summable for any $\lambda > 1$. It follows, by Theorem

[W.30.15.], that $f_\lambda(a)\cdot g_\lambda(a)\cdot\mu(a)$ is summable, because f_λ and g_λ are bounded. Hence, by (0), $(f\,g)_\lambda\,\mu$ is summable.

Take for λ the numbers $1, 2, \ldots, n, \ldots$ We get: $(f\,g)_n\,\mu$ is summable for all $n = 1, 2, \ldots$ Applying Theorem [W.30.22.1.], we deduce that $f\,g\,\mu$ is summable, so the theorem is established.

W.30.24. Theorem. We admit the hypothesis $\boldsymbol{H}_2$. If

1. E is a measurable set of traces,
2. $f(a) > 1$, $g(a) > 1$ for all a,
3. $f(a)\,\mu(a)$, $f^2(a)\,\mu(a)$, $g(a)\,\mu(a)$, $g^2(a)\,\mu(a)$ are (DARS)-summable on E,

then

$$f(a)\,g(a)\,\mu(a)$$

is also summable on E.

Proof. By Theorem [W.30.23.] it suffices to prove, that $f\,g\,\mu$ is admissible. To prove that, take any sequence P_n of complexes, with $\mu(P_n)\to 0$ for $n\to\infty$. We have

$$\Big[\sum_k f(p_{nk})\,g(p_{nk})\,\mu(p_{nk})\Big]^2 \leqq \tfrac{1}{2}\sum_k f^2(p_{nk})\,\mu(p_{nk}) + \tfrac{1}{2}\sum_k g^2(p_{nk})\,\mu(p_{nk}),$$

because for positive numbers a, b we have $a\,b \leqq \frac{1}{2}(a^2+b^2)$.

Now, since f^2, g^2 are both summable, they are admissible. Hence

$$\sum_k f^2(p_{nk})\,\mu(p_{nk}) \to 0 \quad\text{and}\quad \sum_k g^2(p_{nk})\,\mu(p_{nk}) \to 0,$$

which completes the proof.

W.30.25. Till now we have dealt with non negative fields $f(a)$ where $f(a)\geqq 1$. By means of this particular case the general case can be easily settled.

Theorem. We admit Hyp. $\boldsymbol{H}_2$. If

1. E is a measurable set of traces,
2. $f(a)$, $g(a)$ are fields of real numbers, such that $f(a)\,\mu(a)$, $g(a)\,\mu(a)$, $f^2(a)\,\mu(a)$, $g^2(a)\,\mu(a)$ are all (DARS)-summable on E,

then

$$f(a)\cdot g(a)\,\mu(a)$$

is (DARS)-summable on E.

Proof. Define

$$f^+(a) =_{df} \begin{cases} f(a) & \text{for } f(a)\geqq 0,\\ 0 & \text{for } f(a)<0,\end{cases}$$

$$f^-(a) = \begin{cases} 0 & \text{for } f(a)\geqq 0,\\ -f(a) & \text{for } f(a)<0.\end{cases}$$

We define similarly $g^+(a)$ and $g^-(a)$. We have

$$(1)\quad f\cdot g\,\mu = (f^+ - f^-)\cdot(g^+ - g^-)\,\mu = f^-\,g^+\,\mu - f^-\,g^+\,\mu - {} - f^+\,g^-\,\mu + f^-\,g^-\,\mu.$$

All four terms in (1) are products of non negative fields of real numbers. The fields $f^+\mu$, $f^-\mu$, $g^+\mu$, $g^-\mu$ are (DARS)-summable on E, by virtue of Theorem [W.30.18.], because $f\mu$, $g\mu$ are so. Since f^2, g^2 are summable, they are admissible; hence $(f^+)^2\mu$, $(f^-)^2\mu$, $(g^+)^2\mu$, and $(g^-)^2\mu$ are also admissible. It follows that all terms in (1) are admissible (this by means of the inequality $a\,b \leqq \frac{1}{2}(a^2+b^2)$, and by an argument, applied in the proof of the forgoing theorem). Thus to prove our theorem we need to prove that, if $f \geqq 0$, $g \geqq 0$ are summable on E and $f\,g\,\mu$ is admissible, then $f\,g\,\mu$ is summable on E. This has been proved in the case where $f \geqq 1$, $g \geqq 1$, Theorem [W.30.23.]. If $f \geqq 0$, $g \geqq 0$, then $(f+1)$, $(g+1)$ satisfy this inequality,

2) $$(f+1)(g+1)\mu = f\,g\,\mu + f\mu + g\mu + \mu.$$

Since all terms are admissible, it follows that $(f+1)(g+1)\mu$ is admissible. Hence, by Theorem [W.30.23.], $(f+1)(g+1)\mu$ is summable. Hence by (2), since $f\mu$, $g\mu$ and μ are summable, we get that $f\,g\,\mu$ is summable too. The theorem can be considered as established.

W.30.26. Theorem. Under Hyp. $\boldsymbol{H}_2$, if

1. E is a measurable set of traces,
2. $f(a)$, $g(a)$ are fields of real numbers,
3. $f\mu$, $g\mu$, $f^2\mu$, $g^2\mu$ are all (DARS)-summable on E,

then
$$(f+g)\mu \quad \text{and} \quad (f+g)^2\mu$$
are (DARS)-summable on E.

Proof. We have $(f+g)^2\mu = f^2\mu + g^2\mu + 2f\,g\,\mu$. Since all terms are summable, Theorem [W.30.24.], it follows that $(f+g)^2\mu$ is (DARS)-summable on E. Of course $(f+g)\mu$ is also so, because $f\mu$, $g\mu$ are (DARS)-summable on E. This completes the proof.

Chapter W I

Upper and lower summation in the general case. Complete admissibility. Square summability of fields of numbers.

The present chapter can be considered as continuation of [U.] and [W.].

W I.1. In Chapter [W] we have proved several theorems on (DARS)-summation of fields of real numbers, by admitting the hypotheses $\boldsymbol{H}_2$, but almost all the time — with few exceptions — we have worked under the hypothesis, stating that $\boldsymbol{G}$ has no atoms. Since the presence of atoms is related to the presence of point-spectrum of Hermitean operators, the restriction, which forbids atoms, is not convenient, and one must get rid of it some way. The present Chapter W I. is devoted to prove the theorems of the Chapter W. in the general case, where atoms may be available. The hypotheses $\boldsymbol{H}_1$ and $\mathscr{B}$ are rather moderate, so they can be admitted, though $\mathscr{B}$ will be supposed only when really needed.

We shall see, that many theorems proved in Chapter W. can be proved in a general case, though under the additional special new condition, which will be termed "complete admissibility", but which, however, is not too restrictive. We are interested in measurable sets of traces. Now, to the atoms of $\boldsymbol{G}$ there correspond heavy traces; there is a one to one correspondence between atoms of $\boldsymbol{G}$ and the heavy traces.

If α is a heavy trace and A the corresponding atom, then A may be a brick or not. If A is a brick, then every minimal descending sequence $\{a_n\}$ of bricks, which defines α, has the property stating that there exists n_0 such that for all $n \geqq n_0$, we have $a_n = A$.

However, if A is not a brick, the situation is different. The minimal sequences of bricks, defining α, are

$$a_1 \geqq a_2 \geqq \cdots \geqq a_n \geqq \cdots,$$

where

$$A < a_n, \quad (n = 1, 2, \ldots), \quad A = \prod_{n=1}^{\infty} a_n, \quad \lim_{n\to\infty} \mu(a_n) = \mu A > 0.$$

To somata of $\boldsymbol{G}$ there correspond measurable sets of traces. To simplify the language we shall be allowed to identify the somata with the corresponding sets of traces. So if a is a brick which covers the trace α (elusive or heavy), we may write $\alpha \in a$, of course when no dangerous ambiguity would occur. Concerning the proofs of theorems, which are modifications of proofs of theorems in Chapter W., we shall not repeat the proofs in extenso, but we shall confine ourselves to indicate the changes in the proofs. We refer to the literature, given at the end of the book. We shall try to follow the order of lemmas and theorems, as it is in Chapter W.

Section 1 of W I.

Upper and lower summation of fields of numbers in the general case, where the atoms may be present. Preliminaries.

W I.2. We recall the following three theorems, proved in [U.] under hypothesis $\boldsymbol{H}_1$.

W I.2.1. Theorem. If $f(a)$ is a field of real numbers, then the following are equivalent:

I. If S_n is an infinite sequence of complexes with

$$\lim_{u\to\infty} \mu(S_n) =_{df} \lim_{n\to\infty} \sum_k \mu(s_{nk}) = 0,$$

then

$$\lim_{n\to\infty} |f|(S_n) =_{df} \lim_{n\to\infty} \sum_k |f(s_{nk})| = 0;$$

II. if T_n is an infinite sequence of finite or denumerably infinite sets of mutually disjoint bricks with

$$\lim_{n\to\infty} \mu(T_n) =_{df} \lim_{n\to\infty} \sum_k f(t_{nk}) = 0,$$

then

$$\lim_{n\to\infty} |f|(T_n) =_{df} \lim_{n\to\infty} \sum_k |f(t_{nk})| = 0.$$

The property I, or its equivalent II, is called *admissibility of* $f(a)$.

W I.2.2. Theorem. Under hypothesis $\boldsymbol{H}_1$, if $f(a)$ is an admissible field of real numbers, and $p_1, p_2, \ldots, p_n, \ldots$ is any sequence of mutually disjoint bricks, then the sum

$$\sum_{n=1}^{\infty} |f(p_n)|$$

converges.

This is a particular case of Theorem [U.2.3.].

W I.2.3. Theorem. Under hypothesis $\boldsymbol{H}_1$, if $f(a)$ is an admissible field of real numbers, then for every $\varepsilon > 0$ there exists $\delta > 0$, such that, if S is a denumerably infinite or finite collection of mutually disjoint bricks with $\sum_k \mu(s_k) \leqq \delta$, then

$$\sum_k |f(s_k)| \leqq \varepsilon.$$

This has been proved in [U.2.5.].

W I.3. We have proved, under $\boldsymbol{H}_1$, that if a field $f(a)$ of real numbers is admissible, E is a measurable set of traces,

$$P_n = \{p_{n1}, p_{n2}, \ldots\}, \quad (n = 1, 2, \ldots),$$

is a (DARS)-sequence for E, then the sequence of numbers

$$M(P_n) =_{df} \sum_k |f(p_{nk})|$$

is bounded.

This however has been proved under the supposition that $\boldsymbol{G}$ has no atoms. Now, if we peruse the proof in the mentioned theorem, we can notice, that a small modification makes the proof valid also in the case, where $\boldsymbol{G}$ possesses atoms, but provided that E does not contain any heavy trace. We just must use $\mathcal{N}_R$ and $\mathcal{N}_A$ instead of $\mathcal{N}$, and construct the sets $E_{i_1, i_2, \ldots, i_k}$ not for $\boldsymbol{G}$, but for E. Then the following theorem is true:

W I.3.1. Theorem. Under hypothesis $\boldsymbol{H}_1$, if

1. $f(a)$ is a field of real numbers,
2. $f(a)$ is admissible,
3. E is a measurable set of traces, $E > O$,
4. E does not contain any heavy trace,
5. P_n is a (DARS)-sequence for E,

then the sequence of numbers

$$M(P_n) =_{df} \sum_k |f(p_{nk})|$$

is bounded.

W I.4. Theorem. Under $\boldsymbol{H}_1$, if

1. $f(a)$ is a field of real numbers,
2. A, B, C are measurable sets of traces with $A + B = C$, $A \cdot B = O$,
3. there exists a (DARS)-sequence Q_n for C, such that the sequence $|f|(Q_n)$ is not bounded,

then at least for one of the sets A, B there exists a (DARS)-sequence, say P_n, such that $|f|(P_n)$ is not bounded.

Remark. Given Q_n, we just take suitable partial complexes of Q_n, yielding sequences R_n, S_n for A, B respectively. One of them will produce unboundedness.

Proof. If we agree to denote, in general, by $M(T_n)$ the sum $\sum_k |f(t_{nk})|$, where $T_n = \{t_{n1}, t_{n2}, \ldots\}$, then $M(Q_n)$ is not bounded.

Since Q_n is a (DARS)-sequence for C, and $A \leqq C$, there exists a partial complex R_n of Q_n, $(n = 1, 2, \ldots)$, such that R_n is a (DARS)-sequence for A. Hence, $S_n =_{df} Q_n \sim R_n$ is a (DARS)-sequence for B. Now we have

$$R_n \frown S_n = \mathbf{O}, \quad R_n \smile S_n = Q_n, \quad (n = 1, 2, \ldots);$$

hence

$$|f|(R_n) + |f|(S_n) = |f|(Q_n).$$

Since the terms of these sequences are non-negative, and since $|f|(Q_n)$ is not bounded, it follows that at least one of the sequences $M(R_n)$, $M(S_n)$ is not bounded, q.e.d.

W I.4.1. Theorem. Under hypothesis $\boldsymbol{H}_1$, if

1. $f(a)$ is a field of real numbers,
2. A, B are measurable sets of traces,
3. $A \leqq B$,
4. there exists a (DARS)-sequence P_n for A, such that $|f|(P_n)$ is not bounded,

then there exists a (DARS)-sequence Q_n for $\boldsymbol{B}$, such that $|f|(Q_n)$ is not bounded.

Proof. Let $\alpha(n)$ be a subsequence of n, such that

(1) $$|f|(P_{\alpha(1)}) < |f|(P_{\alpha(2)}) < \cdots < |f|(P_{\alpha(n)}) < \cdots \to \infty.$$

$P_{\alpha(n)}$ is a (DARS)-sequence for A.

By Theorem [W.30.5.1.] there exists a subsequence $\beta(n)$ of n and a (DARS)-sequence R_n for $(B - A)$, such that

$$P_{\alpha\beta(n)} \cap R_n = \Theta, \quad (n = 1, 2, \ldots),$$

and where

$$S_n =_{df} P_{\alpha\beta(n)} \cup R_n$$

is a (DARS)-sequence for $\boldsymbol{B}$. We have

$$|f|(S_n) = |f| \cdot (P_{\alpha\beta(n)}) + |f|(R_n), \quad (n = 1, 2, \ldots),$$

and, by (1),

$$|f| \cdot (P_{\alpha\beta(n)}) \to \infty.$$

Since $|f|(R_n) \geqq 0$, it follows that $|f|(S_n) \to \infty$, so the theorem is proved.

W I.5. Lemma. Under hypothesis $\boldsymbol{H}_1$, if

1. E is a measurable set of traces,
2. P_n is a sequence of complexes,
3. α is a heavy trace, A its corresponding atom of $\boldsymbol{G}$,
4. $\alpha \,\overline{\in}\, E$,
5. $|P_n, E|_\mu \to 0$ for $n \to \infty$,

then there exists n_0, such that for all $n \geqq n_0$,

$$A \cdot \operatorname{som} P_n = O.$$

Proof. Suppose the thesis be not true. Then there exists a subsequence $k(n)$ of n, such that

$$A \cdot \operatorname{som} P_{k(n)} \neq O \quad \text{for all} \quad n = 1, 2, \ldots$$

Put

$$Q_n =_{df} P_{k(n)}, \quad (n = 1, 2, \ldots).$$

We have $A \leqq \operatorname{som} Q_n$; hence, if we put

$$Q_n = \{q_{n1}, q_{n2}, \ldots, q_{nk}, \ldots\},$$

there exists a unique k with $A \leqq q_{nk}$. We may suppose that $k = 1$. Thus we have

(1) $$A \leqq q_{n1} \quad \text{for} \quad n = 1, 2, \ldots$$

We have, by hyp. 5,

$$|E, P_n| \to 0;$$

hence

$$|E, Q_n| \to 0;$$

hence

$$\mu(\operatorname{som} Q_n - [E]) \to 0,$$

where $[E]$ is the coat of E, i.e. where

$$\mu(q_{n,1} + q_{n,2} + \cdots - [E]) \to 0 \quad \text{for} \quad n \to \infty.$$

Since, by (1), $A \leqq q_{n1}$, it follows that

$$\mu(A + q_{n2} + q_{n3} + \cdots - [E]) \to 0.$$

Hence $\mu(A - [E]) = 0$, and then $A - [E] = O$, which gives $A \leqq [E]$; but this is a contradiction with hyp. 4. The theorem is proved.

W I.5.1. Lemma. Under Hyp. $\boldsymbol{H}_1$, if

1. E is a measurable set of traces,
2. $A_1, A_2, \ldots, A_s$ are atoms of $\boldsymbol{G}$,
3. E_n does not contain any one of $A_1, \ldots, A_s$,
4. P_n is a sequence of complexes for E,
5. $|P_n, E| \to 0$ for $n \to \infty$,

then there exists n_0, such that if $n \geqq n_0$, then

$$A_1 \operatorname{som} P_n = O, \ \ldots, \ A_s \cdot \operatorname{som} P_n = O.$$

Proof. The theorem follows from Lemma [W I.5.].

W I.5.2. Lemma. Under hypothesis $\boldsymbol{H}_1$, if

1. E is a measurable set of traces,
2. E does not contain any heavy trace,
3. P_n is a (DR)-sequence for E,

then

$$\mathcal{N}(P_n) \to 0 \quad \text{for} \quad n \to \infty.$$

Proof. Hypothesis 3 means, that we have $|E, P_n|_\mu \to 0$ and also $\mathcal{N}_R(P_n) \to 0$ for $n \to \infty$, i.e.

$$\max[\mu(p_{n1} - \beta), \mu(p_{n2} - \beta), \ldots] \to 0 \quad \text{for} \quad n \to \infty,$$

where β is the sum of all atoms of $\boldsymbol{G}$. The thesis $\mathcal{N}(P_n) \to 0$ means:

$$\max[\mu(p_{n1}), \mu(p_{n2}), \ldots] \to 0 \quad \text{for} \quad n \to \infty.$$

First, let us suppose that there is an at most finite number of atoms in $\boldsymbol{G}$. Then by Lemma [W I.5.1.], there exists n_0, such that for all $n \geqq n_0$, $\operatorname{som} P_n$ is disjoint with all atoms, (if available). Hence for $n \geqq n_0$ we have

$$\mu(p_{nk} - \beta) = \mu(p_{nk}),$$

and then

$$\mathcal{N}(P_n) = \mathcal{N}_R(P_n),$$

which completes the proof in our case. Now suppose that there is an infinity of atoms in $\boldsymbol{G}$, say

$$A_1, A_2, \ldots, A_m, \ldots$$

Suppose that $\mathcal{N}(P_n)$ does not tend to 0 for $n \to \infty$. Hence there exists a number $M > 0$ and a subsequence $\alpha(n)$ of n, such that

(1) $$\mathcal{N}(P_{\alpha(n)}) \geqq M \quad \text{for all} \quad n = 1, 2, \ldots$$

Let m be such that

(2) $$\mu(A_{m+1}) + \mu(A_{m+2}) + \cdots \leqq \frac{M}{2};$$

and, by Lemma [W I.5.1.] find n_0, such that, for all $n \geqq n_0$, the atoms $A_1, A_2, \ldots, A_m$ are disjoint with som $P_{\alpha(n)}$. Let $n \geqq n_0$. We have

$$\mu[p_{\alpha(n),k} - \beta] = \mu[p_{\alpha(n),k} - \sum_{s=m+1}^{\infty} A_s].$$

By (1) there exists k' such that

$$\mu[p_{\alpha(n),k'}] \geqq M.$$

For this k' we have

$$\mu[p_{\alpha(n),k'} - \beta] \geqq \mu(p_{\alpha(n),k'}) - \sum_{s=m+1} \mu A_s \geqq M - \frac{M}{2};$$

hence

$$\max[\mu(p_{\alpha(n),1} - \beta), \mu(p_{\alpha(n),2} - \beta), \ldots] \geqq \frac{M}{2} \quad \text{for all} \quad n = 1, 2, \ldots$$

Hence $\mathcal{N}_R(P_{\alpha(n)}) \geqq \frac{M}{2}$ for all $n = 1, 2, \ldots$, which contradicts hyp. 3. The lemma is proved.

W I.6. The above lemma allows to generalize two important Theorems [U.8.] and [U.9.], and without almost any change of proofs:

W I.6.1. Theorem. Under Hyp. $\boldsymbol{H}_1$, if

1. $f(a)$ is a field of real numbers,
2. E is a measurable set of traces,
3. E does not contain any heavy trace,
4. $f(a)$ is admissible,
5. taking account, [W I.3.1.], of boundedness of $f(P_n)$ for every (DARS)-sequence P_n for E, we define for every (DARS)-sequence P_n for E the number

$$X(P_n) =_{df} \limsup_{n\to\infty} \sum_k |f(P_{nk})|;$$

then the supremum

$$\sup_{P_n} X(P_n),$$

taken for all P_n, is finite.

[It follows that, if we put

$$\varphi(P_n) = \limsup_{n\to\infty} \sum_k f(P_{nk}),$$

$$\psi(P_n) = \liminf_{n\to\infty} \sum_k f(P_{nk}),$$

then $\sup\limits_{P_n} \varphi(P_n)$ and $\inf\limits_{P_n} \psi(P_n)$ are also finite.]

W I.6.2. Theorem. Under Hyp. $\boldsymbol{H}_1$ if 1., 2., 3., 4., as before, if we define

$$\Phi(E) =_{df} \sup_{P_n} \varphi(P_n), \quad \Psi(E) =_{df} \inf_{P_n} \psi(P_n),$$

then there exists an (DARS)-sequence T_n for E, such that

$$\lim_{n\to\infty} f(T_n) = \Phi(E).$$

W I.6.3. Theorem. A similar theorem is true for $\Psi(E)$.

W I.7. Lemma. Under Hyp. $\boldsymbol{H}_1$, if

1. α is a heavy trace and A its corresponding atom of $\boldsymbol{G}$,
2. $p_1, p_2, \dots, p_n, \dots$ are bricks with $\alpha \in p_n$, $(n = 1, 2, \dots)$,
3. $\mu(p_n - A) \to 0$ for $n \to \infty$,
4. P_n is the complex, composed of the single brick p_n,

then P_n is a (DARS)-sequence for the set (α), i.e. the set, composed of the single trace α.

Proof. We have $A \leqq p_n$; hence $A - p_n = O$; hence

$$|A, P_n|_\mu = \mu(p_n - A) \to 0;$$

consequently P_n is a (D)-sequence for (α). Let β be the sum of all atoms of $\boldsymbol{G}$. We have $p_n - \beta \leqq p_n - A$; hence

$$\mathcal{N}_R(P_n) = \mu(p_n - \beta) \to 0,$$

so P_n is an (DR)-sequence for (α). We have

$$\mathcal{N}_A(P_n) = \mathcal{N}'(p_n) = \mu(p_n) - \max_{B \leqq p_n} \mu(B),$$

where the maximum is taken for all atoms B, which are contained in p_n. Since $A \leqq p_n$, there exists at least one atom contained in p_n.

Since $\mu(p_n - A) \to 0$, there exists n_0 such that, if $n \geqq n_0$, then the brick p_n cannot contain any other atom B, (outside A), with $\mu B \geqq \mu A$. Indeed if so, we would have $\mu(p_n - A) \geqq \mu(A)$ for an infinity of indices n, and this contradicts hyp. 3. Consequently we have for $n \geqq n_0$:

$$\mathcal{N}_A(P_n) = \mu(p_n) - \mu(A) = \mu(p_n - A) \to 0.$$

Thus we have proved that P_n is a (DAR)-sequence of (α). To prove the ($\boldsymbol{S}$)-property for P_n, it suffices to notice, that the only subsomata of A are A and O. The lemma is established.

W I.7.1. Lemma. Under Hyp. $\boldsymbol{H}_1$ if

1. α is a heavy trace and A the corresponding atom of $\boldsymbol{G}$,
2. P_n is a (DARS)-sequence for (α),

then for sufficiently great n the complex P_n can be written $p_{n1} \cup p_{n2} \cup \cdots$, where $A \leqq p_{n1}$ for $n = 1, 2, \dots$, and $\mu(p_{n2} + p_{n3} + \cdots) \to 0$ for $n \to \infty$, and $\mu(p_{n1} - A) \to 0$, whenever $p_{n2} + p_{n3} + \cdots$ are available.

Proof. We have $|P_n, A|_\mu \to 0$ for $n \to \infty$. We also have

$$A \leqq p_{n1} + p_{n2} + \cdots$$

for sufficiently great n.

Hence, for each of these n, there exists k such that

$$A \leqq p_{n1}.$$

By a slight change of notation, we can have

$$A \leqq p_{n1}.$$

We have

$$\operatorname{som} P_n - A = (p_{n1} - A) + (p_{n2} + p_{n3} + \cdots) \quad \text{for} \quad n \geqq n_0,$$

where n_0 is a suitable index.

Since

(1) $$\mu(\operatorname{som} P_n - A) = \mu(p_{n1} - A) + \mu(p_{n2} + p_{n3} + \cdots),$$

and since

(2) $$\mu(\operatorname{som} P_n - A) \to 0,$$

we get

$$\mu(p_{n2} + p_{n3} + \cdots) \to 0 \quad \text{for} \quad n \to \infty.$$

From (1) and (2) it follows that

$$\mu(p_{n1} - A) \to 0.$$

Thus the lemma is proved.

W I.8. Theorem. Under Hyp. $\boldsymbol{H}_1$, if

1. α is a heavy trace and A the corresponding atom,
2. $f(a)$ is a field of real numbers,
3. $f(a)$ is admissible,
4. P_n is a (DARS)-sequence for (α),
5. $|f|(P_n)$ is not bounded,

then there exists an infinite sequence

$$a_1, a_2, \ldots, a_m, \ldots$$

of bricks, such that

$$\mu(a_m - A) \to 0, \quad a_m \geqq A, \quad (m = 1, 2, \ldots),$$

but $|f(a_m)| \to +\infty$.

Proof. By Lemma [W I.7.1.] there exists an index n_0, such that for every $n \geqq n_0$ we can write $A \leqq p_{n1}$, and

$$\mu(\operatorname{som} P_n - A) = \mu(p_{n1} - A) + \sum_{k \geqq 2} \mu(p_{nk}),$$

where

$$\sum_{k \geqq 2} \mu(p_{nk}) \to 0 \quad \text{for} \quad n \to \infty.$$

Since $f(a)$ is admissible, we can find $n_1 \geqq n_0$, such that for all $n \geqq n_1$

(1) $$\sum_{k \geqq 2} |f(p_{nk})| \leqq 1.$$

Now we have

$$|f|(P_n) = |f(p_{n1})| + \sum_{k \geqq 2} |f(p_{nk})|.$$

Since, by hyp. 5, $|f|(P_n)$ is not bounded, therefore, by (1), the sequence $|f(p_{n1})|$ is not bounded. Hence there exists a subsequence $\lambda(n)$ of n, such that

$$|f(p_{\lambda(n),1})| \to \infty, \quad \text{for} \quad n \to \infty.$$

We have

$$A \leqq p_{\lambda(n),1} \quad \text{for} \quad n \geqq n_1.$$

We also have

$$\mu(p_{\lambda(n),1} - A) \to 0.$$

Indeed, by hyp. 4,

$$\mu(\operatorname{som} P_n - A) \to 0 \quad \text{for} \quad n \to \infty.$$

Hence

$$\mu(p_{\lambda(n)} - A) \to 0.$$

The theorem is proved.

W I.8.1. Theorem. Under Hyp. $\boldsymbol{H}_1$, if

1., 2., 3. as in Theorem [W I.8.],

4. there exists an infinite sequence a_n of bricks, such that $\mu(a_n - A) \to 0$, $A \leqq a_n$, $(n = 1, 2, \ldots)$, and where the sequence $|f(a_n)|$ is not bounded,

then there exists a (DARS)-sequence P_n for (α), such that $|f|(P_n)$ is not bounded.

Proof. By Lemma [W I.7.], if we define P_n as the complex, composed of the brick a_n only, then P_n is (DARS)-sequence for α; so $|f|(P_n)$ is not bounded.

W I.8.2. Theorem. Under Hyp. $\boldsymbol{H}_1$, if

1. $f(a)$ is a field of real numbers,
2. $f(a)$ is admissible,
3. $\alpha_1, \alpha_2, \ldots, \alpha_s$, $(s \geqq 1)$, are heavy traces and $A_1, A_2, \ldots, A_s$ the corresponding atoms,
4. if for every $k = 1, 2, \ldots, s$, $p_1^{(k)}, p_2^{(k)}, \ldots, p_n^{(k)}$, is an infinite sequence of bricks with

$$A_k \leqq p_n^k, \qquad \lim_{n \to \infty} \mu(p_n^k - A_k) = 0,$$

then the sequence $|f(p_n^k)|$ is bounded;

then for every (DARS)-sequence Q_n for

$$(\alpha_1) + (\alpha_2) + \cdots + (\alpha_s)$$

the sequence $|f|(Q_n)$ is bounded.

Proof. Suppose that there is a (DARS)-sequence Q_n for $(\alpha_1)+\cdots+(\alpha_s)$, such that $|f|(Q_n)$ is not bounded. Take a partial complex Q_n^1 of Q_n such that Q_n^1 is a (DARS)-sequence for (α_1). Then $Q_n \sim Q_n^{(1)}$ is a (DARS)-sequence for $(\alpha_2)+\cdots+(\alpha_s)$. Since $|f|(Q_n)$ is not bounded, therefore at least one of the sequences $|f|(Q_n^1)$, $|f|(Q_n \sim Q_n^1)$ is not bounded. The first sequence must be bounded, by virtue of Theorem [W I.8.] and hyp. 4. Hence $|f|(Q_n \sim Q_n^1)$ is not bounded.

Now we can apply to $Q_n \sim Q_n^1$ the analogous argument, getting a (DARS)-sequence for $(\alpha_3)+\cdots+(\alpha_s)$ for which the non-boundedness occurs. Finally we get a (DARS)-sequence for (α_s) with non-boundedness, which contradicts Theorem [W I.8.]. The theorem is proved.

W I.8.3. Theorem. Under Hyp. $\boldsymbol{H}_1$, if

1. $f(a)$ is a field of real numbers,
2. $f(a)$ is admissible,
3. K is the measurable set, composed of any infinite number of heavy traces,
4. for every atom A of K and every sequence a_n of bricks, where $A \leqq a_n$ for all n and $\mu(a_n - A) \to 0$ for $n \to \infty$, the sequence $|f(a_n)|$ is bounded,
5. P_n is a (DARS)-sequence for K,

then the sequence $|f|(P_n)$ is bounded.

Proof. By hyp. 2, we can find $\delta > 0$ such that, if Q is a complex with $\mu Q \leqq \delta$, then

$$|f|(Q) \leqq 1. \tag{1}$$

Since $\sum\limits_n^\infty \mu(A_n)$ is convergent, we can find m, such that

$$\mu(A_{m+1} + A_{m+2} + \cdots) \leqq \frac{\delta}{2}. \tag{2}$$

Now, considering P_n of hyp. 5, we can find a partial complex Q_n of P_n, such that Q_n is a (DARS)-sequence for $(\alpha_{m+1}) + (\alpha_{m+2}) + \cdots$.

Hence

$$\lim_{n\to\infty} |Q_n, A_{m+1} + A_{m+2} + \cdots| = 0.$$

Hence we can find n_0, such that for all $n \geqq n_0$

$$|Q_n, A_{m+1} + A_{m+2} + \cdots|_\mu \leqq \frac{\delta}{2}. \tag{3}$$

From (2) we have

$$|A_{m+1} + A_{m+2} + \cdots, 0| \leqq \frac{\delta}{2};$$

hence by (3),

$$|Q_n, O| < |Q_n, A_{m+1} + \cdots| + |A_{m+1} + \cdots, 1| \leqq \delta,$$

and then

$$\mu Q_n \leqq \delta \quad \text{for} \quad n \geqq n_0. \tag{4}$$

Consequently by (1)

(5) $$\sum_k |f(q_{nk})| \leqq 1 \quad \text{for all} \quad n \geqq n_0.$$

Since $Q_n \subsetneqq P_n$, the sequence P_n is a (DARS)-sequence for K, and Q_n is a (DARS)-sequence for

$$(\alpha_{m+1}) + (\alpha_{m+2}) + \cdots,$$

it follows, that the complementary complex

$$R_n =_{df} P_n - Q_n$$

yields a (DARS)-sequence for

$$(\alpha_1) + (\alpha_2) + \cdots + (\alpha_m).$$

By Theorem [W I.8.2.] and hyp. 4 we see that

(6) $$|f|(R_n)$$

is bounded. From (5) and (6) it follows that $|f|(P_n)$ is bounded.

W I.9. Def. We admit Hyp. $\boldsymbol{H}_1$. Let $f(a)$ be a field of real numbers and K a not empty set of heavy traces. We say that $f(a)$ *satisfies the supplementary condition of admissibility for* K, (or for the set of the corresponding atoms of $\boldsymbol{G}$), whenever the following happens:

If $\alpha \in K$, and A is the corresponding atom, and if $p_1, p_2, \ldots, p_n, \ldots$ is an infinite sequence of bricks, such that $A \leqq p_n$ for $n = 1, 2, \ldots$ and

$$(p_n - A) \to 0 \quad \text{for} \quad n \to \infty,$$

then $|f(p_n)|$ is a bounded sequence.

If K is composed of the single heavy trace α, then we shall also say that $f(a)$ satisfies on α, (on the corresponding atom A), *the supplementary condition of admissibility.*

W I.9.1. Def. We admit the Hyp. $\boldsymbol{H}_1$. Let $f(a)$ be a field of real numbers, and E a measurable set of traces.

We shall say that $f(a)$ *satisfies the condition of complete admissibility on* E, whenever

1. $f(a)$ is admissible,
2. if $K \neq \emptyset$ is the set of all heavy traces, contained in E,

then $f(a)$ satisfies the supplementary condition of admissibility for K.

W I.9.2. Theorem. Under hypothesis $\boldsymbol{H}_1$, if

1. $f(a)$ is a field of real numbers,
2. E is a measurable set of traces, $\mu E > 0$,

then the following are equivalent

I. $f(a)$ is admissible and for every (DARS)-sequence P_n for E and the sequence $|f|(P_n)$ is bounded,

II. $f(a)$ satisfies, on E, the condition of complete admissibility.

Proof. We shall consider three cases:

a) E has no heavy traces,

b) all traces of E are heavy,

c) E contains some heavy traces, but the set of all elusive traces has a positive measure.

Consider the case a) where E does not have any heavy trace. Suppose I takes place; hence $f(a)$ is admissible. Since there are no heavy traces in E, the complete admissibility is just the simple admissibility, so II follows. Suppose that II takes place, i.e. $f(a)$ is admissible, hence, by [W I.3.1.], I follows.

Consider the case b) where all traces are heavy. Suppose II. We apply Theorem [W I.8.2.] and Theorem [W I.8.3.], getting admissibility and the required boundedness. Hence II implies I.

Suppose I. $f(a)$ is admissible. Suppose that there exists a heavy trace α in E, such that $f(a)$ does not satisfy the supplementary condition for α. Hence there exists an infinite sequence a_n of bricks, such that $A \leqq a_n$, and $\lim_{n\to\infty} (a_n - A) = 0$, (where A is the atom corresponding to α), but where

$$|f(a_n)|$$

is not bounded. Then, by Theorem [W I.8.1.], there exists a (DARS)-sequence P_n for (α), where $|f|(P_n)$ is not bounded. Hence by [W I.4.1.] there exists a (DARS)-sequence Q_n for E, such that $|f|(Q_n)$ is not bounded. This, however, contradicts I. Consequently we have proved that I implies II, in our case. Thus we have proved the equivalence of I and II in the cases a) and b).

Let us consider the case c), where E contains some heavy traces, but the set of its elusive traces has a positive measure.

Let I be true, but suppose that II is not true. Since $f(a)$ is admissible, the supplementary condition of admissibility does not holds true on the set K of all heavy traces, contained in E. Hence, by Def. [W I.9.] there exists a heavy trace α, such that for its corresponding atom A, we can find a sequence a_n of bricks with $A_n \leqq a_n$, $\mu(a_n - A) \to 0$ for $n \to \infty$, but such that the sequence $|f(a_n)|$ is not bounded.

The sequence $[a_n]$ is a (DARS)-sequence for (α), but $|f|((a))$ is not bounded. Hence, by Theorem [W I.4.1.], it follows that there exists a (DARS)-sequence P_n for E, such that $|f|(P_n)$ is not bounded. This contradicts I. Thus we have proved that I $\to$ II.

Now suppose II. Let E', E'' be the atomic and the non atomic part of E. We have $E' + E'' = E$, $E' \cdot E'' = O$. The field $f(a)$ satisfies on E' the condition of complete admissibility.

Take a (DARS)-sequence P_n for E, and find, as usually, the partial complexes P'_n, P''_n of P_n, such that $P'_n \cup P''_n = P_n$, $P'_n \cap P''_n = \mathit{O}$.

The sequences P'_n, P''_n are (DARS)-sequences for E', E'' respectively. By Theorem [W I.8.2.] and [W I.8.3.], $|f|(P'_n)$ is bounded, and $|f|(P''_n)$ is bounded. Hence $|f|(P_n)$ is bounded. It follows that II $\to$ I. The theorem is proved.

W I.9.3. Remark. The Theorem [W I.9.2.] summarizes the Theorems [W I.8.1.], [W I.8.2.] and [W I.8.3.].

W I.10. Theorem. Let us admit the Hyp. $\boldsymbol{H}_1$. Let $f(a)$ be a field of real numbers. Suppose that:

1. $f(a)$ is admissible,
2. α is a heavy trace, A its corresponding atom,
3. the supplementary condition of admissibility holds true for (α),
4. for every sequence p_n of bricks, such that

$$A \leqq p_n, \quad (n = 1, 2, \ldots), \quad \lim_{n\to\infty} \mu(p_n - A) = 0,$$

define

$$\varphi(p_n) =_{df} \limsup_{n\to\infty} f(p_n).$$

(We have $\varphi(p_n)$ finite, by hyp. 3.):

Then the set of all $\varphi(p_n)$ is bounded from above.

Proof. We remark that, if

$$(1) \qquad \varphi(p_n) > M > 0,$$

then there exists an index m, such that

$$f(p_m) > M, \quad A \leqq p_m, \quad \mu(p_m - A) \leqq \frac{1}{M}.$$

Indeed, we have $\varphi(p_n) > M + \varepsilon$, where ε is a suitable positive number, i.e.

$$\limsup_{n\to\infty} f(p_n) > M + \varepsilon.$$

Hence we can find a subsequence $\beta(n)$ of n, such that $f(p_{\beta(n)}) > M$ for all n. Now since $A < p_{\beta(n)}$ and

$$\lim(p_{\beta(n)} - A) \to 0,$$

there exists m such that

$$A \leqq p_m, \quad f(p_m) > M, \quad \mu(p_m - A) \leqq \frac{1}{M}.$$

This being done, suppose that our theorem is not true. Hence, the set of all $\varphi(p_n)$ is not bounded from above. Hence we can find an infinite sequence $p_n^1, p_n^2, \ldots$ of sequence of bricks, such that

$$\varphi(p_n^1) > 1, \quad \varphi(p_n^2) > 2, \ldots, \varphi(p_n^m) > m, \ldots$$

and $A \leqq p_n^k$ for all n and k, and

$$\lim_{n\to\infty} \mu(p_n^k - A) = 0 \quad \text{for} \quad k = 1, 2, \ldots$$

Applying the remark, made above, we can find a sequence of indices $m_1, m_2, \ldots$, such that

$$f(p^1_{m_1}) > 1, \quad f(p^2_{m_2}) > 2, \ldots$$

and

$$\mu(p^1_{m_1} - A) \leqq \tfrac{1}{1}, \quad \mu(p^2_{m_2} - A) \leqq \tfrac{1}{2}, \ldots, \qquad A \leqq p^1_{m_1}, \qquad A \leqq p^2_{m_2}, \ldots$$

Consequently, if we put

$$q_n =_{df} p^n_{m_n}, \qquad (n = 1, 2, \ldots),$$

we get

$$\mu(q_n - A) \to 0, \qquad A \leqq q_n \quad \text{and} \quad f(q_n) \to \infty,$$

which contradicts hyp. 3.

W I.10.1. Theorem. Let us admit all hypotheses of Theorem [W I.10.]. Put $\Phi =_{df} \sup_{p_n} \varphi(p_n)$; then there exists a sequence a_n of bricks, such that $A \leqq a_n$ for $n = 1, 2, \ldots$, and

$$\lim_{n\to\infty} \mu(a_n - A) = 0 \quad \text{and} \quad \lim_{n\to\infty} f(a_n) = \Phi.$$

W I.10.1a. Proof. First of all, we shall prove that, if $\Phi - \varphi(p_n) < \varepsilon$, where $\varepsilon > 0$, then there exists an index m, such that

$$A \leqq p_m, \qquad \mu(A - p_m) < \varepsilon \quad \text{and} \quad |\Phi - f(p_m)| < \varepsilon.$$

Indeed, there exists $\eta > 0$, such that

$$(1) \qquad 0 \leqq \Phi - \varphi(p_n) < \varepsilon - \eta.$$

Now, since

$$\varphi(p_n) = \limsup_{n\to\infty} f(p_n),$$

we can find a subsequence $\beta(\dot{n})$ of $\dot{n}$, such that

$$(2) \qquad -\varepsilon < \varphi(p_n) - f(p_{\beta(n)}) < \eta \quad \text{for all} \quad n = 1, 2, \ldots$$

In addition to that, we may choose $\beta(n)$, so as to have for all n:

$$\mu(p_{\beta(n)} - A) < \varepsilon. \quad \text{We have} \quad A \leqq p_{\beta(n)}.$$

From (1) and (2) we get

$$-\varepsilon < \Phi - f(p_{\beta(n)}) < \varepsilon.$$

Putting

$$m =_{df} \beta(1),$$

we get

$$A \leqq p_m, \qquad \mu(p_m - A) < \varepsilon$$

and

$$-\varepsilon < \Phi - f(p_m) < \varepsilon, \quad \text{q.e.d.}$$

W I.10.1 b. Having that, we can proceed to the proof of the Theorem [W I.10.1.]. Since $\Phi = \sup_{\{p_n\}} \varphi(p_n)$, there exists an infinite sequence

$$p_n^{(1)}, p_n^{(2)}, \ldots, p_n^{(s)}, \ldots$$

of sequences of bricks, such that

$$\Phi - \varphi(p_n^{(s)}) < \frac{1}{s}, \quad \text{for} \quad s = 1, 2, \ldots$$

By virtue of the established remark, we can find indices

$$n_{(1)}, n_{(2)}, \ldots, n_{(s)}, \ldots;$$

such that

$$|\Phi - f(p_{n_{(s)}})| < \frac{1}{s}, \qquad A \leqq p_{n_{(s)}}, \qquad \mu(p_{n_{(s)}} - A) < \frac{1}{s}.$$

It follows that

$$\Phi = \lim_{s\to\infty} f(p_{n_{(s)}}),$$

so the theorem is proved.

W I.10.2. Theorem. We admit $\boldsymbol{H}_1$ and the hypotheses 1., 2., 3., 4. of the Theorem [W I.10.]. We define

(1) $$\psi(p_n) =_{df} \lim_{n\to\infty} \inf f(p_n);$$

then the set of all $\psi(p_n)$ is bounded from below. Proof similar to that of Theorem [W I.10.].

W I.10.3. Theorem. We admit $\boldsymbol{H}_1$ and the same hypotheses 1., 2., 3., 4. as before in [W I.10.2.], and the definition (1). Put

$$\Psi =_{df} \inf_{p_n} \psi(p_n).$$

Then there exists a sequence b_n of bricks, such that

$$A \leqq b_n \quad \text{for} \quad n = 1, 2, \ldots, \qquad \lim_{n\to\infty} \mu(b_n - A) = 0$$

and

$$\lim_{n\to\infty} f(b_n) = \Psi.$$

Proof similar to that of [W I.10.1.].

W I.10.4. Theorem. Under Hyp. $\boldsymbol{H}_1$, suppose that $f(a)$ is an admissible field of real numbers. Let α be a heavy trace, on which the supplementary condition of admissibility is satisfied. Let A be the corresponding atom of $\boldsymbol{G}$. Then the set of values

$$\lim_{n\to\infty} \sup f(P_n),$$

where P_n are (DARS)-sequences for (α), coincides with the set of values $\lim_{n\to\infty} \sup f(a_n)$, where a_n are bricks, where

$$A \leqq a_n \quad \text{for} \quad n = 1, 2, \ldots \quad \text{and where} \quad \mu(a_n - A) \to 0.$$

(A similar statement is true for $\lim_{n\to\infty} \inf f(P_n)$.)

Proof. By Lemma [W I.7.] if α is a heavy trace, A its corresponding atom, and

$$p_1, p_2, \ldots, p_n, \ldots$$

is an infinite sequence of bricks with

$$A \leqq p_n, \quad \mu(p_n - A) \to 0,$$

then $\{p_n\}$ is a (DARS)-sequence for (α). By Lemma [W I.7.1.], if P_n is a (DARS)-sequence for (α), then we can find n_0, such that for all $n \geqq n_0$ we have, with a suitable notation,

$$A \leqq p_{n_1}, \quad \mu(p_{n_2} + p_{n_3} + \cdots) \to 0 \quad \text{for} \quad n \to \infty.$$

It follows that

$$|f(p_{n_2})| + |f(p_{n_3})| + \cdots \to 0.$$

Hence

$$\lim_{n\to\infty} f(P_n) = \lim_{n\to\infty} f(p_n).$$

This completes the proof.

W I.10.5. Remark. Under circumstances considered, the numbers Φ and Ψ can be obtained as well by using (DARS)-sequences for (α), as by using sequences of bricks, which enclose A.

W I.10.6. Remark. If we put

$$\Omega =_{df} \max\{|\Phi|, |\Psi|\}$$

under circumstances considered, then there exists a (DARS)-sequence P_n for (α), (and also a sequence a_n of bricks), such that

$$\Omega = \lim_{n\to\infty} |f|(P_n) = \lim_{n\to\infty} |f(a_n)|.$$

Section 2 of W I.

Basic theorems on upper and lower summation for fields of real numbers

W I.11. Theorem. Under Hyp. $\boldsymbol{H}_1$, if

1. the condition $\mathscr{B}$ is admitted,
2. $f(a)$ is a field of real numbers,
3. $f(a)$ is admissible,
4. E, F are measurable sets of traces,
5. E does not contain any heavy trace, (F may),
6. $E \cdot F = O$,
7. P_n, Q_n are (DARS)-sequences for E, F respectively,

then there exist subsequences $s(\dot{n})$, $t(\dot{n})$ of $\dot{n}$ and complexes P'_n, Q'_n such that

1) $P'_n \subseteqq P_{s(n)}$, $Q'_n \subseteqq Q_{t(n)}$, $(n = 1, 2, \ldots)$,
2) $P'_n \frown Q'_n = \mathit{O}$,
3) P'_n, Q'_n are (DARS)-sequences for E, F respectively,
4) $P'_n \smile Q'_n$ is a (DARS)-sequence for $E + F$.

(We can have $s(n) \geqq t(n)$ [or $t(n) \geqq s(n)$], for all n.)

Proof. The proof is a corrolary of the proof of the similar Theorem in [W.2.]. The changes are: we can have $\mathcal{N}(P_\alpha) \to 0$, as needed in [W.2c.] by virtue of hyp. 5 and by using Theorem [W I.5.2.]. Then, of course, $\mathcal{N}_R(P_\alpha) \to 0$ and $\mathcal{N}_A(P_\alpha) \to 0$. This allows to define P'_β and Q'_β as in [W.2d.]. The theorem is established. As its consequences we have:

W I.12. Theorem. Under Hyp. $\boldsymbol{H}_1$, if

1. the condition $\mathcal{B}$ is admitted,
2. $f(a)$ is a field of real numbers,
3. $f(a)$ is admissible,
4. E, F are sets of traces,
5. E does not contain any heavy trace, (F may do),
6. P_n, Q_n are (DARS)-sequences for E, F respectively,
7. $\lim\limits_{n\to\infty} f(P_n) = A$, $\lim\limits_{n\to\infty} f(Q_n) = B$,

then there exist subsequences $s(\dot{n})$, $t(\dot{n})$ of $\dot{n}$, and complexes P'_n, Q'_n, such that

1) $P'_n = P_{s(n)}$, $Q'_n = Q_{t(n)}$, $(n = 1, 2, \ldots)$,
2) $P'_n \frown Q'_n = \mathit{O}$, $(n = 1, 2, \ldots)$,
3) P'_n, Q'_n are (DARS)-sequences for E, F respectively,
4) $P'_n \smile Q'_n$ is a (DARS)-sequence for $E + F$,
5) $\lim\limits_{n\to\infty} f(P'_n) = A$, $\lim\limits_{n\to\infty} f(Q'_n) = B$, $\lim\limits_{n\to\infty} f(P'_n \smile Q'_n) = A + B$.

Proof. The same, as the proof of [W.3.].

W I.13.1. Def. Let $f(a)$ be a field of real numbers, and E a measurable set of traces, (they may be heavy or not). Consider all (DARS)-sequences P_n for E. We define *the upper and the lower sum of $f(a)$ on E* by

$$\overline{S}_E f(a) =_{df} \sup_{P_n} \lim_{n\to\infty} \sup f(P_n),$$

if finite, and

$$\underline{S}_E f(a) =_{df} \inf_{P_n} \lim_{n\to\infty} \inf f(P_n),$$

if finite.

The supremum and the infimum are taken for all (DARS)-sequences for E. Later we shall have existence theorems for $\overline{S}$ and $\underline{S}$. This definition is a generalization of the definition Def. [W.4.].

W I.14. Theorem. Under Hyp. $\boldsymbol{H}_1$ if

1. $f(a)$ is a field of real numbers,
2. E is a measurable set of traces,
3. E does not contain any heavy trace,
4. $f(a)$ is admissible,

then

1) the set of all numbers

$$\chi(P_n) =_{df} \lim_{n\to\infty} \sup |f(P_n|,$$

where P_n varies over the collection of all (DARS)-sequences for E, is bounded,

2) consequently, the set of all numbers

$$\varphi(P_n) =_{df} \lim_{n\to\infty} \sup f(P_n),$$

for varying P_n as before, is bounded,

3) the set

$$\psi(P_n) = \lim_{n\to\infty} \inf f(P_n),$$

for the sum P_n, is bounded,

4) consequently,

$$\sup_{P_n} \lim_{n\to\infty} f(P_n) \quad \text{exists and} \quad \inf_{P_n} \lim_{n\to\infty} \inf f(P_n)$$

exists, where sup and inf are taken for all (DARS)-sequences P_n for E,

5) consequently $\overline{S}_E f(a)$ and $\underline{S}_E f(a)$ exist.

Proof. By Theorem [W I.6.1.]; see also [W I.6.3.].

W I.14a. Our next purpose is to extend this theorem to the case, where are heavy traces. The extension will be made stepwise.

W I.14.1. Theorem. Under Hyp. $\boldsymbol{H}_1$, if

1. $f(a)$ is a field of real numbers,
2. E is a set, composed of a finite, (not null), number of heavy traces,
3. $f(a)$ is completely admissible, Def. [W I.9.],

then

1) the set of all numbers

$$\chi(P_n) =_{df} \lim_{n\to\infty} \sup |f(P_n)|,$$

where P_n varies over the collection of all (DARS)-sequences for E, is bounded,

2) consequently, the sets of all numbers

$$\varphi(P_n) =_{df} \lim_{n\to\infty} \sup f(P_n)$$

and that of all numbers

$$\psi(P_n) =_{df} \lim_{n\to\infty} \inf f(P_n),$$

where P_n varies over all (DARS)-sequences for E, are both bounded,

3) the numbers

$$\sup_{P_n} \limsup f(P_n), \qquad \inf_{P_n} \liminf f(P_n)$$

both exist,

4) consequently $\overline{S}_E f(a)$ and $\underline{S}_E f(a)$ exist.

Proof. Let $E = (\alpha_1) + (\alpha_2) + \cdots + (\alpha_k)$, $n \geqq 1$, where all α_s are heavy traces. Take any (DARS)-sequence P_n for E. By using the fundamental property of (DARS)-sequences and by iterated application of Theorem [W I.8.], we get some sequences

$$P_n^{(1)}, P_n^{(2)}, \ldots, P_n^{(k)}$$

of complexes with properties:

They are (DARS)-sequences for $(\alpha_1), (\alpha_2), \ldots, (\alpha_k)$ respectively for every n; they give disjoint complexes with union P_n.

Now applying Lemma [W I.7.1.], we can find n_0, such that for all $n \geqq n_0$ we can write for the corresponding atoms $A_1, A_2, \ldots, A_k$:

(1) $$A_1 \leqq p_{n1}^{(1)}, A_2 \leqq p_{n1}^{(2)}, \ldots, A_k \leqq p_{n1}^{(k)}, \ldots$$

In addition to that we have

(2) $$\begin{gathered} \mu(p_{n1}^{(1)} - A_1) \to 0, \quad \mu(p_{n1}^{(2)} - A_2) \to 0, \ldots, \\ \mu(p_{n1}^{(k)} - A_k) \to 0 \quad \text{for} \quad n \to \infty, \end{gathered}$$

and

(3) $$\begin{gathered} \lim_{n\to\infty} \mu[p_{n2}^{(1)} + p_{n3}^{(1)} + \cdots] = 0, \ldots, \\ \lim_{n\to\infty} \mu[p_{n2}^{(k)} + p_{n3}^{(k)} + \cdots] = 0. \end{gathered}$$

Since $f(a)$ is admissible, find $\delta > 0$ such that, if for a complex Q we have $\mu Q \leqq \delta$, then $|f|(Q) \leqq \frac{1}{k}$. Now we can find n_1, where $n_1 \geqq n_0$, such that for all $n \geqq n_1$:

(4) $$\mu[p_{n2}^{(s)} + p_{n3}^{(s)} + \cdots] \leqq \delta, \quad s = 1, 2, \ldots, k.$$

Hence we have for $n_1 \geqq n_0$, the inequality $A_s \leqq p_{n1}^{(s)}$, and

$$|f(p_{n2}^{(s)})| + |f(p_{n3}^{(s)})| + \cdots \leqq \frac{1}{k}, \quad (s = 1, \ldots, k),$$

which gives

$$\sum_{s=1}^{k} [f(p_{n2}^{(s)})| + |f(p_{n3}^{(s)})| + \cdots \leqq 1;$$

and hence for $n \geqq n_1$:

(5) $$|f(P_n)| \leqq \sum_{s=1}^{k} |f(p_{n1}^{(s)})| + 1.$$

I say that

(6) $|f(p_{n1}^{(s)})|$ is bounded for $s = 1, 2, \ldots, k$.

Indeed, by (1),

$$A_s \leqq p_{n1}^{(s)} \quad \text{for all} \quad n \geqq n_0;$$

hence for all $n \geqq n_1$. We also have

$$\mu(p_{n1}^{(1)} - A_s) \to 0 \quad \text{for} \quad n \to \infty.$$

Hence by hyp. 3, (6) is true. Consequently

$$\lim_{n\to\infty} \sup |f(P_n)| \leqq \sum_{s=1}^{k} \lim_{n\to\infty} \sup |f(p_{n1}^{(s)})| + 1 \quad \text{for} \quad n \geqq n_1.$$

Since, by Theorem [W I.10.], the set of all numbers $\lim\limits_{n\to\infty} \sup |f(p_{n1}^{(s)})|$ for all (DARS)-sequences for (α_s) is bounded, it follows that the set of all numbers $\lim\limits_{n\to\infty} \sup |f(P_n)|$ with P_n varying over all (DARS)-sequences for E, is bounded from above. The boundedness from above of the set of all numbers $\varphi(P_n) =_{df} \lim\limits_{n\to\infty} \sup f(P_n)$ and the boundedness form below of the set of all numbers

$$\Psi(P_n) =_{df} \lim_{n\to\infty} \inf f(P_n)$$

for variable (DARS)-sequences P_n for s, follows. The existence of $\overline{S} f$, $\underline{S} f$ follows too.

W I.14.2. Theorem. Under Hyp. $\boldsymbol{H}_1$, if

1. $f(a)$ is a field of real numbers,

2. E is measurable with positive measure, composed of finite or infinite number of heavy traces,

3. $f(a)$ is completely admissible on E,

then the thesis of Theorem [W I.14.1.] holds true.

Proof. If E is composed of a finite number of heavy traces, we have the case of Theorem [W I.14.1.]. Let E be composed of an infinite number of heavy traces:

(1) $$\alpha_1, \alpha_2, \ldots, \alpha_k, \ldots$$

As $f(a)$ is admissible, we can find $\delta > 0$, such that, if for a complex R we have $\mu R \leqq \delta$, then $|f|(R) \leqq 1$. Since $\sum\limits_{k=1} \mu(\alpha_k)$ converges, we can find k_0, such that

(2) $$\sum_{k>k_0} \mu(\alpha_k) \leqq \frac{\delta}{2}.$$

Consider the sets

$$E' =_{df} (\alpha_1) + \cdots + (\alpha_{k_0}), \qquad E'' =_{df} (\alpha_{k_0+1}) + (\alpha_{k_0+2}) + \cdots,$$

and take (DARS)-sequences P'_n, P''_n for E', E'' respectively, such that

$$P'_n \cap P''_n = \mathbf{O}, \qquad P'_n \cup P''_n = P_n \quad \text{for} \quad n = 1, 2, \ldots$$

We have

(3) $$|f|(P_n) = |f|(P'_n) + |f|(P''_n).$$

By (2) we have

$$\mu(E'') \leqq \frac{\delta}{2}.$$

Now we can find n_0, such that for all $n \geqq n_0$,

$$|P''_n, E''| \leqq \frac{\delta}{2}.$$

Hence, since

$$|E'', 0| \leqq \frac{\delta}{2},$$

we get

$$|P''_n, O| \leqq \delta, \quad \text{i.e.,} \quad \mu P''_n \leqq \delta,$$

and then

$$|f|(P''_n) \leqq 1.$$

Thus from (3) we get

$$|f|(P_n) \leqq |f|(P'_n) + 1 \quad \text{for all} \quad n \geqq n_0.$$

Hence

$$\lim_{n\to\infty} \sup |f|(P_n) \leqq \lim_{n\to\infty} \sup |f|(P'_n) + 1$$

for all sequences P_n for E. Now, since E' has a finite number of heavy traces, there exists, by Theorem [W I.14.1.] a number $A > 0$, such that $\lim\sup |f|(P'_n) \leqq A$ for all (DARS)-sequences Q'_n for E'; hence for all (DARS)-sequences P'_n, considered for E'. Hence

$$\lim\sup |f|(P_n) \leqq 1 + A,$$

which proves the theorem.

W I.14.3. Theorem. Under Hyp. $\boldsymbol{H}_1$, if

1. $f(a)$ is a field of real numbers,

2. E is a measurable, $(\mu E > 0)$, set of traces, some one of them being heavy, or not,

3. $f(a)$ is completely admissible on E Def. [W I.9.],

then the thesis of Theorem [W I.14.1.] holds true.

Proof. Let $E = E' + E''$, where E'' does not contain any heavy trace and where E' is composed only of heavy traces. Let P_n be a (DARS)-sequence for E. We may decompose $P_n = P'_n \cup P''_n$, where $P'_n \cap P''_n = O$ and where P'_n, P''_n are (DARS)-sequences for E', E'' respectively. We have $|f|(P_n) = |f|(P'_n) + |f|(P''_n)$. We see, by Theorem [W I.14.2.], that the set of numbers $\lim_{n\to\infty} \sup |f|(P'_n)$ is bounded, and by Theorem [W I.6.1.] that $\lim_{n\to\infty} \sup |f|(P''_n)$ is bounded.

Hence the set of all $\lim\sup |f(P'_n) + f(P''_n)|$ is bounded from above, q.e.d.

W I.14.4a. Remark. The above Theorem [W I.14.3.] summarizes the particular cases of Theorems [W I.14.], [W I.14.1.], [W I.14.2.] on boundedness of the set of all $\lim\sup f(P_n)$ for (DARS)-sequences.

It is a generalization of theorem [U.8.], to the most general case of $\boldsymbol{G}$. We are going to extend similarly the Theorem [U.9.]. We shall prove an extension of the theorem Remark [U.9.2.], which has been proved in the case, where $\boldsymbol{G}$ has no atoms.

W I.15. Lemma. Under Hyp. $\boldsymbol{H}_1$, if

1. $f(a)$ is a field of real numbers,
2. E is a set of a finite number $k \geqq 1$, of heavy traces,
3. $f(a)$ is completely admissible on E, Def. [W I.9.1.],
4. $P_n^1, P_n^2, \ldots, P_n^m, \ldots$ are (DARS)-sequences for E, such that the limits

$$\lim_{n\to\infty} f(P_n^1), \ \lim_{n\to\infty} f(P_n^2), \ldots, \ \lim_{n\to\infty} f(P_n^m), \ldots$$

all exist, and where (1) is a sequence of numbers tending to the limit A,

then there exists a (DARS)-sequence S_n for E, such that

$$\lim_{n\to\infty} f(S_n) = A.$$

W I.15a. Proof. Let us fix m for a moment, and consider the (DARS)-sequence P_n^m for E. Since $E = (\alpha_1) + (\alpha_2) + \cdots + (\alpha_k)$, where α_s are heavy traces, we can find k-subcomplexes

$$(1) \qquad P_n^{m1}, P_n^{m2}, \ldots, P_n^{mk} \quad \text{of} \quad P_n^m, \qquad (n = 1, 2, \ldots),$$

such that P_n^{ms} is a (DARS)-sequence for (α_s), the complexes (1) are mutually disjoint and their union is P_n^m. Applying Lemma [W I.7.1.], we can find n_0, (depending on m), such that for all $n \geqq n_0$ we can write

$$(2) \qquad A_1 \leqq p_{n1}^{m1}, A_2 \leqq p_{n1}^{m2}, \ldots, A_k \leqq p_{n1}^{mk}.$$

In addition to that we have

$$(3) \quad \lim_{n\to\infty} \mu(p_{n1}^{m1} - A_1) = \lim_{n\to\infty} \mu(p_{n1}^{m2} - A_a) = \cdots = \lim_{n\to\infty} \mu(p_{n1}^{mk} - A_k) = 0.$$

$$(4) \qquad \mu(p_{n1}^{n1} - A_1) \leqq \frac{1}{m}, \ldots, \mu(p_{n1}^{mk} - A_k) \leqq \frac{1}{m},$$

whatever $m = 1, 2, \ldots$ may be, with $n \geqq n_1(m)$.

We also have

$$(5) \qquad \begin{cases} \lim_{n\to\infty} \mu(p_{n2}^{m1} + p_{n3}^{m1} + \cdots) = 0, \\ \lim_{n\to\infty} \mu(p_{n2}^{m2} + p_{n3}^{m2} + \cdots) = 0, \\ \cdots\cdots\cdots\cdots\cdots\cdots \\ \lim_{n\to\infty} \mu(p_{n2}^{mk} + p_{n3}^{mk} + \cdots) = 0. \end{cases}$$

Of course we can suppose that $n\,1 < n\,2 < \cdots$.

W I.15b. We have supposed that $f(a)$ is admissible. Applying Theorem [W I.2.2.], we can find $\delta_m > 0$, such that for every complex R

with $\mu R \leqq \delta_m$, we have $|f|(R_n) \leqq \frac{1}{m}$. From (5) it follows, that we can find $n_2 \geqq n_2(m) \geqq n_1 \geqq n_0$ such that, for all $n \geqq n_2$ we have

$$(6) \qquad \begin{cases} \mu(p_{n2}^{m1} + p_{n3}^{m1} + \cdots) \leqq \frac{\delta_m}{k}, \\ \cdots\cdots\cdots\cdots\cdots\cdots \\ \mu(p_{n2}^{mk} + p_{n3}^{mk} + \cdots) \leqq \frac{\delta_m}{k}. \end{cases}$$

It follows that for $n \geqq n_2$ we have, whatever m may be,

$$(7) \quad \mu\{(p_{n2}^{m1} + p_{n3}^{m1} + \cdots) + (p_{n2}^{m2} + p_{n3}^{m2} + \cdots) + \cdots + \\ + (p_{n2}^{mk} + p_{n3}^{mk} + \cdots)\} \leqq \delta_m,$$

and hence

$$|f|(p_{n2}^{m1} + \cdots) + |f|(p_{n2}^{m2} + \cdots) + \cdots + |f|(p_{n2}^{mk} + \cdots) \leqq \frac{1}{m},$$

which can be written:

$$(8) \qquad |f|\{p_{n,k+1}^{m} + p_{n,k+2}^{m} + \cdots\} \leqq \frac{1}{m};$$

hence

$$(9) \qquad \left|f(P_n^m) - \sum_{s=1}^{k} f(p_{ns}^m)\right| \leqq \frac{1}{m} \quad \text{for} \quad n \geqq n_2(m).$$

If we define complexes

$$(10) \qquad T_n^m =_{df} p_{n1}^m \cup p_{n2}^m \cup \cdots \cup p_{nk}^m,$$

we can write

$$(11) \qquad T_n^m = p_{n1}^{m1} \cup p_{n1}^{m2} \cup \cdots \cup p_{n1}^{mk};$$

and (9) becomes

$$(12) \quad |f(P_n^m) - f(T_n^m)| \leqq \frac{1}{m} \text{ for } n \geqq n_2(m) \text{ and for all } m = 1, 2, \ldots$$

Since $\lim_{n\to\infty} f(P_n^m)$ exists we can find

$$n_3 = n_3(m) \geqq n_2(m) \geqq n_1(m) \geqq n_0(m),$$

such that for $n \geqq n_3$:

$$\left|\lim_{n\to\infty} f(P_n^m) - f(P^m)\right| \leqq \frac{1}{m}.$$

Hence for $n \geqq n_3$ we get, from (12),

$$(13) \qquad \left|\lim_{n\to\infty} f(P_n^m) - f(T_n^m)\right| \leqq \frac{2}{m} \quad \text{for} \quad n \geqq n_3.$$

Hence

$$(14) \qquad \left|\lim_{n\to\infty} f(P_n^m) - f(T_{n_3(m)}^m)\right| \leqq \frac{2}{m} \quad \text{for all } m.$$

Now we can arrange things so, that

$$n_3(1) < n_3(2) < \cdots \to \infty;$$

so we get from (14) and hyp. 4:

$$\lim_{m\to\infty}\lim_{n\to\infty} f(P_n^m) = \lim_{m\to\infty} f(T_{n_3(m)}^m).$$

It follows

$$(15) \qquad \lim_{m\to\infty} f(T_{n_3(m)}^m) = A.$$

W I.15c. We shall prove, that the sequence of complexes $S_m =_{df} T_{n_3(m)}^m$ is a (DARS)-sequence for E. We have by (11)

$$(15.1) \quad S_m = p_{n_3(m),1}^{m1} \cup p_{n_3(m),1}^{m2} \cup \cdots \cup p_{n_3(m),1}^{mk} \quad \text{for} \quad m = 1, 2, \ldots$$

Consider the sequence

$$(16) \quad [p_{n_3(1),1}^{1s}],\ [p_{n_3(2),1}^{2s}],\ \ldots,\ [p_{n_3(m),1}^{ms}],\ \ldots \qquad (s = 1, \ldots, k).$$

By (2), all these bricks contain A_s, and we have by (7):

$$\mu(p_{n_3(m),1}^{ms} - A_s) \leqq \frac{1}{m} \quad \text{for} \quad m = 1, 2, \ldots$$

Consequently, by Lemma [W I.7.], (16) is a (DARS)-sequence for (α_s). Since the terms in (15.1) are disjoint, we get that S_m is a (DARS)-sequence for E. The lemma is established.

W I.15.1. Lemma. Under hypothesis $\boldsymbol{H}_1$ and 1., 3., 4. of the preceding Lemma [W I.15.], but with changed hypothesis 2 into 2′: E is composed of an infinite (denumerable) number of heavy traces, the thesis of Lemma [W I.15.] holds true.

Proof. Consider the sequence of (DARS)-sequences for E:

$$P_n^1,\ P_n^2,\ \ldots,\ P_n^m,\ \ldots$$

for which the limit of the sequence

$$\lim_{n\to\infty} f(P^1),\ \lim_{n\to\infty} f(P^2),\ \ldots,\ \lim_{n\to\infty} f(P^m),\ \ldots$$

exists and equals A.

Let $\varepsilon > 0$ and find $\delta = \delta(2) > 0$ such that, if for a complex T we have $\mu T < \delta$, then $|f|(T) \leqq \varepsilon$.

Since the series $\sum_{s=1}^{\infty} \mu(\alpha_s)$ converges, we can find N, such that

$$\mu(A_{N+1} + A_{N+2} + \cdots) \leqq \frac{\delta}{2}.$$

We shall make ε and δ tending to 0, and N tending to ∞. Put

$$E' =_{df} (\alpha_1) + \cdots + (\alpha_N),$$

$$E'' =_{df} (\alpha_{N+1}) + (\alpha_{N+2}) + \cdots$$

These sets are measurable. We have $E' \cdot E'' = O$, $E' + E'' = E$. Hence we can find subcomplexes $\bar{P}_n^m$, $\bar{\bar{P}}_n^m$ of P_n^m, such that

$$\bar{P}_n^m \cap \bar{\bar{P}}_n^m = \mathit{O}, \qquad \bar{P}_n^m \cup \bar{\bar{P}}_n^m = P_n^m, \qquad (n = 1, 2, \ldots;\ m = 1, 2, \ldots),$$

and where $\bar{P}_n^m$, $\bar{\bar{P}}_n^m$ are (DARS)-sequences for E', E'' respectively, (this for every m respectively). By hyp. 3 and on account of Theorem [W I.9.2.], the sequences $f(\bar{P}_n^m)$, $f(\bar{\bar{P}}_n^m)$ are bounded for every m.

We can extract from n a subsequence $\alpha_{m(n)}$ (where α_m depends on m), such that $\lim_{n\to\infty} f(\bar{P}_{\alpha_m(n)})$ exists. Since $\lim_{n\to\infty} f(P_{\alpha_m(n)})$ exists, by hyp. 4, it follows that $\lim_{n\to\infty} f(\bar{\bar{P}}_{\alpha_m(n)})$ exists too.

Put

$$\bar{Q}_n^m =_{df} \bar{P}_{\alpha_m(n)}^m, \quad \bar{\bar{Q}}_{\alpha_m(n)}^m =_{df} \bar{\bar{P}}_{\alpha_m(n)}^m$$

for all n and m. We get two sequences

$$(1) \qquad \lim_{n\to\infty} f(\bar{Q}_n^1), \quad \lim_{n\to\infty} f(\bar{Q}_n^2), \ldots,$$

$$(2) \qquad \lim_{n\to\infty} f(\bar{\bar{Q}}_n^1), \quad \lim_{n\to\infty} f(\bar{\bar{Q}}_n^2), \ldots$$

Now $(\bar{Q}_n^m)$ is a (DARS)-sequence for E'. If we consider the set of all numbers

$$\lim_{n\to\infty} \sup f(R_n),$$

where R_n is a (DARS)-sequence for E, we see that this set is bounded. Hence the set of all

$$\lim_{n\to\infty} \sup f(\bar{Q}_n^m) = \lim_{n\to\infty} f(\bar{Q}_n^m)$$

is bounded. Thus the sequence (1) is bounded. It follows, that (2) is also bounded. We can find a subsequence $\beta(m)$ of m, such that the sequence

$$\lim_{n\to\infty} f(\bar{Q}_n^{\beta(1)}), \ \lim_{n\to\infty} f(\bar{Q}_n^{\beta(2)}), \ldots, \ \lim_{n\to\infty} f(\bar{Q}_n^{\beta(m)}), \ldots$$

converges. The sequence

$$\lim_{n\to\infty} f(\bar{\bar{Q}}_n^{\beta(1)}), \ \lim_{n\to\infty} f(\bar{\bar{Q}}_n^{\beta(2)}), \ldots, \ \lim_{n\to\infty} f(\bar{\bar{Q}}_n^{\beta(m)}), \ldots$$

converges too. Put

$$\bar{R}_n^m =_{df} \bar{Q}_n^{\beta(m)}, \quad \bar{\bar{R}}_n^m =_{df} \bar{\bar{Q}}_n^{\beta(m)} \quad \text{for all} \quad n, m.$$

We get two convergent sequences

$$\lim_{n\to\infty} f(\bar{R}_n^1), \ \lim_{n\to\infty} f(\bar{R}_n^2), \ldots, \ \lim_{n\to\infty} f(\bar{R}_n^m), \ldots,$$

$$\lim_{n\to\infty} f(\bar{\bar{R}}_n^1), \ \lim_{n\to\infty} f(\bar{\bar{R}}_n^2), \ldots, \ \lim_{n\to\infty} f(\bar{\bar{R}}_n^m), \ldots,$$

such that their limits exist. Denote their limits by A'_N, A''_N respectively. We have $A'_N + A''_N = A$. $\bar{R}_n^m$; $\bar{\bar{R}}_n^m$ are (DARS)-sequences for E', E'' respectively.

Concerning $\bar{R}_n^m$ and E', we are in the same circumstances, as at the beginning of the proof of Lemma [W I.15.]. We should, in this proof, replace the letter P by $\bar{R}$ and k by N. Indeed, we have now the limits

(3) $$\lim_{n\to\infty} f(\bar{R}_n^1),\ \lim_{n\to\infty} f(\bar{R}_n^2),\ \ldots,\ \lim_{n\to\infty} f(\bar{R}_n^m),\ \ldots$$

and the sequence (3) is convergent. It tends to A'_N. Taking the mentioned proof, but with the present notation, we have the equality

$$E' = (\alpha_1) + (\alpha_2) + \cdots + (\alpha_N),$$

and also N subcomplexes of $\bar{R}_n^m$:

(4) $$r_n^{m1},\ r_n^{m2},\ \ldots,\ r_n^{mN}.$$

We get

$$A_1 \leqq \bar{r}_{n1}^{m1},\ A_2 \leqq \bar{r}_{n1}^{m2},\ \ldots,\ A_N \leqq \bar{r}_{n1}^{mN},$$

and we can find $n_1 = n_1(m)$, such that for all $n \geqq n_1$:

$$\mu(\bar{r}_{n1}^{m1} - A_1) \leqq \frac{1}{m},\ \ldots,\ \mu(\bar{r}_{n1}^{mN} - A_N) \leqq \frac{1}{m}.$$

Applying [W I.2.2.], we can find $\delta_m > 0$, such that for every complex R, with $\mu R \leqq \delta_m$, we have $|f(R)| \leqq \frac{1}{m}$. We can find $n_2 = n_2(m) \geqq n_1$ such that for all $n \geqq n_2$ we have:

$$\mu(\bar{r}_{n2}^{m1} + \bar{r}_{n3}^{m1} + \cdots) \leqq \frac{\delta_m}{N},$$

$$\cdots\cdots\cdots\cdots\cdots\cdots$$

$$\mu(\bar{r}_{n2}^{mN} + \bar{r}_{n3}^{mN} + \cdots) \leqq \frac{\delta_m}{N}.$$

It follows, when we slightly change the notation with reference to $\bar{R}_n^m$:

$$|f|\,\{\bar{r}_{n,\,N+1}^{m} + \bar{r}_{n,\,N+2}^{m} + \cdots\} \leqq \frac{1}{m},$$

i.e.

(5) $$\left|\,|f|(\bar{R}_n^m) - \sum_{s=1}^{N} |f|(\bar{r}_{ns}^m)\right| \leqq \frac{1}{m}.$$

If we define

$$T_n^m =_{df} \bar{r}_{n1}^{m1} \cup \bar{r}_{n1}^{m2} \cup \cdots \cup \bar{r}_{n1}^{mN},$$

we get

$$|f(\bar{R}_n^m) - f(T_n^m)| \leqq \frac{1}{m}$$

for

$$n \geqq n_2(m) \quad \text{and all} \quad m = 1, 2, \ldots$$

Put

$$S_m =_{df} \bar{r}_{n_3(m),1}^{m1} \cup \bar{r}_{n_3(m),1}^{m2} \cup \cdots \cup \bar{r}_{n_3(m),1}^{mN}.$$

We have

$$S_m = T_{n_3(m)}^N,$$

and we prove, that S_m is a (DARS)-sequence for E'. We have

$$\lim_{n\to\infty} f(T_{n_3(m)}^m) = A'_N.$$

The bricks s_{nk} and the number A'_N depend on the choice of N, which in turn, depends on δ, and where δ depends on the chosen $\varepsilon > 0$. Consider the sequence

$$\lim_{n\to\infty} f(\bar{\bar{R}}_n^1),\ \lim_{n\to\infty} f(\bar{\bar{R}}_n^2),\ \ldots,\ \lim_{n\to\infty} f(\bar{\bar{R}}_n^m),\ \ldots,$$

which tends to A''_N

(6) $\bar{\bar{R}}_n^m$ is a (DARS)-sequence for E'', whose measure is $\leqq \frac{\delta}{2}$.

Since $|\bar{\bar{R}}_n^m, E''|_\mu \to 0$, it follows that, starting from a certain index $n_0 = n_0(m)$, we have

$$|\bar{\bar{R}}_n^m, E''| \leqq \frac{\delta}{2}.$$

Since by (4) we have

$$|E'', O| \leqq \frac{\delta}{2},$$

we get

$$\mu\, \bar{\bar{R}}_n^m \leqq \delta.$$

It follows that

$$|f|(\bar{\bar{R}}_n^m) \leqq \varepsilon \quad \text{for all} \quad n \geqq n_0.$$

We have

$$-|f|(\bar{\bar{R}}_n^m) \leqq f(\bar{\bar{R}}_n^m) \leqq |f|(\bar{\bar{R}}_n^m)$$

for all n and m. Hence $|f(\bar{\bar{R}}_n^m)| \leqq \varepsilon$, starting from n_0. Hence $|\lim_{n\to\infty} \bar{\bar{R}}^m| \leqq \varepsilon$ for all m. Consequently $|A''_N| \leqq \varepsilon$. Since $A = A'_N + A''_N$, we have

(7) $$|A - A'_N| \leqq \varepsilon.$$

Let

$$\varepsilon_1 > \varepsilon_2 > \cdots > \varepsilon_p > \cdots \to 0.$$

Find the corresponding δ_p and N_p. We can arrange things, so as to have

$$N_1 < N_2 < \cdots < N_p < \cdots \to \infty.$$

For every N_p we can find, as was shown above, a complex, composed of N_p bricks, say

$$S_p =_{df} \{S_p^1, S_p^2, \ldots, S_p^{N_p}\},$$

which contains the atoms

(8) $$A'_1, A'_2, \ldots, A'_{N_p},$$

where

(9) $$\mu(S_p^k - A'_k) \leqq \frac{1}{p}, \qquad k = 1, 2, \ldots, N_p,$$

and such that

$$|f(S_p - A'_{N_p}| \leqq \varepsilon_p.$$

It follows by (7), i.e. by $|A - A'_{N_p}| \leqq \varepsilon_p$, that

$$|f(S_p) - A| \leqq 2\varepsilon_p.$$

It follows: $\lim_{p\to\infty} f(S_p) = A$.

It remain to prove, that S_p is a (DARS)-sequence for E. It suffices to prove that

$$S_k^k, S_{k+1}^k, \ldots, S_m^k, \ldots$$

is a (DARS)-sequence for A_k. This follows from (8) and (9). Hence, since the finite sum of atoms constitute a μ-dense set in E, therefore S_p is a (DARS)-sequence for E. The theorem is proved.

W I.15.2. Theorem. Under Hyp. $\boldsymbol{H}_1$ and Hyp. $\mathcal{B}$, if

1. $f(a)$ is a field of real numbers,
2. E is a measurable set of traces, (which may be any, heavy or not),
3. $f(a)$ is completely admissible on E,
4. $P_n^1, P_n^2, \ldots, P_n^m, \ldots$ are (DARS)-sequences for E, such that the limits

(1) $$\lim_{n\to\infty} f(P^1), \lim_{n\to\infty} f(P^2), \ldots, \lim_{n\to\infty} f(P^m), \ldots$$

exist, and that the sequens (1) converges to the limit A, then there exists a (DARS)-sequence S_n for E, such that $\lim_{n\to\infty} f(S_n) = A$.

Proof. Put $E = E' + E''$, where $E' \cdot E'' = O$ and where E' is the part, composed of heavy traces only, and E'' that, containing only elusive traces.

The cases, where $E' = O$, or $\mu E'' = 0$, were settled in forgoing theorems. Suppose $E' = O$, $\mu E'' > 0$. Let $\bar{P}_n^m$, $\bar{\bar{P}}_n^m$ be partial complexes of P_n^m with $\bar{P}_n^m \frown \bar{\bar{P}}_n^m = \Theta$ and where $\bar{P}_n^m$, $\bar{\bar{P}}_n^m$ are (DARS)-sequences for E', E'' respectively. The limit

$$\lim_{n\to\infty} [f(\bar{P}_n^m) + f(\bar{\bar{P}}_n^m)]$$

exist, and the sequence $|f(\bar{P}_n^m)|$ is bounded by [W I.9.2.]. Hence we can find a subsequence $\alpha(\dot{n})$ of $\dot{n}$, such that $\lim_{n\to\infty} \bar{P}_{\alpha(n)}^m$ exists. Consequently, since $\lim_{n\to\infty} f(P_n^m)$ exists, the limit $\lim_{n\to\infty} \bar{\bar{P}}_{\alpha(n)}^m$ exists too. Put

$$Q_n^m =_{df} P_{\alpha(n)}^m, \quad \bar{Q}_n^m =_{df} \bar{P}_{\alpha(n)}^m, \quad \bar{\bar{Q}}_n^m =_{df} \bar{\bar{P}}_{\alpha(n)}^m,$$

and consider two sequences:

(2) $$\lim_{n\to\infty} f(\bar{Q}_n^1), \lim_{n\to\infty} f(\bar{Q}_n^2), \ldots, \lim_{n\to\infty} f(\bar{Q}_n^m), \ldots,$$

(3) $$\lim_{n\to\infty} f(\bar{\bar{Q}}_n^1), \lim_{n\to\infty} f(\bar{\bar{Q}}_n^2), \ldots, \lim_{n\to\infty} f(\bar{\bar{Q}}_n^m), \ldots$$

The sequence (2) is bounded. Hence there exists a subsequence $\beta(\dot{n})$ of $\dot{n}$ such that the corresponding subsequences of (2) and (3) have limits, say $\bar{\alpha}$, $\bar{\bar{\alpha}}$. Let us put

$$\bar{R}_n^m =_{df} \bar{Q}_n^{\beta(m)}, \quad \bar{\bar{R}}_n^m =_{df} \bar{\bar{Q}}_n^{\beta(m)}.$$

We have

$$\lim_{m\to\infty}\lim_{n\to\infty}\bar{R}^m_n=\bar{L},\qquad \lim_{m\to\infty}\lim_{n\to\infty}\bar{\bar{R}}^m_n=\bar{\bar{L}}.$$

Applying the Lemma [W I.15.1.], to E' and $\bar{R}^m_n$ and to E'' and $\bar{\bar{R}}^m_n$ we obtain a (DARS)-sequence

$$\bar{S}_n \text{ for } E', \text{ such that } \lim_{n\to\infty} f(\bar{S}_n)=\bar{L},$$

and a (DARS)-sequence

$$\bar{\bar{S}}_n \text{ for } E'', \text{ such that } \lim_{n\to\infty} f(\bar{\bar{S}}_n)=\bar{\bar{L}}.$$

Now, since ***Hyp*** $\mathscr{B}$ is admitted, we can use Theorem [W I.12.]. Indeed E'' does not contain any heavy trace. Thus there exist subsequences $p(\dot{n})$, $q(\dot{n})$ of $\dot{n}$ and subcomplexes $\bar{T}_n$, $\bar{\bar{T}}_n$ of $\bar{S}_{p(n)}$, $\bar{\bar{S}}_{p(n)}$ respectively, $(n=1,2,\ldots)$, such that $\bar{T}_n\cap\bar{\bar{T}}_n=O$, and $\bar{T}_n$, $\bar{\bar{T}}_n$ are (DARS)-sequences for E', E'' respectively. $\bar{T}_n\cup\bar{\bar{T}}_n$ is a (DARS)-sequence for E, and such that

$$\lim_{n\to\infty} f(\bar{T}_n)=\bar{L},\qquad \lim_{n\to\infty} f(\bar{\bar{T}}_n)=\bar{\bar{L}}.$$

It follows that $\lim f(\bar{T}_n\cup\bar{\bar{T}}_n)=L$. The theorem is proved.

W I.16. Theorem. Under Hyp. $\boldsymbol{H}_1$ and $\mathscr{B}$, if

1. $f(a)$ is a field of real numbers,
2. E is a measurable set of traces, (which may be any, heavy or not),
3. $f(a)$ is completely admissible on E,

then

1) the set of all numbers $\chi(P_n)=_{df}\limsup_{n\to\infty}|f(P_n)|$, where P_n varies over all (DARS)-sequences for E, is bounded,

2) the set of all numbers

$$\varphi(P_n)=_{df}\lim_{n\to\infty}\sup f(P_n)$$

and the set of all numbers

$$\psi(P_n)=_{df}\lim_{n\to\infty}\inf f(P_n),$$

where P_n varies over all (DARS)-sequences for E, are both bounded; hence

$$\sup_{P_n}\limsup f(P_n)=\bar{S}_E f(a),$$

and

$$\inf_{P_n}\liminf f(P_n)=\underline{S}_E f(a)$$

both exists,

3) there exist (DARS)-sequences S_n and T_n for E, such that

$$\lim_{n\to\infty} f(S_n)=\bar{S}_E f(a),\qquad \lim_{n\to\infty} f(T_n)=\underline{S}_E f(a).$$

Proof. If P_n is a (DARS)-sequence for E, then by Theorem [W I.9.2.], $|f|(P_n)$ is bounded, so $\lim\sup|f|(P_n)$ exists. If we consider all (DARS)-sequences for E, then by [W I.14.3.], the set of all $\chi(P_n)$ is bounded. Hence the set of all $\varphi(P_n)$ and that of all $\psi(P_n)$ are bounded.

Consequently

$$\sup_{P_n}\chi(P_n),\ \sup_{P_n}\varphi(P_n),\ \inf_{P_n}\psi(P_n)$$

exist and are finite. We can find (DARS)-sequences for E:

$$\bar{R}_n^1,\ \bar{R}_n^2,\ \ldots,\ \bar{R}_n^m,\ \ldots$$

and

$$\bar{\bar{R}}_n^1,\ \bar{\bar{R}}_n^2,\ \ldots,\ \bar{\bar{R}}_n^m,\ \ldots,$$

such that

$$\lim_{m\to\infty}\varphi(\bar{R}_n^m)=\sup_{P_n}\varphi(P_n)=\bar{S}_E f,\qquad \lim_{m\to\infty}\psi(\bar{\bar{R}}_n^m)=\inf_{P_n}\psi(P_n)=\underline{S}_E f.$$

Now

$$\varphi(\bar{R}_n^m)=\lim_{n\to\infty}\sup f(\bar{R}_n^m);$$

hence there exists a subsequence $\alpha(n)$ of n, such that

$$\lim_{n\to\infty}\sup f(\bar{R}_n^m)=\lim_{n\to\infty}f(\bar{R}_{\alpha(n)}^m),$$

and similarly there exists a subsequence $\beta(n)$ of n, such that

$$\lim_{n\to\infty}\inf f(\bar{\bar{R}}_n^m)=\lim_{n\to\infty}f(\bar{\bar{R}}_{\beta(n)}^m).$$

We see, that we are in the conditions of Theorem [W I.15.2.]. Consequently there exist (DARS)-sequences S_n, T_n for E, such that

$$\bar{S}_E f(a)=\lim_{n\to\infty}f(S_n),$$

$$\underline{S}_E f(a)=\lim_{n\to\infty}f(T_n).$$

The theorem is established.

W I.17. Theorem. Under Hyp. $\boldsymbol{H}_1$ and $\mathscr{B}$, if

1. $f(a)$ is a field of real numbers,
2. E is a measurable set of traces, (which may be heavy or not),
3. $f(a)$ is completely admissible,

then

1) $\bar{S}_E f(a)$ is the maximum of all numbers $\lim\limits_{n\to\infty}f(P_n)$, which can be obtained by using (DARS)-sequences for E,

2) $\underline{S}_E f(a)$ is the minimum of all numbers $\lim\limits_{n\to\infty}f(P_n)$ with the same properties.

Proof. By Theorem [W I.16.] there exist (DARS)-sequences S_n, T_n for E, such that

$$\bar{S}_E f(a)=\lim_{n\to\infty}f(S_n),\qquad \underline{S}_E f(a)=\lim_{n\to\infty}f(T_n).$$

Take any (DARS)-sequence P_n for E, such that $\lim_{n\to\infty} f(P_n)$ exists. Denote it by M. The number M belongs to the set of all numbers $\limsup_{n\to\infty} f(Q_n)$, where Q_n varies over all (DARS)-sequences for E.

Indeed

$$M = \lim_{n\to\infty} f(P_n) = \limsup_{n\to\infty} f(P_n).$$

Hence M is not greater than $\sup_{Q_n} \limsup_{n\to\infty} f(Q_n)$. Thus $\bar{S}_E f(a)$ is the required maximum. Similar proof for the second part of the thesis.

W I.18. Theorem. Under Hyp. $\boldsymbol{H}_1$ and Hyp. $\mathscr{B}$, if

1. $f(a)$ is a field of real numbers,
2. E, F are measurable sets of traces, (the traces may be heavy or elusive),
3. $f(a)$ is completely admissible on E,
4. $F \leqq E$,
5. P_n, Q_n are (DARS)-sequences for E, F respectively with $Q_n \subseteqq P_n$ for $n = 1, 2, \ldots$,
6. $\lim_{n\to\infty} f(P_n) = \bar{S}_E f(a)$,
7. either F or $E - F$ does not contain any heavy trace,

then

$$\lim_{n\to\infty} f(Q_n) = \bar{S}_F f(a).$$

Proof. The theorem is a generalization of a similar theorem [W.5.], where G is supposed to not contain any heavy trace. This theorem is essentially based on Theorem [W.3.].

The theorem, extension of Theorem [W.3.], can be proved under the hypothesis, that at least one of the sets E, F does not contain any heavy trace (see Theorem [W I.12.]).

Now, hyp. 7 allows to apply the Theorem [W I.12.], and hence the proof of our Theorem [W I.18.] can be just considered as a repeated proof of [W.5.1.].

W I.19. Theorem. Under Hyp. $\boldsymbol{H}_1$ and $\mathscr{B}$, if

1. $f(a)$ is a field of real numbers,
2. E, F are measurable sets of traces,
3. $E \cdot F = O$,
4. at least one of the sets E, F does not contain any heavy trace,
5. $f(a)$ is completely admissible on $E + F$,

then

$$\text{(1)} \qquad \begin{aligned} \bar{S}_{E+F} f(a) &= \bar{S}_E f(a) + \bar{S}_F f(a), \\ \underline{S}_{E+F} f(a) &= \underline{S}_E f(a) + \underline{S}_F f(a). \end{aligned}$$

Proof follows the same pattern as the proof of Theorem [W.6.], and relies on Theorem [W I.18.].

W I.20. Remark. It is not true, that (1) takes place in the case, where both E and F contain a heavy trace. Indeed, let A, B be two points on the plane with $\mu A > 0$, $\mu B > 0$. There may be available two sequences of bricks

$$a_1 > a_2 > \cdots > A, \quad b_1 > b_2 > \cdots > B$$

with $\prod_n a_n = A$, $\prod_n b_n = B$, and $a_n b_n \neq O$ for all n.

Now, if we have $f(a)$, such that $f(a_n) = f(b_n) = 1$, and for all bricks $f(p) = 0$, it will be impossible to construct a (DARS)-sequence for I and such that there would exist a sequence of complexes, approximating $A + B$ and with measure tending to 2. In the general case we have for $f(a) \geqq 0$ and $E \cdot F = O$:

$$\overline{S}_{A+B} f(a) \leqq \overline{S}_A f(a) + \overline{S}_B f(a),$$

but equality may not take place.

W I.21. Def. To get a more desirable result for upper and lower (DARS)-sum, we shall subject the base $\boldsymbol{B}$ to a farther condition, which we may call *condition of privacy of atoms, or condition* $\mathscr{P}_A$:

If A, B are atoms of $\boldsymbol{G}$, and p_n, q_n are sequences of bricks, such that

$$A \leqq p_n, \quad B \leqq q_n, \quad \lim_{n\to\infty} (p_n - A) = \lim_{n\to\infty} (q_n - B) = O,$$

then there exists n_0, such that for all $n \geqq n_0$ we have $p_n \cdot q_n = O$.

W I.21a. Remark. We shall prove that, if we admit the additional condition $\mathscr{P}_A$, then both sums, the upper and the lower, will be denumerably additive. To prove that, we shall first prove the theorem:

W I.22. Theorem. If

1. $A_1, A_2, \ldots, A_k$ are different atoms, and $\alpha_1, \alpha_2, \ldots, \alpha_k$ the corresponding traces,

2. $E = (\alpha_1) + \cdots + (\alpha_k)$,

3. the hypotheses $\boldsymbol{H}_1$, $\mathscr{B}$, $\mathscr{P}_A$, and complete admissibility are admitted,

then

$$\overline{S}_E f(a) = \sum_{i=1}^{k} \overline{S}_{\alpha(i)} f(a).$$

Proof. By Theorem [W I.17.] we can find a (DARS)-sequence P_n for E, such that

$$\overline{S}_E f(a) = \lim_{n\to\infty} f(P_n).$$

Since P_n is a sequence for E, therefore there exist partial complexes

$$P'_n, P^2_n, \ldots, P^k_n,$$

disjoint with one another, with union $= E$, and where P^s_n is a (DARS)-sequence for (α_s), $(s = 1, 2, \ldots, k)$.

Let us choose a subsequence $\beta(n)$ of n, such that all

$$\lim_{n\to\infty} f(P^s_{\beta(n)})$$

exists for all s. We get

$$\lim f(P_{\beta(n)}) = \bar{S}_E f(a)$$

and

$$\lim_{n\to\infty} f(P^s_{\beta(n)}) \leqq \bar{S}_{(\alpha_s)} f(a).$$

It follows that

(1) $$\bar{S}_E f \leqq \sum_{s=1}^{k} \bar{S}_{(\alpha_s)} f.$$

Let us fix s for a moment, and find a (DARS)-sequence Q^s_n for (α_s), such that

$$\bar{S}_{(\alpha_s)} f = \lim_{n\to\infty} f(Q^s_n),$$

Theorem [W I.17.].

Now by [W I.7.1.], starting from a sufficiently great index n, we have

$$Q^s_n = q^s_{n1} \cup q^s_{n2} \cup \cdots,$$

where

$$A_s \leqq q^s_{n1} \quad \text{for} \quad s = 1, 2, \ldots, k.$$

We also have $\lim \mu(q^s_{n1} - A_s) = 0$; and the sequence of complexes $[q^s_{n1}]$, each composed of a single brick q^s_{n1}, constitutes a (DARS)-sequence for (α_s). In addition to that:

$$\lim_{n\to\infty} \mu(q^s_{n2} + q^s_{n3} + \cdots) \to 0.$$

By virtue of admissibility, we have

$$|f(q^s_{n2})| + |f(q^s_{n3})| + \cdots \to 0.$$

Hence

(2) $$\bar{S}_{(\alpha_s)} f = \lim_{n\to\infty} f(q^s_{n1}).$$

Now we can take account of the condition $\mathscr{P}_A$, by finding n_0, such that for every $n \geqq n_0$, any two bricks

$$q^1_{n1}, q^2_{n1}, \ldots, q^k_{n1}$$

are disjoint. Now, since $[q^s_{n1}]$ is a (DARS)-sequence for (α_s), and since for every n respectively

$$[q^1_{n1}], [q^2_{n1}], \ldots, [q^k_{n1}], \qquad n \geqq n_0$$

are disjoint, we get with

$$R_n =_{df} q^1_{n1} \cup q^2_{n1} \cup \cdots \cup q^k_{n1}$$

a (DARS)-sequence for E. We get

$$\sum_{s=1}^{k} \bar{S}_{(\alpha_s)} f = \lim_{n\to\infty} \sum_{s=1}^{k} f(q^s_{n1}) = \lim_{n\to\infty} f(R_n) \leqq \bar{S}_E f,$$

and by virtue of (1):

$$\bar{S}_E f = \sum_{s=1}^{k} \bar{S}_{(\alpha_s)} f.$$

A similar proof gives the equality for lower sums.

W I.23. Theorem. Under hypothesis $\boldsymbol{H}_1$ and the admissibility we have: if

1. $f(a)$ is a field of real numbers,
2. $E_1, E_2, \ldots, E_m, \ldots$ are measurable sets of traces,
3. $\mu S_m \to 0$ for $m \to \infty$,

then

$$\bar{S}_{E_m} f \to 0 \quad \text{and} \quad \underline{S}_{E_m} f \to 0.$$

Proof. The proof is just the same as for Theorem [W.8.], without any change.

W I.24. Theorem. Under Hyp. $\boldsymbol{H}_1$ and the admissibility we have: if

1. $f(a)$ is a field of real numbers,
2. $\varepsilon > 0$,

then there exists $\delta > 0$, such that if $\mu F \leqq \delta$, then

$$|\bar{S}_F f| \leqq \varepsilon, \quad |\underline{S}_F f| \leqq \varepsilon.$$

Proof. The same as for Theorem [W.8.1.].

W I.25. Theorem. Let us admit the hypotheses $\boldsymbol{H}_1$, $\mathscr{B}$, complete admissibility and $\mathscr{P}_A$. If

1. $f(a)$ is a field of real numbers,
2. E is a set composed of heavy traces only, infinite in number,
3. $E_1, E_2, \ldots, E_m, \ldots$ are disjoint subsets of E, with

$$E_1 + E_2 + \cdots + E_m + \cdots = E,$$

then

$$\bar{S}_E f = \sum_m \bar{S}_{E_m} f \quad \text{and} \quad \underline{S}_E f = \sum_m \underline{S}_{E_m} f.$$

Proof. The proof follows the pattern of Theorem [W.9.].

If we take account of the preceding theorem, we can state the following:

W I.26. Theorem. Let us admit the hypotheses $\boldsymbol{H}_1$, $\mathscr{B}$, $\mathscr{P}_A$ and the complete admissibility. Then if $f(\dot{a})$ is a field of real numbers, and we define for all measurable sets of traces F the set-functions

$$U(F) = \bar{S}_F f(a), \quad L(F) = \underline{S}_F f(a),$$

then they are both denumerably additive. The tribe $\boldsymbol{G}$ may have atoms or not.

Section 3 of W I.

Some theorems on (DARS)-summability of fields of real numbers

W I.27. Theorem. Under Hyp. $\boldsymbol{H}_1$ and $\mathscr{B}$, if

1. $f(a)$ is a field of real numbers,
2. E is a measurable set of traces, $\mu E > 0$,
3. $f(a)$ is (DARS)-summable on E,

then $f(a)$ is completely admissible on E.

Proof. By Theorem [U.2.5.] $f(a)$ is admissible. We must prove that it satisfies the supplementary condition of admissibility, Def. [W I.9.].

Let E' be the subset of E composed of all its heavy traces, and put $E'' = E \sim E'$. Then E'' does not contain any heavy trace.

Let $E' = (\alpha_1) + (\alpha_2) + \cdots$ where α_s are different heavy traces. Since f is (DARS)-summable on E, it follows that f is (DARS)-summable on every (α_s). Hence the set of all numbers, (see [W I.7.1.]), $\limsup_{n\to\infty} p_n^s$, where p_n^s are bricks, such that $A_s \leqq p_n^s$, $\mu(p_n' - A_s) \to 0$, is bounded. Hence the condition [W I.9.] is satisfied and then the complete admissibility of $f(a)$ follows.

W I.28. Theorem. Under Hyp. $\boldsymbol{H}_1$ and $\mathscr{B}$, if

1. $f(a)$ is a field of real numbers,
2. E is a measurable set of traces, with $\mu E > 0$,
3. $f(a)$ is completely admissible on E,
4. for every measurable subset F of E we have $S_F f = 0$,

then $\bar{S}_E |f| = 0$, hence also $\bar{S}_F |f| = 0$ for all measurable sets F where $F \leqq E$.

Proof. Let E'' be the set of all heavy traces in E. It follows, that $E' =_{df} E - E''$ contains elusive traces only. If we suppose that $E' = O$, then the theorem is the consequence of Theorem [W.30.16.].

Suppose that $E = E'$. Then we have

$$E = (\alpha_1) + (\alpha_2) + \cdots,$$

where the number of terms is finite or denumerably infinite. Since $(\alpha_i) \leqq E$, $(i = 1, 2, \ldots)$, we have

$$S_{(\alpha_i)} f = 0;$$

hence if we take any sequence $p_{i1}, p_{i2}, \ldots$ of bricks with $\alpha_i \leqq p_{in}$, $n = 1, 2, \ldots$ and $\mu(p_{in} - \alpha_i) \to 0$, we have

$$f(p_{in}) \to 0.$$

It follows that $|f(p_{in})| \to 0$ for $n \to \infty$, and then, that

$$\bar{S}_{(\alpha_i)} |f| = 0.$$

It follows by Theorem [W I.23.] that

$$\bar{S}_{E'}|f| = 0, \quad \sum_i \bar{S}_{(\alpha i)}|f| = 0.$$

Now let $\mu E' > 0$ and $\mu E'' > 0$. By what has already been proved, we have

$$\bar{S}_{E'}|f| = 0.$$

Concerning E'', we have by Theorem [W.30.16.] $\bar{S}_{E''}|f| = 0$. Consequently

$$\bar{S}_E|f| = 0.$$

W I.29. Theorem. Under Hyp. $\boldsymbol{H}_1$ and $\mathscr{B}$, if

1. $f(a)$ is a field of real numbers,
2. E is a measurable set of traces, with $\mu E > 0$,
3. $f(a)$ is (DARS)-summable on E,

then

$$|f(a)| \text{ is (DARS)-summable on } E.$$

This is a generalization of Theorem [W.30.17].

Proof. Since f is (DARS)-summable on E, it follows that f is summable on E'' and on E', where $E'' = E - E'$, and where E' is the set of all heavy traces in E.

By Theorem [W.30.17.], $|f(a)|$ is summable on E''. The summability of $|f(a)|$ on E' follows from the fact that the sum

$$\sum_\alpha |S_E f(a)| = \sum_\alpha S_E |f(a)|$$

is finite or converging. Since

$$\bar{S}_{E'}|f|, \underline{S}_{E'}|f| \quad \text{and} \quad \bar{S}_{E''}|f|, \underline{S}_{E''}|f|$$

exist by virtue of the complete admissibility, and since

$$\bar{S}_E|f| = \bar{S}_{E'}|f| + \bar{S}_{E''}|f|,$$
$$\underline{S}_E|f| = \underline{S}_{E'}|f| + \underline{S}_{E''}|f|,$$

it follows that

$$\bar{S}_E|f| = \underline{S}_E|f|,$$

so $|f|$ is summable on E.

W I.30. Theorem. The following theorems, taken from [W.30.18.], [W.30.19.], [W.30.19.1.], [W.30.20.], [W.30.21.], [W.30.22.], [W.30.22.1.], [W.30.23.], [W.30.24.], [W.30.25.], [W.30.26.], are true, when we change the condition $\boldsymbol{H}_2$ into $\boldsymbol{H}_1$ and $\mathscr{B}$, and instead of admissibility, we require slightly more: complete admissibility on considered measurable sets of traces.

Proof. The proof consists in splitting the given set E into two disjoint sets E', E'', where E' is composed of heavy traces only, and E'' only of elusive ones. The first part E' is composed of points of a finite or infinite sequence of real numbers, and this part can be mastered by applying known theorems on infinite series.

W I.31. Remark. We leave to the reader the completion of proofs and statements of the corresponding theorems. We do the same in what concerns functions, attached to upper and lower sums.

W I.32. Remark. Analogous theorems for fields of complex numbers can be treated by decomposing

$$f(a) = g(a) + i\,h(a),$$

where g, h are fields of real numbers.

Section 4 of W I.

Theorems on squares of fields of real numbers

W I.40.1. We shall study a notion, which has been used several times, but not explicitely specified and not provided with a suitable term. Now we shall do that:

Def. The scalar field $f(a)$ will be termed *summation-bounded on* E, whenever for every (DARS)-sequence P_n for E the sequence

$$|f|(P_n) = \sum_k |f(p_{nk})|$$

is bounded. E is supposed to be a measurable set of traces.

W I.40.1.1. Theorem. Under Hyp. $\boldsymbol{H}_1$, if

1. E, F are measurable sets of traces, $F \leqq E$,
2. $f(\dot{a})$ is a field of real numbers,
3. f is summation-bounded on E,

then f is summation-bounded on F.

Proof. Take any (DARS)-sequence Q_n for F. By virtue of Theorem [W.30.5.1.] there exists a subsequence $k(n)$ of n and a (DARS)-sequence P_n for $E - F$, such that $Q_{k(n)} \cap P_n = O$, for $n = 1, 2, \ldots$, and such that $Q_{k(n)} \cup P_n$ is a (DARS)-sequence for E. By hyp. 3, the sequence $|f|(Q_{k(n)} \cup P_n)$ is bounded. Hence $|f|(Q_{k(n)}) + |f|(P_n)$ is bounded. Consequently $|f|(P_n)$ is bounded. Since this is true for every (DARS)-sequence for F, the field $f(a)$ is summation bounded on F.

W I.40.1.2. Theorem. Under Hyp. $\boldsymbol{H}_1$, if

1. E, F are measurable sets of traces,
2. $f(a)$ is a field of real numbers,
3. f is summation-bounded on E and on F,

then f is summation-bounded on $E + F$.

Proof. It suffices to prove the theorem in the case, when $E \cdot F = O$. Indeed, by Theorem [W I.1.1.], f is summation-bounded on $(E + F) - E = F - E$, and we have $(F - E) \cdot E = O$.

Having this in mind, suppose that $E \cdot F = O$. Take any (DARS)-sequence P_n for $E + F$. There exist partial complexes Q_n, R_n of P_n, such that $Q_n \frown R_n = O$, $Q_n \smile P_n = R_n$ and where Q_n, R_n are (DARS)-sequences for E, F respectively. Since $|f|(Q_n)$ and $|f|(R_n)$ are both bounded, (by hyp. 3), and since

$$|f|(P_n) = |f|(Q_n) + |f|(R_n),$$

it follows, that $|f|(P_n)$ is bounded.

Since P_n is an arbitrary (DARS)-sequence for $E + F$, the theorem follows.

W I.40.1.2a. Theorem. Under Hyp. $\boldsymbol{H}_1$, if

1. $f(a)$, $g(a)$ are fields of real numbers,
2. E is a measurable set of traces,
3. $0 \leqq g(a) \leqq f(a)$ for all a,
4. $f(a)$ is summation-bounded on E,

then $g(a)$ is summation-bounded on E.

Proof. Let P_n be a (DARS)-sequence for E. We have for some $M > 0$:

$$\sum_k f(p_{nk}) \leqq M \quad \text{for all } n.$$

Since

$$\sum_k |g(p_{nk})| \leqq \sum_k f(p_{nk}),$$

therefore

$$\sum_k |g(p_{nk})| \leqq M, \tag{1}$$

i.e. $g(a)$ summation-bounded on E, because a relation of the kind (1) takes place for all (DARS)-sequences for E.

W I.40.1.3. Remark. Notice, that summation-boundedness, even on *1*, does not imply the admissibility of $f(a)$; see Remark [U.6.]. But the admissibility of $f(a)$ implies the summation-boundedness of $f(a)$ on every measurable set of elusive traces, theorem [W I.9.2.]. To have the summation-boundedness on the whole set E, we must additionly have the condition of supplementary admissibility satisfied, Def. [W I.19.].

Notice that the summation-boundedness on the set (α), composed of the single heavy trace α, is equivalent to the complete admissibility of $f(a)$ on (α). The complete admissibility of $f(a)$ on the set E of traces implies summation-boundedness of $f(a)$ on E.

W I.40.2. Theorem. Under Hyp. $\boldsymbol{H}_1$, if

1. $f(a)$ is a field of real numbers,
2. E, F are measurable sets of traces,
3. $F \leqq E$,
4. $f(a)$ is completely admissible on E,

then $f(a)$ is completely admissible on F.

Proof. By hyp. 4, $f(a)$ is admissible, and f is summation-bounded on every heavy trace α, belonging to E. Hence it is so for every trace belonging to F.

W I.40.2.1. Theorem. Under Hyp. $\boldsymbol{H}_1$, if

1. $f(a)$ is a field of real numbers,
2. E, F are measurable sets of traces,
3. $f(a)$ is completely admissible on E and on F,

then $f(a)$ is completely admissible on $E+F$.

Proof. Follows similar pattern, as the forgoing proof.

W I.40.2.2. Theorem. Under Hyp. $\boldsymbol{H}_1$, if

1. $f(a)$, $g(a)$ are fields of numbers,
2. $0 \leqq g(a) \leqq f(a)$ for all a,
3. E is a measurable set of traces,
4. $f(a)$ is completely admissible on E,

then $g(a)$ is completely admissible on E.

Proof. $f(a)$ is admissible. Let P_n be a sequence of complexes with $\mu(P_n) \to 0$ for $n \to \infty$. Then $\sum\limits_k f(p_{nk}) \to 0$ for $n \to \infty$. By hyp. 2, we have

$$0 \leqq \sum_k g(p_{nk}) \leqq \sum_k f(p_{nk});$$

hence $\sum\limits_k g(p_{nk}) \to 0$ for $n \to \infty$, so $g(a)$ is admissible.

To prove the complete admissibility of g, consider a heavy trace α of E. Denote by A the corresponding atom of $\boldsymbol{G}$. Take any sequence q_n of bricks such that

$$A \leqq q_n, \quad \mu(A - q_n) \to 0 \quad \text{for} \quad n \to \infty.$$

Since $|f(q_n)|$ is bounded, it follows that $|f(q_n)|$ is bounded, so the theorem is proved.

W I.40.3. Remark. We shall study the notions of admissibility and that of summation-boundedness for squares of real fields.

W I.40.4. Theorem. Under Hyp. $\boldsymbol{H}_1$, if $f\mu$, $g\mu$ are admissible, α, β are real constants, then $\alpha f\mu + \beta g\mu$ is admissible.

Proof. By hypothesis

$$S_0 f\mu = S_0 g\mu = 0;$$

hence

$$S_0(\alpha f\mu + \beta g\mu) = 0.$$

By Remark [U.2.6.] theorem follows.

W I.40.4.1. Theorem. Under Hyp. $\boldsymbol{H}_1$, if

1. E is a measurable set of traces,
2. $f\mu$, $g\mu$ are completely admissible on E,
3. α, β are real constants,

then

$\alpha f\mu + \beta g\mu$ is also completely admissible on E.

Proof. A part of the thesis follows from Theorem [W I.40.4.]. The supplementary condition of admissibility can be proved by considering a trace $\alpha \in E$, and applying Remark [W I.40.1.3.].

W I.40.5. Theorem. Under Hyp. $\boldsymbol{H}_1$, if

1. f is a field of real numbers,
2. $|f|^2 \mu$ is admissible,

then

$$|f| \mu \text{ is also admissible.}$$

Proof. Let $P_n = \{p_{n1}, p_{n2}, \ldots, p_{nk}, \ldots\}$ be a sequence of complexes with

$$\sum_k \mu(p_{nk}) \to 0.$$

By hyp. we have

$$\sum_k |f(p_{nk})|^2 \mu(p_{nk}) \to 0.$$

Decompose P_n into two disjoint complexes R_n, S_n where $R_n \cup S_n = P_n$, and where

$$R_n = \{p_{nk} \,|\, |f(p_{nk})| > 1\},$$
$$S_n = \{p_{nk} \,|\, |f(p_{nk})| \leqq 1\}.$$

Put

$$R_n = \{r_{n1}, r_{n2}, \ldots\}, \qquad S_n = \{s_{n1}, s_{n2}, \ldots\}.$$

We have

$$|f(r_{nk})| \leqq f(r_{nk})|^2;$$

hence

$$(1) \quad \sum_k |f(r_{nk})| \mu(r_{nk}) \leqq \sum_k |f(r_{nk})|^2 \mu(r_{nk}) \leqq \sum_k |f(p_{nk})|^2 \mu(p_{nk}) \to 0.$$

On the other hand we have

$$(2) \qquad \sum_k |f(s_{nk})| \mu(r_{nk}) \leqq \sum_k \mu(r_{nk}) \leqq \sum_k \mu(p_{nk}) \to 0.$$

From (1) and (2) we get:

$$\sum_k f(r_{nk}) \mu(r_{nk}) + \sum_k f(s_{nk}) \mu(s_{nk}) \to 0,$$

so $f(a) \mu(a)$ is admissible.

W I.40.5.1. Theorem. Under Hyp. $\boldsymbol{H}_1$, if

1. $f(a)$ is a field of real numbers,
2. E is a measurable set of traces,
3. $|f|^2 \mu$ is completely admissible on E,

then $|f| \mu$ is also completely admissible on E.

Proof. One part of the theorem follows from Theorem [W I.40.5.]. Let A be an atom and α the corresponding heavy trace. Let $\alpha \in E$. Consider any sequence q_n of bricks, such that $A \leqq q_n$, $\mu(q_n - A) \to 0$. Then the sequence of numbers $|f(q_n)|^2 \mu(q_n)$ is bounded, say $< M$, where $M > 0$. For bricks q_n, for which $|f(q_n)| \leqq 1$, we have

$$|f(q_n)| \mu(q_n) \leqq \mu(q_n) \leqq \mu(1)$$

and for bricks q_n, for which $|f(q_n)| > 1$, we have

$$|f(q_n)| < |f(q_n)|^2;$$

hence

$$|f(q_n)| \mu(q_n) < |f(q_n)|^2 \mu(q_n) \leqq M.$$

Thus for all q_n considered we have

$$|f(q_n)| \mu(q_n) \leqq M + 1,$$

so the set of all numbers $|f(q_n)| \mu(q_n)$ is bounded. This being true for any sequence q_n with the property (1); the supplementary condition of admissibility takes place for the given atom. Hence this is true for all atoms $\in E$. Thus the theorem is established.

W I.40.5.2. Remark. The converse of Theorem [W I.40.5.] is not true. Indeed it is possible to have the situation, where $S_1 f(a) \mu(a)$ exists, but $|f(a)|^2 \mu(a)$ is not summation bounded on *1*; hence it is not admissible.

Eg. Take the ordinary interval $(\alpha, \beta\rangle$ and Lebesgues integration. Define $f(\alpha, \beta\rangle = \int_\alpha^\beta k(x)\, dx$, where k is summable but not square summable on $(0, 1\rangle$.

W I.40.5.3. Remark. In the sequel we shall prove only theorems, dealing with, (simple), admissibility. The study of complete admissibility can easily be supplemented.

W I.40.5.4. Theorem. Under Hyp. $\boldsymbol{H}_1$, if

1. $f(a)$ is a field of real numbers,
2. $|f|^2 \mu$ is summation-bounded on *1*,

then $|f| \mu$ is admissible.

Proof. Take any sequence P_n of complexes, such that $\mu(P_n) \to 0$ for $n \to \infty$; P_n is a (DARS)-sequence for the set O.

From hyp. 2, we have, on account of Theorem [W I.40.1.1.] that $|f|^2 \mu$ is summation bounded on O. Hence the sequence

$$(|f|^2 \mu)(P_n) \tag{1}$$

is bounded.

We have

$$\left(\sum_k |f(p_{nk}))| \cdot \mu(p_{nk}\right)^2 \leqq \sum_k |f(p_{nk}) \cdot \sqrt{\mu(p_{nk})}|^2 \cdot \sum_k (\sqrt{\mu(p_{nk})})^2 \leqq$$

$$\leqq \sum_k |f(p_{nk})|^2 \mu(p_{nk}) \cdot \sum_k \mu(p_{nk}).$$

By (1) there exists $M > 0$, such that

$$\sum_k |f(p_{nk})|^2 \mu(p_{nk}) \leqq M.$$

Since $\sum_k \mu(p_{nk}) \to 0$, it follows that

$$|\sum_k | f(p_{nk}) \cdot \mu(p_{nk})|^2 \to 0,$$

which completes the proof.

W I.40.5.5. Remark. The converse theorem is not true. See Remark [W I.40.5.2.]. $|f|\,\mu$ may be admissible, but $|f|^2\,\mu$ not summation-bounded.

W I.40.6. Theorem. Under Hyp. $\boldsymbol{H}_1$, if

1. $f(a)$, $g(a)$ are fields of real numbers,
2. $|f|^2\,\mu$ is summation bounded on *1*,
3. $|g|^2\,\mu$ is admissible,

then

$$|f g| \cdot \mu \text{ is admissible.}$$

Proof. Let P_n be a sequence of complexes with $\mu(P_n) \to 0$ for $n \to \infty$. We have

$$(1) \quad |\sum_k |f(p_{nk})| \cdot |g(p_{nk})| \cdot \mu(p_{nk})|^2 \leqq \leqq \sum_k |f(p_{nk})|^2 \mu(p_{nk}) \cdot \sum_k |g(p_{nk})|^2 \cdot \mu(p_{nk}).$$

By hyp. 2, $|f|^2\,\mu$ is summation-bounded on *1*. Hence by [W I.40.1.1.] it is summation-bounded on *O*. Now P_n is a (DARS)-sequence for *O*; hence

$$\sum_k |f(p_{nk})|^2 \mu(p_{nk}) \leqq M$$

for a suitable $M > 0$. Since $g^2\,\mu$ is admissible, we have

$$\sum_k |g(p_{nk})|^2 \mu(p_{nk}) \to 0 \quad \text{for} \quad n \to \infty.$$

Consequently, by (1),

$$\sum_k |f(p_{nk})| \cdot |g(p_{nk})|^2 \to 0 \quad \text{for} \quad n \to \infty,$$

which completes the proof, because P_n may be any sequence of complexes with $\mu(P_n) \to 0$. As a consequence we have

W I.40.6.1. Theorem. Under Hyp. $\boldsymbol{H}_1$, if

1. $|f|^2\,\mu$, $|g|^2\,\mu$ are both admissible, so is $f g\,\mu$.

Proof. If a scalar field is admissible, it is necessarily summation-bounded on *1*.

W I.40.7. Theorem. Under Hyp. $\boldsymbol{H}_1$, if

1. $|f|^2\,\mu$, $|g|^2\,\mu$ are admissible, so is $(f + g)^2\,\mu$.

Proof. We have

$$|f+g|^2 = |f^2+g^2+2fg| \leqq |f|^2+|g|^2+2|f|\cdot|g|.$$

Now

$$2|f|\cdot|g| \leqq |f|^2+2|g|^2;$$

hence

(1) $$|f+g|^2 \leqq 2|f|^2+2|g|^2.$$

By hypothesis, if $P=\{p_1, p_2, \ldots\}$ is any sequence with $\sum_k \mu(p_k)\to 0$, we have

$$\sum_k |f(p_k)|^2\,\mu(p_k)\to 0$$

and

$$\sum_k |g(p_k)|^2\,\mu(p_k)\to 0.$$

Hence, by (1),

$$\sum_k |f(p_k)+g(p_k)|^2\,\mu(p_k)\to 0, \quad \text{q.e.d.}$$

W I.40.8. Theorem. Under Hyp. $\boldsymbol{H}_1$, if $|f|^2\mu$, $|g|^2\mu$ are admissible, and α, β are complex constants, then

$$|\alpha f+\beta g|^2\,\mu \text{ is admissible.}$$

Proof. This follows from Theorem [W I.40.7.].

W I.40.9. Theorem. Under Hyp. $\boldsymbol{H}_1$, if f is admissible, then

$$|f|^2 \text{ is also admissible.}$$

Proof. Let $P_n=\{p_{n1}, p_{n2}, \ldots\}$ be a sequence of complexes with $\sum_k \mu(p_{nk})\to 0$ for $n\to\infty$. We have

(1) $$\lim_{n\to\infty}\sum_k |f(p_{nk})| = 0.$$

Hence there exists N, such that for all $n \geqq N$

$$\sum_k |f(p_{nk})| < 1.$$

Hence for those n and all k we have

$$|f(p_{nk})| < 1.$$

Hence

$$|f(p_{nk})|^2 \leqq |f(p_{nk})|;$$

consequently from (1) we get

$$\sum_k |f(p_{nk})|^2 \to 0 \quad \text{for} \quad n\to\infty,$$

which completes the proof.

W I.40.9.1. Remark. The theorem can not be inverted. There may exist a field $f(a)$, where $|f|^2$ is admissible and $|f|$ not even summation-bounded, (hence a function not admissible).

Expl. Define

$$f\left(\frac{1}{2}, 1\right\rangle =_{df} \frac{1}{1}, \quad f\left(\frac{1}{2^2}, \frac{1}{2}\right\rangle =_{df} \frac{1}{2},$$

$$f\left(\frac{1}{2^3}, \frac{1}{2^2}\right\rangle =_{df} \frac{1}{3}, \ldots, \quad f\left(\frac{1}{2^{n+1}}, \frac{1}{2^n}\right\rangle =_{df} \frac{1}{n},$$

and put for all other such intervals $(\alpha, \beta\rangle$ of $(0, 1\rangle$:

$$f(\alpha, \beta\rangle =_{df} 0.$$

Since

$$\tfrac{1}{1} + \tfrac{1}{2} + \tfrac{1}{3} + \cdots$$

diverges, therefore f is not admissible.

Hence, if we take a complex, which may be as fine as we like, but with intervals

$$\left(\frac{1}{2^{n+k+1}}, \frac{1}{2^{n+k}}\right\rangle, \left(\frac{1}{2^{n+k}}, \frac{1}{2^{n+k-1}}\right\rangle, \ldots, \left(\frac{1}{2^{n+1}}, \frac{1}{2^n}\right\rangle,$$

we get for the sum a numbers as large as we like.

Nevertheles, since

$$\frac{1}{1^2} + \frac{1}{2^2} + \cdots + \frac{1}{n^2} \cdots$$

converges, it follows that $|f|^2$ is admissible.

W I.40.10. Theorem. Under Hyp. $\boldsymbol{H}_1$, if $|f|^2$, $|g|^2$ are admissible, then fg is also admissible.

Proof. Let $P = \{p_1, p_2, \ldots\}$ be a variable complex with $\sum_k \mu(p_k) \to 0$. We have

$$(1) \qquad \sum_k |f(p_k)|^2 \to 0, \quad \sum_k |g(p_k)|^2 \to 0.$$

Now

$$\left|\sum_k f(p_k)\, g(p_k)\right|^2 \leqq \sum_k |f(p_k)|^2 \cdot \sum_k |g(p_k)|^2;$$

hence from (1) the thesis follows.

W I.40.11. Theorem. Under Hyp. $\boldsymbol{H}_1$, if $|f|^2$, $|g|^2$ are admissible, then $f + g$ is also admissible.

Proof.

$$|f+g|^2 \leqq |f|^2 + |g|^2 + 2|fg| \leqq 2|f|^2 + 2|g|^2.$$

Hence for a variable $P = \{p_1, p_2, \ldots\}$ with $\sum_k \mu(p_k) \to 0$ we get

$$\sum_k |f(p_k) + g(p_k)|^2 \to 0, \quad \text{q.e.d.}$$

W I.40.12. Theorem. Under Hyp. $\boldsymbol{H}_1$, if

1. f is admissible,
2. g is summation-bounded on *1*,

then $f g$ is admissible.

Proof. Let $P_n = \{p_{n1}, p_{n2}, \ldots\}$ be a complex with

$$\sum_k \mu(p_{nk}) \to 0 \quad \text{for} \quad n \to \infty.$$

We have

(1) $$\left|\sum_k f(p_{nk}) \cdot g(p_{nk})\right| \leqq \sum_k |f(p_{nk})| \cdot |g(p_{nk})| \leqq$$
$$\leqq \sum_k |f(p_{nk})| \cdot \sum_k |g(p_{nk})|.$$

Since g is summation-bounded, we have $\sum_l |g(p_{nl})| \leqq M$ for sufficiently great n, and for a suitable $M > 0$. Indeed P_n is a (DARS)-sequence for O, so we can apply Theorem [W I.40.1.1.]. Since f is admissible, $\sum_k |f(p_n)| \to 0$. From (1) follows that $\sum_k |f(p_k) \cdot g(p_k)| \to 0$, so the theorem is established.

W I.40.13. Theorem. Under Hyp. $\boldsymbol{H}_1$, if f, g are admissible, then $f \cdot g$ is admissible.

Proof. This follows from the preceding theorem, since f must be summation-bounded, Theorem [W I.40.12.].

W I.40.14. Theorem. If

1. f, g are summation-bounded on E,
2. α, β complex constants,

then

$$\alpha f + \beta g \text{ is also summation-bounded on } E.$$

Proof. Let $P_n = p_{n1}, p_{n2}, \ldots$ be a (DARS)-sequence for E. We have for some $M > 0$:

$$\sum_k |\alpha f(p_{nk}) + \beta g(p_{nk})| \leqq |\alpha| \sum_k |f(p_{nk})| + |\beta| \sum_k |g(p_{nk})| \leqq$$
$$\leqq |\alpha| M + |\beta| \cdot M,$$

where

$$\sum_k |f(p_{nk})| \leqq M, \quad \text{and} \quad \sum_k |g(p_{nk})| \leqq M \quad \text{for} \quad n = 1, 2, \ldots$$

This proves the theorem.

W I.40.15. Theorem. If f, g are summation-bounded on E, then $f \cdot g$ is also summation-bounded on E.

Proof. Let $P_n = \{p_{n1}, p_{n2}, \ldots\}$ be a (DARS)-sequence for E. We have

$$\sum_k |f(p_{nk})| \leqq M, \quad \sum_k |g(p_{nl})| \leqq M,$$

for a suitable $M > 0$. Since

$$\sum_k |f(p_{nk}) \cdot g(p_{nk})| \leqq \sum_k |f(p_{nk})| \cdot \sum_l |g(p_{nl})|,$$

it follows that

$$\sum_k |f(p_{nk}) \cdot g(p_{nk})| \leqq M^2, \quad \text{q.e.d.}$$

W I.40.16. Theorem. If f is summation-bounded on E, then $|f|^2$ is also summation-bounded on E.

Proof. This follows from the preceding theorem.

W I.40.16.1. Remark. If $|f|^2$ is summation-bounded on E, $|f|$ may not be summation-bounded on E.

W I.40.17. Theorem. Under Hyp. $\boldsymbol{H}_1$, if $|f|^2$, $|g|^2$ are both summation-bounded on E, then $f \cdot g$ is also summation-bounded on E.

Proof. Let $P_n = \{p_{n1}, p_{n2}, \ldots\}$ be a (DARS)-sequence for E. We have

$$\sum_k |f(p_{nk})|^2 \leqq M, \quad \sum_k |g(p_{nk})|^2 \leqq M,$$

for suitable $M > 0$. Now we have

$$\sum_k |f(p_{nk}) \cdot g(p_{nk})|^2 \leqq \sum_k |f(p_{nk})|^2 \cdot \sum_l |g(p_{nl})|^2 \leqq M^2,$$

so the theorem is proved.

W I.40.18. Theorem. Under Hyp. $\boldsymbol{H}_1$, if

1. $f\mu$, $g\mu$ are both summation-bounded on E,
2. α, β are complex constants,

then $(\alpha f + \beta g)\mu$ is also summation-bounded on E.

Proof. Let $P_n = \{p_{n1}, p_{n2}, \ldots\}$ be a (DARS)-sequence for E. Then

$$\sum_k |f(p_{nk})| \mu(p_{nk}) \leqq M, \quad \sum_k |g(p_{nk})| \mu(p_{nk}) \leqq M$$

for a suitable M. Hence

$$\Big|\sum_k (\alpha f(p_{nk}) + \beta g(p_{nk})) \mu(p_{nk})\Big| \leqq |\alpha| \sum_k |f(p_{nk})| \mu(p_{nk}) +$$
$$+ |\beta| \cdot \Big|\sum_k |g(p_{nk})| \mu(p_{nk}) \leqq (|\alpha| + |\beta|) M,$$

which proves the theorem.

W I.40.19. Theorem. Under Hyp. $\boldsymbol{H}_1$, if

1. $|f|^2\mu$, $|g|^2\mu$ are both summation-bounded on E,

then $fg\mu$ is also summation-bounded on E.

Proof. Let $P_n = \{p_{n1}, p_{n2}, \ldots\}$ be a (DARS)-sequence for E. Then

$$\sum_k |f(p_{nk})|^2 \mu(p_{nk}) \leqq M, \qquad \sum_k |g(p_{nk})|^2 \mu(p_{nk}) \leqq M.$$

Now,

$$\left(\sum_k |f(p_{nk})| \cdot |g(p_{nk})| \cdot \mu(p_{nk})\right)^2 \leqq$$

$$\leqq \sum_k |f(p_{nk})|^2 \mu(p_{nk}) \cdot \sum_k |g(p_{nk})|^2 \mu(p_{nk}) \leqq M^2,$$

which proves the theorem.

W I.40.20. Theorem. Under Hyp. $\boldsymbol{H}_1$, if

$$|f|^2 \mu \text{ is summation-bounded on } E,$$

then

$$|f| \mu \text{ is summation-bounded on } E.$$

Proof. We have for (DARS)-sequence P_n for E, as before,

$$\left|\sum_k f(p_{nk}) \mu(p_{nk})\right|^2 \leqq \sum_k |f(p_{nk})|^2 \mu(p_{nk}) \cdot \sum_k \mu(p_{nk}) \leqq$$

$$\leqq \sum_k |f(p_{nk})|^2 \mu(p_{nk}) \cdot \mu(I),$$

which proves the theorem.

W I.40.21. Remark. The converse theorem is not true.

W I.40.21.1. Theorem. Under Hyp. $\boldsymbol{H}_1$, if

1. $f(\dot{a})$ is a field of real numbers,
2. E is a measurable set of traces, which may be any,
3. $S_E |f(\dot{a})| = 0$,

then

$$S_E |f(a)|^2 = 0.$$

Proof. Let P_n be a (DARS)-sequence for E. We have:

(1) $$\sum_k |f(p_{nk})|^2 \leqq \left|\sum_k |f(p_{nk})|\right|^2, \quad \text{because} \quad |f(a)| \geqq 0.$$

Hence, since by hyp. 3, $\sum_k |f(p_{nk})| \to 0$, we get

$$\sum_k |f(p_{nk})|^2 \to 0;$$

hence, by (1),

$$S_E |f(a)|^2 = 0.$$

W I.40.21.2. Remark. The converse theorem is not true.

Expl. If a is one of the intervals obtained by subdivision of $(0, 1\rangle$ into n-equal parts, put

$$f(a) =_{df} \frac{1}{n^{3/4}}$$

and put, for all other subintervals $(\alpha, \beta\rangle$ of $(0, 1\rangle$,

$$f(a) =_{df} 0.$$

These partitions constitute a (DARS)-sequence for 1. We get for the n-th-division

$$f^2(P_n) = n \cdot \left(\frac{1}{n^{3/4}}\right)^2 = \frac{1}{n^{1/2}} \to 0;$$

hence

$$S_E f^2 = 0.$$

On the other hand we have

$$f(P_n) = n \cdot \frac{1}{n^{3/4}} = n^{1/4} \to \infty;$$

hence $S_E f$ does not exist.

W I.40.22. Remark. Let $\mu E > 0$. It may happen that $S_E f(a) > 0$ exists, (is finite), and $S_E f(a)^2 = 0$.

Expl. Put $f(a) = \mu(a)$ where μ is the Lebesguean measure. Let $P_n = \{p_{n1}, p_{n2}, \ldots\}$ be a (DARS)-sequence for E. Then

$$\sum_k \mu(p_{nk}) = \sum_k f(p_{nk}) \to \mu(E) \quad \text{for} \quad n \to \infty.$$

Hence

$$S_E f(a) = \mu(E) > 0.$$

But

$$\sum_k (f(p_{nk}))^2 = \sum_k \mu(p_{nk})^2 \leqq \mathcal{N}(P_n) \cdot \sum_k \mu(p_{nk}) \leqq \mathcal{N}(P_n)\,\mu(1) \to 0.$$

Hence

$$S_E f(a)^2 = 0.$$

W I.40.23. Theorem. Under Hyp. $\boldsymbol{H}_2$, if

1. $f(a)$ is a field of real numbers,
2. E is a measurable set of traces,
3. $S_E |f(a)|^2 \mu(a) = 0$, (hence if $f^2 \mu$ is completely admissible on E),

then

$$S_E |f(a)|\, \mu(a) = 0,$$

hence $f\mu$ is completely admissible on E.

Proof. We may suppose that $f(a) \geqq 0$ for all a.

Let P_n be a (DARS)-sequence for E. We have by Schwarz-lemma:

$$\left[\sum_k f(p_{nk})\,\mu(p_{nk})\right]^2 \leqq \left[\sum_k f(p_{nk}) \sqrt{\mu(p_{nk})}\right]^2 \cdot \sum_k \left[\sqrt{\mu(p_{nk})}\right]^2 \leqq$$

$$\leqq \sum_k f(p_{nk})^2\, \mu(p_{nk}) \cdot \sum_k \mu(p_{nk}) \leqq \sum_k f(p_{nk})^2 \mu(p_{nk}) \cdot [\mu(E) + \sigma_n],$$

where $\sigma_n \to 0$. Hence, by hyp. 3,

$$\lim_{n\to\infty} \left[\sum_k f(p_{nk})\, \mu(p_{nk})\right]^2 = 0,$$

and then $S_E f \mu = 0$, which completes the proof.

W I.40.23.1. Remark. The converse theorem is not true. It may happen that $f(a) \geqq 0$ for every a and $S_I f(a)\,\mu(a) = 0$, but $S_I f(a)^2\,\mu(a)$ does not exist.

Expl. Consider Lebesgue measurable subsets of $(0, 1\rangle$, with Lebesgues measure μ, and take halfopen subintervals of $(0, 1\rangle$ as bricks. Consider the halfopen intervals

$$(1) \qquad \left(\frac{1}{2}, 1\right\rangle, \left(\frac{1}{2^2}, \frac{1}{2}\right\rangle, \left(\frac{1}{2^3}, \frac{1}{2^2}\right\rangle, \ldots, \left(\frac{1}{2^{n+1}}, \frac{1}{2^n}\right\rangle, \ldots$$

Let $f(a)$ have on these intervals the values

$$2^{\frac{1}{2}}, 2^{\frac{2}{2}}, 2^{\frac{2}{3}}, \ldots, 2^{\frac{n+1}{2}}, \ldots$$

respectively, and let $f(a) = 0$ for all other intervals a. We have

$$\mu\left(\frac{1}{2^{n+1}}, \frac{1}{2^n}\right\rangle = \frac{1}{2^{n+1}},$$

for all n, hence

$$(2) \qquad f\left(\frac{1}{2^{n+1}}, \frac{1}{2^n}\right\rangle \mu\left(\frac{1}{2^{n+1}}, \frac{1}{2^n}\right\rangle = 2^{\frac{n+1}{2}} \cdot \frac{1}{2^{n+1}} = \frac{1}{2^{\frac{n+1}{2}}}$$

for all n. We also have

$$\left[f\left(\frac{1}{2^{n+1}}, \frac{1}{2^n}\right\rangle\right]^2 \cdot \mu\left(\frac{1}{2^{n+1}}, \frac{1}{2^n}\right\rangle = 2^{n+1} \cdot \frac{1}{2^{n+1}} = 1$$

for all n. Consider the complex

$$P_n =_{df} Q_1 \cup Q_2 \cup \cdots \cup Q_n \cup \left(\frac{1}{2^{n+1}}, \frac{1}{2^n}\right\rangle \cup$$

$$\cup \left(\frac{1}{2^{n+2}}, \frac{1}{2^{n+1}}\right\rangle \cup \frac{1}{2^{n+1}}, \frac{1}{2^2}\right\rangle,$$

where Q_i are subdivisions of

$$\left(\frac{1}{2^i}, \frac{1}{2^{i-1}}\right\rangle, \qquad (i = 1, 2, \ldots, n).$$

We have

$$(f^2 \mu)(P_n) = 0 + 0 + \cdots + 0 + 1 + \cdots + 1 = n;$$

hence

$$(f^2 \mu)(P_n) \to \infty,$$

so $f^2 \mu$ is not summable. But

$$(f \mu)(P_n) = 0 + 0 + \cdots + 0 + \frac{1}{2^{\frac{n+1}{2}}} + \frac{1}{2^{\frac{2n+2}{2}}} + \cdots + \frac{1}{2^{\frac{2n+1}{2}}} \leqq$$

$$\leqq \frac{1}{2^{\frac{n+1}{2}}}\left[1 + \frac{1}{2^{\frac{1}{2}}} + \frac{1}{2^{\frac{2}{2}}} + \cdots\right] \leqq \frac{1}{2^{\frac{n+1}{2}}} \cdot \frac{1}{1 - \frac{1}{2^{\frac{1}{2}}}} \leqq \frac{1}{2^{\frac{n+1}{2}}}.$$

Hence $f\,\mu(P_n) \to 0$ for $n \to \infty$. It follows that, if $S_I f \mu$ exists, it must be $= 0$. All what we need now, is to prove that $f\,\mu$ is summable. Take any (DARS)-sequence Q_n for $(0, 1\rangle$. To the value of

(3) $$(f\,\mu)\,(Q_n) = \sum_k f(q_{kn})\,\mu(q_{kn}),$$

only intervals (1) can contribute. Since we have here no atoms, therefore $\mathcal{N}(Q_n) \to 0$. To have $\mathcal{N}(Q_n) \leqq \varepsilon$, the only bricks q_{nk}, which matter, are the sufficiently far bricks of (1), i. e. with sufficiently great index. But then $(f\,\mu)\,(Q_n)$ will be $\leqq \dfrac{1}{2^{\frac{n+1}{2}}}$ for sufficiently great n. Hence $(f\,\mu)\,(Q_n) \to 0$. Consequently $f\,\mu$ is summable.

W I.40.24. Theorem. Under Hyp. $\boldsymbol{H}_1$, if

1. $f(a)$ is a field of real numbers,
2. $f(a) \geqq 0$ for all a,
3. E is a measurable set of traces,
4. $S_E f(a)\,\mu(a) = 0$ for (DARS)-summation, (hence $f\,\mu$ completely admissible),
5. $f^2\,\mu$ is admissible,

then
$$S_E f(a)^2\,\mu(a) = 0;$$
hence $f^2\,\mu$ is completely admissible.

Proof. Take a (DARS)-sequence P_n for E, and decompose P_n into two disjoint complexes:
$$R_n =_{df} \{p_{nk} \mid f(p_{nk}) \geqq 1\}, \quad S_n =_{df} \{p_{nk} \mid f(p_{nk}) < 1\}.$$

I say that

(1) $$\lim_{n\to\infty} \mu(R_n) = 0.$$

Indeed, suppose this be not true. Then there exists $\varepsilon > 0$, such that we can select a subsequence $\alpha(n)$ of n, such that $\mu\,R_{\alpha(n)} \geqq \varepsilon$. We have

(2) $$(\mu\,f)\,(R_{\alpha(n)}) = \sum_k f(r_{\alpha(n),k})\,\mu(r_{\alpha(n),k}) \geqq \sum_k \mu(r_{\alpha(n),k}) = \mu\,R_{\alpha(n)} \geqq \varepsilon.$$

Since
$$(\mu\,f)\,(R_{\alpha(n)}) \leqq (\mu\,f)\,(P_{\alpha(n)}) = \sum_k f(p_{\alpha(n),k})\,\mu(p_{\alpha(n),k}) \to 0,$$
for $n \to \infty$, by hyp. 4, we get a contradiction with (2), i.e. with the inequality $(\mu\,f)\,(R_{\alpha(n)}) \geqq \varepsilon$, which is valid for all $n = 1, 2, \ldots$ Thus the equality (1) is established. Since $\lim_{n\to\infty} \mu(R_n) = 0$, we have by the admissibility of $f^2\,\mu$,
$$\lim_{n\to\infty} (f^2\,\mu)\,(R_n) = 0,$$
i.e.

(3) $$\lim_{n\to\infty} \sum_k f^2(r_{nk})\,\mu(r_{nk}) = 0.$$

Concerning the part S_n of P_n, we have

$$\sum_k f(s_{nk})^2 \mu(s_{nk}) \leqq \sum_k f(s_{nk}) \mu(s_{nk}) \leqq \sum_k f(p_{nk}) \mu(p_{nk}).$$

Since, by hyp. 4,

$$\sum_k f(p_{nk}) \mu(p_{nk}) \to 0,$$

we get

(4) $$\lim_{n\to\infty} \sum_k f(s_{nk})^2 \mu(s_{nk}) = 0.$$

If we combine (4) with (3) we get

$$\lim_{n\to\infty} \sum_k f(p_{nk})^2 \mu(p_{nk}) = 0,$$

i.e.

$$S_E f(a)^2 \mu(a) = 0, \quad \text{q.e.d.}$$

W I.40.25. Theorem. Under Hyp. $\boldsymbol{H}_1$ and for (DARS)-sums, if

1. $f(a)$ is a field of real numbers,
2. $f(a) \geqq 0$ for all a,
3. E is a measurable set of traces,
4. $S_E f(a) \mu(a) = 0$,
5. $S_E f(a)^2 \mu(a)$ exists,

then

$$S_E f(a)^2 \mu(a) = 0.$$

Proof. From hyp. 5 it follows that $f^2 \mu$ is admissible, so we are in the conditions of Theorem [W I.40.24.]. The thesis follows.

W I.40.26. Theorem. Under Hyp. $\boldsymbol{H}_1$ and for (DARS)-sums, if

1. $f(a)$ is a field of real numbers,
2. $f(a) \geqq 0$ for all a,
3. E is a measurable set of traces,
4. $S_E f \mu = 0$, (hence $f \mu$ completely additive on E),

then the following are equivalent:

I. $S_E f^2 \mu = 0$,
II. $S_E f^2 \mu$ exists,
III. $f^2 \mu$ is admissible.

Proof. Let I. Then, of course II is true, and then $f^2 \mu$ is admissible. Thus we have I → II → III. Let III. If we take in account hyp. 4, we can apply Theorem [W I.40.24.], getting I. The equivalence is established.

W I.40.27. Theorem. Under Hyp. $\boldsymbol{H}_1$, and for (DARS)-summations, if

1. $f(a)$ is a field of real numbers,
2. $f(a) \geqq 0$ for all a,
3. E is a measurable set of traces,
4. $f(a)^2 \mu(a)$ is admissible,

then the following are equivalent:

I. $S_E f^2 \mu = 0$,

II. $S_E f \mu = 0$.

Proof. Let I. By Theorem [W I.40.23.] we have $S_E f \mu = 0$, i.e. II. Let II. By virtue of hyp. 4., we can apply Theorem [W I.40.24.], getting $S_E f^2 \mu = 0$, i.e. I. The equivalence is established.

W I.40.28. Theorem. Under Hyp. $\boldsymbol{H}_1$ and for (DARS)-sums, if

1. the condition $\mathscr{B}$ is admitted,
2. $f(a)$ is a field of real numbers,
3. E is a measurable set of traces,
4. $f(a)\,\mu(a)$ is completely admissible,
5. $|f(a)|^2\,\mu(a)$ is admissible,

then the following are equivalent:

I. $S_E |f|^2 \mu = 0$,

II. $S_E |f| \mu = 0$,

III. for every measurable set F of traces with $F \leqq E$

$$S_F f \mu = 0.$$

Proof. Let I, i.e.
$$S_E |f|^2 \mu = 0.$$

Then, by Theorem [W I.40.23.],
$$S_E |f| \mu = 0,$$

i.e. II. Let II. i.e.
$$S_E |f| \mu = 0.$$

It follows that
$$S_F |f| \mu = 0 \quad \text{for all} \quad F \leqq E.$$

Hence we have also
$$S_F f \mu = 0, \quad \text{since} \quad |S_F f \mu| \leqq S_F |f| \mu \leqq S_E |f| \mu.$$

Thus we have I $\rightarrow$ II $\rightarrow$ III.

Suppose III. Then, by hyp. 4 and Theorem [W I.28.], $\bar{S}_E |f| \mu = 0$; hence, as
$$0 \leqq \underline{S}_E |f| \mu \leqq \bar{S}_E |f| \mu = 0,$$

it follows

$$S_E |f| \mu = 0. \tag{1}$$

By hyp. 5, (1) and Theorem [W I.40.27.] it follows: $S_E |f|^2 \mu = 0$, i.e. I. Thus the equivalence is establlshed.

Section 5 of W I.

Equivalence of fields of numbers

W I.50.1. We are introducing the following notion of μ-square equivalence of fields of real numbers on A:

Let $f(a)$, $g(a)$ be fields of real numbers. We say that $f(a)$ *is μ-square equivalent to* $g(a)$ *on* A:

$$f(\dot{a}) \approx^2_\mu g(\dot{a}) \text{ on } A,$$

whenever

(1) $$S_A |f(a) - g(a)|^2 \mu(a) = 0.$$

W I.50.2. In [R.7.] we have defined the notion of, (simple), equivalence of fields of real numbers, which in a slightly more general setting, is as follows:

W I.50.2.1. Def. Let A be a measurable set of traces, and $f(\dot{a})$, $g(\dot{a})$ two fields of real numbers, which are (DARS)-summable on A. We say that, on A, $f(a)$ *is equivalent to* $g(a)$ *on* A whenever for every measurable subset F of A,

$$S_F f(a) = S_F g(a).$$

We write:

$$f(a) \approx g(a) \text{ on } A.$$

Usually we shall take the total set W of traces, writing simply $f(a) \approx g(a)$.

W I.50.3. Def. In this Section 5 of W I. we shall admit the following hypotheses, whose assemblage will be called *hypothesis of "comfort"*, and denoted by Hyp. $\mathscr{C}$:

W I.50.3a. Hyp. $\boldsymbol{H}_1$, [U.1.], Hyp. $\mathscr{B}$, hypothesis of admissibility [U 1.], hyp. of complete admissibility.

W I.50.3b. The basic tribe $\boldsymbol{G}$ may have atoms or not. The upper, lower and ordinary summation will be (DARS), and also sequences of complexes for sets of traces will be (DARS)-i.e. completely distinguished.

W I.50.3c. To assure denumerable additivity of upper and lower sums we shall admit the hypothesis $\mathscr{P}_A$ of "privacy of atoms".

W I.50.3d. The considered sets of traces will be always measurable. We shall consider fields of real numbers only, since fields of complex numbers do not seem to involve special difficulties.

W I.50.4. Theorem. The notion of μ-square equivalence on A obeys the basic rules of identity:

1) $f \approx^2_\mu f$ on A,

2) if $f \approx^2_\mu g$ on A, then $g \approx^2_\mu f$ on A,

3) if $f \approx^2_\mu g$ on A, $g \approx^2_\mu h$ on A, then $f \approx^2_\mu h$ on A.

Proof. Only for 3) may be needed.

Let $f \approx^2_\mu g$ and $g \approx^2_\mu h$ on A. We have

(1) $$S_A |f - g|^2 \mu = 0, \quad S_A |g - h|^2 \mu = 0.$$

We have

(2) $$|f - g|^2 = |(f - g) + (g - h)|^2 \leqq |f - g|^2 + 2|f - g| \cdot |g - h| + \\ + |g - h|^2 \leqq 2|f - g|^2 + 2|g - h|^2.$$

By Theorem [W I.27.],

$$|f-g|^2\,\mu, \quad |g-h|^2\,\mu$$

are completely admissible on A. Hence, by Theorem [W I.40.4.1.],

$$\{2|f-g|^2+2|g-h|^2\}\,\mu$$

is also completely admissible on A. Since the fields are non negative, we deduce by Theorem [W I.40.2.2.], that $|f-h|^2\,\mu$ is also completely admissible. Hence by Theorem [W I.16.] the field $|f-h|^2\,\mu$ possesses the lower and upper (DARS)-sums. We have

$$0 \leqq \underline{S}_A\,|f-h|^2\,\mu \leqq \bar{S}_A\,|f-h|^2\,\mu \leqq$$
$$\leqq S_A\,\{2|f-g|^2+2|g-h|^2\}\,\mu = 0,$$

(by (1)). Hence $|f-h|^2\,\mu$ is summable, and we have $S_A\,|f-h|^2\,\mu = 0$, which completes the proof.

W I.50.5. Theorem. Under Hyp. $\mathscr{C}$, if

1. $f(a)$, $g(a)$ are fields of real numbers,
2. α is a real number,
3. A is a measurable set of traces,
4. $f \approx^2_\mu g$ on A,

then

$$(\alpha f) \approx^2_\mu (\alpha g) \text{ on } A.$$

Proof. By definition of μ-square equivalence, we have $S_A\,|f-g|^2\,\mu = 0$. It follows that $S_A\,|\alpha f - \alpha g|^2\,\mu = 0$, which completes the proof.

W I.50.6. Theorem. Under Hyp. $\mathscr{C}$, if

1. $f(a)$, $g(a)$, $f'(a)$, $g'(a)$ are fields of real numbers,
2. A is a measurable set of traces,
3. $f \approx^2_\mu f'$, $g \approx^2_\mu g'$,

then

$$f+g \approx^2_\mu f'+g'.$$

Proof. We have

$$(1) \qquad S_A\,|f-f'|^2\,\mu = 0, \quad S_A\,|g-g'|^2\,\mu = 0.$$

We have

$$|(f+g)-(f'+g')|^2 = |(f-f')+(g-g')|^2 \leqq$$
$$\leqq 2|f-f'|^2+2|g-g'|^2.$$

By arguments, similar to those in the proof of Theorem [W I.50.4.], we get the summability on A of

$$|(f+g)-(f'+g')|^2\,\mu,$$

and then the equality

$$S_A\,|(f+g)-(f'+g')|^2\,\mu = 0,$$

so the theorem is established.

W I.50.7. Remark. If $|f|^2 \mu$ and $|g|^2 \mu$ are both summable on A, then $|f - g|^2 \mu$ is also summable on A. This follows from [W I.30.]. Consequently the sum $S_A |f - g|^2 \mu$ is always meaningful, when we consider square-summable fields.

W I.50.8. Theorem. Under Hyp. $\mathscr{C}$, if

1. $f(a)$, $f'(a)$, $g(a)$, $g'(a)$ are fields of real numbers,
2. α, β are real numbers,
3. A is a measurable set of traces,
4. $f \approx^2_\mu f'$, $g \approx^2_\mu g'$,

then

$$\alpha f + \beta g \approx^2_\mu \alpha f' + \beta g'.$$

Proof. Theorem follows from Theorem [W I.4.] and [W I.5.].

W I.50.9. Theorem. Under Hyp. $\mathscr{C}$, if

1. $f(a)$, $g(a)$ are fields of real numbers,
2. A is a measurable set of traces,
3. $f(a) \approx^2_\mu g(a)$ on A,
4. $A' \leqq A$, A' is a measurable set of traces,

then

$$f(a) \approx^2_\mu g(a) \text{ on } A'.$$

Proof. The proof is immediate.

W I.50.10. Theorem. Under Hyp. $\mathscr{C}$, if

1. A is a measurable set of traces,
2. $f(a)$, $g(a)$ are fields of real numbers,
3. $f\mu$, $g\mu$ are (DARS)-summable on A,
4. $f(a) \approx^2_\mu g(a)$ on A,

then

$$f(a)\,\mu(a) \approx g(a)\,\mu(a) \text{ on } A.$$

Proof. We have

$$S_A |f - g|^2 \mu = 0.$$

By Theorem [W I.40.23.] it follows

(1) $$S_A |f - g|\, \mu = 0.$$

By virtue of Theorem [W I.27.], $|f - g|^2 \mu$ is admissible.

It follows from hyp. 1, that $(f - g)\,\mu$ is completely admissible on A. Applying Theorem [W I.40.28.], we get from (1), $S_E(f - g)\,\mu = 0$ for every measurable subset E of A. Hence, by hyp. 1.,

$$S_E f \mu = S_E g \mu, \quad \text{i.e.} \quad f\mu \approx g\mu \text{ on } A.$$

W I.50.11. Theorem. Under Hyp. $\mathscr{C}$, if

1. $f(a)$, $g(a)$ are fields of real numbers,
2. A is a set of traces,

3. $f^2 \mu$, $g^2 \mu$ are (DARS)-summable on A,
4. $f \mu$, $g \mu$ are (DARS)-summable on A,
5. $f \mu \approx g \mu$ on A,

then

$$f \approx^2_\mu g \text{ on } A.$$

Proof. By hyp. 5,

$$S_E f \mu = S_E g \mu \quad \text{for all measurable} \quad E \leqq A;$$

hence

$$S_E (f - g) \mu = 0 \quad \text{for all measurable} \quad E \leqq A.$$

Hence $f \mu$, $g \mu$ are both completely admissible on A. By Theorem [W I.40.28] we get

$$S_A |f - g| \mu = 0.$$

Hence by Theorem [W I.40.25.], since $|f - g|^2 \mu$ is summable,

$$S_A |f - g|^2 \mu = 0; \quad \text{hence} \quad f \approx^2_\mu g, \quad \text{q.e.d.}$$

W I.50.12. Theorem. Under Hyp. $\mathscr{C}$ and for fields f, g of real numbers, such that $f \mu$, $g \mu$ and $(f^2) \mu$, $(g^2) \mu$ are summable on A, the following are equivalent:

I. $f \mu \approx g \mu$ on A,
II. $f \approx^2_\mu g$ on A.

Proof. This follows from Theorem [W I.10] and [W I.40.11.].

W I.50.13. Remark. In what follows we admit Hyp. $\mathscr{C}$ and consider only fields f of real numbers, such that $f \mu$ and $|f|^2 \mu$ are (DARS)-summable on a measurable A of traces. We call f "*μ-square summable on A*".

W I.50.14. Theorem. Under Hyp. $\mathscr{C}$, if

1. $f(a)$, $g(a)$, $f'(a)$, $g'(a)$ are μ-square summable on A,
2. A is a measurable set of traces,
3. $f(a) \approx^2_\mu f'(a)$ on A, $g(a) \approx^2_\mu g'(a)$ on A,

then

$$f g \mu \approx f' g' \mu \text{ on } A.$$

Proof. First we shall prove that $f g \mu \approx f' g \mu$.

Take a summable subset E of A, with $\mu E > 0$ and choose a (DARS)-sequence P_n for E. Put for $n = 1, 2, \ldots$,

$$N_n = \sum_k f'(p_{nk}) g(p_{nk}) \mu(p_{nk}) - \sum_k f(p_{nk}) g(p_{nk}) \mu(p_{nk}).$$

We have

$$|N_n|^2 = \{\sum_k [f'(p_{nk}) - f(p_{nk})] g(p_{nk}) \mu(p_{nk})\}^2$$

hence by the Cauchy-Schwarz inequality,

$$|N_n|^2 \leqq \sum_k [f'(p_{nk}) - f(p_{nk})]^2 \mu(p_{nk}) \cdot \sum_k |g(p_{nk})|^2 \mu(p_{nk}).$$

By hypothesis, $g(a)$ is square-summable on E. Hence

$$\sum_k |g(p_{nk})|^2 \mu(p_{nk}) \leqq M, \quad \text{for} \quad n = 1, 2, \ldots$$

and for suitable $M > 0$, so

$$|N_n|^2 \leqq \sum_k [f'(p_{nk}) - f(p_{nk})]^2 \mu(p_{nk}) \cdot M.$$

Now, by hyp. 3, $f(a) \approx^2_\mu f'(a)$ on E. Hence $S_E |f' - f|^2 \mu = 0$, so

$$\sum_k |f'(p_{nk}) - f(p_{nk})|^2 \mu(p_{nk}) \to 0 \quad \text{for} \quad n \to \infty.$$

It follows that

$$\lim_{n \to \infty} |N_n|^2 = 0.$$

This being true for all (DARS)-sequences P_n for E, we get

$$S_E(f' - f)\, g\, \mu = 0,$$

and then, since $f \mu$ and $f' \mu$ are summable,

$$S_E f' g \mu = S_E f g \mu.$$

This being true for any $E \leqq A$, we get

$$f' g \mu \approx f g \mu \text{ on } A.$$

Having that we get in a similar way $g' f \mu \approx g f \mu$ on A. Consequently

$$f g \mu \approx f' g' \mu \text{ on } A, \quad \text{q.e.d.}$$

W I.50.15. Remark. In the case where fg and $f'g'$ are both square summable we get, by Theorem [W I.50.11.],

$$f g \approx^2_\mu f' g'.$$

References

(1) RUSSELL, B., and A. N. WHITEHEAD: Principia Mathematicae. Vol. I. Cambridge (England) (1910).

(2) NIKODÝM, O. M.: Sur une propriété de la mesure généralisée des ensembles. (Pewne twierdzenie o mierze uogólnionej.) Prace matematyczno-fiz. Warszawa 1928—1929. Tom 36 (presented at the Congrès des Mathématitien des Pays Slaves à Varsovie 1929).

(3) WAŻEWSKI, T.: C. R. Acad. Sci., Paris 1923 (176), p. 69—70.

(4) FRÉCHET, M.: Sur la distance de deux ensembles mesurables. C. R. Acad. Sci., Paris 1929.

(5) FRÉCHET, M.: Des familles et fonctions additives d'ensembles abstraits. Fund. Math. IV. 1923, p. 391.

(6) MACNEILLE, H.: Partially ordered sets. Trans. Amer. Math. Soc. 42 (1937), p. 416—460.

(7) CARATHÉODORY, C.: Entwurf für eine Algebraisierung des Integralbegriffs. Münch. Sitzber. (1938), p. 27—68.

(8) BIRKHOFF, G.: Lattice theory (revised edition). (1948), New York City Amer. Math. Soc. Colloquium Publ. Vol. 25, XIII and 283 pp.

(9) STONE, M. H.: The theory of representation for Boolean Algebras. Trans. Amer. Math. Soc. 40 (1936), pp. 87ff.

(10) NIKODÝM, O. M.: Un nouvel appareil mathématique pour la théorie des Quanta. Annales de l'Institut Henri Poincaré, tome XI, fascicule II, p. 49—112 (Lectures at the Institut 4, 6, 11, 13, February 1947).

(11) POSSEL, R. DE: Sur la dérivation abstraite des fonctions d'ensemble. J. des Math. 101 (1936), pp. 391—409.

(12) NIKODÝM, O. M.: Critical remarks on some basic notions in Boolean lattices I. Anais da Academia Brasileira da Ciencias. No. 2. Tomo XXIV (1952), 113—136.

(13) NIKODÝM, O. M.: Detto II. Rendiconti del Seminario Matematico della Università di Padova. (1957), Parte II, p. 193—217, vol. XXVII.

(14) NIKODÝM, O. M.: Remarks on the Lebesgue-measure extension device for finitely additive Boolean lattices. Proc. Nat. Acad. Sci. Vol. 37. (1951), pp. 533—537.

(15) NIKODÝM, O. M.: Sur la mesure vectorielle parfaitement additive dans un corps abstrait de Boole. Mém. l'Acad. Roy. Belgique (Classe Sci.). Tome XVII (1938), p. 3—40.

(16) NIKODÝM, O. M.: Sur une généralisation des intégrales de Mr. J. RADON. Fund. Math. XIV. (1929), p. 131—179.

(17) NIKODÝM, O. M.: Sur l'existence d'une mesure parfaitement additive et non séparable. Mém. Acad. Roy. Belgique. XVII, 1939.

(18) NIKODÝM, O. M.: Sur une propriété de la mesure géneralisée des ensembles. Prace Mat.-Fiz. tom 36. Zeszyt II. 1928—1929, p. 65—71.

(19) WILEŃSKI, Fund. Math. XIV, (1929).

(20) FRÉCHET, M.: Sur l'integrale d'une fonctionnelle étendue à un ensemble abstrait. Bull. Soc. Math. France. (1915), t. 43.

(21) RADON, J.: Theorie und Anwendungen der absolut additiven Mengenfunktionen. Sitz.ber. der Math.-Naturwiss.-Klasse der Kais. Akad. Wiss. Wien 1913. Bd. 112, Abt. II, a/2.

(22) SAKS, S.: Theory of the integral. New York (1937).

(23) LEBESGUE, H.: Leçons sur l'intégration. Paris 1905.

(24) MAHARAM, D.: An algebraic characterisation of measure algebras. Ann. Math. vol. 48 (1947), p. 154—167.

(25) WECKEN, F.: Abstrakte Integrale und fastperiodische Funktionen. Math. Z. (1939). Vol. 45, p. 377—404.

(26) STONE, M. H.: Linear Transformations in HILBERT Space and their applications to analysis. Amer. Math. Soc. Colloquium Publ. Vol. 15. New York (1932), VI and 621 pp.

(27) FRÉCHET, M.: Des familles et fonctions additives d'ensembles abstraits. Fund. Math. IV. (1923), p. 329—365.

(28) HOPF, E.: Ergodentheorie. Berlin (1937).

(29) CARTAN, H.: C. R. Acad. Sci., Paris 205 (1937), p. 595.

(30) WAERDEN, VAN DER: Moderne Algebra.

(31) BOURBAKI, N.: Topologie générale. Ch. I. Structures topologiques. Paris (1940).

(32) WALLMAN, H.: Lattices and topological spaces. Ann. Math. 39 [Ultrafilters]. (1938), p. 112—126.

(33) NIKODÝM, O. M.: (Congrès de Bologne). (1928), Vol. VI, Sur le fondement des raisonnements locaux de l'analyse classique.

(34) DUNFORD, N., and J. T. SCHWARTZ, Linear operators Part. I, General theory (1958), 858 pp.

(35) TARSKI, A.: Une contribution à la théorie de la mesure. Fund. Math. 15 (1930) p. 42—56.

(36) NIKODÝM, O. M.: Summation of quasi-vectors on Boolean tribes and its application to quantum theories. I. Mathematically precise theory of P. A. M. DIRAC's Delta-Function. — Rendiconti del Seminario matematico della Università di Padova. Vol. 29. (1959), pp. 1—214.

(37) NIKODÝM, O. M.: Sur les tribus de sous espaces d'un espace de HILBERT-HERMITE. C. R. Acad. Sci., Paris 224 (1947), p. 522—524.

(38) NIKODÝM, O. M.: Tribus et lieux attachés à une classe ordonnée de sous-espaces d'un espace de HILBERT-HERMITE. C. R. Acad. Sci., Paris. 224 (1947), p. 628—630.

(39) NIKODÝM, O. M.: Système général de coordonnées dans l'espace de HILBERT-HERMITE. C. R. Acad. Sci., Paris. 224 (1947), p. 778—780.

(40) NIKODÝM, O. M.: Sur les opérateurs normaux maximaux dans l'espace hilbertien séparable et complet. Notion de „lieux" et ses propriétés. C. R. Acad. Sci., Paris. t. 238. (1954), p. 1373—1375.

(41) NIKODÝM, O. M.: Sur les opérateurs normaux maximaux dans l'espace Hilbertien séparable et complet II. Représentation canonique. C. R. Acad. Sci., Paris. t. 238. (1954), p. 1467—1469.

(42) NIKODÝM, O. M.: Summation of quasi-vectors on Boolean tribes and its application to quantum-theories I. Mathematically precise théory of P. A. M. DIRAC Delta function. Rend. Sem. Mat. della Univ. Padova. Vol. 29 (1959), p. 1—214.

(43) NEUMAN, J. v.: Mathematische Grundlagen der Quantenmechanik. Dover Publ. N. Y. 1943, p. 90.

(44) NAGY, B. v. S.: Spektraldarstellung linearer Transformationen des Hilbertschen Raumes. Ergebn. Math. Grenzgebiete, 5 Bd. Berlin: Springer-Verlag (1942).

(45) NIKODÝM, O. M.: On extensions of a given finitely additive field-valued, non negative measure on a finitely additive Boolean tribe to another tribe, more ample. Rend. Sem. Mat. della Univ. Padova. Vol. 26. (1956), p. 232—327.

(46) BANACH, S.: Teorja operacyj. Warszawa (1931). Théorie des opérations lineaires. (1932).

(47) RIESZ, F.: Über llneare Funktionalgleichungen. Acta Math. 41 (1918), p. 76—98.

(48) NIKODÝM, O. M.: Contribution à la théorie des opérateurs normaux maximaux dans l'espace de HILBERT-HERMITE séparable et complet. J. Math. p. et appl. Tome 36. (1957), p. 129—146.

(49) RIESZ, F.: Über die linearen Transformationen des komplexen Hilbertschen Raumes. Acta sci. math. Szeged 5 (1930).

(50) NEUMANN, J. v.: Allgemeine Eigenschaftstheorie hermitescher Funktionaloperatoren. Math. Ann. 102 (1929), p. 49—131.

(51) HILBERT, D.: Grundzüge einer allgemeinen Theorie der Integralgleichungen. Leipzig 1912.

(52) NIKODÝM, O. M.: Remarques sur les intégrales de STIELTJES en connexion avec celles de M. M. RADON et FRÉCHET [reprints 1939]. Ann. Soc. Pol. Math. Vol. 18 (1945).

(53) CLARKSON, A., and C. R. ADAMS: On definition of bounded variations for functions of two variables. Trans. Amer. Math. Soc. 35 (1933), p. 824—854.

(54) RADON, J.: Theorie und Anwendung der absolut additiven Mengenfunktionen. Sitz.ber. Math. Naturwiss. Kl. Kaiserl. Akad. Wiss. Wien (1913), 112, Abt. II, a/2.

(55) FRIEDRICHS, K.: Beiträge zur Theorie der Spektralschar. I. Mitteilung. Spektralschar auf Intervallen und Spektralzerlegung unitärer Operatoren. Math. Ann. Bd. 110 (1938), p. 54ff.

(56) FRIEDRICHS, K.: Die unitären Invarianten selbstadjugierter Operatoren im Hilbertschen Raume. Jahresber. Deutschen Math. Verein. 45 (1935).

(57) NIKODÝM, O. M.: Sur les êtres fonctionoïdes, une généralisation de la notion de fonction. C. R. Acad. Sci., Paris. 226 (1948), pp. 375; p. 458; p. 541.

(58) JULIA, G.: Sur les projecteurs de l'espace Hilbertien ou unitaire. C. R. Acad. Sci., Paris 1942, p. 456.

(59) JULIA, G.: La théorie des fonctions et la théorie des opérateurs de l'espace Hilbertien. J. Math. 22 (1943).

(60) HELLINGER and TÖPLITZ: Math. Ann. 69 (1910).

(61) DELSARTE: Mém. Sci. math. 59 (1932).

(62) OSGOOD: Non uniform convergence and the integration of series term by term. Amer. J. Math. 19 (1897).

(63) FRÉCHET, M.: Trans. Amer. Math. Soc. 8 (1907).

(64) RIESZ, F.: Acta sci. math. Szeged. 7 (1934).

(65) NIKODÝM, O. M.: Sur les suites convergentes des fonctions parfaitement additive d'ensembles abstraits. Mh. Math. Physik 46 (1933).

(66) NAKANO, H.: Unitärinvarianten im allgemeinen Euklidischen Raum. Math. Ann. Bd. 118 (1941/43), pp. 112—133.

(67) NIKODÝM, O. M.: On Boolean fields of subspaces in an arbitrary HILBERT space. Ann. Soc. Polon. Math. Vol. 17 (1938).

(68) MACNEILLE, H. M.: Extensions of measure. Proc. Nat. Acad. Sci. vol. 24 (1938), p. 188—193.

(69) Bourbaki, N.: Théorie des ensembles (fascicule de résultats). Paris (1939) [Actual. Sci. et Ind. No. 846. Hermann & Co.].
(70) Tukey, J. W.: Convergence and uniformity in topology. Ann. of Math. Studies No. 2, Princeton (1940).
(71) Eberlein, W. F.: A note on the spetral theorem. Bull. Ann. Math. Soc. 52 (1946), p. 328—331.
(72) Smith, K. T.: Sur le théorème spectral. C. R. Acad. Sci., Paris 234 (1952).
(73) Smith, K. T.: Rings of unbounded operators. (Published in the Duke Math. Journ.)
(74) Dieudonné, J., and L. Schwartz: La dualité dans les espaces (*f*) et (*L f*). Ann. Inst. Fourier. Tome 1 (1949), p. 61—101.
(75) Neumann, J. v.: Über adjungierte Funktionaloperatoren. Ann. Math. 33 (1932), p. 294—310.
(76) Weyl, H.: Riemannische Flächen.
(77) Nikodým, O. M.: Tribus de Boole et fonctions mesurables. Tribu spectrale d'une fonction. C. R. Acad. Sci., Paris t. 288 (1949), p. 37—39. — Tribus de Boole et fonctions mesurables. Transformations equimesurables. C. R. Acad. Sci., Paris t. 288 (1949), p. 150—151).
(78) Dirac, P. A. M.: The principles of quantum mechanics (3rd edition). Oxford. At the Clarendon Press (1947).
(79) Schwartz, L.: Theorie des Distributions. Tome 1. Paris: Herman & Co. (1950).
(80) Julia, G.: Introduction Mathématique aux Théories Quantiques (Deuxième Partie), (Gauthier-Villars) (1938).
(81) Nikodým, O. M., and S. Nikodým: Sur l'extension des corps algébraiques abstraits par un procédé généralisé de Cantor.
(82) Nikodým, O. M.: On extension of a given field-valued, non-negative measure on a finitely additive Boolean tribe, to another tribe more ample. Rend. Sem. Mat. Univ. Padova. Vol. 26 (1956), p. 232—327.
(83) Nikodým, O. M.: Contribution to the theory of maximal, normal operators II. An item of operational calculus and its application to resolvent and spectrum. J. reine angew. Math. Bd. 203 (1960), p. 90—100.
(84) Ludwig, G.: Die Grundlagen der Quantenmechanik. Berlin/Göttingen/Heidelberg: Springer-Verlag (Die Grundlehren der Mathematischen Wissenschaften. Bd. 52), pp. XII und 460.
(85) Nikodým, O. M.: Summation of fields of numbers in Boolean tribes I. Auxiliary notions and theorems. Rend. Circolo Mat. Palermo. Serie II., tom IX. Anno 1960, p. 1—57.
(86) Nikodým, O. M.: On upper and lower (DARS)-summation of fields of real numbers in Boolean tribes in the absence of atoms and on (DARS)-summation of fields of numbers admitting two values only. J. Math. p. et appl. (Volume consacré à un hommage postume à J. Peres.
(87) Achieser, N. J., and J. M. Glasman: Theorie der linearen Operatoren im Hilbert-Raum. Berlin: Akademie-Verlag (1954), XII und 396 pp.
(88) Cafiero, Federico: Misura e Integrazione. Edizioni Cremonese. Roma 1959. 451 pp.
(89) Lebesgue, H.: Ann. Éc. Norm. Sup. (3), 27 (1910), p. 380.
(90) La Valleé Poussin, Ch. de: Intégrales de Lebesgue. Fonctions d'ensemble. Classes de Baire. II éd. Paris. Gauthier-Villars (Coll. Borel).
(91) Hahn, H.: Theorie der reellen Funktionen. Leipzig und Berlin 1918.
(92) Hahn, H., and A. Rosenthal: Set functions. Albuquerque (New Mexico) 1948.
(93) Kolmogoroff, A.: Grundbegriffe der Wahrscheinlichkeitsrechnung. Ergebn. Math. Vol. II (1933).

(94) CARATHÉODORY, G.: Vorlesungen über reelle Funktionen. Leipzig und Berlin 1918.

(95) TARSKI, A.: Zur Grundlegung der Booleschen Algebra I. Fund. Math. 24 (1935), p. 177—198.

(96) HALMOS, H.: Measure Theory II. Van Nostrand Comp. Inc. Toronto/New York/ London (1950), XII and 304 pp.

(97) NACHBIN, L.: Une proprieté earactéristique des algèbres Booleiennes. Portugaliae Math. Vol. 6 (1947), p. 115—118.

(98) KAKUTANI, S.: Concrete representation of abstract L-spaces and the mean ergodic theorem. Ann. Math. 42 (1941), p. 523—537.

(99) BOCHNER, S., et R. S. PHILLIPS: Additive set functions and vector lattices. Ann. Math. 42 (1941), p. 316—324.

(100) NIKODÝM, O. M.: Sur les fonctionnelles linéaires. C. R. Acad. Sci., Paris 229 (1949), p. 16—18, p. 169—171, p. 288—289, p. 863—865.

(101) NIKODÝM, O. M.: Summation of fields of numbers in Boolean tribes. Upper and lower summation in the general case. Complete admissibility. Square summability. Equivalence of fields of numbers (IV). J. reine angew. Math. Bd. 214/215 (1964), 84—136 pp.

Alphabetic index

Active (valid) **B**-traces 199
Addition, somatic, of somata 13
Adjoint operation A^* of an operation A 387
Admissible scalar-fields 763
Admissible set of quasi-numbers f_x 766
($\leqq^J$)-algebraic addition; $a \dotplus^J b$ 28
($\leqq^J$)-algebraic subtraction; $a \dot{-} b$ 28
Application (relation, correspondence, mapping) 2
A-sets 541
Atom in a tribe 13
Atom-net-number $\mathcal{N}_A(P)$ 662

Base **B** of tribe (**F**) of figures 175
Base **B** of a tribe 131
Boolean algebra 9
Boolean algebra, formal laws of 9
Boolean lattice (tribe) 8
Borelian extension of a tribe through isomorphism 62
Borelian extension of a tribe within a supertribe, [A I], section 4 59—62
Borelian spectral tribe of N 21
Bricks 131
Brick-coverings 134
B-set s 545

Canonical mapping **C** of **H** 428
Canonical mapping $\boldsymbol{C}^\bullet$ of the space **H** 386
Canonical representation of the normal operator N having a saturated borelian spectral tribe $\mathbf{C} = \mathfrak{M}\,\mathfrak{A}\,\mathfrak{G}^{-1}$ 419
Carathéodory's convex measure $|\,(\dot{a})$ on a tribe [A] 79
Carathéodory's measure theory, [A I], section 7 78—85
Cardinal of a set 19
Cauchys condition 116
Cauchy-Schwarz inequality 169
Chain (linear ordering) 941 footnote
Characteristic function of the trace φ 748
Characteristic of the measure $v(\dot{f})$ 181
Characteristic of the trace 185
Characteristic set M — 837
Characteristic set N — 837
Class J' of sets 243
Coat, inner Q_n — 739
Coat, outer Q_n — 739
Coat $[X]$ of a measurable set X of traces 151
Comparison of the tribe-extension devices **L**, **C**, **N**, [A I], section 10 110—114
Compatibility of spaces 289
($\leqq^J$)-complement; $\mathrm{co}^J a$ 26, 28
J-complement of equivalence class 30
Complex 649
component-density of $\vec{X}$, τ — 718
component of $\vec{X}$, τ — 718
Conditions, absolute for denumerability of measure, [A I], section 3 54—57
Conditions, boundary for bricks 787
Condition of non-overlapping 777
Condition $\mathcal{B}(T, T^*, \mu)$, $\mathcal{B}(T, T^*)$ 101
Condition **F B G** 131
Condition **N** for bricks 191
Condition (propositional function), with one variable $w(\cdot\, x \,\cdot)$, with two variables $v(\cdot\, x, y \,\cdot)$ 1—2
Conditions, relative for denumerability of measure, [A I], section 2 46—53
Condition **S** for plane quarters 191
Condition, supplementary of admissibility on E 895
Confrontation of methods of Lebesgue and of Carathéodory, [A I], section 8 86—95
Convergence, natural somatic, of a_n to a 119
Convergence in μ-square mean 170
Convexity (subadditivity) of $\mu'_e(a')$ 73
coordinate-density of $\vec{X}$, τ — 719
coordinate of $\vec{X}$, τ — 719
Copies P_n of the plane P 421
Corner-measure at infinity 181
Correspondence (relation, mapping, application) 2
Correspondence ($\leqq^J$) 244
Correspondence $\mathfrak{A}(T)$ 160
Correspondence $\mathcal{G}$ 374, 417

Correspondence Γ; $\xi' \cup^{\Gamma} \xi$ 242
Correspondence $\mathfrak{B}$ (related to sub and supertribe) 22
Correspondence $\mathscr{M}$ 249
Correspondence $\boldsymbol{M}$, $X' \cup^{\mathsf{M}} Z$ 371
Correspondence $\mathfrak{S}$ 367
Correspondence $\mathscr{T}$ 606
Correspondence $\mathscr{V}$ 207
Covering of a soma a', in [A I] 70
Covering (in the theory of traces) 147

Diracs equality 751
Diracs function $\delta(E, F)$ 741
Directed sets 143 footnote
Distance $||a, b||_{\mu}$ of somata 115
μ-distance $|E, F|_{\mu}$, $|E, F|$ between somata E, F 39—40
Distance $|P, Q|$, $|E, P|$, $|P, E|$ 649
Distinguished, completely, sequence 676
(D')-distinguished sequence, (general) 676
Distinguished sequence for E 676
Distributive law in a lattice 8
Domain $\mathsf{C}\, R$ of the relation R 2
Double, generalize, Stieltjes' and Radon's integrals 393—401

Eigenspace 484
Eigenvalue 484
Equality of functions, — J^* 245
Equality governing in a theory 2—3, 18
Equality induced (in a tribe by the ideal J, $=^{J}$) 23
Equality-invariance of sets of elements in a theory 18
Equality modulo J 23
Equality; its role in a mathematical theory 18
Equality, specific notion of —, in a mathematical theory 18
Equivalence class $[a]$ corresponding to a, J — 29
Equivalence of fields of numbers 937
Equivalent quasi-vectors, $\vec{f}_x \approx \vec{g}_x$ 697
Existence of two — valued measure on ($\boldsymbol{F}$) 189
Existence, unitary; its meaning 18

Fields $f^{+}(a)$, $f^{-}(a)$ 861
Figures 131
Figures on the plane 175
Filter bases, equivalent 143
Filter base in (T) 143
Filter generated by a filter-base Φ 143
Filter, proper 137
Filter in (T) 136
Finite part π_f of P 537
Fréchets condition 47
Fréchets integral 165
Functions, complex valued, μ-summable of the trace 167
Functions, complex valued, of a trace, μ-square summable 169
Function $\Delta(p, q)$ 726
Function $\Delta''(p, q)$ 735
Function $F(\alpha)$ associated with $f(\alpha')$; $f(\alpha') \cup F(\alpha)$ 247
Function, propositional; (condition) with one variable $w(\cdot\, x\, \cdot)$, with two variables $v(x, y\, \cdot)$ 1—2
Function, simple, of the variable trace 166
Functions of traces, equivalent 170
Fundamental theorem on $\boldsymbol{B}_J$-traces 198
Fundamental theorem on canonical representation of normal operators 425—428
Fundamental theorem on double scale of spaces 366

Generating vector 313
Global elusive and regular, elusive and singular 577
Global heavy and regular, heavy and singular 577

Half-planes, horizontal vertical 174
Half-plane, vertical, $D_v(\alpha)$ 176
Hilbert-Hermite space $\boldsymbol{H}$, H.-H. 252
Homomorphism from (T) into [onto] (T') 19
$(=)$ — $(=')$-Homomorphism from (T) into [onto] (T') 19
Homomorphism, operation and order, $(=)$, $(=')$, from (T) onto [into] (T') 19
Homomorphism of tribes 19
Hopfs condition 57
Hypothesis $\mathscr{B}$ 787
Hypothesis of comfort 937
Hypotheses ***(Hyp Ad)***, ***(Hyp Af)*** 131
Hypotheses ***Hyp*** I, ***Hyp*** II, concerning sets of traces 148
Hypotheses $\boldsymbol{R}$, $\boldsymbol{L}_{\mu}$ 647
Hypothesis $\boldsymbol{R}_e$ 649
Hypothesis $\boldsymbol{S}$ 666

Ideal, denumerably additive 23
Ideal, finitely additive 23

Ideal J in a tribe (T) 22
Ideal, maximal 142
Ideal, proper in (T) 138
Inclusion $(\leqq)$ of spaces 253
Index of α 533
Infinite operations taken from a supertribe 133
Infinite part π_∞ of P 537, 541
Inner coat $[X]_*$ of the set X of traces 149
Inner measure $\mu_e'(a')$ 70
Integral over measurable subset of W 168
Invariant operations with respect to the equality 18
Invariance of a relation in its domain and in its range with respect to the equalities $(\dot{=})$ and $(\ddot{=})$ respectively 18

$\boldsymbol{J}^*$-Equality of functions 245
J-Equality induced in a tribe by the ideal J 23
J-Equivalence class $[a]$ corresponding to a 29
J-Join of equivalence classes 30
Join $\sum_{a \in M} a$, $\bigcup_{a \in M} a$, is meaningful (the join exists) 5
Join $\sum_{n=1}^{\infty} a_n$, $\bigcup_{n=1}^{\infty} a_n$ is meaningful (the join exists) 5
J-Meet of equivalence classes 30
J_N-measurable sets 562
Join in an ordering (sum, union) 4
Julia's theorem 271

Lattice 4
Lattice, complementary 6
Lattice-complement co a of the soma a 6
Lattice, complementary of spaces 256
Lattice, formal laws for addition and multiplication in a lattice 6
Lattice formal laws for 0, 1, co a 7
Lattice, complete 7
Lattice reorganized into a suitable algebra, and conversely 6
Lebesgues device 156 (footnote)
Lebesgues extension device, [A I], section 6 68—76
Lebesgues measure $\mu'(a') = \mu_e'(a')$ 76
Limits, $\operatorname{Lim}_\mu \sup a_n$ $\operatorname{Lim}_\mu \inf a_n$ $\operatorname{Lim}_\mu a_n$ 119
μ-limit of an infinite sequence a_n of somata $\lim_\mu a_n$ 115
Limit, natural somatic 119
Lines, well selected 232
List of theorems and definitions concerning measurable sets of $\boldsymbol{B}_J$-traces 204—205

$\boldsymbol{M}$-Characteristic set 837
Mapping (relation, application, correspondence) 2
Mean value-quasi-vector of f, at x; $\operatorname{val}_x f$ 698
Measurable function 166
μ^*-measurable functions 245
μ'-measurable functions $\Phi(z)$ of the complex variable z 249
Measurable set of $\boldsymbol{B}$-traces 209
Measurable sets, J_N — 562
Γ-measurable soma 79
μ^*-measurable soma 88
μ_e' C-measurable soma 87
μ_e' L-measurable soma 96
Measurable sets, N — 562
Measurable soma, N — 115
μ^*-measure 213
Measure atom 651
Measure of the complex $\mu(P)$, $\mu(p_i)$, $\mu(\operatorname{som} P)$ 649
Measure, denumerably additive on (T) 35
Measure, effective on (T) 35
Measure $\mu(\dot{a})$ finitely additive on (T) 35
Measure-hull, denumerably additive of a tribe, [A I], section 5 62—68
Measure $\mu(X)$ of a measurable set of traces 152
Measure on $\boldsymbol{N}$ 100
Measure-quasi-number 697
μ-measure topology on (T_J) 115
Measure topology on a tribe 39
Metric-space-topology on a tribe 39, 42
Measure, trivial 35
Measure $v(f)$, two-valued on (F) 176
Measure, two valued on a tribe 143
Measure, vector-valued 393
$(\leqq^J)$-Meet; $a \cdot^J b$ 24
Meet of equivalence classes, J — 30
Meet $\prod_{a \in M} a$, $\bigcap_{a \in M} a$ is meaningful (the meet exists) 5
Meet $\prod_{n=1}^{\infty} a_n$, $\bigcap_{n=1}^{\infty} a_n$ is meaningful (the meet exists 5
Minkowski's inequality 169

de Morgan laws 7, 8
Multiplane representation of vectors by functions 377
Multiplane V 421
Multiplicity, constant (uniform), upper, lower 529
Multiplicity of α, lower $\underline{\xi}(\alpha)$ 528
Multiplicity of α, upper $\overline{\xi}(\alpha)$ 527

$\boldsymbol{N}$-Characteristic-set 837
Neighbor traces of a trace 185
Net-number $\mathcal{N}'(P)$ 659
Net-number, reduced $\mathcal{N}_R(P)$ 659
Nikodýms condition 47
Nikodýms device for measure extension, [A I], section 9 96—109
N-measurable sets 562
N-measurable soma 115
Null-element of a lattice, O 6
Null-set of traces 170
Null-soma 0* 100
Number λ, complex, regular with respect to N 485
Numbers, generalized 174
Numbers of the set E, Num$E$ 421
Number of u, Num$u$ 421
Normal (maximal normal) operations 404
Normal (maximal) operator and its canonical representation 402—428
Norm of a vector $\vec{X}$; $\|\vec{X}\|$ 257

Operation (operator) in $\mathbf{H}$ 386
Operations, infinite in a tribe, taken from a supertribe 22
Operation permutable with a projector 391
Operator inverse of N 524
Operator $U f = \varphi f$ with maximal domain, for ordinary functions f 349
Operator $\varphi(N')$ for normal operator N' 461
Operator in the subspace a of $\boldsymbol{H}$ 503
Ordering $(\leq^J)$ induced in a tribe by the ideal J 23
Ordering, linear (L); (chain) 141 (footnote)
Orderings product (intersection, meet of two elements) $a \cdot b$, $a\,b$, $a \frown b$ 3
Ordering admits the product of two elements 4
Orderings sum (union, join) of two elements $a + b$, $a \smile b$ 4
Ordering addition 4
Ordering admits the sum of two elements 4
Orthocomplement co$a$ of the space a 256
Outer coat $[X]^*$ of the set X of traces 149
Outer measure $\mu_e'(a')$ 70

Paucs remark 137
Partitions, nested sequence of 651
Partition of a soma 650
Permutable, strongly, normal operators 622
Permutable, weakly, normal operators 623
Plane quarters 175
Pointed partition 393
Product (meet, intersection) in an ordering 4
Product of two partitions 650
Property Γ of the set ξ' 228
Property $\varphi(\cdot\, x \,\cdot)$ of the unique element 19
Projection of a set E of vectors on the space a, Proj$_a E$, Proj$(a)\,E$ 264
Projection of vector on a space, Proj$_a \vec{X}$, Proj$(a)\,\vec{X}$ 257
(Γ)-property, (R)-property, (S)-property 675
(DARS)-property of sequence 676

$\boldsymbol{Q}_n$-Coat, inner 739
Q_n-Coat, outer 739
Quasi-number with support x 690
Quasi-vector, normal 712
Quasi-vector $\vec{f}_{x,y}$ with support (x_0, E) 726
Quasi-vector with support x 690
Quotient tribe $(T)_J$ 30

Radiation scope of the set E of vectors on the tribe (T) of spaces $\mathcal{M}_T(E)$ 304
Range D R of the relation 2
Rectangles, generalized 174
Rectangle, well selected 232
Relation (correspondence, mapping, application) 1
Relation, empty 2
Relation R generated by the condition $v(\cdot\, x, y \,\cdot)$, is denoted by $\{x, y \mid v(\cdot\, x, y \,\cdot)\}$; $a\,R\,b$ means $v(\cdot\, a, b \,\cdot)$ 2
ᗡ R domain of R, D R range of R 2
Repartition tribe of $f(\dot{z})$ 594

Representation canonical of a normal operator $\boldsymbol{N}$ whose borelian spectral tribe $(\boldsymbol{S}^b)$ is not saturated 420
Representation, space-trace, of a normal operator 417
Representation, trace, of $N(\vec{X})$ 419
Representative of the equivalence class $[A]$ 30
Resolvent of a normal operator 575
Resolvent set of N 486

Saturating sequence of spaces for (T) 328
Saturating sequence of vectors for (T) 328
Saturation compartments 581
Saturation of a tribe of spaces 323
Scalar fields $f(a)$, summation-bounded on E 936
Scalar products of vectors $(\vec{X}, \vec{Y})$ 255
Scale, double, of spaces 336
Scale of repartition of $f(\dot{z})$ 594
Selfadjoint operator in $\boldsymbol{H}$ 403
Selfadjoint operations H', H'' (strongly) permutable 404
Sequence approching M where $M \in \boldsymbol{N}$ 100
Sequences of bricks, descending, minimal 147
Sequence descending of bricks 146
Sequences of bricks, descending, equivalent, $\{a_n\} \sim \{b_n\}$ 147
Sequences, descending where $\{a_n\}$ is included in $\{b_n\}$, $\{a_n\} \leqq \{b_n\}$ 147
Sets, A — 541
Set s, B — 545
Set Ξ of all $\boldsymbol{B}$-traces 215, 217
Set, distinguished, of $\boldsymbol{B}$-traces 220
Sets, indexed 6
Set-number $\mathcal{N}\{p_i\}$, $\mathcal{N}$ P of a partition or complex 651
Set O^*, empty, of $\boldsymbol{B}$-traces 209
Set, ordinary, of traces 217
Set of quasi-vectors, regular on E 690
Set of quasi-vectors, (D')-summable on E (over E) 691
Set of quasi-vectors with support E 690
Set S generated by the condition $w(\cdot x \cdot)$; is denoted by $\{x \mid w(\cdot x \cdot)\}$, $a \in S$ means $w(\cdot a \cdot)$ 2
Set $\boldsymbol{S}$ of spaces 286
Set ξ associated with ξ' 228
Set W^0 of all valid $\boldsymbol{B}$-traces 207
Sets ,,without points'' 12
Side measures at infinity 181
Single-plane representation of vectors by functions 374
Soma of the complex P; som P 649
Soma of an ordering 4
Somata, algebraic addition $a \dotplus b$ of 13
Somata A, B differ from one another in (T) by not more than by σ 96
Somata, disjoint 10
Somata, subtraction of, $a - b$ 9
Space capable of being adjoint to the tribe (T) 293
Spaces (closed subspaces of $\boldsymbol{H}$) 253
Spaces, difference $a - b$ of 260
Space-figures, -bricks, -traces 367, 368
Space of the type Σ, Π, $\Pi\Sigma$, $\Sigma\Pi$ 269
Special sequence for s, induced by $\{Q_n\}$ 739
Special summability on the figure 739
Spectral scale, projector-valued, double of N 409
Space-valued double scale of the normal operator N 409
Spectral scale, projector-valued 404
Spectral scale, space valued 404
Spectral theorem for normal operators 407
Spectral theorem for selfadjoint operations 403
Spectral tribe of N, (simple); $(\boldsymbol{S})$ 411
Spectral tribe of N, borelian; $(\boldsymbol{S}^b)$ 412
Spectrum continuous; discontinuous 486
Spectrum of N 486
Stone's existence theorem 141
Stone's ring 13
Streem, down 143 (footnote)
Stream-sequence 143 (footnote)
Streem, up 143 (footnote)
Strips, vertical, $B_v(\alpha_1 < \alpha_2)$ 177
Strips, vertical; horizontal 174, 175
Subpartition 651
Substratum for subtribes and supertribes 20
Subtribe, completely genuine 22
Supertribe, completely genuine 22
Subtribe, completely genuine, strict 22
Supertribe, completely genuine, strict 22
Subtribe, denumerably genuine 22
Supertribe, denumerably genuine 22
Subtribe, finitely genuine 20
Supertribe, finitely genuine 20
Subtribe, finitely genuine, strict 20

Supertribe, finitely genuine, strict 20
Sum in an ordering (join, union) 4
Sum oscillation, $\Delta_E f(a)$ 857
Sum, special $\underset{W}{S^\bullet} \vec{f}_\tau$ 728
Sum, upper and lower, of fields of real numbers $\overline{S}_E f(a)$, $\underline{S}_E f(a)$ 901
Sum, (DARS), upper and lower $\overline{S}_E f(a)$, $\underline{S}_E f(a)$ 810
Summability on E with respect to a given kind (D') of distinguished sequences 676
Summability on E, with respect to y_0 of the set $\vec{f}_{x,y}$ of quasi-vectors 726
Summation of fields of numbers, admitting two values only 829—838
Symbol $a \perp b$ 257
Symbol $[a \upharpoonleft f(p)]^V$ 425
Symbol R means the domain of the tribe (R) 19
Symbol Point Φ 421
Symbol $\xi^\times$ 218
Symbol Vert ξ 222
Symbol vert α 217
Symmetric difference of somata 13
System (frame) of coordinates, general in $\boldsymbol{H}$ 718—719

Tarski's theorem 144
T-component-density of $\vec{X}$ 718
T-component of $\vec{X}$ 718
T-coordinate-density of $\vec{X}$ 719
T-coordinate of $\vec{X}$ 719
Terrace configuration 537
Terrace mapping $\boldsymbol{C}'$ 553
Theorems, special on Boolean lattice, [A I] 44—119
Topologies on a denumerably additive tribe, [A I], section II 115—129
Trace 147
$\boldsymbol{B}$-traces 149
$\boldsymbol{B}$-trace, ordinary 217
$\boldsymbol{B}$-traces on the plane 176
$\boldsymbol{B}_v$-traces 232
$\boldsymbol{B}_J$-traces 196
Trace x covered by the brick a 147
Trace, elusive 161
Traces, equality of 147
Trace-function, upper and lower, attached to a given field of real numbers 818
Traces, general theory of, [B I] 161
Trace, heavy 161
Traces at infinity 185
Trace, integration, of a function of 165
Traces, measurable function of 165
Traces, measurable sets of 151
Trace, neighborhood of 160
Traces, not at infinity 185
Traces, null set of 159
Trace-representation of normal operators with saturated borelian spectral tribe 411
Trace, representative of 147
Traces, set of, covered by a covering L 174
Traces, the set W of all, 148
Trace-theorem, fundamental 238
Trace-theorem, fundamental generalized 249
Traces, theory of measurability of sets of 149—172
Tribe (Boolean tribe, Boolean lattice) 8
Tribe, completely additive 14
Tribe, denumerably additive 14
Tribe of figures on the plane, [C] 173—205
Tribe $\boldsymbol{F}$ of figures on the plane 175
Tribe, function fitting the 166
Tribe geometrical, of spaces 290
Tribe N 97, 100
Tribe (S*) 370
Tribe of spaces, determined by the scale $s(\dot{Q})$ 344
Tribes $(\mathcal{T})^P$, $(\mathcal{T})^V$ 464
Tribe (T) taken modulo J; (T_J) 29
Tribe, trivial 9

Ultrafilter in (T) 138
$(\leqq^J)$-Union $a +^J b$ 24
Union in an ordering (sum, join) 4
Unit-element of a lattice, 1, 6
Unit-soma 1^*, 100

Vector, generating $\boldsymbol{H}$ with respect to the tribe (T) 321
Vector field on $\boldsymbol{B}$ 675
Vector-field, selected, generated by a given total set of quasi-vectors 69
Vertex of the measure; $v(f)$ 181
Vertex of the trace 185

Wecken's theorem, [A I], section 1 44

Zorn's axiom 142 (footnote)